Applied Superconductivity 1997

Textbooks of related interest:

Physical Methods for Materials Characterisation
P E J Flewitt and R K Wild
Graduate Student Series in Materials Science and Engineering

Symmetry Principles and Magnetic Symmetry in Solid State Physics
S J Joshua
Graduate Student Series in Physics

Handbook of Thin Film Process Technology
Edited by D Glocker and I Shah

Superfluidity and Superconductivity, 3rd edition
D R Tilley and J Tilley
Graduate Student Series in Physics

Superconductivity of Metals and Cuprates
J R Waldram

Other titles of interest:

Nanomaterials: Synthesis, Properties and Applications
Edited by A S Edelstein and R C Cammarata

SQUIDs, the Josephson Effects and Superconducting Electronics
J C Gallop
Measurement Science and Technology Series

Proceedings of interest:

Applied Superconductivity 1995
Institute of Physics Conference Series 148
The Second European Conference on Applied Superconductivity, held in Edinburgh, Scotland, 3–6 July 1995. Volume 1: Plenary Talks and High Current Applications. Volume 2: Small Scale Applications
Edited by D Dew-Hughes

High Temperature Superconductivity
Proceedings of LT-19 Satellite Conference, Cambridge, 13–15 August 1990
Edited by J E Evetts

Magnetism, Magnetic Materials and their Applications
Proceedings of the International Workshop, La Habana, Cuba, 21–29 May 1991
Edited by F Leccabue and J L Sanchez Llamazares

High Temperature Superconductivity
SUSSP 39
Proceedings of the Thirty-Ninth Scottish Universities Summer School in Physics, St Andrews, June 1991
Edited by D P Tunstall and D W Barford

Applied Superconductivity 1997

Proceedings of EUCAS 1997, the Third European Conference on Applied Superconductivity, held in The Netherlands, 30 June–3 July 1997

Volume 2. Large Scale and Power Applications

Edited by H Rogalla and D H A Blank

Institute of Physics Conference Series Number 158
Institute of Physics Publishing, Bristol and Philadelphia

CODEN IPHSAC 158 1–1712 (1997)

British Library Cataloguing in Publication Data

A catalogue record for this book is available from the British Library.

ISBN 0 7503 0485 5 Vol. 1
0 7503 0486 3 Vol. 2
0 7503 0487 1 (2 Vol. set)

Library of Congress Cataloging-in-Publication Data are available

Published by Institute of Physics Publishing, wholly owned by The Institute of Physics, London
Institute of Physics Publishing, Dirac House, Temple Back, Bristol BS1 6BE, UK
US Editorial Office: Institute of Physics Publishing, The Public Ledger Building, Suite 1035, 150 South Independence Mall West, Philadelphia, PA 19106, USA

Printed in the UK by Galliard (Printers) Ltd, Great Yarmouth, Norfolk

Contents

VOLUME 1. SMALL SCALE AND ELECTRONIC APPLICATIONS

Preface xxxvii

Conference organization xxxix

Plenary talks

Wave function symmetry and its influence on superconducting devices
J Mannhart and H Hilgenkamp 1

Materials Aspects for Small Scale Applications

Thallium- and Mercury-containing cuprates in electronic devices
Z G Ivanov and L-G Johansson 7

Preparation and investigation of the ITO/YBCO heterostructures
B Vengalis, V Lisauskas and R Butkute 13

Voltage fluctuations below the critical current in a superconducting BSCCO thin film
C Coccorese, G Jung and B Savo 17

From high temperature superconductivity to nonmetallic behaviour in fully oxygenated $Nd_{1+x}Ba_{2-x}Cu_3O_y$
B Fisher, J Genossar, L Patlagan, G M Reisner and A Knizhnik 21

Critical thickness of YBCO films on CeO_2 buffered sapphire
A G Zaitsev, G Ockenfuss and R Wördenweber 25

Proximity effect in YBCO-$YBCu_{3-x}Co_xO$ bilayer films
E Polturak, G Koren, D Cohen and O Nesher 29

Temperature variation of the microwave power dependence in HTS thin films
L F Cohen, A L Cowie and J C Gallop 33

Memory effects associated with an organised state in the vortex dynamics in $YBa_2Cu_3O_{7-\delta}$ single crystals
S N Gordeev, A P Rassau, P A J de Groot, R Gagnon and L Taillefer 37

Nonlinear surface impedance in 'low' and 'high' T_c-superconductors
J Halbritter 41

Thermal and magnetic limitations of the linear surface resistance of epitaxial HTS films in high microwave films
T Kaiser, C Bauer, W Diete, M A Hein, J Kallscheuer, G Müller and H Piel 45

Study of ramp-type Josephson junctions by HREM
K Verbist, O I Lebedev, G Van Tendeloo, M A J Verhoeven, A J H M Rijnders and D H A Blank 49

Raman and XRD characterisation, microstructure and microwave loss in YBCO 123 thin films
A L Cowie, L F Cohen, G Gibson, Y H Li, J L MacManus-Driscoll and J C Gallop 53

Accuracy and sensitivity of penetration depth measurements in superconducting films using the mutual inductance method
J H Claassen 57

Temperature and orientation dependent dielectric measurements of $LaAlO_3$-substrates
R Schwab, R Spörl, P Severloh, R Heidinger and J Halbritter 61

Modelling the variation of microwave losses induced by DC and RF magnetic fields
J C Gallop, A L Cowie and L F Cohen 65

Microwave properties of dielectrics for HTS devices
C Zuccaro, M Winter, S J Penn, N McN Alford, P Filhol, G Forterre and N Klein 69

Observation of oscillations in the I-V derivative of YBCO thin films
A Verdyan, I Lapsker and J Azoulay 73

DC-pulse characteristic of thin $YBa_2Cu_3O_7$ films
P Lahl and R Wördenweber 77

Coherent motion of vortices in periodic magnetic structures
Y Yuzhelevski, C Camerlingo, M Ginovker, M Guilloux-Viry, R Monaco, A Perrin, B Ya Shapiro and G Jung 81

Magneto-optic and electron beam-induced voltage contrast studies of YBCO thin films
M Zamboni, S A L Foulds, S Koishikawa, K Matsumoto, M Murakami and J S Abell 85

Current density and magnetic field pattern in rectangular YBCO thin films
C Ferdeghini, E Giannini, G Grassano, D Marré, I Pallecchi and A S Siri 89

Electric field effect in (103)/(013) and (110) oriented $YBa_2Cu_3O_7$ thin films
K Herrmann, R Auer, E Brecht and R Scheider 93

Critical current anisotropy of $YBa_2Cu_3O_7$ thin films
J Šouc, K Frölich, V Šmatko, S Takács, F Weiss and G Delabouglise 97

Microwave properties of Nb_3Sn films on sapphire substrates
M Perpeet, W Diete, M A Hein, S Hensen, T Kaiser, G Müller, H Piel and J Pouryamout 101

Doping Ag $YBa_2Cu_3O_{7-\delta}$ thin films in-situ by two-beam pulsed-laser deposition
C K Ong, S Y Xu, Y L Zhou and X Zhang 105

Magnetic characteristics of YBCO thin films on sapphire buffered with CeO_2 substrates
T Nurgaliev, S Miteva, A Spasov, A D Mashtakov, P Komissinski and G A Ovsyannikov 109

Oxygenation mechanism of $YBa_2Cu_3O_{6+x}$ thin films during growth by pulsed laser deposition
J García López, D H A Blank, H Rogalla and J Siejka 113

Wavelength-resolved study of the photodoping effect in $YBa_2Cu_3O_x$
W Markowitsch, C Stockinger, W Lang, W Kula and R Sobolewski 117

Dynamic conductivity of YBCO thin films from low to very high frequencies
I Wilke, G Nakielski, J Kötzler, K O Subke, C Jaekel, F Hüning, H G Roskos and H Kurz 121

Hysteric microwave measurements of YBCO thin films in small swept applied fields
J R Powell, A Porch, M J Lancaster, R G Humphreys and C E Gough 125

$YBa_2Cu_3O_{7-\delta}$/(Sr,Ba)TiO_3 epitaxial combinations for tunable microwave components
Yu A Boikov and Z G Ivanov 129

Ion beam patterning of HTS-thin films
M Kuhn, B Schey, W Biegel, B Stritzker, J Eisenmenger and P Leiderer 133

Properties of $BaTbO_3$ and $La_2CuO_{4+\delta}$ thin films for HTS device applications
R Hojczyk, U Poppe, C L Jia, M I Faley, R Dittmann, C Horstmann, A Engelhardt, G Ockenfuß and K Urban 137

Development of high power magnetron sputtering targets prepared by flame spraying
I Van Driessche, R Carolissen, G Vanhoyland and S Hoste 141

Doping effects in YBCO ceramics
I Nedkov, S Miteva and T Koutzarova 145

Analysis of the photoresponse of Y-Ba-Cu-O thin films on ps to μs timescales
M Zavrtanik, J Demsar, B Podobnik, D Mihailovic and J E Evetts 149

Doping experiments on $YBa_2Cu_3O_{7-x}$ ultrathin films by using electric fields
R Schneider and R Auer 153

Temperature dependent dimensionality of flux dynamics in the superconducting YBCO/PYBCO superlattice
S Y Lang, J T Jeng, H E Horng and H C Yang 157

Passivation for YBCO-Devices using polymerised hexamethyldisilazane
L Mex, K Heinicke, A Krämer, C Francke and J Müller 161

Photoinduced changes of the transport properties in Bi-2201 thin films
C Stockinger, W Markowitsch, W Lang, A Ritzer and D Bäuerle 165

Pinning forces in superconducting Nb/Pd and Nb/CuMn multilayers
C Attanasio, C Coccorese, L Maritato, L Mercaldo, M Salvato, S Prishepa, J M Slaughter and C M Falco 169

Direct bonding of materials to be used in low-temperature electronics
G Kästner, P Kopperschmidt, D Hesse, M Lorenz and U Gösele 173

Thin Film Deposition

Growth and characterization of Re-Ni_2B_2C superconducting thin films
M Iavarone, A Andreone, C Aruta, A Cassinese, F Fontana, F Palomba, M L Russo and R Vaglio 177

In situ RHEED during pulsed laser deposition of complex oxides at high deposition oxygen pressures
G Koster, J Heutink, B L Kropman, A J H M Rijnders, D H A Blank and H Rogalla 181

The growth of $YBa_2Cu_3O_{7-\delta}$ and $PrBa_2Cu_3O_{7-\delta}$ over argon ion milled steps
P J Hirst, M Barnett, N G Chew, J S Abell, M Aindow and R G Humphreys 185

Atomic-absorption controlled reactive evaporation of HTS wafers for microwave applications
V Matijasevic, Z Lu, T Kaplan and C Huang 189

Preparation of Ba-Ca-Cu oxycarbonate superconducting thin films by pulsed laser deposition
U Spreitzer, S Hauser, M Fuchs, G Calestani, A Migliori, H Barowski, T Schauer, A Varlashkin, N Reschauer, T Mayerhöfer and K F Renk 193

Synthesis, characterisation and properties of Tl-2201 thin films
H Q Chen, L-G Johansson, Q-H Hu, D Erts, T Claeson and Z G Ivanov 197

The fabrication of Tl-Ba-Ca-Cu-O thin films on $SrTiO_3$ and $LaAlO_3$ buffered MgO substrates for microwave applications
A P Bramley, A P Jenkins, A J Wilkinson, C R M Grovenor and D Dew-Hughes 201

Structural and superconducting properties of MBE grown $Bi_2Sr_2CuO_{6+\delta}$ thin films
C Attanasio, C Coccorese, T Di Luccio, L Maritato, L Mercaldo, M Salluzzo, M Salvato and S Prishepa 205

Processing of NdBCO thick films on YSZ substrates
T C Shields, A Riley and J S Abell 209

Pulsed oxygen laser deposition of YBCO epitaxial thin films
M R Cimberle, A Diaspro, C Ferdeghini, E Giannini, G Grassano, D Marré, I Pallecchi, M Putti, R Rolandi and A S Siri 213

Structure and electrical properties of the $YBa_2Cu_3O_x$ thin films prepared with pulsed laser deposition in Ar/O_2 atmosphere
P B Mozhaev, A Kühle, G A Ovsyannikov and J L Skov 217

Large area YBCO films deposited by unipolar pulsed magnetron sputtering
J Schneider, J Einfeld, P Lahl, Th Königs, R Kutzner and R Wördenweber 221

Pulsed laser deposition of YBCO on $7 \times 20\,cm^2$
B Schey, W Biegel, M Kuhn, R Klarmann and B Stritzker 225

Temperature restriction for $YBa_2Cu_3O_x$ thin film deposition on CeO_2 buffer layer
A D Mashtakov, I M Kotelyanskii, V A Luzanov, P B Mozhaev and G A Ovsyannikov 229

$YBa_2Cu_3O_7$ films grown on CeO_2/Al_2O_3 substrate by aerosol MOCVD
K Frölich, J Šouc, D Machajdík and F Weiss 233

YBCO thin films prepared on silicon (100) wafers buffered by YSZ and $SrTiO_3$ layers
M Španková, Š Gaži, Š Chromik, I Vávra, P Juhás and Š Beňačka 237

Effect of growth parameters on the grain boundary morphology of $YBa_2Cu_3O_{7-\delta}$ on step edge (001) $LaAlO_3$ substrates
M Gustafsson, E Olsson, H R Yi, M Vaupel and R Wördenweber 241

Helical current flow at the boundary between two tilted anisotropic conductors: 90° YBCO twist boundaries
J R Fletcher, P J King, A Polimeni, R P Campion and S M Morley 245

Growth and characterization of (110) oriented $YBa_2Cu_3O_x$ superconducting thin films
S M Morley, R P Campion and P J King 249

A technology for generation of submicron $YBa_2Cu_3O_{7-x}$ patterns
H Elsner, R P J IJsselsteijn, W Morgenroth, H Roth and H-G Meyer 253

A new $YBa_2Cu_3O_{7-\delta}$ thin film patterning technique for use in multilayer structures
C A J Damen, G Feld, D H A Blank and H Rogalla 257

Passive Devices

Progress, properties and prospects of passive high-temperature superconductive microwave devices in Europe
M A Hein 261

Dielectric dual-mode filter using a high temperature superconducting image plane
S Schornstein, I S Ghosh and N Klein 267

Power dependence of two-dimensional microstrip resonators for cellular filter applications
A P Jenkins, A P Bramley, D Hyland, C R M Grovenor, D Dew-Hughes and D J Edwards 271

High Q HTS shielded composite puck resonators
L Hao, J C Gallop and F Abbas 275

Nonlinear characteristics of HTS planar resonators and filters
I Vendik, O Vendik, T Samoilova, M Gubina, D Kaparkov, S Gevorgian, E Kollberg, A G Zaitsev and R Wördenweber 279

Ag-doped large-area double-sided YBaCuO and GdBaCuO thin films for microwave applications
M Lorenz, H Hochmuth, J Frey, H Börner, J Lenzner, G Lippold, T Kaiser, M A Hein and G Müller 283

Effect of Ag-doping in YBCO on the microwave properties of HTS bandpass filter
J Kim, S-K Han, I-K Yu and K-Y Kang 287

Determination of material and circuit properties using superconducting and normal metal ring resonators
B A Tonkin and M W Hosking 291

Nonlinear microwave losses of large area YBCO thin films
C Zuccaro, N Klein, A G Zaitsev, R Wördenweber, Y Lemaitre and J C Mage 295

High performance $LiNbO_3$ optical modulator with a superconducting transmission line
Y Kanda, K Yoshida and I Uezono 299

A tunable high Q superconducting LC resonator operating at audio frequency
M Bonaldi, P Falferi, R Dolesi, S Vitale and M Cerdonio 303

High Q superconducting resonator for impedence measurements
R Mezzena, P Falferi, M Cerdonio, G Fontana and S Vitale 307

Full-wave analysis of the image hybrid dielectric/HTS resonator
C Sans, J O'Callaghan, D Sancho, R Pous, J Fontcuberta, J-F Liang and G-C Liang 311

Microwave dielectric composite puck resonators
F Abbas, J C Gallop and L Hao 315

High-power high-Q YBaCuO disk resonator filter
M A Hein, B A Aminov, A Baumfalk, H J Chaloupka, F Hill, T Kaiser, S Kolesov, G Müller and H Piel 319

Advanced lumped-element bandpass filters
M Reppel, H J Chaloupka and S Kolesov 323

Investigation of voltage tuneable phase shifters based on HTSC/ferroelectric thin film coplanar waveguides
R A Chakalov, Z G Ivanov, Yu A Boikov, P Larsson, E F Carlsson and S Gevorgian 327

Thin film HTS and ferroelectric slow wave coplanar line resonators
R A Chakalov, Z Wu, L E Davis, T Nurgaliev, Z G Ivanov and A Spasov 331

Fabrication of tuneable microwave resonators using superconducting/ferroelectric heterostructures
I Wooldridge, C W Turner, I D Robertson, S E Schwarz and P A Warburton 335

Experimental study of thin film HTS/ferroelectric CPW phase shifters for microwave applications
E F Carlsson, P K Petrov, R A Chakalov, P Larsson, Z G Ivanov and S Gevorgian 339

Realization and characterization of a millimeter HTS bandpass filter
M Petrizzelli, D Andreone, L Brunetti and C Camerlingo 343

Microwave characteristics of YBCO bandpass filters: closed loop resonators and coupled lines
Y S Ha, S-K Han, K-Y Kang and Y W Park 347

TBCCO-microstrip filters based on dual-mode ring resonators
S Huber, M Manzel, H Bruchlos, F Thrum, M Klinger and A Abramowicz 351

HTS shielded sintered dielectric microwave resonators
S J Penn, A Templeton, X Wang and N McN Alford 355

Aperture coupled high-temperature superconductor microstrip small antennas
H Y Wang, J S Hong and M J Lancaster 359

Characteristics of microstrip-line type microwave high temperature superconducting multiplexer
S-K Han, J Kim, K-Y Kang and Y-S Ha 363

Mixers and Detectors

Pushing the operating range of SIS mixers into the THz regime
T M Klapwijk, P Dieleman and M W M de Graauw 367

SIS junction as a detector at submillimeter wavelength
A Karpov, J Blondel, M Voss and K H Gundlach 373

NbN-MgO-NbN tunnel junctions integrated in aluminium strip lines for terahertz quasiparticle mixers
M Schicke, B Plathner, K H Gundlach, M Aoyagi, S Takada, P Dieleman, J B M Jegers, T M Klapwijk and H van de Stadt 377

Twin-junction SIS mixers for the 230 GHz band
T Noguchi, S-C Shi and H Iwashita 381

Terahertz responses of high-T_c ramp-type junctions on MgO
H Myoren, J Chen, K Nakajima, T Yamashita, D H A Blank and H Rogalla 385

Integrated submm wave receiver with superconductive local oscillator
V P Koshelets, S V Shitov, L V Filippenko, A B Ermakov, W Luinge, J-R Gao and P Lehikoinen 389

Investigation of the response of high-T_c superconducting films to millimeter wave radiation
A Laurinavičius, K Repšas, A R Vaškevičius, V Lisauskas and D Čepelis 393

HTS bolometer on Si_xN_y membrane-preparation and properties
T Heidenblut, B Schwierzi, W Michalke, E Steinbeiß, S Sánchez, M Elwenspoek, M J M E de Nivelle, M P Bruijn, R de Vries, J J Wijnbergen and P A J de Korte 397

Fabrication and characterisation of a Nb diffusion-cooled hot electron bolometer for a 730 GHz waveguide mixer
D Wilms Floet, J-R Gao, W Hulshoff, H van de Stadt, T M Klapwijk and A K Suurling 401

NbN quasioptical phonon cooled hot electron bolometric mixers at THz frequencies
M Kroug, P Yagoubov, G Gol'tsman and E Kollberg 405

Heterodyne Josephson mixing by YBCO double-junction step-edge structure
K Y Constantinian, G A Ovsyannikov, P B Mozhaev, J Mygind and N F Pedersen 409

Experiments on Nb Josephson junctions for optical detection
V Lacquaniti, S Maggi, E Monticone, M Rajteri, M L Rastello and R Steni 413

Submm-wave detector response from array of high-T_c Josephson junctions biased in parallel
K Y Constantinian, G A Ovsyannikov, K Lee, I Iguchi and H Ekstrom 417

Shot noise in NbN SIS junctions suitable for THz radiation detection
P Dieleman, H van de Stadt, T M Klapwijk, M Schicke and K H Gundlach 421

Development of a prototype x-ray detector, using mono crystalline Nb and Ta absorbers
A W Hamster, J M A Webers, T Kachlicki, H Wormeester, J Flokstra, H Rogalla, M L van den Berg, M P Bruijn, F B Kiewiet, O J Luiten and P A J de Korte 425

Digital Applications and Three-Terminal Devices

Digital superconductive electronics: Where does it fit?
F D Bedard 429

Measurement of RSFQ pulse propagation by specially designed shift registers
F-Im Buchholz, W Kessel, M I Khabipov, R Dolata, J Niemeyer and A Yu Kidiyarova-Shevchenko 433

Recent progress in developing HTS quasi-particle injection devices
C W Schneider, R Schneider, R Moerman, G J Gerritsma and H Rogalla 437

Test jigs for superconducting circuits with coplanar waveguides
M Biehl, W Benzing, R Koch, M Neuhaus, T Scherer and W Jutzi 441

Design of an RSFQ autocorrelator and experimental testing of the autocorrelator delay line
P G Litskevitch, A Yu Kidiyarova-Shevechenko, D V Balashov and R Dolata 445

High-Tc superconductor circuit with an upper YBCO groundplane
M Hidaka, H Terai, T Satoh and S Tahara 449

Properties of asymetric, tapered and non-tapered, YBCO Josephson junctions
S P Isaac, E J Tarte, A Moya and M G Blamire 453

Josephson Junctions

Josephson junctions
S Pagano and A Barone 457

Transport mechanisms in HTS junctions
A A Golubov, I V Devyatov, M Yu Kupriyanov, G J Gerritsma and H Rogalla 463

$YBa_2Cu_3O_{7-x}$ Josephson junctions on $NdGaO_3$ bicrystal substrates
Y Y Divin, I M Kotelyanskii, P M Shadrin, O Y Volkov, V V Shirotov, V N Gubankov, H Schulz and U Poppe 467

Electrical field dependence of the quasi-particle transport in $YBa_2Cu_3O_{7-x}$ Josephson junctions in ramp-edge geometry
H Predel, H Burkhardt and M Schilling 471

High-T_c step-edge junctions: problem of reproducibility
I M Kotelyanskii, A D Mashtakov and G A Ovsyannikov 475

Quantum behaviour of Josephson junctions
P Silvestrini, C Granata, B Ruggiero, V G Palmieri and M L Russo 479

Proximity coupling in high T_c Josephson junctions produced by focused electron beam irradiation
W E Booij, A J Pauza, E J Tarte, D F Moore and M G Blamire 483

All refractory light-sensitive Josephson junctions
C Granata and M L Russo 487

Preparation and RBS measurements of Nb/A1-A1O_x/Nb Josephson junctions with very low critical current densities
L Fritzsch, H-J Köhler, F Thrum, G Wende and H-G Meyer 491

Fabrication and measurement of Josephson junctions with hafnium dioxide barrier
H Schulze, F Müller, R Pöpel and J Niemeyer 495

Josephson flux flow in multi-junction stacks: experiment and simulation
N Thyssen, H Kohlstedt, S Sakai and A V Ustinov 499

Characteristics of YBCO grain boundary Josephson junctions on MgO bicrystal substrates
V Štrbík, O Harnack, Š Chromik, M Darula and Š Beňačka 503

Trimming and stability of YBCO Josephson junctions and microbridges
C P Foley, R Driver, S K H Lam, Y Wilson, B Sankrithyan and E E Mitchell 507

Fabrication of YBCO submicron bridges by selective epitaxy growth on a niobium inhibitor
S K H Lam 511

Weak links produced in YBCO thin films using a focused ion beam
M W Denhoff, M Gao, L E Erickson and G Champion 515

Properties of high-T_c Josephson junctions fabricated by electron beam irradiation
M V Pedyash, W E Booij, D F Moore and M G Blamire 519

Properties of YBCP/PBCO/YBCO trilayer structures
Š Chromik, V Štrbík, Š Beňačka, D Machajdík, J Liday, A P Kobzev and P Vogrinčič 523

Microwave properties of intrinsic $Bi_2Sr_2CaCu_2O_y$ Josephson junctions on off-axis substrates
T Ishibashi, K Sato, K Lee and I Iguchi 527

Effective length of annular long Josephson junctions with finite width: theory and experiment
A Wallraff, D Bolkhovsky, V Kurin, N Thyssen and A V Ustinov 531

Anodization-based area reduction process of Nb/Al-AlO_x/Nb Josephson junctions
D Andreone, V Lacquaniti, S Maggi, E Monticone and R Steni 535

Radiation damage of Josephson devices
S Pagano, L Frunzio, R Cristiano, O Mukhanov, A Kirichenko, R Robertazzi, E Mezzetti, L Gozzelino and R Gerbaldo 539

Effect of idle region geometry on the magnetic field curves of Nb/Al-AlO_x/Nb Josephson junctions
V Lacquaniti, S Maggi, E Monticone and R Steni 543

Zero-field resonances in a double-barrier Josephson system with highly transmissive tunnel barriers
I P Nevirkovets, A V Ustinov, E Goldobin, M G Blamire and J E Evetts 547

Phase locked states in stacked Josephson junctions
G Carapella, G Costabile, A Petraglia and N F Pedersen 551

Properties of PdAu barriers of SNS junctions for programmable voltage standards
H Sachse, R Pöpel, T Weimann, F Müller, G Hein and J Niemeyer 555

Spectral study of the Shapiro subharmonic step formation
V K Kornev, A V Arzumanov, K Y Constantinian, A D Mashtakov and G A Ovsyannikov 559

Transport properties and lower critical field of $Bi_2Sr_2Ca_2Cu_3O_{10+\delta}$ Josephson junctions
U Frey, M Blumers, M Basset, J C Martinez and H Adrian 563

Current-phase relation of high-T_c weak links on bicrystal substrates
E Il'ichev, V Zakosarenko, R P J IJsselsteijn, V Schultze, H-G Meyer and H E Hoenig 567

Impact of the symmetry of the superconducting wave function of Josephson tunneling in high-T_c superconductors
Z G Ivanov, E A Stepantsov, F Wenger, T Claeson and P Chaudhari 571

Nonequilibrium properties of electron-beam scribed Josephson junctions in YBCO: Hysteresis and low-temperature behavior
B A Davidson, J E Nordman, B M Hinaus, M S Rzchowski, K Siangchaew and M Libera 575

High-T_c multilayer edge junctions with a Ga-doped YBCO barrier
I Song, E-H Lee, S Park, I Song and G Park 579

C-axis transport properties of $YBa_2Cu_3O_{7-\delta}$/$PrBa_2Cu_3O_{7-\delta}$/$YBa_2Cu_3O_{7-\delta}$ trilayer Josephson junctions
A Schattke, M Engelmann, J C Martinez and H Adrian 583

Oscillators and Volt-Standards

Programmable voltage standards based on HTS Josephson junction arrays
A M Klushin, S I Borovitskii, C Weber, E Sodtke, R Semerad, W Prusseit, V D Gelikonova and H Kohlstedt 587

Transient vs. steady-state dynamics in discrete stacked junction flux-flow oscillators
S Lomatch and E D Rippert 591

Enhanced THz radiation from a-axis oriented YBCO films excited by femtosecond optical pulses
M Hangyo, S Shikii, N Tanichi, T Nagashima, M Tonouchi, M Tani and K Sakai 595

Stack Josephson junctions with sidewall shunt
M Darula, H Kohlstedt, L Amatuni and A M Klushin 599

Transient nonequilibrium superconducting states and ultrashort electromagnetic pulse emission from YBCO by femtosecond laser pulse excitation
M Hangyo, N Wada, M Tonouchi, M Tani and K Sakai 603

Development of superconducting oscillators and mixers using micro-bridges controlled by externals magnetic fields
F Sicking and B Schiek 607

Low phase noise microwave oscillators using temperature compensated high-Q HTS shielded dielectric resonators
I S Ghosh, D Schemion and N Klein 611

Numerical study for higher-order modes in flux-flow oscillators with rf-terminators
S Kohjiro, M Maezawa and A Shoji 615

Shapiro steps of $YBa_2Cu_3O_y$ Josephson junctions in terahertz region
K Lee and I Iguchi 619

Optimization of coherent emission from Josephson junction arrays by variation of the temperature
T Traeuble, M Keck, H Pressler, R Dolata, T Doderer, T Weimann, R P Huebener and J Niemeyer 623

Numerical simulation of Josephson-junction system dynamics in the presence of thermal noise
V K Kornev and A V Arzumanov 627

Improved 10 V Josephson voltage standard arrays
J Kohlmann, F Müller, R Behr, I Y Krasnopolin, R Pöpel and J Niemeyer 631

SQUIDs and SQUID Applications

Smart SQUIDS based on double relaxation oscillation SQUIDs
M J van Duuren, G C S Brons, H E Kattouw, J Flokstra and H Rogalla 635

Magnetic fields characteristics of high-T_c $YBa_2Cu_3O_y$ SQUIDs
H C Yang, J H Lu, H W Yu, D C Jou, H E Horng, J M Wu and S Y Yang 639

$YBa_2Cu_3O_7$ flip-chip SQUID magnetometers with large pick up loops
C Francke, A Krämer, L Mex and J Müller 643

Bias reversal reduction of 1/f noise in sputtered YBCO bicrystal dc-SQUIDs
C Camerlingo, M G Castellano, C Granata, A Monaco, M L Russo, E Sarnelli and G Torrioli 647

HTSC-dc-SQUID gradiometer for a nondestructive testing system
F Schmidl, S Wunderlich, L Dörrer, H Specht, J Heinrich, K-U Barholz, H Schneidewind, U Hübner and P Seidel 651

Non-dissipative noise thermometry measured with an HTS SQUID
L Hao, J C Gallop and R P Reed 655

A 4.2 K superconducting receiver coil and SQUID amplifier for low field MRI
H C Seton, J M S Hutchison and D M Bussell 659

An ultra high quality factor, radio frequency SQUID magnetometer based on a liquid helium cooled quartz resonator
H Prance, R Whiteman, V Schöllmann, R J Prance, T D Clark, J Diggins and J F Ralph 663

SQUIDs with ferromagnetic circuit
S I Bondarenko and A A Shablo 667

A D-SQUID as a possible solution for approaching the classical sensitivity limit
B Chesca 671

Electronic axial gradiometers for unshielded environment operating at 77 K
J Borgmann, H J Krause, G Ockenfuß, J Schubert, A I Braginski and P David 675

Absolute magnetometry by dc-SQUID arrays
P Carelli, K Flacco, M G Castellano, R Leoni and G Torrioli 679

Biepitaxial grain boundary $YBa_2Cu_3O_{7-y}$ thin-film SQUIDs
H E Horng, S Y Yang, W L Lee, H C Yang and J M Wu 683

A multilayer technology for high T_c dc SQUIDs
R P J IJsselsteijn, V Schultze, V Zakosarenko, W Morgenroth and H-G Meyer 687

Modulation properties of series arrays of YBCO dc SQUIDs
S-G Lee, D W Kim, B C Nam, M C Lee and Y Huh 691

YBCO dc SQUIDs on silicon using focused electron beam irradiation
S-J Kim, J Chen, H Myoren, K Nakajima and T Yamashita 695

HTSC SQUIDs on technical substrates
S Linzen, F Schmidl, Y Tian, U Hübner, O Köhler, R Čihař and P Seidel 699

Study of high T_c inductively shunted SQUID properties
M Izquierdo, A B M Jansman, J M Torres Bruna, F J G Roesthuis, J Flokstra and H Rogalla 703

Resonances and hysteresis in inductively shunted high-T_c SQUIDs
A B M Jansman, M Izquierdo, J Flokstra and H Rogalla 707

2-SQUID planar gradiometer on one chip
X H Zeng, H-J Krause, Y Zhang, H R Yi, F Rüders and H Bousack 711

Noise performance of high T_c rf SQUIDs in ambient magnetic fields with and without a flux dam
K E Leslie, L D Macks, C P Foley, J C Macfarlane, K-H Müller and G L Sloggett 715

New type of HTS-SQUID microscope for operation without shielding
Y Tavrin and M Siegel 719

Effect of junction noise in a resistive high T_c SQUID noise thermometer
J C Macfarlane, D Peden, L Hao, J C Gallop and E J Romans 723

Two stage dc SQUID-based amplifier with double transformer coupling scheme
D E Kirichenko, A B Pavolotskij, I G Prokhorova, O V Snigirev, R Mezzena, S Vitale and A V Beljaev 727

Low noise and wide-band 4 GHz amplifier based on dc SQUID
G V Prokopenko, V P Koshelets, S V Shitov, D V Balashov and J Mygind 731

SQUID-amplifier for cryogenic particle detectors based on superconducting phase transition thermometers
O Meier, S Uchaikin, F Pröbst and W Seidel 735

Helmholtz coil systems for the characterization of SQUID sensor heads
A Chwala, F Bauer, V Schultze, R Stolz and H-G Meyer 739

Application of HTS dc SQUIDs with quasiplanar junctions in NDE
M I Faley, U Poppe, K Urban, H-J Krause, M Grüneklee, R Hohmann, H Soltner, H Bousack and A I Braginski 743

HTS SQUID magnetometers and gradiometers fabricated on MgO bicrystal substrates for use in NDE
C Carr, E J Romans, J C Macfarlane, A Eulenburg, C M Pegrum and G B Donaldson 747

Eddy current tomography using rotating magnetic fields for deep SQUID-NDE
A Haller, Y Tavrin, H-J Krause, P David and A I Braginski 751

SQUID-based magnetic microscopy of superconducting thin film samples
S A Gudoshnikov, K E Andreev, N N Ukhansky, L V Matveets, M Mück, J Dechert, C Heiden and O V Snigirev 755

A system for biomagnetic measurement (MCG) with planar single-layer high-T_c CD-SQUID gradiometers in unshielded environment
R Weidl, S Brabetz, F Schmidl, F Klemm, S Wunderlich and P Seidel 759

A seven-channel high-T_c SQUID system for a magnetocardiogram
Y H Lee, I S Kim, H C Kwon, J M Kim, Y K Park and J C Park 763

Corrosion measurements with HTS SQUID gradiometer
R Matthews, S Kumar, D A Taussig, B R Whitecotton, R H Koch, J R Rozen and P Woeltgens 767

Application of SQUID magnetometers in fetal magnetocardiography
A P Rijpma, Y Seppenwoolde, H J M ter Brake, M J Peters and H Rogalla 771

System Aspects and Other Small Scale Applications

Eddy current aircraft testing with mobile HTS-SQUID gradiometer system
H-J Krause, Y Zhang, R Hohmann, M Grüneklee, M I Faley, D Lomparski, M Maus, H Bousack and A I Braginski 775

Recent results of SQUID-based position detectors for gravitational experiments
W Vodel, H Koch, S Nietzsche and J v Zameck Glyscinski 781

Low-noise temperature control using a high T_c superconducting sensor compared to a conventional PRT method
M L C Sing, J P Rice, C Dolabdjian and D Robbes 785

Suppression of superconductivity by injection of spin-polarized current
R J Soulen Jr, M S Osofsky, D B Chrisey, J S Horwitz, D Koller, R M Stroud, J Kim, C R Eddy, J M Byers, B F Woodfield, G M Daly, T W Clinton, M Johnson and R C Y Auyeung 789

Plasma modes in superconducting niobium arrays
F Parage and O Buisson 793

Preparation of regular arrays of antidots in YBa_2Cu_3O7 thin films and observation of vortex lattice matching effects
A Castellanos, P Selders, M Vaupel, R Wördenweber, G Ockenfuss, A van der Hart and K Keck 797

Investigation of the magnetic response of disk-and square-shaped superconducting films
T Nurgaliev 801

Ferroelectric properties of $YBa_2Cu_3O_7/BaTiO_3/Au$ heterostructures
C Schwan, J C Martinez and H Adrien 805

Study of superconductor-insulator-normal-insulator-superconductor (SINIS) junctions
R Leoni, M G Castellano, G Torrioli, P Carelli, A Gerardino and F Melchiorri 809

Four-valve pulse tube refrigerator (FVPTR) with non-metallic regenerator material
J Gerster, L Reißig, I Rühlich, M Thürk and P Seidel 813

Demonstration of magenetic levitation using a cryotiger cooler
C E Gough, R Saini, G Walsh, J P G Price, J S Abell and R Claridge 817

VOLUME 2. LARGE SCALE AND POWER APPLICATIONS

Melt-Textured Materials

YBaCuO large scale melt texturing in a temperature gradient
F N Werfel, U Flögel-Delor and D Wippich 821

Investigation of melt cast processed Bi-2212 conductors and their potential for power applications
P F Herrmann, E Beghin, J Bock, C Cottevieille, A Leriche and T Verhage 825

Improved YBCO bulk material for cryomagnetic applications
P Schätzle, G Krabbes, W Bieger, G Stöver, G Fuchs and S Gruss 829

Top-seeded solution growth of $Sm_1Ba_2Cu_3O_{7-\delta}$ seed crystals for melt texturing of $Y_1Ba_2Cu_3O_{7-\delta}$
Ch Krauns, B Bringmann, C Brandt, M Ullrich, K Heinemann and H C Freyhardt 833

Processing technique for fabrication of advanced YBCO bulk materials for industrial applications
A W Kaiser, H J Bornemann and R Koch 837

Simulation of thermal fields for the optimization of the melt-texturing process of HTSC bulk materials
M Seeßelberg, G J Schmitz, T Wilke and M Ullrich 841

Preparation of top seeded $YBa_2Cu_3O_7$ large single-domain tiles
R Yu, V Gomis, S Piñol, F Sandiumenge, J Mora, M Carrera, J Fontcuberta and X Obradors 845

Study of the influence of the processing rate on the directionally solidified YBaCuO
M Boffa, A DiTrolio, U Gambardella, S Pace, M Polichetti and A Vecchione 849

Dislocation configurations in directionally solidified $NdBa_2Cu_3O_7$
N Vilalta, F Sandiumenge, R Yu and X Obradors 853

Directional solidification in air of $NdBa_2Cu_3O_7$ superconductors with high T_c and J_c
B Martínez, F Sandiumenge, R Yu, N Vilalta, T Puig, V Gomis and X Obradors 857

Effect of hot plastic deformation in oxygen on the critical currents of melt-textured Y-123/Y-211
J Krelaus, A Leenders, L-O Kautschor, M Ullrich and H C Freyhardt 861

Influence of Sm contamination on the superconducting properties of $SmBa_2Cu_3O_{7-\delta}$ seeded melt textured $YBa_2Cu_3O_{7-\delta}$
C D Dewhurst, W Lo, Y H Shi and D A Cardwell 865

Dependence of critical current density on the oxygen content for Y-123 single domain melt-textured samples
V A Murashov, J Horvat, M Ionescu, H K Liu and S X Dou 869

Stress-induced superconducting transition in a melt-processed $YBa_2Cu_3O_{7-\delta}$ bulk superconductor
H-G Lee, I-H Kuk and G-W Hong 873

Texturing of thick film REBCO on metallic substrates by isothermal melt processing
N Zafar, J J Wells, A Crossley and J L MacManus-Driscoll 877

Controlled partial melting of dipcoated Bi-2212/Ag ribbons for combined isobaric and isothermal processing in reduced oxygen atmospheres
A L Crossley and J L MacManus-Driscoll 881

Materials Aspects for Large Scale Applications

Synthesis of $YBa_2Cu_3O_{7-x}$ tapes for high current applications by MOCVD
U Schmatz, F Weiss, T von Papen, L Klippe, O Stadel, G Wahl, D Selbmann, M Krellmann, L Hubert-Pfalzgraf, H Guillon, J Peña and M Vallet-Regi 885

Pulsed current and high voltage measurements on sintered and textured ceramics and BiPbSrCaCuO superconductors
J G Noudem, L Porcar, O Belmont, D Bourgault, J M Barbut, J Beille, P Tixador and R Tournier 889

Longitudinal critical current variation in short samples of Bi-2212/Ag and Bi-2223/Ag monocore tapes and its origin
M Polak, A Polyanskii, W Zhang, J Anderson, E Hellstrom and D Larbalestier 893

Synthesis of $Hg_{1-y}Re_yBa_2Ca_2Cu_3O_{8+x}$ pure phase at normal pressures
S Piñol, A Sin, A Calleja, L Fàbrega, J Fontcuberta, X Obradors and M Segarra 897

Thick film Tl-1223 conductors, processing, microstructure and superconducting properties
J C Moore, P Shiles, C J Eastell, C R M Grovenor and M J Goringe 901

Enhanced grain connectivity and transport critical currents in melt-processed (TlBi)-1223 tapes
J Everett, M D Johnston, G K Perkins, A V Volkozub, J C Moore, C J Eastell, S Fox, D Hyland, C R M Grovenor and A D Caplin 905

High critical currents densities in YBCO films on technical substrates
F García-Moreno, A Usoskin, H C Freyhardt, J Wiesmann, J Dzick, J Hoffmann, K Heinemann and A Issaev 909

Critical current densities in heterogeneous superconductors determined from magnetization measurements
M D Sumption and S Takács 913

Angular dependence of critical current versus magnetic field in Nb_3Sn wires at 9–15T
A Godeke, A Nijhuis, H G Knoopers, B ten Haken, H H J ten Kate and P Bruzzone 917

Influence of Cr plating on the coupling loss in cable-in-conduit conductors
A Nijhuis, H H J ten Kate and P Bruzzone 921

Very high currents and irreversibility fields in bulk $NdBa_2Cu_3O_{7-\delta}$ samples
T Wolf, H Küpfer and A-C Bornarel 925

Homogeneity range of $Sm_{1+x}Ba_{2-x}Cu_3O_z$ solid solutions
D I Grigorashev, E A Trofimenko, N N Oleynikov and Yu D Tretyakov 929

Microstructure and superconducting properties of Hg-1223 HTSC with nonstoichiometric additions
M Reder, J Krelaus, K Heinemann, H C Freyhardt, F Ladenberger and E Schwarzmann 933

On the mechanism of (Bi,Pb)-2223 phase formation
V A Murashov, M Ionescu, G E Murashova, S X Dou, H K Liu and M Apperley 937

Formation of the $Bi_2Sr_2Ca_2Cu_3O_{10}$ phase with and without Pb substitution
J-C Grivel, G Grasso, A Erb and R Flükiger 941

Identification of high pressure phase in Bi–Sr–Ca–Cu–O superconductors by HRTEM and XRD
Y H Li and J L MacManus-Driscoll 945

Electrical characteristics of BSCCO-2223 kA class conductors in the temperature range 77–110K
S P Ashworth, M P Chudzik, B A Glowacki and M T Lanagan 949

Optimisation of the precursor powder for Bi-2223/Ag tapes
J Jiang and J S Abell 953

Preparation of Bi2223 precursor powders by sintering in low oxygen partial pressure
B Zeimetz, G E Murashova, H K Liu and S X Dou 957

Investigation of heat treatment and deformation processes of BPSCCO-2223/Ag conductors
T Fahr, A Hütten, W Pitschke, C Rodig, U Schläfer, M Schubert, P Trinks and K Fischer 961

The mechanism of sausaging in Ag/Bi-2223 multifilament-tapes and the effect of processing parameters
S Kautz, B Fischer, J Müller, J Gierl, B Roas, H-W Neumüller and R F Singer 965

The influence of the oxygen partial pressure on the phase evolution in Ag/Bi-2223 multifilament-tapes
J Müller, B Fischer, J Gierl, S Kautz, H-W Neumüller, B Roas and P Herzog 969

The use of the image method for the calculation of the levitation systems with ideal type-II hard superconductors
A A Kordyuk and V V Nemoshkalenko 973

Comparison of the levitation force measurements of YBCO, BSCCO-2212 and (Tl,Pb)BCCO-1223 at different temperatures
J H Albering, S Gauss, C L Teske and J Bock 977

Critical current homogeneity of Bi-2223 tapes determined by Hall-magnetometry
H-P Schiller, K Grube, B Gemeinder, H Reiner and W Schauer 981

Construction of the current-voltage characteristic in a 12 decade voltage window using magnetisation measurements
J Vanacken, K Rosseel, A S Lagutin, L Trappeniers, M Van Bael, D Dierickx, J Meersschaut, W Boon, F Herlach, V V Moshchalkov and Y Bruynseraede 985

Granularity of the critical current pattern and degradation of macroscopic material properties due to mesoscopic material inhomogeneities
Th Klupsch, P Diko, C Wende, I Sonntag and W Gawalek 989

The peak in the imaginary part of AC susceptibility for a superconductor at different magnetic fields and frequencies
S Takács and F Gömöry 993

Large-area deposition of biaxially textured YSZ buffer layers by using an IBAD technique
J Wiesmann, J Dzick, J Hoffmann, K Heinemann and H C Freyhardt 997

Preparation of biaxially textured YSZ buffer layers on cylindrically curved substrates
J Dzick, J Wiesmann, J Hoffmann, K Heinemann, A Usoskin, F García-Moreno, A Isaev and H C Freyhardt 1001

Critical current of Gd doped $PbMo_6S_8$ synthesised at 1550°C
D N Zheng, A B Sneary and D P Hampshire 1005

Hall probe measurements and analysis of the magnetic field above a multifilamentary superconducting tape
E C L Chesneau, J Kvitkovic, B A Glowacki, M Majoros and P Haldar 1009

Selection of the offset-criterion voltage parameter and its relation to the second differential of a superconductor's voltage-current curve
D W A Willén, W Zhu and J R Cave 1013

Current-voltage characteristics in YBaCuO thin films over more than 13 decades of electric-fields
T Nakamura, Y Hanayama, T Kiss, V S Vysotsky, H Okamoto, T Matsushita, M Takeo, F Irie and K Yamafuji 1017

A modified flux-creep picture
R De Luca, T Di Matteo, S Pace and M Polichetti 1021

Progress in preparing $Tl/Pb(Ba/Sr)_2Ca_2Cu_3O_{8.25+x}$ (Tl-1223) bulk and powder samples in an open system
C L Teske 1025

Flux composition influence at very high temperature on the formation of NdBaCuO materials under quenching
F Auguste, M Liégeois, A Rulmont and R Cloots 1029

Large grained YBCO monoliths for magnetic bearings of a LH_2 tank
M Ullrich, A Leenders, H C Freyhardt, J H Albering and S Gauss 1033

Analysis and modelling of oxygen diffusion in $YBa_2Cu_2O_{7-\delta}$ under non-isothermal conditions
M D Vázquez-Navarro, A Kuršumovic, C Chen and J E Evetts 1037

Materials aspects of substituted Hg-1223 superconductors
S Lee, N Kiryakov, M Kuznetsov, D Emelyanov and Yu D Tretyakov 1041

Rising of the irreversibility line in Re-subsituted Hg-1223 High T_c superconductors
L Fàbrega, B Martínez, A Sin, S Piñol, J Fontcuberta and X Obradors 1045

Sequential processing of composite reaction textured (CRT) bulk $Bi_2Sr_2CaCu_2O_{8+\delta}$ components
P Kosmetatos, A Kuršumovic, D R Watson, R P Baranowski and J E Evetts 1049

Electrical and structural properties of Ag-doped (BiPb)SrCaCuO superconducting system
A Mariño, C E Rojas, H Sánchez and J E Rodríguez 1053

Low field relaxation measurements of induced magnetic moment in Bi(2223)/Ag tape and Bi(2212) single crystal
A Yu Galkin, M Jirsa, L Pust, P Nálevka, M R Koblischka, V M Pan and R Flükiger 1057

Anisotropic magnetoresistance in optimally doped and overdoped $Bi_2Sr_2CaCu_2O_{8+x}$ single crystals
G Heine, W Lang, X L Wang and X Z Wang 1061

Magneto-optic visualization of flux penetration in Bi-2223 tapes after bending
M R Koblischka, T H Johansen and H Bratsberg 1065

Influence of uniaxial pressing on the microstructure and the critical current density of Tl-1223 tapes
T Riepl, S Zachmayer, R Löw, C Reimann, T Schauer and K F Renk 1069

The deposition of oxide buffer layers using a novel laser ablation source
T J Jackson, C H Wang, B A Glowacki, J A Leake, R E Somekh and J E Evetts 1073

Deposition of YSZ, CeO_2, and MgO on amorphous and polycrystalline substrates
M Bauer, J Schwachulla, S Furtner, P Berberich and H Kinder 1077

Growth mechanisms for in-plane aligned YSZ buffer layers deposited on polycrystalline metallic substrates by ion-beam assisted laser deposition
V Betz, B Holzapfel, G Sipos, W Schmidt, N Mattern and L Schultz 1081

Observation and investigations of a nucleation layer in high-J_c YBCO films deposited on polycrystalline substrates
A Usoskin, F García-Moreno, H C Freyhardt, D Jockel, J Wiesmann, K Heinemann, A Issaev, J Dzick and J Hoffmann 1085

Biaxial alignment of YSZ buffer layers on inclined technical substrates for $YBa_2Cu_3O_{7-x}$ tapes
W A J Quinton and F Baudenbacher 1089

Influence of external strains on J_c of YBCO films on thin technical substrates
F García-Moreno, A Usoskin, H C Freyhardt, J Wiesmann, J Dzick, K Heinemann and J Hoffmann 1093

Single source MOCVD of HTSC films onto travelling substrates
O Stadel, L Klippe, G Wahl, S V Samoylenkov, O Y Gorbenko and A R Kaul 1097

Flux Pinning

Critical current density and irreversibility line of 2223 BSCCO tapes enhanced by columnar defects along a part of the tape thickness
R Gerbaldo, G Ghigo, L Gozzelino, E Mezzetti and B Minetti 1101

Critical currents in neutron irradiated Bi- and Tl-based tapes
G W Schulz, C Klein, H W Weber, H-W Neumüller, R E Gladyshevskii and R Flükiger 1105

On the origin of the so-called fishtail effect in single crystals of the RE-123 compounds (RE = Y, Er, Nd)
A Erb, J-Y Genoud, M Dhalle, F Marti, E Walker and R Flükiger 1109

Evidence of a mixed state transition region between the vortex solid and liquid phases in $YBa_2Cu_3O_{7-\delta}$ single crystals
S N Gordeev, A P Rassau, D Bracanovic, P A J de Groot, R Gagnon and L Taillefer 1113

Optimized T_c, B_{irr} and J_c for substituted Y-123 materials
B Dabrowski, K Rogacki, O Chmaissen, J D Jorgensen, J W Koenitzer and K R Poeppelmeier 1117

Current-voltage characteristics in a mixed state of high T_c superconductor
T Kiss, T Nakamura, K Hasegawa, M Inoue, M Takeo, F Irie and K Yamafuji 1121

Flux pinning and critical currents in weakly coupled granular superconductors
A Tuohimaa and J Paasi 1125

The shape-effect and vortex-lattice phase transitions in a BSCCO single crystal
T B Doyle, R Labusch, R A Doyle, T Tamegai and S Ooi 1129

Vortex-lattice melting in high-T_c superconductors
T B Doyle, R Labusch and R A Doyle 1133

Flux pinning and grain coupling in $Bi_2Sr_2CaCu_2O_{8+\delta}$ ceramics: magneto-optical investigations
S-L Huang, M R Koblischka, T H Johansen, H Bratsberg and K Fossheim 1137

Pinning in bulk high-T_c superconductors
M R Koblischka 1141

Scaling of current densities and pinning forces in $NdBa_2Cu_3O_{7-\delta}$
M R Koblischka, A J J van Dalen, T Higuchi, K Sawada, H Kojo, S I Yoo and M Murakami 1145

The influence of Ag doping on the superconducting properties of grain-boundary weak-link in YBCO
A L L Jarvis and T B Doyle 1149

Flux-line localisation by means of segments of correlated defects in YBCO melt-textured
R Gerbaldo, G Ghigo, L Gozzelino, E Mezzetti, B Minetti, R Cherubini and A Wisniewski 1153

Comparative study of J_c, flux pinning, and the influence of irradiation in Bi-2212 ribbon and single crystals
A L Crossley, J Everett, G Wirth, K Kadowaki, C Morgan, C Eastell, C R M Grovenor and A D Caplin 1157

Detailed magnetisation study of inter- and intragranular currents in Ag-sheathed Bi-2223 tape
P Nálevka, M Jirsa, L Pust, A Yu Galkin, M R Koblischka and R Flükiger 1161

Grain connectivity and flux pinning for Bi-2223/Ag tapes obtained by oxide-powder-in-tube method
J Horvat, Y C Guo, R Bhasale, W G Wang, H K Liu and S X Dou 1165

An investigation of the peak effect in the Chevrel phase superconductor tin molybdenum sulphide
I J Daniel, D N Zheng and D P Hampshire 1169

Current Limiters and Cables

Test of 1.2 MVA high-T_c superconducting fault current limiter
W Paul, M Lakner, J Rhyner, P Unternährer, Th Baumann, M Chen, L Widenhorn and A Guérig 1173

Performance of BSCCO cylinders in a prototype of inductive fault current limiter
V Meerovich, V Sokolovsky, J Bock, S Gauss, S Goren and G Jung 1179

Experiments with a 6.6kV/1kA single-phase superconducting fault current limiter
T Yazawa, E Yoneda, S Nomura, K Tsurunaga, M Urata, T Ohkuma, S Honjo, Y Iwata and T Hara 1183

Determination of time constants by direct measurement of magnetic field above superconducting cables
S Takács and N Yanagi 1187

Development of HTS tapes and multistrand conductors for power transmission cables
M Leghissa, J Rieger, J Wiezoreck, H-P Krämer, B Roas, B Fischer, K Fischer and H-W Neumüller 1191

Influence of heating pulses on the quench behavior of a 12 strand Nb_3Sn CICC
V S Vysotsky, M Takayasu, P C Michaels, J H Schultz, J V Minervini, S Jeong and V V Vysotskaia 1195

Transport current distribution in core of multilayer high-T_c superconducting power cable
P I Dolgosheev, V E Sytnikov, G G Svalov, N V Polyakova and D I Belyi 1199

Evaluation of HTS samples for 12.5 kA current leads
M Teng, A Ballarino, R Herzog, A IJspeert, C Timlin, S Harrison and K Smith 1203

Development of a thallium cuprate current lead
D M Pooke, A Mawdsley, J L Tallon and R G Buckley 1207

Fabrication issues in Bi-2212 polycrystalline textured thin rods for current leads
L A Angurel, J C Díez, H Miao, E Martínez, G F de la Fuente and R Navarro 1211

Vapour cooled high T_c superconducting current leads made of Ag sheathed Bi 2223 tapes
B Zeimetz, H K Liu and S X Dou 1215

Development of high-T_c Bi-superconducting current leads for Maglev
E Suzuki and M Kurihara 1219

Properties of Bi (2223)/Ag Au multifilamentary tapes for current leads
W Goldacker, B Ullmann, A Gäbler and R Heller 1223

Application of inductive HTSC current limiters in distribution networks
V Meerovich, V Sokolovsky, S Goren, G Jung, I Vajda, A Szalay and N Gobl 1227

Transition properties of HTS inductive fault current limiter
M Majoros, L Jansak, S Sello and S Zannella 1231

Development of high T_c superconducting elements for a novel design of fault current limiter
A T Rowley, F C R Wroe, M P Saravolac, K Tekletsadki, J Hancox, D R Watson, J E Evetts, A Kuršomovic and A M Campbell 1235

Development of an inductive high-T_c superconducting fault current limiter model
S Zannella, A Arienti, G Giunchi, L Jansak, M Majoros and V Ottoboni 1239

Resistive current limiters with YBCO films
B Gromoll, G Ries, W Schmidt, H-P Krämer, B Seebacher, P Kummeth, S Fischer and H-W Neumüller 1243

Preparation of switching elements for a resistive type HTS fault current limiter
W Schmidt, P Kummeth, R Nies, R Schmid, B Seebacher, H-P Steinrück and H-W Neumüller 1247

Wires and Tapes

Development of Ag sheathed Bi(2223) tapes for power applications with improved microstructure and homogeneity
G Grasso, F Marti, Y Huang, R Passerini and R Flükiger 1251

The effect of silver-alloy sheaths on fabrication, microstructure and critical current density of powder-in-tube processed multifilamentary Bi-(2223) tapes
A Hütten, M Schubert, C Rodig, U Schläfer, P Verges and K Fischer 1255

Critical transport currents of Bi(2223)/Ag tapes under axial and bending strain
B Ullmann, A Gäbler, M Quilitz and W Goldacker 1259

Current distributions in multi-filamentary HTS conductors
A V Volkozub, A D Caplin, H Eckelmann, M Quilitz, R Flükiger, W Goldacker, G Grasso and M D Johnston 1263

Progress in HTS wire and applications development
R Schöttler, G Papst and J Kellers 1267

Third round of the ITER strand bench mark test
H G Knoopers, A Nijhuis, E J G Krooshoop, H H J ten Kate, P Bruzzone, P J Lee and A A Squitieri 1271

The role of vortex melting and inhomogeneities in the transport properties of Nb_3Sn superconducting wires
N Cheggour and D P Hampshire 1275

Biaxially textured substrate tapes of Cu, Ni, alloyed Ni, (Ag) for YBCO films
W Goldacker, B Ullmann, E Brecht and G Linker 1279

Preparation of textured Tl(1223)/Ag superconducting tapes
E Bellingeri, R E Gladyshevskii and R Flükiger 1283

Critical current anisotropy minimum of Tl-1223 superconductors
B A Glowacki 1287

Mechanical endurance of 2223-BSCCO tapes under tensile stress and strain
G C Montanari, I Ghinello, L Gherardi, R Mele and P Caracino 1291

Transverse pressure induced reduction of the critical current in BSCCO/Ag tapes
B ten Haken, A Beuink and H H J ten Kate 1295

Zero field critical current of a 150 m superconducting tape measured in a pancake coil configuration
P Bodin, Z Han, P Vase, M D Bentzon, P Skov-Hansen, R Bruun and J Goul 1299

Limits of application of the BSCCO conductors
B Lehndorff, M Hortig, H-G Kürschner, M Polak, R Wilberg, D Wehler and H Piel 1303

Observation of self-field distribution due to transport currents in Ag-sheathed Bi2223 monofilamentary and multifilamentary tapes
K Kawano and A Oota 1307

Testing the homogeneity of Bi(2223)/Ag tapes by a Hall probe array
P Kováč, V Campbell, D Gregušová, P Eliáš, I Hušek, R Kúdela, S Hasenöhrl and M Ďurica 1311

Hall probe measurements of the magnetic field above current carrying Bi-2223/Ag tapes
J Kvitkovic, E C L Chesneau, M Majoros, B A Glowacki, S P Ashworth, M Ciszek, A M Campbell and J E Evetts 1315

Magnetic flux mapping and current distribution in BSCCO tapes
H-G Kürschner, M Hortig, B Lehndorff and H Piel 1319

Influence of processing and composition on the formation of (Bi,Pb)-2223 tapes
P Haug, D Göhring, M Vogt, A Trautner, W Wischert and S Kemmler-Sack 1323

Multifilamentary composite tapes and round wires based on BiPbSrCaCuO
A D Nikulin, A K Shikov, I I Akimov, F V Popov, D N Rakov, D A Filichev and N I Kozlenkova 1327

Processing and superconducting properties of Ag-Ti-Cu alloy-sheathed BiSrCaCuO tapes
H Miao, M Artigas, G F de la Fuente, F Iriarte and R Navarro 1331

Fabrication and properties of extruded Ag-Mg-Ni sheathed Bi-2223 tapes
L Martini, L Bigoni, F Curcio, R Flükiger, G Grasso, M Migliazza, E Varesi and S Zannella 1335

Bi-2223/Ag- and /Ag-alloy tapes: fabrication and physical properties
T Arndt, B Fischer, H Krauth, M Munz, B Roas and A Szulczyk 1339

Optimization of rolling process for multicore Bi(2223)/Ag tapes made by OPIT technique
P Kováč, I Hušek, L Kopera and W Pachla 1343

Bi 2212 tapes prepared by alternating electro-deposition and heat treatments
F Legendre, P Gendre, L Schmirgeld-Mignot and P Régnier 1347

Fabrication of thin Ag/Bi-2223 mono-core tapes with an overall tape thickness of 40 μm suitable for dissipative power applications
D W A Willén, W Zhu and J R Cave 1351

Evolution of core texturing in the process of Ag/BiSCCO tapes and wires fabrication by OPIT
A Goldgirsh, V Beilin, E Yashchin and M Schieber 1355

The effect of heating rate on Bi-2223 phase formation in composite conductors
A D Nikulin, A K Shikov, D N Rakov, Yu N Belotelova, V E Klepatsky and I I Akimov 1359

Annealing of Ag-clad BiSCCO tapes studied in-situ by high-energy synchrotron x-ray diffraction
T Frello, H F Poulsen, N H Andersen, A Abrahamsen, S Garbe, M D Bentzon and M von Zimmermann 1363

Comparative studies on Bi(2223) phase at the Ag interface and inside the ceramic core in Ag-sheathed composite tapes
W Pachla, H Odelius, U Södervall, P Kováč, I Hušek, H Marciniak and M Wróblewski 1367

AC-Losses

AC losses in Bi-2223 tapes for power applications
L Gherardi, F Gömöry, R Mele and G Coletta 1371

The null calorimetric ac losses measurement method: present state and results
P Dolez, M Aubin, D W A Willén, W Zhu and J R Cave 1377

A new approach to AC loss measurement in HTS conductors
N Chakraborty, A V Volkozub and A D Caplin 1381

Low ac losses in Bi(2223) tapes with oxide barrier
Y B Huang, G Grasso, F Marti, M Dhallé, G Witz, S Clerc, K Kwasnitza and R Flükiger 1385

AC losses of twisted high-T_c superconducting multifilament Bi2223 tapes with a mixed matrix of Ag and $BaZrO_3$
K Kwasnitza, St Clerc, R Flükiger, Y B Huang and G Grasso 1389

Self-field ac losses in dc external fields and non-uniform transverse J_c distribution in Ag sheathed multifilamentary PbBi-2223 tapes
T J Hughes, Y Yang, C Beduz and P Haldar 1393

AC losses in Y-123/hastelloy superconducting tapes
K-H Müller 1397

Computational study of magnetisation losses in HTS composite conductors having different twist-pitch lengths
M Lahtinen, J Paasi, R Mikkonen and J-T Eriksson 1401

Superconducting cables in spatially changing magnetic fields near the end portions of magnetic systems
S Takács 1405

Preisach-type hysteresis modelling in Bi-2223 tapes
D Djukic, M Sjöström and B Dutoit 1409

Numerical calculation of ac losses, field and current distributions for specimens with demagnetisation effects
M Däumling 1413

Self field AC loss measurements of multifilamentary Bi(2223) tapes
B Dutoit, N Nibbio, G Grasso and R Flükiger 1417

Factors affecting the accuracy of transport ac loss measurements on HTSC tapes
H Understrup and P Vase 1421

AC magnetic losses in multifilamentary Ag/Bi-2223 tape carrying DC transport current
M Ciszek, B A Glowacki, S P Ashworth, E C L Chesneau, A M Campbell and J E Evetts 1425

Critical current and self-field loss of BSCCO-2223/Ag tape in bifilar geometry
Y K Huang, J J Rabbers, O A Shevchenko, B ten Haken and H H J ten Kate 1429

Investigations of the AC current loss of twisted and untwisted multifilamentary Ag/ AgMg/ and AgAu/ Bi(2223) tapes
H Eckelmann, M Däumling, M Quilitz and W Goldacker 1433

The influence of the silver sheathed Bi-2223 conductor architecture on the transport ac losses
B A Glowacki, K G Sandemann, E C L Chesneau, M Ciszek, S P Ashworth, A M Campbell and J E Evetts 1437

AC-loss measurements on HTSC cable conductor with transposed BSCCO-tapes
C Rasmussen and S K Olsen 1441

Hysteresis phenomena in the I-V characteristics of Bi(2223)/Ag tapes at 4.2 K
P Kováč, L Cesnak, T Melišek, H Kirchmayr and H Fikis 1445

Magnetic AC loss in multi-filamentary Bi-2223/Ag tapes
M P Oomen, J Rieger and M Leghissa 1449

A novel way of measuring hysteretic losses in multi-filamentary superconductors
M Dhallé, F Marti, G Grasso, Y B Huang, A D Caplin and R Flükiger 1453

Measurements of AC-field losses (50 Hz) on Ag-sheathed $(Bi,Pb)_2Sr_2Ca_2Cu_3O_{10}$ tapes
A Kasztler, H Fikis, P Bauer and H Kirchmayr 1457

AC losses in multifilamentary BiSrCaCuO-2223/Ag tapes studied by transport and magnetic measurements
F Gömöry, L Gherardi, R Mele, D Morin and G Crotti 1461

Comparison of transport and magnetic AC losses in HTS tapes
S P Ashworth, M Ciszek, B A Glowacki, A M Campbell and J E Evetts 1465

Calorimetric AC loss measuring system for HTS carrying AC transport currents
N Magnusson and S Hörnfeldt 1469

Current redistribution and quench development in a multistrand superconducting cable for AC applications at commercial frequency
V S Vysotsky, K Funaki, H Tomiya, M Nakamura and M Takeo 1473

Fusion, SMES, Accelerators and Detectors

Superconductors for fusion magnets: realizations, prototypes and running projects
P Bruzzone 1477

Design considerations of a HTS μ-SMES
R Mikkonen, M Lahtinen, J Lehtonen, J Paasi, B Connor and S S Kalsi 1483

Operation of a small SMES power compensator
K P Juengst, H Salbert and O Simon 1487

Superconducting magnet design issues for a muon collider ring
M A Green 1491

Status of the EU conductor manufacture for the ITER CS and TF model coils
S Conti, S Garrè, S Rossi, A Della Corte, M V Ricci, M Spadoni, G Bevilacqua, R Maix, E Salpietro, H Krauth, A Szulczyk, M Thoener, A Laurenti and C Siano 1495

Ramp rate experiments on an ITER relevant pulsed coil
E P Balsamo, O Cicchelli, M Cuomo, P Gislon, G Pasotti, M V Ricci and M Spadoni 1499

Preliminary test results of a 13 tesla niobium tin dipole
R M Scanlan, R J Benjegerdes, P A Bish, S Caspi, K Chow, D Dell'Orco, D R Diederich, M A Green, R Hannaford, W Harnden, H C Higley, A F Lietzke, A D McInturff, L Morrison, M E Morrison, C E Taylor and J M Van Oort 1503

Motors and Generators

Development of synchronous motors and generators with HTS field windings
G Papst, B B Gamble, A J Rodenbush and R Schöttler 1507

Experimental investigation of a model HTSC alternator
I I Akimov, L I Chubraeva, D A Filichev, L M Fisher, A V Kalinov, D V Sirotko, A K Shikov, V M Soukhov and Igor F Voloshin 1511

Hysteresis electrical motors with bulk melt textured YBCO
L K Kovalev, K V Ilushin, V T Penkin, K L Kovalev, V S Semenikhin, V N Poltavets, A E Larionoff, W Gawalek, T Habisreuther, T Strasser, A K Shikov, E G Kazakov and V V Alexandrov 1515

Prospects for brushless AC machines with HTS rotors
M D McCulloch, K Jim, Y Kawai and D Dew-Hughes 1519

Dynamics of a rotor levitated above a high-T_c superconductor
T Sugiura, Y Uematsu and T Aoyagi 1523

Hysteresis motor with self sustained rotor for high speed applications
X Granados, I Marquez, J Mora, J Fontcuberta, X Obradors, J Pallares and R Bosch 1527

Superconducting bearings in high-speed rotating machinery
T A Coombs, A M Campbell, I Ganney, W Lo, T Twardowski and B Dawson 1531

Superconducting magnetic bearings in a high speed motor
P Stoye, W Gawalek, P Görnert, A Gladun and G Fuchs 1535

Electromagnetic processes in a disc-type cryoalternator
L I Chubraeva and D V Tchoubraev 1539

System Aspects for Large Scale Applications and High Magnetic Fields

Fabrication and transport properties of Bi-2212/Ag multifilamentary coils for high magnetic field generation (II); high field performance test
H Kitaguchi, H Kumakura, K Togano, M Okada, K Tanaka, K Fukushima, K Nomura and J Sato 1543

Behavior of the magnetic flux within an HTS magnetic shielded cylinder for measuring the biomagnetic field
M Itoh, K Mori, S Yoshizawa and S Haseyama 1547

Normal zone propagation studies on a single pancake coil of multifilamentary BSCCO-2223 tape operating at 65K
M Penny, C Beduz, Y Yang, S Manton and R Wroe 1551

Loss reduction in a high quality superconducting coil
O A Shevchenko, H G Knoopers and H H J ten Kate 1555

3-dimensional levitation force measurements in the 15–85K range
S Gauss and J H Albering 1559

The design and fabrication of a 6 tesla EBIT solenoid
M A Green, S M Dardin, R E Marrs, E Magee and S K Mukhergee 1563

Ion-beam switching magnet with HTS coils
D M Pooke, J L Tallon, S S Kalsi, A Szczepanowski, G Snitchler, H Picard, R E Schwall, R Neale, B MacKinnon and R G Buckley 1567

Nonlinear flux diffusion in superimposed ac magnetic field
A L Kasatkin, V M Pan and V V Vysotskii 1571

Trapped field magnets from melt textured YBCO samples
G Fuchs, S Gruss, G Krabbes, P Schätzle, K-H Müller, J Fink and L Schultz 1575

Locally resolved trapped field measurements of melt textured YBCO at low temperatures
T Straßer, T Habisreuther, B Jung, D Litzkendorf, M Wu and W Gawalek 1579

Penetration of magnetic field into melt-textured YBCO-samples
Ch Wenger, G Fuchs, G Krabbes, Th Staiger and A Gladun 1583

Influence of processing parameters on critical currents and irreversibility fields of fast melt processed $YBa_2Cu_3O_7$ with Y_2BaCuO_5 inclusions
K Rosseel, D Dierickx, J Vanacken, L Trappeniers, W Boon, F Herlach, V V Moshchalkov and Y Bruynseraede 1587

Influence of columnar defects on critical currents and irreversibility fields in $(Y_xTm_{1-x})Ba_2Cu_3O_7$ single crystals
L Trappeniers, J Vanacken, K Rosseel, A Yu Didyk, I N Goncharov, L I Leonyuk, W Boon, F Herlach, V V Moshchalkov and Y Bruynseraede 1591

Observation of local variations of stress in fast melt processed $YBa_2Cu_3O_7$ superconductors at Y_2BaCuO_5 inclusions
R Provoost, K Rosseel, D Dierickx, W Boon, V V Moshchalkov, R E Silverans and Y Bruynseraede 1595

Fabrication and transport properties of Bi-2212/Ag multifilamentary tapes and coils for high magnetic field generation
M Okada, K Fukushima, J Sato, K Nomura, H Kitaguchi, H Kumakura, T Kiyoshi, K Togano and H Wada 1599

Fabrication and properties of pancake coils using Bi-2223/Ag tapes
P Verges, K Fischer, A Hütten, T Staiger and G Fuchs 1603

High field performance of Nb_3Al multifilamentary conductors prepared by phase transformation from bcc solid solution
M Kosuge, T Takeuchi, M Yuyama, Y Iijima, K Inoue, H Wada, K Fukuda, G Iwaki, S Sakai, H Moriai, B ten Haken and H H J ten Kate 1607

Some factors affecting the use of internal-tin Nb_3Sn for high field dipole magnets and fusion applications
E Gregory, E Gulko, T Pyon and D Dietderich 1611

Superconducting windings with 'short-circuited' turns
A V Dudarev, A V Gavrilin, Yu A Ilyin, V E Keilin, N Ph Kopeikin, V I Shcherbakov, I O Shugaev and V V Stepanov 1615

Modelling of current transfer in current contacts to anisotropic bulk superconductors
R P Baranowski, D R Watson and J E Evetts 1619

Power dissipation at high current CRT Bi-2212 SC-Ag contacts
A Kuršumovic, J E Evetts, D R Watson, B A Glowacki, S P Ashworth and A M Campbell 1623

Improvements of electric metal superconductor contacts in Bi-2212 polycrystalline textured materials
J M Mayor, L A Angurel, R Navarro and L García-Tabarés 1627

A fast and non-destructive procedure to determine the transport properties of BSCCO/Ag superconducting tapes
J-J Rabbers, W F A Klein Zeggelink, B ten Haken and H H J ten Kate 1631

Contact less characterisation of BSCCO tape
M D Bentzon, D Suchon, P Bodin and P Vase 1635

Computation of critical current through magnetic flux profile measurements
J Amorós, M Carrera, X Granados, J Fontcuberta and X Obradors 1639

Levitation of a small permanent magnet between superconducting $YBa_2Cu_3O_{7-\delta}$ thin films
R Grosser, A Martin, O Kus, E V Pechen and W Schoepe 1643

Modelling of $YBa_2Cu_3O_7$ superconductors for magnetic levitation applications
E Portabella, J Mora, B Martínez, J Fontcuberta and X Obradors 1647

A hybrid magnetic shield employing ferromagnetic iron and HTSC rings
V Meerovich, V Sokolovsky, S Goren, G Jung, G Shter and G S Grader 1651

Possibilities to increase the critical current of solenoids made of anisotropic HTS tapes
J Pitel and P Kováč 1655

Special features of HTS magnet design
J Paasi, M Lahtinen, J Lehtonen, R Mikkonen, B Connor and S S Kalsi 1659

Choice of design margins of superconducting magnets
V E Keilin 1663

Other Large Scale Applications

Simple model and optimization of HTSC bearing
V V Vysotskii and V M Pan 1667

Numerical modelling of high temperature superconducting bearings
F Negrini, P L Ribani, E Varesi and S Zannella 1671

A new type passive magnetic bearing based on high-temperature superconductivity
A V Filatov, O L Poluschenko and L K Kovalev 1675

Magnetic shielding effects found in the superposition of a ferro-magnetic cylinder over a BPSCCO cylinder: Influence of the air gap between the BPSCCO and ferromagnetic cylinders
K Mori, M Itoh and T Minemoto 1679

Liquid phase sintered YBCO hollow cylinders for magnetic shielding
J Plewa, W Jaszczuk, C Seega, C Magerkurth, E Kiefer and H Altenburg 1683

Switching properties of high quality superconductors
O A Shevchenko, H J G Krooshoop and H H J ten Kate 1687

Special Topics

The life and times of Heike Kamerlingh Onnes
R J Soulen Jr 1691

Author index 1697

Inst. Phys. Conf. Ser. No 158
Paper presented at Applied Superconductivity, The Netherlands, 30 June–3 July 1997

YBaCuO Large Scale Melt Texturing in a Temperature Gradient

F.N. Werfel, U. Flögel-Delor, D. Wippich

Adelwitz Technologiezentrum GmbH, Rittergut Adelwitz, D-04886 Adelwitz; Germany

Abstract . To reach an adequate level of prototype device development HTSC bulk materials have to be fabricated into desirable configurations and processed to exhibit current-carrying capabilities of > 30 kA/cm² (77 K, 0 T) as well as mechanical, thermal and electrical stability. YBaCuO blocks, rings and cylinders up to 6 inch diameter were grown in a modified Bridgman furnace using local temperature gradients (ceramo crystal growth - ccg). Without seeding the method is inherently industrial-like exhibiting a well-textured microstructure with cm-size quasicrystals and reduced grain boundary effects. The availability of large monolithic superconductors up to 150 mm diameter in a complicated geometrical shape and a grain alignment optimised for the device geometry lend promise for medium / large scale application as superconducting magnetic bearings (SMB) for high speed rotating machines.

1. Introduction

Melt texture processing which is based on a directional solidification has been most successful avoiding the weak-link problem in bulk YBaCuO superconductors [1,2]. The processing involves a decomposition upon heating and a peritectic-like transformation during cooling: Y123 <------> L035 + Y211 + O_2 . The choice of the temperature profile used in the melt-texturing has a substantial influence on the microstructure, such as the size and the orientation of the 123 domains, the amount and the size of the secondary Y211 phase, trapped CuO and $BaCuO_2$ impurities at domain boundaries, porosity and cracking. A low cooling rate of 1-3 K/h is common to all various methods, to obtain large domains of aligned grains in the sample. Y123 nucleates from the liquid phase directly by a small undercooling below the peritectic temperature of about 1020^0 C. The addition of Ag_2O lowers the nucleation point.
Numerous attempts has been made to analyse and improve the process parameters and materials properties. One of the most prominent technique utilise the top seed nucleation growth of Y123 with seeds of Sm123, Al_2O_3 or MgO. Large domains up to 3-4 cm in size are obtainable [3,4]. The standard levitation against a 0.4 T permanent magnet (∅ 25 mm) with about 80 N reaches 95 % of the theoretical maximum and a 3 cm single domain sample at 77 K can trap more than 600 mT .

However the achievement of uniform alignment over long distances of more than 10 cm has been a problem. Moreover, the domain boundary between seeded grains seem to show strong weak-link behaviour with a high barrier for transport currents.
Consequently the excellent high critical current values of $3\text{-}5\times10^4$ A/cm² of small samples has not been able to transfer to large extensions or even more complicated geometrical structures. Utilising a melt texture ceramo crystal growth (ccg) gradient method we demonstrate that high current YBaCuO monolithic devices for medium /large scale applications can be fabricated. The produced YBaCuO rings and cylinders are stator elements of a superconducting magnetic bearing test stand.

The authors gratefully acknowledge the cooperation with IPHT Jena, FZ Karlsruhe, IFW Dresden, FH Steinfurt

2. Experimental

$Y_{1.4-1.6}Ba_2Cu_3O_{7-\delta}$ precursor powders with 5 wt.% Ag and 0.4 wt.% Pt were prepared by mixing and compacting (2-3 kbar) in blocks, rods, rings and cylinders up to diameters of 15 cm. The green bodies are processed in a modified Bridgman furnace generating different local temperature gradients and fields. The individual processing steps are shown in Fig. 1.

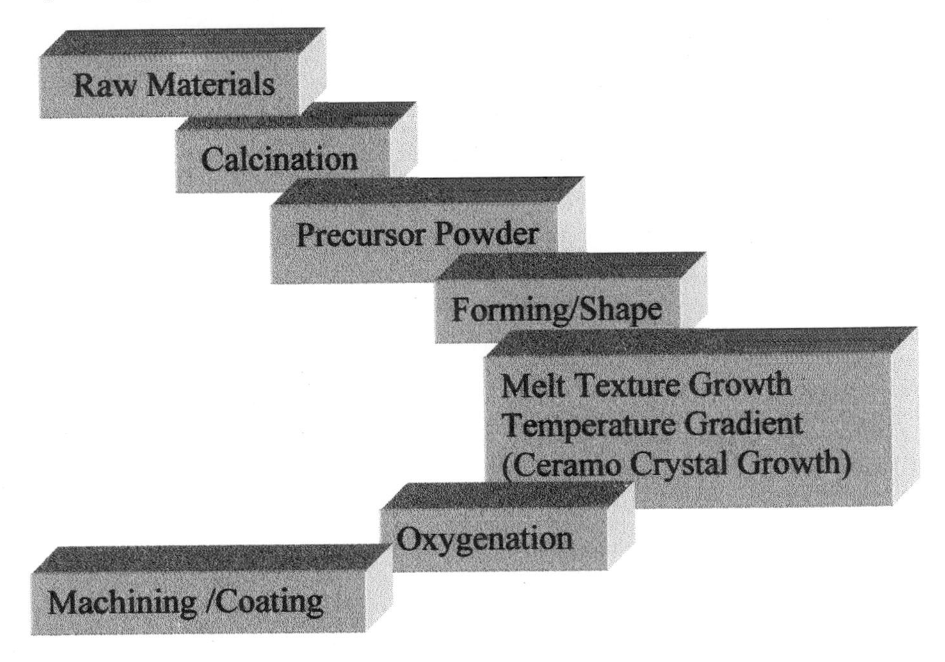

Figure1. Melt texturing ceramo crystal growth (ccg) processing steps

The melt texturing ccg process started with the decomposition at 1100^{0}C and a fast undercooling at the peritectic temperature T_p followed by a cooling down with 1- 3 K / h. The choice of the high temperature level for melting and the duration requires a sensitive adjustment between the heat transport in the compacted bodies maintaining the mechanical stability and the undesired accelerated growing of the Y_2BaCuO_5 (211) phase. An oxygenation at a temperature of 400 - 500 ^{0}C in flowing O_2 for 60-120 hours dependent on the sample size completes the growing procedure.

The key of ccg method is an interplay of the local temperature gradients of typical 60 K/cm and the time dependent cooling down gradient. In a first approximation the cooling rate is determined by the product of the (time dependent) temperature gradient and the growth rate. Because the growing is diffusion determined by a very low kinetics the cooling rate has to be low to maximise the 123 domain formation. When Y123 nucleates from 211 and the liquid the latent heat of the fusion at the liquidus-solidus transformation interface will be removed by convection or heat conduction through the solidified 123. The heat removal can be accelerated or suppressed by a corresponding negative or positive temperature gradient. Unfortunately, the estimated temperature field of the empty furnace at different heater functions will be changed even qualitativèly after furnace loading with the compacted green YBaCuO bodies. Therefore, the amount and position of large monolithic superconductors relative each to another inside the furnace determine the gradient program.

Usually the growing strategy of radial symmetric devices tries to prevent the nucleation of the grains at the sample periphery growing under an statistical angle distribution toward the axis. One way to grow parallel to the sample surface e. g. perpendicular to the sample axis is a high local temperature gradient ratio axial to radial due to the different heater systems. The application of very high temperature gradients in the 100 K/cm region however causes microcracking. Within our growing furnace we have been maintained multidimensional temperature fields giving a cylinder-like texture of the a-b planes (Fig.2).

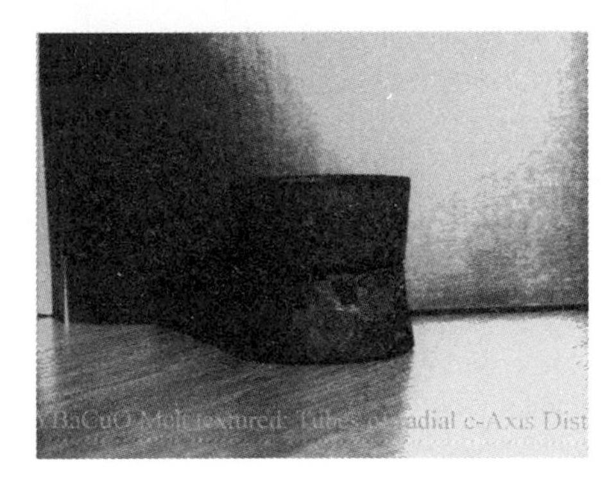

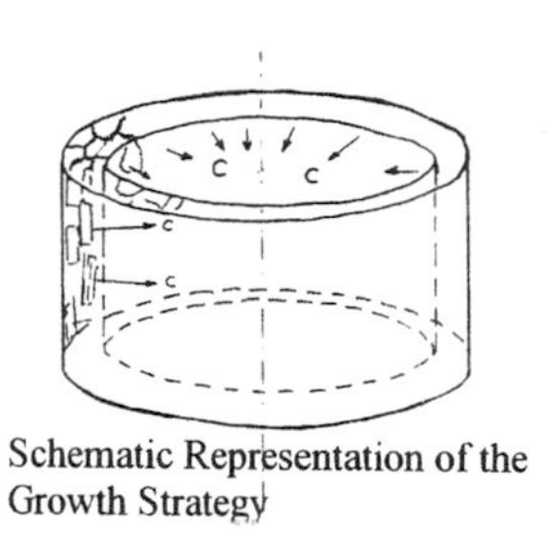

Figure 2. Cylinder-like melt texturing of YBaCuO radial c axes distribution

For the radial geometry texture, a strong positive axial and a reversed radial temperature gradient with respect to the assumed nucleation plane has been applied, setting the hot zone temperature to approximately 1000^0C with a axial temperature gradient of about 60^0 C/cm and a radial gradient of - 20^0C/cm.

3. Results

Using the above described melt textured processing route the materials properties of **large** YBaCuO devices for industrial applications has been estimated by microstructural and magnetic measurements. The results are summarised in Table I.

Table I: YBaCuO Materials Parameter

Superconductor Type	$Y_{1.4-1.6}Ba_2Cu_3O_{7-\delta}$ + Ag_2O + Pt
Method/ Technique	Ceramo Crystal Growth (CCG), modif. Bridgman, without seeding
Temperature Gradients	Spatial: 0-150 K/cm, multidimensional, 1-5 K/ h cooling
Current Density	$3x10^4$ A/cm² (77 K, 0 T, samples >Ø 80 mm)
Standard Levitation	max. 60 - 70 N (0.4 T PM)
Remanent Magnetisation	typ. 350 mT, max. 520 mT
YBaCuO Size	Bulk 30 - 150 mm, Rings <Ø 160 mm
Weight	80 - 2400 g per Superconductor
Density	typ. 96 % (6.4 g/cm³ = 100%)
Domain Size	1-3 cm³, max. 2 x 2 x 3 cm³
Fabricating capacity	8-10 kg YBaCuO Powder per months
Superconductor shape	Rods, Plates, Rings, Cylinders, Tubes

A corresponding standardised levitation and magnetisation measurement [5] of a ccg sample is demonstrated in Fig.3. The trapped flux measurement shows the multigrain character of the sample with about half a dozen domains. The external magnetic field was 0.2 T.

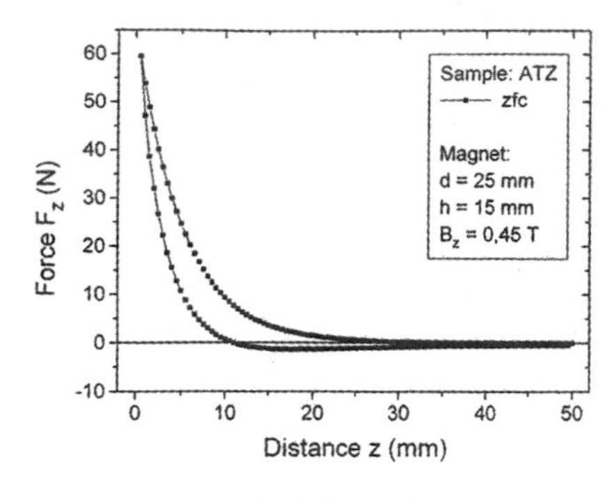

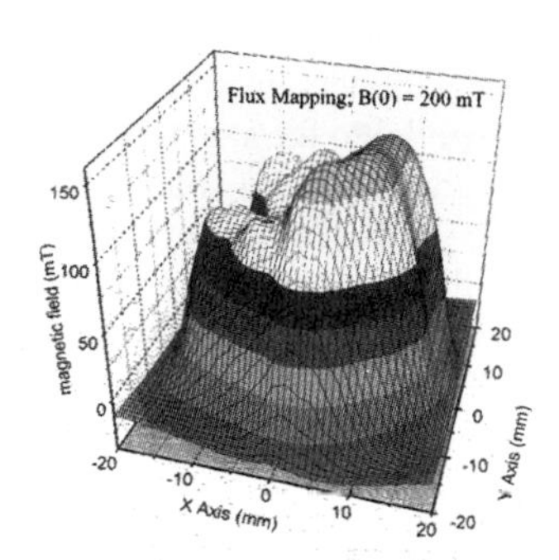

Figure 3. Standard levitation and magnetisation measurement of a ccg multigrain YBaCuO sample

4. Superconducting magnetic bearing (smb) test stand

YBaCuO superconductors of different geometry are assembled together with permanent magnet configurations into a superconducting magnetic bearing device. The properties of different sc bearing configurations are measured in a test stand to adjust on engineering requirements.

Table II: Superconducting magnetic bearing test stand

Rotor weight	10 - 100 N
Rotation speed	max. 120 000 rpm
Cooling system	LN_2, Stirling Cooler 2.5 W/80 K
Armature	3 Phases, 1.2 kW
Rotor bodies	Al/Cu,Ti, Maraging Steel (Hysteresis)
Permanent Magnets	NdFeB, SmCo, armed by CF
Sensors	radial x,y, axial z, Laser 2 μm resolution Force sensors
External Sources	mechanical, magnetic pulses, permanent, single, periodic
Safety system	Stainless Steel container (Vacuum)
Measuring parameters	Radial and axial rotor position, Radial and axial restoring forces Stiffnesses, static and dynamical, Rotor balance , Damping behavior Friction (air, magnetically, eddy currents)

5. Conclusions

Large monolithic YBaCuO superconductors for technical applications are produced on a pre-serial level using a melt texture gradient procedure without seeding. The gradient technology allows a high degree of linear and radial texture of the current carrying a-b planes. The availability of monolithic growth sc rings and cylinders improves the technical performance substantially. On the materials basis prototype superconducting magnetic bearings are constructed and tested.

References

[1] K.Salama, V.Selvamanickam, L.Gao and K. Sun, Appl. Phys. Lett. 54 (1989) 2352
[2] M. Murakami, Modern. Phys. Lett. B4 (1990) 163
[3] G. Krabbes, P. Schätzle, W. Bieger, U. Wiesner, G.Stöver, M.Wu, G. Straßer, Köhler, D. Litzkendorf, K. Fischer, P. Görnert, Physica C 244 (1995) 145
[4] W. Gawalek, T. Habisreuther, T. Straßer, M. Wu, D. Litzkendorf, K. Fischer, P. Görnert, A. Gladun, P. Stoye, P. Verges, K.V. Ilushin, L.K. Kovalev, J. Appl. Supercond. 2 (1995) 465

Inst. Phys. Conf. Ser. No 158
Paper presented at Applied Superconductivity, The Netherlands, 30 June–3 July 1997

Investigation of melt cast processed Bi-2212 conductors and their potential for power applications

P.F. Herrmann[1], E. Beghin[1], J. Bock[2], C. Cottevieille[1] A. Leriche[1] and T. Verhaege[1]

[1]Alcatel Alsthom Recherche, Dept. Electrotrechnique, 91460 Marcoussis, France.
[2]Hoechst AG, 5030 Hürth-Knapsack, Germany.

Abstract - The performances of Melt Cast Processed (MCP) Bi-2212 conductors with integrated silver contacts progressed rapidly during the last years. Nowadays, tubes of a diameter of Ø = 0.3 m and a length up to 0.5 m are fabricated. The critical current in Ø = 70 mm tubes reaches values in the 10 kA range at 77 K and the current density approaches J_c = 10 A/mm^2. As a first application, it has been proven that this interesting current carrying capacity in conjunction with a very low thermal conductivity allows the fabrication of high performance current leads for low temperature applications. Other applications are investigated here with encouraging perspectives, as field windings, power transformers or current limiters. The possibility for operation of this material under electrical fields up to 300 V/m is investigated and the consequences for fault current limitation is worked out.

1. Introduction

Three material options appear today as the most interesting for power applications: silver alloy sheathed Bi-2223 and Bi-2212 conductors, melt cast processed (MCP) Bi-2212 conductors and thick Y-123 films on SS substrates. The MCP conductors appears to be the cheapest conductor option and it reaches 10 kA at 77 K at comparatively low current densities. The centrifugal casting of Bi-2212 tube-conductors uses a similar process [1] as for large stainless steel tubes which are manufactured with a length of 10 m in metallurgical plants. The MCP material has shown in two European Projects on DC & AC current leads [2-4] its superiority for this application where low thermal conductivity and high transport currents are required at the same time. It is also projected to evaluate its potential for power applications as ampere turns for transformer, fault current limiter [5] (FCL) and self limiting power links. High power current limitation has been demonstrated [6] using AC NbTi conductors.

The investigation of the U(I) behaviour up to high electrical fields for evaluation of the material under overload conditions and its current limitation capability is the object of this work. The measurements have been carried out at current densities reaching 30 times the critical current density according to the 1 μV/cm criterion. The currents, necessary for large samples (several 10 kA) would exceed the possibilities of a conventional laboratory. Therefore pulsed measurements have been carried out in small samples of a cross section of 10 mm^2 and a critical current of about 100 A. Transport currents have been generated up to 3000 A allowing to reach electrical fields of 300 V/m. The duration of the current pulse is of the order of about 1 ms.

2. MCP Bi-2212 Samples

Although, the Bi-2212 MCP is a ceramic material, it is comparatively soft and it can be machined using standard tools from classical workshops. MCP tubes can easily be machined in a helix (forming a one layer solenoid) or in any other geometrical form which is achievable from the initial tube. Also, the entire tubes are relatively strong, the material must be strengthened prior for being machined. For the samples which are presented in this work, this was realised by an internal epoxy coating which was introduced in the tube. Using this technique, small samples of length 20 mm and a cross section of about 10 mm^2 have been machined.

Measurement set up: A set of capacitors was charged to voltage levels up to 200 V. The discharge of this capacitor set was triggered by a power electronic control system which also determines the start of the data acquisition of the storing oscilloscope. The capacitor set is discharged through the measurement circuit including the sample during the characteristic discharge time $\tau = (RC)^{-1}$ of about 1 ms. Currents up to 3000 A have been generated.

DC measurements: The room temperature resistivity is found in the range 1.7 - 2.3 10^{-5} Ωm and the DC critical current values are in the range 8 - 10 A/mm^2, the highest J_c value corresponds to the sample with the lowest room temperature resistivity.

3. Overload conditions in MCP conductors

Pulsed measurements: Experimental E(J) curves are separated into three different domains:

1. The sample is said to be in the superconducting state when, for E-values smaller than an arbitrary criterion (usually 1 μV/cm criterion).
2. Above this criterion some dissipation occurs and the sample is said to be in the flux flow state and a resistivity ρ_{ff} results from the depinning of the flux lines [7] :

   $$\rho_{ff}(T) = \rho_n(T) \times B/B_{c2}(T) \tag{1}$$

   Were $\rho_n(T)$ and $B_{c2}(T)$ are the temperature dependent normal resistivity and upper critical field respectively. Under self field conditions, the induction B is proportional to the current, so that ρ_{ff} is also proportional to the current in the sample, as long as the temperature T remains constant.
3. When heating effects become important, a more pronounced resistivity increase is observed. The temperature T increases, so that $B_{c2}(T)$ decreases, and ρ_{ff} evolves rapidly to the normal state resistivity ρ_n. The characteristic signature from this thermally induced response is that the maximum of the electrical field and resistivity are delayed with respect to the maximum value of the current. During the short duration of the current pulse, heating effects can be calculated by the adiabatic approximation:

   $$m\, C_p\, \Delta T = \int \rho(t)\, j^2(t)\, dt \tag{2}$$

   Where m, C_p, ΔT and t are respectively the mass, the heat capacity, the temperature increase and the time.

The signature of the J(t) and the E(t) curves for each domain is shown in figures 1a, 1b, and 1c:

1. The inductive response ($E_{ind} \approx dI/dt$) fig. 1a corresponds to the superconducting state is observed when E_{ind} is large compared to the resistive component $E_{res} = \rho_{ff}J$. From the measurements it is seen that this condition is realised when the current density in the sample remains of the order of 5 x J_c ($\approx$ 50 A/mm^2). The time derivative of the current is maximum at t $\approx$ 0 when the current is switched on which can be seen from the very steep rise of the *"inductive peak"* (10 - 20 μs). It is seen from figure 2 that the maximum electrical field value of the inductive peak is proportional to the maximum current I_{max}. This result is expected from dI/dt ~ U_c, where U_c represents the voltage to which the capacitors have been charged.

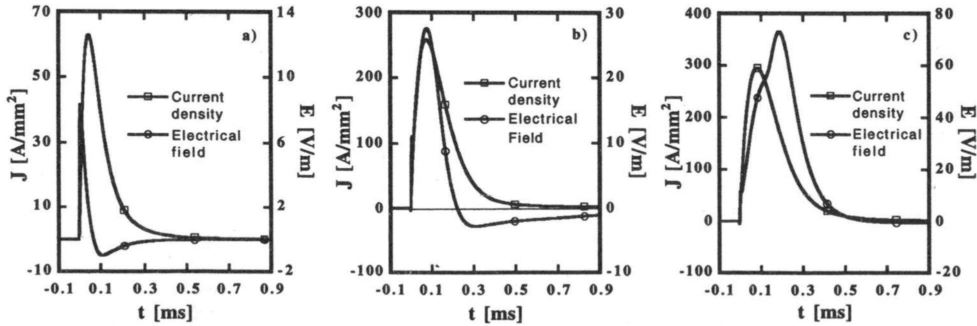

Figure 1: The signature of the three domains of the E(J) curve a) inductive response b) resistive response negligible heating c) development of a thermal induced delayed resistance

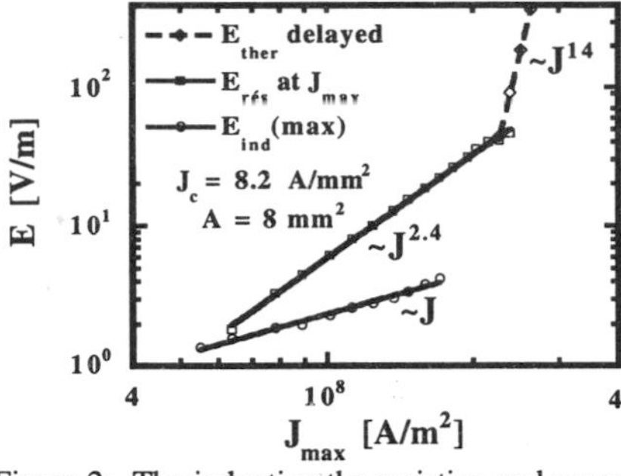

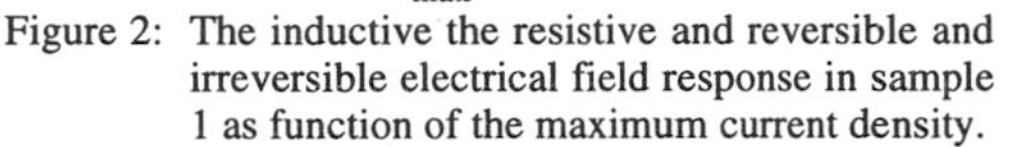
Figure 2: The inductive the resistive and reversible and irreversible electrical field response in sample 1 as function of the maximum current density.

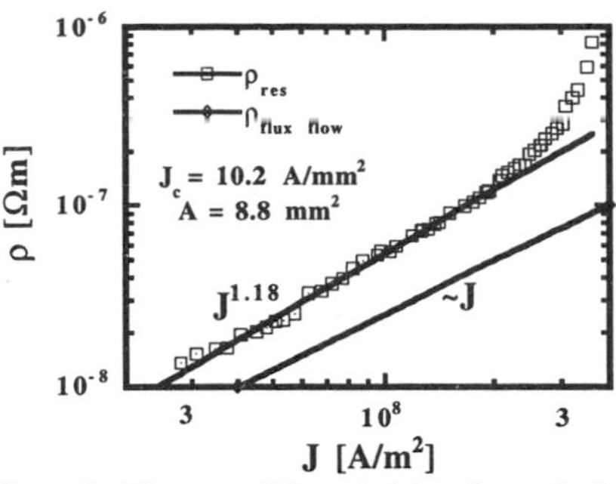

Figure 3: The reversible resistivity of sample 2 and the flux flow resistivity according to equation 1.

2. A *resistive peak* of the electrical field $E = \rho_{ff}(T=77\text{ K}) \times J$ is observed (Fig. 1b) for J_c values in the range 100 - 300 A/mm^2. In order to avoid an inductive contribution to the signal the resistivity is determined at maximum current when dI/dt = 0. The result is shown for two samples in figure 2 and 3 over the whole range at dI/dt = 0. The maximum current is reached 100 μs after the discharge of the capacitors was triggered and a field value of 300 V/m is reached. The fit of the data to $\rho = J^m$ which results in m-values in the 1.1 - 1.4 range is compared to the theoretical flux flow behaviour at T = 77 K (figure 3) according to equation 1 which was calculated with $B_{c2}(77K) = 45$ T using the Abrikosov-Gorkov function with $B_{c2}(T=0) = 200$ T [8] and an extrapolated 77 K normal state resistivity of $\rho_n = 10^{-5}$ Ωm. The experimental resistivity is a factor 2.5 higher than the theoretical $\rho_{ff}(J)$ curve. This could be related to a smaller critical field value in the grain-boundaries which determines the superconducting transport properties of this polycrystalline material. A 77 K critical field value of 20 T instead of 45 T would be in agreement with our measurements.
3. At high current densities (J > 200 A/mm^2) a second maximum in the E(J) for decreasing currents (Fig. 2c) corresponding to a irreversible resistive state $E = \rho(T) \times J$ is observed. The height of this peak is rapidly increasing and it dominates the behaviour of the transition at J values well above 200 A/mm^2. The heating in the sample can be estimated using the assumption that the maximum temperature is reached when the maximum resistivity in the sample is attained and that equation 2 is verified for pulsed measurements. The temperature increase in the sample during the measurement has reached ΔT = 90 K at J_c values of 300 A/mm^2. At still higher currents, the sample was destroyed. The destruction of the sample at relatively low temperature is surprising and it is argued that the destruction is related to mechanical failure (mismatch of heat expansion) and not to burn out. This is supported by the fact that no traces of melting could be found on the broken sample.

4. Fault Current Limitation by MCP Conductors

The above discussed behaviour is of importance for safety and for current limitation. A fault current limiter realised from this material (see fig. 4) would operate in two stages:
1. $T < T_c$: A moderate electrical field is generated instantaneously, due to flux flow and the corresponding heat dissipation increases the temperature. During 1 ms the current is weakly limited resulting in a current peak.
2. After about 1 ms the critical temperature is reached, the normal resistivity of the conductor limits the current to a low value which can be adjusted by optimising the design parameters.

Based on the measured properties, a computer calculation has been realised for the example of a 25 kV 50 Hz power distribution grid. The relevant voltage for current limitation, is between one phase and the ground level which is 14.4 kV$_{rms}$. Under normal steady state operation, the rated current is 1 kA$_{rms}$. In case of a short circuit without FCL this current would rise up to 32 kA$_{rms}$ (i.e. the line inductance equals 2.5 mH). Using a 10 A/mm^2 conductor, this requirement leads to a conductor cross section of 139 mm^2 and to an acceptable

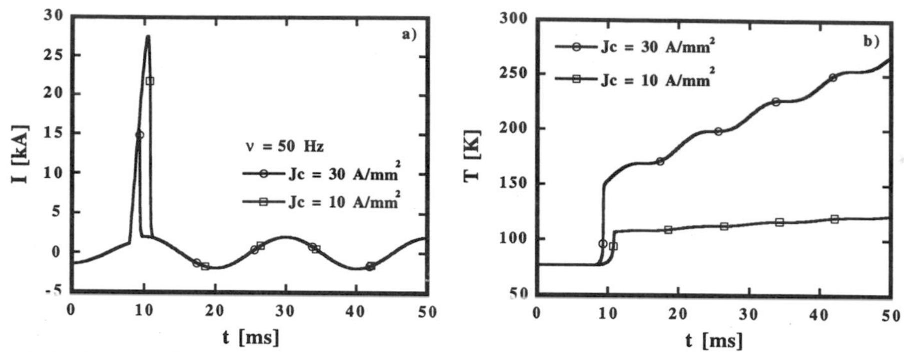

Figure 4: Result of the computer simulation of the current limitation capacity for a fault 2 ms before max current is reached. Time dependent evolution of a) The current b) The temperature

electrical field value of 100 V/m showing that a conductor length of 144 m is necessary. For a 30 A/mm^2 conductor the requirements lead to a cross section which is divided by 3 and an achievable electrical field value of 300 V/m showing that also the length can be reduced by a factor 3.

The short circuit in figure 4 appears at time t = 8 ms, generating a fast current increase, which is practically not influenced by the small flux flow resistance. This regime comes to an end when the critical temperature is attained, at a current of 27.5 kA_{peak} in the first case and only 15 kA_{peak} in the second case, due to more rapid heating. After this transitory phase, the current is strictly limited to 2 kA_{peak}, i.e. less than 1.5 time the rated current, so that small circuit-breakers can be used to break the current after 50 or 100 ms. Due to the lower conductor volume in the second case, the temperature increases more rapidly showing that the system can be optimised with respect to the current peak, the duration of the current limitation which is required and the current level during limitation.

Conclusions

Pulsed Currents measurements have been carried out for the investigation of the E(J) curves of MCP conductors up to electrical fields of 300 V/m and current densities of 300 A/mm^2. Three domains corresponding to the superconducting state, to the resistive flux flow state and to the normal state after heating. The results of these measurements have been used for computer calculation of current limitation in the power grid. A MCP fault current limiter first allows an instantaneous low level current limitation by flux flow. After a delay of typically 1 ms the current is strictly limited to less than 1.5 times the rated current. Despite to the relatively low current density in this MCP material an effective current limitation can be achieved.

References

[1] J. Bock, S. Elschner, P.F. Herrmann and B. Rudolf, "Centrifugal casting of BSCCO 2212 form parts and their first application" Applied Superconductivity IOP Ser. Num. 148 Vol 1, pp 67-72, 1995

[2] P.F. Herrmann, "Current leads for cryogenic applications", To be published in: "Handbook of Applied Superconductivity" IOP publishing London

[3] C. Albrecht, J. Bock, P.F. Herrmann and J.M. Tourre "European development on superconducting oxide-based 1kA (2kA) DC current leads" (Publishable Synthesis Report) Brite EuRam proj. N° BE 4071, 1994

[4] C. Albrecht, J. Bock, P.F. Herrmann, 1996 HTS Current leads for power devices. Synthesis Report from BRITE/EURAM project 7856, contract BRE2 CT 93 0589

[5] "ABB delivers bulk HTS fault current limiter to electric utility" Superconductor Week Vol. 11, N° 4, Feb. 1997

[6] T. Verhaege et al. "Experiments with a high voltage (40 kV) superconducting fault current limiter" Cryogenics Vol. 36, N° 7, pp 521-526, 1996

[7] M. Tinkham "Introduction to Superconductivity" R.E. Krieger Publishing Company Inc. New York 1975

[8] A.I. Glovashkin "Testing HTS compounds in ultrahigh fields by-product of Russian defence conversion" Superconductor Week Vol. 8, N° 8, March 1994

Inst. Phys. Conf. Ser. No 158
Paper presented at Applied Superconductivity, The Netherlands, 30 June–3 July 1997

Improved YBCO bulk material for cryomagnetic applications

P Schätzle, G Krabbes, W Bieger, G Stöver, G Fuchs, S Gruß

Institute of Solid State and Materials Research Dresden, P.O. Box 270016, D-01171 Dresden, Germany

Abstract. Defined oriented single grain YBCO monoliths (0°-10°, 30-45°, 75°-80° c-axis inclination according to the monolith surface) were grown in a melt crystallization process (MCP) by seeding with oriented SmBaCuO seeds. Trapped magnetic fields up to 700 mT with $j_c = 3.5 \times 10^4$ A/cm^2 and a levitation force up to 80 N were realized. Different applications as permanent magnets, performed rings and a levitation bearing demonstrate the high quality of these HTSC bulk material.

1. Introduction

YBCO is the most promising HTSC material for bulk applications. Trapped magnetic field and levitation force are the important properties. Both values are correlated and determined by the microstructure and the grain size of the YBCO sample. Because of this great influence of the microstructure the understanding of the crystallization process becomes more important. The high anisotropy of the Y-123 and the connected superconducting properties make it necessary to produce defined oriented YBCO samples. In the last few years melt texturing has become the most promising method to exceed high YBCO quality bulk material [1]. A further improvement of the magnetic properties was achieved by using a seeding technique at the beginning of the melt texturing process [2, 3, 4]. Several seed materials (MgO, Sm-123, Nd-123, $SrTiO_3$) were used to determine the orientation and number of the growing YBCO grains. The growth of single grains initiated by the seed becomes extremely sensitive to the processing parameters e.g. heating rate, maximum temperature, cooling rate and temperature distribution in the furnace especially in the sample. The composition of the peritectic melt and the volume and mass of the processed YBCO pellet is an important factor for a stabilized growth process. A detailed study of the melt texturing process focussed on the growth of single grain YBCO monoliths with different size and preferred orientation is presented. Several different oriented single grain monoliths were grown defined by SmBaCuO seeds which were 0°, 45° and 90° inclined relating to the c - axis.

2. Experiments

Starting with $YBa_2Cu_3O_{7-\delta}$ (Y-123) powder and an addition of Y_2O_3 and Pt powder which were thoroughly mixed and cylindrical pellets of different diameters (d= 30, 38, 55mm) were prepared by uniaxial pressing. The admixed Y_2O_3 reacts below 940°C according to the following solid state reaction

$$YBa_2Cu_3O_{7-\delta} + n\,Y_2O_3 \rightarrow a\,YBa_2Cu_3O_{7-\delta} + b\,Y_2BaCuO_5 + c\,CuO \qquad (1)$$

[5] which generates an excess of b = 0.38 mol Y_2BaCuO_5 (Y-211) if n = 0.24 mol Y_2O_3 is added (c = 0.16 mol). The generated Y-211 is homogeneously distributed in the pressed pellet and of small size. The amount is optimized to produce small Y-211 inclusion which promote an optimized flux pinning in the Y-123 matrix. The Y-123 phase crystallizes according to the well known reverse of the peritectic reaction at 1020°C ($p(O_2) = 0.21\times10^5$ Pa) [5]. The excess of CuO which is generated during the solid state pre-reaction (1) reduces the crystallization temperature to 1010°C and stabilizes the crystallization process by an enrichment of the melt with copper and changing their viscosity [5].

SmBaCuO seeds were prepared from as-grown SmBaCuO bulk material either by cleaving along the a, b - planes or cut with defined angels of 45° or 30° related to the c-axis.

The pressed YBCO pellets were processed in a modified melt crystallization process (MCP) where the pellets are placed in a tube furnace with an isothermal temperature distribution. The oriented seed is located on the top of the pellet before processing. Heating above the decomposition temperature of Y-123 but below the decomposition temperature of Sm-123 generates the melt suspension. The crystallization takes place during cooling with small rates of 0.5 - 1K/h down to 940°C starting above 1020°C.

After 300 h oxygen annealing between 620 °C and 400°C and polishing the surface X-ray polfigures were measured to determine the orientation of the YBCO monoliths (Seifert diffractometer XRD 3000). A more detailed determination was performed by X-ray micro polfigures with an lateral resolution of about 100µm (mircodiffractometer HI-STAR Siemens). The levitation force F_N (zero field cooled) was measured with an SmCo magnet (d= 25 mm, B_0 = 0.4 T) and the trapped magnetic field (B_z with z = 0.5mm) was detected by scanning with a miniature hall sensor (Lake shore, model HGCA-3020) after magnetization in a superconducting coil with a 2 T field. The microstructure and the chemical composition were examined by SEM and energy dispersive analysis of X-ray (EDAX system).

3. Results and discussion

Beginning at the SmBaCuO seed a YBCO single grain grows over the whole dimension of the monolith. Figure 1 shows a monolith after the melt crystallization process (MCP). Arising at the seed in the middle of the monolith two diagonal lines are obvious. They follow the <110> directions. The angle between them is nearly 90° because of the tetragonal symmetry in which YBCO crystallizes. The single grain was grown from a seed with a, b plane orientation parallel to the monolith surface. X - ray polfigures of the monolith shown in fig. 1 indicates a 10° tilt of the a, b plane corresponding to the edge and an 5 - 8° inclination to the surface. The halfwidth of the (006) peak was in the range of 5°. However, a more detailed x-ray measurement (spot size 100µm) of the orientation in the region of the seed and nearby the edge demonstrate different stages of the crystallization. In the region nearby the seed a sharp orientation exist with tilting and inclined angles less than 2° (halfwidths 1°). In the region of the edges the tilt and inclined angles are changed up to 10 and 15° (halfwidth up to 5 - 10°). In figure 2 the resulting mosaic structure is shown. This structure initiates defects in the Y123 bulk material which influences the flux pinning behaviour.

Fig. 1. Melt textured of a YBCO monolith after the melt crystallization process (MCP)

Fig. 2. Microstructure of a MCP YBCO monolith (top view) (white: Pt - inclusions)

In the case of SmBaCuO seeds with 45° and 90° orientation according to the a, b planes the growth of single grains of nearly the same orientations was achieved. X-ray polfigures of hkl (200/006) reflexes document a resulting inclination of 40° and 75° of the a, b planes. An inclination of 40° reveals x-ray reflexes of the (200) and (006) planes and at an inclination of 75° only (200) reflexes are detectable (Fig. 3).

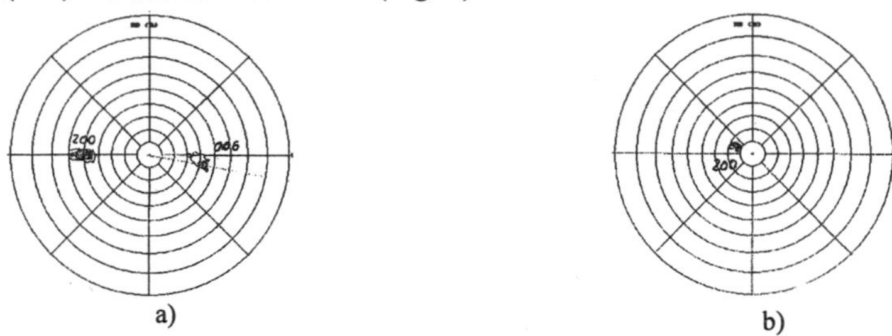

Fig. 3. X-ray polfigures of two different oriented single grain YBCO monoliths, a) 35° inclined, b) 75° inclined

The magnetic properties of MCP YBCO monolith with different dimensions are presented in Table 1 revealing the stability of the process. All monoliths were magnetized with a field orientation parallel to the surface. Because of the anisotropy of B_t and F_N the measured values for the inclined monoliths are low in comparison with the c - axis (0° - 10°) oriented grains, although they reveal the same microstructure. The ratio between trapped magnetic field and monolith mass represents a value from which an optimum monolith dimension for a stabilized growth can be estimated (Tab. 1). A decrease of this ratio is observed if the monolith diameter becomes larger than 38 mm.

diameter [mm]	heigth [mm]	inclination of the c - axis	B_z [mT]	F_N [N]	B_t/m [mT/g]
26	12	0° - 10°	650	60	≈ 17
38	16	0° - 10°	700	75	≈ 12
55	20	0° - 10°	500	80	≈ 4
26	12	30° - 45°	160	-	-
38	16	75° - 80°	230	-	-

Tab. 1. Magnetic properties of single grain YBCO monoliths

3. Properties for applications

MCP YBCO monoliths trap magnetic fields up to 700 mT at 77 K. Figure 4a represents a field profile which is close in agrreement to Bean's model. At lower temperature this value increases to 5.5 T at 30 K limited by the mechanical properties [6].

Permanent magnets of two stacked monoliths reveal fields up to 8.5 T at 51.5 K in a 1 mm gap between them [7] and up to 9 T at 47 K after fixing with resin.

The trapped magnetic field in a YBCO ring prepared by drilling the middle section is shown in figure 4b. The typical cylindrical cone according to Bean's model is deformed and a plateau appears in the middle section. Because of loosing superconducting material the field value is decreased.

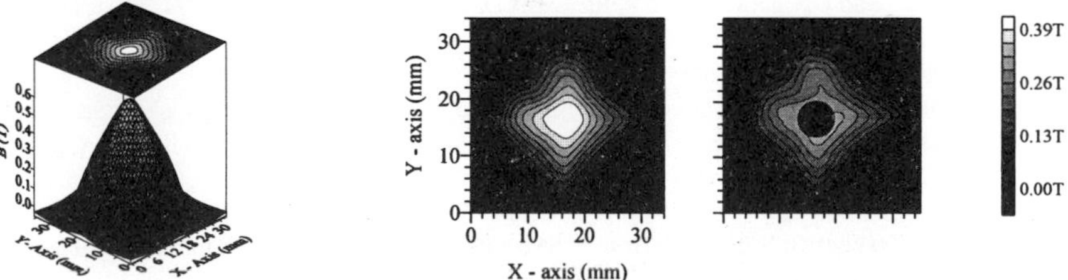

Fig. 4. Trapped magnetic field of a single grain YBCO monolith and after drillling the middle section (●)

An arrangement of six sliced YBCO monoliths was fixed together to form a levitation bearing. A ring of NdFeB permanent magnets (0.4 T) was used to introduce the trapped magnetic field which is shown in figure 5. The stiffness of the bearing was enhanced by performing the magnetization with two NdFeB rings in antiparallel polarization. The trapped field is therefore low in the range of 150 mT but the levitation force of the bearing is up to 80 N with a radial stiffness of 20 N/mm.

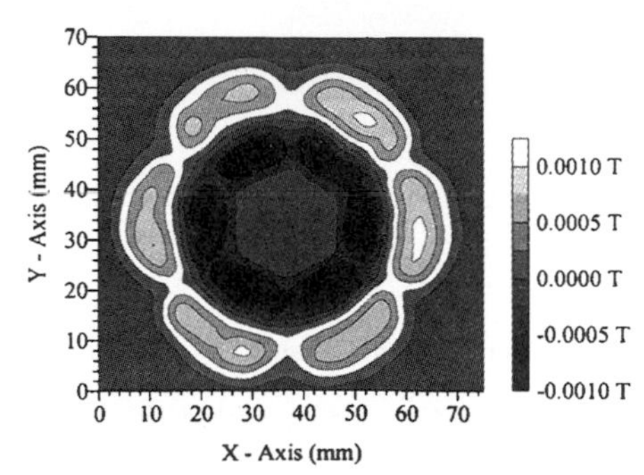

Fig. 5. Trapped magnetic field of a levitation bearing performed with six single grain YBCO monoliths

5. Conclusions

A melt crystallization process (MCP) was established to produce single grained YBCO monliths. Starting from oriented SmBaCuO seeds monoliths of different diameter were grown. The orientation of the grown single grain is controlled by the orientation of the seed. High trapped magnetic fields are realized which allowed to perform different arrangements of the monoliths. Permanent magnets are build by stacking two monoliths. YBCO rings which can be used i. e. electric motors were prepared by drilling. A levitation bearing shows a high levitation force and radial stiffness.

Acknowledgments

The authors wish to thank Mrs. S. Donner for technical assistance, Dr. Schläfer and Dr. Stephan for X-ray measurements. This work was supported by the German Federal Minister of Education and Research under the contracts No. 13 N 69344 and 13 N 6853.

References

[1] Murakami M 1992 *Melt Processed High Temperature Superconductors* (World Scientific Publishing Co. P Ltd)

[2] Meng R L, Gao L, Gautier-Picard P, Rameirez D, Sun Y Y and Chu C W 1994 Physica C **232** 337-346

[3] Morita M, Sawamura M, Takebayashi S, Kimura K, Teshima H, Tanaka M, Miyamoto K and Hashimoto 1994 Physica C **235 - 240** 209 - 212

[4] Schätzle P, Bieger W, Krabbes G, Klosowski J, and Fuchs G 1995 *Inst. Phys. Conf. Ser.* No **148**. 155-158

[5] Krabbes G, Schätzle P, Bieger W, Wiesner U, Stöver G, Wu M, Strasser T, Köhler A, Litzkendorf D, Fisc K, and Görnert P 1995 Physica C **244** 145-152

[6] Fuchs G, Krabbes G, Schätzle P, Stoye P, Staiger T, Müller K - H 1996 Physica C **268** 115-120

[7] Fuchs G, Krabbes G, Schätzle P, Gruß S, Stoye P, Staiger T, Müller K-H, Fink J and Schultz L 1997 Appl. Phys. Lett. **70** (1) 117 - 119

Inst. Phys. Conf. Ser. No 158
Paper presented at Applied Superconductivity, The Netherlands, 30 June–3 July 1997

Top-Seeded Solution Growth of $Sm_1Ba_2Cu_3O_{7-\delta}$ Seed Crystals for Melt Texturing of $Y_1Ba_2Cu_3O_{7-\delta}$

Ch. Krauns[1*], B. Bringmann[1], C. Brandt[1], M. Ullrich[2], K. Heinemann[1], H. C. Freyhardt[1,2]

[1]Institut für Metallphysik, Universität Göttingen, Windausweg 2, D-37073 Göttingen; [2]Zentrum für Funktionswerkstoffe, Windausweg 2, D-37073 Göttingen;

Abstract. $Sm_1Ba_2Cu_3O_{7-\delta}$ (Sm123) seed crystals have been fabricated by a Top-Seeded-Solution-Growth (TSSG) method using $BaZrO_3$ crucibles. Crystals up to a size of 100 mm^2 in the a-b plane have been grown. In this presentation, the crystal growth will be described and the growth parameters will be discussed. Furthermore, the superconducting and structural properties of the Sm123 seed crystals will be presented. These seed crystals have been successfully employed for the melt texturing of $Y_1Ba_2Cu_3O_{7-\delta}$ (Y123) monoliths of diameters up to 50 mm. The superconducting and structural properties of the melt-textured Y123 determined by hall probe measurements and optical microscopy will be presented.

1. Introduction

The growth of single crystals of the high temperature superconductor (HTSC) $Sm_1Ba_2Cu_3O_{7-\delta}$ (Sm123) is interesting both for basic research as well as for the applications of HTSC. Larger good quality crystals are still needed for fundamental research (e. g. for neutron diffraction or latent heat measurements) and in the field of applications these crystals are used as substrates for the homoepitaxial deposition of HTSC thin films [1] and as seeds for the melt texturing of Y123 monoliths. Especially Sm123 single crystals are suitable as seeds, because Sm123 is isostructural to Y123, but its peritectic Temperature T_P in air is approximately 60 °C higher than the T_P of Y123. Therefore, during the melt processing of Y123 a Sm123 seed is more stable than a Y123 seed could be, allowing different treatments. Although the interdiffusion of Sm and Y from the Sm123 seed to the Y123 monoliths during the melt texturing has to be taken into consideration [2]. Furthermore, due to the higher Sm solubility in Ba-Cu-O melts and the not as steep liquidus slopes compared to Y [3] the crystal growth velocities are higher for Sm123 than for Y123 [4].

2. Experiment

Sm123 single crystals are fabricated by a top-seeded solution-growth (TSSG) method [5, 6]. In our case the solvent consists of $Ba_3Cu_5O_8$, while the solute is $Sm_2Ba_1Cu_1O_5$ (Sm211). At the applied temperatures and under the applied oxygen partial pressure (air) Sm123 is the

* corresponding Author; e-mail: ckrauns1@gwdg.de, fax: +49-551-5071750, phone: +49-551-399731

stable solid phase at the surface of the solvent and Sm211 is the stable solid phase at the bottom of the crucible. As the Sm-solubility in Ba-Cu-O varies with temperature [3] , more Sm is solved at the bottom than at the surface. Via convection the Sm saturated solvent of the bottom is transported to the surface and due to the lower temperature the solvent becomes supersaturated. This supersaturation is the driving force for the crystallization in the steady state growth. By choosing a composition of $Sm_2Ba_1Cu_1O_5$ and $Ba_3Cu_5O_8$, the melt does not change during growth according to:

$$2\ Sm_1Ba_2Cu_3O_x\ (s) <\!-\!> Sm_2Ba_1Cu_1O_5\ (s) + Ba_3Cu_5O_8\ (l).$$

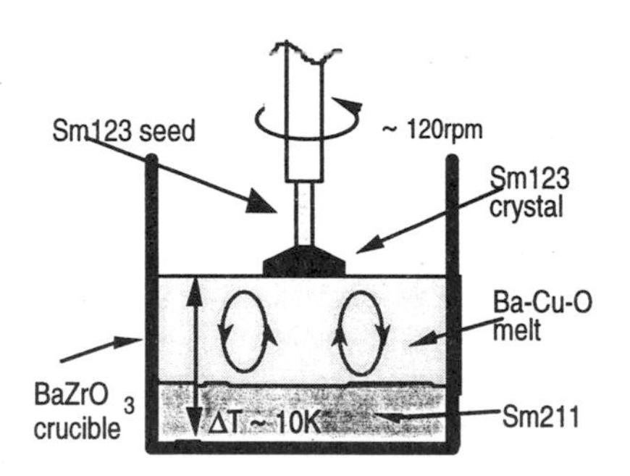

Fig. 1: TSSG of Sm123

Table 1: Process parameters for the crystal growth of Sm123 single crystals

crucible material	$BaZrO_3$
solvent	$Ba_3Cu_5O_8$
solute	$Sm_2Ba_1Cu_1O_5$
surface temperature	1050 °C to 1060 °C
bottom temperature	1065 °C to 175 °C
seed	Sm123 film on MgO
atmosphere	air
pulling velocity	0.01 to 0.2 mm/h
rotation	30 to 120 rpm

The optimal process parameter for the crystal growth are listed in table 1. During the growth time (i. e. several days) the crucible material is exposed to the highly reactive Ba-Cu-O melt, resulting in severe problems of melt contamination respective crystal contamination, crucible corrosion and wetting [7]. Contrary to the growth of Y123 crystals by TSSG [8-10], in which case the problem of crucible contamination can been solved by using Y_2O_3 crucibles, in some case of other RE123 (RE = Rare Earth) HTSC respective RE_2O_3 crucibles (Nd_2O_3 for Nd123 [11] , Sm_2O_3 for Sm123 [5, 6, 12]) can be used. Although, the respective crucible are expensive and difficult to obtain commercially; and for other RE123 phases like Pr123 [13, 14] there are no stable RE_2O_3 crucibles.

Erb et al. reported [15, 16] , that $BaZrO_3$ crucibles are suitable for the growth of RE123 single crystals. Therefore for these crystal growing experiments $BaZrO_3$ crucibles [17] were used. A crucible in crucible technique was necessary, because of occasionally occurring thermal instabilities of the $BaZrO_3$ crucibles. To facilitate the crystal growth of Sm123 on MgO single crystals, beforehand Sm123 films were deposited on the MgO seeds by laser ablation as described in [18].

3. Results and discussions

3.1 TSSG of Sm-123

Sm123 crystals grown by TSSG are shown in figure 2. The Sm123 crystals do often grow in conglomerates as on the surface flowing secondary nucleating crystals attach to the primary seed. Sm123 single crystals were separated from these conglomerates for measurements and seeding experiments The crystals range in size from 2 x 2 mm^2 up to 10 x 10 mm^2 in the a/b plane.

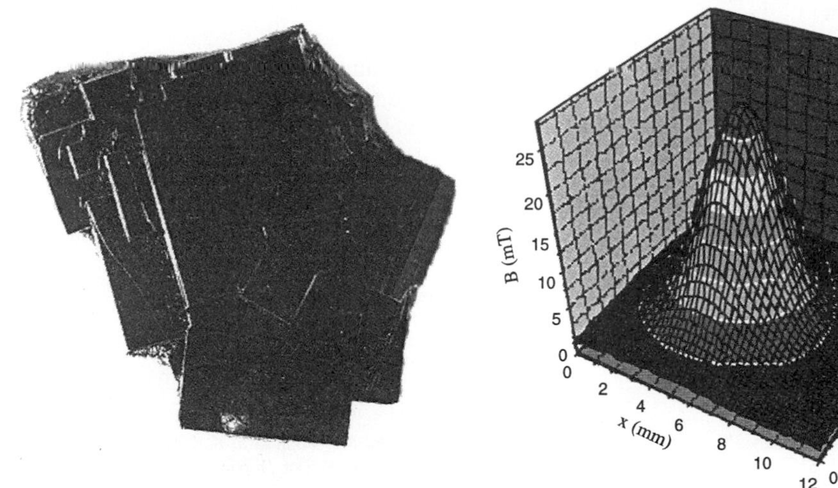

Fig. 2. Optical microscopy of Sm123 single crystals

Fig. 3: Distribution of the remnant magnetization of a Sm123 single crystal at 77 K.

After appropriate oxygenation (i. e. from 500 °C to 300 °C in flowing oxygen for 3 days), T_c has been determined by susceptibility measurements to be approximately 90 K with a ΔT_c of 1 K. EDX-measurements detected no Zr contamination in the Sm123 crystals. Hall probe measurements of the trapped magnetic field at 77 K of the Sm123 single crystals demonstrate their single crystalline structure (figure 3).

3.2 Melt texturing of Y-123 monoliths

Y123 monoliths have been successfully melt textured by both the VGF (Vertical Gradient Freeze) [19, 20] and quasi isothermal batch processes [2], using Sm123 single crystals as iso-structural seeds. The precursors for these melt texturing experiments consists of Y123 plus 25 mol% Y211 and 1.8 mol% PtO_2 additions. In figure 4 a top view of a melt textured Y123 monolith (diameter 32 mm) is shown. With the Sm123 seed is in the middle of the specimen, typical X-like growth pattern can be observed. X-ray measurements of the seed and the grains of the melt textured Y123 monoliths reveal, that the Y123 on the seed of Sm123 grows epitaxially.

Fig. 4: Top view of melt textured Y123 monolith seeded with a Sm123 single crystal

4. Conclusions

Sm123 single crystals up to 100 mm^2 in the a-b plane could be grown by the TSSG method. No contamination of Zr from the crucible material could be detected in the crystals and after appropriate oxygenation T_c is approximately 90 K. Texturing of Y123 monoliths using these Sm123 single crystal seed was successfully performed.

Acknowledgments

This work is supported by the BMBF under the grant numbers 13N6566, 13N6939/9 and 13N6940/9.

References

[1] M. Konishi, K. Hayashi, A. Odagawa, Y. Enomoto, Y. Yamada, M. Nakamura, K. Ohtsu, Y. Kanamori, M. Tagami, S. Koyama and Y. Shiohara 1993, International Symposium on Superconductivity, Hiroshima (Springer-Verlag), 995-998

[2] M. Ullrich, D. Müller, Ch. Krauns, B. Bringmann, A. Leenders, C. Brandt, M. Reder, A. Preusser and H. C. Freyhardt 1997, International Workshop on Superconductivity, Hawai , 76-79

[3] Ch. Krauns, M. Sumida, M. Tagami, Y. Yamada and Y. Shiohara 1994, Z. Phys. B **96**, 207-212

[4] Y. Yamada, Ch. Krauns, M. Nakamura, M. Tagami and Y. Shiohara 1995, J. Mater. Res. **10**, 1601-1610

[5] Ch. Krauns, M. Tagami, M. Nakamura, Y. Yamada and Y. Shiohara 1994, International Symposium on Superconductivity, Kita Kyushu (Springer Verlag), 641-644

[6] Ch. Krauns 1995, Ph.D. thesis, Göttingen

[7] Ch. Krauns, M. Tagami, Y. Yamada, M. Nakamura and Y. Shiohara 1994, J. Mater. Res. **9**, 1513-1518

[8] Y. Yamada, M. Nakagawa, K. Ishige and Y. Shiohara 1992, International Symposium on Superconductivity, Nagoya (Springer-Verlag), 305-308

[9] Y. Yamada and Y. Shiohara 1993, Physica C **182**, 182-188

[10] Y. Yamada 1995, Ph.D. thesis, Tokyo

[11] M. Nakamura, T. Hirayama, Y. Yamada, Y. Ikuhara and Y. Shiohara 1996, Jpn. J. Appl. Phys. **35**, 3882-3886

[12] M. Nakamura, Ch. Krauns and Y. Shiohara 1995, Jpn. J. Appl. Phys. **34**, 6031-6035

[13] M. Tagami, M. Sumida, Ch. Krauns, Y. Yamada, T. Umeda and Y. Shiohara 1993, International Symposium on Superconductivity '93, Hiroshima (Springer), 787-790

[14] M. Tagami, M. Sumida, Ch. Krauns, Y. Yamada, T. Umeda and Y. Shiohara 1994, Physica C **235-240**, 361-362

[15] A. Erb, E. Walker and R. Flükiger 1995, Physica C **245**, 245-251

[16] A. Erb, E. Walker and R. Flükiger 1996, Physica C **258**, 9-20

[17] Reetz GmbH Berlin

[18] S. Sievers, F. Mattheis, H. U. Krebs and H. C. Freyhardt 1995, J. Appl. Phys. **78**, 5545-5548

[19] M. Ullrich 1993, Ph.D. thesis, Göttingen

[20] M. Ullrich and H. C. Freyhardt 1993, IITT International, Gourmay sur Marne , 205-210

Inst. Phys. Conf. Ser. No 158
Paper presented at Applied Superconductivity, The Netherlands, 30 June–3 July 1997

Processing Technique for Fabrication of Advanced YBCO Bulk Materials for Industrial Applications

A W Kaiser, H J Bornemann and R Koch

Forschungszentrum Karlsruhe GmbH, INFP, P. O. Box 3640, 76021 Karlsruhe, Germany

Abstract. We present a continous process for fabrication of semi-finished parts and products based on YBCO. Bulk components are produced powder by metallurgy and melt-texturing. Near net shaping is also used. Oxygenation processes can be monitored in-situ by a macro-thermogravimetric analyzer. Texture of full-size pellets was verified by elastic neutron scattering. Levitation properties under static load levels were analyzed. Flux mapping was used to verify the homogenity of the material andto investigate the field trapping capability. Several frictionless bearing modules have been built ($\Phi \leq 140$mm). Lifting forces are up to 200N per module.

1. Introduction

The compound $Y_1Ba_2Cu_3O_y$ (YBCO) can trap magnetic fields. Levitation resulting from this effect can be utilized in frictionless self-stabilizing magnetic bearings. These can be cooled with liquid nitrogen. Bulk YBCO are produced in a powder metallurgical process based on the melt-texturation process [1, 2]. The desired form and surface finish can be achieved with conventional methods or is reached by near-net-shaping. The oxygen content is adjustable from 6.3 to 6.9.

2. Product range

2.1. Forms available and specification

On a regular basis we are texturing dense (>95%) bulk parts in the form of flat solid cylinders or rings up to 60mm in diameter, or as blocks up to 60mm x 30mm in size. Any thickness up to 25mm can be chosen. Bars with a 10mm x 10mm cross-sectional area can be manufactured up to 90mm in length. Components of more complex geometries can be manufactured as well. Depending on overall size and geometry, these parts are either die-pressed, form pressed, cut from bulk blocks, or assembled from several pieces. New techniques of near-net-shaping have been already tested (Fig. 1).

Figure 1. Examples of bulk YBCO, near-net-shaped before texturing

Compared to conventional sintered YBCO materials, the bulk materials have a number of outstanding features. Screening currents up to 80,000Acm^{-2} in zero field and 20,000Acm^{-2} in B = 1T are reached. A transport current >1800A over 9cm length with 1cm^2 cross-area in self field was obtained using a 3000A dc-power supply. Levitation experiments gave values of 18 Ncm^{-2}.

3. Materials processing

3.1. Characterization

An extensive materials characterization program can be run. Both microscopic and macroscopic material properties are investigated. Levitation properties under static and dynamic load levels are analyzed using a computer-controlled $2\frac{1}{2}$ axis x-y-z motion stage with a three dimensional force sensor unit [3]. A representative measurement is shown in Fig. 2a).

The unit can also be used to measure interaction forces in complete bearing modules. A flux mapping technique is used to check the homogenity of full-size pellets, to evaluate macroscopic screening currents and to investigate the field trapping capability. A typical result is presented in Fig. 2b), showing the remanent magnetization scanned 1mm above the surface of a $\Phi = 33$mm pellet after the external field of 0.7T was reduced to zero.

Using Bean's critical state model [4], the average macroscopic screening current flowing throughout the volume of the pellet is calculated to be on the order of 8000Acm^{-2}. Experiments with $\Phi = 43$mm and 2T were also performed.

The texture of the bulk of full size pellets up to $\Phi = 43$ mm is verified by elastic neutron scattering. Well textured samples show a full width at half maximum (FWHM) of the c-axis distribution (mosaîc spread) as small as 3 and a deviation of the c-axis orientation distribution from the cylinder axis ≤ 2 °.

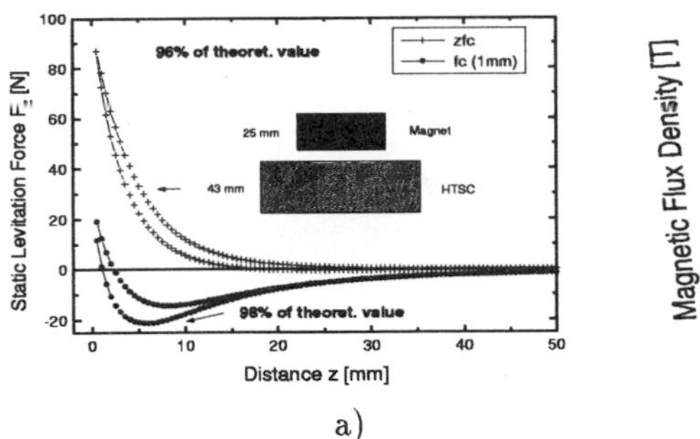

a)

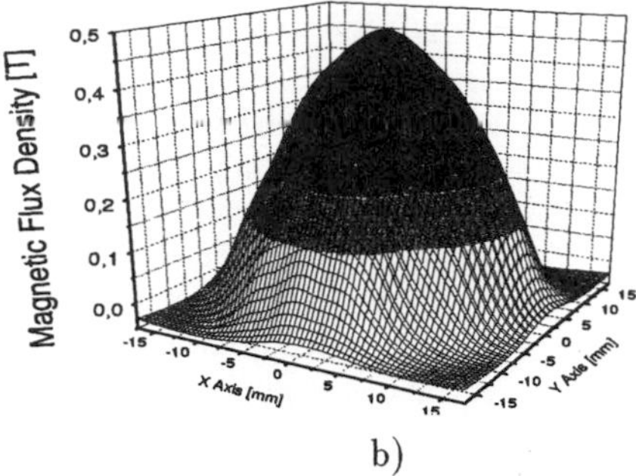

b)

Figure 2. a) Static levitation force vs gap between superconductor and permanent magnet. The small width of the hysteresis loop is evidence for almost complete screening of the magnetic field. The maximum is about 98% of theoretically achievable levitation force, b) Flux map of remanent magnetization of a Φ=43mm pellet at 77K. Applied magnetic field was 0.7T

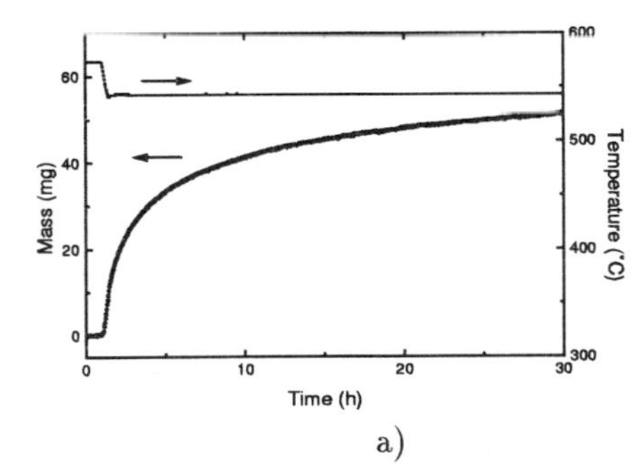

a)

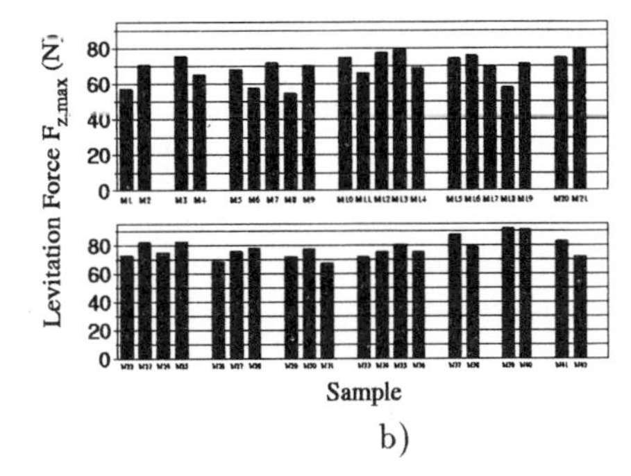

b)

Figure 3. a) Process optimization and process control: Monitoring of the oxygenation process of a 80g pellet, b) Reproducibility of the levitation force for batch-processed pellets Φ=43mm. All samples, quality A and B are included.

3.2. Process control, quality assurance and process reproducibility

Sophisticated in-process test sequences guarantee the high quality standard of the materials, such as in-situ monitoring of the oxygenation process by a macro-thermogravimetric analyzer. Fig. 3a) shows a representative result obtained for a standard size pellet: Φ 32 mm, 80g. In addition to process control, these measurements are also useful for process optimization.

The reproducibility of the process is good. All processed samples are categorized in the two qualities A and B. Here A refers to single domain material, B to two or three domain material. Figure 3b) shows the levitation force of the last batches for samples with a diameter of 43mm. All samples, that means quality A and B are listed. About 80% of processed samples are quality A, 20% are quality B. Quality B samples have a by about 15% lower levitation capability.

4. Applications and bearing manufacturing

Possible applications of bulk YBCO materials include: a) Bearing technology in electrical engineering. Passive, frictionless magnetic bearings could be used in fast rotating electric machinery, such as high speed flywheel systems. b) Magnet technology, such as for example magnetic field generation, control and shielding in electric motors, magnetic separators and magnetic couplings. c) Power engineering, using bulk YBCO for current leads in combination with cryosystems. d) Basic research, using single domain pellets as large quasi-crystals with mosaic spreads $< 3°$ for research, such as neutron spectroscopy.

Using the motion stage, all bulk components intended for use in magnetic bearings are characterized with respect to levitation force under standardized measurement conditions and flux mapping.

Several frictionless bearing modules have been built. Standard sizes range from 40 mm up to 140 mm. Lifting forces are up to 200N for the largest module in combination with a high power, 90mm, permanent magnet ring. Operating temperature is 77K. Depending on the configuration, cooling is provided by either wet cooling, i.e. superconductor is immersed in liquid nitrogen, or conduction cooling, i.e. superconductor is mounted on a cold plate and cooled by conduction only.

5. Summary

In conclusion we have presented a process for fabrication of advanced semi-finished parts and products based on the high-temperature superconductor YBCO. Different geometries and sizes are fabricated on a regular basis. Dedicated quality assurance measures accompany each step of the manufacturing process. With about 80% of the material being single domain, the reproducibility of the process is quite good. The scale-up from semi-automated processing in the laboratory to industrial-type, automated processing will be an essential step towards commercialization of the technology.

As a result superconducting bearings with forces up to 200N can be manufactured.

References

[1] FZK/INFP 1997 *German Patent.*

[2] FZK/INFP 1997 *http:\\www.fzk.infp.de.*

[3] Boegler P and Urban C and Rietschel H and Bornemann H J 1994 *Applied Superconductivity* **2** 315–25

[4] Bean H P 1962 *Phys. Rev. Lett.* **8** 250–3

Inst. Phys. Conf. Ser. No 158
Paper presented at Applied Superconductivity, The Netherlands, 30 June–3 July 1997

Simulation of Thermal Fields for the Optimization of the Melt-Texturing Process of HTSC Bulk Materials

M Seeßelberg†[1], G J Schmitz†, T Wilke‡ and M Ullrich††

† ACCESS e. V., Intzestraße 5, D-52072 Aachen, Germany

‡ Johann Wolfgang Goethe-Universität, Physikalisches Institut, Robert-Mayer-Straße 2-4, D-60054 Frankfurt, Germany

†† Zentrum für Funktionswerkstoffe gGmbH, Windausweg 2, D-37073 Göttingen, Germany

Abstract. Using the software package CASTS (=Computer Aided Solidification TechnologieS), simulations of thermal fields in bulk high T_c superconducting materials during the melt texturing process have been carried out. As examples, simulations of a laser-zone-melting process and of a Vertical Gradient Freeze furnace are presented. These examples demonstrate the efficiency of simulations for optimization of the melt-texturing process.

1. Introduction

Industrial applications of high T_c bulk materials require advanced manufacturing processes with respect to a further improvement of the superconducting properties and reduced process times. Melt growth processes can be improved by an optimal selection of the temperature program for the heaters of the furnace being used. For this purpose, information about the process is required which can hardly be measured experimentally, e.g. the time and the location where the solidification front enters the specimen as well as the position and the shape of the solidification front in time. Many time-consuming experiments can be omitted by using numerical simulations of the temperature fields during the melt texturing process. The software package CASTS (=Computer Aided Solidification TechnologieS) is designed to perform such kind of calculations [1-2]. Besides solving the heat conduction equation this software takes into account the transport of energy from the heaters to the specimen (and vice versa) via infrared radiation as well as heat losses due to convection. The release of latent heat is also considered although in this case its influence is negligible due to the slow growth rates of the solid phase.

Results of simulations of a laser-zone-melting process are presented and simulations of temperature fields in a Vertical Gradient Freeze (VGF) furnace are shown. These examples demonstrate the benefits of simulations of the temperature fields during the melt texturing process.

[1] E-mail: mase@gi.rwth-aachen.de

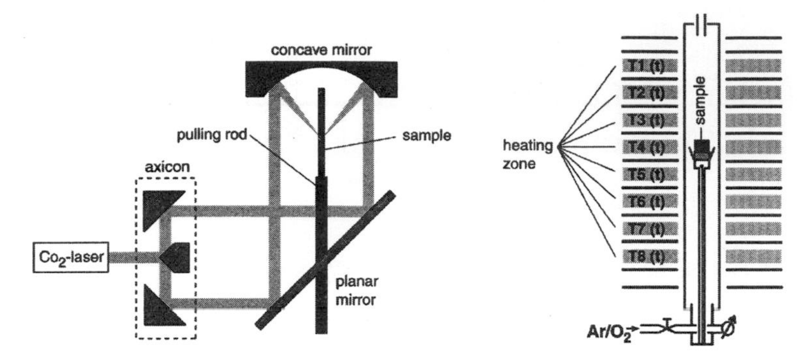

Figure 1. Scheme of the experimental set-up of the laser-zone-melting process (left) and of the Vertical Gradient Freeze furnace (right).

2. Simulation of the laser-zone-melting process

Laser-zone-melting of YBaCuO is a promising process e.g. for the production of HTSC current leads. High production rates of 10 mm/h seem to be within the reach of the method and are possible due to very large temperature gradients in the order of several 100 K/mm during the process. Fig. 1 schematically shows the experimental set-up. The samples are rod-shaped, 4 mm in diameter and up to 70 mm long. To avoid the destruction of the samples by strong thermal stress, a mirror optic is designed which produces a ring-shaped focus on the rod [3-4]. The YBaCuO-rods become more stable and their texture have a better quality if a polycrystalline core remains inside the hot zone. For the formation of such a polycrystalline core, the tuning of the power of the laser and the diameter of the rods is crucial and can be efficiently supported by numerical simulations of the temperature fields inside the rods.

To perform such calculations, the heat transfer coefficient α and the losses of laser power in the mirror system have to be determined. The coefficient α describes the energy loss of the sample to the surrounding atmosphere due to convection and depends on the geometry of the sample. To determine these quantities, the temperature-time curve of a graphit sample in the laser-zone-melting apparatus was measured with a thermocouple. Simulations of the cooling of this sample showed good agreement with the experimentally measured cooling curve when the value $\alpha = 2.5$ mW/(cm^{2}K) was used. Simulations of the temperature during the heating of the sample revealed that only 12 W were absorbed by the specimen although the nominal power of the laser was 30 W.

Simulations of a YBaCuO-rod with a diameter of 4 mm indicate that the optimal result is obtained when a laser power of 16 W is absorbed (Fig. 2). A decreasing laser power leads to an undesired increase of the polycrystalline core. On the other hand, already at an absorbed laser power of 18 W the core vanishes. Experiments with an absorbed laser power of 16 W and 18 W were carried out and the predicted shapes of $Y_2Ba_1Cu_1O_5$-enrichment and of the flux creep zone were observed (Fig. 3). This example demonstrates that simulations are able to predict optimal process parameters. The simulations also reveal that heat flow in the pulling rod influences the temperature field close to the laser focus when the laser-zone-melting process starts at the bottom of the rod. To obtain an optimal result in this stage, the absorbed power must be 18 W.

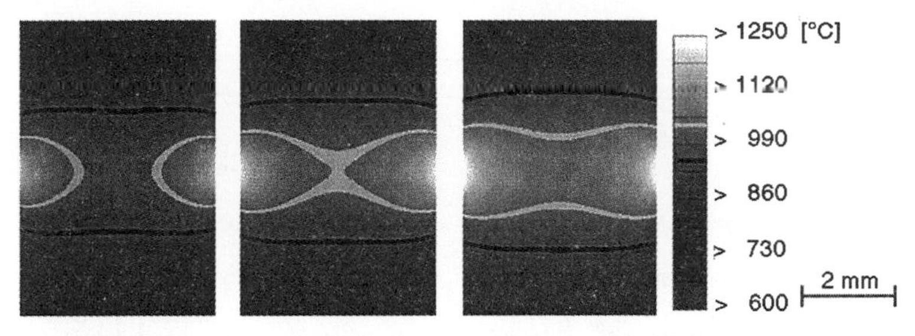

Figure 2. Simulations of the laser-zone-melting process. The diagrams show the temperature field in a section through the center of the sample for an absorbed laser power of 16 W (left), 17 W (middle) and 18 W (right). The bright (dark) line marks the isotherm of 1035 (950) °C.

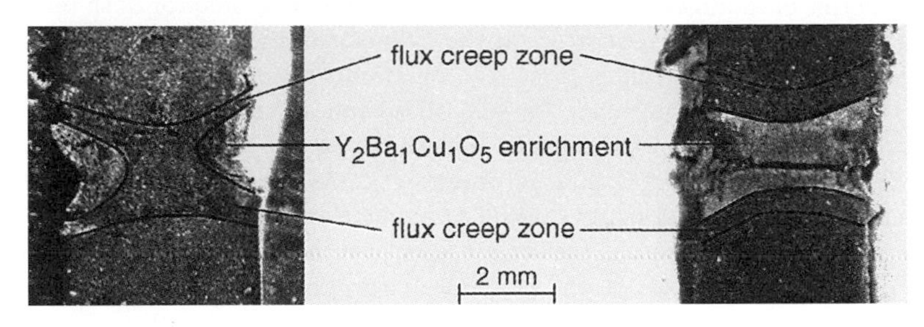

Figure 3. Experimental micrographs of YBaCuO-rods produced with an absorbed laser power of 16 W (left) and 18 W (right).

3. Simulation of a Vertical Gradient Freeze furnace

A Vertical Gradient Freeze (VGF) furnace is a multi-zone furnace designed for directional solidification (Fig. 1). This type of furnace provides better growth rate and temperature control compared to conventional Bridgman methods and nearly all vibrations are avoided because the sample as well as the furnace are kept stationary while a temperature gradient is driven through the samples by independently controlling the different heating zones. A further advantage of the furnace used is that the present set-up allows working in nearly every atmosphere, i.e. any chosen Ar/O_2 ratio. Details of HTSC sample preparation by the VGF method are given in reference [5].

Temperature fields were calculated in cylindrical YBaCuO samples of different sizes without taking into account the influence of the holding device of the sample and of the quartz tube within the furnace. Sample 1 has a height of h=18 mm and a radius of r=22.5 mm; height and radius of sample 2 and 3 are h=18 mm, r=16 mm and h=35 mm, r=5 mm respectively. Calculations have been carried out for two temperature programs P1 and P2 defined by the temperatures $T_1, T_2, \ldots, T_8$ of the eight equidistant heating zones. At the beginning of the process, the eight temperatures of program P1 (P2) amount to T_1=1245 (1090) °C, T_2=1255 (1105) °C, T_3=1180 (1080) °C, T_4 = 1105 (1055) °C, T_5 = 1030 (1030) °C, T_6 = 955 (1005) °C, T_7= 880 (980) °C and T_8 = 805 (955) °C. The cooling rate of each heating zones is set to 1 K/h.

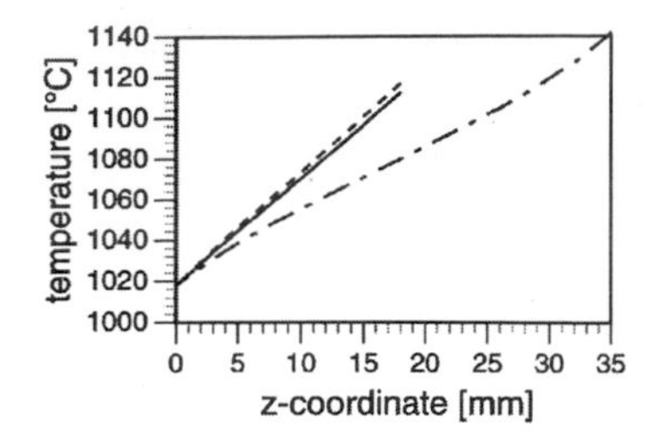

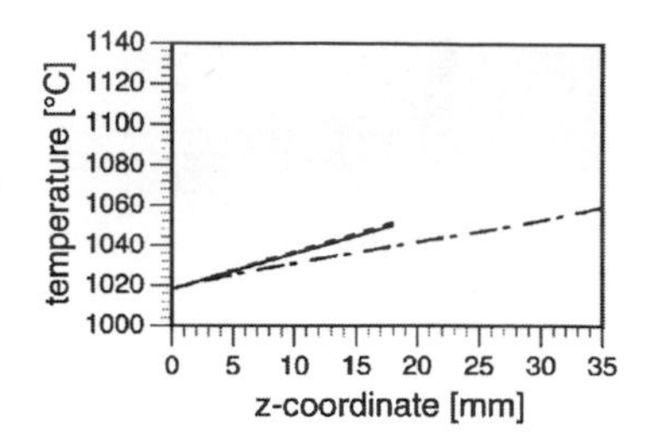

Figure 4. Calculated axial temperature profiles for the temperature programs P1 (left) and P2 (right). Dashed, solid and dashed-dotted lines belong to specimen 1, 2 and 3 respectively.

The isotherms of solidification in the different samples are planar for both temperature programs P1 and P2. The temperature at the cylinder axis increases nearly linear from top to the bottom of the cylinder (Fig. 4). The temperature varies only slightly with respect to the radius of the specimen. On the other hand, the temperature gradient in the specimen decreases with increasing height of the samples. As expected, the temperature gradient in process P1 is larger than in process P2. As a consequence the velocity of the isotherm of solidification in process P2 is about three times larger than in process P1. The calculations also reveal the time t when the isotherm of solidification enters the specimen. Time t depends on the temperature program and on the sample size. Its prediction is of great importance for the reduction of the process time.

4. Conclusions and outlook

Simulations reveal many informations necessary to optimize melt texturing processes which are hardly accessible to experiments. They are not restricted to YBaCuO materials and can be carried out for other HTSC materials and geometries of the specimen.

Acknowledgments

This work is financially supported by the Bundesministerium für Bildung, Wissenschaft, Forschung und Technologie (BMBF) under Grants No. 13N66614 and 13N66593 and by the European Union under contract No. BRE-CT94-1011.

References

[1] Hediger F and Hofmann N 1991 Proc. of the 5th Int. Conf. on Casting, Welding and Advanced Solidification Processes (TMS) 611–619

[2] Seeßelberg M, Schmitz G J, Nestler B and Steinbach I 1996, Proceedings of the ASC '96, Pittsburgh, accepted for publication in IEEE Trans. on Appl. Supercond.

[3] Byer R L et al. 1984 Rev. Sci. Instrum. **55** 1791–1796

[4] Wilke T, Baumgarten C and Assmus W 1997, Proceedings of M^2S-HTSC V, Beijing, accepted for publication in Physica C

[5] Ullrich M and Freyhardt H C 1993 *Superdonducting Materials*, ed. by Etourneau J, Torrance J B and Yamauchi H (France: IITT International, Gournay sur Marne) 205-210

Inst. Phys. Conf. Ser. No 158
Paper presented at Applied Superconductivity, The Netherlands, 30 June–3 July 1997

Preparation of top seeded $YBa_2Cu_3O_7$ large single domain tiles

R.Yu , V.Gomis , S. Piñol , F.Sandiumenge , J.Mora , M.Carrera , J.Fontcuberta , X.Obradors

Institut de Ciència de Materials de Barcelona , C.S.I.C. , Campus Universitat Autònoma de Barcelona , 08193 Bellaterra , Catalunya , Spain

Abstract. Single domain superconductors with diameters up to 40 mm have been prepared by top seeding growth using Nd123 melt textured ceramics as seeds . We show that the use of these seeds is preferred over MgO single crystals to reduce the multinucleation problem. Large levitation forces and critical currents have been measured which demonstrate the single domain character of the samples .

1.Introduction

A large effort is being carried out in the preparation of large single domain tiles of melt textured $YBa_2Cu_3O_7$ superconductors for large scale applications. Top seeding growth appears as a promising technique to avoid multinucleation for the preparation of large single domain tiles, while the vertical Bridgman growth has been shown to be more adapted to long bars with small diameters [1,2] . In this work we present an investigation of the conditions required to prepare single domain tiles and we also present some characterization of the quality of the samples . Complementary investigations of the single domain samples are reported in parallel works [3] .

2.Experimental

Cylindrical samples with diameters between 10 and 40 mm have been prepared with compositions as detailed in Table 1. 1%wt CeO_2 was added to refine the final size of Y_2BaCuO_5 (211) precipitates and to reduce liquid losses [2]. The pellets were pressed uniaxially or through CIP before presintering and the final homogeneity after the melt texturing process was improved in the first case. Single domain bars of $NdBa_2Cu_3O_7$, 7mm in diameter and 100mm in length , were prepared by Bridgman growth [4] . Cylinders were cut to be used as seeds. Commercial MgO single crystals were also tested. Samples were processed at 1.080 °C followed by fast cooling down to 1.020 °C and slow cooling as indicated in Table I, either in an isothermal furnace or in a cylindrical furnace with a temperature gradient. Levitation forces , inductive critical current measurements with a SQUID magnetometer and field dependent local magnetic induction measurements with a Hall microprobe located in the center of the base surface of the cylinder, were performed.

Table 1

Additives	Seed	T_{nucl}	G(°C/cm)	R(°C/h)
8% Y_2O_3	Nd123	~955 °C	--	0.4
8% Y_2O_3	MgO	~ 955 °C	--	0.4
25% 211	MgO	~ 990 °C	--	0.25
25% 211	Nd123	~990 °C	10	1.0
25% 211	MgO	~990 °C	10	1.0

3.Results and discussion

The influence of the precursors , the seed and the thermal cycle on the growth mechanisms have been investigated . In Figure 1 we can observe the heterogeneous nucleation at the seeds which was observed in all the investigated combinations of seeds and precursors but the tendency to multinucleation phenomena seems to differ . Under similar thermal and compositional conditions the use of Nd123 seeds is preferable to MgO , probably because the nucleation energy is smaller due to the higher structural similarity . This is reflected in the relative size of the crystal nucleated in the seed and those homogeneously nucleated. As a consequence the observed levitation forces in multidomain top seeded samples prepared simultaneously are enhanced by about 20 % when Nd123 seeds are used[5] .

The determination of the nucleation temperatures is a difficult problem and our results are consistent with previously reported seed nucleation temperatures reported [6] (a few degrees below T_p) while our homogeneous nucleation temperatures with211 additives appear to be about 10°C lower than those reported by these authors (1000°C) in agreement with previous works [7] . The origin of this discrepancy is probably related to the use of CeO_2 additives instead of PtO_2 or to the higher soaking temperature (1080°C vs 1050°C). An additional advantage in the use of Nd123 seeds instead of Sm123 is the higher peritectic temperature allowing to use

(a)

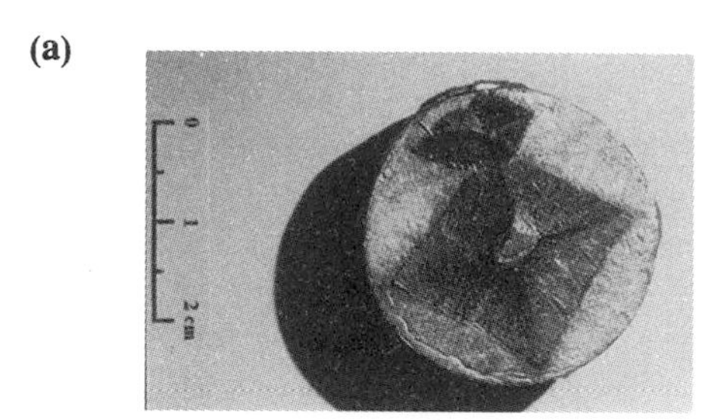

(b)

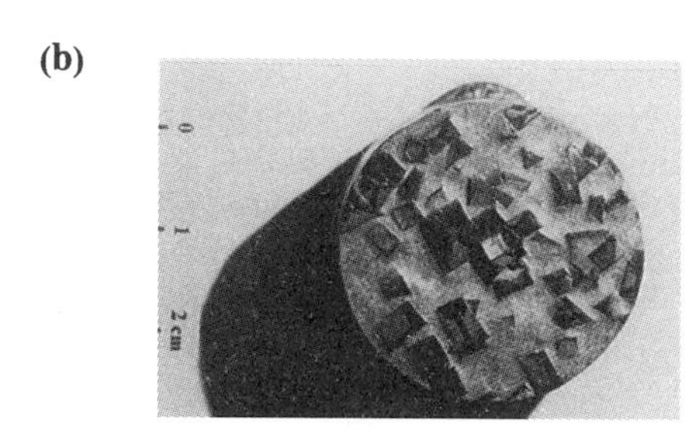

(c)

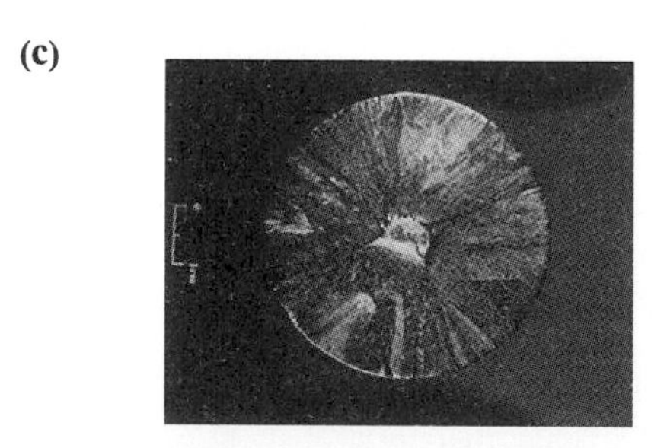

Figure 1 - Optical images of the upper surface in top seeded samples (a) quenched before completing the process (a) sample grown with a MgO seed , (b) sample grown with a Nd123 seed , (c) sample grown up to the completion of the thermal cycle with a Nd123 seed at 1°C/h .

thermal cycles with an isothermal plateau at a higher temperature which increases the solubility of Y ions in the liquid and hence increases the growth rate [8,9]. In this work we have found that cooling rates up to 1 °C/h may be used without perturbing the plane growth front while usually lower cooling rates have been reported [6,10,11]. We also found that the use of Y_2O_3 additives pushes down the nucleation temperatures (Table 1) because the corresponding peritectic temperatures are lower [12].

From the point of view of the understanding of the growth mechanism an appealing feature is a characteristic X-cross at the surface following the principal axes of the (001) plane. These edges develop after a square-like region surrounding the seed which we found to be related to the actual size of the seed but not to its shape. The origin of both the initial square and the X-cross following it is still unclear [6,10]. It has been claimed [10] that these geometrical features result from a small but noticeable 211 particle segregation occuring during the growth of $YBa_2Cu_3O_7$. The anisotropy of solute composition in the growth front would act as a driving force for the particle segregation. The observation of sharp edges then would indicate a good crystallinity of the samples. We have indeed observed, as indicated in Figure 1(c), that in single domain samples prepared at relatively high growth rates (1°C/h) these edges become diffuse and this is then reflected in the critical current distribution which is sligthly depressed at these edges [3].

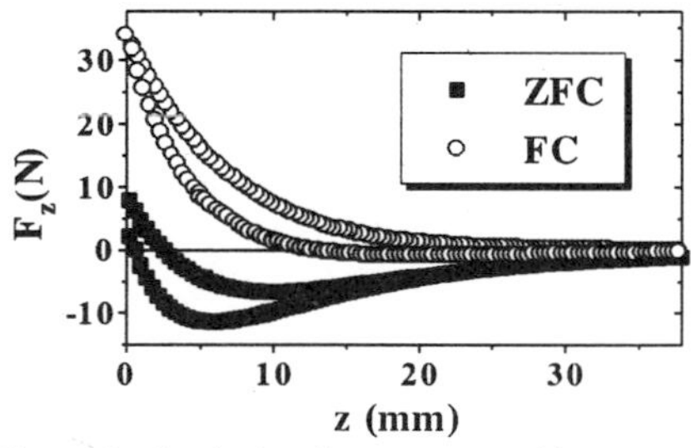

Figure 2 - Levitation forces measured in a sample having a diameter of 40 mm grown with a Nd123 seed at 1°C/h.

The levitation force measurements were performed with a SmCo PM having B_r=0.3T and Φ=25mm. A typical vertical force hysteresis loop is displayed in Figure 2 which corresponds to a sample having a diameter of 40 mm and a thickness of 10 mm. As we have largely discussed in a recent Finite Element modelling analysis of the levitation forces [13], the maximum vertical force measured in these hysteresis loops depends on geometrical parameters and the strength of the permanent magnet. However, when similar diameters are used for both the PM and the SC pellet the levitation force may be normalized to that calculated assuming the Meissner state, i.e. when complete flux exclusion occurs. Further details on the analysis of the levitation force measurements are reported in this conference [3].

Finally, we have tested a new the quality of the samples by means of magnetic remanence profile measurements carried out with a Hall microprobe after a FC process under a field of 1T at 77K. In Figure 3 we show a typical example corresponding to a cylinder with a diameter of 30mm. As it has been demonstrated by several authors this technique appears to be very useful to detect the persistence of grain boundaries in multidomain samples and so it may be validated as a non-destructive tool to characterize the samples. Actually, even in single domain samples a nice correlation between the maximum trapped field and the levitation measured in ZFC processes have been recently demonstrated [15]. This is because, as it may be observed in Figure 3, the single domain samples present some substructure in the remanence profile which may be associated to the existence of subgrain boundaries and lead to a weakening of the overall critical current [3].

In summary large single-domain tiles of $YBa_2Cu_3O_7$ superconductors have been prepared in diameters up to 40 mm by means of top seeding growth using melt textured $NdBa_2Cu_3O_7$ ceramics as seeds. The levitation forces and remanence profile measurements appear to be very useful non-destructive techniques to the quality characterize the quality of single domain samples and, actually, after some modelling, the real critical currents of the bulk sample may be evaluated.

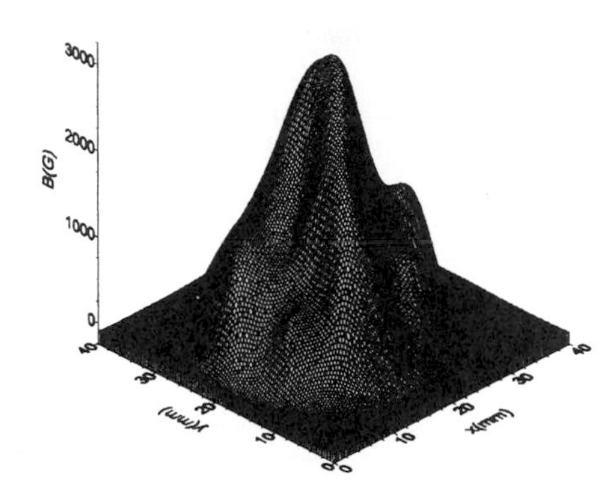

Figure 3 - Magnetic remanence profile at 77K of a cylinder with ϕ=30mm after a FC process under 1T.

Acknowledgements

The financial support of the following projects are acknowledged : CICYT (MAT96-1052) , EC-EURAM (BRE2CT94-1011) , REE (92-2331) and Generalitat de Catalunya (GRQ95-8029) .

References

[1] Meng R L , Gao L , Gautier-Picard P. , Ramirez D. Sun Y Y and Chu C W 1994
Physica C **232** 337
[2] .Piñol S , Sandiumege F , Martínez B , Gomis V , Fontcuberta J , Obradors X , Snoeck E and Roucau C 1994 Appl.Phys.Lett. **65** 1448
[3] Amoros J , Carrera M , Granados X , Fontcuberta J and Obradors X , This Confeence ; Portabella E , Mora J, Fontcuberta J and Obradors X , This Conference
[4] Yu R , Sandiumenge F , Martínez B , Vilalta N and Obradors X (1997) Appl.Phys.Lett. **71**, in press
[5] Yu R, Mora J , Piñol S , Sandiumenge F , Vilalta N , Gomis V , Martínez B , Rodriguez E , Amoros J , Carrera M , Granados X , Camacho D , Fontcuberta J and Obradors X 1997 IEEE Trans. on Applied Supercond. , in press
[6] Gautier-Picard P , Beaugnon E and Tournier R 1997 Physica C **276** 35
[7] Lo W , Cardwell D A , Dewhurst C D and Dung S L 1996 J.Mater.Res. **11** 786
[8] Marinel S, Wang J , Monot I , Delamare M P , Provost J and Desgardin G 1997 Supercond.Sci.Technol. **10** 147
[9] Yu R , Mora J , Vilalta N , Sandiumenge F , Gomis V , Piñol S and Obradors X 1997 Supercond.Sci.Technol. **10** , in press
[10] Honjo S , Cima M J , Flemings M C , Ohkuma T , Shen H , Rigby K and Sung T H 1997 J.Mater.Res. **12** , 880
[11].Chow J C L , Lo W , Dewhurst C D , Leung H T , Cardwell D A and Shi Y H 1997 Supercond.Sci.Technol. **10** 435
[12] Krabbes G , Schatzle P , Bieger W Wiesner U , Stover G , Wu M , Strasser T , Kohler A , Litzkendorf D , Fischer K and Gornert P 1995 Physica C **244** 145
[13] Camacho D , Mora J , Fontcuberta J and Obradors X 1997 J.Appl.Phys. **82** , in press
[14] Strasser T , Habisreuther T , Gawalek W , Wu M , Litzkendorf D , Gornert P , Ilgushin K V and Kovaljov L K 1995 *Applied Superconductivity 1955* , Inst.Phys.Conf.Ser. No 148 , 687
[15] Zhang X H, Parikh A S and Salama K., IEEE Trans. On Applied Supercond. (in press)

Inst. Phys. Conf. Ser. No 158
Paper presented at Applied Superconductivity, The Netherlands, 30 June–3 July 1997

Study of the Influence of the Processing Rate on the Directionally Solidified YBaCuO

M Boffa† A DiTrolio‡ U Gambardella† S Pace† M Polichetti† and A Vecchione†[1]

† Istituto Nazionale per la Fisica della Materia and Dipartimento di Fisica Universitá di Salerno, via Salvador Allende, 84081 Baronissi (SA), Italy.

‡ Istituto Materiali Speciali, CNR Area della Ricerca di Potenza, 85050 Tito Scalo (Pz) Italy.

Abstract.
Directional solidification technique with sample transport in a thermal gradient is used for fabrication of melt textured superconducting YBaCuO bars. Recent developments of the processing for this material are reported. The paper will analyze the influence of the high processing rates (in particular using pulling rate in the furnace up to 4 cm/h) both on the microstructure and on the superconducting properties. Morphological and structural analyses show aligned grains also for samples fabricated with such high pulling rates. From characterizations performed by ac susceptibility measurements and dc magnetization curves results that the onset of superconductivity is 91 K and the critical current density at T = 77 K and at zero magnetic field is about 5000 A/cm^2. These preliminary results indicate that the quality of these Y123 samples appears comparable to (Nd,Sm)-based superconductors grown using similar processing conditions.

1. Introduction

Partial melt processing method [1] has successfully been applied to obtain bulk samples of ceramic superconductors with critical current densities of 10^5 A/cm^2 at 77 K [2, 3]. This fabrication process is based on heating of the Rare Earth (RE)123 solid phase above the peritectic temperature, T_p, where a decomposition into a RE_2BaCuO_5, hereafter named 211, solid and ($BaCuO_2$+CuO) liquid phase takes place. When this partially melted mixture is cooled down through T_p, the formation of a melt textured RE123 phase occurs. Huge RE123 specimens with minimized weak links and large grains aligned with well defined crystallographic orientation are fabricated by using the directional solidification; among various techniques, the directional solidification can be determined

[1] E-mail: VECCHIONE@VAXSA.CSIED.UNISA.IT

by relative motion at a pulling rate R_p between the sample and a thermal gradient G set into the furnace.
The experimental parameters, G and R_p, play a fundamental role to obtain the correct kinetics of solidification of RE123 phase because a stable growth without spurious inclusions is controlled by the cooling rate $R_p \cdot G$. It seems clear that the product $R_p \cdot G$ should be as high as possible to reduce the coarsening of 211 particles during the melting process. On the other hand, as it results from various works [4, 5] an as high as possible G/R_p ratio is also necessary for a stable solid–liquid growth interface in order to obtain a high quality textured samples. In this paper a preliminary study of the effect of the pulling rate R_p on the crystallization mechanism of Y123 grown by the horizontal Bridgman method is reported.

2. Experimental

The details of the preparation of Y123 superconducting bars are described in ref.[6]. Briefly speaking, the samples were fabricated starting from powders of Y_2O_3, $BaCO_3$ and CuO to achieve 50 % Y211 phase in Y123 matrix. The precursor powder was calcined in a furnace at 950°C for 12 hours in air and subsequently oxygenated in extra plateau at 550 °C for 5–7 h. Bars of dimensions 50×5×5 mm^3 were partially melted using the horizontal Bridgman method and heated just above the peritectic temperature T_p. The recrystallization was performed moving the sample in a fixed thermal gradient G=40 °C/cm. The apparatus was designed to obtain R_p values up to 4 cm/h. The samples were prepared with R_p ranging between 0.05 and 4 cm/h using a thermal cycle consisting of a rapid heating up to a temperature $T > T_p$ and of a slow cooling in controlled O_2 atmosphere until the whole sample was passed through the thermal gradient. When this step was over, the samples were cooled down to 550°C at 20 °C/h in air mantaining their temperature at 550 °C under a pure oxygen flow for 48 hours.

3. Results and Discussion

The samples were investigated by various techniques. XRD and SEM with coupled EDS analysis were used for morphological and structural characterizations while superconducting properties were studied by vibrating sample magnetometry and ac susceptibility.
XRD patterns, taken from the face perpendicular to the pulling direction, exhibit an alignment degree D depending on R_p. An estimate can be given by the ratio $D = D_t/D_p$, where D_t and D_p are the ratios of the integrated intensities of the 001 and 111 peaks for a textured sample and for a powdered sample, respectively. In samples grown with $R_p \leq$ 0.1 cm/h showing only 001 peaks, D results larger than 100 as in the case of well oriented Y123. On the other hand, in Y123 fabricated using values of $R_p \geq 1$ cm/h, different peaks from 001 ones are also revealed. However, it has to be pointed out that in these cases a value D = 50 is estimated, which indicates that large textured regions of Y123 phase take place.
Figure 1 shows a typical SEM micrograph of a sample processed at R_p=4 cm/h, i.e. at cooling rate $C = R_p \cdot G$ = 160°C/h. The figure reveals the existence of Y123 texture in

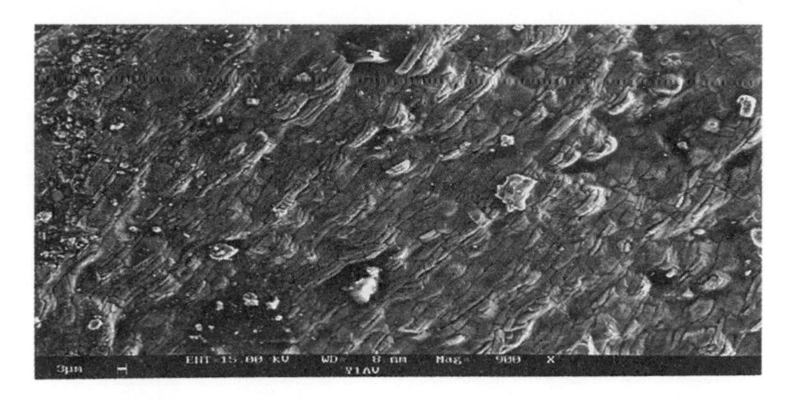

Figure 1. SEM micrograph of the top surface of a sample cooled at 160 °C/h through the peritectic temperature.

domains estimated in about 1 mm in lenght including small roundish in shape and quite uniformly distributed Y211 inclusions. Even if these domains do not extend on typical scale of high quality melt textured superconducting materials, their presence is surprising. Indeed the size of the domains is expected to decrease so rapidly by increasing C, that it has been considered impossible the development of significant single domain above a critical C[5]. These high rates values may suggest that the processing time is long enough to make possible a solidification and growth mechanisms of Y123 material. Also, it was found that the mean size of residual Y211 particles are weakly dependent on R_p [4, 7]. The size of Y211 particles is controlled both by coarsening during the holding time at $T \geq T_p$ and by their dissolution near to solid–liquid growth interface, necessary to supply the yttrium for a correct Y123 growth. For $R_p \sim$1 cm/h in spite of the shortness of the time to dissolve the Y211 particles, the holding time at high temperature is so low that, being the Y211 coarsening inhibited, the yttrium is avaliable to form Y123 phase with high flow. For this reason, the competition between the above mechanisms makes the Y211 size limited to few microns. The initial region of the samples, i.e. the part which first recrystallizes, is strongly disordered and rich of BaCuO+CuO inclusions. This region has a dimension which increases with the processing rate as already observed in other studies [7]. The Y211 phase is dominant to the end of the samples. All these behaviours are expected using a horizontal Bridgman technique as fabrication method. SEM micrograph at higher magnification is reported in Figure 2. From this figure may be identified structures reminding spiral–like or ledge growth generally observed in YBCO thin films [8], single crystal[11] and high quality bulk samples[9, 10]. The Figure 2 also gives evidence for microcracks likely due to thermal contraction resulting from orthorhombic–tetragonal transition during the oxygenation process. This is further corroborated by the observation that the microcracks extend across the identified spiral-like or ledge structure.

The superconducting properties of the samples were investigated by measurements performed by means of a Vibrating Sample Magnetometer. The direction of applied magnetic field was parallel to the sample pulling direction. Ac susceptibility curves measured as function of the temperature on bars processed at the lowest (0.05 cm/h) and the highest (4 cm/h) values of R_p used in this work, show the onset of the supercon-

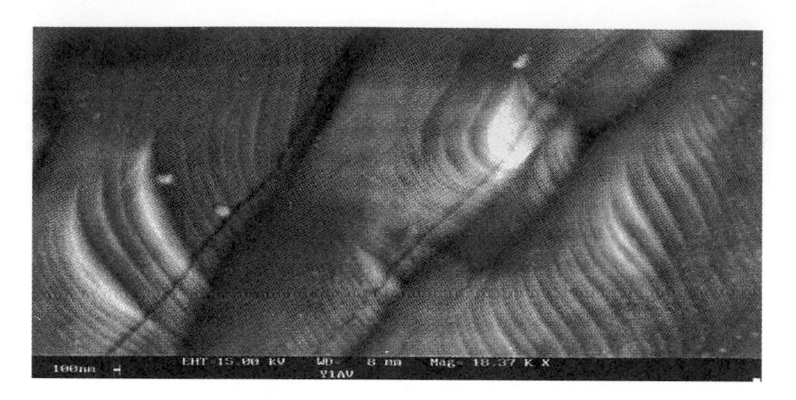

Figure 2. SEM micrograph at higher magnification.

ducting transition at 92 K and 91 K, respectively. It is interesting to note that, from the hysteresis cycles performed at 77 K on all the samples processed at high rates a critical current density ranging between 3000 and 7000 A/cm^2 is evaluated by using Bean's model. Although these values are lower than those obtained in Y123 samples processed at lower R_p, they appear comparable with Nd123 and Sm123 crystallized at the same rate[6, 12]. In summary, these results indicate that partial texturing occurs at high growth rate with fabrication times encouraging in the perspective of potential applications of Y123 superconductors.

References

[1] Jin S, Tiefel T H, Sherwood R C, Davis M E, Van Dover R B, Kammlott G W, Fastnacht R A and Keith H D, 1988 *Appl. Phys. Lett.* **52** 2074–2076.

[2] Murakami M 1992, *Melt Processed High Temperature Superconductors* (Singapore: World Scientific Ltd.) 21–44.

[3] Salama K, Selvamanikam V, and Lee D F, 1992 *Processing and Properties of High T_c Superconductors*, vol.I, (Singapore: World Scientific Ltd.) 155–211.

[4] Izumi T and Shiohara Y, 1992 *J. Mater. Res* **7** 16–23.

[5] Pellerin N, Odier P, Simon P, Chateigner D, 1994 *Physica C* **222** 133–148.

[6] Boffa M, Di Trolio A, Pace S, Saggese A and Vecchione A, Proceedings of Applied Superconductivity Conference, Pittsburgh, USA.

[7] Granados X, Yu R, Martinez B, Pinol S, Sandiumenge F, Vilalta N, Carrera M, Gomis V, Obradors X, 1997 preprint.

[8] Schlom D G, Anselmetti D, Bednorz J G, Gerber C, Mannhart J, 1994 *J. Crystal Growth* **137** 259–267.

[9] Goyal A, Alexander K B, Kroeger D M, Funkenbusch P D, Burns S J, 1993 *Physica C* **210** 197–211.

[10] Jin S, Kammlott G W, Nakahara S, Tiefel T H, Graebner J E, 1991 *Science* **253** 427–429.

[11] Namikawa Y, Shiohara Y, 1996 *Physica C* **268** 1–13.

[12] Salama K, Parikh S and Woolf L, 1996 *Appl. Phys. Lett.* **68** 1993–1995.

Inst. Phys. Conf. Ser. No 158
Paper presented at Applied Superconductivity, The Netherlands, 30 June–3 July 1997

Dislocation configurations in directionally solidified $NdBa_2Cu_3O_7$

N. Vilalta, F. Sandiumenge, R. Yu and X. Obradors

Institut de Ciència de Materials de Barcelona (CSIC), Campus de la Universitat Autònoma de Barcelona, 08193 Bellaterra, Catalunya, Spain.

Abstract. Diffraction contrast imaging is used to characterize the dislocation substructure in $NdBa_2Cu_3O_7$ bars directionally solidified in air (T_c=96K). The dislocation density is higher than that in $YBa_2Cu_3O_7$. In addition, dislocation lines parallel or nearly parallel to the c-axis typically arranged in pile-ups and subgrain beundaries, are commonly found. Microcracks perpendicular to the basal (001) plane are also observed. Most of dislocations glide on the (001) plane, displaying a strong tendency toward dissociation. The extent of the dissociation and the formation of stacking faults strongly depends on the concentration of $Nd_4Ba_2Cu_2O_{10}$ inclusions.

1. Introduction

Comparison between Y123 and Ndl23 materials reveal important differences in J_c(H,T) behavior, which underlying microstructural origin remains almost unexplored. The two signatures of such a differential behavior are: i) A field induced flux pinning process or "fishtail effect" in J_c(H) observed in Ndl23 [1], and ii) An exponential dependence J_c(T) in the low field (≤1T), high temperature (≥40K) regimes, also observed in Ndl23 [2] which contrasts with a softening observed in Y123 correlated with interface pinning [3]. Up to now, point i) has motivated most of microstructural work, focused on the search of any signature of Nd/Ba antisite defect clustering, (see for instance Refs. [4] and [5]). Owing to the dependence of T_c and H_{c2} on the concentration of Nd in Ba sites, such defect could become a flux pinning centre at magnetic fields where the non-substituted matrix is still superconductor [1]. The second aspect has been very recently addressed and points to a reduced efficiency of 422/Ndl23 interfaces in pinning the vortices [2].

The present work shows that Nd123 display new dissociation morphologies and dislocation lines parallel to the c-axis, as well as microcracking along (100). c-axis dislocations may provide strong linear pinning centres perpendicular to the weakly coupled CuO_2 layers. On the other hand, associated long range strain fields may provide areas with depressed T_c and H_{c2} capable to act as field induced flux pinning centres in an analogous way as tentatively proposed for Y123 [6].

2. Experimental

Superconducting Ndl23 bars 150 mm in length and 8 mm in diameter and several 422 additions were directionally solidified in air at T=1120°C [2]. As-solidified samples were submitted to annealing at 950°C in flowing Ar followed by an oxygenation step at 350°C. By this procedure a T_c onset of 96K was obtained. The microstructure of the samples was characterized by diffraction contrast imaging in a TEM operating at 200 kV.

3. Results and discussion

3.1. Dislocation configurations in the (001) plane

Dislocation configurations have been studied in samples containing 5, 15 and 20 % of 422 phase addition. In the samples containing the higher concentration of 422 particles, dislocations present a strong tendency towards dissociation. The density of dislocations and stacking faults is drastically increased at 422/Ndl23 interfaces (Fig. 1). In Fig. 1 a 422 particle is shown on the left side of the image. The precise trace of the particle-matrix interface cannot be clearly observed due to the huge density of stacking faults attached to it. The centre and right hand side of the image also present a high density of [100] or [010] dislocations pinned at the interface. Dislocation lines and stacking faults shown in this micrograph lie on the (001) plane. Fig. 1 clearly shows that perfect [100] (or [010]) dislocations are extensively dissociated leaving stacking faults on segments of their length. A **g.b** analysis [7] of such configurations is consistent with a stacking fault bounded by $\frac{1}{6}[301]$ and $\frac{1}{6}[30\bar{1}]$ partial dislocations (as referred to one twin domain). This configuration is associated with the insertion of a CuO_x layer with height c/6 at the dislocation core [8].

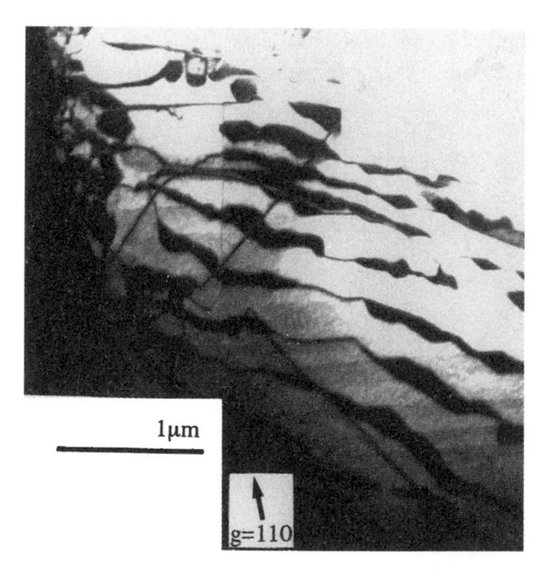

Figure 1: BF TEM micrograph of a Nd123 sample with 20% 422 addition showing a high density of dislocations pinned at a 422 inclusion. Dislocations appear highly dissociated. Dark areas are stacking faults left on segments of their lenght.

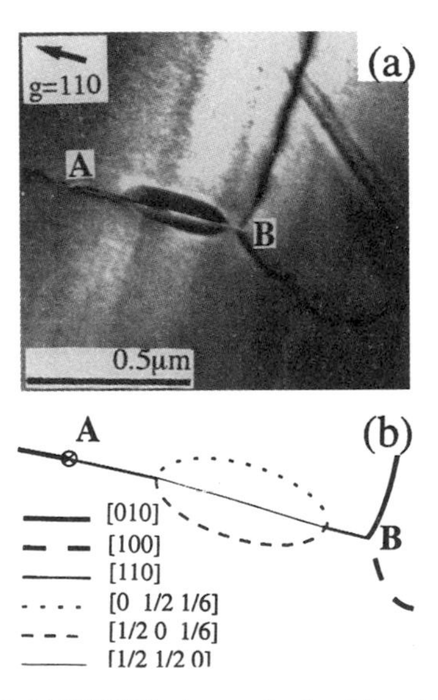

Figure 2: (a) BF TEM micrograph of an interaction configuration. All dislocations lie on the (001) plane except dislocation marked by ⊗ which is parallel to the c-axis. (b) Schematic drawing of the dissociation.

Both perfect and partial dislocation are found to interact with the twin boundaries suggesting that they were introduced in the twinned matrix [9, 10].

Dislocations with Burgers vectors **b**=<110> appear also extensively dissociated as shown in Fig. 2. The reaction may be written

$$[110]\rightarrow\frac{1}{6}[301]+\frac{1}{6}[03\bar{1}]+\frac{1}{6}[110] \qquad (1).$$

The interaction between perfect dislocations at nodes labelled A and B is described by the reaction: [100] + [010] → [110]. It is worth emphasizing that while dislocations joining at B all lie on the (001) plane, in the A node the [100] dislocation is parallel to the c-axis. In samples with lower 422 concentration (5-15% in the precursor) the dislocation density is lower and present a lower tendency towards dissociation. Moreover, the geometry of the dissociated configurations has striking differential features when compared with those found in samples with 20% 422 precursor addition (Fig. 3). The arrowed region in Fig. 3 shows two mutually perpendicular narrow planar defects bounded by partial dislocations associated with a [110] dislocation. A **g**·**b** analysis indicates [7] that the reaction involved here is given by eq. (1); however, the geometrical arrangement is different from that found in Y123 [11]. Dissociated segments are highly elongated in the <100> directions where the in-plane component of the Burgers vector is edge type. Since there is a $\frac{1}{2}[110]$ partial dislocation which transforms Burgers vector $\frac{1}{6}[301]$ into $\frac{1}{6}[03\bar{1}]$ at either side of the dissociation line [see eq. (1)], the resulting configuration consists of two half-loops elongated in perpendicular directions.

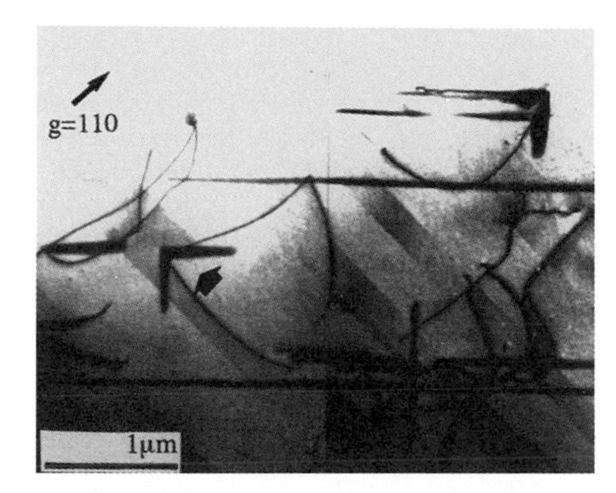

Figure 3: BF TEM micrograph displaying a general view of the microstructure of a sample with 5% 422 addition. Arrowed are two perpendicular narrow stacking faults associated with the dissociation of a [110] dislocation (this dislocation is out of contrast).

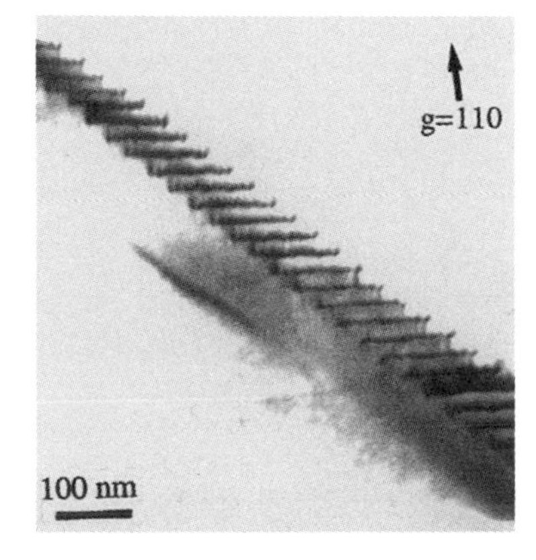

Figure 4: BF TEM micrograph of a (100)/(010) subgrain boundary formed by an array of edge dislocations parallel to the c-axis.

3.2. *Dislocation configurations in the (100), (010) and {110} planes*

Observed dislocation configurations in planes perpendicular to (001) are typically dislocation pile-ups, and subgrain boundaries [7]. Dislocations parallel to the c-axis are typically edge type with **b**=[100]. Figure 4 shows a subgrain boundary composed by an array of parallel [100] (or [010]) dislocations parallel to the c-axis stabilized in the (100)/(010) plane. The estimated line espacing is ~35 nm, which corresponds to a rotation of ≈0.6° about the [001] axis. In consistency with above evidences of dislocation glide on (100)/(010) planes, the present samples also display microcracks parallel to (100)/(010). Figure 5 shows examples of such microcracks intersecting at right angles.

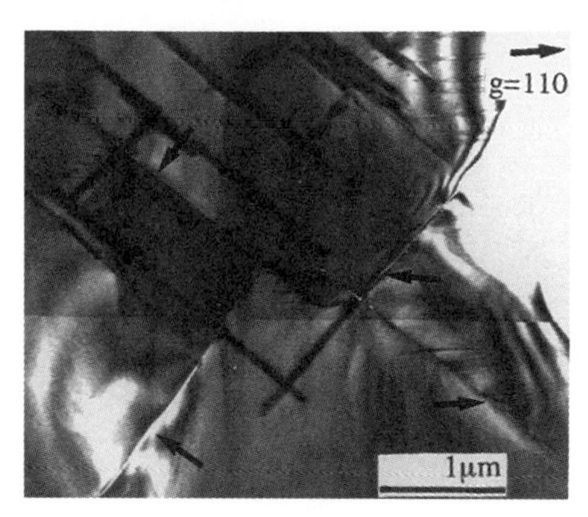

Figure 5: BF TEM micrograph showing mutually perpendicular (100)/(010) microcracks, indicated by arrows.

4. Conclusions

TEM analysis of directionally solidified Nd123 has revealed a higher dislocation density than in similarly processed Y123. Moreover, dislocations display a stronger tendency to dissociate. The resulting morphologies display striking differences which point to an enhanced diffusive regime along dislocation cores and interfaces during the post solidification process. Interestingly, this material contains a high density of dislocation lines parallel to the c-axis which arrangements point to dislocation glide on the (100)/(010) planes. Dislocation substructures and microcracks perpendicular to (001) point to a reduced crystalline anisotropy which correlates with an observed reduced anisotropy of the superconducting properties [12].

References

[1] Murakami M, Sakai N, Higuchi T and Yoo S I 1996 *Supercond. Sci. Technol.* **9** 1015 and refs. therein.
[2] Yu R, Sandiumenge F, Martínez B, Vilalta N and Obradors X 1997 *Appl. Phys. Lett.,* **71** 413
[3] Martínez B, Obradors X, Gou A, Gomis V, Piñol S, Fontcuberta J and Van Tol H 1996 *Phys. Rev. B* **53** 2797.
[4] Ting W, Egi T, Kuroda K, Koshizuka N, and Tanaka S 1997 *Appl. Phys. Lett.* **70** 770
[5] Hirayauna T, Ikuhara Y, Nakamura M, Yamada Y and Shiohara Y 1997 *J. Mater. Res.* **12** 293
[6] Ullrich M, Müller D, Heinemann K, Niel L and Freyhardt H C 1993 *Appl. Phys. Lett* **63** 406
[7] Vilalta N 1997 unpublished work
[8] Rabier J, Tall P D and Denanot M F 1993 *Philos. Mag. A* **67** 1021
[9] Rabier J and Denanot M F 1992 *Philos. Mag. A* **65** 427
[10] Werwerft M, Dijken D K, De Hosson J Th M and Van Der Steen A C 1994 *Phys. Rev. B* **50** 3271
[11] Sandiumenge F, Vilalta N, Piñol S, Martínez B, Obradors X 1995 *Phys. Rev. B* **51** 6645
[12] Martínez B, Sandiumenge F, Yu R, Vilalta N, Puig T, Gomis V and V Obradors X 1997 this conference.

Inst. Phys. Conf. Ser. No 158
Paper presented at Applied Superconductivity, The Netherlands, 30 June–3 July 1997

Directional solidification in air of $NdBa_2Cu_3O_7$ Superconductors with high T_c and J_c

B.Martínez, F. Sandiumenge, R. Yu, N. Vilalta, T. Puig, V. Gomis and X.Obradors

Institut de Ciència de Materials (CSIC); Campus de la U.A.B., E-08193 Bellaterra (SPAIN)

Abstract. Single domain $NdBa_2Cu_3O_{7-x}$ (Nd123) with additions of $Nd_4Ba_2Cu_2O_{10}$ (Nd422) superconducting bars (150 mm in length and 8 mm in diameter) have been fabricated by directional solidification in air. The additions of Nd422 and CeO_2 were found to reduce the loss of liquid phase during melt texturing as well as to refine the Nd422 precipitates. A transition temperature of T_c≈96 K has been detected by using both AC and DC magnetic measurements. This is a very important results that demonstrates that low oxygen partial pressure is not necessary to fabricate Nd123 superconductors with high T and J. This results also make evident that the Nd-Ba substitution is minimized during the in air directional solidification.

1. Introduction

Nd123 has been proposed as a promising alternative to Y123 because of its higher superconducting transition temperature (T_c≈96K) and larger critical current density in high magnetic fields [1]. In addition, it has been reported that the growth rate of Nd123 is about 10 times faster than that of Y123 owing to the higher solubility of Nd in the partial melt, with promising implications on the commercial mass production of high T_c superconductors[2].

Nevertheless, an important drawback of this material is the partial substitution of Ba by Nd cations, the so-called anti-site defects, leading to broad and low superconducting transition temperatures. The use of oxygen-controlled melt growth (OCMG) process with a reduced oxygen atmosphere during solidification [3] has been presented as the only way to overcome this problem. On the other hand, this material usually presents a field-induced pinning process leading to enhanced J_c values at high magnetic fields [3]. It was proposed that the microstructural origin of such behavior lies on the occurrence of nanometric regions having a depressed superconducting order parameter which then become normal at lower magnetic fields than the surrounding matrix [1,4].

2. Experimental

Superconducting Nd123 bars 150 mm in length and 8 mm in diameter were directionally solidified in air at T=1120°C through a thermal gradient of 20°C/cm at a speed of 1mm/h using a vertical Bridgman furnace, including initial additions, 0, 5, 10, 15, and 20 wt% of 422 phase. After solidification the samples were submitted to different thermal treatments in order to optimize T_c. The microstructure of the samples has been analyzed by polarized light microscopy, and using scanning (SEM) and transmission (TEM) electron microscopy while critical currents were measured inductively by using a SQUID magnetometer. The measured 422 particle size distributions are properly described with a log-normal law with a mean particle size of few μm, about two times higher than that achieved in optimized Y123 directionally solidified samples [5]. The superconducting transition temperature of the samples can be optimized up to an onset of 96 K when the as-solidified sample is annealed in Ar at 950°C before the final oxygenation step at 350°C (see Fig. 1). This is a very important point because it makes evident that the use of a reduced oxygen atmosphere [3] is not necessary to obtain Nd123 samples with high T_c and high critical currents.

The OCMG technique is based on the fact that the range of melting points in the $Nd_{1+x}Ba_{2-x}Cu_3O_y$ solid solution is widened as $P(O_2)$ is decreased [6]. Thus, the high T_c composition (x~0), which displays the highest peritectic temperature, is the first to nucleate during the slow cooling process. Similarly, our results may be interpreted considering the fact that the samples are produced by a directional solidification process where the composition having the highest solidification temperature is naturally selected at the single advancing planar growth front, since it is the first to solidify as the bar moves through the temperature gradient. Therefore, unless growth instabilities occur, this directional solidification process allows to obtain the high T_c phase even if the range of melting points in the solid solution is narrow.

3. Results and discussion

Even tough the addition of 422 phase do not seems, in principle, to have a contribution in pinning vortex as relevant as in the case of Y123, an increase of the critical currents and a shift of the irreversibility line towards higher temperatures and magnetic fields (see inset of Fig. 1)is detected as the 422 content is increased from 5% to 15%. The field dependence of J_c at several temperatures in the H||c configuration is depicted in Fig. 2. Together with the already mentioned upward shift of the irreversibility line, a slight enhancement of the fishtail effect is also observed. Such field-induced pinning increment has been explained by considering some degree of Ba substitution by Nd, creating regions with a depressed superconducting order parameter and lower T_c [1]. The observation of an slight enhancement of the fishtail effect, attributed to the spinodal decomposition mechanism proposed for Nd123, which postulates the segregation of anti-site defect clusters dispersed in a superconducting matrix of higher T_c [4], may indicate somehow the interference of the 422 precipitates with this segregation process maybe through the introduction of subtle changes on the microstructure. The existence of such anti-site defects in our samples has been observed thorough contrast variations in TEM [7] and scanning tunnelling microscopy [8] images. Further confidence for this model has been given by Hirayama et al. [9] who detected a sinusoidal variation of the Nd concentration within Nd123 single crystals with a wavelength of a few tens of nanometers.

Regarding to the effectiveness of the 422 interface as flux pinning mechanism, it has already been demonstrated that interface pinning in melt textured Y123/211 composites plays a dominant role in the flux pinning process in the high temperature (T≥40 K) and low field (H≤1 T) regimes [10]. A clear dependence of the zero field critical current $J_c(H=0)$ on the effective Y123/211 interface area, measured through the parameter V/d (V being the effective volume of the non-superconducting inclusions and their mean diameter), has been observed as

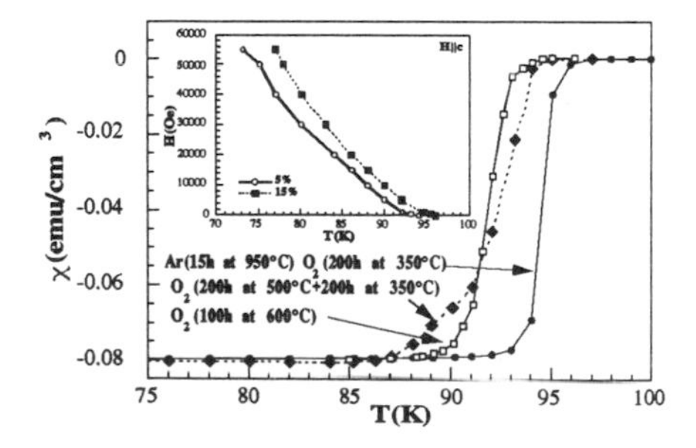

Fig. 1. Low field (H=10 Oe) dc susceptibility of Nd123 samples after different oxygen annealings. Inset: Irreversibility line of two samples with 15% and 5% of Nd422 phase.

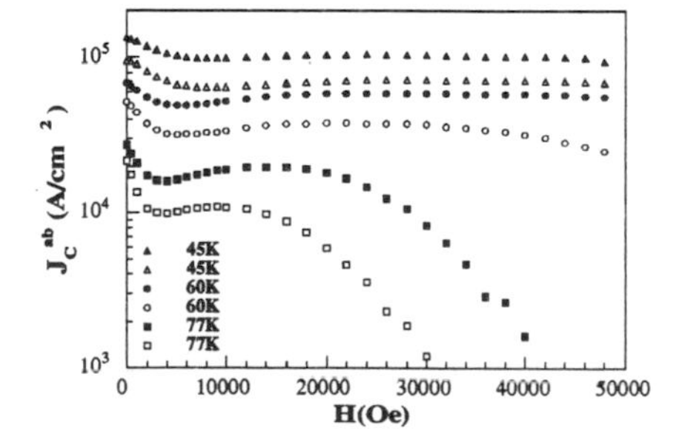

Fig. 2. Field dependence of the critical currents at several temperatures. Full symbols correspond to the sample with 15% of Nd422 phase while open symbols stand for the 5% sample.

well as a $H^{-1/2}$ functional dependence of J_c (H) at high temperatures [11]. A concomitant softening of the temperature dependence of $J_c(T, H\approx0)$ above T≈40K up to T≈75K (see Fig. 3) has also been observed and attributed to Y123/211 interface pinning.

This thermal behavior may be properly described by the expression $J_c(T)=J_c^0 \exp[-3(T/T^*)^2]$ proposed to account for the temperature dependence of the critical current in the case of linear correlated disorder [12]. In the case of Nd123/422 composites the role of the interface pinning has not been extensively explored yet. Unfortunately, the appearance of the fishtail effect makes it difficult to check the validity of the $H^{-1/2}$ dependence in the $J_c(H)$ curves. On the other hand, the thermal dependence of the critical currents in self field conditions exhibits a quasi-exponential behavior up to 80K with no traces of the thermal softening observed in Y123/211 composites [10]. This different behavior is clearly illustrated in Fig. 3, where we show the thermal dependence of the self field critical currents of two Y123/211 samples and a Nd123/422 sample with different V/d parameters, namely V/d≈500 for Nd123, and V/d≈1600 and 160 for Y123. Figure 3 clearly shows that even with a lower value of the V/d parameter the influence of the interface pinning is clearly higher in the Y123-211 system than in the Nd123-422 one. This result points to the fact that interface pinning has only a very limited effect in the flux pinning processes in Nd123/422 composites in strong contrast with what is observed in Y123/211 materials. Nevertheless, it should be mentioned that an increase of the J_c values in the whole range of temperatures and fields is observed (see Fig. 2) when the 422 content of the samples is increased from 5% to 15% nominally.

Since the interface pinning strength is proportional to the gradient of the superconducting order parameter at the interface, it will be highly sensitive to the presence of lattice defects which could smooth out the variation of the superconducting order parameter. In accordance with these considerations, in Y123-211 materials in which interfaces have been observed to be sharp at atomic scale in HRTEM images [13,14] Y123-211 interface pinning do play a dominant role at high temperatures and low fields. Conversely, Nd123/422 interfaces typically contain a layer of defective material. As an example, Fig 4 shows a HRTEM image of a Nd123/422 interface where the Nd123 is slightly deviated (by ≈1°) from the $[010]_{Nd123}$ zone axis and the 422 particle is viewed along $[0\bar{1}1]_{422}$. The interface plane is viewed edge on and corresponds to the $(100)_{422}$ plane. The $(001)_{123}$ planes are inclined by $\alpha=64°$ to the interface. For this orientation it can be observed that the prominent bright matrix planes, spaced by one c-axis length, meet the interface approximately at every three $(011)_{422}$ lattice spacings, i.e. $3d(011)_{422}\sin\alpha\sim d(001)_{123}$. Therefore, both crystals are oriented in such a way that some degree of lattice matching exists at the interface.

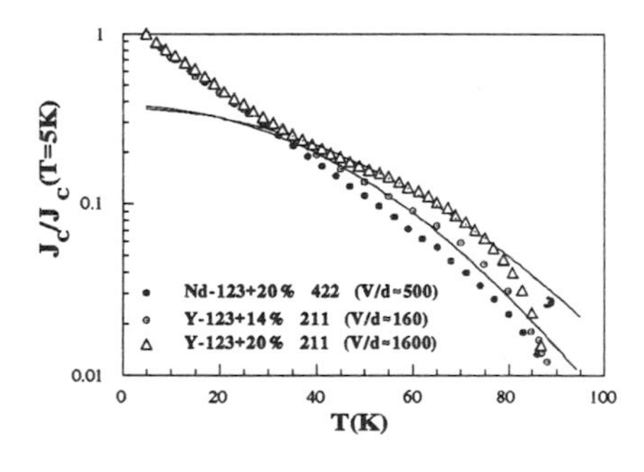

Fig. 3. Temperature dependence of the self-field critical currents in Nd123 and Y123 samples. Note that the thermal softening at high temperatures is very much stronger in Y123-211 composites even if V/d is smaller than in Nd123-422 sample.

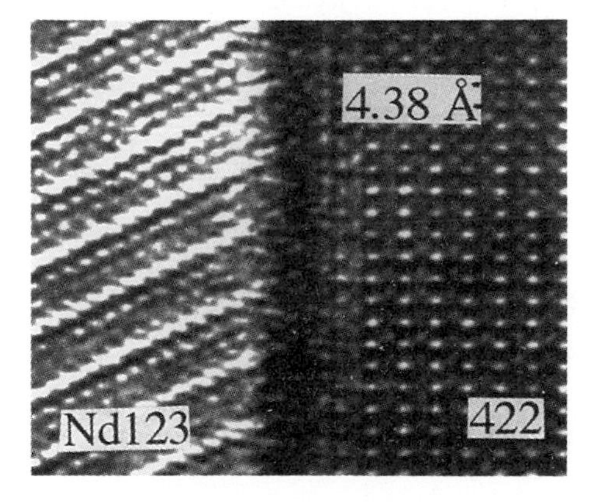

Fig. 4. HRTEM image of a Nd123-422 interface viewed along $[010]_{Nd123}$ and $[0\bar{1}1]_{422}$. The structural width of the interface is about 11 Å.

On the other hand, the interface contains a slab of highly strained material about 11 Å thick. These observations strongly suggest the development of strain relaxation structures which could be interpreted as an array of overlapping interfacial dislocation cores.

Systematic HRTEM analysis has indeed revealed that the development of defective structures is a common feature of Nd123/422 interfaces. We propose that such a distinctive behavior at the interfaces between Nd123-422 and Y123-211 composites lies at the origin of the reduced low field critical currents for similar V/d values and also explains the differences observed in the thermal dependence of J_c between Nd123 and Y123. Both effects suggest a reduced effectiveness of the interface pinning mechanism in the Nd123/422 composites since the broadening of the interface region is directly reflected in a strong decrease of the pinning strength.

In summary we have shown that Nd123 materials with high T_c and critical currents cam be obtained by directional solidification in air avoiding the need of a reduced oxygen atmosphere. At the same time, our results indicate that Nd123-422 interfaces do not play an important role in the pinning process in strong contrast with what has been observed in Y123-211 composites.

Acknowledgements

Acknowledgements: We are grateful to Y. Maniette (Serveis Científico Tècnics de la Universitat de Barcelona) for technical assistance in HRTEM experiments. This work has been supported by CICYT (MAT96-1052), GRQ (95-8029), Programa MIDAS (93-2331) and EC-EURAM (BRE2CT94-1011).

References

[1] M. Murakami, N. Sakai, T. Higuchi and S. I. Yoo, *Supercond. Sci. Technol.* **9**, 1015 (1996).
[2] K. Salama, A. S. Parikh and L. Woolf, *Appl. Phys. Lett.* **68**, 1993 (1996).
[3] S. I. Yoo, N. Sakai, H. Takaichi, T. Higuchi and M. Murakami, *Appl. Phys. Lett.* **65**, 633 (1994).
[4] M. Nakamura, Y. Yamada, T. Hirayama, Y. Ikuhara, Y. Shiohara and S. Tanaka, *Physica C* **259**, 295 (1996).
[5] N. Vilalta, F. Sandiumenge, S. Piñol and X. Obradors, *J. Mater. Res.* **12**, 38 (1997).
[6] S. I. Yoo, N. Sakai, T. Higuchi and M. Murakami, *IEEE Trans. Appl. Supercond.* **5**, 1568 (1995).
[7] T. Egi, J. G. Wen, K. Kuroda, H. Unoki, and N. Koshizuka, *Appl. Phys. Lett* **67**, 2406 (1995).
[8] W. Ting, T. Egi, R. Itti, K. Kuroda, and N. Koshizuka, *Advances in Superconductivity VIII* (Tokyo, Springer 1996) p. 481.
[9] T. Hirayama, Y. Ikuhara, M. Nakamura, Y. Yamada, and Y. Shiohara, *J. Mater. Res.* **12** (1997) Feb. issue.
[10] B. Martínez, X. Obradors, A. Gou, V. Gomis, S. Piñol, J. Fontcuberta and H. Van Tol, *Phys. Rev. B* **53**, 2797 (1996).
[11] M. Murakami, K. Yamaguchi, H. Fujimoto, N. Nakamura, T. Taguchi, N. Koshizuka and S. Tanaka, *Cryogenics 32,* 930 (1992).
[12] D. R. Nelson and V. M. Vinokur, *Phys. Rev. Lett.* **68**, 2398 (1992).
[13] K. Yamaguchi, M. Murakami, H. Fujimoto, S. Gotoh, T. Oyama, Y. Shiohara, N. Koshizuka and S. Tanaka, *J. Mater. Res.* **6**, 1404 (1991).
[14] F. Sandiumenge, S. Piñol, X. Obradors, E. Snoeck and Ch. Roucau, *Phys. Rev. B* **50**, 7032 (1994).

Inst. Phys. Conf. Ser. No 158
Paper presented at Applied Superconductivity, The Netherlands, 30 June–3 July 1997

Effect of hot plastic deformation in oxygen on the critical currents of melt-textured Y-123/Y-211

J Krelaus†[1], A Leenders‡ L-O Kautschor† M Ullrich‡ and H C Freyhardt†‡

† Institut für Metallphysik der Universität Göttingen, Windausweg 2, D-37073 Göttingen, Germany

‡ Zentrum für Funktionswerkstoffe gGmbH, Windausweg 2, D-37073 Göttingen, Germany

Abstract. The influence of an oxygen atmosphere present during high-temperature deformation of melt-textured Y-123/Y-211 on the current density is studied. To this purpose, Bridgman-grown Y-123/Y-211 is uniaxially deformed under compression at $T = 950°C$ with $\dot{\varepsilon} = 2 \cdot 10^{-5}\ s^{-1}$. Current densities are obtained from magnetization measurements applying the anisotropic Bean model for the field parallel/perpendicular to the crystallographic c-axis. For $B \| c$, the current densities of the deformed samples are lower than those of the non-deformed samples at low T and high B. For higher T and for $B \| ab$ they are enhanced. TEM studies show a significant increase of dislocation density with cores in the ab plane introduced by the deformation. The suppression of j at low T is observed also after deformation in air, the enhancement at higher T is a new feature ascribed to the phase stabilizing effect of the oxygen atmosphere during deformation.

1. Introduction

Finely dispersed Y_2BaCuO_5 (Y-211) particles are known to effectively improve the pinning in melt-textured $YBa_2Cu_3O_{7-\delta}$ (Y-123). The interfaces between the Y-123 matrix and the Y-211 inclusions and associated defects are supposed to form strong pinning sites [1]. On the other hand, dislocations introduced by plastic deformation are reported to enhance the current density of melt-textured stoichiometric Y-123 [2]. The combination of both inclusions and dislocations is, therefore, promising for optimizing the current density of Y-123.

Due to its brittleness, plastic deformation of Y-123 requires high temperatures. Deformation experiments at $T = 930°C$ in air with subsequent oxygen (O_2) annealing at 450°C generally depressed current densities for $B \| c$ [3]. The oxygenation step had been observed to result in severely degraded areas in the bulk. Nucleation centres for

[1] E-mail krelaus@umpsun1.gwdg.de

this degradation are supposed to be formed during the deformation in air. In the present work, the effect of a deformation in oxygen on critical currents of Y-123/Y-211 is investigated. The O_2 atmosphere stabilizes the tetragonal Y-123 phase at the deformation temperature and is expected to avoid formation of nuclei for phase degradation.

2. Experimental

Two parallelepipeds are cut from single domained Bridgman grown Y-123/Y-211 [4]. One sample (D, *deformed*) was uniaxially compressed with an angle of 45° to the *ab* plane, $T = 950°$ C, $\dot{\varepsilon} = 2 \cdot 10^{-5}\mathrm{s}^{-1}$ sustaining a pressure of 75 MPa in the steady state creep regime; the other (ND, *non-deformed*) is a reference subjected to all temperature ramps of sample D. The *in-situ* cooled samples have been reoxygenized at 450°C.

Magnetization, M, vs. applied field, B, is measured in a Faraday magnetometer both for $B\|c$ and for $B\|ab$ for $0 \leq B \leq 9$ T and $4.2 \leq T \leq 77$ K. Current densities, j, are calculated applying the anisotropic Bean Model [5]

$$j^{ab,c} = \frac{2\Delta M}{b(1 - b/(3a))} \quad (B\|c), \qquad j^{c,ab} = \frac{2\Delta M}{y} \quad (B\|ab).$$

The first superscript of j denotes the direction of current flow, the second the direction of B. For $B\|ab$, y is the sample length of current flow parallel to ab The second equation is valid for $j^{ab,ab} \gg j^{c,ab}$ which regularly holds for anisotropic HTS materials.

3. Results

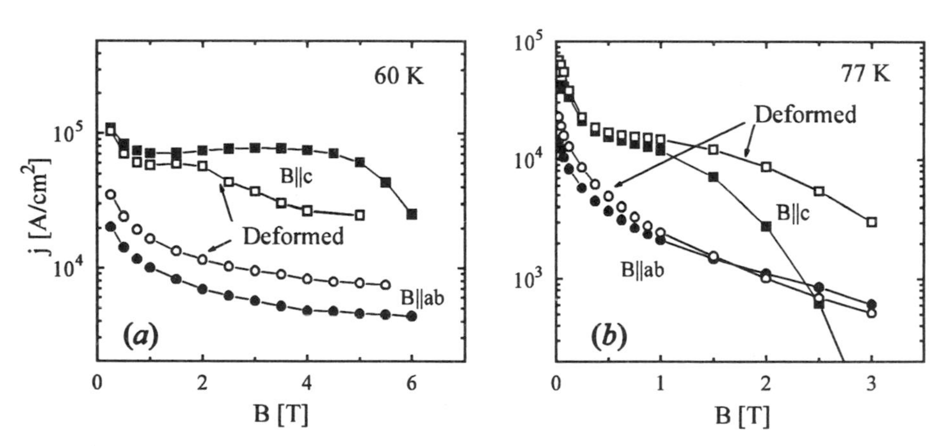

Figure 1. Current densities, j, vs. field, B, (a) at $T = 60$ K and (b) at $T = 77$ K of the deformed and non-deformed samples for $B\|c$ and $B\|ab$

Figure 1 (a) is the j vs. B curve at $T = 60$ K. It compares current densities for deformed (D, open symbols) and non-deformed (ND, full symbols) material, both for $B\|c$ (squares) and $B\|ab$ (circles). For $B\|c$, sample ND exhibits a $j^{ab,c}$ of the order of $1 \cdot 10^5$ A/cm^2. An extended plateau is observed for $B = 2 - 5$ T. The values equal those reported by [3] for the same samples. Deformation reduces $j^{ab,c}$. The suppression is moderate (down to 70 – 90 % of the value for ND) for small applied fields and becomes more drastic (50 % – 25 %) in the high-field regime. The field dependence of $j^{ab,c}$ is modified: the plateau occurring in sample ND is suppressed. For $B\|ab$, $j^{c,ab}$ is about one order of magnitude smaller than for $B\|c$. Neither ND nor D shows a plateau of $j^{c,ab}$. Deformation enhances $j^{c,ab}$ in the whole B regime. The enhancement amounts to factors of about 1.5 – 2. At 77 K, however, a crossover is observed for higher B.

Fig. 1 (b) shows the j vs. B curves at T = 77 K. Deformation increases $j^{ab,c}$, especially for $B > 1$ T where j of sample ND rapidly drops down. For $B\|ab$, the increase of $j^{c,ab}$ is reduced compared to the strong enhancement observed at lower T. At $B = 2$ T, a crossover is observed leading to higher $j^{c,ab}$ of sample ND. Interestingly, also sample ND shows a crossover: at $B \approx 2.5$ T, $j^{ab,c}$ drops below $j^{c,ab}$. This might be explained by intrinsic pinning becoming more dominant at higher T.

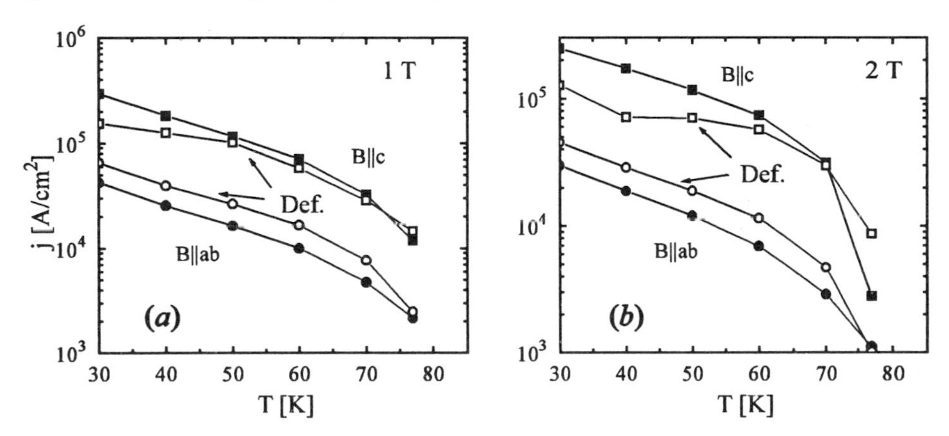

Figure 2. Current densities, j, vs. T, (a) at $B = 1$ T and (b) at $B = 2$ T of the deformed and non-deformed samples for $B\|c$ and $B\|ab$.

Fig. 2 depicts the T dependence of j for both samples at $B = 1$ T and $B = 2$ T. For $B\|c$, the main deformation effect is a reduction of j at low T and a slight enhancement as $T \to T_c$. This effect is more pronounced for $B = 2$ T. For $B\|ab$, j is in general increased. At $T = 77$ K, a crossover arises for higher B.

4. Discussion

The main effect of the plastic deformation is the suppression of $j^{ab,c}$ at low T and for higher B. Current densities at $B = 1$ T are reduced to about 90% (at $T = 60$ K) down to 40% (at $T = 20$ K) of the non-deformed value. At 77 K however, a slight enhancement of j is observed. TEM studies on deformed samples show that mainly dislocations with Burgers vectors $\langle 100\rangle$ and $\langle\bar{0}10\rangle$ are induced. Locally, their density is increased from 10^8 cm^{-2} to 10^{10} cm^{-2}. Moreover, they are observed to be distributed more homogeneously. This suggests to ascribe the $j^{c,ab}$ enhancement to pinning at the additional dislocation cores. Also for $B\|c$, the dislocations could explain the increase of j as $T \to T_c$ if it is considered that extended defects are more effective at higher T

and if point defects are supposed to be the main pinning sites in non-deformed material. Assuming point defects being dominant at low T could also explain the suppression of j at low T and high B. A suggested model is the reduction of point defect density by the dislocation movement during deformation.

Deformation in air [3] leads to partially different results: The current suppression is more evident at higher T. j is reduced to $\approx 40\%$ (at 20 K, 1 T, comparable to our result) down to $\approx 20\%$ (at 77 K, 1 T). As in our results, after deformation in air j drops down more rapidly with increasing B at low and intermediate T. Thus the pinning mechanism in melt-textured Y-123/Y-211 at low T for $B \| c$ is affected in a similar way by deformation both in air and in O_2. As deduced from the $T \rightarrow T_c$ behaviour, an O_2 atmosphere during deformation stabilizes the tetragonal Y-123 phase and avoids formation of degradation nuclei for the subsequent O_2 annealing. After deformation, the superconducting orthorhombic Y-123 can be recovered by oxygenation. TEM studies showing no severely degraded areas support this hypothesis.

5. Summary/Acknowledgement

Deformation of melt-textured Y-123/Y-211 in O_2 has a complex impact on current densities. For $B \| c$, current densities are in general reduced at low or intermediate T. This effect dominates for $B \geq 2T$ where j is reduced to less than 50 % of its starting value. As $T \rightarrow T_c$, deformation enhances j. The crossover is observed for $T \approx 70$ K (for $B \approx 1$ T). For $B \| ab$ ($B \leq 1$ T) current densities are increased by a factor $\approx 1 - 2$. Compared to deformation in air [3], the current density suppression is generally reduced. The O_2 atmosphere suppresses the deformation-induced degradation during subsequent oxygen annealing and is capable to enhance the current densities as $T \rightarrow T_c$. A drastically increased dislocation density is thought to be the reason for this enhancement. The assumption of point defect pinning at low T for $B \| c$ could also explain the reduction of j by a reduction of point defect density due to dislocation movement.

We thank X Obradors and F Sandiumenge from the CSIC, Barcelona, for supplying the melt-textured Y-123/Y-211. This work is supported by the European Union, contract no. BRE-CT94-1011.

References

[1] Murakami M, Fujimoto H, Gotoh S, Yamaguchi K, Koshizuca N, Tanaka S 1991, *Physica* C, **185–189**, 321

[2] Selvamanickam V, Mironova M, Son S, Salama K 1993, *Physica* C, **208**, 238–44

[3] Vilalta N, Sandiumenge F, Rodríguez E, Martínez B, Piñol S, Obradors X and Rabier J 1997, *Philosophical Magazine* B **75**(3) 431–41; Martínez B, Sandiumenge F, Vilalta N, Piñol S, Obradors X and Rabier J 1996, *J. Appl. Phys.* **80**(9) 5515–7

[4] Vilalta N, Sandiumenge F, Piñol S, Obradors X 1997, *J. Mater. Res.* **12**, 38

[5] Sauerzopf F M, Wiesinger H P and Weber H W 1990, *Cryogenics* **30** 650–5

Inst. Phys. Conf. Ser. No 158
Paper presented at Applied Superconductivity, The Netherlands, 30 June–3 July 1997

Influence of Sm contamination on the superconducting properties of $SmBa_2Cu_3O_{7-\delta}$ seeded melt textured $YBa_2Cu_3O_{7-\delta}$

C D Dewhurst, Wai Lo, Y H Shi and D A Cardwell

IRC in Superconductivity, University of Cambridge, Cambridge CB3 OHE, UK.

Abstract. A main processing aim of melt textured $YBa_2Cu_3O_{7-\delta}$ (YBCO) is to develop large grain material with a high J_c which flows over the length scale of the sample, in order to maximise the flux trapping ability of the large YBCO grains. If $SmBa_2Cu_3O_{7-\delta}$ (SmBCO) seeds are used for the solidification process contamination of the 123 matrix occurs at positions close to the seed. We report a study of the magnetic characterisation of a typical large YBCO grain, prepared by SmBCO seeded peritectic solidification, and correlate variations in the transition temperature, T_c, critical current density, J_c, and irreversibility field, B_{IL}, with Sm concentration. The T_c of samples cut from the bulk increases smoothly with sample radius to a distance of the order of mm from the seed position which indicates a variation in Sm concentration at positions away from the seed. The local J_c and B_{IL} are also extremely sensitive to Sm concentration in the vicinity of the seed.

1. Introduction

High temperature superconducting $YBa_2Cu_3O_{7-\delta}$ (YBCO) has significant potential for a variety of permanent magnet-type engineering applications [1-3]. The flux trapping ability of bulk YBCO, which forms the basis of these applications, depends strongly on the magnitude and homogeneity of the critical current density of the material and the length scale over which it flows [4]. Large grain YBCO is typically fabricated by a variety of melt processes based on peritectic solidification and consists of a $YBa_2Cu_3O_{7-\delta}$ (123) phase matrix with Y_2BaCuO (211) inclusions [5,6]. SmBCO seeded YBCO has repeatedly yielded a more controllable growth morphology of the grains although, in contrast to $SrTiO_3$, partial dissolution of the seed at high temperatures results in local Sm contamination of the 123 matrix [6]. The resulting (Y,Sm)BCO solid solution has been observed across the seed/YBCO interface [7] and is likely to influence strongly the local superconducting properties of the material. Other factors, such as the distribution of the non-superconducting 211 inclusions and the effects of platelet and cellular microstructure of the 123 matrix, are also expected to influence the critical current density, J_c, of the bulk sample but should not influence more fundamental parameters such as T_c and B_{IL}. A key parameter in determining the extent of Sm contamination is the distribution of T_c throughout a nominally uniformly oxygenated surface layer of the bulk. A detailed knowledge of the spatial distribution of T_c, J_c and B_{IL} within large YBCO grains is, therefore, not only important in order to understand the net influence such a complex microstructure and Sm contamination has on the superconducting properties, but is also important for predicting the performance of materials for electromagnetic device applications.

2. Experimental

Bulk YBCO samples were fabricated by seeded peritectic solidification, specific details of which are given in Refs. [6,7]. Briefly, a $Y_{1.6}Ba_{2.3}Cu_{3.3}O_{8.5-\delta}$ precursor powder (i.e. equivalent to a final phase composition of $YBa_2Cu_3O_{7-\delta}$ + $0.3Y_2BaCuO_5$) was prepared by spray drying and subsequent calcination. YBCO precursor pellets were prepared by uniaxial die pressing and sintering. Millimeter-sized seeds of average composition $Sm_{1.6}Ba_{2.3}Cu_{3.3}O_{8.5-\delta}$ were placed on the sintered precursor YBCO pellets, melt processed at 1025°C in air and cooled slowly (1°C/h) to 970°C. The resulting bulk samples exhibited the familiar four-faceted growth morphology extending over the entire sample dimensions of diameter 20mm and thickness 6mm. Oxygenation of the sample was performed at a temperature of 550°C for a period of one week.

A slice of width ~2mm was cut from the bulk sample using a diamond cutting wheel and an ethanol lubricant. The slice was chosen such that it incorporated the central seed crystal and the centre of two opposing facets. Twenty seven cubic samples of side ~2mm were subsequently cut from the slice through the entire cross-section, as indicated schematically in Figure 1(a). Sixteen of these samples were chosen for magnetisation measurements in order to determine the local variation of the superconducting properties.

m-H measurements were carried out using an Oxford Instruments vibrating sample magnetometer. Measurements were made with *B//c*-axis (i.e. perpendicular to the sample surface) and at temperatures between 5K and 87K. J_c was estimated from the width of the *m-H* hysteresis loop, Δm, using the Bean model [4]. B_{IL} was determined from the closure of hysteresis in the *m-H* behaviour using a criterion of $\Delta m < 10^{-4}$ emu ($J_c < 5$ A/cm^2). The T_c of each sample was determined from the midpoint of the (sharp) superconducting transition using a Lakeshore *ac* susceptometer in an applied *ac* field of 0.1mT at a frequency of 333Hz. The microstructure of the samples was examined using optical and scanning electron microscopy whilst the cation distribution at the seed/YBCO interface was determined by energy dispersive X-ray (EDX) and electron probe micro-analysis (EPMA) [7].

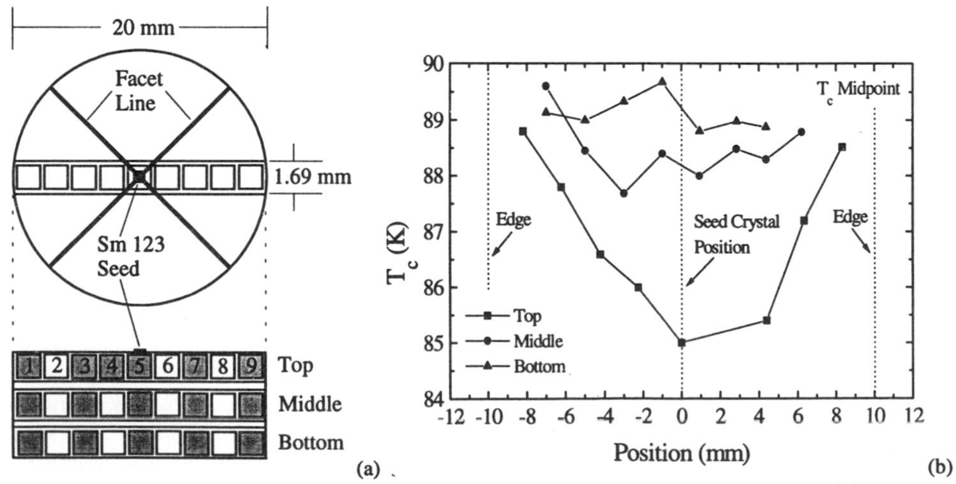

Figure 1. (a) Schematic diagram of a typical four-faceted bulk melt textured $YBa_2Cu_3O_{7-\delta}$ material. Twenty seven cubic samples were cut from the bulk, from a slice through a cross-section of the sample. The shading indicates the samples measured in this report. (b) T_c vs. position for samples taken across the top, middle and bottom sections through the bulk.

3. Results and Discussion

Figure 1(b) shows the distribution of T_c measured for individual samples across the cross-section of a large bulk YBCO sample. It is clear from this figure that the T_c on the top layer of the sample (i.e. that which contains the SmBCO seed) varies systematically and symmetrically with distance from the seed. The positional dependence of T_c is less severe for the middle layer and almost constant for samples at the bottom of the bulk. The changes in T_c can be accounted for by a variation in either composition or oxygen content since T_c, unlike J_c, is not expected to depend on microstructure. Variation in oxygen content is unlikely, however, given that the specimens cut from the bottom of the sample consistently exhibit a higher T_c than those from the top despite being processed under identical oxygenation conditions. On the other hand, EPMA mapping of Sm and Y distributions across the interface between the SmBCO seed and YBCO pellet showed that Sm migrates into the YBCO pellet during processing by both coalescence between Sm-211 and Y-211 particles, and direct Sm diffusion though liquid media in the peritectic state [7]. This results in the formation of superconducting (Y,Sm)BCO solid solutions in the vicinity of the seed crystal. Furthermore, it has been established that the T_c and transition width of these solid solution phases vary with Sm to Y ratio [8]. The T_c of the samples taken in the vicinity of the seed, therefore, depends sensitively upon the ratio of Sm to Y in the superconducting 123 matrix since the superconducting properties of Sm containing RE123 materials is known to be degraded strongly by processing in air [5]. The extent of Sm contamination of the YBCO 123 matrix may be estimated from Figure 1(b) to be about 4 to 5 mm horizontally and about 2mm vertically, presumably indicative of the differing mechanisms of Sm migration along *ab* and *c*-axis directions respectively.

Figure 2(a) shows the lateral variation of J_c at a field of 1T for the samples taken from edge to edge of the top layer of the bulk sample (T1 to T9). J_c is observed to vary symmetrically about the central seed crystal position, and shows a clear suppression in the vicinity of the seed crystal. J_c reaches a maximum towards the edge of the sample at a position of ~4mm from the centre on either side of the sample before decreasing once more towards the sample edge. Several factors may influence the local J_c of the sample including the superconducting properties of the (Sm,Y)BCO solid solution region and microstructural features such as the 211 density, platelet and cellular microstructural features which are known to influence strongly the pinning properties of YBCO [9]. The effects of 211 density and microstructural features on the J_c of typical bulk YBCO samples is discussed further in Refs. [9,10]. It is clear, however, that the large suppression of J_c in the central region below the seed crystal is most likely due to the reduced T_c of the (Sm,Y)BCO solid solution.

Figure 3 shows the loci of the irreversibility lines determined from the closure of hysteresis in the *m-H* behaviour ($J_c < 5A/m^2$) for samples T1, T5 and T8 taken laterally across the top, and M5 and B5 taken through the thickness of the bulk (i.e. at varying distances from the Sm seed crystal). The samples away from the seed (T1, T8, M5 and B5) exhibit the same irreversibility line indicating once more the similar superconducting properties of these samples. The irreversibility line for sample T5, however, can be seen to rise more sharply with decreasing temperature, despite its reduced T_c compared to the other samples. This indicates further the presence of a different superconducting phase containing stronger pinning centres and is consistent with the behaviour expected for Sm containing 123 materials where irreversibility fields of ~13T @77K have been reported [11].

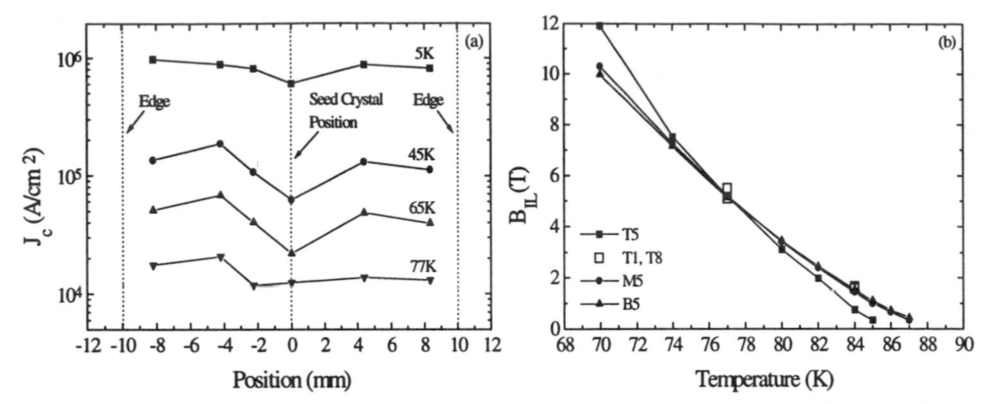

Figure 2. (a) J_c vs. position for samples taken across the top surface of a slice through a bulk $YBa_2Cu_3O_{7-\delta}$ sample. (b) Irreversibility line measured from the closure of hysteresis in the *m-H* curve for samples *T1*, *T5*, *T8*, *M5* and *B5*. The sample closest to the seed crystal (*T5*) exhibits a steeper irreversibility line, despite the reduced T_c, characteristic of $SmBa_2Cu_3O_{7-\delta}$.

4. Conclusions

The local variation of the superconducting properties (i.e. T_c, J_c and B_{IL}) within a bulk melt textured $YBa_2Cu_3O_{7-\delta}$ sample grown by $SmBa_2Cu_3O_{7-\delta}$ seed peritectic solidification have been investigated. These are found to vary symmetrically about the seed crystal indicating bulk growth and similarly symmetrical stochiometric and microstructural properties. The T_c of the nominally uniformly oxygenated samples directly below and adjacent to the seed crystal are suppressed significantly (~5K) by Sm contamination from the seed crystal. The suppression in T_c in the vicinity of the seed is reflected similarly by the locally reduced J_c by almost a factor of 3 relative to the surrounding material although microstructural features, such as 211 distribution are also expected to influence the local J_c distribution. Whilst T_c and J_c are suppressed in the vicinity of the seed the irreversibility line is found to rise more rapidly with decreasing temperature suggesting further the presence of a Sm containing 123 superconducting phase.

References

[1] F. C. Moon and P. Z. Chang, *Appl. Phys. Lett.* **56**, 397 (1990).
[2] W. K. Chu, K. B. Ma, C. K. McMichael and M. A. Lamb, *Appl. Supercond.* **1**, 1259 (1993).
[3] M. Murakami, *Appl. Supercond.* **1**, 1157 (1993).
[4] C. P. Bean, *Rev. Mod. Phys.* **36**, 31 (1964).
[5] M. Murakami, N. Sakai, T. Higuchi and S. I. Yoo, *Supercond. Sci. Technol.* **9**, 1015 (1997).
[6] Wai Lo, D. A. Cardwell, C. D. Dewhurst and S-L. Dung, *J. Mater. Res.* **11**, 786 (1996).
[7] Wai Lo, D. A. Cardwell and P. D. Hunneyball, Submitted to *J. Matter. Res.* (1996).
[8] Wai Lo, S-L. Dung and D. A. Cardwell, *In preperation* (1997).
[9] M. Chopra, S. W. Chan, R. L. Meng and C. W. Chu, *J. Mater. Res.* **11**, 1616 (1996).
[10] Wai Lo, C. D. Dewhurst, Y. H. Shi and D. A. Cardwell, *Proc. International Workshop on Processing and Applications of (RE)BCO Large Grain Materials*, Cambridge UK, 7-9 July (1997).
[11] Th. Wolf, H. Kupfer, H. Wuhl, *Proc. 8th IWCC*, World Scientific (1996).

Inst. Phys. Conf. Ser. No 158
Paper presented at Applied Superconductivity, The Netherlands, 30 June–3 July 1997

Dependence of critical current density on the oxygen content for Y-123 single domain melt-textured samples

V A Murashov [a,b], J Horvat[a], M Ionescu[a], H K Liu[a] and S X Dou[a]

[a] Centre for Superconducting and Electronic Materials, University of Wollongong, NSW 2522 Australia
[b] Moscow Institute of Radio Engineering ,Electronics and Automation, pr. Vernadskogo 78, 117454, Moscow, Russia

Abstract. Large Y - 123 single domain textured sample was produced by slowly cooling Y-123+15%Y-211 precursor seeded with Sm.-123 through the peritectic point. The sample was annealed in flowing O_2 at 380°C for 72 h. An O_2 distribution gradient in the range of 6.64-6.84 was determined in the volume of the sample by measuring "c" parameter of unit cell for the small cleaved sections. The J_c's of this sections were derived from hysteresis loops and T_c's from ac susceptibility. T_c is changing slowly (92→89 K) corresponding x_o=6.84-6.71 and drops sharply for smaller x_o . A maximum J_c of 76000 A/cm^2 (77K, self field) and an irreversibility field H_{irr}=7.2 T was obtained for X_o=6.77 which is close to the point of structural transformation of orthorhombic I → orthorhombic II, where the clustering in the oxygen sublattice is maximal. This short range disordering is providing an additional amount of pinning centres in the presence of the Y-211 inclusions .

1. Introduction

Different kinds of the defects in Y-123 melt-textured samples can form effective pinning centres. This is why the Y-123 texture is one of the most complicated systems for the studying of pinning mechanisms. Strong influence of the small inclusions of Y-211 phase on the value of critical current is well known. However , unlike $YBa_2Cu_3O_x$ single crystals, the information about the influence of oxygen content on the critical parameters is quite scarce. Additional big disadvantage in such study is usually the absence of directly determined absolute data of the oxygen content. Thus study correlates the J_c ,T_c and H_{irr} value with the oxygen content data obtained from P-T-X diagrams without considering the equilibrium state in the experiments. The goal of this work is to show the strong influence of the disordering in oxygen sublattice of Y-123 single domain textured samples on flux pinning in the presence of high concentration of Y-211 inclusions .

2. Experimental

2.1. *Determination of the oxygen content in $YBa_2Cu_3O_x$*

The amount x_o of oxygen was determined from its linear dependence of the "c" lattice parameter in the range x_o=6.40-6.94 with Δx= ±0.01 accuracy. A calibration curve was obtained previously [(b)] by the TG method, using the oxygen loss data from a large mass of Y-123 powder in correlation with "c" parameter determination by XRD. A few direct chemical determinations of O_2 content were undertaken by iodine method, to confirm the accuracy of procedure. Additionally, very thin (20-25 μm) single crystals were annealed in air, in the temperature range 400-580°C for 10 days and their oxygen content determined by the "c" parameter calibration was found to be good agreement with equilibrium data [1,2] taken from P-T-X diagram for these thermodynamic parameters. In this study the reflection (006) was used for the "c" parameter determination.

2.2 *Texturing*

By the use of the single domain textured sample the influence of other sorts of pinning centres (" green phase" inclusions and stacking faults for example) was kept constant for all measurements. Also, we have not taken into consideration the week links such as the grain boundaries between domains with different orientation of "ab" plane to sample surface. Moreover it was shown in preliminary experiments that the oxygen content in poly-domain samples depends on the "ab" plane orientation and changes sharply from one domain to another , even after annealing in the oxygen flow for one week . Single domain samples only, showing a gradual variation of the O_2 content from the periphery to the centre of the bulk were used for the preparation of the cleaved samples with different O_2 concentration.

The melt-textured sample was produced by the use Y-123+15%mass Y-211 ceramic precursor, with size of $30x30x7\ mm^3$ and high porosity substrate of Y-211. This substrate absorbed a part of the liquid of peritectical composition and thus promoted the preservation of a large number of Y-211 pinning centres (<1- 3μm) by texturing. Two fractional Y-211 powder was used in the precursor preparation, where the mass ratio between "1μm" and "5μm" fractions was 1:1. This circumstance let us to increase the part of Y-211 small particles and to avoid the development of microcracks during the texturing and oxygen annealing [3]. The ratio of the mass: Y-123 sample/ Y-211 substrate of 3 provided 40±1% mass of Y-211 phase in the matrix after texturing. The samples were heated up to 1045-1050°C, held for 20 min., cooled at a rate of 90°C/h to 1020°C, held there for 20 min. , cooled at a rate 1°/h to 940°C and than cooled to the room temperature at a rate of 60°C/h. The use of a small seed ($2x1x0.2\ mm^3$) of Sm-123 with (001) surface located in the centre of the bulk provided the development of single domain in the whole volume. A sample of size $26x25x5\ mm^3$ was preoxygenated in O_2 flow for 72 hours at 380°C. The difference of oxygen content from x=6.81 on the sample surface to x=6.64 in the centre of bulk was obtained. The rectangular samples with size $2x2x(0.8\text{-}1.0)\ mm^3$ were cleaved for magnetic measurements at different parts of single domain. The "ab" plane was parallel to the cleavage surface and the change of the oxygen content Δx along the thickness was tested by "c" parameter determination on both sides of the flat. This value was less than 0.02. After first set of magnetic measurements every rectangular sample was cleaved additionally to the thickness of 0.25-0.30 mm to avoid any possible influence of the oxygen gradient on superconducting parameters. The difference of oxygen content of less 0.01 had was determined for samples of this thickness.

3. Results and discussion

The data of T_c, J_c and irreversibility line for the samples with different oxygen content was obtained by an extraction magnetometer. The temperature of superconducting transition changes very slowly (plateau 92→82K) in the range $6.84 \geq x_o \geq 6.71$ but it drops fast with further oxygen decrease . The dependence of J_c on x_o for $T/T_c=0.85$ is shown in Fig.1 for the samples with thickness of 0.25-0.30 mm (curve 1) and 0.8-1.0 mm (curve 2).

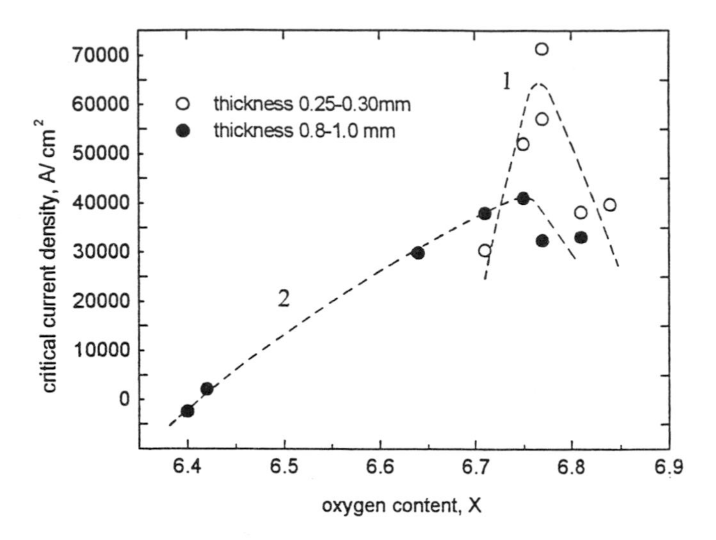

Figure 1. The dependence of Jc on oxygen content x for samples with different thickness, T/Tc = 0.85

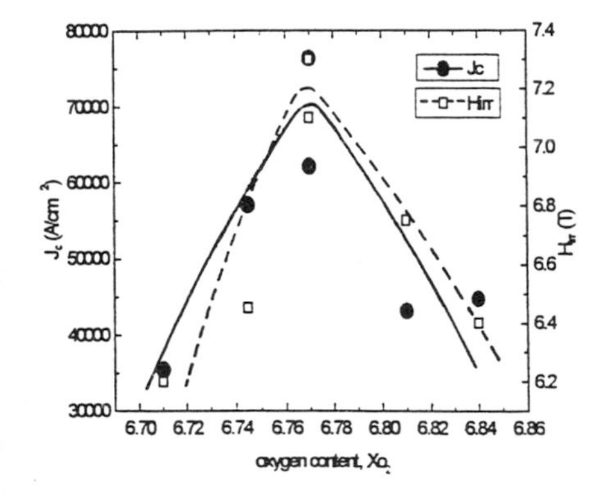

Figure 2: Critical current density Jc and H_{irr} vs. oxygen content , T/Tc = 0.85.

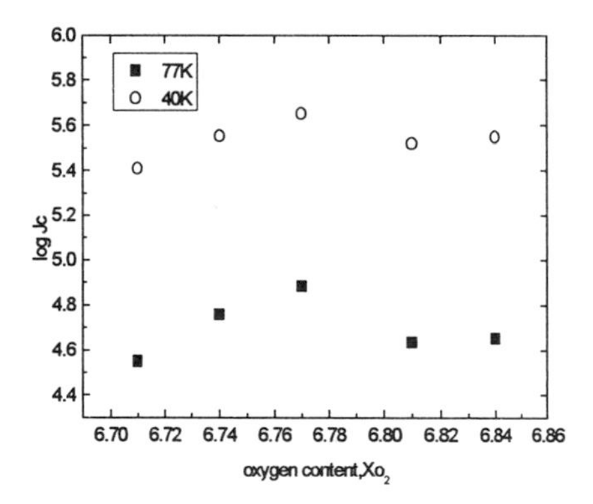

Figure 3: Critical current density Jc vs. content at 40K and 77 K.

One can see the pronounced maximum of the critical current for both curves at x_o= 6.77 and x_o= 6.75, respectively. The maximal value of J_c is very high (76000 A/cm², 77 K, H=0) . The curve 2 shows that the maximum of Jc is obtained by practically the same value of x_o even with a variation of oxygen content along the sample thickness. The average J_c for " thick " samples is smaller (curve 1).The irreversibility field also has a maximum (Fig.2) at the same x_o as critical current density . The same measurements performed at 40 K also give a maximum of J_c for x_o=6.77 (Fig.3).

These results shows a significant influence of the oxygen content on J_c and H_{irr} in a wide temperature range in the presence of large amount of Y_2BaCuO_5 inclusions. The most probable reason for such behaviour of J_c is flux pinning on clusters of oxygen vacancies. This clustering in oxygen sublattice is associated with the possibility of different arrangement of oxygen vacancies in crystal lattice [4]. It is the most pronounced with occurrence of structural transformation of orthorhombic phase I ↔ orthorhombic phase II. In other words, the maximum amount of clusters with both types of short ordering in Y-123 texture is in the range $6.75 < x_o < 6.77$, where this transition occurs in the atmosphere of air. These defects probably give the large peak of J_c, in addition to the influence of flux pinning by Y-211 inclusions.

4. Conclusions

The largest improvement of J_c and H_{irr} has been found for Y-123 single domain melt-textured samples with x_o content corresponding to the structural transition in oxygen sublattice. These data are quite high (J_c= 76000 A/cm^2, 77K, self field; H_{irr}=7.2 T) which proves a strong participation of the point defects in the pinning, even in the presence of the large amount of small Y-211 particles.

Acknowledgements

Authors would like to acknowledge financial support of Metal Manufactures Ltd, Electric Supply Association of Australia, Energy Research and Development Corporation, Department of Energy of NSW, Australian Research Council, Department of Industry, Science and Tourism and Russian National Program of High T_c Superconductivity. Also, the authors are very grateful to A.A.Bush for making available the data for the calibration of "c" parameter versus oxygen content.

References

[1] Kishio K et al. 1987 Jap.J. Appl. Phys. **7** 1228-30
[2] Gallagher P K 1987 Adv. Ceram. Matter. **3B** 632-39
[3] Murashov V A, Schätzle P, Krabbes G, Klosowski J, Wendrock H, Vogel H R, Eversmann K 1996 Physica C **261** 181-88
[4] Goodenough J B 1987 J. Mater.Educ. **9** 619-73

Inst. Phys. Conf. Ser. No 158
Paper presented at Applied Superconductivity, The Netherlands, 30 June–3 July 1997

Stress-induced Superconducting Transition in a Melt-processed $YBa_2Cu_3O_{7-\delta}$ Bulk Superconductor.

Hee-Gyoun Lee, Il-Hyun Kuk and Gye-Won Hong

Superconductivity Research Laboratory, Korea Atomic Energy Research Institute, P.O.Box 105, Yusung-gu, Taejon 305-600, Korea

Abstract Microstructural, crystallographic and magnetic observations were made for a partially oxygenated melt-processed $YBa_2Cu_3O_{7-\delta}$ bulk superconductor. Microstructural observation showed that an alternating microstructure between orthorhombic and tetragonal $YBa_2Cu_3O_{7-\delta}$ phases was developed by partial oxygenation. XRD analysis and the temperature dependence of AC susceptibility indicated the presence of the superconducting phases which were characterized by different oxygen content and various Tc. It is suggested that the superconducting transitions at 35K and 48K are attributed to the stress induced by phase transformation via oxygenation.

1. Introduction

$YBa_2Cu_3O_{7-\delta}$ phase is called an oxygen defect perovskite because there are only seven oxygens per formula unit instead of the nine required for a perfect perovskite. Oxygenation of the $YBa_2Cu_3O_{7-\delta}$ phase induces both a structural and electronic transition with oxygen deficiency, δ. The phase transition from tetragonal to orthorhombic(T-O transition) phase is at $\delta \sim 6.4$[1]. It has been also reported that, in addition to T-O transition, there is an ordered double-cell orthorhombic phase, O_{II}, near $\delta = 0.4 \sim 0.5$, which is characterized by a low Tc of 60K and oxygen ordering [1-3].

For thermodynamic and kinetic reasons, oxygen deficiency, δ, is gradually decreased as the annealing temperature is decreased and annealing period is extended[4,5]. However, it has been reported that the time for full oxygenation of huge $YBa_2Cu_3O_{7-\delta}$ crystal fabricated by a melt process[6] is shorter than that calculated for polycrystalline even if the melt-processed $YBa_2Cu_3O_{7-\delta}$ crystal is absent of open porosity and grain boundaries.

Recently, Lee et al[6,7] showed that inhomogeneous phase transition resulted in planar cracks and the non-uniform oxygenation in a melt-processed $YBa_2Cu_3O_{7-\delta}$ bulk superconductor was induced from the stress effects resulting from oxygenation. The existence of duplicate phase in a partially oxygenated melt-processed $YBa_2Cu_3O_{7-\delta}$ bulk superconductor has been reported from microstructural observations and the investigation of the field dependence of the critical current[8].

In this work, microstructural, crystallographic and magnetic observations were made for a partially oxygenated melt-processed $YBa_2Cu_3O_{7-\delta}$ bulk superconductor. Diamagnetic transitions at 35K and 50K, which might be related with an oxygen ordering, are reported in $YBa_2Cu_3O_{7-\delta}$ superconductor.

2. Experimental Procedure

High-purity (99.9%) Y_2O_3, $BaCO_3$, CuO, and CeO_2 powders were used for present experiment. A starting powder with a nominal composition of Y:Ba:Cu = 1.6 :2.3 :3.3 was prepared with an addition of 0.5 wt% CeO_2. Mixed powder was calcined at 880 ℃ for 24 h three times with intermediate grindings, palletized using cold isostatic press. Pressed green pellets were placed on single crystal MgO substrate and heated to 1050 ℃ with a rate of 100 ℃/h, held for 30 min, downed to 1010 ℃ with a rate of 10 ℃/h, slowly cooled to 970 ℃ with a rate of 1 ℃/h and then quenched in a liquid nitrogen bath.

Oxygenation anneal was performed at 450 ℃ for 5h in air and then air-quenched. The microstructural, crystallographic and superconducting characteristics of the partially oxygenated specimen were investigated by using optically polarized light microscope, X-ray diffraction(XRD) and AC susceptometer. Phase analysis was made by means of XRD pattern using Cu $K_{1\alpha}$ radiation with a fine scan step of 0.01 degree and peak analysis of XRD pattern was made in the range from 2θ = 45° to 50° by using multiple Gaussian distribution function. Both of real part(χ') and imaginary part(χ'') of AC susceptibility were obtained under a various magnetic fields of 0.01, 0.1, 1 and 3 Oe with a frequency of 192 Hz.

3. Results and Discussion

Fig. 1 shows an optically polarized micrograph of a melt-processed $YBa_2Cu_3O_{7-\delta}$ specimen annealed at 450 ℃ for 5h in air after quenching in a liquid nitrogen bath from 970 ℃by interrupting a melt-process. It can be seen that a laminated structure (distinguished by color or contrast in a black and white print) has been developed. It had been reported that this kind of laminated structure was related to a

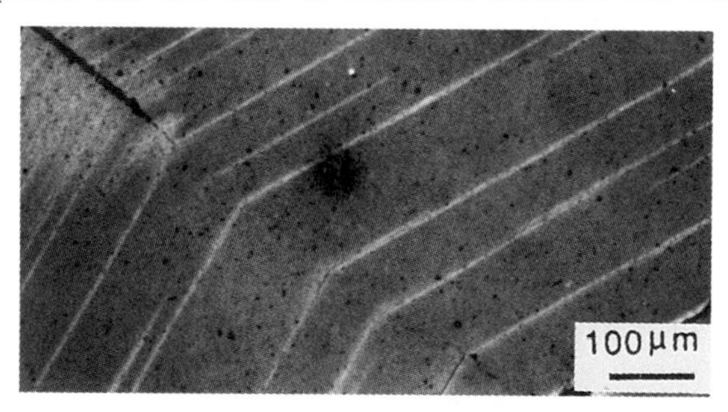

Fig. 1 optically polarized micrograph of a melt-processed $YBa_2Cu_3O_{7-\delta}$ specimen annealed at 450 ℃ for 5h in air after quenching in a liquid nitrogen bath from 970 ℃.

non-uniform phase transition from tetragonal to orthorhombic during low temperature oxygen annealing[6,7]. Lee et al[6,7] showed that the non-uniform oxygenation was induced by non-uniform stress built-up along the c-axis of a melt-processed $YBa_2Cu_3O_{7-\delta}$ crystal. Non-uniform and anisotropic oxygenation led to the T-O phase transition in the same manner; i.e., T-O phase transition occurs non-uniformly and anisotropically accompanying a planar cracks in the $YBa_2Cu_3O_{7-\delta}$ crystal.

Fig. 2 shows real part (χ') and its imaginary part(χ'') of AC susceptibility for a partially oxygenated specimen of Fig. 1. Magnetic field was applied parallel to c-axis of $YBa_2Cu_3O_{7-\delta}$ crystal. Martinez et al[8] reported that, for a melt-processed

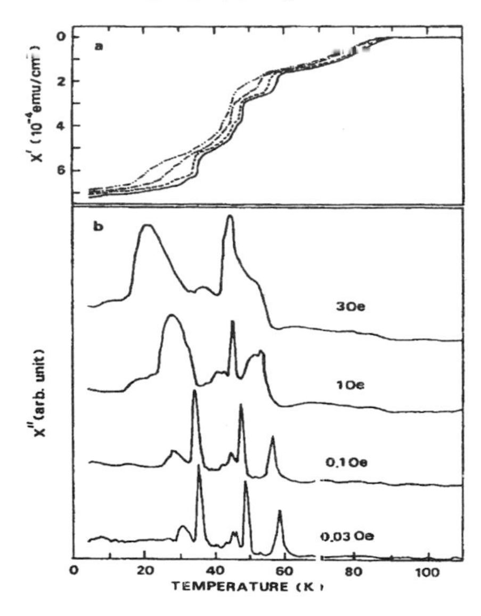

Fig. 2 Real part (χ') and its imaginary part(χ'') of AC susceptibility for a melt-processed $YBa_2Cu_3O_{7-\delta}$ specimen annealed at 450 ℃ for 5h in air after quenching in a liquid nitrogen bath from 970 ℃.

$YBa_2Cu_3O_{7-\delta}$ bulk specimen which was partially oxygenated, the diamagnetic signal from 60K phase was not detected due to the supercurrent from 90K phase when a magnetic field was applied parallel to the basal plane of $YBa_2Cu_3O_{7-\delta}$ crystal. It is very noticeable that apparent diamagnetic transitions existed at various temperatures of 35K, 48K, 58K and 90K and small diamagnetic transitions at 30K and 40K.

It is well known that the structure and the Tc of $YBa_2Cu_3O_{7-\delta}$ is varied with the oxygen deficiency, δ. There is a phase transition from tetragonal to orthorhombic(T-O transition) phase at $\delta \sim 6.4$[1]. In addition to the T-O transition, there appears a transition from O_{II} phase to O_I phase. Tc was 60K and 90K for O_{II} and O_I phase, respectively[1-3]. It is considered that the diamagnetic transitions at 60K and 90K is contributed to the O_{II} and O_I phases, respectively.

Oxygen ordering near $\delta = 6.5$ of the O_{II} phase(60K phase) has been observed by transmission electron microscopy[9] and was revealed to be thermodynamically stable by using the cluster variation method[3]. The diamagnetic transitions at 30K, 35K, 40K and 48K seem to be related with oxygenation of $YBa_2Cu_3O_{7-\delta}$ crystal. Lee et al[6,7] reported that non-uniform oxygenation was induced by stress. Therefore it is considered that the equilibrium oxygen content is varied with stress. We suggest that the diamagnetic transition at 30K, 35K, 40K and 48K is attributed to the nonuniform oxygenation of $YBa_2Cu_3O_{7-\delta}$ crystal induced by stress buildup during low temperature oxygen annealing.

Fig. 3 shows XRD pattern of a melt-processed $YBa_2Cu_3O_{7-\delta}$ bulk specimen which was partially oxygenated by annealing at 450 ℃ for 5h in air. The structure of the specimen was analyzed as a $YBa_2Cu_3O_{7-\delta}$ crystal containing $Y_2Ba_1Cu_1O_5$ but it was difficult to calculate the exact lattice parameter of the crystal because the specimen was consisted of multi-phases magnetically. XRD data showed that 3 main peaks(2θ = 46.04°, 46.19° and 46.55°) and 2 minor peaks(2θ = 46.48°, 46.51° and 46.62°) are present between 2θ = 45.959° and 2θ = 46.63° where (006) peaks of $YBa_2Cu_3O_{7-\delta}$ crystal with δ = 0 and δ = 1 are detected,

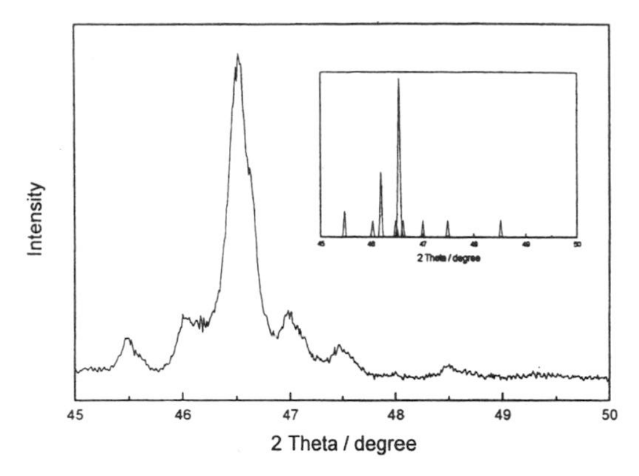

Fig. 3 XRD pattern of a melt-processed $YBa_2Cu_3O_{7-\delta}$ specimen annealed at 450 ℃ for 5h in air after quenching in a liquid nitrogen bath from 970 ℃. Inset is peak analysis of XRD pattern using multiple Gaussian distribution function.

peaks(2θ = 46.04° , 46.19° and 46.55°) and 2 minor peaks(2θ = 46.48° , 46.51° and 46.62°) are present between 2θ = 45.959° and 2θ = 46.63° where (006) peaks of $YBa_2Cu_3O_{7-\delta}$ crystal with $\delta = 0$ and $\delta = 1$ are detected, respectively. It seems that the XRD peaks, which were detected at 2θ = 46.04° and 46.55° , correspond to those from the (006) planes of tetragonal and orthorhombic $YBa_2Cu_3O_{7-\delta}$ crystals with $\delta \sim 1$ and $\delta \sim 0$, respectively.

4. Conclusion

Through the microstructural, crystallographic and magnetic observations of a partially oxygenated melt-processed $YBa_2Cu_3O_{7-\delta}$ bulk superconductor, it is suggested that the superconducting transitions at 35K and 48K are attributed to the stress induced by phase transformation via oxygenation.

Acknowledgment

Authors are grateful to Mr. J.J. Kim for specimen preparation and Mr. Y.I. Kim and Dr. C.S. Kim for XRD measurement and Mr. J. Joenssen and Prof. Dr. K.V. Rao for ac susceptibility measurement.

5. References

[1] B.G. Bagley, L.H. Greene, J-M. Tarascon and G.W. Hul 1987 *Applied Phys. Lett.* **51** 622

[2] R.J. Cava, B. Batlogg, C.H. Chen, E.A. Rietman, S.M. Zahurak and D. Werder 1987 *Nature(London)* **329** 423

[3] L.T. Wille, A. Berera and D. de Fontaine 1988 *Phys. Rev. Lett.* **60** 1065

[4] B.-J. Min, H.-I. Yoo and Y.K. Park 1989 *Materials Letters* **7** 325

[5] Y.K. Park, H.K. Lee, N.M. Hwang, H.K. Kwon, J.C. Park and D.N. Yoon 1988 *J. Amer. Ceram. Soc.* **71** c-297

[6] H.G. Lee, B.H. Ryu, J.J. Kim and G.W. Hong 1996 *Materials Letters* **27** 27

[7] H.G. Lee, I.H. Kuk and G.W. Hong 1997 *J. Materials Science Letters* **16** 80

[8] B. Martinez, V. Gomis, S. Pinol, I. Catalan, J. Fontcuberta and X. Obradors 1993 *Applied Phys. Lett.* **63** 3081

[9] G. Van Tendeloo, H.W. Zandbergen and S. Amelinckx 1987 *Solid State Commun.* **63** 603

Inst. Phys. Conf. Ser. No 158
Paper presented at Applied Superconductivity, The Netherlands, 30 June–3 July 1997

Texturing of thick film REBCO on metallic substrates by isothermal melt processing

N. Zafar, J.J. Wells, A.Crossley and J.L. MacManus-Driscoll

Department of Materials and Centre for High Temperature Superconductivity,
Imperial College, Prince Consort Road, London, SW7 2BP, UK

Abstract. Screen printed thick films of YBCO were melt processed on textured silver substrates in the form of foil and single crystals. Both temperature and oxygen partial pressure were controlled to achieve optimum melting and recrystallisation_times. The best biaxial texturing was achieved on rolled silver foil of texture {110}<110> with a FWHM in azimuthal orientations of ~15°.

1. Introduction

There is great industrial interest in the fabrication of the Tl-based and rare earth (RE) based conductors for applications at 77K, in field. Over the past five years, two main methods for fabricating biaxially textured $YBa_2Cu_3O_{7-x}$ (YBCO) conductors on metallic substrates have been developed. One is ion beam assisted deposition (IBAD) which involves ion assisted fabrication of textured buffer layers on polycrystalline metal substrates, followed by deposition of oriented YBCO of a few microns in thickness. Critical current densities of >1MA/cm^{-2} at 77K have now been achieved over cm lengths [1, 2, 3]. The other technique, developed more recently, is termed "rolling assisted biaxial texturing of substrates" or RABiTs [4]. In the RABiTs method, vapour deposited ceramic buffer layers are still required but they can be grown more rapidly without the need for the slow ion beam processes.

Melt processing of conductors is a viable and potentially more scaleable and cost-effective route for REBCO conductors, although high temperature gradients and slow growth rates are required in order to obtain aligned domains. It is possible that these processing difficulties could be overcome if domains could be multi-seeded from a substrate of the correct texture. This work involves the use of Ag substrate seeding of melt processed YBCO in the absence of an imposed temperature gradient. Ag is the most suitable metallic substrate as far as chemical compatibility goes although it has a large thermal expansion mismatch with YBCO. To overcome the thermal expansion difference thin Ag buffers on another metallic substrate, such as Ni, could potentially be used.

2. Experimental

2.1. Substrate processing

From thin film studies, the $\{110\}\langle 1\bar{1}0\rangle$ Ag substrate texture has been shown to yield biaxal orientation of YBCO, with a high degree of (001) alignment. The {111} and {100} textures, on the other hand, were found to give a number of in-plane orientations of YBCO, and are therefore unsuitable for achieving high critical currents [5]. 1µm polished {110} Ag single crystals were used as substrates as well as Ag foils. With a knowledge of the conditions which lead to the $\{110\}\langle 1\bar{1}2\rangle$ "brass" texture which forms in Ag after extensive cold working [6], we aimed to achieve a $\{110\}\langle 1\bar{1}0\rangle$ texture by a combination of hot rolling and recrystallisation. An ingot of Ag of initial purity 99% was used for thermomechanical deformation. Different reduction pass rates, total reduction rates, rolling temperatures and final recrystallisation temperatures were investigated. For the rolled Ag foils, little attention was paid to forming an atomically smooth surface since we are interested in determining whether it is possible to form a macroscopically, biaxially textured YBCO layer on a macroscopically, textured Ag substrate.

2.2. Precursor deposition

99.99% pure, commercial YBCO powder was used. The average particle size of the powder was 5µm. The powder was calcined at 900°C overnight, in flowing oxygen, before being mixed with a low-temperature-burnout binder. The slurry was screen printed in 5 layers through a mask of dimension 0.5cm x 2cm onto the Ag substrates. The binder from the screen printed track was burnt out at ~700 °C, in air, over a period of 12 hours.

2.3. Melt processing

The burnt-out, screen printed YBCO/ Ag samples were processed under conditions of reduced temperature and pressure, using a coulometric titration system which has been described earlier [7]. The processing schedule was designed to follow the stability diagram for YBCO [7]. Samples were melt processed at ~945°C (in the absence of a temperature gradient) near 3×10^{-3} atm., and the pO_2 was controlled to bring the samples above the peritectic for partial melting and below it, again, for recrystallisation. The complete melt processing cycle time was <70 hours.

Substrate and thick film sample charaterisation were undertaken using x-ray Bragg Bretano analysis, and x-ray pole figure analysis, optical and scanning electron microscopy. In the pole figure analyses of the YBCO films, sample translation was employed to ensure that the area of the tape being sampled was equivalent for all parts of the pole figure. At least 50 domains were sampled in each analysis.

3. Results and Discussion

For the hot rolling experiments, the optimum conditions which led to the $\{110\}\langle 1\bar{1}0\rangle$ orientation were: hot rolling temperature - 600°C, total reduction ~98.5%, recrystallisation temperature - 800°C. A pole figure of a typical thermomechanically deformed and recrystallised foil is shown in Figure 1. The pole figure shows the presence of a single $\{110\}\langle 1\bar{1}0\rangle$ texture with a FWHM in ϕ of ~15 °. The best results of biaxial texture in YBCO were achieved for tapes processed using the hot rolled and recrystallised silver foil. A FWHM in ϕ of the main domain set of ~ 15° was achieved as shown in Figure 2 (~50 domains sampled). Very good out-of-plane texture was achieved both on the foil and single crystals. Figure 3 shows the XRD of a typical melt textured tape on a $\{110\}\langle 1\bar{1}0\rangle$ textured

Ag substrate and shows near complete 'c' axis texturing. Note that the intensity is plotted on a 'square' scale to highlight the presence of second phases and misoriented grains. Y_2BaCuO_5 is always present in melt textured YBCO domains as a result of the nature of the YBCO crystal growth process from the melt.

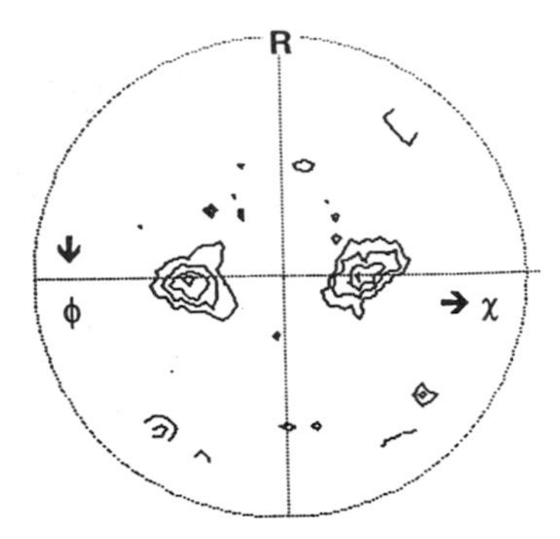

Figure 1 : {111} x-ray pole figure of Ag foil after hot rolling and recrystallisation. The rolling direction is indicated as 'R'. ϕ is the angle of sample rotation around the substrate normal. χ is the angle of sample rotation around the axis parallel both to the substrate plane and the rolling direction.

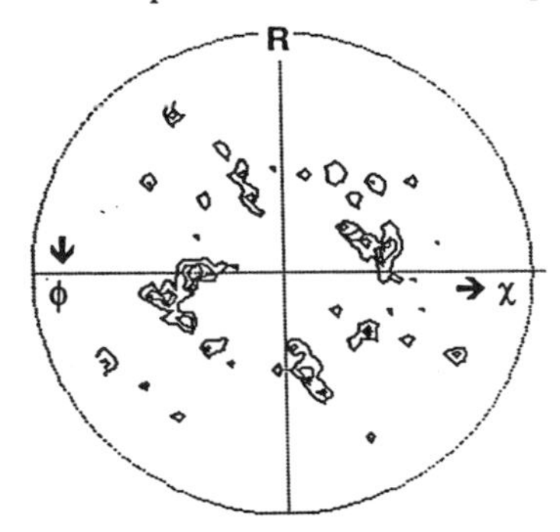

Figure 2: {103} Pole figure for 30 μm YBCO melt processed tape on $\{110\}\langle 1\bar{1}0\rangle$ Ag foil (of ~15° FWHM in-plane texture). Approximately fifty domains were sampled. A predominant {001}<010> YBCO texture is observed.

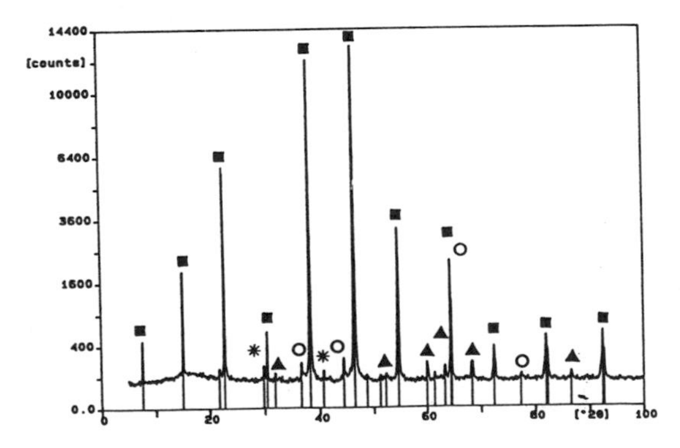

Figure 3: X-ray diffractogram of 30 μm YBCO thick film melt processed on $\{110\}\langle 1\bar{1}0\rangle$ Ag. Closed squares - (001) peaks, triangles- {101} peaks, circles - Ag, and asterices- Y_2BaCuO_5

An optical micrograph of the surface of a melt processed tape on rolled Ag substrate is shown in Figure 4. Typically the domains were between 200 and 500μm in size, and in cross-section were observed to be of the tape thickness, indicative of seeding and growth from the substrate.

Figure 4: Surface of melt processed YBCO on Ag

4. Conclusions

A ~15° in-plane spread of grain orientations was achieved for 30μm YBCO melt processed_ at ~945°C (in the absence of a temperature gradient), at ~3x10^{-3} atm, on {110}<110> Ag with a similar degree of in-plane misorientation.

Acknowledgments

The authors are very grateful to the Royal Society for funding N.Z, the Engineering and Physical Science Research Council of the U.K. for funding J.W.W, to National Grid PLC for provision of research funds, and to IBM for provision of equipment.

References

[1] Y. Iijima, N. Tanabe, O. Kohno, and Y, Ikeno, Appl. Phys. Lett. 60, 769, 1992.

[2] X.D. Wu, S.R. Foltyn, P. Arendt, J. Townsend, et al. Appl. Phys. Lett. 65 (15), 1961, 1994.

[3] H.C. Freyhardt, J. Hoffman, J. Wiesmann, J. Dzick, K. Heinemann, A. Isaev, F. Garcia-Moreno, S. Sievers, and A. Usokin, "Proceedings of the 1996 Applied Superconductivity Conference", Pittsburgh, U.S.A., August 25-30, 1996.

[4] A. Goyal, D.P. Norton, J.D. Budai, M. Paranthaman, E.D. Specht, D.M. Kroeger, D.K. Christen, Q. He, B. Saffian, F.A. List, D.F. Lee, P.M. Martin, C.E. Klabunde, E. Hartfield, and V.K. Sikka, Appl. Phys. Lett. 69 (912), 1795, 1996.

[5] J.D. Budai, R.T. Young, and B.S. Chao, Appl. Phys. Lett. 62 (15) 1836, 1993.

[6] I.L. Dillamore and W.T. Roberts, Metallurgical Reviews 10(39), 271, 1965.

[7] J. L. MacManus-Driscoll, J. C. Bravman and R. Beyers, Physica C 241,401, 1995.

Inst. Phys. Conf. Ser. No 158
Paper presented at Applied Superconductivity, The Netherlands, 30 June–3 July 1997

Controlled partial melting of dipcoated Bi-2212/Ag ribbons for combined isobaric and isothermal processing in reduced oxygen atmospheres

A.L. Crossley and J.L. MacManus-Driscoll.

Centre for High Temperature Superconductivity, Imperial College of Science, Technology and Medicine, Prince Consort Road, London SW7 2BP, U.K

Abstract. A detailed study has been made of the control and optimisation of partial melting of dipcoated Bi-2212/Ag ribbons using reduced oxygen partial pressures. A coulometric titration technique has been employed to vary the oxygen partial pressure in a region of the phase diagram corresponding to binary melting and the amount of partial melting has been quantified. It has been shown that complete melting of the Bi-2212 for short time periods is required for achieving optimum ribbon microstructure. The information has been used to process ribbons using either or both isothermal and isobaric methods. The combined method was found to be most effective Fully connected ribbons with only minor quantities of second phases have been obtained.

1. Introduction

It has been found that melt processing is necessary for producing practical conductors of $Bi_2Sr_2Ca_1Cu_2O_{8+\delta}$ on silver (Bi-2212/Ag). During melt processing the material is heated above the decomposition temperature of Bi-2212 in order to produce sufficient liquid such that upon solidification the Bi-2212 phase is dense. The phase assemblages during and after partial melting are complex since Bi-2212 melts incongruently. Conventional melt processing in air often results in large alkaline earth cuprate (AEC) phases and copper-free phases which can cause grain misalignment, obstruct current paths and lower the critical current density (J_c)[1].

One approach to improving phase purity and grain alignment has involved processing in different oxygen partial pressures. By choosing a region of the stability diagram which will produce secondary phases that do not grow as fast as the AEC phase in air, it may be possible to both minimise residual liquid and reduce grain misalignment[2]. In addition the optimum amount of partial melting at the peritectic has not been studied in detail for any pO_2 range and this is an important quantity for microstructural optimisation.

The following study will be restricted to a region of the phase diagram corresponding to binary melting where the $pO_2 \sim 6x10^{-3}$atm and the temperature $\sim$ 830°C. This region is preferable because only one precipitated crystalline phase, $(Ca_{1-x}Sr_x)_2CuO_3$, has been observed and the phase grows more slowly than the secondary phases which form in air. We have quantified the amount of partial melting necessary for tape optimisation and have used reduced pO_2's to give phase pure tapes.

2. Experimental

3cm lengths of dipcoated $Bi_{2.2}Sr_2Ca_1Cu_2O_{8+\delta}Ag_{0.1}$ 'green' ribbons from Oxford Instruments Inc were used for this study. A burnout procedure of 8h at 800°C was conducted in air. A 1g pellet of $Bi_{2.2}Sr_2Ca_1Cu_2O_{8+\delta}Ag_{0.1}$ was also wrapped in Ag and placed above the ribbon. The ribbon contains on average only 100mg of superconductor. The 1g pellet was included in the system to ensure that a sufficient mass of superconductor was available to give only small changes in oxygen stochiometry in the ribbon sample as oxygen was titrated into or out of the system.

The samples were installed into a coulometric titration (CT) system for the processing and phase studies. The CT system consisted of a quartz tube in which the oxygen partial pressure was controlled and measured using an yttria stabilised zirconia (YSZ) electrolyte. The sample was installed in the tube and was heated independently from the YSZ. Temperature control was to within ± 2°C and oxygen partial pressure was to better than $\pm 10^{-6}$atm.The system was then sealed, flushed and back filled with Ar, then heated to 830°C.

Isothermal studies were conducted at 830°C near $6x10^{-3}$atm and isobaric processing studies near 830°C and at $6x10^{-3}$atm. The combined processing route consisted of isothermal processing up to point of optimum partial melting followed by isobaric processing for recrystallisation.

3. Result and discussion

Figure 1 shows the oxygen partial pressure plateau for a 1g Bi-2212 pellet with and without 0.1mol Ag additions. As the oxygen is removed, the Bi-2212 sample clearly shows an oxygen partial pressure plateau at $5.75x10^{-3}$atm. The plateau for the sample containing Ag is very narrow and can be seen to start at $6x10^{-3}$atm with a change in gradient. Ag is known to influence the melting point of Bi-2212 and this shows the onset of melting beginning at a higher pO_2. The change in size and shape of the plateau suggests that the Ag inhibits oxygen release upon melting. Ag droplets have been observed by other workers[3] and have been

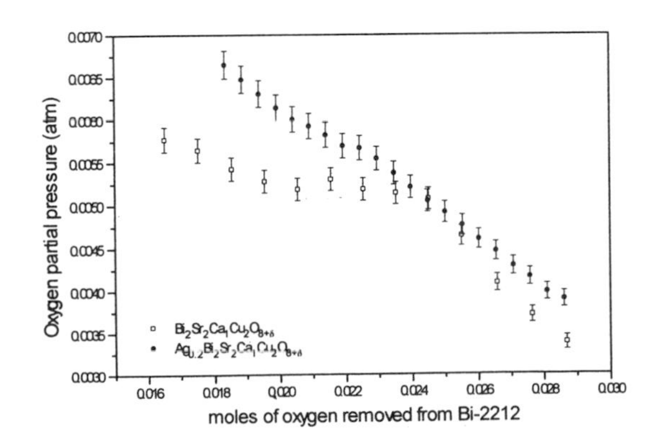

Figure 1 Bi-2212 oxygen partial pressure plateau with and without Ag

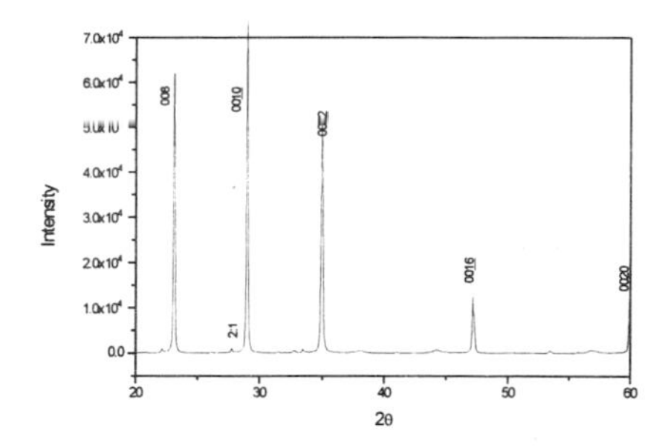

Figure 2 XRD of surface of ribbon processed by combined isothermal and isobaric technique

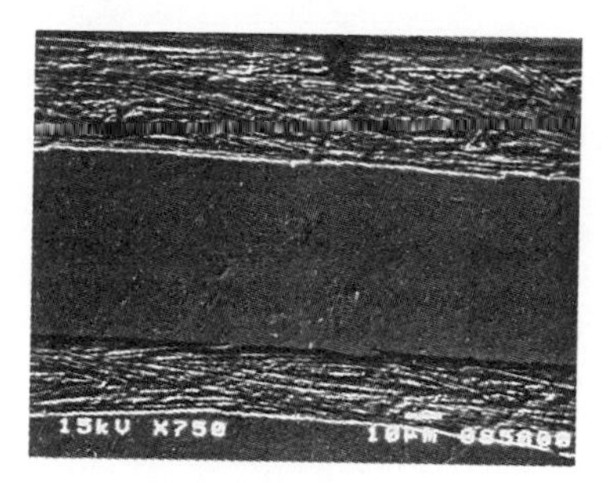

Figure 3 SEM of etched cross section of Bi-2212 tape after combined processing

found within the Bi-2212, but have not been incorporated into the crystalline structure. A possible explanation for the difference in observed plateau is that as the Bi-2212 begins to melt, Ag dissolves in the liquid phase. Oxygen has a higher solubility in the Ag incorporated in the liquid and thus the overall oxygen release is suppressed.

The microstructures of ribbons processed isothermally at different points along the Bi-2212+Ag 'plateau' differed markedly. A phase pure microstructure could only be obtained by processing just at the end of the 'plateau' corresponding to complete partial melting of Bi-2212 but not before or after that point. For optimally processed ribbons a combined approach proved practical.

The optimised ribbon, corresponding to melting at the end of the plateau, was characterised by XRD and WDS, and the magnetic properties measured using a vibrating sample magnetometer (VSM). The secondary phases were 2:1AEC and liquid, with no evidence of any other AEC phases or of a Cu-free phase. From XRD (figure 2) it is observed that only 00ℓ peaks are present with some minor peaks of 2:1AEC. An SEM micrograph of a polished and etched longitudinal cross-section of the ribbon shown in figure 3. Only Bi-2212 plates are apparent in the ribbon and the plates are highly aligned with no evident porosity.

Ribbons processed by other researchers [4],[5] under isothermal conditions and at reduced pO_2's, showed copper free phases and other minor phases which was not the case here. The critical current density at 10K is shown in figure 4.

Jc measured for our combined isobaric and isothermal processed ribbons are of similar values to ones obtained for processing in air. Since the microstructure is very clean, one might expect higher critical current densities. However it is probable that the grains were not of optimum size. For the same processing times at lower pO_2's, smaller plates were observed than are usually observed in ribbons processed in air because of the reduced kinetics at 830°C as compared to at 885°C (melting point in air). Therefore in order to further increase J_c above values obtained in air, longer crystal growth times are required. (i.e. longer annealing times at below the maximum processing temperatures).

4. Summary

The results of our CT processing study show that the Bi-2212 needs to be completely partially melted at the peritectic but not beyond this point. This corresponds to a complete

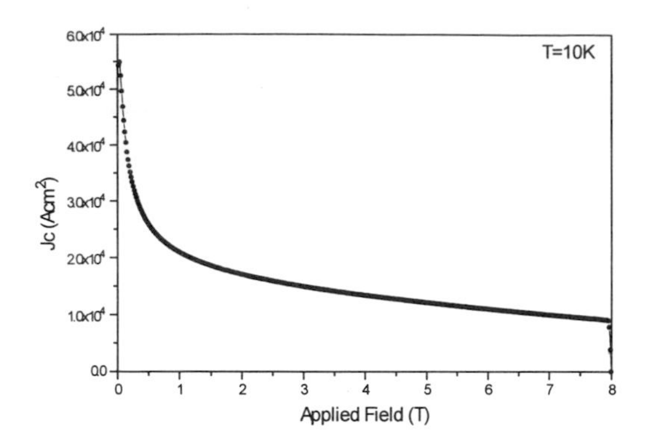

Figure 4 Jc-H at 10°K behaviour of tape processed using combined approach.

decomposition of the sample on the stability line. Any further below the stability line and large 2:1AEC secondary phases grow. For the low oxygen partial pressures and temperatures (pO_2~$6x10^{-3}$atm, T~830°C) used in this study a binary melting reaction was observed . The crystalline product of the Bi-2212 decomposition grows very slowly in the melt. This allowed a highly aligned material with excellent phase purity to be obtained. A combination of isothermal and isobaric processing has been found to be the best route for obtaining highly aligned, near phase pure tapes.

Acknowledgements

The authors would like to thank Oxford Instruments and the EPSRC for the CASE award for A.L.Crossley. They also thank Lisa Cowie, and Ian McDougall for providing the Bi-2212 material, and IBM for donation of laboratory equipment.

References

[1] Feng Y, Hautanen K.E, High Y.E, Larbelestier D.C, Ray II R, Hellstrom E.E, Babcock S.E, 1992 Physica C **192** 293-305

[2] MacManus-Driscoll J.L, Bravman J.C, Savoy R.J, Gorman G, Beyers R.B, 1994 *J. Am. Ceram. Soc 77 9 2305*

[3] Ray R.D, Hellstrom E.E, 1995 *Supercond. Sci. Technology*. **8** 430

[4] Holesinger T.G, Johnson J.M, Coulter J.Y, Safar H, Phillips D.S, Bingert J.F, Bingham B.L, Maley M.P, Smith J.L, Peterson D.E, 1995 *Physica C* **253** 182-190

[5] Funahashi R, Matsubara J, Ogura T Ueno K, Ishikawa H, 1997 *Preprint to Physica C*

Inst. Phys. Conf. Ser. No 158
Paper presented at Applied Superconductivity, The Netherlands, 30 June–3 July 1997

Synthesis of $YBa_2Cu_3O_{7-x}$ tapes for high current applications by MOCVD

U.Schmatz[a], F.Weiss[a], T.von Papen[a], L.Klippe[b], O.Stadel[b], G.Wahl[b], D.Selbmann[c], M.Krellmann[c], L.Hubert-Pfalzgraf[d], H.Guillon[d], J.Peña[e], M.Vallet-Regi[e]

[a]LMGP-ENSPG-INPG, 38402 St Martin d'Hères, France, [b]IOPW, 38108 Braunschweig, Germany, [c]IFW, 01171 Dresden, Germany, [d]LCM, 06108 Nice, France, [e]Univ. Complutense, 28040 Madrid, Spain

Abstract. For high current applications (power lines, current limiter), $YBa_2Cu_3O_{7-x}$ films of high superconducting qualities in sufficient thickness are necessary on suitable substrates (tapes). In order to achieve these desired properties, high growth rates have to be combined with epitaxial film growth on long lengths. Optimisation and results for films grown with high growth rates by single-source MOCVD techniques using traditional as well as new precursors in band coating reactors are presented. Compatibility with technical substrates was achieved by depositing suitable buffer layers.

1. Introduction

The growth of films by MOCVD generally allows to obtain high growth rates, and therefore is an attractive deposition technique for an industrial scale-up. However, in thermal MOCVD of $YBa_2Cu_3O_{7-x}$, problems in deposition reproducibility are encountered in the use of the Barium β-diketonate thd precursor, which is thermally instable under evaporation conditions. The search for more stable Barium precursors yielded fluorinated β-diketonate precursors. However, the presence of Fluorine in the deposition is not desirable for $YBa_2Cu_3O_{7-x}$ phase formation. In a different approach, we have developed single source feeding techniques which allow to keep the precursor reservoir at ambient conditions, and transport only the precursor immediately used. In these instantaneous-evaporation systems, it is also possible to use precursors that are not suitable for classical thermal evaporation due to their thermal decomposition characteristics.

For high current applications, suitable technical substrates have to be used for $YBa_2Cu_3O_{7-x}$ deposition. In order to obtain high critical current densities, the possibility to grow in-plane oriented $YBa_2Cu_3O_{7-x}$ has to be provided by the substrate. For cables and magnetic windings, substrates should be in tape form. As tapes generally are polycrystalline materials, in-plane orientation has to be induced by a buffer layer deposited by Ion Beam Assisted Deposition or by inclined deposition. Mechanically treated, rolling-assisted biaxially textured metal substrates also have shown their potential as a support for $YBa_2Cu_3O_{7-x}$ film deposition, if suitable buffer layers permit to transfer their orientation and serve as an effective diffusion barrier (RABiTS)[1].

2. Single liquid source MOCVD

The main problem encountered in thermal MOCVD of $YBa_2Cu_3O_{7-x}$ at high growth rates is the thermal instability of the Barium precursor commonly used, « Ba(thd)$_2$ ». This product's structural and physical properties depend on the preparation technique chosen. For high growth rates, as large quantities of precursor have to be evaporated, the precursor has to be heated to high evaporation temperatures, close to the decomposition temperature. As the precursors physical properties change with time (due to decomposition or oligomerisation), gas phase composition is not constant. For more reliable precursor feeding, it is advantageous to use a technique where only the precursor actually consumed in the reaction zone is transported and the precursor reservoir is left under inert conditions. This can be achieved by introducing the precursors as liquid into the evaporation zone. In the case of solid precursors, dissolving the precursors in an appropriate ratio in an organic solvent yields a single source solution, which can then be introduced into the reaction chamber. Different possibilities of introducing the source solution into the reactor have been studied: the source solution can be introduced as an aerosol, as micro-droplets directly injected, or transported with a band.

In Aerosol Assisted MOCVD (AAMOCVD)[2], the source solution is transported as fine droplets of micronic size. This aerosol is generally generated by ultrasonic excitation, either by a piezoelectric transducer, or by an ultrasonic nozzle. The aerosol is transported into the heated evaporation zone, where solvent and precursors are evaporated. This gas phase is then transported to the reaction zone, where CVD reaction can take place. As the solvent recipient is exposed to the operating pressure of the reactor, this has to be chosen above the vapour pressure of the solvent. Diethylenglycol dimethyl ether has been used due to its physical properties which allow a ultrasonic pulverisation and due to the good solubility of β-diketonate precursors.

In Injection Assisted MOCVD[3], source solution droplets of typically microliter size are injected into a heated evaporation zone by computer controlled electric microvalves with typical opening times in the microseconds range. As droplet size is considerably larger than in the case of AAMOCVD, for complete evaporation, a sufficiently long evaporation time has to be assured. As the solution reservoir is pressurised, it is possible to use solvents with vapour pressures higher than the operating pressure and therefore it is possible to use lower operating pressures than in AAMOCVD. Solvents typically used in this process are monoglyme, hexane and tetrahydrofurane, which are more volatile than diglyme and therefore evaporate very easily when injected into the reactor.

For evaporation from a band, the source solution is applied to a porous band (flow is controlled either by a mass flow controller, or the solution is applied spot by spot using microvalves[4], [5], [6]). This band passes through a zone heated to a temperature where the solvent is evaporated but not the precursors. By this means, the solvent can be eliminated from the process. The precursors remain on the band, which moves into the evaporation zone. Here the temperature chosen allows the evaporation of the precursors, which are then transported by a carrier gas to the reaction zone. The main difference of this process to AAMOCVD or Injection Assisted MOCVD is the fact that no solvent vapours are present in the reaction zone.

The commonly used hydrated Barium thd precursor presents the disadvantage of varying properties from one batch to another and of a poor long-time stability. However, air stable, fluorine-free Barium β-diketonate compounds exist, where the Barium's coordination sphere has been saturated by an additional ligand. These products present an evaporation behaviour which is not suitable for thermal evaporation: in heating the precursor, the adduct ligand first dissociates from the molecule, yielding a non-constant gas phase composition with time. In the case of the liquid feeding/instantaneous evaporation techniques, where precursor transport is activated by other means than their evaporation, their use is possible as only the precursor immediately used is heated and totally evaporated. We have used the Barium thd precursor with various adducts: ortho-phenantroline, bi-pyridyl, and triglyme. With these stabilised precursors, we have obtained more easily reproducible results for films deposited by AAMOCVD or Injection Assisted MOCVD.

3. Results

Liquid source MOCVD techniques have been used for $YBa_2Cu_3O_{7-x}$ film deposition on the typical monocrystalline substrates used for high temperature superconductor deposition as $SrTiO_3$, $LaAlO_3$, and MgO. Superconducting films with state-of-the-art properties ($T_c > 90K$, $j_c > 10^6$ A/cm^2) have been obtained by any of the liquid source processes presented. As film composition can easily be varied in these techniques by simply adjusting the source solution composition, we have been able to obtain a variety of film compositions. Films with a slight excess in copper generally present the best superconducting properties, the excess copper is observed as precipitates of copper oxide on the film's surface.

On rolling-assisted textured metallic substrates (Fe-Cr alloy), in-plane oriented CeO_2 films have been obtained by heteroepitaxy. Figure 1 presents the φ scan on the 111 line of a CeO_2 film deposited by AAMOCVD on a metallic substrate, with FWHM of 10°. However, surface roughness of these films still has to be improved in order to provide the possibility to obtain high quality $YBa_2Cu_3O_{7-x}$ films on these substrates.

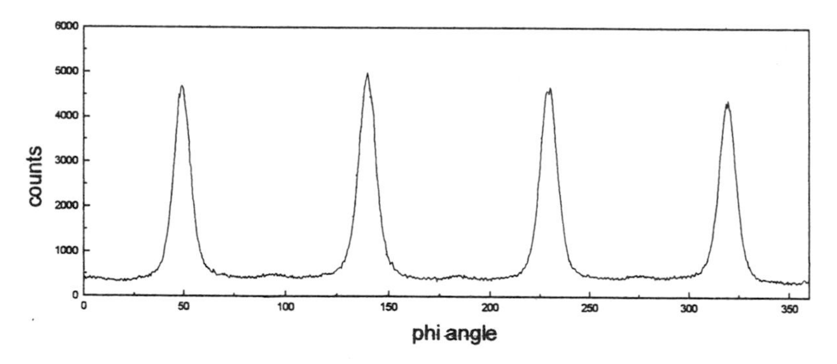

Figure 1: φ scan of a CeO_2 film deposited by AAMOCVD on an oriented metallic substrate

$YBa_2Cu_3O_{7-x}$ films of 0.8μm thickness have been deposited in a band coating reactor, running in band coating mode (substrate speed: 4 cm/h, growth rate: 4 μm/h). On monocrystalline $SrTiO_3$ substrates, state-of-the-art superconducting characteristics ($T_c > 90$ K, j_c (77 K, 0T) > 2 10^6 A/cm^2) have been obtained. The film shows a homogeneous surface in SEM and purely c-axis oriented in-plane epitaxied $YBa_2Cu_3O_{7-x}$ as is evidenced by X-ray diffraction. On polycrystalline substrates, as fibre texture is observed, critical current density is lower by 3 orders of magnitude (Figure 2).

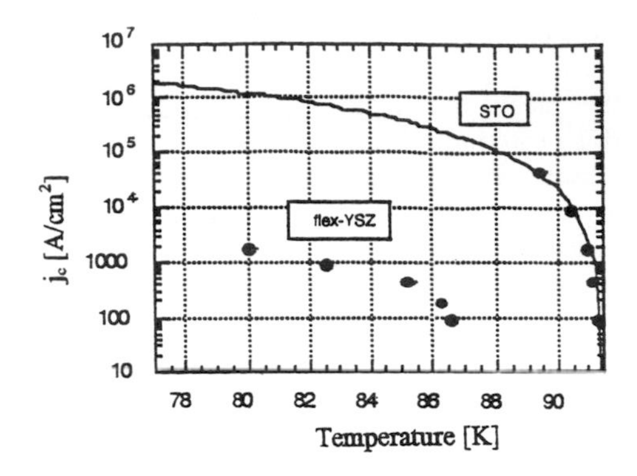

Figure 2: critical current densities obtained on moving 1*1cm^2 substrates in the band coating reactor. STO = $SrTiO_3$, flex-YSZ = flexible YSZ tape

4. Discussion and Conclusions

With the different single liquid source techniques established, we have obtained films with state-of-the-art superconducting properties in static mode reactors. In the recently constructed band coating reactor, on a moving monocrystalline substrate, the same properties have been obtained. This shows the possibility to cover suitable moving tape substrates with high quality films by liquid source MOCVD techniques.

The possibility of MOCVD to grow in-plane oriented buffer layers on textured metallic substrates by heteroepitaxy has been demonstrated. For further optimised film surface quality, these buffer layers are suitable for $YBa_2Cu_3O_{7-x}$ deposition.

The possible combination of buffer layer and superconductor deposition in a MOCVD reactor using two deposition zones is a promising concept for industrial scale-up of $YBa_2Cu_3O_{7-x}$ band coating.

Acknowledgements

This work has been supported by the European Commission under BRITE-EURAM BR2-CT94-0742.

References

[1] A.Goyal et al., Appl.Phys.Lett. 69 (1996) 1795-1797
[2] F.Weiss et al., J.de Physique IV, C3 (1993) 321-328
[3] F.Felten et al., J.de Physique IV C5 (1995) 1079-1086
[4] A.R.Kaul and B.V.Selzenev, J.de Physique IV, C3 (1993) 375-378
[5] L.Klippe and G.Wahl, Applied Superconductivity (1995), 611-614
[6] S.Pignard et al., J.de Physique C1 (1997) 483

Inst. Phys. Conf. Ser. No 158
Paper presented at Applied Superconductivity, The Netherlands, 30 June–3 July 1997

Pulsed current and high voltage measurements on sintered and textured ceramics BiPbSrCaCuO superconductors

J.G. Noudem[a,b], L. Porcar[a,b], O. Belmont[b,c], D. Bourgault[a], J. M. Barbut[b], J. Beille[c], P. Tixador[d] and R. Tournier[a]

[a]EPM-Matformag, [c]Laboraroire Louis Néel, [d]CRTBT/LEG, CNRS, B.P. 166, 38042 Grenoble cedex 09, France
[b]Schneider Electric S.A, A3, rue Volta, 38050 Grenoble cedex 09, France

Abstract. High pulsed currents with a duration varying from 1.25 to 20 ms have been applied at 77 K to sintered and textured samples. For I > Ic, two types of behaviour are observed : When the average temperature of the sample is lower than Tc, a critical current is recovered at the end of the pulse. Above Tc, samples remain normal after the pulse because the sample temperature is too high leading to a maximum instantaneous applied power up to 140 GW/m^3 in the textured sample. V(I) curves for sintered and textured samples are compared. At 15 times Ic, the measured electric fields are 1000 V/m and 400 V/m for textured and sintered samples respectively.

1. Introduction

One of the potential applications of high-Tc-superconductors is as Fault Current Limiters (FCL). In resistive FCL, the transition from the superconducting materials are frequently studied using d.c., a.c. or pulsed currents [1], but the measurements at high pulsed current and high voltage are very scarce. Bi:2223 superconducting sintered and textured bars have been fabricated.

The current limiting performance of bulk Bi:2223 superconductors are described. The transition to the conducting normal state is investigated at 77 K by applying pulse currents far above critical current up to 20 ms duration. We observe two regimes following the average temperature reached by the sample is below or above Tc that we discussed in this work.

2. Experiments

As starting material we used commercially available 2223 bismuth powder ($Bi_{1.8}Pb_{0.4}Sr_2Ca_{2.2}Cu_3O_{10.3+x}$, grade 1, d_{50} = 3.1 µm, Hoechst AG). The powder was pressed under a uniaxial pressure of 1 GPa into pellets of 20 mm in diameter and 6 mm in thickness. These pellets underwent the thermal and stress cycles in air required for texturation as describes previously report [2]. Long bar samples were pressed into rods ($\Phi = 5$ mm, L = 120 mm) with a cold isostatic pressure (1 GPa) and sintered at 845 °C during 90 hours in air. Contacts for current and voltage measurements were made by silver paint. All measurements were carried out using the ohmic standard four-point method. The critical current density at 77 K of each sample was deduced from d.c measurements using 1 μV/cm criterion. For sintered and textured samples, Jc are respectively 300 A/cm^2 and 1000 A/cm^2.

The pulsed current is produced by capacity discharge. The sample is immersed in liquid nitrogen at 77 K. Damping sinusoidal current up to 20 KA with a varying duration (1.25 to 20 ms) could be applied.

3. Results and discussion

Typical current-voltage pulsed waveforms applied to the textured sample are shown in figure 2. The peak value of the pulse (20 ms) current has been adjusted around 300 A. Since this magnitude is much higher than the critical current of 90 A, a voltage is measured and shows : (i) the phase transition between the superconducting and resistive state for I > Ic (ii) that the sample returns to the superconducting state when the pulse current falls. Indeed the superconducting state is recovered using applied currents of up to 10 times Ic. An explanation of this recovery is that the sample temperature is low with respect to Tc. The recovery of the superconducting state during the current fall is dependent on the temperature attained by the sample, in agreement with the critical current line Ic (T).

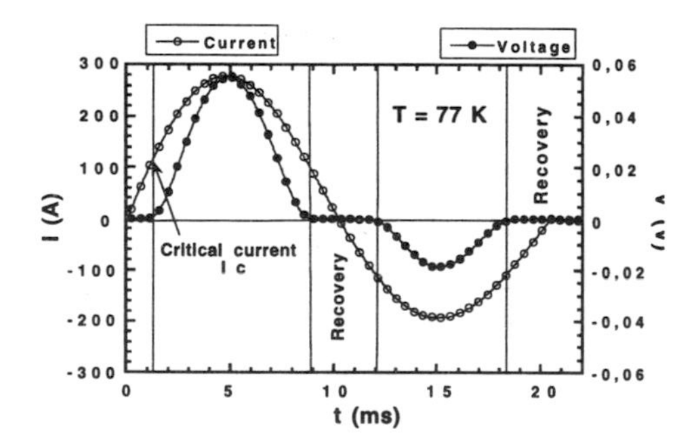

Figure 2 : Current and voltage waveforms given by a textured sample

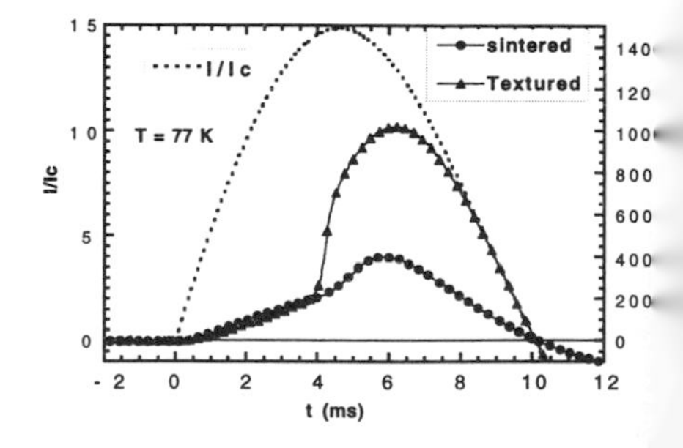

Figure 3 : Electric fields measured for sintered and textured samples for I up to 15 Ic

Figure 3 shows the variation of electrical field during a high pulse current (15 Ic) for the textured and sintered bars. The maximum current is 1350 A for a textured sample, corresponding to a current density of 15000 A/cm^2. The voltage signal reaches 6 V and 33 V between the voltage taps on textured and sintered bars respectively. The deduced electric field is higher for the textured sample (1000 V/m) than for the sintered bar (400 V/m). The transition in the normal state of the textured sample occurs faster due its higher critical current density (Jc) which is releated to the temperature increase velocity dT/dt function of ρJ^2 where ρ is the resistivity and J current density (see Table 1). These results suggest that a reduction of the fault current to about one quater of the unlimited peak value is possible in the first half period.

In table 1, the ability to limite the current of samples with the various instantaneous power per unit volume ρJ^2 at 15 times Ic for textured and sintered samples are compared. The analysis of these values shows that the maximum instantaneous power per unit volume between the two voltage taps applied to the textured and sintered samples are 140 GW/m^3 and 16 GW/m^3 respectively and the maximum energy per unit volume transmitted to the textured sample after a 20 ms pulse is 925 MJ/m^3.

This result demonstrates that textured samples could be a candidate for a resistive FCL application also these ceramics samples are mechanically strong and very homogeneous as is shown by their resistance to repeated high energy dissipation furthermore the measurements are highly reproducible.

Table. 1 : Sintered and textured bars characteristic at I = 15 Ic during 20 ms pulse.

superconductor	Jc (A/cm^2)	$\rho J = E$ (V/m)	$\rho J^2 = E.J$ (GW/m^3)	dT/dt (K/ms)
sintered	300	400	16	10
textured	1000	1000	140	55

The temperature and current dependence of the mean resistivity ρ (T,I) of all the textured samples are plotted on figure 4. The average temperature was calculated using the specific heat and the approximate relationship applicable to an adiabatic and homogeneous system :

$$\int UIdt = \int VC_p dT$$

C_p is the specific heat and varies as C_p = 0.0155 T ($J/K/cm^3$) for Bi:2223 [3], V : total volume of the sample (cm^3), I : pulse current amplitude (A), U : voltage (Volt), t : time (s), T : Temperature (K). The resistivity variation is plotted for different values of the pulse current up to 15 Ic with a duration of 20 ms. Two distinct behaviours are observed when the average temperature is above and below Tc.

(i) When the mean temperature of the sample is lower than Tc, a critical current is recovered at the end of the pulse. At 77 K, the sample is fully superconducting and the resistivity increases slowly with thermally activated flux motion. The slope above 0.07 $m\Omega$.cm is identified as the flux flow resistivity associated with the resistance of the weakly linked intergrain medium.

(ii) Above Tc, the sample remains normal after the pulse. The resistivity has similar values for different pulse currents. In the normal state, a good agreement is obtained between the classical resistivity measurement on small samples (I<<Ic) and the reconstituted plot using high pulse current and voltage.

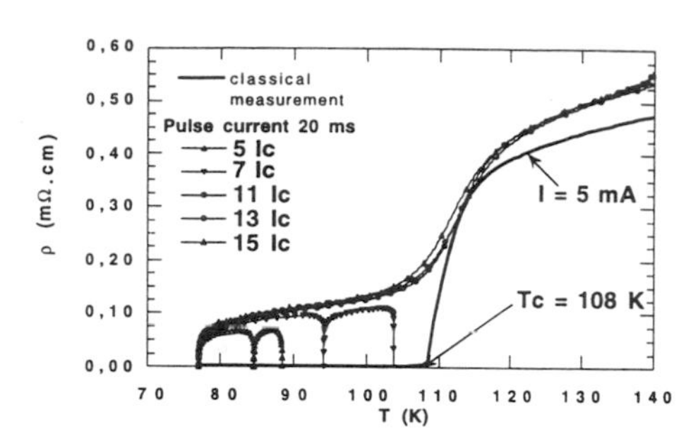

Figure 4 : Resistivities versus temperature at different currents for a textured sample.

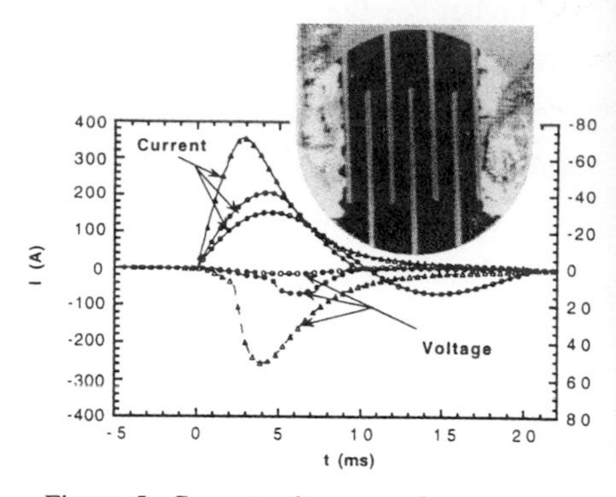

Figure. 5 : Current-voltage waveforms for textured meander sample

It is important for FCL applications to develop long length conductors. Textured meander samples have been obtained using a wire saw. The sample in figure 5 has a total length of 7.8 cm and a cross-section of 1x1.3 mm^2. Figure 5 shows that, when the pulsed current increases, the voltage also increases and reaches a maximum of 50 V at 350 A.

4. Conclusion

The superconductor to normal transition of sintered and textured Bi:2223 ceramics samples submitted to high pulsed currents has been studied. We have demonstrated that long samples can be made. Two regimes are observed in the ρ (T,I) diagram : superconducting recovery state when the mean temperature of the sample is lower than Tc and an purely resistive state above Tc. This work shows the ability of the bulk Bi:2223 to support a high current and voltage and high power per unit volume up to 140 GW/m^3. The textured samples have better electrical characteristics than the sintered samples, the maximum electric fields measured were 1000V/m and 400V/m respectively. The high voltage transition of the textured samples is highly favorable for the development of a current limiter device.

References

[1] L. Porcar, D. Bourgault, J.M. Barbut, M. Barrault, P. Germi and R. Tournier : Physica C **275** (1997) 293.

[2] J.G. Noudem, J. Beille, E. Beaugnon, D. Bourgault, D. Chateigner, P. Germi, M. Pernet, A. Sulpice and R. Tournier : Supercond. Sci. Technol. **8** (1995) 558.

[3] M. Ausloos, M. Benhaddou and R. Cloots : Physica C **235-240** (1994) 1767.

Inst. Phys. Conf. Ser. No 158
Paper presented at Applied Superconductivity, The Netherlands, 30 June–3 July 1997

Longitudinal Critical Current Variation in Short Samples of Bi-2212/Ag and Bi-2223/Ag Monocore Tapes and Its Origin

M Polak*, A Polyanskii, W Zhang, J Anderson, E Hellstrom, D Larbalestier**

Applied Superconductivity Center, University of Wisconsin, Madison, USA
* also Institute of Electrical Engineering, Slovak Academy of Sciences, Bratislava, Slovakia,
** also Institute of Solid State Physics, Russian Academy of Sciences, Chernogolovka, Russia

Abstract. Variations in local longitudinal critical currents in short samples of Bi-2223 and Bi-2212 monocore tapes were studied. The sample critical current and local values of the critical current in some sections of the Bi-2212 tapes increased considerably if the silver was removed from the test section. This effect was smaller in Bi-2223 tapes. Current transfer voltages associated with inhomogeneities are responsible for the observed behavior. Magneto-optical experiments support the results of the transport measurements.

1. Introduction

Many interesting and important experiments are made with short (test section ~ 10 mm) samples that are supposed to be mostly homogeneous. The standard assumption concerning representative samples is that they are homogeneous. However, a systematic study on 22 short samples of powder-in-tube Bi-2223/Ag [1] showed surprisingly large variations: I_c values scattered from 3.5 A to 28 A. Large differences between I_c measured in a 30 mm long test section of a Bi-2223/Ag tape and "local" values (~ 1.7 mm long sections) was also found using a "sliding" contacts technique [2]. To determine the degree of variation of longitudinal critical current in short samples of Bi-2212/Ag and Bi-2223/Ag we measured "local" V-I curves and estimated "local" critical currents using potential taps with spacing that ranged from 1 to 2.5 mm, which were attached to the test section (10 mm typically) of short samples. To investigate the role of silver on the V-I curves and local critical currents, we measured the samples both with and without the silver sheath. The test sections of the samples were also observed using magneto-optical imaging to study the flux penetration of the external magnetic field applied perpendicular to the broad tape plane.

2. Experimental

Samples from Bi-2223/Ag monocore tape (A), as well as from Bi-2212/Ag (B), were prepared. The powder in tube (PIT) tape A, with nominal composition $(Bi,Pb)_2Sr_2Ca_2Cu_3O_x$, was prepared by standard procedures described previously [3] in unit lengths of about 100 mm. The tape dimensions were 2.8x0.15 mm and the core cross-section was 1.2×10^{-3} cm^2. The critical current density (1μV/cm, 77 K, self field) of the tape core was measured on 10 mm long test sections, and was found to range from 1.25×10^4 A/cm^2 to 1.6×10^4 A/cm^2. The PIT tape B with the composition $Bi_{2.1}Sr_2CaCu_{1.95}O_x$ was prepared by melt processing [4] in unit lengths of about 120 mm. The tape dimensions were 2.9x0.15 mm and the core cross-section was 1×10^{-3}cm^2. The critical current densities (1μV/cm, 4.2 K, self field) in the tape core varied from 1.5 to 2.5×10^5 A/cm^2 when measured in 10 mm test sections. Measurements with transport current were made on ~30 mm samples. Several potential wires (diameter ~ 0.1mm) with a spacing that ranged from 1 to 2.5 mm were attached to the ~ 10 mm long test section of the sample using a conductive silver paint. The ends of the tape were soldered to the Ag-foil current leads. On these samples we measured "local" V-I curves and "local" critical currents I_{cl} (between the neighboring potential taps), as well as V-I curves and I_c for the whole test section. After finishing the measurements on the original silver sheathed samples we proceeded to either modify for MO observations, or measure "local" V-I curves without the silver sheath. In the first case we cut off the test section, removed the potential wires, and removed the upper silver sheath by chemical etching. The indicator film for MO observations was put directly on to the remaining BSCCO core. In the second case the sample was attached to a special bridge, and we then removed the silver from the test section except for a of small island of silver on the upper side of the tape, to which potential wires were connected. Then, the V-I measurements were done by the same method that for the silver sheathed samples.

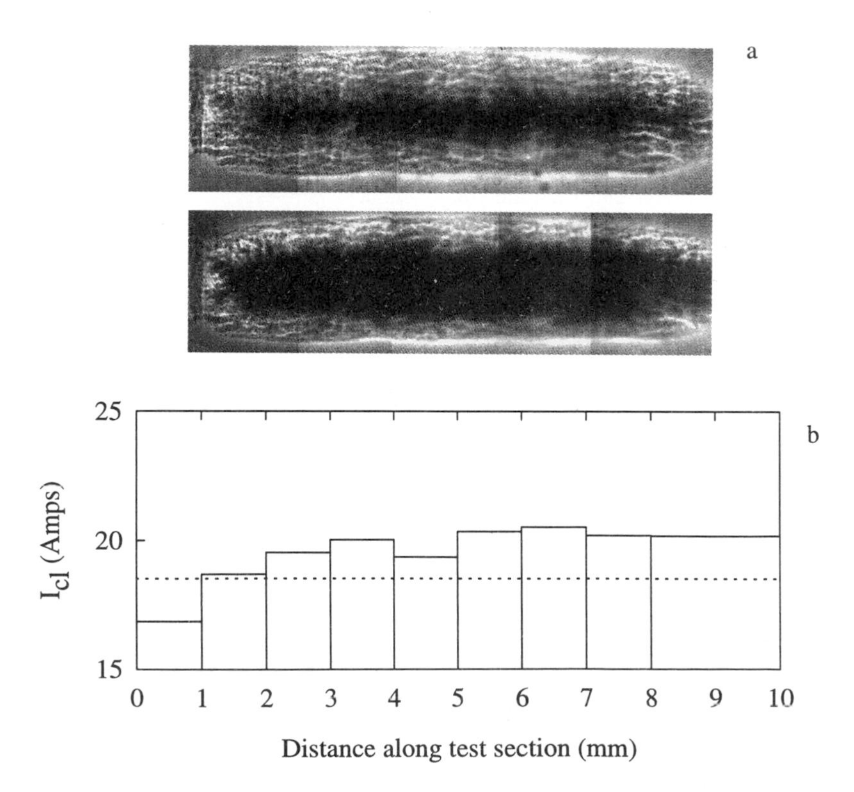

Fig.1 a: Magneto-optical image of the sample A/1 at 8mT, 77K (top picture) and 8mT, 40 K (bottom picture), b: local critical currents I_{cl} (full lines) and I_c of the whole test section (dashed line).

3. Results and discussion

Fig.1 shows the results for the sample A/1 (from the Bi-2223/Ag tape A). The plot in Fig.1b shows the local critical currents, I_{cl}, in the test sample section, as well as the critical current of the whole test section I_c, both at the mean electric field E=1μV/cm. The "local" critical currents range from 91% of I_c to 111% of I_c. The MO image of the nearly whole test section was taken when the external magnetic field was 8 mT at 77 K (top) and 40 K (bottom) as shown in Fig. 1a. Excluding the ends of the sample no remarkable variation of the width of the areas penetrated by magnetic field (bright areas along the core edges) can be observed. One can see a slight narrowing of the screened (dark) core part on the right side of the sample. Core thickness variation was also measured and the observed differences were within 10 % of the mean thickness value. We believe that the variation of I_{cl} is due to differences in core thickness and to local regions of poor grain connectivity. The measurements of the local values, I_{cl} on another sample, A/2, showed that the critical currents of some sections as well as the whole test section slightly increased (~1.5%) when we removed the silver sheath.

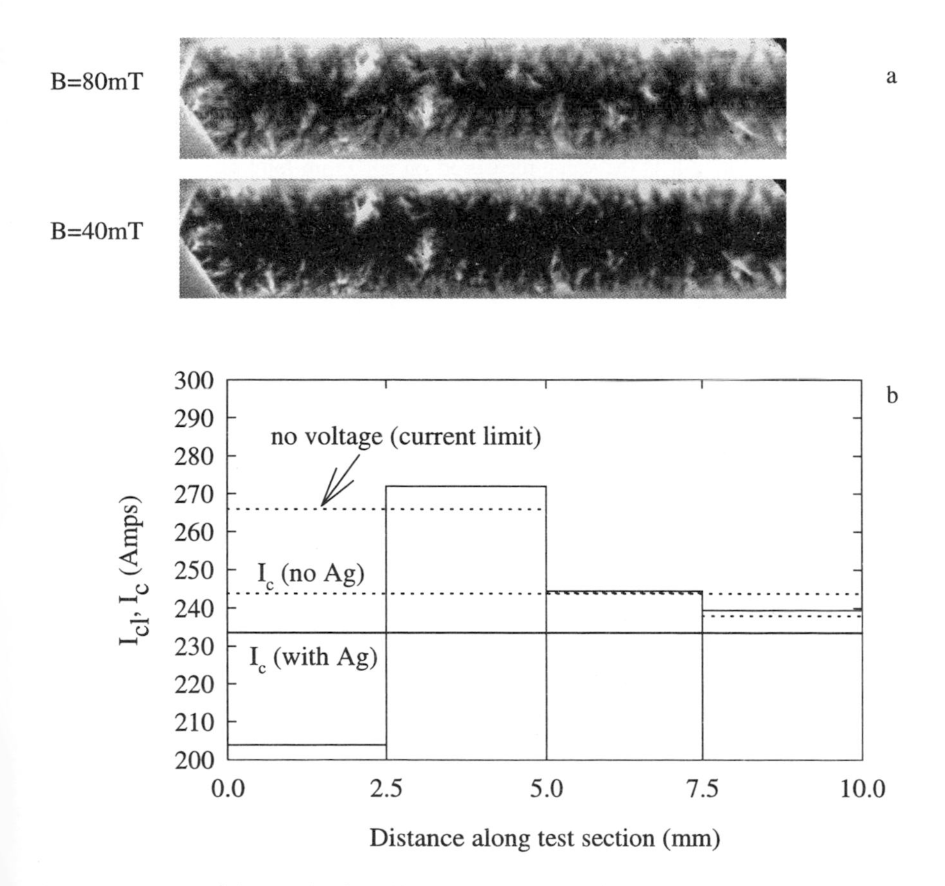

Fig.2. a: MO image of the sample B/1 at 40 K, b: I_{cl} measured with Ag sheath (full lines) and without it (dotted lines). Inset: V-I curves with Ag (full line) and without Ag (dotted line) for the subsection 0-2.5 mm.

Similar measurements were done on the Bi-2212 sample B/1. In the Bi-2212 samples much more important differences in I_{cl} with and without Ag were found (Fig.2b). In the inset of Fig.2b we see V-I curves for the subsection from x=0 to 2.5 mm with silver and without silver. This local V-I curve has a linear section (the section between 2.5 mm and 5 mm behaved similarly), while usual nonlinear curves were measured for other subsections. The linear sections are due to the current transfer in the vicinity of localized defects. After we removed the silver, these linear (ohmic) voltages disappeared (see inset). As a consequence, I_{cl} and I_c of this particular section increased from 203.9 A to more then 266 A (current limit with respect to the heating of the weakest sections) after Ag was removed. The MO image shown in Fig.2a shows clearly that the structure of the penetrated flux is different from that for Bi-2223 sample.

From this experiment we deduced 2 important consequences of inhomogeneous samples. First, the voltage measured in the test section of Ag-sheathed samples has a component due to the flux motion, but also an important component due to the current transfer in the vicinity of a defect, which disappears if silver is removed. Second, the local critical current I_{cl} can be higher if the measured sample section does not have a silver sheath. Thus, the true inhomogeneity of the BSCCO core can be measured on bare samples only.

4. Conclusions

The "local" critical current measured in 1 mm long subsections of a Bi-2223 sample varied from (90-110)% of I_c, where I_c is the value measured on a 10 mm long section. In short Bi-2212/Ag samples, I_{cl} varied from 95% I_c to 112.5% I_c. In Bi-2223/Ag samples the I_{cl} variation is probably due to the core thickness variation and a variation of the grain connectivity, while in 2212 samples, macropores cause the variation of I_{cl}. Some of the local V-I curves have ohmic components which disappeared after the silver sheath was etched off. Thus, the voltage measured on silver sheathed tapes also has components which are not associated with the fluxoid movements. These voltages influence the shape of the measured curves and I_c considerably, in Bi-2212/Ag tapes in particular. The differences between Bi-2223 and Bi-2212 was also demonstrated by MO imaging of the field penetration into the Bi-based tapes, which revealed a substantial difference between the fine structure of the penetrated magnetic flux. We conclude that to study the homogeneity of the core, the local critical currents must be measured on bare samples only.

References

[1] Osamura K Nonaka S, Matsuno K, Ito H, Sakai A, Ochiai S 1996 Proc. of the 8th Int. Workshop on Critical Currents in Superconductors, Ed. T. Matsushita and K. Yamafuji, World Scientific

[2] Cave J R, Willen D W A, Nadi R, Cameron C, Zhu W 1995 IEEE Trans. on Appl. Superc. **5** 1294-1297

[3] Dorris S E, Prorok B C, Lanagan M T, Sinha S and Poeppel R B, 1993 Physica C **212** 66

[4] Polak M, Zhang W, Polyanskii A, Hellstrom E E and Larbalestier D C, presented at the Applied Superconductivity Conference 1996, Pittsburgh, to be published in IEEE Trans. on Appl. Superconductivity

[5] Polak M, Zhang W, Parrell J, Cai X Y, Polyanskii A, Hellstrom E E, Larbalestier D C and Majoros M, to be published in Superc. Sci. Technol.

Inst. Phys. Conf. Ser. No 158
Paper presented at Applied Superconductivity, The Netherlands, 30 June–3 July 1997

Synthesis of $Hg_{1-y}Re_yBa_2Ca_2Cu_3O_{8+x}$ pure phase at normal pressures

S.Piñol, A.Sin, A.Calleja, L. Fàbrega, J.Fontcuberta, X.Obradors, M.Segarra*.

Institut de Ciència de Materials de Barcelona (CSIC). Campus de la UAB. E-08193-Bellaterra. Spain.
(*) Dept d'Engenieria Química i Metal.lurgia. Universitat de Barcelona. Martí i Franqués, 1. E-08028 Barcelona. Spain.

Abstract. We have prepared ceramics of the superconducting phase $Hg_{0.8}Re_{0.2}Ba_2Ca_2Cu_3O_{8+x}$ with a critical temperature onset of about 138K and transitions widths of 8K (10%-90%). The superconductors were prepared using precursor powders of $Ba_2Ca_2Cu_3O_7$ mixed with stoichiometric additions of ReO_3 and HgO. We have utilized the vacuum sealed quartz tube technique for the preparation of the superconducting phase.

1. Introduction.

Mercury based copper oxide superconductors, $HgBa_2Ca_{n-1}Cu_nO_y$ (n= 1, 2, 3,...) have been extensively studied in the last years, because their high superconducting transition temperature. The member of this series with three CuO_2 layers (n=3), is the most interesting in high temperature superconductivity, because it has the highest critical temperature (136 K) recorded at ambient pressure [1,2]. However, the synthesis of unsubstituted high-purity $HgBa_2Ca_2Cu_3O_{8+x}$ remains a serious challenge. One of the reasons is that the preparation of these materials seems to be very delicate and requires extraordinary care during all steps of the sample preparation. Many synthesis procedures have been reported for these compounds with various success. The undoped Hg-1223 cuprate has been obtained as an almost single phase only under pressures of several tens of kilobars [3,4]. Several groups have tried to obtain undoped Hg-1223 at normal pressures by the evacuated quartz tube method, but the purity of the phase was not more than 75% [5, 6, 7]. This difficulty arises from the oxygen-deficient character of these compounds, specially at the mercury layers (HgO_δ), which favors the instability of the structure. The solid HgO precursor decomposes at low temperature (450°C) into mercury and oxygen. So, the formation of the superconducting phases is due to the reactions between solid-solid and solid-vapor phases. Further incorporation of oxygen and thus stabilization of the superconducting phase can be achieved by higher-valence cation doping. It leads to a higher purity and sometimes to almost single phase Hg-1223. Recently, it was found that some chemical substitutions play the same role as the applied high pressure. In the last three years, different new substitutions have been studied to stabilize these compounds. One of the most efficient doping cations is rhenium; it does not reduce superconducting critical temperature and it has been argued that the irreversibility line of the substituted (Hg-Re)-1223 compound reaches 15 T at 77 K thus exceeding that of $YBa_2Cu_3O_7$.

In this paper we present the synthesis of $Hg_{1-y}Re_yBa_2Ca_2Cu_3O_{8+x}$ ($0.1<y<0.25$) superconducting ceramics by using the sealed quartz tube technique. The conditions required to obtain single phase superconducting oxides are explored.

2. Experimental.

Starting powders for the synthesis of $Hg_{1-y}Re_yBa_2Ca_2Cu_3O_{8+x}$ were obtained by mixing multiphase $Ba_2Ca_2Cu_3O_{7+x}$ 99.9% with HgO and ReO_3. The mixtures were carried out in two steps. Firstly, the $Ba_2Ca_2Cu_3O_{7+x}$ was homogenised with ReO_3 in an agate mortar and annealed at 850°C and 930°C for 15h in oxygen atmosphere with an intermediate regrinding. Then, the powders were homogenised with stoichiometric additions of HgO and pressed up to $10Tm/cm^2$. Finally, the pellets were encapsulated in an evacuated quartz tube (10^{-2} torr) and annealed at different rates and temperatures in order to optimise the purity of the Hg-1223 phase. The purity of the phase and cell parameters were determined by X-ray diffractometry; sample composition was determined by Energy Dispersive X-ray Spectroscopy. Superconducting properties were characterised by AC Susceptibility Measurements.

3. Results and discussion.

In a first series of experiments, we have explored the loss of weight of green ceramics with $Hg_{0.9}Re_{0.1}Ba_2Ca_2Cu_3O_{8+x}$ compositions. We have encapsulated in quartz ampoules of $5cm^3$ volume, pellets of 300 mg and we have introduced rapidly the ampoules inside the isothermal zone of a preheated furnace (flash technique) at 850°C for different times. All the encapsulated samples were cooled down to room temperature by quenching in air. Figure 1 shows the weight loss, due to HgO evaporation, of the different pellets as a function of the annealing time. At the first stage, the unreacted HgO sublimates to the gass phase and in a second stage, it is incorporated again into the pellets in a fast (few minutes) vapour-solid reaction. Prolonged annealing at 850°C lead again to the complete loss of mercury from the pellets. Drops of liquid mercury were found in the ampoules when weight losses were important. No superconducting phases were formed in these conditions and $CaHgO_2$ crystals were identified by EDX analysis. In another series of experiments, we have explored the effect of the rate of heating. We have heated the encapsulated samples at a heating rate of 150°C/h until 850°C , kept at this temperature during different times and then cooling the samples down to room temperature at 150°C/h. No weight losses were detected for annealing times up to 3h.

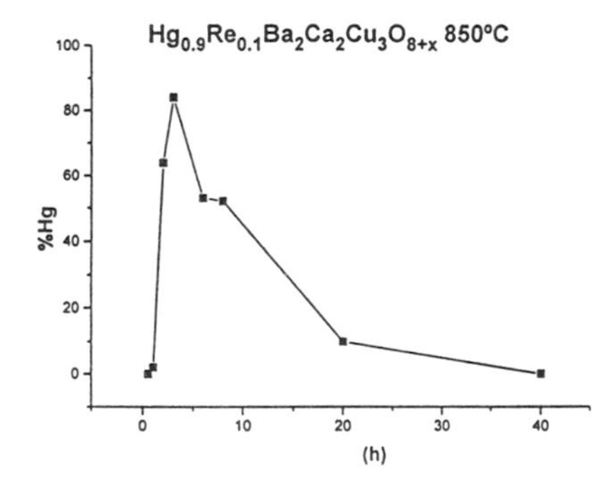

Fig. 1. Weight loss due to the HgO evaporation from the pellets heated to 850°C (flash technique).

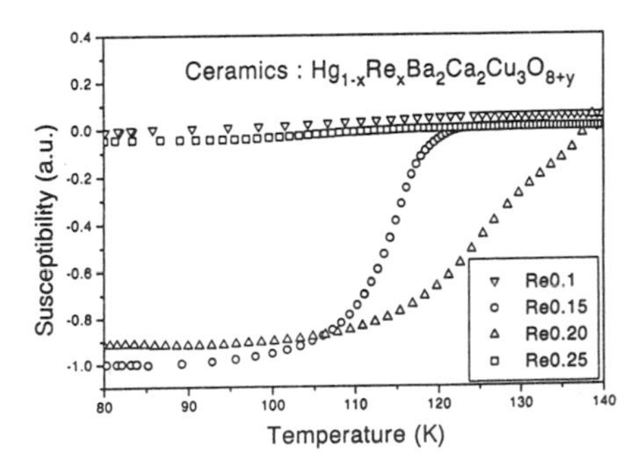

Fig.2. Magnetic susceptibility for samples having different Re composition, annealed at 900°C.

However, no superconducting phases were detected by magnetic susceptibility measurements and EDX analysis. Only $CaHgO_2$ and $BaCuO_2$ were observed by EDX analysis. The same series of experiments were repeated at 900°C and a sharp superconducting transition at 125K was found by magnetic susceptibility measurements. According to the X-ray powder pattern it corresponds to the (Hg Re) 1212. No (Hg Re) 1223 phase was identified.

In a new series of experiments we have explored the phase formation and stability using different starting rhenium compositions (y= 0.10, 0.15, 0.20 and 0.25). The rate of heating was kept constant at 150°C/h and the annealing temperature at 900°C. Only a narrow Re composition close to 0.20 was found to be appropriate to obtain an almost pure (Hg-Re)-1223 phase, as shown in Figure 2. The previous experiments indicate that the pressure generated by the HgO decomposition in the quartz tube is a key parameter controlling the phase formation and purity. Consequently, it is to be expected that the initial mass of material may have some influence on these synthesis. When exploring this influence, we have observed that the pressure inside the capsule decreases when increasing the rhenium concentration (y). Using a quartz capsule ($5cm^3$) of 8 mm internal diameter and 1 mm wall thickness we can prepare 3.2 g of $Hg_{1-y}Re_yBa_2Ca_2Cu_3O_{8+x}$ with y= 0.18 and 0.20. For higher (y=0.22) and lower (y=0.16) Re concentrations, the tube was not resistant enough for the vapour pressure of the compounds. To obtain the y=0.16 sample we should decrease the mass of the pellet until 1.6 g. This result means that there is a minimum in the partial pressure of the $Hg_{1-y}Re_yBa_2Ca_2Cu_3O_{8+x}$ compounds close to the rhenium composition x=0.18, and this pressure is close to the quartz tube resistance (~15 atm. for a 1mm thickness tube).

Next, in order to investigate the effects of longer annealings on the sample purity and reactivity, we have studied the influence of a post annealing process at 900°C. Samples doped with Re y=0.20 and 1.9g in a $5cm^3$ capsule were selected. After 3 hours of annealing, the samples show a porous microstructure formed mainly by 1223 grains weakly connected. Prolonged annealing, do not increase the purity of the phase, but at 6h and 9h some decomposition of the 1223 phase giving mainly $CaHgO_2$ and $BaCuO_2$ is detected by X-ray diffraction. Fig. 3 shows the magnetic susceptibility signal of a pellet (a) and a powder (b). After grinding and mixing the prereacted powder with an added amount of HgO to compensate the weigth loss, the mixture is encapsulated again and we have repeated the same annealing at 900°C. We have found that the purity of the 1223 phase increases considerably

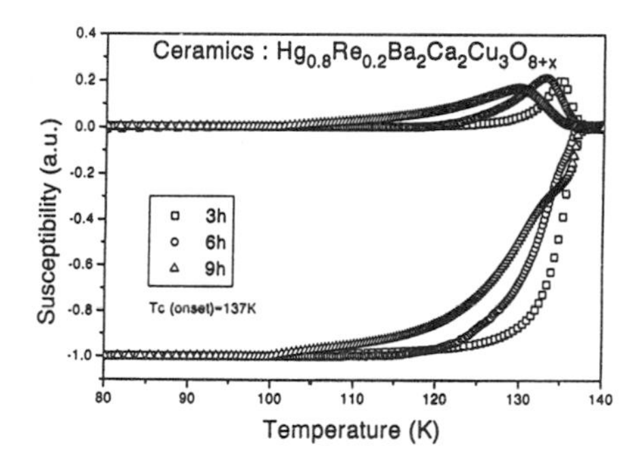

Fig. 3 a. Magnetic susceptibility for a ceramic sample.

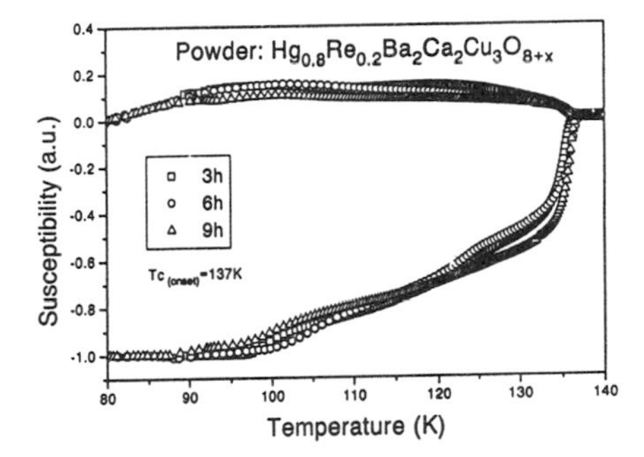

Fig. 3 b. Magnetic susceptibility for powdered samples.

(see Fig. 4) and the impurities such as $CaHgO_2$ and $BaCuO_2$ decrease. Further homogenisation does not result in a significant change of sample purity. On the contrary, after 3 cycles a decomposition of the phase is detected by X-ray diffraction. When we have repeated the same experiment with Re=0.18, we have seen that the impurities are eliminated almost completely, indicating that this composition is the optimal in our experimental conditions.

Finally, we have studied the influence of the mass on the purity of the phase. We have encapsulated 3.3 g of $Hg_{0.8}Re_{0.2}Ba_2Ca_2Cu_3O_{8+x}$ in a 5 cm^3 ampoule and we have repeated the same thermal treatment. The purity of the phase was increased indicating that the higher pressure due to the additional mass leads to an amelioration of the purity as confirmed by X-ray diffraction pattern and magnetic measurements (Fig. 5). The parameters of the cells were calculated by high precision lattice constant determination. The crystal plane (110) gives rise to a single peak at about $2\Theta=32.8°$, corresponding to the tetragonal I4/mmm structure. We have studied samples with compositions Re= 0.16, 0.18 and 0.20. The cell was found to be tetragonal in all of them and no differences were found in the values of the parameters which were: a=b= 3.857 Å ± 0.002 and c= 15.69Å ± 0.01. The value of a(=b) is similar to that found by other authors for undoped and doped compounds with Tl and Pb, but thc c parameter is shorter due to the stronger character electropositive of the Re cation.

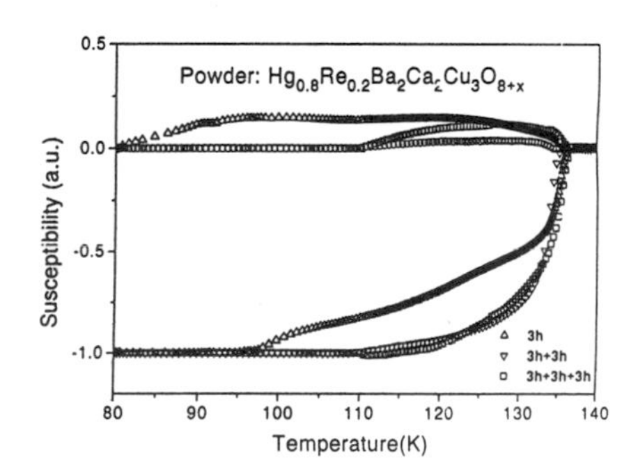

Fig. 4. Magnetic susceptibility of a sample encapsulated and annealed several cycles.

Fig. 5. Magnetic susceptibility of samples with different mass annealed at the same conditions.

Acknowledgments

This work has been supported by the CICYT project: MAT96-1052-C04-04.

References

[1]-S.N.Putilin, E.V.Antipov, O.Chmaissem and M.Marezio. Nature 362 (1993) 226.

[2]- A.Schilling, M.Cantoni, J.D. Guo and H.R.Ott. Nature 363 (1993) 56.

[3]- E.V.antipov, S.M. Loureiro, C.Chaillout, J.J.Capponi, P.Bordet, J.L.Tholence, S.N.Putilin and M.Marezio. Physica C 215 (1993) 1.

[4]-M.Hirabayashi, K.Tokiwa, H.Ozawa, y.Noguchi, M.Tokumoto and H.Ihara. Physica C 219 (1994)6.

[5]- R.L.Meng, L.Beauvais, X.N.Zhang, Z.J. huang, Y.Y.Sun, Y.y.Xue and c.w.Chu. Physica C 216 (1993) 21.

[6]- A.Schilling, O.Jeaudupeux, J.D.Guo and H.R.Ott. Physica C 216 (1993) 6.

[7]-H.M.Shao, L.J.Shen, J.C.Shen, X.Y.Hua, P.F.Yuau and X.X. Yao. Physica C 232 (1994) 5.

Inst. Phys. Conf. Ser. No 158
Paper presented at Applied Superconductivity, The Netherlands, 30 June–3 July 1997

Thick film Tl-1223 conductors; processing, microstructure and superconducting properties

J.C. Moore, P. Shiles, C.J. Eastell, C.R.M. Grovenor and M.J. Goringe

Department of Materials, University of Oxford, Parks Road, Oxford, OX1 3PH, UK.

Abstract. Tl-1223 thick films have been fabricated using 2 techniques; electro-phoretic deposition (EPD) of $Tl_{0.78}Bi_{0.22}(Sr_{0.8}Ba_{0.2})_2Ca_2Cu_3O_x$ powder and spray pyrolysis of a precursor film followed by thalliation. A partial melt process was used for EPD tapes to improve grain connectivity and increase the grain size and some c-axis alignment was achieved. For spray pyrolised tapes, Sr substitution was found to be essential to stabilize the 1223 phase and good c-axis alignment was observed.

1. Introduction

The Tl-1223 phase has a high irreversibility line making it an ideal candidate for high current applications at 77K. However, despite extensive investigation, Tl-1223 tapes made using the powder-in-tube method have demonstrated poor critical current properties in magnetic fields due to a lack of grain alignment [1]. Thick film fabrication techniques offer an alternative route for the fabrication of long length low cost conductors. Tl-based superconducting films with good c-axis alignment and promising properties have been produced using standard deposition processes such as electrodeposition [2], electro-phoretic deposition [3], dip-coating [4] and most commonly spray pyrolysis [5, 6]. Recently, spray pyrolyzed films with biaxial texture have been fabricated on textured silver substrates with J_c values of $10^4 Acm^{-2}$ (77K, 1T) [6]. However, it is not yet clear which of these deposition processes is the most suitable for the production of long lengths of Tl-1223 conductor.

With the aim of developing a continuous process, we have investigated two approaches to thick film fabrication of Tl-1223 tapes: (i) electro-phoretic deposition (EPD) of a Tl-1223 coating followed by a partial melt process in a thallium rich atmosphere and (ii) the deposition of a non-thallium precursor by spray pyrolysis followed by a thalliation step to form the superconducting phase.

2. Tape characterization

Tapes fabricated by either EPD or spray pyrolysis were heated in an alumina crucible containing Tl-2212 source powder which was sealed with Ag foil. The crucible was heated in air in a single zone box furnace fitted with an inconel insert. The deposited coatings and

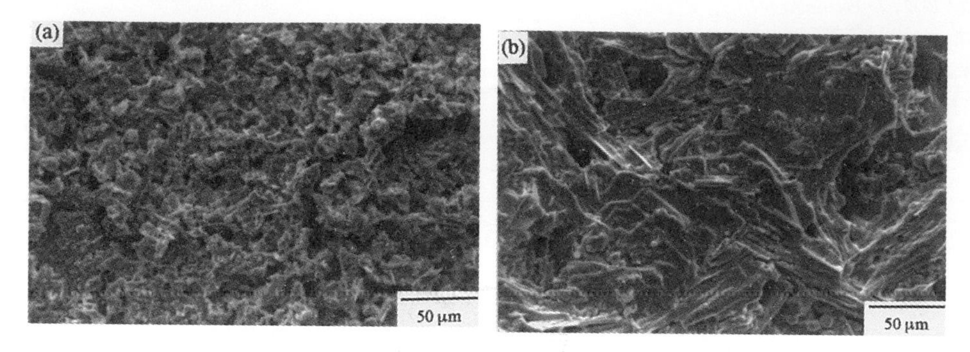

Figure 1 SEM micrographs of EPD tapes fabricated using melt temperatures of (a) 920°C and (b) 925°C.

fully reacted tapes were studied using XRD and SEM. T_c and I_c values were determined using a standard 4 pt technique for the tapes with the most promising microstructures.

3. Electro-phoretic deposition

Electro-phoretically deposited tapes (with 25-50μm thick coatings) were fabricated using similar apparatus and conditions to those previously described for Bi-2212 [7] using superconducting powder of composition $Tl_{0.78}Bi_{0.22}(Sr_{0.8}Ba_{0.2})_2Ca_2Cu_3O_x$. We have not yet attempted to optimize the deposition conditions for Tl-1223 powder and, so far, our research has concentrated on the development of a two step melt process adapted from our previous work on PIT tapes [8]. We have shown in PIT tapes that partially melting the Tl-1223 powder is essential to produce good inter-grain connectivity and improved properties [9, 10].

We have observed that the phase reactions during melt processing of electro-phoretic tapes are very similar to those observed in PIT tapes. XRD has shown the decomposition of the 1223 phase to 1212 and SrCaCu oxides at the partial melt temperature followed by the re-formation of the 1223 phase during the subsequent slow cool and annealing steps. SEM studies have shown an identical trend to that observed earlier in PIT tapes of increasing grain size with increased partial melt temperatures, Figure 1. A typical platy microstructure with a grain size of 50-60μm has been obtained for a melt temperature of around 925°C. Higher temperatures lead to extensive decomposition of the 1223 phase which cannot then be re-formed in the second stage of the process. Figure 2 shows the effect of the annealing time at 850°C on I_c values for tapes melted at 925°C. Promising I_c values of 5-6A at 77K have been measured for times of 3-3.5 hours. XRD analysis shows that the tapes also contain the 1212

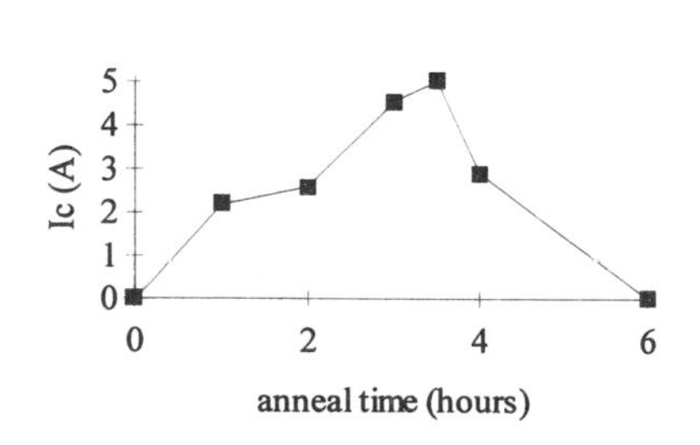

Figure 2 Diagram showing the effect of annealing time at 850°C on Ic values at 77K.

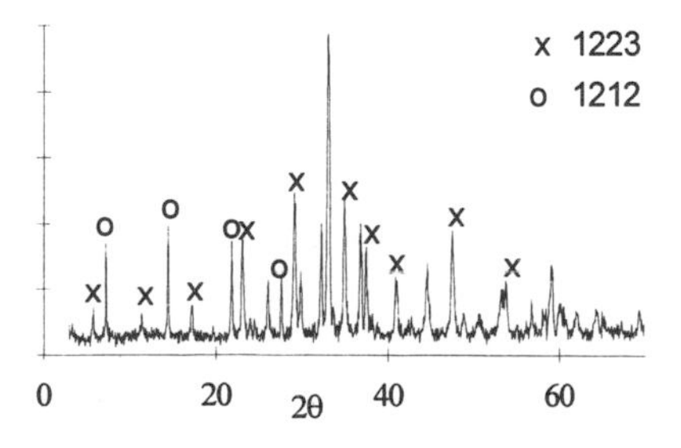

Figure 3 XRD pattern of a typical tape showing some c-axis alignment of 1212 and 1223 phases.

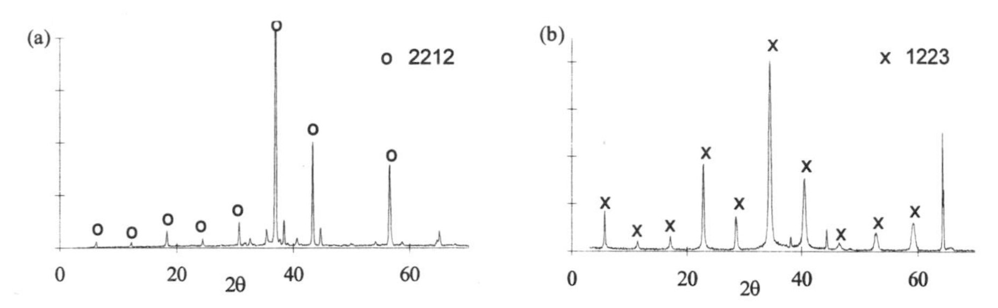

Figure 4 XRD patterns for spray pyrolyzed tapes of precursor composition (a) $Ba_2Ca_2Cu_3O_x$ and (b) $SrBaCa_2Cu_3O_x$ showing the majority phases 2212 and 1223 respectively.

phase and some c-axis grain alignment, as indicated by the (00*l*) peaks labelled in Figure 3, so that further refinement of the heat treatment conditions is required. Although these tapes have a greater degree of alignment than seen in melt processed or in-situ reacted PIT tapes, these patterns indicate that a 25μm coating is too thick to achieve good texture.

4. Spray pyrolysis deposition

Tapes were fabricated using similar apparatus and deposition conditions to those previously described [11]. We have developed this work to study the formation of the 1223 phase by investigating the effect of substituting elements as well as the effect of silver and silver alloy substrates. Tapes have been fabricated with different Sr/Ba ratios and Bi compositions and typical thalliation conditions are 865°C for 0.5 hours with 0.2g 2212 source powder. Figure 4 shows XRD patterns for tapes of precursor composition $Ba_2Ca_2Cu_3O_x$ and $SrBaCa_2Cu_3O_x$ with both tapes having good c-axis alignment. For a precursor composition of $Ba_2Ca_2Cu_3O_x$, the majority phase is 2212 with some 2223. We did not observe the formation of the 1223 phase for a wide range of thalliation conditions using this precursor composition. For a Sr/Ba ratio of 1, the 1223 phase is easily formed without requiring a significant increase in the reaction temperature. Initial experiments investigating the addition of Bi to the precursor film have shown a slight increase in T_c but no significant change in microstructure in contrast to the more dramatic effects seen in powder samples.

Previous studies have shown [5], and we have also found, that the microstructure and thickness of the precursor coating has a significant influence on the microstructure and

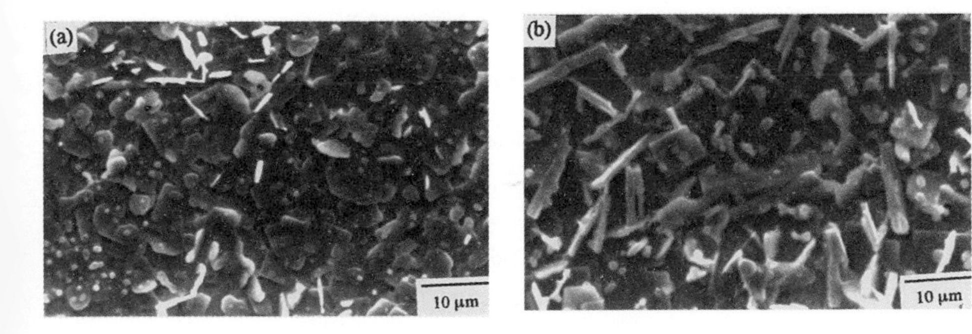

Figure 5 SEM micrographs of typical spray deposited films of composition $TlSrBaCa_2Cu_3O_x$ of approximate thickness (a) 5μm and (b) 10μm.

degree of c-axis alignment seen in the fully reacted tape. Figure 5 shows typical microstructures from tapes of composition $TlSrBaCa_2Cu_3O_x$ of approximate thickness 5μm and 10μm showing an increase in the number of misorientated grains in the thicker tape. The micrographs also show round Ca(Tl,Ag) oxide particles seen on the surface of all the tapes. Particles of similar appearance are also visible in spray pyrolyzed films made by other groups. J_c values of 2-4000 Acm^{-2} (77K) have been measured so far.

Silver alloy substrates would be advantageous in a continuous process due to their improved mechanical strength so we have also studied the effect of silver and Ag(0.25%Ni0.25%Mg) substrate materials. The properties of tapes fabricated on Ag(NiMg) alloy substrates have been found to be extremely poor with T_c depressed by up to 20K. Initial TEM studies have revealed the presence of small CuNiMg oxide particles throughout the superconducting matrix suggesting that there may be a deleterious reaction between the silver alloy and the superconducting phase leading to the observed decrease in T_c.

5. Conclusions

Electro-phoretically deposited Tl-1223 tapes have been fabricated using a partial melt process. The reaction processes in this partial melt process have been found to be similar to those previously observed in PIT tapes and a platy microstructure with a typical grain size of 50-60 μm has been achieved. Promising I_c values of 5-6A and some c-axis texture has been obtained in approximately 25 μm thick coatings. Good c-axis texture has been achieved in spray pyrolyzed tapes, as it is easier to deposit superconducting layers of less than 10μm in thickness with good inter-grain connectivity by spray pyrolysis. We have found that for our processing conditions the substitution of Sr for Ba is essential in forming the 1223 phase.

References

[1] e.g. Fox S, Moore JC, Jenkins R, Grovenor CRM, Boffa V, Bruzzese R and Jones H 1996 *Physica C* **257** 332-340

[2] Bhattacharya RN and Paranthaman M 1995 *Physica C* **251** 105-110

[3] Negishi H, Koura N, Idemoto Y and Ishikawa M 1996 *Jpn. J. Appl. Phys.* **35** 4302-4306

[4] Selvamanickam V, Pfaffenbach K, Kirchoff D, Gardner D, Hazelton DW and Haldar P to appear in IEEE Trans. Appl. Supercond. 1997

[5] Specht ED, Goyal A, Kroeger DM, Mogro-Campero A, Bednarczyk PJ, Tkaczyk JE and DeLuca JA 1996 *Physica C* **270** 91-96

[6] Okada M, Sato J, Higashiyama K, Yaegashi Y, Nagano M, Kumakura H and Togano T 1997 *Superlattices and Microstructures* **21A** 105-110

[7] Ming Yang, Goringe MJ, Grovenor CRM, Jenkins R and Jones H 1994 *Supercond. Sci. Technol.* **7** 378-388

[8] Moore JC, Hyland DMC, Salter CJ, Eastell CJ, Fox S, Grovenor CRM and Goringe MJ 1997 *Physica C* **276** 202-217

[9] Everett J see this volume

[10] Eastell CJ, Moore JC, Fox S, Grovenor CRM and Goringe MJ submitted to J. Superconductivity

[11] Su LY, Grovenor CRM, Goringe MJ, Dewhurst CD, Cardwell DA, Jenkins R and Jones H 1994 *Physica C* **229** 70-78

Acknowledgements This work was funded by EPSRC grant number K83405.

Inst. Phys. Conf. Ser. No 158
Paper presented at Applied Superconductivity, The Netherlands, 30 June–3 July 1997

Enhanced grain connectivity and transport critical currents in melt-processed (TlBi)-1223 tapes

J Everett, M D Johnston, G K Perkins, A V Volkozub, J C Moore†, C J Eastell†, S Fox†, D Hyland†, C R M Grovenor† and A D Caplin

Center for High Temperature Superconductivity, Blackett Lab., Imperial College, London SW7 2BZ, UK.
†Department of Materials, University of Oxford, Parks Road, Oxford OX1 3PH, UK

Abstract. We report detailed transport current-voltage (*I-V*) studies of *in situ* and melt-processed (TlBi)-1223/Ag tapes. Although their low field critical current density J_c performance is limited by grain boundary "weak-links", there is a plateau region in their $J_c(B)$ characteristics at high fields. In this plateau region, the melt-processed tape shows a transport J_c that is enhanced significantly over that of the *in situ* processed tapes.

1. Introduction

Unfortunately, unlike BSCCO, the thallium cuprates appear not to achieve good grain texture during tape fabrication [1]. Despite this, optimisation of processing techniques has produced Tl tapes with self-field critical current densities J_c's at 77 K of ~ 2×10^4 A/cm^2, although these J_c values are depressed rapidly in an applied magnetic field [2, 3].

The importance of achieving good grain texture in the Tl-1223 phase is highlighted by the high J_c's (~ 7×10^4 A/cm^2 at 77 K and 0 T) of strongly *c*-axis textured Tl-1223 thick films [4]. Here, the grain connectivity is close to optimal showing, that it is good granular alignment that is crucial for achieving the high J_c's. On the other hand, for (TlBi)-1223/Ag tapes produced by the powder-in-tube (PIT) method, it has been suggested that granular alignment will become the decisive factor in improving transport J_c only after extensive and close intergranular connectivity has been established [2]. Thus a combination of enhanced granular connectivity and alignment is needed to improve J_c and reduce grain boundary "weak-link" behaviour in (TlBi)-1223/Ag tapes. In this paper we show how the intergranular connectivity and transport J_c of a melt-processed (TlBi)-1223/Ag tape, are enhanced significantly over those of *in situ* processed (TlBi)-1223/Ag tapes.

2. Experimental

We have studied two (TlBi)-1223/Ag tapes prepared by *in situ* reaction of Tl- and Bi-oxides with a SrCaCuO precursor. The superconductor core thickness t of tape IS75 is ~75 μm and its cross-sectional area A ~10^{-3} cm^2; for tape IS150, t ~150 μm, A ~3×10^{-3} cm^2. A third tape studied was prepared using a melt-processing technique, with t ~95 μm and A ~1.5×10^{-3}cm^2.

Full details of tape fabrication and microstructure are given elsewhere [5, 6]. Tape IS75 has a denser core than tape IS150. However, they have a similar microstructure, consisting of well-defined platelets typically 1-2 μm in length. Neither IS75 nor IS150 show any granular alignment. Furthermore, their microstructure reveals crystalline and amorphous secondary phases. Amorphous phases are of particular importance in the *in situ* processed samples, as they tend to coat a high proportion of their grain boundaries. Another feature of these tapes is the existence of a few regions where there are clusters of four or five grains that result in small, but well-connected dense areas of microstructure.

The microstructure of the melt-processed tape is markedly different from the *in situ* processed tapes. It consists of densely packed regions (colonies) ~ 60-70 μm in length, of less well defined Tl-1223 grains. Grain boundaries appear to be well-connected, having large areas of contact, although there is no long-range granular alignment. While the microstructure of the tape reveals no amorphous phases, it does show that the melt-process leads to a range of crystalline secondary phases that are intimately mixed with the superconducting grains. It is important to note that no attempts were made to maximise J_c in this tape, by further pressing or optimisation of the rolling process.

Transport *I-V* (or electric field-current density *E-J*) characteristics were taken in a pulsed current transport rig. We subtracted the parallel current contribution of the Ag sheath using $I_{core} = I - V/R_n$, where I_{core} is the current in the core, and R_n the sheath resistance.

3. Results and discussion

Fig. 1 shows the *E-J* characteristics of IS75, IS150 and MP at 77 K and 40 K. At 77 K, the *I-V*'s of all three tapes display the "weak-link" signatures well known for polycrystalline YBCO: linear dissipation above a sharply defined critical current [7]; However, IS150 has a

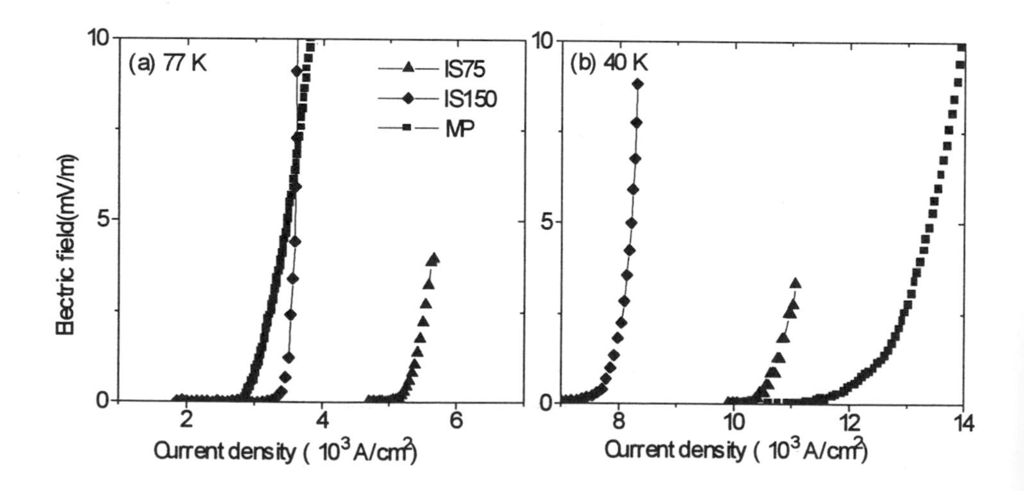

Figure 1. Transport *E-J* characteristics of the (TlBi)-1223/Ag tapes IS75, IS150 and MP in zero applied field at (a) 77 K and (b) 40 K. The curves have been corrected for the parallel current contribution in the sheath.

slightly sharper critical current than IS75 and MP. At 40 K, the onset of dissipation in MP becomes very rounded and is far less sharp than in both of the *in situ* processed tapes. This indicates that the melt-process improves the severe grain boundary "weak-link" problems that occur in (TlBi)-1223/Ag tapes fabricated using an *in situ* processing route [8].

Fig. 2 shows the transport $J_c(B)$ curves obtained for IS75, IS150 and MP at 77 K and 40 K. At 40 K, all three tapes display an initial rapid suppression of J_c, followed by a high field plateau in J_c that starts from ~ 50 mT; at ~ 500 mT, J_c has fallen from the zero field value of ~ 9900 A/cm^2, ~ 7700 A/cm^2 and ~ 11400 A/cm^2 by factors ~ 25 in IS75, ~ 34 in IS150 and ~ 6 in MP respectively. Suppression of J_c by an applied field is slightly faster at 77 K. Here, the field dependence of J_c for IS75 and MP is roughly the same and better than for IS150 out to ~100 mT. However, at 500 mT, J_c has fallen from ~ 5100 A/cm^2 to ~20 A/cm^2 in IS75, ~ 3400 A/cm^2 to ~ 2 A/cm^2 in IS150 and ~3000 A/cm^2 to ~ 80 A/cm^2 in MP. At 0.1 mV/m, the electric field criterion used to evaluate J_c, the parallel current path in the sheath is negligible in all three tapes at 40 K. On the other hand at 77 K, the sheath current becomes dominant in IS150 above ~ 300 mT [9]. It is evident from these data that there is a significant improvement in both the field dependence and high field plateau value of transport J_c in the melt-processed tape compared with the *in situ* processed tapes.

The precipitous low field drop in J_c is consistent with the model of dissipation at Josephson junction grain boundary "weak-links" [10]. However, this model fails to explain the relatively field independent $J_c(B)$ characteristics of the three tapes at higher fields. Recent measurements of the anisotropy of J_c(transport) with respect to applied field direction for the three tapes, showed that as the field is increased their critical current anisotropy decreases. Interestingly, it also showed that in the plateau region of their $J_c(B)$

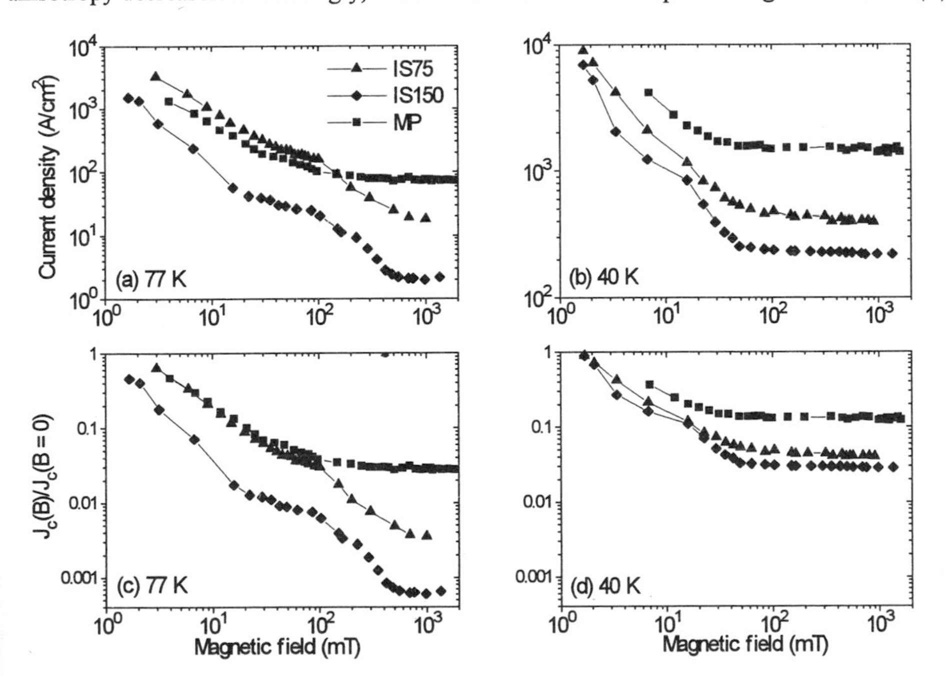

Figure 2. Transport $J_c(B)$ characteristics for tapes IS75, IS150 and MP at (a) 77 K and (b) 40 K, evaluated at 0.1 mV/m and corrected for sheath contribution. Field dependencies of the normalised J_c for three tapes, (c) 77 K, (d) 40 K. The magnetic field was applied perpendicular to the plane of the tapes.

characteristics the response was isotropic [8], indicating the absence of any grain texturing with respect to tape plane. It is important to note that this behaviour is completely different from BSCCO-2223/Ag PIT tapes, where the critical current anisotropy *increases* with field. This is indicative of different patterns of current flow in the two systems at high fields [8, 9].

A recent study of the magnetic behaviour the three (TlBi)-1223/Ag tapes showed that for MP, at 40 K, the dominant magnetic screening currents initially flow coherently around the whole sample, while at higher fields the current pattern breaks up into islands corresponding to ~ colony size (~ 100 μm) [8, 9]. For IS75 and IS150 on the other hand, full coherence is not reached even at low fields, and it is estimated that the screening current loops flow on a scale corresponding to a grain size or to small clusters of well-connected grains (~ 10 μm). Furthermore, a comparison between the transport $J_c(B)$ and magnetic $J_c(B)$ behaviour of the three tapes showed that their magnetic intragranular current densities were one to orders of magnitude higher than their intergranular transport current densities.

4. Conclusions

Although signatures of "weak-link" behaviour are still evident in the (TlBi)-1223/Ag melt-processed tape, it should not detract from the potential of the processing method. Transport studies described here show that, the grain connectivity and transport $J_c(B)$ performance of this conductor is improved significantly over that of the (TlBi)-1223/Ag *in situ* processed tapes. The coating of grain boundaries by amorphous phases is abundant in the *in situ* processed samples, and it is likely that the improvements achieved in the melt-processed tape are due largely to the absence of amorphous material from its microstructure. Encouragingly, intragranular pinning in the *in situ* and melt-processed tapes is strong enough to suggest that the achieved enhancement in granular connectivity will boost the prospects of (TlBi)-1223 being used for high field applications at both low and high temperatures. Recent advances made in the synthesis and performance of strongly *c*-axis textured Tl-1223 films indicate that the goal now should be to obtain the large-scale grain alignment required to alleviate the "weak-link" behaviour in the (TlBi)-1223/Ag phase.

Acknowledgements

Work supported by UK EPSRC. J Everett thanks The National Grid Company plc for financial support for this work through the CASE Award scheme.

References

[1] Eastell C J, Moore J C, Fox S, Grovenor C R M and Goringe M J 1997 *Submitted to Journ. Supercond.*

[2] Ren Z F, Wang J H, Miller D J and Goretta K C 1994 *Physica C* **229** 137

[3] Selvamanickam V *et al.* 1996 *Physica C* **260** 313

[4] Parilla P A *et al.* 1995 *IEEE Trans. Appl. Supercond.* **5** 1958

[5] Moore J C, Fox S, Salter C J, He A Q and Grovenor C R M 1995 *Inst. Phys. Conf. Ser.* **148** 483

[6] Moore J C *et al.* 1997 *Physica C* **276** 202

[7] Everett J *et al.* 1996 *Supercond. Sci. Technol.* **9** 1077

[8] Everett J *et al.* 1997 *to be published*

[9] Everett J *et al.* 1997 *to be published in IEEE Trans. Appl. Supercond.* **7**

[10] Clem J R 1988 *Physica C* **153-155** 50

Inst. Phys. Conf. Ser. No 158
Paper presented at Applied Superconductivity, The Netherlands, 30 June–3 July 1997

High critical currents densities in YBCO films on technical substrates

F García-Moreno[1], A Usoskin[1], H C Freyhardt[1,2], J Wiesmann[2], J Dzick[2], J Hoffmann[2], K Heinemann[2] and A Issaev[2]

[1]Zentrum für Funktionswerkstoffe Gött. gGmbH, Windausweg 2, D-37073 Göttingen, Germany
[2]Institut für Metallphysik, Univ. of Göttingen, Windausweg 2, D-37073 Göttingen, Germany

Abstract. **$YBa_2Cu_3O_{7-x}$ (YBCO) films with high-quality in-plane alignment have been grown on buffered Ni foils and yttria-stabilized zirconia (YSZ) ceramic ribbons using a modified pulsed-laser-deposition technique based on a quasi-equilibrium substrate heating and variable azimuth scanning of the target with the laser beam. Critical current densities, J_c, up to 2×10^6 A/cm^2 (77 K, 0 T) have been found in 600-800 nm-thick YBCO films. A considerable narrowing by 6-8° of the Φ-scan-X-ray peaks for YBCO films in comparison with biaxially textured YSZ buffer layers was observed for both substrate types.**

1. Introduction

Recent progress in deposition of high temperature superconducting (HTSC) films on technical polycrystalline substrates demonstrates a realistic prospect of reaching critical currents, which are nearly as high as for films on single crystalline substrates. Here new results are reported for $YBa_2Cu_3O_{7-x}$ (YBCO) films prepared by a modified pulsed-laser-deposition (PLD) method on thin (0.12 mm) and flexible Ni foils and yttria-stabilized zirconia (YSZ) ceramic ribbons with critical current densities, J_c, up to 2×10^6 A/cm^2 (77 K, 0 T), which is the highest J_c value obtained so far on technical substrates.

2. Substrates

Ni foils of 99.98% purity were electro-chemically polished to reach a mirror-like surface with a roughness of better than 0.02 μm. Polycrystalline YSZ ceramic ribbons exhibited a similar surface quality provided by a conventional mechanical polishing. YSZ buffer layers prepared by ion-beam-assisted-deposition (IBAD) were employed to meet requirements needed for a high-quality in-plane texture for YBCO films as well as to suppress an oxidation in the case of Ni foils.

3. IBAD buffer layer

The buffer layers were prepared with a dual-ion-beam system employing two ion sources of the Kaufman type. The method was similar to that one used by Iijima [1] and has been described in detail elsewhere [2]. The assisting ion beam with an energy of 300 eV was directed to the substrate under an incident angle of 55°. The ion density at the substrates amounted to about 250 $\mu A \cdot cm^{-2}$. The temperature of the substrate (bonded to the holder without additional cooling) increased during deposition from room temperature at the beginning to about 150°C after 3 hours. With a deposition rate of ~ 0.08 nm/s a film thickness of d = 900 nm was reached within this time. The in-plane textures of buffer layers were characterised by Φ-scans of the (111)-peaks measured employing a 4-circle X-ray diffractometer (with Cu-K_α radiation). These investigations were performed with IBAD-YSZ layers on single crystalline silicon, covered with its intrinsic surface oxide layer, as well as with polycrystalline ceramic and metallic substrates. As was shown recently [2], the texture improves with increasing thickness of the buffer layer. This improvement can be approximated by an exponential decay function. A saturation of texture level corresponds to 7° for an 8 µm-thick YSZ film. The YSZ layers for the YBCO deposition were grown to a film thickness of 900 nm and, thus, a full-width-half-maximum (FWHM) value of 17° was achieved (Fig. 1). The real FWHM at the YSZ surface are expected to be better than the measured ones because of an integration over the X-ray penetration depth.

4. High Tc films

For the preparation of the HTSC films a PLD method (308 nm, 1 J, 30 ns, 0.5 mbar of O_2) was employed. A quasi-equilibrium radiation-heating technique [3] based on the heating of the substrate in a high-temperature „black cavity“ combined with a short periodic exposition of the substrate surface to the synchronised laser plume was used to maintain a precise (better than ± 1°C) substrate temperature of ~ 760°C during film deposition. It is known that even for rotating YBCO targets an increasing surface roughness develops with increasing ablation time. The plasma plume becomes unstable and the deposition rate decreases. Periodic variations of the azimuth of the incident laser beam during YBCO target ablation were used to stabilize the deposition rate to a level of ± 2 % which seems to be much better than the ± 20 % usually reached by a conventional target scanning.

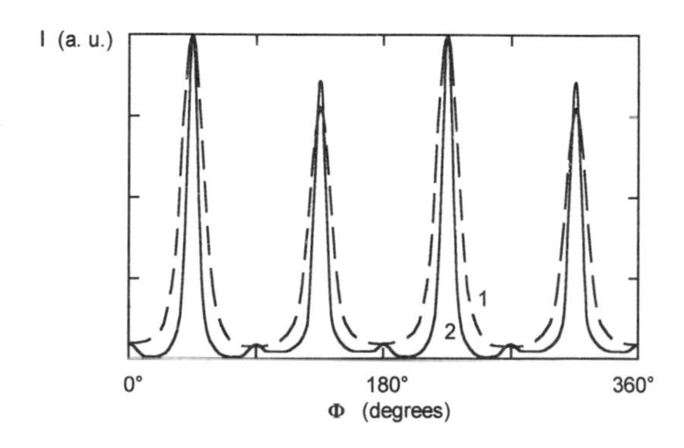

Fig 1. Φ-scans of IBAD-YSZ buffer layer (1) and of YBCO film (2) grown on YSZ ribbon with IBAD-YSZ buffer. 1 - d = 900 nm, χ = 55°, FWHM = 17.4°; 2 - d = 600 nm, χ = 45°, FWHM = 9.7°.

The critical current densities, J_c, of the films were measured by a dc method as well as by an inductive ac (100 kHz) technique based on the effect of magnetic shielding. Both of these methods yield similar results. Typical temperature dependences of J_c for the (600-1000) nm thick YBCO films of different quality on Ni foils in comparison with film on $SrTiO_3$ are shown in Fig. 2.

Table 1. HTSC and structural parameters of YBCO films on metallic and ceramic substrates

Substrate/buffer	T_c [K]	ΔT_c [K]	FWHM buffer	FWHM YBCO	J_c [A/cm^2]
YSZ / -	87	1	-	> 60°	1.7×10^5
YSZ / IBAD	88.5	1	17.4°	9.7°	1.2×10^6
Ni / polycr. YSZ	83	4	90°	-	1×10^4
Ni / IBAD YSZ	89.5	0.5	18.7°	10.9°	2×10^6

From X-ray analysis performed by Φ-scans of the (103) peak the quality of the in-plane texture of the YBCO films was determined (see Fig. 1). The observed FWHMs of ~ 10° confirm the high degree of alignment of the YBCO crystallites. These FWHM values are considerable better than the respective ones for the buffer layer (see Table 1).

The Φ-scan of Fig. 1 for YBCO film exhibits the four peaks of the well known 45° epitaxy relationship between the YBCO and the YSZ cell and also a small contribution of a 0° relationship. The relative fraction of the grains with the 0° and 45° relationship can be increased up to ~ 10 % on Ni foils and up to ~ 30 % on ceramic YSZ by varying deposition rate and substrate temperature during YBCO deposition. Contrary to this, for YSZ buffers with FWHM > 20° only the 45° relationship was found and, therefore, only for the high-level textured buffers the alternative growth orientations can be observed.

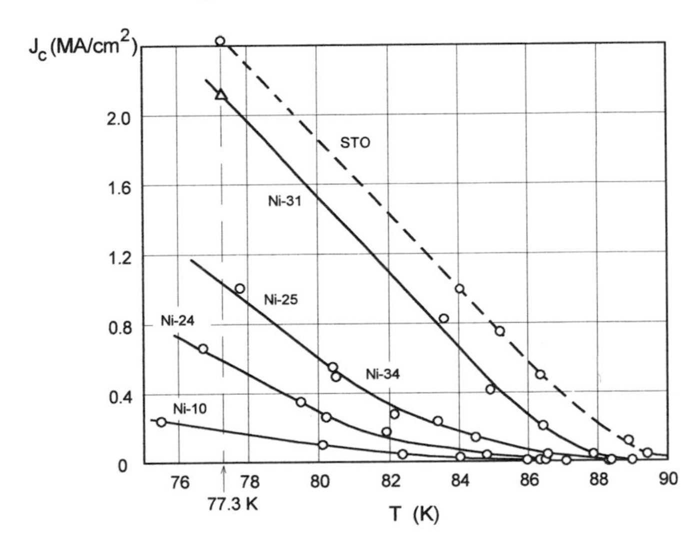

Fig. 2. Temperature dependences of J_c for YBCO films of various quality on Ni foils (Ni-) and on $SrTiO_3$ (STO).

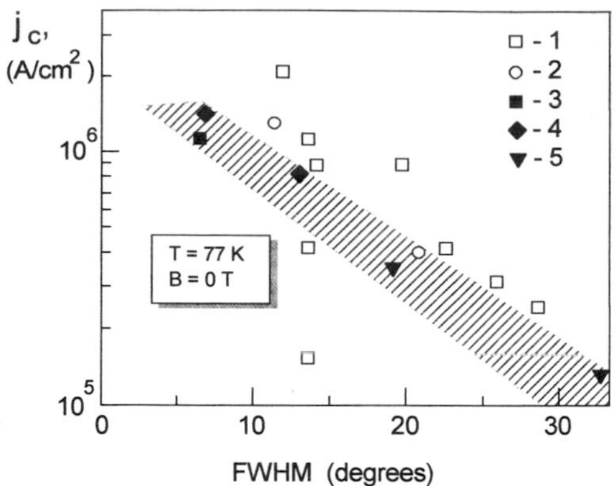

Fig. 3. Critical currents densities of YBCO films deposited on Ni/IBAD-YSZ (1) and YSZ-ribbon/IBAD-YZS (2) versus FWHM. Data (3) - (5) correspond to the results obtained for technical substrates in references [4-6] respectively.

From figure 3 one can conclude that for a given FWHM of the YBCO film (a) our best J_c values are significantly higher than the presently known ones [4-6] , and (b) they can be varied within one order of magnitude. The latter indicates that not only the degree of in-plane texture is responsible for the HTSC critical properties, but also other microstructural parameters of the HTSC film, i. e. imperfections of grain boundaries and microcrystallite links, etc., which are mainly determined by the conditions used for YBCO film deposition.

The main results obtained for metallic and polycrystalline ceramics are summarized in Table 1. The typical well reproducible values of J_c are about a factor of 2 less than the highest values observed.

Acknowledgement

This work was performed within a development project of Kabelmetal electro GmbH which is supported by German BMBF (Project No. 13N6924/6), and was also supported by Siemens AG together with German BMBF (Project No. 13 N6482).

References

[1] Y. Iijima Y, Tanabe N, Kohno O and Ikeno Y *Appl. Phys. Lett.* **60** (1992) 769

[2] Wiesmann J, Heinemann K and Freyhardt H C *Nucl. Instr. Meth. B* **120** (1996) 290

[3] Usoskin A, Freyhardt H C, García-Moreno F, Sievers S, Popova O, Heinemann K, Hoffmann J, Wiesmann J and Isaev A *Applied Superconductivity 1995* IOP Conf. Series **148** (1995) 499

[4] Foltyn S *High Tc Update* **9/10** (1995) No. 1

[5] Wu X D, Foltyn R, Arendt P N, Blumethal W R, Campbell I H, Cotton J D, Coulter J Y, Hults W L, Maley M P, Safar H F and Smith J L *Appl. Phys. Lett.* **67** (1995) 2397

[6] Iijima Y, Onabe K, Futaki N, Tanabe N, Sadakata N, Kohno O and Ikeno Y *J. Appl. Phys.* **74** (1993) 1905

Inst. Phys. Conf. Ser. No 158
Paper presented at Applied Superconductivity, The Netherlands, 30 June–3 July 1997

Critical current densities in heterogeneous superconductors determined from magnetization measurements

Mike D. Sumption*, Silvester Takács**

*Materials Science and Engineering, The Ohio State University, Columbus, Ohio, USA
**Institute of Electrical Engineering, Slovak Academy of Sciences, 842 39 Bratislava, Slovakia

Abstract. The basic mechanisms determining the critical current densities of proximity-coupled heterogeneous superconducting structures in different regions of the magnetization curve are summarized. We show that besides the usually considered criteria (pinning force equal to the Lorentz force and quasi-particle velocity equal to the depairing velocity of the Cooper pairs, respectively), two additional conditions can determine the critical current density in these structures: the shearing of the flux line lattice in the "weak" superconducting volume (without moving the flux line lattice in the "strong" superconductor) and the elastic deformations (bowing out) of flux line portions in this volume (with pinned portions in the "strong" superconductor). The basic calculation principles of the corresponding critical current densities are given and the results are compared with magnetization measurements on NbTi/Cu fine filament (FF) superconductors. Some comments are given concerning the possible applications to other superconducting structures, like the high temperature superconductors.

1. Introduction

Practical superconductors for AC applications should have small filaments to reduce the hysteresis losses in changing magnetic fields or due to fluctuations. The matrix between them must be normal metal (N), otherwise the stability of the conductor would be decreased. Additionally, the filaments (S) are twisted in order to reduce coupling-related losses. The lower limit for the thickness between the filaments is determined by proximity effect (PE), which makes the matrix "partially" superconducting. If the coupling between filaments is too strong, they behave like a coupled structure with larger diameter. The whole bundle is a heterogeneous mixture of S and N parts, resulting in anisotropic electromagnetic properties. The flux lines (FLs) penetrating such structures are anisotropic, too. The pinning properties depend on the direction of FLs and the direction of the Lorentz force, acting on them at applying the current.

In superconductors and superconducting structures, generally three kinds of critical currents are described. The *depairing* current destroys the Cooper pairs and is the maximum current which can flow without resistivity. However, in the presence of magnetic field the FLs have to be pinned by lattice imperfections and/or surfaces, otherwise the viscous flow of the flux line cores would destroy the superconductivity. The corresponding critical current density is therefore determined by the volume *pinning* force. Lastly, the *Josephson* critical current is the maximum current which can flow between superconductors joined by normal or insulator material. This current is supposed to determine the critical current density in the network models of weakly coupled superconducting grains, e.g. also in some high T_c superconductors. However, the need to pin S-N-S currents in these structures is generally ignored.

2. Currents in twisted and untwisted samples

In transport measurements (current flowing mainly along the filaments), as well as generally for *twisted* filaments, the flux lines (FLs) have to cross the filaments, the critical current density is therefore determined by pinning the flux lines. There is very little difference in FLs which are entering or leaving the superconducting strands. The magnetization curve is nearly symmetric at higher fields. In magnetization measurements of *untwisted* samples, the current flows in different directions, crossing also the filaments through normal — or partially superconducting — regions. The definition of the critical current J_c and the so-called depinning field B_D (where the superconductivity in the normal metal essentially vanishes) is then very difficult, mainly due to some uncertainties in boundary conditions on the interfaces [1]. Typical magnetization curves of twisted and untwisted composites in different regions are shown schematically in figure 1. After the FLs enter into the normal regions (between $B_{c1,n}$ and $B_{c1,s}$), they penetrate the filaments at the fields above $B_{c1,s}$. At this point of the magnetization curve, the currents between S and N can have still a small contribution due to the shear mechanism. The J_c values from different parts of the magnetization curves [2,3] are given in figure 2 for NbTi/Cu composite with 700 nm filament diameter and 140 nm interfilamentary spacing.

The *depairing current density* in the PE induced normal metal (S') can be defined in analogy with superconductors, including the anisotropic nature in S' [4]

$$J_{Dn} = J_{Ds}(\varepsilon^*_\lambda)^2 \varepsilon^*_\xi \exp(-\alpha K_n d_n)/(\varepsilon_\lambda \varepsilon_\xi)^{1/2}$$

where J_{Ds} is the depairing current density in S [5], and we have also included the *anisotropy* (ε) and *heterogeneity* (ε^*) factors, e. g. $\varepsilon_\lambda = \lambda_1/\lambda_2$ (index 1 perpendicular and 2 parallel to the filaments), $\varepsilon_\xi^* = \xi_s/\xi_2$ (the parameters in S are isotropic). In this way, the problem of boundary conditions can be replaced by the problem of finding those factors which can be measured experimentally. Although the functional form of the exponential field dependence is in good agreement with the experiments (figure 2), the slope is different from the theoretical values [4, 5] by a factor of about 7. It is interesting to compare this result with the Josephson coupling current for anisotropic superconductors where $J_D = \varepsilon^2 J_{Ds}\, \xi/d_n$, having the same power of the anisotropy factor (which is supposed to be only one).

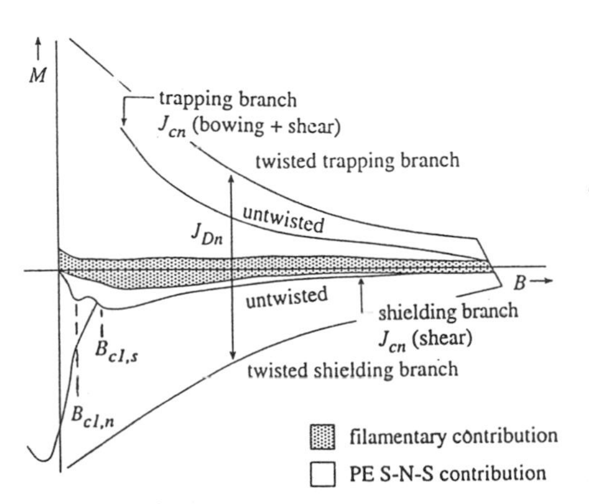

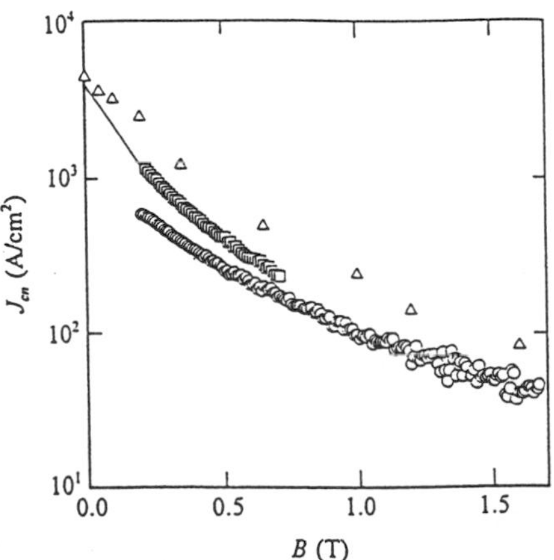

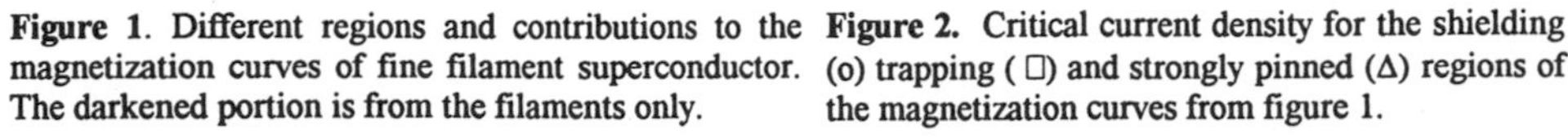

Figure 1. Different regions and contributions to the magnetization curves of fine filament superconductor. The darkened portion is from the filaments only.

Figure 2. Critical current density for the shielding (o) trapping (□) and strongly pinned (Δ) regions of the magnetization curves from figure 1.

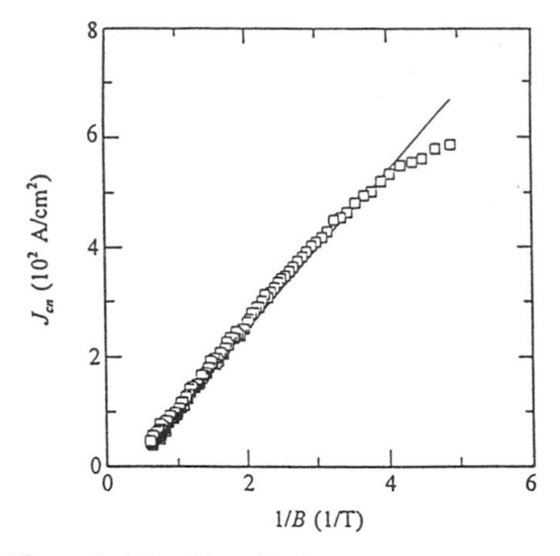

Figure 3. A fit of the critical current density vs 1/B for the shielding branch of an untwisted composite.

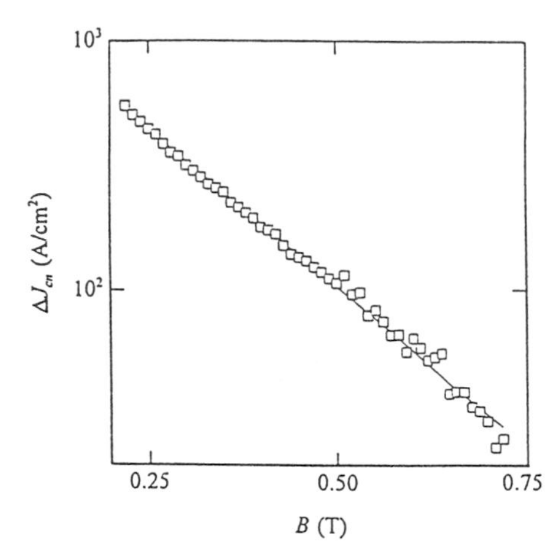

Figure 4. Field dependence of the difference between the trapping and shielding critical current densities.

The possibility of *shearing* the flux line lattice was considered already by Kramer [6] in calculating his well-known scaling model. We use here the model of Pruymboom et al [7], leading to $J_c = 0.1GB_{c2}(1-b)^2 / 8\mu_0\kappa^2 W$ where G is a geometrical factor, B_{c2} the upper critical field, κ the Ginzburg-Landau parameter, W the width of the layer in which the FLs are moving and $b = B/B_{c2}$. This should be modified in our case by replacing B_{c2} by B_D and the averaging of the order parameter instead in S'. The result is

$$J_c = \frac{0.1GB_D}{8\pi\mu_0\kappa^2 W\alpha} \sim 1/B$$

where, $\alpha = 2\pi B/ K_n\Phi_0$, K_n is the inverse correlation length in N. Figure 3 illustrates the qualitative 1/B dependence of J_c. The inverse of the x-intercept gives $B_D = 3.5$ T. The resulting value for the lower critical magnetic field B_{c1} is either 2.7 mT or 15.6 mT, depending upon whether we use the theoretical or experimental value for α [2]. The measured value of B_{c1} is between these (5.0 mT).

After reversing the field, a very small number of FLs which are going only through S', leave the composite very easily (they are very weakly pinned). Those FLs which are to a large extent existing in S', but are partially also in the filaments, can be deformed in S' (*bow out*) due to the Lorentz force. The only force acting against the Lorentz force is due to the change of elastic energy [8] (increased length and curvature). Both of these energies are proportional to the square of the deviation a from the equilibrium position. The relative increase in the self energy is $\delta l/l = (8/3)\,(a/l)^2$, where l is the length of the (unpinned) FL. The numeric factor calculated by including the energy due to the curvature is $p = [8/3 + 16\,C/(\ln\kappa + 1.0)]$,where C is changing from about + 0.5 to - 0.4 at $l = 2\,\lambda$ to $l = 0.4\,\lambda$. Analogous calculations can be performed for pancake like structures. The correction factors are of the same order. The change of the self energy is compensating the Lorentz force, approximately equal J_cB. Using the de Gennes boundary conditions [9] for the change of the order parameter across the boundary, we obtain $\Delta_N/\Delta_S \approx 1/8$. The critical current density due to this mechanism is then reached when the displacement of the FLs a is about the FL spacing. This condition leads to

$$J_{cn} = \frac{\varepsilon_\lambda B_{cs}^2 \xi_n^2 p}{2\mu_0(8l)^2 aB} \exp(-B/B_0)[\ln(\kappa)+0.5]$$

where the anisotropy factor is $\varepsilon_\lambda \approx 10^{-2}$. We obtain the critical current density by subtracting the shielding from the trapping branch (figure 4). From the slope, we have $B_0 = 0.3$ T. Taking $B_{cs} = 0.2$ T, $\xi = 300$ nm, $\kappa = 50$, and $a = 60$ nm at $B = 0.5$ T, we get from the pre-exponential factor $l \approx 3.3$ µm, which seems reasonable, being some number of filament diameters.

3. Conclusions

We have shown that pinning is required for PE coupling currents in FF strands, and that this pinning takes two forms in untwisted strands. The first one is the shear mechanism, when FLs shear around the lattice in the filaments. This determines the critical current density in the *shielding branch* of the magnetization curve. In the *trapping branch*, partial FL movement is possible, in analogy with (but distinct from) the pancake vortices in high Tc superconductors. This mechanism involves the FL deformation (bowing). For twisted composites, however, the PE depairing current can be reached. In addition, we have calculated the maximum depairing current and its field dependence including the properties of both parts of the composite. Comparing with the experimental results, qualitative agreement is obtained concerning the field dependence of these critical current densities in NbTi/Cu composites. In this way, the different anisotropy coefficients for the composite can also be determined.

Fine filament composites have a very well defined geometries. Therefore, their study can give us a deeper insight into some basic mechanisms of FL motion, applicable to many different structures, including high T_c superconductors, artificial pinning centres (APC) and superlattices. The structure of fine filament superconductors is somewhat analogous to high T_c superconductors at the atomic level, where it is supposed that superconductivity is caused by the CO layers between the "metallic" parts. However, an even closer analogy appears to pinning properties of FLs close to the grain boundaries, which should be determining the critical current density (mainly in higher fields), rather than the "pure" Josephson currents. For materials with APC, a critical point in calculating J_c is the knowledge of the order parameter through the pin. The results are very sensitive to the *boundary conditions* used in those calculations. At present there is an uncertainty in this direction and even some hints for the incorrectness of present assumptions [1, 4]. It should be possible to test these boundary condition assumptions using the considerations outlined above.

Acknowledgments

The authors appreciate fruitful discussions with E H Brandt and also E W Collings. This work was partially supported by the NRC (Grant 31020100) and the Slovak Grant Agency VEGA.

References

[1] Sumption M D and Collings E W *IEEE Trans. Appl. Supercond.* **AS-7** (1997) in press
[2] Sumption M D and Collings E W *Cryogenics* **34** (1994) 491
[3] Sumption M D *Proximity effect magnetization and energy loss in multifilamentary composites: Influence of strand design and sample geometry* (Thesis, Ohio University, Athens) (1996)
[4] Sumption M D and Takács S submitted to *Physica C*
[5] Takács S *Czech. J. Phys. B.* **36** (1986) 524
[6] Kramer J *J. Appl. Phys.* **44** (1973) 1360
[7] Pruymboom A, Kes P, van der Drift E and Radelaar S *Cryogenics* **29** (1989) 232
[8] Brandt E H *Phys. Rev. B* **34** (1986) 6514
[9] Deutscher G and de Gennes P G *Superconductivity* Ed. R D Parks (New York: Marcel Dekker) (1969)

Inst. Phys. Conf. Ser. No 158
Paper presented at Applied Superconductivity, The Netherlands, 30 June–3 July 1997

Angular Dependence of Critical Current versus Magnetic Field in Nb_3Sn Wires at 9-15T

Arno Godeke[1], Arend Nijhuis[1], Hennie G. Knoopers[1], Bennie ten Haken[1], Herman H. J. ten Kate[1] and Pierluigi Bruzzone[2]

[1]Low Temperature Division, University of Twente, The Netherlands, [2]ITER Joint Central Team, Naka-gun, Japan.

Abstract. In most applications of Nb_3Sn superconductors the transverse magnetic field component is dominating the transport properties. However for particular applications (as for example joints in high field magnets) it is proposed to position the superconductor in longitudinal field aiming for an enhancement of the critical current. Since the critical current is normally measured in transverse fields, very little is known about the influence of reducing the angle between the current and magnetic field in Nb_3Sn. In the framework of the ITER Nb_3Sn strand development program, the critical current of two types of Nb_3Sn wires is investigated as a function of the angle between the transport current and the applied field. An increase in the critical current occurs when the angle is reduced. For comparison, the enhancement is also verified with magnetic measurements on the AC-loss using a sample in different orientations of the applied field. The observed angular dependence is compared to the present flux-pinning models for the critical current in type II superconductors. An anisotropic maximum pinning force is observed in magnetic fields from 9 to 15 tesla.

1. Introduction

The critical current density (J_c) in type II superconductors is mainly determined by the pinning of the flux-line lattice on various structural imperfections. It is well known that NbTi wires exhibit a strong dependence of the maximum pinning force on the angle between the transport current and the applied magnetic field (B). This is due to the cold work of the wire during the production process that shapes the α-phase precipitations. In many Nb_3Sn applications, the superconductors are exposed to fields with longitudinal components. Nevertheless, in comparison to NbTi, little is known about the influence of reducing the angle between the current and field in Nb_3Sn. An anisotropy in the maximum pinning force in superconducting Nb_3Sn strands has also been recently reported in [1].

In order to study this anisotropy, the critical current is measured as a function of the angle between strand axis and magnetic field for two strands of the ITER benchmark series [2]. The critical pinning force density F_p is determined from $J_c(B)$; F_p is equal to the critical value of the Lorentz force $J_c \times B$ per unit volume [3]. As only the perpendicular field component gives rise to a Lorentz force that acts on the flux line lattice, the pinning force can be written as $F_p(B,\phi)=J_c(B)\cdot B\cdot sin(\phi)$, in which ϕ is the angle between the current and the applied field. For small angles between J and B, J_c will no longer be determined by the critical balance between F_p and the Lorentz force, but by the threshold of flux line cutting instability, as summarised in [4].

The magnetisation loss versus the angle of the applied field is measured in order to verify the I_c measurements. The anisotropy can be defined as the axial critical current density, J_{cz}, divided by the azimutal critical current density, $J_{c\theta}$. In the magnetisation measurements, the AC as well as the DC fields are oriented in the same direction.

Accordingly, the orientation of the screening currents will vary, from parallel to the wire axis (J_{cz}, for $\phi = 90°$) up to perpendicular to the wire axis ($J_{c\theta}$, for $\phi = 0°$). In the critical current measurements the angle of the field is varied with respect to the strand axis, whereas the current remains in the direction of the strand axis. From the existing formulas for the energy loss per cycle for perpendicular and parallel-applied fields, it can be deducted that [5]:

$$\alpha = \frac{J_{c\theta}(B_{\parallel})}{J_{cz}(B_{\perp})} = \frac{4 \cdot d_z \cdot Q_{h(0°)}}{\pi \cdot d_\theta \cdot Q_{h(90°)}} = \frac{4 \cdot d_z}{\pi \cdot d_\theta} \cdot q_h(0°)\,, \quad (1)$$

in which Q_h is the loss per cycle, d_θ is the effective filament diameter for field with parallel orientation and d_z for perpendicular direction. If a normalised pinning force is deducted from the resistive measurements; $f_p(\phi) = F_p(\phi)/F_p(90°)$, a correlation can be made between the loss- and the critical current measurements.

2. Test set-up.

For the critical current measurements versus the angle of the applied field, the sample is mounted on a stainless steel holder, which can be rotated inside the 60 mm bore of a 16 T superconducting solenoid. The samples are straight sections of 58 mm length, with voltage taps in the centre on a distance of 10 mm. The electromagnetic loss in superconductors without transport current consists of hysteresis and coupling loss [6]. The hysteresis loss strongly depends on the amplitude of the applied field and is independent of the ramp rate for low frequencies. The loss measurements are performed in a magnetometer at liquid He temperature. The field is taken sinusoidal with a frequency from $f = 2.5$ mHz up to 20 mHz. The test set-up for magnetisation measurements is described in [7]. The preparation procedure of the sample for the hysteresis loss measurements is as follows. A heat-treated strand is cut into pieces which a length of 20 mm, which is more than 2 times the twist pitch of the strands. In total 57 of such strand pieces are used to prepare the hysteresis loss sample. The angle of the strand pieces is varied from 90 to 0 degrees. The hysteresis loss is obtained by extrapolation of the linear $Q_{tot}(f)$ curves down to zero frequency.

3. Sample material.

Two strands from the 3rd ITER benchmark action are selected, manufactured by different production techniques. One strand is a ‘bronze’ type strand manufactured by Vacuumschmelze (GE) and the other strand is from the ‘internal Sn’ type, manufactured by Europa Metalli (I) [2]. The electric field criterion to determine J_c is 5 μV/cm for the Vacuumschmelze sample and 10 μV/cm for the Euro Metalli sample. The I_c measurements are carried out at five field amplitudes (9, 11, 12, 13 and 15 T) and six different angles ϕ, including the perpendicular and parallel orientation (90°, 72°, 54°, 36°, 18° and 0°). Notice that the strand axis is not identical with the current direction in the filaments because the wire samples are twisted ($L_p \approx 10$ mm).

4. The critical current versus angle of applied field

The critical current measurements for different angles and fields on both conductors are summarised in Figure 1 and Figure 2. The calculated results of $f_p(\phi)$ versus the angle between field and wire are shown in Figure 3 and Figure 4 for both conductors and all applied field values. A similar $f_p(\phi)$ is observed for field values between 9 and 15 T. The largest deviation occurs at 9 T in the bronze sample. The reduction of f_p for small angles in the internal Sn strand is more severe than in the bronze material.

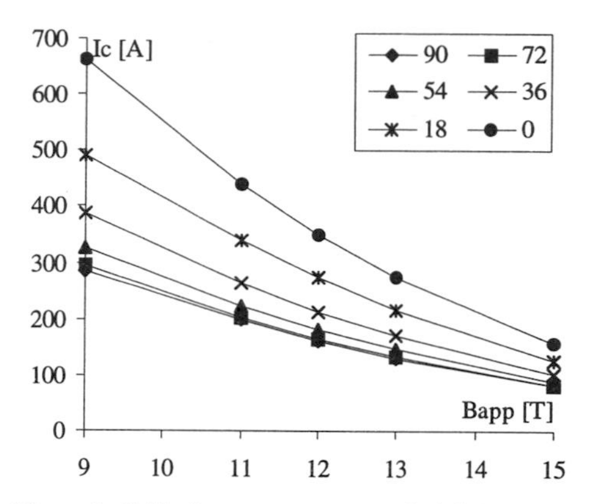

Figure 1. Critical current versus applied field and angle for the internal Sn-strand

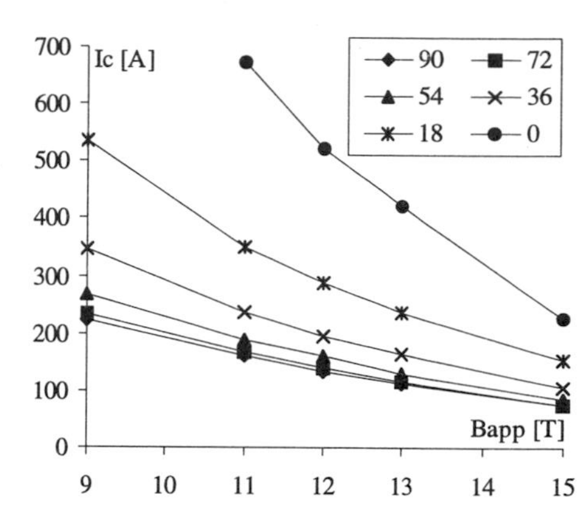

Figure 2. Critical current versus applied field and angle for the bronze-strand

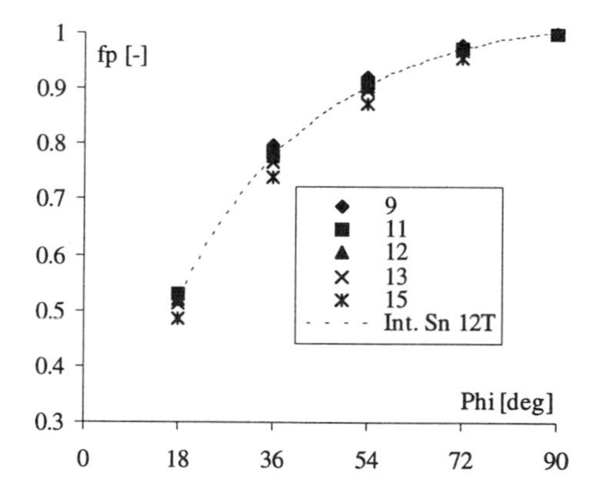

Figure 3: Normalised pinning force versus the angle and the applied field for the internal Sn-strand.

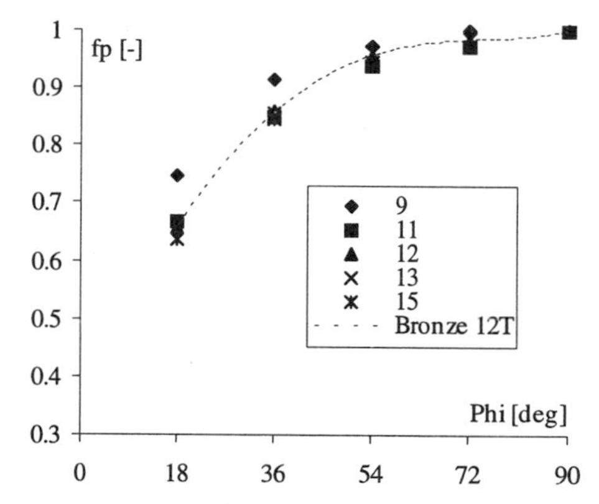

Figure 4: Normalised pinning force versus the angle and the applied field for the bronze-strand.

5. The magnetisation versus the angle of the applied field.

The magnetisation is measured for different angles with amplitudes of 2 and 3 T without a DC background field and once with an amplitude of 0.5 T and a background field of 1 T. The results of the magnetization measurements for both conductors are summarized in Figure 5. In the assumption that the effective strand diameters, d_z and d_θ are similar for both directions, values of α, deduced from the results of the magnetisation measurements, are calculated to be 0.85 and 0.38 for the bronze and the internal Sn conductor respectively. It can be seen in Figure 3 and Figure 4 that the different samples behave qualitatively as expected.

An other way to analyse the data is to plot the normalised pinning force and hysteresis loss together as a function of the angle between the strand axis and the applied field. As can be seen in Figure 5 the data for the internal Sn sample coincides perfectly, whereas the data for the bronze sample deviates approximately by a factor 1.3. The relative difference between the α values, as deduced from the AC-loss measurements, is much larger than the relative difference in $f_p(\phi)$ obtained from the I_c measurements. For the internal Sn sample, the α, corresponds well with the tendency that is observed in the $f_p(\phi)$ determination with J_c. In the bronze sample, the α value is high compared to the $f_p(\phi)$ values for lower angles. This can be explained by an increased effective filament diameter in the magnetisation loss at parallel field or by a different method of pinning for $J_{c\theta}$ (radial pinning) and J_{cz} (transverse pinning). For both analyses it is clear that the twistpitch of the samples can give rise to a certain error.

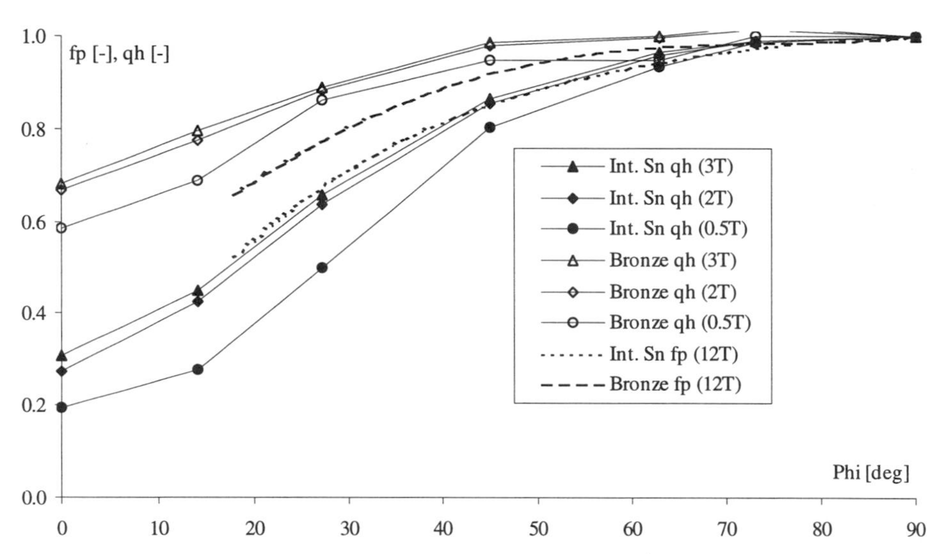

Figure 5: The normalised hysteresis loss versus the field orientation with respect to the strand axis in comparison to the normalised pinning force deduced from the critical current measurements.

6. Conclusions

The measurements on strands manufactured by two different production techniques clearly show an anisotropy in the critical current. The critical current rises strongly at reduction of the angle between the strand axis and the applied field. Both the J_c and Q_h determination show an angular dependence in f_p that is more pronounced in the internal Sn wire compared to the bronze sample.

References

[1] Schild, T., Cziazinsky, J-L. Duchateau, presented at the ASC'96 in Pittsburgh, 1996.
[2] Knoopers, H.G.M., et al, 1997 *presented on EUCAS'97.*
[3] Wilson, M.N., Superconducting magnets, Oxford University Press, Oxford New York, 1997.
[4] Evetts, J., Concise Encyclopedia of Magnetic & SC Materials, Pergamon Press, 1992.
[5] Roovers, A., Phd Thesis University of Twente, 1989.
[6] Campbell, A.M. 1982 *Cryogenics* ***22*** 3-16.
[7] Verweij, A.P., L.E. Erikson, H.H.J. ten Kate, presented at the IISSC, San Fransisco, 1993.

Inst. Phys. Conf. Ser. No 158
Paper presented at Applied Superconductivity, The Netherlands, 30 June–3 July 1997

Influence of Cr Plating on the Coupling Loss in Cable-in-Conduit Conductors

Arend Nijhuis[1], Herman H. J. ten Kate[1] and Pierluigi Bruzzone[2]

[1]Low Temperature Division, University of Twente, the Netherlands, [2]ITER Joint Central Team, Naka-gun, Japan.

Abstract. The coupling current loss of Nb3Sn based Cable-In-Conduit (CIC) conductors can be controlled by the interstrand resistance (R_c) with a Cr coating on the strand surface. The R_c determines the amount of coupling loss that is a driving factor for the design of large cabled conductors operating under a time varying field, like the conductors for the ITER fusion project. Nine CIC specimens have been compared which are fully identical except for the Cr plating of the strand surface. The coupling losses of the 48 strands CIC specimens, can change up to a factor of four depending on the vendor of the Cr plating. The measurement results of coupling losses backed by direct interstrand R_c measurement, show a clear correlation between loss and Cr vendor, however, the quantitative correlation has a margin of error. In addition, the statistical piece-to-piece scattering in coupling loss, the possible influence of the Cr vendor on the value of the RRR and the dependence of the sample length on the R_c were studied. Moreover the strand crossovers are examined by micrography and surface roughness tests.

1. Introduction

The coupling loss in CIC conductors depends on many parameters, including the cable void fraction, twist pitches, the inter- and intra-strand resistance, the history of cabling, jacketing and the load at the strand crossovers [1,2]. The ITER strands are coated with a thin layer of Cr primarily to avoid diffusion bonding, which may occur at the strand crossovers during the heat treatment. Each Home Team has selected different vendors for applying the Cr coating on the strands. It is assumed that the electrical R_c at the crossovers substantially depends on the proprietary methods used by the different vendors to plate the strands.

The coupling losses and R_c's of nine identical CIC specimens (1x3x4x4) have been compared. The only difference between the specimens are the strand surfaces, which are coated by seven different Cr vendors. The test results of the AC loss and R_c on conductor specimens prepared from strands plated by different Cr vendors confirmed the role of the Cr plating process parameters in the R_c of the Nb3Sn CIC conductors [3]. A number of additional tests, carried out on the same conductor sections, are carried out for a better understanding of the R_c at the Cr plated crossovers and the influence on the inter-strand coupling loss ($n\tau$). The electromagnetic loss in superconductors without transport current consists of hysteresis and coupling loss [4]. The coupling loss per volume strand per cycle increases linearly with frequency f and the square of the amplitude of the magnetic field B_a:

$$Q_{cpl} = \frac{\pi \cdot B_a^2 \cdot \omega \cdot v \cdot n \cdot \tau}{\mu_0} \qquad [\text{J/m}^3\cdot\text{cycle}]. \qquad (1)$$

The applied field is $B_a.\sin(2\pi ft)$ and the $n\tau$ is derived from the initial slope of the loss curve. The τ for a strand is given by (2), where L_p is the twist pitch and $\sigma_\perp$ is the effective electrical conductivity in the transverse direction. The coupling loss is measured by magnetisation, without and in a 1 T background DC field and a field sweep of ΔB=0.4 T.

$$\tau = \frac{\mu_0}{2} \cdot \left(\frac{L_p}{2\pi}\right)^2 \cdot \sigma_\perp, \qquad [\text{s}]. \qquad (2)$$

2. Results

The results of the magnetisation and R_c measurements are presented in [3] and some of the results are summarised in Table 1. The strands investigated are labeled corresponding to their home team. Among the Cr plated conductors with 36.5 % void fraction, the range of loss is up to a factor of four. Micro-graphs of the strand cross-sections are taken of all cable types in order to check the bronze layer, diffusion barrier and strand layout in relation to the spread in $n\tau$ among different cable types. No significant differences in the strand layouts are noticed. In the case of US-1 the bronze layer is relatively thin compared to the others, but this does not correspond to the high R_c found for US-1. The opposite is found for the JA-1 cable, a low R_c in combination with a relatively thick bronze layer. This illustrates that the surface condition of the strand is far dominant for R_c with respect to the variations in the internal geometry of the strands.

To assess the range of statistical, piece-to-piece variation in the coupling loss, due to inhomogeneity of the R_c, the AC loss test is repeated for three more sections of two CIC types: RF-3 at 36.5% and RF-3 at 32.5% void fraction. The variation in $n\tau$ is within a range with a standard deviation of 16 %. A variation in the cabling pitch also causes a spread in $n\tau$ among different CIC conductor sections of the same Cr vendor type. For that reason, the cabling pitch of the last cabling stage is determined for all cable types. The average cabling pitch of the last stage amounts to 95 mm and the standard deviation is $\pm$ 5 %.

The R_c is also measured on 560 mm long specimens and some scattering (30% in the worst case) has been observed in the same specimen for strand pairs with equivalent position [3]. To assess the extent of the inhomogeneity of the R_c, three specimens are cut to shorter lengths and the R_c of the same strand pairs are measured. The specimens are cut down first to 380 mm (four pitches), then to 190 mm (two pitches), 145 mm and eventually to 95 mm (one pitch). The three selected specimens are RF-2, RF-3 and EU-1. The R_c for all measured strand combinations shows a maximum variation within 30 % along the length of a cable. The actual scattering in the product of $n\tau$ and the R_c is given by the standard deviation of 36 % without DC field and 28 % in a 1 T field. As the variation in the twist pitch amounts to 5 % then the scattering in the $R_c \cdot n\tau$ can be attributed to the variation of the R_c along the length of the cable.

In [3] a correlation in the $n\tau$ results and R_c is obtained in relative units, looking at the ratio between the performance of different conductors. In order to verify the correlation for the absolute units, the measured trans-conductance is related to the coupling loss time constant using the formula for the decay time constant (2). Calculation of τ results in an $n\tau$ of 610 ms for 0 T (n=2) and an $n\tau$ of 93 ms for 1 T. In the case of an applied DC field the measured $n\tau$ is 430 ms; more than 4 times the calculated value. If no background field is present the RF-1 specimen behaves like a monolithic conductor obeying the well-known time constant model (2) with the twist pitch of the last cabling stage. For an applied DC field the average of all the 'measured' $n\tau$ values is almost 4 times the average of the 'calculated' results. Without DC field the average of the 'measured' $n\tau$ values is 2.5 times the average of the calculated results.

The RRR values have been measured by the Bochvar Institute on bare specimens of the five batches of strand (distinguished by shading in Table 1) from which the sections for the Cr vendors action originally have been cut. The scattering was relatively small (190-225). The RRR is also measured on two strands from each of the nine CIC's (vf=36.5 %) to assess the

Table 1. Summary data measured on the CIC conductors with different Cr plating (void fraction 36.5 %).

	B_{dc}=1 T		B_{dc}=0 T					Contact Area	non-contact
Sample :	$n\tau$ [ms]	$R_{c\text{-Stage 3}}$ [nΩm]	$n\tau$ [ms]	$R_{c\text{-Stage 3}}$ [nΩm]	d_{Cr} [μm]	RRR	L_p [mm]	R_a [nm]	R_a [nm]
RF-1	430	3.9	740	0.54	0	213	94	508	511
RF-2	47	27	58	21	2.2	119	96	139	164
RF-3	29	46	36	34	2.1	117	93	327	468
JA-1	108	11	127	6.0	2.5	96	106	319	192
JA-2	67	21	69	11	2.5	96	95	206	247
JA-3	76	22	83	12	3.1	95	90	174	315
US-1	37	30	38	24	2.0	128	92	223	347
EU-1	71	17	75	10	1.9	99	89	317	343
EU-2	38	21	37	15	1.8	107	96	150	330
Average	91	23	124	16	2.3	107	95	263	324
Stand.dev.	115	11	206	10	0.4	13	5	119	115
%	126	47	167	60	19	12	5	45	35

influence of the different plating processes on the RRR degradation (Table 1). The average RRR value of 213 found for specimen RF-1 is in good agreement with the 210 measured by Bochvar. The average of all RRR values after Cr coating is 107 with a standard deviation of 12 %. This means that the RRR is reduced drastically with 50 % after applying the Cr coating and cabling. Notice that the pairs of strands coming from the same billet produced by Bochvar have more or less corresponding RRR values. Only JA-3 and US-1 are an exception.

Evidence of sticking contacts at strand crossovers has actually been found after removing the jacket and opening cable JA-1, the Cr plated specimen with the highest loss. The strongest bonding is found in the bare stranded cable RF-1.

From each of the nine CIC (36.5% void fraction), a section is used for metalographic investigations of the Cr plating at the strand crossovers by using SEM technique. Some typical results of the scans are shown in Figure 1. A large number of cracks in all directions divide the coating into small Cr scales of approximately 10 μm or smaller. Material analyses made from the material in between the Cr scales prove that in some cases Cu is present in these channels. On few locations the copper is even stowed upwards to the surface (especially at the boundary of a crossover area). The effective area of the Cu to Cu contacts between strands is extremely small and thereby its role in the R_c can be neglected. Notwithstanding the cracks in the Cr layer, the plating is still quite effective for all Cr-layer types. The surfaces of the RF-2 and RF-3 layers are very rough and considering the scattering in particle size and the large degree of crumbling it seems that the layer growth during the plating process does not result into a solid crystalline structure. The results of the plating procedures and the properties of the Cr layer of JA-1, JA-2 JA-3 and US-1 seem to be more or less similar. The surface is relatively smooth and the layer growth during the plating process is solid and homogeneous.

The clear differences in crystal growth and surface profile can explain at least for a considerable part the difference in R_c among the chromium layer vendors. The extreme roughness of the surface explains at least partly how such a thin Cr layer of ~2 μm causes a relatively high inter-strand contact resistance.

The surface roughness is determined for each different Cr layer at a strand-to-strand contact area and at a non-contact area. After the Cr coating, the wire surface has become less rough. This implies that the final surface roughness is highly determined by the original condition and profile of the strand surface before plating. The change of the surface profile from the hard Cr scales due to deformation at the strand-to-strand contact crossovers is relatively small.

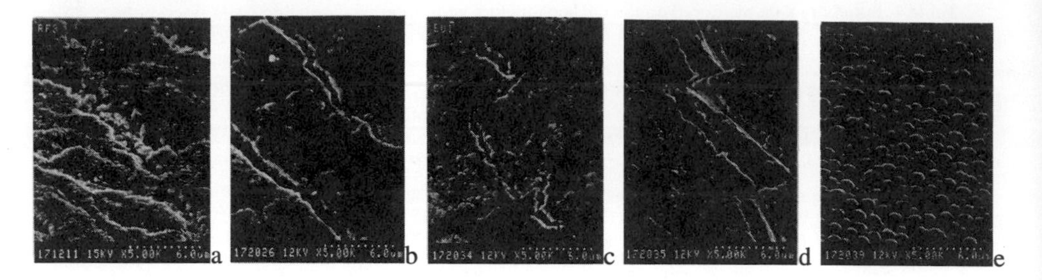

Figure 1. SEM photographs from magnified crossover areas between two strands (5000 x) of specimen (a) RF-3, (b) JA-3, (c) EU-1, (d) EU-2 with Cu stowed upwards between the Cr scales, (e) EU-2, non-contact area showing the small and undamaged spheres typically for EU-2.

The thickness of the Cr layer is obtained by using the difference in weight, Δm, of a piece of strand before and after etching the Cr layer in a solution of HCl. The correlation between $n\tau$ and d_{Cr} is very poor so the Cr layer thickness does not play a significant role in the R_c. In fact none of the parameters from Table 1 could improve the correlation between $n\tau$ and R_c. It is suggested that the R_c primarily depends on the surface condition of the strand but also on the bulk properties such as crystalline structure and degree of crumbling, determined by the used coating process. The Cr coating is also important because it acts as a substrate for surface oxidation [5]. This also could be an important effect responsible for the increase of R_c after bending or electromagnetic loading [2]. It is not clear which effect, mechanical aspects (changing contact surface and pressure) or oxidation has the largest impact on the increase of the R_c after mechanical or electromagnetic loading.

3. Conclusions

The large differences in coupling loss time constants (up to a factor of four), found for cables manufactured with strands covered by different Cr vendors, is not due to the statistic variations introduced during the cabling process (inhomogeneity of the R_c or deviations in the cabling pitch). Nor can it be clearly attributed to the RRR value or the thickness of the Cr layer or the bronze barrier. The differences can be attributed to the varying properties of the Cr layers supplied by different vendors, suggesting that the surface layer condition of the strand (mechanical aspects and oxidation) is far dominant for R_c

The RRR of the bare specimen is 213 and the average RRR after Cr coating is 107.

A large number of cracks divides the coating into a fine structure of islands. In some cases, Cu is present in the crack channels but the plating is still quite effective for all Cr-layer types. The clear differences in crystal growth and surface properties between layers from different vendors can explain at least for a considerable part the difference in R_c among the chromium layer vendors.

References

[1] Nijhuis, A., H.H.J ten Kate, P. Bruzzone, and L. Bottura 1996 *IEEE Trans on Mag.* ***32***, 2743-2746.
[2] Nijhuis, A., H.H.J. ten Kate, and P. Bruzzone 1996 paper presented at *ASC-96 Pittsburg*.
[3] Bruzzone, P., A. Nijhuis, and H.H.J. ten Kate 1997 *Proceedings of the ICEC/ICMC-16* 1243-1248.
[4] Campbell, A.M. 1982 *Cryogenics* ***22*** 3-16.
[5] Kimura, A. et.al. 1996 *Cryogenics* ***36*** 681-690.

Inst. Phys. Conf. Ser. No 158
Paper presented at Applied Superconductivity, The Netherlands, 30 June–3 July 1997

Very High Currents and Irreversibility Fields in Bulk $NdBa_2Cu_3O_{7-\delta}$ Samples

T Wolf, H Küpfer, A-C Bornarel

Forschungszentrum Karlsruhe, Institut für Technische Physik,
Postfach 3640, D-76021 Karlsruhe, Germany

Abstract. Single crystals and melt-textured samples of $NdBa_2Cu_3O_{7-\delta}$ exhibiting T_c values up to 96 K were prepared. Current densities, j, and irreversibility fields, B_{irr}, were optimized by Y doping and by introducing a peak effect caused by oxygen deficiency. At 77 K in pure crystals, B_{irr}, values parallel to the c axis of up to 13.4 T were obtained whereas, in melt-textured samples, the highest B_{irr} value was found to be 11.8 T. Critical current densities at 77 K and 2.5 T were $8.3 \cdot 10^4$ A/cm^2 and $1.2 \cdot 10^5$ A/cm^2 in pure and Y doped single crystals, respectively. To obtain currents that high also in melt textured samples, strongly interacting defects must be avoided which raise j only at low fields. Current densities of up to 5.5 10^4 A/cm^2 at 77 K and 3 T in addition to a B_{irr} of 8 T were obtained. The correlation and interference between the current density and B_{irr} in the presence of weakly and strongly interacting defects, such as small clusters of oxygen vacancies and 422 precipitates, respectively, as well as the influence of the twin structure are analyzed and discussed.

1. Introduction

It is well known that most rare earth elements can replace Y in the $YBa_2Cu_3O_{7-x}$ (Y-123) structure to form a stable superconducting compound. Meanwhile, the Nd-123 compound has been shown to exhibit the highest transition temperature, T_c, in this family. This should result in increased irreversibility fields and currents at 77 K, which moves Nd-123 into the range of interest for technical applications. Moreover, the large Nd ion tends to substitute for Ba in the unit cell leading to a solid solution with reduced T_c values. Recent measurements of the critical current as a function of the magnetic field showed a peak effect which was attributed to such a cation disorder [1]. However, the same peak effect was observed in Y-123. Here, the magnetization versus H curve was explained by the presence of impurity point defects, point-like oxygen vacancies, and twins [2].

As Y-123 exhibits no cation disorder at least down to a level of 1000 ppm, we are not convinced that the Nd/Ba substitution is responsible for the peak effect in Nd-123. This investigation was carried out to clarify whether Nd-123 can be used for technical applications, and to correlate the pinning behavior of Nd-123 with its defect structure. As this cannot be determined by studying sintered or melt-textured samples only, also single crystals were grown and characterized.

In order to be certain to prepare a stoichiometric Nd-123 phase without any Nd/Ba disorder one must avoid Nd-123 + Nd-422 mixtures and, instead, start from low Nd mixtures, and also reduce the oxygen partial pressure during sample synthesis.

2. Experimental

2.1 Crystal Growth

Nd-123 single crystals were grown from BaO/CuO fluxes by slow cooling. Calcination of the high-purity (5 N) Nd_2O_3, $BaCO_3$, and CuO powder mixtures took place in a Y-stabilized ZrO_2 crucible in an air-tight chamber furnace at 1010 to 1030°C. The furnace atmosphere was 50-60 mbar of air/950-940 mbar of Ar. After a complete melt had formed, the furnace was allowed to cool slowly so that the crystal growth process could begin. At 950-960°C, which was thc cnd of the cooling ramp, the remaining liquid flux was decanted within the furnace to separate the crystals from the flux. Isometric crystals additionally exhibiting (101)-faces, with dimensions of up to 6 x 6 x 3 mm^3, were obtained. After cooling to room temperature, selected crystals were further annealed in a tubular furnace under an oxygen flow of 1 bar. This process occured at 600-380°C and took about 800 h. Another oxidation treatment in the range of 480-280°C and at 175 bar oxygen followed which took about another 300 h. The crystals were measured in this highly oxidized state and after various reduction steps in 1 bar of oxygen in the temperature range of 310-430°C. The oxygen content of each state was obtained by the isotherms of Kishio et al [3].

2.2 Melt Texture

For the preparation of melt textured Nd-123 pseudo-crystals we used either pure, vacuum synthesized Nd-123 powders (TA1), or Nd-123 + 5% CuO + 5% BaO mixtures (TB series) or Nd-123 + 5% CuO + 4% BaO, 1% SrO mixtures (TC1). After milling, the powders were pressed uniaxially at 3 t for 30 s into to green pellets with a diameter of 15 mm and thicknesses of 1 to 5 mm. Texturing took place in an atmosphere of 80 mbar of air and 920 mbar of Ar. After fast heating to 1050°C and a holding time of 0.5 h, the pellets were quickly cooled to 1030-1020°C and then, at rates of 1.3°C/h to 1.5°C/h to approx. 900°C where the furnace was shut off. Oxygen annealing was performed in an oxygen flow of 1 bar or in a static oxygen atmosphere of up to 400 bar, similar to the treatment of single crystals. The oxygen contents of the samples were determined from Kishio's isotherms [3].

2.3 Superconductivity Characterization

The magnetic moment, m, of the crystals and the melt textured samples produced by an induced shielding current was measured with a vibrating sample magnetometer. Measurements were performed as a function of the applied magnetic field, B, the temperature, T, and the angle, ϕ, between the B direction and the c axis of the samples at a constant sweep rate, dB/dt, of 10^{-2} T/s, which corresponds to an electric field, E, at the sample surface of about 10^{-2} μV/cm. The current density, j, was determined from the hysteresis width, Δm, by the extended Bean model. The E-criteria for determining the irreversibility field, B_{irr}, were the same as for the determination of j. T_c values were determined from the midpoint of the diamagnetic signal at $5 \cdot 10^{-4}$ T.

3. Results and Discussion

3.1 *Critical Currents*

The current density versus B for a single crystal is shown in Fig. 1 at various temperatures. It passes through two shallow peaks, PE1 and PE2, which result from point-like defects and the twin structure, respectively. The PE2 peaks become more and more pronounced with rising temperature, due to the higher thermal activation of the point-like defects which, compared to twins, therefore become the less dominating pinning defects. This correlation is supported by the very similar j(B,T) dependence of a pure twinned Y-123 crystal where PE2 vanishes with increasing angle, ϕ, between B and the c axis of the crystal [4].

The nature of PE1 was resolved after small reduction steps which introduced oxygen vacancies into the Cu chains without changing the level of Nd/Ba substitution. Fig. 2 shows that increasing the oxygen deficiency from $\delta = 0.04$ to 0.11 (XA1 to XA6) caused PE1 to rise to $j = 8.3 \cdot 10^4$ A/cm^2 and fall again, whereas PE2 already disappeared at low δ values. This behavior was also observed in pure Y-123 crystals and can be explained by the increasing concentration of pointlike defects which reduce the influence of the twin structure. Simultaneously, B_{irr} (77 K) decreased from 13.4 T to 7.8 T, while T_c remained constant or decreased only by 1 K after the last reduction step.

Further possibilities to introduce pointlike defects into the crystals and, in this way, increase the peak effect, PE1, exist by substituting Nd by Y, Ba by Nd, or Ba by Sr. While our Nd/Ba substitutions are not yet conclusive the Y/Nd substitution led to a j value of $1.2 \cdot 10^5$ A/cm at 77 K and 2 T, which is the highest critical current so far observed in Nd-123 material.

In contrast to single crystals, melt-textured samples contain uncorrelated defects, such as precipitates, pores or microcracks. This gives rise to relatively high critical currents at low fields and reduces the influences of twins and point-like defects. In other words, 422 precipitates reduce both B_{irr} and PE1. Therefore, the amount of second phases in melt-textured samples should be kept as small as possible. Fig. 3 shows the results of three melt-textured samples after high-pressure oxidation. Keeping their 422 content below 10% allowed the PE1 peak-effect to be observed. The influence of two subsequent oxygen reduction steps of TB1 is shown in Fig. 3. Again, the same behavior can be observed as in single crystals. Increasing oxygen deficiency, δ, causes B_{irr} to decrease and PE1 to increase

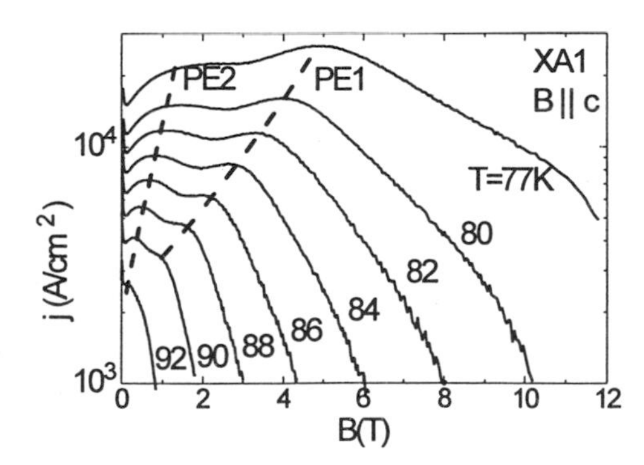

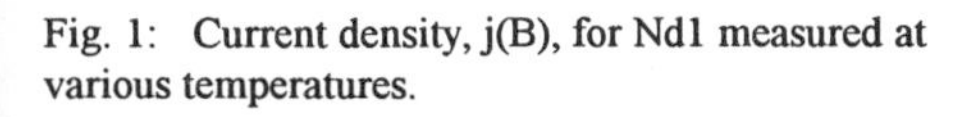
Fig. 1: Current density, j(B), for Nd1 measured at various temperatures.

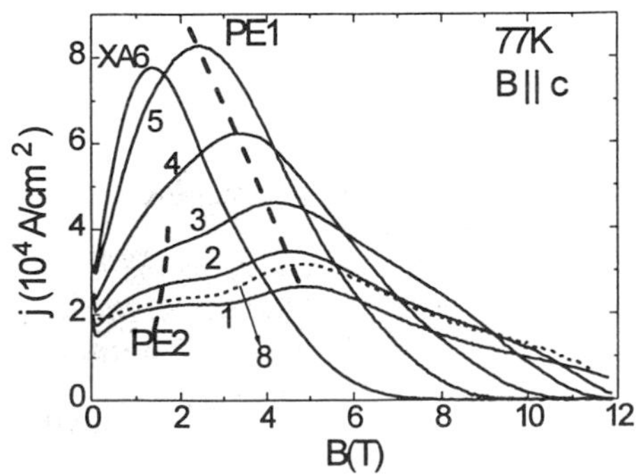

Fig. 2: Current density, j(B), at various oxygen deficiencies in a twinned crystal.

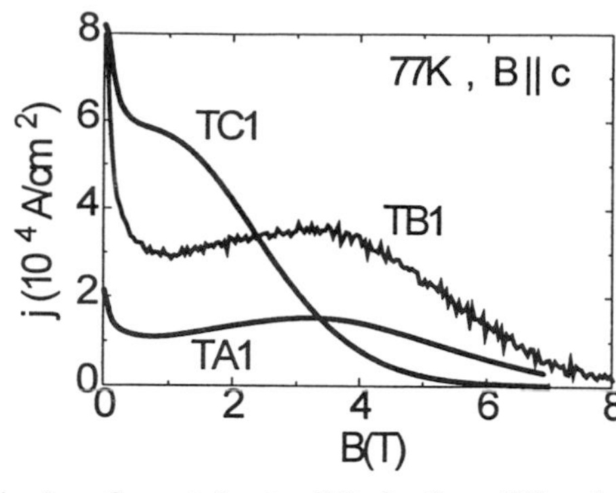

Fig. 3: Current density, j(B), for three differently prepared Nd-123 melt-textured samples after high-pressure oxydation. TA1 is the purest sample while TC1 contains the largest amount of second phases and other impurities.

Fig. 4: Current density, j(B), after high-pressure oxidation (TB1) and after two subsequent oxygen reduction steps, TB2 and TB3.

first and then decrease again. Due to the faster oxidation kinetics the intervals for δ are larger than in single crystals. The best compromise so far between the influence of precipitates and point-like defects led to values of $j = 5{\cdot}10^4$ A/cm^2 at 77 K and B = 3.5 T. B_{irr} values of up to 11.8 T were observed in very pure, nearly fully oxidized Nd-123 melt-textured samples.

4. Conclusion

Nd-123 single crystals and melt textured samples with a little of Nd/Ba substitution and 422 second phase content were prepared from low Nd starting mixtures at low oxygen pressures. After proper oxidation, the samples exhibited T_c values of up to 95.5 K. At 77 K, B || c the highest irreversibility fields so far of all HTSC materials of 13.4 T and 11.8 T were measured in the single crystals and melt-textured samples prepared, respectively. Variation of the oxygen and Y contents of the samples allowed critical currents of $1.2{\cdot}10^5$ A/cm^2 at 77 K/2 T to be obtained for single crystals, and of $5{\cdot}10^4$ A/cm^2 at 77 K/3.5 T for melt-textured samples. The j versus B curves were explained by the interaction between twins, point-like defects, and second phases. No influence on the critical current of Nd/Ba substitution was observed.

References

[1] Yoo S I, Murakami M, Sakai N, Higuchi T and Tanaka S 1994 *Jpn. J. Appl. Phys.* **33** L1000 - L1003

[2] Wolf T, Küpfer H and Wühl H 1996 *Proc. 8th Int. Workshop on Critical Curents in Superconductors* (Singapore: World Scientific) 411-414

[3] Kishio K, Hasegawa T, Suzuki K, Kitazawa K and Fueki K 1989 *Mat. Res. Soc. Symp. Proc.* **156** 91-97

[4] Wolf T, Bornarel A-C, Küpfer H, Meier-Hirmer R and Obst B to be published in *Phys. Rev. B*

Inst. Phys. Conf. Ser. No 158
Paper presented at Applied Superconductivity, The Netherlands, 30 June–3 July 1997

Homogeneity range of $Sm_{1+x}Ba_{2-x}Cu_3O_z$ solid solutions

D I Grigorashev, E A Trofimenko, N N Oleynikov and Yu D Tretyakov

Inorganic Chemistry Chair, Department of Chemistry, Moscow State University, V-234 Vorob'evy Hills, Moscow 119899, Russia

Abstract. The region of $Sm_{1+x}Ba_{2-x}Cu_3O_z$ solid solutions (Sm123ss) formation was studied at different po_2 and T. It was found that the maximum substitution degree x_{max} is 0.75 for po_2=1 atm (T=1000°C) and ~0.4 for $po_2=3.2\cdot10^{-4}$ atm (in nitrogen, T=850°C). Calculations of lattice parameters showed that lengths of both **a** (**b**) and **c** axis linearly decrease with the increase of x. According to ac-susceptibility measurements the T_c value of $SmBa_2Cu_3O_z$ (x=0) is 88K and drops abruptly with the increase of x.

1. Introduction

A new approach to the enhancement of flux pinning in RE-Ba-Cu-O superconductors was proposed by Murakami et al [1,2]. It is based on the use of rare-earth (RE) analogues of HTSC-phase $YBa_2Cu_3O_z$ with the possibility of heterovalent substitution of Ba for RE (RE=La, Nd, Sm, Eu, Gd). Upon specific heat treatment it is possible to obtain samples with compositional microfluctuations that may act as effective pinning centers [1]. But in order to understand the real mechanism of pinning center formation in RE123ss it is important to know the stability range, crystal structure and properties of these compounds.

Different authors have been studied phase diagram for RE123ss [3-7], but their data are contradictory: perhaps the use of different precursors and cooling regimes affect the results of measurements.

Recently the results of Nd123ss studies were published by Goodilin et al [8]. It was found that the character of $Nd^{3+}\rightarrow Ba^{2+}$ substitution changes at different temperatures and po_2. It seems interesting to obtain the same results for other RE-elements.

The aim of the present work was to study the formation region of $Sm_{1+x}Ba_{2-x}Cu_3O_z$ solid solutions at different conditions (T, po_2) and to study their superconductig properties.

2. Experimental details

The samples of $Sm_{1+x}Ba_{2-x}Cu_3O_z$ solid solutions with different x were synthesized by convenient ceramic route by using carbon-free precursors: Sm_2O_3, $Ba(NO_3)_2$ and CuO. After mixing, nitrate decomposition and calcination the samples were ball-milled in acetone and pressed into pellets at P=700 MPa. Then the samples were placed on single-crystalline MgO substrates and inserted at 850-1050°C into the preliminary calibrated tubular furnace (Naber Labotherm). Isothermal annealing was carried out in oxygen and nitrogen (P_{O2}=3.2·10^{-4} atm) for 20-200 hrs with subsequent quenching on air.

In order to study the phase composition and lattice parameters of the obtained samples XRD-analysis in Guinier chamber (Cu Kα-radiation, Ge internal standard) was used.

Differential thermal and thermogravimetric analyses were perfomed in oxygen by Sinku-Riko analyzer (specimen weight~150 mg, heating rate 2°C/h).

Superconducting properties of solid solutions prepared at 950°C at P_{O2}=1 atm and oxygenated (25 hrs - 400°C, flowing oxygen) were studied by means of ac-susceptibility measurements by APD Cryogenics setup (H_a=1 Oe).

3. Results and discussion

Basing on the results of XRD and thermal analysis the stability range for Sm123ss was determined for different P_{O2} (fig.1). One may see a wide homogeneity region of Sm123ss in oxygen: $0 \leq x \leq 0.75$ at T=1000°C. A decrease of the temperature to below 900°C leads to an abrupt narrowing of the Sm123ss stability region with x_{max}=0.4 at T=850°C. Upon decrease of p_{O2} the stability range for Sm123ss becomes narrower and in nitrogen x_{max} was found to be approximately 0.4 (T=850°C). With increasing temperature the homogeneity region of Sm123ss shrinks.

According to the XRD-data a further increase of the Sm content leads to secondary phases like Sm_2CuO_4 and CuO. At $T \geq 1000$°C and P_{O2}=1 atm the formation of liquid phase is observed. Perhaps this proceeds due to melting of eutectic compositions with inclusions of CuO.

In order to determine the high-temperature boundary of the Sm123ss stability we performed DTA/TG analysis of singlephase samples prepared at 950°C in oxygen (fig.2). We observed the existence of two superpositional peaks on the DTA-curve at T~1050-

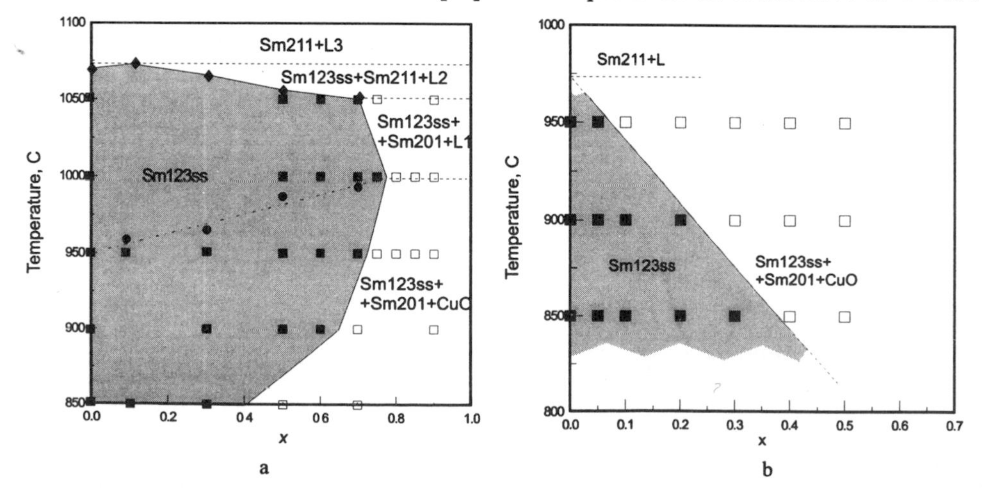

Fig. 1. Homogeneity range of $Sm_{1+x}Ba_{2-x}Cu_3O_z$ solid solutions in oxygen (a) and nitrogen (b): ■ - singlephase samples; □ - multiphase samples; ♦ - decomposition points of Sm123ss according to DTA measurements; ● - position of "low-temperature" peak on DTA-curves.

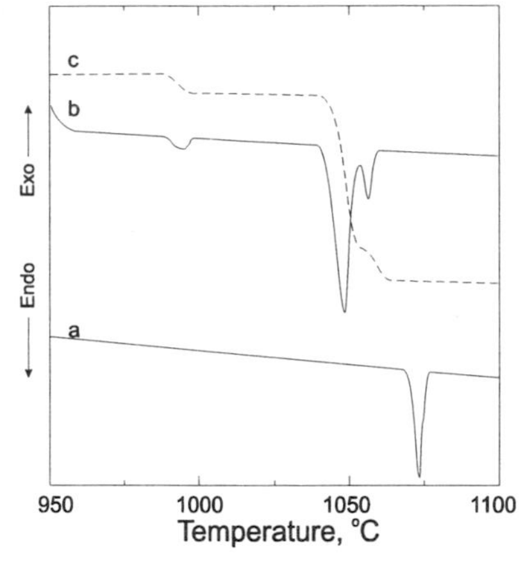

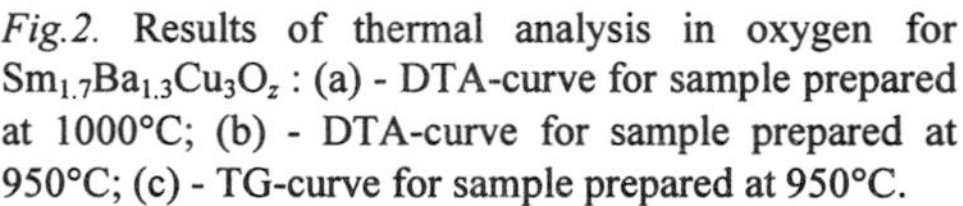

Fig.2. Results of thermal analysis in oxygen for $Sm_{1.7}Ba_{1.3}Cu_3O_z$: (a) - DTA-curve for sample prepared at 1000°C; (b) - DTA-curve for sample prepared at 950°C; (c) - TG-curve for sample prepared at 950°C.

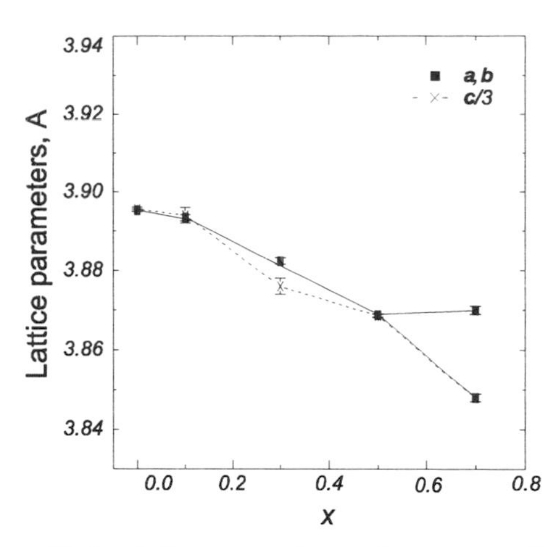

Fig.3. Lattice parameters of as–quenched $Sm_{1+x}Ba_{2-x}Cu_3O_z$ solid solutions obtained in oxygen at 950°C.

1075°C accompanied by a two-step weight loss. Perhaps this effect originates due to two-step "melting" of Sm123ss: firstly proceeds the decomposition of solid solution with higher substitution degree x_1 to solid solution with lower x_2 ($x_1>x_2$); the second stage may be assigned to the melting of the latter Sm123ss(x_2). As it can be expected from this supposition upon increase of x the temperature gap (ΔT) between the superpositional peaks increases: $\Delta T\approx10$°C for x=0.3 and 30°C for x=0.7.

It is interesting to note the principal change of thermal behavior for Sm123ss synthesized at T=1000°C. There was observed a single endothermic peak on DTA-curves with a constant temperature ($T\approx1075$°C) for all substitution range $0\leq x\leq0.7$ (fig.2.a).

Moreover it should be noted that the existence of the "low-temperature" peak at T~950-1000°C is accompanied by weight loss for samples prepared at 950°C (figs.1,2).

This phenomenon can be explained by the supposition of non-equilibrium state of the obtained samples and different concentration of antistructure defects in solid solutions synthesized at temperatures below and above the "low-temperature" peak [9].

Calculations of the lattice parameters show that both **c** and **a**(**b**) axis shrink with the increase of the Sm→Ba substitution degree (fig.3). The tetragonal structure of Sm123ss undergoes orthorhombic distortion at x>0.5. Probably such a distortion originates due to an ordering process of Sm^{3+} ions on Ba^{2+} sites. Partially it is confirmed by observations of RE^{3+} tendency to form couples via oxygen of four-coordinated copper in RE123ss[10]. Also one should note the equality of **a** and **c**/3 parameters values in all substitution region range i.e. the crystal structure of Sm123ss may be described as consisting of three cubic blocks.

Ac-susceptibility measurements allow us to determine T_c of the obtained solid solutions. It was found that for "pure" Sm123 (x=0) the T_c=88K (fig.4). With the increase of Sm content in Sm123ss T_c drops down and for x=0.3 it is only 33K. One may suppose that upon the increase of the degree of heterovalent substitution significant changes in the charge carrier density occur that lead to strong degradation of T_c. Also it should be noted that results obtained in the present work are in contradiction with the results obtained for Nd123ss by Goodilin et al [8]. Despite of little difference between Sm^{3+} and Nd^{3+} there was observed a very low T_c for Nd123ss samples synthesized in oxygen: even for "pure" Nd123 T_c was 45K.

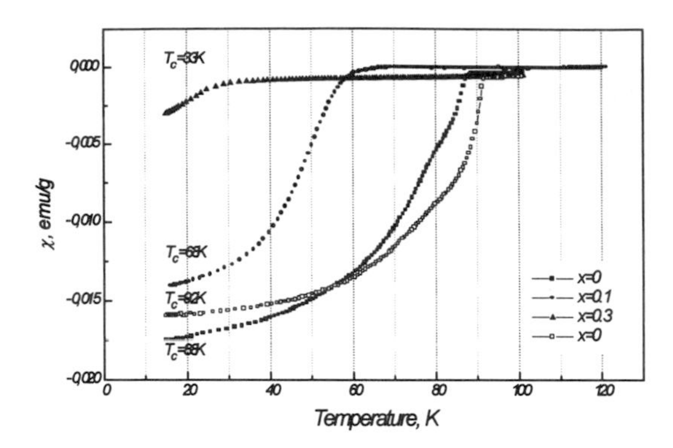

Fig.4. Ac-susceptibility measurements of fully oxygenated $Sm_{1+x}Ba_{2-x}Cu_3O_z$ samples prepared at 950°C and quenched on air (solid symbols) and into liquid nitrogen (open symbols).

It seems that such a difference occurs due to various methods of samples quenching. In [8] samples were quenched in the furnace by a flow of gaseous nitrogen. Hence, the cooling rate in that case is much smaller than in our experiments (samples were quenched in air). This is also confirmed by experiments with samples quenched into liquid nitrogen: here T_c for $SmBa_2Cu_3O_z$ rises up to 92 K. But in order to understand the observed phenomena additional experiments should be done.

4. Conclusions

1. The homogeneity range of $Sm_{1+x}Ba_{2-x}Cu_3O_z$ solid solutions at different temperatures and oxygen partial pressures was studied. The maximum substitution degree was found to be x_{max}=0.75 at po_2=1 atm and 0.4 at $po_2=3.2\cdot10^{-4}$ atm.
2. Upon increase of x in $Sm_{1+x}Ba_{2-x}Cu_3O_z$ a decrease of the lattice parameters was found. At x>0.5 orthorhombic distortion of Sm123ss unit cell proceeds probably due to ordering of antistructure defects.
3. The T_c of "pure" Sm123 (x=0) is 88K and drops rapidly with the increase of x. That may be the consequence of changes in charge carrier density upon heterovalent $Sm^{3+}\rightarrow Ba^{2+}$ substitution.

References

[1] Murakami M, Yoo S-I, Higuchi T, Sakai N 1994 *Jpn. J. Appl. Phys.* **33** L715-7
[2] Yoo S-I, Murakami M, Sakai N, et al 1994 *Jpn. J. Appl. Phys.* **33** L1000-3
[3] Akinaga H, Katon H, Takita K, et al 1988 *Jpn. J. Appl. Phys.* **27**(II) L610-2
[4] Wong-Ng W, Paretzkin B, Fuller E R 1990 *J. of Solid State Chem.* **85** 117-32
[5] Li S, Hayri A, Ramanujachary K V, et al 1988 *Phys. Rev. B* **38** 2450-4
[6] Fomichev D V, D'yachenko O G, Mironov A V, Antipov E V 1994 *Physica C* **225** 25-33
[7] Lindemer T B, Specht E D, Martin P M, et al 1995 *Physica C* **225** 65-75
[8] Goodilin E A, Oleynikov N N, Antipov E V, et al 1996 *Physica C* **272** 65-78
[9] Trofimenko E A, Grigorashev D I, Oleynikov N N, et al 1997 *Doklady Akademii Nauk* to be published
[10] Kramer M J, Karion A, Dennis K W, et al 1994 *J. Electron. Mater.* **23** 1117-20

Inst. Phys. Conf. Ser. No 158
Paper presented at Applied Superconductivity, The Netherlands, 30 June–3 July 1997

Microstructure and superconducting properties of Hg-1223 HTSC with nonstoichiometric additions

M. Reder†, J. Krelaus†, K. Heinemann†, H.C. Freyhardt†, F. Ladenberger‡ and E. Schwarzmann‡

† Institut für Metallphysik, Hospitalstr. 3-7, 37073 Göttingen, Germany

‡ Institut für Anorganische Chemie, Tammannstr. 4, 37077 Göttingen, Germany

Abstract. The influence of nonstoichiometric additions of ReO_3 in $HgBa_2Ca_2Cu_3O_{8+\delta}$ (Hg-1223) has been examined with respect to microstructure and superconducting properties. Grain size and shape depend significantly on the Re content whereas the intragranular defect structure does not change. EDX examinations suggest the existance of a limit of solubility of Re in Hg-1223. The critical temperature, T_c, of the as prepared (Hg,Re)-1223 samples is above 130 K and could not be improved by O_2-/N_2- annealing. A maximum of magnetical hysteresis versus Re content has been observed for 15% Re which might be caused simply by a grain size maximum.

1. Introduction

Hg-1223 is the HTSC with the highest T_c reported so far (T_c = 135 K at p =1 bar [1], rising up to 164 K under higher pressure [2]). Unfortunately, up to now the critical current densities of the Hg-12(n-1)n system is not very high. Because of this, it is very important to examine possibilities to optimize microstructural defects of the material and to improve the grain connectivity. Of special interest are effects of non-stoichiometric additions and the control of stacking faults in the Hg-1223. Stacking faults are always present in Hg-1223 and can be described as layers of $HgBa_2CaCu_2O_{6+\delta}$ (Hg-1212) or $HgBa_2Ca_3Cu_4O_{10+\delta}$ (Hg-1234). In the present work the influence of Re-addition on microstructure and superconducting properties of Hg-1223 are being discussed. Further, the effect of additional O_2 and N_2 annealing on T_c is presented.

2. Experimental

(Hg,Re)-1223 has been synthesized using HgO, ReO_3, BaO, CaO and CuO as starting materials. The oxides were mixed and filled in gold tubes which were sealed using a hydrogen torch. The samples were then annealed in N_2 for 24 h at 840°C under a pressure of 600–800 bar. To produce a series with varying Re content the oxides were mixed with the ratio $Hg_{1-x}Re_xBa_2Ca_2Cu_3O_{8+\delta}$ and x=0.05, 0.10, 0.15, 0.20, 0.25 and 0.30. Pieces of each sample were annealed at 300°C in O_2 respectively N_2 . Microstructural characterization of the samples was performed by SEM on fractured surfaces as well as on polished surfaces. The phase composition was examined by EDX in an analytical SEM. The intragranular defects were visualized by TEM using samples prepared by grinding and ion-milling at 4 kV. The magnetization measurements were carried out in a Faraday magnetometer and T_c was determined from ac-susceptibility measurements.

3. Microstructure

Already the macroscopic appearance of the samples differs significantly with Re content as indicated in table 1. For a better understanding of these differences microstructure and phase composition have to be known.

Table 1. Grain size and macroscopic appearance of $Hg_{1-x}Re_xBa_2Ca_2Cu_3O_{8+\delta}$

Re content x	Macroscopic appearance	Grain size "a" [μm]	Grain size "c" [μm]	a/c
0.05	brittle, porous	12.9 (±0.8)	2.3 (±0.3)	5.6 (±0.7)
0.10	hard, porous	18.8 (±1.5)	4.3 (±0.6)	4.4 (±0.8)
0.15	very hard, dense	17.7 (±1.3)	4.8 (±0.6)	3.6 (±0.5)
0.20	very hard, dense	16.1 (±1.5)	4.1 (±0.5)	3.9 (±0.6)
0.25	hard, porous	11.3 (±0.5)	1.2 (±0.1)	9.7 (±0.9)
0.30	brittle, porous	12.9 (±0.6)	1.6 (±0.1)	7.9 (±0.7)

3.1. SEM

SEM examinations of the Re doped samples reveal differences in grain size, grain shape and porosity. The samples with x=0.15 and x=0.20 appear to be quite dense whereas the others contain many pores. From several SEM micrographs for each Re content x the mean grain size has been determined distinguishing between the short and the long extension of the platelike grains called "c" and "a" (see table 1). A maximum in grain size and a minimum in the a/c ratio are observed for $0.10 \leq x \leq 0.20$.

3.2. EDX

To clarify the question whether the Re has been incorporated into the Hg-1223 phase energy dispersive X-ray (EDX) spectra of Hg-1223 grains were examined. A plot of x(EDX) versus x(starting material) shows that the initial slope flattens at higher x (see fig. 1a). The change of the slope occurs between x=0.10 and x=0.20. The error bars in figure 1a) are thought to demonstrate the resolution limit of the EDX system: in a Re-free Hg-1223 sample x was determined to 0.035 which represents the resolution of the EDX system. Figure 1a) leads to the hypothesis that between x=0.10 and x=0.20 the solubility

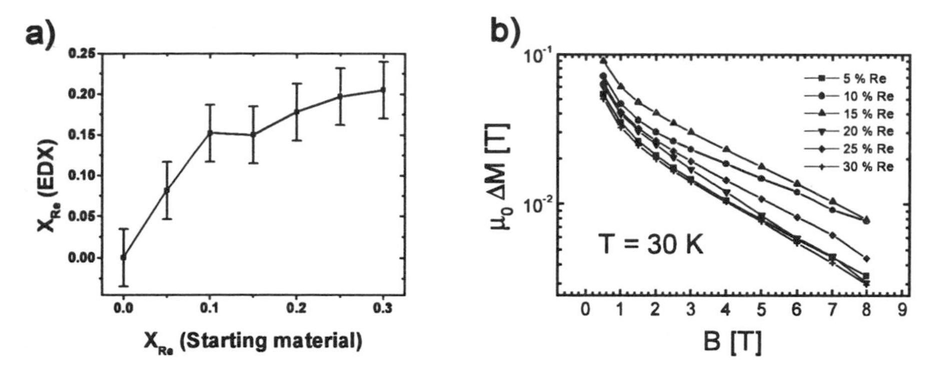

Figure 1. a) x of the starting material versus the x value determined in (Hg,Re)-1223 grains by EDX. b) Magnetic hysteresis at 30 K

limit of Re in Hg-1223 is reached and higher amounts of Re remain in the material as secondary phases. This is supported by the occurence of Re-rich regions which have been observed for $x \geq 0.20$ in polished samples by EDX mapping as well as on fractured sample surfaces by X-ray analyses in spot mode. Structural considerations of Shimoyama and coworkers lead to the assumption of a solubility limit of Re in $Hg_{1-x}Re_xSr_2Ca_2Cu_3O_{8+\delta}$ at x=0.25 [3]. Our experiments show that this limit might be reached at lower x for $Hg_{1-x}Re_xBa_2Ca_2Cu_3O_{8+\delta}$.

3.3. TEM

The analysis of the intragranular defect structure by TEM does not show significant changes with varying x. In all samples examined a great number of stacking faults were present. These stacking faults are arranged almost periodically in some grains whereas they show no periodic arrangement or are not found at all in other grains. Superstructures of Re arranged periodically on Hg sites like described for $Hg_{1-x}Re_xSr_2Ca_2Cu_3O_{8+\delta}$ by [3] and [4] were not found.

4. Superconductivity

4.1. Critical temperature

The critical temperature, T_c, of the (Hg,Re)-1223 samples was determined to be above 130 K in all our samples (see table 2). This is different from pure Hg-1223 where as prepared samples normally show a T_c of about 125 K and optimized T_c of 135 K is only reached by additional annealing in O_2 (typically 24 h at 300°C). In (Hg,Re)-1223 additional annealing in O_2 or N_2 did not significantly enhance of T_c. On the contrary, in the samples with x=0.15 and x=0.30 O_2 annealing resulted in a decrease of T_c to about 130 K (possibly because of O_2 overdoping). Probably Re can be incorporated in the HgO planes and cause a higher O_2 content in these planes due to a higher valence than Hg [3]. This additional O_2 is thought to optimize T_c. Similar results have been reported for Hg-1212 [5].

Table 2. Critical temperatures ($T_{C,Onset}$)

Re content x	T_c [K], as prepared	T_c [K], N_2 annealed	T_c [K], O_2 annealed
0	125	not measured	135
0.05	135	135.6	135.5
0.10	133.8	135.3	135.3
0.15	134.7	133.1	130.7
0.20	133.3	133.8	134.3
0.25	134.6	134.6	132.5
0.30	134.3	133.7	130.1

4.2. Magnetic hysteresis

The magnetic hysteresis ΔM of the (Hg,Re)-1223 samples at 30 K exhibit a maximum for x=0.15 for all applied fields from 0.5 to 8 T (see fig. 1b). This effect might be caused by enhanced bulk pinning due to the Re incorporated in the Hg-1223. Reasons for enhanced bulk pinning could be additional pinning centers and/or a lowered anisotropy of j_c. A comparison with the grain size distribution determined in 3.1 leads to the insight that the maximum of ΔM might also be a grain size effect.

5. Conclusion

By Re doping of Hg-1223 the grain size as well as the shape of the (Hg,Re)-1223 grains can be modified. We assume the solubility limit for Re in Hg-1223 to be between x=0.10 and 0.20 .

In Re doped Hg-1223 no additional O_2 annealing is necessary to reach the optimal value of T_c. A maximal ΔM signal for x=0.15 might be explained due to grain size effects.

Acknowledgments

This work is supported by the German BMBF under grant number 13N 6622.

References

[1] S.N. Putilin, E.V. Antipov, O. Chmaissem, M. Marezio *Nature* **362** (1993) 226

[2] C.W. Chu, L. Gao, F. Chen, Z.J. Huang, R.L. Meng, Y.Y. Xue *Nature* **365** (1993) 323

[3] O. Chmaissem, J.D. Jorgensen, K. Yamaura, Z. Hiroi, M. Takano, J. Shimoyama, K. Kishio *Phys. Rev. B* **53** (1995) 14647

[4] K. Yamaura, J. Shimoyama, S. Hahakura, Z. Hiroi, M. Takano, K. Kishio *Physica C* **247** (1995) 351

[5] C. Wolters, K.M Amm, Y.R. Sun, J. Schwartz *Physica C* **267** (1996) 164

Inst. Phys. Conf. Ser. No 158
Paper presented at Applied Superconductivity, The Netherlands, 30 June–3 July 1997

On the mechanism of (Bi,Pb)-2223 phase formation

V A Murashov[a,b], M Ionescu [a], G E Murashova[a], S X Dou[a], H K Liu[a] and M Apperley[c]

[a]Centre of Superconducting and Electronic Materials, University of Wollongong, NSW 2522, Australia
[b]Moscow Institute of Radio Engineering, Electronics and Automation, pr. Vernadskogo 78, 117454, Moscow, Russia
[c]MM Cables, 1 Heathcote Rd, Liverpool, NSW 2170, Australia

Abstract. Some aspects of the Bi-2212→Bi-2223 transformation within the Bi_2O_3-SrO-CaO-CuO system were investigated using a quasi solid state reaction. A 2212 crystal self melt grown, having the approximate dimensions (2x2x1.5)mm^3 was sandwiched between two pellets of powder precursor with nominal composition $Bi_{1.85}Pb_{0.35}Sr_{1.95}Ca_{2.05}Cu_{3.05}O_x$ and sintered in air in the temperature range 855 870°C, for up to 500h, with quenching after each 100h of reaction. The phase transformation kinetics were monitored using optical and electronic microscopy on the surface of 2212 crystal, corresponding to the (a, b) crystallographic plane. After 100h, a small amount of liquid was observed, and the nucleation of (Ca, Sr)-cuprates was found. After 200h, the crystal surface was further etched preferentiall along the a and b axes, with the formation of the 2223 phase under the nuclei of the (Ca, Sr)-cuprates.

1. Introduction

All aspects of the 2212→2223 transformation are very important in the Bi-2223/Ag tape manufacturing process. According to the phase diagram [1], above 850°C the Bi_2O_3-SrO-CaO-CuO system contains a liquid component. Its composition and quantity depends strongly on the position, relative to the surface of the 2223 homogeneity volume, where the transformation of secondary phases into 2223 can occur. The presence of a high-temperature liquid phase can strongly influence the kinetic of the conversion reaction, the type and amount of secondary phases present in the tape and the size of 2223 crystals. Information regarding the nature of the kinetics of the interaction between liquid phase and 2212 phase is absent in the literature at the present time. This work reports on the study of the interaction between the liquid phase of a specific composition and the 2212 phase.

2. Experimental

Bi-2212 crystals were grown using self flux in an ZrO_2 crucible, in a vertical tube furnace with an axial temperature gradient of approximately ~60°C/cm. The load was heated in the temperature range 700-980°C, and cooled at a rate of 3.5°C/h. The typical size of the crystals obtained in this way was 6x6 mm^2 in the (a,b) plane, and 0.10-0.15 mm along the c axes.

A Bi-2212 probe crystal, grown as described above, having a size of 2x2x0.15 mm^3 was pressed between two pellets formed out of Bi-2223 precursor powder, with nominal composition $Bi_{1.85}Pb_{0.35}Sr_{1.95}Ca_{2.05}Cu_{3.05}O_x$ and a phase assemblage adequate to the formation of 2223 phase, in which 2212 is the predominant phase. The pellets were sintered in air, in the temperature range of 855-870°C for up to 500h with periodic air quenching after every 100h of sintering. Above this temperature, the 2212 crystal starts to melt and on the other hand the amount of liquid phase generated by the powder precursor is high, hampering the experiment. The kinetic of 2212→2223 phase transformation was monitored by optical and SEM-EDS microscopy on the broad face of the 2212 crystal , which corresponded to the (a,b) crystallographic plane.

3. Results and discussion

The interaction between the liquid phase and the (a,b) plane of the 2212 probe crystal was visible at all temperatures within 855-870°C range, but was more pronounced at 870°C. At this temperature, the small amount of liquid in the system proved to be a very important transformation parameter. The entire surface of the crystal was covered by a liquid film after the first 100 h, and nuclei of secondary phases were formed, having an approximate diameter of 0.3μm.

After 200 h, the size of the secondary phase crystals increased to between 1-3μm, displaying an isomorphic growth habit. The EDS analyses of these crystals revealed a composition rich in Ca and Sr and poor in Bi, identified as (Ca, Sr)-cuprates.

The surface of the 2212 probe crystal showed tracks resembling chemical etching, directioned along the "a" and "b" crystallographic axes. The etch pits originated under the secondary phase crystals, where their depth reached a maximal. The solute was transferred to the areas adjacent to the etch pits, where the composition already corresponded to the 2223 phase.

After 300 h, the surface of the 2212 crystal (Figure 1) typically looked like that seen in an epitaxial Bi-2223 film, but grown from a large number of nucleation centres.

After 400 h, the surface transformation of the 2212 crystal was complete, and a further reaction time only resulted in an increase of the thickness of the 2223 film. The number of secondary phases precipitated on the surface of the probe crystal decreased and their relative size increased. It was noted that the kinetics of the liquid-assisted interaction described above was different for certain orientations of the (a,b) plane of the 2212 probe crystal, relative to its surface. When the (a,b) plane had an orientation other than perpendicular to the normal of the crystal surface, where the conversion reaction takes place, the size of the surface nucleated secondary phases were much smaller. This is illustratedin the right hand side of Fig. 1, for which the surface of the crystal did not correspond with the (a,b) plane, and where the boundary between two crystallographic orientations can be seen. In this area, the size of the (Sr, Ca)-cuprate crystals was small, but their number was high. Where the conversion reaction 2212→2223 was limited, no etching pattern of the 2212 crystal surface was possible to identify.

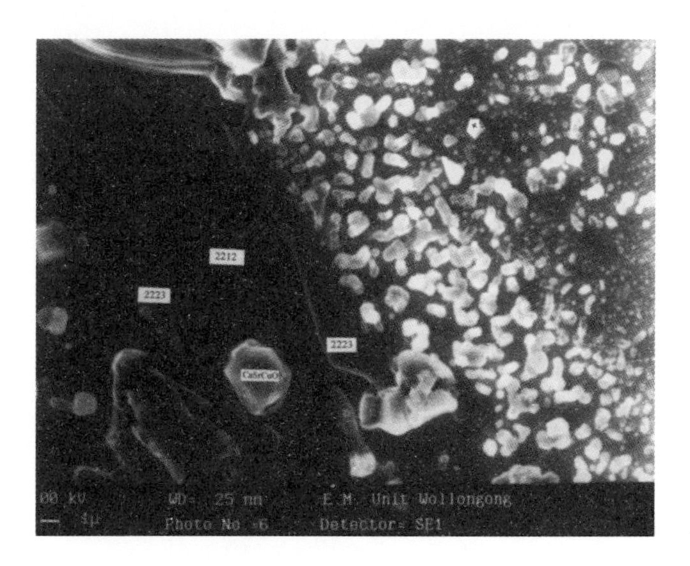

Figure 1: The etching pattern and the (Sr, Ca)-cuprate crystals on the surface of (a,b) plane of a 2212 crystal after reaction at 870°C for 300h

4. Conclusion

The peritectic liquid phase, present in the Bi_2O_3-SrO-CaO-CuO system, having a nomin al composition of $Bi_{1.85}Pb_{0.35}Sr_{1.95}Ca_{2.05}Cu_{3.05}O_x$, reacted with 2212 crystals via (Ca, Sr)-cuprates to form the 2223 phase. In air, the reaction rate reaches a maximum at 870°C. Chemical etching patterns and nuclei of (Ca, Sr)-cuprates were observed on the surface of the 2212 crystals, which were thought to promote the 2212→2223 transformation reaction. Under these conditions, the formation of the 2223 phase takes place at the interface between 2212 and (Ca, Sr)-cuprate crystals through a liquid-assisted film growth mechanism. The size of (Ca, Sr)-cuprate crystals was found to be an important factor in the 2212→2223 transformation process. The speed of the transformation reaction was found to be different along different crystallographic directions in the 2212 crystal.

Acknowledgments

The authors are gratefull for their financial support to Metal Manufactures Ltd, the Energy Research and Development Corporation, the Department of Industry, Science and Tourism and Russian National Program of High Tc Superconductivity.

References

[1] Majewcki P, Kaesche S, Aldinger F 1996 Adv. Mater. **8** n 9 762-5

Inst. Phys. Conf. Ser. No 158
Paper presented at Applied Superconductivity, The Netherlands, 30 June–3 July 1997

Formation of the $Bi_2Sr_2Ca_2Cu_3O_{10}$ phase with and without Pb substitution

J.-C. Grivel, G. Grasso, A. Erb and R. Flükiger

University of Geneva, dpt. of condensed matter physics, 24 quai Ernest Ansermet, CH-1211 Genève 4, Switzerland

Abstract. The $(Bi,Pb)_2Sr_2Ca_2Cu_3O_{10+\delta}$ phase has been synthesised both in the Pb-doped and the Pb-free systems using calcined oxide and carbonate mixtures as starting reagents. X-ray powder diffraction and electrical resistivity measurements were used to characterise the final products. The Pb-free samples show a lower degree of crystalline perfection. Zero resistance is achieved in a single step in Pb-doped specimens, whereas Pb-free samples exhibit more complicated transitions although they appear as nearly single Bi(2223) phase in the X-ray diffraction patterns.

1. Introduction

Owing to the large potential of the $(Bi,Pb)_2Sr_2Ca_2Cu_3O_{10+\delta}$ (Bi(2223)) phase for a possible use in superconducting devices operated at liquid nitrogen temperature, much effort has been devoted to the study of the synthesis and formation mechanism of this compound. In the case of the Pb-doped Bi(2223) phase, numerous reports describing the reactions leading to the formation of high purity Bi(2223) ceramics have been published [1]. On the contrary, the Pb-free Bi(2223) compound has been much less studied, owing to the great difficulties encountered in the achievement of samples with large amounts of the Bi(2223) phase without Pb substitution. Some papers report on the successful synthesis of the Pb-free Bi(2223) phase either from non-stoechiometric nominal compositions [2,3] or from mixtures of selected phases [4]. In this contribution, we report on the synthesis of Pb-free Bi(2223) ceramic samples with a nominal composition close to the ideal $Bi_2Sr_2Ca_2Cu_3O_{10+\delta}$ stoechiometry, by use of precursors consisting of calcined oxide and carbonate powders. The quality of the final product is compared with that of Pb-doped samples.

2. Experimental details

The starting reagents for the sample synthesis were high purity (99.999%) Bi_2O_3, PbO, $SrCO_3$, $CaCO_3$ and CuO powders that were mixed in appropriate proportions in a planetary ball mill with isopropanol. The nominal compositions were $Bi_{1.72}Pb_{0.34}Sr_{1.83}Ca_{1.97}Cu_{3.13}O_{10+\delta}$ for the Pb-doped phase and $Bi_{2.06}Sr_{1.83}Ca_{1.97}Cu_{3.13}O_{10+\delta}$ in the Pb-free case. After milling, the isopropanol was evaporated by heating under constant stirring. The powder was calcined first at 750°C for 50h, then pelletised under a pressure of 100 MPa, calcined at 820°C for 50h for the Pb-doped and the Pb-free samples and finally at 860°C for 50h with an intermediate grinding and re-pelletising for the Pb-free precursor powders only. 1.0 g of precursor powders were formed into flat pellets under a pressure of 0.3 GPa. For the thermal treatments, the pellets were located in an Al_2O_3 crucible with a covering plate of the same material. The Pb-doped pellets were sintered in air at 849°C for 160h. The Pb-free samples were repeatedly heat treated in air at 870°C, with intermediate grinding and re-pelletising after 22h, 61h, 100h and 161h total sintering time. The pellets were placed on a $BaZrO_3$ substrate to avoid a reaction with the alumina crucible.

X-ray diffraction (XRD) patterns were recorded on powdered samples in a θ-2θ diffractometer using Ni-filtered Cu Kα radiation. XRD peak profiles were fitted to a pseudo-Voigt function and the effective domain size (D_{eff}) was evaluated as described elsewhere [1].

3. Results and discussion

Typical XRD patterns of Pb-doped and Pb-free Bi(2223) samples are shown in figure 1. A striking feature is the difference in the shape of the reflections. Without Pb, the Bi(2223) peaks of (*hkl*) indices with $l \neq 0$ are significantly larger than in the Pb-doped phase. Table 1 contains the values of the mean coherent domain size (D_{eff}) evaluated from the peak shape for various reflections. It appears that in general, the D_{eff} values are significantly larger in the Pb-doped Bi(2223) phase.

The (200) reflection however yields similar values for both kind of samples. The origin of this phenomenon may be found in the nature of the stacking faults present in the Bi(2223) crystallites. As observed by TEM, stacking faults in this phase are mainly located along the (00*l*) direction. In the case of the (*hk0*) reflections, the X-ray beam is diffracted within the a-b planes and narrower peaks result from the longer range order present inside the Bi(2223) planar layers.

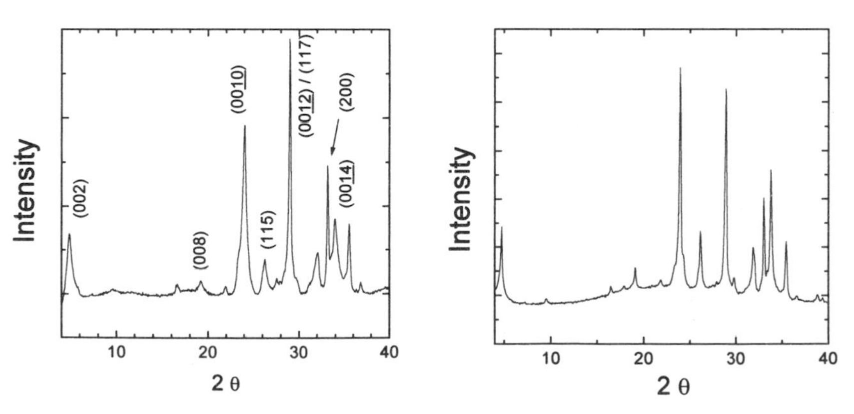

Figure 1 : XRD patterns of Pb-free (left) and a Pb-doped (right) Bi(2223) phase samples.

Table 1 : D_{eff} values for some reflections from the Pb-doped and Pb-free Bi(2223) phase

reflection :	*(0010)*	*(115)*	*(0014)*	*(200)*
Pb-doped :	60 nm	45 nm	35 nm	≈400nm
Pb-free :	25 nm	13 nm	8 nm	≈400 nm

The lesser degree of crystalline perfection achieved in the Pb-free Bi(2223) phase is likely to play a role in the transport properties of ceramic samples. Figure 2 depicts resistive measurements performed on Pb-doped and Pb-free Bi(2223) ceramics. For the Pb doped sample, a single sharp transition is observed, with a $T_{c(R=0)}$ of 106K. In the Pb-free case however, two resistive drops appear around 110K and 80K respectively. Resistance only vanishes after a tail, which extends to lower temperatures as the number and length of heat treatments increases.

A detail of the corresponding XRD patterns is shown in figure 3 in order to illustrate the development of the Bi(2223) phase. Although after 22h sintering, the amount of the Bi(2223) compound is still quite small, the resistive drop around 110K is evident. Further grinding and sintering steps result in the formation of a large proportion of the Bi(2223) phase (≥80% of sample volume). However, this does not result in the achievement of a single resistive transition, as is the case for Pb doped samples. This phenomenon may result from the low crystalline perfection of the Bi(2223) phase. Low quality grain boundaries main also play a role in this effect.

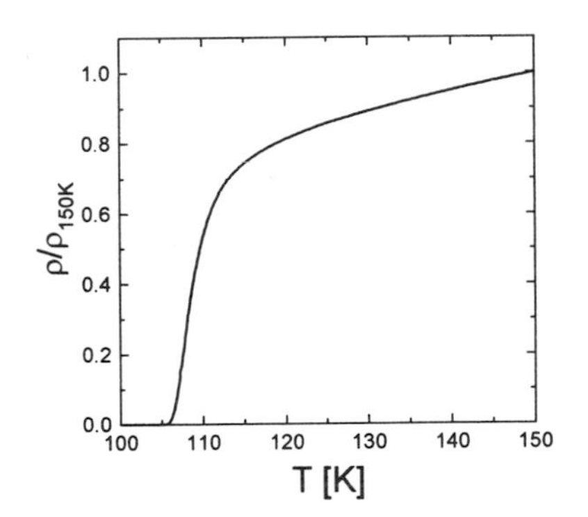

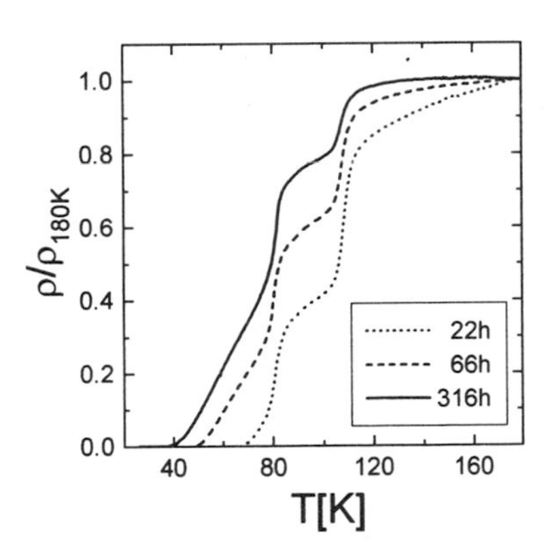

Figure 2: Temperature dependence of the electrical resistivity measured on a Pb-doped ceramic after 160h sintering at 849°C (left) and Pb-free samples sintered for various total durations at 870°C.

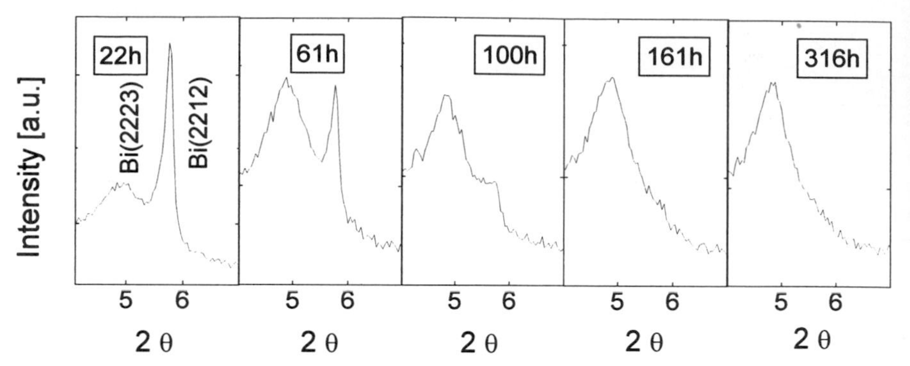

Figure 3: detail of XRD patterns recorded on Pb-free samples, showing the evolution of the Bi(2212) and Bi(2223) (*002*) reflections as a function of heat treatment time. (sintering at 870°C). The samples were ground and pelletised between each heat treatment.

4. Conclusions

Starting from classical precursors, we have succeeded in forming nearly single phase Bi(2223) ceramics with and without Pb substitution. The XRD patterns recorded on the powdered samples show that the Pb-doped Bi(2223) phase tends to present a longer crystalline coherence along the direction perpendicular to the basal plane of the structure than the Pb-free phase. The poor crystallinity of the Pb-free Bi(2223) phase may play a role in the uncomplete resistive transitions of the ceramic samples.

Acknowledgements

This work was supported by the Swiss National Foundation (PNR 30), the Priority Program « Materials Research and Engineering » (PPM) and the Brite Euram II Project no BRE2 CT92 0229 (OFES BR060).

References

[1] Grivel and Flükiger 1996 *Supercond. Sci. Technol.* **9** 555-564 (and references therein)

[2] Herkert W, Neumueller H-W and Wilhelm M 1989 *Solid State Commun.* **69** 183

[3] Majewski P, Hettich B, Schultze K and Petzow G 1991 *Adv. Materials* **3** 488

[4] Sastry P V P S S and West A R 1994 *J. Mater. Chem.* **4** 647

Inst. Phys. Conf. Ser. No 158
Paper presented at Applied Superconductivity, The Netherlands, 30 June–3 July 1997

Identification of High Pressure Phase in Bi-Sr-Ca-Cu-O Superconductors by HRTEM and XRD

Y. H. Li and J. L. MacManus-Driscoll

Department of Materials, Imperial College, London SW7 2BP, UK.

Abstract. The composition and structure of the main decomposition product of Bi-2212 after annealing under high oxygen pressures was determined by WDS, TEM and XRD analyses. The high pressure decomposition reaction for Bi-2212 can be written as:
$Bi_{2.1}Sr_{1.95}Ca_{0.95}Cu_2O_{8+\delta} + O_2$ --> $(0.692/\alpha)(Bi_{2\alpha}Sr_{2.48\alpha}Ca_{1.19\alpha}O_y) + (1.968)CuO +$
$(0.323/\beta(Bi_{2\beta}Sr_{0.53\beta}Ca_{0.39\beta}O_x) + (0.0064)Bi_{11}Sr_9Cu_5O_y$
The highly oxidised phase $Bi_2Sr_{2.48}Ca_{1.19}O_y$ has the same composition as a less oxidised phase whose solid solution region was found to be centred on the Bi:(Sr+Ca) 1:2 or (Bi+Cu):(Sr+Ca) 9:(11+5) positions. The average atomic positions of the high pressure phase have been determined by high resolution images and image simulations along the [110] direction with different sample thickness. In addition, experimental XRD has been compared with calculated XRD data to elucidate the structure of the phase.

1. Introduction

Among the various high Tc oxide systems discovered, the Bi-Sr-Ca-Cu-O system has been investigated extensively due to its potential for practical applications in the form of wires and tapes [1-3]. Many of the studies reported in the literature are aimed at understanding the reaction path-ways for formation of the individual phases and optimising the process parameters for the synthesis of phase-pure materials with the correct microstructure [4-6]. However, the understanding is incomplete and requires further detailed studies of the thermodynamics of these compounds.

2. Experimental

Commercial phase pure Bi-2212 powder was used for the annealing study. The powder was annealed in 60 bar O_2 at 800 °C for 12 hrs followed by rapid cooling to room temperature. Subsequently the decomposed products were characterised using several techniques. The upper stability line of Bi-2212 was determined by reforming Bi-2212 from the decomposed products using a coulometric titrate technique, as reported previously [7].

In order to determine the solid solution range of the alkaline earth bismuthate 2:3:1 (~ 9:11:5) which has been reported to be a major decomposed product of Bi-2212 under oxidising conditions, several samples of composition near 2:3:1 and 9:11:5 with small fractions of Cu were made by solid state reaction in air at 800 °C. The compositions of several points in each phase mixture were measured by Wavelength Dispersive Spectrometry (WDS).

3. Results and Discussion

The upper stability line for Bi-2212 determined by coulometric titration is shown in Fig.1. The kinetics of Bi-2212 decomposition over the temperature range studied were very slow; e.g., at 700°C, after a titration of a few coulombs of charge into the system, oxygen equilibration took on the order of 2 days.

The X-ray diffraction pattern of the Bi-2212 sample annealed under high oxygen pressure shows a major phase, CuO, and an unknown phase. Three phases were found in the decomposed sample from microprobe analysis: Bi:(Sr+Ca) = 1:2, Bi:(Sr+Ca) = 2:1 and CuO. Using the average phase composition of several measured points, the high pressure decomposition reaction for Bi-2212 can be written as:

$Bi_{2.1}Sr_{1.95}Ca_{0.95}Cu_2O_{8+\delta} + O_2 \dashrightarrow (0.692/\alpha)(Bi_{2\alpha}Sr_{2.48\alpha}Ca_{1.19\alpha}O_y) + (1.968)CuO +$
$(0.323/\beta(Bi_{2\beta}Sr_{0.53\beta}Ca_{0.39\beta}O_x) + (0.0064)Bi_{11}Sr_9Cu_5O_y$

The XRD diffractogram of the 2:3:1 (or 9:11:5) phase formed in air corresponds to the (monoclinic P_c) 2:3:1 phase. The compositions of the highly oxidised phase both made by decomposition of Bi-2212 (open circles), and of the phase formed for compositions near 2:3:1 (or 9:11:5) in air (closed circles) at 800 °C are shown in the $1/2Bi_2O_3$-SrO-CaO pseudo-ternary of Fig.2. For the high pressure phase the compositions are also close to 9:11:5 but are slightly more Bi rich than the phase which was formed in air. If the small Cu concentration found in both the higher and lower phases is taken into account and Cu is assumed to be in the Sr or Ca position, the Bi:(Sr+Ca+Cu) ratio is closer to 1:2. On the other hand, if Cu is presumed to substitute on the Bi site, the (Bi+Cu):(Sr+Ca) ratio moves toward 1:1.78 (9:11+5). This is possible in the oxidised phase since Bi^{5+} ions are likely to be stable and the ionic radius of Bi^{5+}, 0.074nm, is close to Cu^{2+}, 0.072nm. The solid solution range of the phase formed in air is close to 9:11:5 but extends towards the 2:3:1 composition.

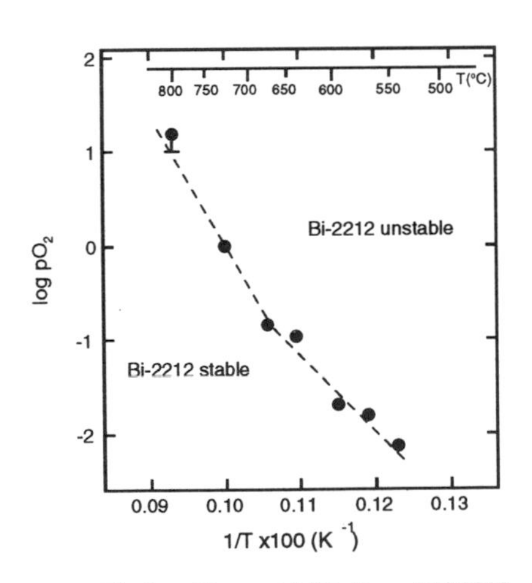

Fig.1. Upper stability line of Bi-2212.

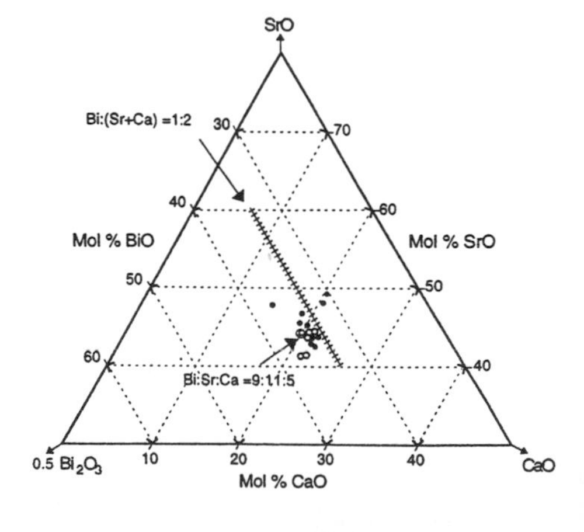

Fig.2. Compositions plotted on a ternary triangle.

In the TEM study of the high pressure phase, a series of electron beam diffraction patterns obtained by tilting around c*-axis of the structure confirm that it is orthorhombic as shown in Fig. 3, which shows the three diffraction patterns obtained when tilting around c*-axis which were identified as belonging to [010], [120] and [110].

Many possible structures were built based on that it is a perovskite structure with Sr ions occupying A-sites and Bi and Ca ions occupying B-sites. The formula for this structure is $Sr_3(CaBi_2)O_9$. The frame of the structure is shown in Fig. 4. Since only Sr ions occupy the A-sites (indicated by 'A'), we suppose that they occupy all the A-sites equally. As both Bi and Ca ions occupy B-sites, the B-sites were indicated by B_1, B_2, B_3 and B_4 respectively to identify their different positions. There are possibilities for Bi and Ca ions to occupy different B-sites. In addition, Ca and Sr can inter-substitute. Since in the diffraction pattern of Fig.3, every other spot along c*-axis is quite weak, this suggests that the atomic plane at the B_1 and B_2 positions should be not very different from the atomic plane at the B_3 and B_4 positions. Therefore, we supposed, at first, that Bi ions take up B_1 and B_4 positions and Ca ions take up B_2 and B_3 positions. In order to keep an overall Bi:Sr:Ca ratio of 2:3:1, only Bi positions are fully occupied. The occupancy for Sr is 0.75 and for Ca is 0.5. The XRD intensities were calculated based on this structure using 'Lazy' and compared with the experimental XRD data. Since the oxygen occupancy causes little change in the XRD intensities due to its comparatively low scattering factor relative to those of the other elements present, no attempt was made to determine the occupancy of the oxygen positions and so it was taken to be unity.

The experimental data shows that the principal XRD peaks of the phase occur at 2θ values (relative intensities) of 29.9 (100), 43 (36.4), 29.6 (33.6) and 53.8 (31.1). However, even though the calculated intensities based on the above structure gave similar intensities at the same positions as above, i.e. 29.9 (100), 43.0 (39.3), 29.6 (25.9) and 53.8 (28.9), it gave two strong peaks at 2θ values of 18 (53.8) and 18.3 (51.8), while the experimental data showed only two small peaks at the same positions, i.e. 18 (7.6) and 18.3 (6.3). These two peaks correspond to (011) and (101) reflections. In the (011) and (101) planes in the structure are the planes which are alternatively occupied by either Bi ions or Ca ions. Since all Bi positions are fully occupied and all Ca positions are half empty, it is possible to move some Bi ions to the Ca positions, giving an intensity reduction of the (011) and (101) peaks. Therefore, the occupancy of the Bi positions was reduced from 1 to 0.75 and some Bi ions were added on the Ca positions to give an occupancy of 0.25. The calculated XRD intensities based on this structure gave two small peaks at 2θ values of 18 (9.6) and 18.3 (9.3) which correspond to (011) and (101) peaks. The major peaks calculated based on this new structure occur at 2θ values of 29.9 (100), 43 (37), 29.6 (25.9) and 53.8 (28.9), which is a good match to the experimental data shown above bearing in mind that we did not take into account of oxygen vacancies and associated ion position shifts and possible peak overlap with the $Bi_2(Sr,Ca)_1O_y$ phase present in the decomposed phase structure.

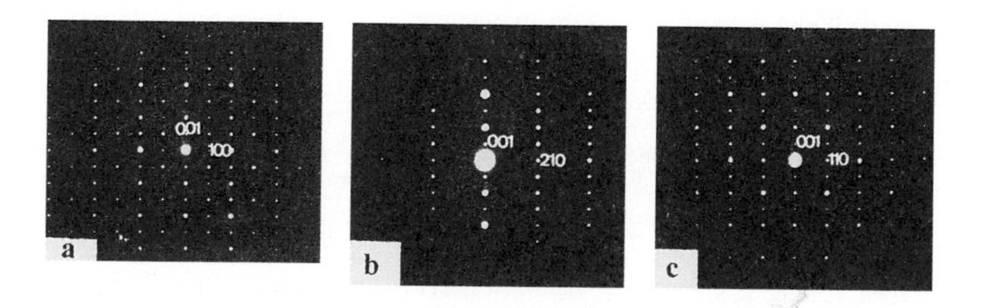

Fig.3. Three diffraction patterns obtained when tilting around c*-axis.

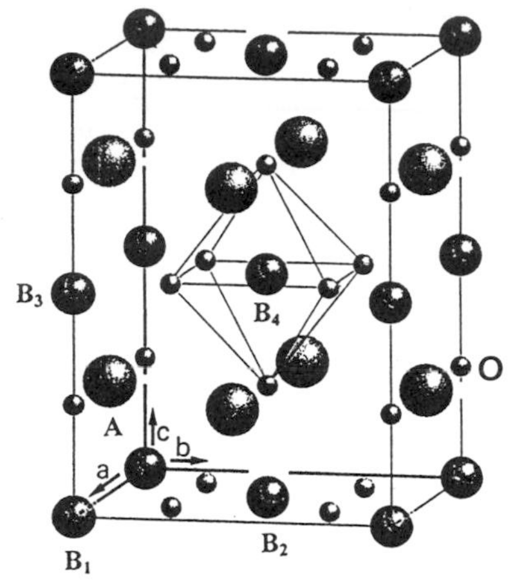

Fig.4. The frame of the proposed high pressure phase structure.

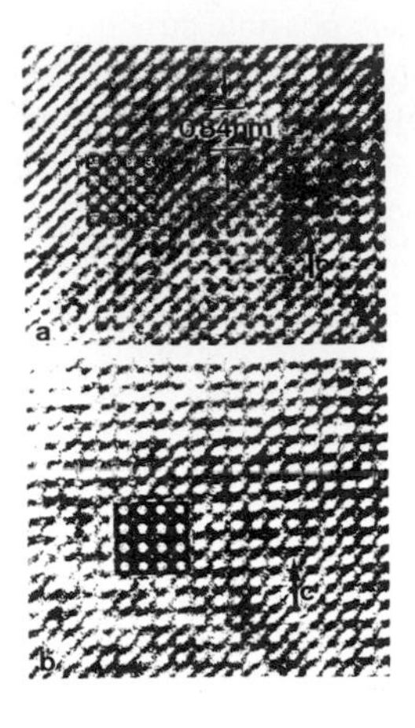

Fig.5. High-resolution pictures and the Simulated images (inserts).

High-resolution TEM pictures of the high pressure phase along [110] direction with different sample thickness are shown in Fig.5. The defocus for the all image simulations is 70nm and the sample thicknesses are 1.3nm and 3nm respectively in Fig.5a and Fig.5b. It can be seen that the simulated images match the experimental ones well.

4. Conclusion

From an average of several WDS analysis points, the high pressure decomposition reaction for Bi-2212 can be written as:

$Bi_{2.1}Sr_{1.95}Ca_{0.95}Cu_2O_{8+\delta} + O_2 \dashrightarrow (0.692/\alpha)(Bi_{2\alpha}Sr_{2.48\alpha}Ca_{1.19\alpha}O_y) + (1.968)CuO +$
$(0.323/\beta(Bi_{2\beta}Sr_{0.53\beta}Ca_{0.39\beta}O_x) + (0.0064)Bi_{11}Sr_9Cu_5O_y$

The highly oxidised phase $Bi_2Sr_{2.48}Ca_{1.19}O_y$ has the same composition as the less oxidised phase whose solid solution region was found to be centred on the Bi:(Sr+Ca) 1:2 or (Bi+Cu):(Sr+Ca) 9:(11+5) positions. The average atomic positions of the high pressure phase have been determined by high-resolution images and image simulations along the [110] direction with different sample thickness. In addition, experimental XRD has been compared with calculated XRD data for possible structures of the phase. The average structure was confirmed to be a perovskite-like structure with Sr occupying A-sites and Bi and Ca occupying B-sites. The average occupancy of Bi and Sr positions is 0.75 and other positions were occupied by both Ca and Bi ions with an occupancy of 0.5 for Ca and 0.25 for Bi.

References

[1]. M. Ueyama, T. Hikata, T. Kato and K. Sato, Jpn. *J. Appl. Phys.* 30 (1991) L1384.
[2]. Q. Li, K. Brodersen, H. A. Hjuler and T. Freltoft, *Physica C* 217 (1993) 360.
[3]. D. C. Larbalestier, X. Y. Cai, Y. Feng, H. Edelman, A. Umezawa, G. N. Riley and W. L. Carter, *Physica C* 221 (1994) 299.
[4]. M. R. De Guire, N. P. Bansal, D. E. Ferrel, V. Finan, C. J. Kim, B. J. Hills and C. J. Allen, *Physica C* 179 (1991) 333.
[5]. Y. E. High, Y. Feng, Y. S. Sung, E. E. Hellstrom and D. C. Larbalestier, *Physica C* 220 (1994) 81.
[6]. W. Lo and B. A. Glowacki, *Physica C* 193 (1992) 253.
[7]. J. L. MacManus-Driscoll, Y. H. Li and Z. Yi, *J. Am. Ceram. Soc.* 80 (1997) 807.

Inst. Phys. Conf. Ser. No 158
Paper presented at Applied Superconductivity, The Netherlands, 30 June–3 July 1997

Electrical characteristics of BSCCO-2223 kA class conductors in the temperature range 77-110K

S.P.Ashworth, M.P.Chudzik(+), B.A.Glowacki(*), M.T.Lanagan(+)

IRC in Superconductivity, Cambridge University, Cambridge, UK, CB3 OHE
(+) Argonne National Laboratory, Argonne, Ill. USA
(*) also at Department of Materials Science and Metallurgy, Cambridge University

Abstract: Monoliths capable of carrying very high currents (up to 1kA at liquid nitrogen temperatures) have been produced by cold pressing BSCCO-2223 powders. These conductors have rectangular cross sections of order 1cm^2, are up to 10cm long and have low resistance current injection contacts. In this paper we present data on the transport critical current and V-I characteristics of these conductors as a function of temperature in liquid nitrogen between 77K and 110K and as a function of magnetic field to 0.8T. These measurements utilise transport currents approaching 1kA and consequently require liquid cooling at all temperatures, this is achieved using a pressurised liquid nitrogen cryostat. The critical current dependence on temperature and V-I characteristics are shown to be very different to those of, for example, BSCCO-2223 OPIT tapes. The effect of the production processes of these two conductor forms and the subsequent microstructure are discussed and related to the different electrical characteristics.

1.Introduction

There is considerable interest in the use of hts as for example current leads into superconducting magnets. In this application the conductor can be quite short (<15cm perhaps), but needs to carry high currents (over 100A). The current lead, whilst not usually exposed to the full magnetic field of the magnet, will none the less experience a 'stray' magnetic field, which could be considerable. One particularly attractive form of conductor for this application are the cold pressed BSCCO-2223 bars considered in this paper. These bars do not have any metal cladding or support so the thermal conductivity is low, the critical temperature is over 100K consequently cooling and stability problems are somewhat alleviated. In application as a current lead from for example the 77==K region to the 4.2K region of a magnet the usefulness of the material will be determined by its behaviour at the high temperature end, in particular its response to magnetic fields.

2. Experiment

These measurements utilise a pressurised liquid nitrogen cryostat which enables the liquid nitrogen bath temperature to be stabilised in the range 75-124K with an accuracy of±0.05K [1]. This is accomplished by controlling the nitrogen pressure using a PID controller and pneumatic valves. In these measurements sufficient heat leak is allowed in the system to ensure that the liquid nitrogen is always boiling (it is also possible to conduct measurements in sub-cooled liquid). In this boiling liquid the heat transfer from the superconductor and current injection contacts is highly efficient. An 800A current generating 100μV/cm (i.e. above the critical current) along the superconductor produces a surface heat flux of 40mW/cm^2 to be removed by the liquid coolant to avoid sample heating. The temperature rise for such a heat flux will be

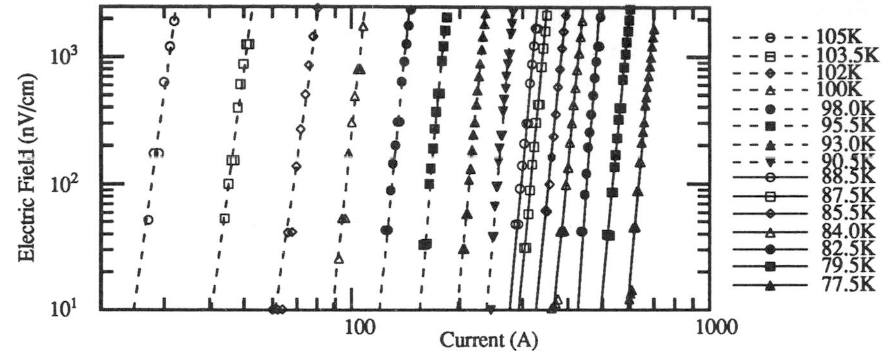

Figure 1: Electric field generated by a transport current along a BSCCO-2223 bar at various temperatures

below 0.05K in boiling liquid nitrogen. It is to be expected that heat fluxes of up to 1W/cm^2 are feasible before the onset of film boiling. A previously calibrated Pt resistance thermometer was used to accurately measure temperatures, and to ensure that the liquid nitrogen was indeed at the saturation temperature. The current was swept from zero to approximately 120% of the critical current over a period of 300seconds and the voltages recorded. Magnetic fields were applied perpendicular to the face of the bar by an electromagnet.

The BSCCO-2223 bars used in these experiments were produced by cold pressing of the superconductor powder and are described more fully in [2]. The bars are up to 10cm long and typically 1cm x 5mm in cross section. Low resistance current contacts are achieved by sputtered silver and then low melting point solder to copper braid. This production technique does not introduce an overall grain alignment. The BSSCO-2223 bar data are compared with high quality PIT BSCCO-2223 multifilamentary tape produced by Intermagnetics General Corporation.

3.Results and Discussion

In Figure 1 we show typical electric field-current data on linear scales for a BSCCO-2223 bar at temperatures in the range 77-102K and zero applied field. The critical current of this sample at 77K, defined using the 1μV/cm criterion, was approximately 700A. This corresponds to a critical current density of 1500A/cm^2 at 77K. It is interesting to note that at the critical current the magnetic field at the surface of the sample is over 30mT. From data such as this we can obtain the critical current variation with temperature, as shown in Figure 2, which also shows the effect of small magnetic fields on the critical current density, these are field cooled measurements. The critical current becomes negligible at a temperature of 103K in zero field, well below the maximum T_c of the BSCCO-2223 superconductor. The processing of these conductors has been optimised for maximum critical current at 77K, it is apparent that this is not simply a case of maximising the T_c as might be the case in a less granular sample. Although the zero field data seems to be quite linear with temperature, there is apparently some slight curvature. The 13mT and 25mT applied field data emphasises and confirms the presence of this curvature.

Of more technological interest is the behaviour of these bars in zero field cooled situation, as might be experienced in the 'stray' field of a magnet. Figure 3 shows the reduction in critical current density at 77.4K as a field (applied perpendicular to the face of the sample) is increased to 0.8T and subsequently decreased. In an application were the bars act as current leads between the 4.2K and 77K regions of a magnet, the behaviour at 77K will limit the transported current. In the inset to Figure 3 the same data is compared to the behaviour of a

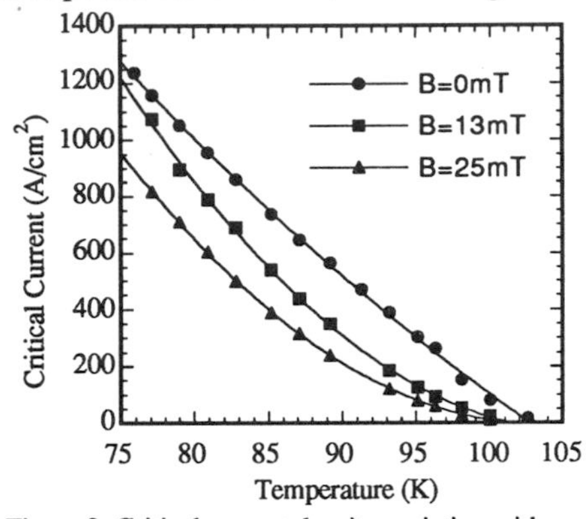

Figure 2: Critical current density variation with temperature for a BSCCO-2223 bar. Data is shown for zero applied field and for two applied field cooled values.

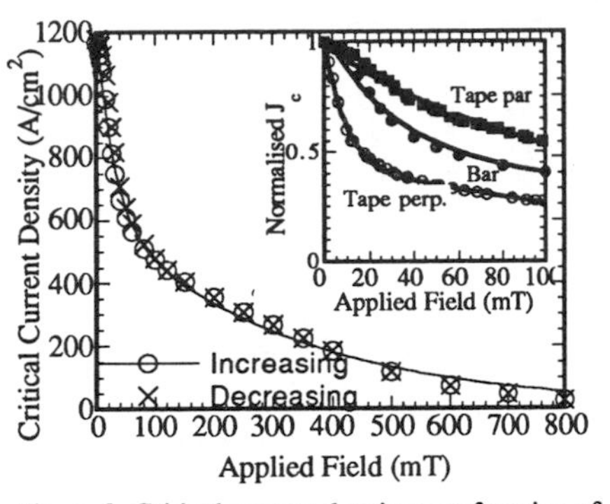

Figure 3: Critical current density as a function of applied field (zero field cooled) for a BSCCO-2223 bar at 77.4K. Inset compares the behaviour of tape and bar. The lines are double exponential fit (see text).

PIT tape, these conductors typically show a very rapid decrease in J_C with fields applied perpendicular to the tape face and significantly slower decrease if the field is parallel to the tape face. The behaviour of the bar for field perpendicular to the face is intermediate between the two tape curves. This can be understood in terms of the microstructure of the two conductors, in high quality tape considerable effort is directed toward producing a high degree of alignment of the BSCCO-2223 material within the tape. On the other hand the BSCCO bars have no significant alignment of the material within the conductor - which, whilst giving a low critical current density compared to the tape, significantly reduces the design problems in applications. There is not such an imperative to ensure that the grains in the bar are oriented parallel to all stray fields.

The reduction in J_C with applied field of BSCCO-2223 PIT tapes can be well described using a double exponential equation;

$$I_c = I_{c1}e^{-H/H_1} + I_{c2}e^{-H/H_2} \quad (1)$$

In this form the two exponential terms are often assumed to refer to weak and strong links between grains. Fitting this equation to the data for the 2223 bar also produces an excellent fit up to 0.8T, yielding values of 327mT and 29mT for the two characteristic fields. These values should be compared with those for the PIT tape data of figure 4, 194mT and 9mT. It is difficult to see how the 'weak' links in the bar have a higher characteristic field than those in the highly aligned PIT tape, this implies that equation (1) should simply be viewed as providing a four parameter fit, with no deeper meaning.

It is interesting to make a more complete comparison of the PIT tape and the pressed bar. The E-J characteristics of conductors are usually very enlightening, in Figure 4 we compare the E-J curves of the two materials at various temperatures. To aid this comparison the current densities have been normalised by the value at 77.4K. In the region 77K-93K the J_C's of the two conductors change in a very similar way (the curves at each temperature cross near to the 1μV/cm line), this behaviour would not continue to temperatures near to T_0. It is also apparent that the gradient of the two sets of curves are very different, the lines on Figure 4 represent a power law fit to the data - the index (n value) for the high quality PIT tape vary between 28 and 30 (decreasing with increasing temperature), whilst the bar varies between 14 and 16. These power law lines are reasonable fits to the data, but closer examination of Figure 4 reveals that there is actually a downward curvature in all these lines. Recall that all this E-J data is taken in zero applied field, we would not expect to see a true power law behaviour in BSCCO-2223 without the application of a magnetic field. Further comparison between tape and

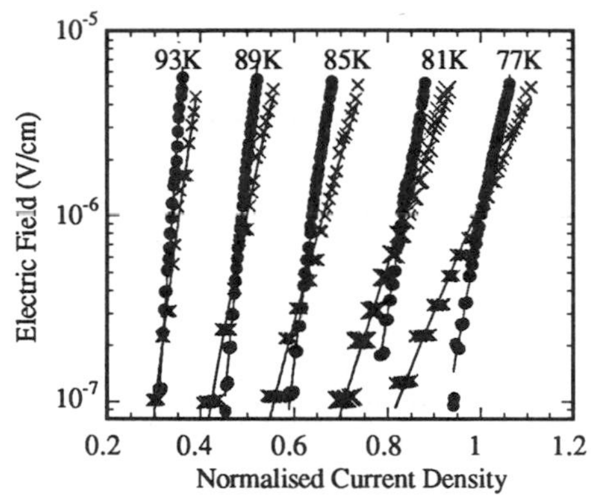

Figure 4: Electric field as a function of normalised current density ($I_c(T)/I_c(77.4K)$) for a BSCCO -2223 bar (crosses) and PIT tape (open circles) for a number of temperatures.

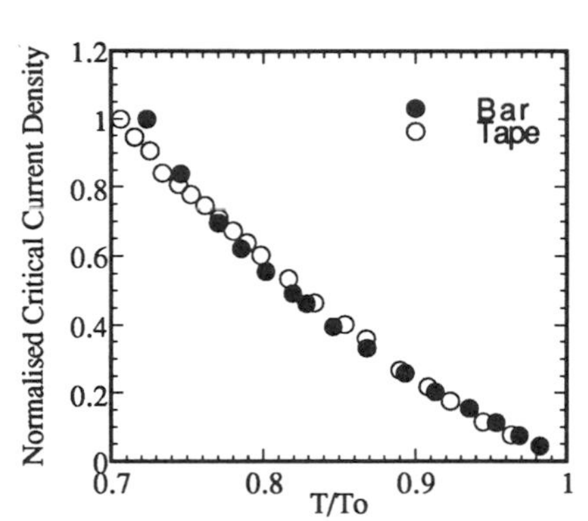

Figure 5: Comparison of critical current densities of BSCCO-2223 bar and PIT tape. T_o is the temperature at which the critical current becomes zero. J_c are normalised to unity at 77.4K

bar is shown in Figure 5 in which we show the reduction in J_C with temperature for both tape and bar, the temperature in this case is normalised by the temperature at which the critical current goes to zero (107K for the bar and 109K for the tape). It is apparent that on these axes the behaviour of the two conductors is almost indistinguishable. Attempts to fit the data of Figure 5 with functions such as $(1-T/T_C)$ or $(1-T/T_C)^2$ are not particularly successful.

3.Conclusions

The cold pressed BSCCO-2223 bars discussed here are shown to have critical currents up to 700A at 77K with complete cryo-stability in liquid nitrogen. No thermal run away occurs up to 150% of the critical current. On exposure to a magnetic field perpendicular to the bar face, the reduction in critical current density is significantly slower than for a BSCCO-2223 PIT tape in a similar configuration, this can be attributed to a lack of grain alignment in the bar. This is an extremely useful property if the bars are to be used as current leads in a magnet. The E-J characteristics of the bars are less steep than for the PIT tape, in a PIT tape this is usually taken as evidence that the processing of the material has not been fully optimised.

Acknowledgements

We would like to thank Intermagnetics General Corporation for supplying BSCCO-2223 OPIT tape.

References

[1] BA Glowacki, EA Robinson, SP Ashworth Cryogenics 37 p173-175 (1997)

[2] MP Chudzik, BJ Pulzin, JJ Picciolo, RL Thayer, BL Fisher, P Kostic IEEE Trans. on Applied Supercond. 6&7 p2102 (1997)

[3] PN Mikheenko, J Horvat, QY Hu, M Ionescu, SX Dou Supercond. Sci. and Tech. 6

Inst. Phys. Conf. Ser. No 158
Paper presented at Applied Superconductivity, The Netherlands, 30 June–3 July 1997

Optimisation of the Precursor Powder for Bi-2223/Ag Tapes

J Jiang and J S Abell

School of Metallurgy and Materials, University of Birmingham, Birmingham, B15 2TT, UK

Abstract. The effects of precursor powder calcination condition and particle size on the properties of the Bi-2223 tapes have been studied. It has been found that both microstructure and critical current density J_c depend strongly on the powder calcination condition and particle size distribution. A small particle size promotes the formation of the Bi-2223 phase and coarse powder leads to $(Ca,Sr)_2CuO_3$ and $(Ca,Sr)_2PbO_4$ remaining in the final tape. Too fine a powder may cause the formation of large alkaline-earth cuprate grains. Different calcination conditions resulted in different phase assemblages in the precursors, which determined the reactivity of the precursor, i.e. the formation rate of the Bi-2223 phase. The J_c tends to increase when the amount and size of non-superconducting phase decrease in the final tapes. It is suggested that for high J_c values a better precursor is a mixture of 2212, CuO and $(Ca,Sr)_2CuO_3$ with appropriate amount of $(Ca,Sr)_2PbO_4$ in the one-powder process.

1. Introduction

The fabrication of high critical current density J_c conductors by the oxide-powder-in-tube technique employing the Bi-2223 phase superconductor has been found to depend on a large number of processing parameters[1-4]. Among these are the precursor powder stoichiometry, phase assemblage, particle size, and thermomechanical processing conditions. Although several kinds of processing techniques have been used to prepare the precursor powder, which include the conventional solid state reaction, co-precipitation, co-decomposition, freeze drying and spray drying, the phase assemblage and particle size depend on the subsequent powder heat treatment[4-5]. In this work, the dependence of J_c and microstructure on the precursor powder calcination conditions and the precursor particle size have been investigated.

2. Experimental

Powders were prepared by the solid state reaction method[6]. The nominal composition of the precursor powder was $Bi_{1.8}Pb_{0.4}Sr_{2.0}Ca_{2.2}Cu_3O_x$. The powder calcination conditions are given in Table 1. Powders SS1, SS2, and SS3 were designed to explore the effect of calcination temperature, while powders SS2, SS4, SS5, and SS6 were designed to explore the effect of calcination period and intermediate grinding. Powders SS-5 and SS-6 were calcined twice with intermediate grindings in an agate motar. The corresponding tapes were designed TS1-6. In order to explore the effects of powder particle size powder SS-2 was ground further. After grinding for 3 hours, 4 hours, and 5 hours, the powder was taken as powder SP1, SP2 and SP3, respectively, and corresponding tapes TP1-3. The particle size distributions of powders were analysed by using a Coulter LS Particle Size Analyser. The powders were processed by the powder-in-tube method with the same processing parameters[6]. The tapes were sintered four times in air at 836°C for 60 hours with intermediate pressing with a pressure of 2GPa. X-ray (Cu-K_{α}) diffraction was used to identify the phases. Microstructure was examined on polished cross sections of the tapes by SEM and EDX analysis.

3. Results and discussion

3.1 Effects of precursor powder calcination

The phase assemblages of the calcined precursor powders based on the XRD and EDX analysis are summarised in Table 1. Increasing the calcination temperature from 810 to 830°C, extending the calcination time from 12 to 48 hours and applying intermediate grinding resulted in a decrease in the relative amount of $(Ca,Sr)_2PbO_4$, and CuO, and an increase in $(Ca,Sr)_2CuO_3$. From Table 1 for TS1, TS2, and TS3, we note that increasing the calcination temperature of the precursor led to the faster formation of the 2223 phase in the tapes. The increased formation rate is believed to be due to the formation of lead-doped Bi-2212 and $(Ca,Sr)_2CuO_3$ phases in the precursor. It seems that a mixture of Ca_2CuO_3, CuO and (Bi,Pb)-2212 may be the best choice for the preparation of the 2223 phase. However, the highest conversion rate during the first sintering does not correspond to the highest critical current density in the final tape. The highest critical current density (J_c=2.9x10^4A/cm^2 and I_c=32A, 77K, self-field) was achieved in tape TS2 after 4 cycles of thermomechanical treatment.

This means that there exist other factors which exert a strong influence over the transport properties. This can be interpreted in terms of the liquid phase formation associated with decomposition of $(Ca,Sr)_2PbO_4$, since microcracks were more often observed in tape TS-3 as shown in Fig1.(c). It is believed that lead released from Ca_2PbO_4 can react with pre-existing CuO to form a eutectic liquid[7]. This liquid can help the diffusion of atoms which are needed for the formation of the 2223 phase, and on the other hand, it can heal cracks resulting from the intermediate mechanical deformation. The powder in TS3 contained more $(Ca,Sr)_2CuO_3$, therefore, correspondingly less Ca_2PbO_4. Thus, the extent of liquid phase formation in TS-3 is reduced, thereby limiting the healing of microcracks in this tape; a lower J_c is subsequently observed. So, an appropriate amount of $(Ca,Sr)_2PbO_4$ in the precursor powder is important for a high J_c value. Fig.1 shows the micrographs of longitudinal cross-sections of the final tapes. The grey matrix was the Bi-2223. The black regions were mainly $(Ca,Sr)_2CuO_3$, and the white were $(Ca,Sr)_2PbO_4$ or Pb-rich 3221phase[8]. For the tape TS2, the amount of the second phases was less and the size smaller (about 1-2μm). Thus, the presence of large impurity phase grains, especially in tapes TS3 and TS6, is another limiting factor for the high J_c value.

Table 1 Calcination conditions and phase assemblages of the precursor powders, and conversion ratios and critical current densities of the corresponding tapes

powder ID	temperature (°C)	time (hour)	phase assemblage	tape ID	fractional conversion to 2223*	J_c** (kA/cm^2)
SS1	810	24	2212,$(Ca,Sr)_2PbO_4$,CuO,some $(Ca,Sr)_2CuO_3$	TS1	42.2%	24
SS2	820	24	2212,$(Ca,Sr)_2PbO_4$,CuO,$(Ca,Sr)_2CuO_3$	TS2	64.5%	29
SS3	830	24	2212,$(Ca,Sr)_2PbO_4$,some CuO, $Ca,Sr)_2CuO_3$	TS3	86.9%	22
SS4	820	12	2212,$(Ca,Sr)_2PbO_4$,CuO,some $(Ca,Sr)_2CuO_3$	TS4	60.8%	26
SS5	820	2x12	2212,$(Ca,Sr)_2PbO_4$,CuO,$(Ca,Sr)_2CuO_3$	TS5	66.7%	20
SS6	820	2x24	2212,$(Ca,Sr)_2PbO_4$,some CuO,$(Ca,Sr)_2CuO_3$	TS6	81.3%	18

*: Fractional conversion from 2212 to 2223 was calculated by using the (008) and (0010) XRD reflection intensities of the 2212 and 2223, respectively, for the tapes after the first sintering at 836°C for 60 hours.
**: Critical current densities of the fully processed tapes

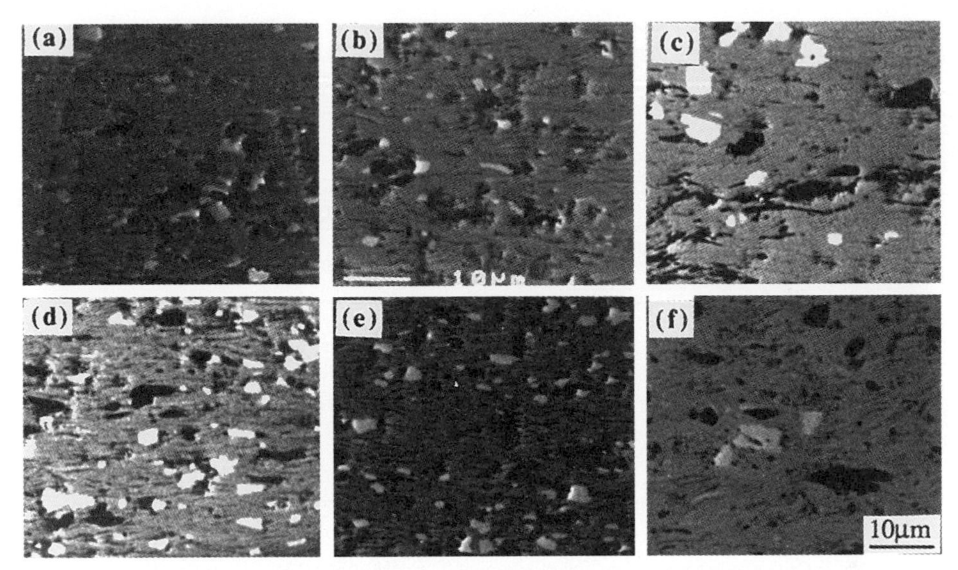

Fig.1 Backscattered electron images of cross sections of the fully processed tapes. (a):TS1, (b): TS2, (c): TS3, (d): TS4, (e): TS5, (f): TS6

3.2. Effects of precursor powder particle size

The average particle sizes by volume of powder SP1-3 are listed in Table 2. A mixture with the finest and most homogeneously distributed particles was achieved in powder SP3. Particles of CuO and Ca_2CuO_3 with size of larger than 5μm were often observed in the polished cross-section of as-rolled tape TP1. Fig.2 shows the dependence on sintering temperature of critical current (I_c) of the tape sintered for 60 hours. The I_c of tape TP3 was the most sensitive to sintering temperature between 834°C and 842°C. Optimised J_c values of these three tapes were obtained after sintering four times at 836°C for 60 hours with intermediate pressing. J_c values for the fully processed tapes are given in Table 2. The highest J_c value was achieved in tape TP-2 although tape TP-3 exhibited the highest formation rate of the 2223 phase.

Fig.3 shows the backscattered images of the polished cross sections. The grey matrix was the 2223 phase. Tape TP3 appeared to have a larger average 2223 grain size. A large number of black and white particles were observed in tape TP1, while fewer black particles were seen in tape TP2. Coarse black particles were present in TP3. EDX analysis revealed that the black particles in tape TP1 and TP2 were $(Ca,Sr)_2CuO_3$, and $(Ca,Sr)_2CuO_3$ or $(Sr,Ca)_{14}Cu_{24}O_{41}$ in tape TP-3. Coarse white particles, $(Ca,Sr)_2PbO_4$, or 3321 phases, were present in TP1, the size and quantity of which were reduced significantly in tape TP-2, and were absent in TP3. For tape TP1 containing the coarse precursor SP1, the Ca_2PbO_4 and Ca_2CuO_3 can not be consumed completely and they grow during the annealing. By extension of the powder grinding time, the average precursor particle size was decreased, and the amount of Ca_2PbO_4 and Ca_2CuO_3 was limited significantly in the final tapes, thus leading to an increased J_c value in tape TP2. Further decreasing the particle size of powder SP-2 led to powder SP-3. However, the J_c did not increase further in tape TP3, while larger 2223 grains and $(Sr,Ca)_{14}Cu_{24}O_{41}$ and Ca_2CuO_3 phases were observed. This indicates that an optimised average particle size is important. It is suggested that an excess amount of liquid phase is formed due to the very fine particles present in tape TP3, and that the growth of large alkaline-earth cuprate grains is related to the presence of the excess amount of liquid phase[7].

Table 2 Grinding time and average particle size of precursor powder, and fractional conversion to 2223 and critical current densities of corresponding tapes

Powder ID	Grinding time(hours)	Average particle size by volume (μm)	Tape ID	Fractional conversion to 2223(%)*	J_c** (kA/cm^2)
SP1	3	6.8	TP1	47.6	21
SP2	4	5.5	TP2	59.0	29
SP3	5	4.3	TP3	64.5	24

*: Fractional conversion from 2212 to 2223 was calculated by using the (008) and (0010) XRD reflection intensities of the 2212 and 2223, respectively, for the tapes after the first sintering at 836°C for 60 hours.
**: Critical current densities of the fully processed tapes

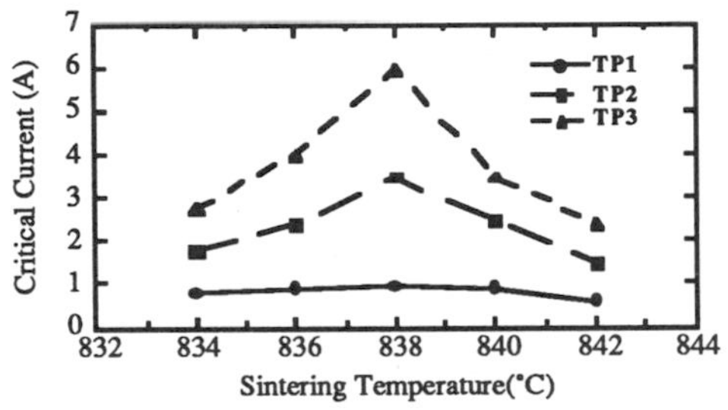

Fig.2 Critical current as a function of sintering temperature for tapes TP1, TP2 TP3 after sintering once in air for 60 hours

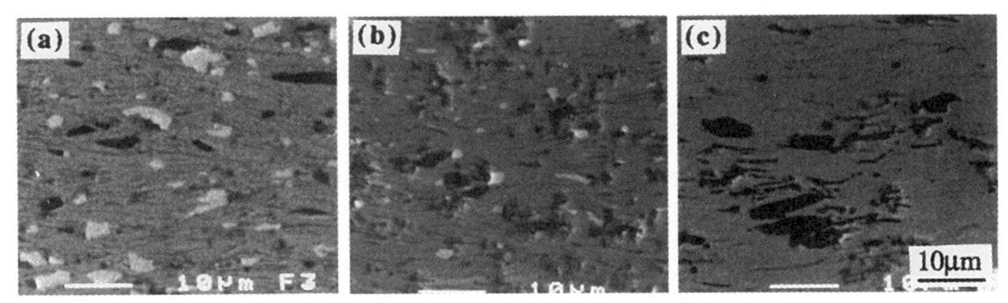

Fig.3 SEM images of cross sections of the fully processed tapes. (a): TP-1, (b): TP-2, and (c) TP-3

4. Conclusion

In order to limit the amount and size of the secondary phases in the final tapes, and to improve the J_c, the phase assemblages and particle size of the precursor powder must be carefully controlled during the powder processing. It is clear that an optimised phase assemblage in the precursor powder is important for high J_c values in the Bi-2223 tapes, and a mixture of 2212, CuO and $(Ca,Sr)_2CuO_3$ with an appropriate amount of $(Ca,Sr)_2PbO_4$ to promote liquid phase formation may be the best precursor combination.

References

[1] Haldar P and Motowidlo L R 1992 *JOM*, **44**(10) 54
[2] Dou S X and Liu H K 1993 *Supercond. Sci. Technol.* **6** 297
[3] Li Q, Brodersen K, Hjuler H A and Freltoft T 1993 *Physca C* **217** 360
[4] Grasso G, Jeremie A and Flükiger R 1995 *Supercond. Sci. Technol.* **8** 827
[5] Huang Y T, Shy D S, Chen L J 1995 *Physica C* **254** 159
[6] Jiang J, and Abell J S, to be published in *Physica C*
[7] Wong-Ng W, et al. 1992 *Am. Ceram. Soc. Bull.* **71** 1261
[8] Dou S X, Liu H K, Zhang Y L, and Blant W M 1991 *Supercond. Sci. Technol.* **4** 203

Inst. Phys. Conf. Ser. No 158
Paper presented at Applied Superconductivity, The Netherlands, 30 June–3 July 1997

Preparation of Bi2223 precursor powders by sintering in low oxygen partial pressure

B. Zeimetz, G. E. Murashova, H. K. Liu, S. X. Dou

Centre for Superconducting and Electronic Materials, University of Wollongong, Wollongong NSW 2522, Australia

Abstract. Precursor powders for Bi2223 powder in tube tapes were prepared by spray drying and subsequent calcination and grinding. If a low oxygen partial pressure is used in the final sintering stage, the reactivity of this powder during the tape sintering (in air) is strongly increased. However, critical currents are lower compared to our reference tapes. We have performed quenching experiments on tapes at early sintering stages, to investigate the speed and the nature of 2223 formation. There is direct evidence for growing of 2223 on 2212 'templates'.

1. Introduction

Within the efforts to gain a more comprehensive understanding of the complex chemical phase relations in the Bi2223 system (e.g. [1 ,2] and Ref.s therein) and in particular the nature of the Bi2223 phase formation(e. g. [3 ,4]), the focus has been on the role of the lead, and of the lead containing phases.

It has been claimed [5] that the Bi2223 phase forms in a two step process from Bi2212 through the reactions

$$Ca_2PbO_4 + Bi2212 \Rightarrow Bi(Pb)2212 + SP \quad (1)$$

$$Bi(Pb)2212 + SP \Rightarrow Bi2223 \quad (2)$$

From this it follows that the lead distribution in the precursor, i.e. content of Ca_2PbO_4 and Bi(PB)2212 will play a crucial role. Ca_2PbO_4 is considered as vital, because it provides a liquid phase to speed up the 2223 formation [5][6]. On the other hand, if the 2223 phase grows from 2212 only if lead is incorporated, the presence of Ca_2PbO_4 would require an additional, unnecessary step (i. e. reaction (1)).

Dorris et al. [7], and Luo et al. [8] have investigated this problem systematically by mixing powders with variation of lead containing phases $Bi_{2-X}Pb_XSr_2CaCu_2O_{8+d}$ and Ca_2PbO_4, but constant overall stochiometry. They found that, the more Pb is incorporated into 2212, the faster is the conversion into 2223, and the higher is the final critical current I_c of the tape.

Similar work has been done by Jeremie et al [9]. However, in their work the reactivity of the powder with Pb2212 is slightly *reduced* compared to their reference 'standard' powder (inferred from fig. 4 in [9]), while tape critical current densities were about equal.

Instead of using the 'two powder process'[10] as done in the works discussed above, the Pb can also be 'shifted' according to equation (1) by sintering the *powder* in a low oxygen atmosphere [11]. This is due to the fact that the Pb solubility of Bi2212 increases with decreasing oxygen pressure [12].

In the present work, we use this effect to produce powders with variing Pb distribution, and compare their performance in Ag sheathed PIT tapes. While we could confirm that the reactivity of a powder with high amount of Pb-2212 is strongly increased, we found that the I_c

performance of such a tape is *worse* compared to our reference sample. We attribute this to inhibition of Bi2212 recrystallization in the early stages of tape sintering.

2. Experimental

A Bi(Pb)2223 precursor powder with nominal stochiometry $Bi_{1.85}Pb_{0.35}Sr_{1.9}Ca_{2.03}Cu_{3.05}O_{10+X}$ was prepared by dissolution of nitrates in nitric acid and solidification in a spray dry machine. The powder was then heat treated at 800 °C for 24 h, ground in an agate mortar, heat treated at 810 °C for 12 h, and again ground. After this step the powder was divided into two batches, which we will refer to as powders 'A' and 'B'. Powder A was sintered in air, while powder B was sintered in a low pO_2 gas mixture.

Each powder was then ground briefly and filled into a silver tube. The tubes were deformed into tapes [13] with identical deformation schedules.. Short (40 mm) samples of the tapes were heat treated in air. Some were 'quenched' (50 °C/min) after 1, 10, 20, 35, 50 h, and investigated by X-ray diffraction and Electron Microscopy.

Other samples were heat treated twice at 840 °C in air for 50 h, with an intermediate 'sandwich rolling' [14], and their critical currents (1 μV/cm) determined at 77 K in self field.

3. Results and Discussion

Fig. 1 shows XRD patterns of precursor powders A and B. The main differences are the supression of the Ca_2PbO_4 peaks in powder B, and the width increase of the Bi2212 200/020 peak at 33.0°. The latter indicates that the lead content in the Bi2212 phase has increased [12]. This was also confirmed with EDX measurements in SEM samples.

The phase evolution during the tape sintering was examined with XRD, Fig. 2. While in tape A (from powder A) the Bi2223 formation commences only after 20 hours, in tape B (containing powder B) Bi2223 can be observed after just 1 hour of sintering, and the transformation into Bi2223 is essentially complete after 35 h.

However, the critical currents of samples from powder/tape B are only around half of those from tape A (typically $I_c(A) = 24$ A, $I_c(B) = 14$ A).

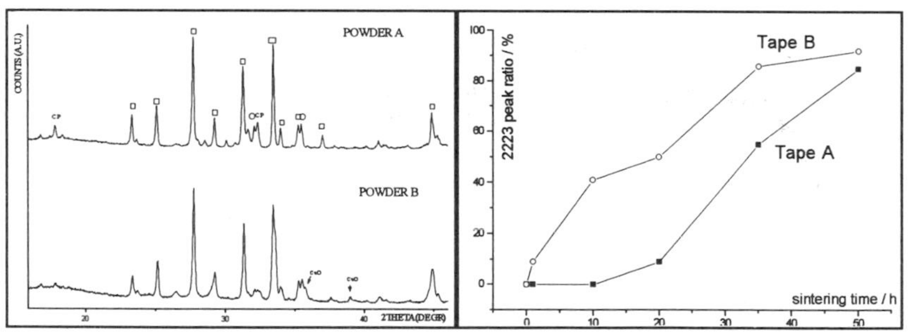

Fig. 1: XRD patterns of powders A and B; □ = Bi2212; O = $(Sr,Ca)_2Cu_2O_Y$ CP = Ca_2PbO_4

Fig. 2: Bi2223 formation rate; expressed as ratio of Bi2223(00<u>10</u>) and Bi2212(008) XRD peak heights $h_{2223}/(h_{2223}+h_{2212})$

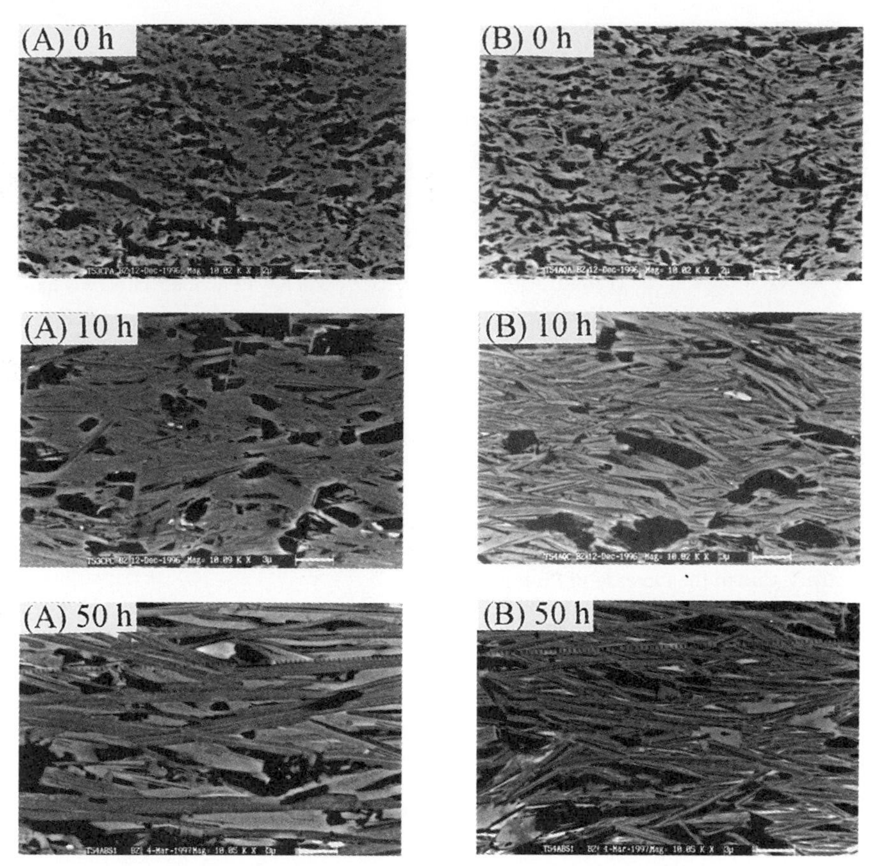

Fig. 3: SEM micrographs of samples from tapes A and B, sintered for 0, 10 and 50 hours

The micrographs of tapes A and B in Fig. 3 confirm the XRD results (fig. 2), i.e. in tape B the Bi2223 phase forms much faster compared to A. However, the Bi2223 grains are much larger (longer and thicker) in tape A compared to B. This is the reason for the much higher critical currents found in tape A, as the number of grain boundaries is strongly reduced.

It can also be seen in Fig. 3 that the Bi2212 phase is recrystallizing in the first few hours of sintering. This effect has been studied in detail by Thurston et al [15], who performed transmission XRD measurements of tapes. They found that the Bi2212 phase is realigning (with c-axis along the tape) during early sintering stages, and the Bi2223 phase formation starts after an initial 'delay period', very similar to our tape A. In our tape B, in contrast, the Bi2212 recrystallization is inhibited due to the Bi2223 growth.

We conclude that the 'better' Bi2212 morphology (i.e. larger Bi2212 grains with higher degree of alignment) of tape A leads to a better Bi2223 morphology. This argument implicitly assumes that the Bi2223 formation is governed by a direct growth on Bi2212 'templates'. While the nature of Bi2223 formation [3,4] is still controversial and we don't want to enter the discussion here, we show micrographs in Fig. 4 to support our argument.

Our main conclusion is, that the initial 'delay' of Bi2223 formation, probably due to reaction (1), is beneficial, because it allows the regrowth and alignment of the Bi2212, leading to an improved Bi2223 morphology.

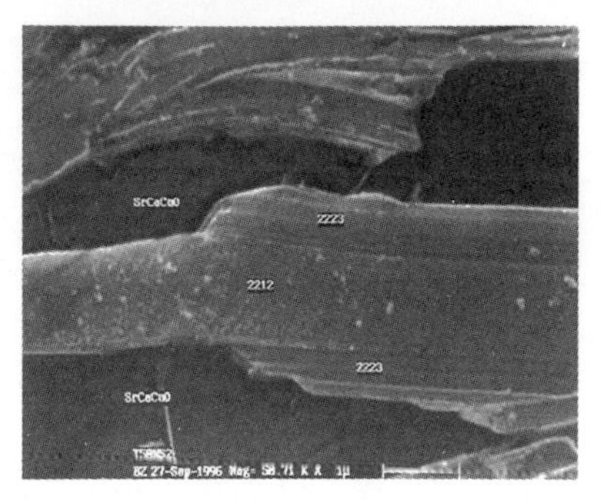

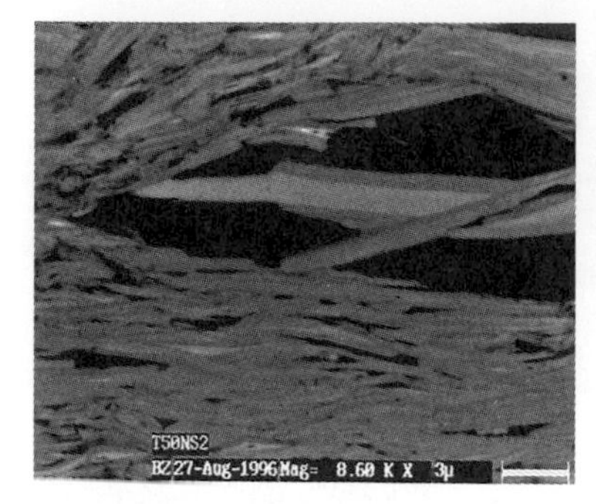

Fig. 4: a) SEM image of a sintered tape; showing Bi2223 grains growing on both sides of a Bi2212 'template'; b) backscattering electron image of the same area

4. Conclusions

While we could confirm that

- sintering a Bi2223 precursor powder in a low oxygen partial pressure strongly increases its reactivity, due to the increase of Pb-rich Bi2212,

we also found that

- a higher reactivity of the precursor not necessarily leads to a higher critical current of the final tape,

because

- an initial delay in Bi2223 formation is beneficial in allowing the regrowth and realignment of the Bi2212 phase

and

- the Bi2212 morphology in the tape determines the final Bi2223 morphology.

Acknowledgements

We gratefully acknowledge financial support of the ARC, Metal Manufacturers Ltd., Commonwealth Department of Industry, Science and Tourism, ERDC and Energy Supply Association Ltd.

References

[1] S. Kaesche, P. Majewski, F. Aldinger, J. Electr. Mater. 24 (1995) 1829
[2] P. Majewski, S. Kaesche, F. Aldinger, subm. to J. Am. Cer. Soc. (1996)
[3] Q. Y. Hu, H. K. Liu, S. X. Dou, Physica C 250 (1995) 7
[4] J.-C. Grivel, R. Flukiger, Supercond. Sc. Techn. 9 (1996) 555
[5] J.-C. Grivel, R. Flukiger, J. of Alloys and Compounds, 1996 (acc. for publ.)
[6] S. E. Dorris, M. A. Pitz, J. T. Dawley, D. J. Trapp, J. Electr. Mater. 24 (1995) 1835
[7] S. E. Dorris, B. C. Prorok, M. T. Lanagan, N. B. Browning, M. R. Hagen, J. A. Parrell, Y. Feng, A. Umezawa, D. C. Larbalestier, Physica C 223 (1994) 163
[8] J. S. Luo, S. E. Dorris, A. K. Fischer, J. S. LeBoy, V. A. Maroni, Y. Feng, D. C. Larbalestier, Supercond. Sc. Techn. 9 (1996) 412
[9] A. Jeremie, R. Flukiger, Physica C 267 (1996) 10
[10] S. E. Dorris, B. C. Prorok, M. T. Lanagan, S. Sinha, R. B. Poeppel, Physica C 212 (1993) 66
[11] Y. Idemoto, S. Ichikawa, K. Fueki, Physica C 181 (1991) 171
[12] Y. Iwai, Y. Hoshi, H. Saito, M. Takata, Physica C 170 (1990) 319
[13] S. X. Dou, H. K. Liu, Supercond. Sci. Techn. 6 (1993) 297
[14] W. G. Wang, H. K. Liu, Y. C. Guo, B. Bain, S. X. Dou, Appl. Supercond. 3 (1995) 599
[15] T. R. Thurston, U. Wildgruber, N. Jisrawi, P. Haldar, M. Suenaga, Y. Wang, J. Appl. Phys. 79 (1996) 3122

Inst. Phys. Conf. Ser. No 158
Paper presented at Applied Superconductivity, The Netherlands, 30 June–3 July 1997

Investigation of Heat Treatment and Deformation Processes of BPSCCO-2223 /Ag conductors

T. Fahr, A. Hütten, W. Pitschke, C. Rodig, U. Schläfer, M. Schubert, P. Trinks and K. Fischer

Institute of Solid State and Materials Research Dresden, P. O. Box 27 00 16, D-01171 Dresden

Abstract. This paper presents results of thermoanalytical and XRD investigations at room temperature and in a high temperature chamber concerning the early stage of the transformation of precursors for the fabrication of Bi,Pb-2223 / Ag tapes. The occurence of two parallel reactions is suggested at the beginning of reaction annealing of the conductors, (i) the formation of Bi,Pb-2212 from the Pb free phase and (ii) the formation of a liquid phase accelerating the Bi,Pb-2223 phase formation. High temperature XRD experiments show the appearence of a preferential orientation of the 2212 phase in the period of heating the tape to the reaction temperature. A second part shows the dependence of the microhardness and sausaging in monofilamentary tapes on rolling parameters as well as results of the production of a multifilamentary tape with 99 filaments and a length of 550 m, where a critical current density of 25,8 kA /cm^2 was achieved.

1. Introduction

The oxide-powder-in-tube (OPIT) technique is the most employed method in fabricating Bi,Pb-2223 / Ag tapes on an industrial scale for instance for power transmission cables. The performance of such devices strongly depends on the microstructure of the ceramic core material consisting of platelike Bi,Pb-2223 crystals. The evolution of the microstructure is influenced by the processing conditions, because the phase composition in the Bi,Pb-2223 phase system only very slowly reaches the equilibrium state. The crystal alignment of the 2223 phase is influenced by the texture of the 2212 phase at the beginning of the reaction annealing of the tapes [1]. So it is usefull to know the chemical reactions occuring in the early stage of the precursor transformation [2]. On the other hand, well aligned and c-axis oriented Bi,Pb-2223 crystals can only be produced if the filament is free of sausaging. This paper gives results of thermoanalytical investigations of precursor powders and tape samples that are discussed as the occurence of two independent reactions at the beginning of precursor transformation. First results of high temperature XRD experiments are discussed with respect to the texture development of the precursor phases. Finally, the effect of rolling parameters on the evolution of the microhardness in monofilamentary tapes and the beginning of sausaging are introduced.

2. Experimental Methods

Monofilamentary tapes have been fabricated with precursors of the nominal chemical composition $Bi_{1,8}Pb_{0,4}Sr_2Ca_{2,1}Cu_3O_x$ and pure silver as metallic component of the composite. DSC measurements of powder samples that were air quenched after various annealing times and of green tapes were performed with a „NETZSCH DSC 404“. XRD measurements were carried out with Co-K_α . For high temperature experiments in reflection mode a Buehler chamber was used with a controlled atmosphere and a platinum heating strip where a multifilamentary tape sample was mounted with the Ag on top of the tape removed. All thermal treatments were carried out in an atmosphere of N_2 / 8% O_2 . The microhardness of the filaments in mono core tapes was determined as Vickers hardness HV 0,01 at transverse cross sections of the tapes.

3. Results and discussion

3.1. Thermoanalytical investigations and XRD

Fig. 1 shows DSC traces of precursor powder samples annealed for various periods of time at 815°C and subsequently air quenched. The trace of the precursor state (0h) shows only one broad endothermic peak in the temperature range between 850-880°C. This effect corresponds to the melting of the complete phase system. The traces of powders quenched after 1-5h of annealing show a smaller peak infront of the melting peak. XRD measurements reveal the reversible [3] formation of the orthorhombic Bi,Pb-2212 phase (Fig.2) form tetragonal Bi-2212, a main precursor phase, in this early stage of the reaction. In this period almost no 2223 phase can be detected. The XRD pattern of the sample 1h at 835°C annealed, again shows the tetragonal Bi-2212 phase after air quenching. This heat treatment occured above the onset temperature (825°C) of the small endothermic peak in DSC traces of samples showing the orthorhombic 2212 phase after quenching. So we attribute the prepeak (DSC scans 1-5h 815°C) to the decomposition of Bi,Pb-2212. The very weak endothermic effect at 815-825°C in samples annealed for at least 7h at 815°C we have discussed as the

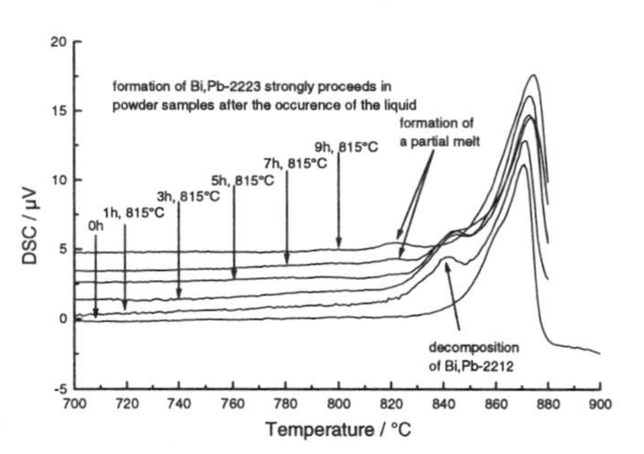

Fig.1 DSC scans of precursors air quenched after various annealing times at 815°C

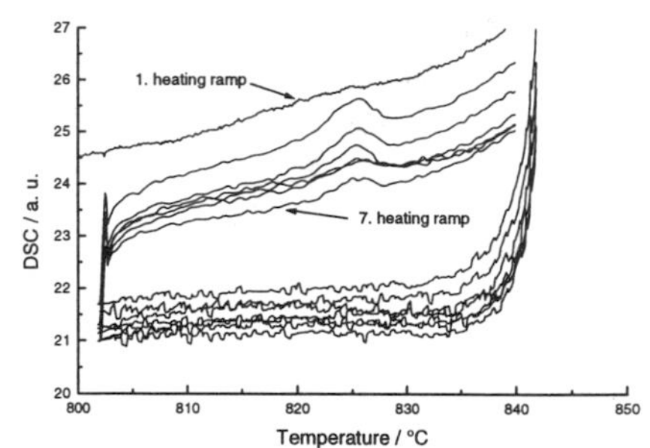

Fig.3 DSC curves of a tape sample with Ag sheath material (800-840°C, 2 K/min)

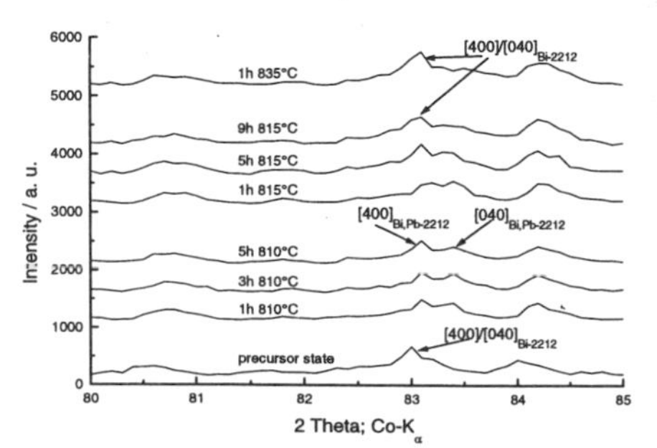

Fig.2 Formation of Bi,Pb-2212 from Bi-2212 in the early stage of reaction annealing at 815°C

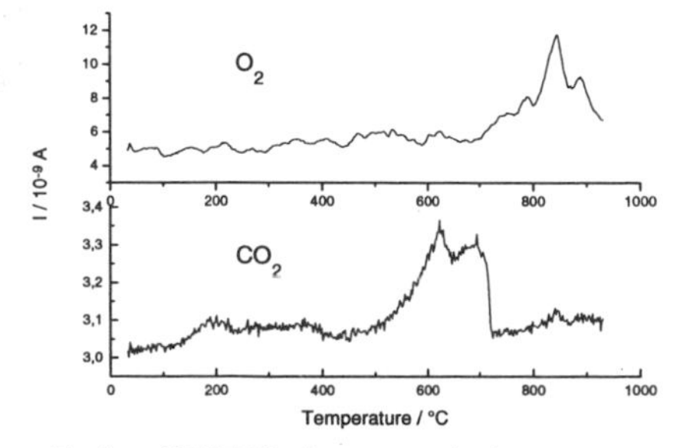

Fig.4 DTA/MS of a precursor in Ar

formation of a liquid phase in the ceramic material, which strongly accelerates the formation of the 2223 phase [4]. It shall be pointed out, that the liquid also occurs in the powder alone, not only in contact with silver, which now is shown in Fig.3. DSC scans of tape samples show relatively weak endothermic reactions in the temperature range of the tape processing compared to powder samples. Because the tape contained an amount of 2223 phase of about 35% after the DSC measurement with the temperature swept from 800-840°C, the endothermic peak should be attributed to the formation of the partial melt. The liquid is not present immediately after the processing temperature of tapes is reached but yet in the secound heating ramp. Now, the quantity of heat detected decreases from one heating ramp to the next. It is not clear so far, if this is caused by a fast decrease of the amount of liquid in the filament or by the formation of a bad heat conducting solid phase at the ceramic / silver interface as a product of the wetting of silver by the partial melt.

The results of a mass spectroscopic detection of gaseous species during heating the precursor powder (Fig.4) shows the release of oxygen in the range of the processing temperature. In order to distinguish the oxygen evolved by the powder from that of the annealing atmosphere, these measurements had to be done in Argon. In this sense only qualitative conclusions can be drawn. The oxygen is released in two steps, at around 800°C and above 825°C in Ar. While the secound effect is correlated to the complete reduction of the precursor to be seen by metallic copper after the experiment, the first small effect should be attributed to the reduction of Pb(IV) phases such as $(Sr,Ca)_2PbO_4$. This can be correlated to XRD results that revealed the formation of the Pb containing Bi,Pb-2212 phase in the early stage of annealing. The Pb is incorporated into 2212 in the bivalent state, which occurs over the decomposition of Pb(IV) phases with oxygen as by-product. Additionally, the release of CO_2 occurs as a result of an incomplete calcination of the precursor, which can produce bubbles during the tape processing.

High temperature XRD experiments show that in the period of heating the multifilamentary tape to the processing temperature the crystals of the 2212 phase undergo a preferential orientation (Fig.5) to be seen at the strong [001] reflexes. This texture obviously is adjusted in the period before the transformation conditions to the 2223 phase are reached. Immediately after the temperature is constant, Ca_2PbO_4 and CuO decompose and form the liquid phase from which 2223 slowly crystallizes.

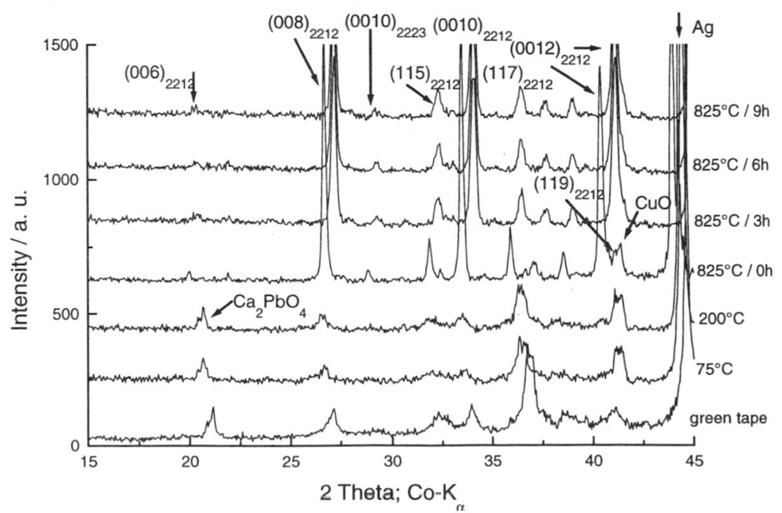

Fig.5 HTXRD on multifilamentary tapes in N_2 / 8% O_2 (Co-K_α)

3.2. Deformation parameters and sausaging

In Fig. 6 the evolution of the microhardness of the oxide core in monofilamentary tapes is presented in dependence on the diameter of the rolls employed. While the deformation with rolls of a diameter of 80 mm causes a relatively strong densification of the ceramic

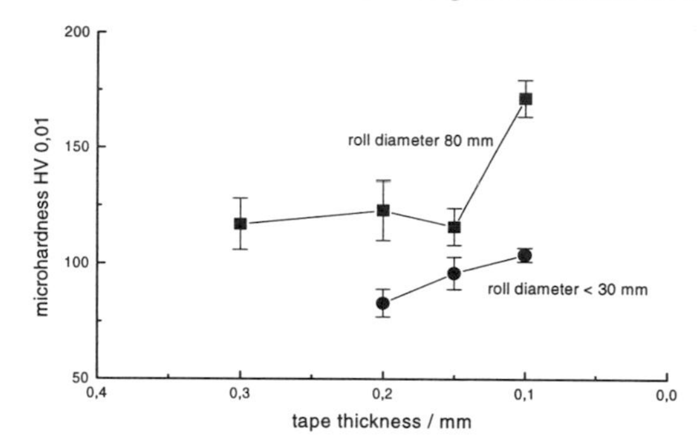

Fig.6 Evolution of the microhardness HV 0,01 of the core of monofilamentary tapes in dependence on rolling parameters

material, the use of rolls with a diameter lower than 30 mm leads to a weaker densification of the core down to a tape thickness of 100 μm.
The tape with a thickness of 100 μm deformed with 80-mm-rolls has shown sausaging indicated by the kink in the upper curve, while that prepared with small rolls was free of sausaging. These results could be verified at multifilamentary tapes as well [5].

3.3. Fabrication of a long length tape

Finally, the results of the fabrication of a multifilamentary tape with a length of 550 m are presented:
tape length: 550 m; tape width: 0,237 mm; tape thickness: 3,62 mm;
critical current values measured over the whole length of the tape (77 K, 0T):
$I_C = 67$ A $j_C = 25,8 \pm 2,2$ kA / cm^2 $j_E = 7,8 \pm 0,7$ kA / cm^2

Acknowledgements

This work was performed within the SIEMENS joint project "HTS for Power Engineering", supported by the German Federal Minister of Research under contract 13N6481

References

[1] G. Grasso, A. Perin, R. Flükiger; Physica C **250** (1995), 43
[2] J. S. Luo, S. E. Dorris, A. K. Fischer, J. S. LeBoy, V. A. Maroni, Y. Feng, D. C. Larbalestier; Supercond. Sci. Technol. **9** (1996), 412
[3] J. C. Grivel, R. E. Gladishewskii, E. Walker, R. Flükiger; Physica C **274** (1997), 66
[4] K. Fischer, T. Fahr, A. Hütten, E. Müller, M. Schubert, D. Schläfer, R. Wenzel; IEEE Trans. Appl. Supercond. **7** (1997), 2223
[5] A. Hütten et al.; oral presentation at EUCAS 1997

Inst. Phys. Conf. Ser. No 158
Paper presented at Applied Superconductivity, The Netherlands, 30 June–3 July 1997

The mechanism of sausaging in Ag/Bi-2223 multifilament-tapes and the effect of processing parameters

S. Kautz[1,3], B. Fischer[2], J. Müller[1,2], J. Gierl[3], B. Roas[1], H.-W. Neumüller[1], R. F. Singer[3]

[1] Siemens AG, Corporate Technology, Erlangen, Germany. [2] ISKP, University of Bonn, Bonn, Germany. [3] Department of Materials Science, University of Erlangen, Erlangen, Germany.

Abstract. The mechanical deformation of Ag/Bi-2223 multifilament-tapes by flat rolling is limited by the effect of sausaging. In-situ observations of tape rolling show that sausaging is the result of discrete shearing-processes which can be explained by a simple model. The shearing is caused by a shear-band-formation in the silver-matrix which depends mainly on the kind of deformation, degree of deformation, the structure and mechanical properties of the Ag matrix and the precursor. Repeated shearing of the silver-matrix in periodic distances and different directions leads to a necking in the ceramic filaments which causes the phenomenon of periodic sausaging.
Processing parameters like initial wire size, annealing procedures or number of filaments affect the beginning and the periodicity of sausaging by shear-band-formation. Experimetal results support this new sausaging model and make it possible to predict optimum mechanical deformation parameters for producing Ag/Bi2223 multifilament tapes without sausaging.

1. Introduction

The processing of Ag/Bi2223 mulifilament-tapes presents a large number of interdependent variables, which influence the Jc and the homogenity of the final tape. Future applications of high temperature superconducting wires and tapes depend on the production of long length high quality material. In order to achieve this a detailed understanding of the mechanical deformation - one of the most important processes in fabricating high-Tc superconducting wires and tapes - is required. For this composite of two very different materials, high temperature superconducting ceramic powder cores in a metallic silver matrix, a demanding treatment is essential. If the tape rolling is not done carefully, a wavy Ag/BSCCO interface, called sausaging, develops, which leads to a reduced effective current path and, more seriously, disturbs the desired texture of the superconducting grains. This paper is a contribution to the discussion about inhomogenous plastic deformation of Ag/Bi2223 multifilament-tapes and presents a new model of sausaging-formation.

2. Experimentals

Ag/Bi2223 multifilament composite conductors are produced by the powder-in-tube (PIT) technique[1,2]. The oxid powder with a stoichiometry of $Bi_{1,8}Pb_{0,4}Sr_{2,0}Ca_{2,1}Cu_{3,0}O_x$ is isostatically pressed and filled in pure Ag tubes. The formation of monofilamentary wires is carried out by conventional deformation processing of the rods via swaging and drawing. Fabrication of multifilamentary composites requires the additional process of bundling the monofilamentary wires into a Ag tube. The multifilamentary bundel is then deformed to a round wire which is transformed into a tape by flat rolling. For investigating the sausaging-formation an in-situ rolling experiment is carried out. A pre-rolled multifilament tape which tends to build sausaging by the next rolling reduction is the source material. This tape is rolled only for a part of the tape length. Then the rollers are stopped and devided during the rolling process, so that the tape can be taken out. Longitudinal cross-sections of the tape show then the transition from the non-sausaging part of the tape (mark A in Fig.1) to the sausaging part (mark E in Fig.1). So the formation of sausaging in the rolling gap - the deformation zone - can be investigated in detail.

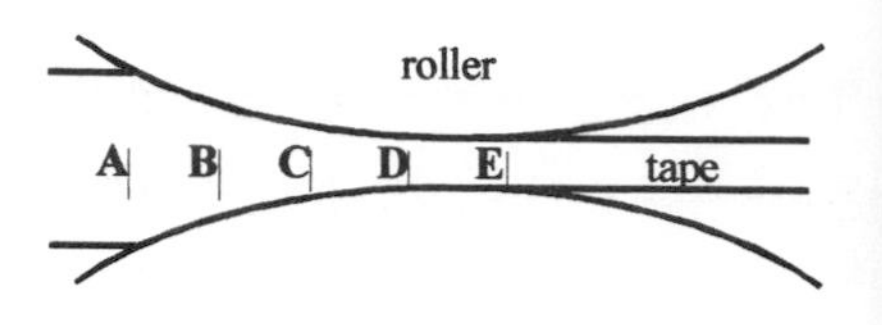

Fig. 1: Scheme drawing of an insitu rolling experiment with position marks A - E

3. Results

3.1. Sausaging-formation

The result of this insitu rolling experiment is shown in Figure 2, where the development of sausaging by shearing pro-cesses is simple. This observations lead to a new model of sausaging-formation.The shearing is caused by a shear-band-formation in the silver-matrix

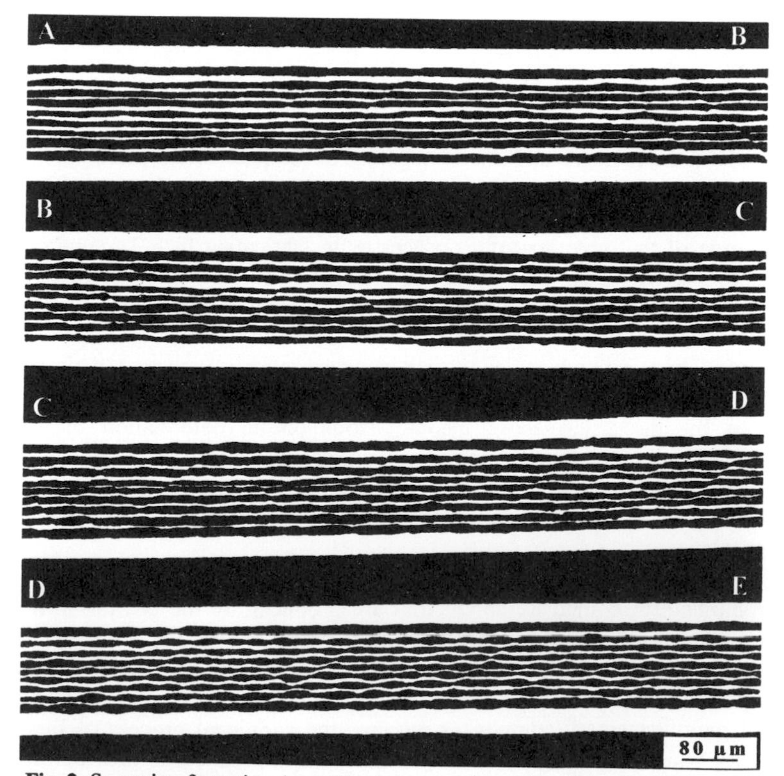

Fig. 2: Sausaging-formation; longitidinal cross-sections taken from the rolling gap positions marked A - E in fig. 1

which depends mainly on the kind and degree of deformation and the structure and mechanical properties of the Ag matrix. Shear-band-formation is a typical reaction in face centered cubic metals, which are de-formed very strongly by rolling [3]. But there is another important prerequisite for this descrete deformation process. A critical powder core density, which depends on the processing parameters and the properties of the powder, has to be exceeded, before the deformation of the tape is concentrated on the silver martix. If this critical powder core density is exceeded some more rolling leads to the shearing processes in the Ag matrix. The shearing does not stop at the filaments. They were also cut by the shearing forces. Repeated shearing in periodic distances and different directions build the necking in the ceramic filaments which causes the phenomenon of periodic sausaging.

3.2. The effect of processing parameters

Processing parameters like wire size before rolling, annealing procedures or number of filaments affect the beginning and the periodicity of sausaging.

The first signs of sausaging formation are the shearing-bands which can be seen in zone A - B of Fig. 2. This situation is defined as the beginning of sausaging formation. Another interesting parameter is the periodicity of sausaging specified by the necking-distance along the filaments in the middle of the Ag/Bi2223 multifilament-tape, measured when sausaging is completely developed, like in zone D-E of Fig. 2. The wires discussed contain the same percentage mass of powder, are processed by the same rollers and the same deformation degree per rolling step.

One of the examined processing parameters is the initial wire size: As drawn and annealed (350°C, 30min.) wires with different diameters were rolled to tapes as described in section 2. Figure 4 and 5 show that a smaller wire diameter does not allow such a high total area reduction as a bigger one, before sausaging begins. Therefore the necking-distance increases when the diameter rises.

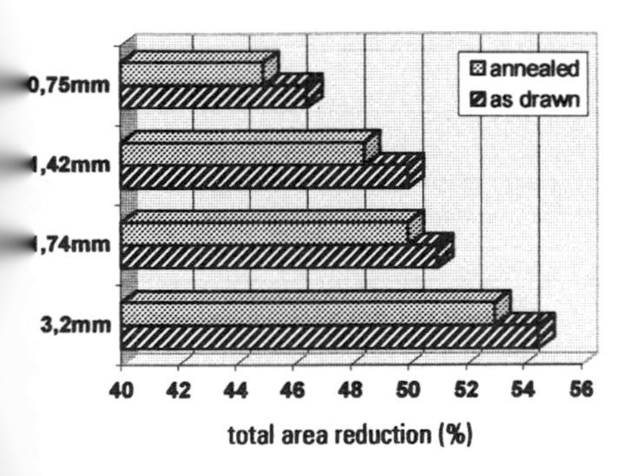

. 4: Start of sausaging-formation at a total area reduction depending on the wire diameter before rolling (55 Fil.).

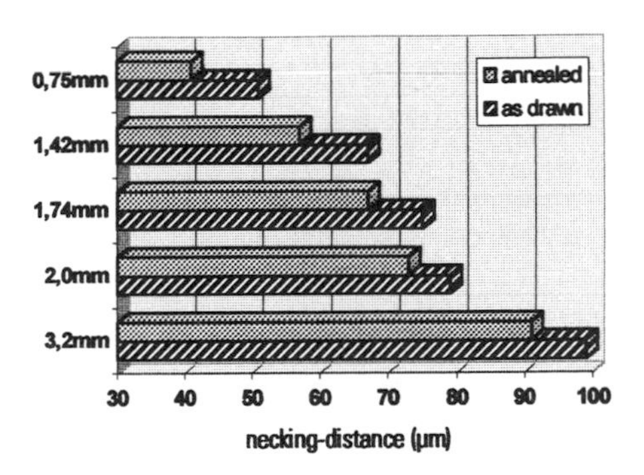

Fig. 5: Different necking-distances caused by different wire diameters before rolling (55 Fil.).

Both phenomenon can be explained by the degree of freedom of the deformation in the compound. If there is enough space for the powder to glide and rotate, the mobility of the particles enables a higher degree of deformation. Also an increasing amount of Ag between the filments is a good prerequisite for a homogeneous deformation. The dependence of sausaging development in tapes with a different number of filaments can explained in the same way. Figure 6 and 7 show that the necking-distance decreases and the shearing-process starts at a smaller total area reduction when the number of filaments increases.

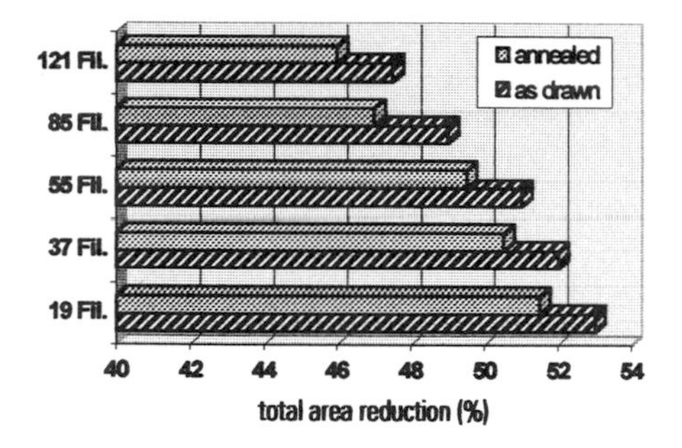

Fig. 6: Start of sausaging-formation at a total area reduction depending on the number of filaments in the wire (1,75 mm dia.).

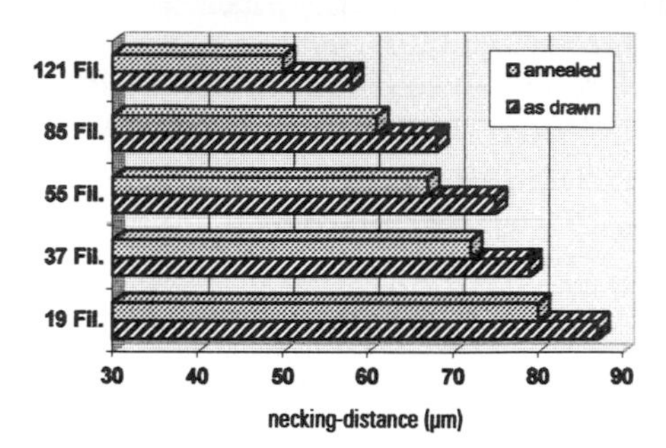

Fig. 7: Different necking-distances caused by a different number of filaments in the wire (1,75 mm diam.).

Another interesting effect is the early beginning of sausaging-formation and the reduction of the necking-distance by using an annealed (350°C, 30 min.) wire instead of an as drawn (figures 5-8). The annealing process changes the structure and mechanical properties of the silver matrix in a way which accelerates the mechanism of shear-band-formation. Also the increasing difference of the mechanical properties between silver matrix and precursor filament play an important role.

The reasons, correlations and the achievments of sausaging-formation and the effects of processing parameters are still investigated and will be published soon.

Acknowledgement

The authors wish to acknowledge their co-workers from Vacuumschmelze GmbH, Siemens AG and collaborating partners within the BMBF-programme „Superconductivity for Power Engeneering“. This work has been supported by the German Federal Ministry for Education, Science, Research and Technology BMBF under grant no. 13N6481.

References

[1] Heine K 1989 Appl. Phys. Lett. **55** 2441

[2] Tenbrink J 1991 IEEE Trans. Magn. **27** 1239

[3] Aernoudt E 1980 Verformung und Bruch-Conference, Leoben, Austria, Oct. 16.-17. 1980

Inst. Phys. Conf. Ser. No 158
Paper presented at Applied Superconductivity, The Netherlands, 30 June–3 July 1997

The influence of the oxygen partial pressure on the phase evolution in Ag/Bi-2223 multifilament-tapes

J. Müller[1,2], B. Fischer[1], J. Gierl[3], S. Kautz[1,3], H.-W. Neumüller[1], B. Roas[1], P. Herzog[2]

[1] Siemens AG, Corporate Technologie, Erlangen, Germany. [2] ISKP University of Bonn, Germany.
[3] Department of materials science, University of Erlangen, Germany.

Abstract. For a further increase of the j_c-level in Bi-2223 multifilament tapes different features like phase evolution and origin of different phases have to be understood and controlled. In this contribution the influence of the oxygen partial pressure of the sintering atmosphere on the Bi-2223-phase evolution in tapes has been studied. The precursor powder inserted in the silver tube had a nominal stoichiometry of Bi 1.8, Pb 0.4, Sr 2.0, Ca 2.1, Cu 3.0 and was calcined in air. The sintering atmosphere was nitrogen with oxygen contents of 0,001 % to 100 % O_2. The optimized sintering temperature declines sharply with decreasing oxygen partial pressure. Plumbat compounds seem to be necessary for a further increase of the critical current.

1 Introduction

The importance of the high-T_c superconducting phase $(Bi,Pb)_2Sr_2Ca_2Cu_3O_{10}$ for producing polycrystalline, bulk materials and tapes is now well established. The unique properties of this phase and their parameters have been discussed extensively in the literature [1,2,3]. One important parameter to control the phase formation is the oxygen content of the sintering atmosphere which affects the superconducting properties [4]. According to Endo et al. [5], the decrease in oxygen pressure lowers the melting point of the $Bi_{0.8}$ $Pb_{0.2}$ $Sr_{0.8}$ $Ca_{1.0}$ $Cu_{2.0}$ O_y. The purpose of the present paper is, first, to determine the dependence of the Bi(Pb)2223 phase evolution in Ag-sheated tapes on the oxygen partial pressure and temperature. Second, to find out which parameter determines mainly the dependence from the oxygen partial pressure and the temperature.

2 Experimental

Bi-2223 powder with an overall composition of $Bi_{1.8}$ $Pb_{0.4}$ $Sr_{2.0}$ $Ca_{2.1}$ $Cu_{3.0}$ O_x was packed into a Ag-tube and processed to a 55-filament-tape by established techniques [6,7]. The grade of filling of the Ag-sheated tape was 25 ± 5 %. The samples used in the experiment were typically 3.5 mm wide, 0.27 mm thick and 45 cm long. The unreacted Ag-sheated Bi(Pb)-2223 tapes were annealed in different oxygen atmospheres (0,001 % - 100 % O_2 in N_2 at 1013 Pa) the temperature was varied from 450°C up to 950°C with sintering times between 0.1 and 160 h. The sample temperature was monitored using a calibrated Pt/Pt-Rh thermocouple positioned close to the samples and was stabilized within ± 1°C. The critical current (I_c) was measured at 77 K and 0 T with the 1 μVcm^{-1} criterion. X-ray diffraction (XRD) measurements were made on peeled sections (after sintering) of each sample using a Siemens D500 diffractometer with Cu Kα radiation. The fractional conversion of the Bi(Pb)-

2223 was calculated from the XRD intensity ratios of the (0010) peak of the Bi(Pb)-2223 and (008) peak of Bi(Pb)-2212.

3 Results

In the samples studied in this paper the microstructure of the grains consists mainly of an intergrowth of the Bi(Pb)-2212 and the Bi(Pb)-2223 phase. Under different oxidizing conditions during sintering the formation rate of the Bi(Pb)-2223 varies significantly and as a consequence the volume fraction of Bi(Pb)-2223 vs. Bi(Pb)-2212 differs for samples heat treated for a certain amount of time. To study the phase evolution of Bi(Pb)-2223 in Ag-sheated tapes the dependence of the sintering temperature on the oxygen content of the atmosphere was determined (fig. 1) to show exactly where this phase appears.

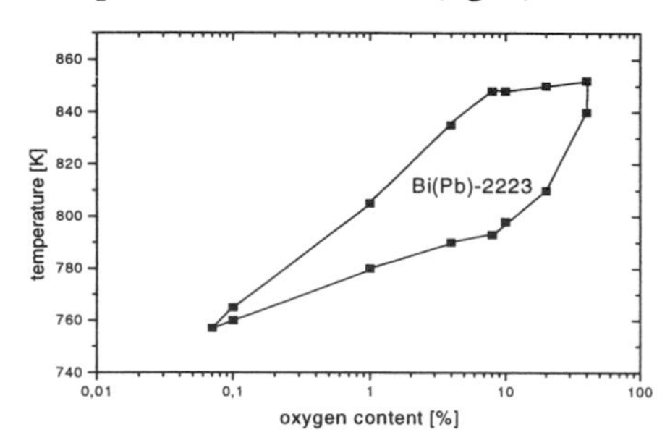

Figure 1. Dependence of the phase stability of Bi(Pb)-2223 in Ag-sheated tapes on temperature and oxygen partial pressure

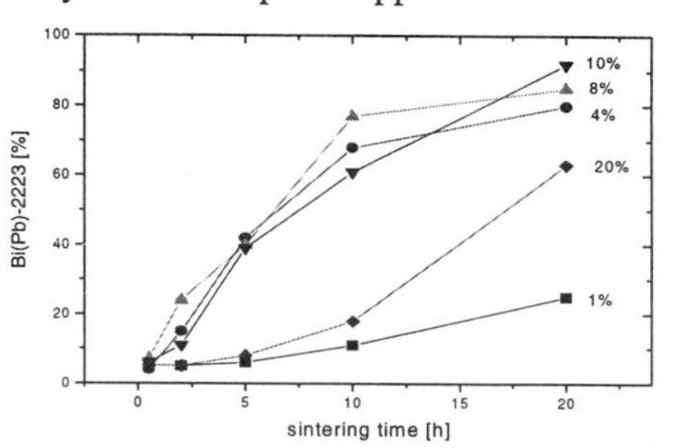

Figure 2. Bi(Pb)-2223 fraction after a sintering time of 0.1h up to 20h in 1%, 4%, 8%, 10% and 20% oxygen content

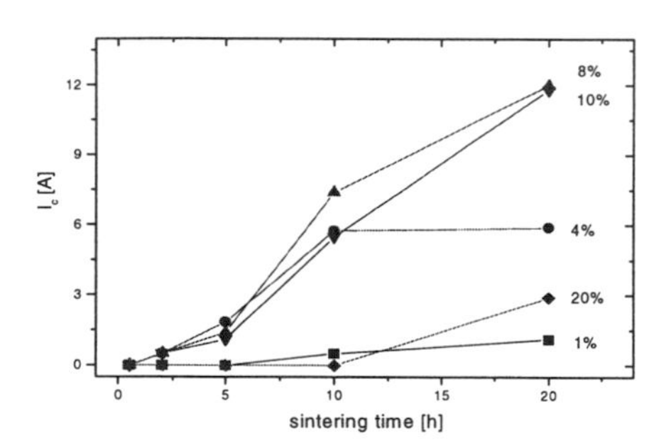

Figure 3. Critical current after a sintering time of 0.1h up to 20h in 1%, 4%, 8%, 10% and 20% oxygen content

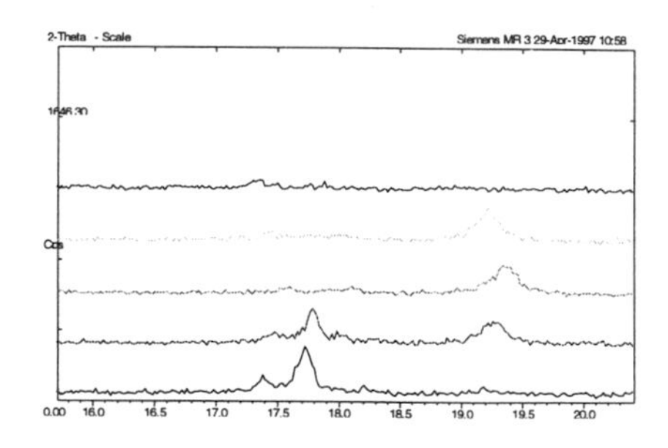

Figure 4. Part of the XRD-diagramm of the samples annealed in different oxygen partial pressures for 10h (from bottom to top: 20%, 10%, 8%, 4% and 1% oxygen)

The Bi(Pb)-2223 phase is stable in Ag-sheated tapes for oxygen partial pressures of about 0.07 % up to 40 % O_2 and a temperature range between 755°C and 850°C. Outside of this region no Bi(Pb)-2223 phase could be stabilized for a variation of the sintering time between 0.1 and 160h. The temperature for stabilizing the Bi(Pb)-2223 phase decreases with decreasing oxygen content of the sintering atmosphere. Besides the stability region the dependence of the critical current density and the Bi(Pb)-2223 - fraction rate on the sintering time have been studied for different oxygen partial pressures at the optimized (T,p_{O2}) values

The sintering temperature for each oxygen partial pressure (T, p_{O2}) was determined from the fastest rate of convertion from Bi(Pb)-2212 to Bi(Pb)-2223 with XRD-measurements and the maximum critical current values. In fig. 2 data for the variation of the volume fraction of Bi(Pb)-2223 as function of the sintering time are displayed. The Bi(Pb)-2223 phase development after 10h is faster in 8%, 10 % and 4 % than in 1 % and 20 % oxygen content. A further increase of the sintering time to 20h lead to nearly the same result. The phase fraction of the samples annealed in 20 % O_2 increased to about 70% phase fraction while the 1 % O_2 annealed samples remained at 20 % Bi(Pb)-2223 phase fraction. A comparison with fig. 3 shows that after 10h annealing no significant increase of the critical current (Ic) could be observed in low oxygen partial pressures like 1 % and 4 % O_2. A further increase in the critical current could however be observed in samples annealed in 8 % up to 20 % O_2. The results from fig. 2+3 were compared with parts from the XRD-measurements from the samples after a sintering time of 10h (fig. 4). A benefit effect of the plumbat compounds on the Bi(Pb)-2223 phase evolution was discussed in the literature [8]. A comparison of the evolution of the plumbat compounds (fig. 4) with the measured phase fraction (fig. 2) and critical current data (fig. 3) was choosen to understand and explain the results. In order to have no interference of XRD-peaks we chose the range from 17° to 18.5° to observe the plumbat compounds. The following different phases appear in fig 4.: Sr_2PbO_4 (17,4°), Ca_2PbO_4 (17,8 and 18,2°), $Pb_3Bi_{0.5}Sr_2Ca_2Cu_1O_x$, the so called „3321"-Phase (18°) identified from Khaled et al. [8] and Wang et al. [3] and the „right" peak at 19,2°, the 008 of the Bi(Pb)-2223 phase. In fig. 4 there are obviously no plumbat compounds (17,5-18,2°) present in the samples annealed in 1 %, 4 % and 8 % oxygen content whereas they are present for 10 % and 20 %. An increase of the sintering time to 20 h caused plumbat compounds also appear for 8 % O_2. We believe that there is a direct correlation between the development of the critical current and the content of plumbat compounds during the sintering time. Especially the appearence of plumbat compounds for an oxygen content of 8 % after 20 h sintering supports this assumption. An increase of the sintering time up to 80 h lead to the same critical current for 8 % and 20 % with a maximum to 10 % O_2. The maximum of the critical current is not reached in a sintering atmosphere of 20 % oxygen. This could rely on too many plumbat compounds being formated in this atmosphere. The appearence of plumbat compounds depends above all on the oxygen content and the temperature of the sintering atmosphere. The first appearence of plumbat compounds, with regard to the oxygen content, was determined from Idemoto et al. [9] for Bi(Pb)-2223 bulk material with an overall stoichiometry of $Bi_{1.74}\,Pb_{0.27}\,Sr_{1.86}\,Ca_{2.09}\,Cu_{3.04}\,O_x$ at about 5-6 % O_2 and 700 °C. To determine the chemical stability of Pb^{2+} and Pb^{4+} for different temperatures and oxygen partial pressures in Ag-sheated Bi(Pb)-2223-tapes we annealed samples between 450°C and 950 °C in 0.001 up to 100 % oxygen for totally 10h. XRD measurements on the peeled samples led to the following stable region for plumbat compounds (fig 5.). The result of the experiments with the Ag-sheated Bi(Pb)-2223 tapes is nearly the same as found by Idemoto et al. [9] for bulk material in 1991. Plumbat compounds are stable below 650 °C up to 5-6 % oxygen. An increase up to 100 % causes the appearence temperature to fall down to 450 °C. Above 650 °C in low oxygen partial pressures the plumbat compounds disappear after a certain time. This unstable region exists up to a temperature of 850 °C where the melting of this compound begins. In oxygen partial pressures above 10 % the plumbat compounds form until they melt at about 920 °C.

Fig 5. also explains why the plumbat compounds appear later for 8 % than for 10 % oxygen content of the sintering atmosphere, because for 10 % the formation of plumbat compounds allready begins at the sintering temperatures of about 600 °C, 50 °C below the formation temperature in 8 % O_2. The consequence for the Bi(Pb)-2223 phase evolution is that if

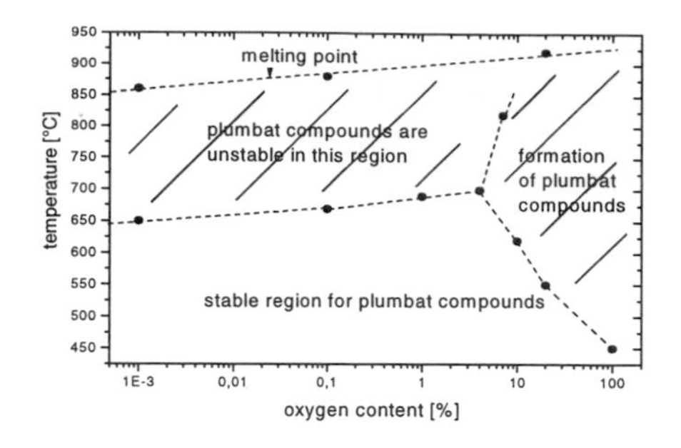

Figure 5. Dependence of the stability of plumbat compounds in Bi(Pb)-2223 tapes on the temperature and oxygen content of the sintering atmosphere

plumbat compounds are required for a better phase evolution of Bi(Pb)-2223 in Ag-sheated tapes, oxygen contents above 6-7 % oxygen are required for sintering temperatures above 800 °C.

4 Conclusions

The oxygen content in the sintering atmosphere (1013 Pa) has a distinct influence on the formation of the high-Tc Bi(Pb)-2223 phase in Ag-sheated tapes. The formation temperature decreases with decreasing oxygen partial pressure. Below an oxygen partial pessure of about 0.07 to 0.1% O_2 (at 1013 Pa) and a temperature of 760 °C no Bi(Pb)-2223 could be stabilized. The upper limit of the stbility region is at about 40 % O_2 and a temperature of about 850 °C. The rate of Bi(Pb)-2223 formation and the critical current show a maximum for oxygen contents of about 8 % to 10 % O_2. The formation rate and therefore the critical current is directly correlated with the content of plumbat compouds observed with XRD. It seems to be a fact that a small content of plumbat compounds is necessary for a Bi(Pb)-2223 phase evolution after a certin amount of sintering time.

Acknowledgement

The authors wish to acknowledge their collaboration partners within the BMBF-program „Superconductivity for Power Engeneering“: This work has been supported by tge German Federal Ministry of Education, Science, Research and Technology BMBF under grant no. 13N6481

References

[1] O. Eibl, Supercond. Sci. Technol. **8** (1995) 833
[2] I.Van Driessche et al., Supercond. Sci. Technol. **9** (1996) 843
[3] W.G. Wang et al., Supercond. Sci. Technol. **9** (1996) 881
[4] Y. Idcmoto and K. Fueki, Physica C 168 (1990) 167
[5] U. Endo, S. Koyama, T. Kawai , Jpn. J. Appl. Phys. 27 (1988) L 1476
[6] K. Heine, J. Tenbrink, M. Thöner, Appl. Phys. Lett. 55 (1989) 2441
[7] J. Tenbrink, M. Wilhelm, K. Heine and H. Krauth, IEEE Trans. Magn. 27 (1991) 1239
[8] Khaled et al., J. Mat. Sci. Elec. 7 (1996) 261
[9] Y. Idemoto et al., Physica C 181 (1991) 171

Inst. Phys. Conf. Ser. No 158
Paper presented at Applied Superconductivity, The Netherlands, 30 June–3 July 1997

The use of the image method for the calculation of the levitation systems with ideal type-II hard superconductors

A A Kordyuk and V V Nemoshkalenko

Institute of Metal Physics, Kiev 252680, Ukraine

Abstract. The development of the melt textured and single crystal HTS with a very high pinning makes these materials very close to ideal type-II hard superconductors. We have applied the image method for calculation of the static and dynamic parameters of the system where a permanent magnet levitates above such a superconductor. We have found the exact solution of the levitation forces and elasticity for the point magnetic dipole over a flat ideal hard superconductor. We have checked the obtained results for melt textured HTS samples and found a range of magnetic fields where the experimental and theoretical results are in a full agreement.

1. Introduction

The investigation of the magnetic levitation was strongly stimulated by the discovering of the high temperature superconductors (HTS) with an opportunity to use a liquid nitrogen. A lot of works described the properties of different systems with levitation then. Many interesting results about macroscopic magnetic properties of the HTS were obtained [1–3] but the interest in these systems for large scale applications apears only with the development of the melt-textured HTS technology [4]. The levitation systems with such a HTS are actively studied last time [5,6].

In previous papers [7–9] we described the elastic properties of the permanent magnet (PM)—HTS system. We had shown that in such a system the granular HTS at the liquid nitrogen temperatures may be considered as a set of small isolated superconducting grains. It was true for the elastic properties of such a system and for its dampings determined by a.c. energy losses in the HTS sample. The melt-textured large grain HTS samples very differ from granular ones in levitation properties. Firstly, they have very strong pinning resulting in the absence of the effect of the PM rise above HTS sample at its cooling [7]. Secondly, the small isolated grains approximation can not work for large grains.

2. Advanced mirror image method

In this paper the absolutely hard superconductor approach is used. The sense of this approach is to use the surface shielding currents to calculate the magnetic field distribution outside the superconductor and to obtain from it the elastic properties of the PM—HTS

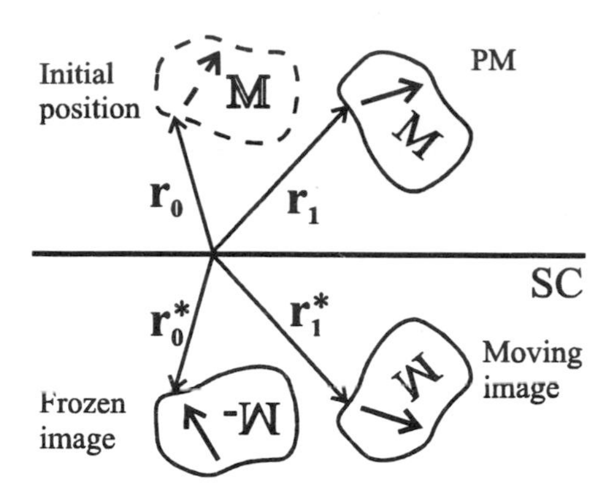

Fig. 1. The advanced mirror image method illustration

system. The magnetic field inside such ideal superconductor $\mathbf{B}(\mathbf{r})$ does not change with PM displacements. The feasibility of this approximation is determined by the condition $d \ll L$, where d is the field penetration depth and L is the character system dimension (firstly the distance between PM and HTS). With such approximation this problem has the exact analytical solution for the case of a magnetic dipole over a flat superconductor in the field cooled (FC) case.

To describe the FC behavior of the PM the advanced mirror image method was applied. The method is illustrated by Fig. 1. Its distinction from the usual one, which is applied to the type-I superconductors, is in the using of the frozen PM image that creates the same magnetic field distribution outside the HTS as the frozen magnetic flux does. From the uniqueness theorem, the magnetic field distribution in an area with no induced currents is uniquely determined by the normal field component on its boundary. In other words, the distribution of this component determines the PM—HTS interaction, and in turn is determined by the PM initial position (FC position). For the PM with initial position $\mathbf{r}_0 = (x_0, y_0, z_0)$ and volume magnetization $\mathbf{M}$ (see Fig. 1, z axis is vertical) that generates the magnetic field $\mathbf{H}(\mathbf{r} - \mathbf{r}_0, \mathbf{M})$ the normal magnetic field component on HTS surface $\rho = (x, y, 0)$ is equal to the same component of its reverse image with $\mathbf{r}_0^* = (x_0, y_0, -z_0)$ and $-\mathbf{M}^*$ (the operation * maps any vector symmetrically about ρ-surface)

$$H_z(\rho - \mathbf{r}_0, \mathbf{M}) = H_z(\rho - \mathbf{r}_0^*, -\mathbf{M}^*). \tag{1}$$

It is required that $H_z(\rho)$ should be unchanged at any PM displacements $\delta\mathbf{r} = \mathbf{r}_1 - \mathbf{r}_0$ (and rotations $\mathbf{M} \to \mathbf{M}_1$) from the initial position. To do this the presence of another image with $\mathbf{M}_1^*$ is required. This image moves with PM to $\mathbf{r}_1^*$ position ($H_z(\rho - \mathbf{r}_1, \mathbf{M}_1) + H_z(\rho - \mathbf{r}_1^*, \mathbf{M}_1^*) = 0$). Thus, the interaction between the PM and shielding current can be described by the interaction of the PM with net field of two images

$$\mathbf{H}_{\mathrm{im}}(\mathbf{r}) = \mathbf{H}(\mathbf{r} - \mathbf{r}_0^*, -\mathbf{M}^*) + \mathbf{H}(\mathbf{r} - \mathbf{r}_1^*, \mathbf{M}_1^*), \tag{2}$$

and for the field outside and inside the superconductor we can write:

$$\mathbf{B}(\mathbf{r}) = \begin{cases} \mathbf{H}(\mathbf{r} - \mathbf{r}_0, \mathbf{M}) + \mathbf{H}_{im}(\mathbf{r}), & z > 0 \\ \mathbf{H}(\mathbf{r} - \mathbf{r}_0, \mathbf{M}), & z < 0 \end{cases} \tag{3}$$

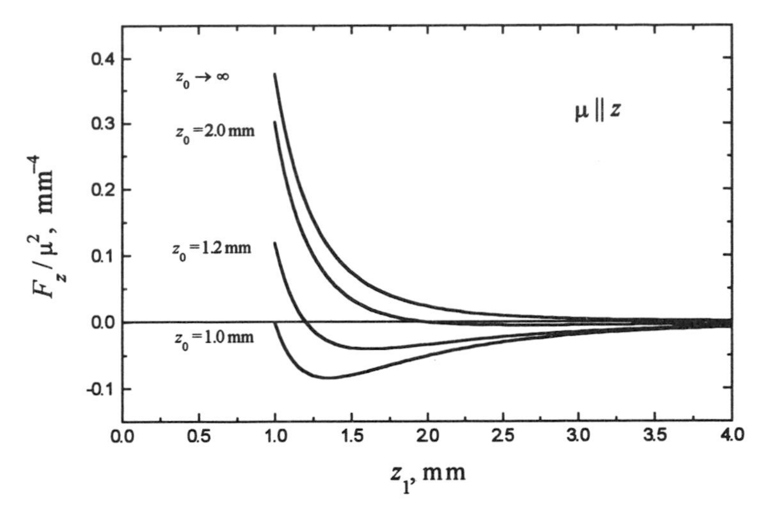

Fig. 2. The levitation force acting on the magnetic dipole with $\mu \parallel z$ for the different its FC positions z_0 *vs* its distance to the superconducting surface z_1.

The above relations are true for any shape of PM with any **M** distribuion in the PM volume V. For the case of the point magnetic dipole with the magnetic moment $\mu_0 = \int_V \mathbf{M}\,dV$ the analytical solution can be obtained.

The force acting on the magnetic dipole

$$\mathbf{F}(\mathbf{r}_1) = (\mu_1 \cdot \nabla)\,\mathbf{H}_{im} = \left(-(\mu_1 \cdot \nabla)\,\nabla\left(\frac{(-\mu_0{}^* \cdot (\mathbf{r} - \mathbf{r}_0{}^*))}{|\mathbf{r} - \mathbf{r}_0{}^*|^3} + \frac{(\mu_1{}^* \cdot (\mathbf{r} - \mathbf{r}_1{}^*))}{|\mathbf{r} - \mathbf{r}_1{}^*|^3}\right)\right)_{\mathbf{r} = \mathbf{r}_1} \tag{4}$$

can be calculated symbolically for any PM displacement and μ direction. For example, for $\mathbf{r}_1 = (x_0, y_0, z_1)$ and $\mu \parallel z$

$$F_z(z_1) = 6\,\mu^2\,[(2z_1)^{-4} - (z_1 + z_0)^{-4}]. \tag{5}$$

The $F_z(z_1)$ dependencies for different z_0 are presented in Fig. 2.

The elastic properties of PM—HTS system $\omega = \omega_0[1 + \gamma\,(A/z_0)^2]$ (oscillation frequencies ω_0 and nonlinearities γ, A is the PM amplitude) can be obtained from the expansion of $\mathbf{F}(\delta\mathbf{r})$ on $\delta = \delta s / z_0$ (for any mode $s = x, y, z$):

$$F_s(\delta) = -m\,\mu^2 z_0^{-4}(k_0\delta + k_1\delta^2 + k_2\delta^3) + O(\delta^4), \tag{6}$$

$$\omega_0 = \mu\sqrt{\frac{k_0}{m z_0^5}}, \quad \gamma = \frac{3k_2}{8k_0} - \frac{5}{12}\left(\frac{k_1}{k_0}\right)^2, \tag{7}$$

where m is PM mass. The values of the coefficients k and γ are presented in the Table I.

For the melt-textured HTS where the energy losses have predominantly hysteretic nature the feasibility of such approximation can be determined from critical state model (the

Table I. The values of the coefficients k and γ are calculated from (7) and (8).

μ direction	Mode s	k_0	k_1	k_2	γ
$\mu \parallel z$	$x = y$	3 / 8	0	–45 / 128	–45 / 128
	z	3 / 4	–45 / 16	105 / 16	–165 / 64
$\mu \parallel x$	x	9 / 32	0	–75 / 256	–25 / 64
	y	3 / 32	0	–15 / 256	–15 / 64
	z	3 / 8	–45 / 32	105 / 32	–165 / 64

thickness of the layer carrying the critical current J_c must be well less than PM—HTS distance) and for the above configuration

$$z_0 >> d = \frac{c}{4\pi} \frac{h(\rho, s)}{J_c}, \tag{8}$$

where $h = 2\,dH_\rho/dz$. This condition is much more stronger than it is necessary to validate the using of the described approach for melt-textured HTS. Even for $J_c \sim 10^4$ A/cm² and for $h \sim 100$ Oe the penetration depth $d \sim 0.1$ mm. The experimental values of resonance frequencies that we obtained for PM over the single domain HTS sample system ($m = 0.021$ g, $\mu = 1.6$ g cm³, $z_0 = 2.5$ mm) are in complete agreement with theoretical ones [10].

3. Conclusions

The obtained results can be usefull to examine the quality of single crystal HTS samples. If the melt-textured sample has more than one domain the resonance frequencies and levitation forces are appreciably reduced. Thus, the elastic properties of the PM—HTS system can be used to obtain the information about "granularity" of such HTS samples but to determine from this the number of domains per sample the additional investigations are needed.

The exact analitical solutions that were obtained for the point magnetic dipole over HTS surface can be used as a limit case to check the correctness of the numerical calculations of the elasticity in arbitrary PM–HTS systems.

References

[1] Moon F C, Yanoviak M M and Ware R 1988 *Appl. Phys. Lett.* **52** 1534

[2] Brandt E H 1988 *Appl. Phys. Lett.* **53** 1554

[3] Grosser R, Jäger J, Betz J, and Schoepe W 1995 *Appl. Phys. Lett.* **67** 2400

[4] Salama K, Selvamanickam V, Gao L and Sun K 1989 *Appl. Phys. Lett.* **54** 2352

[5] Yang Z J, Hull J R, Mulcahy T M and Rossing T D 1995 *J. Appl. Phys.* **78** 2097

[6] Sugiura T and Fujimori H 1996 *IEEE Trans. Magn.* **32** 1066

[7] Kordyuk A A and Nemoshkalenko V V 1996 *Appl. Phys. Lett.* **68** 126

[8] Nemoshkalenko V V and Kordyuk A A 1995 *Low Temp. Phys.* **21** 791

[9] Kordyuk A A and Nemoshkalenko V V 1996 *Journal of Supercond.* **9** 77

[10] Kordyuk A A and Nemoshkalenko V V 1997 *Phys. Metal. and Adv. Technol.* to be published

Inst. Phys. Conf. Ser. No 158
Paper presented at Applied Superconductivity, The Netherlands, 30 June–3 July 1997

Comparison of the Levitation Force Measurements of YBCO, BSCCO-2212 and (Tl,Pb)BCCO-1223 at Different Temperatures

J. H. Albering[1], S. Gauss[1], C. L. Teske[2], Joachim Bock[1]

[1]Hoechst AG, Corporate Research & Technology, G864, D-65926 Frankfurt am Main, Germany
[2]Institut f. Anorganische Chemie, Christian-Albrechts-Universität, Otto-Hahn-Platz 6/7
D-24098 Kiel, Germany

Abstract. Levitation force measurements on various HTS materials such as sintered, MMTG and single-domain YBCO, MCP-BSCCO-2212 and (Tl,Pb)BCCO-1223 were performed in the temperature range of 15 - 85 K. The dependence of the levitation forces on domain (crystallite) size and the sample thickness have been studied as a function of temperature.

1. Introduction

Measurements of the levitation forces between superconducting bulk parts and permanent magnets give basic information for the design and optimization of bearings for flywheels, transport systems or liquid gas containers. Specially for the liquid hydrogen technology a number of applications may be realized (LH_2-containers with superconductor-magnet bearings or pumps for cryo fluids). The main advantage of such LH_2-application would be the inherent already present cooling. Thus, no additional expensive cryosystem has to be installed to operate the bearing or pump system. While the properties of various YBCO materials are well investigated for LN_2 temperature [1-3, and literature quoted therein], the knowledge about the levitation properties of YBCO and other HTS ceramics such as BSCCO-2212 or the TlBCCO-1223 phases at lower temperatures is relatively poor [4]. The present publication gives a comparison of the temperature dependent levitation properties of various HTS materials.

2. Sample Preparation

Levitation measurements were performed on various HTS materials such as single-domain, multi-domain and sintered YBCO, melt-cast-processed-(MCP)-BSCCO-2212 and sintered (Tl,Pb)BCCO-1223. The single domain YBCO samples were obtained by melt-processing top-seeded (Nd-123) YBCO bulk parts of different geometries. The samples containing YBCO-123, YBCO-211 (or Y_2O_3) and PtO_2 were heat treated by the standard methods [5, 6]. Large single-crystalline samples with domain sizes up to 35 mm diameter were obtained by this method. The multi-domain samples were processed by melt-texturing cold-isostatically pressed pellets of the powder mixtures described above and YBCO powders with silver additives [7]. Samples with different grain sizes were produced by modification of the synthesis conditions. BSCCO-2212 plates were obtained by the MCP process as published before [8]. The samples casted by that method were either of cylindrical shape or had the form of large rectangular plates (up to 20 cm length). Levitation measurements were performed on flat cylindrical or square samples cut from the large bulk parts. Pellets of the (Tl,Pb)BCCO-1223 materials were prepared by uniaxially pressing of the respective ceramic precurcor powders followed by several heat treatments [9]. The size of the samples used for the levitation measurements varied between 28 - 35 mm in diameter and 1-15 mm thickness. The dependence of the levitation forces of YBCO and BSCCO-2212 on the sample thickness were performed on one sample each, which was thinned mechanically between the measurements.

3. Experimental Set up

For the investigation of the levitation properties at temperatures lower than 77 K a new test bench had to be developed. The samples are fixed in vacuum on a two-stage cryocooler. A magnet mounted on a tripod, which is connected to three force sensors (for the measurement of forces F_x, F_y and F_z in three dimensions) can be moved 3-dimensionally above the sample. In the standard measuring mode the magnet system is moved only in z-direction towards the superconductor. Due to the thermal isolation of the cryocooler the minimum superconductor (SC)-permanent magnet (PM) distance is limited to 4.6 mm. Samples up to 38 mm diameter and 20 mm thickness can be mounted in our experimental set up. The magnet ($SmCo_5$) used for most of the measurements presented in this publication has a cylindrical shape (25 mm diameter, 15 mm height) and a flux density of 0.34 T. The construction of the apparatus and the possible measurement modes are described in more detail elsewhere [10].

4. Results

The levitation forces of YBCO (grain sizes ranging from 0.1-32 mm), BSCCO-2212 and (Tl,Pb)BCCO-1223 have been measured in the temperature range between 15-85 K. The characteristic zero-field-cooling (zfc) levitation vs. distance curves for some selected materials and temperatures are shown in Fig. 1.

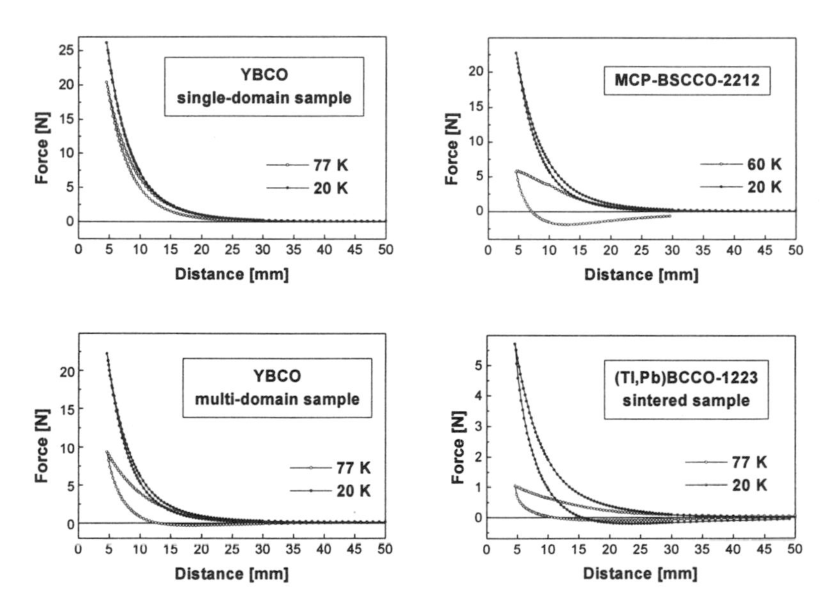

Fig. 1 Hysteresis loops in the levitation force measurements of YBCO, BSCCO-2212 and (Tl,Pb)BCCO-1223 at various temperatures.

Due to the set up of our test bench the minimum SC-PM distance is limited to 4.6 mm. In order to obtain data for the maximum levitation forces, comparable with those of other research groups, the curves can be fitted by an exponential decay function and extrapolated to a distance of 0.1 mm. Fig. 2 shows the results of such a fitting routine for the data measured at 20 K with a $SmCo_5$-magnet system supported by iron flux collector poles.

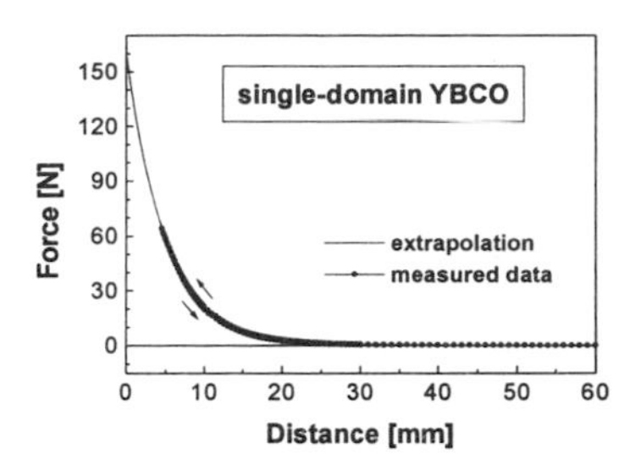

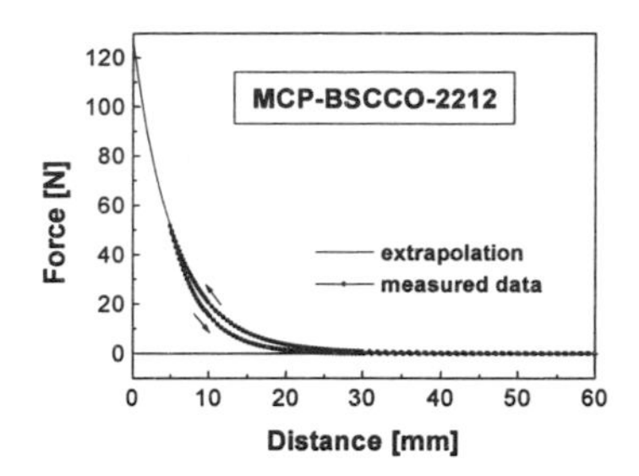

Fig. 2 Levitation force measurements of single-domain YBCO and MCP-BSCCO-2212 at 20 K using a magnet with iron collector poles. The measured data (dots) were fitted and extrapolated to zero distance by an exponential decay function.

Fig. 3 shows the temperature dependence of the maximum levitation force values for several HTS materials (taken from zfc measurements). The maximum levitation forces for single-domain YBCO were found to be 20 N at 77 K and 26 N at 15 K (4.6 mm distance, 0.34 T magnet). Multi-domain YBCO samples show significantly lower levitation forces at 77 K (up to 10 N at 4.6 mm distance), but the forces measured at 15-40 K are in the same range like those of the single-domain samples. This can be ascribed to the disappearence of the weak links between the YBCO-grains at low temperatures. A similar behaviour was observed for the BSCCO-2212 material (F_{max} = 22 N, 15 K, 4.6 mm distance) . In opposite to YBCO the increase of the levitation force at lower temperatures for BSCCO is due to the strong increase of j_c. For sintered YBCO and (Tl,Pb)BCCO-1123 samples both effects can be ob-served simultaniously (F_{max} = 6-7 N, 15 K, 4.6 mm distance). In the temperature range from 40-80 K the increasing forces can be correlated with the increasing j_c of the materials, while the low temperature measurements are governed by the disappearence of the weak link effect.

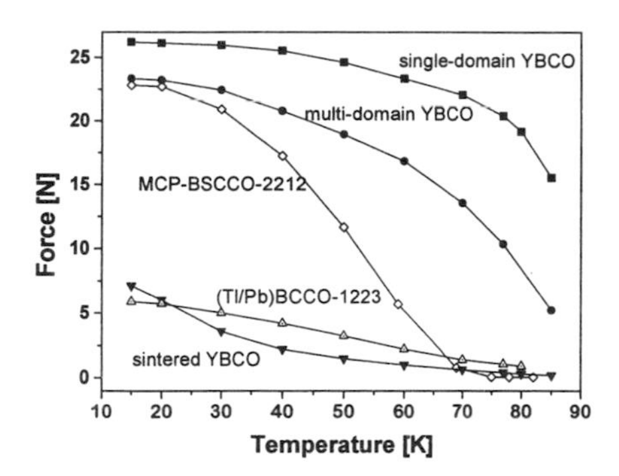

Fig. 3 Temperature dependent levitation forces of various HTS materials. The data represent the maximum force values of zero-field-cooling levitation measurements at a superconductor-magnet distance of 5 mm.

The dependence of the levitation forces for YBCO pellets of various crystallite sizes (0.1-32 mm) but the same sample size is shown in Fig. 4. Obviously, there is a continous transition of the levitation properties from sintered to single-domain samples. With increasing grain size the effect of the temperature dependence on j_c and the flux pinning become the dominating parameters for the levitation forces, while the properties of the samples with small crystallites and a great number of grain boundaries are limited by the temperature dependence of the quality of the intergrain contacts.

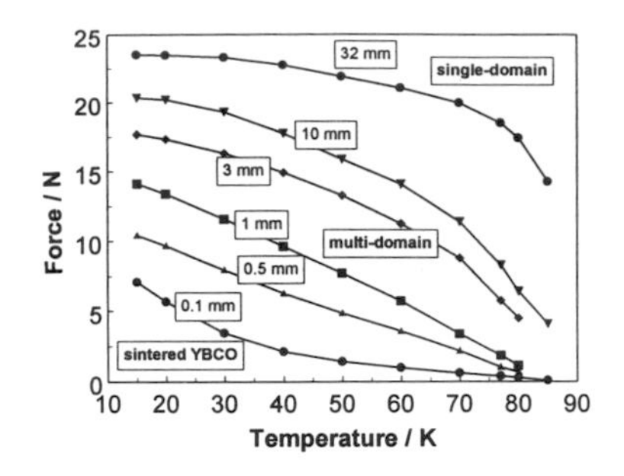

Fig. 4 Influence of the crystal size on the temperature dependent levitation forces of YBCO. Overall sample geometries are similar.

The levitation measurements on a MMTG-YBCO and a MCP-BSCCO-sample as a function of the sample thickness revealed that there is only a weak dependence of the SC-PM forces above a „*critical thickness*". Below that value (~5 mm for multi-domain YBCO, ~3 mm for BSCCO-2212) the forces decrease dramatically with decreasing thickness, due to the penetration of magnetic flux through the samples.

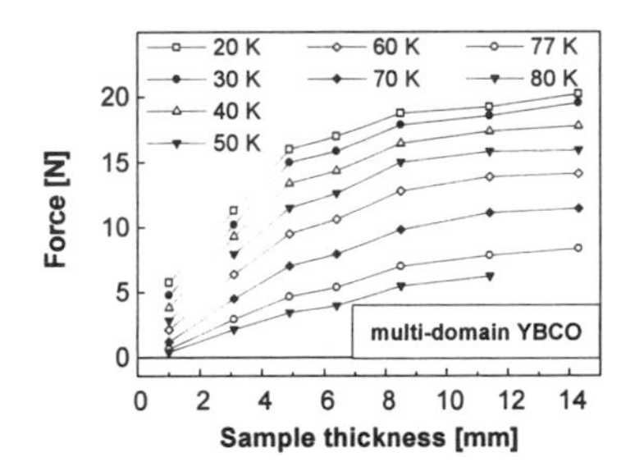

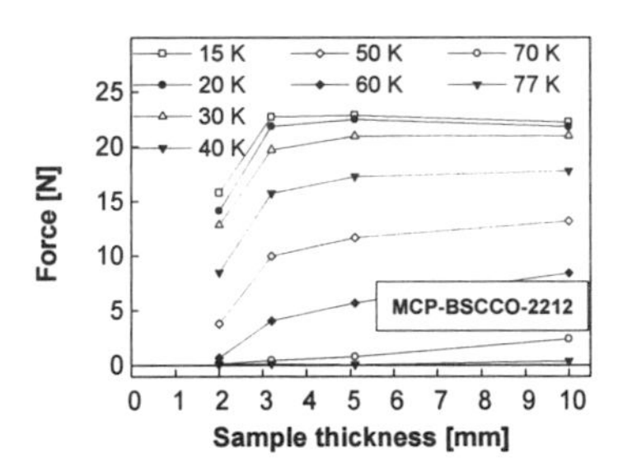

Fig. 5 Dependence of the levitation force on the sample thickness (5 mm distance SC-magnet) for multi-domain YBCO and MCP-BSCCO-2212.

5. Conclusions

A comparison of the levitation properties of different YBCO materials, MCP-BSCCO-2212 and (Tl,Pb)BCCO-1223 was presented for different temperatures. The weak properties and the temperature dependence of jc can be analysed by the levitation forces at low temperatures. At 20 K multigrain YBCO and MCP-BSCCO exhibit similar levitation forces than single-domain YBCO.

Acknowledgements

The authors would like to thank H. May for supply of two magnet systems. This work was partially funded by the german BMBF under contract 13N6939/9.

References

[1] P. Boegler, C. Urban, H. Rietschel, H. J. Bornemann, Appl. Supercond. **2** (1994) 315.
[2] H. Teshima, M. Morita, M. Hashimoto, Physica **C 269** (1996) 15.
[3] A. Sanchez, C. Navau, Physica **C 275** (1997) 322.
[4] S. Gauss, S. Elschner, 4. Intern. Symposium on Magnetic Bearings, Zürich, 1994.
[5] M. Murakami, Supercond. Sci. Technol. **5** (1992) 185.
[6] M. Murakami, N. Sakai, T. Higuchi, S. I. Yoo, Supercond. Sci. Technol. **9** (1996) 1015.
[7] S. Gauss, S. Elschner, Advances in Croygenic Engineering (Materials) **38** (1992) 907.
[8] S. Elschner, J. Bock, M. Brand, S. Gauss, Appl. Supercond., Proc. Eucas, Göttingen (1993) 285.
[9] C. L. Teske, J. H. Albering, S. Gauss, Proceedings, Eucas, Eindhoven (1997).
[10] S. Gauss, J. H. Albering, Proceedings Eucas, Eindhoven (1997).

Inst. Phys. Conf. Ser. No 158
Paper presented at Applied Superconductivity, The Netherlands, 30 June–3 July 1997

Critical current homogeneity of Bi-2223 tapes determined by Hall-magnetometry.

H-P Schiller, K Grube, B Gemeinder, H Reiner, W Schauer

Forschungszentrum Karlsruhe, Institut für Technische Physik, D-76021 Karlsruhe, Germany

Abstract. An inductive method has been used to determine the critical current density j_c for Y-123 films and Bi-2223/Ag tapes in dependence on magnetic field B, temperature T, and electric field criterion E. Miniature Hall sensors record the magnetic field B_s produced by screening currents in the sample which flow in response to an external field sweep dB_e/dt. For ring shaped YBaCuO films j_c and the E(j) characteristic can easily be determined from B_s and the ring geometry assuming j_c to be constant over the ring width. - The spatially inhomogeneous current distribution in multifilamentary BiSrCaCuO tape conductors impedes a quantitative measurement of j_c using the Hall-magnetometry method. Rather it offers a qualitative test of the j_c homogeneity along and across the tape direction using an array of 15 Hall sensors to measure the B_s field distribution. Quantitative results for the critical current were obtained by multi-contact transport measurements. Tapes were investigated at 77 K by pulling them through the field B_e in the midplane gap of a copper split coil solenoid. The operating capability of this test procedure was demonstrated by measuring the j_c homogeneity of a 17 m long Bi-2223/Ag tape.

1. Introduction

It is well known that the critical current density of long length Bi-2223/Ag tapes is considerably degraded compared to short sample results. For the latter values close to $7 \cdot 10^4$ A/cm² can be achieved at 77 K / 0 T / 1 μV/cm, whereas the current density at 1 km tape length is reduced by a factor of about four to 1.5 or $2 \cdot 10^4$ A/cm² [1]. In addition, the current density values actually achieved have been estimated to be only about 5 - 10 % of the intrinsic current carrying capacity of the BiSrCaCuO material. This hidden potential is due to inherent limitations, restricted grain connectivity, microstructural defects and technical shortcomings. Exploring the current limiting mechanisms is thus a main task in view of technical applications. - In our contribution we describe a fundamental step in this direction: testing long length Bi-2223 tapes with respect to their critical current homogeneity. We use a contactless, inductive method which measures the magnetic field distribution produced by induced screening currents.

2. Experimental

The experimental procedure to determine j_c and E(J) is described in detail in [2, 3] for c-axis YBaCuO films. Screening currents are induced in the sample by an external magnetic field sweep dB_e/dt. Currents flow with the critical current density j_c according to the critical state model and produce a sample field B_s which is measured by a miniature Hall sensor (active area 50 μm × 50 μm). B_s is the z-component of the sample field, perpendicular to the sample plane, at the position z_0 of the Hall sensor (some tenths of a mm above the sample). When measuring B_s the external field B_e is compensated by a second Hall sensor. The magnetic field profile parallel to the sample surface at $z = z_0$, $B_s(x,y,z_0)$, is uniquely related to the two-dimensional screening current distribution in the sample. - The determination of j_c from B_s is straightforward for a ring sample with the ring width small compared to its radius and $j_c = const$ over the width. In this case the splitting of the hysteresis loop $B_s(B_e)$ is directly proportional to $j_c(B_e)$. As shown previously [4] the assumption $j_c = const$ is a good approximation also for disk shaped and rectangular samples, therefore $j_c = G \cdot B_s$ is generally valid where G is some geometrical function which can be computer calculated. - The external field $B_e = B_0 + B_r(t)$ may be composed of a background field B_0 ($\leq$10 T) and the ramp field $B_r(t)$ ($\leq$140 mT) which induces the screening currents [2]. The electric field is determined by the sweep rate: $E \propto dB_e/dt = dB_r/dt$. The E-field contribution from the sample field change dB_s/dt is about two orders of magnitude less and can be neglected, if the measurements are performed in the state of full field penetration. In the case of constant B_e (relaxation experiment), however, dB_s/dt is the only contribution which determines the electric field criterion.

Besides measurements on films, Bi-2223/Ag tapes were investigated at 77 K (LN_2) by pulling the tape through the static field in the midplane gap of a copper split coil solenoid (Fig. 1). The Hall probe is placed beneath the coil winding where the field gradient dB_e/dr is approximately constant and the central coil field of 140 mT maximum is reduced by about 40 %. When passing the coil gap the tape magnetization or B_s goes through a hysteresis loop $B_s(B_e)$ from B_e=0 to the maximum B_e at the coil center and back to $B_e = 0$. Reversing the pulling direction means measuring B_s at the same B_e but on the opposite loop branch due to the asymmetric Hall sensor position. The width of the loop is determined by the irreversible part of the magnetization. The variation of B_s along the tape length is a measure of the inhomogeneous flow of the shielding currents. For the 55 filament tape investigated the magnetization (B_s) has its maximum in the virgin run at $B_e \approx 6$ mT. For $B_e \geq 20$ mT the hysteresis loops $B_s(B_e)$ of subsequent runs coincide with the virgin curve. The tape speed $v = dr/dt$ determines the electric field $E \propto (dB_e/dr) \cdot v$ and via the E(j) relation the critical current density j_c and B_s. To measure the lateral B_s distribution a linear Hall sensor array

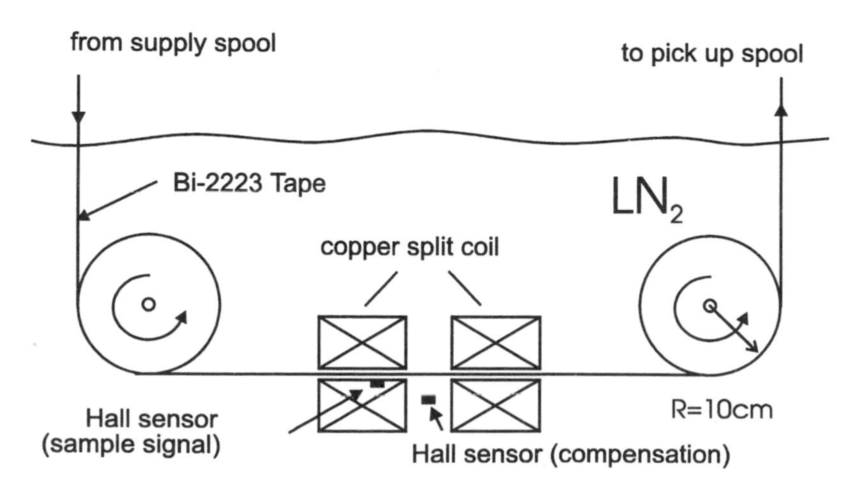

Fig.1 Schematic view of the rig to measure the critical homogeneity by pulling the Bi-2223 tape through the split coil gap.

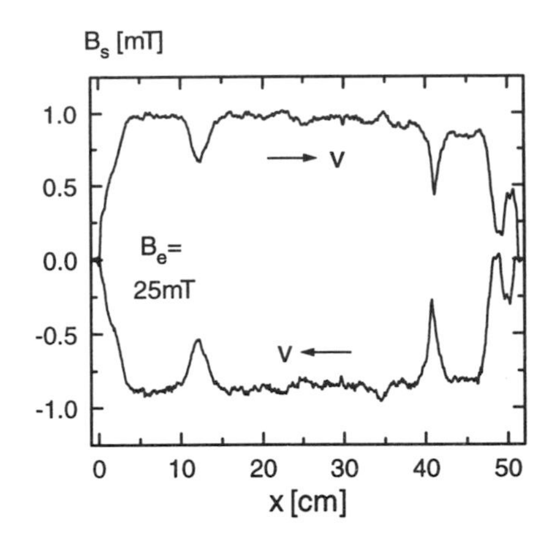

Fig.2 Hall field signal B_s for forward and backward speed direction of the Bi-2223 tape (v = 3 mm/s).

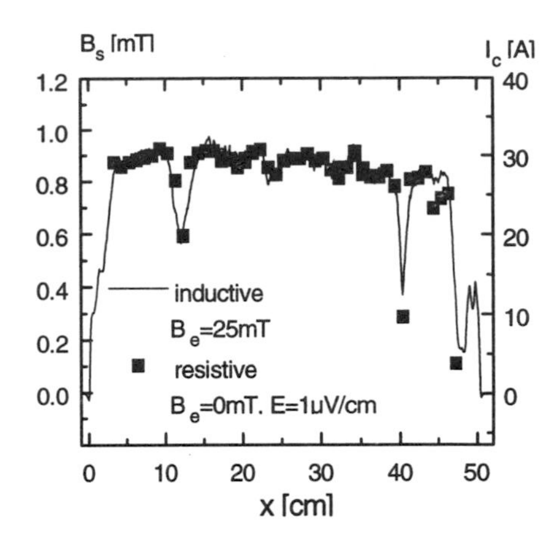

Fig.3 Inductive B_s signal and resistive critical current I_c (■) for the same 50 cm tape section as shown in Fig.2.

consisting of 15 Hall sensors 50 μm × 50 μm each has been used. - As the multifilamentary structure of the tape conductor with an inhomogeneous j_c distribution for different filaments and even within each filament impedes the direct conversion $B_s \rightarrow j_c$ the measured B_s values along and across the tape serve as a quality test of the j_c homogeneity. Quantitative results for the critical current I_c were obtained from additional resistive measurements. For every neighboring pair (1 cm distance) of 50 voltage taps I_c was determined at 1 μV/cm (Fig.3).

3. Results and discussion

Figure 2 shows the B_s signal along the initial 0.5 m of a 17 m long, 55 filament Bi-2223/Ag tape. The modulation of B_s reflects inhomogeneities of the current flow in the tape, obviously due to structural defects. For the opposite pulling direction the reverse signal is quite symmetric with respect to the $B_s = 0$ axis. Thus the contribution of the reversible magnetization to the signal is negligible.

To correlate the variation of B_s along the tape length x with the critical current I_c multicontact resistive measurements have been performed for the same tape section. Figure 3 shows that the B_s variation is exactly reproduced by the resistively determined I_c. Both distributions $B_s(x)$ and $I_c(x)$ coincide if the scaling coefficient β, $B_s = 1$ mT $\leftrightarrow I_c = \beta$ A, is properly adjusted (β = 33 in Fig.3). This proves that the sample field B_s is really suited to reveal the I_c length inhomogeneities. Both types of measurement probe the same intergranular current system in the accessible external field range up to 85 mT.

Tests on a two meter section of the Bi-2223 tape are presented in Fig. 4. The sample field B_s shows a periodic collapse every ~10 cm which can be correlated with the periodic appearance of bubbles at the tape surface. The degradation is a real I_c reduction as again demonstrated by resistive control measurements reproducing the same periodic structure. Additional 'long wave length' (~1 m) periodic I_c fluctuations were observed which may be attributed to a temperature inhomogeneity during the heat treatment of the tape bobbin. The severe degradation of the critical current, at some positions up to 50 %, is mainly caused by a mechanical maltreatment, e.g. small radius bending with opposite curvature.

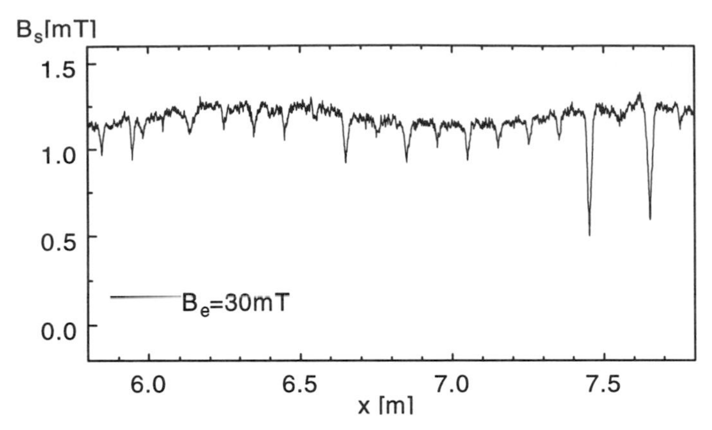

Fig.4 Long length test of the tape current homogeneity showing periodic defects attributed to the formation of bubbles at the tape surface.

Lateral measurements of B_s using a linear Hall probe array of 15 sensors provide further information of the current distribution in the tape. The B_s profile shows the same V-shape as previously found for a single filament Bi-2223 conductor [5].

4. Summary

An inductive method is used to characterize continuously the j_c homogeneity of long length Bi-2223 tapes at 77 K. Shielding currents are induced in the tape when pulling it through the field of a copper split coil. They produce a sample field B_s which is measured by a Hall sensor. The variation of B_s and the resistively determined critical current distribution along the tape length coincide. Thus both intergrain transport and shielding current are affected in the same way by structural defects at the external field accessible (< 100 mT). Therefore recording the sample field serves as an easy and fast critical current homogeneity test.

References

[1] Flükiger R, Grasso G, Grivel J C, Hensel B, Marti F, Huang Y, Perin A 1996 *Critical Currents in Superconductors* (World Scientific, Singapore) 69-74

[2] Schauer W, Windte V, Dickgießer M, Yin D, Neumann C, Maurer W, Reiner H 1994 *Adv. in Cryogen. Engineering* **40** 393-400

[3] Komarkov D, Neumann C, Schiller H P, Gemeinder B, Schauer W 1995 *Applied Superconductivity EUCAS 1995, Edinburgh* (IOP Publishing LTD, Bristol) 959-62

[4] Polák M, Windte V, Schauer W, Reiner H, Gurevich A, Maurer W, Wühl H 1992 *Supercond. Sci. Techn* **5** S403-6

[5] Paasi J A J, Lahtinen J 1993 *Physica C* **216** 380-2

Inst. Phys. Conf. Ser. No 158
Paper presented at Applied Superconductivity, The Netherlands, 30 June–3 July 1997

Construction of the Current-Voltage Characteristic in a 12 Decade Voltage Window using Magnetisation Measurements.

J Vanacken, K Rosseel, A S Lagutin, L Trappeniers, M Van Bael, D Dierickx, J Meersschaut*, W Boon, F Herlach, V V Moshchalkov and Y Bruynseraede

Laboratorium voor Vaste-Stoffysica en Magnetisme, *Instituut voor Kern en Stralingsfysica, Katholieke Universiteit Leuven, Celestijnenlaan 200 D, B-3001 Leuven, Belgium.

Abstract. Pulsed Field Magnetisation Measurements (PFMM) in a field range up to 50 Tesla are compared with SQUID and VSM magnetisation measurements in lower fields. Experiments on fast melt processed $YBa_2Cu_3O_7$ with Y_2BaCuO_5 inclusions show high values of the "magnetic" critical current ($j_C >$ 10^9 A/m^2 at T=20 K and $\mu_0 H$~4 T). The j_C values are quite different for the pulsed field, the SQUID and VSM magnetisation measurements since the voltage criterion associated with these different magnetisation measurement techniques extends over a 12 decade window, thus allowing us to construct the *E-j* characteristic. The PFMM seem to be most suitable to determine the "real" critical current which is defined as the current density j inside the sample reaching its limiting value j_c.

1. Introduction

The critical parameters of the high temperature superconductors (HTSC) are very often only known only in the d.c. range (H < 10T) of magnetic fields and therefore the high field values (H > 30T) of these parameters must be found from extrapolations. These extrapolations involve some uncertainty. Therefore, a direct measurement of the superconducting critical parameters in high fields is still of great importance. To generate high fields, however, pulsed field techniques are needed. In this way, fields up to μ_0H=60 Tesla or more can be obtained, with a typical duration of ~ 20 msec. One particular feature in measuring superconductors under these conditions is that not only the field magnitude but also the applied magnetic field sweep rate (dH/dt) is a function of time t. The latter is very important, because shielding currents in these materials are sensitive to the dH/dt which determines the variation of the voltage V in the voltage-current (I-V) curves [1].

2. Experimental techniques

Sample preparation method: The precursor powders, Y_2O_3 and $Ba_2Cu_3O_x$, corresponding to 80wt% $YBa_2Cu_3O_{7-x}$ and 20wt% Y_2BaCuO_5 nominal composition are prepared by freeze drying. These powders are cold isostatically pressed at 1200 MPa and put on a high quality $BaZrO_3$ substrate. The fast melt processing of these samples consists of a heating up to

1030°C, followed by a stabilisation at this temperature for about 2 hours. At this temperature the Y_2O_3 and $Ba_2Cu_3O_x$ react to form the 211 phase + liquid. A fast cooling of the melt with a rate of 120°/h, results in the growth of the 123 phase. The fast cooling rate prevents total decomposition of submicron 211 particles and reduces the coarsening of larger 211 inclusions, thus producing 211 particles with a mean size well below 1 μm. The as-grown samples consist of randomly ordered crystalline domains of about 0.5 mm. [2]

Magnetisation measurements: Three experiments using different techniques for measuring magnetisation (PFMM, VSM and SQUID) have been carried out on the same fast melt processed sample, which was machined into the shape of a cylinder, with radius R=8.3 10^{-4} m, and the volume V=1.62 10^{-9} m^3. Thc PFMM were performed in pulsed magnetic fields up to 50T with a constant pulse duration τ_0 of about 20 ms and at a temperature T=20K [3]. For pulsed field conditions a high measuring sensitivity (better than 10^{-2} emu in fields above 30 T) is achieved using an inductive magnetic sensor which has been especially designed for this purpose. For the steady field magnetisation measurements commercial magnetometers were used: a SQUID from Quantum Design (MPMS) and a VSM from O.I. (MagLab). All techniques have been calibrated using the same Ni-sample.

3. Model for Constructing the Current-Voltage Characteristic

The construction of the E(J) - relation is based on the Maxwell equations. We can estimate the electrical field E from: $\left|rot\ \vec{E}\right| = \frac{E}{R} = -\frac{\partial B}{\partial t} = -\mu_0 \frac{\partial (H+M)}{\partial t}$, resulting in

$$E = -R\mu_0 \frac{\partial (H+M)}{\partial t} \tag{1}$$

where E is the electrical field, B the magnetic induction, M the magnetisation, H the magnetic field, μ_0 = 1.2566 10^{-6} mkg/C^2, and R the radius of the sample. From the magnetic moment induced by a constant current density J_c inside a cylindrical sample we obtain: $m = \int_0^R j(r)\pi r^2\ h\ dr = \frac{1}{3}\pi\ h\ J_c\ R^3 = M\left(\pi\ R^2 h\right) = M\ V$, resulting in

$$J = J_c = \frac{3\ M}{R} \tag{2}$$

The above equations are calculated in cylindrical coördinates (ρ,θ,z) using $J_c(H)$ is constant (Bean model), and |rot $\boldsymbol{E}$| ~ E/ρ, which is satisfied if $\boldsymbol{E} = E_\varphi(\rho)\ \mathbf{e}_\varphi$, and $E_\varphi(\rho)/\rho \gg \partial E_\varphi(\rho)/\partial\rho$, a condition which is usually met, and in fact is an electrical analogue of the Bean model. In all the following derivations, the above formalism (1), (2) is used to calculate E and J.

4. Results and Discussion

From figures 1.a, 1.c and 1.e, it can be seen that the time-windows in which all the three experiments are performed are essentially different. The PFMM measurement is carried out in ~ 5-10 ms, the VSM measurements in ~ 2 s and the SQUID measurements in 2 minutes (times between recording two successive measuring points). The change in *dH/dt* is larger than the change in *dM/dt* for both the PFMM and VSM technique, in contrary to that for the

SQUID technique where *dM/dt* >> *dH/dt*. As a result, for the PFMM and VSM technique, both *dM/dt* and *dH/dt* contribute to the induced electrical field, whereas for SQUID only *dM/dt* does. All experiments shown in figure 2 are for fixed field $\mu_0 H = 4$ Tesla and for a fixed temperature T=20 K. The data are grouped into 3 parts, PFMM (raising & lowering branch), VSM and SQUID. Combining the E and J data gives the E-j relation as shown in Fig. 2. Thanks to the combination of the three different techniques used for magnetic measurements the induced electrical field can be changed over 12 decades, — from $E = 10^{-11}$ V/m to $E = 10^{1}$ V/m.

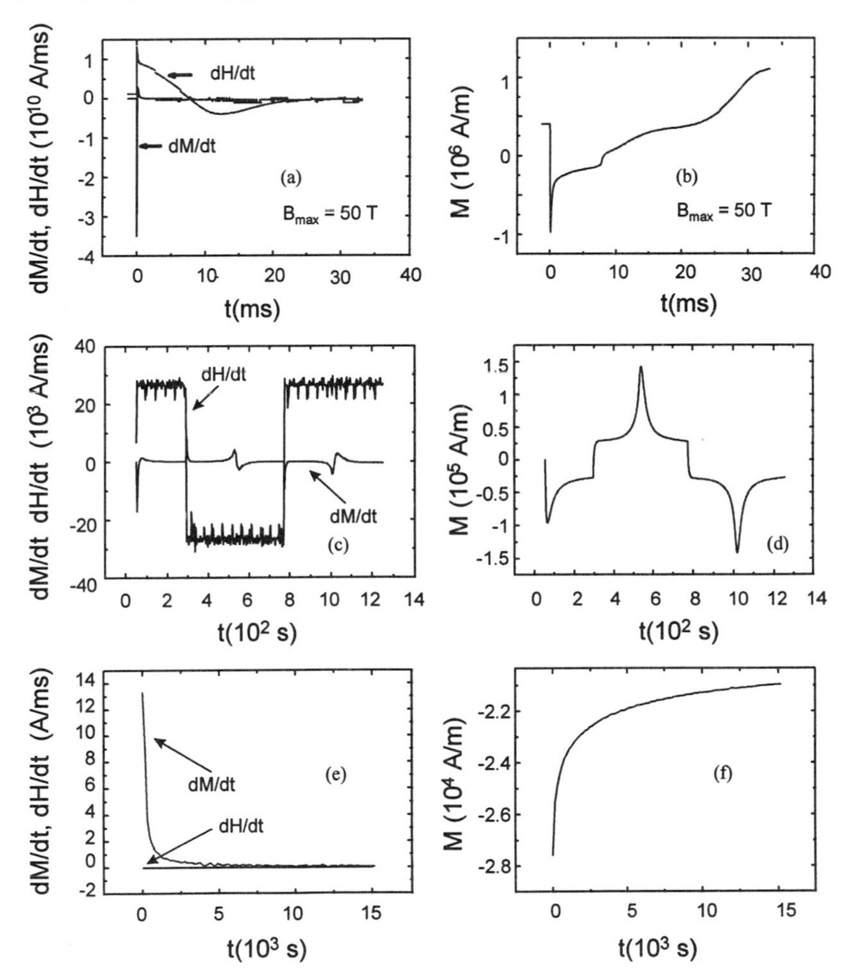

Figure 1. The time-derivatives of the magnetisation and field versus time obtained using the PFMM (a), VSM (c)and SQUID (e) technique. The magnetisation versus time for the PFMM (b), VSM (d)and SQUID (f) technique. The data in the first column will be used to determine the induced electrical field, whereas the data in the second column will determine the current density.

Although the electrical field changes over 12 decades, the current density is changed only by (1-2) decades, —from $J = 9\ 10^{7}$ A/m to $J = 1.7\ 10^{9}$ A/m. Therefore we can conclude that the E-J transition is rather sharp.

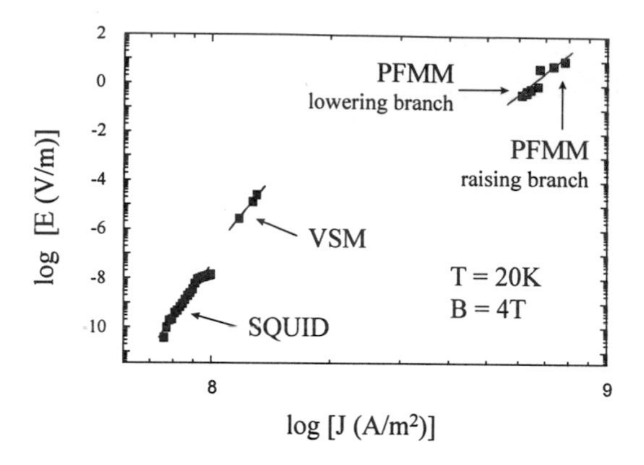

Figure 2. The E-J characteristic calculated using Eq.(1), (2) from the data presented in fig. 1. All data are for an external field $\mu_0 H$ = 4 Tesla, and at a temperature T=20K.

From the presented data, it is also clear that the *E-J* relation for the different techniques shows different slopes *dE/dj*. The SQUID and VSM data fall more or less on a straight line, indicating a power law in the E-J relationship. The high E values which are obtained by the PFMM technique, clearly don't fit onto the same straight line. This could be related to the fact that the *E-J* transition is usually sharper at low *E* and it becomes more smeared out at high *E* values. For such high E- values, we are close to the flux flow state, but still in the giant flux creep regime. Using the criterion given by Eq.3, we can estimate the upper values for μ_0 dH/dt within the framework of this flux creep model [4]:

$$j(H, {}^{dH}\!/_{dt}) = j_c \left[1 - \frac{kT}{U_c} \ln\left(\frac{v_0 \, H}{R \, {}^{dH}\!/_{dt}} \right) \right] \rightarrow j_c \tag{3}$$

The above limit $j \rightarrow j_c$ is valid when $dH/dt \rightarrow v_0 H / R$. Taking v_0 = 1 m/s, R= 8.3 10^{-4} m and H= 4 10^7 A/m, we find an upper limit for $dH/dt \sim$ 5 10^{10} A/ms. From figure 1.a, it becomes clear that we are indeed just below that value. In this way, it can be stated that the PFMM measures the current density, which is the closest one to the real critical current density j_c.

Acknowledgements

This work is supported by the FWO-Vlaanderen, the Flemish GOA and the Belgian IUAP programs. J.V., D.D. and A.L. are Postdoctoral Fellows supported by the K. U. Leuven Onderzoeksraad. L. T. and K. R. are Research Fellows supported by the I.W.T.

References

[1] Keller C, Kupfer H, Meier-Hirmer R, Wiech H, Selvamanickan R and Salama K 1990 *Cryogenics* **30** 410.
[2] Dierickx D and Van der Biest O 1995 *Eur. J. Solid State Inorg. Chem.* **32** 711.
[3] Lagutin A S, Vanacken J, Harrison N and Herlach F 1995 *Rev.Sci.Inst.* **66** 4267.
[4] Schnack H, Activation energies from magnetization relaxation & Vice versa in high-Tc, Ph. D. thesis, Acad. Pers B.V. Amsterdam, V.U. Amsterdam 1995.

Inst. Phys. Conf. Ser. No 158
Paper presented at Applied Superconductivity, The Netherlands, 30 June–3 July 1997

Granularity of the Critical Current Pattern and Degradation of Macroscopic Material Properties due to Mesoscopic Material Inhomogeneities

Th Klupsch†[1], P Diko‡, C Wende†, I Sonntag†, and W Gawalek†

† Institut für Physikalische Hochtechnologie Jena, PF 100239, D-07702 Jena, Germany

‡ Institute of Experimental Physics, Slovak Academy of Sciences, Watsonova 47, 04353 Kosice, Slovak Republic

Abstract. Because the defect structure of superconducting materials for cryomagnetic applications (e.g., melt textured YBCO) is characterized by length scales of several orders of magnitude, the critical current density must be considered, in general, as a space- dependent function $j_c = j_c(\boldsymbol{r})$ on a mesoscopic length scale L. It is shown that any regular critical current pattern for $j_c(\boldsymbol{r}) = const$ is unstable against such mesoscopic fluctuations of $j_c(\boldsymbol{r})$. In particular, there appears a granular current pattern due to the formation of local current loops on the characteristic length scale L while, at the same time, macroscopic material properties (remanent magnetic momentum, trapped induction) show a degradation. Analytical results within a consistent model description are presented.

1. Introduction

The melt-textured YBCO and RE-substituted YBCO materials are promising candidates for cryomagnetic applications. As well known, the process of melt-texturing enables the preparation of samples consisting of large quasi single-crystalline domains of the superconducting 123-matrix with specified mesoscopic defects in the μm-range (see, e.g.,[1], [2]). It has been proved that the inclusions of the 211-phase widely govern other details of the complex mesoscopic defect structure in two characteristic length scales represented by the mean diameter D_{211} and the mean next-neighbouring distance d_{211} of these 211-islands [3]. If any region with the diameter L_1 (say $L_1 \approx 3d_{211}$) is representative of all local properties of the complex mesoscopic defect structure, the defect structure is called homogenous if these local properties are not changed on length scales $\gg L_1$.

[1] E-mail: klupsch@main.ipht-jena.de

It is accepted that the 211-inclusions support the flux line pinning by creating additional microscopic defects within the surrounding matrix due to thermal microstress relaxation [1],[2]. But details (e.g., the creation of point-like or correlated pinning centres) are not known. We will assume that the critical current density j_c which is governed by the statistical properties of the defect structure can be defined, as a local quantity, on regions with the characteristic length dimension L_1. Then it is expected that j_c becomes enhanced in 211-high concentration regions, compared with the 123-YBCO matrix. So far, only few experimental investigations are known which qualitatively prove such a prediction [4].

During the past years it was shown by detailed polarization optical studies that also for the quasi single-crystalline domains of the melt-textured YBCO material, the mesoscopic defect structure is by no means homogenous (see, e.g., [5]) but can significantly change on distances of few d_{211}. But this means that on length scales $L > L_1 > d_{211}$, j_c becomes a random, space-dependent function $j_c = j_c(\boldsymbol{r})$. The effect of other structural deficiencies such as low angle grain boundaries (see,e.g.,[6]) may be included in such a description (sharply dropping down of $j_c(\boldsymbol{r})$). The existence of space-dependent fluctuations of j_c for melt-textured YBCO was recently proved directly by local susceptometric methods [7].

2. Persistent Current Pattern Deformation

The remanent state of an axially magnetized, single-domain cylinder (radius R_0, height l) with the (mean) crystallographic c-axis parallel to the cylinder axis will be considered. A cylinder coordinate system (r, φ, z) is used where the top cylinder surface is at $z = 0$. The persistent current flow $\boldsymbol{j}(\boldsymbol{r})$ is in the crystallographic (a,b)-direction only. Then the (two-dimensional) isotropic critical state model can be used where $\boldsymbol{j}(\boldsymbol{r})$ is given by $\mathrm{div}\boldsymbol{j}(\boldsymbol{r}) = 0, |\boldsymbol{j}(\boldsymbol{r})| = j_c(\boldsymbol{r})$ together with the condition of continuous and vanishing normal component of $\boldsymbol{j}(\boldsymbol{r})$ at inner boundaries and at the sample surface, respectively. The $\boldsymbol{B}$-dependence of $j_c(\boldsymbol{r})$ may be neglected. If $j_c(\boldsymbol{r}) = j_0 = const$, we arrive at the well-known, azimuthal, persistent current flow of the Bean model $\boldsymbol{j}(\boldsymbol{r}) = \boldsymbol{j_0}(\boldsymbol{r}) = j_0\boldsymbol{e}_\varphi$.

Assuming now that in the sheet between z and $z + \delta z$, there may be a restricted region $\Omega(z)$ on the characteristic length scale L with an enhanced, local critical current density $j_c(\boldsymbol{r}) > j_0$. Then an additional, local current loop system $\boldsymbol{j_1}(\boldsymbol{r})$ confined in $\Omega(z)$ is formed so that the local, persistent current flow between z and $z + \delta z$ would be $\boldsymbol{j}(\boldsymbol{r}) = \boldsymbol{j_0}(\boldsymbol{r}) + \boldsymbol{j_1}(\boldsymbol{r})$. To get estimations we study a simplified but exactly solvable and consistent model where the global current flow between z and $z + \delta z$ near $\Omega(z)$ is assumed to be homogenous, $\boldsymbol{j_0}(\boldsymbol{r}) = const, |\boldsymbol{j_0}(\boldsymbol{r})| = j_0(z)$, and $|\boldsymbol{j}(\boldsymbol{r})| = j_1 = const$ within $\Omega(z)$. Furthermore, $\Omega(z)$ may be approached by an arbitrary oriented, lateral rectangle. Note that from our model, the current flow beyond $\Omega(z)$ is not changed in any way (criterion of consistency). Therefore, the sample sheet between z and $z + \delta z$ may be subdivided into non-overlapping regtangles $\Omega_\alpha(z), \alpha = 1, 2, \ldots$ where in each of those, the constant, lateral current density $j_c(\boldsymbol{r}) = j_\alpha(z) \geq j_0(z)$ may be given. Thus, any function $j_c(\boldsymbol{r}) \geq j_0(z)$ in the sheet δz may be approached, following this principle. Then the model solution described above is straightforwardly generalized to $\boldsymbol{j}(\boldsymbol{r}) = \boldsymbol{j_0}(\boldsymbol{r}) + \sum_\alpha \boldsymbol{j_\alpha}(\boldsymbol{r})$ where the $\boldsymbol{j_\alpha}(\boldsymbol{r})$ represent closed current loop systems confined in $\Omega_\alpha(z)$ ("granularity").

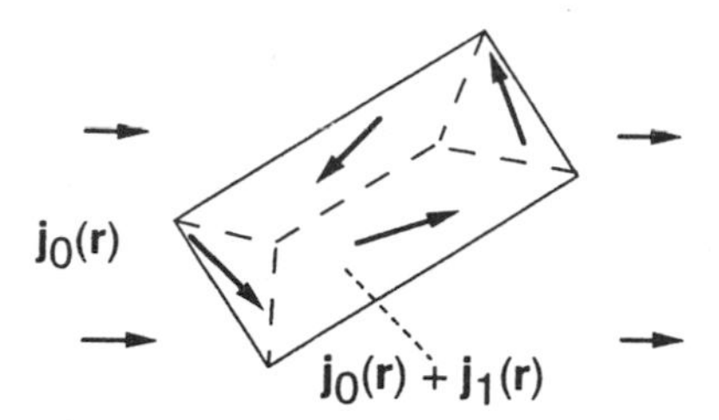

Figure 1. Current pattern near a rectangular region with enhanced j_c.

The decribed procedure to find the $\Omega_\alpha(z)$ is not yet unique. As well known, any critical current pattern in the remanent, critical state becomes uniquely determined by the magnetization conditions. For the fully penetrated, remanent critical state, a subdivision into a minimum number of rectangular regions $\Omega_\alpha(z)$ has to be carried out so that for each of those a maximum surface to perimeter ratio is reached. For this case, the electrical field strength parallel to $\boldsymbol{j_\alpha(r)}$ becomes a maximum during magnetization. Finally, our model can be extended taking into account (i) involved $\Omega_\alpha(z)$, (ii) localized "holes" where $j_c(\boldsymbol{r}) < j_0$, and (iii) the formation of current rings with radii $R \gg L$ due to percolation in rectangular regions $\Omega_\alpha(z)$ connected together. For the cases (ii),(iii), $\boldsymbol{j_0(r)}$ splits up into separated, quasi concentric rings which may be approached by the effective global current $\boldsymbol{j_0^{(eff)}(r)} \approx j_0^{(eff)}\boldsymbol{e_\varphi}$ where $j_0^{(eff)} \approx j_0$. The crucial point is that any local enhancement of $j_c(\boldsymbol{r})$ in spatial dimensions L within any ring (which may not be avoided) does not contribute to $j_0^{(eff)}$, but it again creates local current loops. Of course, a more precise analysis can only be given within the dynamical theory where instead of j_c, the nonlinear, space-dependent conductivity is the material property of interest. For principle considerations, the analysis given so far should be sufficient.

3. Degradation of Macroscopic Properties

Two characteristic quantities describing macroscopic sample properties will be discussed: the remanent magnetic momentum m and the peak value of the remanent induction B_p measured at r=0, 0< $z \ll R_0$.We find $m \approx m_0(1 + \beta(<\delta j_\alpha>/j_0)(<\delta R_\alpha>/R_0)), B_p \approx B_{p0}(1 + (\beta/3)(<\delta j_\alpha>/j_0)(<\delta R_\alpha>/R_0))$, where $m_0 = (\pi/3)j_0R_0^3l, B_{p0} = (\mu_0 j_0/2)l\times$ Arsinh(R_0/l) are the "nondisturbed" contributions from $\boldsymbol{j_0(r)}$ only. $<\cdots>$ means averaging on α and z. β is the total volume fraction of all local current loop systems. δR_α and δj_α are an effective radius and ring current density of the αth local current system given by $\delta j_\alpha \approx j_\alpha(1-(j_0/j_\alpha)^2), \delta R_\alpha^3 \approx (3/4\pi)\delta a_\alpha^2\delta b_\alpha(1-\delta a_\alpha/3\delta b_\alpha)$, respectively, with δa_α,δb_α as the lateral length of $\Omega_\alpha(z)$ (where $\delta b_\alpha \geq \delta a_\alpha$).

There appear the two governing, geometrical parameters β and $\delta R =<\delta R_\alpha>$. They can be estimated using polarization microscopy on polished surfaces combined with image processing, looking for characteristic dimensions of subgrains as well as of high defect concentration regions. In Fig.2, an example of a mesoscopic defect structure of a melt-textured YBCO-sample with Ce-addition ($j_c \approx 3\cdot 10^4 Acm^{-2}$ at $77K$ from VSM) prepared with a relatively large starting mean 123-powder particle size of about $400\mu m$ is shown. We find an averaged size of sub-domains and 211-high concentration regions of about $50\mu m$ and $400\mu m$, respectively, and $D_{211} \approx 1\mu m$, The volume fraction

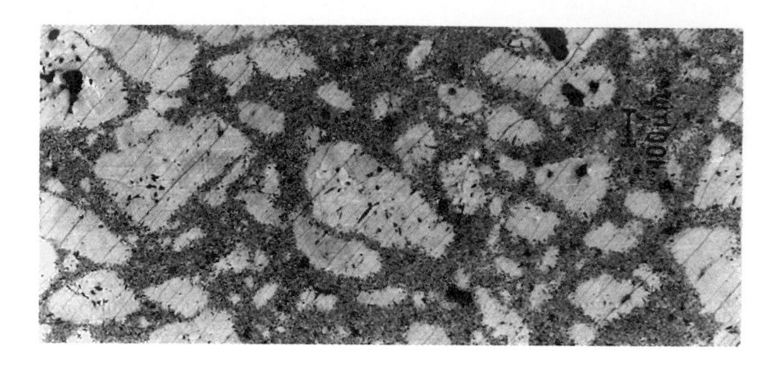

Figure 2. Micrograph of 211-high concentration regions (dark) and the nearly defect-free 123-matrix (bright) in Ce-doped melt-textured YBCO.

of the 211- particles is about 0.4. Also, there are extended 211-low concentration regions which may be considered as belonging to the 123-matrix. We estimate $\beta \geq 0.4$ and $\delta R/R_0 < 1 \cdot 10^{-1}$. This means that for the example considered, the wanted enhancement of the critical current density in the 211-high defect concentration regions contributes with an efficiency of few percent to m and B_p.

4. Conclusion

It was shown that the macroscopic properties of melt-textured YBCO samples sensitively depend on long-range fluctuations of the mesoscopic defect structure. Of course, the governing geometrical parameter δR may change on a broad scale depending on the preparation conditions. To avoid a degradation of the material properties the preparation conditions should be optimized to get δR as large as possible to approach the ideal case of a homogenous defect structure.

This work was supported through BMBF contract No. 13N 6646, Grant Agency of Slovak Academy of Sciences (Project No. 2/1323/94) and the DFG project YBCO-HRPLM.

References

[1] Murakami M 1992 *Melt Processed High Temperature Superconductors* (Singapore: World Scientific)

[2] Sandiumenge F, Pinol S, Obradors X, Snoek E and Roucau C, 1994 *Phys.Rev.B* **50** 7032

[3] Diko P, Gawalek W, Habisreuther T, Klupsch Th and Görnert P, 1995 *Phys.Rev.B* **52** 13658

[4] Endo A, Chaughan H S and Shiohara Y 1996 *Physica* **C273** 107

[5] Diko P, Todt V R, Miller D J and Goretta K C, 1997 *Superlattices and Microstructures* **21** 403

[6] Mironova D, Du G, Rusakova I and Salama K, 1996 *Physica* **C271** 15

[7] Zeisberger M, Campbell A M, Wai Lo and Cardwell D A, 1996 Contribution to the ASC 96 Conference and to be published

Inst. Phys. Conf. Ser. No 158
Paper presented at Applied Superconductivity, The Netherlands, 30 June–3 July 1997

The peak in the imaginary part of AC susceptibility for a superconductor at different magnetic fields and frequencies

Silvester Takács, Fedor Gömöry

Institute of Electrical Engineering, Slovak Academy of Sciences, 842 39 Bratislava, Slovakia

Abstract. The position of the peak in the imaginary part of AC susceptibility is calculated at different DC magnetic fields and in dependence on the amplitude and frequency of the AC magnetic field. The calculations are based on the results of the linear diffusion theory for the complex penetration depth, as well as on the non-linear theory starting from the force equilibrium equation for the flux lines including all forces and effects acting on them (pinning, flux flow, flux creep, reversible motion of the flux line lattice). By comparing the theoretical results with the experiments, we show that the linear theories of flux diffusion - choosing reasonable values for their material parameters - are not sufficient to describe properly the basic mechanisms of the flux penetration into superconductors with considerable effect of flux creep and flux flow. On the other side, the results of the non-linear theory are in qualitative agreement with the experiments.

1. Introduction

One of the most measured parameters of high T_c superconductors is the temperature T_p, at which the imaginary part in the complex AC susceptibility peaks [1, 2] — in dependence on the external parameters (DC field value, frequency and amplitude of the AC field). It is often used also for determining the irreversibility line (IL), although this procedure could be strictly valid only in the zero frequency limit. Namely, there should be some region with active superconducting shielding already above the temperature T_p and some irreversibility should be present. This was demonstrated [3] by the fact that the flux flow regime is established already above T_p. In spite of these limitations for determining the IL (one can only hardly avoid tails in different parameters at approaching the critical temperature T_c) [2 - 5], the parameter is very interesting for the experiments. It can be easily measured by many methods and its position is approximately given by penetrating the magnetic field just into the centre of the sample [1].

We have shown previously, that the experimental results on the frequency dependence of the AC susceptibility in different materials (YBCO, BSCCO) [3, 6] cannot support the position of T_p as the depinning line, as calculated from the linear theory for the flux diffusion [7]. In this contribution, we compare our experimental results with more precise calculations of the linear theory for the complex flux penetration depth [8, 9], including all effects on the flux line lattice (pinning, flux creep — FC, flux flow — FF, reversible flux line motion — RFLM [10]), as well as with the non-linear theory from the force equilibrium equation, based on our previous results [3, 10, 11].

2. Penetration depth and penetration field

The complex penetration depth λ_{eff} calculated in the linear theory is given by [8]

$$\lambda_{eff}^2 = \lambda^2 + \lambda_C^2 \frac{1 - i/\omega\tau}{1 + i\omega\tau_0}$$

where λ is the Ginzburg-Landau and λ_C the Campbell's penetration depth, $\lambda_C = B/(\mu_0\alpha)^{3/2} = (B_{DC}d/\mu_0 J_c)^{1/2}$, d - the interaction distance, α - Labusch's parameter, T_c, J_c - critical temperature and current density, T - temperature, $t = T/T_c$, $\tau = \tau_0 \exp(u) = (\eta/\alpha) \exp(u)$, $u = U/k_BT$, U - the pinning barrier, k_B - Boltzmann's constant, $\eta = B_{c2}\Phi_0/\rho_n$ is the flux line viscosity, $\rho_f = \rho_n B/B_{c2}$, ρ_n - the normal state resistivity, B_{c2} - the upper critical field, Φ_0 - flux quantum, B - the total field, B_{DC} - DC field component, supposed to be much larger than the AC amplitude b, $f = 2\pi/\omega$ (= 10^n) - frequency of the applied field, μ_0 - permeability of the vacuum, i - the imaginary unit. Similar expression was obtained by Coffey and Clem [9].

After some rearrangements, we obtain at penetrating the slab of thickness $2D$, supposing $\lambda << D$, for the result of Brandt [8] (a) and Coffey and Clem [9] (b), respectively:

$$\lambda_C^2[(1 + (1/\omega\tau)^2)/(1 + (\omega\tau_0)^2)]^{1/2} = D^2 \quad \text{(a)}$$

$$(\rho_f/\mu_0\omega)[(\varepsilon^2 + (1/\omega\tau')^2)/(1 + (\omega\tau')^2)]^{1/2} = D^2 \quad \text{(b)}$$

with $\tau' = (\eta/\alpha)(I_0^2(\nu) - 1)/I_1(\nu)I_0(\nu)$, $\varepsilon = 1/I_0^2(\nu)$, I being the Bessel functions with the imaginary argument, $\nu = u/2$.

The first attempt to calculate the penetration field from forces on flux lines was made by Müller [12], with main emphasis on FC. The result was a nearly linear dependence of T_p on ln (f). We started with FF, including later pinning, FC, with some modifications also RFLM [3, 11] for a slab of thickness D. Our approach was used for different materials [13]. With Kim's field dependence for J_c, we succeeded then in solving the force equilibrium equation in the one-dimensional case [11]: $(1/\mu_0)dB/dx = P/(B + B_0) + C + Q/B$ (the sign of the constants depend on the field changing mode). The creep factor is given by [12] $C \sim \ln(f)$, P and B_0 are the parameters of Kim's model and $Q = (8/3)B_{c2}DB_p f/\rho_n$ [11]. Taking $B_0 = 0$, which is well satisfied for YBCO [6], we obtain the flux distribution equation

$$B - \frac{P+Q}{C}\ln\left|B + \frac{P+Q}{C}\right| = \mu_0 Cx + \text{const.}$$

and for the penetration field $B_p = \beta_p(P + Q)/C$,

$$\beta_p - \ln|\beta_p + 1| = \mu_0 C^2 D/(P + Q)$$

This equation can be used for the determination of the penetration temperature T_p.

3. Comparison with experiments

The experimental results for $t_p = T_p/T_c$ are given in figure 1 for different fields. Using known and/or supposed temperature dependencies of the parameters $\nu \sim (1 - t)^{3/2}/B$, $B_{c2} \sim (1 - t)^2$, $J_c \sim (1 - t)$ [7 - 13], we obtain in the low frequency limit $(1 - t) \sim 1/\omega$ in both linear theories

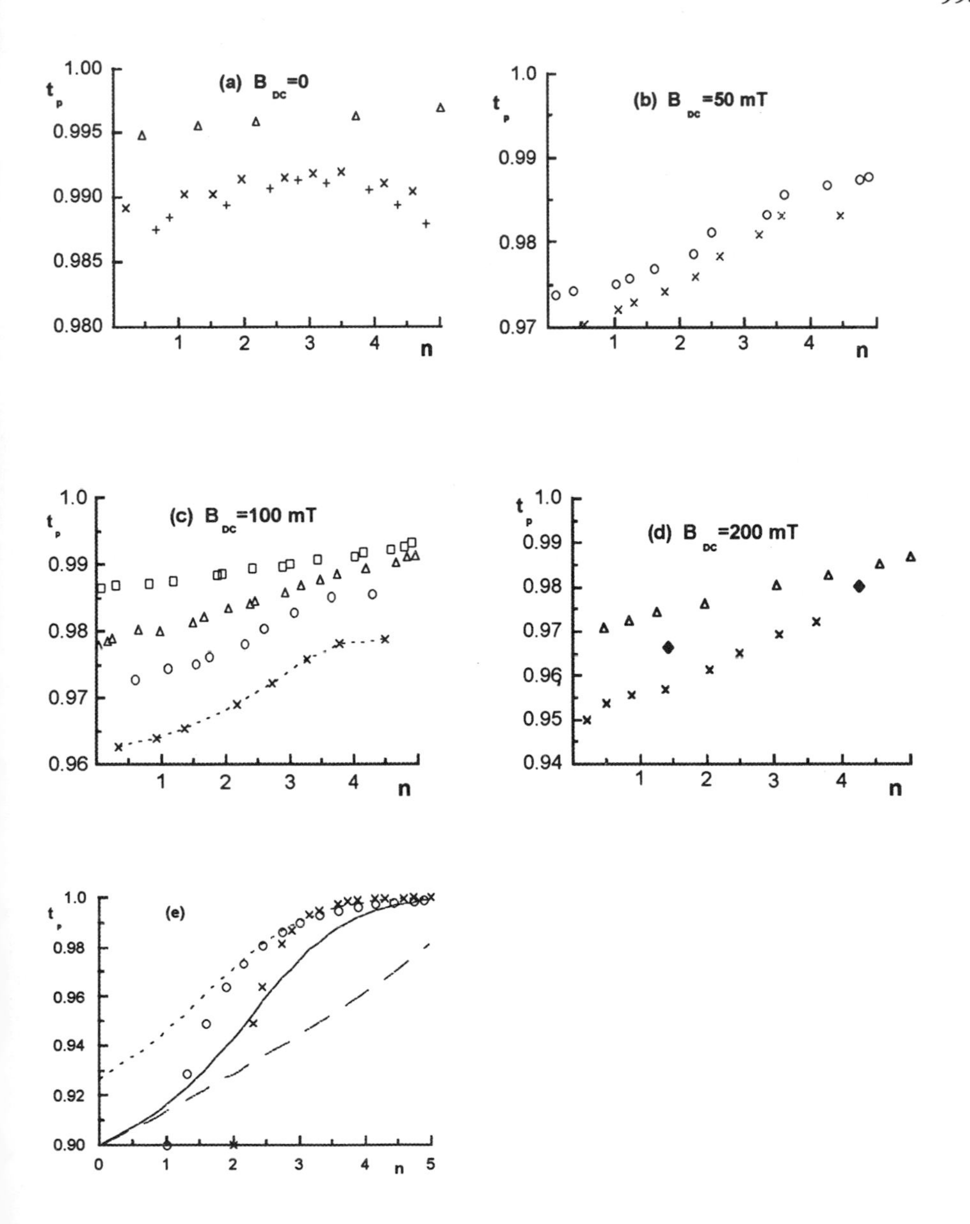

Figure 1. The measured frequency dependence ($f = 10^n$) of the peak temperature $t_p = T_p/T_c$ at different DC fields (a - d) and AC field amplitudes (open squares - 0.1 mT, Δ - 1 mT, ∇ - 1.5 mT, o - 2 mT, ♦ - 2.5 mT, • - 4 mT, × – 5 mT, + – 6.5 mT. The measured material was melt-grown YBCO with T_c = 92 K. The theoretical curve (e) gives the results for the flux creep (dashed line), the extreme flux flow model with no pinning (dotted line) and the inclusion of all forces on the flux lines (solid line). The values for the $1/f$ (×) and $1/f^{1/2}$ dependencies (o) are given as well.

for $\nu \ll 1$. The same dependence holds in the high frequency limit for any values of ν. As it is seen in figure 1, there is no region of this frequency dependence in experiments (see also [3, 12, 13]). In the low frequency limit and $\nu \gg 1$, we have $(1-t)^k \exp[-\nu_0(1-t)^{3/2}] \sim \omega$ with $k = \pm 1/2$ for the results of Coffey and Clem [9] and Brandt [8], respectively. Without going into details, this is the only region which comes close to the experimental results. However, we could find rather hardly the parameters leading to T_p values comparable with the experiments in the given frequency range. We need more precise values for high T_c superconductors to decide whether the linear theory is sufficient in this region.

4. Conclusions

As given in figure 1e, the inclusion of pinning, flux creep and flux flow into the force equilibrium equation leads to qualitative agreement with the experiments, with the exception of zero DC field. Due to the strong non-linearity in the latter case, much more exact treatment is possibly needed. In addition, one can assume the normal fluid component to be then more active, as in the most cases (mainly at low AC amplitudes) T is very close to T_c. This may cause the decrease of T_p with increasing frequency. Another important reason for this behaviour could be the temperature different from T_c, at which pinning vanishes. To decide, how exactly our simplified treatment describes the experiments quantitatively, more knowledge of the temperature dependencies of the parameters is needed. The role of RFLM on the position of the peak is not clear, although there is some evidence that this effect is weakened by stronger FF [3]. However, it has a considerable effect on the absolute value of the peak [14].

The linear theory fails to describe the experiments in many limiting cases. It seems to come close to the experimental results only in the high frequency region, supposing high values of the pinning potential $\nu \gg 1$. This case should be examined more precisely.

Acknowledgment

This work was partially supported by the Slovak Grant Agency for Science VEGA.

References

[1] Clem J R *Physica* **C153-155** (1988) 50

[2] Nicolo M and Goldfarb R B *Phys. Rev.* **B39** (1989) 6615

[3] Takács S and Gömöry F *Cryogenics* **33** (1993) 133

[4] Loegel B, Mehdaoui A, Bolmont D, Danesi P, Bourgault D and Tournier R *Physica* **C210** (1993) 432

[5] Fabbricatore P, Gemme G, Moreschi P, R. Musenich R, R Parodi R and Zhang B *Phys. Rev.* **B50** (1994) 3189

[6] Takács S, Gömöry F and Lobotka P *IEEE Trans. Magn.* **MAG-27** (1991) 2206

[7] Clem J R, Kerchner H R and Sekula T S *Phys. Rev.* **B14** (1976) 1893

[8] Brandt E H *Physica* **C195** (1992) 1

[9] Coffey M W and Clem J R *Phys. Rev. Letters* **67** (1991) 386

[10] Campbell A M *J. Phys.* **C4** (1971) 3186

[11] Takács S, Gömöry F, Pevala A and Lobotka P *Supercond. Sci. Technol.* **5** (1992) S452

[12] Müller K H *Physica* **C168** (1990) 585

[13] Lee C W and Kao Y H *Physica* **C256** (1996) 183

[14] Matsushita T, Otabe E S and Ni B *Physica* **C182** (1991) 95

Inst. Phys. Conf. Ser. No 158
Paper presented at Applied Superconductivity, The Netherlands, 30 June–3 July 1997

Large-area deposition of biaxially textured YSZ buffer layers by using an IBAD technique

J Wiesmann, J Dzick, J Hoffmann, K Heinemann, and H C Freyhardt

Institut für Metallphysik, Universität Göttingen, Hospitalstr. 3/7, 37073 Göttingen, Germany

Abstract. On large-area ceramic substrates up to 10 x 10 cm^2 biaxially textured YSZ-films were deposited using an ion-beam-assisted deposition method (IBAD) with two Kaufman ion sources with diameters of 11 cm. The alignment is characterized by the full width at half maximum (FWHM) of a (111) ϕ scan, determined with a four-circle diffractometer. The best YSZ films on small substrates (1 cm^2) show remarkably good in-plane alignment with a FWHM of only 7°. In general the texture improves with increasing film thickness. For 10 x 10 cm^2 substrates the texture varies laterally within a range of 15° to 20° FWHM. This homogeneous distribution of texture was reached by employing four consecutive rotations by 90° around the substrate normal. The YSZ films enable a deposition of YBCO films with a $j_c > 10^6$ A/cm^2 on the total area. First experiments on an area of 20 x 20 cm^2 show in-plane texture better than 35° and on a circle of 20 cm in diameter better than 25°, respectively. Therefore, the substrates had to be rotated and shifted during deposition. The deposition rate of the film volume reached 12 nm·m²/h, i.e. 300 nm per hour were deposited on 20 x 20 cm^2.

1. Introduction

Many high-current applications of $YBa_2Cu_3O_{7-x}$ thin films (YBCO) require a deposition on large-area polycrystalline substrates. For this a buffer layer between the substrate and the YBCO film is necessary to protect them from interdiffusion and to transmit a biaxial c-orientated texture to the YBCO for preventing large angle grain boundaries. In 1991 Iijima et al. found a method for the preparation of the, up to now, most suitable buffer material on non-singlecrystalline materials [1]. They deposited biaxially textured cubic YSZ films (yttria stabilised zirconia: ZrO_2 with 8 mol% Y_2O_3) by using an ion-beam-assisted deposition technique (IBAD). These biaxially aligned YSZ films grow with the <100> direction parallel to the substrate normal and the <111> direction parallel to the assisting ion beam and allow an epitaxial growth of YBCO films with the (011) plane in YBCO parallel to the (001) plane in YSZ, i.e. a 45° epitaxy relation exists between YBCO and YSZ.

We used a similar IBAD process for the preparation of YSZ and CeO_2 films [2] and developed this method further for the deposition on planar, large-area substrates up to 10 x 10 cm^2 [3] and cylindrical substrates [4]. With our highly textured YSZ films the following critical current densities j_c (at 77 K, 0 T) in YBCO films on non-singlecrystalline substrate materials were achieved: On polycrystalline metals (Ni) j_c reached $2 \cdot 10^6$ A/cm^2 [5], on cylindrical curved Ni foils j_c reached $0.25 \cdot 10^6$ A/cm^2 [4] and on 10 x 10 cm^2 ceramics j_c exceeds 10^6 A/cm^2 [6].

2. Experimental

For the deposition of YSZ a dual-ion-beam equipment was employed with two ion sources of the Kaufman type with diameters of 11 cm. It is similar to the one used by Iijima et al.[1] and has been described in detail elsewhere [3]. The assisting ion beam with an energy of 300 eV was directed onto the substrate holder with an angle of 55° to the substrate normal. The ion density at the substrates amounted to about 250 $\mu A/cm^2$. The temperature of the uncooled substrate holder increased from room temperature at the beginning of the experiment to about 150°C after 3 hours. The investigations were mainly performed on silicon single crystals covered with the native oxide and on polycrystalline ceramic substrates of a size up to 10 x 10 cm^2. The orientation and microstructure of these substrates have no remarkable influence on the quality of the YSZ layers. For some depositions a computer controlled rotation by 90° around the substrate normal was employed to homogenize the in-plane texture and the film thickness. For substrates with areas larger than 10 x 10 cm^2 an additional shift of the holder in the substrate plane was performed. The deposition rate reached about 0.08 nm/s resulting in a total film thickness of 900 nm after 3 hours. ϕ scans of the (111) peaks, which were measured with a 4-circle diffractometer (Cu-radiation: $\lambda = 0.154$ nm) at $\chi = 55°$, characterize the in-plane texture.

3. Results

Recently we could demonstrate that the in-plane texture, characterized by the FWHM - given in degrees - of a (111) ϕ scan, improves with the film thickness d in µm (figure 1). This improvement can be described numerically with an exponential decay. Our measured FWHM values in (111) ϕ scans could be fitted by the following function:

$$\mathrm{FWHM} = 7° + 29° \times \exp(-d\,[\mu m] / 0.7)$$

A saturation of the texture is observed at 7°. This value was in fact achieved in a 5.5 µm thick YSZ film. The improvement of the in-plane alignment demonstrates that after the nucleation phase a further growth selection takes place even in films with a thickness of several 100 nm. The real texture at the surface of the YSZ film is obviously better than the value measured by x-rays because of their penetration depth > 1 µm.

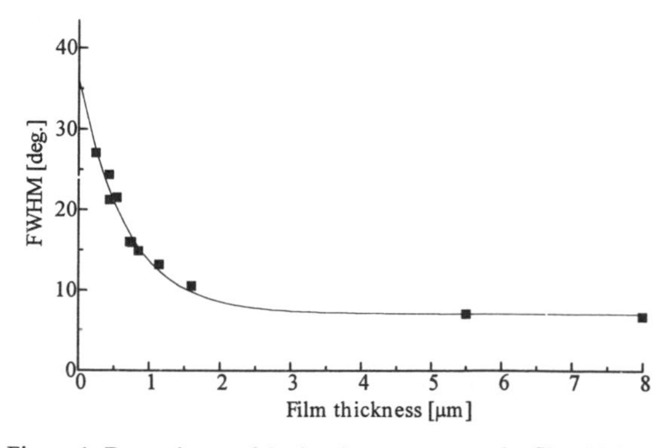

Figure 1: Dependence of the in-plane texture on the film thickness

Therefore, YBCO layers epitaxially grown on YSZ typically exhibit a FWHM, which is 5° to 8° less than the one of the YSZ layer on the substrate [6]. An in-plane texture of 20° FWHM of the YSZ film and thus a film thickness of 600 nm - under optimized deposition conditions - is necessary for a film growth of YBCO with j_c above 10^6 A/cm^2.

In the following YSZ was always deposited with the same deposition parameters. The films without any substrate movement had a thickness of 900 nm after a deposition time of 3 hours. The quality of the in-plane texture reached a FWHM of 15°. The same value can be reached during the same deposition time, but with four consecutive rotations in one step by 90°, each one after a quarter of the total deposition time. Thus one of the <111> axes always lies parallel to the assisting ion beam during the deposition. The film growth is independent on 90° rotations, which make a homogenization of the main properties on large areas possible [3]. On 10 x 10 cm^2 a texture could be achieved in a range of 15° to 21° for the FWHM. The local deposition rate d varies locally and temporally on large area substrates during their movement, and thus d does not represent the total deposition speed easily. Therefore, the information of a volume deposition rate d_v is more useful. On 10 x 10 cm^2 d_v reached 3 nm·m^2/h. This means, that after one hour 300 nm are deposited onto 0.1 m^2.

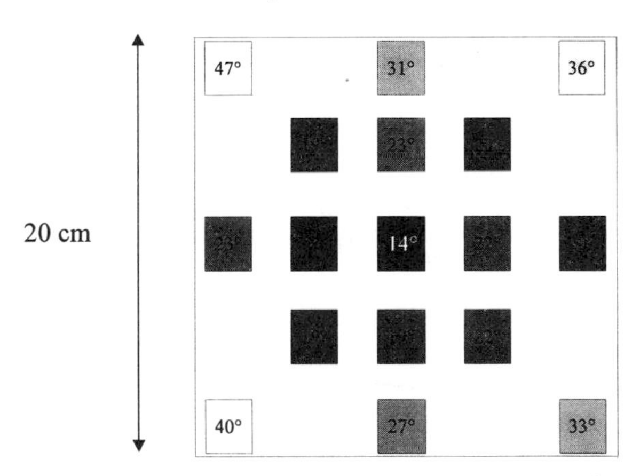

Figure 2: Deposition with four consecutive rotations by 90°: FWHM of (111) ϕ scans on 20 x 20 cm^2

First experiments were performed with small substrates (25 pieces) mounted on an area of 20 x 20 cm^2 with a spacing of 5 cm to the next. The value of d_v reached 12 nm·m^2/h. Without substrate movement the in-plane alignment is very inhomogeneous, but it can be improved by four consecutive rotations by 90° (figure 2). Only at the edges the FWHM of the in-plane alignment drops very fast.

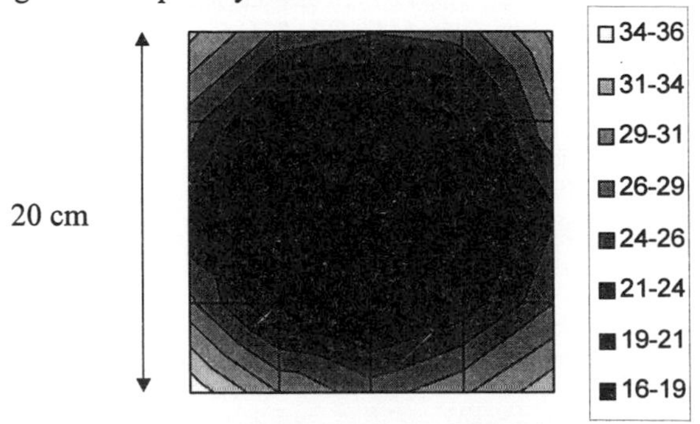

Figure 3: Deposition with rotation and shifting: FWHM of (111) ϕ scans on 20 x 20 cm^2

Consequently a shift of the substrate in both directions of the substrate plane was employed in addition to the rotation for a further homogenization. The achieved in-plane texture is shown in figure 3. The texture varies in a range from 15° to 35° for the FWHM, on a circular area with a diameter of 20 cm it is better than 26°. In the nearest future the substrate movement will be optimized with respect to the texture quality and the total deposition time.

4. Conclusion

The ion-beam-assisted-deposition technique enables the growth of biaxially textured YSZ layers on large areas up to 20 x 20 cm^2. The quality of the in-plane texture, characterized by the FWHM of a (111) ϕ scan, improves with increasing film thickness and saturates at a FWHM of 7°. The in-plane texture can be homogenized on large areas by employing four consecutive rotations by 90° around the substrate normal and in addition by a shifting in the substrate plane during the deposition. Thus on 20 x 20 cm^2 - filled with small substrates - a FWHM better than 26° was obtained on a circular area with a diameter of 20 cm. The volume deposition rate reached 12 $nm \cdot m^2/h$. The results demonstrate the feasibility of a deposition of biaxially aligned YSZ layers on non-singlecrystalline substrates even up to 20 x 20 cm^2 with large volume deposition rates and high in-plane alignment, that allows YBCO film growth with critical current densities around 10^6 A/cm^2. YBCO films were already deposited with a j_c better than 10^6 A/cm^2 on polycrystalline ceramic substrates of a size of 10 x 10 cm^2, which were covered with our biaxially textured YSZ layers [6].

Acknowledgement

This work was supported by the German BMBF and the Siemens AG under grant number 13N6482. M Bauer, R Semerad, and H Kinder (TU München) are gratefully acknowledged for providing the results on the YBCO films deposited on the YSZ layers.

References

[1] Iijima Y, Tanabe N, Ikeno Y, and Kohno O *Physica C* **185** (1991) 1959

[2] Wiesmann J, Hoffmann J, Usoskin A, García-Moreno F, Heinemann K, and Freyhardt H C *Applied Superconductivity 1995, IOP Conf. Series* **148** (1995) 503

[3] Wiesmann J, Heinemann K, and Freyhardt H C *Nucl. Instrum. Meth. B* **120** (1996) 290

[4] Hoffmann J, Dzick J, Wiesmann J, Heinemann K, García-Moreno F, and Freyhardt H C *J. Mater. Res.* **12** (1997) 593

[5] García-Moreno F, Usoskin A, Freyhardt H C, Wiesmann J, Dzick J, Issaev A, Hoffmann J and Heinemann K *High T_c Updates* **10** (1996) 22

[6] Kinder H, to be published in the *Proceedings of the EUCAS 1997*

Inst. Phys. Conf. Ser. No 158
Paper presented at Applied Superconductivity, The Netherlands, 30 June–3 July 1997

Preparation of biaxially textured YSZ buffer layers on cylindrically curved substrates

J Dzick[1], J Wiesmann[1], J Hoffmann[1], K Heinemann[1], A Usoskin[2], F García-Moreno[2], A Isaev[1] and H C Freyhardt[1,2]

[1] Institut für Metallphysik, Hospitalstr. 3/7, 37073 Göttingen, Germany.
[2] Zentrum für Funktionswerkstoffe gGmbH, Windausweg 2, 37073 Göttingen, Germany.

Abstract. Biaxially textured yttria stabilized zirkonia (YSZ) buffer layers are prepared on cylindrically curved substrates with a radius of curvature of 6 mm. The YSZ films are deposited by an ion-beam-assisted deposition process (IBAD) with two Kaufman ion sources. Because of the curved geometry of the substrates only one quarter of a circle in maximum can be coated at the same time. The angle between the assisting ion beam and the local substrate normal varies continuously in the deposition area. A homogeneous YSZ film can only be obtained by a continuous rotation of the substrate during the deposition. The quality of the in-plane alignment depends on the total film thickness, the rotation velocity and the width of the deposition area. A slow movement along the rotation axis during deposition has no influence on the growing film. Our best observed in-plane texture is around 24° full width at half maximum (FWHM). First results of YBCO ($Y_1Ba_2Cu_3O_{7-x}$) films deposited on curved Ni substrates with YSZ buffer layers exhibit a j_c of 0.25 $MAcm^{-2}$ at 77 K and 0 T.

1. Introduction

High critical current densities in YBCO thin films require a reduction of the content of large angle grain boundaries, what could be achieve by growing these films epitaxially on appropriate single crystals. For technical applications the use of polycrystalline instead of single-crystalline substrates is possible if some additional textured buffer layer is deposited onto the substrates to facilitate epitaxial growth of a well-orientated YBCO film.

In 1991 Iijima et al. prepared YSZ films with an IBAD process on polycrystalline metallic substrates [1]. These films have a (001) texture parallel to the substrate normal and an in-plane alignment of the grains, which can be characterized by the FWHM of the (111) peaks measured with a ϕ scan at χ=55°. While YBCO films can be grown epitaxially on such a YSZ buffer layer with a low content of large angle grain boundaries, the YSZ film also protects the superconducting film from interdiffusion of substrate components [2]. Our best YBCO films on planar polycrystalline Ni substrates with a YSZ layer exhibit a j_c of 2 $MAcm^{-2}$ at 77 K and 0 T [3].

Beside YSZ buffer layers on planar substrates YSZ films were prepared on cylindrical substrates [4]. The investigations were set off by the question if it is possible to deposit textured YBCO films on cylindrical substrates and therefore if it is possible to prepare an epitaxial surface on cylindrically curved substrates. Such substrates (e.g. tubes) with a YBCO layer can be used for various applications, e.g. rf conductor cables or resonators.

2. Preparation

The construction of the dual ion beam equipment is shown in figure 1 (left drawing). One Kaufman ion source (sputter source) is directed on the YSZ target (ZrO_2 with 8 mol% Y_2O_3). It uses Xenon atoms for sputtering. The sputtered material is deposited on the opposite substrate. The beam of the second Kaufman source is directed onto the growing film. This source supplies Argon atoms with an energy of 300 eV. The special deposition parameters are described elsewhere [4]. Both ion sources do have a diameter of 2.5 cm. Electropolished Ni foils (5 mm × 10 mm × 0.3 mm) are used as substrates. These are cleaned with ion beam etching before the deposition starts.

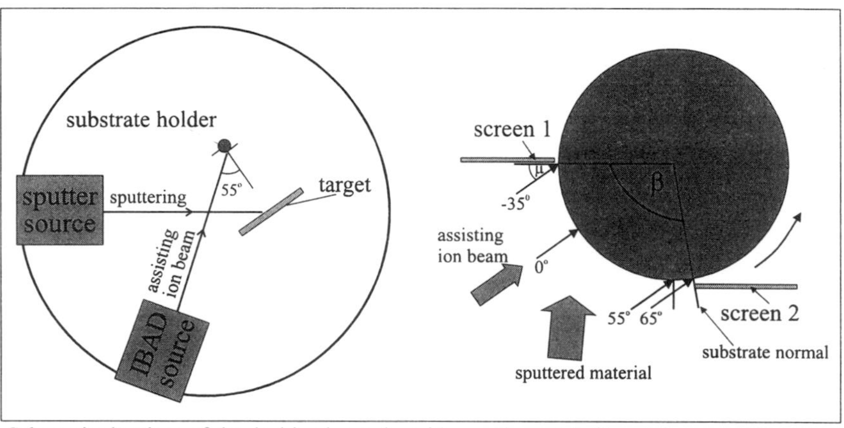

Fig. 1. Schematic drawings of the dual-ion-beam-deposition equipment (left) and the deposition geometry for cylindrical substrates (right). The deposition window is determined by screen 1 and 2 while μ indicates the angle between the assisting ion beam and the substrate normal. The angle β describes the angle of opening of the deposition window.

The coating of a cylindrical substrate is quite different to the deposition on planar substrates. While on planar substrates the angle of incidence of the assisting ion beam is always around the optimum of 55° to the substrate normal, the curvature of cylindrical substrates produces a continuous variation of the angle of incidence as it is shown in figure 1 (right drawing). Therefore optimal IBAD conditions occur just in a small area in the whole deposition window and the preparation of a homogeneous YSZ film requires a rotation of the substrate during the deposition.

The two screens (shown in figure 1) constrain the angle of opening (β) of the deposition window. Screen 2 cuts off the deposition area where the angle of incidence becomes larger than 65°, that is known to cause an undesired (111) orientated growth of the YSZ film [5]. Screen 1 is variable in its position so that the angle of opening of the deposition window can be adjusted to 40°, 60° or 100°.

3. Results and discussion

The stationary coating (this means without rotation) of small Ni substrates allows to determine how the YSZ film grows under different angles of incidence (μ) of the assisting ion beam. In figure 2 the fraction of (001) growth (left axis) and the fraction of in-plane alignment (right axis) is shown for different angles of incidence. From -35° up to 65°

pronounced (001) growth is observed. For larger angles a sudden change to (111) growth is found. The region of in-plane aligned growth is restricted for it is observed under angles of incidence from 25° up to 65°. The best in-plane alignment is found under optimum IBAD conditions ($\mu = 55°$).

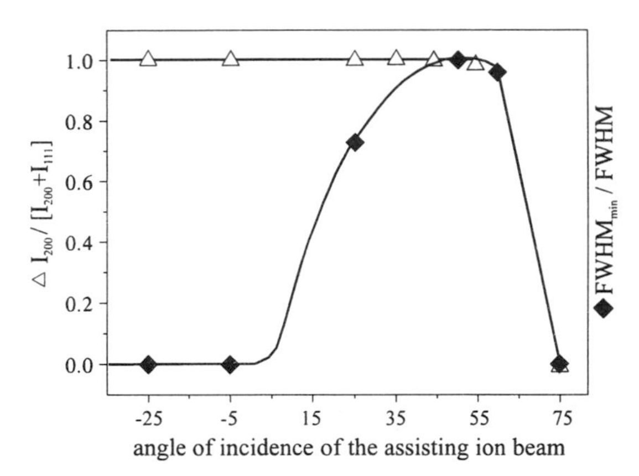

Fig. 2. Dependence of the in-plane alignment (FWHM) and the (001) texture on the angle of incidence (μ) of the assisting ion beam with respect to the local substrate normal.

A rotation during the deposition homogenizes the texture of a YSZ film. The in-plane alignment improves with increasing film thickness (figure 3) similar to the behaviour found for YSZ films on planar substrates [6]. Up to now the best in-plane alignment with a FWHM of 25° was observed in a 500 nm thick YSZ film (figure 3). For a constant film thickness the increase of the angle of opening of the deposition window leads to a decline of the in-plane alignment of the film (figure 3). This happens due to an increased share of non-optimum IBAD conditions that causes growth disturbances. An additional axial movement by 5 mm/min during deposition has no effect on the in-plane alignment.

The rotation velocity (10°/min - 1000°/min) does not influence the in-plane alignment if a deposition window with an angle of opening not larger than 60° is used. For larger deposition windows the in-plane alignment deteriorates with increasing rotation velocity (figure 4).

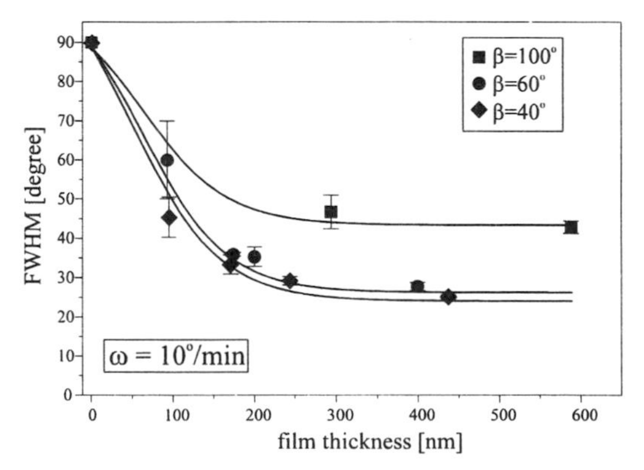

Fig. 3. Dependence of the FWHM of the in-plane alignment on the film thickness for different angles of opening β of the deposition window (see also figure 1).

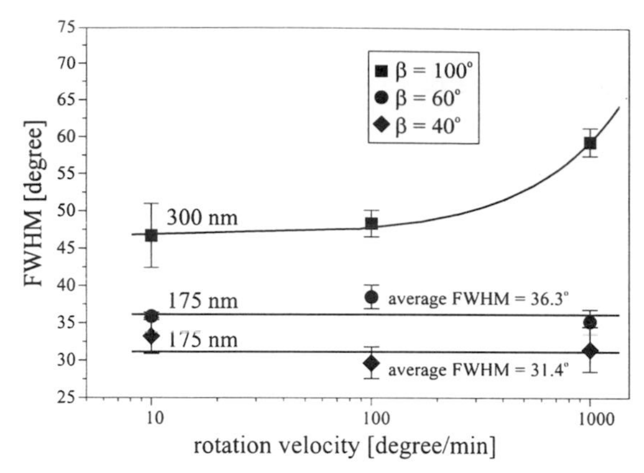

Fig. 4. Dependence of the FWHM of the in-plane alignment on the rotation velocity for different angles of opening β of the deposition window (see also figure 1).

YBCO films were deposited by laser ablation on cylindrically curved Ni foils with YSZ buffer layers. Critical current densities up to 0.25 $MAcm^{-2}$ at 77 K and O T were obtained. Therefore, the coating of cylindrical substrates with YSZ films and superconducting YBCO films seems to be attractive for technical applications.

In summary, it is possible to deposit YSZ films with a biaxial alignment on cylindrical substrates. The IBAD process can be applied on long tubes as an additional movement along the axis of the tube does not affect the in-plane alignment. The quality of the in-plane alignment of the YSZ films is sufficient to produce YBCO films with critical current densities well above 0.1 $MAcm^{-2}$ at 77 K and 0 T [7].

Acknowledgement

This work was performed within a development project of kabelmetal electro GmbH which is supported by the German BMBF (Project No. 13 N 6924/6).

References

[1] Iijima Y, Tanabe N, Ikeno Y and Kohno O *Physica C* **185** (1991) 1959
[2] Simon T *Ph.D. Thesis* University of Göttingen (1995)
[3] García-Moreno F, Usoskin A, Freyhardt H C, Wiesmann J, Dzick J, Isaev A, Hoffmann J and Heinemann K *HIGH Tc UPDATES* **10** (1996) No. 22
[4] Hoffmann J, Dzick J, Wiesmann J, Heinemann K, García-Moreno F and Freyhardt H C *J. Mater. Res.* **12** (1997) 593
[5] Wiesmann J *Diploma Thesis* University of Göttingen (1994)
[6] Wiesmann J, Heinemann K and Freyhardt H C *Nucl. Instrum. Meth. B* **120** (1996) 290
[7] Wiesmann J, Hoffmann J, Usoskin A, García-Moreno F, Heinemann K and Freyhardt H C *Applied Superconductivity 1995* IOP Conf. Series **148** (1996) 290

Inst. Phys. Conf. Ser. No 158
Paper presented at Applied Superconductivity, The Netherlands, 30 June–3 July 1997

Critical current of Gd doped $PbMo_6S_8$ synthesised at 1500 °C

D N Zheng, A B Sneary and D P Hampshire

Department of Physics, University of Durham, Durham DH1 3LE, UK

Abstract. We have made a series of samples with nominal composition $Pb_{1-x}Gd_xMo_6S_8$ ($0 \leq x \leq 0.3$) using a hot isostatic pressing process. A high reaction temperature of 1500 °C has been used to incorporate Gd into the lattice. The critical current density J_c of these samples has been measured in order to investigate possible flux pinning due to the magnetic Gd ions. Adding Gd markedly decreases the critical temperature T_c and increases normal state resistivity. We attribute these changes to a decrease in carrier concentration. Also J_c and the irreversibility field decrease progressively with increasing Gd level. The pinning force density curves of the doped samples were found to be similar to the undoped sample, indicating that J_c is limited by the same mechanism. The results suggest that impurity phases, degraded grain boundaries or a reduction in the critical parameters are responsible for low J_c and B_{irr}.

1. Introduction

Chevrel phase materials $PbMo_6S_8$ and $SnMo_6S_8$ are promising candidates for high field applications due to their high B_{c2} values [1, 2]. Study of flux pinning and looking for ways of increasing the critical current density J_c in these materials are of significant importance. In the Chevrel phase superconductor $PbMo_6S_8$, a substantial amount of magnetic Gd ions can be doped without reducing its T_c greatly. It has been suggested that the Gd ions could interact with flux lines and this interaction will affect flux pinning and possibly lead to an increase of the J_c of the material. In fact, it has been demonstrated that ferromagnetic particles dispersed in samples of superconducting Hg-13%In alloy can increase critical current [3]. Preliminary calculations for the Gd doped $PbMo_6S_8$ system suggest that a J_c value as high as 10^{10} A/m^2 can be achieved by pinning flux lines with Gd ions. In order to investigate this possible pinning mechanism experimentally, we have made a series of samples with nominal composition $Pb_{1-x}Gd_xMo_6S_8$ and have measured J_c and other properties of the samples.

2. Experimental

A solid state reaction method has been used for synthesising the ceramic samples. It has been found that at a reaction temperature between 1000-1100 °C most Gd was not incorporated into the lattice, and forms secondary phases instead. In order to overcome this problem, a higher reaction temperature (1500 °C) was used. The samples were further subjected to a hot isostatic press treatment at 2000 bar and 900 °C for 6 hours to increase sample density and improve the connections at grain boundaries. X-ray diffraction and electron microscopy have been employed for the characterization of the samples. Ac susceptibility, magnetoresistivity and dc magnetic hysteresis measurements have been carried out to investigate flux pinning and critical current properties of the samples.

3. Results and Discussion

Phase purity has been checked by powder X-ray diffraction. At low doping levels (x<0.06), the samples are almost single phase within the resolution of the technique. While at high doping levels, impurity phases such as Mo_2S_3 and Gd_2S_3 are found. The microprobe analysis undertaken on the x=0.1 and 0.3 samples confirmed the existence of the secondary phases, but also showed that the Chevrel phase is the majority phase. As revealed by the energy-dispersive x-ray spectroscopy (EDX), the Chevrel phase grains contained an amount of Gd approximately at the level of doping.

The resistivity data are shown in Fig.1 for the samples. The T_c is reduced progressively with increasing Gd doping level. However, up to a doping level of x=0.1, the decrease of T_c is not significant. Two reasons are believed to be responsible for the decrease of T_c. Firstly, the charge transfer from the Pb site to the conducting Mo_6S_8 clusters is altered due to the substitution of Gd^{3+} ions for the Pb^{2+} ions. Secondly, the unit cell volume is decreased after Gd doping. These two factors change the density of states at the Fermi surface and thus lead to a decrease of T_c. Hall measurements suggest that in $PbMo_6S_8$ holes are the dominant charge carriers [4]. The substitution of Gd^{3+} for Pb^{2+} means a reduction of holes and thus charge carriers. Therefore we would expect the reduction of charge carriers to result in an increase of resistivity. The experimental data show that this is indeed the case. Doping could also reduce the mean free path which can contribute to the increase of the resistivity. However, since the conduction is taking place in Mo_6S_8 clusters which are far from the Pb and Gd ions, we suggest that the decrease of carrier concentration due to charge transfer is the dominant effect. In Table I, we summarize the results obtained on these samples. We note that the T_c and resistivity of x=0.06 and x=0.1 samples do not follow the trend from the nominal composition, rather the data suggest that the real doping level in the x=0.06 sample is higher than in the x=0.1 one.

Magnetic hysteresis curves have been measured for all the samples. For the Gd doped samples, there is a paramagnetic contribution which increases systematically with increasing the doping level. Assuming that the Gd ions are free of magnetic correlations (we have seen previously that free Gd which is not substituted into the grains forms Gd_2S_3 and other secondary phases which contribute little to the paramagnetic background [5]), we can compare the paramagnetic moment measured to the theoretical value given by the standard theory. This gives an estimated doping level of 0.27 for the x=0.3 sample, which is consistent with EDX results.

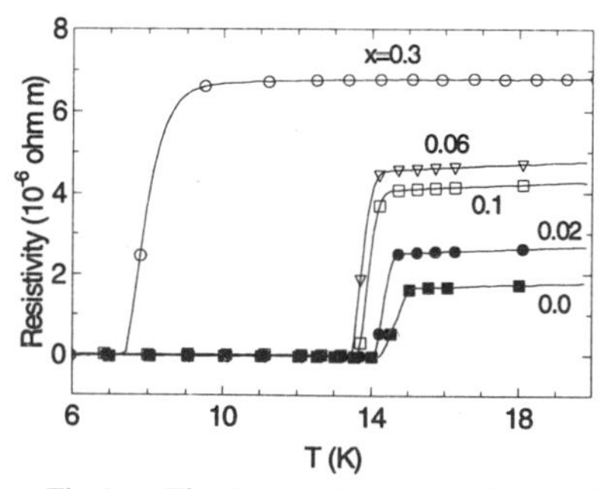

Fig.1. The temperature dependence of resistivity for the $Pb_{1-x}Gd_xMo_6S_8$ samples.

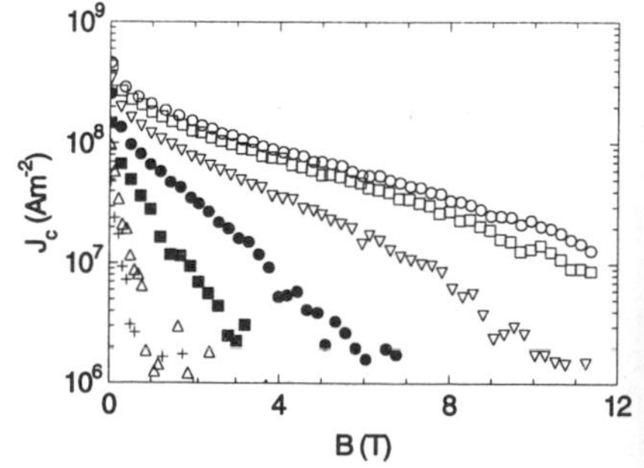

Fig.2. The field dependence of J_c for the x=0.1 sample at 4.2, 5.2, 7.1, 8.5, 9.9, 11.1 and 12.2 K (from top to bottom).

Table I. Critical temperature T_c, normal state resistivity ρ (at 15 K) and critical current density of the $Pb_{1-x}Gd_xMo_6S_8$ samples.

	T_c (K)	ρ (μΩm)	J_c (10^8 Am^{-2}) 2 T	6 T	10 T
x=0.00	14.3	1.6	4.1	1.8	0.9
x=0.02	14.1	2.4	3.8	1.5	0.75
x=0.06	13.6	4.7	1.2	0.49	0.22
x=0.10	13.7	4.2	1.5	0.57	0.2
x=0.30	7.4	6.7	0.0036	0	0

J_c is calculated from the hysteresis data using the Bean model. The values at 4.2 K are shown in Table I. Figure 2 shows the field dependence of J_c for the x=0.1 sample. The overall trend is that J_c decreases as more and more Gd is added. In Fig.3, $J_c^{1/2}B^{1/4}$ is plotted against B for the x=0.1 sample. The linear behaviour suggests that the pinning force density F_P ($=J_c \times B$) can be described by the Kramer relation $F_P \propto b^{1/2}(1-b)^2$. Although J_c changes significantly with different Gd doping level, we found that F_P curves for the samples with doping levels up to x=0.1 follow the Kramer relation. Since this relation has been seen in Nb_3Sn in which grain boundaries are believed to be major pinning centres. The F_P curves observed suggest that J_c is determined by grain boundary pinning. As evidence to support this, we have found that the J_c of the $PbMo_6S_8$ sample is low compared to the values previously reported for the samples reacted at 1000 °C. The high reaction temperature results in a large grain size, thus reducing the number of grain boundaries.

For the x=0.3 sample, the curves in the Kramer plot diverge from linear behaviour at low field (below $0.4B_{irr}$) and show a trend to saturation. At high fields, the curves are still linear.

Alden and Livingston have proposed that the pinning force curve due to pinning by magnetic particles is of the form $b^{1/2}(1-b)$ [3], which shows a negative curvature in the Kramer plot (i.e, $J_c^{1/2}B^{1/4}$ versus B) as b approaches 1. For the data shown in Fig.3, however, the Kramer relation appears to be more appropriate.

By performing linear extrapolations of the data for each temperature to $J_c^{1/2}B^{1/4}=0$, the irreversibility field can be estimated. The data are shown in Fig.4 and show a systematic decrease of B_{irr} with increasing doping level. There could be two reasons which cause the decrease of B_{irr}. First, T_c and other intrinsic parameters such as critical fields and penetration depth are changed by doping Gd in $PbMo_6S_8$.

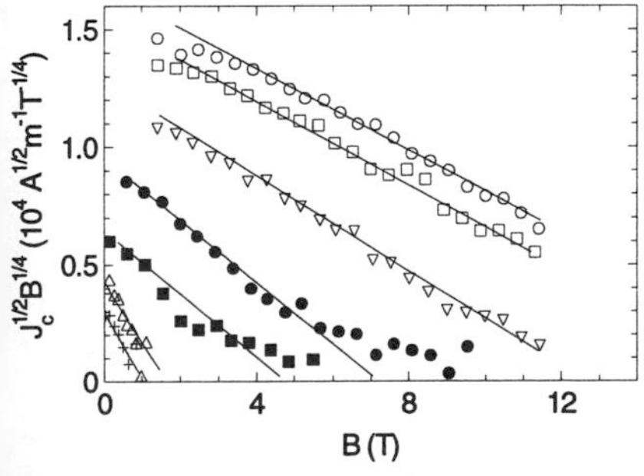

Fig.3. Kramer plot for the x=0.1 sample. Temperature is (from top): 4.2, 5.2, 7.1, 8.5, 9.9, 11.1 and 12.2 K

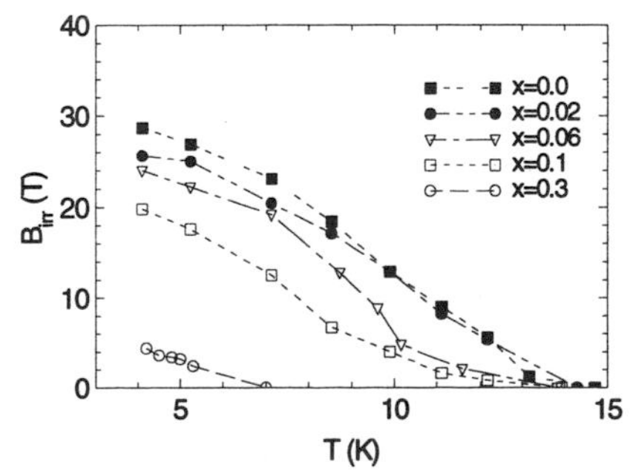

Fig.4. The temperature dependence of the irreversibility field for the $Pb_{1-x}Gd_xMo_6S_8$ samples.

Secondly impurity phases and the degradation of grain boundaries can also decrease B_{irr}, and probably play a more important role. The T_c and resistivity data have indicated that the amount of Gd doped into grains in the x=0.1 samples is actually less than that of the x=0.06 sample. This also suggests that there are more impurity phases presented in the former sample. Also, the existence of impurity phases could affect the properties of grain boundaries and make the sample more granular. As shown in Fig.4, the $B_{irr}(T)$ line of the x=0.1 sample is lower than the x=0.06 sample even though the former sample has a higher T_c value. This suggests that the low irreversibility line observed is mainly due to the existence of impurity phases and degraded grain boundaries.

4. Conclusions

We have made a series of Chevrel phase $Pb_{1-x}Gd_xMo_6S_8$ samples using a hot isostatic pressing process and have measured the critical current density and other properties of the samples. We found that a reaction temperature as high as 1500 °C is needed to incorporate Gd into the grains of Chevrel phase. X-ray diffraction data show that small amounts of impurity phases are present in the samples at doping level with $x \geq 0.06$. Adding Gd decreases T_c and increases normal state resistivity. This is attributed to the change of carrier concentration. Both J_c and B_{irr} are reduced whereas the pinning force function is found to be similar between the samples, suggesting one dominant mechanism operating in the samples. The results show that the decrease of B_{irr} in the doped samples is related to impurity phases and degraded grain boundaries.

Acknowledgements

We thank C. Eastell and M. Goringe (University of Oxford) for performing EDX measurements; C. Lehmann for help in X-ray diffraction measurement; M. Decroux (University of Geneva), N. Cheggour and I.J. Daniel for valuable discussions. This work is supported by the Engineering and Physical Sciences Research Council, UK.

References

[1] Odermatt R, Fischer Ø, Jones H and Bongi G 1974 *J. Phys. C: Solid State Phys.* **7** L13.
[2] Foner S, McNiff Jr E J and Alexander E J 1974 *Phys. lett.* **49A** 269.
[3] Alden T H and Livingston J D 1966 *J. Appl. Phys.* **37** 3551.
[4] Meul H W 1986 *Helv. Phys. Acta* **59** 417.
[5] Zheng D N and Hampshire D P 1995 *Inst. Phys. conf. ser.* **148** 255.

Inst. Phys. Conf. Ser. No 158
Paper presented at Applied Superconductivity, The Netherlands, 30 June–3 July 1997

Hall probe measurements and analysis of the magnetic field above a multifilamentary superconducting tape

E C L Chesneau[1], J Kvitkovic[2], B A Glowacki[1], M Majoros[2], P Haldar[3]

[1]IRC in Superconductivity and Department of Materials Science, University of Cambridge, U.K.
[2]Institute of Electrical Engineering, Slovak Academy of Sciences, Slovak Republic.
[3]Intermagnetic General Corporation, Latham, New York, 12110, U.S.A.

Abstract. When characterising superconducting tapes it is useful to know the current flow inside a sample since its magnitude and any local variations are a good indicator of the uniformity and quality of a tape. Hall probe measurements are of use for this since they are non-intrusive and the sample preparation for measurements is relatively simple. Here we present preliminary results of Hall probe measurements made above a sample of multifilamentary superconducting tape. These measurements have a high spatial resolution as a result of a small Hall probe size. Also, all three components of the magnetic field can be measured not just the vertical component as is usual. Several methods exist for the back calculation of two-dimensional currents from magnetic fields, we compare three such methods.

1. Introduction

In order to optimise the critical current density, J_c, of Ag/Bi-2223 tape the effect of many processing parameters must be considered and quantified [1,2]. To do this it is common to use measurements of the overall J_c of the tape, often in conjunction with microstructural analysis. However, for several of the factors influencing J_c, such as the tape cross-section and the silver-to-superconductor ratio, it would also be useful to know where the current is flowing within the tape and not just the overall value of the current flow. This could lead to a more accurate identification of the effect of processing parameters, for example in producing regions of locally high J_c, and thus further optimisation of the overall J_c. Magneto-optic imaging [3,4] and slicing [5,6] experiments have been used to determine the current flow within tapes. However, both of these methods are intrusive so, for some applications, analysis of measurements of the magnetic field surrounding a tape may prove to be more practical. Here we compare 3 methods; gradient, convolution and matrix inversion; for the back calculation of the underlying current from Hall probe measurements of the magnetic field above a tape.

2. Experimental

A multifilamentary tape (IGC, 37 filaments) was used for the measurements; the cross-section is shown in Figure 1. The tape had a self field critical current at 77K of 24.2A [7]. For the first two methods of analysis a square scan (20x20 points) was made of the field component perpendicular to the tape plane whilst an external field of 2.5mT was applied

Figure 1 Cross-section of the multifilamentary tape showing significant amount of intergrowth between the filament. The filaments distort along the length of the tape which can lead to coalescence of the filaments and some difficulty in distinguishing all 37 filaments at any point. The total width of the tape is 3.4mm.

normal to the tape plane; the step size was 150µm. For the third method of analysis, a line scan of all 3 components of field was made after applying and then removing a field of 20mT; the step size was 100µm. The scans were made 150µm above the tape surface using 50µmx50µm Hall probes. Further details of the experimental arrangement are given in reference [8].

3. Analysis

3.1 Gradient method

We assume that the sample can be treated as a horizontal 2-dimensional or thin sheet carrying a quasi-static current with the measurement probe lying on the surface or very close to it. In this arrangement only H_z, the vertical component of magnetic field, is significant and the Maxwell equation;

$$\nabla \times \mathbf{H} = \mathbf{J}$$

where $\boldsymbol{J}$ is the local current density, can be reduced to;

$$\frac{\partial H_z}{\partial x} = J_y \qquad \text{and} \qquad -\frac{\partial H_z}{\partial y} = J_x$$

Thus, only one component of the magnetic field need be measured to identify 2 components of current. The 2-dimensional approximation is reasonable since the superconductor within the tape has a high aspect ratio, of approximately 21, and the alignment of grains along the tape will result in the vertical currents being much less than the horizontal. However, as the probe is moved further away from the sample, the results will become less accurate; in both the absolute current density values and the spatial distribution of the current.

3.2 Deconvolution method

Allowances can be made for the effect of a separation of the probe and the sample by using the method of Roth *et al* [9]. A brief outline of the method is given here but a detailed explanation can be found in reference [9]. The problem is approached with the Biot-Savart equation;

$$\mathbf{B}(\mathbf{r}) = \frac{\mu_o}{4\pi} \int_{-\infty}^{+\infty} \int_{-\infty}^{+\infty} \int_{-\infty}^{+\infty} \frac{\mathbf{J}(\mathbf{r}').(\mathbf{r}-\mathbf{r}')}{|\mathbf{r}-\mathbf{r}'|^3} d^3\mathbf{r}'$$

where co-ordinates with a dash represent those in the current plane and those without represent those in measurement plane. As before we assume a 2-dimensional current carrying sheet and in addition; a square scan i.e. n by n measurements, which covers all of the current carrying regions. Thus the divergence of the current density is zero throughout the scan area and so J_x and J_y are related. Under these conditions the Biot-Savart equation can be simplified to;

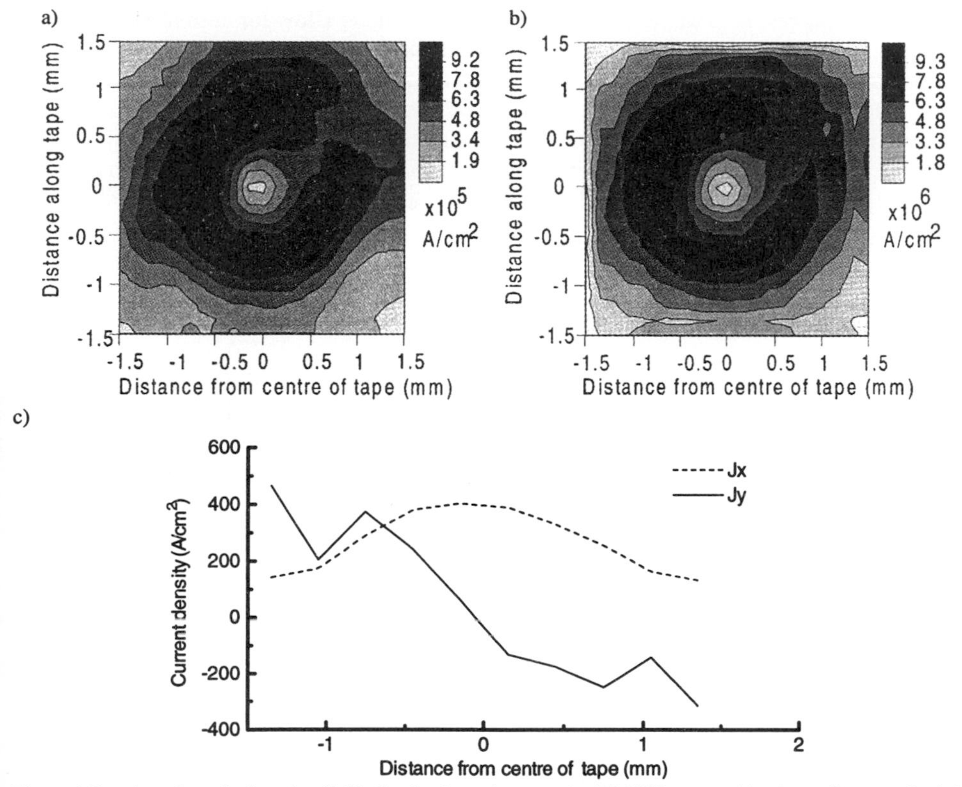

Figure 2 Results of analysing the field distribution above a Ag/Bi-2223 tape with a) gradient method b) convolution method c) matrix inversion method.

$$B_z(x,y) = \frac{\mu_o tz}{4\pi} \int_{-\infty}^{+\infty} \int_{-\infty}^{+\infty} \frac{J_x(x',y').(y-y') - J_y(x',y').(x-x')}{\left((x-x')^2 + (y-y')^2 + z^2\right)^{3/2}} \, dx'dy'$$

where t is the thickness of the conductor. This equation can then be split into two similar terms each of which is the convolution of a current distribution with a weighting function. Thus we can Fourier transform the measured magnetic field, $B_z(x,y)$ and divide by the transform of the filter function to obtain the transform of the current distribution J_x or J_y.

The spatial resolution of this method is limited by the height of the probe above the sample which must be sufficient to maintain the thin sheet approximation. Its use is also restricted by the need for a square scan within which the current is constrained to flow.

3.3 Matrix Inversion method

This is based on methods developed for use with biomagnetic measurements [10]. The field at any measurement point can be expressed as a linear function of the current density at every point, this function will vary between measurement points depending on the experimental arrangement. Thus, the magnetic field to current relationship can be represented as a matrix **A** with the resulting equation as $\mathbf{m} = \mathbf{Ac}$, where **m** is a vector of all measurements and $\boldsymbol{c}$ is a vector of the current values. In reference [10] the current is taken as due to point sources, here we have expanded the technique to allow for current flow through a volume; a distinction which is expressed via **A**, and have only considered 2 components of current. Singular value decomposition (SVD) was then used to invert **A** to find **c**.

This method can be easily expanded to consider a 3-dimensional current distribution however in this case the current is no longer a unique function of the field. This can be overcome by using methods such as linear estimators [10], pseudo inverses and maximum statistical entropy [11] to invert **A**, but significant assumptions must be made about the underlying current.

4. Results and conclusions

The results of analysing the field distributions by all 3 methods are shown in Figure 2. As expected the gradient method finds a similar distribution to the convolution method but finds a maximum current density that is lower (0.1kA/cm^2 compared with 1.3kA/cm^2). The matrix inversion method also finds a low current density, but this may be due to the method not being fully optimised yet. Both of these are lower than the engineering density of the tape; 9.4kA/cm^2. This may imply the presence of significant magnetic relaxation during the time scale of the experiment, as found previously [12,13].

Interestingly there is little sign of the current flow being constrained by the filaments. This may be due to the significant amount of intergrowth between filaments providing good quality current links across the width of the tape [14].

In conclusion; we have illustrated the use of three methods for the back calculation of currents from magnetic fields. For near 2-dimensional tapes, unless the exact value of current density is required, the gradient method is the simplest to implement and sufficiently accurate. For more exact values of current density, the convolution method can be used but this has more restrictions on its use. The most complicated to implement is the inversion method but this is intrinsically the most flexible in terms of describing the experimental arrangement.

References

[1] Willis J O, Ray II R D, Bingert J F, Phillips D S, Beckman K J, Smith M G, Sebring R J, Smith P A, Bingham B L, Coulter J Y and Peterson D E 1997 *Physica C* **278** 1-10

[2] Li S, Bredehöft, Hu Q Y, Liu H K, Dou S X and Gao W 1997 *Physica C* **275** 259-265

[3] Parrell J A, Polyanskii A A, Pashitski A E and Larbalestier D C 1996 *Supercond. Sci. Technol.* **9** 393-398

[4] Welp U, Gunter D O, Crabtree G W, Luo J S, Maroni V A, Carter W L, Vlasko-Vlasov V K and Nikitenko V I 1995 *Appl. Phys. Lett.* **66** 1270-1272

[5] Grasso G, Hensel, Jeremie A and Flükiger R 1995 *Physica C* **241** 45-52

[6] Larbalestier D C, Cai X Y, Feng Y, Edelman H, Umezawa A, Riley G N and Carter W L 1994 *Physica C* **221** 299-303

[7] Ciszek M, Glowacki B A, Campbell A M, Ashworth S P, Liang W Y, Haldar P and Selvamanickam V *IEEE Trans. Appl. Superconductivity* **7** *(in press)*

[8] Kvitkovic J and Majoros M J. 1996 *Magnetism and Magnetic Materials* **158** 440-441

[9] Roth B J, Sepulveda N G and Wikswo Jr J P 1989 *J. Appl. Phys* **65** 361-372

[10] Smith W E, Dallas W J, Kullmann W H and Schlitt H A 1990 *Appl. Optics* **29** 658-666

[11] Clarke C J S and Janday B S 1989 *Inverse Problems* **5** 483-500

[12] Paasi J, Polák M, Lahtinen M, Plechácek and Söderlund L 1992 *Cryogenics* **32** 1076-1083

[13] Glowacki B A, Noji H and Oota A 1996 *J. Appl. Phys.* **80** **2311**-2316

[14] Ashworth S P, Glowacki B A, Chesneau E C L, Ciszek M and Haldar P *IEEE Trans. Appl. Superconductivity* **7** *(in press)*

Inst. Phys. Conf. Ser. No 158
Paper presented at Applied Superconductivity, The Netherlands, 30 June–3 July 1997

Selection of the offset-criterion voltage parameter and its relation to the second differential of a superconductor's voltage-current curve

Dag W. A. Willén, Wen Zhu and Julian R. Cave

IREQ Hydro-Québec, 1800, boul. Lionel-Boulet, Varennes (Québec) J3X-1S1 Canada

Abstract: A relatively rapid characterisation method that gives valuable information about the superconducting transport behaviour is the calculation of the second differential of the superconductors E-I curve. This curve can be interpreted as a distribution of critical current values in a superconducting sample, as resulting from factors such as sausaging, cracks, phase content, and the distribution of magnetic flux-pinning forces. Here, we present measurements on samples of various qualities and with various metallic sheath materials, and the effect these factors have on the shape of the E-I curve and its second differential. The relation to the offset critical current criterion is discussed in terms of intrinsic material properties and the selection of the criterion voltage parameter, E_c(offset).

1. Introduction

One of the main tasks in the development of high-temperature superconductors for industrial applications is to improve their current carrying capacity. This requires characterisation methods that accurately evaluate the benefits of new fabrication procedures.

The commonly used 1 μVcm^{-1} (onset) criterion for the critical current, I_c, is dominated by the weakest regions or mechanisms in a sample, often induced by the presence of cracks, voids, or non-superconducting phases. The advantages of a fabrication process in terms of properties of "good" regions will be hidden in such a measure. It is therefore advantageous to study the *extended* E-I curves (Fig. 1), up to a few times the value of I_c(onset).

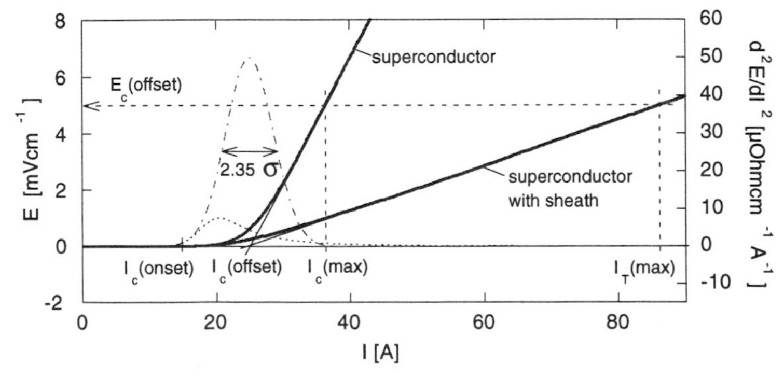

Fig. 1. Schematic of a superconductor's E-I curve (full lines), its second differential (dotted lines), and the offset criterion for the critical current for a superconductor with and without a conductive sheath.

The second differential of the E-I curve can be interpreted as a distribution of critical currents [1-10] determined by a distribution of pinning potentials [1, 4, 8, 9] in uniform conductors, and with increased width due to non-uniformity of the superconductor [2-5]. The second differential is often bell-shaped and is in some cases well approximated by a Gaussian distribution [1, 3].

The intercept of the tangent of the E-I curve with the I axis has been named the "offset" criterion of the critical current [11]. The offset criterion gives a zero I_c-value for normal conductors [11], is more reproducible than the onset criterion in process evaluation [12], and is insensitive to sheath material [13]. Depending on the choice of the voltage parameter, E_c(offset), from which the tangent is drawn, the resulting I_c(offset) will correspond to the average value of I_c, $\langle I_c \rangle$, or a lower value [1, 4, 11] (Fig. 1). However, sample heating can distort the result for high currents and voltages. It is therefore desirable to have some guidance in how to select the parameter E_c(offset) in terms of intrinsic parameters, in order to cover a sufficient portion of the resistive transition with a minimal amount of distortion due to heating.

2. Measured and intrinsic I_c distributions

Noise in the measured E-I data can introduce a significant error into the calculation of the second differential, requiring heavy smoothing [4, 6]. The amount of smoothing can be reduced by an appropriate selection of the discrete measurement points (E_n, I_n) [6]. A simple estimate of the maximum error is $\delta(d^2E/dI^2)_n \sim E_n/(I_n-I_{n-1})^2$, using the assumptions that the error in current, $\delta(I_n)$, is small compared to the step length, (I_n-I_{n-1}), and that the error in voltage, $\delta(E_n)$, is proportional to the voltage, E_n. Here, the current step length was increased continuously from ~60 mA at 0 A to ~ 600 mA at 60 A in order to reduce this error. In addition, some smoothing was performed before each differentiation.

For Ag-sheathed Bi-2223 tapes, the I_c-distribution extends into a voltage and current regime where the shunting of current into the silver sheath is substantial [14-16]. The current flowing in the superconductor, I_{sc}, is most readily approximated by $I_{sc}=I_m - E_m/R_{Ag}$, where I_m and E_m are the measured current and electric field, and R_{Ag} is the resistance of the silver sheath per unit length. The *intrinsic second differential* of the superconductor and the *measured second differential* of the composite conductor are then related by the equation

$$(d^2E(I_{sc})/dI_{sc}^2) = (d^2E(I_m)/dI_m^2)\ \{1-(dE(I_m)/dI_m)/R_{Ag}\}^{-3}\ , \qquad \text{(Eq. 1)}$$

and its inverse,

$$(d^2E(I_m)/dI_m^2) = (d^2E(I_{sc})/dI_{sc}^2)\ \{1+(dE(I_{sc})/dI_{sc})/R_{Ag}\}^{-3}\ . \qquad \text{(Eq. 2)}$$

Here, $dE(I_m)/dI_m$ and $dE(I_{sc})/dI_{sc}$ denote the first differentials. With Eq. 1, the intrinsic second differential of the superconductor can be calculated from a measurement of a sheathed conductor, and Eq. 2 can be used to fit measured data with a theoretical intrinsic critical current distribution.

Figure 2 shows the measured values of a silver-sheathed Bi-2223 sample without (a) and with (b) an additional soldered brass strip. Equation 1 has been used to derive the same intrinsic second differential from the two cases, and the resulting distribution is illustrated using a Gaussian fit (c) of the form

$$(d^2E(I_{sc})/dI_{sc}^2) = R_f(\sigma(2\pi)^{1/2})^{-1}\exp\{-(\langle I_c\rangle-I_{sc})^2/2\sigma^2\}. \qquad \text{(Eq. 3)}$$

Then, Eq. 2 has been used to recreate the second differentials (d) and (e) (solid lines). It is

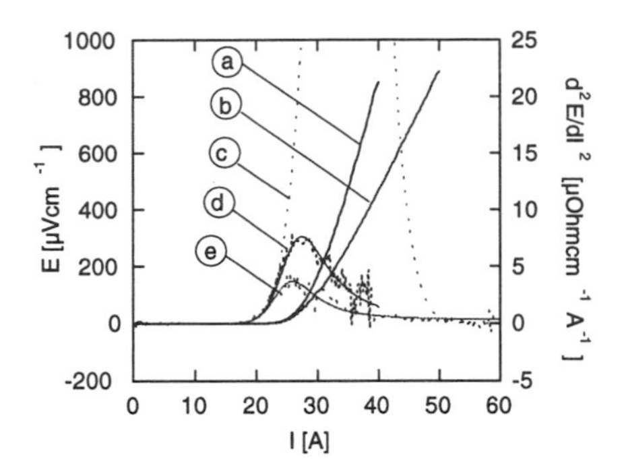

Fig. 2. E-I data for a Ag/Bi-2223 tape (a), with an additional soldered brass strip (b). The intrinsic (d^2E/dI^2) is fitted with a Gaussian (c). Eq. 2 has then been used to calculate (d) and (e).

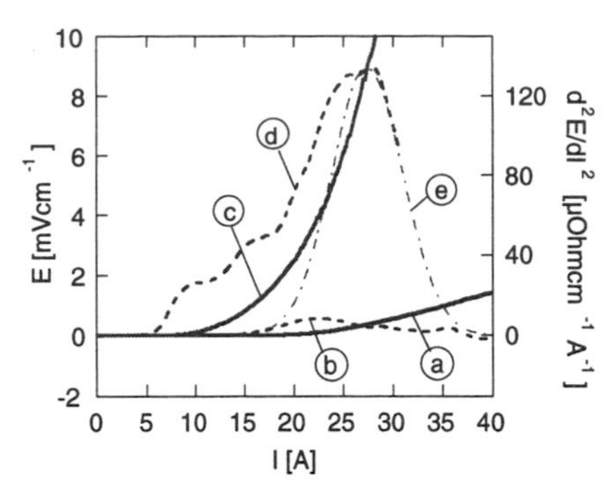

Fig. 3. E-I data for a Bi-2223 sample with Ag sheath (a, b) and after the sheath was etched off (c, d). The intrinsic second differential extends to the high values indicated by the Gaussian (e).

seen that the measured distributions have an asymmetric bell shape, although the intrinsic superconductor distribution is well fitted by a symmetric Gaussian.

A more direct way to obtain the intrinsic d^2E/dI^2 distribution is to physically remove the silver sheath [16]. Figure 3 shows transport measurements on a Ag/Bi-2223 sample before (a, b) and after (c, d) the silver sheath was etched off. Some damage occurred in the sample during etching and cool down, as seen in the second differential at low current (d). An estimation of the undamaged intrinsic distribution is shown in (e).

3. Selection of the offset criterion voltage parameter

Figure 2 does not cover the complete transition of the superconductor, in spite of the relatively high electric field, and therefore the offset criterion would indicate a value of I_c(offset) smaller than $<I_c>$ even if measured from the highest voltage. In Fig. 3 the distribution extends to above 30 A and 10 $mVcm^{-1}$. To reach this high voltage would be problematic for the sheathed conductor due to heating.

The shunting of current into the sheath and the power dissipation is expected to be reduced for a high-resistivity sheath material. Figure 4 shows the E-I characteristic of an "all-rolled" AgAu/Bi-2223 sample. The measurement in 0.9 T magnetic field yields R_{AgAu}=722 $\mu\Omega cm^{-1}$ (a), and in self-field the slope is R_T=670 $\mu\Omega cm^{-1}$ (b) for two ramp speeds of the current. Thus, the flux-flow resistance of the superconductor is approximately R_f=9.3 $m\Omega cm^{-1}$. The offset-criterion with E_c(offset)≅15 $mVcm^{-1}$ gives I_c(offset)=$<I_c>$=29 A. The intrinsic distribution was modelled by a Gaussian using these parameters. A value of σ=2.2 A recreates the E-I curve well using Eq. 2 and integration twice.

From Fig. 5, we deduce that an increased distribution width will result in a higher value of the electric field at the completion of the resistive transition, E(max). For a symmetric distribution, E(max) is given by (noting that R_f is resistance per unit length)

$$E(max) = (\Delta I_c R_f)/2 = (\Delta J_c A_{sc} \rho_f /A_{sc})/2 = (\Delta J_c \rho_f)/2 , \qquad \text{(Eq. 4)}$$

where ρ_f is the flux-flow resistivity of the superconductor, A_{sc} the core cross-sectional area, and ΔJ_c the absolute spread in I_c/A_{sc}. Thus, E(max) is independent of sample dimension and voltage gauge length, and is a suitable upper limit for the selection of E_c(offset). For the tape in Fig. 4, this implies a maximum value of E(max) ≅ $6\sigma R_F/2$ ≅ 60 $mVcm^{-1}$. This is,

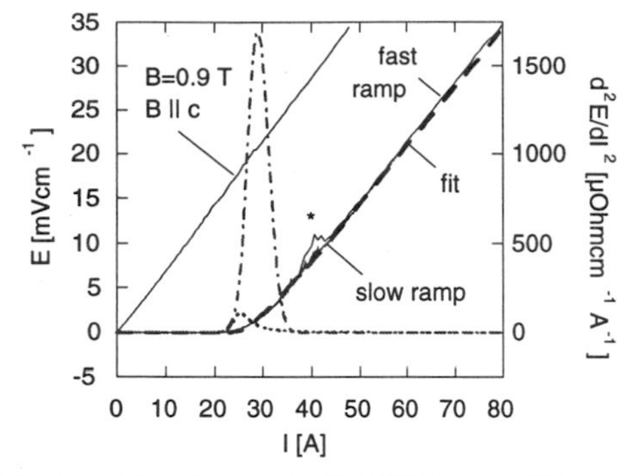

Fig. 4. E-I data for a AgAu/ Bi-2223 tape in 0.9 T and in self-field. As in Fig. 2, a Gaussian has been scaled using Eq. 2 and integrated twice to recreate the voltage trace. * - heating effects.

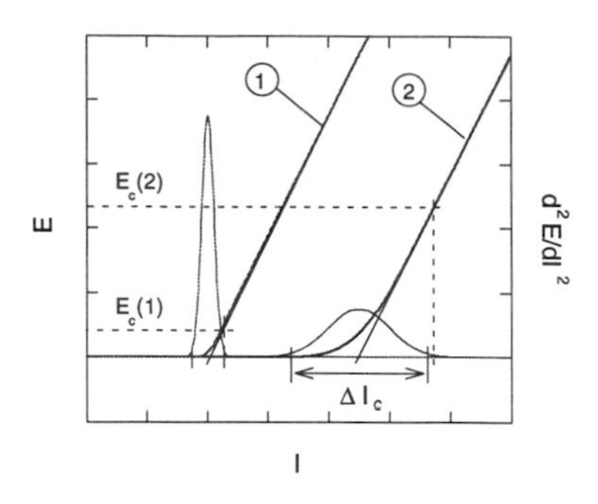

Fig. 5. Schematic of the influence of the transition width, ΔI_c, on the selection of the offset criterion voltage parameter for unsheathed superconductors with the same flux-flow resistance, R_f.

however, an impracticably high voltage and in fact much lower values of E_c(offset), in the range of 5-15 $mVcm^{-1}$, appear to give good estimates of $\langle I_c \rangle$.

4. Conclusion

Extended E-I curves have been measured for Bi-2223 samples with pure silver sheath, with the sheath etched off, and with resistive AgAu-alloy sheath.

The presence of a conductive matrix skews the measured second differential, (d^2E/dI^2), compared to a more symmetric *intrinsic critical current distribution*. The latter can be determined by using a modified second-differential function.

Provided that a proper voltage parameter, E_c(offset), is used, the offset criterion for the critical current is a simple means to extract the intrinsic average critical current of the superconductor core material. In the case of a symmetric I_c distribution, an upper limit for this voltage parameter is given by the expression $E_c(\text{offset})=(\Delta J_c\ \rho_f)/2$ and is independent of sample dimensions but proportional to the transition width. For the samples presented here, values of E_c(offset) of 5-15 $mVcm^{-1}$ give good results.

References:

[1] J Baixeras and G Fournet 1967 *J. Phys. Chem. Solids* **28** p1541
[2] R G Jones, E H Rhoderick and A C Rose-Innes 1967 *Physics letters* **24A** p318
[3] E Yu Klimenko and A E Trenin 1985 *Cryogenics* **25** p27
[4] W H Warnes and D C Larbalestier 1986 *Cryogenics* **26** p643
[5] C J G Plummer and J E Evetts 1987 *IEEE Transaction on Magnetics* **MAG-23** p1179
[6] L F Goodrich, A N Srivastava, M Yuyama and H Wada 1993 *IEEE Ttrans. Appl. Supercond.* **3** p1265
[7] Osvaldo F Schilling, Katsuzo Aihara and Shin-pei Matsuda 1992 *Physica C* **201** p397
[8] I Kusevic, E Babic, M Prester, S X Dou and H K Liu 1993 *Solid State Communication* **88** p241
[9] B Brown, J M Roberts and J Talc 1997 *Physical Review B* **55** p55
[10] J Cave, D Willén, R Nadi, D Cameron and W Zhu 1994 *IEEE Tans. Appl. Supercon.* **5** p1294
[11] J W Ekin 1989 *Appl. Phys. Lett.* **55** p905
[12] D Willén, C Richer, P Critchlow, M Goyette, R Nadi, et al. 1995 *Supercond. Sci. Technol.* **8** p347
[13] J R Cave, H D Ramsbottom, D Willen et al. 1996 *Proc. 8 th IWCC* p279 (World Sientific, 1996)
[14] D N Matthews, K-H Müller, C Andrikidis, H K Liu and S X Dou 1994 *Physica C* **229** p403
[15] C M Friend et Al. 1995 *Inst. Phys. Conf. Ser. No. 148* p431
[16] M Polak, W Zhang, E E Hellstrom and D C Larbalestier, 1995 *Inst. Phys. Conf. Ser. No. 148* p427

Inst. Phys. Conf. Ser. No 158
Paper presented at Applied Superconductivity, The Netherlands, 30 June–3 July 1997

Curret-voltage characteristics in YBaCuO thin films over more than 13 decades of electric-fields

T Nakamura[1], Y Hanayama[1], T Kiss[1], V Vysotsky[1], H Okamoto[2], T Matsushita[1,3], M Takeo[1], F Irie[2] and K Yamafuji[1]

[1] Graduate School of Information Science and Electrical Engineering (ISEE), Kyushu University, Fukuoka 812-81, Japan.
[2] Kyushu Electric Power Co. Inc., Fukuoka 815, Japan.
[3] Department of Computer Science and Electronics, Kyushu Institute of Technology, 680-4 Kawazu, Iizuka 820, Japan

Abstract. Combining the four-probe method and the magnetization method, we have measured electric-field vs. current-density (*E-J*) characteristics in $YBa_2Cu_3O_{7-\delta}$ thin films over more than 13 decades of electric-fields. It has been shown that the *E-J* curves obtained by the both methods lie on a continuous line each other. Furthermore, it has been pointed out that the scaling of *E-J* curves is valid only in a limited range of electric-field. That is to say, S-like-shape ln*E*-ln*J* curve can be observed even in the so-called glass regime.

1. Introduction

The phase transition between vortex-glass state and vortex-liquid state has been predicted theoretically by Fisher [1] in weakly pinned high-T_c superconductors, then it was demonstrated by Koch et al. [2] for the first time in their scaling analysis of transport characteristics in $Y_1Ba_2Cu_3O_{7-\delta}$(YBaCuO) thin films. As is well known, a large number of scaling characteristics have also been observed in the other high T_c cuprates. However, the scaling is usually discussed only in a limited electric-field region as narrow as 3 to 4 decades. Therefore, the measurement in a wider range of electric-field is necessary for a study on a dynamic state of flux line lattice in high T_c superconductors in detail. These measurements are also important from a practical point of view, because the electric-field varies complicatedly in a wide range depending on various operation conditions.

The *E-J* characteristics have usually been obtained by the conventional four-probe method. However, such contact method has limitation of sensitivity, which is usually no better than 0.1μV/cm. On the other hand, contactless magnetization methods have been used to obtain *E-J* characteristics in the lower electric-field region[3,4].

In this study, we measured *E-J* characteristics over more than 13 decades of electric-fields in a YBaCuO thin film combining the four-probe method and the magnetization method, and showed that *E-J* curves were scaled only in a limited range of electric-field.

2. Experimental

The sample used for the measurement is a c-axis oriented YBaCuO superconducting thin film,

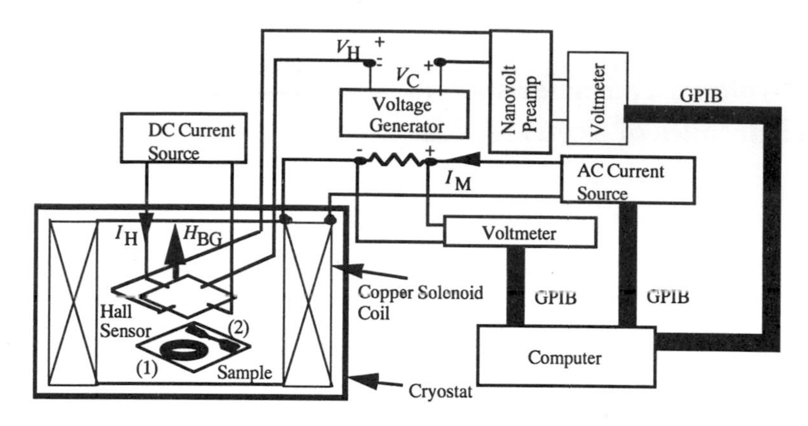

Fig.1 Schematic diagram of magnetization measurement. V_H is Hall voltage, and V_C is voltage to cancel the background field H_{BG}, which is applied by a permanent magnet (not shown). External sweep magnetic-field is applied by a copper solenoid coil; magnet constant, k_m=25.6 mT/A. Sample (1) is a ring for the magnetization measurement, and sample (2) is a bridge having the same line width of the ring for the four-probe measurement.

200nm in thickness deposited on a $SrTiO_3$ substrate (10x10 mm^2) by the excimer laser ablation method. In order to cancel the geometric effect of the film, we adopt a ring shape; 3mm in diameter, $2r$, and 200 μm in line width, w, for the magnetization measurement. A bridge having the same line width of the ring is also used for the four-probe measurement. These two samples are made from the same film by a typical chemical etching technique. The magnetization measurement is realized by monitoring the magnetic flux density in the center of the ring, which is generated by the shielding current flowing through the ring due to the sweeped external magnetic field. The magnetic flux density is measured by a Hall sensor; Hall coefficient, s_H=0.1V/T at the nominal bias current of 150mA. By measuring the magnetization curves as a function of the external magnetic field sweep rate, *E-J* curves can be obtained as follows.

$$E = \frac{r}{2}\frac{dB_e}{dt} \tag{1}$$

$$J = \frac{2rB_s}{\mu_0 wt} \tag{2}$$

where, B_e is the external magnetic field , and μ_0 is the magnetic permeability for vacuum. We assume $r>>w$, and $dB_e/dt>>dB_s/dt$, with B_s denoting the sample field due to the shielding current.

The magnetic relaxation is also measured to estimate the lower electric-field range of *E-J* curves. The external field B_e is applied perpendicular to the film surface using a copper solenoid coil; magnet constant, k_m=25.6 mT/A, and a Sm-Co permanent magnet is also used for the background field.

3. Results and Discussion

To check the validity of the system, we first measured the *E-J* characteristics by four-probe method and magnetization method in self-field at 77.3K. The data obtained by the four-probe method are in the range of 10^{-4}V/m to 10^0V/m, whereas the data by the magnetization method

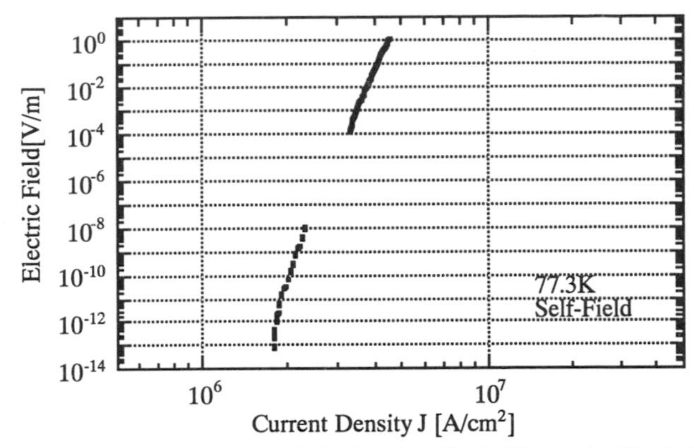

Fig.2 Electric-field(E) vs. current-density(J) characteristics in YBaCuO thin film obtained by the four-probe method (solid line) and the magnetization method (dotted line) in self-field at 77.3K.

are in the range of 10^{-13}V/m to 10^{-8}V/m. As can be seen in Fig.2, both of them; the four-probe method(solid line) and the magnetization method(dotted line), lie on a same line, then the E-J curve over more than 13 decades of electric-field is obtained. The critical current density of the film is $3.2\text{x}10^6$A/cm^2 at the 1μV/cm criterion, and the critical temperature is 91K.

By using the same system, we obtained the temperature dependence of E-J curves at a magnetic-field of 0.52T as shown in Fig.3. If we limit the electric-field range to about 4 to 5 decades as a typical measurement, we can see that the ln$E$-ln$J$ curves change from convex to concave as the temperature is increased in high- and low-electric-field region, respectively. However, as we can see at 61.2K for example, the gradient of ln$E$-ln$J$ curve in the low electric-field region is clearly smaller than that of the high electric-field region. Namely, even in the so-called glass region determined by the four-probe method, concave ln$E$-ln$J$ curves are observed in the low electric-field region obtained by the magnetization method. This result means that the scaling is valid only in a limited electric-field region.

Actually, if we limit the electric-field range, the scaling is valid for the transport data and the magnetization data as shown in Fig.4(a) and 4(b), respectively. However, the scaling

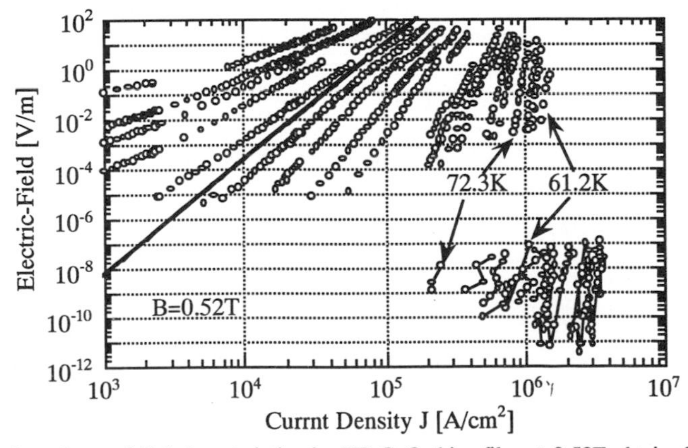

Fig.3 Temperature dependence of E-J characteristics in YBaCuO thim film at 0.52T obtained by four-probe method (top) and magnetization method (bottom). Temperatures were 61.2, 66.5, 68.3, 72.3, 76.0, 77.7, 80.7, 81.6, 82.7, 83.9, 85.0, 85.8, 86.7, 87.6, 88.2, 88.8, 89.3, 89.7, 90.0, 90.4 K for the four-probe method, and 25.5, 30.4, 33.4, 42.4, 47.3, 51.5, 56.6, 61.2, 63.8, 66.5, 72.3 K for the magnetization method. T_g(=88.4K) shown by the solid line is determined by the scaling analysis for the data obtained by the four-probe measurement.

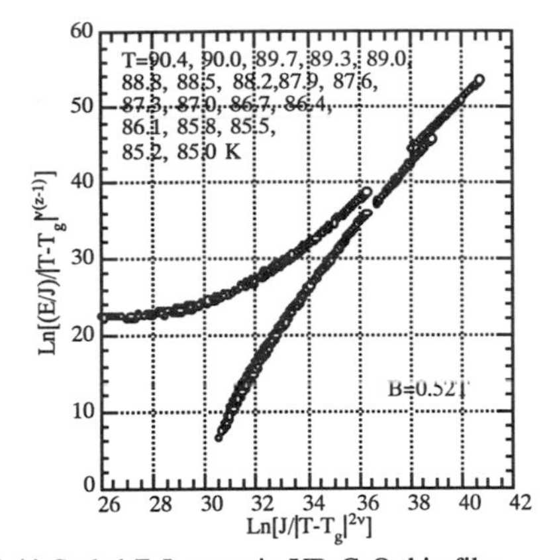

Fig.4 (a) Scaled *E-J* curves in YBaCuO thin film at 0.52T obtained by the four-probe method. The corresponding electric-field is in the range of 10^{-5}V/m to 10^{0}V/m. The scaling parameters are T_g=88.4K, z=8.2, and ν =1.4.

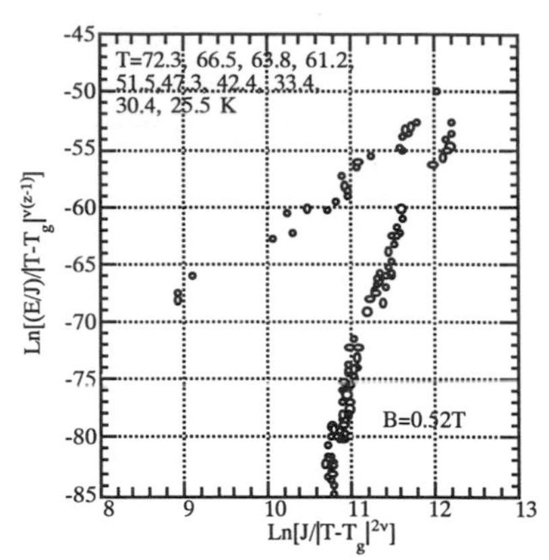

Fig.4 (b) Scaled *E-J* curves in YBaCuO thin film at 0.52T obtained by the magnetization method. The corresponding electric-field is in the range of 5×10^{-12}V/m to 10^{-7}V/m. The scaling parameters are T_g=56.6K, z=22.5, and ν =0.6.

parameters are totally different from each other. Namely, as the electric-field range is decreased, the parameter T_g decreases and z increases drastically. For the data by the four-probe method where the corresponding electric-field is in the range of 10^{-5}V/m to 10^{0}V/m, the transition temperature, T_g, is 88.4K, dynamic critical index, z, is 8.2 and static critical index, ν, is 1.4. On the otherhand, for the data by the magnetization method where the electric-field is in the range of 5×10^{-12}V/m to 10^{-7}V/m, T_g = 56.6K, z = 22.5 and ν =0.6 although there is some ambiguities in determing the T_g.

4. Conclusion

Combining the four-probe method and the magnetization method, we measured the *E-J* characteristics in a wide range of electric-field. It was shown that *E-J* curves obtained by the both methods agree quantitatively with each other. Even in the so-called glass region determined by the four-probe method, concave ln*E*-ln*J* curves are observed in the low electric-field region obtained by the magnetization method. Namely, S-like shape ln*E*-ln*J* curves are observed. This means that the scaling is valid only in a limited electric-field region. If we limit the electric-field region to about 4 to 5 decades, the scaling is valid in high- and low-electric-field region, respectively. The scaling parameters, however, are totally different from each other. That is, the transition temperature decreases and the dynamic critical index increases drastically in the low electric-field regime.

References

[1] Fisher M P A 1989 *Phys. Rev. Lett.* **62** 1415-18
[2] Koch R H, Foglietti V, Gallagher W J, Koren G, Gupta A, and Fisher M P A 1989 *Phys. Rev. Lett.* **63** 1511-14.
[3] Polak M, Windte V, Schauer W, Reiner J, Gurevich A, and Wuhl H 1991 *Physica C* **14** 14-22
[4]Charalambous M, Koch R H, Masselink T, Doany T, Feild C, and Holtzberg F, 1995 *Phys. Rev. Lett.* **75** 2578-81

Inst. Phys. Conf. Ser. No 158
Paper presented at Applied Superconductivity, The Netherlands, 30 June–3 July 1997

A modified flux-creep picture

R. De Luca[1], T. Di Matteo, S. Pace and M. Polichetti

INFN, INFM, Dipartimento di Fisica, Universita' degli Studi di Salerno, Via S. Allende, I-84081 Baronissi (SA), ITALY

Abstract. We study the creep problem in type-II superconducting materials for which the characteristic pinning center dimension l is larger than the coherence length ξ. The average value of the escape time of flux quanta from these pinning centers is calculated by adopting an appropriate expression for the pinning potential U. The attempt frequency ν in the Arrhenius formula is seen to depend explicitly on the current density J and the electric field E_{creep} due to flux-creep, calculated by this approach, qualitatively agrees with the experimental $E-J$ curves.

1. Introduction

Experimental current-voltage characteristics of high-T_c superconductors require a revision of the classical flux-creep theory [1]. Indeed, there are features of the E-J curves obtained at a fixed temperature T and at a fixed applied field H that do not find an explanation in the realm of the traditional Anderson-Kim model [2]. As an example, the overall curvature of a single logE-logJ curve goes from negative to positive as the temperature increases. Moreover, magnetic relaxation measurements [3] and transport measurements [4] show that, when a classical flux-creep picture is invoked, a divergent dependence of the pinning energy barrier on the current density J is necessary in order to justify the sharp decrease of dissipative phenomena in the low voltage limit. Other approaches have been proposed to explain these peculiarities [5]. However, in these models the flux creep problem is presented by means of different flux lattice regimes and a diverging energy barrier height ΔU for vanishingly small current densities J and for low temperature is seen as indicative of a vortex glass state.

In the present work we propose a slightly modified classical flux-creep approach to the problem of relaxation phenomena of magnetic states in type-II superconductors in the presence of large pinning centers. We start by revisiting the derivation of the Arrhenius

[1] **E-mail: DELUCA@vaxsa.csied.unisa.it.**

formula for the mean escape time τ of a single fluxon from a pinning center of characteristic dimension l much greater than the coherence length ξ. Then, by introducing an appropriate non-diverging pinning potential U, under the same basic assumptions made in the standard derivation, we notice that the attempt frequency ν in[B the Arrhenius formula depends on the current density J. Finally the E-J curves are calculated.

2. Escape time from a potential well

The coherence length of high-T_c superconductors is of the order of Å, so that it may well result smaller than the characteristic dimensions of defects. In this way, the interaction potential between a vortex and a pinning site may be written as follows:

$$U(x)/U_0 = tanh((x-l)/\xi) - tanh((x+l)/\xi) - \epsilon_0(J/J_0)x/\xi \tag{1}$$

where $2l$ is the size of the pinning center, U_0 is the depth of the potential well for $J = 0$, and $\epsilon_0 = \Phi_0 J_0 L/U_0$, Φ_0 being the elementary flux quantum, L the vortex line length and J_0 the critical current of the superconductor at $T = 0K$. The coherence length ξ and the potential barrier height U_0 depend on the temperature, so that we may set [6]:

$$U_0(t) = u_0\sqrt{1-t^4} \tag{2}$$

$$\xi_0(t) = \frac{\xi_0}{\sqrt{1-t}} \tag{3}$$

where t is the reduced temperature and u_0 and ξ_0 are the zero temperature values of the potential well depth and of the coherence length, respectively.

We can calculate the value of the potential U at the local minimum x_{min} and at the adjacent maximum x_{max}, so that the barrier height ΔU can be written as:

$$\Delta U = U(x_{max}) - U(x_{min}) \tag{4}$$

In Fig.1a and Fig.1b we report the potential well shape in configuration space and the barrier height ΔU as a function of the the current density J for different values of the temperature, respectively, as derived from Eqs.(1) through (4).

In Fig.1a we notice how the temperature dependence of the coherence length induces a smoothing of the potential well shape. We may recall that the Arrhenius formula is valid in the limit of small thermal energies with respect to potential barrier heights. In Fig.1b we draw a horizontal line $\Delta U = 5k_BT$ for $T = 0.8T_c$, which intersects the corresponding ΔU vs. J curve at a point P. This point, which may be considered to be characteristic of the limit of application of the Arrhenius formula as derived with the aid of classical stochastic methods, moves toward higher values of J as the temperature decreases.

Furthermore, the solution of the differential equation for the average escape time τ calls for a perfectly reflecting potential at the top and for a perfectly absorbant potential at the bottom of the well itself. In order to take account of the presence of other pinning sites in the superconducting medium, one needs to opportunely change the problem's boundary conditions. However, even if these conditions are met, a numerical solution of the integral form of the solution for τ is required when $\Delta U \simeq k_BT$. This problem will be tackled in a forthcoming paper.

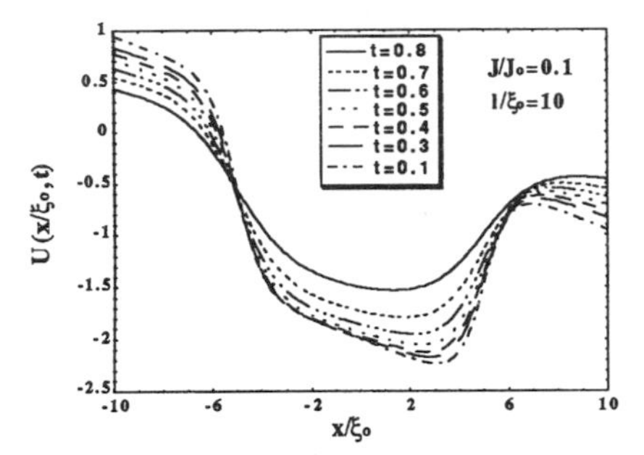

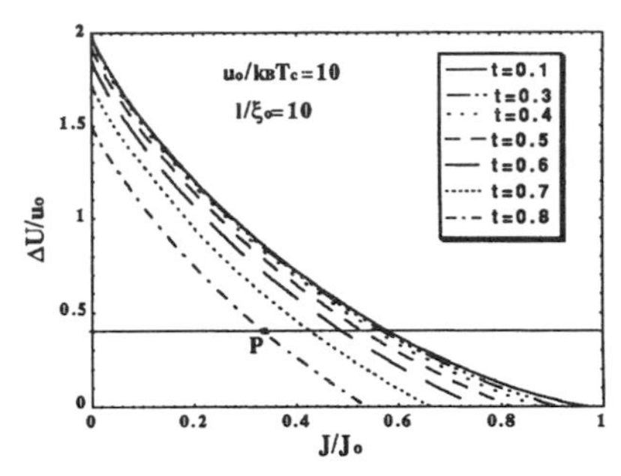

Figure 1. a) Pinning potential in a type-II superconductor in the presence of large pinning sites as a function of the spatial variable x at various temperatures. b) Energy barrier height as a function of the current density J at various temperatures.

In the present work, instead, we start by the usual Arrhenius formula for the average value of the escape time from a potential well:

$$1/\tau = \nu(J) \ \ \exp(-\Delta U/k_B T) \tag{5}$$

where

$$\nu(J) = \sqrt{U''(x_{min})U''(x_{max})}/2\pi\beta \tag{6}$$

is the attempt frequency, which is here calculated in terms of the damping constant β and of the geometric mean of the curvature of the pinning potential at x_{min} and x_{max} [7]. In this way, the electric field E_{creep} can be written in the following way:

$$E_{creep} = E_0(\nu(J)/\nu_0)exp(-\Delta U/k_B T) \tag{7}$$

with E_0 and ν_0 being normalizing constants. The ν vs. J dependence is shown in Fig.2a, while the E_{creep} vs. J curves are shown in Fig.2b for various values of the reduced temperature t.

From Fig.2a we notice that the attempt frequency goes to zero for vanishing values of the current density. In this way, the electric field E_{creep} goes faster to zero in the $J = 0$ limit when compared to the usual case found in the literature, where the attempt frequency is taken as independent of J. Therefore, the E_{creep} vs. J curves do reproduce the characteristic features of the experimental findings. Indeed, from Fig.2b we also notice that, for decreasing temperature values, the $logE_{creep}$ vs. $logJ$ plots go from positive to negative curvature.

3. Conclusions

Starting from the well known Arrhenius formula we interpret, in the realm of a unified non-traditional flux-creep picture, the E-J curves of high-T_c superconductors. The analysis is carried out by first noticing that in this class of superconductors the coherence length may be small when compared to the characteristic pinning site dimensions.

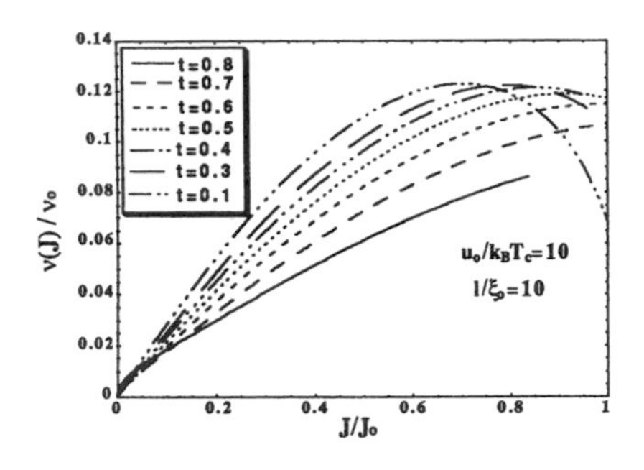

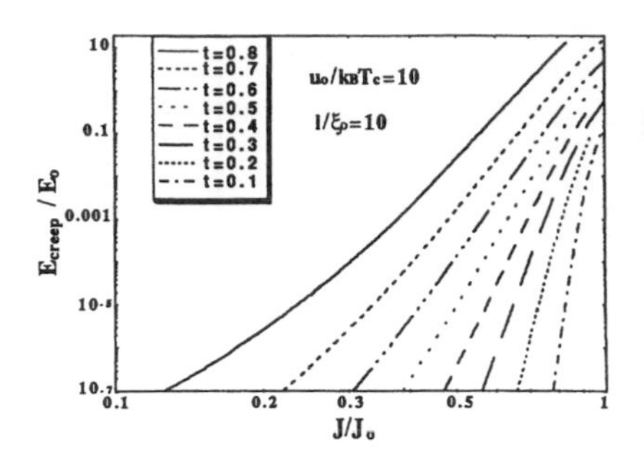

Figure 2. a) Current density dependence of the attempt frequency at various temperatures. b) Electric field due to flux creep as a function of the current density J at various temperatures.

Second, by revisiting the formal derivation of the Arrhenius formula and by abiding to the assumptions there made, we find that the attempt frequency must depend on the current density J. In particular, the ν vs. J dependence is such that $\nu \to 0$ as $J \to 0$. In this way, the dissipative phenomena at low voltages are seen to vanish without invoking a diverging energy barrier height and under a condition which is valid even for low fields. Therefore the current voltage characteristics for these systems can be derived in an unitary fashion. The resulting E-J curves derived for different temperatures are in qualitative agreement with experiments and well reproduce the observed features in the experimental voltage-current characteristics.

Acknowledgments

The authors would like to thank prof. S. De Martino and prof. S. De Siena for helpful discussions.

References

[1] Huse D A, Fisher M P A and Fisher D S 1992 *Nature* **358** 553–559

[2] Anderson P W and Kim Y B 1964 *Rev. Mod. Phys.* 39–43

[3] Maley M P, Willis J O, Lessure H and McHenry M E 1990 *Phys. Rev. B* **42** 2639–2642

[4] Zeldov E, Amer N M, Koren G, Gupta A, McElfresh M W and Gambino R J 1990 *Appl. Phys. Lett.* **56** 680–682

[5] Blatter G, Feigel' man M V, Geshkenbein V B, Larkin A I and Vinokur V M 1996 *Rev. Mod. Phys.* **66** 1125–1388

[6] Tinkham M 1975 *Introduction to Superconductivity* (New York: McGraw-Hill)

[7] Gardiner C W 1985 *Handbook of Stochastic Methods* (Berlin: Springer-Verlag)

Inst. Phys. Conf. Ser. No 158
Paper presented at Applied Superconductivity, The Netherlands, 30 June–3 July 1997

Progress in Preparing Tl/Pb(Ba/Sr)$_2$Ca$_2$Cu$_3$O$_{8.25+x}$ (Tl-1223) Bulk and Powder Samples in an Open System

Christoph L. Teske

Institut für Anorganische Chemie, Christian-Albrechts-Universität zu Kiel, Otto-Hahn-Platz 6/7, D-24098 Kiel;

Jörg H. Albering and Stephan Gauss

Hoechst AG, CRT-TP, HTSL-Project, G 864, D-65926 Frankfurt a. M.

Abstract. The preparation of Tl-1223 in an open system on a 100g scale is described. New results on the lead containing material are reported. Bulk parts and PIT-tapes with different compositions were prepared. The samples are characterised by R/T-, AC-curves and Rietveld refinements. Critical current densities of up to 8.3 kA/cm^2 were obtained.

1. Introduction

The Tl-1223 type HTS-compounds meanwhile have become a more and more interesting material for technical applications (thin films for electronic devices and flexible tapes and cables for magnets and motors). Among the relevant HTS-ceramics (YBCO and B(P)SCCO) they distinguish themselves by higher critical temperatures ($112 < T_c < 120$K) and higher current densities (j_c) in terms of magnetic field dependence at 77 K. The preparation of these Tl-containing HTS-ceramics at elevated temperatures (T > 500-600°C) is restricted due to the high volatility of the toxic Tl_2O. Thus, usually sealed containers (quartz ampoules) and gold foils are used [1-3]. This laboratory method does not yield sufficient amounts of material for practical applications (e.g. PIT-wires) and therefore is not appropriate for commercial use. We already reported on a scaleable process using a quasi open system with flowing gas atmosphere for the preparation of bulk and powder samples on a 100 g scale [4]. Meanwhile we have tested several nominal compositions to show that our method produces competitive materials. Here we report on the recent results with lead containing samples.

2. Experimental

2.1 General remarks

We are using starting mixtures consisting of Tl_2O_3/PbO and precursors (precalcinated oxocuprates of Ca, Sr, and Ba). In order to achieve a high phase content of Tl-1223 and to minimize the Tl-losses by fitting the reaction conditions, each composition primarily has to be investigated by thermal analysis (TG/DTA in flowing atmosphere). The optimal temperature for the solid state reaction ranges between 880 and 905 °C in case of air and must be elevated for about 30 K for pure oxygen atmosphere. The latter is lowering the mean Tl_2O ($+O_2$) losses (3-5%) below 2%. On the other hand pure oxygen atmosphere increases the tendency to form e.g. $BaPbO_3$ as an additional impurity phase. For the preparation of larger quantities (by a multistage process) the conditions of the DTA/TG-experiment are simulated using a vertically mounted tube furnace [5] (heating rate 6-10 K/min; holding time at $T_{max} \approx 12$ min). Each batch consists of 3-4 pellets (∅ = 40 mm, mass ≈ 33 g, or of several bars (3-4 × 5 × 35-40 mm). Between each of the 3-4 heat treatments the samples are reground, sieved and a surplus of Tl_2O_3 in the magnitude of the expected loss is added. The reground and sieved bulk material can also be used for the production of Powder In Tube (PIT) superconducting wires. The biggest sample produced in our laboratory so far was more than 340g.

2.2 Experiments with lead containing starting mixtures

A starting mixture with the nominal („Hitachi" [6]) composition of $(Tl/Pb)Ca_2(Sr/Ba)_2Cu_3O_{8+x}$ was used (heating rate 10 K/min, $T_{max} \approx 900$ °C, air). The microstructure analysis of the obtained relatively dense HTS- ceramic showed an average grain size of ≈ 20 μm for the Tl-1223 crystals and a small amount of impurity phases. The crystal structure was refined by Rietveld analysis [7] of the X-ray powder data. $(Ca/Sr)_2CuO_3$ and $BaPbO_3$ were detected as impurities (see Figure 1.). During our previous investigations on lead free Tl-1223 we experienced that a surplus of Calcium seemed to have a positive effect on T_c.

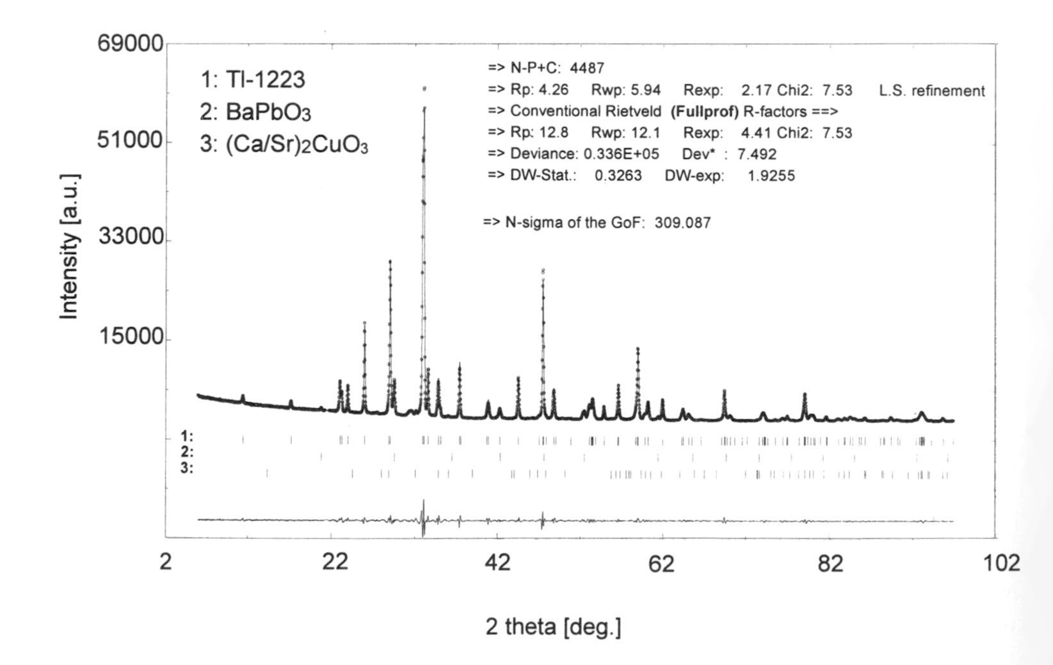

Fig.1. Observed, calculated and difference XRD-intensity profiles for Tl-1223. Nominal composition: $(Tl/Pb)Ca_2Sr_{1.6}Ba_{0.4}Cu_3O_{8+x}$. *Phase 1*: 89.5; *2*: 5.7; *3*: 4.8 %Cont. of the total integrated intensity (100x).

It is known from the literature [8], that Ca substitutes partially for Ba or Sr. Samples of the nominal composition $(Pb/Tl)Ca_{2+x}Sr_{1.6}Ba_{0.4-x}Cu_3O_{8.5+\delta}$ ($0 < x < 0.4$) were prepared and simultaneously heated in the furnace, similar to the method mentioned above.

3. Results and Discussion

The critical temperature obtained for $(Tl/Pb)Ca_2Sr_{1.6}Ba_{0.4}Cu_3O_{8+x}$ was $T_c \approx 117$ K. Silver tubes (ca. 0.2 × 5 cm) were filled with the reground material and 30-40 cm long mono filamentary tapes were produced by rolling (filled cross section ≈ 0.2-0.3 mm²). After annealing critical current densities of $j_C = 8.2\ kA/cm^2$ (1μV criterion) were achieved. The investigated series with Ca rich samples showed a dependence of T_c on x (see Figure 2.). A maximum $T_c \approx 119$ K was achieved at $x \approx 0.2$. With increasing x it became more and more difficult to produce tolerable pure material. At last a new sample ($x = 0.2$) was prepared in the above described manner. Figure 3. shows the resistance and the AC-signal in terms of temperature. A comparison of the refined lattice parameters (a = 3.82500(5), c = 15.43482(31) for $(Tl/Pb)Ca_2Sr_{1.6}Ba_{0.4}Cu_3O_{8+x}$ and a = 3.82304(12), c = 15.35688(72) [Å] for $(Pb/Tl)\ Ca_{2.2}\ Sr_{1.6}\ Ba_{0.2}\ Cu_3O_{8.5+\delta}$) supports the assumption that Ca substitutes for Ba. PIT-wires with this composition were investigated as well. Although the intrinsic properties seemed to be satisfying, only comparatively low critical currents were obtained. The maximum of the achieved critical current density in those tapes was 5-6 kA/cm², which may be due to impurities and low grain alignment. The investigation by the Rietveld powder XRD method resulted in a sample containing a remarkable amount of $(Ca/Sr)_2CuO_3$ (see Figure 4.).

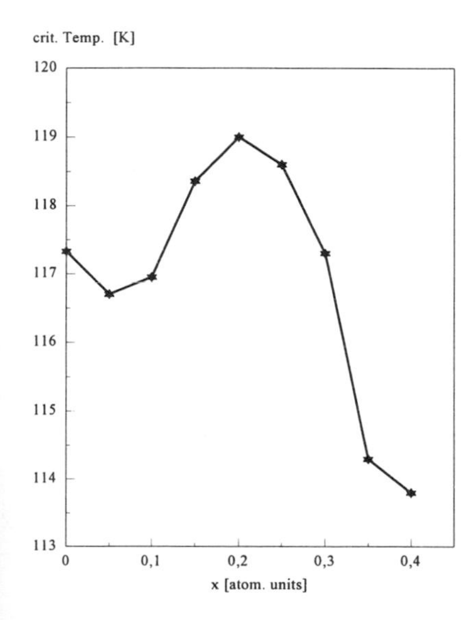

Fig. 2. Critical temperature T_c in terms of x for $(Pb/Tl)Ca_{2+x}Sr_{1.6}Ba_{0.4-x}Cu_3O_{8.5+\delta}$.

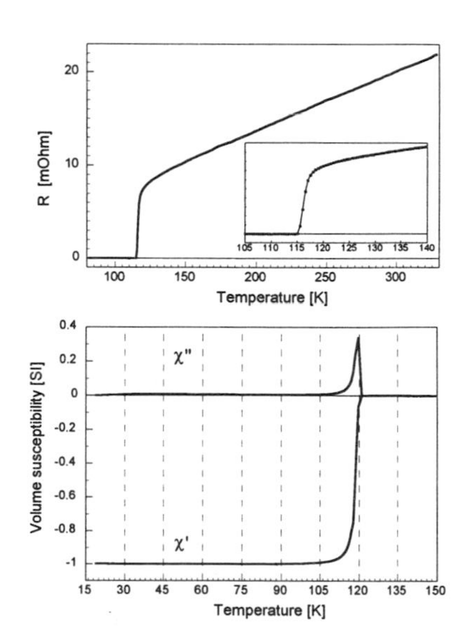

Fig. 3. Resistance and AC-signal in terms of temperature for the nominal composition: $(Pb/Tl)Ca_{2.2}Sr_{1.6}Ba_{0.2}Cu_3O_{8.5+\delta}$.

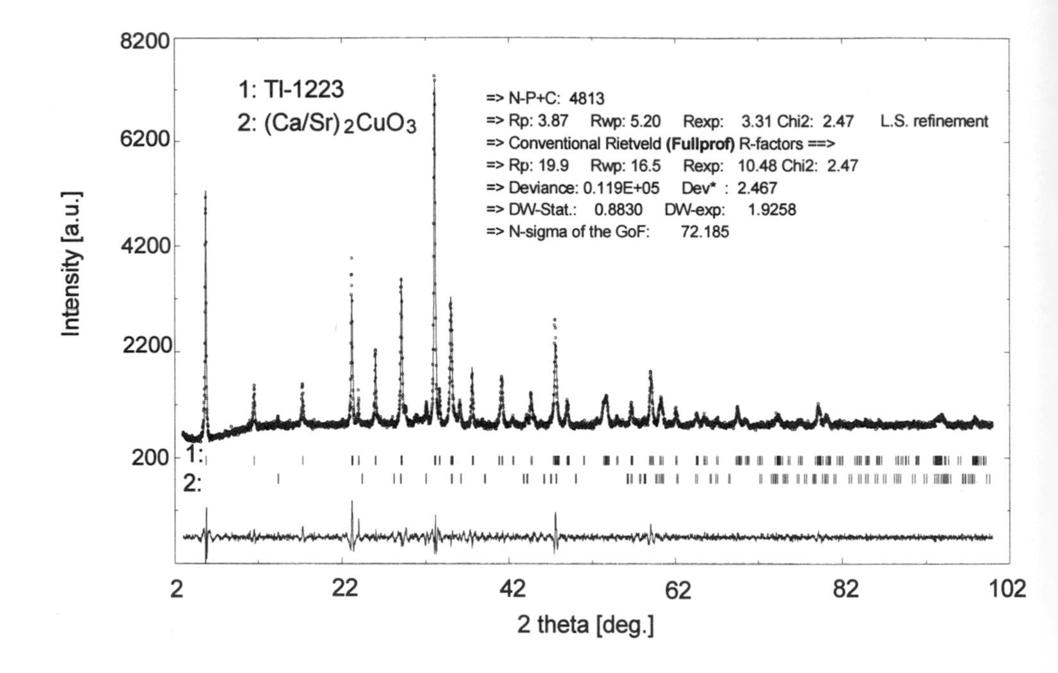

Fig.4. Observed, calculated and difference XRD-intensity profiles for Tl-1223. Nominal composition: $(Tl/Pb)Ca_{2.2}Sr_{1.6}Ba_{0.2}Cu_3O_{8+x}$. *Phase 1*: 81.1; *2*: 18.9 %Cont. of the total integrated intensity (100x).

4. Conclusions

The samples prepared on a large scale by our method in an open system show very similar properties to those known from the literature. Especially $(Tl/Pb)Ca_2Sr_{1.6}Ba_{0.4}Cu_3O_{8+x}$ can be produced in larger amounts with appropriate quality. To improve the properties of Tl-1223 a lot more of chemical investigations must be done. Our method is qualified to yield sufficient amounts of bulk (and powder) material for further investigations on different additives (such as Bi_2O_3 and Li_2O) to improve phase purity and the alignment of the ceramic particles.

References

[1] M. A. Subramanian, C. Tonardi, J. Gopalakrishnan, P. Gai, J. Calabrese, T. Askew, R. Filippen, A. Sleight 1988 *Science* **242**, 249

[2] S. Parkin, V. Lee, A. Nazzal, R. Savoy, R. Beyers 1988 *Phys. Rev. Lett.* **60**, 2539

[3] R. Sugise, M. Hirabayashi, N. Tereda, M. Jo, T. Shomomura, H. Ihara 1988 *Jap. J. Appl. Phys*. **24/9**, 1709

[4] Chr. L. Teske, Hk Müller-Buschbaum, Chr. Lang and S. Elschner 1995 *Appl. Supercondutivity*, Vol. **1** 99

[5] Chr. L. Teske, J. H. Albering and S. Gauss 1997 *J. Superconductivity* to be published

[6] T. Kamo, T. Doi, A. Soeta, T. Yuasa, N. Inoue, K. Irmara and S.-P. Matsuda 1991 *Appl. Phys. Lett.* **59** 3186

[7] H Rietveld 1969 *J. Appl .Crystallogr*. **2**, 65

[8] R. E. Gladyshevshii, Ph. Galez, K. Lebbou, J. Allem and, R. Abraham, M. Couach, R. Flükiger, J.-L. Jorda, M. Th. Cohen-Adad 1996. *Physica* **C 267** 93

Inst. Phys. Conf. Ser. No 158
Paper presented at Applied Superconductivity, The Netherlands, 30 June–3 July 1997

Flux composition influence at very high temperature on the formation of NdBaCuO materials under quenching

F Auguste, M Liégeois, A Rulmont and R Cloots

SUPRAS, Institute of Chemistry, B6, University of Liège, B-4000 Liège, Belgium

M Ausloos

SUPRAS, Institut of Physics, B5, University of Liège, B-4000 Liège, Belgium

Abstract. We have investigated the influence of the flux chemical composition on the formation of Nd-123 phase. The flux was monitored by changing the proportions of BaO and CuO in the initial mixture. Neodynium was provided by the partial dissolution of Nd_2O_3 in the flux raised at very high temperature (1350°C). Quenched specimens have been produced in order to investigate the microstructure and chemical composition of the various phases. Optical polarized light microscopy, scanning electron microscopy and energy dispersive X-ray microanalysis were performed. The microstructures show a very high dependence on the flux initial composition. A discussion based on the ternary Nd_2O_3 -BaO-CuO phase diagram system shows that a 4 BaO + 12 CuO precursor is promising for the formation of single crystals of the 123 phase.

1. Introduction

High-Tc superconducting crystals are difficult to grow because these complex chemical compounds melt incongruently (i.e., decompose with partial melting). SUPRAS has developed complex low-melt solvents (with several components) and other advanced techniques to mitigate these problems [1-3]. Among them is a magnetically melt textured growth method [1]. In order to optimize the texturing, an appropriate seed on the top or in the initial powder may be very influential. $Y_1Ba_2Cu_3O_x$-like crystals usually grow as thin plates with reasonably well-defined crystal morphology. Obtaining a uniform optimum oxygen distribution in the crystals is also difficult. The bigger the crystal, the larger the problem. They are also annoyingly very brittle. Therefore crystals with a well defined degree of perfection, as indicated by different measurements are of interest. Moreover the magnetic and electrical properties are extremely sensitive to the crystal perfection, i.e. to very little intracrystal flux pinning and so-called weak links. Although much work has already been done in the field [4-6], it is important to re-investigate the "123" phase diagram with respect to the microstructure itself, and in "new" systems.

2. Experimental

Six ceramic superconductors with a variable composition $Nd_2Ba_xCu_yO_{10-z}$ intended to contain Nd-123 as a major phase were synthesized according to the "thermal cycle" shown in

Fig. 1, starting form a mixture of powders of various oxides in appropriate proportions. An appropriate BaO-CuO mixture was quickly raised from room temperature to 800°C at a 150°C/h rate. The mixture was maintained at such a temperature during 48 hours, then removed at room temperature for an addition of Nd_2O_3 powder, in a proportion such that the mixture corresponded to a specific point (Q_i) in the ternary phase diagram Nd_2O_3-BaO-CuO (Table 1). The new mixture was put back into the oven at 900°C and raised to 1350°C at a 150°C/h rate. After a 1 to 10 minutes at that temperature the sample was quenched between copper plates in air at room temperature or in oil if the mixture was too viscous. The samples were inserted in epoxy resin for observation with various microscopic techniques. The samples were polished thereafter. Samples Q_3 and Q_6 are unfortunately damaged and could not be further observed. Optical polarized light microscopy (PLM), scanning electron microscopy (SEM) and energy dispersive X-ray (EDX) microanalysis were performed on the four other samples.

3. Results

The Q1 sample (Fig. 2) results from an initial precursor 4 BaO + 6 CuO. An array of approximately equal (ca. 0.020 mm x 0. 001 mm) size 123 needles is observed with side branches. The chemical composition is characterized by the presence of Al on Cu sites. The sample formula is better written $Nd_{1+x}Ba_{2-x} (Cu_{1-y}Al_y)_3O_{7-z}$ with $x<0.5$ and $y<0.25$. The dark phase in Fig. 2 corresponds to the non reacted liquid phase containing an excess of BaO and Cu_2O.

Table I. Definition of samples (Qi) with respect to the inital "flux precursor" x BaO + y CuO.

sample	x	y
Q1	4	6
Q2	8	6
Q3	8	1
Q4	4	12
Q5	1	12
Q6	7	18

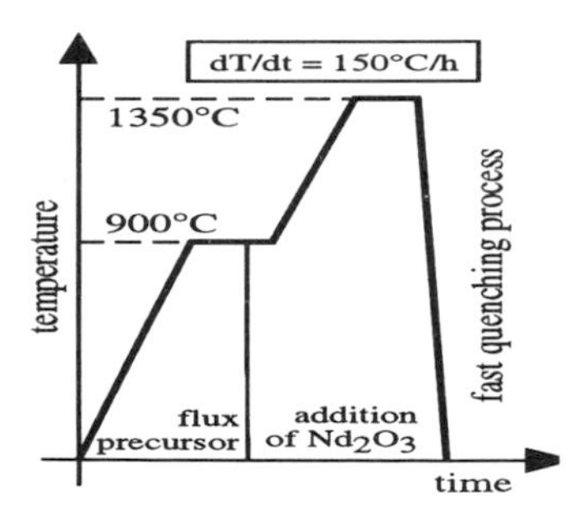

Fig. 1 (a) Thermal cycle used in this investigation.

The Q_2 sample (Fig. 3) results from an initial precursor 8 BaO + 6 CuO. An array of approximately equal (ca. 0.10 mm x 0. 01 mm) size disconnected 287 needles is observed with Al replacing Cu on almost all (i.e. 92%) Cu sites. Another type of smaller and less elongated (ca. 0.010 mm x 0.002 mm x 0.002 mm) grains appears to be made of $NdBa_{2+<x>}Cu_3O_{7-v}$ with $<x> \approx 0.3$ on the tie line between Nd-123 and Nd-163 in the phase diagram [4]. They appear whiter on the scanning electron micrograph. The to-be-superconducting ceramics thus appear to be made of a mixture of a 123 solid solution and Al phases. There is much unreacted liquid phase containing an excess of BaO and Cu_2O.

The Q4 sample (Fig. 4) results from an initial precursor 4 BaO + 12 CuO. A square array of approximately equal (ca. 0.100 mm x 0.010 mm) size disconnected 123 needles is observed, together with other smaller and less elongated Cu2O grains. The sample formula can be written $Nd_{1+x}Ba_{2-x}Cu_3O_{7-z}$, with $x \approx 0.2$.There is some Al in the 123 grains. The non reacted liquid phase contains an excess of BaO and Cu2O.

The Q5 sample (Fig. 5) results from an initial precursor BaO + 12 CuO. An array of approximately equal (ca. 0.03 mm x 0. 001 mm) size disconnected Nd_2CuO_4 needles is found, each one surrounded by 123 $Nd_{1+x}Ba_{2-x}Cu_3O_{7-w}$ ($x \approx 0.36$) micrograins as a minority phase. Small Cu_2O grains are seen between the needles.

Fig. 2 SEM of Q_1 sample. Needles of the 123 phase composition are observed and are embedded in a medium containing an excess of BaO and Cu_2O.

Fig. 3 SEM of Q2 sample. Grains with the 287 composition are observed. Smaller white grains are visible on the top of the 287 grains. Their chemical composition lies on the tie-line between Nd-123 and Nd-163 in the phase diagram.

4. Discussion

As expected the initial flux composition is very relevant for the production of Nd-123 phase. The present investigations have allowed us to observe stable $Nd_{1+x}Ba_{2-x}Cu_3O_{7-z}$, or rather $Nd(Ba_{1-y}Nd_y)_2Cu_3O_{7-z}$ solid solutions. Several phases contain Al, meaning that some flux compositions are more reactive than others with the container, i.e. Ba deficient fluxes (like Q4 and Q5) are more inert towards Al_2O_3. Thus the starting chemical composition close to Q4, instead of the usual Q6 seems of interest for getting 123 single crystals in Al_2O_3. Notice that the melting temperature is higher and the flux viscosity greater than in usual conditions. The more or less symmetric composition, i.e. Q2, with respect to the stoichiometric 123 in the phase diagram is not so promising for the production of the 123 phase.

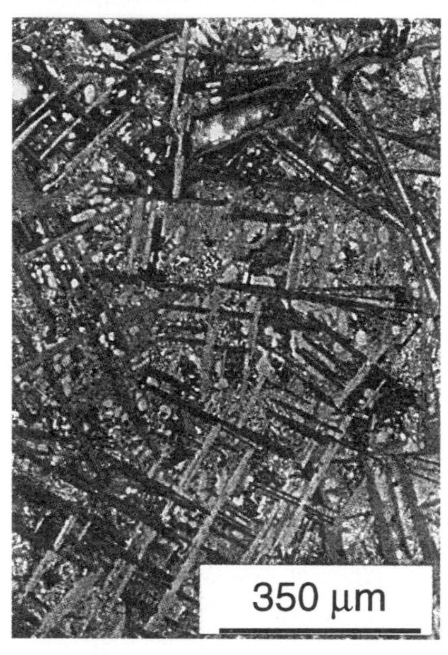

Fig. 4 PLM of Q4 sample. Well-defined, interconnected 123 needles are observed. Plenty of Cu_2O particles are distributed in the matrix.

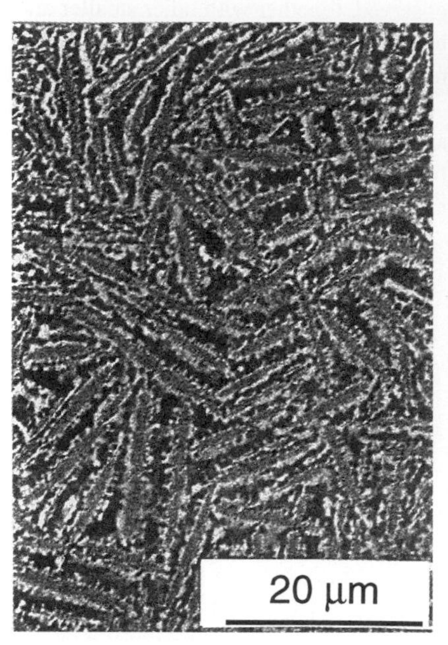

Fig. 5 SEM of Q5 sample. Nd_2CuO_4 elongated grains are clearly observed, and are surrounded by the 123 phase as the minor phase.

Acknowledgements

Part of this work has been financially supported through grant ARC 94/99-174. Thanks to Prof. H.W. Vanderschueren for allowing us to use the SEM and EDX at the Measurement and Instrumentation in Electronics Laboratory (MIEL)

References

[1] Cloots R, Vandewalle N, Ausloos M 1994 Appl. Phys. Lett. **65** 3386-3388

[2] Ausloos M, Bougrine H, Cloots R, Rulmont A, Gilabert A, Laval J Y 1995 Inst. Phys. Conf. Ser. **148** 103-106

[3] Stassen S, Rulmont A, Ausloos M, Cloots R 1996 Physica C **270** 135-143

[4] Bieger W, Krabbes G, Schätze P, Zelemina L, Wiesner U, Verges P, Klosowski J 1996 Physica C **257** 46-52

[5] Aselage T, Keefer K 1988 J. Mater. Res. **3** 1279-91

[6] Kore P, Broaten O, Kjekshus A 1992 Acta Chem. Scan. **46** 805-840.

Inst. Phys. Conf. Ser. No 158
Paper presented at Applied Superconductivity, The Netherlands, 30 June–3 July 1997

Large grained YBCO monoliths for magnetic bearings of a LH_2 tank

M. Ullrich†[1], A. Leenders†, H.C. Freyhardt†‡, J.H. Albering§, and S. Gauss§

† Zentrum für Funktionswerkstoffe gGmbH Göttingen, Windausweg 2, D-37073 Göttingen, Germany
‡ Institut für Metallphysik, Windausweg 2, D-37073 Göttingen, Germany
§ Hoechst AG, D-65926 Frankfurt am Main, Germany

Abstract. YBCO monoliths used in a PM/SC arrangement of a magnetic bearing of a cryogenic LH_2 vessel were melt-textured by VGF, MTG and TSMG methods. The differences in j_c between large-grained and single-domain samples decreases strongly with decreasing temperature. It seems that at a distinct temperature weak links due to grain boundaries become ineffective leading to a strong increase of the measured magnetic moment of the large-grained samples. Simultaneously, the measured levitation forces of the large-grained samples strongly increase with decreasing temperature. At low temperatures the levitation forces of large-grained and single-domain samples are nearly identical.

1. Introduction

Hydrogen can become a highly attractive fuel for the future. Storage of liquid hydrogen (LH_2) is most attractive with respect to volume as well as mass density compared to conventional storage systems. The performance of a LH_2 vessel will be substantially improved if magnetic bearings are used. Due to the inherent cooling, i.e. the fact that no additional effort has to be undertaken to cool the superconductors (SC) below the superconducting transition temperature, a bearing consisting of permanent magnets (PM) and SC is highly competitive compared to conventional magnetic bearings consisting of PM arrangements. For a stable levitation in more than two directions a conventional magnetic bearing needs to be electronically controlled whereas for the SC/PM arrangements this is not required.

In this contribution the influence of decreasing the temperature from the boiling point of liquid nitrogen, i.e. 77 K, down to the boiling point of liquid hydrogen, i.e. 20 K, on the critical current density and, of course, on the levitation force of the SC is discussed for different sample qualities, i.e. for samples with and without grain boundaries.

[1] E-mail: ullrich@umpsun1.gwdg.de.

2. Experimental

YBCO monoliths were prepared by standard melt-texture growth (MTG) as well as by top-seeded-melt growth (TSMG) [1] by placing top-seeded-solution grown $Sm_1Ba_2Cu_3O_x$ (Sm-123) [2] seeds onto the top surface of the precursors ($Y_1Ba_2Cu_3O_{7-\delta}$ + ~25 mol% Y_2BaCuO_5 + 1.8 mol% PtO_2, $\phi \approx 32$ mm, height ≈ 15 mm, Hoechst AG). Furthermore, some samples were melt textured by using the vertical-gradient-freeze (VGF) method, i.e. by moving a vertical temperature gradient of 25 K/cm electronically through the samples [3]. Magnetization measurements were performed by cutting small rectangular specimens from the melt-textured samples and placing them in a Faraday balance. The critical current density was determined by applying the anisotropic Bean critical state model [4] and assuming superconducting screening currents to flow over the total sample dimensions, i.e. the external sample dimensions were used as the radii of this screening currents. The demagnetization factor was determined from the virgin magnetization curves. The structure and single crystalline quality of the samples was checked by X-ray diffraction.

The levitation forces of the melt-textured cylinders were measured quasi statically at 77 K as well as in the temperature range from 20 - 85 K using an experimental set-up explained in detail in [5]. A $SmCo_5$ magnet was used in both cases ($\phi = 25$ mm, h = 20 mm). The quality of the melt-textured samples was also checked by scanning the remnant induction using a hall probe, i.e. determining the number of the magnetic domains of the sample.

3. Results and Discussion

In Fig. 1a the critical current density, j_c^S, of a melt-textured-quasi-single crystal is compared to the heuristic parameter, j_c^M, of a melt-grown sample with grain boundaries. This heuristic parameter was determined by assuming that the superconducting screening currents flow over the total external sample dimensions, which is, of course, wrong for a sample which is composed of several grains. Thus, j_c^M has no distinct physical meaning. This parameter, however, helps to understand the influence of decreasing temperature on the measured magnetic moment of the sample and, thus, on the weak link nature of the grain boundaries. Whereas at high temperature j_c^M is much lower and, moreover, exhibits an even stronger field dependence than j_c^S, this difference is highly reduced at lower temperatures. Fig. 1b and 1c show the temperature dependence of j_c^S as well as j_c^M and the ratio of $j_c^S(T)$ to $j_c^M(T)$ for B = 2 T, respectively. It becomes obvious from this figures that at low temperatures the values of $j_c^S(T)$ and $j_c^M(T)$ are nearly comparable. Only very weak links (large cracks, etc.) seem to reduce the effective radii of the screening currents. At higher temperatures, e.g. for B = 2 T for temperatures above 40 K, the difference between $j_c^S(T)$ and $j_c^M(T)$ increases significantly. This can be explained by josephson $S\tilde{S}S$/SNS/SIS junctions at grain boundaries which become normal conducting with increasing temperature, thus, decreasing the effective radii of the screening currents.

Measurements of the zero-field-cooled (ZFC) levitation force do also support that melt-textured-large-grained samples, i.e. samples with grain boundaries, exhibit at low temperatures nearly the performance of single-domain samples. Fig. 2 shows a ZFC levitation force measurement of such a sample with grain boundaries. A strong increase of the levitation force with decreasing temperature is observed. At 30.1 K such a sample

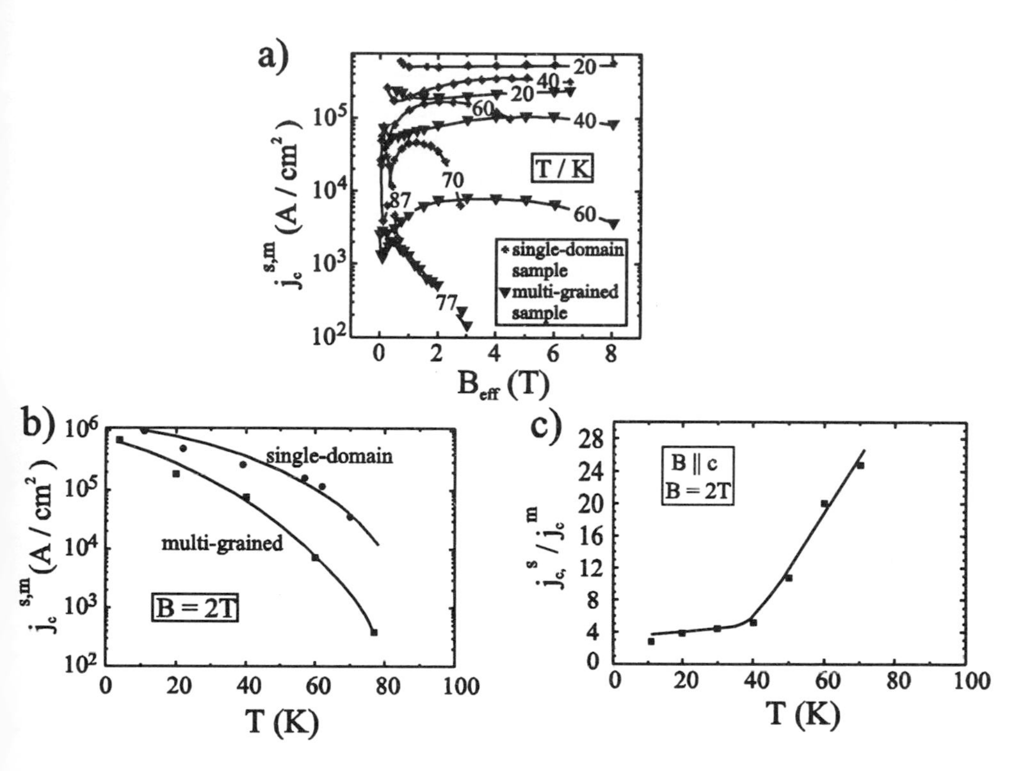

Figure 1. (a) Comparison of the field dependence of the single-domain (j_c^S) and a melt-textured sample containing grain boundaries (j_c^M). Note that in the case of the single-domain sample j_c^S denotes the intra as well as the intergranular critical current density whereas the heuristic parameter, j_c^M, of the large-grained sample with grain boundaries has no distinct physical meaning. (b) Critical current density, j_c^S, and heuristic parameter, j_c^M, versus temperature, T, for B = 2 T. (c) Critical current density, j_c^S, of the single-domain sample divided by the heuristic parameter, j_c^M, versus temperature, T, for B = 2 T.

can levitate 75,5 N (at d = 0.5 mm) which is about 86.8 % of the calculated theoretical limit [6].

Even single-domain samples do not exhibit much higher levitation forces. The dependence of the levitation force on the temperature of single-domain samples, however, is not as pronounced [7].

Thus, for applications at high temperatures, e.g. at 77 K, single-domain samples prepared by TSMG must necessarily be applied. At lower temperatures, i.e. at the working temperature of a cryogenic LH_2 vessel, even melt-textured samples with grain boundaries which can easily and economically be prepared without seeding can be applied due to the improvement of the coupling at the grain boundaries with decreasing temperature.

4. Conclusions

It was shown that the difference of the parameter j_c^M and j_c^S of large-grained and single-domain samples is strongly reduced with decreasing temperature. It seems that at a dis-

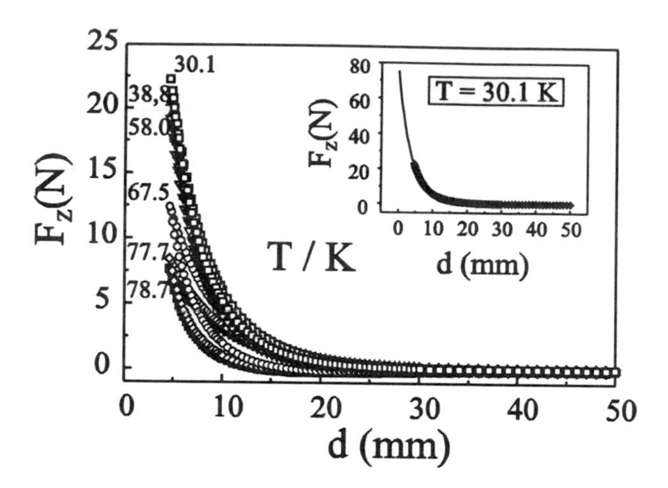

Figure 2. ZFC axial levitation force, F_z, versus distance, d, for temperatures, T = 78.7, 77.7, 67.5, 58.0, 38.8, and 30.1 K. Note that due to the experimental set-up the minimal distance between the melt-textured sample and the $SmCo_5$ magnet amounts to about 5 mm. The inset shows an extrapolation to d = 0.5 mm.

tinct temperature weak links at grain boundaries, i.e. josephson $S\tilde{S}S$/SNS/SIS junctions, become superconducting leading to a strong increase of the measured magnetic moment of large-grained samples. Consequently, the measured levitation force of large-grained samples does also strongly increase with decreasing temperature. At low temperatures the levitation forces of large-grained and single-domain samples are nearly identical.

Thus, even large-grained-melt-textured YBCO monoliths can be used at liquid hydrogen temperature – without negativ impact on the performance – for a magnetic bearing of a LH_2 vessel consisting of a PM/SC arrangement.

Acknowledgments

This work is supported by the BMBF under grant numbers 13N6939/9 and 13N6940/9.

References

[1] Ullrich M, Müller D, Krauns Ch, Bringmann B, Leenders A, Brandt C, Reder M, Preusser A and Freyhardt HC 1997 *Proc. of the ISTEC/MRS International Workshop on Superconductivity, Big Island, Hawaii, USA pp 76-79*

[2] Krauns Ch, Bringmann B, Brandt C, Ullrich M, Heinemann K and Freyhardt HC *this conference*

[3] Ullrich M and Freyhardt HC 1993 *Superconductir ȷ Materials* ed J Etourneau et al (Gournay sur Marne, France, IITT International) pp 205-210

[4] Wiesinger HP, Sauerzopf FM and Weber HW 1992 *Physica C* **203** 121

[5] Gauss S and Albering JH *this conference*

[6] Canders WR, May H and Palka R 1997 *Proc. of the 8th International Symposium on Non-Linear Electromagnetic Systems (ISEM), Braunschweig, Germany*

[7] Albering JH, Gauss S, Teske C and Bock J *this conference*

Inst. Phys. Conf. Ser. No 158
Paper presented at Applied Superconductivity, The Netherlands, 30 June–3 July 1997

Analysis and Modelling of Oxygen Diffusion in $YBa_2Cu_3O_{7-\delta}$ under Non-isothermal Conditions

M.D. Vázquez-Navarro, A. Kuršumovic, C. Chen[1] and J.E. Evetts.

Department of Materials Science and IRC in Superconductivity, University of Cambridge, Pembroke St., Cambridge CB2 3QZ, UK,
[1]University of Oxford, Department of Physics, Parks Rd., Oxford OX1 3PU, UK

Abstract. Thermogravimetric analysis (TGA) has been used to measure oxygen diffusion in $YBa_2Cu_3O_{7-\delta}$ (YBCO) powder and single crystals in a pure oxygen atmosphere. Measurements were made up to 700 °C and the experiments were carried out at constant heating rate. A computational model that reproduces the weight change of an YBCO sample under non-isothermal conditions has been developed. Constant ramp rate experiments avoid the problem of initial transients encountered in isothermal experiments and can be analysed quantitatively for the diffusion constant and activation energy.

1. Introduction

Numerous studies have shown that the properties of $YBa_2Cu_3O_{7-\delta}$ (YBCO) are strongly dependent on the oxygen content. Although the equilibrium oxygen stoichiometry depends on temperature and partial oxygen pressure, the actual oxygen concentration distribution in a sample depends on the value of the diffusion coefficient and the annealing history of the sample. Knowledge of the required annealing schedule to produce homogeneous specimens with high transition temperatures in the shortest possible time is important to optimise the fabrication process. Because of its significance the chemical diffusion coefficient has been studied by a large number of groups. However, none of these studies agree on a single value for the diffusion coefficient, the scatter being sometimes as large as seven orders of magnitude.

Oxygenation studies have mostly been carried out at constant temperature by studying the relaxation behaviour of properties directly dependent on the oxygen content, such as the electrical resistivity [1] or weight (thermogravimetry) [2], and other techniques such as internal friction [3] or differential scanning calorimetry [4]. Oxygenation under non-isothermal conditions has not been studied in such detail due to the fact that the material is constantly in a non-equilibrium state, since the equilibrium oxygen concentration varies continuously with changing temperature [5]. The study reported here uses TGA to follow the oxygenation behaviour of YBCO powder and single crystal samples under non-isothermal conditions. There are no analytical expressions for this type of diffusion, so a computational model has been developed that reproduces the weight change of an YBCO sample under non-isothermal conditions.

2. Experimental procedure

YBCO powder samples were prepared using Seattle Superconducting Corporation YBCO orthorhombic powder, with an average particle size of 5 μm in diameter and a phase purity >99%. The single crystals were grown by the flux growth method and consisted of platelets that were thin in the *c*-direction and had areas that ranged from 0.04 mm^2 to 1 mm^2 in the *ab* plane.

A Perkin-Elmer System 7 low temperature (≤ 1000°C) thermogravimetric unit was used to carry out the weight change measurements. The heating rates used were 1, 30, 50, 90, 110 and 130°C/min and the temperatures to which the samples were heated were 450°C and 700°C. A null run under identical conditions without a sample was subtracted from each experiment to compensate for buoyancy and aerodynamic forces. The nominal sample weight was 20-30 mg, the gas flow was ≈ 60ml/min and the gas used 100% O_2.

Using the analogy between the heat conduction and diffusion processes the computer model developed was based on finite element programmes written to study the heat transfer in steel specimens heated locally by concentrated solar energy [6, 7]. It considers the diffusion process to occur only along the *ab* plane, since the value of the diffusion coefficient along the *c* direction is ≈ 10^6 times smaller than that on the *ab* plane [8]. The powder particles are taken to be spherical of radius *r*, and in order to carry out the calculations they are divided into parallel sections of thickness *dh* and each of these sections is divided into concentric rings separated by a distance *dr*, as it can be seen in Fig. 1 a) and b). The programme calculates the rate of transfer of oxygen through each one of the rings in each section with respect to the volume of each ring and integrates this value over the whole volume of the sphere.

Taking Fick's first law of diffusion: $F = D\frac{\partial C}{\partial x}$ (valid for rings where *dr* is small compared to the radius of the sphere), where *F* is the rate of transfer per unit area of section, *C* is the concentration of diffusing substance, *x* is the distance considered normal to the section and *D* is the diffusion coefficient, we can calculate the rate of transfer of material through a chosen element of the sample.

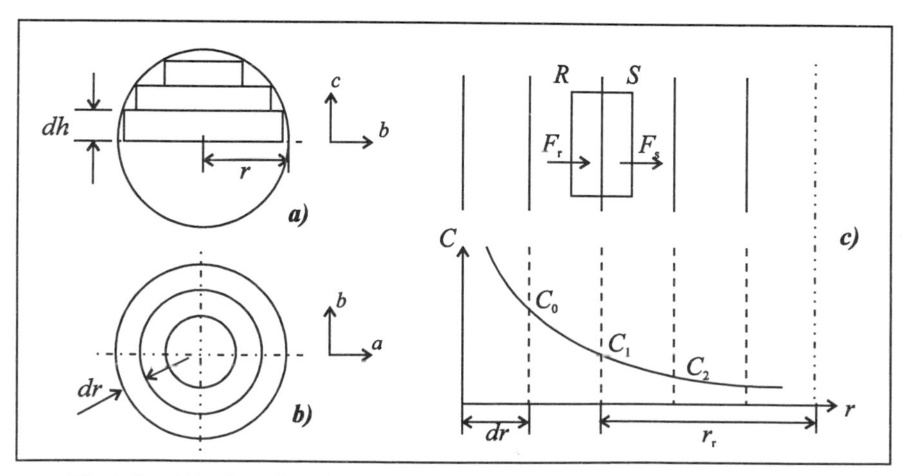

Fig. 1 a) and b) schematics of divisions of sphere in model, **c)** Fick's first law of diffusion

In a short time step, τ, the net amount of oxygen accumulated in the element shown in Fig.1 c) is equal to the amount that flows in through R minus the amount that flows out through S, this is:

$$q_{net} = q_r - q_s = -\frac{D\tau}{dr}(2\pi r_r dh)(2C_1 - C_0 - C_2)$$

This relationship enables us to calculate the variation in accumulation or loss of oxygen in the theoretical powder particle as diffusion time increases.

The calculation must be stable throughout the whole range of temperatures. Increasing the time increment, τ, decreases the computing time but if too long time increments are used the calculation becomes catastrophically unstable. The optimum time increment depends on the size of the element dr. If dr is very small, the time step needs to be very small too for the calculation to converge. This increases considerably the computing time. However, dr cannot be too large or the integration of the weight gain over the whole area will carry a considerable error. Calculations with very small steps (of the order of 10^{-6} cm) have been carried out and they all converge. To optimise computing time and ensure convergence in the calculations the values of dr taken ranged between 1/20 and 1/30 of the particle radius and the time steps varied between 0.001s to 0.01 s.

3. Results and discussion

The experiments involved a variable temperature ramp followed by an isothermal anneal. For oxygenation that takes place before the isothermal stage the equilibrium stoichiometry is constantly changing, which makes the analysis non-trivial.

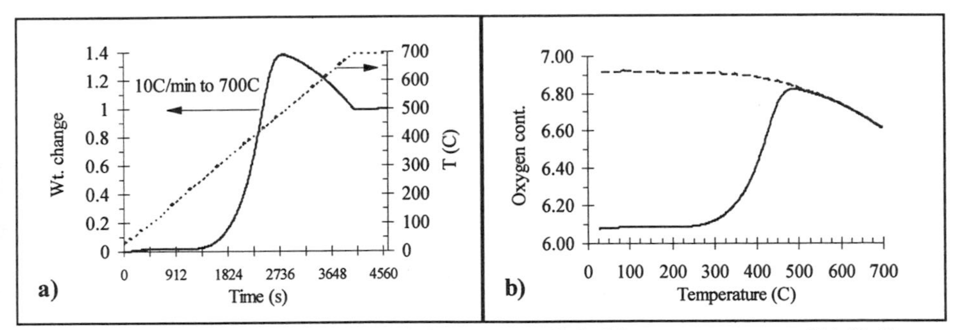

Fig. 2 a) Normalised weight change at 10 °C /min to 700 °C, (dashed line = temperature profile), b) Change in oxygen content at 10 °C/min to 700 °C with T, (dashed line = experimental data for equilibrium oxygen content at each temperature).

A typical TGA trace for 10 °C/min ramp to 700 °C is shown in Fig. 2 a). The maximum is associated with the achievement of the equilibrium concentration of oxygen for that particular temperature throughout the sample. From then on, the change in weight is equivalent to that under equilibrium conditions. In Fig. 2 b) the data is replotted as a function of oxygen content and temperature.

Experiments at different ramp rates show that the temperature at which the sample reaches equilibrium increases with increasing ramp rate, since the sample has less time at each temperature and therefore the oxygen absorption is correspondingly smaller (Fig. 3 a)).

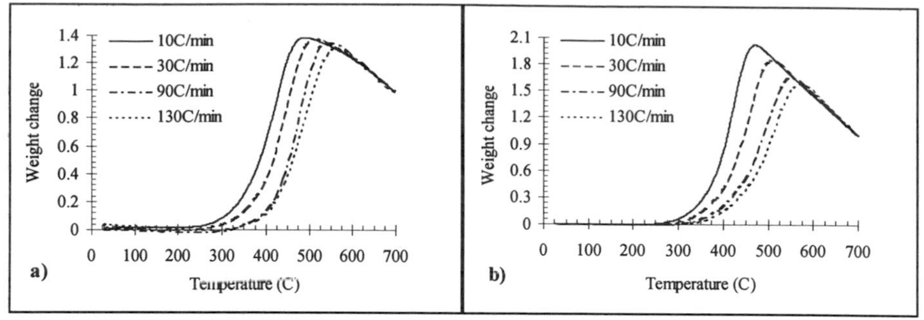

Fig. 3 a) Normalised weight change at different ramp rates to 700 °C (experimental), b) Normalised weight change at different ramp rates to 700 °C (theoretical).

All of these non-equilibrium diffusion experiments have been reproduced using the model (see Fig. 3 b)). By iterating the value of the diffusion coefficient used in the analysis an accurate determination can be made of the actual diffusion coefficient of oxygen in YBCO as a function of temperature. The diffusivity at 450 °C was $6.47(\pm 0.4)*10^{-11}$ cm^2/s. It has been observed in the experimental data that there is a slight spread in the final equilibrium values of the curve from the turnover point (maximum) to the final temperature, and this is thought to be due to the size distribution of the particles in the powder sample. The model is currently being used to study the effect of a combination of different particle sizes on the final shape of the curve.

4. Conclusions

Oxygen diffusion under non-isothermal conditions is being studied by analysing experimental thermogravimetric data with the help of a computer model. The analysis avoids for the first time the problems of the initial transient encountered in isothermal studies and yields accurate values for oxygen diffusivity in YBCO.

Acknowledgements

We thank J. Ruiz for valuable discussions about the model. This work is supported by an EPSRC CASE studentship with EA Technology.

References

[1] LaGraff J A and Payne D A 1993 *Physica C* **212** 478-486.

[2] Kishio K Suzuki K Hasegawa T Yamamoto T and Kitazawa K 1989 *J. Solid State Chem.* **82** 192-202.

[3] Xie X M Chen T G and Wu Z L 1989 *Phys. Rev. B* **40** 4549-4556.

[4] Glowacki B A Highmore R J Peters K F Greer A L Evetts J E 1988 *Supercond. Sci. and Techn.* **1** 7-11.

[5] Gallagher P K 1987 *Adv. Ceram. Mater.* **2** 632-639.

[6] Ruiz J Fernández B J Belló J M 1989 *2nd Seminar of Surface Engineering with High Energy Beams Sci. and Tech.* 161-172.

[7] Rodríguez Donoso G P Ruiz J Fernández B J Vázquez-Vaamonde A J 1995 *Mat. and Design* **16** 163-166.

[8] Rothman S J Routbort J L Welp U Baker J E 1991 *Phys. Rev. B* **44** 2326-2333.

Inst. Phys. Conf. Ser. No 158
Paper presented at Applied Superconductivity, The Netherlands, 30 June–3 July 1997

Materials aspects of substituted Hg-1223 superconductors

S. Lee, N. Kiryakov, M. Kuznetsov, D. Emelyanov and Yu.Tretyakov

Department of Chemistry, Moscow State University, 119899 Moscow, Russia

Abstract. Hg-1223 phase with substitutions of Pb for Hg and Sr for Ba were synthesized by ampoule technique. The substitution and preparation of homogeneous precursor powders by spray and freeze drying methods allowed us to reduce the impurities phases, improve the chemical stability of the precursors and final samples. and to get the samples with the T_c of 110-130 K without additional oxygen annealing. Fabrication of the ceramic, fibers and thick films confirmed the advantages of the substituted Hg-1223 in comparison with the conventional one.

1. Introduction

$HgBa_2Ca_2Cu_3O_z$ phase (Hg-1223) is a promising candidate for practical application because of high critical temperature (T_c=135K) and critical current density (j_c) [1-2]. For utilization of this phase for materials preparation several problems should be solved related with the preparation of homogeneous and active precursor powders, impurity phases formation, degradation of the precursors and final samples, necessity of high pressure equipment and glove box with the inert atmosphere, optimization of superconducting properties, etc.

Chemical substitution can be very effective pathway to overcome above problems. At present, different substitutions in the Hg-1223 (i.e. Hg by Pb, Tl or Re) used for preparation of thin [3] and thick films [4], fibers [5], tapes [6] etc. Recently, the substitution of Ba for Sr and synthesis of Ba-free HgRe-1223 was reported [7]. However high-pressure method that used for preparation of Sr-containing Hg-1223 [8] makes impossible their utilization for practical application.

In this paper we report the synthesis of the Hg-1223 samples through the homogeneous precursors with the substitution of Hg by Pb and Ba by Sr by ampoule method and their utilization for materials preparation.

2. Experimental

Spray drying of nitrates solution was performed using laboratory spray-dryer (Buchi) at t=200°C. In the freeze drying method, the solution was sprayed into liquid nitrogen. The frozen droplets were placed into a freeze drier (SMH-15, Usi Froid) under vacuum < 10^{-3} Torr at t=-40°C for 24 hours and slowly warmed to room temperature. Thermal decomposition of the nitrates salt mixtures were carried under vacuum at the heating rate of 100°C/h to the temperature of 600°C for 1 h. The precursor were mixed in agate mortar with

HgO. For preparation of the ceramic samples precursor mixture was compacted into the pellets. Heat treatment was carried in evacuated quartz ampoules at the temperatures of 850-870°C for 5-15 hours. Additional details of the synthesis were described previously [9]. The powder with the particle size <5 micron, containing Hg-1223 as the major phase was obtained by ball milling in absolute hexane for 30 min. Materials inks for screen-printing and extrusion were prepared by mixing of the powder with the organic binder. For thick film preparation track of 10 x 10 mm with thickness of 40 micron were screen printed onto MgO substrate. Fiber samples were obtained by cold isostatic pressing or extrusion of powder mixed with organic binder The sintering process of thick film or fiber was carried out in evacuated quartz ampoule at 850-870°C for 5-15 h. together with the reactant pellet of the same nominal composition.

3. Results and Discussion

Application of the spray and freeze drying methods and further thermal decomposition of the nitrates mixture under vacuum allowed us to avoid melting and to obtain very homogeneous oxides precursors with the particle size less than 1 micron. Further utilization of these precursors for synthesis of Hg-1223 showed their advantages in comparison with the conventional methods of powders preparation.

Recently we found that optimal substitution of Hg by Pb for the $Hg_{1-x}Pb_xBa_{2-y}Sr_y$ $Ca_2Cu_3O_z$ phase is x =0.2-0.3 [10-12]. For such samples the minimal amount of impurities observed and T_c of 110-130 K was obtained depending of Sr content. For example $Hg_{0.7}Pb_{0..3}Ba_{1.5}Sr_{0..5}Ca_2Cu_3O_y$ sample has rather high T_c of 128 K. Moreover the maximum Tc observed in as-prepared samples in contrast to the $HgBa_2Ca_2Cu_3O_z$ phase (fig.1). We also found that for the substituted samples irreversibility line lies at a higher field than that for the pure Hg-1223 phase (fig.2)

For $Hg_{0.7}Pb_{0..2}Ba_{0.5}Sr_{1.5}Ca_2Cu_3O_y$ and $Hg_{0.7}Pb_{0..3}Sr_2Ca_2Cu_3O_y$ compositions the substitution of Ba by Sr significantly improve their chemical stability to the degradation. It is possible to perform all synthetic procedure without glove bag, required for other Hg-based phases. For the $Hg_{0.7}Pb_{0..2}Ba_{0.5}Sr_{1.5}Ca_2Cu_3O_y$ composition the T_c value was estimated as 122-125 K for as-prepared samples which is higher than T_c of "Hitachi composition" $Tl_{0.5}Pb_{0.5}Sr_{1.6}Ba_{0.4}Ca_2Cu_3O_y$ (115-118 K) which is widely used for the preparation of Tl-based materials. Annealing in the reduced atmosphere for optimization of oxygen content increases the T_c of the substituted Hg-based samples.

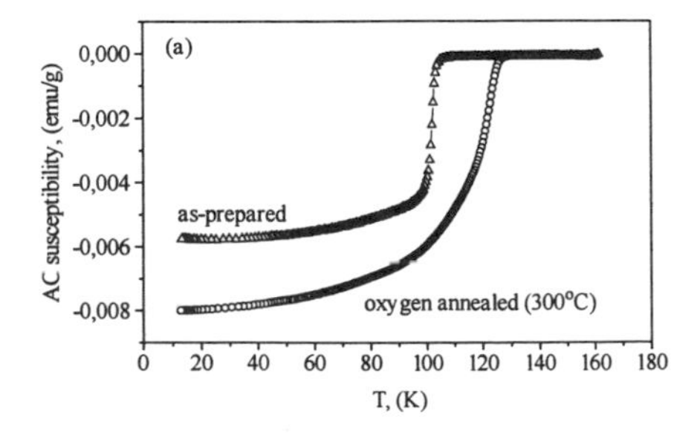

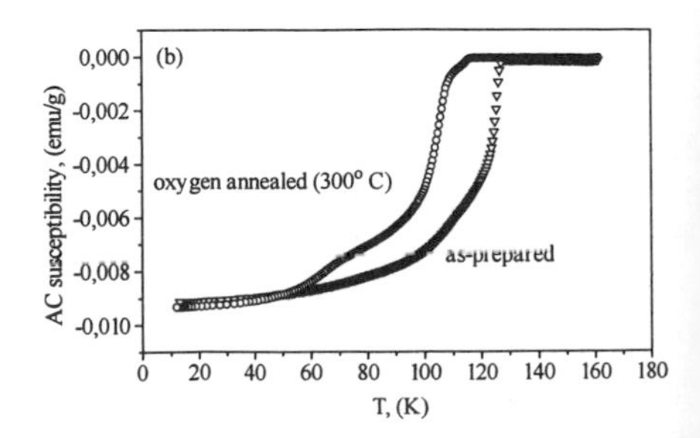

Fig.1. AC susceptibility data for as- prepared samples and after oxygen annealing at 300°C for 30 h: a) $HgBa_2Sr_2Ca_2Cu_3O_y$, b) $Hg_{0.7}Pb_{0.3}Ba_{1.5}Sr_{0.5}Ca_2Cu_3O_y$

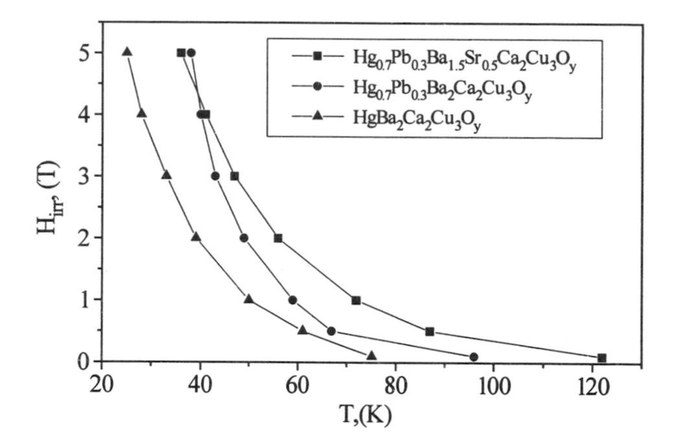

Fig.2 Irreversibility line of Hg-1223 with the substitution of Hg by Pb and Ba by Sr measured on grain-aligned powders (H//c)

Another advantage of the Pb and Sr-contained samples is the prevention of $BaCuO_2$ impurity formation. We found that this impurity phase forms the layer at the grain boundaries and drastically reduce the transport Jc value in Hg-1223. In contrast, the impurities of SrCa-cuprates observed in the Sr-containing samples form the small particles between the crystallites that do not affect so much on the quality of the contacts. Ac susceptibility and transport j_c measurements for the $Hg_{0.7}Pb_{0..2}Ba_{0.5}Sr_{1.5}Ca_2Cu_3O_y$ samples are shown in fig.3.

The further experiments showed that powders of such compositions allow their utilization with the organic binders without serious degradation observed for the conventional Hg-1223. This fact is important for the preparation of thick films, tapes and fibers materials. As an example, the XRD pattern of thick film on MgO substrate is shown in fig. 4. Another experiments for the preparation of the thin films and wires are in progress.

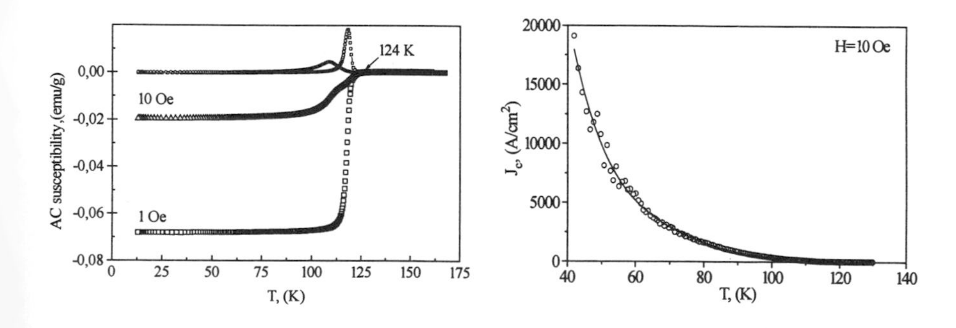

Fig.3. a) AC susceptibility data for the ceramic sample with the composition $Hg_{0.7}Pb_{0.3}Ba_{0.5}Sr_{1.5}Ca_2Cu_3O_y$ in different magnetic fields; b) temperature dependence of critical current density for this sample in the magnetic field H=10 Oe.

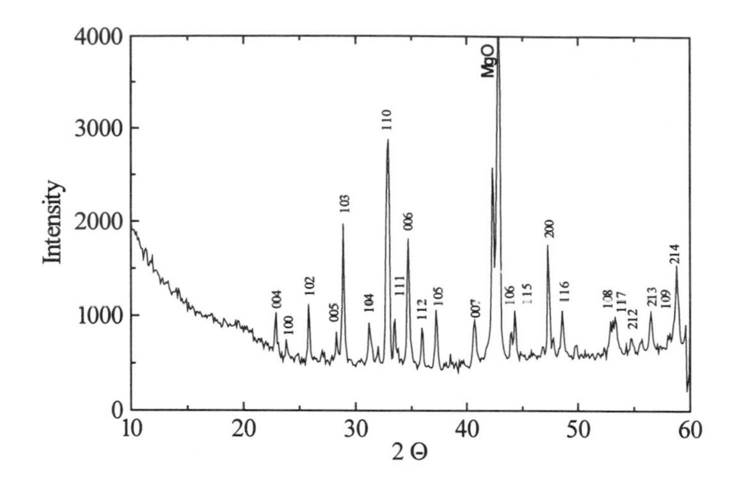

Fig.4 XRD pattern of the $Hg_{0.7}Pb_{0.3}Ba_{0.5}Sr_{1.5}Ca_2Cu_3O_y$ thick film on MgO substrate after sintering at 870°C for 15 hours.

Acknowledgements

The work was supported by Russian Scientific Council for Superconductivity and INTAS project N 94-1399. We would like to thank V.Lennikov, M.S.Kim and Sung-Ik Lee for investigation of superconducting properties of our samples and discussion.

References

[1] Schilling A, Cantoni M, Guo J D, Ott H R 1993 *Nature* **363** 56-58
[2] Krusin-Elbaum L, Tsuei C C, Gupta A 1995 *Nature.* **373** 679-681
[3] Brazdeikis A, Flodstrom A S, Bryntse I. 1996 *Physica C* **265** 1-4
[4] Singh H K, Saxena A K, Srivastava O 1996 *Physica C* **262** 7-12
[5] Goto T 1995 *Physica C* **247** 133-136
[6] Meng R L, Hickey B, Wang Y Q, Sun Y Y, Gao L, Xue Y Y, Chu C W 1996 *Appl.Phys.Lett.* **68** 3177-3179
[7] Chmaissem O, Jorgensen J D, Yamaura K, Hiroi Z, Takano M, Shimoyama J, Kishio K 1996 *Phys.Rev.B.* **53** 14647-14654.
[8] Hahakura S, Shimoyama J, Shiino O, Hasegawa T, Kitazawa K, Kishio K 1994 *Physica C* **235-240** 915-916
[9] Lee S, Mun, M O, Bae M K, Lee S I 1994 *J.Mater.Chem.* **4** 991-992
[10] Lee S, Shlyakhtin O A, Mun, M O, Bae M K, Lee S I 1995 *Supercond.Sci.Technol.* **8** 60-64
[11] Lee S, Kuznetsov M, Kiryakov N, Emelyanov D, Tretyakov Y 1997 *Proc.of Intern. Conf. on Materials. and Mechanisms of Supercond. (M^2S-HTSC-V), 27 Feb-2 Mar. Beijing (China)* to be published in Physica C.
[12] Lee S, Kiryakov N, Plesenkova O, Emelyanov D, Tretyakov Y 1996 *Russ.J.Inorg.Chem.* to be published.

Inst. Phys. Conf. Ser. No 158
Paper presented at Applied Superconductivity, The Netherlands, 30 June–3 July 1997

Rising of the irreversibility line in Re-substituted Hg-1223 High T_c Superconductors

L.Fàbrega, B.Martínez, A.Sin, S.Piñol, J.Fontcuberta, X.Obradors

Institut de Ciència de Materials (CSIC); Campus de la U.A.B., E-08193 Bellaterra (SPAIN)

Abstract. Transport and magnetization measurements as a function of field and temperature, carried out on as-prepared and oxygen annealed Hg-1223 superconductors with 18% Re, indicate that this substitution increases the irreversibility line due to surface barriers, and also enhances bulk pinning at low temperatures ($T \leq 40K$). Oxygen annealing slightly decreases T_c and has second order effects on the enhancement of the superconductor irreversibility.

1.Introduction

Mercury cuprates have roused considerable interest since their late discovery [1], because they are the highest T_c superconductors ($T_c \approx 135K$ for the 1223 phase). These compounds have, however, two important drawbacks: first, their unstability and difficult synthesis. Second, their irreversibility line lies very low, because of a very weak bulk pinning. In fact, the magnetic irreversibility has been shown to be mainly due to surface barrier effects down to rather low temperatures [2]. The ineffectiveness of pinning centers is thought to be caused by the high anisotropy of these cuprates: studies in single crystals indicate that their anisotropy ratio is $\gamma \equiv (m_c/m_{ab})^{1/2} \approx 50$ [3], close to that of the isostructural Bi-series.

An important advance was realized when Shimoyama et al. [4] showed that a partial substitution of Hg by high valency ions Cr or Re favours the stabilisation of these phases and allows their obtention at atmospheric pressure. Re-substituted phases display similar T_c and higher irreversibility lines [4-7]. The latter have been associated to a reduction of the anisotropy, due to the remarkable shortening of the c-axis length and a metallization of the interlayers [7], which has been argued to be caused by the metallic character of the ReO_6. In fact, the first measurements on grain-aligned samples appear to confirm the anisotropy reduction and provide values of $\gamma \approx 8$ [8], which should result in a significant increase of the vortex stiffness and therefore improve the flux pinning effectiveness.

There are, though, several questions to be addressed. First, it has been shown that the rise of the irreversibility line is strongly dependent on the oxygen content of the sample, which in turn does not significantly alter T_c. And second, as we already mentioned, magnetic irreversibility in unsubstituted samples is dominated by surface barriers, down to very low temperatures. It is therefore necessary to stablish the exact effect of the Re-substitution and oxygen incorporation on the superconductor anisotropy, and analyze their repercussion on the surface barriers and bulk pinning, and consequently on the irreversibility line.

In this paper we address these important issues, by measuring hysteresis cycles and resistance versus temperature in as-prepared and oxygen annealed $Hg_{0.82}Re_{0.18}Ba_2Ca_2Cu_3O_{8+\delta}$ samples. We follow the evolution of the critical temperature and the irreversibility line, and we complete the study by measuring the temperature and field dependent magnetization for a grain-aligned, as prepared sample (H//c), in order to look for the origin of the magnetic irreversibility and relate it to the Re and oxygen effects. In as-prepared samples, we observe an increase of the irreversibility line, as reported by other groups, and a significant enhancement of bulk pinning at low temperatures, as compared to unsubstituted samples. This becomes the dominant contribution to the irreversible magnetization for T≤40K; above this temperature, the irreversibility associated to the flux penetration through surface barriers overcomes bulk pinning. The irreversibility line shifts towards higher temperatures after oxygen annealing, but this effect is much less important than that of the Re-substitution.

2. Experimental

Single-phased, ceramic $Hg_{0.82}Re_{0.18}Ba_2Ca_2Cu_3O_{8+\delta}$ was synthesized following the procedure described in ref. [9]. Some as-prepared samples were post annealed in 1 bar of oxygen at 300°C, for 10 to 40h.

An as-prepared sample was ground and the powder was dispersed in an insulating epoxy down to a volume concentration of 3%, and then oriented by the application of a magnetic field. X-ray diffraction confirmed a good orientation of the superconducting grains, with a c-axis spread of 3.5°, determined from the full width at half maximum of the rocking curve of the (0 0 6) peak.

Temperature dependent resistance measurements were carried out under applied fields H≤70kOe for polycrystalline bars, by using the standard four-probe method and with currents between 0.1mA and 2mA. Magnetization hysteresis cycles up to 55 kOe were measured with a Quantum Design SQUID magnetometer, on powder and a grain oriented bar (H//c) of dimensions $2x2x5mm^3$, the longest one being parallel to the c-axis.

3. Results and discussion

The effect of oxygen annealing on T_c has been analyzed by measuring the low field (H=10 Oe) susceptibility and the resistance at zero magnetic field, for powders as-prepared (P0) and oxygen annealed during 10, 20 and 40h (P1,P2 and P4, respectively). Both sets of data reveal a unique superconducting transition of similar width for all samples, with the onset at

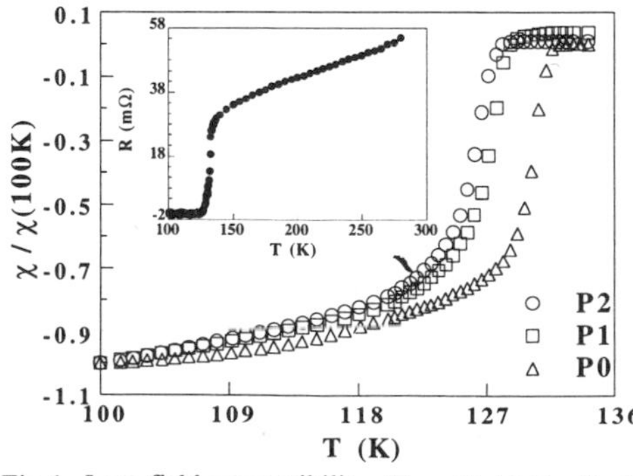

Fig.1. Low field susceptibility (Zero Field Cooling) for P0, P1 and P2 samples. Inset: resistive transition of P0 at zero applied field.

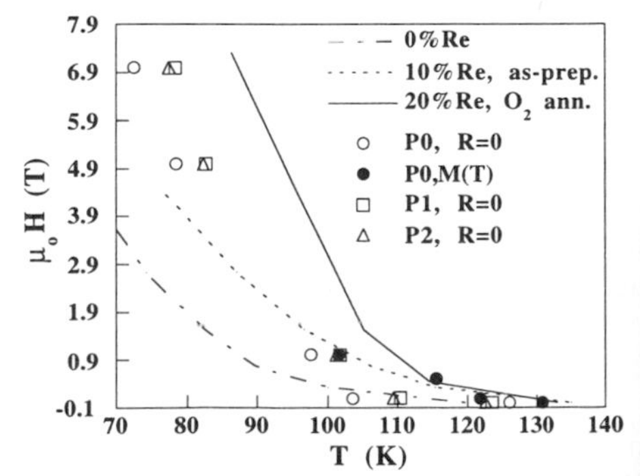

Fig.2. IL (R=0 criterion) for P0, P1 and P2; also shown is the low field IL from M(T) data for P0. The three lines are IL's for several powders, from ref. [6].

temperatures T_{on}=133K, 128.6, 128.1 and 128.0 K, respectively (Fig.1). The as-prepared powder has a resistance more than twice higher than the annealed ones; all of them display, though, a similar value $d\ln R/T \approx 0.005 K^{-1}$.

Fig.2 displays the irreversibility line (IL) for P0, P1 and P2, determined from the R(T) data, using the zero-resistance criterion. Also shown in this figure are the IL's reported by Kishio et al. for Hg-1223 powders with different Re and oxygen contents [6]. Upon comparison with them we see that our as-prepared samples display an IL in between that of as-prepared samples with 10% Re and that of the 1 bar O_2-annealed powder with 20% Re. The IL lies in any case well above that of the unsubstituted compound. On the other hand, oxygen annealing of our samples moves the IL slightly higher. We can conclude, therefore, that the effect of Re-substitution -which, in turn, involves the incorporation of a certain amount of oxygen- is far more important than that of our post-annealings. It can also be appreciated that the little effect of O_2 appears to saturate: indeed, the IL's for P1 and P2 are very close. A similar behaviour has been observed for the critical temperature and the normal state resistance. Such a weak effect of oxygen is in contrast with the results reported by Kishio et al. [6]. These authors annealed the samples at higher O_2 pressure.

In order to analyze in more detail the effects of Re and oxygen on the superconductor irreversibility and its origin, we have performed measurements of isothermal hysteresis cycles for the powder and grain oriented samples. Fig.3 displays the M(H) curves for T=20, 40,50 and 70K for the as-prepared, grain-oriented sample (H//c). At the higher temperatures the field decreasing branch is flat and close to zero; similar hysteresis cycles, considered a signature of the existence of surface barriers [10], have been currently reported for the Hg-phases, with and without Re [2,8]. They denote that bulk pinning is irrelevant and that magnetic irreversibility is governed by the thermally activated penetration of flux lines over the surface barriers. The dominant contribution of surface barriers in our samples is further confirmed by the dependence of the irreversible magnetization $M_{irr}(H)$ with the inverse of magnetic field, as predicted by theory [11].

At lower temperatures (T≤40K), a significant contribution to M_{irr} from bulk pinning develops, as indicated by the increasing symmetry of the hysteresis cycles [12]; accordingly, the irreversibility field experiences a stepper increase than at higher temperature, and rises from H_{irr}≈17 kOe at 50K to H_{irr}=46 kOe at 40K (see Fig.4). Such a dominant contribution of bulk pinning has not been reported for the unsubstituted 1223-phases, where surface barriers appear to be the main contribution to irreversibility, at least down to 15K [2], and therefore bulk pinning is irrelevant for almost the whole phase diagram of the superconductor.

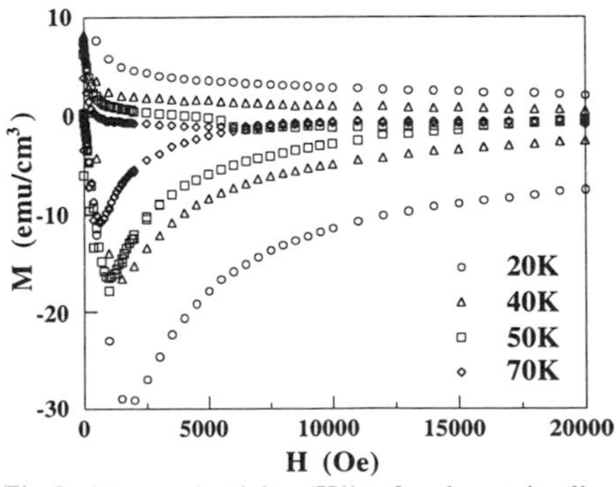

Fig.3. Hysteresis cicles (H//c) for the grain-aligned as-prepared sample, at T=20,40,50,60 and 70K.

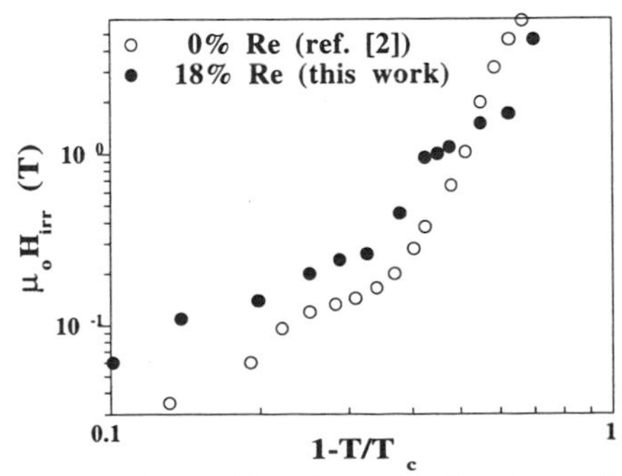

Fig.4. IL from M(H) for grain-aligned samples (H//c): ours (18%Re), and one without Re (ref.[2]).

In Fig.4 we compare the irreversibility line (H//c) obtained from M(H) data for our oriented sample with the equivalent line for a sample without Re [2]. At high temperatures we see that, as already observed in Fig.2, the IL for the Re-substituted sample lies higher than that for the pure 1223 compound; the difference between both, however, diminishes when decreasing the temperature, and they cross at T≈50K; this is the temperature below which bulk pinning in our sample becomes relevant. Below 50K, the magnetic irreversibility is smaller for the Re-substituted sample. This result appears to be in sharp contradiction with the observed enhancement of bulk pinning at low temperatures. A possible explanation could be a Re-induced change in the magnetic irreversibility $M_{irr}(T)$ associated to surface barriers, producing a reduction of M_{irr} at T<50K, with favours the observation of bulk pinning.

4. Conclusions

Our results indicate that the enhancement of the irreversibility line associated to the Re-substitution is due to both an increase of surface barrier effects at high temperatures and an improvement of bulk pinning, which is clearly appreciated at low temperatures and moderate/low fields. Bulk pinning, however, displays a fast decrease with temperature and magnetic field, so that at high temperatures and fields surface barriers continue to dominate the superconductor irreversibility. The increase of bulk pinning may be the result of the anisotropy reduction due to the metallization of the interlayers, as already suggested. This metallisation might also account for a change in the irreversible magnetization due to surface barriers, as our results seem to point out.

Oxygen annealing increases the irreversible magnetization and slightly moves higher the irreversibility line. These effects, though, are second-order as compared to those induced by the Re-substitution, because of a bad oxygenation during our O_2 annealings or because of different oxygen contents of the distinct as-prepared samples.

Acknowledgements

We would like to acknowledge finantial support by the CICYT (MAT 96-1052 and MAT 94-1924) and the Generalitat de Catalunya (GRQ95-8029).

References

[1] A.Schilling, M.Cantoni, J.D.Guo, H.R.Ott 1993, *Nature* **363**, 56
[2] Y.C.Kim et al. 1995 *Phys. Rev.* B **52**, 4438
[3] G.Le Bras et al. 1996 *Czechoslovak J. of Phys.* **46(S3)**, 1769
[4] J.Shimoyama et al. 1994 *Physica C* **224**, 1; K.Yamaura et al. 1995 *Physica C* **246**, 351
[5] O.Chmaissen et al. 1996*Phys. Rev. B* **53**, 14647
[6] K.Kishio et al. 1996 *J. of Low Temp. Phys.* **105**, 1359
[7] J.L.Tallon et al. 1996 *J. of Low Temp. Phys.* **105**, 1379
[8] H.Yamasaki et al. 1997, *Physica C* **274**, 213
[9] S.Piñol, A.Sin, A.Calleja, J.Fontcuberta, X.Obradors, F.Espiell 1997 *J. of Superconductivity* (in press)
[10] L.Burlachkov et al. 1994 *Phys. Rev. B* **50**, 16770
[11] J.R.Clem 1974 *Procceedings of the 13th Conference on Low Temperature Physics* (Eds. K.D.Timmerhaus, W.J. O'Sullivan and E.F.Hammel; Plenum, NY) **3**, 102
[12] L.Fàbrega et al., submitted to *Physica C*

Inst. Phys. Conf. Ser. No 158
Paper presented at Applied Superconductivity, The Netherlands, 30 June–3 July 1997

Sequential processing of Composite Reaction Textured (CRT) bulk $Bi_2Sr_2CaCu_2O_{8+\delta}$ components

P Kosmetatos, A Kuršumovic, D R Watson, R P Baranowski and J E Evetts

Department of Materials Science and Metallurgy and IRC in Superconductivity,
University of Cambridge, Pembroke St., Cambridge CB2 3QZ, UK

Abstract. A critical parameter for the commercial viability of any processing technique for high critical current bulk applications is the ability to adapt it to other than simple monolithic artefacts. In particular, the ability to react large components in sequential stages as well as to join previously reacted components must be demonstrated. The CRT technique is a multi-seeding process whereby an aligned 10wt% distribution of high aspect ratio MgO fibres in a Bi-2212 matrix drives the nucleation and texturing process in a partial melting stage. Components thus produced exhibit 10^5 Acm^{-2} at 10K and 12T and self field limited J_c of $1\text{-}4\times10^3$ Acm^{-2} at 77K. We report here the development of heating cycles that allow the repeated post-reaction melting of such components whilst retaining their physical shape, microstructural texture and critical current carrying characteristics. Long lengths of precursor material have also been reacted in stages and the ability to form joints between components has been demonstrated.

1. Introduction

A principal deficiency of standard melt texturing techniques for bulk superconductors is their dependence on a single growth front. This in turn limits the overall speed of the process and hence the ability to scale it up. The Composite Reaction Texturing (CRT) technique employs multiple nucleating sites, and therefore its effectiveness is in principle independent of the size and shape of the artefact. CRT for Bi-2212 utilises an aligned 10wt% distribution of high aspect ratio (5 μm thick, 200 μm long) MgO fibres to drive nucleation and can produce components reaching 10^5 Acm^{-2} at 10K and 12T and self field limited J_c of $1\text{-}4\times10^3$ Acm^{-2} at 77K [1].

Like every other technique that aspires to ultimate commercial viability, CRT must be adapted to components of a complex shape and an appreciable size. Its versatility in the manufacture of essentially any desired shape has already been demonstrated [2]. Practical size limitations, invariably due to size of the furnace available, will be removed if the hitherto static process can be developed into a continuous one.

The reaction of long lengths of green tape in segments inevitably results in the formation of a Heat Affected Zone (HAZ) between the still-green segments and those already reacted. It is therefore necessary to investigate: (a) whether the already reacted area in a HAZ will be adversely affected by reprocessing; (b) whether the stress involved in densifying a part of the sample while keeping the rest unreacted will be such as to fracture the component; (c) whether the unreacted volume will subsequently densify and texture properly. Furthermore, the ability to re-react a CRT ceramic will be very useful in cases of initially poorly textured samples or for recovering failed components.

Finally, it is essential to develop a reliable joining process for CRT artefacts, since it is much more efficient to manufacture simple components that can subsequently be joined in complex shapes instead of large, monolithic structures. Joints between previously reacted CRT components can be made using a lower melting point Li-doped Bi-2212 compound as a solder [3], but it is still worthwhile investigating whether even this complication can be eliminated.

2. Experimental

CRT Bi-2212 components of 5-20 cm in length and 1 mm in thickness have been produced by doctor-blading a slurry consisting of Bi-2212 powder supplied by Hoechst AG, 5wt% Ag powder, MgO fibres manufactured in-house, a binder (PVB) and a solvent (cyclohexanone). The tapes thus produced are subsequently warm-pressed into a laminate and the organics are burned off. During the consequent melt texturing the samples are kept at the maximum temperature of 887°C for 20 minutes and then cooled slowly at 3°C/h. The process must take place in flowing O_2 to ensure proper densification. Samples produced in such a way however do not have optimum oxygen stochiometery and hence optimum T_c [5]. It is therefore necessary to anneal the samples after the conclusion of the reaction to achieve the maximum possible J_c that the microstructure of the sample will allow. Fig. 1 shows the variation of J_c with respect to annealing temperature for various samples in an atmosphere of 2%O_2 in Ar. All samples were quenched to ambient from the temperature indicated. It is clear that optimum stoichiometry for this atmosphere is reached around 600°C.

Components thus reacted were subsequently subjected to a series of remelting-retexturing cycles utilising exactly the same thermal routes as before. Furthermore, a component which failed locally after being subjected to a very high current (about 10 times its J_c) was also reprocessed under the same conditions.

In order to react a long green artefact in segments, a three-zone furnace was set up to have a narrow hot zone and a sharp temperature gradient of about 15 K/cm on one side. The green sample was then placed in such a way as to keep a large part of it unreacted. The samples were then re-reacted in their entire length by eliminating this temperature gradient.

For the joining studies a series of previously reacted samples were placed in an open-sided die in a vertical furnace and press joined along the c-axis by applying a static load of up to 2 MPa. The thermal route used was identical to normal CRT, i.e. the samples were remelted, retextured and annealed as above.

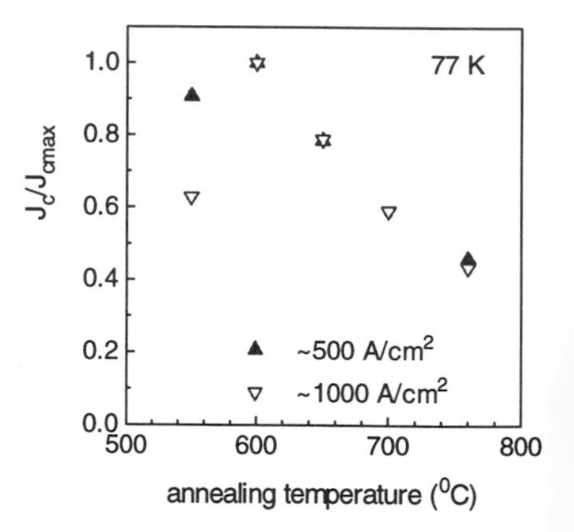

Fig. 1 Variation of Jc with respect to annealing temperature in 2% O_2 in Ar (all samples quenched)

3. Results and discussion

3.1. Repeated remelting of reacted CRT components

Fig. 2 shows that transport J_c in reprocessed CRT samples is initially robust but generally begins to deteriorate appreciably after the third reprocessing cycle. This deterioration is directly associated with loss of connectivity in the sample due to the formation of large voids in the microstructure (Fig. 3) . The effect, known as bubbling, has been frequently reported [5] and is cumulative with each consequent reprocessing. Bubbling is still not entirely understood and is attributed to, among other causes, CO_2 evolved by the burning of impurities or ambient O_2 trapped in the sample. It is usually not a problem in normal CRT because of the relatively low heating rates involved and the more open porosity in the green components compared with powder compacts. During reprocessing however, it appears that the time above the melting point allowed to the sample is not sufficient to fully heal existing pores, possibly because of the large average size of the textured grains compared with the powder particles in the green sample. It is likely that a more prolonged stay at 887°C will ameliorate the problem.

The recovery of a previously failed sample after a single reprocessing cycle is notable. The sample, with original J_c of about 550 A/cm^2, carried no supercurrent after being locally melted by the passage of an very high current pulse but was restored to 450 A/cm^2 after reprocessing. This could be of significance when contemplating the field serviceability of commercial applications.

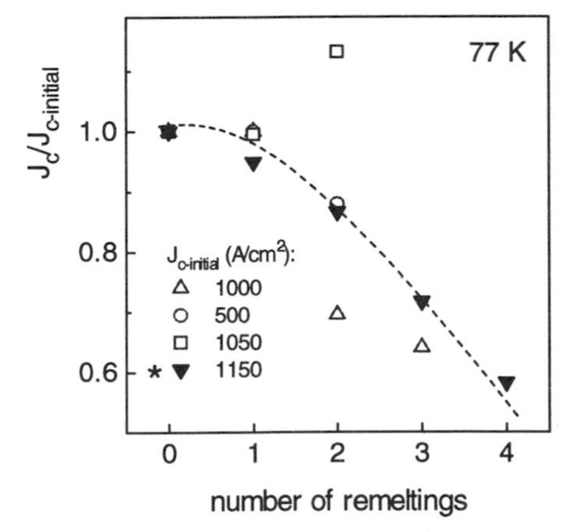

Fig. 2 Self-limited transport J_c variation as a function of reprocessing cycles. Samples quenched from 600°C except: * furnace cooled from 760°C

Fig. 3 Backscattered SEM micrograph of a four times reprocessed sample displaying extensive porosity due to bubbling.

3.2. *Reaction of green components in linear segments*

When subjected to reaction in stages, the reacted segment of the CRT artefact predictably shrunk by as much as 50% in thickness and up to 20% in width while the green segment remained unchanged. In no case did this differential shrinkage lead to fracture of the components. After reprocessing however, the pronounced HAZ invariably failed to densify and texture, resulting in the samples rapidly losing connectivity in its vicinity or even turn resistive (Fig. 4). This retention of porosity (Fig. 5) is attributed to the large particle size in the HAZ, due to the sintering of the unreacted particles during the initial reaction. It remains to be seen whether extended dwell times above the melting point will overcome this problem.

3.3. *Joining of previously reacted CRT components*

Joining along the c-axis by reprocessing the components in contact with each other has been shown to be feasible. In the case of no external load a useful and strong joint was formed despite the fact that densification was not fully achieved. Such joints retained more than 80% of the original J_c. The use of an

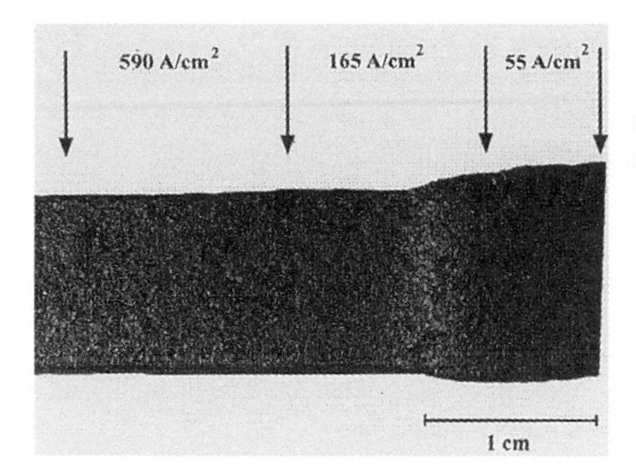

Fig. 4 Top view of CRT sample reacted in stages, showing an undensified HAZ (right). The critical currents corresponding to each segment of the sample are shown

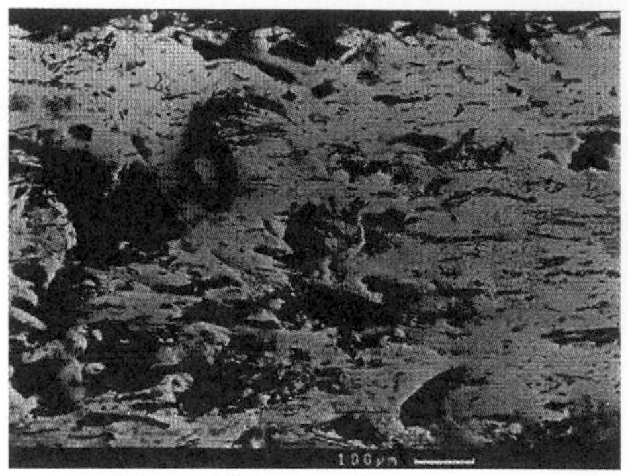

Fig. 5 Backscattered SEM micrograph of the cross section of the HAZ displaying extensive porosity towards the segment reacted only once (left of picture)

external load appears to cause more problems than it solves and is invariably deleterious to transport properties (J_c dropping to 10% at 0.75 MPa and to zero above 1.25 MPa). Despite aiding densification, external pressure inflicts great stresses on the MgO fibres, usually fracturing them. In the open-sided dies we have employed, the pressure has also led to the expulsion of most of the liquid from the sides of the die.

4. Conclusion

The properties of CRT components have been shown to be very robust to a number of remelting cycles up to the point where microstructure deterioration through bubbling severely affects their connectivity. Remelting and retexturing can also be employed to recover a major portion of the original properties of failed components. Long lengths of green samples can be reacted in stages with no danger of fracturing the preform, but current thermal routes do not appear to allow proper densification of the HAZ involved. Useful joints between previously reacted components have been manufactured without recourse to a solder, but excessive loads during press joining appear to cause insurmountable problems of liquid expulsion or adhesion to the die.

Acknowledgements

Research supported by an EPSRC CASE award in collaboration with Cryogenics Ltd. The authors are grateful to Dr. B. A. Glowacki of Cambridge University for supplying the furnace employed in the press-joining experiments.

References

[1] Soylu B. et al., 1992 *Appl. Phys. Lett.* **60** 3183–3185
[2] Watson D. R. et al, 1995 *Super. Sci. Tech.* **8** 311–316
[3] Soylu B. et al., 1995 *Trans. Appl. Superconductivity* **5** 1463-1466
[4] Quilitz M. and Maier J., 1996 *J. Superconductivity* **9** 121–127
[5] Haughan T. et al., 1996 *Physica C* **266** 183-190

Inst. Phys. Conf. Ser. No 158
Paper presented at Applied Superconductivity, The Netherlands, 30 June–3 July 1997

Electrical And Structural Properties Of Ag-Doped (BiPb)SrCaCuO Superconducting System

A. Mariño, C. E. Rojas, H. Sánchez and J. E. Rodríguez

Department of Physics Universidad Nacional de Colombia. Bogotá, Colombia

Abstract. Bulk samples of $(Bi_{0.8}Pb_{0.2})_2Sr_2Ca_3(Cu_{1-x}Ag_x)_4O_{12+\delta}$ for $0 \leq x \leq 0.02$ have been prepared by the solid state reaction method. For low nominal silver (Ag) doping $x \leq 0.01$, critical current density (Jc) increased with increasing Ag content. This improvement can be associated with microstructural changes. The formation and structure of the high Tc phase (110K) was not affected for low silver doping ($x \leq 0.01$). Large nominal Ag doping, $x \geq 0.02$, however inhibits the formation of thc 110 K phase and decreases Jc.

1. Introduction

The mechanism of the superconductivity of high Tc superconductors (HTSC) is not yet established. It is, however, generally accepted that the carriers responsible for the conductivity are holes or electrons introduced by cation doping or changing the oxygen content into the CuO_2 planes. The existence of these two dimensional CuO_2 planes has been regarded as essential ingredient for the occurrence of the superconductivity. The substitution of copper with other ions like transition-metal elements is very useful to know the effects on the superconducting properties of these high Tc copper oxides.

Earlier studies have reported an improvement in electrical properties of YBCO and (2212) and (2223) BSCCO systems by silver, niobium and tungsten doping.[1-3].

In this work we report the effect of Ag-doping on the electrical and structural properties of the 2234 BSCCO system. The 2234 BSCCO compound is characterized mainly by the presence of the high Tc Phase (110K). The possible formation of the 2234 phase, however, could be responsible for some characteristics observed in Si-doped 2234 BSCCO compound [4].

2. Experimental

Bulk samples of nominal composition $(Bi_{0.8}Pb_{0.2})_2Sr_2Ca_3(Cu_{1-x}Ag_x)_4O_{12+\delta}$ with $0 \leq x \leq 0.02$ were prepared by solid state reaction method using high purity powders of Bi_2O_3, PbO, $SrCO_3$, $CaCO_3$, CuO and $AgNO_3$. The materials were ground thoroughly, preheated at 800°C for 20h and then ground and pressed (9 tons/cm^2) in the form of small pellets. Thereafter the pellets were sintered at 850°C for 140h in air and cooled with the furnace turned off.

The analysis of the phases present in the samples and the determination of the lattice constants were carried out by powder X- ray diffraction using Cu-Kα radiation. Electrical resistance was measured with the standard four probe technique and the critical current density Jc was recorded by the resistive method with the 10μV/cm criterion. Magnetic susceptibility was measured using an a.c. Hartsshorn- type bridge.

3. Results and Discussion

Figure 1 shows the temperature dependence of the resistivity for several x values ($0.001 < x < 0.02$).

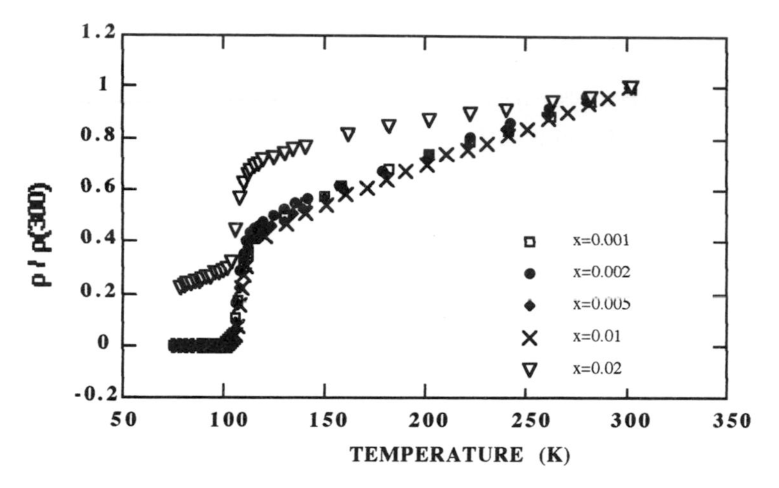

Fig. 1. Normalized resistivity vs. Temperature for different Ag-concentrations.

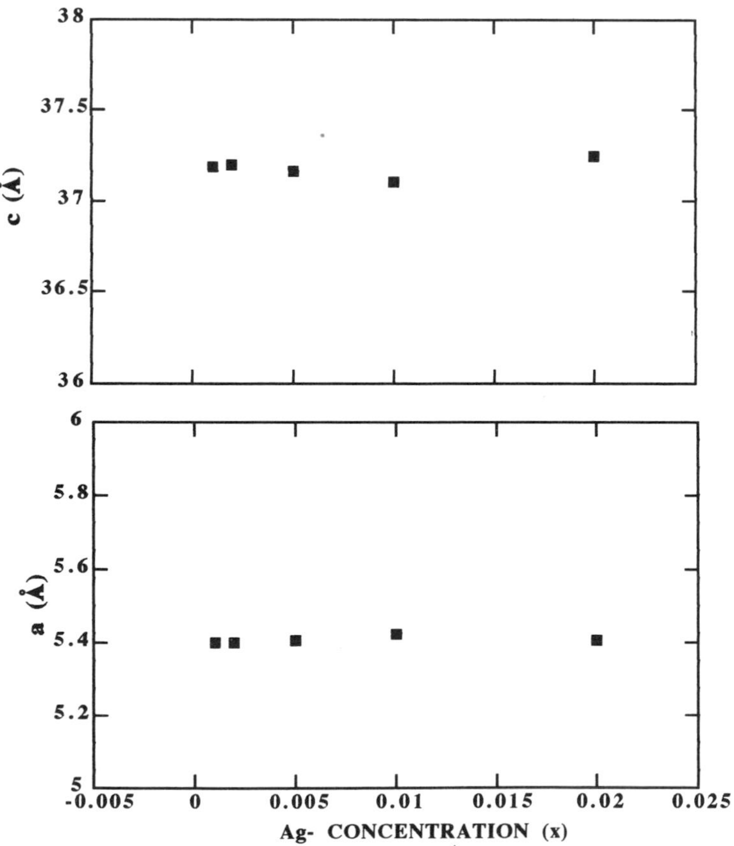

Fig. 2. Unit cell dimensions of the 110K phase vs. X.

The room temperature resistivity increases with Ag- doping and for high concentrations ($x > 0.02$) the zero resistance phenomenon disappears above the boiling point of liquid nitrogen.

The behaviors of both residual resistivity (ρ_0) and transition width (Δt), as a function of Ag-doping, correlate well. The lowest ρ_0 and Δt values correspond to the samples with $0.005 < x < 0.01$ (see Table 1).

TABLE 1

Ag-concentration (x)	ρ0 (mΩ-cm)	ρ (300K) (mΩ-cm)	Δt (K)
0.001	0.20	46.1	6.0
0.002	0.21	52.6	6.2
0.005	0.16	62.0	5.0
0.01	0.10	60.3	3.5
0.02	0.60	61.0	5.8

Residual resistivity (ρ_0), room resistivity (ρ_{300}) and transition width (Δt), as a function of Ag-doping.

The results of magnetic susceptibility are coincident with the resistivity curves. An increment of the diamagnetic signal was observed for samples with $0.005 < x < 0.01$. Larger Ag- concentrations ($x > 0.01$) decrease the diamagnetic signal indicating a reduction of the high Tc superconducting volume.

The X-ray diffraction patterns of samples with $x \leq 0.01$ show that the high Tc phase (110K) is predominant. Larger Ag-concentrations ($x > 0.01$), however, displayed multiphase samples, in that case both the low Tc (80K) phase and the Ag diffractive peak increase by increasing x. Then low Ag-doping doesn't affect the formation of the high Tc phase.

The lattice parameters of the 110K phase did not change with Ag-concentration as observed in Fig. 2. This indicates that Ag-doping doesn't affect the structure of the high Tc phase.

Fig.3 shows the critical current density Jc(77K) vs. x. An enhancement of Jc from $\sim 10 A/cm^2$ to $\sim 80 A/cm^2$ as the nominal Ag- doping increased from $x = 0.0$ to $x \sim 0.005$ was observed. Further increase of the Ag content produced a corresponding decrease in Jc. It is believed that the microstructural changes produced by Ag-doping result in a corresponding increase of Jc.

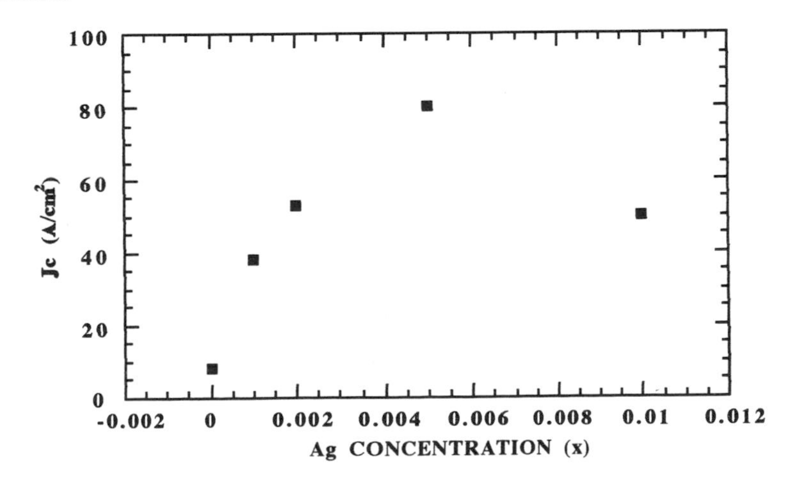

Fig. 3. Critical current density at 77K for different Ag concentrations.

As observed with X-ray diffraction, part of the Ag segregates on the BSCCO grain boundaries. This effect can either increase the contact surface between grains by decreasing the grain size or change the weak link type.[5, 6].

Reploted data of Jc as a function of $(1 - T/Tc)^{\alpha}$ for a sample with x=0.005 are displayed in fig.4. The straight line represents the least square fit of the experimental data to the power law $(1 - T/Tc)^2$, close to Tc. The nearly square law ($\alpha \sim 2$) suggests that the weak link are mainly superconducting-metal-superconducting (SNS) type. The same behavior was observed for all Ag-doped samples.

At higher Ag-concentrations (x>0.01) the increase of the low Tc phase and the formation of non superconducting phases reduces Jc, as shown in fig.3.

It is worthwhile to note that it was observed a decrease of Tc by increasing Ag-concentration, which, could be related with disorder induced in the CuO_2 planes. This fact is under current analysis.

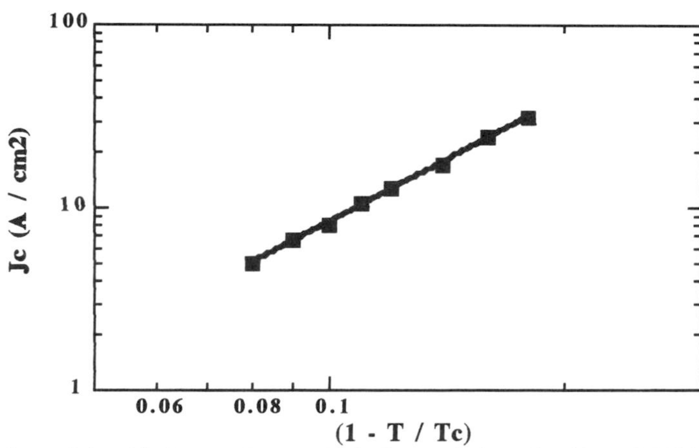

Fig. 4 . Dependence of the critical current Jc on (1- T/Tc) for a typical sample with x = 0.005. The straight line is a plot of Jc α $(1 - T/Tc)^2$.

4. Conclusions

The Ag-doping affects the superconducting properties of the (BiPb)SrCaCuO system but doesn't affect the structure of the high Tc phase. For x < 0.01 the Jc increases. Larger doping levels (x > 0.01) inhibit the formation of the high Tc phase and decrease Jc. The enhancement of the critical current density can be attributed either the increase of the contact surface between grains or the increasing of SNS weak link type.

References

[1] Y Matsumoto et al. 1988 Mater Res Bull **23** 1241
[2] M Kuwabara and N Kusaka 1988 Jpn J Appl Phys **27** L1504
[3] R Yufang et al 1990 Sol State Comm **75** 625-627
[4] A Sanchez II Sanchez and A Marino 1996 Czech J Phys **46** 1487-1488
[5] S Ravi and U Seshu Bai 1992 Sol State Comm **83** 117-121
[6] J Joo J P Singh and R B Poeppel 1992 J Appl Phys **71** 2351

Inst. Phys. Conf. Ser. No 158
Paper presented at Applied Superconductivity, The Netherlands, 30 June–3 July 1997

Low field relaxation measurements of induced magnetic moment in Bi(2223)/Ag tape and Bi(2212) single crystal

Alexandr Yu. Galkin[1], Miloš Jirsa[1], Ladislav Půst[1], Petr Nálevka[1], Michael R. Koblischka[2], Vladimir M. Pan[3], and René Flükiger[4]

[1]Institute of Physics, AV ČR, Na Slovance 2, CZ-180 40 Praha 8, the Czech Republic
[2]Department of Solid State Physics, Norwegian University of Science and Technology, N-7034 Trondheim, Norway
[3]Institute for Metal Physics, NASU, Vernadsky Blvd. 36, Kiev 252142, Ukraine
[4]Département de Physique de la Matière Condensée (DPMC), Université de Genève, 24, Quai Ernest-Ansermet, CH-1211 Genève 4, Switzerland

Abstract. Relaxation of the induced magnetic moment and of the associated critical current was studied in detail at low external fields on a Bi(2223)/Ag tape and a Bi(2212) single crystal. In the low-field range, both samples exhibited a large variation of the relaxation rate $R = -\mathrm{d}|m|/\mathrm{d}\ln t$ and the normalized relaxation rate $S = -\mathrm{d}\ln|m|/\mathrm{d}\ln t$. The maxima of the $R(B_e)$ and $S(B_e)$ dependencies are correlated with the shape of the magnetic hysteresis loop. Above 0.5 T, two relaxation regimes were recognized. The crossover from the fast to the slow relaxation regime is discussed.

1. Introduction

The Bi-based superconducting tapes represent the most perspective material for technical applications. One of the important characteristics is the relaxation behavior. Due to the low effective activation energy, a significant role of surface effects, and the inhomogeneous local field distribution at low external fields, study of the dissipative mechanisms in the Bi-based superconductors is a rather complex task. In granular samples like the Ag-sheathed Bi(2223)/Ag tapes the situation is further complicated by presence of the intra- and intergranular currents. To be able to consider effects of the intergranular currents on the total relaxation behavior of the studied Bi(2223)/Ag tape, the relaxation in a Bi(2212) single crystal was also investigated. The aim of this paper is to describe quantitatively relaxation measurements at low fields and to find some correlation between relaxation characteristics of both samples.

2. Experiment and Discussion

The monofilamentary Ag-sheathed tape of $(Bi,Pb)_2Sr_2Ca_2Cu_3O_{10}$ [Bi(2223)/Ag with dimensions of $5\times2.8\times0.032$ mm^3 and the $Bi_2Sr_2CaCu_2O_x$ single crystal [Bi(2212)] with dimensions of $2.6\times1.1\times0.035$ mm^3 were investigated for this paper. The tape was prepared by the standard powder-in-tube method followed by rolling, the crystal was grown using a directional solidification method employing the stationary crucible in a strong temperature gradient [1].

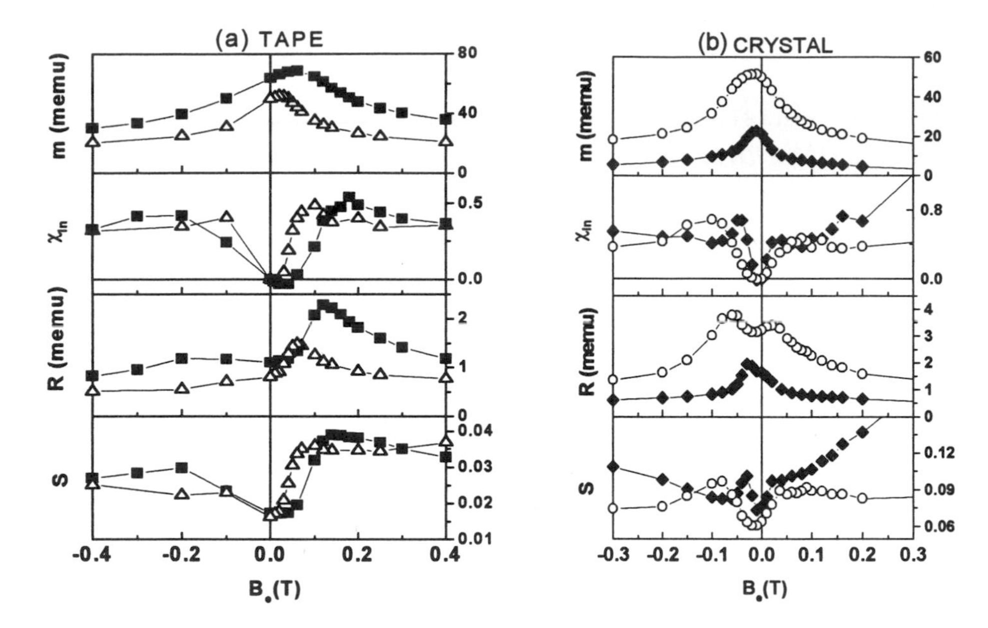

Figure 1. Comparison of the field dependencies of magnetic moment m, logarithmic susceptibility χ_{ln}, relaxation rate $R= - d|m|/dln t$ and normalized relaxation rate $S= - dln|m|/dln t$ for (a) Bi(2223)/Ag tape at 10 K (squares) and 20 K (triangles), and (b) Bi(2212) single crystal at 10 K (circles) and at 15 K (diamonds)

Magnetic hysteresis loops (MHL) and conventional relaxation of the induced magnetic moment at a constant magnetic field were measured by means of a QD SQUID magnetometer. To ensure the full critical state of the vortex system, the target field was reached from a large field ($|B_e| \geq 3$T). The applied field was always perpendicular to the sample plane. Conventional relaxation was measured at different fields, each for 1000 seconds after the field stop. To establish shift t_0 of the time scale origin, the experimental $m(\ln t)$ dependencies were fitted by the simple logarithmic function $A + B \ln(t+t_0)$. After the time correction for t_0, the relaxation rate $R=-d|m|/d\ln t$ and the normalised relaxation rate $S=-d\ln|m|/d\ln t$ were evaluated.

In Fig. 1(a) the field dependencies of the magnetic moment m, the logarithmic susceptibility $\chi_{ln}=d\ln|m|/d\ln|B_e|$, R, and S are compared for Bi(2223)/Ag at 10 K and 20 K. In Fig. 1(b) the comparison is also made for a Bi(2212) single crystal at 10 K and 15 K. Similar peaks appear at low fields on the $R(B_e)$, $S(B_e)$, and $\chi_{ln}(B_e)$ dependencies in both samples. For the Bi(2223)/Ag tape at 10 K, there is always a big peak lying at a positive (descending) field and a smaller peak positioned at a negative (ascending) field. For the Bi(2212) single crystal at 10 K the situation is just opposite: the bigger peaks lie at negative fields while the smaller ones occur at positive fields. With increasing temperature the big peaks shift towards B_e=0 while the smaller ones almost disappear. Only one, central maximum is observed on the $m(B_e)$ dependence. It appears for the Bi(2223)/Ag tape at a positive field, for the Bi(2212) single crystal at a negative field. With increasing temperature, the maximum shifts towards B_e=0 for both samples.

The significant feature of the maxims of the $\chi_{ln}(B_e)$, $S(B_e)$, and $R(B_e)$ dependencies is that, for the given sample, they consistently shift with temperature. This indicates that the shapes of $R(B_e)$, $S(B_e)$, and $\chi_{ln}(B_e)$ dependencies are correlated. Perkins et al. [2] found from the scaling properties that in $TmBa_2Cu_3O_7$ single crystal S is a linear function of χ_{ln} at high

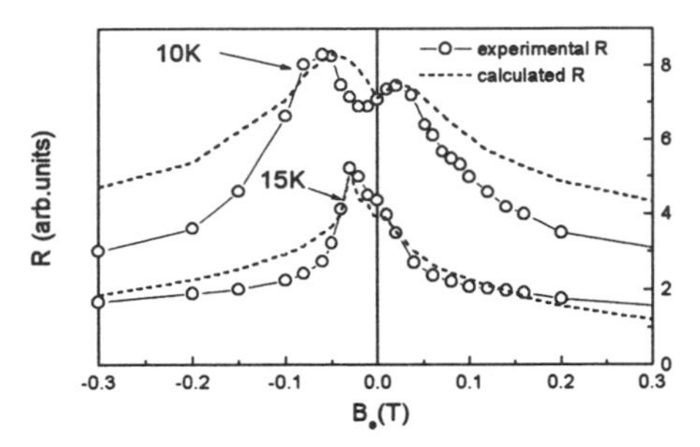

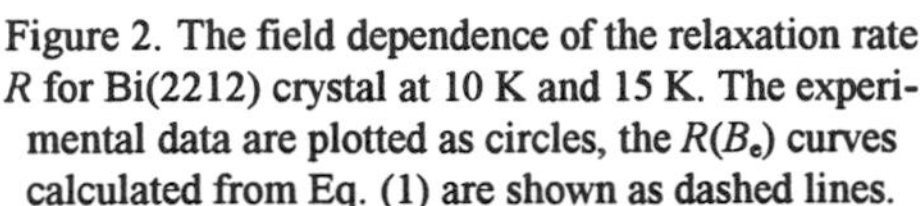

Figure 2. The field dependence of the relaxation rate R for Bi(2212) crystal at 10 K and 15 K. The experimental data are plotted as circles, the $R(B_e)$ curves calculated from Eq. (1) are shown as dashed lines.

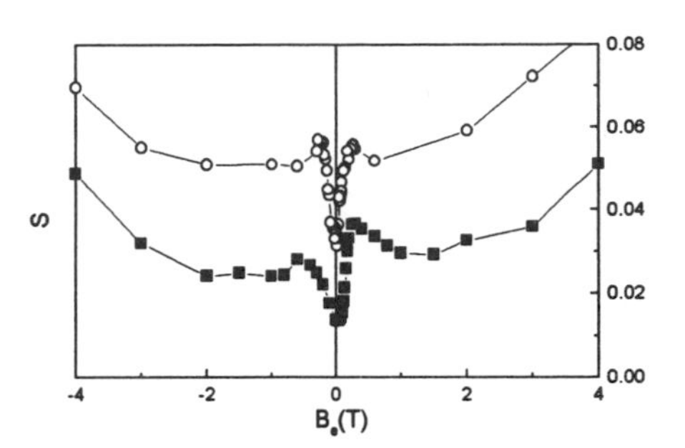

Figure 3. The field dependence of normalized relaxation rate S for Bi(2223)/Ag tape (squares) and Bi(2212) single crystal (circles), both at T=5 K

fields and temperatures. Our experiments show that this relationship holds also at low fields.

To describe the relaxation process at low fields, we used the equation of motion deduced with taking into account the backward jumps of vortices [3,4]. For a slab magnetized in the plane one obtains

$$dj / d\ln t = \frac{B_e \sinh(\sigma j / j_c) + \mu_0 l j \cosh(\sigma j / j_c)}{(B_e \sigma / j_c + \mu_0 l) \cosh(\sigma j / j_c) + \mu_0 l j \sigma / j_c \sinh(\sigma j / j_c)} \qquad (1)$$

where j is the current density, $j = -m/(a^2 bc)$, (a, b, c) are dimensions of the sample, $2l$ is the distance between two adjacent pinning sites, and $\sigma = U_0/kT$. The simplest, Kim-Anderson, form of the $U(j)$ dependence is assumed, $U = U_0(1 - j/j_c)$, with field independent j_c and U_0. Even though this simple model does not take into account the stray field effects that are important at low fields [2,5,6], it describes surprisingly well the experimentally observed relaxation processes at low fields. This is documented by the fit of the experimental $R(B_e)$ curves for the single crystal at 10 K and 15 K as shown in Fig. 2. The calculated curves have two maxims similar to the experimental ones.

The dramatic change of the normalized relaxation rate S with field in the low field region (0-1.5 T for the tape and 0-0.5 T for the crystal) (see Figs. 1 and 3) seems to be just due to normalization to m. It is a strong function of field in this region which reflects in the shape of the $S(B_e)$ dependence. The strong field dependence of m and consequently also of S at fields below the penetration field, B_p, has been also attributed to the self-field effects [2,5,6]. In the intermediate field range, S changes more gradually and develops a shallow minimum. Such a behavior has been also observed in other types of high-T_c specimens [6]. The high-field part of $S(B_e)$ rises steeply and tends towards S=1 at B_{irr} (irreversible field).

During relaxation measurements in the Bi(2223)/Ag tape we observed a clear crossover between two relaxation regimes on the $m(\ln t)$ curve (see Fig. 4). The similar crossover has been also observed in BSCCO [7] and YBCO [8] single crystals and was ascribed to transition from surface to bulk pinning [9]. Paasi et al [10] attributed the abrupt change in the $m(\ln t)$ slope observed in a Bi(2223)/Ag tape at low fields to the transition from a fast intergranular to a slow intragranular current relaxation.

Bending of granular Bi-based tapes into a small radius was proposed as a method how separate effects of inter- and intragranular currents [11]. After bending, the intergranular currents are believed to vanish while the intragranular ones to be only slightly affected. In order to study the role of the intergranular currents in the relaxation process, we bent the

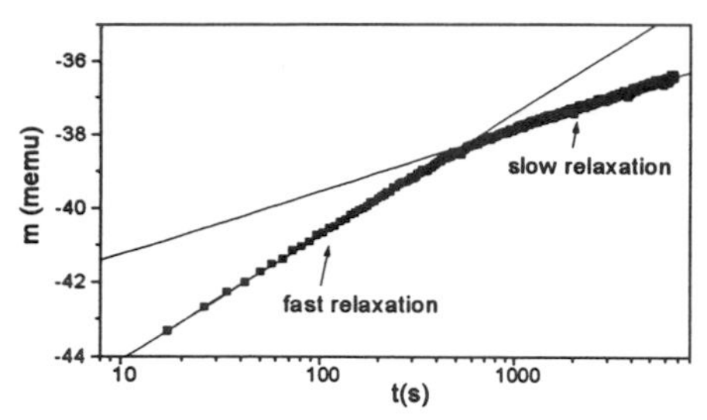

Figure 4. Observed crossover between two regimes of the magnetic moment relaxation at T=5 K and B=3 T

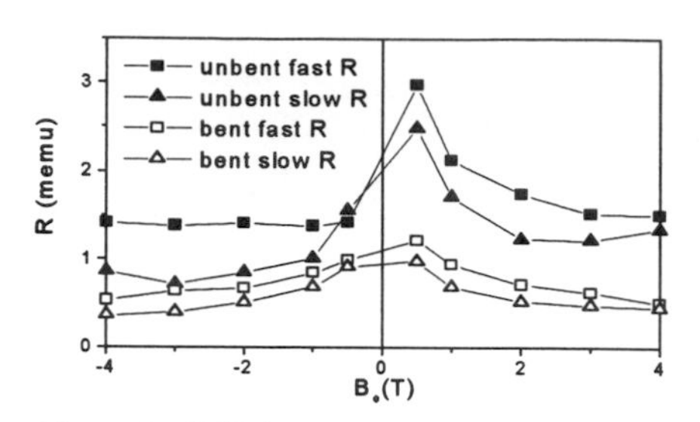

Figure 5. Field dependence of relaxation rate R for two relaxation regimes: fast and slow for the unbent and bent Bi(2223)/Ag tape at T=5 K.

Bi(2223)/Ag sample into a small radius [12]. The field dependence of the relaxation rate R is presented in Figure 5 for the slow and fast relaxation regime before and after the bending. It is clearly seen that the bending significantly reduces asymmetry of $R(B_e)$. The other important feature is the reduction of the difference between R for the slow and fast relaxation.

3. Conclusions

We studied the low field relaxation processes in a Bi(2223)/Ag tape and a Bi(2212) single crystal. A strong correlation between the relaxation rate $R= - d|m|/d\ln t$, the normalized relaxation rate $S= - d\ln|m|/d\ln t$, and the logarithmic susceptibility $\chi_{\ln}=d\ln|m|/d\ln|B_e|$ was found. The shape of the $R(B_e)$ dependence is qualitatively described by means of the flux-creep equation derived for a slab magnetized along the plane, taking into account the backward jumps of vortices and field-independent U_0 and j_c. The crossover observed on the $m(\ln t)$ dependence in the Bi(2223)/Ag tape might indicate two time windows at which different current contributions dominate the relaxation process.

Acknowledgements

This work was supported by GA AVČR under the contract A1010512.

References

[1] M. H. Ionescu, C. C. Sorrell, S. X. Dou, and R. Ramer, J. Supercond. **1**, 81 (1994).
[2] G. K. Perkins, L. F. Cohen, A. A. Zhukov, A. D. Caplin, Phys. Rev. B. **51**, 8513 (1995).
[3] H. G. Schnack, R. Griessen, J. G. Lensink, C. J. van der Beek, and P. H. Kes, Physica C**197**, 337 (1992).
[4] M. Jirsa, L. Půst, H. G. Schnack, and R. Griessen, Physica C **207**, 85 (1993).
[5] M. Jirsa, L. Půst, D. Dlouhý, Phys. Rev. B **55**, 3276 (1997).
[6] M. Däumling and D. C. Larbalestier, Phys. Rev. B **40**, 9350 (1989).
[7] N. Chikumoto, M. Konczykowski, N. Motohira, and A. P. Malozemoff, Phys. Rev. Lett. **96**, 1260 (1992)
[8] S. T. Weir et al., Phys. Rev. B **43**, 3034 (1991).
[9] L. Burlachkov, Phys. Rev. B **47**, 8056 (1993).
[10] J. Paasi, M. Polák, M. Lahtinen, V. Plecháček, and L. Söderlund, Cryogenics **32**, 1076 (1996).
[11] K.-H. Müller, C. Andrikidis, H. K. Liu, S. X. Dou, Phys. Rev. B **50**, 10218 (1994).
[12] P. Nálevka et al., to be published in Proceedings of EUCAS'97 Veldhoven, Netherlands.

Inst. Phys. Conf. Ser. No 158
Paper presented at Applied Superconductivity, The Netherlands, 30 June–3 July 1997

Anisotropic magnetoresistance in optimally doped and overdoped $Bi_2Sr_2CaCu_2O_{8+x}$ single crystals

G Heine[a], W Lang[a,b], X L Wang[c], X Z Wang[d]

[a]Ludwig Boltzmann Institut für Festkörperphysik, Kopernikusgasse 15, A-1060 Wien, Austria, and Institut für Materialphysik der Universität Wien, Strudlhofgasse 4, A-1090 Wien, Austria.

[b]Department of Electrical Engineering and Laboratory of Laser Energetics, University of Rochester, Rochester, NY 14627, USA.

[c]Center for Superconducting and Electronic Materials, University of Wollongong, Wollongong, NSW 2500, Australia.

[d]Department of Physics, Simon Fraser University, Burnaby B.C., V5A 1S6, Canada.

Abstract. We report on measurements of the in-plane and out-of-plane resistivity and the magnetoresistance (MR) of $Bi_2Sr_2CaCu_2O_{8+x}$ single crystals with different oxygen contents. The in-plane resistivity obeys a metallic-like temperature dependence with positive MR. The out-of-plane resistivity shows an activated behaviour above T_c with a metallic region at higher temperatures and negative MR. In-plane and out-of-plane measurements were analysed in the framework of a recent model for thermodynamic order parameter fluctuations. Within this model the positive in-plane MR above T_c is caused by the suppression of the fluctuation conductivity enhancement, whereas the negative out-of-plane MR arises from the reduction of the fluctuation-induced pseudogap in the one-electron density of states by the magnetic field. Using this theoretical background, and considering the MR of normal-state quasiparticles, our experimental data can be explained excellently over a wide temperature range.

1. Introduction

One of the most puzzling features of the high temperature superconductors is the different temperature dependence of the in-plane and out-of-plane resistivity and the opposite sign of the in-plane and out-of-plane magnetoresistance (MR), respectively. It is a well known fact, that thermodynamic fluctuations play a major role in the cuprates near the superconducting phase transition. The hierarchy of the different fluctuation effects is responsible for the sign of these corrections to the normal-state conductivity tensor. The "direct" or Aslamazov-Larkin (AL) contribution stems from the acceleration of Cooper pairs, created in a thermodynamic non-equilibrium, and results in an enhancement of the conductivity. Another contribution is caused by the reduction of the single-electron density of states (DOS) due to fluctuation pairing, thus, diminishing the normal-state conductivity. Generally, the AL effect dominates the in-plane properties, whereas the DOS contribution might prevail for transport along the c-axis. Both contributions are reduced in a magnetic field, resulting in positive in-plane and negative out-of-plane MR, an observation which hardly can be explained by conventional anisotropic transport models.

2. Experimental methods

Single crystals of $Bi_2Sr_2CaCu_2O_{8+x}$ were grown by a self-flux method and a segregation growth technique [1]. The oxygen content was adjusted by annealing in flowing oxygen at 500°C to obtain overdoped samples (L6, L12, $T_c \approx 80$ K) and in flowing argon at 600°C for the optimally doped crystals (C1, C2) with maximum $T_c \approx 90$ K.

Electrical contacts were established on evaporated silver pads using thin Au wires. Two opposite faces of the crystal were plunged into silver paint to form current contacts for the in-plane measurements. Ring-shaped current electrodes with centered voltage pads were used for the out-of-plane measurements. The MR measurements were performed in a superconducting magnet providing fields up to ±13 T. The temperature measurement took place in zero field using a calibrated platinum resistor. During the magnetic field sweep, the temperature was held constant by means of a capacitance sensor resulting in a temperature stability better than ±30 mK. Lock-in technique at a frequency of 17.4 Hz was applied for the transport measurements.

3. Results and discussion

The temperature variation of the anisotropic resistivity in the oxygen-overdoped samples is shown in Figure 1. The in-plane resistivity exhibits metallic behaviour from T_c to 300 K and $\rho_{ab}(300\text{ K}) \approx 200$ μΩcm. The out-of-plane resistivity shows $d\rho_c/dT > 0$ only above 170 K. Below this temperature, ρ_c follows a thermally activated behaviour. In the normal state $\rho_c(T) = a\exp(\Delta/T) + bT + c$, with $\Delta = 332$ K. The zero-resistance $T_{c0} = 79$ K in both crystals. Figure 2 presents the corresponding resistivity data of the optimally doped samples. The in-plane resistivity also is metallic, but, as expected, with a higher room temperature value $\rho_{ab}(300\text{ K}) = 260$ μΩcm. The out-of-plane resistivity exhibits the activated behaviour ($\Delta = 384$ K) in the entire temperature range of our measurement.

In Figure 3, the magnetoconductivities (MC) $\Delta\sigma_{B,xx} = 1/\rho_{ab}(B) - 1/\rho_{ab}(0)$ and $\Delta\sigma_{B,zz} = 1/\rho_c(B) - 1/\rho_c(0)$ are shown for convenient comparison with theoretical results. Note that corresponding MC and MR have opposite sign. Thus, in our samples, the in-plane MR is positive, while the out-of-plane MR is negative. A recent theory for order parameter fluctuations in layered superconductors by Dorin *et al.* [2] provides the theoretical

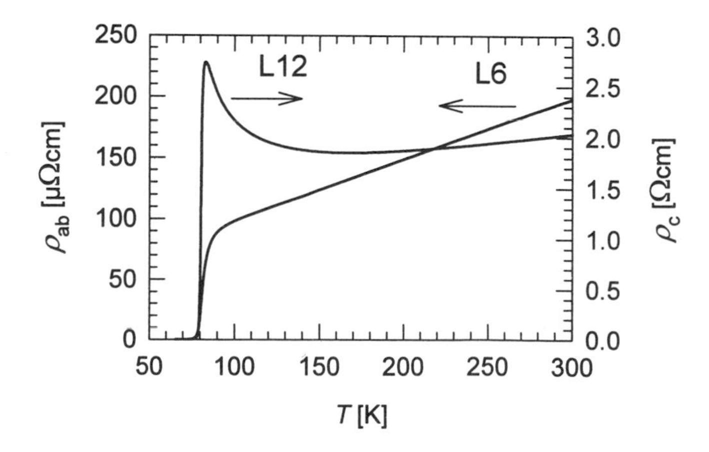

Figure 1: In-plane (sample L6) and out-of-plane (sample L12) resistivity of the oxygen-overdoped $Bi_2Sr_2CaCu_2O_{8+x}$ crystals.

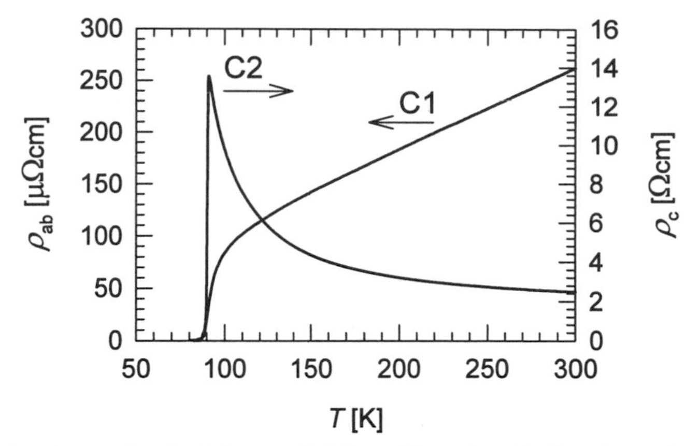

Figure 2: In-plane and out-of-plane resistivity of the optimally doped samples C1 and C2, respectively.

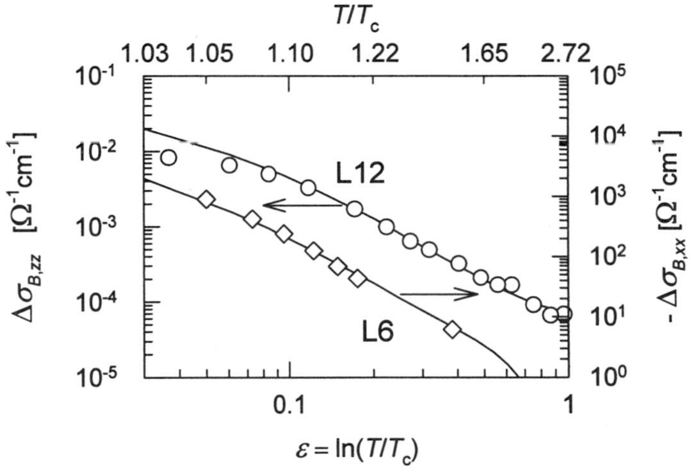

Figure 3: In-plane (L6) and out-of-plane (L12) magnetoconductivities of the overdoped samples in a magnetic field of 12 T. The lines represent the theoretical fits. For details see the text.

interpretation of our results. Since the magnetic field interacts with the charge carriers via orbit and spin, one has to consider the orbital and the Zeeman contributions to the AL and the DOS process. The spin effect, not included in [2], was calculated by applying a renormalization to the reduced temperature $\varepsilon = \ln(T/T_c)$ according to [3]. Additionally, the MR of the normal-state quasiparticles [4] has to be considered in the analysis of the in-plane MR at higher temperatures. This normal-state contribution to the MC is estimated from the Hall angle θ_H, obtained on similar crystals [5], using the relation $\Delta\sigma^N_{B,xx} \approx -A\tan^2\theta_H/\rho_{ab}$ [6]. A depends on the non-sphericity of the Fermi surface and is of the order of one. The parameters, used to fit our in-plane data, were the Fermi velocity $v_F = 1.8\times10^7$ cm/s, the interlayer coupling energy $J/k_B = 8.3$ K, the transport scattering time $\tau = 1.4\times10^{-14}$ s, and $A = 0.7$.

On the other hand, the c-axis MC is positive from T_c to our experimental limit at 210 K, indicating a dominance of the DOS contribution. To evaluate the data, we included the orbital and Zeeman DOS, as well as the orbital and Zeeman AL contributions. The

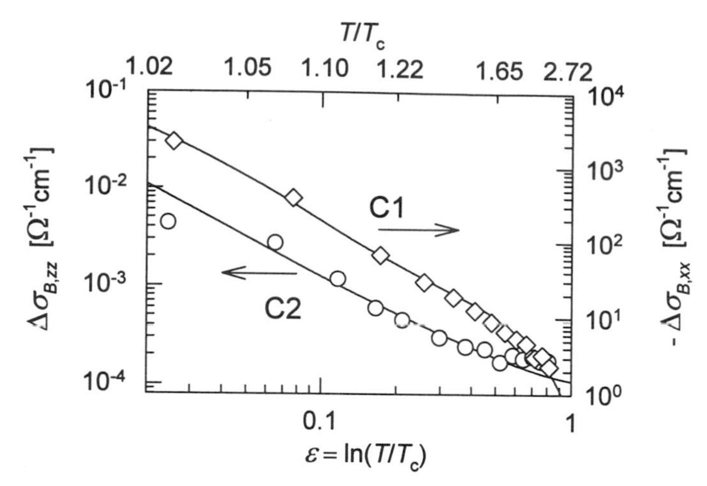

Figure 4: In-plane (C1) and out-of-plane (C2) magnetoconductivities of the optimally doped samples at $B = 12$ T. The lines represent the theoretical fits.

resulting parameters were $v_F = 1.8\times10^7$ cm/s, $J/k_B = 6.3$ K and $\tau = 1.1\times10^{-14}$ s. A sign change of the MC arising from the different polarities of the competing AL and DOS contributions was not observed, in contrast to $YBa_2Cu_3O_7$ single crystals [7].

Figure 4 shows our MC results for the optimally doped samples. The parameters to fit the data were obtained by the same procedure as for the overdoped crystals. The analysis of the in-plane MC yields $v_F = 1.8\times10^7$ cm/s, $J/k_B = 4$ K, $\tau = 1.0\times10^{-14}$ s, $A = 2.7$, and the out-of-plane measurements were fit by $v_F = 0.8\times10^7$ cm/s and the previous values for J and τ.

Finally, we compare the physical parameters obtained above with the quantities deduced from normal-state transport measurements. The Fermi velocity estimated from Hall measurements on similiar crystals [5] is $v_F = 1.4\times10^7$ cm/s, similar to our values from the MC analysis. Our results for τ range from 1.0×10^{-14} to 1.4×10^{-14} s, which also is in good agreement with a typical carrier mobility $\mu = R_H/\rho_{ab} \approx 5$ cm^2/Vs at room temperature inferred from the Hall angle [5]. The rather low value of J, which slightly increases in the overdoped samples, corresponds to the normal-state conductivity anisotropy [2] $\sigma_{ab}/\sigma_c = v_F^2\,\hbar^2/(J^2 s^2)$ [$s = 1.54$ nm is the interlayer spacing], observed in our crystals at about 110 K, i.e., in the fluctuation regime.

Acknowledgements

This work was supported by the Fonds zur Förderung der wissenschaftlichen Forschung, Austria.

References

[1] Wang X L, Ai Z P, Shang S X, Wang H, Jiang M, Wang X Z, Bäuerle D 1994 *J.Cryst.Growth* **139** 86-8
[2] Dorin V V, Klemm R A, Varlamov A A, Buzdin A I and Livanov D V 1993 *Phys.Rev. B* **48** 12951-65
[3] Aronov A G, Hikami S and Larkin A I 1989 *Phys. Rev. Lett.* **62** 965-8
[4] Yan Y F, Matl P, Harris J M and Ong N P 1995 *Phys. Rev. B* **52** R751-4
[5] Forro L, Mandrus D, Kendziora C, Mihaly L and Reeder R 1990, *Rhys. Rev. B* **42** 8704-6
[6] Sekirnjak C, Lang W, Proyer S and Schwab P 1995 *Physica C* **243** 60-8
[7] Axnäs J, Holm W, Eltsev Y and Rapp Ö 1996 *Phys. Rev. Lett.* **77** 2280-3

Inst. Phys. Conf. Ser. No 158
Paper presented at Applied Superconductivity, The Netherlands, 30 June–3 July 1997

Magneto-optic visualization of flux penetration in Bi-2223 tapes after bending

M. R. Koblischka, T. H. Johansen, and H. Bratsberg

Department of Physics, University of Oslo, P.O. Box 1048, Blindern, 0316 Oslo 3, Norway.

Abstract. The effect of bending on silver-sheathed Bi-2223 tape is investigated by means of magneto-optical (MO) visualization of flux distributions. One single piece of rolled tape is used throughout all experiments starting from as-prepared, and subsequently a bending strain between 2.25 % and 9 % is applied to the tape. Flux patterns are visualized using an intact tape, i. e. the visualization is done while keeping the silver sheath intact after removing the bending strain.

1. Introduction

Bending of Ag-sheathed $Bi_2Sr_2Ca_2Cu_3\,O_{10+\delta}$ (Bi-2223) tapes is an important issue for practical use, e. g. for the preparation of coils and cables. Several authors have shown that bending (and subsequent straightening) of the tapes decreases the intergranular critical current density vastly [1, 2]. The bending strain produces cracks in the superconducting core and breaks the connectivity between the grains. Magneto-optical (MO) imaging is an ideal tool to study flux penetration and current flow in high-T_c superconductors [3], and also in samples of technical interest [4, 5].

The aim of this paper is to provide MO flux patterns of as-prepared and bent tapes for various applied fields. We mainly focus on how the flux entry and exit takes place. Therefore, we present a complete set of images for each bending step, showing the field increase, decrease, and the corresponding remanent state at $T = 50$ K. In contrast to earlier work [1], we have chosen large bending strains $\epsilon = 2.25$ to 9 % (bending diameters 4 – 1 mm) in order to reach a state where, following the assumptions made in Ref. [6], the connections between the grains should be completely broken. The crossover into such a state where a sample becomes truely granular should be clearly visible in our magneto-optical images. This situation plays an important role for magnetic measurements on Bi-2223 tapes as bending to a small diameter would then allow to separate the inter- and intragranular contributions to the magnetic moments measured.

2. Experimental

The MO imaging is based on the Faraday effect in a magneto-optical active layer. Here we have used a Bi-doped YIG film with in-plane anisotropy with a thickness of 4 μm, half of which corresponding to the spatial resolution of our experiment. In order to obtain images with a relatively high contrast, an indicator film with a high field sensitivity (~ 0.1 mT) was used. The

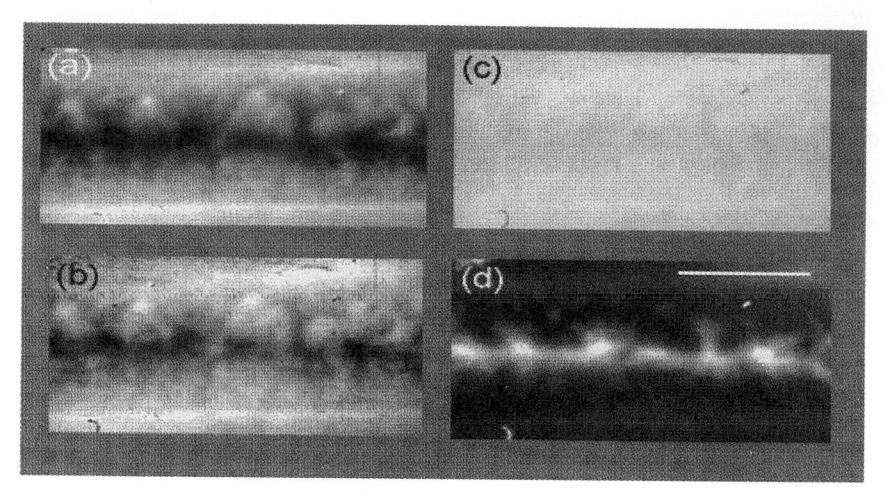

Figure 1. Flux patterns after zero-field cooling of an as-prepared tape. $T = 50$ K and (a): 15 mT, (b): 30 mT, (c): 45 mT, and (d): remanent state. The marker is 2 mm long.

flux line lattice is imaged as bright areas, whereas the flux-free Meissner area stays dark. The images presented here are, therefore, maps of the z-component of the local magnetic field, B_z. The images are recorded using an 8-bit Kodak DCS 420 CCD digital camera and subsequently transferred to a computer for processing. In the MO apparatus the sample was mounted on the cold finger of an optical Helium flow cryostat [7]. The magnetic field was applied perpendicular to the tape plane using a copper solenoid coil with a maximum field of ± 120 mT.

All measurements were performed using monofilamentary silver-sheathed (Pb,Bi)-2223 tapes prepared by the 'powder-in-tube' method with subsequent drawing and rolling [8]. For the MO measurements, we used a single piece of tape measuring externally (including the silver sheath) 8.2 mm in length, 3.8 mm in width and 90 μm in thickness. The silver sheath is 20 μm thick. The tape is bent to 4 mm diameter and then down to 1 mm diameter in steps of 1 mm using a brass rod of the corresponding size. This procedure yields bending strains $\epsilon = t/d$, where t is the total tape thickness and d denotes the bending diameter, ranging from 2.25 to 9 % [9]. The MO indicator film is laid on top of the tape and carefully centered. This enables the MO images to be compared directly to each other.

3. Results and Discussion

Fig. 1 presents flux patterns of the as-prepared tape at 50 K. It is remarkable that near the edges the field pattern is quite uniform. This reflects clearly the better orientation and grain growth along the silver sheath. When the flux reaches the wider part of the tape core, some defects due to foreign phases appear altering the current flow slightly. In the remanent state the pinned flux is confined to the sample centre. This so-called d^+-line of the currents is a clear fingerprint of a homogeneous, thin superconductor. The d^+-lines appear where the currents flowing along the sample edges have to perform a sharp turn and in the sample centre (see the definitions given in Refs. [10, 3]). Furthermore, negative vortices appear along the edges of the tape core (d^--lines). This is the consequence of a large demagnetization factor.

Fig. 2 presents flux distributions after bending the tape to a strain $\epsilon = 2.25$ %, followed by re-straightening. The flux patterns are now completely altered as the flux starts to penetrate the sample through a network of channels created by the bending procedure. These crack

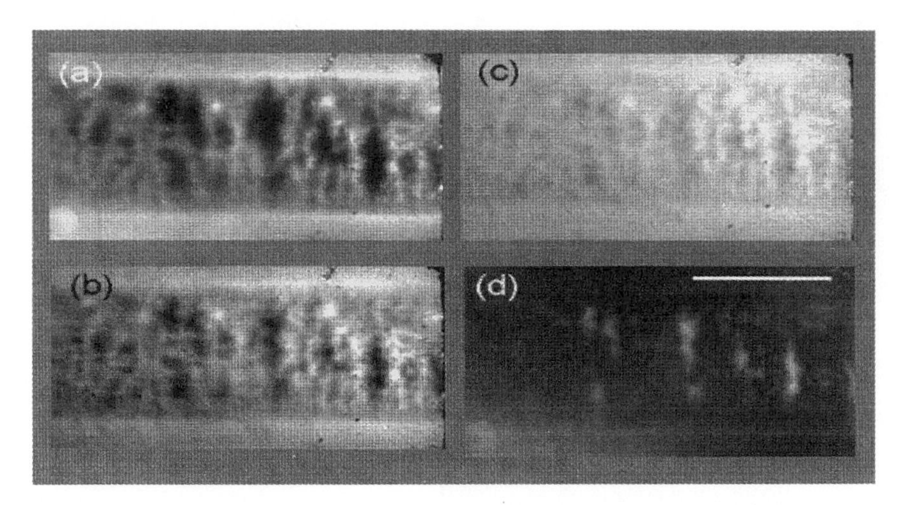

Figure 2. Flux patterns Bi-2223 tape after application of a bending strain of 2.25 %. $T = 50$ K and (a): 15 mT, (b): 30 mT, (c): 45 mT, and (d): remanent state. The marker is 2 mm long.

have changed the corresponding current flow considerably as the currents form now several small loops enclosing grain clusters with still intact coupling between the Bi-2223 grains. In the remanent state (d) the flux is pinned within these clusters, and the d^+-line has vanished. However, some transport current is still flowing throughout the sample as indicated by the field overshoot along the sample edges.

In Fig. 3, the tape is shown after applying a bending strain of 9 %. The basic flux pattern is similar to that of fig. 2. The images clearly demonstrate that both intra- and intergranular currents are affected by the bending procedure as the achieved intensities (grey values)are vastly reduced. A field of 12 mT is now sufficient to fully penetrate the sample. Note that image (c), taken in reducing field, is practically the inverse to image (a). In the remanent state, only a small amount of flux stays pinned within the grains. This shows that the intragranular pinning is quite small at this elevated temperature.

The crack network caused by bending was found to appear already at small bending strains in Ref. [1], so we may conclude that increasing the bending strain does not change the crack pattern, but the efficiency of the cracks to reduce the current density increases with strain magnitude. Furthermore, the homogeneity of the as-prepared tape plays an important role. Furthermore, clear fingerprints of still flowing *intergranular* currents are visible: The field overshoots at the edges of the tape core are clearly observed. The field distribution *outside* the tape core is identical to a homogeneous, thin superconducting strip. These two observations demonstrate that there are still intergranular currents flowing in the strongly bent sample; a true granular sample would not show these signatures [3, 11]. Our observations demonstrate the capability of the magneto-optical imaging for non-destructive testing of superconducting tapes. Using the same piece of tape during the entire bending experiment is only possible if the flux patterns can be visualized *while keeping the silver sheathing intact* as demonstrated in this work.

To conclude, it is shown that bending creates a network of cracks running perpendicular to the rolling direction. The presence of these cracks changes the flux patterns drastically: Flux enters the sample only through the cracks, and on field reversal, flux penetrates the grains (clusters), and leaves from the cracks. Bending the tape to a very small diameter of 1 mm ($\epsilon \sim 9$ %) does not remove the intergranular currents completely. The grains are still coupled

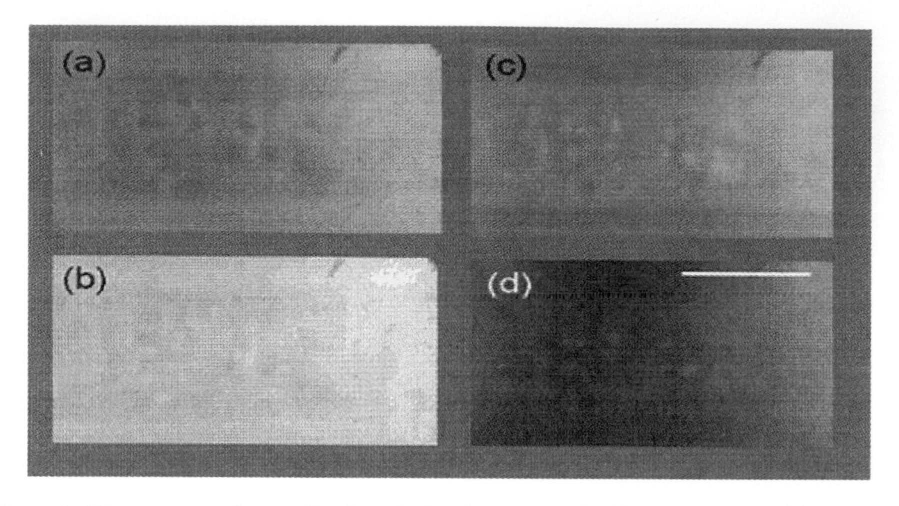

Figure 3. Flux patterns after application of a bending strain of 9 %. $T = 50$ K and (a): 6 mT, (b): 12 mT, (c): 6 mT (reducing field), and (d): remanent state. The marker is 2 mm long.

together and form small clusters.

Acknowledgments

We would like to thank Prof. Flükiger (University of Geneva) and P. Bodin, M. Bentzon (NST, Brøndby) for the (Pb,Bi)-2223 tapes. We also acknowledge valuable discussions with L. Půst, and P. Nálevka (Institute of Physics, Academy of Sciences of the Czech Republic). This work is supported by The Research Council of Norway.

References

[1] Polak M, Parrell J A, Polyanskii A A, Pashitski A E and Larbalestier D C 1997 *Appl. Phys. Lett.* **70** 1034

[2] Yau J and Savvides P 1994 *Appl. Phys. Lett.* **65** 1454; Suenaga M, Fukumoto Y, Haldar P, Thurston T R and Wildgruber U 1995 *Appl. Phys. Lett.* **67** 3025; Kováč P, Kopera L, Hušek I and Cesnak L 1996 *Supercond. Sci. Technol.* **9** 792

[3] Koblischka M R and Wijngaarden R J 1995 *Supercond. Sci. Technol.* **8** 199

[4] Koblischka M R 1995 *IOP Conf. Ser.* **275** 271

[5] Pashitski A E, Gurevich A, Polyanskii A A, Larbalestier D C, Goyal A, Specht E D, Kroeger D M, DeLuca J A and Tkaczyk J E 1997 *Science* **275** 367

[6] Müller K-H, Andrikis C and Guo Y C 1997 *Phys. Rev. B* **55** 630

[7] Johansen T H, Baziljevich M, Bratsberg H, Galperin Y, Lindelof P E, Shen Y and Vase P 199 *Phys. Rev. B* **54** 16264

[8] Grasso G, Perin A, Hensel B and Flükiger R 1993 *Physica C* **217** 335

[9] Koblischka M R, Johansen T H and Bratsberg H 1997 *Supercond. Sci. Technol.* (in press)

[10] Schuster T, Indenbom M V, Koblischka M R, Kuhn H and Kronmüller H 1994 *Phys. Rev.* B **4** 3443

[11] Koblischka M R, Schuster T and Kronmüller H 1994 *Physica C* **219** 205; Moser N, Koblischka R and Kronmüller H 1990 *J. Less Common Metals* **164-165** 1308

Inst. Phys. Conf. Ser. No 158
Paper presented at Applied Superconductivity, The Netherlands, 30 June–3 July 1997

Influence of uniaxial pressing on the microstructure and the critical current density of Tl-1223 tapes

T Riepl[1], S Zachmayer[1], R Löw[1], C Reimann[2], T Schauer[1] and KF Renk[1]

[1] Institut für Experimentelle und Angewandte Physik, Universität Regensburg, D-93040 Regensburg.
[2] Physikalisches Institut III, Universität Erlangen-Nürnberg, D-91058 Erlangen.

Abstract. Silver sheathed Tl-1223 tapes were produced by the powder-in-tube process starting with superconducting powders of the nominal compositions $Tl_{0.89}Bi_{0.25}Ba_{0.4}Sr_{1.6}Ca_2Cu_3O_x$ and $Tl_{0.79}Pb_{0.35}Ba_{0.4}Sr_{1.6}Ca_2Cu_3O_x$. Ag tubes, filled with the powders, were formed to tapes by drawing and rolling and heat treated for 6 hours at 840°C in flowing oxygen. Some of the tapes were pressed uniaxially at pressures up to 3.4 GPa before the heat treatment.
For the tapes made of Bi substituted Tl-1223 we received following results. SEM investigations of pressed and unpressed tapes showed that the pressing step increased the density of the superconducting core and reduced the average grain size. Pressed tapes showed cracks parallel to the current direction. The pressing step increased the critical current density J_c at 77 K from 8 kA/cm^2 (unpressed) to 17 kA/cm^2 (pressed). J_c decreased rapidly in applied magnetic fields, indicating that weak links limited the critical current. At 0.1 T the critical current of unpressed tapes dropped to 2 % of I_c(0 T), while that of pressed tapes dropped only to 10 % of I_c(0 T), indicating that uniaxial pressing caused not only a higher number of percolative current paths but also a higher fraction of strong linked grain boundaries.
Comparable results were obtained for tapes made of Pb substituted Tl-1223. The critical current density at 77 K was 5 kA/cm^2 in unpressed tapes and 18 kA/cm^2 in pressed tapes. In an attempt to produce long-length pressed tapes we have already manufactured a tape of 1.2 m length with a critical current density of 12 kA/cm^2 over the full length.

1. Introduction

During the last years, superconducting Tl monolayer cuprates (Tl-1223) have been of great interest because of their high critical transition temperature (T_c) up to 124 K [1] and their high irreversibility field at liquid nitrogen temperature [2,3]. Many attempts have been made to prepare tapes from this material by the powder-in-tube process [4-9], most of them involving a pressing step. In this paper, we examine the influence of uniaxial pressing on the critical current density and the microstructure of powder-in-tube tapes, prepared by an ex-situ process.

2. Experimental

Superconducting pellets of Tl-1223 were synthesized following a conventional ceramic route. The pellets were pulverized and a small amount of Tl_2O_3 (0.02 formula weight) was added. Silver tubes of 6 mm outside diameter and 1 mm wall thickness were filled with the powder and formed to tapes by drawing and rolling. Some of the tapes were pressed uniaxially at pressures up to 3.4 GPa, while others remained unpressed. Finally the tapes were heat treated for 6 hours at 840°C in flowing oxygen. We studied two different pellet compositions: $Tl_{0.85}Bi_{0.25}Sr_{1.6}Ba_{0.4}Ca_2Cu_3O_x$ ((Tl,Bi)-1223) and $Tl_{0.75}Pb_{0.35}Sr_{1.6}Ba_{0.4}Ca_2Cu_3O_x$ ((Tl,Pb)-1223).

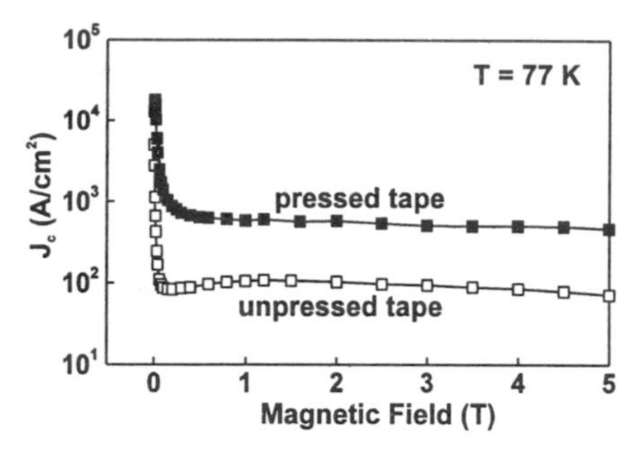

Fig. 1. Transport J_c of a pressed and an unpressed tape in a magnetic field perpendicular to the tape face.

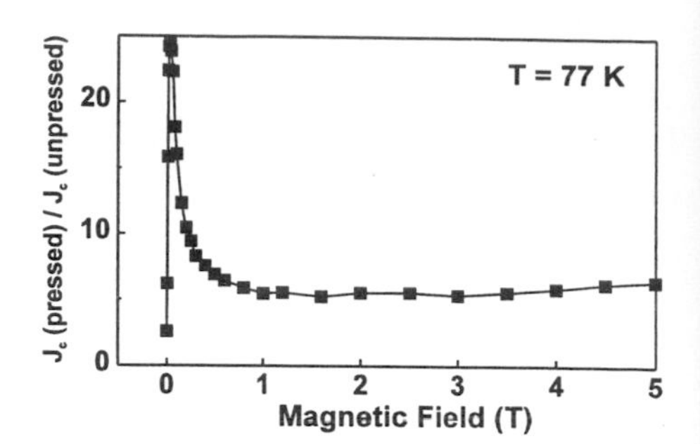

Fig. 2. Ratio of the critical current density in a pressed and an unpressed tape versus applied magnetic field.

3. Results and discussion

3.1 Examination of (Tl,Bi)-1223 tapes

The critical current density J_c at 77 K in a magnetic field perpendicular to the tape face of an unpressed sample and a sample that was pressed at 3.4 GPa is shown in fig. 1. The pressed tape has a J_c of 17 kA/cm^2 in zero field. The superconducting current is maintained up to a magnetic field of 5 T, but due to weak links J_c decreases rapidly in a small applied field. This is in good agreement to the results obtained by other groups [5,8]. The unpressed tape has a lower J_c in zero field (8 kA/cm^2) and performs very poor in magnetic fields with a minimum of J_c around 20 mT. The ratio of J_c in both tapes is shown in fig. 2. It rises sharply from around 2 for zero field to 24 around 20 mT. In magnetic fields higher than 1 T the ratio is in the range of 5 to 7. This indicates that pressing not only increases J_c in zero field, but also improves the performance in a magnetic field, reducing the weak link behaviour. The percentage of strong linked grains rises by a factor of 3.

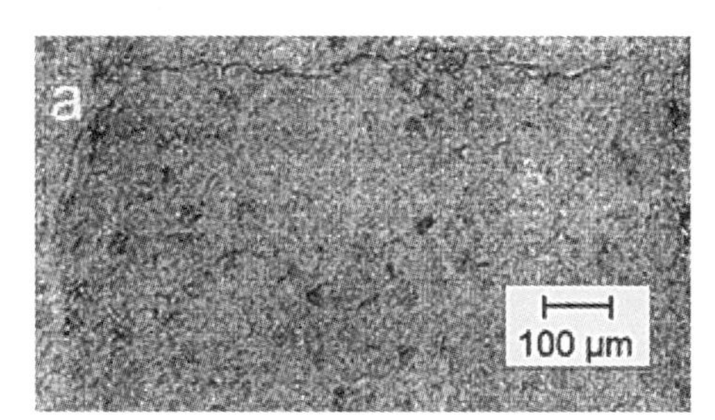

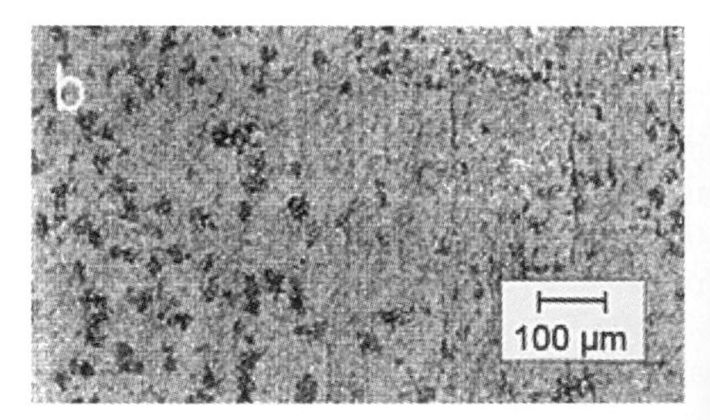

Fig. 3. Surface of the superconductor in (a) an unpressed and (b) a pressed tape after removing the silver sheath.

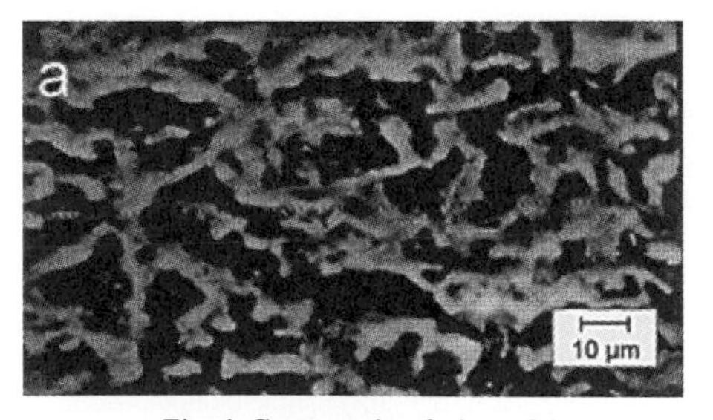

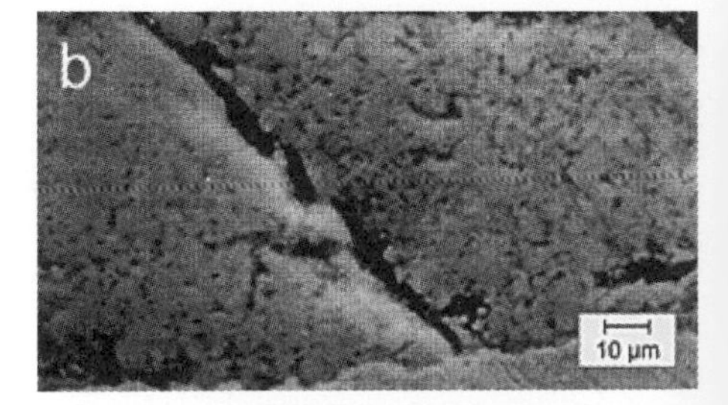

Fig. 4. Cross sectional view of the superconductor in (a) an unpressed and (b) a pressed tape.

Table 1. Structural data of unpressed and pressed (Tl,Bi)-1223 tapes.

Property	unpressed tape	pressed tape
tape dimensions	0.2 mm × 1.1 mm	0.08 mm × 2.3 mm
superconducting core fraction	35 %	30 %
ω-scan: FWHM	29°	33°
average grain size	4.2 μm	2.8 μm
maximum grain size	16 μm	12 μm
density of exclusions	260/mm²	60/mm²

After removing the silver sheath, further investigations were made. Fig. 3 shows the surface of the superconductor in both tapes. The pressed tape has black inclusions (Ba, Ca and Cu rich phases according to EDX analysis) and also cracks in vertical direction (parallel to the current in the tape), while the unpressed tape is quite homogeneous with less clear inclusions and without cracks. Higher magnifications showed that the superconducting grains are smaller in the pressed tape. The maximum and average grain size, as well as the density of exclusions are given in table 1. XRD analysis was performed on both tapes. θ-2θ scans indicated that they consisted mainly of Tl-1223 without evidence of other phases. ω scans of the (006) Bragg peak revealed that there is no texture, but a preferred orientation of the c axis perpendicular to the tape surface. The FWHM was slightly higher in the pressed tape (table 1), indicating that the pressing caused no c axis alignment as shown for bulk samples in reference [12], but even reduced the preferred orientation of the c axis.

Cross sections were investigated by SEM (fig. 4). In the unpressed tape, the superconductor is very porous, while it is very dense in the pressed tape. The cross section of the pressed tape shows also one of the cracks that run parallel to the current direction. The superconducting core fraction was 35 % for the unpressed tape and 30 % for the pressed tape (table 1). The density of the ceramic core was increased by 35 % during the pressing.

3.2 Comparison of tapes made of (Tl,Bi)-1223 and (Tl,Pb)-1223

For tapes made of the lead substituted material we obtained nearly the same results. Fig. 5 shows the critical current density of tapes made of (Tl,Bi)-1223 and (Tl,Pb)-1223 at 77 K in zero field versus applied pressure during pressing. Unpressed tapes of (Tl,Pb)-1223 had a lower J_c than unpressed tapes of (Tl,Bi)-1223. The results were about equal for both materials if a pressure around 1 GPa was applied. For higher pressures, J_c of (Tl,Pb)-1223 tapes was considerably higher than in the case of (Tl,Bi)-1223 tapes. This means that pressing has a stronger influence on the lead substituted material than on the bismuth substituted material. The average increase of J_c was 2.4 (kA/cm^2)/GPa for (Tl,Bi)-1223 tapes and 5.4 (kA/cm^2)/GPa for (Tl,Pb)-1223 tapes. The influence of pressing on the structure of

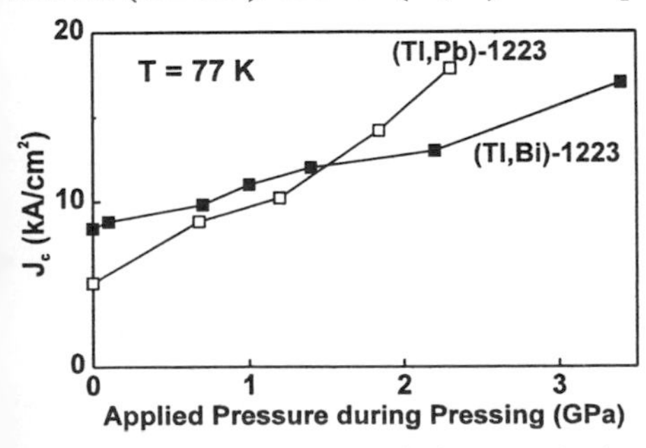

Fig. 5. J_c (77 K, 0 T) versus applied pressure for the materials (Tl,Pb)-1223 and (Tl,Bi)-1223.

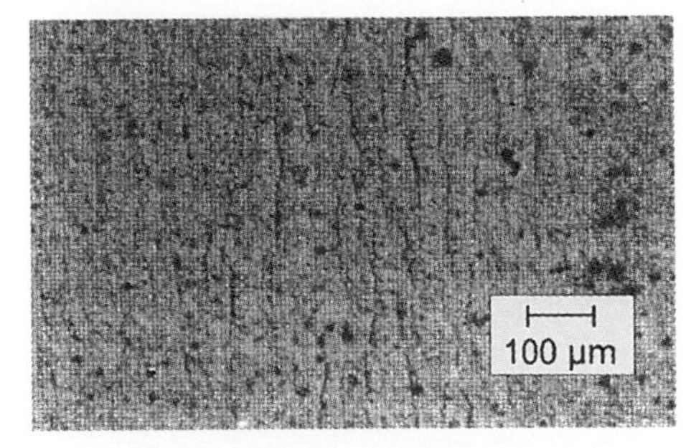

Fig. 6. Surface of the superconductor in a pressed (Tl,Pb)-1223 tape.

Table 2. Critical current densities of long Tl-1223 tapes.

Preparation	Length (m)	Core J_c (kA/cm^2)	Effective J_c (kA/cm^2)	Reference
rolled monofilament	24	10.0	1.6	[10]
pressed 19-filament	1.3	9.2	1.8	[11]
rolled 37-filament	1.3	12.0	2.4	[11]
pressed monofilament	1.2	12.0	3.4	this work

the tapes was almost identical for both materials. The main difference was a higher tendency to the formation of cracks in the lead substituted material (compare figures 3(b) and 6).

3.3 Long tapes

We have prepared long samples by sequential pressing, using dies with a length of 14 cm and a slight curvature. The tapes were moved forward in steps of 3 cm and pressed with a force of 250 kN. The critical current density of these tapes was 12 kA/cm^2. The sequentially pressed tapes were almost homogeneous over the total length and showed a performance in magnetic fields similar to that of short pressed tapes. The superconducting core fraction of 30 % led to a high effective J_c (I_c /total cross section) of 3.4 kA/cm^2 over the total length. To our knowledge this is the highest effective J_c obtained so far for long Tl-1223 tapes (table 2).

4. Conclusions

Our investigations show, that pressing reduces the average grain size and also the degree of c axis alignment. However, pressing also densifies the superconducting core of Tl-1223 tapes. This increases J_c of the tapes in zero field and additionally improves the behaviour in magnetic fields. This is true for both examined materials, (Tl,Bi)-1223 and (Tl,Pb)-1223. These improvements can be transferred to long tapes by applying sequential pressing.

Acknowledgements

The measurements in high magnetic fields were performed at Siemens AG Corporate Technology. This work was supported by the Bayerische Forschungsstiftung through the Bayerischer Forschungsverbund Hochtemperatur-Supraleiter (FORSUPRA).

References

[1] R.S. Liu, S.F. Hu, D.A. Jefferson and P.P. Edwars, *Physica C* **198**, 318 (1992).
[2] T. Nabatame, Y. Saito, K. Aihara, T. Kamo and S.P. Matsuda, *Jpn. J. Appl. Phys.* **32**, L484 (1993).
[3] W. Mexner, J. Hoffmann, S. Heede, K. Heinemann, H.C. Freyhardt, F. Ladenberger, E. Schwarzmann, *Z. Phys. B* **101**, 181 (1996).
[4] S. Wu, B.A. Glowacki and W.Y. Liang, "Improvement of Critical Current Density and Microstructure of Tl-1223 Tapes by a Two-Powder Method", presented at ASC, Pittsburgh 1996.
[5] V. Selvamanickam, T. Finkle, K. Pfaffenbach, P.Haldar, E.J. Peterson, K.V. Salazaar, E.P. Roth, J.E. Tkaczyk, *Physica C* **260**, 313 (1996).
[6] J.C. Moore, S. Fox, C.J. Salter, A.Q. He and C.R.M. Grovenor, "Study of Tl-1223 for use in wire and tape fabrication", in Applied Superconductivity 1995, Vol. 1, ed. by D. Dew-Hughes, Bristol: Institute of Physics Publishing (1995).
[7] R.E. Gladyshevsky, A. Perin, B. Hensel and R. Flükiger: "Preparation and physical charcterisation of Tl(1223) tapes with j_c (77K,0T) > 10 kA/cm^2)", in Applied Superconductivity 1995, Vol. 1, ed. by D. Dew-Hughes, Bristol: Institute of Physics Publishing (1995).
[8] Z.F. Ren, C.A. Wang, D.J. Miller and K.C. Goretta, *Physica C* **247**, 163 (1995).
[9] M. Okada, K. Tanaka and T. Kamo, *Jpn. J. Appl. Phys.* **32**, 2634 (1993).
[10] Z.F. Ren and J.H. Wang, *Appl. Phys. Lett.* **61**, 1715 (1992).
[11] V. Selvamanickam, K. Pfaffenbach, D. Kirchhoff, M. Gardner, D.W. Hazelton and P. Haldar "Development of Tl-1223 and Y-123 Conductors for 77 K HTS Applications", presented at ASC Pittsburgh 1996.
[12] Z.F. Ren, J.H. Wang, D.J. Miller and K.C. Goretta, *Physica C* **229**, 137 (1994).

Inst. Phys. Conf. Ser. No 158
Paper presented at Applied Superconductivity, The Netherlands, 30 June–3 July 1997

The deposition of oxide buffer layers using a novel laser ablation source

T J Jackson, C H Wang, B A Glowacki, J A Leake, R E Somekh and J E Evetts

Department of Materials Science and Metallurgy, University of Cambridge, Pembroke Street, Cambridge, CB2 3QZ, UK

Abstract. The growth of single crystal/highly textured films of oxides on metals has become of considerable interest recently as a building block to the exploitation of conductors of YBCO based HTS materials. This study describes work in which oxide films are deposited by pulsed laser ablation onto previously deposited, single crystal metal thin films. The metal films are sputter deposited onto single crystal oxide substrates in a linked, but separate, UHV chamber. The samples are then transferred to the laser ablation chamber. Different levels of oxidation are used in the laser ablation process. In its most novel form, the oxide films are grown using a sub-millisecond oxygen pulse synchronised with the ablating laser pulse. In this way the overall oxygen exposure of the metal is minimised and controlled. Other background gases have been introduced to change the environment of the growing oxide films. Epitaxial (002) $SrTiO_3$ has been grown directly on (002) Ni, with a 0.48^0 rocking curve; this work is being extended to YSZ and CeO_2 where there is a greater tendency to grow with a (111) orientation.

1. Introduction

To date, most work on buffer layer oxides has focussed on deposition of the buffer layer onto polycrystalline or textured metal substrates [1, 2, 3]. In this work, single crystal metal layers sputtered onto single crystal oxide substrates are used as the base layer, in order to explore some of the issues fundamental to the growth of the buffer layer oxide. One of the key issues is the supply of oxygen to the oxide material being deposited. For the deposition of many oxide materials, a second source of oxygen in addition to the oxygen derived from the target is required to achieve the desired structure and stoichiometry. This necessarily results in exposure of the metal surface to oxygen too, leading to oxidation of the surface or even of the body of the metal if the diffusivity of oxygen is sufficiently high at the temperatures involved.

In the present work, we are interested in growing buffer layers of c-axis, or (002), oriented strontium titanate ($SrTiO_3$), yttrium stabilised zirconia (YSZ) and cerium oxide (CeO_2) on (002) Ni. The presence of (111) NiO at the oxide/metal interface inhibits the nucleation of the buffer layer oxide with (001) orientation [2]. It is important, therefore, to minimise the amount of oxygen to which the metal surface is exposed. In this work, Ni films are first deposited in ultra high vacuum conditions by dc planar magnetron sputtering onto (001) $LaAlO_3$ and MgO substrates. The samples are then transferred into a linked laser ablation chamber without breaking UHV conditions for deposition of the buffer layer oxide.

The ablation may be carried out under a variety of conditions designed to ensure sufficient oxygen supply to the buffer layer oxide being deposited while minimising the partial pressure of oxygen in the deposition system. Depositions may be carried out under UHV or in flowing argon or forming gas (4% H_2 in Ar). The most novel mode of operation is with a sub-millisecond oxygen pulse, aimed towards the target and synchronised with the firing of the laser. The pulsed oxygen source has been described in detail elsewhere [4] so will be considered here only briefly. Two simultaneous electronic trigger pulses are produced by a driver unit. One trigger pulse opens the pulsed oxygen valve. The other trigger pulse passes through an electronic delay unit before firing the laser. The delay is set by using a gated photodiode to measure the brightness of the ablation plume. With the optimum delay time the oxygen pulse meets the ablation plume under the target, causing a visible increase in the brightness of the plume due to enhanced oxidation. Figure 1 shows the variation in the (integrated) brightness of the ablation plume from a $SrTiO_3$ target with delay time, for delay times from 0.1 to 300 ms. The duration of the oxygen pulse in these experiments was 201 μs. The optimum delay time, at which the plume is brightest, is 1.1 ms. Such pulses repeated at 1.1 Hz result in a rise in the average, background pressure in the laser ablation chamber of 0.02 Pa.

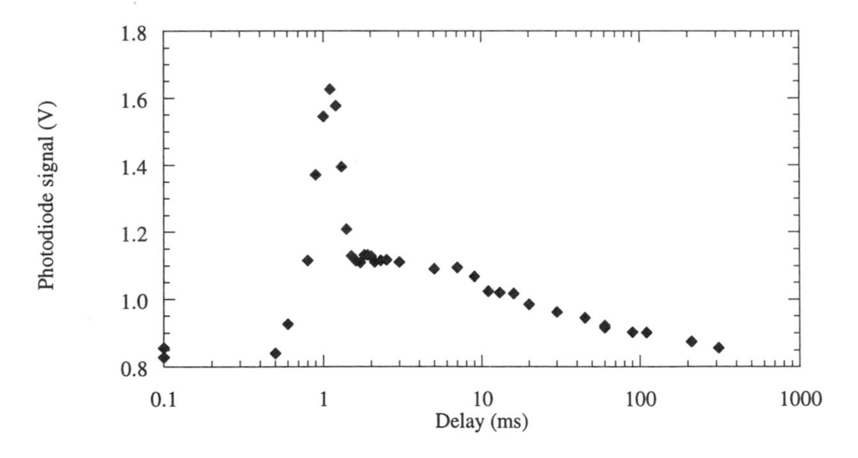

Figure 1 A gated photodiode is used to measure the brightness of the ablation plume. The target for these measurements was $SrTiO_3$. The delay time (x-axis) is the delay between opening the pulsed oxygen valve and firing the laser. The duration of the oxygen pulse was 0.2 ms. The brightness at 0.1 ms delay is the same as that recorded in UHV conditions (no oxygen pulse). The plume is brightest when the oxygen pulse meets the ablation plume beneath the target, at a delay time of 1.1 ms. The oxygen pulse broadens to approximately 2 ms full-width-half-maximum in transit to the target. The brightness at long delay times is thought to be due the presence of a slow component in the oxygen pulse which is formed by collisions and turbulence in the pulse.

Calculations of the Gibbs free energy of formation of bulk ZrO_2, CeO_2 and $SrTiO_3$ [5] show that these oxides are all stable under UHV at the temperatures of interest for thin film deposition. Many oxygen deficient phases are also stable. While NiO is stable under UHV, under thermal equilibrium conditions it is expected to be reduced by the presence of Ce, Zr, Sr or Ti. Thus it may be possible to form an initial "template" layer of the desired buffer layer oxide by laser deposition in UHV. Alternatively a low partial pressure of oxygen may be achieved by deposition in either argon or in forming gas, which presents reducing conditions for NiO if the partial pressures are controlled carefully [5]. We have shown previously that, in the deposition of MgO onto Nb, the niobium is protected from oxygen in the MgO ablation plume if the ablation is performed in an argon ambient [4].

2. Deposition of the buffer layer oxide

In accordance with the arguments presented above, $SrTiO_3$ was deposited onto (001) Ni films with the following conditions of gas supply. The first 20 nm was grown with 660 laser pulses at a repetition rate of 1.1 Hz in a flowing argon ambient of 7 Pa cm (1 Pa of argon with the substrates 7 cm below the target). The following 80 nm were grown with the synchronised oxygen pulses and no argon ambient. The pulses used were 201 μs in duration and also at 1.1 Hz. The substrates were heated to 700 ±20 ^{0}C and the laser fluence was set at 1.5 J cm^{-2}, at which fluence each laser pulse removes approximately 5 nm over the 4x4 mm^2 ablation area.

X-Ray diffraction using the θ-2θ geometry with separate θ drive was first used to investigate the crystalline properties of the films. An in-plane epitaxial ratio I(002)/I(110) of 413 was measured for $SrTiO_3$ on the Ni film described above, with ΔΩ(002) of 0.48^0, compared with ΔΩ(002) of 0.46^0 for the nickel film. Out of plane phi scans of Ni [111] and $SrTiO_3$ [110] are shown in Figures 2a and 2b respectively. The raw data is presented. Figure 2a shows a 4-fold symmetry coincident with that of the MgO substrate [111] peak (not shown), which evidences the single crystal nature of the Ni film. The symmetry in figure 2b for the $SrTiO_3$ [110] phi scan is less clear cut, confirming that the epitaxy is less well developed.

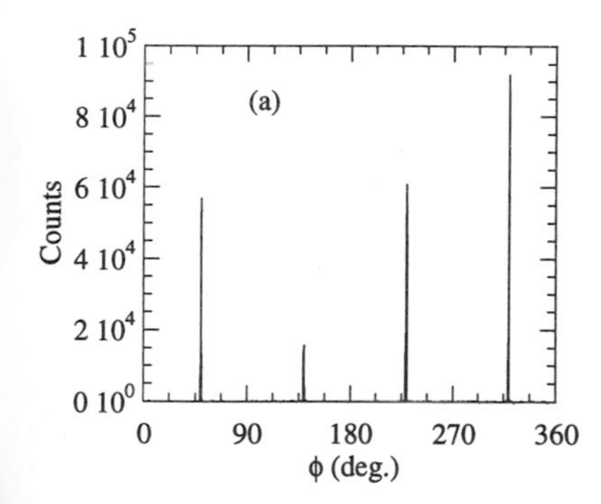

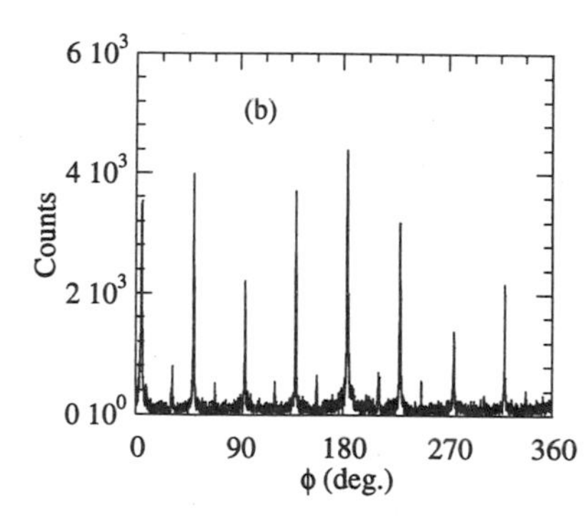

Figure 2: X-Ray diffraction phi scans (a) Ni [111] and (b) $SrTiO_3$ [110] showing the degree of epitaxy in the two layers. The peaks are uneven in height because of the offset between the Ni [002] direction and the substrate normal. The offset, which cannot be corrected for in our diffractometer, is 0.98^0.

More information can be gained from study of deposition conditions further removed from optimum. For example, when the initial template layer was grown in UHV, not in argon, the in-plane epitaxial ratio was reduced by two orders of magnitude. "UHV" refers to initial outgassing rates for oxygen and water of 5x10^{-9} Pa s^{-1} and 2x10^{-8} Pa s^{-1} respectively. The presence of NiO was not detected by X-Ray diffraction in these $SrTiO_3$/Ni bilayers.

The presence of NiO has been detected after the deposition of YSZ and CeO_2 under conditions similar to those described above. The diffusivity of atomic oxygen in YSZ is of the order of 10^{-6} cm^2 s^{-1} at 1000 K, compared with 10^{-9} cm^2 s^{-1} in CeO_2 and 10^{-10} cm^2 s^{-1} in $SrTiO_3$ [6]. For example, YSZ was deposited by two different routes. In the first experiment, a 40 nm thick initial template layer was deposited in an argon ambient of 7 Pa cm, followed by a further 80 nm deposited with the synchronised oxygen pulses. X-Ray diffraction showed that this procedure, similar to that used successfully for $SrTiO_3$, was insufficient to prevent oxidation of the nickel at the Ni/YSZ interface. In the second experiment, the full thickness of the YSZ layer was deposited under UHV conditions, in which case no NiO was detected. In both of these experiments, the YSZ showed a tendency to grow with the (111) orientation, irrespective of the different conditions of gas supply during deposition. X-Ray measurements on the YSZ films deposited entirely under UHV conditions showed an in-plane epitaxial ratio I(002)/I(111) equal to 1, and rocking curve widths $\Delta\Omega(002)$ and $\Delta\Omega(111)$ of 0.70^0. A low oxygen ambient may be required to nucleate the (001) orientation more fully; this may be realised by the use of the pulsed oxygen in a forming gas ambient. The importance of the thickness of the initial template layer is also being investigated. In the case of CeO_2, deposition of the initial template layer in UHV led to the formation of NiO at the Ni/CeO_2 interface. The CeO_2 grew with (111) orientation only, with a rocking curve width $\Delta\Omega(111)$ of 0.62^0.

3. Conclusions

Fine tuning of the deposition conditions is necessary for the preparation of epitaxial buffer layer oxides on single crystal metal surfaces. The conditions appear to be strongly material dependent. The issue of sufficient oxygen supply to nucleate the correct phase and orientation of the buffer layer while maintaining the integrity of the metal surface is particularly complex, but is extremely important for conductor development. The multi-gas approach taken in this work provides a means by which this issue may be investigated.

Acknowledgments

This work is supported by the UK Engineering and Physical Sciences Research Council, grant number GR/H/97420.

References

[1] Reade R P, Berdahi P, Russo R E and Garrison S M 1992 *Applied Physics* Letters **61** 2231-3

[2] Norton D P, Goyal A, Budai J D, Christen D K, Kroeger D M, Specht E D, He Q, Saffian B, Paranthaman M, Klabunde C E, Lee D F, Sales B C and List F A 1996 *Science* **242** 755-7

[3] Hawsey R and Peterson D P 1996 *Superconductor Industry* Fall 1996, 23-9

[4] Wang C H, Jackson T J, Somekh R E, Leake J A and Evetts J E 1997 *Measurement Science and Technology* (in press)

[5] Jackson T J (to be published)

[6] Tidrow S C, Wilber W D, Tauber A, Schauer S N, EckartD W, Finnegan R D and Pfeffer R L 1995 *Journa of Materials Research* **10** 1622-34

Inst. Phys. Conf. Ser. No 158
Paper presented at Applied Superconductivity, The Netherlands, 30 June–3 July 1997

Deposition of YSZ, CeO_2, and MgO on amorphous and polycrystalline substrates

M Bauer, J Schwachulla, S Furtner, P Berberich, and H Kinder

TU München, Physik Department E10, 85747 Garching, Germany

Abstract. We deposited YSZ, CeO_2 and MgO thin films on SiO_2/Si, Ni alloy and poly-YSZ substrates by e-gun evaporation. To achieve a biaxial texture, the substrate was inclined with respect to the deposition direction.
Columnar growth was found using Scanning Electron Microscopy (SEM) of fracture cross sections. Pole figure measurements show varying normal and inplane texture. YSZ films were (100) or (111) oriented depending on the substrate temperature with only slight inplane orientation. In case of CeO_2 (111) oriented growth was found with an inplane texture of 26° FWHM in the best case. MgO films were about (211) oriented with the best inplane texture of 22° FWHM.
YBCO thin films deposited on poly-YSZ substrates with MgO buffer layers showed epitaxial growth. Nevertheless, the j_c was not improved, compared to the j_c of films with no inplane orientation because of the (211) orientation of the buffer layer.

1. Introduction

For the deposition of high quality high T_c superconductor films such as $Y_1Ba_2Cu_3O_7$ (YBCO) on polycrystalline or amorphous substrates like metal tapes biaxially oriented buffer layers have to be used. Several methods to deposit these buffer layers have been published recently. A very simple and fast method is the inclined substrate deposition (ISD) using pulsed laser deposition [1]. Very high deposition rates being 0.5 μm/min were published for ISD. Therefore, this deposition technique is suitable for the deposition on large areas as for example long metal tapes. Electron-beam evaporation is a very well known deposition method that also has the capability to deposit with a very high rate. In this paper we report on biaxially aligned yttria stabilised zirconium oxide (YSZ), CeO_2, and MgO thin films that were deposited with ISD using e-beam evaporation.

2. Experiments

We used a standard e-beam evaporation source. The target material were pieces of sintered YSZ (11 wt.% Y_2O_3), CeO_2 or MgO respectively. The substrate was clamped to a Cu heater block or radiation heated by a small oven surrounding the substrate. It could be inclined by an angle α ranging from 0° up to 70° to deposit under these angles. The deposition rate for ISD in the direction perpendicular to the substrate surface was 250...500 nm/min measured by a quartz crystal monitor with a deposited film thickness between 0.4 μm and 1.0 μm.

Oxygen was introduced so that the oxygen partial pressure at the substrate position was 4×10^{-4} mbar with a background pressure below 10^{-6} mbar. Silicon with 300 nm thermally grown amorphous SiO_2 was used as a substrate for most of the films. Additionally, some films were deposited onto polished Ni alloy (Hastelloy C®) and polycrystalline YSZ substrates. These films showed similar pole figures. For the SEM fracture cross sections the SiO_2/Si substrates were broken parallel to the deposition direction and a thin Au layer was deposited to avoid charging of the substrate. In order to investigate the influence of the substrate, we deposited thin films on inclined ($\alpha > 0$) as well as not inclined ($\alpha = 0$) substrates.

3. Results for YSZ buffer layers

YSZ films deposited with $\alpha = 0$ have two different orientations measured by X-ray Θ–2Θ diffraction. At $T_{substrate} < 250°C$ the films grow (100) oriented changing to a (111) orientation at higher substrate temperatures up to 550°C.

Nearly the same orientation is found in films deposited with an inclined substrate. In case of the low substrate temperature polefigure measurements reveal a (100) orientation that is slightly inclined towards the deposition direction. Furthermore, the film is only very poorly inplane oriented. On the other hand, films deposited at high substrate temperatures are (111) oriented (figure 1) with the [111] axis being slightly tilted by an angle of 16° towards the deposition direction, too. The inplane texture of the YSZ [-111] peak is 50° FWHM.

A fracture cross section of a YSZ film deposited at high substrate temperature is shown in figure 2. Growth columns that are tilted towards the deposition direction can be seen clearly, which is the case for all films investigated. However, the tilt angle is always smaller than the substrate inclination angle α. This is due to self shadowing effects and was reported to be the case for many other deposited materials, too [2].

4. Results for CeO_2 buffer layers

The orientation of CeO_2 films deposited on a not inclined substrate is different to that of YSZ films. At substrate temperatures below 200°C the films grow (111) oriented changing to a (110) orientation at medium substrate temperatures. At substrate temperatures above 450°C up to 600°C a (100) oriented growth is observed.

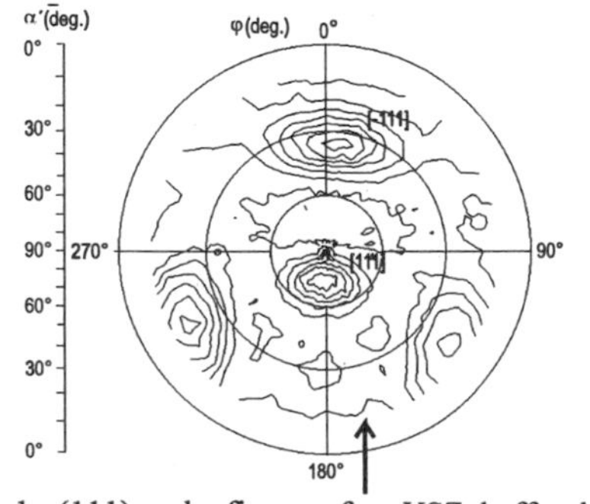

Fig. 1: (111) pole figure of a YSZ buffer layer deposited at 400°C and α = 55°. The arrow indicates the deposition direction.

Fig 2: SEM image of the fracture cross section of a YSZ buffer layer deposited at 400°C and α =55°.

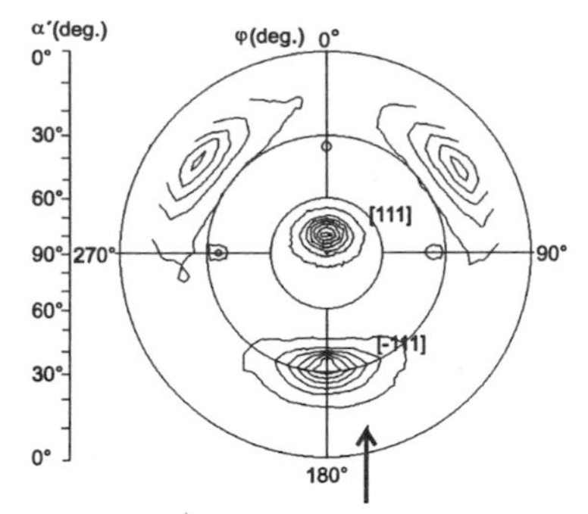

Fig. 3: (111) pole figure of a CeO_2 buffer layer deposited at 160°C and α = 55°.

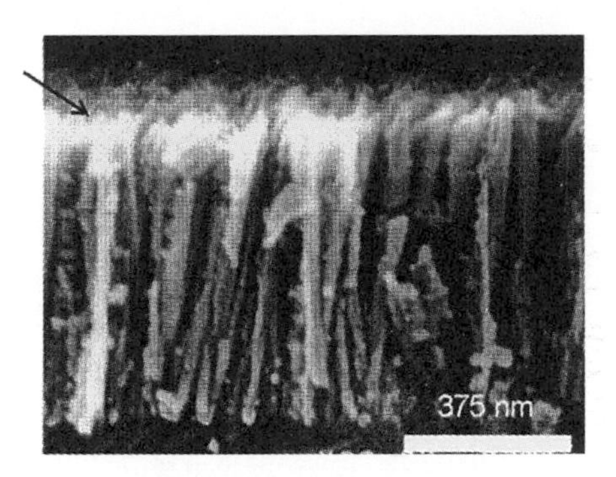

Fig. 4: SEM image of the fracture cross section of a CeO_2 buffer layer deposited at 160°C and α =55°.

In contrast to the results for YSZ, the orientation changes as the substrates are inclined. A typical pole figure of CeO_2 (111) is shown in figure 3. The film is (111) oriented with an inplane texture of 24° FWHM. The fracture cross section of this film is shown in figure 4. It can be seen clearly that CeO_2 grows in a columnar growth mode but in this case the columns are not inclined towards the evaporation source, but even slightly against it. The surface of the film consists of planes with the normal direction about parallel to the deposition direction. In comparison with the polefigure it is obvious that these are the (-111) crystal planes or vicinal (-111) planes.

5. Results for MgO buffer layers and YBCO on MgO buffer layers

MgO films deposited on not inclined substrates are predominantly (111) oriented at all temperatures. As the substrate is inclined, the orientation changes to a tilted (100) orientation with inplane texture. A typical pole figure of MgO (200) is shown in figure 5. The centre (100) peak is tilted towards the deposition direction by an angle β = 40°. This is about equivalent to the (211) orientation where β equals 35°. The FWHM Ω of the [010] and [001] peak respectively is 22° which is the best value measured so far. A fracture cross section (figure 6) of this film again reveals columnar growth with vertical columns even at a deposition angle α of 55°. The top of the columns is the (100) or vicinal (100) crystal plane oriented towards the deposition source. The value of Ω decreases with increasing substrate inclination whereas the tilt angle β increases. If the substrate temperature is increased above 500°C, Ω starts to decrease.

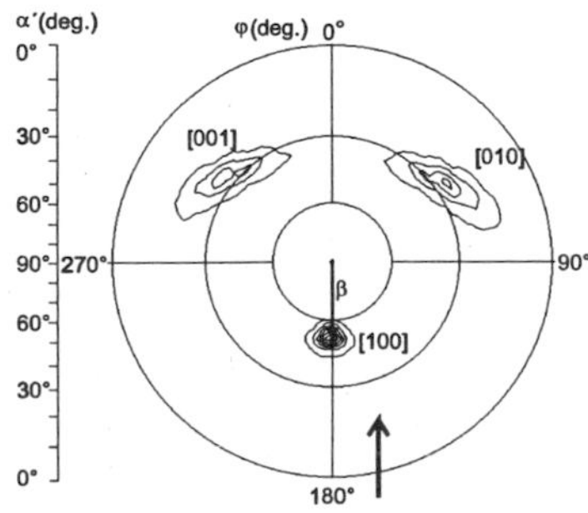

Fig. 5: (200) pole figure of a MgO buffer layer deposited at 300°C and α = 55°.

Fig. 6: SEM image of the fracture cross section of a MgO buffer layer deposited at 160°C and α =55°.

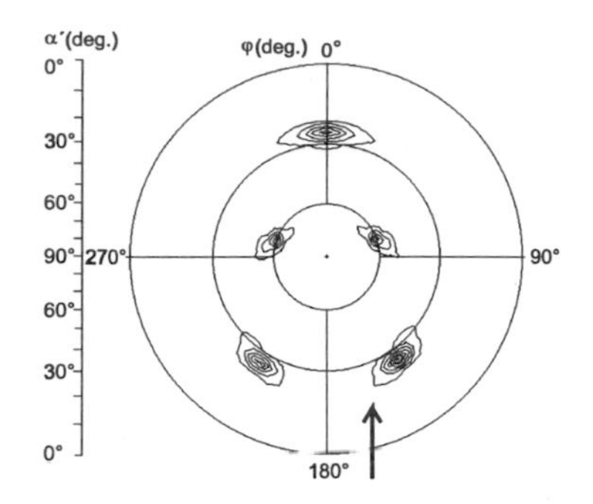

Fig. 7: (103) pole figure of a YBCO film deposited on ISD MgO/poly-YSZ substrate.

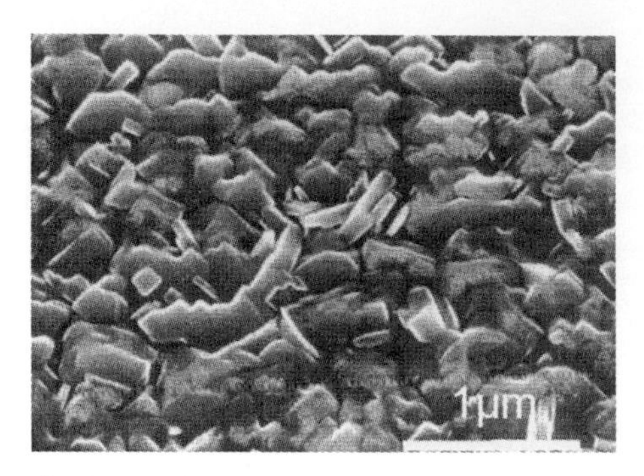

Fig. 8: Top View SEM image of the YBCO film on ISD MgO/poly-YSZ substrate.

Both CeO_2 and MgO show a similar growth mode for ISD. It can be characterised by columnar growth, where the top of the columns is about a low indexed crystal plane oriented towards the deposition source. In case of CeO_2 this is the (111) crystal plane, in case of MgO the (100) plane.

In contrast to YSZ at high temperatures and CeO_2, which have an orientation unsuitable for YBCO film growth, our ISD MgO films can be used as buffer layers for YBCO film growth. After an epitaxial 500 nm thick MgO film was deposited onto the ISD MgO at 700° C and low deposition rates, the 500 nm thick YBCO film was grown using thermal coeporation that is described elsewhere [3]. The YBCO (103) pole figure (figure 7) clearly shows epitaxial growth on the MgO buffer layer with an inplane texture of 20° FWHM. In spite of this inplane orientation the j_c of 20 kA/cm² at 77 K (T_c = 84 K) is not higher than the j_c of c-axis YBCO films with no inplane texture. This is most likely due to the large tilt angle β that results in a poor connectivity of the YBCO grains and also a large amount of small a-axis oriented grains visible in the top view SEM image (figure 8).

6. Conclusion

We have shown that it is possible to grow biaxially oriented YSZ, CeO_2, and MgO thin films using e-beam evaporation on inclined substrates. YSZ films at low substrate temperatures show only a poor biaxial texture. At high substrate temperatures YSZ films are (111) orientated as well as all CeO_2 films. Therefore, these films are not suitable as buffer layers for HTSC thin films.

On the contrary, MgO is about (211) textured with an inplane FWHM as small as 22°. It is possible to grow YBCO epitaxially on these buffer layers. Due to the (211) texture, the j_c of the films is not increased in comparison to not textured films. Further improvements have to be done.

References

[1] Hasegawa K, Yoshida N, Fujino K, Mukai H, Hayashi K, Sato K, Ohkuma T, Honjyo S, Ishii H and Hara T 1996 *Proc. of the ICEC 16, May 1996, Kitakyushu, Japan*

[2] Leamy H J and Gilmer G H 1980 *Current topics in Material Science, Vol. 6, ed. E. Kaldis (North-Holland, Amsterdam)* 309-344

[3] Berberich P, Utz B, Prusseit W and Kinder H, *Physica C* **219** 497-504

Inst. Phys. Conf. Ser. No 158
Paper presented at Applied Superconductivity, The Netherlands, 30 June–3 July 1997

Growth mechanisms for in-plane aligned YSZ buffer layers deposited on polycrystalline metallic substrates by ion-beam assisted laser deposition

V Betz[1], B Holzapfel[1], G Sipos[2], W Schmidt[2], N Mattern[1] and L Schultz[1]

[1]IFW Dresden, Helmholtzstr. 20, D-01069 Dresden, Germany
[2]Siemens AG, Research Laboratories, P.O. Box 32 20, 91050 Erlangen, Germany

Abstract. Highly textured yttria stabilized zirconia (YSZ) is deposited on mechanically polished Ni-based substrates using ion-beam assisted pulsed laser deposition. With pulsed laser deposition in off-axis geometry a thin intermediate CeO_2 layer and an $YBa_2Cu_3O_{7-d}$ (YBCO) layer are deposited on the biaxially aligned YSZ films. Depending on deposition parameters, in-plane orientations of 9° and 4.4° full-width at half maximum are reached for YSZ and YBCO, respectively, resulting in a critical current density of 2.3×10^6 A/cm². We report on superconducting properties of the YBCO films and the influence of deposition parameters on YSZ film quality. For YSZ films deposited at room temperature, we found a small decrease in deposition rate due to the ion bombardment of less than 10%, suggesting the aligning process is not due to anisotropic sputtering. From specially deposited YSZ films we found that film grains cannot change their respective in-plane growth direction. The successful biaxial alignment of a novel buffer layer material (Pr_6O_{11}) is shown.

1. Introduction

Large scale applications based on high temperature superconductors such as transmission cables or high field magnets require the preparation of long wires and tapes with high critical currents [1]. For the $Bi_2Sr_2Ca_2Cu_3O_{10+\delta}$ system this can be achieved by the oxide-powder-in-tube (OPIT) process, but high field applications of Bi-based superconductors are limited to low temperatures (< 35 K) due to their high anisotropy [2] resulting in very low irreversibility fields at high temperatures. $YBa_2Cu_3O_{7-x}$ (YBCO) thin films prepared on single crystal substrates show high critical current densities > 10^6 A/cm² at 77 K with a weak magnetic field dependence and a low J_c-anisotropy [3]. However, YBCO films on polycrystalline metal substrates for long tape applications show large angle grain boundaries leading to a strong weak-link behaviour drastically reducing critical current densities in magnetic fields [4]. The growth of biaxially oriented buffer layers on polycrystalline metal substrates by ion-beam assisted deposition methods is one promising route to achieve biaxially textured YBCO films with a subsequently high critical current density [5-9].

In this paper, we present yttria stabilised zirconia (YSZ) and Pr_6O_{11} films grown by ion-beam assisted pulsed laser deposition (IBALD) with and without seedlayers produced by an intermediate substrate rotation during film deposition. From this experiments, new findings are derived towards the still not well understood [10] underlying film-growth mechanisms in the ion-beam assisted deposition process. On YSZ films, YBCO is deposited with an intermediate thin CeO_2 layer to avoid YBCO misorientations due to the high lattice mismatch between YSZ and YBCO [11].

2. Film growth

YSZ and Pr_6O_{11} buffer layers were deposited at room temperature by IBALD in a deposition chamber similar to a previously described [12]. A KrF-excimer laser is used to ablate a stoichiometric YSZ (9.5 mol% Y_2O_3) or Pr_6O_{11} target applying an energy density of 2-5 J/cm² and a laser repetition rate of 5 to 15 Hz. The target-to-substrate distance was set to 90 mm. Deposition rates were measured without assisting ion-beam using a quartz monitor mounted at substrate position. The resputter rate of the ion-beam during deposition was determined from the difference of expected and measured film thickness. An RF plasma source with a mean beam energy of 300 eV and 50 μA/cm² ion flux density was used as assisting source. The ion-beam divergence of the source is measured to 5° FWHM using a two-stage faraday-cup assembly [13]. A mixture of argon and oxygen (ratio 2.2:1) was used as ion gas. Due to ion gas flow, the chamber base pressure of 1×10^{-6} mbar increased to 3.7×10^{-4} mbar during film deposition. Both, YBCO and CeO_2 films were deposited by off-axis PLD [14] at 760°C in 0.4 mbar O_2-atmosphere. Some of the YBCO films have been prepared by standard PLD under similar conditions.

3. Results

3.1. IBALD film growth

All grown YSZ and Pr_6O_{11} films show a columnar microstructure [15] with an exclusive (001)-orientation and a certain in-plane alignment, mainly dependent on film thickness and ion-to-atom ratio. The degree of in-plane orientation strongly increases with film thickness up to about 1 μm. Since measured with x-ray analysis, the value is related to the whole film thickness, the surface orientation is somewhat better and is represented by the orientation of a heteroepitaxially on the YSZ film grown CeO_2 layer (Fig. 1 a)).

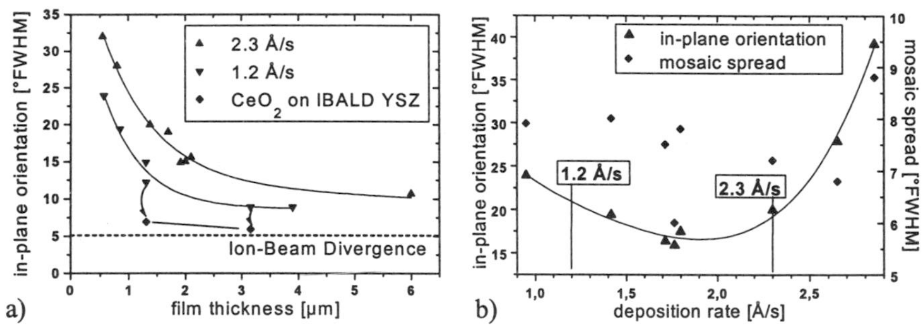

Fig. 1: Dependence of the achieved in-plane orientation on a) film thickness and b) deposition rate. For all experiments shown in figure 1 b) the same deposition time was used.

For a growth rate of 1.2 Å/s, the degree of in-plane orientation approaches 5° FWHM (fig. 1 a), the value of the internal ion-beam divergence [16]. A variation of the deposition rate, i.e. the ion-to-atom ratio, leads to a minimum of the in-plane orientation width at about 1.9 Å/s for the used ion-beam parameters. This rate is the optimum to achieve a maximum orientational order within a certain time. A slower growth rate only increases the time needed to achieve a desired orientation, whilst a faster growth rate limits the best achievable orientation (fig. 1a). The best achieved in-plane orientations are 8.9° FWHM for a 3.2 μm YSZ film, corresponding to 4.4° FWHM in the subsequently deposited YBCO. For Pr_6O_{11}, 16° FWHM at about 1 μm film thickness were obtained (fig. 2).

For a study of the growth mechanisms during the IBALD process, YSZ films were grown with a 45° substrate rotation after 90 nm, 180 nm, and 550 nm film growth. After rotation, the films were grown to a final thickness of 1100 nm. The in-plane orientational order was measured with x-ray analysis in standard geometry and under gracing incidence. The smaller penetration depth under gracing incidence allows for an evaluation of the surface-near film orientation. Figure 3 shows the respective ϕ-scans for the grown YSZ films.

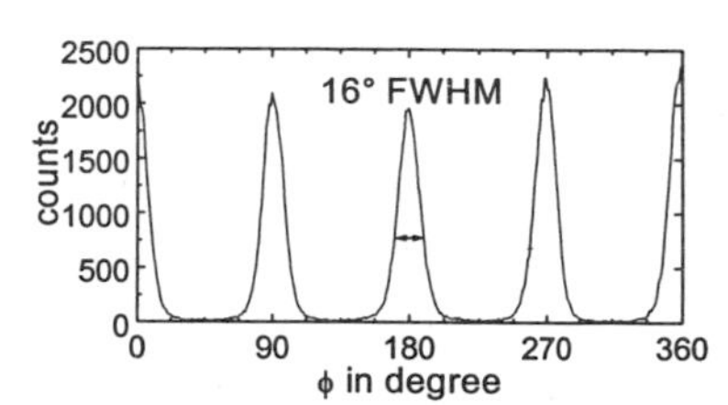

Fig. 2: Pr_6O_{11} film grown by IBALD. x-ray ϕ-scan for ψ=55°.

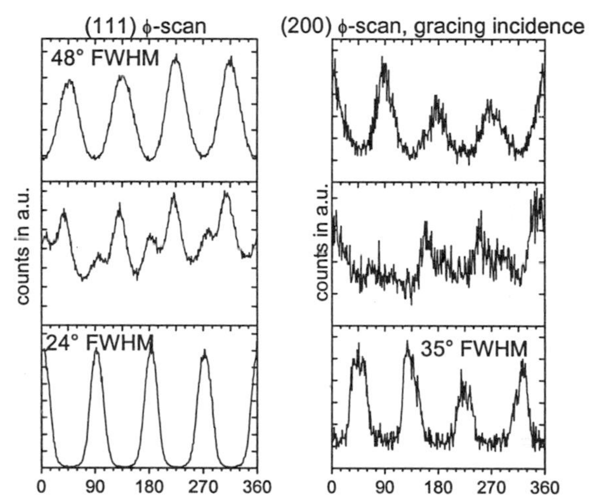

Fig. 3: X-ray analysis of 1100 nm thick YSZ films grown by IBALD with an intermediate 45° substrate rotation at 90 nm (a), 180 nm (b), and 550 nm (c). Left: ϕ-scans of the YSZ (111) pole for ψ=55°. Right: ϕ-scans of the YSZ (200) pole under gracing incidence. Ion-beam incidence is along ϕ=0° before, and ϕ=45° after substrate rotation.

After 90 nm of film growth, the orientational order of the film is only weak, and crystallites of all in-plane orientations can be found. After rotation of the substrate, a different set of crystallites is selected by the ion-beam and starts to dominate the texture. Due to the seedlayer, the reached in-plane alignment of 48° FWHM (fig. 3 a) is not as good as the 16° FWHM (11° FWHM for gracing incidence) obtained in a YSZ film deposited without substrate rotation. If the substrate rotation takes place after 180 nm of film growth, the initial film orientation is visible in the x-ray pattern (small peaks). Still, a small number of crystallites are present under 45° misorientation with respect to the initial ion-beam direction. Those are selected by the new ion-beam direction and try to dominate the film surface orientation (fig. 3 b, comparison of left and right pattern). After 550 nm film growth, the in-plane texture is already well developed, and there are no crystallites to be selected by the ion-beam after substrate rotation. The film orientation therefore corresponds to the initial ion-beam direction (fig. 3c). The homoepitaxial growth of the columns is retained. No new nucleation sites are created during IBALD film growth in all conducted experiments. Due to a lack of ion-beam assistance along the YSZ (111) direction, the in-plane orientation decreases during the film growth after substrate rotation (fig. 3 c). Although the resputter rate due to the assisting ion-beam was found to be less than 10%, this study is coherent with the phenomenological findings of the growth and extinction model proposed by Sonnenberg [15]. Further studies to clarify the driving force for the growth selection are under way.

3.2. YBCO films

On the grown YBCO films, T_c was determined from AC-susceptibility measurements and varied between 88.5 K and 89.5 K for all samples. J_c was determined by magnetisation measurements and transport measurements across a 15 μm wide bridge (fig. 4). The best value of $2.3 \times 10^6 A/cm^2$ was achieved for an YBCO film which had an in-plane orientation of 4.4° FWHM. This film was deposited in standard PLD geometry and had a thickness of 250 nm.

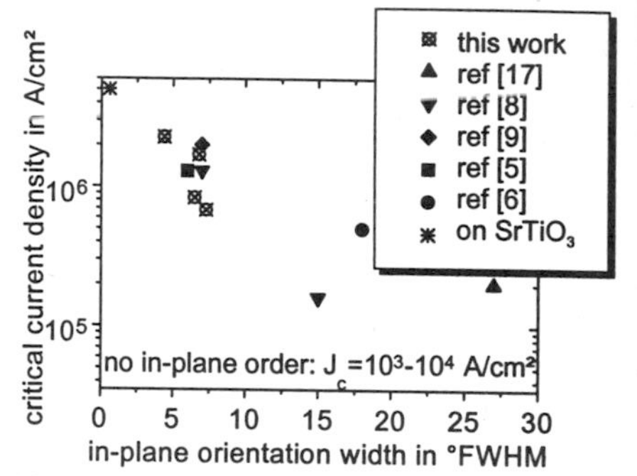

Fig. 4: Dependence of the critical current density on the film in-plane orientation.

4. Conclusions

In summary, we have successfully demonstrated the growth of biaxially aligned YSZ and Pr_6O_{11} films. In-plane orientations of 8.9° and 16° FWHM were achieved, respectively. Growth studies on IBALD YSZ showed the aligning process to be growth controlled rather than nucleation controlled. No new nucleation sites are formed during IBALD. On YSZ deposited YBCO films showed in-plane orientations of up to 4.4° FWHM, resulting in a J_c of $2.3 \times 10^6 A/cm^2$.

References

[1] D C Larbalestier and M P Maley, *MRS Bull.* **XVIII**, 50 (1993).

[2] P Schmitt, P Kummeth, L Schultz, and G Saemann-Ischenko, *Phys. Rev. Lett.* **267** (1991).

[3] B Roas, L Schultz and G Saemann-Ischenko, *Phys. Rev. Lett.* **64**, 479 (1990).

[4] D Dimos, P Chaudhari, J Mannhart, and F K LeGoues, *Phys. Rev. Lett.* **61**, 219 (1988).

[5] X D Wu, S R Foltyn, P N Arendt, W R Blumenthal, I H Campbell, J D Cotton, J Y Coulter, W L Hults, M P Maley, H F Safar, and J L Smith, *Appl. Phys. Lett.* **67**, 2397 (1995).

[6] Y Iijima, K Onabe, N Futaki, N Tanabe, N Sadakata, and O Kohno, Y Ikeno, *J. Appl. Phys.* **74**, 1905 (1993).

[7] R P Reade, P Berdahl, and R E Russo, S M Garrison, *Appl. Phys. Lett.* **61**, 2231 (1992).

[8] A Knierim, R Auer, J Geerk, G Linker, O Meyer, H Reiner, R Schneider, *Appl. Phys. Lett.* **70**, 661 (1997).

[9] H C Freyhardt, J Dzick, K Heinemann, J Hoffmann, F Garcia Moreno, A Usoskin, J Wiesmann, *IEEE Trans. Appl. Supercond.*, in press (1997).

[10] R M Bradley, J M E Harper, and D A Smith, *J. Appl. Phys.* **60**, 4160 (1986).

[11] S. M Garrison, N Newman, B F Cole, K Char, and R W Barton, *Appl. Phys. Lett.* **58**, 2168 (1991).

[12] V Betz, B Holzapfel and L Schultz, *Thin Solid Films*, in press (1997).

[13] Ch Huth, H-C Scheer, B Schneemann, and H-P Stoll, *J. Vac. Sc. Technol. A* **8**, 4001 (1990).

[14] B Holzapfel, B Roas, L Schultz, P Bauer, and G Saemann-Ischenko, *Appl. Phys. Lett.* **61**, 3178 (1992).

[15] N Sonnenberg, A S Longo, M J Cima, B P Chang, K G Ressler, P C McIntyre, and Y P Liu, *J. Appl. Phys.* **74**, 1027 (1993).

[16] V Betz, B Holzapfel, and L Schultz, *IEEE Trans. Appl. Supercond.*, in press (1997).

[17] F Yang, E Narumi, S Patel, D T Shaw, *Physica C* **244**, 299 (1995).

Inst. Phys. Conf. Ser. No 158
Paper presented at Applied Superconductivity, The Netherlands, 30 June–3 July 1997

Observation and investigations of a nucleation layer in high-J_c YBCO films deposited on polycrystalline substrates

A Usoskin[1], F García-Moreno[1], H C Freyhardt[1,2], D Jockel[1], J Wiesmann[2], K Heinemann[2], A Issaev[2], J Dzick[2] and J Hoffmann[2]

[1]Zentrum für Funktionswerkstoffe Gött. gGmbH, Windausweg 2, D-37073 Göttingen, Germany
[2]Institut für Metallphysik, Univ. of Göttingen, Windausweg 2, D-37073 Göttingen, Germany

Abstract. A relatively high thickness threshold of (200-400) nm was found in the dependences of J_c versus film thickness and was interpreted (and confirmed by TEM cross-sectional analysis) in terms of a nucleation layer which appears at the initial stage of film growth and does not possess a high J_c . It was shown that in this nucleation stage misoriented YBCO crystallites grow at the surface of unbuffered as well as buffered polycrystalline substrates. Nevertheless, during further film growth these unfavourably oriented grains are overgrown leading to an excellent in-plane orientation of the YBCO layer. A competition between these differently oriented nucleation crystallites as well as the suppression of the unfavourable growth is described in the frame of a model taking into account a kinetics of surface material transfer and its dependence on efficiency of adatom trapping by alternatively oriented nucleation centers. The true critical current densities which can be carried by the top part of the HTSC film with a higher microstructure perfection are estimated, at least, to (3-4)×10^6 A/cm^2.

1. Introduction

Wide fields of investigations and applications of HTSC films are connected now with $YBa_2Cu_3O_{7-x}$ (YBCO) films deposited on polycrystalline technical substrates which allow a significant increase of film surface dimensions [1, 2]. High J_c of ~ 2 MA/cm^2 recently achieved in YBCO films grown by pulsed laser deposition on Ni foils [3] demonstrates a realistic prospects of their technical use.

One can find a number of differences in structural and conducting properties of YBCO films grown on polycrystalline substrates with or without pre-textured buffer layer in comparison with the same films grown on single crystalline substrates. One of them, a considerably higher effect of the nucleation layer on the superconducting parameters, was pointed out in the course of our previous study [2].

The present study is aimed to the investigation of the kinetics of seeding, growth and overgrowth of a nucleation layer during YBCO film deposition and of the influence of these structural modifications on the critical current density.

2. Experimental details

YBCO films were prepared by pulsed-laser-deposition employing an excimer laser Lambda-3308 (0.5 J, 308 nm, 25 ns, 300 Hz) and a quasi-equilibrium heating technique described elsewhere [2]. 0.1-0.2 mm thick yttria-stabilized zirconia (YSZ) ceramic ribbons as well as polycrystalline Ni foils preliminary covered with ~ 300 nm thick YSZ buffer layer obtained by ion-beam-assisted-deposition (IBAD) were used as substrates. Unbuffered YSZ ribbons were also employed for the experiments.

The critical current densities, J_c , and transition temperatures, T_c , of the films were measured by a dc method as well as by an inductive ac (100 kHz) technique based on the effect of magnetic shielding. Both of these methods yield similar results.

3. Results and discussion

Typical dependences of J_c on thickness t for YBCO films on Ni foils and YSZ ribbons in comparison with film on $SrTiO_3$ are shown in Fig. 1. For the investigated polycrystalline substrates, thickness thresholds in the $J_c(t)$ dependences were found in the range of (200-400) nm which are significantly larger than the ~ 10 nm observed for single-crystalline $SrTiO_3$.

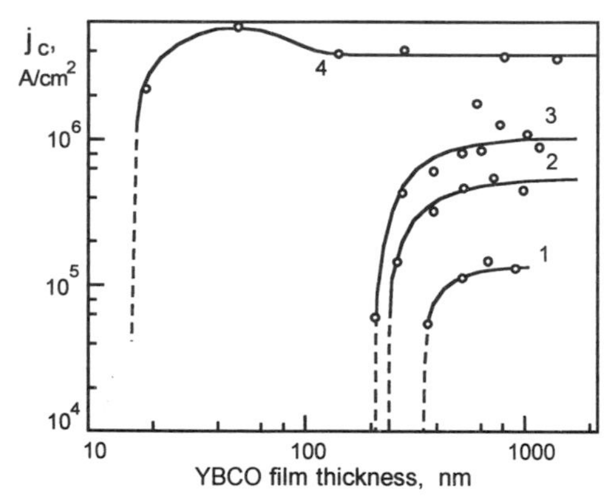

Fig. 1. Dependences of J_c on the thickness t for YBCO films deposited on YSZ ribbons (1), YSZ ribbons with IBAD buffer (2), Ni foils with IBAD buffer (3) and single crystalline $SrTiO_3$ (4).

Such a type of J_c threshold dependences can be successfully described by the following function:

$$J_c(t) = J_{c\infty}(1 + t_d / t)^{-1} \quad (1)$$

where $J_{c\infty}$ is the J_c of the film with $t >> t_d$, t_d is the thickness of the nucleation layer which appears in the beginning of film growth and does not possess a considerable J_c .

A cross-sectional TEM view of YBCO film grown on polycrystalline substrate with an IBAD buffer layer is shown in Fig. 2. The film possess a columnar structure. Each of the columnar crystallites acquires an orientation where the c axis is parallel to the film normal and the a- and b- directions are determined by the buffer texture. Nevertheless, at the first stage of growth, for the thicknesses < 300 nm, one can also find differently shaped crystallites which are randomly oriented. A set of shaped pores seen in the photo corresponds to the fraction of such crystallites which fall out in the course of sample preparations (grinding, etching, etc.)

Fig. 2. Cross-sectional view of YBCO film deposited on YSZ ceramic ribbon with IBAD buffer.

for TEM analysis. Thus, the observed structure inhomogeneity confirms the nucleation (or „dead") layer model which was determined from $J_c(t)$ investigations.

To understand how the deteriorated nucleation structure of YBCO can be self-improved during the further film growth the following model calculations were performed. Let us consider a pair of alternatively oriented nucleation centers at the stage when they get in a direct contact. It is possible to show that during their development they acquire shapes similar to those shown in Fig. 3 a, and can be described by a dependence

$$h(r) = h_0 \left\{ 1 + \ln \left[\left(1 - \frac{\pi}{2} \frac{1+\xi^2}{\xi^2} \left(\frac{r}{L} \right)^2 \right) \left(1 - \frac{\pi}{2} \frac{1+\xi^2}{(1+\xi)^2} \right)^{-1} \right] \right\} \qquad (2)$$

following from the consideration of the kinetics of material transfer of impinging molecular/ion flow over the substrate to the surfaces of nuclei. In equation (2) h and r denote the vertical and radial coordinates of the growing surface (see Fig. 3 a), L the spacing between nucleation centers, h_0 the height of contacting surfaces, and ξ ratio of adatom trapping

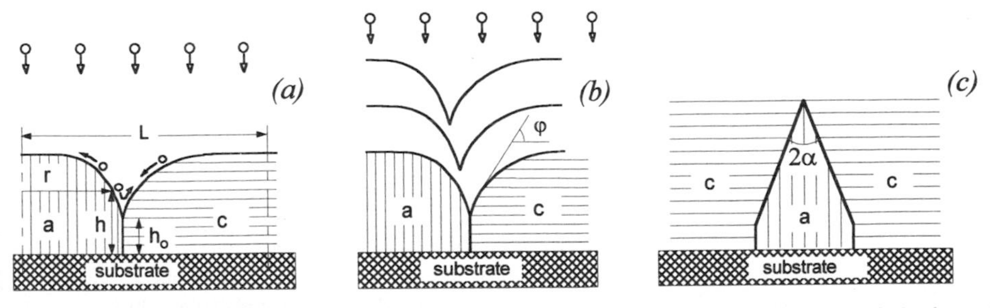

Fig. 3. Schematic views of inter-nuclear valley and adatom diffusion over its surface (*a*), of „horizontal" movement of the valley during further deposition (*b*), and of the final shape of misoriented (a-) crystallite when overgrown by c-oriented surround (*c*). a and c indicate the growth orientations of YBCO nucleation centers.

probabilities for two differently oriented crystalline planes (with c- and a-orientations in the case of simultaneous a- and c-growth, see Fig. 3 a).

As a result, near the contact area of a- and c- grains the film surface acquires the shape of a valley (Fig. 3 a, b). Because of the intrinsic anisotropy of the trapping rates [4] for diffusing adatoms (see Fig. 3 a), the deposition rate at the surfaces which are more near to a- (or b-) crystalline planes is significantly higher (by a factor 3 .. 10) [5]. Obviously, this must cause a „horizontal" shift of the valley as indicated in Fig. 3 b. The shift is directed towards the center of the misoriented crystallite, and due to that it finally acquires a triangular form in the cross-sectional view (see Fig. 3 c). The angle, 2α, at the top of the overgrown a-crystallite can be estimated by a consideration of the relation of the „horizontal"/"vertical" movement rates of the valley depending on the relation of the trapping probabilities ξ, as

$$\alpha = \mathrm{a}\tan\left[\frac{1}{\tan\varphi}\,\frac{\tan\varphi - 1}{\tan\varphi + 1}\,\frac{1-\xi}{1+\xi}\right] \tag{3}$$

where φ is the angle determining the depth of the valley (see Fig. 3 b). The 2α values for the range of $0.1<\xi<0.33$ [4, 5] were amounted to 10°-16° from Eq. (3), while the experimental numbers following from the shape the shaped pores in Fig. 2 correspond to 20°-25°. The calculated values are considered to be in a satisfactory agreement with experiment, especially, taking into account that certain simplifications were implicated in deriving Eq. (3).

In summary, one can conclude that in YBCO films grown on polycrystalline substrates a (200-400) nm-thick „dead" layer appears during film deposition. This nucleation layer practically does not contribute to the superconducting current flow. It partly consists of misoriented YBCO crystallites. Nevertheless, a sufficient number of the YBCO crystallites in this layer acquires a proper growth orientation determined by textured buffer layer. These crystallites serves as seeds for the further growth of a uniform top layer of the HTSC film which is capable of high superconducting current transport reaching J_c levels estimated to $(3\text{-}4)\times10^6$ A/cm^2. A mechanism causing such a selective overgrowth is determined by a kinetics of surface material transfer and its dependence on efficiency of adatom trapping by alternatively oriented crystalline planes of nucleation centers.

Acknowledgement

This work was performed within a development project of Kabelmetal electro GmbH which is supported by German BMBF (Project No. 13N6924/6), and was also supported by Siemens AG together with German BMBF (Project No. 13 N6482).

References

[1] Iijima Y, Onabe K, Futaki N, Tanabe N, Sadakata N, Kohno O and Ikeno Y *J. Appl. Phys.* **74** (1993) 1905

[2] Usoskin A, Freyhardt H C, García-Moreno F, Sievers S, Popova O, Heinemann K, Hoffmann J, Wiesmann J and Isaev A *Applied Superconductivity 1995* IOP Conf. Series **148** (1995) 499

[3] Garcìa-Moreno F, Usoskin A and Freyhardt H C *High Tc Update* **10** (1996) No. 22

[4] Goyal A, Alexander K, Kroeger D, Funkenbusch P and Burns S *Physica C* **210** (1993) 197

[5] Endo A, Chauhan H, Nakamura Y and Shoihara Y *J. Mater. Res.* **11** (1996) 1114

Inst. Phys. Conf. Ser. No 158
Paper presented at Applied Superconductivity, The Netherlands, 30 June–3 July 1997

Biaxial alignment of YSZ buffer layers on inclined technical substrates for $YBa_2Cu_3O_{7-x}$ tapes

W A J Quinton and F Baudenbacher

IRC in Superconductivity, Madingley Road, Cambridge, CB3 0HE, UK

Abstract. In-plane aligned $YBa_2Cu_3O_{7-x}$ (YBCO) films were deposited on Hastelloy substrates using biaxially aligned yttria stabilised zirconia (YSZ) buffer layers. The YSZ films were fabricated by pulsed laser deposition (PLD) on inclined substrates. The orientation and the in-plane alignment of the YSZ buffer layer depends strongly on the deposition conditions. The YSZ growth direction perpendicular to the substrate plane can be adjusted between the (111) and the (002) directions by varying the substrate temperature. In-plane alignment of either orientation can be achieved by simply inclining the substrate towards the plume axis, with azimuthal full-width-half-maxima of the (111) reflections in pole figures of typically 24° under optimised conditions. The inclination does not significantly affect the growth orientation perpendicular to the substrate surface. The subsequent deposition of YBCO on (002) oriented biaxially aligned YSZ buffer layers yielded epitaxial YBCO films as shown by x-ray θ-2θ scans and pole figures. Initial critical current measurements on wet chemically etched tracks resulted in values of 1.0×10^5 A/cm^2 in zero field at 77 K.

1. Introduction

YBCO films on flexible metallic tapes may be the high temperature superconducting wire technology that enables high field applications at liquid nitrogen temperatures. In contrast to Bi- and Tl- based superconductors, YBCO has an intrinsically higher irreversibility field (B_{irr}) enabling the operation at higher temperatures and magnetic fields [1]. The critical current limiting factor in YBCO wires and tapes fabricated by powder in tube methods are mainly large angle grain boundaries acting as weak links [2]. Thin film techniques have been successfully employed to achieve the necessary c-axis orientation and in-plane alignment avoiding large angle grain boundaries.

The key is to use textured substrates – with buffer layers if interface reactions and interdiffussion processes prevent the growth of high quality YBCO films [3]. In order to coat flexible metal tapes, additional novel process steps have to be introduced to fabricate biaxially aligned buffer layers suitable for YBCO growth. Currently there are two techniques under investigation. The first to be reported was Ion Beam Assisted Deposition (IBAD) [4,5], whereby an additional ion beam, under an oblique angle of incidence, is used to impart the desired orientation to the buffer layer. The biaxial alignment arises from a growth rate differentiation caused by impinging ions and the shadowing of slower growing grains [6]. The differentiation is thought to result from an orientation dependent damage threshold and sputtering yield. The ion bombardment causes the removal of a predominant fraction of deposited material leading to a relatively slow buffer layer deposition process.

The second approach uses thermo-mechanically processed biaxially textured nickel substrates (RABiTS) to achieve aligned buffer layer growth. The high temperatures and oxygen environment required for the deposition of oxide buffer layers causes a detrimental oxidation of the nickel surface requiring additional noble metal layers or processing in a reducing atmosphere [7,8]. The process complexity in both cases may prevent the scalability to technical applications requiring long lengths.

However, Hasegawa *et al.* reported on a method to form biaxially aligned YSZ films on Hastelloy without IBAD or RABiTS by simply inclining the substrate at elevated temperatures [9]. In this paper we follow a similar approach and describe a new technique to grow biaxially aligned YSZ buffer layers on Hastelloy substrates at room temperature. The technique can easily be scaled up since no elaborate substrate heating system is required.

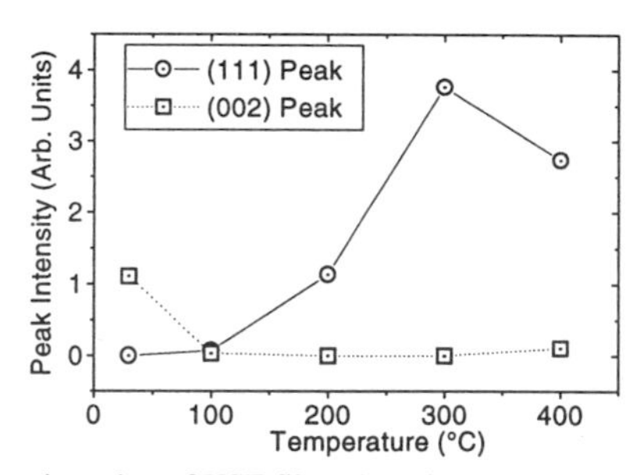

Figure 1. The crystallographic orientation of YSZ films deposited using the conventional on-axis geometry, with an oxygen pressure of 0.001 mbar, as a function of substrate temperature.

2. YSZ deposition

The YSZ films were prepared by pulsed laser deposition (PLD) using a KrF excimer laser (λ=248 nm) at a pulse frequency of 10 Hz. The Hastelloy C substrates were electrochemically polished and mounted on a our rotatable substrate heater. The crystallographic alignment and the quality of our YSZ films was studied by x-ray diffraction measurements as a function of oxygen pressure, substrate temperature, laser energy density, target-to-substrate distance, and inclination of the substrate relative to the plume axis.

In our first optimisation run we varied the oxygen pressure at a substrate temperature of 400 °C in a conventional PLD geometry with the substrate and target surfaces parallel. We observed (002) and (111) mixed orientations at higher pressures, becoming an almost pure (111) texture for films deposited at pressures of 0.001 mbar. None of the films deposited in this arrangement was in-plane aligned.

The variation of substrate temperature at the optimum oxygen pressure in the parallel configuration shows a change in the growth orientation perpendicular to the substrate surface as shown in Fig. 1. The uniaxial orientation changes from (111) to (002) at substrate temperatures below 100 °C, giving a pure (002) texture at room temperature. It was then possible to achieve biaxial alignment simply by inclining the substrate either at room temperature or at 400 °C, depending on the required orientation. Fig. 2 shows an x-ray pole figure of a YSZ film on a Hastelloy substrate deposited at room temperature. The inclination angle between the substrate normal and the plume axis was 55°. Notice that the crystallographic orientation perpendicular to the substrate surface is tilted by approximately 15° due to the effect of this inclination.

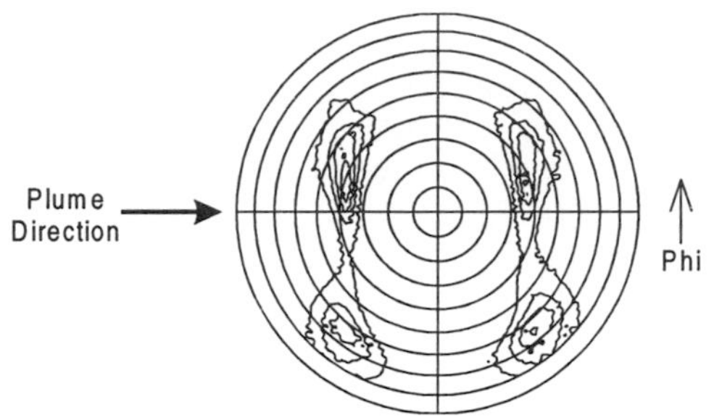

Figure 2. Pole figure showing the YSZ (111) reflections from a film deposited at room temperature onto a metallic substrate inclined by 55°.

At this temperature, pressure and inclination, other deposition parameters were varied. The target-to-substrate distance was varied between 45 and 90 mm and found to have no noticeable affect on the in-plane alignment of the YSZ films. However, it was found that there is an optimum laser energy density, for which the best alignment is attained. For each YSZ film deposited using varying energy densities, the azimuthal full-width-half-maxima (FWHM) of the (111) reflections were measured, and these can be seen in Fig. 3.

3. Discussion

It is thought that the alignment mechanism is most probably similar to that of IBAD. In this case, however, the alignment occurs due to ions and high energy particles originating in the laser plume. In both cases, the required angle between the substrates' normal and their respective ion sources, is 55°, which is angle between the (111) and (002) directions.

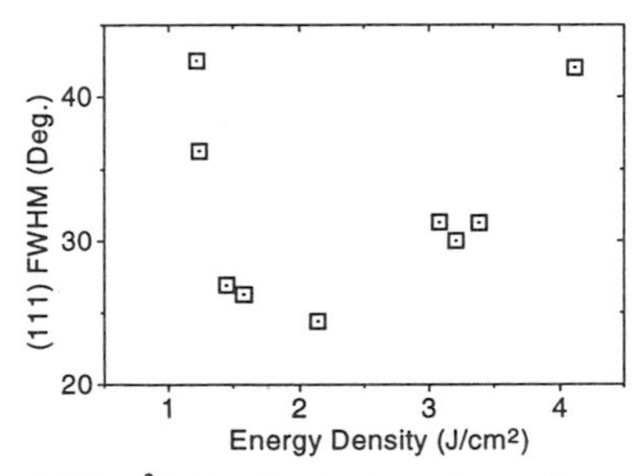

Figure 3. A laser fluence of around 2 J/cm^2 is found to be the optimum for producing well aligned YSZ films. The y-axis shows the average full-width-half-maxima of the four YSZ (111) peaks.

At low laser energy densities, the momentum of the bombarding particles may be below the damage threshold of unfavourable growth orientations. At higher laser fluences, however, the increase in the ion density within the plasma, leads to strong interactions, so increasing the angular spread of impacting particles. This results in a greater spread in the in-plane crystal orientation. A similar behaviour has been observed with IBAD [10] at much higher ion energies.

4. YBCO films

YBCO films were deposited, using PLD in the conventional on-axis geometry, onto these (002) oriented, biaxially aligned YSZ buffer layers. X-ray diffraction measurements in the Bragg-Brentano-geometry show these subsequent superconducting layers to be purely c-axis orientated. Pole figures show an improved FWHM of around 18°, and that the 15° tilt of the YSZ (002) orientation is not transferred to the YBCO layer.

The superconducting properties were measured in four point geometry on photolithographically patterned and wet etched tracks. Zero point resistivities were up to 90 K, with transition widths of 2 K. Critical current densities were around 1.0×10^5 A/cm^2.

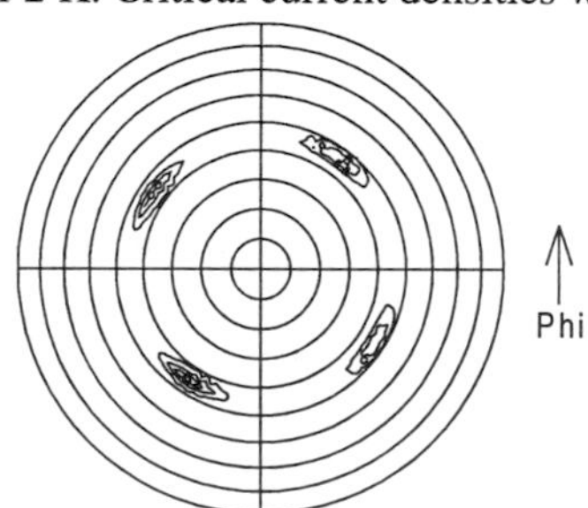

Figure 4. Pole figure of a YBCO film fabricated onto a YSZ buffer layer deposited onto an inclined metallic substrate. The reflections shown are those of the YBCO (103) orientation.

5. Conclusions

Highly aligned YSZ buffer layers can be deposited at room temperature on metallic substrates, without the need for a secondary ion beam or substrate texturing. In-plane alignment is achieved simply by inclining the substrate by 55° away from the plume axis. Subsequently deposited YBCO films are shown to have an improved biaxial alignment.

References

[1] Iye Y, Tamegai T, Takeya H and Takei H 1987 *Jpn. J. Appl. Phys. Pt. 2* **26** L1057-9

[2] Dimos D, Chaudhari P, Mannhart J and LeGoues F K 1988 *Phys. Rev. Lett.* **61** 219-22

[3] Ockenfuß G, Baudenbacher F, Prusseit-Elffroth W, Hirata K, Berberich P and Kinder H 1991 *Physica C* **180** 30-33

[4] Iijima Y, Tanabe N, Kohno O and Ikeno Y 1992 *Appl. Phys. Lett.* **60** 769-71

[5] Reade R P, Berdahl P, Russo R E and Garrison S M 1992 *Appl. Phys. Lett.* **61** 2231-3

[6] Sonnenberg N, Longo A S, Cima M J, Chang B P, Ressler K G, McIntyre P C and Liu Y P 1993 *J. Appl. Phys.* **74** 1027-34

[7] Goyal A, Norton D P, Budai J D, Paranthman M, Specht E D, Kroeger D M, Christen D K, He Q, Saffian B, List F A, Lee D F, Martin P M, Klabunde C E, Hartfield E and Sikka V K 1996 *Appl. Phys. Lett.* **69** 1795-7

[8] He Q, Christen D K, Budai J D, Specht E D, Lee D F, Goyal A, Norton D P, Paranthaman M, List F A and Kroeger D M, Physica C 275 (1997) 155-61

[9] Hasegawa K, Yoshida N, Fujino K, Mukai H, Hayashi K, Sato K, Ohkuma T, Honjyo S, Ishii H and Hara T 1996. To be published in *Proc. of the 9th Int. Symp. on Supercond. (ISS '96), Sapporo, Japan.*

[10]Ensinger W, Nucl. Instr. and Meth. 1995 **B 106** 142-6

Inst. Phys. Conf. Ser. No 158
Paper presented at Applied Superconductivity, The Netherlands, 30 June–3 July 1997

Influence of external strains on J_c of YBCO films on thin technical substrates

F García-Moreno[1], A Usoskin[1], H C Freyhardt[1,2], J Wiesmann[2], J Dzick[2], K Heinemann[2] and J Hoffmann[2]

[1]Zentrum für Funktionswerkstoffe Gött. gGmbH, Windausweg 2, D-37073 Göttingen, Germany
[2]Institut für Metallphysik, Univ. of Göttingen, Windausweg 2, D-37073 Göttingen, Germany

Abstract. A study of critical currents J_c of biaxially textured $YBa_2Cu_3O_{7-x}$ (YBCO) films deposited on thin (0.12 mm) polycrystalline Ni foils and YSZ ribbons under compressive and tensile strains, caused by substrate bending, has been performed by employing a special device which allows both to vary *in situ* the bending radius in liquid N_2 and to measure J_c. A significant increase of J_c in YBCO films under compressive strains has been observed. J_c was found to be reversible in the range of strains between, at least, ± 0.5 %. Within these limits, J_c is almost constant for a tensile strain, but shows a maximum under compression. J_c at this maximum can be up to 2 times higher than the initial J_c of the as-deposited films. No considerable changes of T_c were found in the range of strains from -1.4 % (compression) to +0.6 % (tension). A mechanical deterioration of YBCO film was observed for the tensile strains of > 0.6 %.

1. Introduction

High mechanical flexibility of high-T_c film/technical substrate architectures represents one of the important requirements which one should meet to employ the films in the numerous technical fields which are widely discussed now [1-2]. Nevertheless, the problem of physical mechanisms limiting this flexibility because of J_c deterioration in high-T_c films was not studied sufficiently. Obviously, results obtained on varying of T_c under uniaxial strains (see, e.g. [3]) do not introduce enough understanding to the problem of J_c behaviour.

Here we report on investigations of critical currents J_c of biaxially textured $YBa_2Cu_3O_{7-x}$ (YBCO) films deposited on thin (0.12 mm) polycrystalline Ni foils and YSZ tapes under compressive and tensile strains caused by substrate bending.

2. Film preparation

High critical-current YBCO films were deposited exploiting a pulsed-laser-deposition (PLD) method under the following conditions: 308 nm wavelength, 2.7 J/cm^2 energy density on the target, 0.5 J pulsed energy, (8-12) Hz pulse repetition rate, 0.5 mbar oxygen pressure, 760° C substrate temperature. Two modifications of the deposition process were employed to improve the PLD method: a 2-way scanning of the target with the laser beam [4] and a quasi-equilibrium radiative substrate heating [5], which allow a significant increase of stability of the deposition rate and of the substrate temperature.

Two types of polycrystalline technical substrates, 0.125 mm-thick Ni foils and yttria-stabilized zirconia (YSZ) ribbons of the same thickness, were used for YBCO film deposition. Ni foils of 99.98% purity were electro-chemically polished to reach a mirror-like surface with a roughness of better than 0.02 μm. Polycrystalline YSZ ceramic ribbons exhibited a similar surface quality provided by a conventional mechanical polishing. YSZ buffer layers prepared by ion-beam-assisted-deposition (IBAD) [6] were employed to meet requirements needed for a high-quality in-plane texture for YBCO films as well as to suppress an oxidation in the case of Ni foils. 600-800 nm-thick YBCO films with critical current densities of (0.2-2) MA/cm^2 (77 K, 0 T) have been used for the present experiments.

3. Bending experiments and discussion

A study of critical current densities J_c of 600-800 nm-thick YBCO films under compressive and tensile strains, caused by substrate bending, has been performed by employing two special devices. In the first method the substrate with the film was placed between hard cylinders and a concave surface (see Fig. 1 a), so that the film could be bent to the known radius which can be varied step by step. The relation between the bending radius r and the strain δ in the film is given by the ratio $|\delta| = t/2r$, where t denotes the thickness of the substrate ($t \ll r$). The critical current density, J_c , and critical temperature T_c of the films were measured in this case by an inductive shielding method [4] for different radii. In the course of the measurements it was necessary to maintain precisely a constant distance between the film the field coil which had to be located almost at the concave side of the sample. This first method was used only for the Ni foils because of the mechanical properties of the YSZ ribbons which does not allow

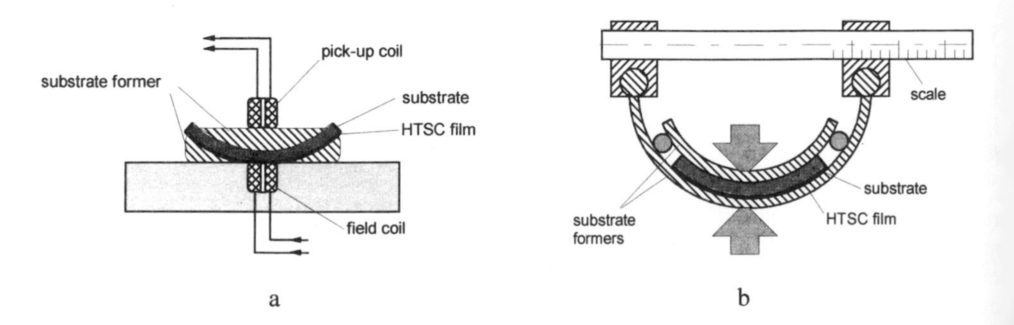

Fig. 1. Schematic view of the devices used for J_c , T_c measurements of bent YBCO films on polycrystalline ribbons. a - inductive ac method with a step-by-step variation of the bending radius, b - 4-probe dc method with smooth change of bending radius *in sity*.

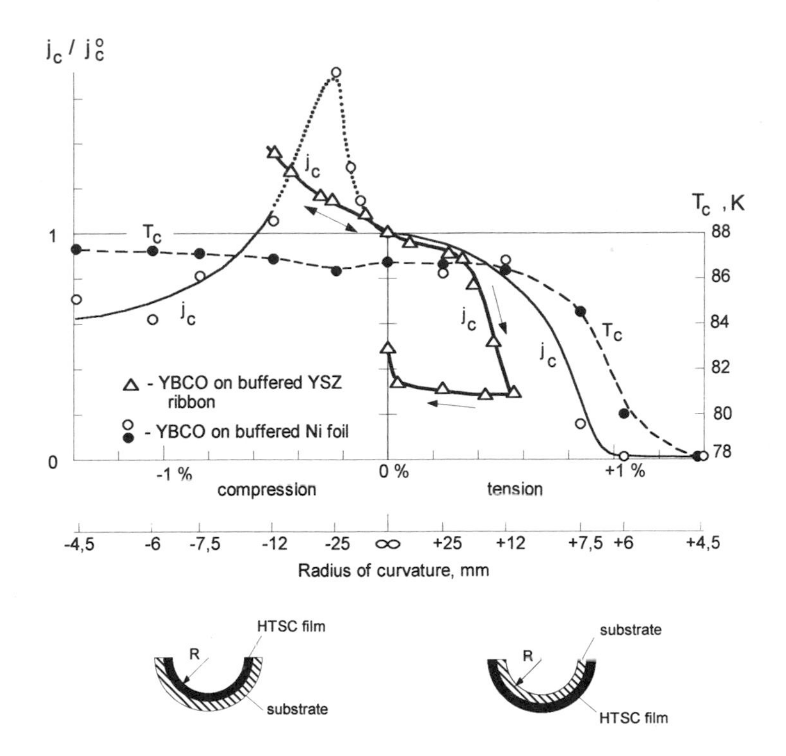

Fig. 2. J_c and T_c dependences on compressive and tensile strains and bending radius for YBCO films deposited on 0.125 mm thick YSZ and Ni ribbon with IBAD buffer layer. J_c° corresponds to ~ 10^5 A/cm^2.

considerable plastic deformations. In the second method (see Fig. 1 b), the samples were inserted between two springy steel tapes. The method allows both to vary *in situ* at 77 K the bending radius and to measure J_c by a 4-probe dc technique. As it was found, both of these methods yield similar results.

The dependences of J_c and T_c on tensile and compressive strains are shown in Fig. 2. J_c was found to be reversible in the range of strains between, at least, ± 0.5 %. Within these limits, J_c is almost constant for a tensile strain, but shows in the case of Ni substrates a maximum under compression. J_c at this maximum can be up to 2 times higher than the initial J_c of the as-deposited films. No considerable changes of T_c were found in the range of strains from -1.4 % (compression) to +0.6 % (tension) for YBCO films deposited on Ni foils. The bending radii < 12 mm correspond to the limit of the mechanical stability of YSZ ribbons, and because of that the measurements were not possible to continue at the strains higher than 0.5 % and lower than -0.5 % for this type of brittle YSZ substrates.

The structural analysis of the films and substrates was performed by scanning electron microscopy (SEM). A mechanical deterioration of YBCO film was observed for tensile strains larger than 0.5 %. In this case a set of cracks can be observed (see Fig. 3. a). The cracks are formed perpendicular to the direction of the tension. The longitudinal dimension of cracks can exceed 10 μm, the width is of (50-200) nm, and the spacing corresponds to (2-5) μm. Obviously, these cracks are responsible for the observed quick drop of J_c at tensile strains above 0.5 % (see Fig. 2). Due to the uniaxial deformation and, as a result, to the uniaxial arrangement of the cracks, an anisotropy of J_c should be expected in such films.

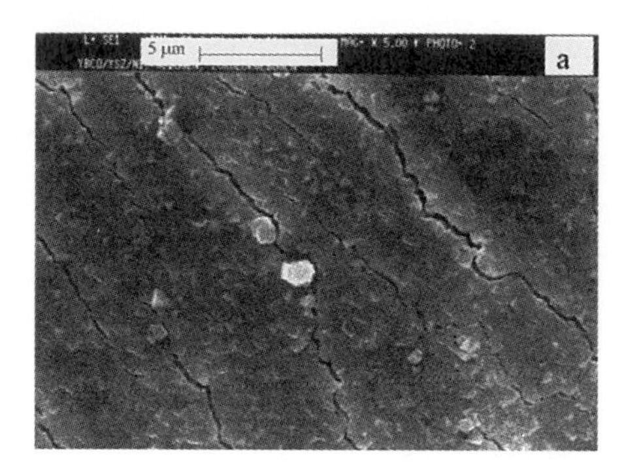

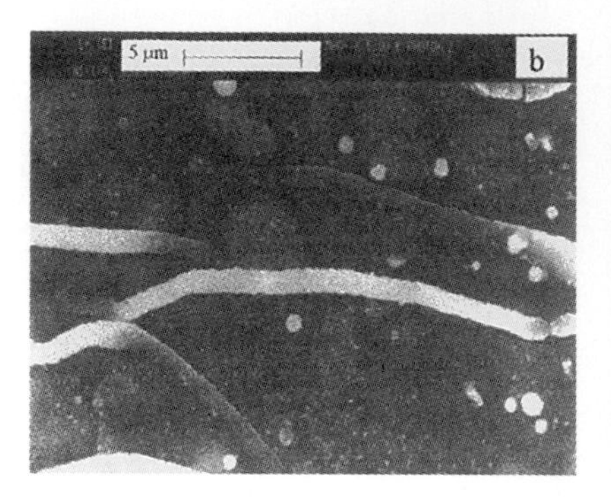

Fig. 3. SEM view of cracks in YBCO film caused by ~ 1 % tensile strain (a), and of blistering caused by ~ 2 % compressive strain (b). 800 nm-thick YBCO film is deposited on buffered 0.12 mm-thick Ni foil.

For compressive strains smaller than 1 % no deterioration can be found by SEM technique, but it does not mean that the cracks of smaller width are not developed. Some cracks in the film become visible for compressive strains ≥ 0.8 % when the substrate is bent back to the original planar state. At the higher compressive strains > 2 % blisters with characteristic dimensions of 5-10 µm appear (see Fig. 3 b).

The most attractive peculiarity found in the course of the present investigations is the maximum of J_c in the films under compressive strain. Such a behaviour of J_c can be caused by variations either (a) of *inter*-grain properties (i. e. of weak links between crystallites) or (b) of *intra*-grain properties. Measurements of the YBCO film conductivity in the normal state versus tensile and compressive strains were performed to distinguish which of these two mechanisms is responsible for the observed effect. As a first result, no considerable variations exceeding 2 % were found for the electrical conductivity. This seems to indicate that the J_c variations might not be connected with a modification of inter-crystallite links, but rather with changes of intra-crystallite properties (e. g. with variations of YBCO electron/phonon spectra) under uniaxial deformation.

Acknowledgement

This work was performed within a development project of Kabelmetal electro GmbH which is supported by German BMBF (Project No. 13N6924/6), and was also supported by Siemens AG together with German BMBF (Project No. 13 N6482).

References

[1] Vaglio R *Applied Superconductivity 1995* IOP Conf. Series **148** (1995) 781

[2] Komarek P *Verhanl. DPG* (Münster, 1997) **6** No. 32 (1997) 1010

[3] Welp U *Phys. Rev. Lett.* **69** (1992) 2130

[4] Usoskin A, Freyhardt H C, García-Moreno F, Sievers S, Popova O, Heinemann K, Hoffmann J, Wiesmann J and Isaev A *Applied Superconductivity 1995* IOP Conf. Series **148** (1995) 499

[5] Freyhardt H C, Usoskin A and Neuhaus W, *German Patent* No. P 42 28 573.9-45 (1992)

[6] Wiesmann J, Heinemann K and Freyhardt H C *Nucl. Instr. Meth. B* **120** (1996) 290

Inst. Phys. Conf. Ser. No 158
Paper presented at Applied Superconductivity, The Netherlands, 30 June–3 July 1997

Single Source MOCVD of HTSC films onto travelling substrates

O Stadel, L Klippe, G Wahl

Institut für Oberflächentechnik und Plasmatechnische Werkstoffentwicklung (IOPW), Technische Universität Braunschweig, Bienroder Weg 53, 38108 Braunschweig, Germany

S V Samoylenkov, O Y Gorbenko, A R Kaul

Department of Chemistry, Moscow State University, 119899 Moscow, Russia

Abstract. By two different single source MOCVD techniques superconducting Y-123 and Lu-123 films were grown on various substrates. $YBa_2Cu_3O_{7-\delta}$ layers were deposited on polycrystalline YSZ tapes (10 cm length) and (100)-$SrTiO_3$ short samples in a MOCVD system developed for continuous coating of tape substrates. EDX analysis of the tape showed a homogeneous distribution of $YBa_2Cu_3O_{7-\delta}$ film stoichiometry and film thickness. XRD figures exhibit strong preferential c-axis oriented film growth. The samples were superconducting at T_c > 80 K (tape) and T_c > 90 K ($SrTiO_3$).

With another MOCVD-single source system $LuBa_2Cu_3O_{7-\delta}$ films were obtained on $LaAlO_3$ and $SrTiO_3$ crystals with $j_c(77\ K) = 1.5\text{-}2.5\cdot10^6\ A/cm^2$ and T_c = 87-90 K. In addition onto both sides of $LaAlO_3$ substrates $LuBa_2Cu_3O_{7-\delta}$ films of almost identical quality were deposited.

1. Introduction

The development of high-quality HTSC films on long tapes promises industrial applications like superconducting fault current limiters and magnets. CVD is known as a suitable technique for large scale applications. In conventional multi source CVD degradation of Ba-precursor ($Ba(thd)_2$) causes problems of stability of deposition stoichiometry. Using single source techniques constant film stoichiometry can be maintained over long process periods [1, 2].

For deposition on tapes a MOCVD system was developed utilizing a band evaporator [3, 4] and a small hot wall reactor. In another single source system using powder flash evaporation [5, 6] double sided deposition of $LuBa_2Cu_3O_{7-\delta}$-films onto (100)-$LaAlO_3$ substrates were carried out. Possible applications for these films can be microwave devices, such as resonators.

2. Experimental

The CVD-system developed for the deposition on tapes consisted of three main parts, the mass-flow control unit, the band evaporator and the reactor. The precursors ($Y(thd)_3$, $Ba(thd)_2$ and $Cu(thd)_2$) were dissolved in diglyme (diethylene-glycol-dimethyl-ether). The solution was continually supplied by a mass-flow controller for liquids (BRONKHORST HI-TEC). The operation of the band evaporator, which separates solvent and precursors has been reported elsewhere [3, 4]. In the reactor at 850°C $YBa_2Cu_3O_{7-\delta}$ was deposited onto the substrate, which was fixed to a substrate holder (Inconel). This substrate holder was lead

Table 1 Experimental conditions for deposition on tapes

Deposition temperature	850°C
Heated pipes	270°C
Precursors	Y-, Ba-, Cu-$(thd)_n$
Solution	0.025 mol/l metal-thd in diglyme (Y:Ba:Cu=1:2.6:2.7)
Deposition rate	3-15 cm/h (thickness 0.5μm)
Total pressure	500 Pa
Oxygen partial pressure	450 Pa

through the deposition chamber, a short cylindrical hot wall reactor (∅: 4 cm, height: 6 cm). The travelling speed of the linear motion was approximately 10 cm/min. After deposition the sample was annealed at 460°C for 30-60 min in pure oxygen (10^5 Pa). In order to obtain superconducting films with high critical current densities also (100)-$SrTiO_3$ single crystals were coated with $YBa_2Cu_3O_{7-\delta}$. In the reactor configuration used it was possible to minimize gas phase reactions, while the temperature of the moved substrate remained constant during deposition. To demonstrate the capabilities of this combination of small hot wall reactor and evaporating system, yttrium stabilized zirconia was deposited from thd-precursors at growth rates of 0.5-1.5 μm/min [7].

Using powder flash evaporation MOCVD, $LuBa_2Cu_3O_{7-\delta}$-films were prepared in another hot wall reactor [5]. Previously the deposition process was optimized to obtain $LuBa_2Cu_3O_{7-\delta}$-films with high superconducting characteristics on various substrates. Single crystalline $LaAlO_3$ substrates (0.5 mm thick) were used for the deposition on both sides. After the first side was covered with $LuBa_2Cu_3O_{7-\delta}$ the sample was turned around and during the subsequent deposition on the back side the $LuBa_2Cu_3O_{7-\delta}$-film was protected by a thin $NdGaO_3$-plate.

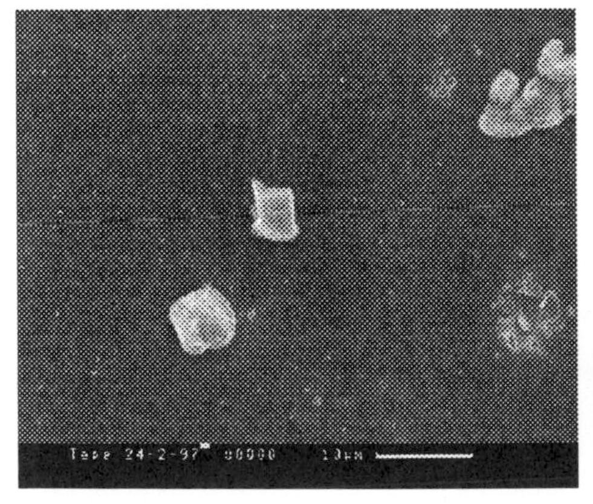

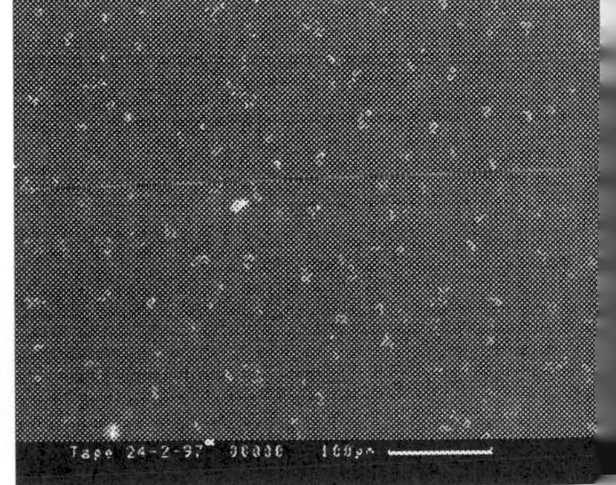

Fig. 1.SEM images of YBCO-film (thickness 0.5 μm) on polycrystalline YSZ tape

3. Results

Morphology, structure and stoichiometry of the $RBa_2Cu_3O_{7-\delta}$-films (R = Y, Lu) were analyzed by SEM, XRD and EDX. Superconducting properties were characterized by resistive and ac-magnetization measurements. Fig. 1 shows the surface of a $YBa_2Cu_3O_{7-\delta}$ film (thickness 0.5 μm), which was deposited onto a moved polycrystalline YSZ-substrate. The surface was very smooth with a small density of precipitations. Over the whole length no changes of morphology occured. X-ray scans (θ-2θ, Cu-K_α) were performed at different sites along the tape axis in steps of 1 cm (fig. 2). Apart from the peaks of the polycrystalline YSZ substrate, only (001) of $YBa_2Cu_3O_{7-\delta}$ reflexes were found. EDX analysis (Cu-K_α, Ba-L_α, Zr-L_α, Zr-K_α) of the 10 cm tape showed deviations of stoichiometry and film thickness only in the range of measuring accuracy (±5%).

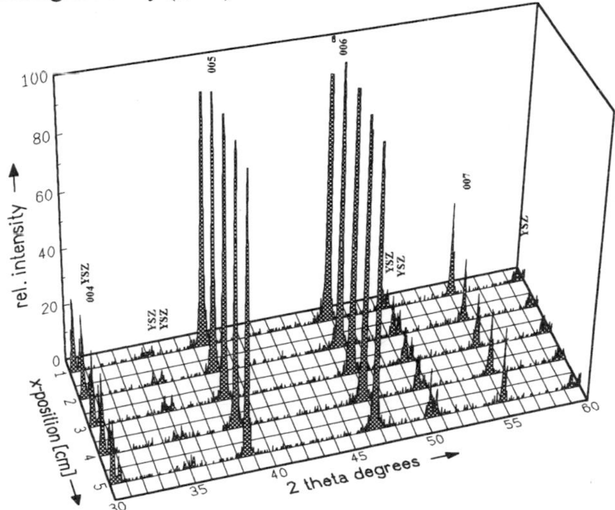

Fig. 2. XRD patterns (Cu-K_α) of the YBCO film measured in steps of 1 cm along the tape axis.

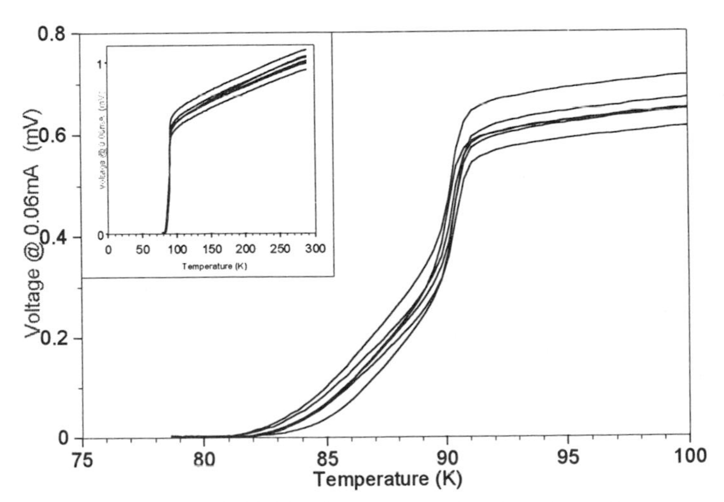

Fig. 3. Resistive measurement of a YBCO film on a tape substrate in steps of 1 cm.

The transition temperature of the $YBa_2Cu_3O_{7-\delta}$ film was determined by resistive measurements (figure 3). For this, contacts were fixed on the $YBa_2Cu_3O_{7-\delta}$ film with distances of 1 cm between each other. All plots showed similar behaviour with $T_{c,onset}$ at 90 K and $T_{c,zero}$ near 82 K. The critical current density j_c measured by ac-Magnetization was less than 10^3 A/cm^2. With the same experimental conditions $YBa_2Cu_3O_{7-\delta}$ was deposited onto moved (100)-$SrTiO_3$-substrates. Values of critical current density reached $2 \cdot 10^6$ A/cm^2 (77 K) with $T_c = 91$ K.

$LuBa_2Cu_3O_{7-\delta}$ films deposited by powder flash evaporation MOCVD on (100)-$SrTiO_3$, (100)-$LaAlO_3$, (100)-YSZ and ceria buffered R-sapphire exhibit $T_c = 87$-90 K and j_c(77 K) – 1.5-2.5$\cdot 10^6$ A/cm^2. In addition $LuBa_2Cu_3O_{7-\delta}$ films were prepared on both sides of $LaAlO_3$ single crystals. The sharp ($\Delta T_c = 1.5$ K) superconducting transition at 86.5 K demonstrates the almost identical quality of the films on both sides.

4. Discussion

In a continuous MOCVD process $YBa_2Cu_3O_{7-\delta}$ films were deposited onto moved substrates. Over the whole length (10 cm) of a polycrystalline YSZ tape almost identical properties were measured. The sample was superconducting with $T_c > 80$ K, but critical currents at 77 K were below 10^3 A/m². Films deposited under the same experimental conditions onto a moved (100)-$SrTiO_3$ substrate reached a critical current density of $j_c(77\ K) = 2 \cdot 10^6$ A/cm^2. Further investigations will be concerned with deposition on textured tapes in order to obtain higher values of the critical current.

$LuBa_2Cu_3O_{7-\delta}$ films with high values of critical current density were prepared on various single crystalline substrates. Double-sided deposition of high quality $LuBa_2Cu_3O_{7-\delta}$ films on $LaAlO_3$ yielded almost identical properties on both sides.

Achknowledgements

The authors acknowledge financial support from the EU (Brite EuRam II project no. 7887), from the german BMBF (13 N 6947/5) and the Volkswagen foundation.
We further like to thank Dr. Neumüller and Dr. Schmidt (Siemens AG, Erlangen) for colaboration.

References

[1] Feng-Chi Y, Yi-Yuan X, Ji-Ping C, Guo-Wen Y and Bin-Ji C 1996 *Supercond. Sci. Technol.* **9** 991.
[2] Onabe K, Nagaya S, Shimonosono T, Iijima Y, Sadakata N, Saito T, Kohno O 1996 *Proc. of ICEC16/ICMC: PSI-e2·52, Kitakyushu, Japan.*
[3] Klippe L and Wahl G 1995 *EUCAS* 611.
[4] Klippe L and Wahl G 1996 *Proc. E-MRS Spring Meeting Strasbourg, in print.*
[5] Samoylenkov S, Gorbenko O, Graboy I, Kaul A, Tretyakov Y, 1996 *J. Mater Chem.* **6 (4)** 623.
[6] Samoylenkov S, Gorbenko O, Kaul A, 1996 *Physica C* **267** 74-78.
[7] Wahl G et al. 1997 *Proc. E-MRS Spring Meeting Strasbourg, to be published.*

Inst. Phys. Conf. Ser. No 158
Paper presented at Applied Superconductivity, The Netherlands, 30 June–3 July 1997

Critical current density and irreversibility line of 2223 BSCCO tapes enhanced by columnar defects along a part of the tape thickness.

R.Gerbaldo, G.Ghigo, L.Gozzelino, E.Mezzetti and B.Minetti

Istituto Nazionale di Fisica della Materia - U.d.R Torino-Politecnico; Istituto Nazionale di Fisica Nucleare - Sezione di Torino; Politecnico di Torino, Torino, Italy.

P.Caracino and L.Gherardi

Pirelli Cavi S.p.A, Milano, Italy.

Abstract. This paper analyses the implantation of columnar defects on a surface layer of Ag/BSCCO-2223 tapes. Columnar-defect length was ~ 5% of the whole specimen. Large shift of the IL was detected. The IL shape shows features characteristic of strong vortex localisation in high temperature regimes. 3D-2D cross-over field shifts towards higher fields. Furthermore, the experiment detects a significant enhancement of the J_c and the hardness to the dropping off of the perfomance in field. IL and J_c anisotropy of the irradiated samples is reduced.

1. Introduction

Tape shaped BSCCO-2223 superconductors sheathed in a silver cladding are of interest for a variety of magnetic, coil and multistrand conductor applications. There is an interest in pushing the J_c values higher at high field and temperature, where so far a dropping off in performance is observed. Above the self-field, where weak link behaviour is dominant, flux creep plays a dominant role in limiting J_c's. Heavy-ion irradiation producing columnar tracks with radial dimension (5-10 nm) [1] close to the coherence length is ideal for core pinning of vortices. Substantial J_c enhancements were obtained by means of tracks either extending themself along the full sample thickness [2] or extending at least for the 50% of the entire thickness [3]. The specifity of our irradiations is to introduce only segments of columnar defects produced by 0.25 GeV Au ions and implanted on the surface. The length of the columnar defect is about 5% of the entire thickness of the specimen. Nevertheless, strong shift of the IL's towards higher fields and temperatures were systematically observed. The magnitude order of the shift is comparable with that obtained with splayed columnar defects from 0.8 GeV p^+ irradiation, crossing the whole sample [4]. It will be shown that in our case the track density plays a significant role in determining the IL shape. We also report substantial J_c enhancement over the entire range of temperature and fields. In particular, the defect-induced hardness to the drop off in-field performances at high temperature must be outlined. These results, obtained by means of a surface treatment, give some hints on possible Bi-2223 technology concerning tape-shaped materials.

2. Experimental

The samples, all monofilamentary Ag/BSCCO-2223 tapes of total width 3 mm and total thickness 0.2 mm, were prepared by the powder in tube technique described elsewhere [5]. The thickness of the superconducting core was 100 μm. Because of the limited penetration depth of Au ions in silver, it was necessary to prepare the samples by removing the silver on one side of the tape.

The specimens were irradiated with 0.25 GeV Au^{16+} ions at the 15 MV Tandem-XTU facility of the I.N.F.N.-Laboratori Nazionali di Legnaro (Padova, Italy). The samples were mounted in vacuum and irradiated perpendicularly to the tape surface, at room temperature with a beam collimated to 7 mm in diameter. The dose rate was less then $8\cdot10^8$ ions/cm^2·s to avoid sample heating. Au ion fluences of 0.5, 1, 1.5 and $2\cdot10^{11}$ ions/cm^2 were used (the fluence values are affected by an error of about 10%). These fluences correspond to dose equivalent fields, B_ϕ, of 1, 2, 3 and 4T, respectively. With B_ϕ we intend the field which would be ideally required to fill each track with a flux quantum ($B_\phi = n\phi_0$, where ϕ_0 is the flux quantum and n the track density). TRIM code simulations show that only in a surface layer of about 5 μm the energy released in the scattering processes overcomes the threshold of 20 MeV/μm, which is considered to be the lowest limit for the production of continuous tracks in high temperature superconductors. Au ions implanted at about 14 μm.

The samples, before and after irradiation, were characterized by means of a.c. and d.c. magnetic measurements using a LakeShore 7225 magnetometer/susceptometer with applied magnetic fields up to 4000 kA/m. One sample was fully characterised before irradiation while a coarse characterisation of other ones was made in order to check the specimen homogeneity. The IL's were obtained by the onset of the third harmonic of the a.c. susceptibility and J_c values were obtained from isothermal hysteresis loops.

3. Results and discussion

In Fig.1 the IL's for twin samples irradiated with the 4 different fluences above mentioned are shown. It appears that the introduction of columnar defects of limited length can lead to strong shift of IL towards higher magnetic fields and temperatures.

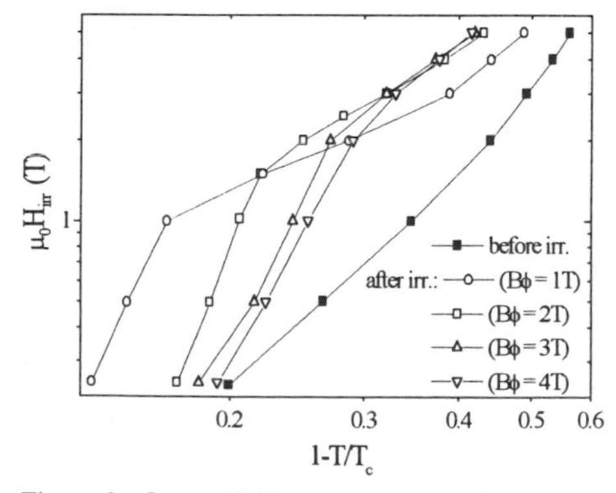

Figure 1 - Irreversibility lines before and after 0.25 GeV Au ion irradiation for different fluences.

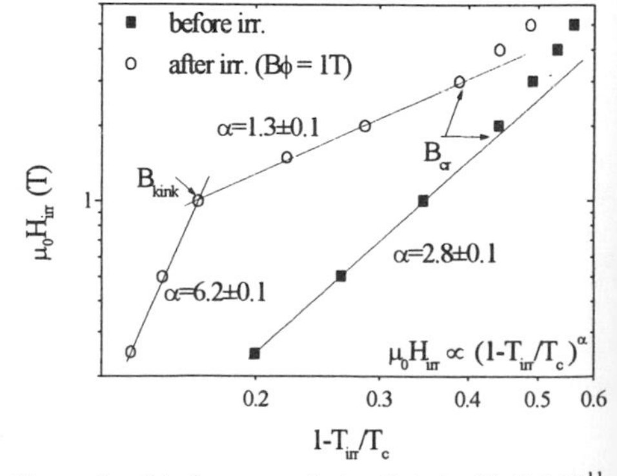

Figure 2 - IL's for a sample irradiated with $0.5\cdot10^{11}$ Au-ions/cm^2 (B_ϕ = 1T). For the IL after irradiation, the field, B_{kink}, where strong vortex localisation sets up, and the 3D-2D cross-over field, B_{cr}, are shown.

The magnitude order of the shift is comparable with that obtained with splayed columnar defects from 0.8 GeV proton irradiation [4].The shape of the after irradiation IL's at higher temperature is characterised by the setup of strong vortex localisation, whose signature is a kink and a steeper rise of the curves below the kink. The typical trend si shown in Fig.2. This behaviour indicates that at higher temperature, where the localisation length diverges, the limited length of our columnar defects does not dramatically influence the performance. Nevertheless, the position of the kink is shifted towards lower reduced temperature, T/T_c, and reduced field, B/B_ϕ, as the density of tracks increases (Fig.3).

We speculate that, in case of defects of limited penetration length, L_p, the dynamic regime we are dealing with, when the kink point is approached from above, is mainly driven by the track density. The flux lines in BSCCO-2223 are expected to be coupled over a very long length (a few microns), as well as in Bi-2212 superconductors [6]. The pinning energy, assumed to be linearly dependent on the longitudinal correlation length along the c-axis, L_c, reaches a saturation value in correspondence of $L_c(T) \approx L_p$. As the track density increases, the repulsion energy between vortices increases gradually shifting the ratio between the two energies towards larger contributions of the vortex repulsions. Such a mechanism should be taken into account for a correct planning of the magnetic fields where performances are to be optimised.

Before irradiation, the samples present the usual 3D-2D dimensional cross-over. In the low field region the IL can be fitted by the three-dimensional-like power law $H_{irr} \propto (1-T/T_c)^\alpha$ with $\alpha = 2.8 \pm 0.1$ for H parallel to the ion tracks. In the high field region ($\mu_0 H_{irr} > 2$ T) a deviation from the power law starts to appear due to the breakdown of the interlayer coupling. Above the cross-over field, B_{cr}, the IL can be fitted by the two-dimensional-like exponential function $H_{irr} \propto \exp(-T/T_0)$ [7] with $T_0 = 14.2 \pm 0.8$ K. After irradiation, the expected increase of B_{cr} was observed for the sample irradiated at the lowest fluences (Fig.2), while for the highest fluences up to 5 T no cross-over phenomena were found.

Au-ion irradiation caused a marked reduction of IL anisotropy as well as of J_c anisotropy. While (as previously discussed) the IL's measured for H parallel to the ion tracks were shifted to higher temperature and field, the IL's measured for H parallel to the tape surface remained almost unchanged. The whole effect is the abatement of the IL anisotropy after irradiation (Fig.4).

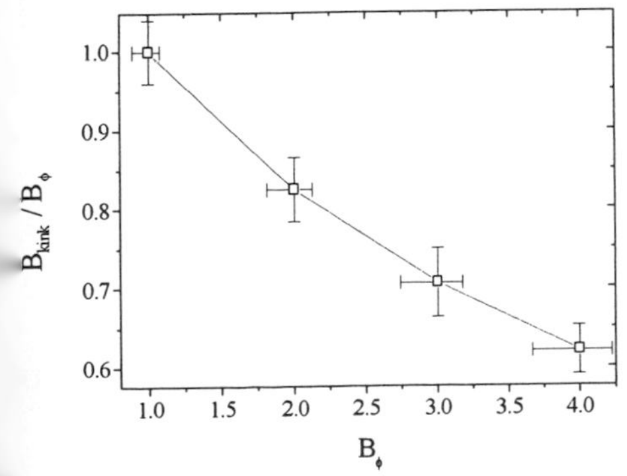

Figure 3 - Ratio between the cross-over field to strong vortex localisation regime (B_{kink}) and the dose equivalent field (B_ϕ) vs B_ϕ

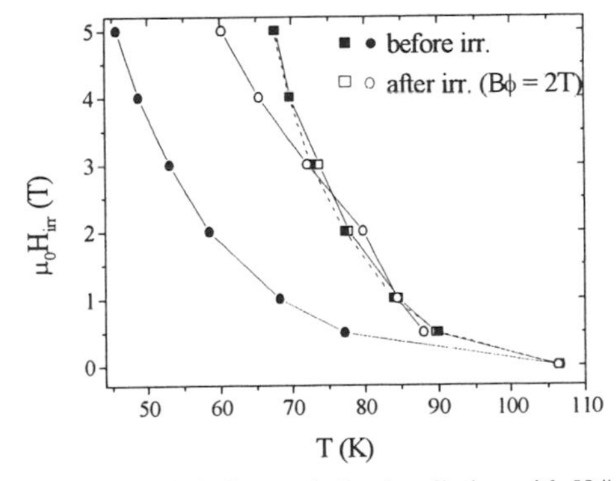

Figure 4 - IL's before and after irradiation with H || ion tracks (circle) and H || tape surface (square) for a sample irradiated with $1 \cdot 10^{11}$ Au-ions/cm^2 (B_ϕ = 2T).

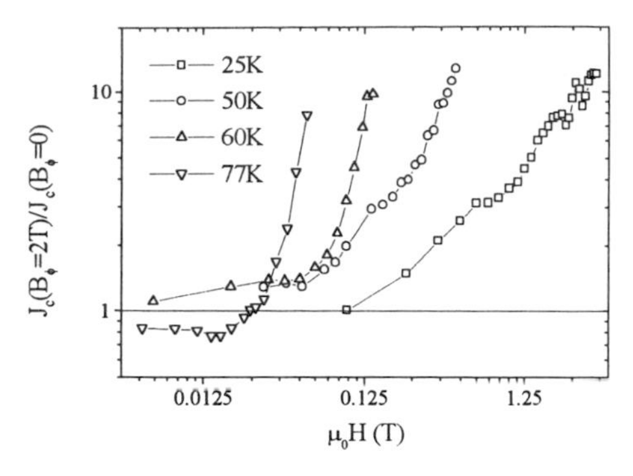

Figure 5 - J_c enhancement after irradiation for the sample irradiated with $1\cdot10^{11}$ Au-ions/cm^2 (B_ϕ = 2T).

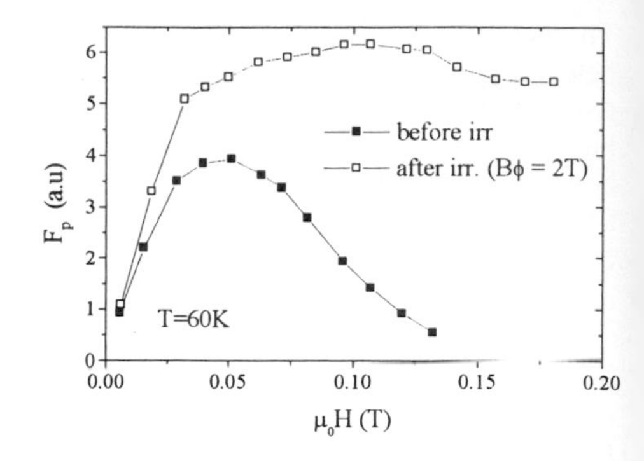

Figure 6 - Pinning force before and after irradiation for the sample irradiated with $1\cdot10^{11}$ Au-ions/cm^2 (B_ϕ = 2T).

Significant J_c enhancement in the full range of the investigated temperatures were observed. The enhancement ratios increase with the magnetic field and become prectically "infinite" in regions above the IL of the unirradiated sample (Fig.5). In particular, it is worthwhile to note the defect induced hardness with respect to the drop-off of the J_c with the field at slightly higher temperature. This trend is also put in evidence by noticeable indipendence of the pinning force on field (Fig.6).

4. Conclusions

Surface columnar defects strongly shift IL of BSCCO-2223 tapes towards higher fields and temperatures. A kink in the after irradiation IL shows as a signature of *strong* vortex localisation at higher temperature. The position of the kink is shifted towards lower reduced temperature, T/T_c, and reduced field, B/B_ϕ, as the density of tracks increases. The 3D-2D dimensional cross-over in the IL in Au ion irradiated samples is shifted towards higher fields. The J_c enhancements are characterised by an induced hardness to the in-field dropping off at higher temperature. IL anisotropy of the irradiated samples is reduced. The obtained results give some hints for further improvement of the Bi-2223 technology, concerning tape shaped materials, when an appropriate *surface* pinning is introduced.

References

[1] Zhu Y., Buhdani R.C., Cai Z.X., Welch D.O., Suenaga M., Yoshizaki R., Ikeda H., 1993 *Phil.Mag.Lett.* **67** 125

[2] Kummeth P., Struller C., Neumuller H.W., Saemann-Ischenko G., and Eibl O. 1994 *Critical Currents in Superconductors*, ed. by H.W.Weber, World Scientific, 311

[3] Civale L., Marwick A.D., Wheeler R., Kirk M.A., Carter W.L., Riley G.N. and Malozemoff A.P. 1993 *Physica* **C 208** 137

[4], Krusin-Elbaum L., Thompson J.R., Wheeler R., Marwick A.D., Kim Y.C., Christen D.K., Li C., Patel S., Shaw D.T., Lisowski P. and Ullmann J. 1994 *Appl. Phys. Lett.* **64** 3331

[5] Gherardi L., Caracino P., Coletta G. and Spreafico S. 1996 *Materials Science and Engineering* **B 39** 66

[6] Kiuchi M, Yamato Y., Matsushita T. 1996 *Physica* **C 269** 242

[7] Song Y.S., Hirabayashi M., Ihara H. and Tokumoto M. 1994 *Physical Review* **B 50** 16644

Inst. Phys. Conf. Ser. No 158
Paper presented at Applied Superconductivity, The Netherlands, 30 June–3 July 1997

Critical Currents in Neutron Irradiated Bi- and Tl-based Tapes

G.W.Schulz[1], C.Klein[1], H.W.Weber[1], H.W.Neumüller[2], R.E.Gladyshevskii[3], and R.Flükiger[3]

[1] Atominstitut der Österreichischen Universitäten, A-1020 Vienna, Austria
[2] Siemens AG, Zentralabteilung Forschung und Entwicklung, D-91050 Erlangen, Germany
[3] DPMC, Université de Genève, CH-1211 Genève

Abstract Experimental results on the magnetic field behavior, including the angular dependence, of the critical current densities in silver-sheathed tapes are presented. The experiments consist of transport measurements in a wide temperature range and in external magnetic fields up to 6 T. In Bi-2223 enhancements of the transport critical current densities J_c are observed at higher magnetic fields after irradiation with fast neutrons, in particular for H||c. This is attributed to an improvement of the flux pinning capability by the neutron induced defects, but the weak link structure is somewhat damaged as evidenced by the small degradation of J_c at low fields. The angular dependence of J_c demonstrates that the "pinning dominated" angular range is extended from H||c towards H||ab and that the overall anisotropy of J_c decreases. These findings are compared with the behavior of various Tl-1223 tapes prior to and following neutron irradiation.

1. Introduction

High critical currents are the key requirement for practical applications of Bi-2223 and Tl-1223 tapes. The microstructure of these superconducting tapes consists of small platelet-like grains. The mechanisms of current flow and its path through the tape are still under discussion [1] [2] [3] [4]. There are two fundamental limits for J_c. The connectivity between the grains which limits the *inter*granular current, and the pinning mechanism within the grains which limits the *intra*granular current. In this paper we report on experiments on silver sheathed Bi-2223 and Tl-1223 tapes, which were irradiated by fast neutrons. Results on a cross-over from the weak link dominated region to a pinning dominated region are presented.

2. Experimental

Bi-2223 silver sheathed tapes with 19 and with 55 filaments and a Tl-1223 monocore tape were investigated. The 19 filament tape (MK4/1) was prepared in 1995, the 55 filament tape (S7A3) and the Tl-1223 tape were prepared in 1997. All tapes were produced by the powder in tube (PIT) technique. The critical currents of the samples were determined from transport measurements using an electrical field criterion of 1μV/cm or 5 μV/cm. A superconducting split pair magnet was used to apply external magnetic fields up to 6 T.

Since the sample holder can be rotated, the angle between the tape surface and the external magnetic field can be varied from 0° (H||ab) to 90° (H||c). The irradiation was made in the central core position of the TRIGA Mark-II reactor in Vienna. At full reactor power, the flux density of fast neutrons amounts to $7.6 \cdot 10^{16}$ $m^{-2}s^{-1}$ (E>0.1 MeV). The samples were irradiated to a fluence of $2 \cdot 10^{21}$ m^{-2} at ambient temperature (~ 50° C).

3. Results on Bi-2223

To investigate the influence of the weak link structure on the current flow, we measured the critical current in increasing and subsequently in decreasing magnetic fields for H||ab and H||c. At 77 K, the difference in J_c between increasing and decreasing external magnetic field is for both Bi-based samples quite low. At lower temperatures the hysteresis of the 19 filament tape (MK4/1) increases, whereas it is still low in the 55 filament tape (S7A3). This clearly shows that MK4/1 is more weak link dominated than S7A3. These weak links are one of the limiting factors for the critical current. Another limitation is the pinning of the flux lines in the grains. The addition of artificial pinning centers can help to clarify the type of limitation in subsequent measurements of the critical current.

Neutron irradiation was used to produce artificial pinning centers, i.e. mainly spherical defect cascades with a diameter of 5 nm including the strain field, which is the ideal size for pinning [5]. At a fluence of $2 \cdot 10^{21}$ m^{-2} the defect density amounts to $1 \cdot 10^{22}$ m^{-3}.

The behavior of J_c at 77 K as a function of field is shown in Fig. 1. J_c decreases slightly after irradiation in the low field region, since some weak links were obviously damaged during irradiation. This effect is much smaller in S7A3, since less weak links are present. The increase in the number of pinning centers in the grains cannot compensate for this effect at low fields. J_c increases after neutron irradiation at higher fields as an effect of

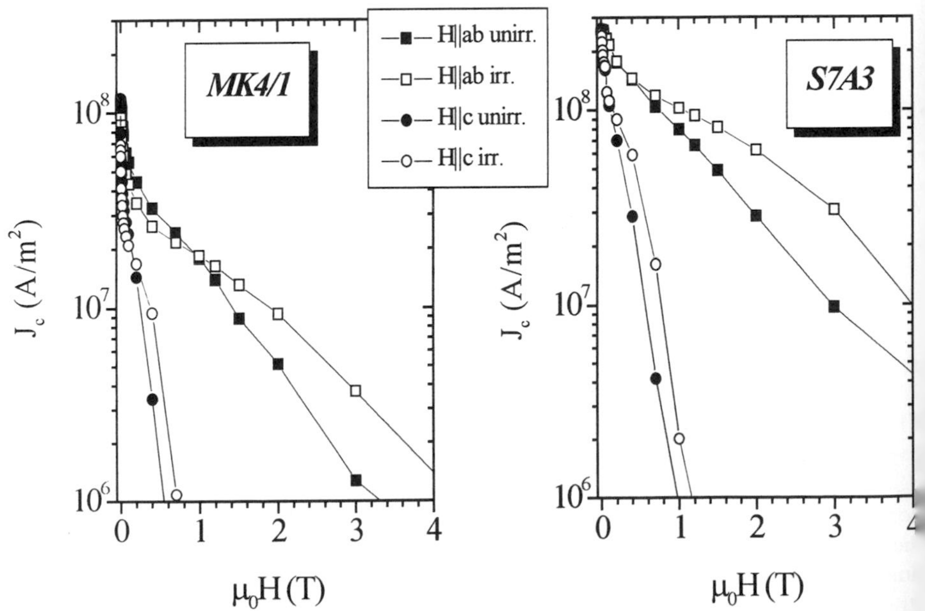

Fig.1: J_c as a function of the applied magnetic field at 77 K. The samples were irradiated to a fluence of $2 \cdot 10^{21}$ m^{-2}.

the new pinning centers.

The cross-over between the unirradiated and the irradiated state separates the weak link dominated regime from the pinning dominated regime. The cross-over for H||ab occurs at 900mT (MK4/1) and at 400 mT (S7A3), respectively. The smaller weak link dominated range for the latter sample can also be seen for H||c, where the cross-over occurs at 10 mT in contrast to 150 mT in MK4/1. With decreasing temperature all cross-overs shifts to higher magnetic fields.

We also investigated the anisotropy of the critical current by measuring J_c as a function of the orientation of the tape with respect to the external magnetic field. 0° corresponds to H||ab and 90° to H||c. The effect of the pinning centers is also found in the rotation measurements [Fig. 2]. The intersection shifts to lower angles with increasing field. The anisotropy decreases after irradiation due to the isotropically distributed artificial defects.

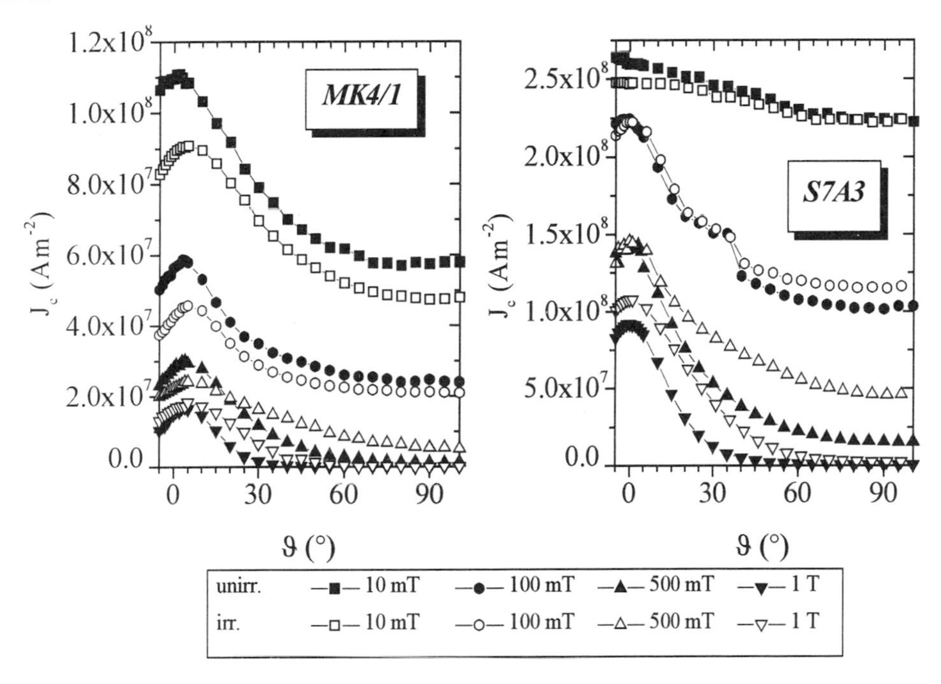

Fig. 2: Rotation measurement at 77 K. The samples were irradiated to a fluence of $2 \cdot 10^{21}$ m^{-2}.

4. Results on Tl-1223

Tl-12223 is in principle more suitable than Bi-2223 because of its higher T_c and its higher irreversibility line. However, J_c is considerably lower, because the grains in the tape are not well aligned and not well connected and the critical current is completely weak link dominated. This can be seen from the hysteresis of J_c in external magnetic fields and from the dramatic decrease of J_c with increasing field. Further evidence for the weak link dominated behavior is provided by the field [Fig. 3] and the angular dependence of J_c after irradiation. J_c decreases at all fields, since pinning was not the limiting factor for the critical

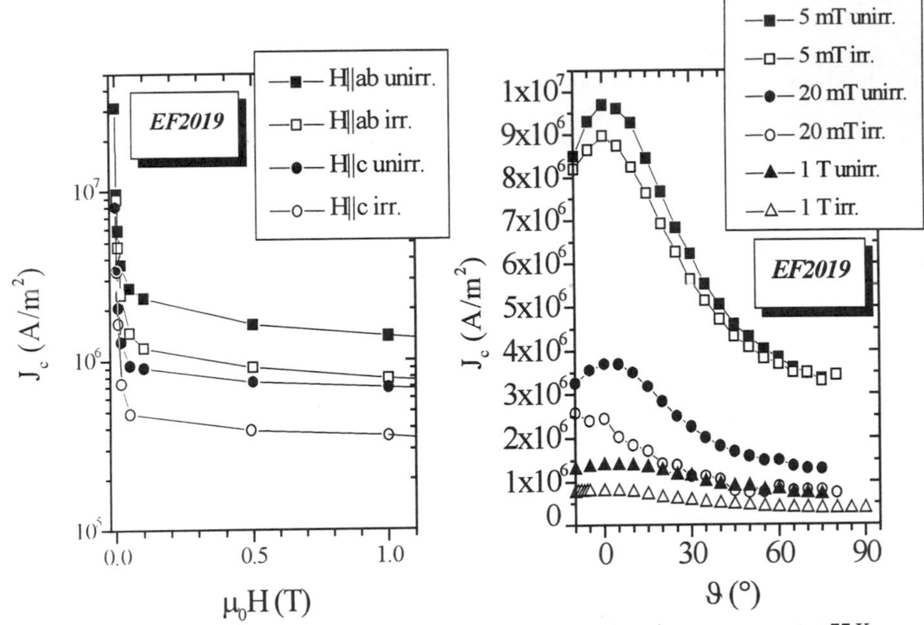

Fig. 3: J_c as a function of the applied magnetic field at 77 K. The sample was irradiated to a fluence of $2\cdot10^{21}$ m^{-2}.

Fig. 4: Rotation measurement at 77 K.

current. In the rotation measurements [Fig. 4], no intersection between the unirradiated and irradiated state, as reported before for Bi-2223, is found.

5. Summary

The critical current density J_c of Bi-2223 and of Tl-1223 tapes was measured before and after irradiation with fast neutrons. The transition from the weak link dominated regime to the pinning dominated regime shifts to lower fields, when the hysteresis of J_c becomes lower. This transition could not be found in Tl-1223, because the tapes are completely weak link dominated.

Acknowledgment

This work is supported in part by the EU Brite Euram program under contract no. BRE2-CT93-0531.

References

[1] L.N.Bulaevskii, L.L.Daemen, M.P.Maley, and J.Y.Coulter, Phys. Rev. B **48** 13798 (1993)
[2] Q.Y.Hu, H.W.Weber, S.X.Dou, H.K.Liu, and H.W.Neumüller, J. Alloys Comp. **195** 515 (1993)
[3] B.Hensel, G.Grasso, and R.Flükiger, Phys. Rev. B **51** 15456 (1995)
[4] A.P.Malozemoff, G.N.Riley, Jr., S.Fleshler, and Q.Li, in Proc. SPA'97, Xi'an, China
[5] M.C.Frischherz, M.A.Kirk, J.Farmer, L.R.Greenwood, and H.W.Weber, Physica C **232** 309 (1994)

Inst. Phys. Conf. Ser. No 158
Paper presented at Applied Superconductivity, The Netherlands, 30 June–3 July 1997

On the origin of the so-called fishtail effect in single crystals of the RE - 123 compounds (RE = Y, Er, Nd)

Andreas Erb, Jean-Yves Genoud, Marc Dhalle, Frank Marti, Eric Walker and René Flükiger,

Département de Physique de la Matière Condensée, Université de Genève, 24, quai Ernest Ansermet, 1211 Genève 4 Switzerland

Abstract. We report on experiments performed on twinned crystals grown in the recently developed non reactive crucible material $BaZrO_3$. Due to the very high purity (5 N) the experiments are not obscured by residual impurity effects [1,2]. In $YBa_2Cu_3O_{7-\delta}$ the so-called fishtail effect in the magnetisation curves can be suppressed and re-established by appropriate annealing procedures with or without changing the overall oxygen content. Thus, only a locally altered distribution of the oxygen vacancies e.g. a clustering of the oxygen deficient regions must be responsible for this anomaly [3]. For other rare earth systems additional complications occur due to inhomogeneities in the metal sublattice. Again, eliminating these microstructural inhomogeneities leads to the absence of the fishtail anomaly. Combining the influences of different origin on the critical currents gives way to optimised properties.

1. Introduction

The origin of the sample dependent anomaly in the irreversible magnetisation hysteresis curves in single crystals of the 123- high T_c superconductors, often referred to as fishtail effect is still controversially discussed. The understanding of this feature however is important for both application purposes as well as for fundamental research.

2. Experimental

We measured the irreversible magnetisation of single crystals of 123 compounds with different rare earth elements and mixtures of rare earth elements. For this purpose the single crystals have been grown in $BaZrO_3$ crucibles, which avoids contamination with metallic impurities. Starting materials were of 99.999 at. % purity, the purity of the resulting crystals has been found to be better than 99.995 at. % as determined by wet chemical analysis (ICP-MS). To study the influence of the substitution of Y by rare earth elements (Nd , Er) on the pinning properties a series of $YBa_2Cu_3O_{7-\delta}$-, $NdBa_2Cu_3O_{7-\delta}$-, $ErBa_2Cu_3O_{7-\delta}$ - and of $Nd_{0.41}Er_{0.59}Ba_2Cu_3O_{7-\delta}$ single crystals has been prepared. In the case of pure $YBa_2Cu_3O_{7-\delta}$ the anomalous increase in the irreversible magnetisation curves can be suppressed by an annealing which leads to homogenous distribution of the oxygen vacancies, eg. a high temperature high pressure annealing to avoid the formation of oxygen vacancy clusters. An other way to suppress the anomaly is to fully oxygenate the $YBa_2Cu_3O_{7-\delta}$ crystals to a state of O_7. Both these treatments remove structural inhomogeneities of the oxygen sublattice big enough to pin the vortices.

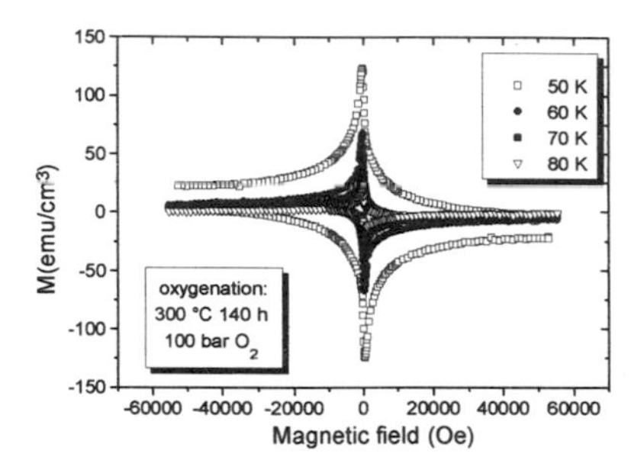

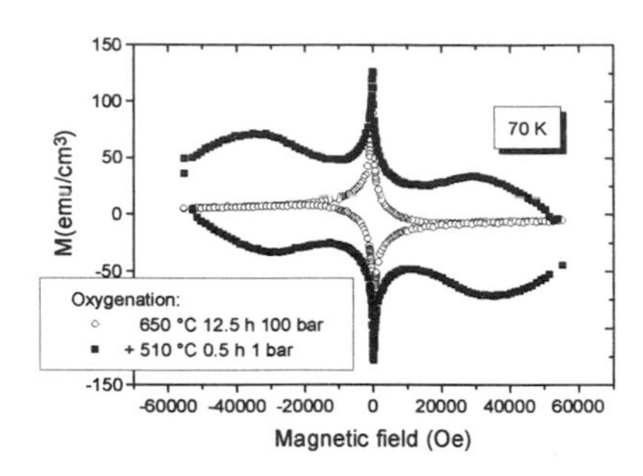

Fig. 1 : Magnetisation curves for pure $YBa_2Cu_3O_{7-\delta}$ single crystals after different heat treatments resulting in a complete oxygenation $YBa_2Cu_3O_{7.0}$ (left) and optimal doping (right).

For pure $YBa_2Cu_3O_{7-\delta}$ it has therefore been concluded that the existence of a fishtail anomaly is due to an inhomogenous distribution of oxygen vacancies, leading to the formation of clusters with lower H_{c2} and hence to an increase of the pinning on increasing magnetic field [3]. This explanation has already been proposed by Vargaz et al. [4], however experimental proof was lacking due to the lack of sufficiently pure samples. Another way to explain the fishtail effect using models of a crossover of different pinning regimes like for example in Ref. [5], assume that the fishtail effect is an intrinsic property of $YBa_2Cu_3O_{7-\delta}$. However, it is hard to understand how these scenarios should work in pure samples, where the effect can be switched on and off according to the annealing regimes as we have shown above.

3. Results and Discussion

3.1 $NdBa_2Cu_3O_{7-\delta}$

Microstructural inhomogeneity can be avoided or favoured by appropriate sample preparation methods. For melt-textured material normally high critical currents are desired and efforts have been undertaken to deliberately include such local inhomogeneities by doping with Pt and other metals. The problem of doping with other metals is that the size of the precipitates of the foreign phases supposed to provide pinning are often too large to be compared to the coherence length ξ. The light rare earth 123-compounds, like Nd and Sm , however show an intrinsic tendency to produce a local variation in the metallic sublattice upon slow cooling as it has been shown by Nakamura et al [6]. Again this local variation on the Ba and RE site can be suppressed by appropriate preparation and annealing methods.

A major disadvantage of the $NdBa_2Cu_3O_{7-\delta}$ system when compared to the $YBa_2Cu_3O_{7-\delta}$ is that the oxygenation temperature for obtaining T_c values above 90 K is very low as it can be seen from Fig. 2, where the T_c values for different rare earth 123 compounds as a function of annealing temperature are given [2]. In the case of the $NdBa_2Cu_3O_{7-\delta}$ the maximum values for T_c are found after annealing procedures at temperatures which are about 200°C lower than in the case of $YBa_2Cu_3O_{7-\delta}$ or $ErBa_2Cu_3O_{7-\delta}$. Using the values for the diffusion coefficient for oxygen in the 123 compounds [7] one finds that the time required for a homogenous oxygenation is about 100 times longer for a massive sample of $NdBa_2Cu_3O_{7-\delta}$ than it is for $YBa_2Cu_3O_{7-\delta}$. For practical applications, however, this circumstance results in a serious prolongation of the time needed for the production of optimized samples for levitation devices.

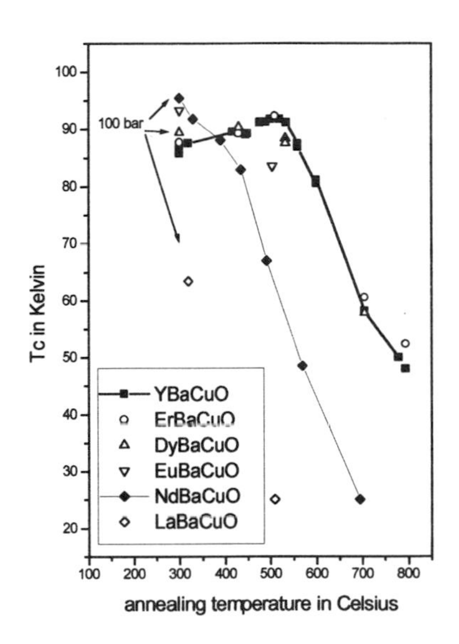

Fig. 2 T_c - values for different RE - 123 single crystals in dependence of the annealing temperature for 1 bar oxygen atmosphere (except the 300 °C treatment)

For application purposes the Nd System is superior to the pure YBCO System since it combines the local inhomogeneities of both, oxygen vacancy clusters like in the case of $YBa_2Cu_3O_{7-\delta}$ and - as an additional source for strong pinning - the local variation in the Nd- and Ba - sublattice. Both mechanisms contribute to the magnetisation curves obtained after a normal oxygenation procedure at 320 ° C following slow cooling, as it can be seen in Fig 3, producing a broad maximum in the J_c curves with a structure suggesting 2 maxima. Again, eliminating the local inhomogeneity by an appropriate annealing leads to a complete absence of the fishtail effect(also shown in Fig.3 right)

The high temperature annealing (700 °C 20 h 100 bar O_2 which is suitable to suppress both the oxygen clustering as well as the spinodal decomposition of the Nd 123 system lowers T_c to only about 50 K in the case of the Nd system. Hence, the oxygen content has to be raised by a further annealing treatment at low temperature to increase the transition temperature.

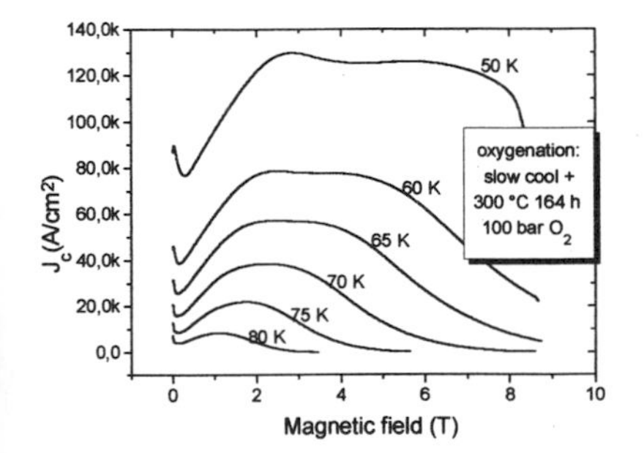

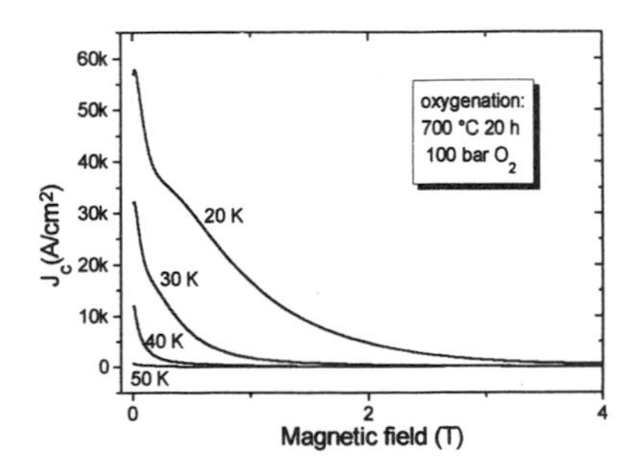

Fig. 3 Critical currents for a $NdBa_2Cu_3O_{7-\delta}$ single crystal (AE159G) after different annealing regimes.

The result of such an annealing at 320 °C 100 bar, which rises T_c again back to a value of 92 K is compared with the original state after the slow cool low temperature annealing (dashed lines) in Fig. 4. Evidently the low oxygenation temperature does not allow the metal atoms to change site and to re-establish the state with a local inhomogeneity of the metallic sublattice, since the high J_c values at higher fields do not re-establish. This is explained by the fact that the site change of metallic ions requires a higher activation energy than the diffusion of oxygen.

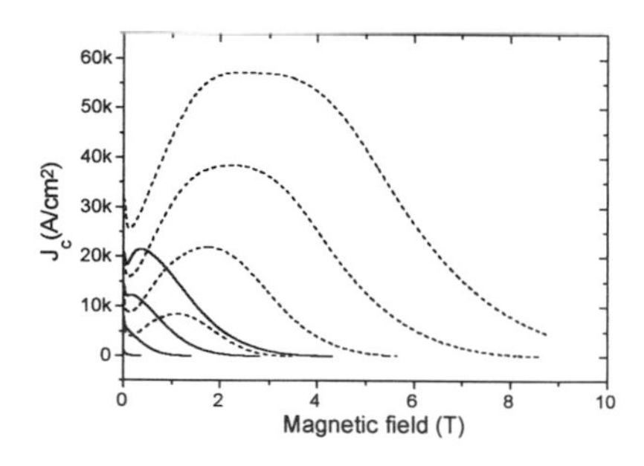

Fig.4 : $NdBa_2Cu_3O_{7-\delta}$ single crystal AE159G after different annealing regimes. J_c at 65, 70, 75 and 80 K(see text)

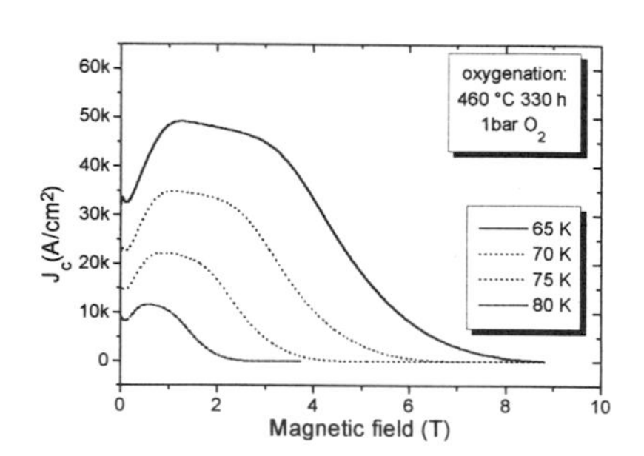

Fig. 5 : $Nd_{0.41}Er_{0.59}Ba_2Cu_3O_{7-\delta}$ single crystal, Magnetisation curves

3.2 $Nd_{0.41}Er_{0.59}Ba_2Cu_3O_{7-\delta}$

As mentioned above the major disadvantage of the $NdBa_2Cu_3O_{7-\delta}$ system for practical application lies in the very low annealing temperatures required for achieving the highest T_c values. This can be partly overcome by addition of a heavier RE - element. Doping the Nd system with Er leads to an pronounced increase of T_c for the same annealing temperature when compared to the pure Nd system as it can be already expected from Fig.2. For a $Nd_{0.41}Er_{0.59}Ba_2Cu_3O_{7-\delta}$ single crystal we obtain a T_c of 89 K after oxygenation at 460 °C. While the J_c values (Fig. 5) are comparable to those for the pure $NdBa_2Cu_3O_{7-\delta}$ single crystal (Fig. 4) the diffusion coefficients are raised by more than an order of magnitude, hence avoiding the major disadvantage of the pure $NdBa_2Cu_3O_{7-\delta}$ system.

References

[1] A. Erb, E. Walker, R. Flükiger, Physica C 245 (1995) 245; A. Erb, E. Walker, R. Flükiger, Physica C 258 (1996) 9

[2] A. Erb et al. 1997 Proc. of the M^2HTSC-V conference Beijing, Peoples Republic of China, to be published in Physica C

[3] A. Erb, J.-Y. Genoud et al. J. Low. Temp. Phys. 105, 1023 (1996)

[4] J. L. Vargas and D. C. Larbalestier, Appl. Phys. Lett. 60, 1741 (1992)

[5] Y. Abulafia et al., Phys. Rev. Lett. 77, 1596 (1996)

[6] M. Nakamura, Y. Yamada, T. Hirayama; Y. Ikuhara, Y. Shiohara, S. Tanaka, Physica C 259 (1996) 295

[7] A. Erb , B. Greb, G. Müller-Vogt, Physica C 259 (1996) 83

Inst. Phys. Conf. Ser. No 158
Paper presented at Applied Superconductivity, The Netherlands, 30 June–3 July 1997

Evidence of a Mixed State Transition Region Between the Vortex Solid and Liquid Phases in $YBa_2Cu_3O_{7-\delta}$ single crystals*

S.N. Gordeev[a], A.P. Rassau[a], D. Bracanovic[a], P.A.J. de Groot[a], R. Gagnon[b], L.Taillefer[b]

[a]Department of Physics, University of Southampton, Highfield, Southampton, SO17 1BJ, United Kingdom.

[b]Department of Physics, McGill University, Montreal (Quebec), H3A 2T8, Canada

Abstract. Transport and a.c. susceptibility measurements have been performed on detwinned $YBa_2Cu_3O_{7-\delta}$ single crystals in the vicinity of the melting line. We have extracted threshold current (I_v) versus temperature dependences from a set of $V(I)$ curves using three different voltage criteria. These dependences showed a sharp rise over a transition region in the vicinity of T_m, the width of which was independent of the particular voltage criterion. A similar sharp transition was observed in the out of phase component of the a.c. susceptibility, the width of which was independent of the applied a.c. field. We propose that within the transition region, the vortex solid and liquid phases coexist. From a comparison of the extracted $I_v(T)$ curves with a low current ohmic $R(T)$ dependence we estimate that the resistivity falls to zero when between 10 and 30 % of the vortices are in the solid phase. We explain this observation in terms of vortex liquid flowing in channels between solid domains.

1. Introduction

Recent calorimetric [1] and magnetisation [2] studies have revealed that, in clean detwinned $YBa_2Cu_3O_{7-\delta}$ single crystals, the phase transition from the vortex liquid to the vortex solid state is first order. Theory predicts [3] that on increasing T above the melting temperature T_m, the shear modulus C_{66} of the vortex lattice should show a sharp decrease thereby leading to an increase the mobility of the vortices. This effect has been observed using a number of different techniques. In transport measurements it is seen as a sharp rise in the resistivity at T_m. In magnetic measurements the transition manifests itself as a jump in the reversible magnetisation which has been associated with an increase in the vortex density. However, point-like disorder is known to push the melting line down to lower temperatures [4], thus a broadening of the phase transition is expected in samples where the point disorder has some spatial variation. Very recently, Fendrich *et al.* [5] demonstrated that the increase in the reversible magnetisation at T_m actually occurs over a temperature range of about 0.2 K. This implies that between the liquid and solid phases there exists a transition region (TR) in which both phases coexist. In this paper we present results of transport and a.c. susceptibility studies

* This work is part of a project supported by EPSRC (UK). S.G. acknowledges support from RFFR (Russia), grant No. 96-02-18376a.

on clean detwinned $YBa_2Cu_3O_{7-\delta}$ single crystals. We demonstrate that both methods give further evidence for the existence of a transition region between the liquid and solid states.

2. Experimental

The experiments described in this paper were performed on two detwinned $YBa_2Cu_3O_{7-\delta}$ single crystals from the same batch [6]. The crystal used in transport measurements had dimensions 2.10 mm × 0.75 mm × 85 μm. Current contacts were painted onto the sides of the sample providing a uniform current flow along the *ab*-plane. The transport measurements were performed with a SQUID picovoltmeter using standard lock-in techniques. The probing current ($J//ab$) was modulated with a square wave of frequency f_m = 68Hz. All transport measurements were performed at a field of 2 T (with $H//c$). During measurements, the temperature stability of the system was ~ 1 mK. The crystal used in the a.c. susceptibility measurements had dimensions 1.69 mm × 1.08 mm × 91 μm. These measurements were performed using a coaxial mutual inductance system with superimposed a.c. and d.c. fields.

3. Results and discussion

Shown in Fig. 1 are three sets of voltage-current curves measured at B = 2 T. The first set of curves in the sequence (fine lines) run from 90.0 K to 89.2 K separated by a temperature interval of 0.1 K. These curves can be seen to be ohmic and thus correspond to the vortex liquid state. The next set of curves (bold lines) encompass the temperatures between 89.18 K and 88.92 K with a temperature step of 20 mK. Within this transition region the shape of the *V-I* curves evolves rapidly with decreasing temperature. The low-current section of each of these curves is ohmic whereas the high-current part shows a pronounced negative curvature. The final set of curves (fine lines) consist of temperatures in the range 88.9 K to 87.0 K again with a temperature step of 0.1K. These curves are highly nonohmic which is typical of the vortex solid state.

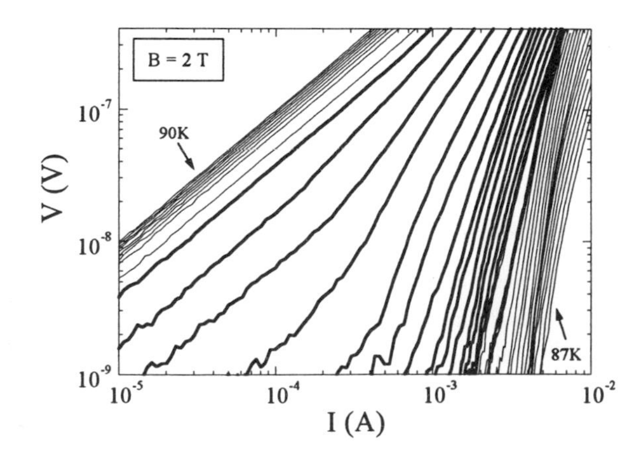

Figure 1. Current-voltage curves measured at B = 2 T. The thin lines are separated by a temperature interval of 0.1 K whilst the thick lines have a separation of 0.02 K.

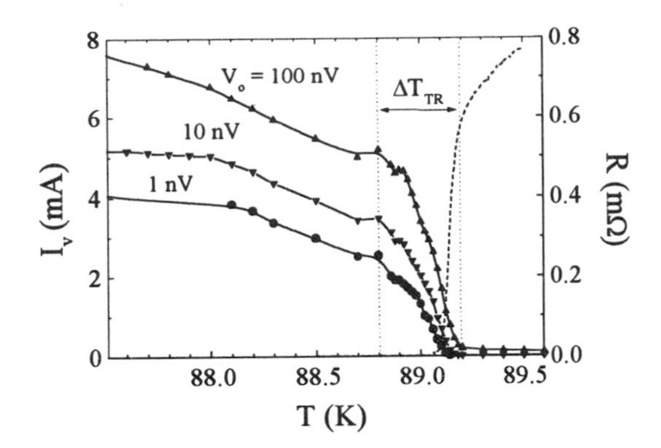

Figure 2. Threshold current versus temperature plots at $B = 2$ T extracted from Fig. 1 using three different voltage criteria (V_o). The dashed line shows the ohmic $R(T)$ dependence for $I = 0.1$ mA.

Clear evidence of a transition region is seen in Fig. 2 where the $I_v(T)$ curves, extracted from Fig.1, are shown. The transport current I_v which induces the fixed voltage V is comparatively high in the solid state and has very low value in the liquid state. We have found that the temperature width of the transition region (ΔT_{TR}) over which I_v drops is sample and field dependent. For the data shown above $\Delta T_{TR} = 0.40$ K but for different samples (at $B = 2$ T) we observed that the ΔT_{TR} values varied over the approximate range 0.1 K - 0.4 K.

There are two mechanisms which could contribute to the rapid change in the current over the TR. The first of these is the softening of the C_{66} modulus across the vortex system as a whole. The second possible contribution is due to the fact that even slight spatial variations of the melting temperature (as a result of sample inhomogeneity) could lead to a broadening of the transition. As shown in Fig. 2, the width of transition region is independent of the particular voltage criterion used thus the second contribution seems to be more important.

Fig. 3 shows the temperature dependence of the out of phase component of the a.c. susceptibility χ'' and its derivative $d\chi''/dT$ for several different a.c. fields at d.c. field $B = 2.2$ T in the vicinity of T_m. These data demonstrate that the a.c. dissipation increases

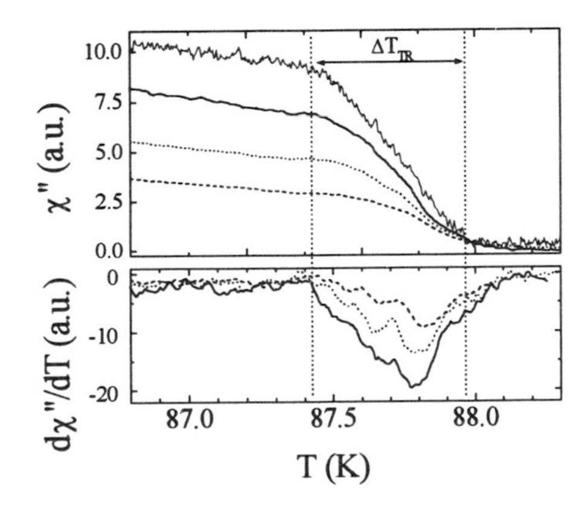

Figure 3. Temperature dependence of the out of phase component of the a.c. susceptibility and its derivative at d.c. field $B = 2.2$ T for a.c. fields of: 25 μT (normal line), 50 μT (bold lines), 100 μT (dotted lines) and 200 μT (dashed lines)

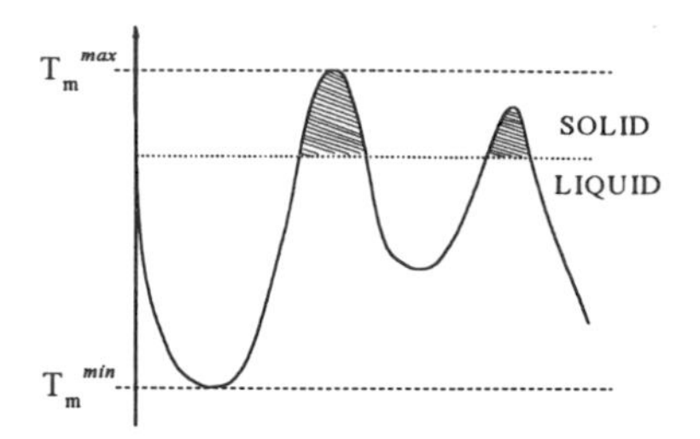

Figure 4. An example of how the local melting temperature could vary across the width of a sample.

strongly across the transition region as the temperature decreases. The shape of the $\chi''(T)$ curves is very similar to the $I_v(T)$ dependences presented in Fig.2. An increase of the amplitude of the applied a.c. field leads to an increase of the density of shielding currents induced in the sample and also to an increase of the electric field within the sample. As shown in Fig. 3, it does not affect the temperature width of the transition region. This is clearly seen from the temperature dependences of $d\chi''/dT$. Even the positions of the small peaks on these dependences remain the same for different a.c. fields. All these observations seem confirm that the structure of the vortex system within TR is determined by small spatial variations of the melting temperature.

We propose the following explanation for the existence of a transition region at T_m. A slight degree of inhomogeneity is expected even in high quality samples (this could, for example, arise due to variations in the oxygen content). Thus due to this inhomogeneity or as a result of surface barriers, there should be a weak spatial variation in the local melting temperature. The width of the transition region ΔT_{TR} depends on the range of variation of the local melting temperature. Within the transition region, vortices solidify at points in the sample where the local melting temperature is at its highest (see Fig.4). Thus in the TR the vortex liquid and liquid phases coexist. When a transport current is applied the dissipation which arises is mainly due to the flow of the vortex liquid within the channels between the solid domains. Thus the *V-I* curves at low currents are expected to be ohmic. At higher currents the Lorentz force becomes large enough to move vortices within the solid domains. At this point the dissipation increases at the same time becoming nonohmic. As T decreases the width of the channels also decreases the ohmic resistance falling to zero upon closing of the last channel between solid domains. We have found that for different samples the temperature width of the ohmic resistivity drop varies between 0.1 and 0.3 of ΔT_{TR} (see Fig. 2). This implies that the channels close when the proportion of vortices in the solid phase is between 10 % and 30 % of the total.

In summary, we have observed a transition region at the melting line in both transport and a.c. susceptibility measurements. We explain the transport properties in this region in terms of a model of coexistent vortex liquid and solid phases.

References

[1] Schilling A *et al.* 1996 Nature **382** 791-793
[2] Liang R *et al.* 1996 *Phys. Rev. Lett.* **76** 835-838
[3] Larkin A I *et al.* 1995 *Phys. Rev. Lett.* **75** 2992-2995
[4] Giamarchi T and Le Doussal P 1997 *Phys. Rev. B* **55** 6577-6583
[5] Fendrich J A *et al.* 1996 *Phys. Rev. Lett.* **77** 2073-2076
[6] Gagnon R *et al.* 1994 *Phys. Rev. B* **50** 3458-3462

Inst. Phys. Conf. Ser. No 158
Paper presented at Applied Superconductivity, The Netherlands, 30 June–3 July 1997

Optimized T_c, B_{irr} and J_c for Substituted Y-123 Materials

B. Dabrowski[a], K. Rogacki[a], O. Chmaissem[b], J.D. Jorgensen[b], J.W. Koenitzer[c] and K.R. Poeppelmeier[c]

[a]Physics Department, Northern Illinois University, DeKalb, IL 60115, USA
[b]Materials Science Division, Argonne National Laboratory, Argonne, IL 60439, USA
[c]Department of Chemistry, Northwestern University, Evanston, IL 60208, USA

Abstract. Several Y123 related compounds doped with the transition elements for Cu were synthesized to study the relationship between fundamental properties of the blocking layer and T_c, B_{irr} and J_c. Synthesis conditions and compositions were optimized to obtain the highest T_c's for $YSr_2Cu_{3-x}M_xO_z$ and $YBaSrCu_{3-x}M_xO_z$. For $YSr_2Cu_{3-x}M_xO_z$ compounds, using elevated oxygen pressure annealing at ~ 650 C, the oxygen content was increased to z ~ 7.30 and T_c's were increased to ~ 77 K for M = Re, W and Mo (x = 0.2) by substitution of small amounts of Ca for Y. For the $YBaSrCu_{3-x}M_xO_z$ compounds, the highest transitions, T_c = 87 K, were obtained for M = Mo with x ~ 0.075 after high oxygen pressure anneal at 600 C. The Mo-substituted material is orthorhombic for x < 0.1 and tetragonal for x > 0.1. Oxygen content is above 7 and Mo substitutes on the Cu-"chain" site. The scaled irreversibility fields are better than for Y123.

1. Introduction

$YBa_2Cu_3O_7$ (Y123) compound is of special importance for the power and magnet applications of high temperature superconductors because it retains high critical current, J_c, in intense magnetic fields and displays large irreversibility field, B_{irr}, at 77 K. High J_c and B_{irr} are caused by the strong intrinsic flux pinning by the short and metallic blocking-layer (the CuO chains). Short and metallic blocking layer provides a strong coupling between the two-CuO_2-planes which is mediated through the relatively thin intermediate region of the structure consisting of the blocking layer, d(block), and twice the distance between the Cu(plane) and O(apical), d(apical)[1,2]. Recently, we have investigated how B_{irr} and J_c of the compounds with two-CuO_2-planes depend on structural features of the blocking layer[1]. We report here the synthetic, structural and superconducting properties of the Y123 materials containing smaller Sr in place of a larger Ba in the intermediate region of the structure and transition elements substituted in place of Cu in the blocking layer.

2. Optimized Synthesis

Polycrystalline samples of $Y_{1-y}Ca_ySr_2Cu_{3-x}M_xO_z$ and $YBaSrCu_{3-x}M_xO_z$ (M = transition element) were synthesized from stoichiometric mixtures of the oxides and carbonates in air. Samples were fired several times at increasing temperatures, checked for phase purity with x-ray diffraction, annealed under several oxygen pressures and temperatures, and T_c measured.

The high pressure annealings were done for selected samples in pure oxygen (220 atm. O_2) at 400-900 C followed by slow cooling (0.2 deg/min.) to room temperature. Our goal was to maximize the superconducting T_c by determining the optimum synthesis and annealing conditions of temperature and oxygen pressure.

Pure $YSr_2Cu_3O_7$ compound with Sr replacing Ba can be prepared only at very high oxygen pressures [5,6]. This material was stabilized at atmospheric pressure by the partial substitution for copper in the blocking layer with small transition or post-transition metals[7,8]. The highest superconducting T_c was observed at a minimum amount of substitution for Cu required to form the single-phase material. T_c is related to the formal valence of the substituted ion; the higher valences corresponding to the higher observed T_c's. We have achieved the highest $T_c = 77$ K with M = W and additional substitution of Ca for Y in $Y_{1-y}Ca_ySr_2Cu_{3-x}M_xO_z$ (x = 0.2, y = 0.1 and z =7.3)[1]. Slightly lower T_c's, ~75 K, were obtained for Mo and Re. The optimum preparation conditions were determined to be, in the first step, firing in air at 990 C, followed by high oxygen pressure anneal at 600 C.

The Sr substituted material for Ba, $YBaSrCu_3O_z$, can be prepared in air with maximum T_c = 82 K[9]. We have found that substitution of transition elements for Cu decreases T_c for samples synthesized in air. However, in many cases T_c's of 80 K could be recovered after the high oxygen pressure anneal. The largest increases of T_c to 87 K were observed for Mo substitution[10]. This is the first instance of an improved superconducting properties of Y123 material with substitution of transition element for copper. Figure 1 shows normalized resistivity for the single phase samples of $YBaSrCu_{3-x}Mo_xO_z$ with x = 0, 0.1 and 0.2 after the anneal at 1 atm. and at high oxygen pressure. T_c's depend on both the doping level x and oxygen content z. The highest transitions were obtained for x = 0.05 - 0.1.

3. Structure and Oxygen Content

Neutron powder diffraction data were obtained for ~ 3g samples at room temperature. Rietveld structural refinements were carried out using the GSAS code[11] over the range 0.5 < d < 4.0 Å. The patterns were indexed using a tetragonal cell, space group P4/mmm for $YSr_2Cu_{3-x}M_xO_z$ (M = W, Mo, Re). A model of the structures with a split chain-oxygen position consisting of a mirror-site (0, 0.5, 0) and off-mirror site (x, 0.5, 0) gave best fit.

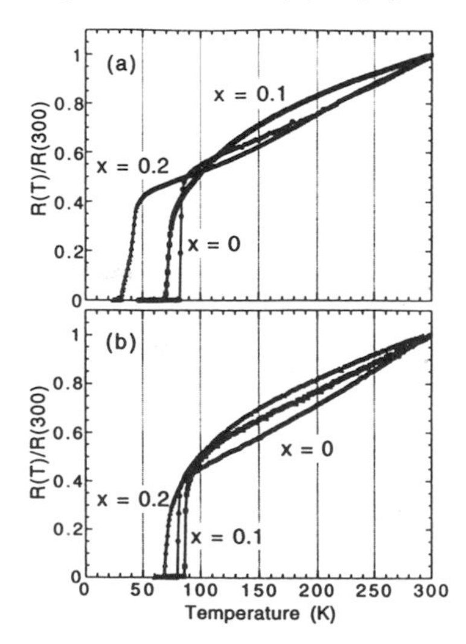

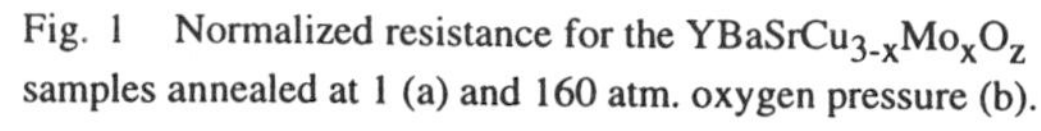
Fig. 1 Normalized resistance for the $YBaSrCu_{3-x}Mo_xO_z$ samples annealed at 1 (a) and 160 atm. oxygen pressure (b).

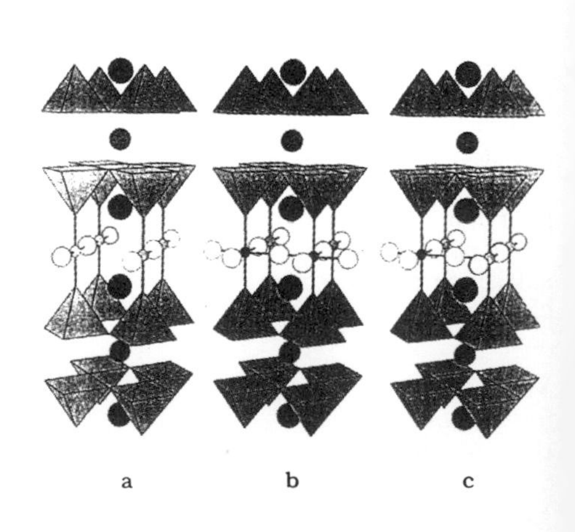

Fig. 2 Model defect structures for Y123 (a), dimmers of corner shared Mo and W octahedra (b), and isolated Re octahedra (c).

The substituted cation M was found to occupy solely the Cu-"chain" site in the blocking layer. The oxygen stoichiometries of all samples, z = 7.30(3), were obtained from the oxygen occupancies. The refined oxygen stoichiometries are approximately equal to the oxygen contents found by replacing chain Cu^{3+} with M cations in their highest allowed oxidation states consistent with the preparation conditions (Mo^{6+}, W^{6+}, and Re^{7+}). For Mo and W substitutions, the maximum observed T_c occurred at doping levels of x = 0.20 while for Re, at a doping level of x = 0.15. Therefore, three oxide anions are added to the chain layer for every 2 Mo or W cations substituted, while two oxide anions are added for every Re cation. Since the substituted cations exist in an octahedral environment in the blocking layer, the following model accounts for the observed oxygen stoichiometries. In the case of Mo and W, randomly distributed dimmers of corner shared octahedra give the observed 2/3 (Mo, W)/O ratio, while, in the Re sample, isolated Re octahedra are consistent with the observed 1/2 Re/O ratio. The model structures compared to Y123 are shown in Figure 2.

Refinements for $YBaSrCu_{3-x}Mo_xO_z$ showed that the x = 0.05 sample crystallizes in an orthorhombic space group Pmmm while the x≥0.1 samples crystallize in the tetragonal space group P4/mmm. The substituted Mo was found to occupy solely the Cu-chain site; Ba and Sr were found to share the Ba site. The apical oxygen atom had a large thermal factor owing to the multiple arraignments of the Sr and Ba atoms, and the Cu and Mo atoms. The chain-oxygen atoms in the blocking layer were found to occupy two independent sites (0, 0.5, 0) and (x, 0.5, 0). The total refined oxygen stoichiometries 7.07(4), 7.16(3), and 7.30(3) for x = 0.05, 0.1, 0.2, respectively indicate that the ratio of nominal Mo to the amount of excess oxygen in the blocking layer is very close to 2:3 in all three samples. Thus, the model of randomly distributed dimmers of corner shared MoO_6 octahedra in the blocking layer accounts again for the observed oxygen stoichiometries in a similar manner to $YSr_2Cu_{3-x}Mo_xO_y$.

4. Superconducting Properties

To study the superconducting properties, the ac susceptibility (χ) and dc magnetization were measured for solid pieces and powdered samples in applied field from 0 to 7 T with a Quantum Design PPMS system. The transition onsets, T_{co}, of the real part of χ decrease slowly with increasing magnetic field. Because the temperature at which χ reaches 3%, $T_c(3\%)$, is close to the temperature at which the resistivity attains a zero value, the 3% value was chosen to define the irreversibility lines from χ measurements. The irreversibility fields $B_c(3\%)$ were obtained from χ plots for several $Y_{1-y}Ca_ySr_2Cu_{3-x}M_xO_z$ compounds. The highest $B_c(3\%)$ field was observed for the $Y_{0.9}Ca_{0.1}Sr_2Cu_{2.8}W_{0.2}O_z$ sample.

The dc magnetization measurements were performed to determine the irreversibility field, B_{irr}, and the upper critical field, B_{c2}. Large differences were observed for measured critical fields B_{c0} and B_{irr}, similar to those observed for other high temperature superconductors. The large difference indicates a substantial influence of thermally activated processes on weak pinning down to 30 K. Smaller difference was found between $B_c(3\%)$ and B_{irr} indicating that χ measurements can be conveniently used to ascertain B_{irr}.

The critical current density, J_c, was determined from magnetization loops for powdered samples by applying the Bean formula. The grain diameter was used as a scaling length because the samples showed granular behavior in ac measurements at dc fields larger than 1 T. Figure 3 shows the critical current at 5, 20 and 30 K for the $Y_{0.9}Ca_{0.1}Sr_2Cu_{2.8}W_{0.2}O_z$ sample. At 5 K the calculation gives $J_c \sim 5x10^5$ and $2x10^5$ A/cm^2 for fields of 1 and 7 T, respectively, showing only weak dependence on the applied field. However, J_c drops much more rapidly with increasing magnetic field at elevated temperatures. Similar ac susceptibility and dc magnetization measurements were performed for the $YBaSrCu_{3-x}Mo_xO_z$ samples with x = 0, 0.05 and 0.1 annealed at high oxygen pressure. Figure 4 shows the irreversibility fields vs. scaled temperature obtained from ac and dc measurements and compared with B_{irr} for the single-crystals of several compounds with different T_c's. When plotted on the reduced temperature scale, the Sr- for Ba and Mo-substituted materials have scaled irreversibility fields better than pure Y123, however at 77 K,

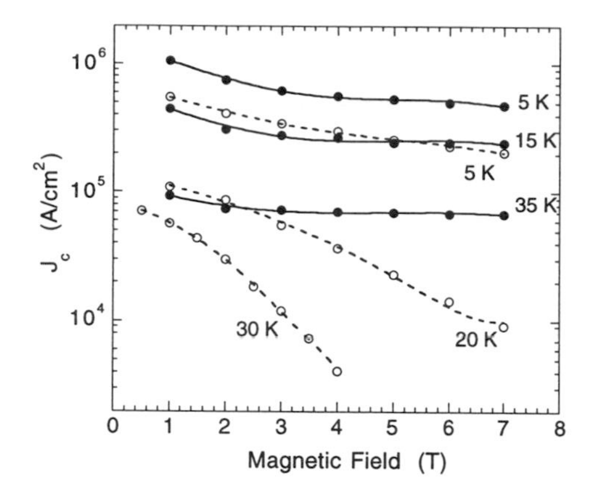

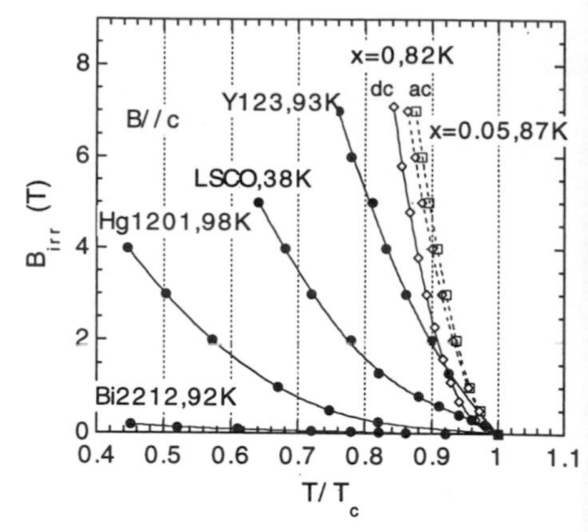

Fig. 3 Critical current for the $Y_{0.9}Ca_{0.1}Sr_2Cu_{2.8}W_{0.2}O_z$ sample at 5, 20 and 30 K (open symbols) and $YBaSrCu_{2.9}Mo_{0.1}O_z$ at 5, 15 and 35 K (filled symbols).

Fig. 4. B_{irr} vs. reduced temperature for the $YBaSrCu_{3-x}Mo_xO_z$. B_{irr} for the single-crystals of several compounds is also shown.

pure Y123 has larger B_{irr}. Thus, the intrinsic flux pinning may be significantly modified by the Sr- for Ba and Mo-substitution in the intermediate region of the structure. Also the critical current densities, J_c , shown on Fig. 3 are remarkably large even at temperatures much above 30 K. At 35 K, the estimated J_c is about 10^5 A/cm^2 at 1 T and shows very weak dependence on the applied field. The $YBaSrCu_{3-x}Mo_xO_z$ materials seem to be interesting for applications at temperatures as high as 70 K where J_c of is about 10^3 A/cm^2 at 3 T.

In conclusion, we have optimized the synthesis conditions and compositions for chemically stabilized $YSr_2Cu_{3-x}M_xO_z$ materials. The highest T_c (~ 77 K), irreversibility field, and critical current density were achieved for the Ca-doped $Y_{0.9}Ca_{0.1}Sr_2Cu_{2.8}W_{0.2}O_z$ sample. Similar optimization of superconducting properties was performed for the $YBaSrCu_{3-x}M_xO_z$ compositions. Mo substitution produces improved materials. Structural and superconducting properties indicate that physical properties of the intermediate region in addition to its thickness determine the effective coupling between the superconducting CuO_2 planes and that it is possible to influence the coupling by the proper choice of substituted metallic ions, formation of complex defects and by modifying the oxygen content.

This work was supported by the National Science Foundation Science and Technology Center for Superconductivity under grant #DMR 91-20000 (BD, KR, OC, JWK, KRP) and the U.S. Department of Energy, BES - Materials Sciences under contract No. W-31-109-ENG-38 (JDJ).

References

[1] Dabrowski B et al. 1997 *Physica C* ***277*** *24*
[2] Shaked H et al. 1994 *Crystal Structures of the High-Tc Superconducting Copper-Oxides, Elsevier Science*
[3] Wagner J L et al. 1993 *Physica C* ***210*** *447*
[4] Jorgensen J D at al. 1996 *Recent Developments in High Temperature Superconductivity, Springler-Verlag, Berlin p. 1-15*
[5] Okai B 1990 *Jap. J. Appl. Phys.* ***29*** *L2180*
[6] Chandrachood M R 1992 *Physica C* ***194*** *205*
[7] Sunshine S A 1989 *Chem. Mater.* ***1*** *331*
[8] Den T and Kobayashi T 1992 *Physica C* ***196*** *141*
[9] Veal B W et al. 1989 *Appl. Phys. Lett.* ***51*** *279*
[10] Dabrowski B et al. (unpublished)
[11] Larson A C and Von Dreele R B 1985-1990 *General Structure Analysis System University of California, Berkeley*

Inst. Phys. Conf. Ser. No 158
Paper presented at Applied Superconductivity, The Netherlands, 30 June–3 July 1997

Current-voltage characteristics in a mixed state of high T_c superconductor

T Kiss, T Nakamura, K Hasegawa, M Inoue, M Takeo, F Irie* and K Yamafuji

Graduate School of Inform. Sci. and Electrical Eng., Kyushu Univ., Fukuoka 812-81, Japan.
*Kyushu Electric Power Co., Inc., Fukuoka 815, Japan.

Abstract. Depinning properties in a random pin medium under the influence of thermal agitation have been studied by the Monte-Carlo method. It has been shown that the depinning probability is scaled on a power function in the vicinity of the depinning threshold, that is the minimum value of critical current density. The simulation results have been compared with the measurements in a $Y_1Ba_2Cu_3O_{7-\delta}$ thin film, then the scaling based on the statistic feature of critical current density is confirmed. The scaling is consistent with that of the well-known glass-liquid transition within the measured electric field region. A new analytical expression describing the current-voltage curves has also been derived, then the origin of the extremely wide scaling regime is discussed.

1. Introduction

Electric field (E) vs. current density (J) characteristics are essential for a practical application of high T_c superconductors (HTSC). The phase diagram of vortex state in HTSC complicatedly depends on materials, however, the observed E-J characteristics seem to have common features such as (1) ln$E$-ln$J$ curves are convex at a temperature, T, below the so called glass-liquid transition temperature, T_{GL}, whereas they are concave at $T>T_{GL}$, (2) the E-J curves are scaled around T_{GL} with the aid of dynamic critical index, z, and static critical index, ν, [1] then (3) the scaling region is much wider than the theoretical prediction [2, 3]. From a practical side, the scaling is important since we can predict the E-J curves over wide range of T and magnetic field, B, by use of the scaling feature. Furthermore, if the relationship among pinning strength, material quality and the scaling properties is clarified, it will be important as an evaluation criterion of HTSC. However, little is known so far about the influence of flux pinning on the scaling.

In this paper, we investigate stochastic property of pinned fluxoids in a random medium by the Monte-Carlo technique. Then, the dynamic behavior of the fluxoids in a mixed state of HTSC are studied. The results are compared with the measurements in a $Y_1Ba_2Cu_3O_{7-\delta}$ (YBaCuO) thin film.

2. Monte-Carlo Study

Assuming random point pins where elementary pinning strength has a random Gaussian distribution, we carried out a Monte-Carlo simulation [4]. The configuration of pinned fluxoids are calculated based on a simplified Labush equation in the presence of thermal agitation. As the temperature is increased, pinning potential is shallowed due to thermal fluctuation, then the critical current density of flux bundle, J_c^{FB}, is decreased. Here, J_c^{FB} decreased due to the thermal agitation is denoted by $\hat{J}_c^{FB}$. Around the transition temperature, T_{GL}, the clusters with $\hat{J}_c^{FB} = 0$ have a fractal distribution as shown in Fig. 1. Namely, the area of the unpinned cluster, A_{cl}, is proportional to a power of its radius, R; *i.e.* $A_{cl}(\hat{J}_c^{FB} = 0) \propto R^{D_F}$ with fractal dimension $D_F \cong 1.7$ [5]. This fact indicates that the correlation length of the unpinned clusters diverges as T approaches T_{GL}.

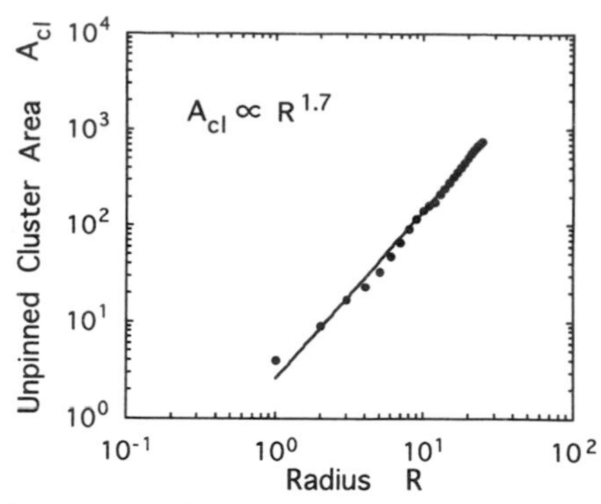

Fig. 1 Fractal distribution of the unpinned clusters around $T \sim T_{GL}$ obtained by the Monte-Carlo simulation, where A_{cl} is the unpinned cluster area inside the circle of radius R.

Figure 2(a) shows critical current density in flux bundle row, $\hat{J}_c^{FR}$, in the direction of Lorentz force under the influence of thermal agitation, where the temperature is represented by the dispersion of Gaussian thermal fluctuation, σ_{th}. As temperature approaches T_{GL}, the minimum value of $\hat{J}_c^{FR}$ denoted by J_{cm} approaches 0 as

$$J_{cm} = J_T \left|1 - T / T_{GL}(B)\right|^{2\nu} \tag{1.a}$$

as shown in Fig. 2(b), where J_T is constant. This scaling feature reflects the divergence of the correlation length among unpinned clusters [5].

Note that the scaling shown in eq. (1.a) is essentially the same kind of phenomena as the well-known scaling of macroscopic pinning force [6] in the vicinity of the irreversibility field because eq. (1.a) can be rewritten as a function of B as follows.

$$J_{cm} = J_T \left|1 - T / T_{GL}(B)\right|^{2\nu} \cong J_B \left|1 - B / B_{GL}(T)\right|^{2\nu}, \qquad \text{with} \quad J_B \cong \left(\left|\frac{\partial T_{GL}}{\partial B}\right| \frac{B_{GL}}{T_{GL}} \right)^{2\nu} J_T. \tag{1.b}$$

where B_{GL} is the transition field.

Furthermore, as shown in Fig. 2(c), the simulation results show that the $\hat{J}_c^{FR}$ distribution is scaled in the from of the Weibull function defined by eq. (2).

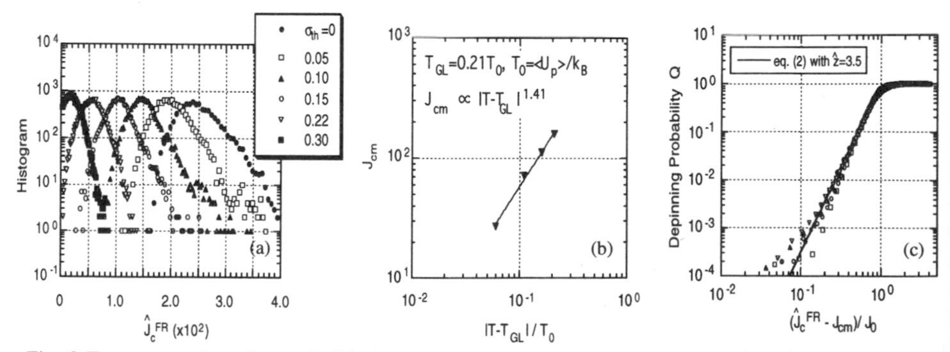

Fig. 2 Temperature dependence of critical current density in flux bundle row obtained by the Monte-Carlo simulation, where σ_{th} is the strength of thermal fluctuation, (b) temperature dependence of the depinning threshold, J_{cm}, and (c) scaled depinning probability function.

$$Q(j_c)=\left\{1-\exp\left[-(j_c-j_{cm})^{\hat{z}}\right]\right\}S(j_c-j_{cm}), \tag{2}$$

with $j_c=\hat{J}_c^{FR}/J_0$, $j_{cm}=J_{cm}/J_0$, $\hat{z}=(z+2-D)/(D-1)$,

where z is the dynamic exponent, J_0 is scaling parameter, D (=3 in the present case) is the dimension of fluxoids, and $S(j)$ is the step function defined by $S(j)$=0 at j<0 and $S(j)$=1 at j>0.

3. *E-J* characteristics

Based on the results in the preceding section, *E-J* relationship is obtained as follows in the vicinity of J_{cm}.

$$E(J)=\rho_{FF}\int_0^j Q(j_c)dj_c$$

$$\cong \frac{\rho_{FF}}{\hat{z}+1}J\left(\frac{J}{J_0(T_{GL})}\right)^{\hat{z}}\left\{1+\beta(J_{cm}/J)^{\frac{1}{2\nu}}\right\}^{-\hat{z}}\left\{1-(J_{cm}/J)\right\}^{\hat{z}+1} \qquad for \quad T<T_{GL}, (3.a)$$

$$\cong \frac{\rho_{FF}}{\hat{z}+1}J\left(\frac{J}{J_0(T_{GL})}\right)^{\hat{z}} \qquad for \quad T=T_{GL}, \quad (3.b)$$

$$\cong \frac{\rho_{FF}}{\hat{z}+1}|J_{cm}|\left(\frac{|J_{cm}|}{J_0(T_{GL})}\right)^{\hat{z}}\left(1-\gamma(\rho_{lin}/\rho)^{\frac{1}{2\nu\hat{z}}}\right)^{-\hat{z}}\left\{(1+J/|J_{cm}|)^{\hat{z}+1}-1\right\} \quad for \quad T>T_{GL}, \quad (3.c)$$

with $\rho\equiv E/J$ and $\rho_{lin}\equiv\lim_{J\to 0}\frac{dE}{dJ}\cong\rho_{FF}\left(\frac{|J_{cm}|}{J_0(T_{GL})}\right)^{\hat{z}}$,

where ρ_{FF} is the uniform flux flow resistivity, β and γ are the numerical parameters determine the temperature dependence of J_0 at $T<T_{GL}$ and $T>T_{GL}$, respectively. Note that J_0 is nearly proportional to the average value of $\hat{J}^{FR}$; therefore, we assume here that $J_0(T)$ depends linearly on T around $J_0(T_{GL})$ for simplicity. Then, $J_0(T)$ can be approximately expressed in the scalable form by J or ρ as

$$J_0(T)\cong J_0(T_{GL})\left[1+\beta(J_{cm}/J)^{1/(2\nu)}\right] \qquad for \quad T<T_{GL}, \tag{4.a}$$

$$J_0(T)\cong J_0(T_{GL})\left[1-\gamma(\rho_{lin}/\rho)^{1/(2\nu\hat{z})}\right] \qquad for \quad T>T_{GL}. \tag{4.b}$$

It can be seen that eqs. (3.a) to (3.c) satisfy the scaling predicted by Fisher [1]. However, these equations indicate that the scaling region can be very wide because of the temperature dependence of $J_0(T)$ [5].

4. Comparison with the measurements in a YBaCuO thin film

T- and *B*-dependent *E-J* curves measured in a laser deposited YBaCuO thin film are shown in Figs. 3(a) and 3(b), respectively. Those *E-J* curves are collapsed on the same line by normalizing by the nondimensional current density $(J-J_{cm})/J_0$ as shown in Fig. 3(c). This result is consistent with that of the Monte-Carlo simulation shown in Fig. 2(c). Moreover, $J_{cm}(B,T)$ agrees with eq. (2), and the value of ratio $\ln(J_B/J_T)$=-7.5 for $T<T_{GL}$, -8.2 for $T>T_{GL}$ are consistent with the value, $\ln[(|\partial T_{GL}/\partial B|\cdot B_{GL}/T_{GL})^{2\nu}]=-7.7$, estimated from the observed transition line $T_{GL}(B)$. The solid lines in Fig. 3 are eq. (3), where T is replaced by B in the case of B-dependence. The parameters are shown in the caption, where $Y_0\equiv\ln\left[\rho_{FF}(\hat{z}+1)^{-1}J_0(T_{GL})^{-\hat{z}}\right]$.

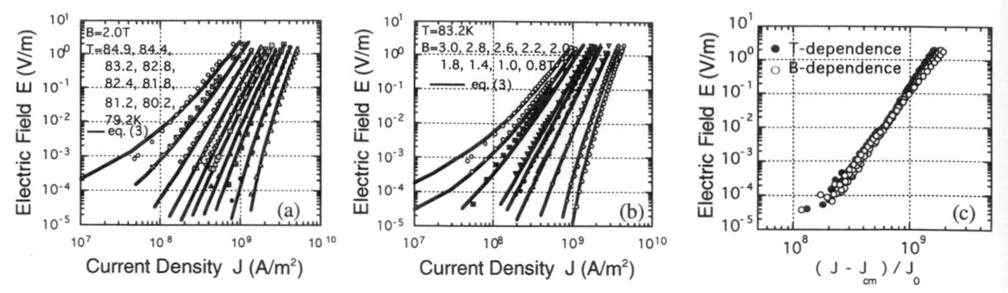

Fig. 3 Comparison with measurements. (a) T-dependent E-J curves. The solid lines are eq. (3) with the numerical parameters; z=7.7, ν=1.4, T_{GL}=83.2K, Y_0=-92.2, $\ln J_T$=29.5 for $T<T_{GL}$, $\ln J_T$=28.5 for $T>T_{GL}$, β=2.9, γ=2.1 obtained by the scaling analysis. (b) B-dependent E-J curves. The numerical parameters are z=7.7, ν=1.4, T_{GL}=2.0T, Y_0=-92.3, $\ln J_B$=22.0 for $B<B_{GL}$, $\ln J_B$=20.3 for $B>B_{GL}$, β=2.0 and γ=1.6. (c) Scaled E-J curves as in the same way obtsined by the Monte-Carlo simulation shown in Fig. 2(c)

Isotherm E-J curves are also shown in Fig. 4, where the solid lines are obtained from eq. (3) with the same parameters shown in Fig. 3, whereas the broken lines indicate $\beta=\gamma=0$ which is assumed to be corresponding to the critical regime obtained for a constant pinning potential. This result indicates that the remarkably wide scaling region is strongly influenced by the temperature dependence of average pinning strength, that is $J_0(T)$.

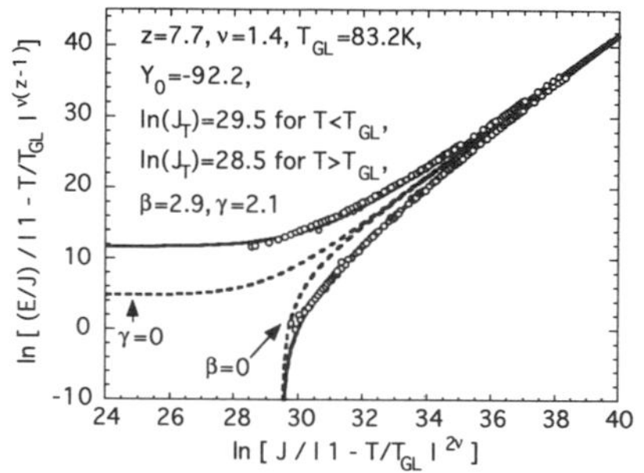

Fig. 4 Temperature dependence of the average pinning strength and the scaling curve. The dots are the data. The solid lines are eq. (3) for β=2.9 and γ=2.1, whereas the broken lines are for $\beta=\gamma=0$.

5. Conclusion

Statistic property of $\hat{j}_c^{FR}$, which is the critical current density of a flux bundle row in a random pin medium under the influence thermal agitation, has been studied by the Monte-Carlo method. The results show a new type of scaling, namely the E-J curves are collapsed on a universal curve by normalizing by the nondimensional current density $(J-J_{cm})/J_0$, where J_{cm} is the minimum value of $\hat{j}_c^{FR}$ and J_0 is the parameter representing the width of $\hat{j}_c^{FR}$ distribution. The measurement results in a YBaCuO film support the scaling. It has been shown that the scaling based on the $\hat{j}_c^{FR}$ distribution is consistent with the well-known scaling of glass-liquid transition within the measured range. Moreover, it can be seen that the scaling is the same kind of phenomena with that of macroscopic pinning force if the $\hat{j}_c^{FR}$ distribution is considered. New analytical expression for the E-J curves has also been derived. It has been shown that the scaling region can be very wide since J_0 is varied with temperature approximately in the scalable form.

References

[1] Fisher M P A 1989 *Phys. Rev. Lett.* **62** 1415-8
[2] Blatter G *et al.* 1994 *Rev. Mod. Phys.* **66** 1125-388
[3] Yamasaki H *et al.* 1995 *IEEE Tarans. Appl. Supercon.* **5** 1888-91
[4] Kiss T *et al.* 1996 *Proc. 8th IWCC* 39-44; Yamafuji K and Kiss T 1996 *Physica C* **258** 197-212
[5] Yamafuji K and Kiss T, 1997 to be published in *Physica C*
[6] Awaji S *et al.* 1996 *Proc. 8th IWCC* 183-6

Inst. Phys. Conf. Ser. No 158
Paper presented at Applied Superconductivity, The Netherlands, 30 June–3 July 1997

Flux pinning and critical currents in weakly coupled granular superconductors

Antti Tuohimaa and Jaakko Paasi

Laboratory of Electricity and Magnetism, Tampere University of Technology, P.O. Box 692, Fin-33101 Tampere, Finland.

Abstract. We present a computational study of intergranular flux pinning and critical current density in granular superconductors. The intergranular current system was modelled as a two dimensional Josephson junction array, which consists of superconducting grains connected via overdamped short Josephson junctions. Special attention was paid to the influence of non-superconducting intergranular defects of μm-size on the flux pinning in the array, with the aim to find out sizes and shapes of defects that lead to high overall critical current density of the array. It was observed that an array with long but narrow defects in the direction of the transport current can carry considerably higher overall current densities than an array with square defects.

1. Introduction

It is frequently assumed that the flux pinning in grain surfaces and in long grain boundary (GB) Josephson junctions (JJ) determine the intergranular (transport) critical currents of granular superconductors. For magnetic fields higher than the Josephson lower critical field of a GB the magnetic flux penetrates into the GB. If the GB is long, the flux can be trapped in the JJ [1-2]. Instead, if the GB is short, the magnetic flux passes through the GB into the non-superconducting defect which can be for example a void between the grains or well connected grain blocks. The penetrated flux induces a supercurrent on the surfaces of the grains around such a region, thus forming a superconducting current loop. The grains in question can still be in the Meissner state. The current loop prevents free intergranular flux motion. Hence the defect acts as a pinning center for the flux in the space between the superconducting grains. In granular superconductors e.g. in Bi-2223, only a fraction of the GBs are long and there exist a great number of non-superconducting defects, of μm-size, such as voids between the grains and second phase particles, which have a considerable impact on the critical current at low magnetic field region, where the $1/H$ dependence of the critical current density J_C is present [3]. On the contrary, at high fields the J_C decreases exponentially and it is limited by intragranular pinning.

In this paper we concentrate on influence of the μm-size intergranular defects on the flux pinning. Often the J_C of a JJ differs from the maximum Josephson current density of the junction. In granular material an additional flux redistribution phenomena exist due to the coupling of adjacent current loops and junctions. Therefore we use a JJ array in order to model the intergranular flux pinning and critical current density in granular superconductors.

2. Model

We used an idealized JJ array model, which is shown in Fig. 1(a). It presents a cross-section of an array that is infinite in the direction of the magnetic field and in which the defects are rectangular and homogeneously distributed. The length of all JJs is W and the maximum Josephson current per unit length along field direction is $I_W=J_J W$, where J_J is maximum Josephson current density. The defects have dimensions D and C to the direction of y and x, respectively, leading to lattice constants $a_y=D+W$ and $a_x=C+W$.

In the equivalent circuit model corresponding to Fig. 1(a) the grains are located at the nodes. Between the grains there exist ideal short JJs which are shunted by a resistance R. The junctions are characterised by gauge-invariant phase differences $\upsilon_{i,j}$ and $\theta_{i,j}$ for the vertical and horizontal junctions, respectively. All the junctions are taken to be identical and we assume they have overdamped dynamics. For the $\theta_{i,j}$ the dynamic equation is [4],

$$\frac{d\theta_{i,j}}{dt}=\frac{R}{L}\left\{\frac{2\pi\phi_{i,j}}{\phi_0}-\frac{2\pi\phi_{i,j-1}}{\phi_0}-\beta\sin\theta_{i,j}\right\} \tag{1}$$

where ϕ_0 is the flux quantum, β is the SQUID-parameter $\beta=2\pi LI_W/\phi_0$, L is the self inductance of the loop per unit length along field direction and the magnetic flux $\phi_{i,j}$ of the loop (i,j) is obtained from the flux quantization condition as $\phi_{i,j}=\upsilon_{i,j}+\theta_{i,j+1}-\upsilon_{i+1,j}-\theta_{i,j}$. An analogous equation holds also for the vertical junctions. Let us assume that the array is infinite in the direction of x-axis and that the array is cooled under the zero field condition, thus all the vertical phases differences are zero and the equations reduce to that of one-dimensional JJ array. In the external magnetic field H the boundary condition is given by $\frac{d\theta_{i,1}}{dt}=\frac{R}{L}\left\{-\frac{2\pi\phi_e}{\phi_0}+\frac{2\pi\phi_{i,1}}{\phi_0}-\beta\sin\theta_{i,1}\right\}$, where $\phi_e=\mu_0 H a_x a_y$.

Stationary solutions of the system depend only on the value of β and, via the boundary conditions, on the external magnetic flux ϕ_e. The parameter β includes all the parameters of the model: C, D, W and J_J. In the Meissner state the magnetic flux density inside the grains falls off exponentially with the distance from the grain surface up to $W/2$ with characteristic length λ_L. In this work we set λ_L=0.15 μm. Now, taking into account that the magnetic flux density in the defect is constant in our case, the LI_W is

$$LI_W=\mu_0 I_W\left\{CD+2(C+D)\lambda_L(1-e^{-W/2\lambda_L})+2\pi\lambda_L\left[\lambda_L-(\lambda_L+W/2)e^{-W/2\lambda_L}\right]\right\} \tag{2}$$

The terms on the right hand side corresbond with the grey coloured area shown in Fig 1(a).

The array J_C is derived from the Ampere's law and it is

$$J_{Cx,ij}=\frac{1}{\mu_0}\frac{|\Delta_y B_{ij}|}{a_y}=\frac{1}{\mu_0}\frac{|\Delta_y\phi_{ij}|}{(a_y)^2 a_x} \tag{3}$$

where $\Delta_y\phi_{ij}=\phi_{i,j}-\phi_{i,j-1}$. In the given form the $J_{Cx,ij}$ is a local variable. If we are interested in overall J_{Cx} we have to average the $J_{Cx,ij}$ over the array taking into account the oscillation of the $J_{Cx,ij}$ as a function of ϕ_e. This was done by calculating the magnetic flux difference between adjacent loops $\Delta_y\phi_{ij}=\phi_{i,j}-\phi_{i,j-1}$ as a function β, where $\Delta_y\phi$ with a given β is an average of 100 different ϕ_e values, which are chosen such that magnetic flux is fully penetrated in the array and ϕ_es are equally distributed over one period of $\phi_{ij}(\phi_e)$ oscillation which is ϕ_0. The averaging was done because of the $\phi_{ij}(\phi_e)$ oscillation of a JJ-loop [1]. The resulting $\Delta_y\phi(\beta)$ is given in Fig. 1(b).

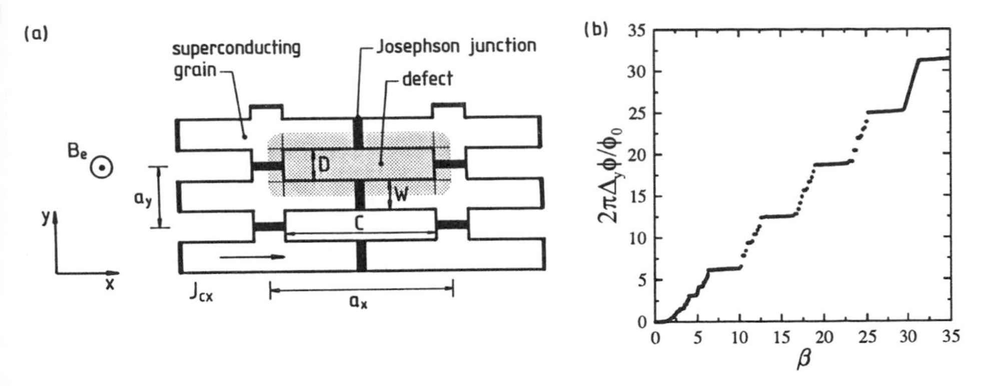

Figure 1. (a) The infinite JJ array model. (b) The normalized average flux difference between adjacent loops as a function of β.

The influence of the defect shapes (C,D) and the JJ lengths (W) on the critical current density of the array was studied by using the following procedure. First, the values of $\beta(J_J,W,D,C)$ were calculated for the given (J_J,W,D,C) combination. Then, the $\Delta_y\phi(\beta)$ was extracted from Fig. 1(b). Finally, the J_C was obtained by using eq. (3).

3. Results

The influence of J_J,W,C and D on the array J_C was studied systematically. A large number of computation runs were done for $J_C(J_J,W,D,C)$. Here we give an outline to the main results with the aim to show the general tendencies in the results. Because $j_C=J_C/J_J$ was not sensitive to small changes in J_J we fixed J_J=1 GA/m^2. The results are shown for the j_C by carefully selected plots where we have fixed either W,C or D and the j_C is presented as a function of the two other variables.

At first we present the $j_C(W,C)$-surface, Fig. 2(a), for D=0.1 μm which is a representative value for narrow voids or non-stoichiometric layers between the grains. The results show the longer C the higher j_C. In real samples the defects cannot be very long and therefore we set C=10 μm for the study of $j_C(W,D)$, Fig. 2(b). From the figure we can see that the best j_C -values are obtained when $W\approx$ 0.5 μm. Hence we choose W=0.5 μm for the $j_C(C,D)$-surface shown in Fig. 2(c). According to the results the best j_C was obtained when C>>D. The longer C-side the better. Favourable D depends on the values of J_J and W. For the J_J=1 GA/m^2 the best results are obtained when D<<1 μm and $W\approx$ 0.5 μm, provided that C>>D. Then the array J_C can be over 50 % of J_J. On the other hand, values j_C>30 % are obtained in a great variety of (J_J,W,D,C)- combinations. If these values with long but narrow defects are compared to the j_C=20 % obtained by square lattice arrays with parameters D=C=1.2 μm and W=0.8 μm, Fig. 2(d), which presents the optimized j_C -value square arrays when J_J=1 GA/m^2, we can conclude that long but narrow defects enable stronger flux pinning than square defects leading to higher overall array J_C.

The favourable lattice structure of the model array is quite similar to the microstructure of the longitudinal midsection of textured Bi-2223 tapes [5]. This suggests that non-stoichiometric layers or voids between adjacent grains can be effective pinning centers providing high j_C in well oriented materials.

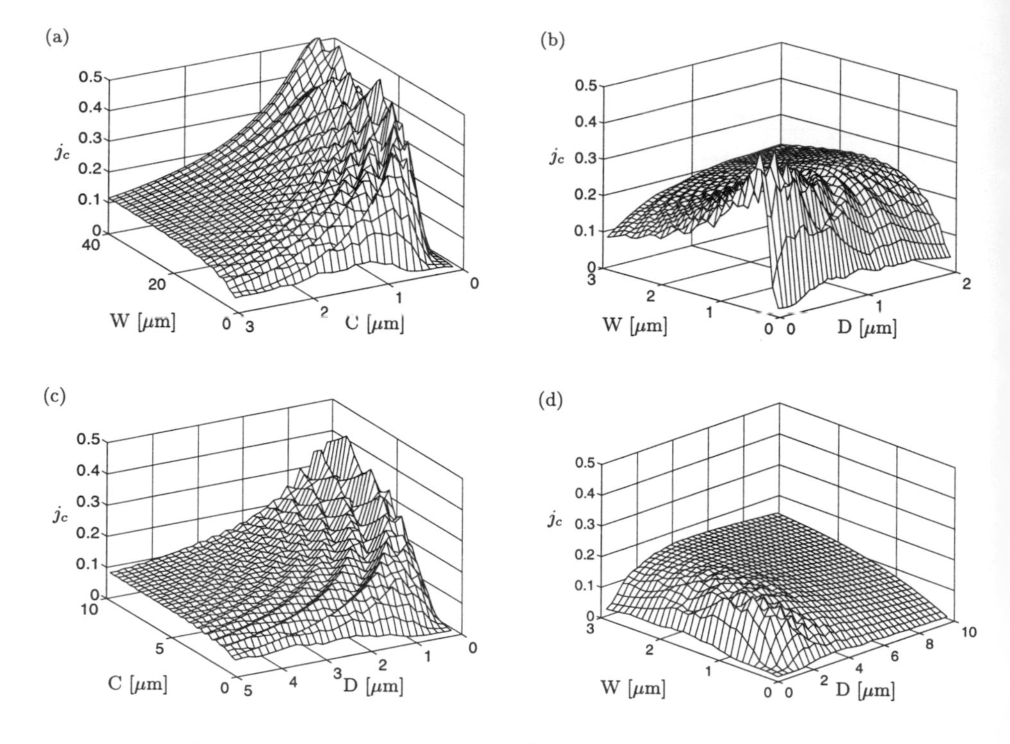

Figure 2. The j_C-surfaces for J_J=1 GA/m^2 (a) D=0.1 μm, (b) C=10 μm, (c) W=0.5 μm and (d) C=D. For the legend see the text.

Therefore, when considering the flux pinning in granular superconductors, especially in low magnetic fields, we should not only take into account the flux pinning in superconducting grains and grain boundaries but also the pinning due to intergranular defects of μm-size.

4. Conclusion

We have studied the influence of μm-size defects on intergranular flux pinning and critical current density of granular superconductor. A granular superconductor was modelled as a JJ array where self-inductances are taken into account. The results show that an array with long and narrow defects in the direction of the trasport current enable higher critical current densities than an array with square defects.

References

[1] Likharev K K 1986 *Dynamics of Josephson junctions and circuits* (Philadelphia: Gordon and Breach)
[2] Fehrenbacher R, Geshkenbein V B and Blatter G 1992 *Phys. Rev. B* **45** 5450-5466
[3] Edelman H S, Parrel J A and Larbalestier D C 1997 *J. Appl. Phys.* **81** 2296-2301
[4] Paasi J A J and Tuohimaa A H 1996 *Physica C* **259** 10-26
[5] Hensel B, Grasso G, Flükriger R 1995 *Phys. Rev. B* **51** 15456-15473

Inst. Phys. Conf. Ser. No 158
Paper presented at Applied Superconductivity, The Netherlands, 30 June–3 July 1997

The shape-effect and vortex-lattice phase transitions in a BSCCO single crystal

T. B. Doyle [a], R. Labusch [b], R. A. Doyle [c], T. Tamegai [d] and S. Ooi [d]

[a]Department of Physics, University of Natal, Durban 4001, South Africa.
[b]Institut für Angewandte. Physik, Der Technischen Universität. Clausthal, 38678 Clausthal-Zellerfeld, Germany.
[c]IRC in Superconductivity, University of Cambridge, CB3 0HE, England.
[d]Department of Applied Physics, Univ. of Tokyo, Hongo, Bunkyo-ku, Tokyo, 113, Japan.

Abstract. Iterative numerical calculation, based on a rigorous theoretical treatment that explicitly includes the equilibrium constitutive $B[H^{rev}]$ and the critical current density $J_c(B)$ functions, for the magnetisation behaviour in disc geometry, are compared with isothermal, quasi-static, experimental data on a disc-shaped BSSCO single crystal in perpendicular applied field.

1. Introduction

Analysis of the magnetisation behaviour in platelet shaped single-crystals of high-T_c superconductors is of particular importance in the study of vortex-lattice pinning mechanisms, phase behaviour, and, in particular, of the physical processes which determine the 'irreversibility line" [1]. In this non-ellipsoidal geometry the magnetisation behaviour shows a pronounced "geometry" or "shape" effect which is evident by a delay in the applied field H^o for initial vortex penetration into the zero field cooled (ZFC) state and by an "intrinsic" hysteresis in the M-H^o behaviour for $H^o < 0.5H_{c1}$. To obtain the vortex-pinning and vortex-lattice phase behaviour from experimental data in strip, platelet or disc specimens the contributions from the geometry effect and from other possibly significant effects, such as surface-barriers and anisotropy, need to be known. This has only recently been made possible with the development of a theory for the geometry effect [2-5].

In the present work the results of iterative numerical calculation, based on a rigorous theoretical treatment of the geometry effect [3,5], are compared with quasi-static, isothermal M-H^o data obtained for a disc-shaped BSCCO single-crystal and are discussed in terms of the above considerations. The treatment explicitly includes the equilibrium constitutive $B[H^{rev}]$ and the critical current density $J_c(B)$ behaviours. In the context of the present conference proceedings this, and the following paper [7], are of relevance in regard to vortex-pinning behaviour and to the origin of the "irreversibility line" in the high-T_c systems.

2. Experimental

The specimen preparation and measurement procedure are described in detail elsewhere [5]. The sample used here is a slightly over-doped BSCCO single-crystal, with T_c = 86K, which was cleaved and polished to produce a disc, with the c-axis normal to its plane, of thickness

$d = 0.01 \pm 0.002$ mm and radius $R = 0.5 \pm 0.01$ mm. Isothermal magnetisation measurements were made using a low-field Cryogenic Consultants SQUID magnetometer and the remnant magnetic field was reduced to below 0.1 Oe before each isothermal run.

3. Results and discussion

Figures 1(a) and (b) show a self-consistent scaling of the experimental M-H^o data onto the "ideal" (i.e. zero-pinning) calculated results for the geometry effect behaviour (solid and dashed curves), normalised by H_{c1} and for the BSCCO specimen of aspect ratio $2R/d \cong 100$.

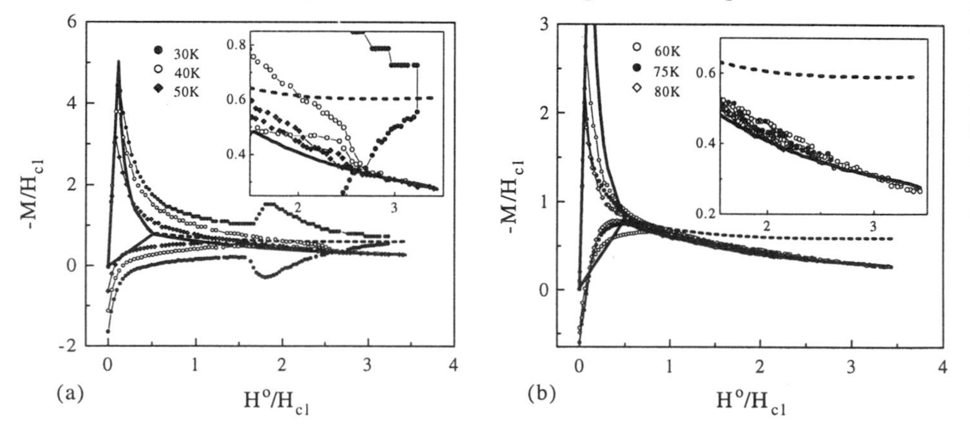

Figures 1(a) and (b). Normalised isothermal M-H^o data at temperatures as indicated. The dashed and solid curves are the calculated "ideal" behaviours for the "solid" and "liquid" vortex-lattice respectively (see text).

The dashed curves in Figures 1(a) and (b) give the "ideal" behaviour for temperatures $T < 40$K and, where appropriate, at fields below the "melting" transition [6] ($H_m/H_{c1} = 2.52$ - present case). The solid curves give the asymptotic behaviour above the "melting" transition. We propose that the $B[H^{rev}]$ relations for these curves correspond, respectively, to "solid" and "liquid" vortex-lattice phases. Explicit, semi-empirical expressions for these relations are given elsewhere [5,7]. It is apparent from these figures that the geometry effect, for the "ideal" pin-free case, extends only up to $H_{or}/H_{c1} \cong 0.5$ (large κ limit [4,5]). At higher fields the "ideal" $M(H^o)$ behaviour is reversible. (For H^o increasing in the range $H^p < H^o < H_{or}$, where H^p is the applied field for initial vortex penetration, penetrating vortices collect at the centre of the disc in a "pool" that grows in size to fill the disc at $H^o = H_{or}$ [2-5]). Values for $H_{c1}(T)$ and for $h^p(T) = H^p/H_{c1}$, as obtained from the fits of Figure 1, are shown in Fig. 3(b). The relative lack of scatter in these results gives some confidence in the fits. Over the temperature range measured, i.e. 30 - 80K, $H_{c1}(T)$ varies as $(1-T/T_c)$ and linear extrapolation yields $T_c = 86 \pm 1$K and $H_{c1}(0) = 460 \pm 5$ Oe.

It will be noticed from Figure 1 that for $T > 40$K: (i) the field for initial flux penetration H^p does not scale with H_{c1} (see also Figure 3), and (ii) the "melting" transition which is well defined at $T = 40$K, gets progressively weaker with increasing temperature. This behaviour is not observed in specimens with smaller aspect ratios (see for example [8] where $H^p(T)$ is generally found to vary approximately as $(1-T/T_c)$, (i.e. to scale with $H_{c1}(T)$) and the "melting" step is clearly discernible from $T \cong 40$K to near T_c. It has been proposed and argued elsewhere [5] that a decrease in $H^p(T)/H_{c1}(T)$, for $T > 40$K, in large aspect ratio low-pinning strength BSCCO specimens, may be a consequence of line-tension anisotropy and vortex de-coupling for the vortices penetrating the specimen rims at a non-zero angle relative to the c-axis. The dashed curve fit to the $h^p(T)$ data in Figure 3(b) is given by $h^p(T) - h^p(T_c) = 0.19(1-T/T_c)^2$. This temperature dependence has been predicted for the vortex de-coupling field [1] and indirectly supports the above proposal. The apparent progressive decrease in the magnitude of the melting transition step above 40K is proposed to be related to the relatively very small thickness of the specimen and to be due to some combination of experimental noise and to enhanced effects of specimen inhomogeneity.

In Figures 2(a) and (b) the results of calculated fits to the experimental M-H^o data are shown for $T = 30$ and 40K respectively. The M-H^o data at $T = 30$K shows the well-known "arrowhead" effect [9] (onset at $H_{ah}/H_{c1} = 1.52$ - present case). For this fit we have used $B[H^{rev}]$, as determined for the "solid" vortex-phase [5,7], and $J_c(B)$ as follows: (i) For $H^o < H_{ah}$; $J_c(B) = J_{co}(1+B/B_o)^{-3}$; with $J_{co} = 8.7\times10^7\text{Am}^{-2}$, $B_o = 2.5B_{c1}$ and, (ii) for $H^o > H_{ah}$; $J_c(b) = J_{co}\, b \exp(-b^{-3/2})$; $b = B/B_{c1}$ with $J_{co} = 3.3\times10^8\text{Am}^{-2}$. These pinning behaviours correspond to predicted 3D collective pinning theory [10] for the large-bundle and the small-bundle regimes, respectively.

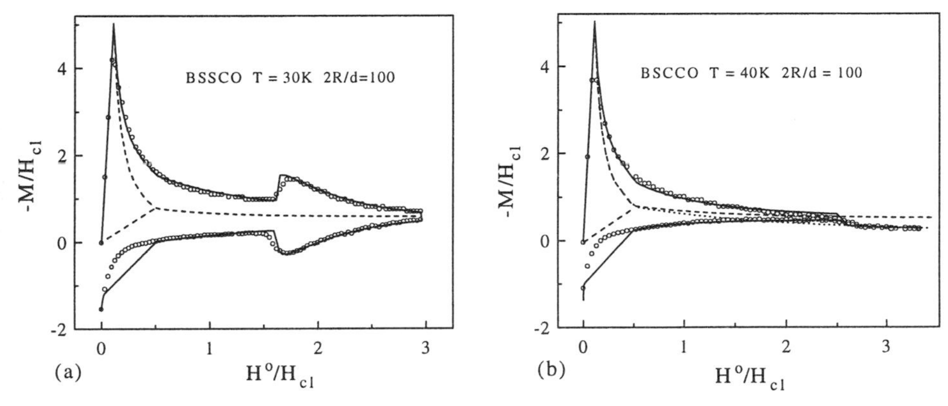

Figures 2(a) and (b). As for Figure 1 at $T = 30$K and 40K with the dashed and dotted curves corresponding to the solid and liquid phases respectively. The solid curves are calculated fits to the data as described in the text.

The M-H^o data a $T = 40$K (Figure 2(b)) shows the well-known "step-effect" or "magnetic-jump" (onset at $H^o = H_m$) which has been associated with a first-order "melting" (B increasing) transition [2]. This "step" is modelled here on the assumption that it corresponds to a first-order transition between the equilibrium $B[H^{rev}]$ behaviours for the "solid" and "liquid" phases, as identified above, at some critical value of the local induction $B = B_m$. A more detailed treatment of this "melting" transition is given elsewhere in these proceedings [7]. For this fit we have used: (i) for $B < B_m$; $B[H^{rev}]$, for the "solid" phase [5,7], and $J_c(B) = J_{co}(1+B/B_o)^{-3}$ with $J_{co} = 6.5\times10^7$ Am^{-2} and $B_o = 1.5B_{c1}$, and (ii) for $B > B_m$; $B[H^{rev}]$ for the "liquid" phase, and $J_c(B) = 0$. The "melting" transition occurs at $B_m/B_{c1} = 2.52$ with a discontinuous change in M of the order $\Delta M/H_{c1} = +0.2$.

Fits to the M-H^o data for T $\geq$ 50K were done following the same method as for $T = 40$K but with $h^p(T)$ pre-set in the calculations to the experimentally found values (see Figure 3(b)). The fit for $T = 50$K is shown in Figure 3(a). The results for $J_{co}(T)$ in the low-field 3D regime, at the various measurement temperatures, are shown in Figure 3(b). The exponentially decreasing temperature dependence of J_{co} on T (dotted curve) is consistent with a (collective) creep limited mechanism for flux pinning.

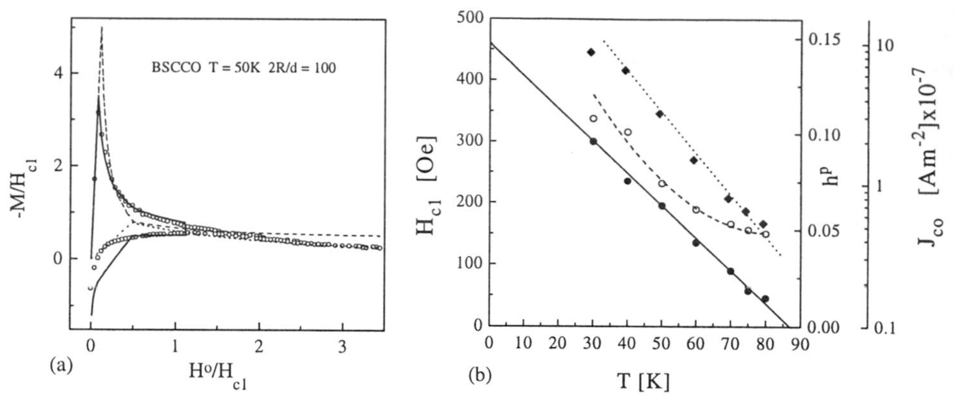

Figure. 3(a) As for Fig. 2 at $T = 50$K. (b) $H_{c1}(T)$ (•), $h^p(T)$ (o) and $J_{co}(T)$ (♦) from the fits (see text).

In conclusion, the generally good agreement between the calculated and the experimental M-H^o data presented here establishes with some certainty that the geometry effect dominates the magnetisation behaviour in disc (and platelet) geometry in this particular system, at least for $T > 30$K. There is no apparent indication of any surface-barrier effect on $h^p(T)$, which for $T < 40$K, is in close agreement with the geometry effect prediction. The small deviation from theory in the decreasing H^o branch may be due to a surface-barrier effect but may also be consequent on the mechanism of vortex escape through the specimen rims [5]. In relation to the M-H^o hysteresis (and the "irreversibility line") the results of the present investigation establish that any hysteresis for $H^o > 0.5H_{c1}$ is not determined by the geometry effect and is well described by bulk pinning.

Acknowledgements

The Foundation for Research Development and the University of Natal, South Africa are thanked for financial support. Nigel Hussey is thanked for help running the SQUID.

References

[1] For a general review: E. H. Brandt, *Rep. Prog. in Phys.* **58** (1995) 1465.
[2] E. Zeldov et al, *Phys. Rev. B* **49** (1994) 9802.
[3] T. B. Doyle and R. Labusch, Proc. CMOS'1996. Publ. in *Journ. Low Temp. Phys.* **105**, Nos. 5/6 (1996) 1207.
[4] M. Benkraouda and J. R. Clem, *Phys. Rev. B* **53** (1996) 5716.
[5] T. B. Doyle, R. Labusch and R. A. Doyle. Submitted to *Physica C*.
[6] H. Safar et al, *Phys. Rev. Lett.* **68** (1992) 2672.
[7] T. B. Doyle, R. Labusch, R. A. Doyle. These Proceedings.
[8] T. Hanaguri et al, *Physica C* **256** (1996) 111.
[9] N. Chikumoto et al, *Phys. Rev. Lett.* **69** (1992) 1260, *Physica C* 185-189 (1991) 2201.
[10] G. Blatter et al, *Rev. Mod. Phys.* **66** (1995) 1125.

Inst. Phys. Conf. Ser. No 158
Paper presented at Applied Superconductivity, The Netherlands, 30 June–3 July 1997

Vortex-lattice melting in high-T_c superconductors

T. B. Doyle [a], R. Labusch [b] and R. A. Doyle [c]

[a]Department of Physics, University of Natal, Durban 4001, South Africa.
[b]Institut für Angewandte Physik, Der Technischen Universität Clausthal, 38678 Clausthal-Zellerfeld, Germany.
[c]IRC in Superconductivity, University of Cambridge, CB3 0HE, England.

Abstract The magnetisation "step" associated with the vortex-lattice "melting" transition in high-T_c superconductors is investigated using a rigorous treatment for the magnetisation behaviour of a disc-shaped specimen. The analysis yields the flux density, field and magnetisation profiles in the specimen and predicts an intermediate state with the expected behaviour at the solid/liquid interface. Calculations are also made for comparison with experimental magnetisation data on a BSCCO single-crystal disc specimen.

1. Introduction

The large anisotropy and significant thermal energy, in relation to typical energies associated with the vortex-lattice in the high-T_c superconductors, give rise to a rich vortex-lattice phase behaviour, which includes a liquid and a solid phase, in these systems [1]. The detailed nature of the transition, or transformation, between these phases has been the subject of much recent research [2-4] and controversy [5]. Theoretical treatments [6-7], supported by Monte Carlo simulations [8] predict that, in pin-free systems, the transition is first-order. In the presence of strong pinning the transition is thought to be continuous, or second-order, from the liquid to a vortex-glass or Bose glass [9]. Various experimental investigations, on platelet high-T_c single-crystal specimens in perpendicular applied field, including neutron scattering, muon-spin rotation, global and local magnetisation measurements, and calorimetric measurements (see [10] for a general review) show discontinuous behaviours that are generally interpreted to be consistent with a first-order transition. The situation in platelet or disc geometry is, however, complicated by inhomogeneous demagnetisation fields with the consequence of inhomogeneous flux density $\boldsymbol{B}(\boldsymbol{r})$ even in the absence of flux pinning. This leads to a spatial localisation of the melting transition and hence also to the expectation of an "intermediate-state". In this geometry, in increasing applied field, a "melted" vortex-lattice zone is first nucleated near the middle of the specimen where, on account of the "geometry-effect" [11], the flux density profile has a flat maximum. With further increase in applied field, this zone, which may break down into a fine-scale intermediate state, grows rapidly to fill the specimen [3]. For a first-order magnetic transition the Clausius-Clapeyron relation between a discontinuous change in magnetisation ΔM and an entropy change ΔS per vortex per layer is given [12] by:

$$\Delta M = -(B / s\phi_o)(dT / dH_m)\Delta S \qquad (1)$$

where s is the CuO double-layer spacing and (dT/dH_m) is the slope of the melt phase-boundary. In local magnetic field measurements (see for example [3]) it is usual to measure $\Delta B \cong \Delta M(1\text{-}D)$ or $\Delta(B\text{-}\mu_o H^o)$, where H^o is the applied field, and D is an effective demagnetisation coefficient.

2. Results and discussion

In the present work we use the results of a rigorous theoretical treatment [13,14] to calculate, by iterative numerical method, the internal field, flux density and magnetisation profiles, and hence also the global magnetisation, near the melting transition for a disc-shaped specimen in perpendicular applied field. We also use results from the previous paper in these proceedings [11]. In particular, the melting step is modelled as a first-order transition, at some critical induction $B = B_m$, between two distinct, constitutive equilibrium $B[H^{rev}]$ behaviours, corresponding to the "solid" and "liquid" phases, respectively. An approximate semi-empirical expression is used for the inverse $B[H^{rev}]$ relation [13], namely:

$$H^{rev}[B] = [B + \gamma(B_{c2} - B) + B_\gamma(1-\gamma)\exp(\frac{-\alpha B}{B_\gamma})] / \mu_o, \text{ where } \quad \gamma = [1.16(2\kappa^2 - 1)]^{-1},$$

$$B_\gamma = \frac{(1-\gamma B_{c2})}{(1-\gamma)}$$ and $B_{c2} = \beta\kappa^2 \ln\kappa$. The parameters α, β and κ (GL parameter) are obtained empirically by fitting to the $M\text{-}H^o$ data for a disc specimen [11,13]. The treatment does not include a term for the solid/liquid interface energy, which should be relatively very small, and implicitly assumes cylindrical symmetry. It does not, therefore, allow for the possibility of topological structure along the solid/liquid interface or for any fine-scale intermediate-state phases in the solid or liquid zones (but see below).

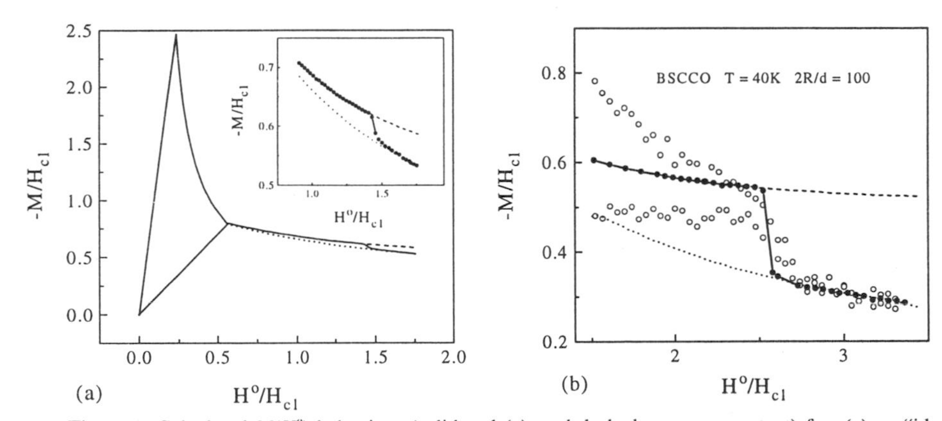

Figure 1. Calculated $M(H^o)$ behaviour (solid and (•), and dashed curves - see text) for: (a) an "ideal" specimen with κ = 70 and $2R/d$ = 20 (The inset is a magnified plot of the "melting-step"), and (b) the BSCCO crystal (data points (o)) with $2R/d$ = 100 [11] and calculated behaviour (see text).

In Figure 1(a) the calculated $M(H^o)$ behaviour (solid curve) is shown for a "ideal" (i.e. pin-free) disc specimen with κ = 70 and aspect ratio $2R/d$ = 20. The parameters for this calculation have been chosen to give clear, unambiguous results but are not atypical for BSCCO platelet specimens with the c-axis and the applied field normal to their plane. Figure 1(b) shows a calculated fit (solid curve and points (•)) for comparison with the melting transition in the very large aspect ratio BSCCO ($\kappa \cong 70$, $2R/d = 100$) specimen of the previous paper [11]. For the dashed and dotted curves in these figures, corresponding to "solid" and "liquid" phases respectively [11], we have used the following parameters: (i) For the "ideal" specimen: $\kappa = 70$, $\alpha = 1.8B_{c1}$, $\beta = 5.3$ (solid) and $\beta = 4$ (liquid) with the transition at $B_m = 1.4B_{c1}$, and (ii) for the BSCCO specimen: as above but with $\beta = 2$ (liquid) and with the transition at $B_m = 2.5B_{c1}$. The magnitude of the global magnetisation steps, from inspection of Figures 1(a) and (b), are $\Delta M \cong 0.044H_{c1}$ (ideal) and $\Delta M \cong 0.2H_{c1}$ (BSCCO).

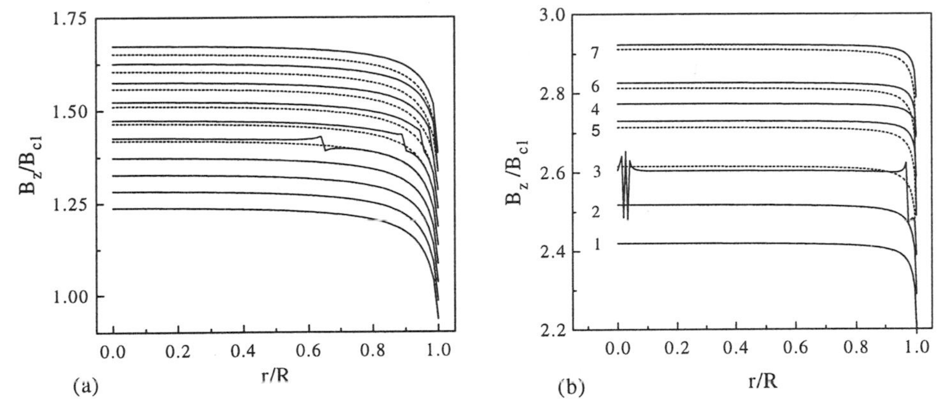

Figure 2(a) and (b). Calculated, reduced flux density profiles at constant H^o, H^o increasing, near the melting transition, for the "ideal"($2R/d$ = 20) and the BSCCO ($2R/d$ = 100) specimens, respectively (solid curves). The dashed curves show the behaviour without the melting transition (see text).

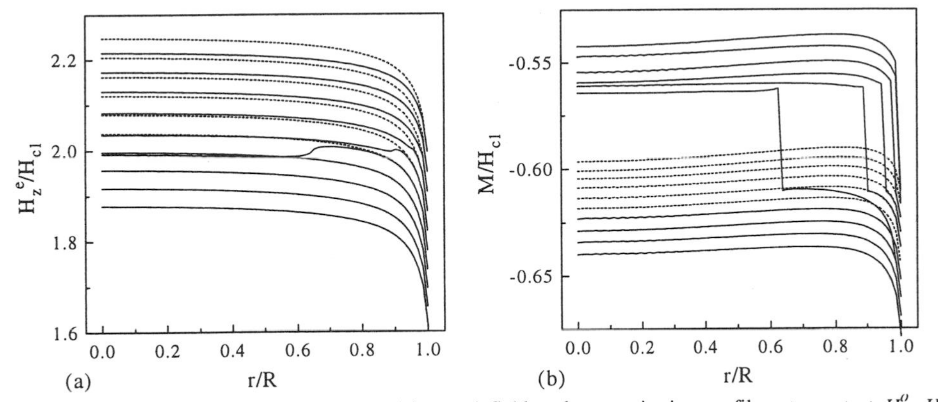

Figure 3(a) and (b). Calculated, reduced internal field and magnetisation profiles at constant H^o, H^o increasing, near the melting transition, for the "ideal" specimen, respectively (see Fig.1 and text).

In Figures 2(a) and (b) the calculated, reduced flux density $B_z(\rho)$; $\rho = r/R$, $z \parallel \boldsymbol{H}^o$, profiles, at constant H^o with H^o increasing in approximately equal steps, are shown for the "ideal" and for the BSCCO specimens, respectively (solid curves). The dashed curves in these figures show the calculated behaviour in the absence of a melting transition. The apparent stability (uniformity) of $B_z(\rho)$ in the "melted" region, $\rho < \rho_m$, where $\rho_m \equiv r_m/R$ is the position of the melting step, is consequent on the choice, in the calculations, of a large step in H^o between successive profiles. When this step is reduced the calculated $B_z(\rho)$ in this regime become increasingly unstable and oscillates between the equilibrium values for the two phases. This is evident in the $B_z(\rho)$ profiles of Figure 2(b) for the much larger aspect ratio BSCCO specimen where it is also more difficult to obtain stable solutions at small ρ_m. The nature of this instability is suggestive that, in reality, the "melted" zone should break down into an intermediate state with a scale determined by the solid/liquid interface energy. This problem and the detailed shape of the melting step profiles (which are determined by the current sheet at the solid/melt interface), with particular reference to the local micro-Hall effect measurements [3,5] will be treated in detail elsewhere. Note also that, for the BSCCO specimen, as a result of its larger aspect ratio, the $B_z(\rho)$ profiles are much flatter than for the "ideal" specimen, and also that the close proximity of the large "melting step" to the edge of the specimen leads to a reversal of the profiles 4 and 5 (but not, however, of the $M(\rho)$ profiles).

Figures 3(a) and (b) show the calculated reduced, internal field and magnetisation profiles, corresponding to the flux density profiles for the "ideal" specimen of Figure 2(a). It is apparent from Figure 3(b) that the magnitude of the local step, ΔM, is independent of position ρ_m as expected. The global magnetisation M during the melting transition can therefore, be expressed as $M(H^o) \cong [M(H^o)]_{sol} - \Delta M \rho_m^2(H^o)$ where $[M(H^o)]_{sol}$ is the equilibrium solid phase behaviour (dashed curve in Figure 1(a)). Since $[M(H^o)]_{sol}$ is approximately constant over the transition the total change in the global M is, therefore approximately equal to the local ΔM. For the BSCCO specimen, using the results of Figure 1(b) i. e. $\Delta M/H_{c1} \cong 0.2$ and $H_{c1} \cong 230$ Oe [11], we obtain $4\pi\Delta M \cong 46$Oe and, hence from Eq.1, $\Delta S/k \cong 0.15$, which is close to other measured values [3,4].

Acknowledgements

T. B. Doyle wishes to thank Eli Zeldov for helpful discussion, and the Foundation for Research Development (South Africa) for financial assistance.

References

[1] G. Blatter et al, *Rev. Mod. Phys.* **66** (1995) 1125.
[2] H. Safar et al, *Phys. Rev. Lett.* **68** (1992) 2672.
[3] E. Zeldov et al, *Nature* 375 (1995) 373.
[4] U. Welp et al, *Phys. Rev. Lett.* **76** (1996) 4809.
[5] D. E. Farrell et al, *Phys. Rev. B* **53** (1996) 11807.
[6] E. Brezin et al, *Phys. Rev. B* **31** (1985) 7124.
[7] D. R. Nelson et al, *Phys. Rev. Lett.* **60** (1988) 1973.
[8] D. S. Fisher et al, *Phys. Rev. Lett.* **62** (1991) 1415.
[9] D. R. Nelson and V. M. Vinokur, *Phys. Rev. Lett.* **68** (1992) 2398.
[10] E. H. Brandt, *Rep. Prog. Phys.* **58** (1995) 1465.
[11] T.B. Doyle et al, previous paper in these Proceedings.
[12] R. E. Hetzel, A. Sudbo and D. A. Huse, *Phys. Rev. Lett.* **69** (1992) 518.
[13] R. Labusch and T. B. Doyle, and T. B. Doyle, R. Labusch and R. A. Doyle, submitted to *Physica C.*
[14] T. B. Doyle and R. Labusch, *Proc Int. Conf. on the Physics and Chemistry of Molecular and Oxide Superconductors*, Aug. 1996, Karlsruhe, Germany. Publ in *J. Low Temp. Phys.* **105**, Nos. 5/6 (1996) 1207.

Inst. Phys. Conf. Ser. No 158
Paper presented at Applied Superconductivity, The Netherlands, 30 June–3 July 1997

Flux Pinning and Grain Coupling in $Bi_2Sr_2CaCu_2O_{8+\delta}$ ceramics: Magneto-optical Investigations.

Sun-Li Huang† , M R Koblischka†‡ , T H Johansen‡ , H Bratsberg‡ and K Fossheim†

† Department of Physics, NTNU Trondheim, N-7034 Trondheim, Norway

‡ Department of Physics, University of Oslo, P.O. Box 1048, Blindern, 0316 Oslo, Norway

Abstract. Efforts are being made to improve the flux pinning and grain coupling in bulk Bi-2212 ceramics, prepared by a partial melting process. For this purpose, the penetration of vortices is visualized by means of the magneto-optical Faraday effect technique at various temperatures ranging from 18 K up to 77 K. Our Bi-2212 ceramic samples are found to show a strong coupling between the grains, which is manifested by the observation of a flux front propagating into the sample on increasing field. To investigate the pinning properties, we studied remanent states obtained after applying external magnetic fields up to 160 mT.

1. Introduction

Despite the large progress made with the development of large, bulk $YBa_2Cu_3O_{7-\delta}$ superconductors, also bulk $Bi_2Sr_2CaCu_2O_{8+\delta}$ (Bi-2212) ceramics may be expected to play an important role for future applications. However, there is still some work to be done to improve the pinning capabilities of the Bi-2212 system. It is especially important to understand the properties of flux penetration and to control the trapping in bulk Bi-2212 samples, as the samples consist typically of small, platelet-like grains. Magneto-optical (MO) visualization of flux (for a recent review on these techniques, see Ref. [1]) can be an important tool to elucidate these properties, as investigations can be carried out on various scales ranging from a whole sample down to individual grains [2, 3]. Most MO investigations carried out using Bi-based superconductors focus on Bi-2223 tapes [1, 4], and only one experiment is carried out on Bi-2212 ceramics [3].

2. Experimental

Bi-2212 ceramics are prepared using a partial melting processing technique [5]. The transition temperature is 94.5 K, showing a sharp transition. The grain size is about 60

Figure 1. Flux patterns after zero-field cooling at $T = 18$ K, showing the entire sample and the propagation of a flux front in increasing field. The scale bar is 1 mm.

μm (SEM, polarization microscopy). The magneto-optic (MO) imaging is based on the Faraday effect in a magneto-optical active layer. Here we have used a Bi-doped YIG film with in-plane anisotropy with a thickness of 4 μm.

In this way, the flux line lattice is imaged as bright areas, whereas the flux-free Meissner area stays dark. The images presented here are, therefore, maps of the z-component of the local magnetic field, B_z. The images are recorded using an 8-bit Kodak DCS 420 CCD digital camera and subsequently transferred to a computer for processing. In the MO apparatus the sample was mounted on the cold finger of an optical Helium flow cryostat [6]. The magnetic field was applied perpendicular to the tape plane using a copper solenoid coil with a maximum field of ±120 mT, remanent states can be generated up to ± 180 mT.

3. Results and Discussion

Flux patterns of the Bi-2212 ceramic are presented in fig. 1 at $T = 18$ K using a small magnification, in order to verify the existence of a flux front during flux penetration in increasing external field. A complete shielding against the applied field is observed up to 7 mT. On further increase of the field flux enters the sample preferentially through some 'easy' channels, then the remaining bulk is penetrated. On field reversal, even a fingerprint of the discontinuity lines of the currents [7] can be observed. Furthermore, MO observations can be carried out up to 77 K, where Bi-2212 single crystals are clearly above the irreversibility line [8].

Figure 2 presents various remanent states at $T = 18$ K using a higher magnification. An increasing amount of trapped flux can be seen inside the grains on increasing the field. These areas of trapped flux are found to be equal or larger than the grain size

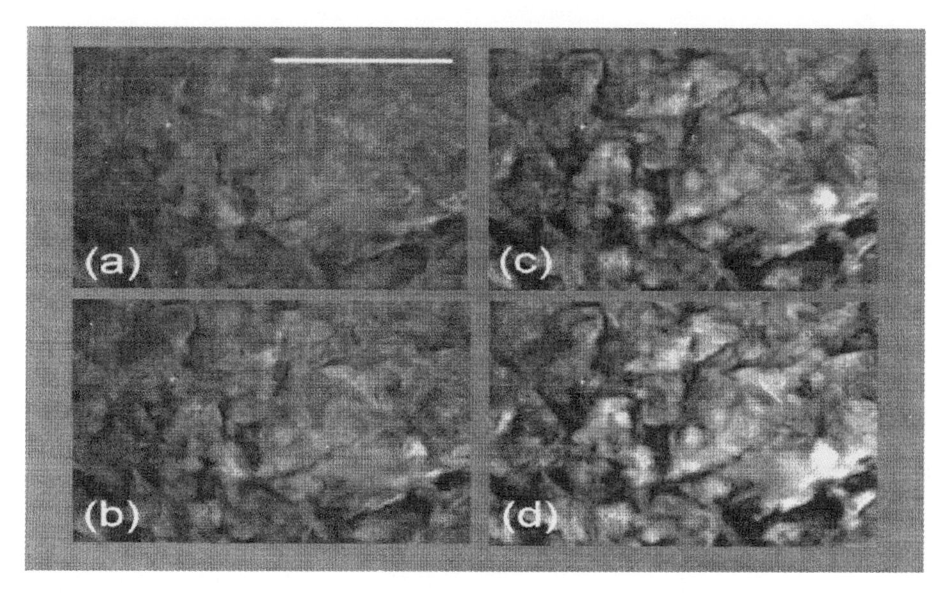

Figure 2. Various remanent states ($\mu_0 H_a = 0$ T) at $T = 18$ K after applying 60 mT (a), 90 mT (b), 120 mT (c), 150 mT (d). The marker is 200 μm.

determined by SEM; so some well-coupled clusters must be present in the sample. To obtain *local* critical current densities by means of MO imaging we use images of the remanent state, taken at various positions throughout the sample. In the remanent state, flux is pinned *inside* the grains (clusters), and the vortices have left the 'easy' channels between the grains. This implies that only *intragranular* currents are flowing within the grains. Therefore, investigations of remanent states avoid the problems which arise when analysing a state with an applied field [2, 3], e. g. the flux amplification between well shielding grains [9]. After calibration, flux density profiles are obtained from the detected light intensity in the images. The *local* current densities (fig. 3) are obtained from the measured gradients of B_z; $j_c = 1/\mu_0 \times \Delta B_z/\Delta x$ where x is in the observation plane. This method assumes that the Bean model is valid, i. e. that the aspect ratio of a grain is $\approx$1. As the thickness of a grain (cluster) is a priori unknown, the obtained data for j_c are a lower limit. Furthermore, as shown in Ref. [2, 3], the *local* current densities depend strongly on the grain size, r_G. The grain size determined by SEM is about 60 μm for the sample under study. Figure 3 shows that the largest currents are obtained for grains close to this value; these are the 'true' intragranular currents. Grains larger than 60 μ are presumably grain clusters, where coupling between different grains (with different orientation) is involved. This causes a reduction of the local j_c. In Ref. [2], it was shown for $YBa_2Cu_3O_{7-\delta}$ ceramics that $j_c \sim 1/r_G$. This seems to be also valid for the Bi-2212 ceramics studied here.

As conclusion, we have found that partial melting processing of Bi-2212 ceramics produces good connections between the grains so that a flux front can be observed during initial penetration. The MO imaging allows to determine the local current density *inside* individual grains (clusters) from images of the remanent state as a function of the grain (cluster) size. It is hoped that the *local* investigations of flux distributions may contribute

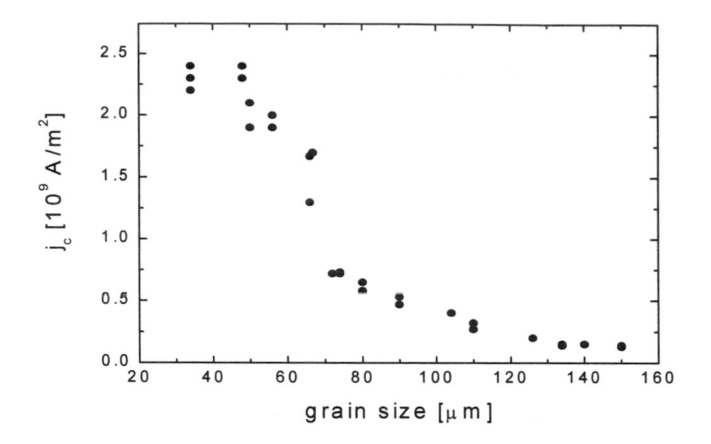

Figure 3. *Local* critical current density as a function of the grain (cluster) size; data from remanent state after applying a field of 150 mT, $T = 18$ K.

to a further development of the pinning properties of Bi-based bulk superconductors.

Acknowledgements

We gratefully acknowledge financial support by The Research Council of Norway.

References

[1] Koblischka M R and Wijngaarden R J 1995 *Supercond. Sci. Technol.* **8** 199

[2] Moser N, Koblischka M R, and Kronmüller H 1990 *J. Less Common Metals* **164-165** 1308

[3] Koblischka M R, Schuster T, and Kronmüller H 1994 *Physica C* **219** 205

[4] Polak M et al 1997 *Appl. Phys. Lett.* **70** 1034; Pashitski A E *et al.* 1995 *Physica C* **246** 133

[5] Huang S L *et al.* 1995 *Supercond. Sci. Technol.* **8** 32

[6] Johansen T H et al 1996 *Phys. Rev. B* **54** 16264

[7] Schuster T *et al.* 1994 *Phys. Rev. B* **49** 3443

[8] Koblischka M R *et al.* 1995 *Physica C* **249** 339; Indenbom M V *et al.* 1994 *Physica C* **222** 203

[9] Waysand G 1988 *Europhys. Lett.* **5** 73; Hodgdon M L, Navarro R, and Campbell L J 1991 *Europhys. Lett.* **16** 677

Inst. Phys. Conf. Ser. No 158
Paper presented at Applied Superconductivity, The Netherlands, 30 June–3 July 1997

Pinning in bulk high-T_c superconductors

M R Koblischka

Department of Physics, NTNU Trondheim, N-7034 Trondheim, Norway and Department of Physics, University of Oslo, N-0316 Oslo, Norway.

Abstract. The volume pinning forces, F_p, of several high-T_c superconductors are compared to each other using literature data and to those obtained in $NdBa_2Cu_3O_{7-\delta}$ samples. It is shown that the pinning at small normal inclusions which is dominant in most high-T_c systems, leads to a peak at around 0.3 when plotting the normalized volume pinning forces, $F_p/F_{p,max}$, versus the reduced field defined by $h = H/H_{irr}$. In contrast to this, data of $NdBa_2Cu_3O_{7-\delta}$ show the peak at $\approx$0.5, thus indicating a strong pinning at composition fluctuations providing a scatter in T_c ($\Delta\kappa/\delta T_c$-pinning). It is shown that this kind of pinning is ideal for applications at $T = 77$ K.

1. Introduction

Flux pinning is one of the crucial problems in the development of technical high-T_c superconductors, especially because of the high operating temperature aimed for in applications. Moreover, the pinning at temperatures around 77 K is of great importance for the applications, but several high-T_c materials still suffer from quite poor pinning properties at this temperature, mainly the Bi- or Tl-based superconductors. The newly developed light rare earth (LRE) superconductors [1] present a novel approach to this problem. In these materials, the fishtail or peak effect shows a peak at large H_a, thus providing a high critical current density at $T = 77$ K and above. Furthermore, flux creep is considerably reduced in the peak region.

The study of volume pinning forces, $F_p = j_c \times B$, in conventional superconductors [2] showed that a scaling law $F_p = A(h)^p(1-h)^q$ holds, with A being a numerical parameter, h is the reduced field defined by $h = H_a/H_{c2}$ (H_a denotes the applied field), and p and q are parameters describing the actual pinning mechanism. The position of the peak is given by $h_0 = p/(p+q)$. Several pinning functions are described by Dew-Hughes [2] using a direct summation model. Such a scaling of the volume pinning forces holds also in high-T_c materials as found by several authors [3]. In this paper, these scalings are compared using literature data and measurements on the newly developed $NdBa_2Cu_3O_{7-\delta}$ (NdBCO) samples showing outstanding pinning properties [4].

2. Results and Discussion

To describe the pinning in high-T_c materials, one can safely assume that core pinning is dominant due to the large κ values. This leaves two different sources of pinning; either by non-superconducting (normal) particles embedded in the superconducting matrix leading to a scatter of the electron mean free path (δl-pinning) or pinning provided by spatial variations of the Ginzburg parameter associated with fluctuations in the transition temperature, T_c (δT_c-pinning) [5]. Pinning is different for various sizes of pinning sites compared to the flux line spacing $d = 2/\sqrt{3}(\Phi_0/B)^{0.5}$. The interaction volume, V, of point pins is ξ^3, whereas volume pins $V \sim d^3$. Based on experiments dealing with high-pressure oxygen loading in $YBa_2Cu_3O_{7-\delta}$ (YBCO) single crystals prepared in $BaZrO_3$ crucibles [6], one can conclude that oxygen vacancies are the main pinning source in YBCO, being also responsible for the formation of the so-called fishtail shape of magnetization loops [7].

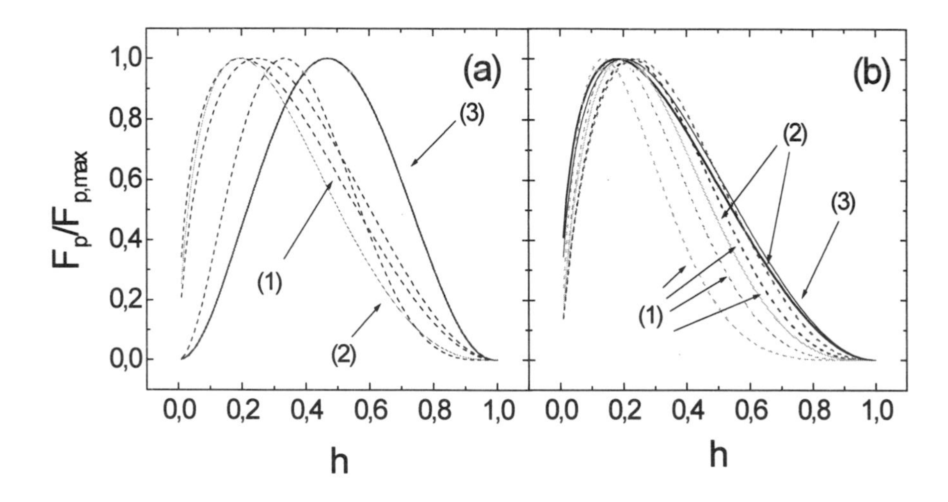

Figure 1. Compiled literature data of F_p-plots. In (a) F_p-data of YBCO samples are shown: (1) – YBCO single crystals [3] and melt-processed YBCO [8], (2) – irradiated YBCO single crystal [15], (3) – NdBCO single crystal and OCMG [4]. In (b), data of Bi-based high-T_c superconductors are shown: (1) – Bi-2212 wires and tapes [9], (2) – Bi-2223 tapes [10], (3) – Bi-2223 thin films [11].

A comparison of the data of various high-T_c materials becomes possible using plots of F_p, normalized by its maximum value, $F_{p,max}$, versus the reduced field, h. The experiments show that the appropriate scaling field for the high-T_c materials is the irreversibility field, H_{irr}, instead of the upper critical field, H_{c2}.

Fig. 1 (a) presents the compiled literature data of YBCO, together with the measurements on NdBCO; the literature data were extracted by using the peak position, h_0, or the parameters p and q. Conventional superconductors exhibit a variety of peak positions (see e. g. [2] and references therein). For high-T_c materials, the peak is generally found at low h. For various YBCO samples the peak in the F_p diagram is found at $h_0 \approx 0.3$ [3, 8]. This pinning is mostly ascribed to oxygen vacancies [3], or, in the case of melt-processed YBCO, to finely dispersed $YBa_2Cu_2O_5$ (Y-211) inclusions [8]. At high h, all authors have found a typical tail being caused by flux creep which is dominant at high fields/high temperatures. Data are typically only available at temperatures above 60 K, where both $F_{p,max}$ and H_{irr} can be directly accessed.

Figure 1 (b) presents the F_p-diagrams of Bi-based superconductors (both Bi-2212 [9] and Bi-2223 [10]). The peaks are found at $h_0 \approx 0.2$, but the tail at high h is much more pronounced. The pinning is strong at low temperatures, and is decreasing rapidly when reaching the irreversibility line being located relatively low. This means that flux creep effects are dominating the high field part of the diagram, so the direct summation model cannot be applied here. To overcome this, several models were developed to describe the pinning with flux creep theory [9]. A similar behaviour is also obtained for most Tl- or Hg-based superconductors, reflecting the grade of anisotropy.

In literature, several attempts are made to improve the pinning by adding small sized ($V \sim \xi^3$) non-superconducting particles into the superconductors. Such particles can be Y-211 [8], Ca_2CuO_3 [12], carbon nanotubes [13], nano-sized gold particles etc. These particles provide effective pinning at low temperatures, but the F_p-diagrams do not show a shift towards higher h as the basic pinning mechanism remains unchanged. Irradiation procedures like heavy-ion irradiation were shown to be very effective to increase the critical current densities [14]. Quite recently, it was shown that the columnar tracks loose their pinning efficiency at temperatures above 60 K and the improvement of pinning is only very small at temperatures like 77 K. Also in this case, the basic pinning mechanism is unchanged. This was e. g. shown by Xiao et al. [15].

In (LRE)BCO, the pinning originates from a different source, as there is an exchange between the Ba and Nd atoms possible leading to composition fluctuations. This provides variations in T_c ($\Delta\kappa$-pinning [2] or δT_c-pinning [5]), shifting the peak towards higher values. Correspondingly, the peak is found at $h_0 \approx 0.5$. Furthermore, as this pinning is active mainly at high T/high h, the flux creep, which is normally dominant in this region, is considerably reduced. In Ref. [16] strong evidence for δl-pinning was presented for YBCO thin films. Several authors found evidence for δT_c-pinning in e.g. Pr-doped YBCO and (K,Ba)BiO_3 single crystals [17]. The composition fluctuations in NdBCO provide variations in T_c, but with the advantage that the overall T_c is not reduced as in the case of Pr-doping. As shown in Ref. [18], the physics of flux pinning in NdBCO is strongly related to that of the Pr-doped YBCO.

These observations lead straightforwardly to the conclusion that the ideal pinning center for high-T_c materials at ≈ 77 K is a superconducting one with at least a size of the order of $\xi^2 d$ providing a scatter in T_c ($p = 2$, $q = 1$). Such a pinning mechanism would especially be important for the Bi-/Tl-based high-T_c materials, where strong bulk pinning vanishes above 60 K and is then mainly governed by geometrical barriers [19]. Small normal conducting pinning sites ($p = 1$, $q = 2$) will always cause the peak in F_p at low h. However, the additional presence of strong non-superconducting pinning sites within the samples should lead to broad peaks in the F_p-diagrams, so $j_s(T, H_a)$ could be improved considerably. This means that ideally both types of pinning should be present in a high-T_c superconductor.

Acknowledgements

I would like to thank M. Murakami (SRL-ISTEC, Div. VII, Tokyo) for the hospitality during my stay in Tokyo where the measurements on the NdBCO samples were performed.

References

[1] for a recent review, see Murakami M et al 1996 *Supercond. Sci. Technol.* **9** 1015

[2] Dew-Hughes D 1974 *Philos. Mag.* **30** 293

[3] Civale L et al 1991 *Phys. Rev. B* **43** 13732; Klein L et al 1994 *Phys. Rev. B* **49** 4403; Satchell J S et al 1988 *Nature* **334** 331; Pan V M et al 1989 *Cryogenics* **29** 392; Wördenweber R et al 1990 *Cryogenics* **30** 458; Li J N et al 1990 Physica C **169** 81; Nishizaki T et al 1991 *Physica C* **181** 223.

[4] Koblischka M R et al 1996 *Phys. Rev. B* **54** R6893

[5] Blatter G et al 1994 *Rev. Mod. Phys.* **66** 1125

[6] Erb A et al 1996 *J. Low Temp. Phys.* **105** 1023

[7] Jirsa M et al 1997 *Phys. Rev. B* **55** 3276

[8] Murakami M in *Melt Processed High Temperature Superconductors*, ed. M. Murakami (World Scientific, Singapore, 1992)

[9] Löhle J et al 1992 *J. Appl. Phys.* **72** 1030; Fabbricatore P et al 1996 *Phys. Rev. B* **54** 12543

[10] Kiuchi M et al 1996 *Physica C* **260** 177; Martini L et al 1994 *Physica C* **235-240** 3033

[11] Yamasaki H et al 1993 *Phys. Rev. Lett.* **70** 3331

[12] Majewski P et al 1994 *Appl. Supercond.* **2** 93

[13] Fossheim K et al 1995 *Physica C* **248** 195

[14] L. Krusin-Elbaum et al 1996 *Phys. Rev. Lett.* **76** 2563

[15] Xiao Z L, Häring J, and Ziemann P 1994 *Physica C* **235-240** 2979

[16] Griessen R et al 1994 *Phys. Rev. Lett.* **72** 1910

[17] Wen H H et al 1995 *Physica C* **251** 371; Harneit W et al 1996 *Europhys. Lett.* **36** 141

[18] Blackstead H A and Dow J D 1997 *Appl. Phys. Lett.* **70** 1891

[19] Zeldov E et al 1995 *Europhys. Lett.* **30** 367; 1994 *Phys. Rev. Lett.* **73** 1428

Inst. Phys. Conf. Ser. No 158
Paper presented at Applied Superconductivity, The Netherlands, 30 June–3 July 1997

Scaling of current densities and pinning forces in $NdBa_2Cu_3O_{7-\delta}$

M R Koblischka[1], A J J van Dalen[2], T Higuchi, K Sawada, H Kojo, S I Yoo, and M Murakami

Superconductivity Research Laboratory, International Superconductivity Technology Center, 1-16-25, Shibaura, Minato-ku, Tokyo 105, Japan.

Abstract. Critical current densities, j_s, and volume pinning forces, F_p, are obtained in a wide temperature ($5 < T < 92$ K) and field range ($0 < \mu_0 H_a < 9$ T) on two different Nd-$Ba_2Cu_3O_{7-\delta}$ samples (a single crystal and a melt-processed one) by SQUID magnetometry. It is shown that the critical current densities of both samples can be scaled well around and above the fishtail peak. Above 60 K, a good scaling of the normalized volume pinning force, $F_p/F_{p,\max}$ (versus the reduced field, $h = H_a/H_{\rm irr}$) can be established. Experimental evidence for strong pinning at superconducting volume/surface defects ($\Delta\kappa$- or δT_c-pinning) is given. These defects are ascribed to spatial composition fluctuations found in light rare earth superconductors.

1. Introduction

Recently, the development of the light rare earth superconductors of the 123-type containing Nd, Sm, Eu, Gd instead of Y presented a novel approach to increase the current densities at elevated temperatures. These superconductors are characterized by an enhanced transition temperature, T_c, ranging between 92 and 96 K and the presence of composition fluctuations as the light rare earth atoms may substitute on the Ba sites [1]. These composition fluctuations are technologically interesting as they may give rise to additional flux pinning [2] by a scatter of T_c or κ (δT_c- [3] or $\Delta\kappa$-pinning [4]).

Here, we present a scaling analysis of the current densities and pinning forces for a $NdBa_2Cu_3O_{7-\delta}$ (NdBCO) single crystal and a melt-processed sample.

2. Experimental

Single crystals of NdBCO are grown by a flux growth method in controlled oxygen atmosphere as described in Ref. [5]. The crystal is twinned, and has the shape of a thin platelet with dimensions $0.79 \times 0.63 \times 0.08$ mm^3. The sample shows a sharp transition with a $T_{c,\rm onset}$ of 93.8 K. The melt processed sample is prepared via the oxygen-controlled melt growth process (OCMG) described in Ref. [1]. Additionally, the sample contains fine $Nd_4Ba_2Cu_2O_{10}$ (Nd-422) inclusions with an average size of 1.7 μm. Finally, this sample is cut into a cubic shape

[1] present address: Norwegian University of Science and Technology, N-7034 Trondheim, Norway
[2] present address: Argonne National Laboratory, Materials Science Division, Argonne, Illinois USA

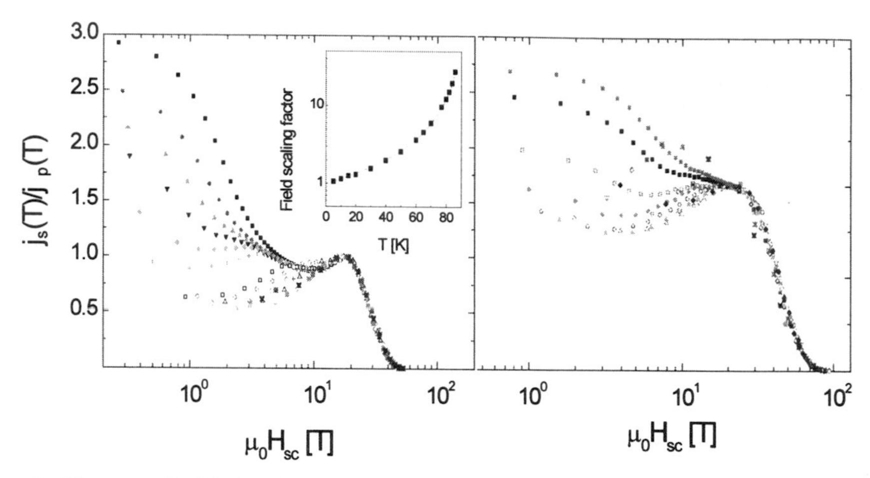

Figure 1. The normalized induced current densities j_s/j_p plotted versus $\mu_0 H_{sc}$ of both samples, (left) – single crystal, (right) – OCMG. All data between 5 and 92 K are shown. The fishtail peaks are found at constant positions of 18 T (single crystal) and 22 T (OCMG). The inset presents the obtained field scaling factors as a function of temperature.

with dimensions 1.92 $\times$ 1.86 $\times$ 0.3 mm^3. $T_{c,onset}$ of this sample is determined to 94.7 K. Magnetization loops are obtained using a Quantum Design MPMS-7 SQUID equipped with a 7 T superconducting magnet. In order to avoid field inhomogeneities, the scan length is set to 15 mm. The magnetic field is applied parallel to the c axis of the sample.

Both samples show a pronounced fishtail effect [6, 7, 8], and the single crystal shows even two peaks at 40 K $\leq T \leq$ 74 K. To clarify the origin of these peaks, we followed the scaling approach described in Ref. [8]. The scaling was started from high temperatures, adding the low temperatures curves one after another. In this way, field scaling factors are obtained as a function of temperature as presented in the inset to fig. 1 (left). The line indicates the fit using $H_{sc} = H/(1 - T/T^*)^p$ with T^* and p as fit parameters, yielding $p = 1.27$ and $T^* = 92.9$ K. T^* is within the error margins equal to the independently determined $T_{c,onset}$. The scaling is working well at fields $H_a \geq H_p$. For $H_a < H_p$, the presence of the intermediate peak (30 K $\leq T \leq$ 74 K) and of the central peak (5 K $\leq T \leq$ 30 K) is disturbing the scaling. After scaling, the fishtail peak is found at a constant position $\mu_0 H_{sc}$ = 18 T. On the right side of figure 1, the scaling of j_s of the OCMG sample is shown. Also in this case, a good scaling is obtained above H_p. As this sample is much thicker than the single crystal studied here, the effects of the low field peak are more pronounced. Here, the fishtail peak is found at $\mu_0 H_{sc}$ = 22 T.

In Fig. 2, the plots of the normalized volume pinning force, $F_p/F_{p,\max}$ versus the reduced field, $h = H_a/H_{irr}$, are shown for both samples. To determine h, only the experimentally obtained data were used; the corrections necessary were always within the experimental error. For $T <$ 70 K, h was treated as a free parameter, but following the temperature dependence $(1 - T/T^*)^p$ with $T^* = T_c$ and $p \approx 1.2$ [9]. In both cases, the scaling is very good. The position of the peak, h_0, is determined to be 0.48 (single crystal) and 0.42 (OCMG). The data were then fitted to the functional dependence given by $F_p/F_{p,\max} = A(h)^p(1-h)^q$, with A = numerical parameter, p and q are describing the actual pinning mechanism. The position of the maximum in the F_p-plot, h_0, is given by $p/p+q$. Best fits to our data yield $A = 20.31 \pm 3$, p =

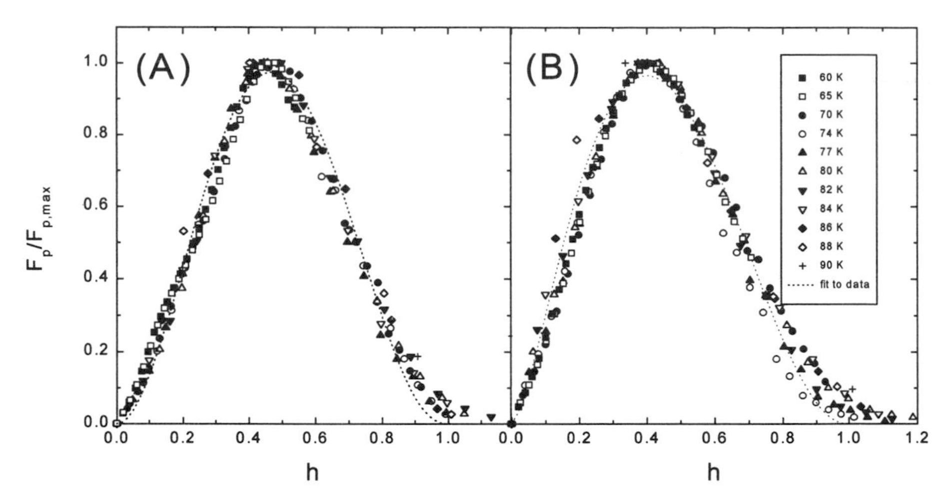

Figure 2. Plots of the scaled volume pinning forces, $F_p/F_{p,max}$, versus the reduced field, $h = H_a/H_{\rm irr}$, for the single crystal (A) and the OCMG sample (B) at 60 K $\geq T \geq$ 90 K.

2.08±0.09, q = 2.35 ±0.11 (single crystal) and A = 11.7±1.7, p =1.48 ±0.08, b = 2.23±0.12 (OCMG). Literature data of $YBa_2Cu_3O_{7-\delta}$ (YBCO) samples are typically yielding $p = 2$ and $q = 4$, implying a peak at $h_0 = 0.33$ [10]. For melt-processed YBCO with fine Y-211 inclusions similar values are found [11]. It is important to point out that the intermediate peak in the $j_s(\mu_0 H_a)$ data of the single crystal form a small shoulder at $h \approx 0.15$; indicating indeed an anomalous enhancement of the pinning force. The peak position at ≈0.5 is, following the model of Dew-Hughes [4] a strong indication of pinning provided by superconducting defects with an interaction volume of $\xi^2 d$ where d is the intervortex spacing given by $d = 2/\sqrt{3}(\Phi_0/B)^{0.5}$. This pinning is ascribed to the composition fluctuations (providing a scatter in T_c) found in the NdBCO system [12]. The high peak position is also the reason for considerably suppressed flux creep found at elevated temperatures as described in Ref. [13].

The analogy of the current scaling and the scaling of the volume pinning forces can be used to deduce an estimation for the irreversibility field at 0 K, $H_{\rm irr}(0)$, as both scalings differ only by this factor. For the melt-processed sample we obtain in this way $H_{\rm irr}(0)$ = 49 T and for the single crystal 38.5 T.

In conclusion, we have obtained a good scaling of the critical current densities in NdBCO samples above the fishtail peak. The volume pinning forces yield a very good scaling in the temperature range between 60 K and 92 K. Experimental evidence for δT_c-pinning at the composition fluctuations is given.

Acknowledgments

This work was partially supported by New Energy and Industrial Technology Development Organization (NEDO) for the R & D of Industrial Science and Technology Frontier Program. M.K. and A.D. are grateful for support from Japanese Science and Technology Agency (STA).

References

[1] for a recent review, see Murakami M et al 1996 *Supercond. Sci. Technol.* **9** 1015

[2] Murakami M et al 1994 *Jpn. J. Appl. Phys.* **33** L715; Yoo S I et al 1994 *Appl. Phys. Lett.* **65** 633; Murakami M et al 1994 *Physica C* **235-240** 2781

[3] Blatter G et al 1994 *Rev. Mod. Phys.* **66** 1125

[4] Dew-Hughes D 1974 *Philos. Mag.* **30** 293

[5] Sawada K et al 1995 4th Euro Ceramics Vol. 6 ed. A. Barone, D. Fiorani, and A. Tampieri, (Gruppo Editoriale Faenza Editrice) p. 293.

[6] Zhukov A A et al 1995 *Phys. Rev. B* **51** 12704

[7] Jirsa M and Půst L 1997 *Physica C* (to be published)

[8] Jirsa M et al 1997 *Phys. Rev. B* **55** 3276

[9] Koblischka M R et al 1996 *Phys. Rev. B* **54** R6893

[10] Civale L et al 1991 *Phys. Rev. B* **43** 13732; Klein L et al 1994 *Phys. Rev. B* **49** 4403; Satchell J S et al 1988 *Nature* **334** 331; Nishizaki T et al 1991 *Physica C* **181** 223.

[11] Murakami M in *Melt Processed High Temperature Superconductors*, ed. M. Murakami (World Scientific, Singapore, 1992).

[12] Koblischka M R presented at the M^2S-HTSC V conference, Beijing, 28.2.–4.3.1997, to be published in Physica C; Koblischka M R, this conference.

[13] van Dalen A J J et al 1996 *Supercond. Sci. Technol.* **9** 659

Inst. Phys. Conf. Ser. No 158
Paper presented at Applied Superconductivity, The Netherlands, 30 June–3 July 1997

The influence of Ag doping on the superconducting properties of grain-boundary weak-link in YBCO

A L L Jarvis† and T B Doyle‡

† Department of Electronic Engineering, University of Natal, Durban, South Africa
‡ Department of Physics, University of Natal, Durban, South Africa

Abstract. Low-field, isothermal magnetisation measurements on polycrystalline YBCO specimens doped with varying concentrations of Ag are analysed using a critical state model (CSM). The inter-granular critical current density $J_c(H)$ has a dependence which has been obtained from a percolation model which assumes a random spatial distribution of current in an individual grain-boundary 'weak-link' Josephson junction (GBJ). For this case there exists a field-independent critical component of $J_c(H)$ which depends on the GBJ microstructure. This component increases when the Ag concentration is increased in the range 0 wt% - 5 wt%. A decrease is observed at a concentration of 10 wt%. This behaviour is consistent with the prevailing ideas in respect of the effect of Ag doping on the microstructure of the GBJ.

1. Introduction

The introduction of Ag in the high-Tc superconductor systems is know to improve critical current density J_c [1]. The objective of this investigation is to clarify the role of Ag doping in polycrystalline YBCO. Isothermal zero-field-cooled (ZFC) magnetisation measurements (M-H_a) are made on control and doped Ag specimens. These measurements are analysed using a CSM which yields various superconducting parameters for the GBJs. The change in these parameters with Ag doping is investigated.

2. Theory

2.1. Critical state model

In application to polycrystalline granular superconductors, the CSM is specifically formulated for the 'Josephson' regime $H < H_{c1g}$, where H_{c1g} is the intra-granular lower critical field. In this regime J_c is an intrinsic property of the 'weak-link'. The model uses an inter-granular critical current-field dependence, $J_c(H)$ which has been obtained from a percolation model, for a random GBJ 3D network, in which the current flow in an individual GBJ is assumed to be spatially random[2, 3]. The percolation model yields a universal expression of the form

$$J_c(H) = AJ_{c0}\left[\exp(-\frac{H}{H_0}) + \delta\right] \tag{1}$$

where $J_{c0} \equiv J_c(H = 0, T), H_0 = H_0(T)$ and $\delta = \delta(T)$ are characteristics of the GBJ network and $A \approx 1.25$ is a constant. H_0 is the field corresponding to a single flux quantum, ϕ_0, in the GBJ. A non-zero value of the parameter δ turns out to be necessary to account for the low-field magnetisation behaviour and for the transport critical current behaviour which manifests a plateau at intermediate H in granular material [4].

The CSM is fitted with characteristic parameters $J_{c0}(t)$, $H_0(t)$, $\delta(t)$ and permeability $\mu_e(t)$ (where $t = T/T_c$) as fitting parameters to the experimental isothermal low field M-H_a data.

2.2. Temperature dependence of characteristic CSM parameters

The temperature variation of each of characteristic fitting parameters is compared with theory and appropriate model predictions. These latter fits yield, *inter alia*, values for various characteristic lengths of the average GBJ, see table 1.

The effective permeability μ_e allows for the partial screening of the intra-granular material and for a non-superconducting volume fraction in the material. We use a columnar grain approximation model [5] for which μ_e is given by, $\mu_e(t) = f_n + f_s[2I_0(\Lambda)/I_1(\Lambda)]$, where f_n and $f_s \approx 1 - f_n$ are, respectively, the normal and superconducting volume fractions, I_0 and I_1 are modified Bessel functions, and $\Lambda = \Lambda(t) \equiv r_{eg}/\lambda_g(t)$. Here r_{eg} is the effective average grain radius and $\lambda_g(t)$ as the mean inter-granular penetration length. In the present work we obtain $\lambda_g(t)$, from the reversible part of magnetisation M_{rev}, versus applied magnetic field H_a in the intermediate-field domain $H_{c1g} << H_a << H_{c2g}$ using the method of Kogan [6]. The $\lambda_g(t)$ data is fitted to $\mu_e(t)$ data, with the use of the expression for $\mu_e(t)$, which yields r_{eg} and f_n for the different specimens. The predicted r_{eg} of Ag doped specimens compare favourably with results obtained from polarized optical micrographs. The characteristic field H_0 is given by, $H_0(t) = \eta B_0/\mu_e(t) \cong \{(\eta\phi_o)\} / \{(\mu_e(t)[2\lambda_g(t) + d]a_o)\}$ [7], where B_0 is the induction in the plane of an average junction corresponding to a single flux quantum ϕ_o in the junction, $\eta = O(1)$ is a geometrical/morphological constant which allows for the superconducting anisotropy, grain-shape anisotropy and average GBJ orientation, and d is the effective junction-barrier width.

The behaviour of the zero-field critical current density $J_{c0}(t)$ may be modelled on various weak-link junction systems. In this paper the Ambegaokar-Baratoff model for a SIS junction, corrected for order parameter suppression at the junction interfaces [8], and a direct (dilute) pinning summation model [9] is used and the following expression is obtained: $J_{c0}(t) = J_{c0I}\frac{\Delta_i(t)}{\Delta_i(0)}\tanh[\Delta_i(t)/2k_BtT_c]$. Here $J_{c0I} = [4\alpha/\pi a_o^2][\pi\Delta_i(0)/eR_N]$ where $\alpha = O(1)$, R_N is the normal state junction resistance, Δ_i is the suppressed gap parameter at the junction boundaries given by $\Delta_i = \Delta_\infty(t)\tanh[\omega(t)]$ with $\omega(t) = x_o/\sqrt{2}\xi(t)$, where x_o is the suppression range for the gap parameter near the junction boundaries, $\Delta_\infty(t)$ is the gap parameter far from the junction boundaries and $\xi(t)$ is the effective coherence length in the superconducting material.

The reduced field-independent critical current for the GBJ with a random microstructure and critical current density distribution is given by $\delta = [4a/S]^{1/2}\psi$ where S is the average junction area and a and ψ are, respectively, the average area of the structural inhomogeneties and a rms relative fluctuation of critical current density in the average GBJ [10]. There is an implicit temperature dependence, which can arise from changes in local superconducting properties in GBJ i.e. lower critical temperature superconducting phase behaviour [7].

3. Experiment

All YBCO specimens were prepared from a precursor YBCO powder. Bulk doped and control specimens were pelletized at a pressure of 64 MPa, then sintered for 3 hours at 950°C, cooled at 0.2°C/min to 300°C, annealed in an oxygen atmosphere for 20 hours at 300°C and then cooled down to room temperature at 0.2°C/min. Silver doping (2%, 5%, 10% by weight) was achieved by appropiate addtion of Ag_2O to the precursor powder. ZFC magnetisation measurements were made using a vibrating sample magnetometer on bars($10 \times 2 \times 2$ mm^3) cut from the parent specimens,

4. Results and discussion

Figure 1(a) shows typical CSM fits to low-field M-H_a data of Ag-doped YBCO specimens at 80.5K. The resulting CSM fitting parameters, as a function of temperature, with appropriate model fits, are shown in figures 1(b)-(d). An increase of J_{c0} with increasing Ag-doping is observed (inset of figure 1(d)) and is due to the decrease in the resistivity R_nA/d, (see table 1) in the intergranular material. This decrease has been attributed to a proximity effect resulting from the presence of Ag in the GBJs [11]. $\delta(t)$ was found to be approximately constant for each Ag concentration (see table 1).

Table 1. Various characteristic lengths and fitting parameters obtained from appropriate model fits.

Ag%	$r_{eg}(\mu m)$	f_n	$\mu_e(0)$	η	d(nm)	$\omega(0)$	$J_{c0I}(A/cm^2)$	$R_nA(p\Omega.m^2)$	δ
0	1.97(2)	0.06(5)	0.27(7)	0.32(1)	0.13(2)	1.87(9)	3380	7.00(8)	0.038(5)
2	2.57(5)	0.20(4)	0.36(5)	0.31(5)	0.13(5)	1.41(5)	29200	0.75(7)	0.014(3)
5	4.53(7)	0.34(3)	0.43(3)	0.59(3)	0.09(9)	2.54(7)	31000	0.79(3)	0.009(2)
10	3.18(6)	0.28(9)	0.40(2)	0.47(4)	0.10(4)	1.81(4)	53900	0.44(1)	0.005(1)

$J_c(H)$ (equation 1) increases with increasing Ag-doping in the range 0 wt% to 10 wt% in agreement with results published elsewhere [1]. The field-independent component δJ_{c0} of $J_c(H)$ shows an increase with Ag-doping only in the range 0 wt% to 5 wt% (see figure 1(d)). This suggests an increase in microstructure fluctuations in GBJs in this concentration range. When YBCO is doped with 10 wt% Ag a decrease is observed. This suggests that Ag doping at this concentration has a different effect on the GBJs resulting in a more homogenous microstructure. This latter effect could be due to the possibility that Ag 'scavenges' the grain-boundaries thereby absorbing impurities [12] and leaving 'cleaner' Ag-doped GBJs. This may help to explain the results established elsewhere, that 10 wt% Ag is the optimum doping concentration to increase inter-granular critical current in polycrystalline YBCO[1].

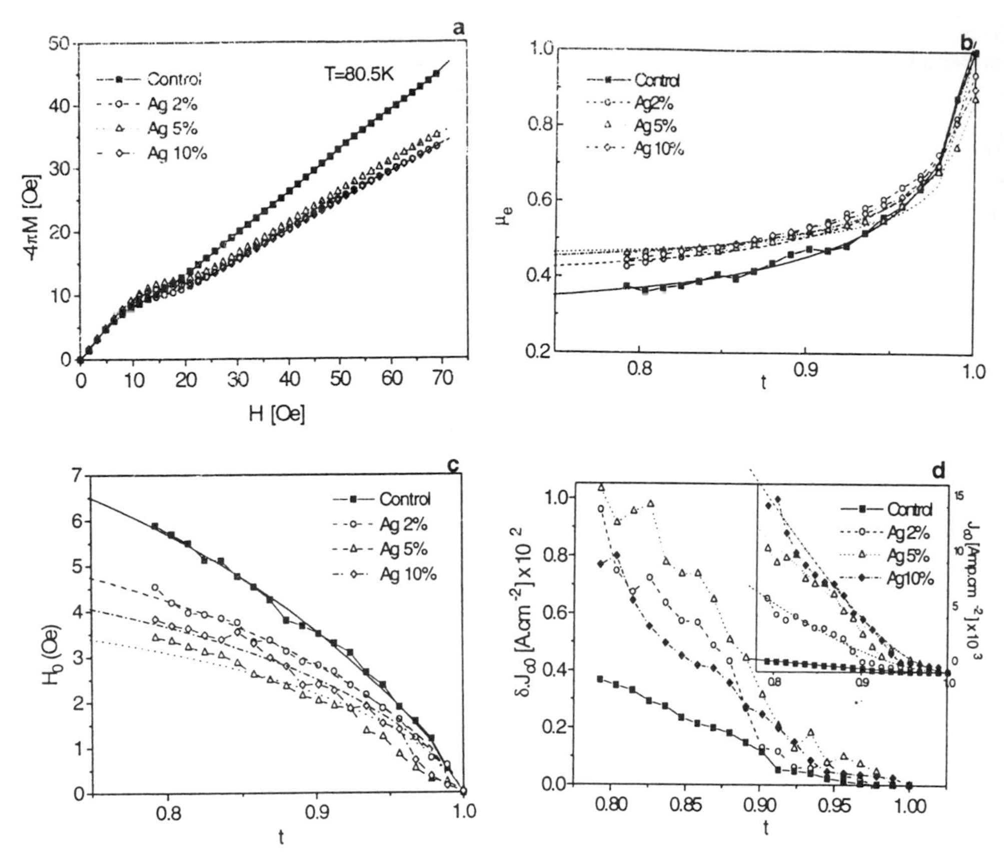

Figure 1. (a) CSM fits(lines) of low-field M-H_a data(symbols) and (b)-(d) CSM fitting parameters behaviour as a function of temperature with appropriate model fits.

References

[1] Goto S and Shiiki K 1995 *Jpn. J. Appl. Phys.* **34** 4760–4764

[2] Meilikhov E Z 1990 *Supercond.* **3** 1110–1117

[3] Meilikhov E Z and Doyle T B 1994 *Proc. 7th Int. Critical Currents Workshop*

[4] Ekin J W, Hart H R and Gaddipati A R 1990 *J. Appl. Phys.* **68** 2285–2290

[5] Clem J R 1988 *Physica. C* **153-155** 50–55

[6] Kogan V G, Fang M M and Sreeparna Mitra 1988 *Phys. Rev. B* **38** 11958–11961

[7] Doyle T B, Doyle R A, Minkov D, Stepankin V N and Yakovets U P 1994 *Physica C* **233** 253–262

[8] Deutscher G and Chaudhari P 1991 *Phys. Rev. B* **44** 4664–4671

[9] Doyle T B and Doyle R A 1993 *Phys. Rev. B* **47** 8111–8118

[10] Barone A and Paterno G 1982 *Physics and Applications of the Josephson Effect* (New York: John Wiley)

[11] Jung J, Cheng S C and Frank J P 1990 *Phys. Rev. B* **42** 6181–6195

[12] Deslandes F, Raveau B, Dubots P and Legat D 1989 *Solid State Commum.* **71** 407–410

Inst. Phys. Conf. Ser. No 158
Paper presented at Applied Superconductivity, The Netherlands, 30 June–3 July 1997

Flux-line localisation by means of segments of correlated defects in YBCO melt-textured

R Gerbaldo[1], G Ghigo[1], L Gozzelino[1], E Mezzetti[1], B Minetti[1], R Cherubini[2] and A Wisniewski[3]

[1] Istituto Nazionale di Fisica della Materia - Politecnico di Torino. Istituto Nazionale di Fisica Nucleare - Sezione di Torino. Politecnico di Torino, Torino, Italy

[2] Istituto Nazionale di Fisica Nucleare - Laboratori Nazionali di Legnaro, Legnaro (PD), Italy

[3] Dept. of Physics, Polish Academy of Science, Warsaw, Poland

Abstract. YBCO melt textured samples were studied after Au ion irradiation of two opposite surface layers of the samples and after proton irradiation. Correlated disorder introduced by *surface* Au ion irradiation is effective at high reduced temperatures and leads to the occurrence of a kink in the irreversibility line and to the shift of this line towards higher fields and temperatures. Critical current enhancement *increasing* with temperature and field was observed. Random disorder due to proton irradiation does not change the irreversibility line and provides critical current enhancement *decreasing* with temperature and field.

1. Introduction

Flux line (FL) localisation is one of the most challenging problems of HTS materials, both from the point of view of a basic understanding of the physics involved, and from the point of view of applications. The dynamics of the FL lattice in the region below the irreversibility line (IL), the region of applicability of these materials, is variously affected by the interaction with different types of static disorder. Randomly distributed point-like defects act collectively and promote FL wandering [1]. On the contrary, in the presence of *correlated disorder* (i.e. in the presence of defects with linear or planar structure) pinning is stronger, less sensitive to thermal fluctuations and leads to FL localisation [2]. When both types of disorder are present, the competition between wandering and localisation results in complex FL dynamics [3]. Irradiation experiments provide a useful tool to introduce in the material extrinsic defects in a rather controlled manner. It has been experimentally confirmed (e.g. Ref. [4]) that the creation of columnar defects by means of high energy heavy ions in the whole volume of a crystal provides FL localisation. The aim of this paper is to check if the ion irradiation of surface layers on the opposite sides of bulk samples can also lead to effects of FL localisation. This kind of irradiation affects less than 13% of the sample volume, hence it is not obvious whether it could result in significant changes of the IL and J_c. In order to outline effects of FL localisation we compare the influence of ion-induced *correlated disorder* with the influence of *random disorder* induced by p^+ irradiation producing randomly distributed point-like defects. All the investigated samples come from the same batch and don't significantly differ in pre-irradiation characteristic, what enables us to do a systematic study.

2. Experimental details

The YBCO melt-textured samples were prepared at the Politecnico di Torino and present a uniform dispersion of non superconducting Y_2BaCuO_5 inclusions within large, well aligned YBCO grains [5]. The samples were cut 0.20, 0.27 and 0.50 mm thick. The measurements were performed by means of a Lake Shore 7225 susceptometer-magnetometer. The values of J_c were obtained from hysteresis loops, according to the modified Bean model, and the $\chi''(T)$ peaks values were used to define the IL. Irradiations with 0.24 GeV Au ion and 3.5 MeV proton were performed at room temperature and in vacuum at the 15 MV Tandem XTU accelerator and at the 7 MV Van de Graaff CN accelerator, respectively, of the I.N.F.N.-Laboratori Nazionali di Legnaro (Padova, Italy). The samples were irradiated perpendicularly to the ab-planes on both the surfaces. High energy heavy ions produce linear tracks of amorphous material. By means of TRIM simulations we estimated that 0.24 GeV ^{197}Au ions produce columnar defects about 5 μm long in our YBCO samples, affecting only surface layers of the samples (5-13 % of the total volume). The irradiation fluence ($1.25 \cdot 10^{11}$ ions/cm^2) corresponds to a dose equivalent field (the field corresponding to equal density of vortices and columnar pins) B_ϕ = 2.5 T. In the case of 3.5 MeV protons, defects are point-like (less than 1 nm wide). The implantation depth is about 70 μm (51% of the volume affected). The fluence was $\Phi=1.86 \cdot 10^{16}$ p/cm^2.

3. Critical current density and irreversibility line

The introduction of extrinsic defects by means of irradiation results in enhancement of the J_c, but this effect is very different depending on the type of induced disorder. The pinning mechanism in unirradiated and in proton irradiated samples is qualitatively the same, i.e. pinning by randomly distributed small defects. As a consequence one can expect that proton irradiation results in two opposite effects: in enhancement of the J_c (more pins interacting with FL's) but also in enhancement of FL wandering (due to the increased density of metastable states). In other words, the FL lattice becomes more sensitive to fluctuations. This is what we experimentally observed in our samples: we obtained significant increase of

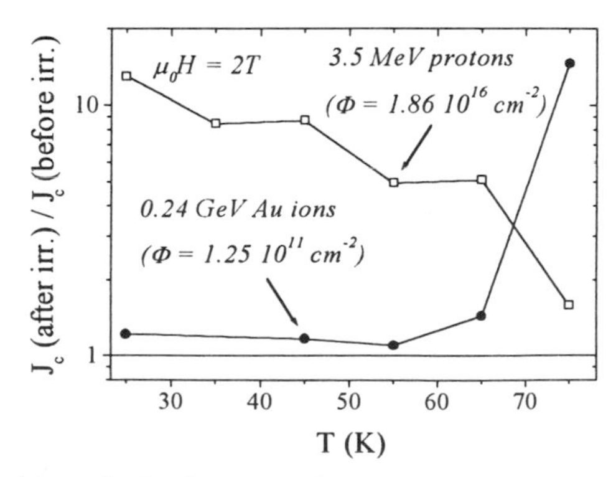

Figure 1 - J_c enhancement factors vs temperature at μ_0H = 2 T for one of the samples irradiated with 3.5 MeV p^+ and for one of the samples irradiated with 0.24 GeV Au ion.

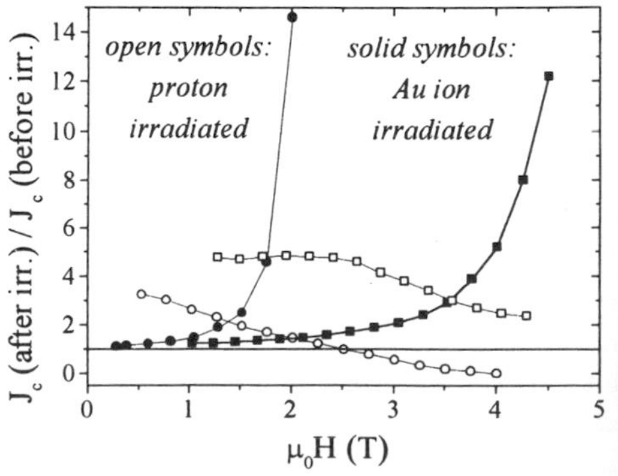

Figure 2 - J_c enhancement factors vs field at T=65K (squares) and 75K (circles), for a sample irradiated with 3.5 MeV p^+ and for a sample irradiated with 0.24 GeV Au ion.

J_c (see also Ref. [6]), but this enhancement *decreases* with increasing temperature (Figure 1) and magnetic field (Figure 2). In Figure 2 a crossover to a lowered J_c at 2.5 T and 75 K is also visible. Moreover, studies of magnetisation relaxation show that the J_c decrease with time is steeper after irradiation [7]. On the other hand, the characteristic feature of correlated disorder is FL localisation. If the irradiation of surface layers of the samples with Au ions leads to global localisation effects, the $J_c(H,T,t)$ behaviour should drastically change. The presence of both types of disorder, *random* (pre-existent) and *correlated* (ion induced), determines a complex behaviour. FL wandering and FL localisation are competing effects and the latter can prevail only above an unbinding temperature T_u [3]. In fact, localisation in columnar defects provides a greater stability of the FL lattice vs. thermal fluctuation and internal interactions [2]. This picture is nicely confirmed by our experimental results: effects of FL localisation in Au irradiated samples emerge as J_c enhancements which *increases* with increasing temperature (Figure 2) and magnetic field (Figure 3). Magnetisation relaxation measurements also show that the creep rate is decreased [8]. The temperature dependence of the enhancement factor indicates that the unbinding transition temperature for $\mu_0H = 2T$ is about 65 K. The result of proton irradiation demonstrates that pinning from random disorder decreases with the temperature because the smaller defects lose its effectiveness. The somehow symmetric curves in indicate that in fact the onset of localisation shows up when pinning from random disorder is weaker.

The position of the irreversibility line in our samples can be well described by the power law $H_{irr}(T) \propto (1-T/T_c)^{\alpha}$, with values of the exponent α ranging from 1.3 to 1.5, see for example the inset of Figure 4. The increase of random disorder, provided by the addition of proton-induced point-like pins, does not influence the position of the IL, although below it a significant J_c enhancement is observed (inset of Figure 3). This result is in accordance with other literature findings [9]. The effects of Au ion irradiation are very different, i.e. a shift of the IL towards higher fields and temperatures and the appearance of a "kink", corresponding to a change of slope, near the dose equivalent field. The shift of the IL at 5 T ranges from 2 to 4 K, depending on the sample (see Figure 3). This is remarkable because the irradiation affected only a limited part of the samples. The specific IL feature of heavy ion irradiated samples is the occurrence of the kink (Figure 4). The value of the exponent α in a power law $H_{irr} \propto (1-T/T_c)^{\alpha}$ varies from 1.7 to 1.8 (depending on samples) below the kink (i.e. higher temperatures) and from 1.1 to 1.3 above the kink (i.e. lower temperatures). The kinks occur

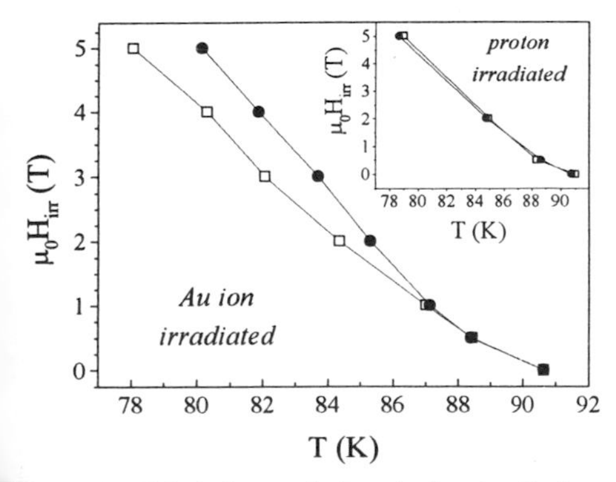

Figure 3 - IL's before and after Au ion irradiation. The inset shows IL's of a proton irradiated sample. (Open symbols: before irradiation; solid symbols: after irradiation).

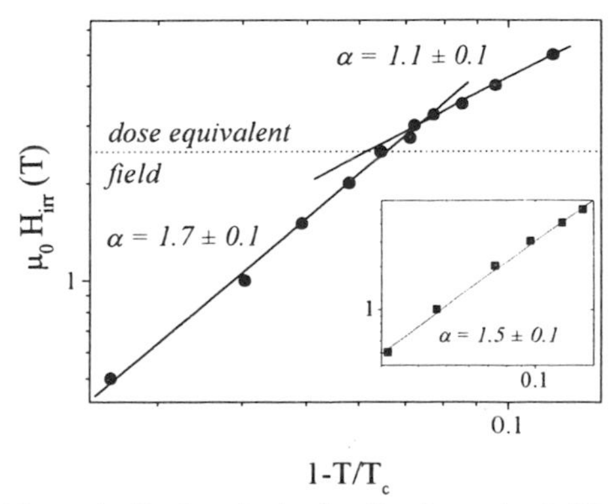

Figure 4 - IL of an Au-ion irradiated sample. Solid lines are power law fits $H_{irr} = A(1-T/T_c)^{\alpha}$. In the inset analogue data and fit are shown for one of the unirradiated sample.

at fields B* ranging from 2.6 ± 0.3 T to 3.0 ± 0.3 T (depending on samples), i.e. close to the dose equivalent field $B_\Phi = 2.5 \pm 0.3$ T. According to Nelson and Vinokur [2] , the presence of columnar defects leads to the formation of a Bose-glass phase, with vortex lines localised in the pins. At low fields all vortices are pinned by columns and the IL is shifted to higher temperatures from the melting line of a clean lattice. When the vortices outnumber the pins, the vortex array becomes polycrystalline and the IL is expected to approach the melting line. Krusin-Elbaum et al. [4], who studied YBCO single crystals irradiated with 1 GeV Au ions, associated the field B* with an "accommodation field" below which the vortices are pinned independently. For $B<B^*$ the IL becomes more upwardly curved, as expressed by larger effective exponents α. Above B* the observed linear behaviour of the IL is consistent with the entanglement transition in the vortex liquid. The occurrence of the same behaviour for our surface irradiated samples indicates that FL's can be localised also if they are not pinned in their whole length by the columnar tracks. Moreover, we observed B* values close to the dose equivalent field, while for ion irradiated YBCO single crystals [4] and for YBCO thin films [10] B* approximately equal to $0.5B_\Phi$ and $0.7B_\Phi$, respectively, were reported. This could be attributed to a higher degree of FL entanglement in our case, due to the absence of ion tracks in the inner part of the sample.

4. Conclusions

The comparison between radiation-induced correlated and random disorder allowed us to outline FL localisation effects which emerge after surface Au ion implantation, although it affects only less then 13 % of the total sample volume. The importance of this result is due to the effectiveness of columnar surface defects at high temperatures and fields, where the FL pinning provided by random (e.g. proton-induced) defects lose its strength. Correlated disorder leads to the occurrence of a kink in the irreversibility line and to the shift of this line towards higher fields and temperatures. Enhancement of critical current density, which increases with increasing temperature, field and relaxation time, was also observed. Random disorder due to proton irradiation does not change the irreversibility line and only provides a J_c enhancement which decreases with increasing temperature, field and relaxation time.

References

[1] Blatter G, Feigel'man M V, Geshkenbein V B, Larkin A I and Vinokur M V 1994 *Rev. Mod. Phys.* **66** 1125.

[2] Nelson D R and Vinokur V M 1993 *Phys. Rev. B***48** 13060.

[3] Balents L and Kardar M 1994 *Phys. Rev. B***49** 13030.

[4] Krusin-Elbaum L, Civale L, Blatter G, Marwick A D, Holtzberg F H and Feild C 1994 *PRL* **72** 1914.

[5] Abbattista F, Mezzetti E et al. 1995 *Applied Superconductivity 1995* vol. 148, D. Dew-Hughes Eds., Bristol: IOP Publishing, p. 543.

[6] Mezzetti E, Colombo S, Gerbaldo R, Ghigo G, Gozzelino L, Minetti B and Cherubini R 1996 *Phys. Rev. B* **54** 3633.

[7] Mezzetti E et al., to be published.

[8] Mezzetti E, Gerbaldo R, Ghigo G, Gozzelino L, Minetti B and Cherubini R, *IEEE Trans. Appl. Superc.*, in print.

[9] Civale L, Marwick A D, McElfresh M W, Worthington T K, Malozemov A P, Holtzberg F H, Thompson J R and Kirk M A 1990 *Phys. Rev. Lett.***65** 1164.

[10] Nakielski G, Rickertsen A, Steinborn T, Wiesner J, Wirth G, Jansen A G M and Kotzler J 1996 *Phys. Rev. Lett.* **76** 2567.

Inst. Phys. Conf. Ser. No 158
Paper presented at Applied Superconductivity, The Netherlands, 30 June–3 July 1997

Comparative study of J_c, flux pinning, and the influence of irradiation in Bi-2212 ribbon and single crystals

A.L. Crossley,[1] J. Everett,[1] G. Wirth,[3] K. Kadowaki,[4] C. Morgan,[2] C. Eastell,[2] C.R.M. Grovenor,[2] and A.D. Caplin[1]

[1]Centre for High Temp. Superconductivity, Imperial College, London SW7 2BP, UK.

[2]Department of Materials, University of Oxford, Oxford OX1 3PH, UK.

[3]Gesellschaft fur Schwerionenforschung, 64291 Darmstadt, Germany.

[4]National Research Institute for Metals, 1-2-1, Sengen, Tsukuba-Shi, Ibaraji 305, Japan.

Abstract. In the BiSrCaCuO (BSCCO) superconductors there has been considerable success in overcoming weak link problems, and at high temperatures intra-grain flux motion is thought to be the dominant dissipation mechanism. Polycrystalline BSCCO tapes have critical current densities (J_c) at high temperatures that are large compared with those in BSCCO single crystals. In order to further improve J_c in the tapes, it is essential to identify the additional flux pinning mechanisms in them. The microstructural and electrical homogeneity of $Bi_2Sr_2CaCu_2O_{8+d}$ ribbon (Bi-2212/Ag) provides a good basis for comparison with of Bi-2212 single crystals, in order to investigate and highlight the pinning in polycrystalline ribbons. We report a detailed study on the effect of irradiation on J_c in both Bi-2212 single crystals and Bi-2212/Ag ribbons using magnetisation measurements. Information regarding the intra-granular flux pinning of the grains within the ribbon has been obtained.

1. Introduction

The small coherence length and large penetration depth in high temperature superconducting materials lead to relatively low pinning energies. The effective pinning potential in the BSCCO phases is further reduced by their highly anisotropic layered structure, which causes flux to penetrate as 2D pancake vortices rather than extended 3D vortices. Above ~30K in fields greater than ~0.1T, flux motion becomes a severe current limiting mechanism. It is the dominant current limiting mechanism in tapes at high temperatures [1] (in contrast to weak-link problems at grain boundaries). To increase J_c it is necessary to pin pancake vortices and prevent flux motion. A 3D flux line can be pinned along its length by point defects, but because pinning energies within the CuO layers are relatively small, such defects are less effective for pancake vortices. Alternatively, it may be possible to decrease the superconducting anisotropy by making the Bi-O insulating layers more metallic. Much attention is being given to enhance J_c, with a range of approaches, such as addition of MgO particles as point defects [2] and oxygen overdoping to reduce anisotropy [3].

For the reasons outlined above, J_c in clean single crystal Bi-2212 is very small above 40K, yet a relatively high J_c is found in polycrystalline BSCCO tapes [4]. Thus in addition to a high degree of grain alignment in the tapes (to ensure good interconnection between grains), their high J_c requires that intragranular pinning be large compared to single crystals.

The effects of ion beam irradiation on $Bi_2Sr_2Ca_2Cu_3O_{10+\delta}$ (Bi-2223) tapes have been studied extensively [5], and it has been shown that columnar defects created by neutron and heavy ion irradiation improve the J_c-field performance substantially.

In this study we use magnetisation measurements to estimate the intragrain J_c within the crystallites of Bi-2212 in a high J_c ribbon before and after irradiation, and compare it to those of virgin and irradiated clean Bi-2212 single crystals.

2. Experimental

The Bi-2212 ribbons were prepared by dip-coating, followed by burnout of the organic binder, and a melt processing schedule of melting for 0.1h at 880C in air, recrystallisation at 50C/hr, followed by an 18h post anneal at 840C. The single crystals were grown by a travelling floating zone technique [6]. Microstructural analysis was carried out using x-ray diffraction (XRD) in θ-2θ mode and transmission electron microscopy (TEM).

Samples were irradiated with energetic U ions at 1×10^{11}ions/cm^2, to dose-equivalent fields (B_ϕ) of 0.5T and 2T. The magnetisation measurements were made with a vibrating sample magnetometer (VSM). A constant sweep rate of 0.02T/s was used for the ribbon samples with the applied field normal to the ribbon.

For a polycrystalline sample, the magnetisation is the sum of inter- and intra-granular contributions. Their relative weight can be assessed using the "length-scale" technique [7]; when the length-scale becomes much smaller than the macroscopic sample dimension, it can be inferred that the latter is dominant, i.e. intra-granular contributions. The J_c of the ribbon crystallites was then calculated using the Bean model, so that $J_c=\Delta M/3a$, where ΔM is the irreversible magnetisation and a the length scale of current flow, here estimated to be 20μm, the average grain radius in the *ab* plane. In order to ensure that the single crystal and the ribbon crystallite J_c's were comparable in terms of electric field (E), they were standardised using the relation $E \sim J^n$, with n of order 10 [8].

3. Results

XRD revealed the ribbons to be highly aligned in the *c*-direction and contained only small amounts of secondary phases. TEM of the virgin ribbon showed the crystallites to contain a large number of point-like defects, which are between 1 and 3 unit cells in the *c*-direction and extend even further in the *ab* plane.

The length scale data for the ribbon (Figure 1) show that at 40K, in fields above 0.1T the magnetisation is dominated completely by intra-granular screening currents; in contrast the 10K data shows the sample is fully connected to fields above 1T. Thus the contribution to J_c from the inter-granular currents at 40K above ~0.1T is negligible, and the J_c of the ribbon crystallites can be obtained.

The J_c-H behaviour of the virgin ribbon crystallites and the virgin single crystal are quite different (Figure 2). At 0.1T the J_c within the crystallites is nearly two orders of magnitude larger than in the single crystal, thus confirming that the pinning is a great deal stronger in the former than the latter; also the pinning is sustained to considerably higher fields in the crystallites than in the single crystal.

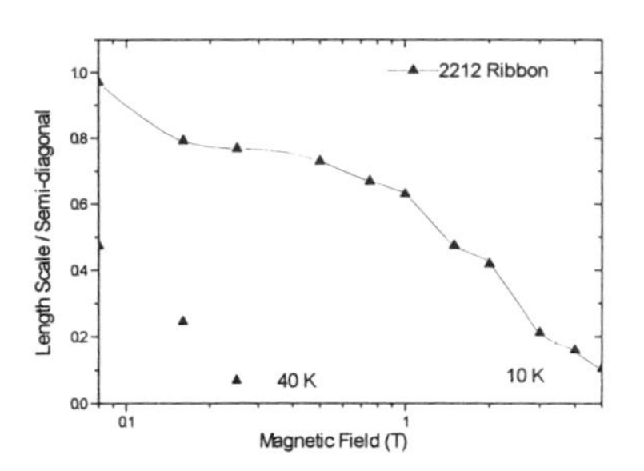

Figure 2 ''Length scale'' behaviour of the Bi2212/Ag ribbon at 10K and 40K

It is well established that irradiation enhances the pinning dramatically in Bi-2212 single crystals, up to fields of order B_ϕ. We see that at low fields, J_c increases by two orders of magnitude, and also remains field independent up to fields approaching B_ϕ. In contrast, irradiation increases J_c in the ribbon crystallites only slightly, although there is a more significant improvement in the field-dependence. It is noteworthy that irradiated ribbon data and the (corrected) irradiated crystal data run almost parallel to each other, there being a factor of about 5 between them. Since the pinning in both is dominated by the same defect (irradiation columns) at the same density, the same J_c should be expected. However, we have two major uncertainties: the value of a to adopt for calculation of J_c of the ribbon crystallites, and also that the ribbon is not completely dense;

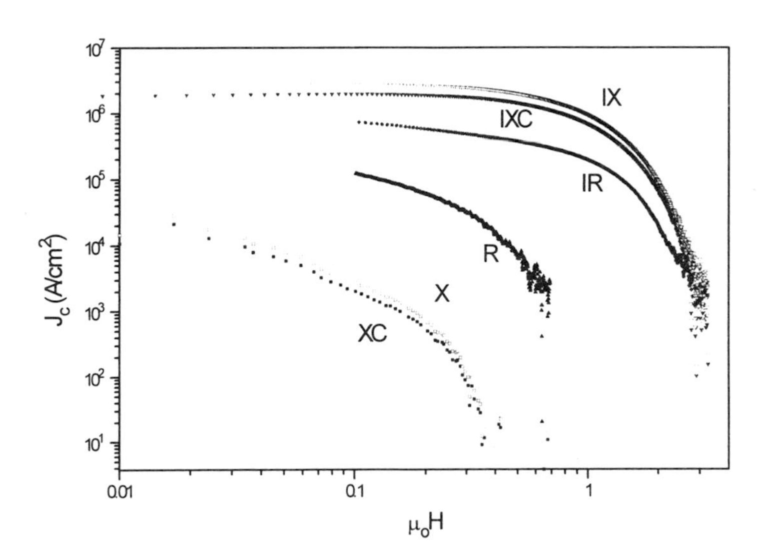

Figure 1. Shows intra-granular J_c versus H at 40K for the virgin (R) and irradiated (IR) ribbon (length scale equal to average grain size of 20μm), and virgin (X) and irradiated (IX) single crystal (data corrected for sweep rates are indicated as XC and IXC). The samples were irradiated to a matching field of 2T.

consequently the discrepancy in the J_c's for the crystal and the ribbon is not untoward.

4. Conclusion

It is clear that the crystallites in the virgin ribbon contain a high density of effective pinning centres, whose behaviour with field is similar to that of irradiation columns. A high density of point like defects were observed within the crystallites of the ribbon and it is possible that they are responsible for the higher than expected critical current observed in Bi-2212 tapes.

Further microstructural analysis is necessary to reveal the exact nature of the pinning sites in melt processed Bi-2212/Ag.

Acknowledgements

The authors would like to thank the EPSRC and Oxford Instruments plc. for the CASE award for ALC.

References

[1] Dhallé M, Cuthbert M, Johnston M D, Everett J, Flükiger R, Dou S X, Goldacker W, Beales T and Caplin A D, 1997 *Supercond. Sci. Technol.* **10** 1 21-31
[2] Wei W, Sun Y, Schwartz J, Gooretta K, Balachandran U, and Bhargave A 1996 *Preprint.*
[3] Kotaka Y, Kimura T, Shimoyama J, Kitazawa K, Yamajuji K, Kishio K, and Pooke D 1995 *Physica C* **235-240** 1529-1530
[4] Frank H, Lemmens P, Lethen J, Wiesener J, Wirth G, Wagner P, Adrian H, and Guntherodt 1994, *Physica C* **235-240** 2739-2740
[5] Dou S X,and Liu H K 1993 *Supercond. Sci, Technol.* **6** 297-314
[6] Mochiku Tand Kadowaki K, *Trans. Mats Res. Soc. Japan* 1993 **19** 349
[7] Angadi M A, Caplin A D, Laverty J R and Shen Z X, 1991 *Physica C* **177** 267
[8] Dhallé M, Cuthbert M N, Thomas T, Perkins G K, Caplin A D, Yang M, and Goringe M J 1995, *IEEE Trans. Appl. Supercond.* **5** 1317

Inst. Phys. Conf. Ser. No 158
Paper presented at Applied Superconductivity, The Netherlands, 30 June–3 July 1997

Detailed magnetisation study of inter- and intragranular currents in Ag-sheathed Bi-2223 tape

Petr Nálevka[1,2], Miloš Jirsa[1], Ladislav Půst[1], Alexandr Yu. Galkin[1], Michael R. Koblischka[3], and René Flükiger[4]

[1]Institute of Physics ASCR, Na Slovance 2, CZ-180 40 Praha 8, Czech Republic.
[2]KIPL, FJFI CVUT, V Holešovičkách 2, CZ-180 00 Praha 8, Czech Republic.
[3]Department of Solid State Physics, Norwegian University of Sciences and Technology, N-7034 Trondheim, Norway.
[4]Université de Genève, 20, Rue d'Ecole de Médecine, CH-1211 Genève 4, Switzerland

Abstract. Magnetic hysteresis loops of a rolled Bi-2223/Ag tape were measured at various temperatures and fields up to ± 5 T. In order to separate the inter- and intragranular current contributions, the tape was bent to a small diameter and measured again after straightening. At low temperatures, a significant intergranular current hysteresis was observed for the increasing and decreasing field. Shapes of the experimental hysteresis loops are discussed in detail along with a deduced relationship between intergranular and intragranular currents.

1. Introduction

Ag-sheathed tapes of $(Bi,Pb)_2Sr_2Ca_2Cu_3O_{10+\delta}$ (Bi-2223/Ag) are known as favourite candidates for high power technical applications. Due to the granular nature of this superconducting material, two types of the critical currents are induced in the tapes: the intergranular currents and the intragranular ones. In spite of an enormous experimental and theoretical effort, relations between the inter- and intragranular currents are not fully understood yet.

Bending of an Ag-sheathed tape to a small diameter was recently suggested as a way how to separate the intergranular currents from intragranular ones [1].

The aim of the present paper is to study the inter- and intragranular currents in a Bi-2223/Ag tape at elevated temperatures by means of the bending method.

2. Experiment and discussion

The studied Ag-sheathed Bi-2223 tape was prepared by the standard powder-in-tube technique followed by rolling.

All measurements were performed in a Quantum Design's SQUID system MPMS-5S. The external magnetic field was always applied along the normal to the tape plane. We measured magnetic hysteresis loops (MHL) at temperatures from 5 K to 77 K and at field up to B = ± 5 T. After bending the sample to diameter 1.25 mm and straightening it again, the measurement was repeated.

Magnetic hysteresis loops measured before and after the bending are plotted in figures 1(a) and (b). The most significant feature of the curves is the drop of magnetic moment after

the tape bending. Also the shape of the MHL is changed. The bending removes two anomalies observed originally in the tape:
(i) Maximum of the magnetic moment lies no more at positive external fields and
(ii) The MHLs become symmetrical at high fields and low temperatures.

The MHL maximum lying at positive fields contradicts the conventional scheme based on the assumption that the induced critical current reaches its maximum at zero magnetic field. As the internal magnetic field should lag behind the external one [2], the MHL maximum should lie at negative external fields on the descending field branch. Simple models of the intergranular current transport [3,4] do not account for this effect. Recently, Müller et al. [5] analysed position of the MHL peak in Bi-2223/Ag tapes and attributed the anomalous peak position to demagnetising effects (stray fields) arising from intragranular currents.

The asymmetry of the magnetic moment at low temperatures before the tape bending contradicts again the conventional ideas. It is characterized by the significantly higher (in absolute value) magnetic moment in the decreasing field than in the increasing field, in contrast to the traditional picture of a symmetric irreversible magnetic moment (due to flux pinning on defects) superimposed onto a negative equilibrium reversible moment. The simple model of Bean-Livingston surface barrier predicts just an opposite effect, i.e. the current in the decreasing field to be constant and much smaller than that in the increasing field [6,7].

Our present experiments show that both the above-mentioned anomalies are associated with intergranular currents and reflect the fact that at low temperatures these currents are limited by grain boundaries. After the tape bending, the MHL shape changes significantly and the anomalies disappear (fig. 1(b)). The MHL maximum shifts to the proper (negative) side of the field scale and no critical current suppression is observed during the field increase. As documented in figures 1(b) and 2, after the bending each MHL consists of a vertically symmetric irreversible contribution due to intrinsic pinning [2] and of an equilibrium reversible background. The intragranular currents rapidly decay with increasing temperature and field.

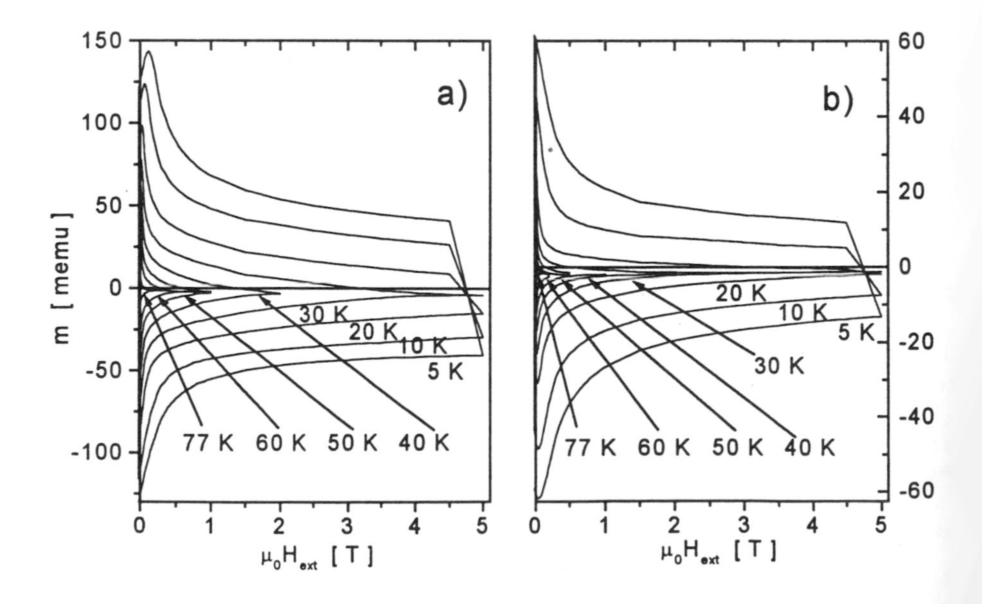

Fig. 1. Hysteresis loops measured in the Bi-2223/Ag tape before (a) and after (b) bending.

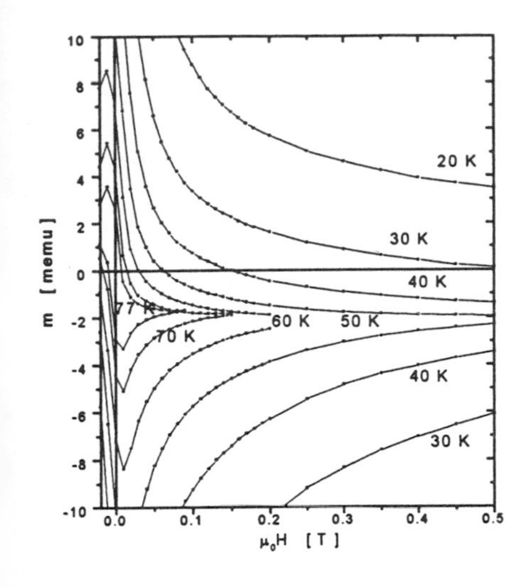

Fig. 2. Low-field parts of the MHLs of the bent tape for temperatures above 30 K.

In figure 3, we present hysteresis loops obtained by subtracting the hysteresis loops measured on the bent tape from those measured before the bending. These MHLs correspond to intergranular currents only. The two peculiarities observed on the MHLs before the tape bending are present on the "differential" MHLs, too. We conclude therefore that the intergranular currents are responsible both for the anomalous peak position and for the magnetic moment asymmetry in the decreasing and increasing field observed at low temperatures.

Below 30 K, the intergranular currents in decreasing field follow an $H^{-1/3}$ dependence (in agreement with Ref. [8]) (see inset in fig. 2). The intergranular currents in increasing field are practically field-independent at 5 K, however, with increasing temperature they become $\propto H^{-1/3}$. We emphasize that the field-

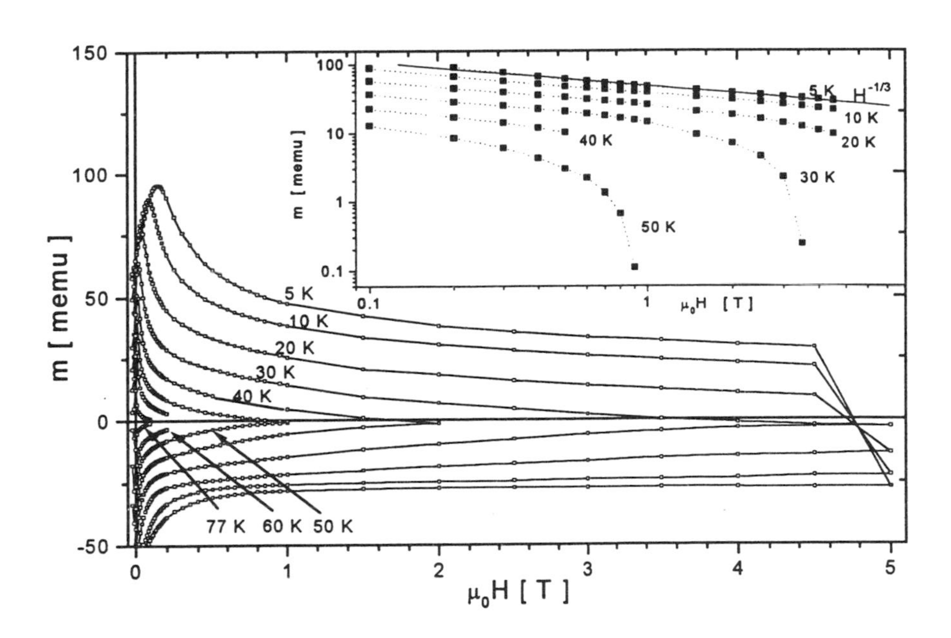

Fig. 3. Hysteresis loops related to intergranular currents. These curves were obtained by subtracting the MHLs measured after the bending from those measure before. Inset: Descending field branches (dotted) for low temperatures in a log-log scale.

independent intergranular currents in increasing field at 5 K egree with the results obtained by transport measurements [9] where the transport currents were found to be field-independent up to high fields.

The stray fields from the intragranular currents can result in an intergranular current hysteresis [5] but the magnitude of such an effect is much smaller than that observed in our sample at low temperatures. Another source of hysteresis may be Bean-Livingston surface barrier at grain boundaries [10]. The associated surface currents together with the inhomogeneous Josephson junctions may in principle lead to a field-independent current in the increasing field. However, the effects observed in Bi-2223/Ag tapes have different character from those attributed to Bean-Livingston surface barrier as observed on clean Bi-based single crystals [7]. Also the predicted $j_c \propto H^{-1}$ dependence is far from the observed $j_c \propto H^{-1/3}$. As discussed in [11-13], geometrical barrier can give rise to an asymmetry observed in the Bi-2223/Ag tape. The thermally activated vortices have enough energy to overcome such a barrier above 20 K. Thus, at higher temperatures the MHLs associated with the intergranular currents become symmetric.

3. Conclusions

Magnetic hysteresis loops measured on a Bi-2223/Ag tape before and after the tape bending to a small diameter were examined in detail. Two significant anomalies were observed on the MHLs before the bending, (i) the central peak on the MHL positioned at positive (descending) fields and (ii) the vertical asymmetry of the MHLs at temperatures below 20 K. Both these effects were shown to be connected with the intergranular currents. While the anomalous peak position could be explained by demagnetising effects of intragranular currents [5], the MHL asymmetry and especially the associated field independence of critical currents in increasing fields can hardly be due to a demagnetisation shift of the MHL branches [5]. Neither can they be attributed to Bean-Livingston barrier. The effect might be probably due to some kind of a geometrical barrier [11-13] leading to constitution of additional surface currents at grain edges.

Acknowledgements

This work was supported by GA AVČR under contract A1010512.

References

[1] K.-H. Müller, C. Andrikidis, H. K. Liu, S. X. Dou: Phys. Rev. B **50** (1994) 10218
[2] C. P. Bean: Phys. Rev. Lett. **8** (1962) 250; Rev. Mod. Phys. **36** (1964) 31
[3] L. N. Bulaevskii, L. L. Daemen, M. P. Maley, J. Y. Coulter: Phys. Rev. B **48** (1993) 13798
[4] B. Hensel, J.-C. Grivel, A. Jeremie, A. Perin, A. Pollini, R. Flükiger: Physica C **205** (1993) 329
[5] K.-H. Müller, C. Andrikidis, Y. C. Guo: Phys. Rev. B **55** (1997) 630
[6] C. P. Bean and J. D. Livingston: Phys. Rev. Lett. **12** (1964) 14
[7] N. Chikumoto, M. Konczykowski, N. Motohira, K. Kishio: Phys. Rev. Lett. **69** (1992) 1260
[8] M. Dhallé, M. Cuthbert, M. D. Johnston, J. Everett, R. Flükiger, S. X. Dou, W. Goldacker, T. Beales, A. D. Caplin: Supercond. Sci. Technol **10** (1996) 21
[9] B. Hensel, G. Grasso, R. Flükiger: Phys. Rev. B **51** (1995) 15456
[10] A. I. Dyachenko: Physica C **213** (1993) 167
[11] J. Provost, E. Paumier, A. Fortini: J. Phys. F **4** (1974) 439
[12] J. R. Clem, R. P. Huebener, D. E. Gallus: J. Low Temp. Phys. **12** (1973) 449
[13] E. Zeldov, A. I. Larkin, V. B. Geshkenbein, M. Konczykowski, D. Majer, B. Khaykovich, V. M. Vinokur, H. Shtrikman: Phys. Rev. Lett. **73** (1994) 1428

Inst. Phys. Conf. Ser. No 158
Paper presented at Applied Superconductivity, The Netherlands, 30 June–3 July 1997

Grain connectivity and flux pinning for Bi-2223/Ag tapes obtained by oxide-powder-in-tube method

J Horvat, Y C Guo, R Bhasale, W G Wang, H K Liu and S X Dou

Centre for Superconducting and Electronic Materials, University of Wollongong, NSW 2522, Australia, e-mail: jhorvat@uow.edu.au

Abstract. Bi-2223/Ag superconducting tapes were prepared with oxide-powder-in-tube (OPIT) method, with intermediate deformation between sintering stages. Critical current was measured by a standard four-probe method, at 77K. Magnetic field was applied perpendicularly to the tape plane. Critical current density J_c, quality of the links between the grains and the field of the peak of the pinning force density (H^*) were obtained from these measurements for a number of samples of different quality. It was obtained that J_c increases with the quality of the links between the grains, for any value of H^*. However, there is a peak of Jc at about H^*=180mT, which becomes increasingly pronounced as the quality of the links increases. This peak is an artifact of OPIT: intermediate deformation is beneficial for the improvement of flux pinning and H^*, however excessive deformation can be detrimental for the formation of the links.

1. Introduction

Oxide-powder-in-tube-method (OPIT) is a convenient way of improving grain connectivity and flux pinning of Bi-based superconductors. Grain connectivity is improved because of alignment of the grains by intermediate mechanical deformation and subsequent sintering. Numerous reports claim that dislocations introduced upon mechanical deformation are responsible for increase in pinning [1,2]. On the other hand, it has been clearly shown that such dislocations can be easily annealed-out by short sintering [3], casting a doubt on dislocations as origin of the pinning. Nevertheless, increased density of dislocations and improved flux pinning have been observed in samples after mechanical deformation and prolonged sintering [1,4]. A way out from this controversy was found in experimental results which give a strong indication that micro- and nano-cracks, introduced by mechanical deformation and imperfectly healed upon subsequent sintering, are responsible for increased pinning [5].

Therefore, the larger the deformation, the larger increase of flux pinning is obtained. However, grain connectivity is also influenced by the deformation. It is well known that critical current density (J_c) initially increases with the degree of mechanical deformation, reaches a maximum and then decreases [6]. This is usually ascribed to cracks which cannot be healed by subsequent sintering. Consequently, one has to find a compromise between improvement of flux pinning and optimal grain connectivity, to obtain the tapes with large J_c at elevated fields.

In this paper we present dependence of J_c on flux pinning and quality of the inter-grain links for a number of Bi-2223/Ag tapes. These results are a characteristic of samples prepared by OPIT method, using preparation procedures commonly found in literature. Improved procedures, which usually do not get published, may give quantitatively different results.

2. Experimental details

Bi-2223/Ag tapes were prepared by standard oxide-powder-in-tube method [7]. A large number of samples of different quality, prepared under different conditions, was studied. The powder composition, degree of mechanical deformation, number of deformation steps and sintering temperature were varied. The details of the preparation procedures of the samples, as well as the effect of different preparation procedures on the relevant properties of the samples, can be found elsewhere [5,8,10]. In this report, some relationships between commonly used characterization parameters are studied. Experimental results obtained for these samples will give us some general characteristics of OPIT method.

Critical current was measured by standard four-probe method at 77K, with magnetic field perpendicular to the tape face. Field of the peak of pinning force density (H^*) was used as a measure of flux pinning. It has been shown earlier that values of H^* are in qualitative agreement with measurements of irreversibility field, obtained from magnetic hysteresis loop [5]. J_c was obtained by dividing critical current by cross-sectional area of superconducting core.

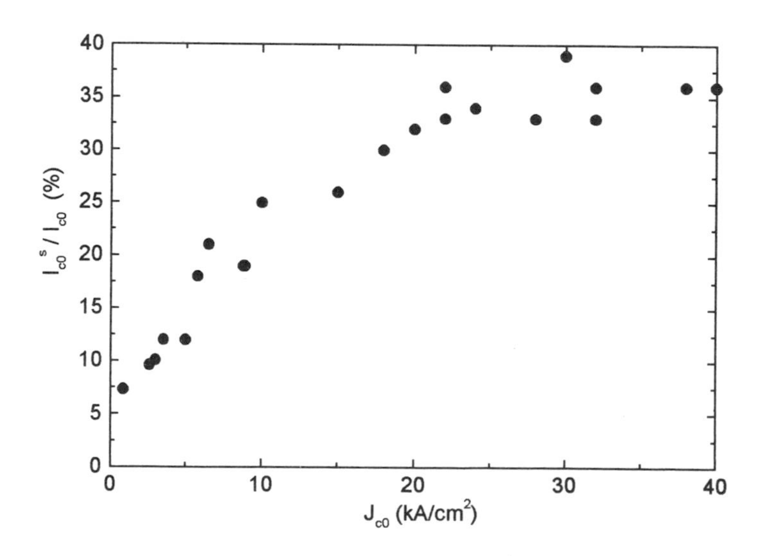

Figure 1: I_{c0}^s/I_{c0}, representing the quality of the links between the grains, vs. J_{c0} for a number of differently prepared samples.

Quality of the links between the grains was monitored by measuring the ratio of critical current through strong links in total critical current, in zero field (I_{c0}^s/I_{c0}) (Fig.1)

[8, 9]. The value of I_{c0}^s/I_{c0} was obtained by extrapolating the field dependence of normalized critical current (I_c/I_{c0}) from elevated fields (H>100mT) to zero field [8]. Since at elevated fields weak links do not conduct the current, the employed part of I_c/I_{c0} vs. H reflects the conductance of strongly linked grains only. More details on the procedure for obtaining I_{c0}^s/I_{c0} and its physical meaning can be found elsewhere [8,9]. Larger values of I_{c0}^s/I_{c0} (above 20%) indicate good grain connectivity. It has been shown for a number of different samples that I_{c0}^s/I_{c0} can be reliably used for this purpose [8]. I_{c0}^s increases with J_{c0} and saturates at about 40% for J_{c0}>20 kA/cm^2 (Fig. 1).

3. Experimental results and discussion

Figure 2 shows J_c, normalized to maximum J_c of the samples measured, vs. H^* and I_{c0}^s/I_{c0}. There are two main features of the plot: J_c increases with I_{c0}^s/I_{c0} and there is a ridge along I_{c0}^s/I_{c0}-axis centered at about H^*=180mT. The ridge is not noticeable for I_{c0}^s/I_{c0} < 15%, however it becomes increasingly pronounced for larger values of I_{c0}^s/I_{c0} (Fig.2).

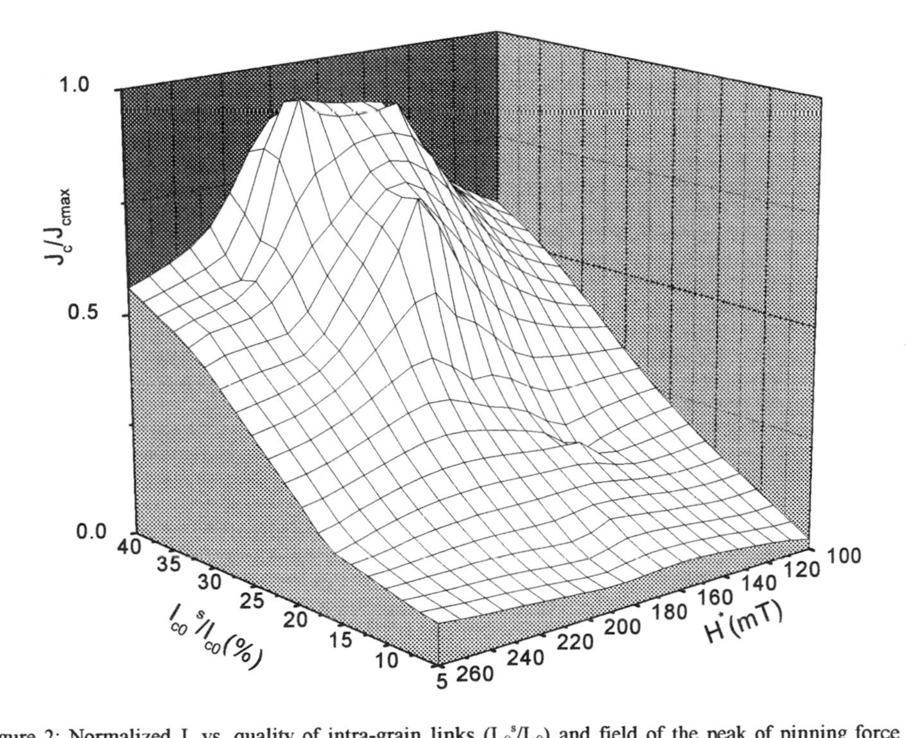

Figure 2: Normalized J_c vs. quality of intra-grain links (I_{c0}^s/I_{c0}) and field of the peak of pinning force density (H^*) for a number of Bi-2223/Ag tapes of different quality.

The increase of J_c with I_{c0}^s/I_{c0} is not surprising, since larger values of I_{c0}^s/I_{c0} indicate higher quality of the links. The ridge is a consequence of improvement of H^* and J_c with intermediate pressing and subsequent sintering. With pressing and sintering, one aligns grains improving grain connectivity and quality of the links. In addition to this, one introduces new pinning centres in the tapes through healing of micro-cracks [5]. However, too large deformation introduces large cracks that cannot be healed and grain connectivity and J_c worsen [6], even though *quality* of the links is not affected to a large extent. On the other hand, H^* continues increasing with large deformation. Combination of these two effects gives the ridge in Fig.2.

An interesting feature is apparent absence of the ridge for the tapes of poor quality, for $I_{c0}^s/I_{c0}<15\%$. This probably occurs because of poor grain alignment and connectivity of these tapes. The effect of large cracks introduced by excessive deformation is probably masked out by substantial improvement of packing density and grain alignment. Namely, even a small improvement of the grain alignment gives proportionally larger improvement of J_c for these tapes than for the ones with better initial grain alignment.

4. Conclusions

Recently introduced parameter for assessment of quality of links between the grains [8, 9], I_{c0}^s/I_{c0}, was employed for characterization of tapes prepared by OPIT method. Relationship between J_c, flux pinning and grain connectivity, which is a characteristical feature of OPIT method, was clearly revealed. Dependence of J_c on flux pinning exhibits a peak, which is the most pronounced for tapes with high grain connectivity. This is a consequence of deteriorating grain connectivity with excessive mechanical deformation, which on the other hand only further improves the pinning.

Acknowledgments

Authors would like to acknowledge financial support of Metal Manufactures Ltd, Electric Supply Association of Australia, Energy Research and Development Corporation, Department of Energy of New South Wales, Australian Research Council and Department of Industry, Science and Tourism.

References

[1] Dou S X, Guo Y C and Liu H K 1992 *Physica C* **194** 343
[2] Grindatto D P, Daumling M, Grasso G, Nissen H-U and Flükiger R 1996 *Physica C* **260** 25.
[3] Yang G, Shang P, Sutton S D, Jones I P, Abell J S and Gough C E 1993 *Phys. Rev. B* **48** 4054.
[4] Miller D J, Sengupta S, Hettinger J D, Shi D, Gray K E, Nash A S and Goretta K C 1992 *Appl. Phys. Lett.* **61** 2823.
[5] Horvat J, Bhasale R, Guo Y C, Liu H K and Dou S X 1997 *Supercond. Sci. Technol.* **10** 409.
[6] Guo Y C, Liu H K and Dou S X 1993 *Mater. Sci. Eng. B* **23** 58.
[7] Dou S X and Liu H K 1993 *Supercond. Sci. Technol.* **6** 297.
[8] Horvat J, Wang W G, Bhasale R, Guo Y C, Liu H K and Dou S X 1997 *Physica C* **275** 327.
[9] Horvat J, Dou S X, Liu H K and Bhasale R 1996 *Physica C* **271** 51.

Inst. Phys. Conf. Ser. No 158
Paper presented at Applied Superconductivity, The Netherlands, 30 June–3 July 1997

An Investigation of the Peak Effect in the Chevrel Phase Superconductor Tin Molybdenum Sulphide.

I J Daniel D N Zheng and D P Hampshire.

Department of Physics, University of Durham, Durham, DH1 3LE, U.K.

Abstract Chevrel Phase superconductors have great potential for high field applications due to their high upper critical fields. In this study the unusual field dependence for the critical current density (J_c) in $Sn_1Mo_6S_8$ was investigated. The samples were fabricated using a Hot Isostatic Press operating at 2000 bar and were characterised for phase composition using x-ray powder diffraction and electromagnetically using resistivity, susceptibility and dc. magnetisation as a function of field and temperature. J_c and the pinning force were calculated from the dc. magnetisation measurements. The peak effect was found in high fields and characterised in detail. This pinning mechanism could provide a route for achieving the required current density in Chevrel Phase materials for the next generation of high field magnets.

1. Introduction

The Chevrel Phase superconductor $Sn_1Mo_6S_8$ has previously been seen to exhibit a peak in the pinning force in the region close to the irreversibility field (B_{irr}) [1]. This peak effect may be important for high field magnet applications where strong flux pinning at high fields is critical for acheiving the current densities required. Kramer investigated the variation with microstructure of the peak effect [2,3] and found it to be a general phenomena among superconducting materials. He proposed a parameterisation of the pinning force of the form

$$F_P = \alpha . B_{irr}^{n}(T)\ . b^{1/2} . (1-b)^2 \qquad (1)$$

There has been little work reported on the temperature dependence of the peak effect in $Sn_1Mo_6S_8$. In this work we parameterise the field and temperature dependence of the pinning force using Kramer's functional form.

2. Experimental

2.1 *Sample Fabrication*

Molybdenum powder (< 8μm mesh, 99.95%) and sulphur chips (99.999%) were reacted to form MoS_2 which was sealed in a molybdenum crucible with Sn(II)S (<8μm mesh, 99. 5%), and reacted at 1100°C for 24 hours. The samples were then wrapped in molybdenum foil and sealed under vacuum in stainless steel tubes. These billets were then hot isostatically pressed at 800°C and 2 kbar argon gas for 8 hours. All powder preparation was completed in a glove box with less than 10 ppm oxygen and water.

2.2 Measurements

Samples were characterised using x-ray powder diffraction and found to be predominantly single-phase with low levels of molydenum as a second phase impurity. Variable temperature ac. susceptibility and ac. resistivity measurements were completed in dc. fields up to 14 T. Dc. magnetisation measurements were made at temperatures from 4.2K to above T_c in fields up to 11.6 T.

3. Results

3.1 Dc. Magnetisation measurements

The magnetisation data (Figure 1) clearly show a non monotonic decrease in magnetic moment with increasing applied magnetic field. Using Bean's critical state model [4], the size of the hysteresis was used to calculate the critical current density (Figure 2) giving a value of 1×10^8 A.m^{-2} at 10 T and 4.2K. The J_c data were replotted using standard Kramer plots ($J_c^{1/2}B^{1/4}$ vs B) and the temperature dependence of the irreversibility field was determined from the high field data. These values (figure 4) are consistent with those determined from the onset of the susceptibility transition. In figure 5, a scaling analysis of the pinning force data is provided by fitting it to a curve of the form of equation 1.

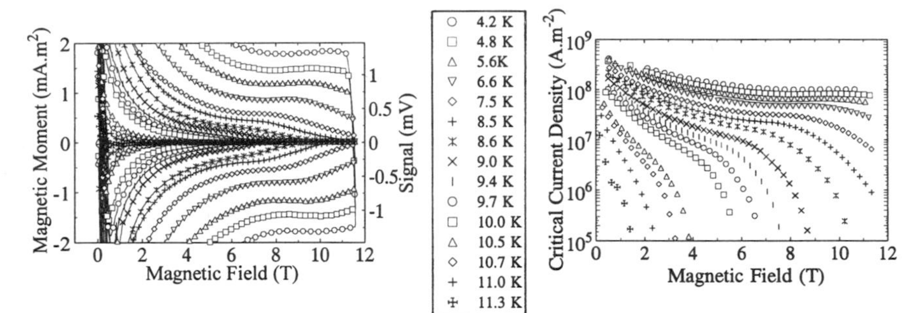

Figure 1 Magnetisation curves at various temperatures

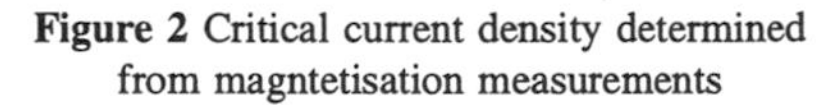

Figure 2 Critical current density determined from magntetisation measurements

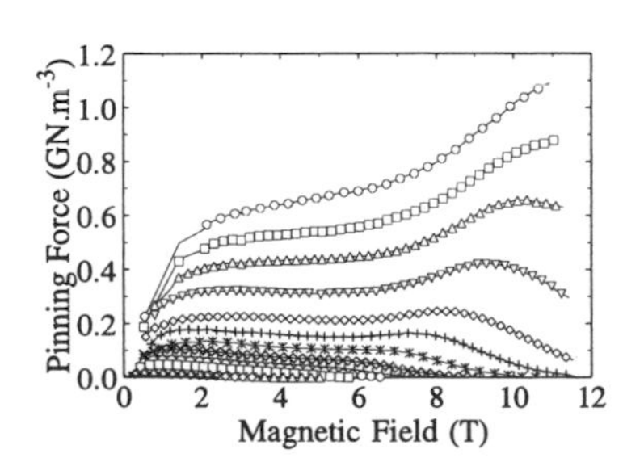

Figure 3 Volume Pinning Force as a function of temperature. For legend see Figures 1 and 2

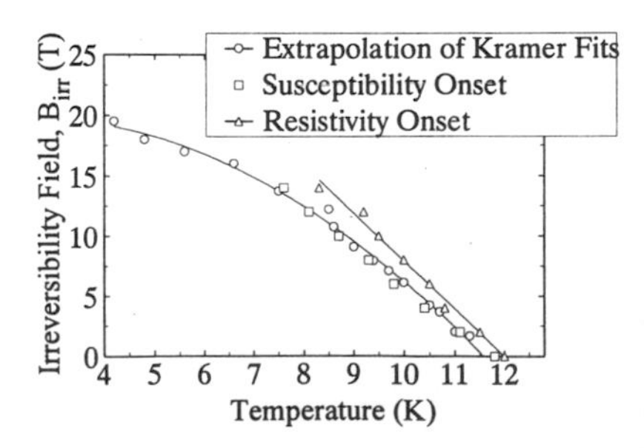

Figure 4 Irreversibility Fields as determined by Kram Extrapolation and Susceptibility Onset

The volume pinning force was normalised by a prefactor $\alpha_H.(B_{irr})^n$ which gives a single Universal Scaling law for $b > 0.75$. The data in figure 5 clearly show a lack of scaling that suggests that more than one pinning force mechanism operates across the whole field range. In figure 6, a second mechanism has been identified by normalising the data at low fields using a second prefactor $\alpha_L.(B_{irr})^n$ that gives a Universal Scaling law for $b < 0.15$. Figures 5 and 6 demonstrate that at each temperature, two mechanisms operate - one at low reduced field, the other at high reduced field which is conventionally called the peak effect. The cross-over from the low reduced field mechanism to the high reduced field mechanism is systematic and is strongly temperature dependent.

The values of n in the prefactors, $\alpha_H.(B_{irr})^n$ and $\alpha_L.(B_{irr})^n$, used to normalise the data in figures 5 and 6 are found from the straight line fits in figure 7. The value of 3.1 ± 0.1 was determined for n for the high field pinning mechanism and 2.4 ± 0.1 was found for the low field pinning mechanism.

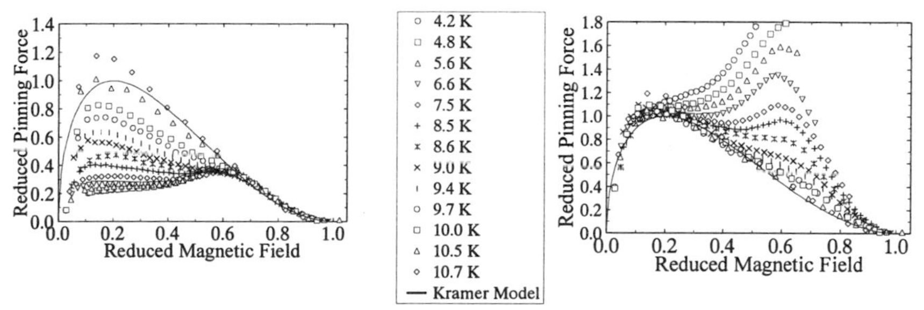

Figure 5 Reduced Pinning Force Curve using $B/B_{irr} > 0.7$ for Reduction

Figure 6 Reduced Pinning Force Curve using $B/B_{irr} < 0.15$ for Reduction

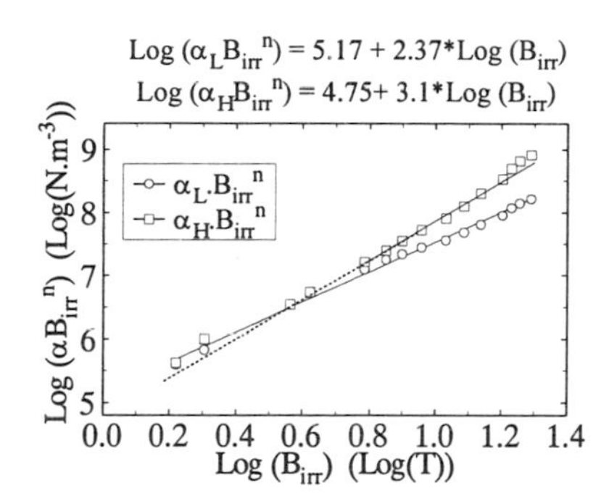

Figure 7 Log log plot of $\alpha_L.B_{irr}^n$ and $\alpha_H.B_{irr}^n$ against B_{irr}

3.3 *Ac. susceptibility and resistivity measurements*

The critical temperature was 11.8K determined by the onset of the susceptibility transition in zero field, and the field dependence of this onset is shown in Figure 4. The normal state resistivity was found to be $440\mu\Omega$.cm at room temp and $15\mu\Omega$.cm just above T_c at $T = 12K$.

4. Discussion

The peak effect was first studied in detail by Kramer who examined the effect of microstructure on the volume pinning force for various materials and found that the pinning appeared to be optimised in high fields. Kramer proposed that the relatively low pinning strength at low field could be overcome by improving the microstructure. In this paper we have completed a detailed investigation of the scaling properties of $Sn_1Mo_6S_8$ as a function of temperature and magnetic field. Evidence for two separate mechanisms has been produced which operate in low reduced field and in high reduced field and have a functional form given by:

$$b < 0.15 \qquad F^*_{pL} = \alpha_L B^{2.4}_{irr}(T).b^{1/2}.(1-b)^2$$

$$b > 0.6 \qquad F^*_{pH} = \alpha_H B^{3.1}_{irr}(T).b^{1/2}.(1-b)^2$$

where $\alpha_L = 1.5 \times 10^5$ N.m^{-3}.T$^{-2.4}$ and $\alpha_H = 5.6 \times 10^4$ N.m^{-3}.T^{-3}

Both these equations have the same field dependence $b^{1/2}(1-b)^2$ which is often associated with grain boundary pinning. We suggest that these two mechanisms could result if the grain size were bimodal. Alternatively, there may be a mechanism that is not simple grain boundary pinning which nevertheless gives the Kramer field dependence. Finally we note that if the high field pinning mechanism could be continued into the low field regime then the critical current density achieved at 4.2 K would increase by factor 5 at intermediate low fields as can be seen for figures 5 and 6.

5. Conclusion

The peak effect in $Sn_1Mo_6S_8$ has been studied in detail using variable temperature high field magnetisation measurements. Evidence has been found for two pinning mechanisms operating in this material - one at low reduced field and one at high reduced field. The functional form of these two mechanisms has been derived. We note that extending the high field mechanism throughout the entire superconducting phase may provide a means to increase J_c in Chevrel phase materials to those required for high field magnet applications.

Acknowledgements

We would like to thank Dr N Cheggour and H A Hamid for their most useful discussion in the production of this paper and Dr C Lehmann for the xray powder diffraction work. This work was funded by the E.P.S.R.C.

References

[1] Bonney L A, Willis T C and Larbalestier D C 1995 *J. Appl. Phys.* **77**, 6377
[2] Kramer E J 1975 *J. Elec. Mat.* **4** 839
[3] Kramer E J 1973 *J. Appl. Phys.* **44** 1360
[4] Bean C P 1964 *Rev. Mod. Phys.* **36** 553

Inst. Phys. Conf. Ser. No 158
Paper presented at Applied Superconductivity, The Netherlands, 30 June–3 July 1997

Test of 1.2 MVA High-T_c Superconducting Fault Current Limiter

W. Paul[1], M. Lakner[1], J. Rhyner[1], P. Unternährer[1], Th. Baumann[1], M. Chen[1], L. Widenhorn[2], A. Guérig[3]

[1]ABB Corporate Research, CH-5405 Baden, Switzerland. [2]ABB High Voltage Technologies, CH-8050 Zürich, Switzerland. [3]Nordostschweizerische Kraftwerke AG, CH-5401 Baden, Switzerland

Abstract. A 3-phase superconducting fault current limiter with a rated power of 1.2 MVA has been built, tested, and installed in a power plant. The device is based on the "shielded iron core concept". The superconducting part consists of a stack of rings made of Bi-2212 ceramic. They were fabricated by a special partial melt process and have a diameter of 38 cm, a height of 8 cm, and a thickness of 1.8 mm. The current-voltage characteristic obeys a power law $U \sim I^{\alpha}$ with $\alpha \approx 5$. The critical current density defined by the 1 μV/cm criterion is about 1400 A/cm^2. The nominal current and voltage of the device are 70 A and 10.5 kV, respectively. In 3-phase short circuit tests with a prospective fault current of 60 kA the current was limited to about 700 A in the first half wave. After 50 ms the limited current was below 250 A. The test results are in excellent agreement with detailed simulations of both the normal operation and the behaviour under fault conditions.
The current limiter has been installed in the auxiliary line of a hydropower plant for an one year endurance test. It is the first superconducting device tested in a power plant under actual operating conditions.

1. Introduction

The world-wide increasing demand for electrical power leads to a growing interconnection of electrical networks. As a consequence, the short-circuit currents (I_{sc}) increase. They can be more than 20 times higher than the nominal current (I_N) and thus lead to high mechanical and thermal stresses, both of which are proportional to I_{sc}^2. All electrical equipment has to be designed to withstand these stresses. It is obvious, from both technical and economical points of view, that a device that can reduce I_{sc} is desirable. Within all discussed and already realised current limiters, only the Superconducting Fault Current Limiter (SCFCL) offers ideal electrical behaviour. Most SCFCL are based on the transition from the superconducting to normal conducting state. In case of a short, the critical current of the superconductor will be exceeded, leading to an increase of its resistance which will limit I_{sc}. In the past several SCFCL have been built based on Low Temperature Superconductor (LTS) [1,2,3]. However, due to the high cooling costs, they have never entered the market. With the advent of High Temperature Superconductors (HTS) SCFCL based on these new materials is being developed. The activities concentrate on two concepts. One is the so called "resistive" SCFCL where the HTS is directly connected in the line which has to be protected [2,3,4]. Such a device is very compact, but needs long length conductor and current leads connecting the HTS with the power line at room temperature, which lead to high thermal losses. Most models of resistive HTS SCFCL are based on thin films of YBCO. The second concept is the so called "shielded iron core" SCFCL, where the HTS is coupled inductively in the line. This type of device needs no current leads, but is very large and heavy because of the iron core [5,6]. We have built and tested a 1.2 MVA 3-phase prototype of the second type of SCFCL.

2. ABB concept of "shielded iron core" SCFCL

The "shielded iron core" SCFCL mainly consists of a normal conducting coil, a superconducting tube, a cryostat, and an iron core, all of which are concentrically arranged (Fig. 1). The device is essentially a transformer in which the secondary winding is the superconducting tube. The primary winding (coil) is connected in series with the line which has to be protected. Only the superconducting tube is cooled with liquid nitrogen (77 K).

Fig. 2 shows the equivalent circuit of the device. R_1, R_2 are the resistances of the coil and HTS-tube (i.e. 1-turn coil) respectively, L_S is the stray inductance, L_{11}, L_{22}, L_{12} are the inductances of and between the primary and secondary sides, and I_1, I_2 are the currents in the primary and secondary sides. The equivalent circuit is exact, even in the presence of non-linearity and temperature dependence [5].

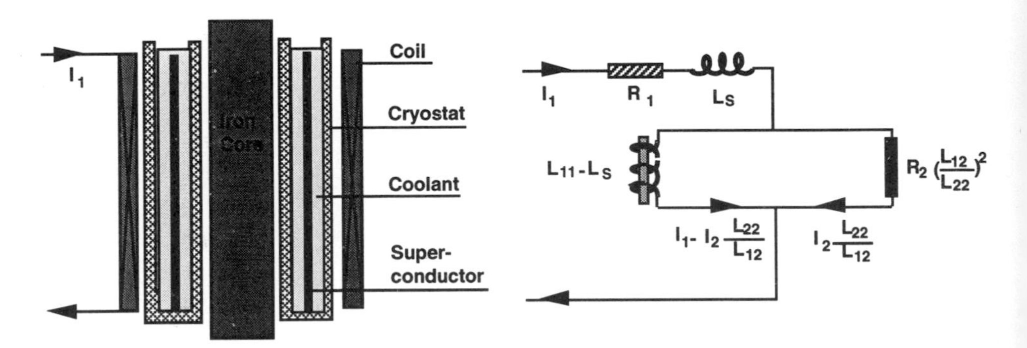

Fig. 1 Scheme of ABB SCFCL

Fig. 2 The exact equivalent circuit

3. 1.2 MVA prototype

3.1. Coils

Two sets of coils have been used. Set I has 250 turns and Set II has 280 turns. Their DC-resistances are about 200 and 250 mΩ, respectively. The height and radius of both coils are $h \approx 140$ cm and $r_{pr} \approx 24$ cm, respectively. Set II which is used for the prototype installed in the power plant is wound on a glass-fibre epoxy tube, which is grounded inside via a conducting varnish with 1 kΩ area resistance.

3.2. Iron-core

The iron is not magnetised in normal operation, the core can thus be made of rather thick and cheap steel. We used 4 mm thick construction steel. The height and radius of the core are $h_{co} = 190$ cm and $r_{co} = 17$ cm, respectively. The effective permeability of the open core is about $\mu = 15$. The total mass of one core is about 1300 kg. For mechanical reasons the three cores were joined via a steel frame.

3.3. Cryostat

Since the cryostat is exposed to a magnetic AC-field even in normal operation, it should be made of non-magnetic and electrically insulating material. For our cryostat we used glass-fibre reinforced epoxy with a vacuum multi-layer insulation, which was cut into stripes to avoid closed current loops. The thermal losses per cryostat are about 10 W at 77 K.

3.4. Superconductor

We produced HTS-rings with 8 cm height by processing Bi-2212 powder in cylindrical Ag-moulds which were rotating in a furnace [7]. The rings have radius r_{sc} = 19 cm and wall thickness d = 1.8 mm. The Bi-2212 material has the voltage-current characteristic $V \sim I^{\alpha}$ with $\alpha \approx 5$ [5,8]. The critical current density, as defined by the 1μV/cm criterion, is about 1400 A/cm^2. In order to avoid hot spots and to reduce the tensile stress on the ceramic, the rings were reinforced with a steel bandage. It served both as electrical bypass and mechanical reinforcement. To further enhance the mechanical stability an additional glass-fibre epoxy composite was applied. Two such HTS rings are shown in Fig. 3 where the one in the background is still contained in the Ag mould and the other one has the Ag mould removed and is mechanically reinforced. The superconducting tubes (HTS-tubes) finally used in the SCFCL were built from a stack of 16 reinforced rings which were supported by a glass-fibre epoxy composite housing with a mechanical damping around each individual ring in order to prevent them from direct mechanical impact. The 3 HTS-tubes were mechanically suspended from the top in the cryostats. The total height of the stack was 140 cm.

Fig. 4 shows the 3-phase FCL, as installed in the hydropower plant. The three vertical cylinders correspond to the 3 phases of the SCFCL. Also visible is the steel frame which mechanically integrates the three iron-cores, the automatic refill system and part of the LN_2 storage tank.

Fig. 3 Bi-2212 ceramic rings produced by a partial melt process

Fig. 4 The 3-phase 1.2 MVA FCL as installed in the auxiliary line of a power plant

3.5. Nominal Current

The critical current of a tube as defined by the 1μV/cm criterion is about I_{c2} = 32 kA. However, due to the smooth I(V) curve of the Bi-2212 the superconductor stays thermally stable up to about 1.5 times that value. For the maximum operation current we chose I_{2max} =

28 kA, to allow for inhomogeneity in j_c and some overload capacity. The nominal current of our device thus was $I_N = I_{2max}/(n\sqrt{2}) \approx 70$ A and 80 A, depending on the number of turns of the coils. Below I_N the impedance of the SCFCL is mainly given by $\omega L_S = 0.9$ and 1.5 Ω, respectively for the two different sets of coils described earlier. Tests on stacks with one HTS ring missing (i.e. 15 instead of 16 rings) showed no significant change in I_{c2} and ωL_S.

4. Tests and discussion

4.1. Short circuit tests

Short circuit tests with a prospective fault current of about 60 kA were performed via a circuit comprised of a generator, which was excited up to 10.5 kV phase-phase voltage, the SCFCL, an inductive load, R_L, and a circuit breaker in series (see Fig. 5). A short was simulated by bypassing the load. After 50 ms the circuit was opened by the breaker. About 100 tests were performed using coils of Set I (I_N = 80 A). Most tests were single phase tests where one phase of the current limiter in series with load and breaker was connected between two phases of the generator. The limitation was always smooth without large over-voltages. The Bi-2212 first limits the current by its I(V) characteristic at 77 K to about 7 times I_N, then it warms up and thus further reduces the current to about 3 times I_N after 50 ms. Simulations [5] of both the normal operation and the behaviour under fault conditions are in good agreement with the experiments.

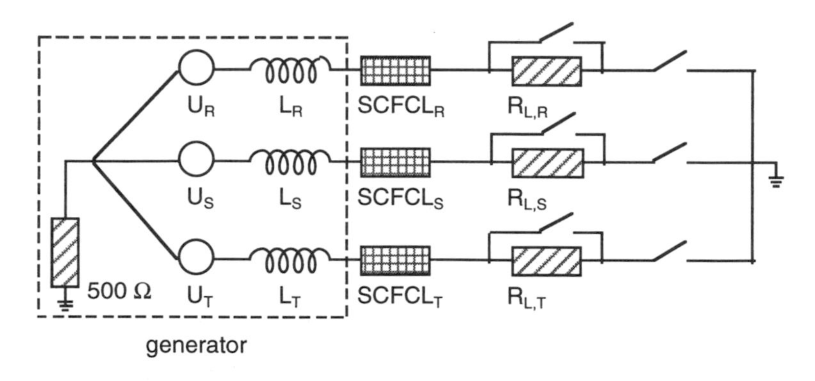

Fig. 5 Scheme of the three-phase short circuit test

During the fault the temperature of the superconductor rises locally well above room temperature, leading to high thermo-mechanical stresses. The magnetic field during the fault leads to a mechanical pressure on the outer surface of the HTS-tube of about 0.2 bar. The axial forces on the rings at both ends of the stack are in the order of 1500 N. These high thermo-mechanical and magnetic forces would cause damage to rings if they were not sufficiently reinforced.

Fig. 6 shows the time evolution of the circuit current and the voltage drop over the SCFCL (n = 250, $I_N = 80\ A_{rms}$) for a 3-phase short at a phase-phase voltage of 9.5 kV. The double-peaks in phase S right after the initiation of the short are due to a time delay between the 3 switches.

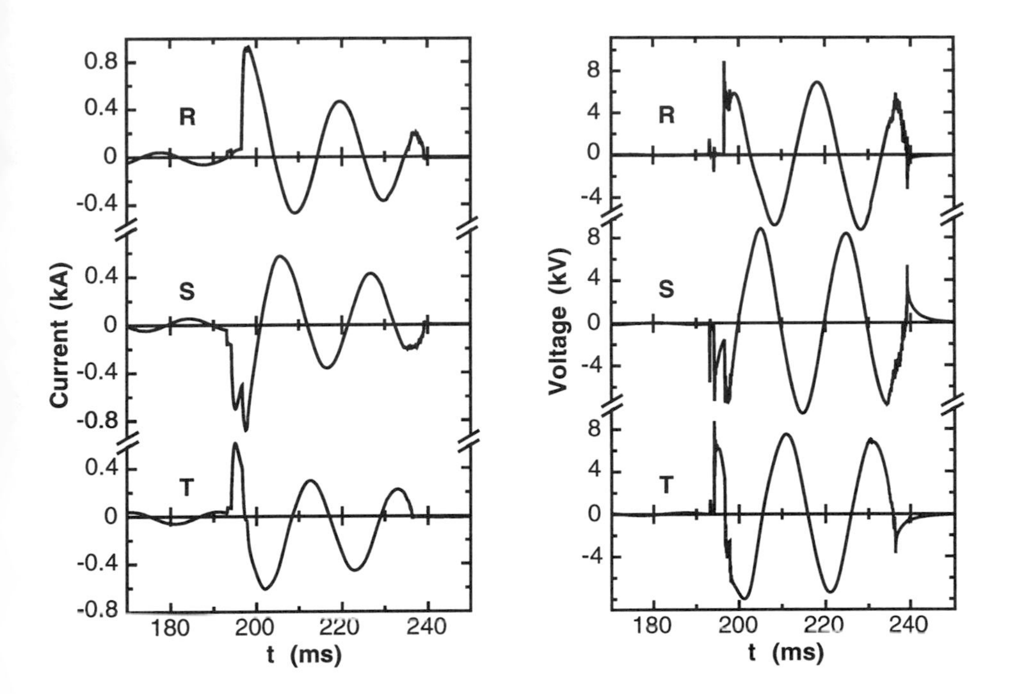

Fig. 6 Three-phase short circuit test results, evolution of circuit current (left) and voltage drop over the SCFCL (right)

4.2. Endurance test in power plant

After several dielectric tests (90 kV BIL and 50 kV AC-test) the prototype (coil Set II) was installed in the Swiss hydropower plant "Kraftwerk am Löntsch" (see Fig. 4). The device is connected in the auxiliary line, as shown in Fig. 7. The cooling is provided by a 2000 l LN_2-storage tank from which the 3 cryostats are automatically refilled. An endurance test was started in November 1996 in order to gain experience with the cooling system and possible fatigue ageing of the HTS due to 100 Hz magnetic forces. After 6 months of operation we did not face any major problems. The losses of the device are about 80 l LN_2/day (cryostats: 15 l, transmission lines: 35 l, tank: 30 l). At an average current of less than 10 A, the AC-losses are negligible. They are $\propto I^3$ and are about 40 W at $I_N = 70$ A. As can be seen, the losses are mainly due to those of the transmission lines and the tank. Therefore, the cooling would be more efficient by employing a closed cycle system.

5. Conclusion

A SCFCL prototype based on the "shielded iron core" concept has been built and tested. The three phase device with a rated power of 1.2 MVA utilises stacks of Bi-2212 HTS rings (38 cm in diameter) which were fabricated by a partial melt process.

The prototype has been installed in a hydropower plant and is at present subject to an endurance test. The installation demonstrates the technical feasibility of HTS current limiters and marks a major step towards a SCFCL product.

We believe that our Bi-2212 based approach can be scaled up to the 10 MVA-region where we see first commercial applications.

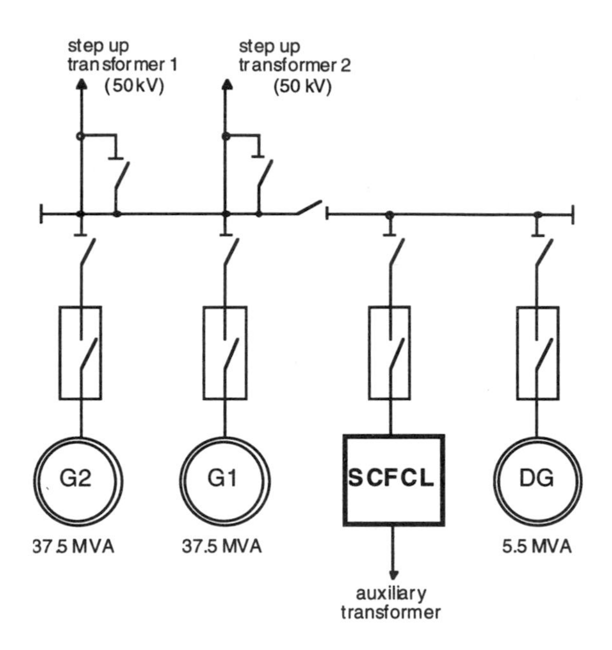

Fig. 7 Scheme of hydropower plant, “KW am Löntsch”, Switzerland

Acknowledgements

We thank F. Platter, V. Mallick, L. Gauckler and his group for helpful discussions and we appreciate technical assistance from R. Weder and M. Hoidis. Financial support from the National Energy Research Foundation (NEFF), the Swiss Utilities Study Fund (PSEL) and the Swiss National Fund (NFP30) is gratefully acknowledged.

References

[1] B.P. Raju, K.C. Parton and T.C. Bertram, "Fault Current Limiter Reactor with Superconducting DC Bias Winding", paper no. 23-03 CICRE (1982).

[2] T. Verhaege et al,, "Experiments with a high voltage (40 kV) superconducting fault crrent limiter", Cryogenics., vol. 36, no.7 , 521-526 (1996).

[3] S. Nomura, T. Yazawa, K. Tasaki, E. S. Yoneda, K. Yamamoto, M. TAkahashi, Y. Yamada, H. Nakagnome, H. Maeda, T. Ohkuma, M. Nakade, and T. Hara, “Development in Superconducting Fault Current Limiter in TEPCO and TOSHIBA", Presented IEA International Workshop on Fault Current Limiters". (IEA stands for International Energy Agency), held in Jerusalem, Israel, in April, 1995

[4] G. Ries, B. Gromoll, H. W. Neumüller, W. Schmidt, H. P. Krämer and S. Fischer, “Development of Resistive HTSC Fault Current Limiters”, Inst. Phys. Conf. Ser. No. 148 (IOP Publishing Ltd) (EUCAS’95) pp. 635-8.

[5] W. Paul, Th. Baumann, J. Rhyner, and F. Platter "Test of 100 kW High-Tc Superconducting Fault Current Limiter", IEEE Trans. Appl. Supercond., vol. 5, no.2 , 1059-62 (1995).

[6] D.W.A. Willen and J.R. Cave, "Short Circuit Test Performance of Inductive High Tc Superconducting Fault Current Limiter", IEEE Trans. Appl. Supercond., vol. 5, no.2, 1047-50 (1995).

[7] M. Chen, Th. Baumann, P. Unternährer and W. Paul, “Fabrication and Characterisation of Superconducting Rings for Fault Current Limiter Application”, presented at M2S-HTSC-V, Beijing, Feb. 97, to appear in Physica C.

[8] W. Paul and J.P.Meier, "Inductive measurements of voltage-current characteristics between 10^{-12}V/cm and 10^{-2}V/cm in rings of Bi-2212 ceramic", Physica C, vol. 205, 240-246 (1993).

Inst. Phys. Conf. Ser. No 158
Paper presented at Applied Superconductivity, The Netherlands, 30 June–3 July 1997

Performance of BSCCO cylinders in a prototype of inductive fault current limiter

V Meerovich[1], V Sokolovsky[1], J Bock[2], S Gauss[2], S Goren[1], G Jung[1]

[1]Physics Dept., Ben-Gurion University, P. O. Box 653, Beer-Sheva, 84105, Israel
[2]Hoechst AG, Corporate Research&Technology, D-50354 Frankfurt/Main, Germany

Abstract. Operation of inductive fault current limiter at high levels of short circuit currents results in creation of hot thermal domains in superconducting rings, capable of causing their mechanical destruction. A method of preventing local overheating by means of employing superconducting with smeared S-N transition has been investigated. It was found that such elements can be successfully employed in inductive current limiters. Investigations of operation of models of inductive current limiter prove that the most important parameter of a superconducting element designated to operate in such devices is the flux relaxation rate and its dependence on the ac current amplitude.

1. Introduction

The principle of operation of a superconducting inductive fault current limiter (FCL) is based on magnetic coupling between closed superconducting coil and a protected circuit. Inductive FCL consists of primary normal metal coil coupled via a ferromagnetic core to a secondary superconducting coil assembled in the form of a set of superconducting rings or cylinders [1]. The transition of an active element to the dissipative state under the influence of the fault current flow causes an increase of the FCL impedance to the level capable of limiting the fault current. The feasibility of FCL operation at liquid nitrogen temperatures has been demonstrated in several small-scale models employing high-T_C superconducting (HTSC) cylinders. However, with the increasing scale of the device, i.e., with increasing fault current levels severe technical problems have been encountered. The major problem consists in formation of overheated local thermal domains in the superconducting element frequently manifesting themselves as an excess local boiling of liquid nitrogen on the cylinder surface. At high power levels these domains are capable of causing destruction of the ring leading to fatal failure of the device [2].
Thermal domain formation in active HTSC elements can be prevented by increasing the homogeneity of superconducting cylinders; stabilization by normal conductor inclusions (composite superconductors), shunting of the entire FCL by an external resistor; adding an extra coil closed by a resistive load to the FCL, or employing superconducting elements with smeared S-N transition. Superconducting BSCCO cylinders with required smeared S-N transition can be obtained by centrifugal casting procedure developed by Hoechst AG [3]. In this paper we present the results of investigations of performance of such superconducting elements in a small scale model of inductive FCL.

2. FCL operation in short-circuit experiment

Typical oscilloscope traces of voltage across the primary coil in the model are shown in Fig. 1 for various levels of short-circuit current. Instead of abrupt changes in the device impedance, seen previously in the FCL models employing YBCO and BSCCO superconductors with sharp S-N transition [1,2], we observe a slow increase of the impedance over a wide range of currents. In a marked difference to the previously employed materials, it is impossible to determine the sharp value of the FCL activation current neither from the material dc I-V characteristic, nor from the FCL ac I-V curve, see Fig. 6. A significant increase of the device impedance occurs now at the currents almost an order of magnitude higher then the critical current determined by 1 μV/cm criterion. However, in the same time no effects related to the formation of thermal domains and hot spots were observed.

This unconventional performance of centrifugal casting BSCCO cylinders, with respect to behavior of hitherto investigated HTSC elements, raises several questions concerning the nature of physical processes responsible for their characteristics. We are also facing the problem of redefining the value of the FCL "activation current" and establishing new criteria for the evaluation of the physical parameters of superconducting elements relevant to the performance of FCL. We have addressed these problems in extended investigations of static and dynamic magnetic properties of centrifugal casting BSCCO cylinders.

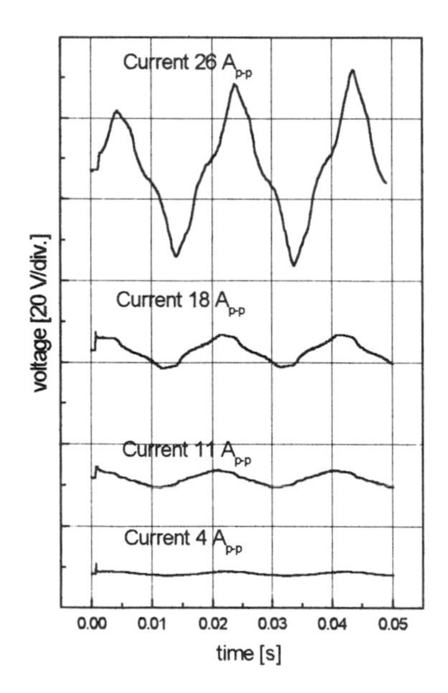

Fig. 1. Voltage drop across FCL at various levels of fault current. 1μV/cm criterion dc critical current corresponds to 1.2 A current in the primary coil.

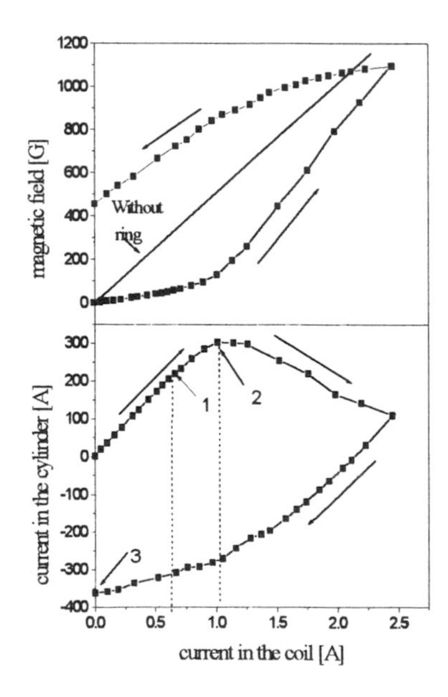

Fig. 2. Quasi-static magnetic field linked to FCL core and calculated circulating current in the cylinder as a function of dc primary coil current..

3. Static and dynamic magnetic characteristics of BSCCO cylinders

The static magnetic properties were evaluated by measuring the primary current induced magnetic field in FCL model by means of a cryogenic Hall sensor [4]. Fig. 2a shows typical behavior of the magnetic field penetrating the ferromagnetic core as a function of the primary coil current, while current circulating in the superconducting cylinder, as calculated from this characteristic, is shown in Fig. 2b. Three characteristic points can be seen in Fig. 2. Point #1 corresponds to the onset of noticeable DC dissipation in the superconductor, point #2 marks the maximum current in the cylinder, and point #3 corresponds to the trapped flux. At each current flowing above point #1 we observe slow growth of magnetic field with the growth rate decreasing in time. The observed field growth is due to the relaxation of superconducting currents circulating in the cylinder [5]. In such conditions one needs to determine a criterion at which quasi-static conditions are reached. We have adopted the criterion of no measurable change of the magnetic field ($\Delta B < 0.01$ G) during 5 minutes time interval.

Fig. 3 shows the response of magnetic field in the ferromagnetic core to a current step in the primary coil. The observed magnetic field decay can be well fitted by the logarithmic expression describing the classical creep flux in hard superconductors [5]. The experimentally determined dependence of the logarithmic creep rate on the height of the current step is illustrated in Fig. 4.

Fig. 5 shows the initial part of FCL field-current dynamic characteristic as obtained for various amplitudes of 50 Hz current. The current at which one observes an abrupt increase of the magnetic field marks the activation of the FCL. Thus determined activation current increases with increasing amplitude of the primary coil current. Clearly the noticeable response of the FCL impedance can be detected only when magnetic relaxation rate becomes of the same order of magnitude as the rate of change of the primary current. In fact, despite a pronounced increase of the relaxation rate with increasing current, as shown in Fig. 4, at currents more then 10 times higher than dc dissipation onset (point #1) the decay rate still sufficiently exceeds the characteristic time of the 50 Hz process (0.005 s).

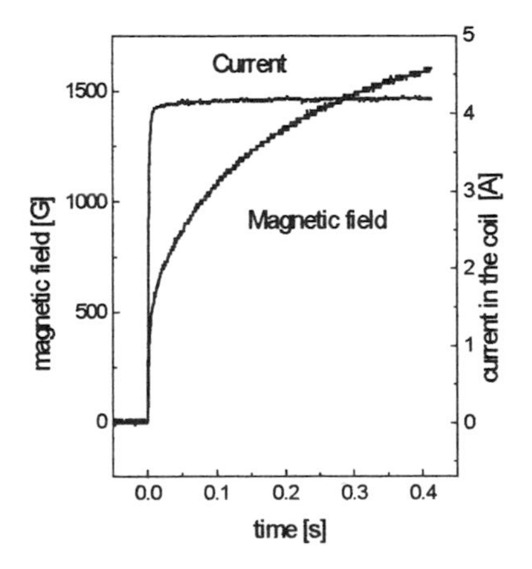

Fig. 3. Magnetic response of the cylinder. to the step current excitation in the primary coil.

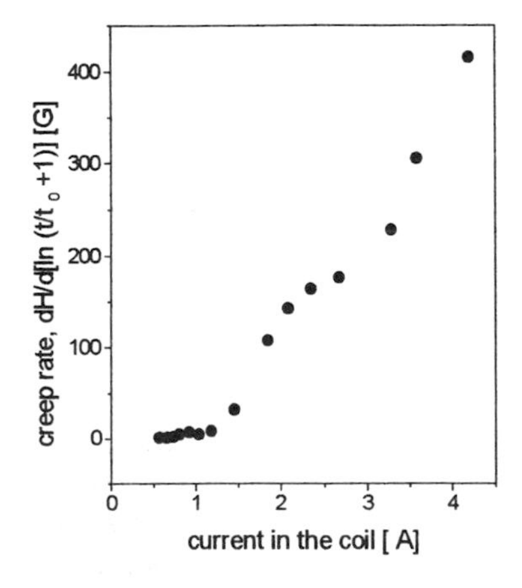

Fig. 4. Magnetic relaxation rate in BSCCO cylinder as a function of the height of current step.

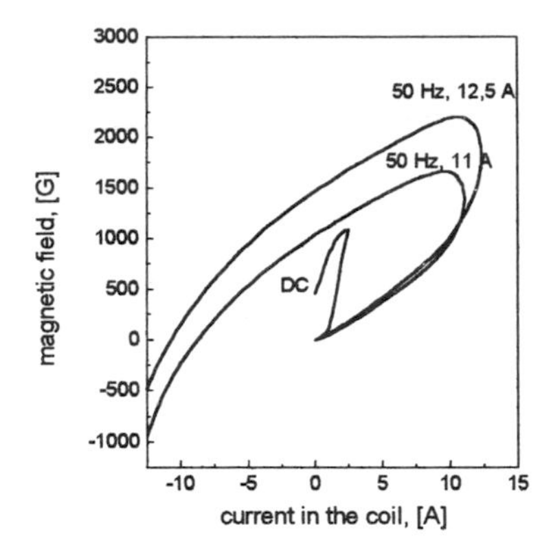

Fig. 5. 50 Hz dynamic magnetic response for various current amplitudes in comparison to DC response.

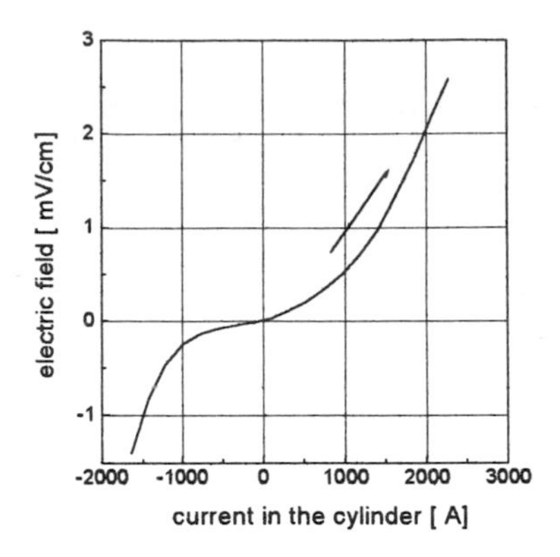

Fig. 6. AC electric field in HTSC cylinder vs. coil current, as evaluated from FCL oscilloscope traces.

4. Discussion

The smeared S-N transition and the corresponding smooth I-V characteristic of the investigated cylinders, see Fig. 6, prevent the formation of thermal domains during the fault event, even at very high levels of short circuit currents. The cylinders with such characteristics can efficiently operate as active elements of an inductive FCL without causing problems due to local overheating. However, the activation current of the FCL employing these cylinders can not be clearly determined from the static characteristics. The value of the activation current depends on the relation between magnetic relaxation rate in the superconductor and the rate at which current in the primary coil changes. With increasing fault current level the FCL activation current increases.

We conclude that for inductive FCL applications the proper figure of merit of a superconducting element with a smeared S-N transition should be the magnetic relaxation rate and not the critical current. The latter can be used only for materials characterized by sharp dissipation onset.

Acknowledgments

This research was supported by a Grant from the G.I.F., German-Israeli Foundation for Scientific Research and Development.

References

[1] Sokolovsky V, Meerovich V, Grader V and Shter G 1993 *Physica* **C 209** 277-281

[2] Meerovich V, Sokolovsky V, Jung G and Goren S 1995 *Proc. EUCAS'95 , IOP. Conf. Ser. No. 148* 603

[3] Bock J, Elschner S, Herrmann P F, Rudolf B 1995 *Proc. EUCAS'95 , IOP Conf. Ser. No. 148* 67-72

[4] Goren S, Jung G, Meerovich V, Sokolovsky V, Skoletsky I, Shter G, Grader G and Homjakov V V 1994 *Proc. EUCAS '93 October 4-8, 1993, Gottingen, Germany, Applied Superconductivity* 749-752

[5] Yeshurun Y, Malozemoff A P and Shaulov A 1996 *Rev. Mod. Phys.* **68** (3) 911- 949

Inst. Phys. Conf. Ser. No 158
Paper presented at Applied Superconductivity, The Netherlands, 30 June–3 July 1997

Experiments with a 6.6kV/1kA Single-phase Superconducting Fault Current Limiter

Takashi Yazawa, Eriko Yoneda, Shunji Nomura, Kazuyuki Tsurunaga, Masami Urata, Takeshi Ohkuma*, Shoichi Honjo*, Yoshihiro Iwata* and Tsukushi Hara*

Toshiba Co., 4-1 Ukishima-cho, Kawasaki, 210 JAPAN
*Tokyo Electric Power Co., 4-1 Egasaki-cho, Tsurumi,-ku, Yokohama 230, JAPAN

Abstract. The authors have developed a 6.6kV/1kA single-phase superconducting fault current limiter. The apparatus demonstrates the integration of all the essential elements; a vacuum vessel, a helium vessel, a current limiting coil wound with low-T_c conductors, current leads, and refrigerators. All the thermal losses into the cryostat are designed to be compensated by the refrigerators. In non-fault conditions, a low thermal loss is achieved by reducing the AC loss of the limiting coil and the conduction loss through Bi2223 current leads. The fault conditions was successfully investigated with a short circuit generator. When the rated voltage of 6.6 kV was applied, a fault current of 56 kA was limited to 5.2 kA in less than 300 μs. The released energy up to 72 kJ was completely stored in the closed vessel.

1. Introduction

Fault current limiters[1]-[3] restrict fault currents in transmission or distribution lines, which has a merit of omitting a capacity renewal for equipment such as circuit breakers. Also demands for self-limiting interconnection by fault current limiters will be surely increased for effective electric power use. Tokyo Electric Power Co. and Toshiba Co. are collaborating in a research on resistive fault current limiters, using NbTi cable, operating passively with a low response time. Aiming towards the application for the high voltage line, the group has investigated 6.6kV/1kA single phase fault current limiter as a preliminary step. The apparatus is the integration of all the necessary elements for a superconducting apparatus.

2. 6.6kV/1kA superconducting fault current limiter

Figure 1 shows a schematic drawing for the 6.6kV/1kA single-phase superconducting fault current limiter, Fig.2 shows the photograph of its outlook and table 1 shows the specifications.. The apparatus is mainly comprised of a vacuum vessel, a current limiting coil in a liquid helium vessel, a pair of current leads and two refrigerators. A Gifford-McMahon (GM) refrigerator is used for the 80K stage and a GM refrigerator with a Joule-Thomson (JT) loop for the 4K stage. All the evaporated gas in the liquid helium vessel is designed to be compensated by the 4K refrigerator in both the fault and the non-fault conditions.

The conductor length wound in the limiting coil is determined by the enthalpy protection requirement in fault conditions. The temperature rise during current limitation is given by the following equation;

$$\int_0^{\Delta t} \frac{V^2}{R} dt = S\ell \int_{4.2[K]}^{T_f} C dT \qquad (1)$$

where V is the rated voltage, Δt is the period for current limitation, R is the resistance, S is cross-sectional area, ℓ is conductor length, C is the specific heat of the conductor. The final temperature T_f is designed to be 200 K, which defines the length according to the equation (1) and hence the resistance.

The conductors used are multi-strand conductors made of ultra-fine NbTi filaments in a high-resistive Cu30%Ni matrix. In the strand, the filament diameter is reduced to 0.14 μm and the twist pitch is 1.1 mm for the reduction of the hysteresis and coupling losses. The strands are twisted to a (6+1) x (6+1) type two stage cable. The strand twist pitches are designed to decrease the saturated region in a strand for self-field condition[4].

Two identical conductors were wound in opposite directions on four nested bobbins, one conductor around the 1st and the 4th bobbins in series, the other around the 2nd and 3rd. The bobbins are made of fiber reinforced plastics. A composite of high modulus alumina and high strength polyester is used for the fiber to increase the coil stability[5].

The current leads are located in the vacuum and made of phosphorus deoxidised copper for the high temperature region over 80K and of Bi2223 for the low temperature region below 80K. The Bi2223 leads are ϕ34mm in diameter and hollow shaped, which reduces self-field losses compared to the bulk leads with the same cross-sectional area. The high temperature ends of the Bi2223 leads are cooled by the nitrogen heat pipes connected between the thermal anchor for the leads and the cooling stage for the GM refrigerator.

The helium vessel is made of stainless steel to endure the pressure increase in the fault condition. A cylindrical shaped shield made of OFHC copper is set inside the vessel for the eddy current loss reduction. According to the numerical analysis, the loss is estimated to be below 0.1 W at the rated coil current of 1kA.

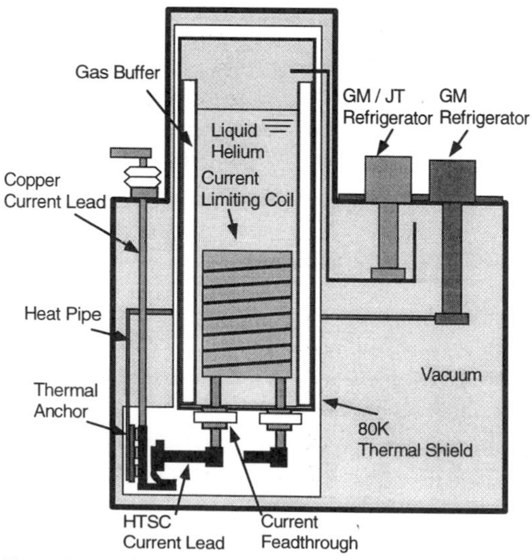

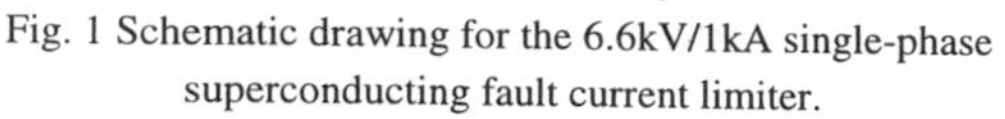

Fig. 1 Schematic drawing for the 6.6kV/1kA single-phase superconducting fault current limiter.

Fig. 2 Photograph of the 6.6kV/1kA single-phase superconducting fault current limiter.

Table 1 Specifications of the 6.6kV/1kA single-phase superconducting fault current limiter

Size	φ1200×1800mm
Rated voltage and current	6.6 kV / 1kA
Impedance for non-fault condition	20 μH (6.3mΩ)
Impedance for fault condition	6.0Ω
Refrigerators	GM Ref. for 80K stage GM/JT Ref. for 4.2K stage

3. Experiment and discussion

A fault current limiting experiment was carried out with a short circuit generator. The experiment simulated the distribution line use. The peak values of the fault currents without the current limiter were set by an applied voltage, a reactor and a resistor in the circuit.

Figure 3 shows one typical experimental result. In this experiment, a voltage of 6.6 kV was applied, and the fault current was set to be 56.6 kA. The fault current limiter successfully limited the fault current to 5.2 kA in 300 μs. Figure 4 shows the relation of the peak of the limited current and the generated resistance against the applied voltage. In all the experiments, the fault currents were successfully limited to 5 kA with generated resistance of 2 to 4 Ω. This figure also shows the consumed energy. The generated vapours were fully stored inside the closed vessel.

The experimental results presented above show that the obtained resistance is at most 70 % of the fully normal value, which is not preferable for the reduction of the energy consumed and the recovery time.

Figure 5 shows the quench current dependency of the coil on the current sweep rate. These data were obtained by the means of a capacitor bank circuit. The data in the figure were obtained with several LC resonance frequencies as a parameter. Below the sweep rate of 20 MA/s, the quench current has a positive correlation on the sweep rate. Above 20 MA/s, however, the quench current indicates a almost constant value. This characteristics is explained by the following mechanism. At a relatively low current I_{dis}, mechanical disturbance occurs at some surface around the cable. The quench current I_q is then determined by the sweep rate and the thermal diffusion time τ in a strand, which is described in the following equation;

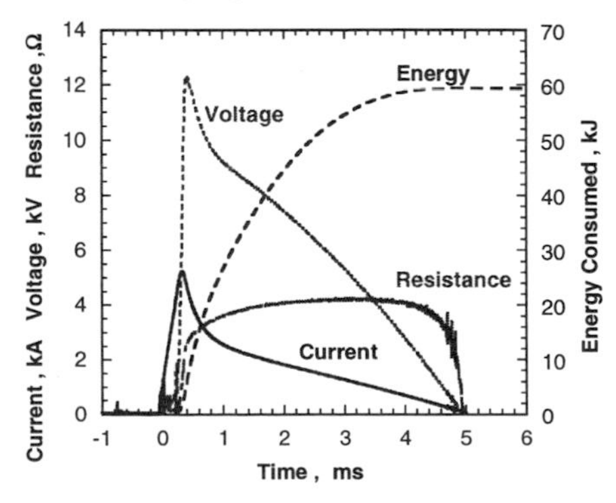

Fig. 3 A typical experimental result showing time-dependent current, voltage, resistance and energy consumed.

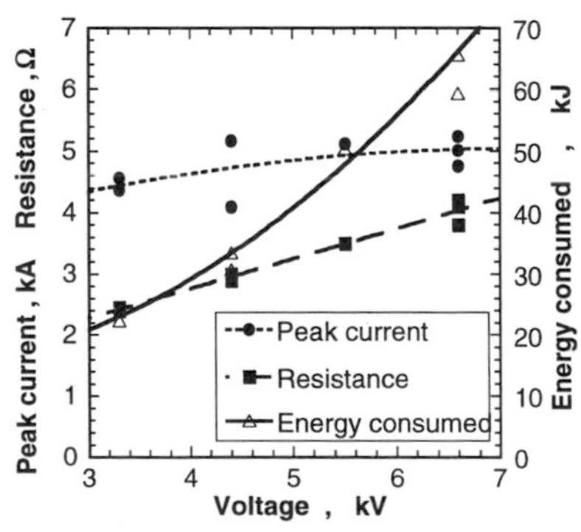

Fig. 4 The dependence of the peak current, the generated resistance and the energy consumed as the function of applied voltage.

$$I_q = I_{dis} + \tau \frac{dI}{dt} \qquad (2)$$

The results below 20 MA/s are well described by I_{dis} of 2.5 kA and τ of 60 μs. The value τ is described as the thermal diffusion time constant given by the following equation;

$$\tau = \frac{4a^2}{\pi^2 D_\theta} \qquad (3)$$

where a is the strand radius and D_θ is thermal diffusivity of the strand. Above the sweep rate of 20 MA/s, another electro-magnetic factor such as the AC losses or unbalanced current distributions among strands are considered to be dominating the quench behaviour.

Figure 6 shows a direct plot of the coil voltage as a function of the current, which also indicates the quench characteristics mentioned above. At a low sweep rate, the normal voltage appears for a low current and increases gradually. This behaviour causes a non-full quench over the whole length. On the other hand, at a high sweep rate, a sharp normal transition occurs. Eliminating mechanical disturbance at low current is expected to result in a coherent quench over the entire range of sweep rates.

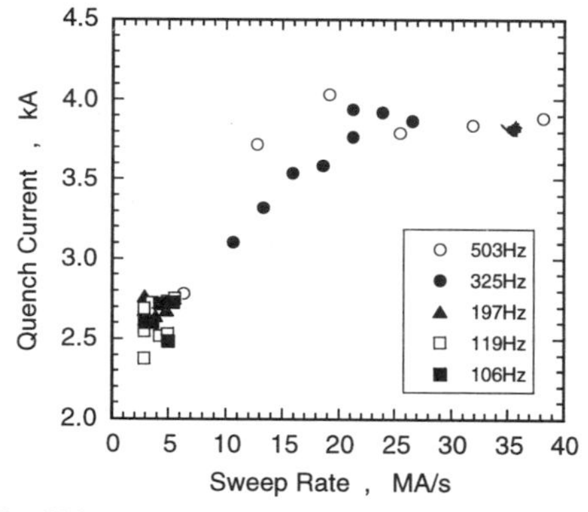

Fig. 5 The dependence of the quench current upon the sweep rate.

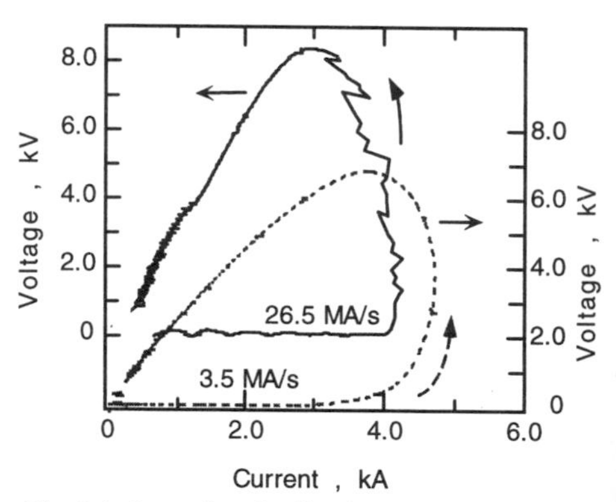

Fig. 6 A direct plot of coil voltage as a function of the current.

4. Conclusion

A resistive type 6.6kV/1kA single-phase superconducting fault current limiter is developed. In experiments with a short circuit generator, a fault current of over 56 kA was limited to 5.2 kA in less than 300 μs. Dealing with the improvements of quench characteristics and electric insulation, we continue to investigate the fault current limiter system for applications at a higher voltage level.

References

[1] T. Hara, et. al, *IEEE PES*, 1992
[2] T. Verhaege et. al , *Cryogenics*, 1996, p.521
[3] D.W.A.Willen and J. Cave, IEEE *Trans Appl Super*, vol.5, 1995, p.1047
[4] N. Amemiya, et.al., *IEEE Trans. Appl Super*, vol.5, 1995, p.984
[5] K.Tasaki, et.al., *IEEE Trans. Mag*., vol. 32, 1996, p. 2407

Inst. Phys. Conf. Ser. No 158
Paper presented at Applied Superconductivity, The Netherlands, 30 June–3 July 1997

Determination of time constants by direct measurement of magnetic field above superconducting cables

Silvester Takács*, Nagato Yanagi**

*Institute of Electrical Engineering, Slovak Academy of Sciences, 842 39 Bratislava, Slovakia
**National Institute for Fusion Science, Toki, Gifu 509-52, Japan

Abstract. The calculations of magnetic field and of corresponding magnetization for slabs in exponentially decreasing and linearly increasing applied magnetic fields are used to evaluate the time constants from experimental measurements of magnetic fields above superconducting cables. The measurements are performed with Hall probes on different places of cables for the helical coils of the LHD project and subtracted from the magnetic field far from the sample. The demagnetization factor of the sample and the finite distance of Hall probes from the induced currents are included by a general shape factor. Both regimes of field changes can be used, if the time constant or the field rate of the applied field differs up to two orders of magnitude from the time constant of cable. The exponential decrease is more suitable, as the time constant and the shape factor can be determined independently. The importance of the finite sample length, of inhomogeneous applied fields and current distribution, as well as of the flux creep on determining the time constant can be demonstrated, too.

1. Introduction

There is a number of methods to determine the time constants of superconducting composites and/or cables. The differences in the obtained results, even on the same samples, could be often very large [1]. The differences caused by the performed methods include in the most cases the problem of compensation, i. e. subtracting the unwanted part of the signal from the induced field, current or voltage. However, the obtaining of a simple "exponential" function should be considered as unessential with improved computerization of measurements.

The superconducting cables for large scale applications, like the Large Helical Device (LHD) project [2], should carry large transport currents. They are made by multistage process from filaments, filament bundles, strands, subcables, etc. The measurement are due to the interconnection of the components and other effects, like the end effects [3], not always easy.

In the most cable structures, the coupling losses between strands and/or subcables are the most important ones, therefore the measured time constants are determined by this type of losses in our superconducting cables. At evaluating the time constants from the measured AC losses, the existence of different time constants causes an additional problem for flat cables [4]. The time constant derived from the low frequency limit can be very different from the time constant determined from the maximum of the losses/cycle. The latter one should determine the actual rate for the decrease of the induced currents.

To measure the time constant in a way independent of AC losses, we applied a new method measuring the “inner” magnetic field above the cable surface directly by Hall probes. The theoretical results for the simplified model of one-dimensional flux diffusion in slabs are compared with the measurements. These included the full sample of the helical conductor for

LHD (KISO-32), the sample without the aluminum stabilizer (Al cut-off) and the aluminum stabilizer itself. Both the "perpendicular" (applied field perpendicular to the wide side of the cable or conductor) and "parallel" geometries were used. The applied field was exponentially discharged, $B_e = B_0 \exp(-t/\tau_0)$, or linearly increased with the slope B_0/t_0.

2. Solution of the flux diffusion equation

Generally, the magnetic field distribution in flat cables can be calculated in the low frequency region only. We simplify the calculations by considering a slab of thickness $2w$ ($-w \le x \le w$) in parallel field and solve the one-dimensional flux diffusion equation

$$\frac{\partial^2 B}{\partial x^2} = \frac{1}{D}\frac{\partial B}{\partial t} \tag{1}$$

where D is the diffusion constant. The general solution for an *exponential field change* is [5]

$$B(x,t) = \frac{4}{\pi} B_0 G \sum_{k=0}^{\infty} f_k(x) \frac{\tau/\tau_0}{(2k+1)^2 - \tau/\tau_0}\left(e^{-t/\tau_0} - e^{-t/\tau_k}\right) + B_0 e^{-t/\tau_0} \tag{2}$$

with $f_k(x) = (-1)^k \cos[(2k+1)\pi x/2w]/(2k+1)$, $\tau_k = \tau/(2k+1)^2 = 4w^2/D\pi^2(2k+1)^2$. The shape factor G includes the geometrical effects of the demagnetization factor, the finite distance of the Hall probes from the currents, etc. It can be seen that the *adjustable parameters* are G and τ as the "basic" *time constant* of the structure, whereas τ_k determine the relaxation of the induced higher Fourier components. At comparing with the measurements, the local value of B, its difference to the applied field (magnetization M), as well as some averaged values and other derived quantities (e.g. calculating the voltage in a pick-up coil) can be useful. We use here the magnetization at the cable center: $M = B(0) - B_e$.

In the *linear regime*, $B_e = B_0\, t/t_0$, the general solution of the diffusion equation (1) is

$$B(x,t) = B_0 t/t_0 - \frac{4}{\pi}\frac{B_0}{t_0} G \sum_{k=0}^{\infty} f_k(x)\tau_k\left(1 - e^{-t/\tau_k}\right) \tag{3}$$

As for the exponential field decay, the *adjustable parameter* τ should be determined by the parameters of the cable only, its value is the same as previously for the exponential field decay.

These results give us also the possibility to determine the time constant of the induced currents. The Hall probe in the perpendicular geometry measures the field in the cable center:

$$B(0) = B_e - \frac{4}{\pi} G \frac{\tau}{t_0} \sum_{k=0}^{\infty} \frac{(-1)^k}{(k+1)^3}\left(1 - e^{-t/\tau_k}\right) = B_e - B_0 G \frac{\tau}{t_0}\left\{\frac{\pi^2}{8} - \frac{4}{\pi}\sum_{k=0}^{\infty}\frac{(-1)^k}{(k+1)^3} e^{-t/\tau_k}\right\} \tag{4}$$

This formula can be used after switching on the applied field, but due to the nonlinearity in the starting stage it has little practical consequences. However, the relative magnetization at the centre for about $t > 5\tau$ has a constant contribution $M/B_0 \approx -\pi^2 G\tau/8t_0$. Thus, we can determine the product $G\tau$, no more both parameters independently. We are able to compare the results on the same cable (different places, different magnetic field distribution), as shown in the following. For $t \to \infty$, we obtain a quadratic field distribution of the magnetization,

$$B_\infty(x) = \frac{B_0}{t_0} t - \frac{4}{\pi}\frac{B_0}{t_0} G\tau \sum_{k=0}^{\infty} \frac{(-1)^k}{(k+1)^3} = \frac{B_0}{t_0} t - \frac{\pi^2}{8}\frac{B_0}{t_0} G\tau\left[1 - \left(\frac{x}{w}\right)^2\right]$$

3. Characterization of samples and experimental results

Previously, the time constants of LHD samples were determined from frequency dependence of the normalized losses per cycle [6]. The B-type [7] measurements (measuring the field, usually perpendicular to the Hall probe) are restricted to the cases where the differences of the signals from the Hall probes to an equivalent position without the sample are sufficient. As the calculations show, this can be assured in our experiments, if the discharging or generally the ramp rate is within about two orders of magnitude of time constants of measured samples. A Rutherford-type flat cable with 15 NbTi/Cu strands was measured which should carry the nominal current of 13 kA in the helical coils of LHD at maximum magnetic field of 6.9 T in Phase I operation [2]. Pure aluminum stabilizer is used with copper casing. The time constant of this highly conducting aluminum stabilizer lies in the same range as that for the coupling currents between the strands, causing some disadvantages for determining the time constants.

The effective dimensions of the superconducting part are 13.2 mm (width) × 3.13 mm (thickness), to which Al (width 12.4 mm, thickness 5.07 mm) is attached. The Hall probes were attached on various positions of 660 mm and 220 mm long samples (five and two times the cabling pitch). The split coil was discharged from the central magnetic field of 3 T. The local magnetic field on the conductor samples as well as in the "vacuum" were measured by Hall probes during the field decay. In the linear regime, the peak field of 3 T was reached after 100 s and 200 s. Figure 1 shows the temporal evolution of the measured field (Hall probe B2) at exponential discharge of the applied field (time constant τ_0 = 14 s, B5), as well as the difference between these two signals. The corresponding magnetization at the centre is due to the screening currents. The time constant τ and G could be determined very precisely.

In spite of the simple model, very good agreement was obtained with the experimental results. This enables us to consider the obtained τ value to be the right decay time of coupling currents in the cable. The fitting was equally good for other cases: cables with and without Al, the Al stabilizer itself, both in magnetic fields perpendicular and parallel to the wide side of the cable. Especially, the time constant of 0.75 s for aluminum in perpendicular field, which corresponds to $\rho_{Al} \approx 2.6 \times 10^{-11}$ Ωm at 3 T, supports the measurements and the model well.

The measurements at *linear* field increase support the usefulness of our approach too, although the determination of the time constant is possible only by knowing the value for the shape factor. We expect at low field changes that the flux can nearly fully penetrated into the loops. Therefore $G \approx 1$ in the perpendicular geometry and we can compare the results with exponential field change and at different places on the cable (Table 1). The magnetization decreases slowly, which is probably caused by flux creep.

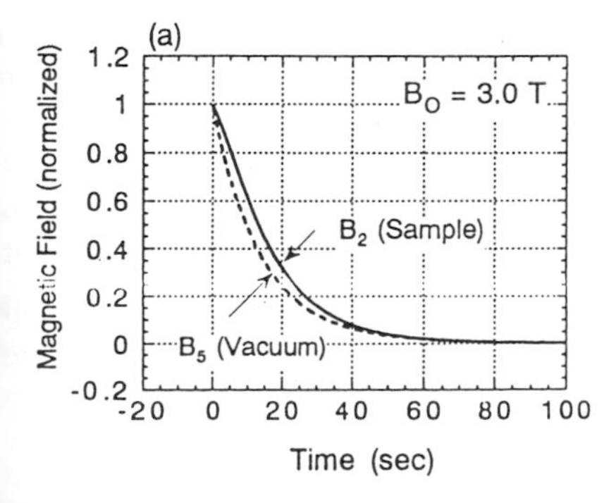

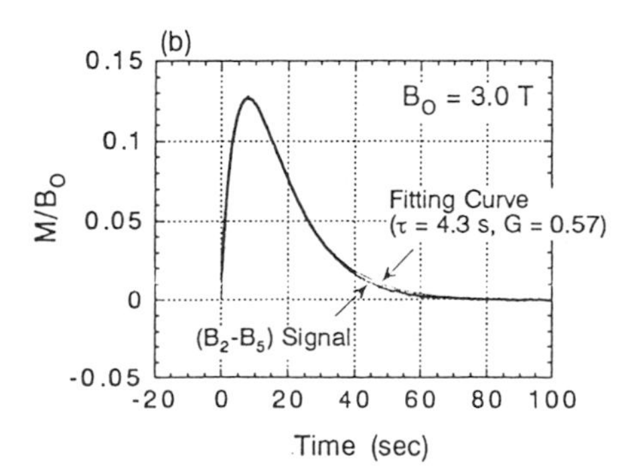

Figure 1. Normalized magnetic fields measured with the Hall probes during an exponential field decay (a) and the difference between the "inner" field and the "vacuum" field (b).

Table 1. Time constants (in s) of the cable by different field changes

regime	Ia	Ib	IIa	IIb	III
cable center	5.5	4.9	6.5	4.9	4.9
close to end	13.8	10.6	21.1	13	12.2

a - from the maximum magnetization, b - at the end of the linear field increase (already influenced by flux creep), I - $\dot{B}_e = 0.03$ T/s, II - $\dot{B}_e = 0.015$ T/s; III - from the exponential field decay

4. Conclusions

The direct measurement of magnetic fields on superconducting cables and normal conductors with Hall probes appears to be very advantageous and can be used for determining the time constant without requiring any additional assumptions. The results for exponential field decay are very consistent, the shape factor and the time constant can be determined independently. The excellent fitting of the experimental results with two independent parameters was possible in the *whole time interval* after discharging (linear increase of the magnetization, reaching the maximum, decreasing phase up to exponential behavior at large times) is remarkable. Some methods use only the last stage of this regime. One has to have in mind that during the whole interval we were starting with high magnetic fields, going through intermediate ones up to low fields, at very different field rates. Namely, there are at least four parameters determining the basic shape of *M(t)* curve (starting slope, position and height of the maximum, exponential behavior), the excellent fitting of curves with two parameters shows the good applicability of our procedure, as well as the appropriateness of our model. Therefore, these results can be taken as "decisive" for obtaining the *right value* of the time constant.

The *analytic results* for the flux diffusion can be used for other types of measurements (determining the voltages in pick-up coils, current distribution, especially in inhomogeneous fields), as it was here successfully applied to full cables, to cables with aluminum cut-off, as well as to Al stabilizer itself, in fields both perpendicular and parallel to the wide side of cable.

The time constant in *inhomogeneous field* is higher [5] than in the homogeneous one, the length dependence is in accordance with theory [8]. The time constant at the cable centre is not length dependent. Evidently, the smaller loss creation in shorter samples is mainly close to the ends, without changing the induced currents at the centre considerably.

Acknowledgments

The authors appreciate fruitful discussions with O Motojima and S Imagawa, the general support by the late J Yamamoto, and the many staffs in the LHD group for supporting the experiments. S T would like to thank for partial support by the Japan Society for the Promotion of Science and by the Slovak Grant Agency VEGA.

References

[1] Wada T and Tachikawa K *Adv. Cryog. Eng. Mater.* **38** (1992) 731
[2] Motojima O, Akaishi K, Fujii K *et al* *Fusion Eng. Design* **20** (1993) 3
[3] Takács S, Yanagi N and Yamamoto J *IEEE Trans. Appl. Supercond.* **AS-5** (1995) 2
[4] Takács S and Yamamoto J *Adv. Cryog. Eng. Mater.* **42** (1996) 1233
[5] Yanagi N, Takács S, Mito M, Takahata K, Iwamoto A and Yamamoto J *Proc. ICEC16/ICMC* (London: Elsevier) (1996); *Cryogenics* in press
[6] Sumiyoshi F, Kawabata S, Fukushima K *et al* *IEEE Trans. Appl. Magn.* **30** (1994) 2491
[7] Cross RW and Goldfarb RW *IEEE Trans. Magn.* **27** (1991) 1796
[8] Takács S *Supercond. Sci. Technol.* **9** (1996) 137

Inst. Phys. Conf. Ser. No 158
Paper presented at Applied Superconductivity, The Netherlands, 30 June–3 July 1997

Development of HTS tapes and multistrand conductors for power transmission cables

M. Leghissa, J. Rieger, J. Wiezoreck, H.-P. Krämer, B. Roas, B. Fischer, K. Fischer* and H.-W. Neumüller

Siemens AG, Corporate Technology, P.O.Box 3220, 91050 Erlangen, Germany
*Institute for Solid State and Materials Research, P.O.Box 270016, 01171 Dresden, Germany

Abstract. Significant progress has been achieved in the performance of machine-stranded HTS cable conductors. Results on a 10 m long conductor yielding AC losses of less than 1 W/m operating at 2000 A_{rms} will be presented. This development has been facilitated by the improvement of 2223 BPSCCO multifilamentary tape performance, i.e., a reproducible production of long lengths (> 100 m) within the 20 kA/cm^2 range as well as short-sample critical current densities up to 41 kA/cm^2.

1. Introduction

The progress in the HTS cable development requires the fabrication and test of cable conductors with continuously improving performance in current capability and conductor length [1,2,3,4]. The recent improvement in the performance at Siemens - demonstrated by the product of critical current and conductor length - is shown in Fig. 2. Target values for a commercial product (transmission cables with power ratings ≥300 MVA) are in the range of $I_c \times L \approx 500$ kAm. A major requirement for HTS cables is therefore the availability of HTS tapes of long lengths, typically 100 m to 1000 m.

2. Development of 2223 BPSCCO/Ag-sheathed multifilamentary tapes

For the fabrication of the 10 m / 4 layer cable conductor a total of 2 km multifilamentary 2223 BPSCCO/Ag tapes of high quality have been fabricated with a typical batch length of 90 m - 110 m. Originally a critical current of 30 A has been specified. However, due to the improvement of our preparation process this value has been exceeded to an average 40 A. As shown in Fig. 2 j_c-values for long-lengths samples are 20 kA/cm^2. HTS tapes have been fabricated with lengths up to 600 m. For a 600 m-batch a performance of 40 A / 18 kA/cm^2 has been achieved.

Raising the critical current density is an essential requirement for an economic breakthrough of the HTS technology in large scale applications. By optimising the deformation step and the thermomechanical treatment we have enhanced the critical current density in short samples to 41 kA/cm^2 and even 39 kA/cm^2 over a length of 0.7 m. Taking

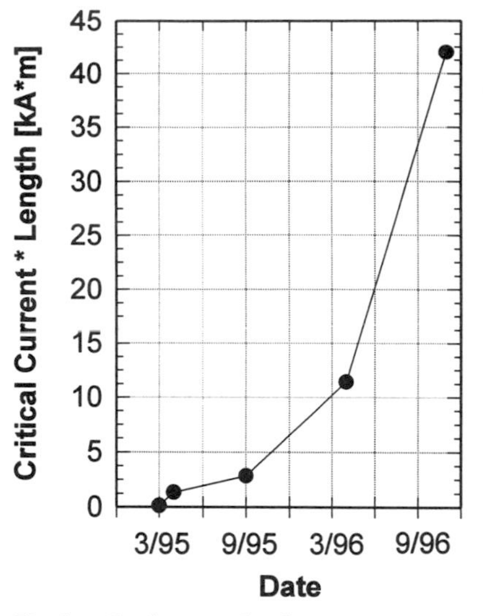

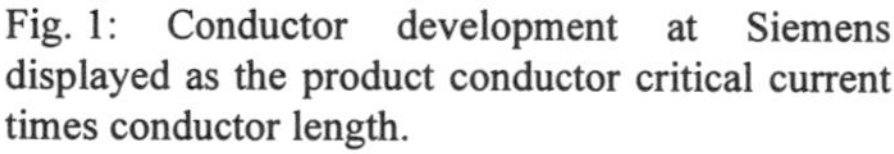
Fig. 1: Conductor development at Siemens displayed as the product conductor critical current times conductor length.

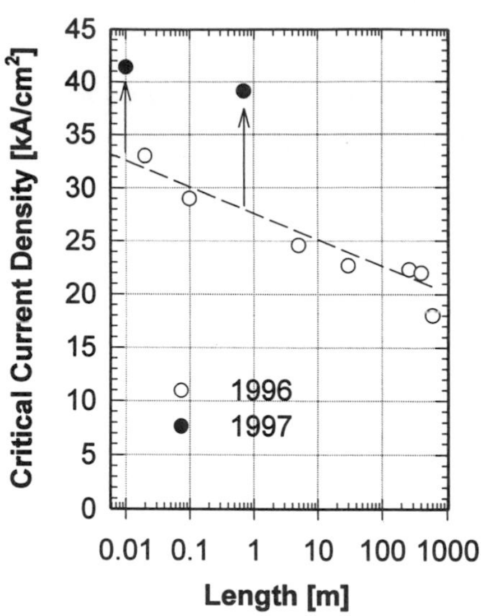

Fig. 2: Critical current densities of multi-filamentary 2223 BPSCCO/Ag-tapes versus sample length at 77 K.

into account that magneto-optical investigations already revealed j_c-values of 80 kA/cm^2 in the central filaments of our non-optimised tapes [5] - and recently values up to 60 kA/cm^2 have been reported elsewhere [6] - we believe that the "standard" powder-in-tube technology is capable of reaching the target of 100 kA/cm^2.

Efforts have been also made to improve the mechanical stability and to lower the AC losses of HTS tapes. The critical tensile stress at room temperature has been raised from 40 MPa for a pure Ag-sheath to 100 MPa using a high-strength sheath material. Losses in twisted AC conductors have been lowered to 0.6 W/(kAm) at 77 K / 0.1 T.

3. Fabrication and test of a 10 m HTS cable conductor

3.1. Conductor fabrication

Using a total of 2 km 2223 BPSCCO tape a flexible 4 layer (145 tapes) machine-strande cable conductor has been made using a special low-loss design with individual layer isolated from each other. The cable conductor has a total length of 10.5 m . The layers ar soldered together at the conductor ends. The conductor has been equipped with severa temperature sensors, voltage contacts, axial field coils and Rogowski coils between th layers in order to study DC and AC losses, temperature increase and current distributio between the individual layers.

3.2. Cooling system

For testing the conductor a cooling system with a closed LN2 cycle has been constructec The closed cooling cycle is pressurised in order to increase the boiling point of nitroger Using a heat exchanger the LN2 is cooled to 77.2 K. Using a pressure of 2 bar a boiling poir

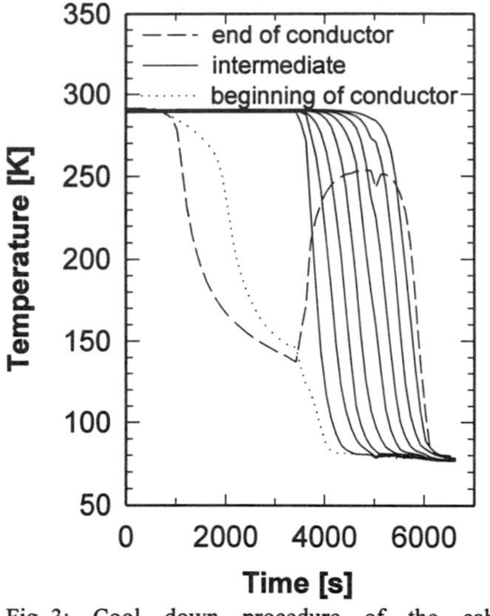

Fig. 3: Cool down procedure of the cable conductor. The curves show the conductor temperature at different positions. Pumping within the LN2-cycle is started at t=3400 s

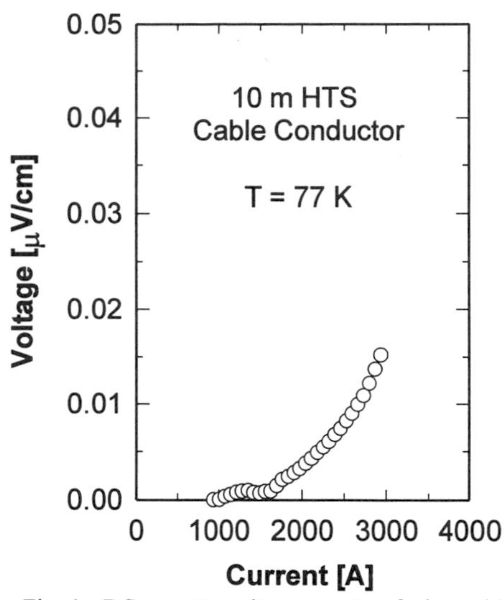

Fig. 4: DC current-voltage curve of the cable conductor. At 3000 A the capacity of our power supply is exceeded. From extrapolation of the I-V curve we expect a critical current in the order of 5 kA.

of 83.5 K enables a temperature gradient of up to 5.3 K along the conductor while keeping the nitrogen in the liquid phase.

Fig. 3 shows the cool down behaviour. First both ends of the conductor are cooled via heat conduction to the nitrogen bath. Than the pump for the LN2 cycle is switched on (t_0=3400 s), leading to a temperature decrease starting at the LN2 inlet and an initial temperature increase at the outlet end of the conductor. After 2600 s the system has reached the final temperature of 77 K, yielding a cool down time of approximately 4 min/m.

3.3 DC current-voltage curve

First a DC test of the conductor has been performed in order to determine the critical current. As displayed in Fig. 4 at a current of 3000 A - the capacity of our DC power supply - a voltage of 0.016 μV/cm has been measured, indicating a critical current of ≥ 5 kA. An upper imit of 5.8 kA is given by the sum of the individual tapes. From I-V-measurements at lifferent positions we conclude that the performance of the conductor is very homogeneous vithout a strong degradation due to the winding, handling and cool down.

3.4. AC performance and loss measurements

The conductor is connected via a transformer to a generator supplying current in the requency range between 35 Hz and 60 Hz. In Fig. 5 the temperature increase of the onductor is plotted for operation at 3000 A_{rms} (55 Hz). After approximately 500 s quilibrium is reached, giving a proof that a stable operation at high currents is possible.. The largest temperature increase (ΔT=1.2 K, depending on the LN2 mass flow) is observed t the end of conductor.Finally the AC losses have been measured electrically as a function

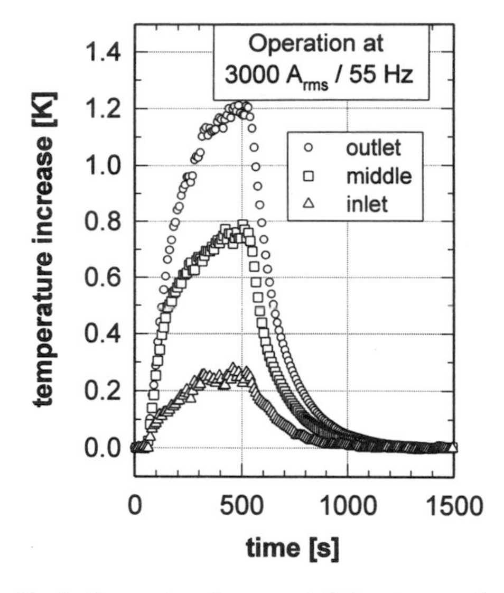

Fig. 5: Temperature increase at different parts of the conductor during operation at 3000 A_{rms}. The highest increase is at the end of the conductor. At t=500 s the current has been switched off.

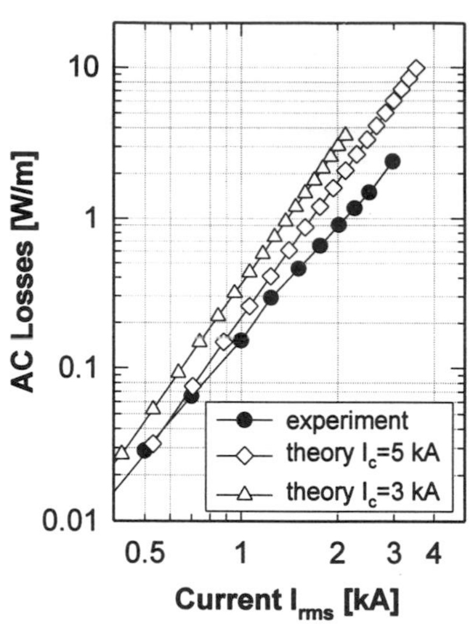

Fig. 6: Electrical measurement of the AC losses versus conductor current at 55 Hz. For comparison calculations using the Bean model are shown.

of current amplitude and frequency. As shown in Fig. 6 the amplitude dependence shows nearly a cubic dependence. Moreover the frequency dependence (not shown here) is linear, indicating hysteretic losses.

At an operating current of 2000 A_{rms} the losses at 55 Hz are 0.9 W/m, which is the lowest value for a HTS cable conductor reported so far. For comparison Fig. 6 shows calculations of the losses using the Bean model for a superconduting cylinder [7]. Our conductor shows clearly lower losses due to the special low-loss design with an optimised current distribution between the layers. Using the Rogowski coils we have been able to measure the current within the individual layers, yielding a value of I_i/I_{total}=0.25±0.04 of the total current for the layers i=1-4.

Acknowledgements

The authors wish to acknowledge their co-workers from Vacuumschmelze GmbH, Siemens AG and collaborating partners within the BMBF-programme "Superconductivity for Power Engineering" supported by the German BMBF under grant no. 13N6481.

References

[1] Scudiere J et al. 1996 Adv. in Cryogenic Engineering **42** 899 (New York, Plenum)
[2] Leghissa M et al. 1997 IEEE Trans. Appl. Supercond., **7** 355
[3] Shibata T 1996 Adv. in Supercond. VIII Vol. 2 1303 (Tokyo, Springer)
[4] Miura O et al. 1996 Croygencis **36** 589
[5] Th. Schuster et al. 1996 Appl. Phys. Lett. **69** 1954
[6] A. Malozemoff et al. 1997 M2S- Conference, Beijing, China, Feb. 28 - March 3
[7] G. Vallego and P. Metra 1995 Supercond. Sci. Technol. **8** 476

Inst. Phys. Conf. Ser. No 158
Paper presented at Applied Superconductivity, The Netherlands, 30 June–3 July 1997

Influence of heating pulses on the quench behavior of a 12 strand Nb_3Sn CICC

V S Vysotsky

ISSSP RSC "Kurchatov Institute", Moscow, Russia; Present address: Kyushu University-36, Graduate School of ISEE, Fukuoka 812-81, Japan

M Takayasu, P C Michael, J H Schultz and J V Minervini

Plasma Science and Fusion Center, MIT, Cambridge, MA 02139, USA

S Jeong

Korea Advanced Institute of Science and Technology 373-1, Taejon 305-701, Korea

V V Vysotskaia

Baikov Institute of Metallurgy, Moscow, Russia

Abstract. During several field ramps, sequential heat pulses were applied to one strand or to one triplet of the cable in the experiment with direct measurements of the current in each strand of a CICC. Despite the extra thermal disturbance introduced into the cable, the quench field in some cases was higher in the presence of heat pulses than during tests at the same conditions without heating. This paper compares experimentally derived strand current distribution data obtained during field ramps with and without heat pulses. The results indicate that the heating changes the current distribution among strands in a way that improves RRL behavior of the CICC. Thus, small heat disturbances artificially introduced into a CICC, may improve the cable's RRL behavior.

1. Introduction

We studied the Ramp Rate Limitation (RRL) phenomenon by direct measurement of the current in each strand of a chrome coated 12 strand Nb_3Sn CICC [1,2]. The sample was tested at fixed DC currents in linearly increasing external magnetic field. The sample demonstrated a strong RRL phenomenon. The strand measurements indicate that the main cause of the sample's preliminary quench was severe current non-uniformity that occurred in the CICC [2,3]. This preliminary quench starts from one single strand that reaches its critical value before the rest of the strands in the cable. The quench then propagates to several other strands, whose currents are next closest to the critical value [3]; this cascade process leads to entire cable quench. However, the results also show several single strand quenches that recover rather than propagate to the rest of the cable. Thus, *a certain current distribution must be present within the CICC to enable a single strand quench to propagate*.

Some of the cable strands in our experiments were instrumented with resistive heaters [1,2] that were located in a well cooled area outside the sample's conduit. During several field ramps, sequential heat pulses were applied to one strand or to one triplet. An unexpected result was observed in which the quench field was higher in the presence of the applied heat pulses than at the same conditions without heating. The present paper compares the experimental data on current distribution among strands during field ramps with and without heating pulses. The influence of heat pulses on the current distribution and quench behavior of a CICC is discussed.

2. Experimental data

A detailed description of the experimental procedure of our sample is presented in [1,2]. We tested a two-stage, 12 strands (3×4) CICC wound in a 8 turns, 2 layers coil. Current sensors mounted near the helium inlet to the inner layer of the coil are indicated by the code letter "I" - the code letter "O" - is used to designate sensors at the end of the outer coil layer. The four triplets are designated by the code letters "A" through "D", and each strand in a triplet by a number from 1 to 3. Sensor location "ID-2" thus indicates: the inlet end of the second strand in triplet D. The same strand on opposite side of the sample is designated "OD-2". Each strand in the sample is instrumented with both Hall probes and pick-up coils, unfortunately, data acquisition system limitations did not allow us to monitor all signals simultaneously.

Three sets of heat pulse experiments were performed at the same DC transport current of 2000A. The first, baseline case was performed without application of heat pulses to the strands. A second set was performed by applying a heat pulse train to strand OA-1 only, while the third set was performed with the heat pulses applied to the entire OA triplet. The heater power was about ~2.5 W per strand applied to an ~1 cm strand length. The pulse duration was 0.8 ~ 0.9 s with 0.7 s pause between pulses. The heating pulses started at ~3 T during the field ramp and continuec until sample quenched. The experiments were performed at three different field ramp rates The dependence of the quench field on ramp rate is shown in Figure 1. One can see that i spite of applied heating, the quench field at low dB/dt is higher than without heating.

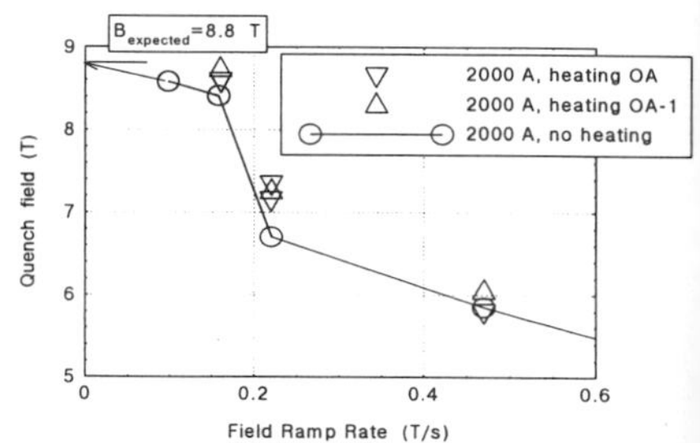

Figure 1. Ramp Rate Limitation in the model sample, with and without applied heat pulses.

To determine why the applied heating improves RRL behavior we compared the stranc current vs. field traces during field ramps with and without heating. In [3] we showed tha quench in our sample was initiated by the quench of the strand IB-1; the quench the propagates to strands IA-1, IA-2, IA-3 and IC-1 and ID-2. These strands appear to be ke participants in the quench development process during the heat pulse experiments as well.

A set of current vs. field traces is shown in Figure 2 for IB-1 strand during heat puls experiments at three different ramp rates. With and without heating. From Figure 2, one ca see that currents of strand IB-1 *became less after heat pulses were applied.* Thus, the quenc initiated by the strand IB-1 started later in case of heating. For strands IB-2 and IB-3, th currents are unchanged by the heat pulses. Although the heat pulses were applied to the O triplet, we did not observe any change in the IA triplet currents at any field ramps.

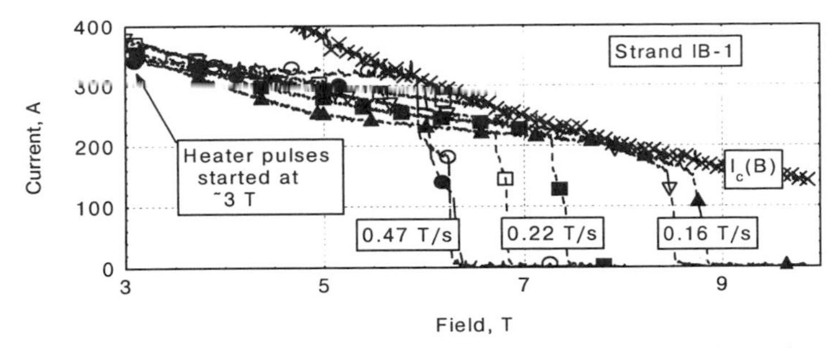

Figure 2. Current traces for the strand IB-1 at different field ramps with and without heat pulses applied to the OA triplet. The cases with heating are shown by closed symbols. The heating from field level ~2-3 T reduces the current in strand IB-1 that usually initiates the quench of entire cable [3]. This strand current reduction leads to quench at higher field.

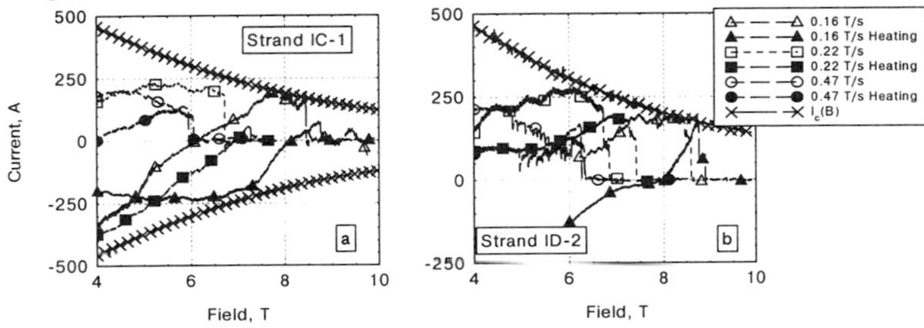

Figure 3. a) current traces for strand IC-1, and b) current traces for strand ID-2. The applied heat pulses significantly reduce the currents in the strands IC-1 and ID-2; this make cable more stable and promotes a higher quench field.

Current vs. field traces for strands IC-1 and ID-2 at three ramp rates are shown in Figures 3a and 3b both with and without the applied heat pulses. For these strands, heating in the triplet OA caused a significant change of the strand currents; they became much less than in unheated case, especially at low dB/dt.

Traces for strand OA-1 at two different field ramp rates are shown in Figure 4. Heating pulses were applied just to this strand. At 0.16 T/s ramp rate, current drop-recovery events are clearly seen starting at a field of about 6.2 T. They appear due to small normal spots in the strand/subcable generated by the heater. Such behavior was typical during other field ramps at low dB/dt; current drop and recovery events in the OA triplet always appeared at ~6.2 T. For dB/dt=0.47 T/s the current traces for strand OA-1 are very different from the cases of low dB/dt and we simply could not observe any drop-recovery events (DRE) because in this case the quench occurs at fields below 6.2 T. We also note that the application of heat pulses to strand OA-1 change the slope of the current vs. field trace for this strand at low dB/dt, and that strand OA-1 current reaches the $I_c(B)$ curve more smoothly. Although the current in OA-1 reaches $I_c(B)$

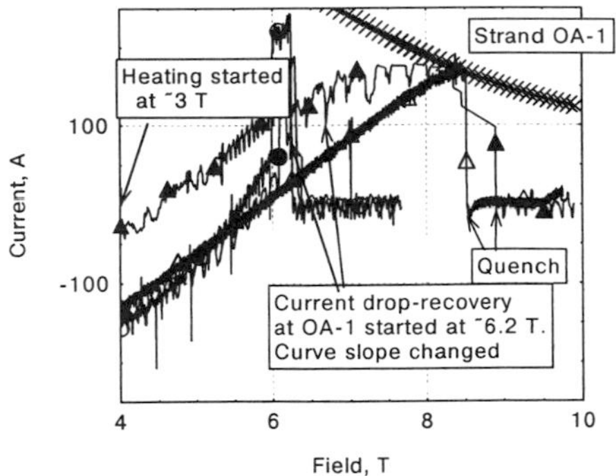

Figure 4. Current vs. field traces in the strand OA-1. Heating was applied only to this strand. Current drop-recovery events begin at the field ~6.2 T due to the applied heating. The heat pulses reduce the current slope at low ramp rates and hence it touches $I_c(B)$ curve more smoothly. Symbols - see Figure 3.

earlier in the heated case, the strand does not quench at this moment because of its "smooth" approach to the critical current transition.

3. Discussion

In [3] we showed that due to severe non-uniformity of current distribution among strands in a CICC one strand could reach its critical current before the entire cable reaches the strand average I_c. Under certain condition the quench of this strand may lead to a quench of the entire cable. The necessary condition seems to be that some of the strands inside a CICC also have currents sufficiently near their critical values so that they are also destabilized when first strand quenches. The current maldistribution inside of CICC is caused by changing field that induces current through closed current loops formed either through the termination joints or through interstrand resistances [1,2,3]. Thus, during field ramp the *current in the strands may change in a very sophisticated way* depending on dB/dt, the length of the cable, and the interstrand and joint resistances [1,2,3].

The results in this paper indicate that the application of a *small heat pulse* to one strand in a CICC can *induce sufficient changes to the current distribution inside CICC* to moderate RRL behavior especially at low ramp rates. In our case, the heat pulses generate normal spots in the heated strands that lead to current redistribution that reduces the current in the strand IB-1 enough to shift the quench in this strand to higher field levels. It drastically changes the currents in the strands IC-1 and ID-2 that are important for quench to develop to entire cable. Similarly, as seen in Fig. 4, the small normal spot that forms in strand OA-1 during heating delays the induced current rise due to field changes. That is why current in strand OA-1 touches $I_c(B)$ curve more smoothly when it is heated and reduces the probability of quenching in this strand. For cases in which quench due to RRL phenomena precludes the onset of DRE (such as was observed in Fig. 4 at 0.47 T/s), there is almost no change of the quench field with heating. Thus, artificially induced DRE may improve the RRL behavior of the CICC. The action of the small artificial heating is similar to action of partial drop-recovery events that may happens inside CICC due to intrinsic reasons [4]. Small disturbances inside CICC that do not lead to entire cable quench improve current distribution and an RRL behavior. Similar results were received with full size CICC in [5].

4. Summary

Small heat pulses applied to one strand or one triplet in a CICC can change the current distribution inside the cable and thus reduce the currents in strands that trend to reach their critical value first. This improves the RRL behavior of the entire cable. Contrary to intuitive reasoning, these small, sequential heat pulses do not reduce stability, but may actually improve current distribution inside CICC and increase its stability.

References

[1] Vysotsky V.S., et al, *Proceedings of ICEC-16/ICMC,* (1997) 1215-1226
[2] Vysotsky V.S. et al, (1997) *Cryogenics* **37** in press
[3] Vysotsky V.S. et al, (1997) *IEEE Trans. on Applied Superconductivity* **7** in press
[4] Takayasu M. et al, (1997) *IEEE Trans. on Applied Superconductivity* **7** in press
[5] Kozumi N. et al, (1994) *Cryogenics* **34** 1015-1022

Inst. Phys. Conf. Ser. No 158
Paper presented at Applied Superconductivity, The Netherlands, 30 June–3 July 1997

Transport current distribution in core of multilayer high-T_c superconducting power cable

P.I.Dolgosheev, V.E.Sytnikov, G.G.Svalov, N.V.Polyakova *
D.I. Belyi **

*Superconducting Wire & Cable Department, JSC "VNIIKP"Moscow, RUSSIA
**Cabix Consulting , Moscow, RUSSIA.

Abstract. Power transmission cables are expected to be one of the earliest commercial applications of high-Tc superconductors. Many firms have attempted to increase the current carrying capacity of cable by enlarging the number of superconductor layers , but such attempts have not been successful. This difficulty is connected with the current distribution effect and the main objective of our work was to meet the challenge. The theoretical investigations of current distribution among the HTS-cable layers are executed and the recommendations on cable core design are presented.

1. Introduction

The use of HTS cables in power systems necessitates an increase of their current carrying capacity as compared to a level achieved by now. Most of the firms have selected the way to solve a problem of building up the current carrying capacity of the cable by increasing a number of layers of superconductive tapes to 4-10 [1-7]. At the moment currents of 1 to 12 kA were obtained at short sample lengths (around 7 m long) [1-3]. However, in testing 50 m models, the achieved values of critical current are not in excess of 2 kA [4,5]. One of the primary causes of decrease of current carrying capacity is a non-uniform distribution of current between the layers and the impact of the length of experimental sample.

2. Current distribution between superconducting layers

We have presented certain principles of designing cores of superconducting cables based on our experience in the field of low-temperature superconducting cables [8,9] and wires [10-12].The goal is to create the cable core with the current distribution, which is proximate to homogeneous distribution of current between core layers for maximum cable current. The value of the core total current, when the current reaches its critical value in one of the layers, is assume to be a maximum current of the cable core. A mathematical model for analysis of the distribution of currents between the core layers is based on our publications [11,12] and has been updated with regard to a specific nature of HTS conductors and cores. Authors are hopeful of publishing a detailed description of the theory of HTS multilayer core in 1997-98.

Listed below are the results of analysis of endless multilayer core for traditional design. A thickness of superconducting tape is assume to be much less that the layer radius. A direction of tape twisting in a layer is alternating from layer to layer and twist pitches are the same in any layer. Lets consider a conductor with a central element whose outer diameter equals 39.5 mm; the conductor is made of HTS tapes 0.3 mm thick provided with electric insulation, 0.1 mm per side. To demonstrate the above , Fig. 1 shows a design relations of maximum current of conductors (Icore) consisting of one to ten layers standardized to a current at a smooth superconducting cylinder (Itube) whose diameter is equal to the diameter of the outer layer. As is obvious from the analysis, there is no way to provide a uniform distribution of transport current between the layers without using a special decisions for current supply. The examination of superconductors in the form of two to ten layers evidence that in a superconductive state , when a active voltage drop along the conductor is not observed, in essence, a transport current flows in two outer layers only, and a total current in the underlying layers approaches zero.

We have developed a program of optimization of design parameters of the cable core (pitches and directions of twisting) with the preset diameter of layers and current carrying capacity of the superconductor in each layers which would provide a maximum value of the current at the most rational use of the superconductor cross-section. Positive results were obtained for the cores consisting of two to ten layers. Table 1 lists the results of optimization of four-layer core for all eight possible models. The parameters of the core have been assumed to be the same as for Fig. 1. In addition, it was assumed that the gaps between the tapes are zero, and a linear critical current per unit of width of the tape is equal to 60A/cm. In Table 1: i_i is the current fraction in i - layer. The criteria is a maximum accessible current of the core. It is easily seen that for a classic design with the alternating direction of twisting at $\gamma_1 = \gamma_3 = 1,\ \gamma_2 = \gamma_4 = -1$ (option 7), variation of twisting pitches alone does not lead to a positive result. Out of eight possible options of design of four-layer core, maximum value Imax is provided by option 1 and 3. In option 1 all the layers are twisted to one side, while in option 3, two inner layers are twisted to one side , and two outer layers - to the other. Twist pitches have to be different for layers, their optimization are not subject of the paper.

A design of the cable core balanced out against a maximum value of current Imax is harmonic in reference to current leads both for electrically insulated and uninsulated layers. Critical current in the layers are achieved simultaneously and one would expect more steep characteristics V-I that was observed in publications [1,4] which is connected with the non-availability of transverse currents between the layers in a balanced design of the core.

3. Time dependence of transport current distribution in core layers (influence of sample length)

Fig. 2 illustrates the results of calculation of four-layer cores 1 m long (Fig. 2a) and 50 m lor (Fig. 2b and 2c) from electrically insulated layers at a variation of the current input speed. T following two core design have been examined: option A, prototype of option 7 shown in Ta le 1, but with similar pitches of tape twisting in the layers equal to 0.5 m (Fig. 2a and 2b) ar option B which is option 3 shown in Table 1 (Fig. 2c). The values of electrical resistance current leads connected in series with each layer is set equal to 10^{-9} Ohms. A common curre variation law is taken to be exponential, i.e. $I_o(t) = I_o(1 - e^{-t/\tau})$, where τ is a time constan

Variation of values Io and τ makes possible to also apply the simulation to line and sinusoidal (1/4 of a period) common current alteration laws. A value of the outer layer critical current for both options of the cores makes around 780 A. The superconductor cross-sectior for both options of the cables ensures Io=3000A.

A maximum speed of the common current input at time instant t=0 equals Io/ τ a comes to 10 A/s (Fig. 2b) ; 150 A/s (Fig. 2a) and 2*10^6 A/s (Fig. 2c). As indicated in Fig.2a

Table 1. The results of the total maximum current optimization (I max).

Variant	1	2	3	4	5	6	7	8
γ_i	$\gamma_{1,2}=1$ $\gamma_{3,4}=1$	$\gamma_{1,2,3}=1$ $\gamma_4=-1$	$\gamma_{1,2}=1$ $\gamma_{3,4}=-1$	$\gamma_1=1$ $\gamma_{2,3,4}=-1$	$\gamma_{1,3,4}=1$ $\gamma_2=-1$	$\gamma_{1,4}=1$ $\gamma_{2,3}=-1$	$\gamma_{1,3}=1$ $\gamma_{2,4}=-1$	$\gamma_{1,2,4}=1$ $\gamma_3=-1$
i_1	0.2491	-0.0004	0.2432	0.2231	-0.1436	-0.2351	0.2000	-0.1908
i_2	0.2376	0.2367	0.2575	- 0.2848	0.3529	0.3489	-0.3007	0.0774
i_3	0.2597	0.3983	0.2621	0.5609	0.4388	0.4647	0.5716	0.5736
i_4	0.2532	0.3650	0.2372	0.5008	0.3519	0.4215	0.5282	0.5399
I_{Max}	2950.4	1967.4	2964.1	1364.9	1725.2	1690.1	1373.3	1366.1
I_{max}/I_{tube}	3.640	2.427	3.657	1.684	2.128	2.086	1.694	1.685

for option A at a sample length of 1m and Imax = 150 A/s , the layer current become closer in value approximately in 120 s and the common expected current Imax will come to at least 3 kA. At Imax >150 A/s , the outer layer current will achieve its critical value as soon as after t>30 s. With the increase of the sample length to 50 m (Fig. 2b) the outer layer current will achieve its critical value after 160 s even at Imax = 10 A/s. While the common current will amount to as low as 1240 A.

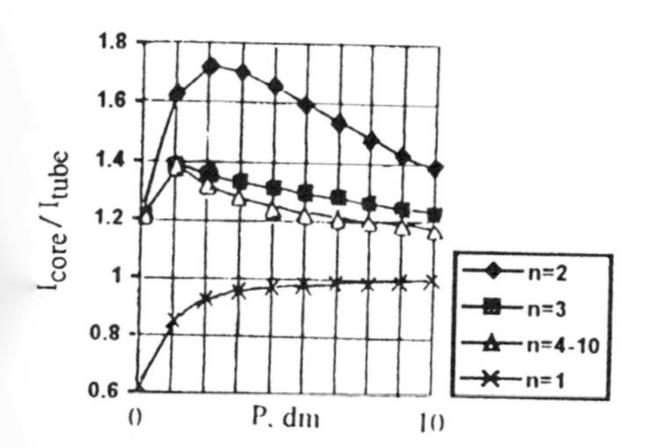

Figure 1. I_{core}/I_{tube} versus the twisting pitch (P) of layers

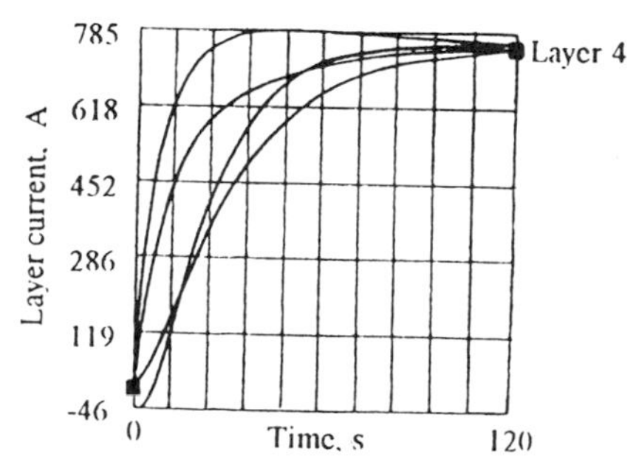

Figure 2a.

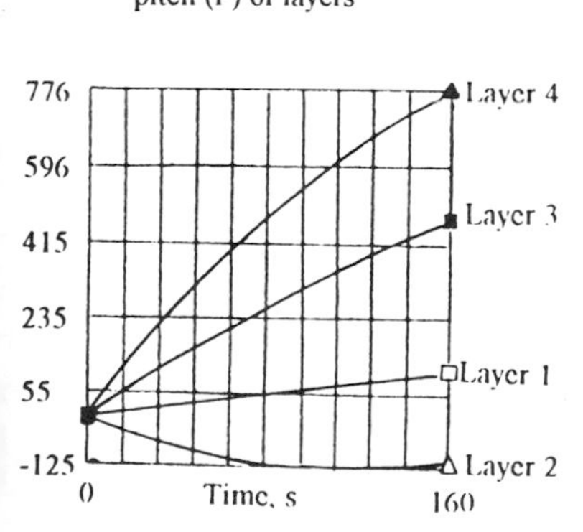

Figure 2b.

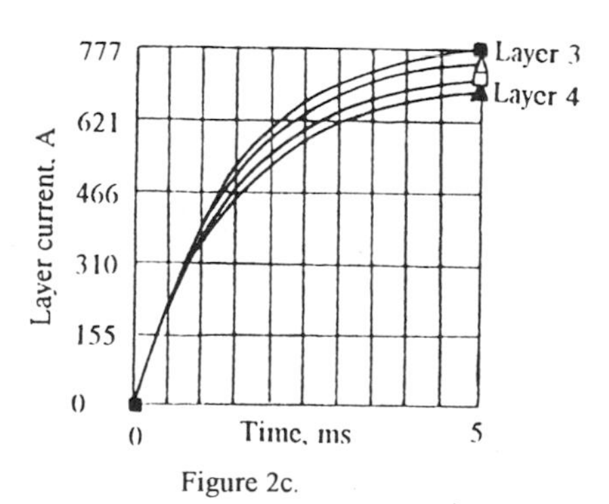

Figure 2c.

Figure 2. Time dependence of the current distribution in the core layers

The dependencies illustrated in Fig. 2a and 2b provide a quantitative and qualitative assessment of the reduction of current carrying capacity realized in the experiment at transition from short model, 1 m long, to 50 m models. For the core design balanced out as per Imax (option B), a distribution of currents between the layers essentially does not depend on Imax or on the sample length. The currents at all the layers attain their critical values simultaneously at l = 50 m and Imax = $2*10^6$ A/s , within 0.005 s; for both cases the common current equals ≈ 3 kA. The analysis performed at Imax = $2*10^6$ A/s , revealed that option B of four layer cable is the only one suitable for use on alternating current at a frequency of 50 - 60 Hz.

For electrically uninsulated layers (irrespective of the values of resistance at current leads), distribution of currents between the layers Ii at the cable section far removed from the current leads corresponds to data listed in Table 1.

4. Conclusion

- The alternation of twisting direction of the tapes from layer to layer results in non effective utilization of superconductor cross-section. The useful utilization of superconductor can be provided with simultaneously variation of the directions and the pitches on all layers.
- The experiments on the short samples truly represent the electromagnetic processes on real length core only for a balanced out design core and for uniformly soldering of layers on current leads.
- For the balanced cable core the coupling losses are tended to zero and the electrical insulation between layers is not necessity.
- It is desirable to execute the experiments for detailed investigations of current distributions between layers on a short samples of multilayer core.
- The long length core modeling may be accomplished by a variation of a current input speed.

Authors are ready to render the every assistance on the design of the balanced cable cores and on the organization and the execution of its experimental investigations.

References

[1] J.J Cannon J R., Minot M. J., Buczek D., Vellego G., Metra P.. 1995 IEEE Trans. on Appl. Supercon., v.5, N 2

[2] Sinha V.. ,1996 Southwire Users Meeting Atlanta, USA, Report III-A

[3] Hara T., Ishii H., Honjo S.. 1996 ICEC 16/ ICMC, Kitakyushu, Japan, , part 2, p. 963.

[4] Fujikama J., Saga N., Ohmatsu K et al.. 1996 ICEC 16/ ICMC,Kitakyushu, Japan, part 2, 975.

[5] Mukoyama S., Miyoshi K.,Tsubouti H. et al.. 1996 ICEC16/ ICMC, Kitakyushu, Japan, 2, . 979.

[6] Leghissa M., Fischer B., Roas B., Jenovelis A., Kantz S., Neumuller W.- H. 1996 Appl. Supercond. Conf. , , Pittsburgh USA, rep. LE-4.

[7] Lue J. W, Lubell M.S., Jones E.C., Demko J.A., Kroeger D.M., Martin P.M.. Appl. Supercond. Conf. - 96. rep.LOB -4.

[8] Morgan G. H. ., Forsyth E.B.. Preprint BNL 19848, 1976 p. 13.

[9] Peshkov I.B.,. Svalov G.G, Dolgosheev P.I. et al.. 1979 IEEE Trans. on Magn., , v. 15, N 1.

[10] Turck. B.. 1974 Cryogenics, August, p. 448.

[11] Sytnikov V.E.. Svalov G.G.,.Meshchanov G.I, Dolgosheev P.I. 1983 -Cryogenics. v. 23, N2, p. 77.

[12] Sytnikov V.E., Svalov G.G., Peshkov I.B.. 1989 Cryogenics.,v. 29, N 10, p. 971.

Inst. Phys. Conf. Ser. No 158
Paper presented at Applied Superconductivity, The Netherlands, 30 June–3 July 1997

Evaluation of HTS Samples for 12.5 kA Current Leads

Marc Teng, Amalia Ballarino, Robert Herzog, Albert IJspeert
CERN, 1211 Geneva 23, Switzerland

Ciaran Timlin, Stephen Harrison , Kevin Smith
Oxford Instruments Ltd, Old Station Way, Eynsham, Witney, Oxon OX8 1TL, England

Abstract. The Large Hadron Collider will require the conduction of more than 3400 kA between room temperature and 1.8 K. To reduce the heat load into the liquid He, a development programme was started on current leads with HTS sections. This paper reports on the evaluation of HTS samples for 12.5 kA from a number of suppliers. The tests were performed in a purpose built assembly, with the main criteria being current carrying capacity, heat conduction and quench behaviour. Results show the viability of HTS components in high-current leads, although further engineering efforts are required for some aspects.

1. Introduction

Leads conducting high currents from ambient to cryogenic temperatures represent a large fraction of the heat load in many cryogenic systems. Conventional vapour cooled leads with a theoretical heat load of 1.04 W/kA[1] would imply an overall heat load of 3.5 kW at 4.2 K to conduct the required 3400 kA for LHC[2]. Aiming to reduce the 4.2 K heat load, a project was launched at CERN in 1995 to study the use of High Temperature Superconducting (HTS) materials for the coldest part of the lead, up to about 70 K. HTS materials have a low thermal conductivity but they are perfect electrical conductors in this temperature range. The heat from the conventional upper section of a complete lead loads the cryogenic system mainly at higher temperatures, leading to substantial savings in liquid He.

A dedicated test facility was designed and built by Oxford Instruments in close collaboration with CERN. In April 1996 a first set of HTS samples carried 13 kA from 70 K to 4.2 K. To encourage development of HTS for high current leads, companies were invited to supply prototype samples, complying with the following requirements:

1. To conduct 12.5 kA from 75 to 4.2 K with a heat load of at most 1.3 W at 4.2 K. This is about 10% of the heat load of a conventional vapour cooled lead.

2. To withstand a quench caused by a temperature increase at the upper end of the sample and to carry nominal current for 60 s in the quenched state. This requirement corresponds to an emergency de-excitation.

2. Test Facility

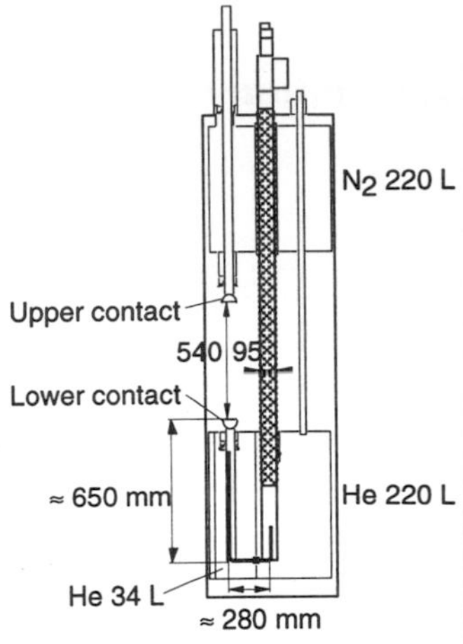

A special cryostat for testing 12.5 kA leads has been designed and built by Oxford Instruments (Fig. 1) [2][3].The samples are installed in vacuum between terminals at 4.5 K and 70 ± 5 K, respectively. Heat conduction at the cold terminal is measured by monitoring the He boil-off. The temperature at the warm terminal can be controlled by pumping on liquid nitrogen.

Voltages, current, temperatures and the He boil-off are recorded by a data acquisition system built with Labview. A second computer was used to control the power converter, either by entering commands manually, or in control-loop mode. This mode enables the simulation of a shunt parallel to the sample, by continuously adjusting the current to a value calculated from the momentary sample voltage and the set values of current and shunt resistance.

Fig. 1: Test facility

Such a shunt might be necessary in real operation to allow the discharge of the magnets, even when the HTS has become normal conducting. Control-loop mode drives the current down as voltage builds up across the sample. The advantage of this arrangement is that quench tests can be performed with different shunt values without physical interference, saving thermal cycles.

3. Experimental results

The performance of the samples was evaluated by current cycling, while observing the heat conducted into liquid He and by quench testing. For the latter the current was set to nominal and the top end temperature increased until the lead quenched. Table 1 summarises the test results of the 7 samples, received from 4 manufacturers. In addition it features extrapolated values for the thermal conduction (Q') with the appropriate physical shunt in parallel.

3.1 BSCCO tapes

American Superconductor Corporation (ASC, Westborough MA) provided a sample made of BSCCO-2223 PIT ribbon stacks. The silver matrix is doped with gold to reduce the thermal conductivity. The test of the original sample revealed a high resistance in the lower connection (HTS-LTS-bolted contact) of ~250 nΩ. After modification the joint resistances improved to 1.9 nΩ (HTS-LTS), but the heat conduction increased slightly. Current carrying tests showed a low overall heat load of 1.49 W at full current. During the quench tests the sample carried the full current, without shunt emulation for 42 s measured from the time the sample voltage increased above 5 mV. After this period it suffered damage.

3.2 YBCO modules

The EURUS (EURUS Technologies Inc, Tallahassee FL) sample consists of 72 parallel modules, each containing YBCO filaments sealed in a resin and surrounded by a steel spiral which serves both to shield the external magnetic field and as a shunt. Wires of 160 mm in length are connected to both ends of the modules. For the cold end superconducting wires are included.

Table 1

Sample		BSCCO tapes	BSCCO tapes mod.	YBCO modules	YBCO bars	BSCCO tube (70)	BSCCO tube (113)	BSCCO tube (Ag)
HTS material		BSCCO-2223	BSCCO-2223	YBCO	YBCO	BSCCO-2212	BSCCO-2212	BSCCO-2212
Shape		powder in Ag tapes	powder in Ag tapes	filaments in resin	4 Bulk bars in parallel	Bulk tube	Bulk tube	Ag clad bulk tube
Manufacturer		ASC	ASC	Eurus	Haldor Topsøe A/S	Hoechst	Hoechst	Hoechst
HTS length	[mm]	520	520	200	285	200	95	200
$\mathbf{I_{max}}$	[kA]	12.5	12.8	9.5	11	6.5	13	12.7
$\mathbf{R_{shunt,max}}$	[μΩ]	100	100	50	25	25	25	∞
Q'(I = 0 A)	[W]	0.67	1.19	3.04	0.41	0.32	1.34	>16
$\mathbf{Q'_{sh}(I = 0\ A)}$	[W]	0.97	1.49	5.44	5.21	5.12	6.14	>16
$\mathbf{Q'_{sh}(I = 12.5\ kA)}$	[W]	N.A.	1.79	7.59	6.58	5.90	7.7	>17
$\mathbf{Q'_{shunt}}$	[W]	0.3	0.3	2.4	4.8	4.8	4.8	0
$\mathbf{V_{max}}$	[mV]	520	950	140		265	53	132
$\mathbf{R_{max}}$	[μΩ]		90	75		500	40	11
$\mathbf{R_{lower\ connection}}$	[nΩ]	N.A.	1.9	20	4.7	70	600	8.9
$\mathbf{P_{lower\ connection}}$	[W]	N.A.	0.312	1.80	0.57	2.96	94	1.39
$\mathbf{R_{upper\ connection}}$	[nΩ]	N.A.	24.5	19	350	4500	565	454
$\mathbf{P_{upper\ connection}}$	[W]	N.A.	3.8	1.7	40	100	88	70

I_{max}: maximum current attained; Q'(I=0), Q'_{sh}(I=0,12.5 kA): heat conduction at zero current and corrected for appropriate shunt resistances and optimised connections at zero current and 12.5 kA; Q'_{shunt}: heat conduction of an appropriate shunt;
V_{max}, R_{max}: maximum voltage and resistance during quench test.

At 9.5 kA heating in the upper connection wires caused an increase of the temperature and a quench of the HTS. Because of the limited maximum current the quench test was performed at 7 kA. It lasted 72 seconds, with the current reduced by the action of the simulated shunt. After quench the sample was cooled down again and ramped up to 7 kA, showing no degradation.

3.3 YBCO bars

The Topsøe A/S (Lyngby, DK) sample consisted of 4 bars of bulk YBCO in parallel, each 285 mm long and supported in a tube. End caps on their top ends are connected via braids to the upper terminal. The heat load by conduction was 0.41 W at I = 0. A bottom joint resistance of 4.7 nΩ helped to keep the heat load down to 1.78 W at 11 kA. With a shunt comparable to the BSCCO bulk, this results in 5.21 W and 6.58 W respectively. However, the resistances of the braids dissipated about 40 W, such that the top temperature of the HTS rose from 68 to 73 K, upon which the current was stopped manually. The shunt emulation current control was inactive. When trying to ramp up anew, the sample appeared broken. The braid resistance was about 230 nΩ at 4 kA, increasing to 350 nΩ at 11 kA.

3.4 BSCCO tubes

First tests and commissioning of the test facility were conducted with two BSCCO-2212 tubes in parallel from Hoechst (Frankfurt, DE)[3] . During processing of these tubes silver contacts are integrated on both ends to facilitate solder connections to end caps. The end caps are connected via braids to the upper terminal and directly to the bottom terminal respectively. Earlier experience showed that quenches caused local heating which the low thermal conductivity prevented from spreading.

This usually results in a break of the sample close to the warm end[4]. The wall thickness of all these tubes is between 7 and 8 mm; the cross-section is determined by the diameter.

The first new sample (BSCCO tube (70)), a single 70 mm diameter tube was wrapped with a copper foil with indium to make a good thermal contact, prolonging the warm end cap by 2 cm to improve the spreading of heat generated during a quench. This measure had proved successful on smaller samples [4]. The sample survived two quench tests with 1 and 10 μΩ simulated shunt resistances. With a shunt resistance of 100 μΩ the sample broke.

A second test on a 113 mm diameter, 95 mm long tube (BSCCO tube (113)) proved that the full current could be carried by a single sample. However, the bottom solder joint resistance increased reversibly from 35 to 600 nΩ above 6 kA.

For a third test a 70 mm diameter tube was clad with silver (BSCCO tube (Ag)) in an attempt to improve quench behaviour. A 200 μm thick sheet of silver covered the entire tube. This silver conducted more than 10 W thermally during the test. The calculated resistance of the silver was about 10 μΩ. This sample carried the full 12.5 kA current at 77 K, although it is rated for only 7 kA at 77 K. It was quenched at 12.5 kA, conducting current for 44 s with the voltage above 5 mV. After this test, the sample was cooled down and the current was ramped up to 6 kA showing no degradation. The bottom solder contact had a resistance of 10 nΩ, lower than on previous tests with this material.

4. Conclusions

The encouraging result of this test series is that all the materials were able to conduct high current with low heat conduction. The limitations encountered stem mainly from too much heat generation in the normal conducting parts of the connections at both ends of the HTS. Localised heating during quench is an intrinsic problem of bulk material, but measures can be taken to alleviate the associated breakage problem, as shown with the BSCCO-2212(Ag) sample. Moreover, the mechanical fragility of the bulk material would necessitate the packaging to be carefully designed, for a current lead application.

Heat generation at the warm connection of some samples (BSCCO-2212(70), YBCO filaments, YBCO bars) led to a rapid temperature rise at high current and consequently to a quench of the HTS material. This limited the maximum current or the time the maximum current could be maintained during the experiments. Heat generation at the cold connection of other samples (BSCCO-2223, BSCCO-2212(70,113)) increased the heat load into the liquid helium to levels much higher than acceptable. Careful dimensioning and preparation of the connections results in acceptably low resistances, as shown by some of the samples.

The modified BSCCO-2223 sample came closest to satisfying the heat load requirements. In particular, the current density in the HTS material is sufficient for this purpose, the limitation regarding burn-out being determined by the characteristics of the matrix and/or the shunt.

References

[1] M N Wilson 1983 *Superconducting Magnets* (Oxford: Clarendon)
[2] A Ballarino, A IJspeert, M Teng, U Wagner, S Harrison, K Smith, L Cowey 1996 *ICEC* **16** proceedings 1143-6 (Elsevier)
[3] S Harrison, D Jenkins, K Smith, M Townsend, A Ijspeert, A Ballarino, M Teng 1996 *EPAC96* proceedings 2296 (Bristol: Institute of Physics)
[4] A Ballarino, A IJspeert 1996 *ICEC* **16** 1147-50

Inst. Phys. Conf. Ser. No 158
Paper presented at Applied Superconductivity, The Netherlands, 30 June–3 July 1997

Development of a thallium cuprate current lead

D M Pooke[1], A Mawdsley, J L Tallon, and R G Buckley

New Zealand Institute for Industrial Research, PO Box 31310, Lower Hutt, New Zealand

Abstract. We report the utilisation of the Tl-1223 phase in a superconducting current lead. This compound offers the advantages of a higher transition temperature and higher magnetic irreversibility line than the more commonly used Bi-2212 and Bi-2223 phases. A processing path was developed centring on simple sintering steps to offer a rapid, cost-effective route to production of component bars, yet providing industry-competitive current densities. A stoichiometry was selected for high T_c and good phase purity. The processing route provided a dense sintered product displaying critical current densities $J_c \sim 2500A/cm^2$ at T=77K in 100A-class leads, with significant J_c even at T=110K. This high temperature operation may make the Tl-phase a compelling choice in applications sensitive to refrigeration load. A 750A prototype compound current lead has been constructed from component bars, and tested non-destructively to 900A. Measurements on individual bars has revealed a need for further development of the current contacts.

1. Introduction

High-T_c superconducting (HTS) current leads have the capacity to reduce the refrigeration costs of cryogenic devices, particularly those operating at or near 4K. This is due to the low thermal conductivity and zero dc electrical resistance of HTS materials, allowing them to carry large electrical currents without transferring heat into the cooled environment [1]. Energy efficiency is an important issue in the commercialisation of cryogenic technologies; for example, one Watt of heat generated at 77K requires about 20W of refrigeration to expel this heat to room temperature. The benefits of HTS current leads will broaden if they are able to operate at the highest possible temperature, and if their performance can be maintained in magnetic fields. With a target operating temperature of 110K, the commonly used Bi-2212 or -2223 phases are ruled out. The Tl- and Hg- phases, however, display both higher Tc's and irreversibility fields, and so could be investigated. Given the ease of processing the Tl materials compared to the Hg compounds, we have concentrated on the Tl system, and the outcome of development to a prototype stage (Fig. 1) is reported in this paper.

2. Composition and Process Development

While Tl-2223 has the highest T_c of the Tl- phases, the processing window is smaller and we found it more difficult to achieve high currents with this material. Several compositions of the Tl-1223 phase were trialled instead; we settled on a composition often reported in the

[1] E-mail: d.pooke@irl.cri.nz

Figure 1. Prototype current lead, with 8 Tl-1223 bars, tested to 900A

literature [2], $Tl_{0.6}Pb_{0.4}Ba_{0.4}Sr_{1.6}Ca_2Cu_3O_{10}$, because it yielded promising J_c's at an early stage, and had a Tc above 115K.

Process development was restricted to simple solid-state reaction and sintering steps: in the first instance it was of interest to find the performance limits of untextured material, as this offers an economic route to sample fabrication. Processing was carried out in Au bags to reduce Tl loss. Ag paint was applied to the ends of the bars and sintered at 800°C for the formation of current contacts. The aim of the processing schedule was to obtain small densely packed grains of good phase purity, as it was observed that large grain growth was detrimental to J_c. Grain growth was notable at high temperatures, in compositions high in Ba (in which a melt phase can be accessed at lower temperatures), and also with the addition of Ag as reported in other works [3].

The performance of the bars shown in Fig. 2 is typical of the better bars produced. The improvement from the "intermediate" to the "prototype" bar resulted from the introduction of a post-anneal step of 60h at 915°C. This was found to raise T_c by around 2K, and to make the T_c less sensitive to subsequent annealing (for example, for the Ag contacts). The $J_c \sim 2500 A/cm^2$ at 77K in self-field is remarkable for untextured ceramic, and likewise the reduction in field is less than a factor of 10 to 0.5Tesla. In fact J_c (77K, $\mu_0H=0.1T$) compares well with values found for many textured ceramics indicating the existence of a network of strong links between grains. The ability to carry current at temperatures as high as 110K is a distinguishing feature of these bars. The properties of the best bars are summarised in Table 1.

3. Prototype current lead

A current lead of around 750A capacity was targeted. This would consist of eight element bars arranged symmetrically around a tufnel former. The end caps were machined in Cu and plated first with Ni, then with a Au layer to facilitate forming a low resistance contact to the

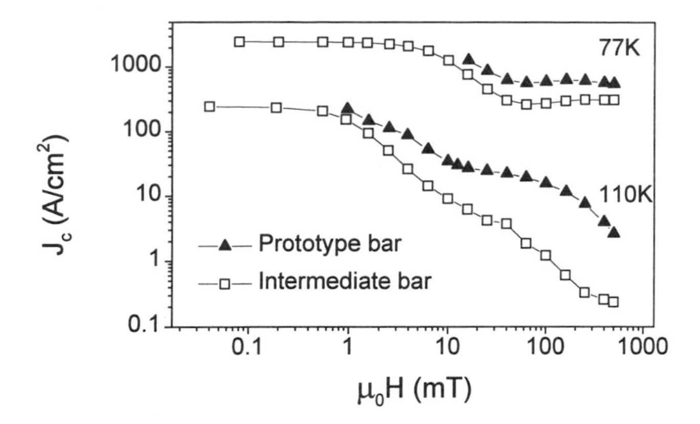

Figure 2. Field dependence of critical current at 77K and 110K for a bar produced at an intermediate stage of the development programme, compared with a bar prepared for the prototype lead. The low-field J_c of the latter exceeded the supply capacity of the current source at the time of measurement.

Table 1. Properties of bars in the prototype current lead

Property	Value
Size	20mmx10mm^2
Density	~88% theoretical
T_c	typically 115K
J_c at:	A/cm^2
77K, self-field	>2500
77K, 0.01T	1900
110K, self-field	270
110K, 0.01T	35
Estimated heat leak (assuming 1W/m-K):	
1000A, self-field, 77-4K	~300mW
200A, self-field, 110-60K	~200mW
200A, 0.01T, 110-60K	~1.4W
Contact resistance at 0.015, 77K	0.3μΩ
Contact heating at 77K, 200A	~10mW
Repeated plunging into liquid nitrogen:	no deterioration in J_c

ceramic bars. Each bar was soldered into these end caps with a low melting-point In-Ag solder. Torsional rigidity of the structure was provided by the tufnel core; one end cap was free to slide vertically on the core to allow for differences in the thermal expansion between the bars and the core. Additional Cu end-connectors provided for connection to a current supply.

Before assembly the J_c and T_c of each bar was measured. Some variability in bars produced in different batches was evident (Fig. 3a), but bars within a batch were consistent in performance

The completed composite lead was immersed in liquid nitrogen for evaluation. The device survived applied currents up to 900A. Critical current was measured, as shown in Fig. 3b, with voltage probes attached to each leg. It is evident that some legs did not share the current load, even at 900A and at the highest applied field. Absolute current capacity of

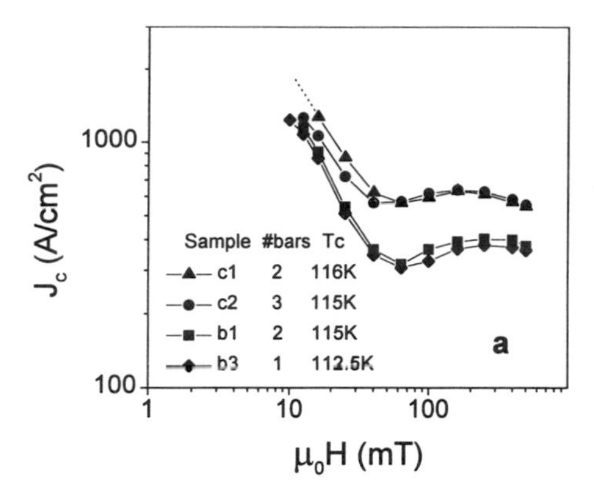

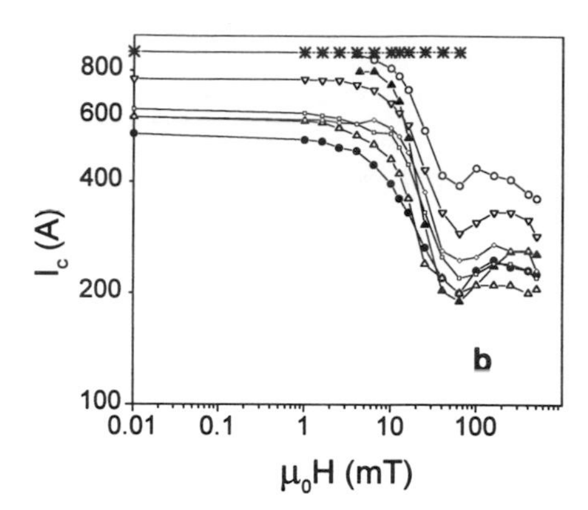

Figure 3. (a) The 77K field dependence of J_c of bars making up the prototype current lead. The number of bars from each batch is indicated in the legend. Bars from within a batch behave identically. (b) J_c (1 μV/cm criterion) of each leg of the current lead as a function of field. Variation between legs is caused by non-uniformity of contact resistances.

this prototype is thus limited by the poor current sharing between legs, presumably because of inconsistent contact resistances. Nevertheless, the device showed no degradation despite being overdriven to 900A, and displayed excellent mechanical properties, with no breakages of bars despite the thermal shocks imposed by direct plunging into liquid nitrogen. These results represent a successful demonstration of the technology.

4. Conclusions

We have developed a prototype current lead consisting of bars of untextured Tl-1223. The materials processing route is simple yet the critical currents achieved are comparable to many reported for textured samples. Operation at temperatures as high as 110K is a remarkable feature of the bars, offering a level of performance not found in other current leads. Issues of material uniformity, and of current contacts remain to be addressed.

Acknowledgements

The authors thank American Superconductor Corporation for their support in this project.

References

[1] Hull J R 1993 *IEEE Trans. on Appl. Supercond.* **3** 869-74
[2] Mair M, Künig W T and Gritzner G 1995 *Supercond. Sci. Technol.* **8** 894-9
[3] Park C, Bayya S S, Sriram D, and Snyder R L 1995 *Appl. Supercond.* **3** 139-146

Inst. Phys. Conf. Ser. No 158
Paper presented at Applied Superconductivity, The Netherlands, 30 June–3 July 1997

Fabrication issues in Bi-2212 polycrystalline textured thin rods for current leads

L.A. Angurel, J.C. Díez, H. Miao, E. Martínez, G.F. de la Fuente and R. Navarro

Departamento de Ciencia y Tecnología de Materiales y Fluidos, ICMA, CSIC-Universidad de Zaragoza, María de Luna, 3 50015-Zaragoza, Spain

Abstract. Up to 10 cm long Bi-2212 polycrystalline textured thin rods have been fabricated using a laser float zone melting system. The reproducible high values of the transport critical current, together with a low thermal conductivity and low resistance electrical contacts, have enable the development of current leads based on these materials. The influence of different growth conditions and fabrication parameters on the final properties of the textured samples is analysed. On 1 mm diameter rods, optimum critical currents densities at 77 K of 5.5 kA/cm^2 have been obtained.

1. Introduction

One of the most promising applications of Bi-Sr-Ca-Cu-O (BSCCO) bulk textured materials is the development of current leads. BSCCO materials, with appropriate texture, combine high critical current densities, J_c, and low thermal conductivities. Until now, commercially available current leads have been fabricated using BSCCO materials with major components of the Bi-2212 and Bi-2223 phases textured by mechanical deformation (BSCCO/Ag sheathed tapes) and Melt Casting Process (MCP) [1]. Using this last technique, large scale rings, rods and tubes are obtained reaching $J_c(77\ K) \approx 4\ kA/cm^2$.

The Laser Floating Zone technique (LFZ) [2] is an alternative texturing process adequate for the fabrication of cylindrical current leads. In this technique, a melting zone is induced on a long precursor (ceramic or powder pressed rod) by focusing a power laser beam. The initial use of a seed to start the growth and the stationary motion of the melted zone at controlled speed by control of the seed pulling and precursor feeding define the process. Due to the high thermal gradients induced at the solidification interface growth rates up to 50-100 mm/h are possible. The anisotropic BSCCO crystallites becomes textured with its *a-b* planes near parallel to the growth direction.

Up to this moment, best results with the LFZ technique have been reached on Bi-2212 textured materials. In spite of the higher T_c, BSCCO materials with the cathions stoichiometry 2223 are less adequate because the Bi-2223 phase can not be obtained directly from melting. Only after appropriate sintering process it is possible to recover this phase, but losing the induced texture [3]. In this work, the main fabrication parameters which control the final characteristics of LFZ Bi-2212 textured thin rods optimized for their application as current leads are reviewed.

2. Processing issues

The main parameters of the LFZ fabrication process of Bi-2212 rods with diameters between 1 and 3 mm for current leads are:

i) The characteristics of the precursor. It has been observed a great correlation between the precursor stoichiometry and the final T_c of the LFZ textured sample. Considering that in most applications, at least a part of the sample is going to work about 77 K, T_c should be increased as much as possible because the proximity of the working temperature would greatly reduce the J_c value. On samples grown from precursor stoichiometries that lead to T_c>90 K, the textured rods after annealing may recover these values leading to J_c(77 K) performances of technological interest. For a given stoichiometry the texture of the sample does not depend on the degree of reaction in the precursor; i.e., the same results are obtained for precursors formed by mixtures of unreacted oxides, by fully reacted Bi-2212 powders (isostatically pressed) or by ceramic bars [4].

ii) The Laser source. Two different continuous wave lasers have been used: a CO_2 (λ=10.6 μm) laser and a Nd:YAG (λ=1.06 μm). Due to the differences in the penetration depth, the second one should be more suitable in the processing of thicker samples [5].

iii) The growth process. Fixing the above conditions, the growth atmosphere, diameter and speed control determine the final texture and microstructure of the BSCCO rods. Along the fabrication, a controlled artificial dry air atmosphere has been used in the growth chamber. The BSCCO grains tend to form colonies piling up the platelet single crystals in the *c*-direction [6]. Lower growth rates and smaller diameters yields higher colonies size and better alignment. Parallel to the *a-b* planes, the colonies may reach dimensions of 2000x200 μm^2 on Bi-2212 rods of 1 mm of diameter and grown at 5 mm/h, which are reduced to 30x15 μm^2 for 70 mm/h. In the first case, the colonies have dimensions comparable to the sample and, in consequence, the cylindrical symmetry is lost. From these facts, it could be expected that the lower the growth rate the higher should be I_c, but this is not the case because it has also been observed that low growth speeds induce the appearance of cracks between the colonies which reduce I_c. As conclusion, the optimum growth rates correspond to intermediate values between 15 and 30 mm/h. The particular value depends on the focusing optical system that has been used to induce the melting zone, as well as of the rod diameter.

iv) The annealing process. For the used diameters, unless for very low growth rates, as-grown rods do not show superconductivity at 77 K. The primary solidification phase is the Bi-free $(SrCa)CuO_2$ oxide, appearing two Bi containing phases, one with composition close to the Bi-2201 phase and another to the Bi-2212. Upon short annealing times, for instance 12 h at 845 °C in air, the $(SrCa)CuO_2$ grains are reabsorbed and the more stable 2212 phase becomes majoritary. Most of the non-superconducting phases disappear and only small traces of $(SrCa)_2CuO_3$ and $(SrCa)_{14}Cu_{24}O_{41}$ appear at the surface of the sample. I_c(77 K) increases with the annealing time reaching a saturation value for long times between 60 and 100 h at 845 °C in air. This value is believed to be determined by the self-field at the surface of the sample. On samples of 1 mm of diameter an optimum value of I_c(77 K)= 46 A has been reached which reduces 2.5 times for growth rates of 70 mm/h. Upon increasing the rod diameters to 1.65 and 1.85 mm although the optimum I_c(77 K) increases to 75 and 90 A, respectively, J_c decreases.

v) The fabrication of contacts. Appropriate metallic low resistance contacts with the rods is a very important technological issue because of the possible energy losses caused by the flow of high intensities. Reproducible resistance contacts below 1 μΩ have been obtained with a silver painting process and copper metallic ends soldered with tin alloy as explained elsewhere [7]. These low values of the resistance contact have allowed the characterization of these samples at temperatures between 60 K and 77 K, range in which I_c>200 A have been reached.

3. Physical properties and performances

The homogeneity of the superconducting properties has been checked on a 10 cm rod by placing a series of voltage contacts at 1 cm intervals. The values of I_c(77 K) deduced with the usual 1 μV/cm criteria have been represented in Fig. 1a. It has been observed that the differences between different sections of the sample is lower than a 20 %. Only when the final end of the measured sample is very close to the metallic chuck that fixes the precursor, there is a systematic change with improved performances in this region. This is caused by the increased thermal gradient in this position.

The irreversibility line on these textured samples, determined from magnetic measurements, follows a dependence $B_{irr}(T)=B_0(1/t-1)^2$ ($t=T/T_c$), which compared with the single crystal behaviour is shifted towards higher temperatures and fields. This shows that additional intergranular pinning mechanisms play an important role as it has been confirmed by the analysis of the distribution of pinning energies derived from magnetisation relaxation measurements [8]. Annealing time also shifts the irreversibility line enlarging the region of applicability but reaching saturation simultaneously with I_c(77 K). It should be noted that, during the annealing, the I–V(77 K) characteristics for intensities above I_c may be scaled one upon the other using this value, showing that the pinning mechanisms are not modified in the annealing process and there are only improvements of the intergranular strength connectivity [6].

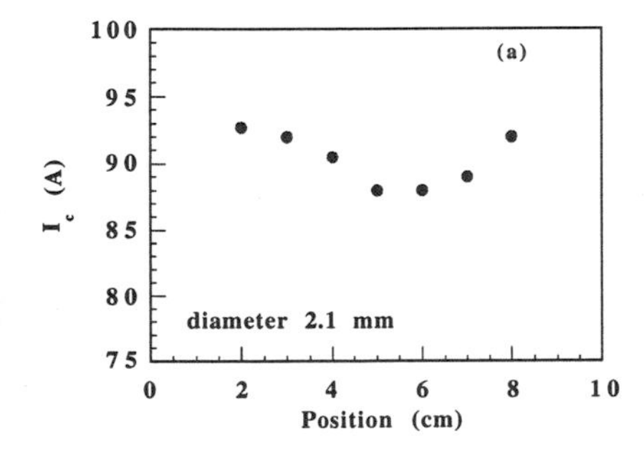

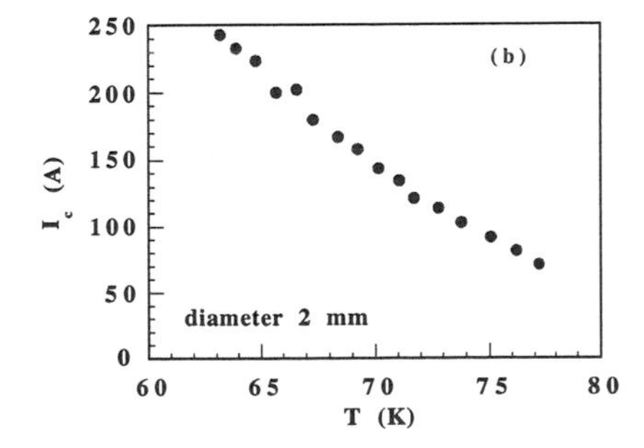

Figure 1: (a) Homogeneity of I_c in a sample with a legth of 10 cm. (b) Temperature dependence of I_c in a sample with a diameter of 2 mm.

The proximity of T_c and the working temperatures produces an important dependence of the transport $I_c(T)$ as is displayed in Fig. 1b. It has been observed that when T_c>90 K, and in consequence the values of I_c(77 K) are improved, I_c(65 K) ≈ 3 I_c(77 K), and the same relation has been determined between I_c(50 K) and I_c(65 K) from magnetic measurements. This temperature dependence is strongly related with the irreversibility line considering that the pinning mechanisms are associated to the surface of the superconducting grains.

A major problem associated with applications of Bi-2212 based materials at high temperatures is the strong field reduction of $I_c(H)$. At 77 K an exponential dependence, $I_c(H)=I_c(0)\exp(-H/H_0)$, has been obtained. The characteristic field, H_0 is higher in samples with better transport properties, although always below 200 Oe [6]. In the low temperatures regime, however, this field dependence losses sensibility changing to a potential one. The crossover from one to the other behavior take places at a temperature which in optimum cases are about 60 K.

From the performances on these Bi-2212 rods where no special cautions have been taken as well as considering the fabrication characteristics, it is concluded the very real possibility of its use in applications as current leads where only low magnetic fields would be present. However, additional effort should be performed concerning the thermal and mechanical stabilization by using appropriate composites of these materials.

Acknowledgments

The financial support of the Spanish Ministry of Education and Science (CICYT 95-0921-C02-01 and 02) and R.E.E. is acknowledged.

References

[1] Elschner S, Bock J, Brommer G and Herrmann PF 1996 *IEEE Trans. on Magnetics* **32** 2724-2727

[2] Angurel L A, De la Fuente G F, Badía A, Larrea A, Diez J C, Peña J I, Martínez E and Navarro R 1997 *Studies of High Temperature Superconductors* vol 21 (New York: Nova Sciece Publishers, Inc.) p 1

[3] Miao H, Díez J C, Angurel L A, Peña J I and De la Fuente G F 1997 *Solid State Ionics* (in press)

[4] Miao H, Díez J C, Angurel L A and De la Fuente G F 1997 *to be submitted*

[5] De la Fuente G F, Díez J C, Angurel L A, Peña J I, Sotelo A and Navarro R 1995 *Adv. Mater.* **7** 853-856

[6] Díez J C, Martínez E, Angurel L A, Peña J I, De la Fuente G F and Navarro R 1997 *to be submited*

[7] Mayor J M, Angurel L A, Navarro R and García-Tabarés L 1997 *in this conference*

[8] Martínez E, Angurel L A, Díez J C, Lera F and Navarro R 1996 *Physica C* **271** 133-146

Inst. Phys. Conf. Ser. No 158
Paper presented at Applied Superconductivity, The Netherlands, 30 June–3 July 1997

Vapour Cooled High Tc Superconducting Current Leads made of Ag sheathed Bi 2223 tapes

B. Zeimetz, H. K. Liu, S. X. Dou

Centre for Superconducting and Electronic Materials, University of Wollongong, Wollongong NSW 2522, Australia

Abstract. We have developed 1-2 kA current leads for a Superconducting Magnetic Energy Storage (SMES) System. Our design combines the high efficiency of vapour cooling with the engineering advantages of silver sheathed tapes. We report on design details of the leads and preliminary performance measurements.

1. Introduction

Current leads have become the first commercially available product made of bulk High Temperature Superconductors (HTS), due to the fact that the Critical Current Densities J_c required are not very high, and the flexibility of the conductors is not a critical issue. The crucial parameter of a current lead is the heat leak per current Q/I, which has to be minimized [1][2]. For a conduction cooled HTS lead it is given by

$$Q / I_c = 1 / (J_c * L) * \int_{T\,min}^{T\,max} \kappa(T) dt$$

While the length L is usually determined by the size of the cryogenic system (and consideration of cryogenic stability [1]), one has to maximize the Critical Current Density J_c and/or to minimize the thermal conductivity $\kappa(T)$.

Numerous efforts to produce efficient HTS current leads have basically diverged into two directions: One can either produce bulk rods or tubes of HTS such as YBaCuO(123) [3][4][5], Bi2212 [6][7] or Bi2223 [4][7][8], with a rather low J_c, but very low $\kappa(T)$ [9]. Another possibility is to utilize silver sheathed HTS tapes [10][11][12], with high J_c and high $\kappa(T)$ (of the silver sheath). In the latter case the thermal conductivity can be drastically reduced by using alloyed silver [12].

However, the comparison must not be limited to physical properties Jc vs. κ, but include some engineering considerations. Important advantages of tapes are:

- low resistance contacting (joining) is achieved by conventional soldering (while it is a major problem for 'bulk' HTS)
- tapes can be aligned such that the influence of an external magnetic field on J_c is minimized
- silver works as mechanical and chemical protection
- danger of quenching is strongly reduced due to high thermal/electrical conductivity of silver

So far, the Current Leads based on HTS tapes have been designed as Conduction Cooled leads (i.e. they are only cooled at the lower end), because e. g. the tapes were cast into resin for the purpose of mechanical support [11]. In this work we present a design allowing Vapour Cooling, which enables a much higher cooling efficiency [1][2].

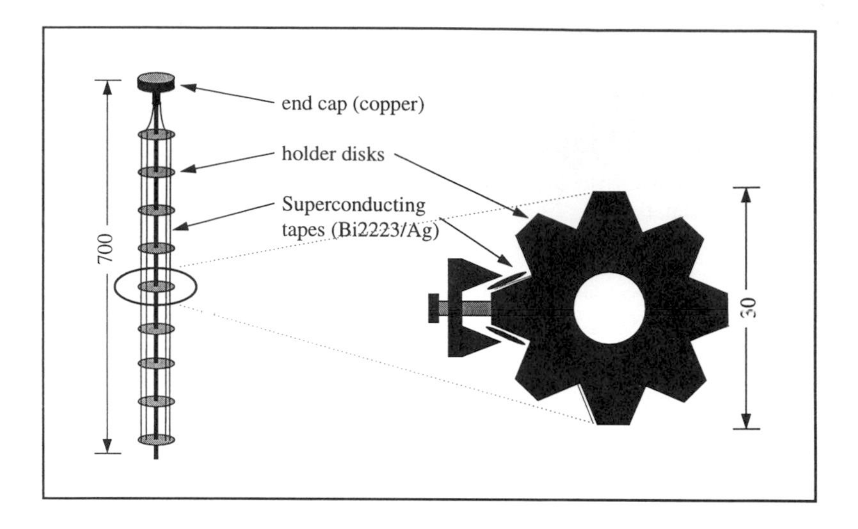

Fig. 1: Schematic sketch of current lead (left; without tube) and holder disks (right) for clamping of HTS tapes

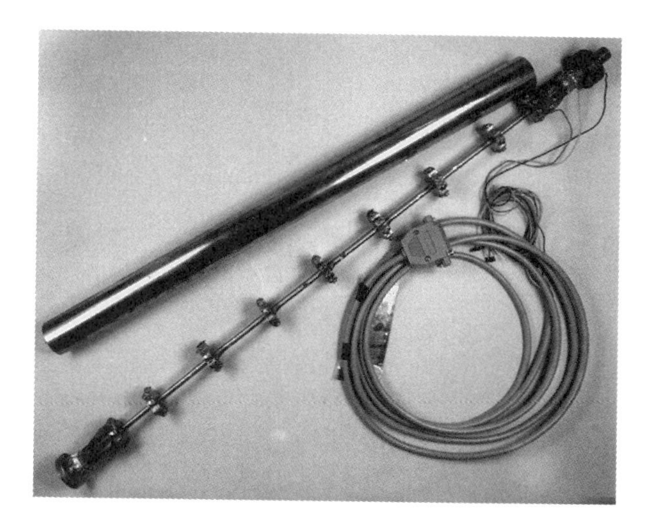

Fig. 2: Photograph of current lead with steel tube, voltage leads and copper end caps

2. Current Lead Design

Fig. 1 shows a drawing, and Fig. 2 a photograph of our lead. A steel rod is at the core of the lead, with steel disks along its length. The tapes are attached to the disks with steel clamps.

The entire construction is enclosed in a steel tube, which serves as mechanical protection as well as a guide for helium vapour. Hence, the lead operates with Vapour Cooling of the tapes along the entire lenght.

The clamps at the ends of the tapes are made of copper instead of steel. They are simultaneously serving as electrical connection, and the tapes can be soldered to them. This results in very low contact resistance. The end caps of the lead are made of copper too. With four threads at each cap, they can be screwed to other components of the cryogenic system.

Our design with screwed clamps ensures maximum flexibility; the lead can be optimized (in terms of current capacity) for any given configuration by addition, removal or exchange of individual tapes.

Furthermore, the number of tapes can be varied along the length, with more tapes at the high temperature end (taking into account the strong $J_c(T)$ dependance), as was shown by other groups [11]. The actual temperature distribution along the lead depends on the specifications of the cryogenic system, and in particular on the design of the normal conducting part of the current lead.

3. HTS Tape Performance

While the leads have not yet been tested at full capacity, we report on preliminary performance tests on individual tapes.

To maximize the critical current at minimum heat leak we follow several strategies.

1. Alloyed silver material (currently being tested) is to be used as sheath material, reducing the thermal conductivity by 1-2 orders of magnitude.
2. We specialize in producing tapes with very 'thin' sheath (typical cross section ratio 50%, compared to 75% for 'usual' tapes).
3. Thin sheaths tend to crack during rolling, therefore such tapes cannot be rolled to very small thickness, which is desirable for high current densities J_c. The relatively low J_c of thick tapes (typically 300 μm) can be strongly improved by adding silver particles with appropiate size to the core, as we have reported elsewhere [13].

Fig. 3a shows Critical Current versus Temperature for a tape, and Fig. 3b Magnetic Field dependance of Critical Currrent at 77 K. The tape has a a cross section of 300 μm x 4 mm, with a fill factor of around 50 %, and 25 w% silver added to the core.

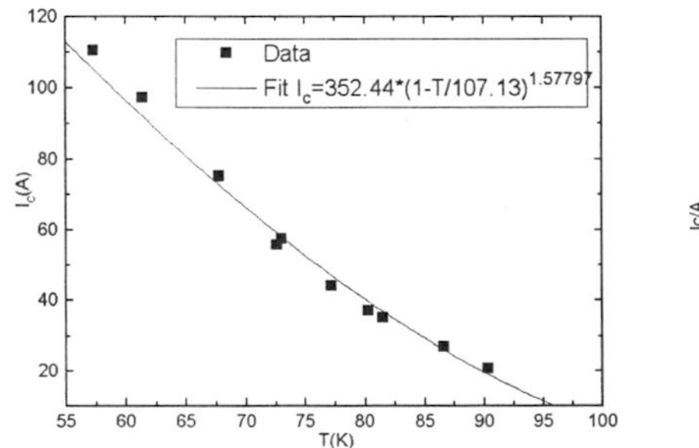

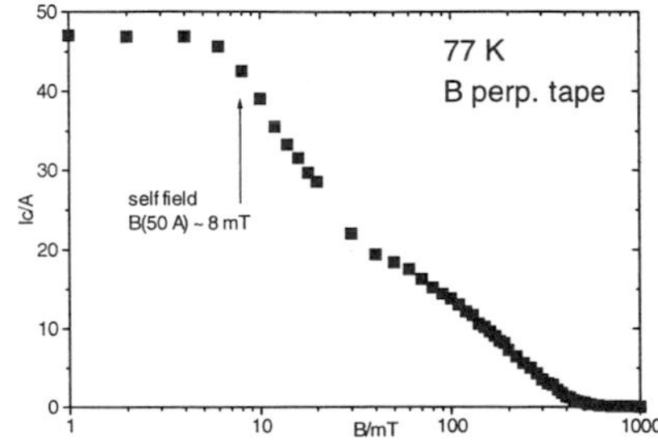

Fig. 3: a) $I_c(T)$ in self field for a single tape (300 μm x 4 mm; 1 μV/cm)
b) $I_c(B)$ characteristics at 77 K of a similar tape; the plateau at low fields indicates a strong self field effect

4. Conclusions

Our current lead design shows two distinctive features:
- vapour cooling of HTS tapes
- maximum flexibilty of current capacity (and optimum heat leak per current Q/I), due to the fact that tapes can be added or changed according to demand.

The maximum capacity of the other parts of the leads (copper connections etc.) is designed for 1000 - 2000 A; the actual capacity will be tested in the near future.

Acknowledgments

The authors gratefully acknowledge the financial support of the Australian Research Council, Metal Manufacturers Ltd., the Commonwealth Department of Industry, Science and Tourism, the Energy Research and Development Corporation, and the Energy Supply Association of Australia Ltd.
We are very grateful to E. Babic and I. Kusevic of the University of Zagreb for extensive $I_c(B,T)$ measurements.

References

[1] M. N. Wilson, Superconducting Magnets, Clarendon Press Oxford 1983
[2] J. R. Hull, Cryogenics 29 (1989) 1116
[3] R. C. Niemann, Y. S. Cha, J. R. Hull, W. E. Buckles, B. R. Weber, S. T. Yang, Proc, of CEC/ICMC 95, Columbus, July 17-21, 1995
[4] D. Ponnusamy, Z. Li, K. Ravi-Chandar, IEEE Trans. Appl. Supercond. 5 (1995) 769
[5] N. Alford, T. W. Button, S. J. Penn, P. A. Smith, IEEE Trans. Appl. Supercond. 5 (1995) 809
[6] J. Bock, H. Bestgen, S. Elschner, E. Preisner, IEEE Trans. Appl. Supercond. 3 (1993) 1659
[7] Herrmann, P. F. et al., IEEE Trans. Appl. Supercond. 3 (1993) 876
[8] T. Hasebe, T. Tsuboi, K. Jikihara, S. Yasuhara, J. Sakaraba, M. Ishihara, Y. Yamada, IEEE Trans. Appl. Supercond. 5 (1995) 821
[9] C. Uher, J. Supercond. 3 (1990) 337
[10] R.C. Niemann, Y. S. Cha, J. R. Hull, C. M. Rey, K. D. Dixon, IEEE Trans. Appl. Supercond. 5 (1995) 789
[11] T. Kato, K. Sato, T. Masuda, T. Shibata, Y. Hosoda, S. Isojima, S. Terai, T. Kishida, E. Haraguchi, in: T. Fujita, Y. Shiohara (Ed.s), Advances in Superconductivity IV, Vol. 2, Springer, Tokyo 1994 (Proc. of ISS 1993)
[12] H. Fujishiro, I. Manabu, K. Noto, M. Matsukawa, T. Sasaoka, K. Nomura, J. Sato, S. Kuma, IEEE Trans. Magn. 30 (1994) 1645
[13] B. Zeimetz, H. K. Liu, S. X. Dou, Supercond. Sci. Techn. 9 (1996) 888

Inst. Phys. Conf. Ser. No 158
Paper presented at Applied Superconductivity, The Netherlands, 30 June–3 July 1997

Development of High-Tc Bi-Superconductor Current Leads for Maglev

Eiji Suzuki and Minoru Kurihara

Railway Technical Research Institute, Kokubunji–City, Tokyo 185, Japan

Abstract. We have developed high–Tc current leads having a 600 A current capacity with no cooling gas for maglev system. These leads composed of high–Tc superconductor with Bi–2223 wires sheathed in Ag–Au alloy could realize low heat leakage property and less consumption of liquid helium during energization and de–energization. We have further investigated the characteristics of these leads combined with an actual superconducting coil.

1. Introduction

The current leads for maglev superconducting magnets were conventionally cooled by a large amount of helium gas to reduce the rise of temperature in themselves during energization and de–energization. And this operation has caused a large disturbance in the refrigerating system of maglev. Further the heat leakage through these leads made of copper was not negligible. Therefore we developed the low heat leakage current leads using high–Tc superconductor with Bi–2223 wires sheathed in Ag–Au alloy which had a 600 A transmission capacity even without cooling gas.

We have at first investigated the current carrying capacity and the low heat leakage properties of these leads working at the temperature between about 5 K and 80 K and next examined the characteristics of these leads in the practical use combined with an actual superconducting coil.

2. Performance of high-Tc current leads

2.1. Structure of current leads

These high–Tc current leads are composed of 11 pieces of Ag–Au alloy sheathing Bi–2223 wires of a 430 mm length which are so stacked that the critical current will be more than 600 A. They are fixed on an FRP by an epoxy resin and copper terminals are provided at both ends. Table 1 lists the specific items of these leads.

Table 1 specific items of high–Tc current leads

superconductor	Bi2223	thickness × width × length(each)	
material of sheath	Ag–10at%Au	=0.45 × 6 × 430mm	
wire	61 multi–core wire	number of lamination	11 sheets

2.2. Current-carrying test

A test was done to see whether these leads of which both cold ends were joined by the short NbTi wire could carry or not the operating current in a practical use according to a pattern in which the ramp rate was 5 A/s and the time holding 600 A was two minutes. The current leads are surrounded with the FRP pipes in order to prevent them from being cooled by the vaporized helium gas simulating the state with no cooling gas [1].

We observed no difference between the temperatures in these leads at 0 A and 600 A. No voltage was also generated through each lead conductor in carrying a current of 600 A except a sufficiently small one due to the joint resistance at the cold copper terminals (0.115mV). The heat leakages into liquid helium were 0.25 W/leads at 0 A, and 0.33 W /leads at the operating current of 600 A, which corresponded respectively to about half of those in the conventional copper leads [2].

These leads which need no cooling gas can eliminate any disturbance to the refrigerator and simplify the refrigerating system of maglev.

3. Test with the leads combined with a superconducting coil

3.1. Main features of the combined superconducting coil

For the purpose of investigating the characteristics of these current leads in a rather strong magnetic field and their behaviour in an abnormal state (i. e. quenching) of a superconducting magnet, we tested them as combined with an actual superconducting coil. The superconducting coil having a persistent current switch (PCS) with which we combined these leads is one shaped like a race track having a width of 1070 mm and a height of 500 mm similar to the one used in the Yamanashi Maglev Test Line and is wound in 1400 turns with a superconducting wire of NbTi having a rectangular cross section of about 2 mm × 1 mm. The magnetomotive force is 700 kA (the current of 500 A), the self inductance is 2.72 H and the storage energy is 340 kJ.

3.2. Results of the combined tests

Through the results of the tests combining these current leads with a superconducting coil, we could get the following facts [3].

(1) We could recognize no occurrence of voltages and no rise of temperatures in the

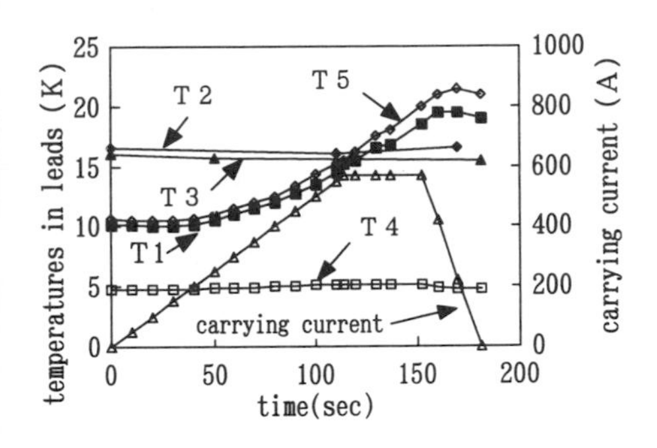

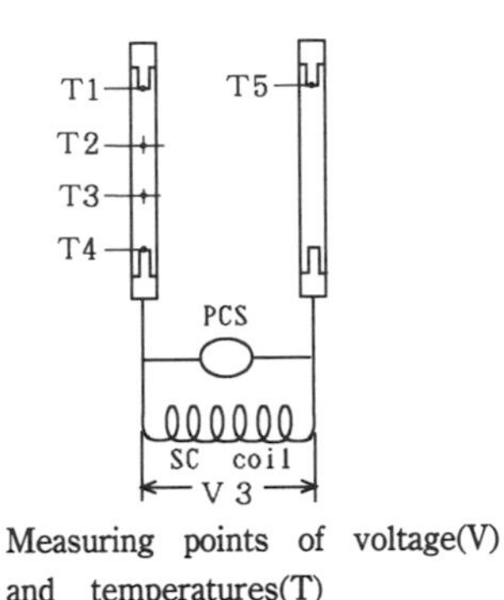

Measuring points of voltage(V) and temperatures(T)

Figure 1 Variation of temperatures in the leads when the SC coil is energized 800 kA

lead conductors when we carried ultimately a current of 600 A through the closed PCS at the respective ramp rates of 5, 10 and 20 A/s.

(2) It was proved that the high−Tc current leads are able to successfully operate without breaking their superconductivity during the energization up to a magnetomotive force of 800 kA (the current of 571 A). The strength of magnetic field which affects the lower parts of the current leads is about 0.33 T.

The variation of temperatures in the leads is shown in Figure 1 when the energization was executed. In this diagram, we recognize that only the temperatures at the nearest points (T1, T5) to the hot ends rise by about 10～12 degrees in these operations. The initial temperatures of these points were lower than those of the middle parts because the temperatures of the hot ends were apt to vary by the test condition.

It can be seen that the mean volume of liquid helium consumed during the energization or de−energization in the case of the high−Tc current leads is below half that in the conventional leads made of copper meshes.

(3) We performed four tests diminishing the magnetic strength by opening PCS artificially under the persistent current mode of a magnetomotive force of 700 kA. The current leads could stably operate with a smooth decay of current every time.

(4) Tests of inducing a coil quenching

The tests of inducing the quenching of the superconducting coil energized up to 700 kA were executed three times. In this case, we could induce the superconducting

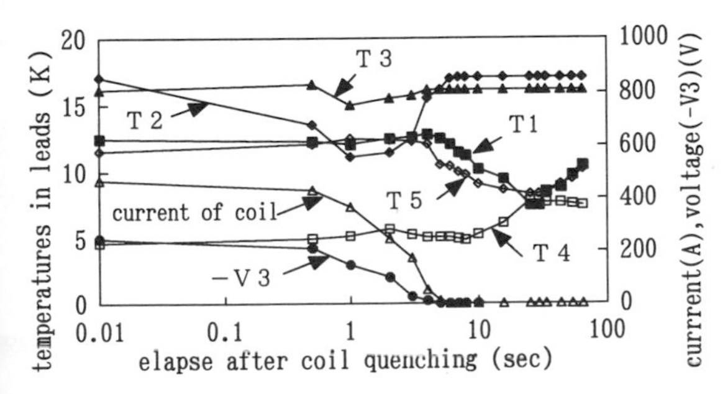

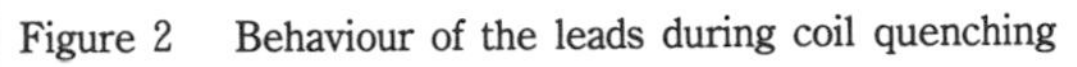
Figure 2 Behaviour of the leads during coil quenching

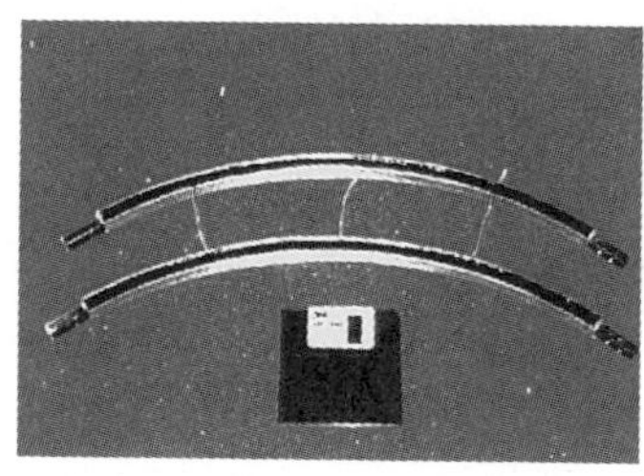
Figure 3 Warped current leads incorporated with an SC coil

coil to quench because of the decaying rate of current accelerated by changing the resistance joining the both ends of the coil from 0.06 Ω to 0.56 Ω.

Figure 2 shows the immediate behaviour of the leads during the coil quenching. Although the temperature at each point of the leads varies complicatedly, the generation of voltages through the lead conductors did not take place and the stability of superconductivity in the high-Tc current leads could be ensured. The maximum voltage (V3) generated between both ends of the superconducting coil was −275 V.

4. Vibration test of warped current leads

On the basis of the above mentioned results, we developed warped high-Tc current leads incorporated with a superconducting coil as shown in Figure 3 and confirmed the capacity of carrying a current of 571 A in the strong magnetic field of 1 T without cooling gas.

Further we submitted these warped leads to a severe vibration test carrying a current of 500 A through the closed PCS while the superconducting coil was vibrated within the range of frequencies 80∼400 Hz by an oil actuator. It was proved that the leads could endure without quenching the same acceleration of 200 $m/s^2{}_{p-p}$ as the superconducting coil suffers in actual running.

5. Conclusions

We have developed high-Tc Bi-superconductor current leads with a 600 A transmission capacity which need no cooling gas and showed their low heat leakage properties. According to the results of tests, the heat leakage and the consumption of liquid helium can be reduced to about half that in the case of the conventional copper leads. We also energized a magnetmotive force of 800 kA by combining these leads with the actual superconducting coil and induced a quenching of the coil to simulate the abnormal state of maglev in running.

Further a vibration test with carrying a current of 500 A was executed using another leads of warped shape under the magnetic field of 1 T. In every case these current leads could successfully operate without breaking their superconductivity even under no gas-cooling.

This research is subsidized by the Ministry of Transport, Japan.

References

[1] T. Kato et al., Proceedings of the 9th Int. Symposium on Superconductivity (ISS' 96), p184, Oct. 1996

[2] M. Kurihara et al., the 55th Meeting on Cryogenics and Superconductivity, E3-15, p270, Nov 1996

[3] E. Suzuki et al., the 55th Meeting on Cryogenics and Superconductivity, E3-16, p271, Nov. 199

Inst. Phys. Conf. Ser. No 158
Paper presented at Applied Superconductivity, The Netherlands, 30 June–3 July 1997

Properties of Bi (2223)/Ag Au Multifilamentary Tapes for Current Leads

W Goldacker, B Ullmann, A Gäbler, R Heller

Forschungszentrum Karlsruhe, Institut für Technische Physik,
Postfach 3640, D-76021 Karlsruhe, Germany

Abstract. BSCCO(2223) tapes for the application in current leads for superconducting coils need replacement of the pure Ag sheath through a low thermal conduction material as AgAu alloy to achieve reduced thermal losses compared to conventional current leads. 7-filamentary BSCCO tapes with Ag92Au8 alloy sheaths were prepared and critical current densities J_c up to 11 $kAcm^{-1}$ at 77 K were achieved. The $J_c(B,T)$ behavior indicates a nearly pure Bi(2223) phase, but compared to Ag sheathed tapes a stronger J_c degradation with fields presumably due to a less good texture of the BSCCO phase. The mechanical data characterized by I_c vs. axial tensile stress investigation and a special distortion experiment demonstrate that AgAu sheaths are mechanically weaker than Ag sheaths with an unsufficient performance for technical applications. These tape properties will be discussed with respect to the application aspects in a 1 kA current lead.

1. Introduction

For Bi(2223)/Ag tapes, the choice of modified sheath materials or combination of two different materials is very restricted, since the oxygen permeability of pure Ag is a sensitive boundary condition for the formation of the BSCCO phase during the final heat treatment of the tape [1]. But modifications of the sheath for mechanical strengthening, enhanced electrical resistivity and low thermal conductivity are necessary for the different applications in coils, power cables and current leads. Important is, that the material properties are improved at the operation temperature of the devices. For the application of BSCCO tapes in current leads for superconducting coils, a low thermal conductivity of the tape is required for the temperature range 4.2 K - 77 K. The new sheath should, if possible, replace the pure Ag completely, which requires chemical digestibility with the superconductor. As best choice, AgAu alloys were proposed [2] which can be used with up to ≈ 15 at.% Au content. Alloying Ag with Au reduces the thermal conductivity of Ag as function of Au content up to a factor of 10^{-3} at T = 20 K for > 11 at.% Au content and a factor of < 10^{-1} at T = 77 K. The addition of Au to the Ag sheath has no significant chemical consequence for the reaction of BSCOO in direct contact with the sheath [3]. Current leads out of this material will surely not be used in every device for cost reasons, but especially for large coils and devices, as the coils for the ITER fusion reactor, these costs can be covered by reduced cooling and refrigeration costs. The presented paper reports on the preparation and characterization of multifilamentary BSCCO/AgAu tapes, which were applied in a 1 kA prototype current lead [4]. Of special interest were the mechanical properties of the tape.

2. Experimental

For the preparation of the BSCCO/AgAu tapes 8 at.% Au content was used as compromise between costs and improved material properties. To insure sufficient mechanical stability and avoiding the usual but in this case more risky multifilamentary bundling process, a 7 filamentary conductor was produced, starting from a 25 mm diam. AgAu rod, drilled with seven 5.5 mm diam. holes of 120 mm length. The precursor (Merck, pyrolized nitrate solution) was filled directly and compacted by a piston and closed with Ag rods. Deformation was made by swaging with intermediate annealings (300 °C, 20 min). The 1.8 mm thick wire was rolled to the tape by means of a four roll machine (side limitation). The final tape dimensions were 2.4 x 0.19 - 0.21 mm after a two stage heat treatment at 818 °C/25 hrs. in 8% O_2/Ar. The bending radius during annealing was 90 mm. About 280 m tape were produced. The superconductor content was finally 18% due to the powder filling method without CIP compaction.. Characterization of the tapes was performed investigating phase content (X ray), resistivity, critical temperature T_c, critical current density $J_c(T,B)$ and especially the J_c degradation with axially applied strain and under distortion around the tape axis.

Fig. 1: Cross section of 7-filamentary BSCCO(2223)/AgAu tape

3. Results

The deformation behavior of the BSCCO/Au tapes was quite good applying 10% reduction steps and intermediate annealings. The filaments showed slight sausaging (see fig. 1) sensitively depending on the tape rolling procedure. The critical parameter of the final mechanical-thermal treatment was the rolling step suffering from the softness of the annealed AgAu. Therefore the achieved critical current varied from 6.5 to 9.5 A, corresponding with maximum critical current density J_c = 11000 Acm^{-2}. The resistive T_c measurement in fig. 2 for the first and second annealing step gives a final T_c = 108 K indicating no chemical reaction with the matrix. Also the corresponding $I_c(T)$ curves show that after the second annealing mainly (2223) phase was obtained as verified by X ray powder diffraction (> 95% (2223) phase), too. The critical current reached 30 A at T = 25 K for an average tape (see Fig. 3). The dependence of the critical current on magnetic field (see fig. 3) shows a stronger degradation compared to standard Ag sheathed tapes. The reason for this behavior may result from a worse texture due to the non ideal filament cross section and a too thick filament size for optimised currents. A bad texture quality, as 10 - 12° grain misalignment around the c-axis (FWHM) determined from the orientation dependence of Ic(B), leads consequently to contributions of the in plane field component, for which I_c degrades much

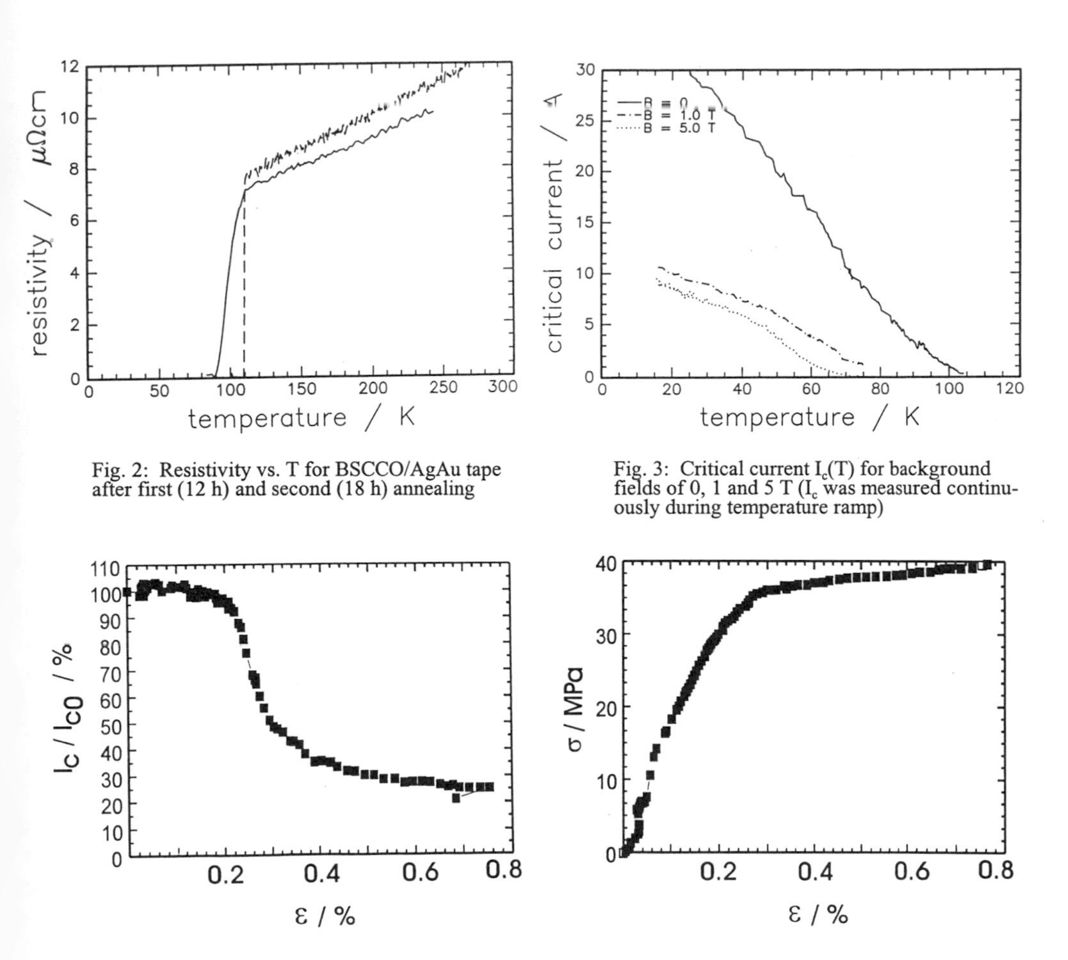

Fig. 2: Resistivity vs. T for BSCCO/AgAu tape after first (12 h) and second (18 h) annealing

Fig. 3: Critical current $I_c(T)$ for background fields of 0, 1 and 5 T (I_c was measured continuously during temperature ramp)

Fig. 4: Critical current I_c and axial stress vs. axial applied tensile strain of a BSCCO/AgAu tape

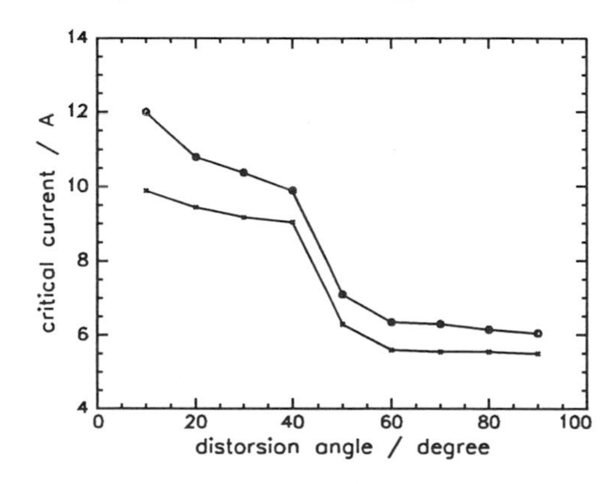

Fig. 5: Critical current as function of longitudinal tape distortion of a 2 tape staple (tape 1 and tape 2) of 110 mm length

stronger with magnetic field. In fig. 4 the degradation of the critical current with axially applied strain is given measured on a precision, gauged strain rig. Irreversible I_c degradations occur for strains $\varepsilon > 0.18\%$ corresponding to a stress of < 35 MPa both values being 10 - 15% smaller compared to Ag sheathed tapes ($\sigma = 40$ MPa, $\varepsilon \geq 0.2\%$) [5]. The Young's modulus E = 16 GPa, is also smaller than E = 20 GPa in Ag sheathed tapes. Therefore AgAu behaves significantly softer than Ag. Applying the BSSCO/AgAu tapes in the current lead design, proposed in ref. [4], requires a distortion of the tape staples around their longitudinal axis with a twist pitch of one meter (twist pitch = length of 360° turn), a design to prevent damage from thermal stresses during the cooling cycles. Fig. 5 gives the result from a distortion experiment made with a two tape staple of 110 mm length, where both tapes were measured separately. Strong irreversible I_c degradations occur for distortion angles > 40° corresponding to a twist pitch of 0.9 - 1 m. This proves that applied in the current lead device the BSCCO tapes have only a very small mechanical reserve for additional applied strains which justifies the lead design preventing thermal stresses [6].

4. Conclusions

The application of low thermal conductive Ag92Au8 material in 7-filamentary BSCCO(2223) tapes was demonstrated and current densities of $J_c = 11$ kAcm^{-2} obtained which have potential for further significant improvement, since reference Ag-sheathed tapes from the same precursor batch reached $J_c > 30$ kAcm^{-2}. Reasons for the limited J_c are the too small filament number and a too large tape size and filament thickness resulting from compromises in the tape preparation technique due to cost limitation. Significant improvement are expected for higher filament numbers of 19 or 37 filaments and applying the usual bundling technique of hexagonal monocores. For technical applications with advantages against conventional Cu current leads, increased current densities are necessary, but as important higher tolerable stresses of ≈ 100 MPa are required for save operation and to enhance the mechanical reserve for occurring strains. This can only be achieved by mechanically reinforced tapes, using a stronger second component beside the very soft AgAu in the sheath.

Acknowledgements

This work has been performed within the frame of the European Fusion Technology Programme. The valuable experimental assistance of H. Orschulko, S. Zimmer and R. Bussjaeger should be acknowledged.

References

[1] Quilitz M and Goldacker W, to be publ. in *Superc. Science Techn.*

[2] Fujishiro H, Ikebe M, Noto K, Matsukwa M, Sasaoka T, Nomura K, Sato J, Kuma S, MT13 Conf., Victoria, Canada 1993

[3] Nomura K, Sasaoka T, Sato J, Kuma S, Kamakura H, Togano H and Tomita N, *Appl. Phys. Lett.* **69** (1 1994, p. 112

[4] Heller R, Friesinger G, Goldacker W, Kathol H, Ullmann B , Fuchs A M, Jakob B, Pasztor G, Vécsey and Wesche R, pres. Appl. Superc. Conf. Pittsburgh P.A. USA, Aug. 25th - 30th, 1996, to be published in *IEEE Trans. on Appl. Superc.*

[5] Ullmann B, Gäbler A, Quilitz M, Goldacker W, presentation 14 BR-4, EUCAS, Eindhoven, June 30th July 3rd 1997, to be published in Proc.

Inst. Phys. Conf. Ser. No 158
Paper presented at Applied Superconductivity, The Netherlands, 30 June–3 July 1997

Application of inductive HTSC current limiters in distribution networks

V Meerovich[1], V Sokolovsky[1], S Goren[1], G Jung[1], I Vajda[2], A Szalay[3], N Gobl[3]

[1]Physics Department, Ben-Gurion University, P.O.B. 653, 84105 Beer- Sheva, Israel
[2]Department of Electrical Machines, Technical University of Budapest, Budapest, Hungary
[3]Metalltech Ltd., Budapest, Hungary

Abstract. The application of an inductive HTSC fault current limiter in distribution network is considered. The basic principles of choice of the device parameters are formulated. The limiter parameters are calculated for two designs of the device. The first of them is the design employing a closed magnetic system with an air gap, the second is based on a magnetic system consisting of only a rod without yokes.

1. Introduction

The successes in technology of superconductors make the creation of a superconducting fault current limiter (FCL) the reality in the very near future. One promising design is an inductive limiter based on high temperature superconductors (HTSC) [1,2]. The FCL consists of a primary normal metal coil coupled via a ferromagnetic core to a secondary short-circuited superconducting coil assembled in the form of a set of HTSC rings or cylinders.

Some power FCL prototypes have been built and successfully tested [2]. Large HTSC rings and cylinders with high critical current density are available to build FCLs for distribution networks of 6-15 kV (50-70 cm in diameter). An actual problem is to find the effective location of the device in a power system and to determine the FCL parameters required for its successful operation. In the present paper, we consider the sample of the FCL application in a power system and show the method of the determination of the FCL parameters for an actual location in the network.

2. Utility requirements to FCL parameters

A real inductive superconducting FCL has a non-zero impedance under normal network operation determined by the leakage fluxes of the coils and the resistance of the primary coil. The impedance rapidly increases (before the first fault current peak) under a fault and the FCL should return to the initial state in a short time after a breaker interrupts fault current. The impedances under normal and fault conditions, the activation current, the activation and recovery times are the basic "external" parameters of an FCL. The acceptable values of these parameters lie within the limits defined by an electric power network. To determine FCL parameters under fault conditions we consider the diagram of currents as shown in Fig. 1.

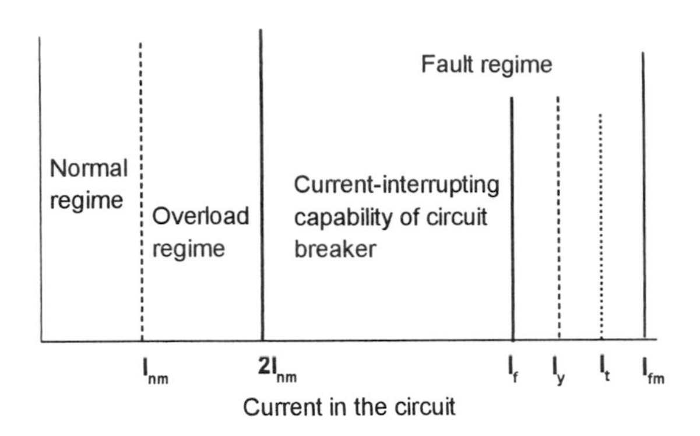

Fig. 1 Diagram of currents in power network.

FCL must limit the steady-state component of a fault current below the current-interrupting capability $\mathbf{I_f}$ of the circuit breakers and the first peak - below the admissible value $\mathbf{I_y}$. For the FCL impedance $\mathbf{z_p}$, the limited steady-state current is expressed as:

$$\mathbf{I_L} = \mathbf{U_f} /(\mathbf{z_k}+\mathbf{z_p}) < \mathbf{I_f} \tag{1}$$

where $\mathbf{U_f}=\mathbf{U_n}/\sqrt{3}$ is the phase voltage; $\mathbf{z_k}$ is the circuit impedance under fault condition.

The value of the maximum fault current is determined by the FCL impedance $\mathbf{z_p}$ and the activation current $\mathbf{I_a}$. Previous experiments [1,2] have shown two possible operation regimes: a slow transition extending for several periods and a quick transition during the first half-period. They refer to different relations between the heating time of the superconductor up to the critical temperature $\mathbf{t_h}$ and the time of transient current attenuation in a superconductor $\mathbf{t_e}$. Using the mathematical model of the FCL [3], we have estimated the maximum current in the circuit containing the FCL in two limiting cases: $\mathbf{t_e} << \mathbf{t_h}$ and $\mathbf{t_e} >> \mathbf{t_h}$:

$$\mathbf{max(I)} \approx \sqrt{2}\cdot\mathbf{I_L}\left(1+\frac{\mathbf{z_k}+\mathbf{z_\sigma}/2}{\mathbf{z_k}+\mathbf{z_p}}\right)+\mathbf{I_a}\left(1-\frac{\mathbf{z_k}+\mathbf{z_\sigma}/2}{\mathbf{z_k}+\mathbf{z_p}}\right)<\mathbf{I_y}, \tag{2}$$

$$\mathbf{max(I)} \approx 2\sqrt{2}\cdot\mathbf{I_L}-\mathbf{I_a} < \mathbf{I_y} \tag{3}$$

where $\mathbf{z_\sigma}$ is the FCL impedance when the secondary coil is in the superconducting state. In the first case, the temperature was assumed to be constant during the first half-period. In the second case, we considered an abrupt resistance change of the superconductor to the normal state resistance value at the activation current.

Fig. 1 shows that the activation current has to fall within the limits between the amplitudes of the admissible overload current $2\mathbf{I_{nm}}$ and the current capability of circuit breakers. On other hand, a steady-state limited fault current must also be higher than the maximum current of the admissible overload. Otherwise, the control equipment does not "see" a fault condition. From this follows that the current of activation has to be in the range:

$$2\cdot\sqrt{2}\cdot\mathbf{I_{nm}}\cdot\mathbf{U_f}/(\mathbf{U_f} - 2\cdot\mathbf{z_p}\cdot\mathbf{I_{nm}}) < \mathbf{I_a} < \mathbf{I_f}\cdot\sqrt{2}\,. \tag{4}$$

The expressions (1)-(4) determine the range of the FCL external parameters.

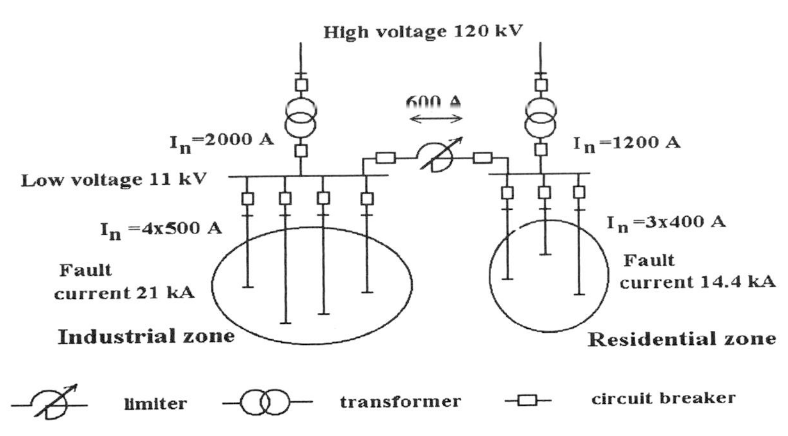

Fig. 2 Schematic of two distribution substations with connecting tie.

3. FCL application in a distribution network

As an example, we consider a distribution network with the configuration and parameters shown in Fig. 2. The network consists of two separate substations: the first supplies an industrial zone and the second supplies a residential area. The development of the regions requires to build new lines of high voltage and new substations. Another way of increasing the power supply is the one by interconnecting the buses of the substations. This connection is possible because the consumption of power in these regions is separated in time. In industrial zones the consumption reaches the maximum value during a work-day, in the residential areas the consumption peaks occur during the morning and evening hours. However, the connection of the substations leads to the increase of fault currents up to 35.5 kA which is more than the current-interrupting capabilities of the circuit-breakers installed in this network. Thus, the connection would need the replacement of all breakers and other equipment and devices of the substations.

To solve the problem of fault currents in interconnected system, one can install FCLs. The installation of FCLs in an every feeder branch requires a lot of devices, rebuilding of both substations and increase in the substation area. To use FCLs between transformers and buses, they must be designed for large currents and high nominal power. Moreover, this FCL location may be problematic because the FCL may influence the normal regime of the circuit. The most effective FCL location is the connecting tie (Fig.2). Since the weakest place of the system is the circuit-breakers of the residential area (current-interrupting capability of 20 kA), the fault current in the connection must be limited to the value less than 5.5 kA. For this, the impedance of the current limiter under fault condition must be more than 0.9 Ohm. The activation current can be taken as high as 3 kA.

As an example of practical performance of the device, the parameters were calculated for two various designs of the magnetic system (Fig.3) [4]. The first is an open magnetic system including only a rod without yokes. The second design is based on a closed magnetic system with air gaps. Fig. 4 shows the core diameter and the number of turns as functions of the limited current, in the region determined by expressions (1)-(4). With the increase of the limiter impedance, the number of turns in the primary coil increases but the diameter of the

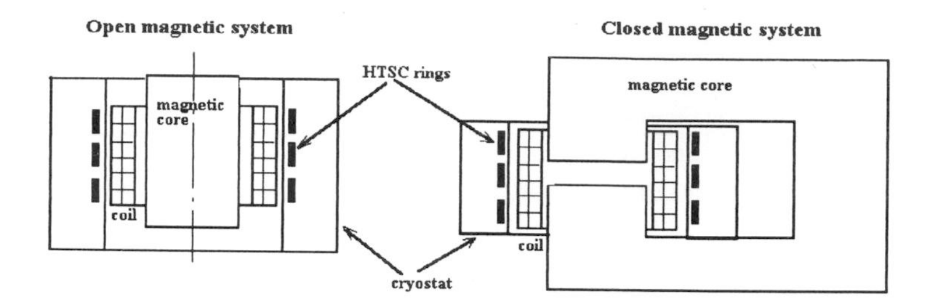

Fig. 3 Schematic of FCL design.

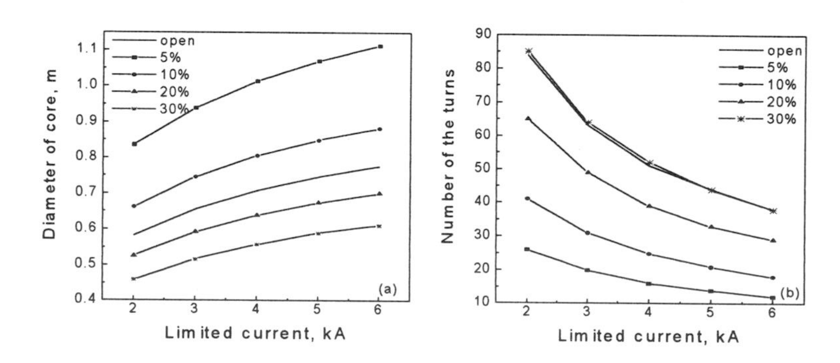

Fig. 4 (a) Core diameter and (b) number of turns in the coil as functions of limited fault current for various percent ratios of air gap/core diameter (closed magnetic system) and for open magnetic system.

magnetic core decreases. The weight of the device decreases, but the critical current of the superconducting rings increases proportional to the number of turns. The comparison of two magnetic systems shows that, at small air gaps, the number of turns for the closed magnetic system is less than that for the open magnetic system and the diameter is greater. With the increase of the air gap these relationships reverse. It should be noted that even at equal core diameters the weight of the closed magnetic system is more than four times larger than the weight of the open magnetic system. The question is what is simpler to fabricate: the rings with high critical current and with smaller diameter or the rings with lower critical current.

We conclude that the case investigated above presents the example of the most effective application of FCLs in distribution substations: it allows one to change the electric network without rebuilding substations.

Acknowledgements

This work was supported by the Ministry of Science of Israel and the National Research Fund, National Scientific Fund of the Hungarian Academy of Sciences: contracts T022513 and T019857.

References

[1] Meerovich V, Sokolovsky V, Jung G and Goren S 1995 *Proc. EUCAS'95 , Inst. Phys. Conf. Ser.* **148** 603

[2] Giese R F 1995 *Fault Current Limiters - A Second Look* (Argonne National Laboratory:Report)

[3] Meerovich V, Sokolovsky V, Jung G and Goren S 1995 *Proc. Eighteenth convention of IEEE* Report **1.5.4**

[4] Vershinin Y N, Meerovich V M, Naumkin I E, Novikov N L and Sokolovsky V L **1989** *Electrical Technology USSR (UK)* **1** 1

Inst. Phys. Conf. Ser. No 158
Paper presented at Applied Superconductivity, The Netherlands, 30 June–3 July 1997

Transition properties of HTS inductive fault current limiter

M Majoros,[1] L Jansak,[1] S Sello,[2] S Zannella[2]

[1] Institute of Electrical Engineering, Slovak Academy of Sciences, Bratislava, Slovak Republic
[2] CISE SpA, Segrate (Milan), Italy

Abstract. Transition properties of an inductive fault current limiter consisting of a primary copper winding, a high - T_c superconducting ring and an iron core were analyzed. A mathematical model taking into account the dynamic current - voltage characteristic of the superconductor as well as the non linearity of the iron core magnetic characteristic was adopted. Obtained results were compared with experimental data. The possible mechanism of self - oscillations observed in circuits with HTS fault current limiters is proposed.

1. Introduction

High temperature superconductors (HTS) have been considered for current limiting devices based on the fast transition from the superconducting state to the resistive one when the current overcomes the critical value. Recent developments have already shown the feasibility to realize inductive fault current limiters using HTS in the form of hollow cylinders or stacked rings [1]. Recently, self-oscillations in circuits with high-T_c superconducting current limiters were observed [2,3]. In this work the limiting properties of an inductive fault current limiter consisting of a primary copper coil with N_1=400 turns, a HTS cylinder and an iron core are analysed. A mathematical model [4] taking into account the full dynamic current-voltage characteristic of HTS as well as the non linear behaviour of the iron core was adopted to evaluate its transition properties. The possible mechanism of self-oscillations is proposed.

2. Mathematical model

The circuit diagram of the limiter shown in Fig. 1 can be described by the following equations

$$R_{10}I_1 + \frac{d\Phi_1(I_1)}{dt} + \frac{d\Phi_{12}(I_2)}{dt} = V_o\,sin(\omega t) \qquad (1a)$$

Table 1: Parameters a,b,c in eq. (4) of the non linear primary, secondary and mutual fluxes.

	a	*b*	*c*
$\Phi_1(I_1)$	5.2	7.9	74
$\Phi_{12}(I_2)$	4.68	3160	74
$\Phi_2(I_2)$	0.013	3160	74
$\Phi_{21}(I_1)$	0.0117	7.9	74

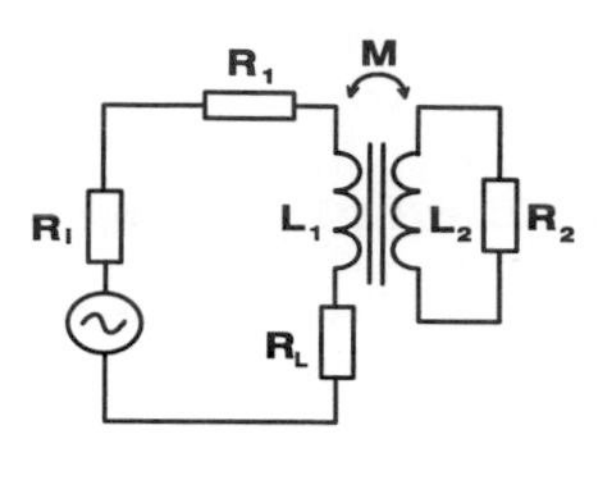

Fig.1: Circuit diagram of the fault current limiter.

$$R_2(I_2,B,T)I_2+\frac{d\Phi_2(I_2)}{dt}+\frac{d\Phi_{21}(I_1)}{dt}=0 \tag{1b}$$

where $V_o \sin\omega t$ is the voltage produced by the power supply, $\omega=2\pi f$, where f is the frequency, $R_{10}=R_1+R_i+R_L$ is the total primary circuit resistance, (R_i - internal resistance of the power supply, R_1 - resistance of the primary coil, R_L - load resistance). I_1, I_2 are the primary and secondary currents, respectively. R_2 (I_2, B, T) is the non linear resistance of the superconductor which depends on the current I_2 and also on magnetic field B and temperature T. $\Phi_1(I_1)$ is the magnetic flux of the primary coil produced by primary current and $\Phi_2(I_2)$ is the magnetic flux of the secondary ring produced by the secondary current. $\Phi_{12}(I_2)$ is the magnetic flux of the primary coil generated by the secondary current and $\Phi_{21}(I_1)$ is the magnetic flux of the secondary ring generated by the primary current. Introducing the differential self and mutual inductances $L_1(I_1)=d\Phi_1(I_1)/dI_1$, $M_{12}(I_2)=d\Phi_{12}(I_2)/dI_2$, $L_2(I_2)=d\Phi_2(I_2)/dI_2$ and $M_{21}(I_1)=d\Phi_{21}(I_1)/dI_1$ the equations (1) can be written in the compact form:

$$R_{10}I_1+L_1(I_1)\frac{dI_1}{dt}+M_{12}(I_2)\frac{dI_2}{dt}=V_o\sin(\omega t) \tag{2a}$$

$$R_2(I_2,B,T)I_2+L_2(I_2)\frac{dI_2}{dt}+M_{21}(I_1)\frac{dI_1}{dt}=0 \tag{2b}$$

In addition to $R_2(I_2,B,T)$, the equations (1), (2) contain another non linear terms Φ_i, Φ_{ji} and L_i, M_{ij} (i,j=1,2) which depend on the iron core B(H) characteristic. In the present model [4] we suppose $R_2(I_2,B,T)$ to be a dynamic characteristic which can be approximated by the following Fermi-like function:

$$R_2(I_2)=R_{2n}/\left(\exp\left[\frac{I_{2c}-|I_2|}{\Delta I_2}\right]+1\right) \tag{3}$$

where R_{2n} is the HTS normal state resistance, I_{2c} is the critical current of HTS ring corresponding to the resistance $R_{2n}/2$ and ΔI_2 is a parameter characterising the width of the transition. The non linear behaviour of the iron core, neglecting its hysteresis, was fitted by the following function

$$\Phi(I)=aI/(b+c|I|) \tag{4}$$

with a,b,c as the fitting parameters. Because of the highly non linear functions (3), (4) the solution of an initial value problem for the system of two coupled non linear ordinary differential equations (1), (2) belongs to so called stiff problems and special numerical procedures must be adopted [4]. We used a third-order semi-explicit Runge-Kutta

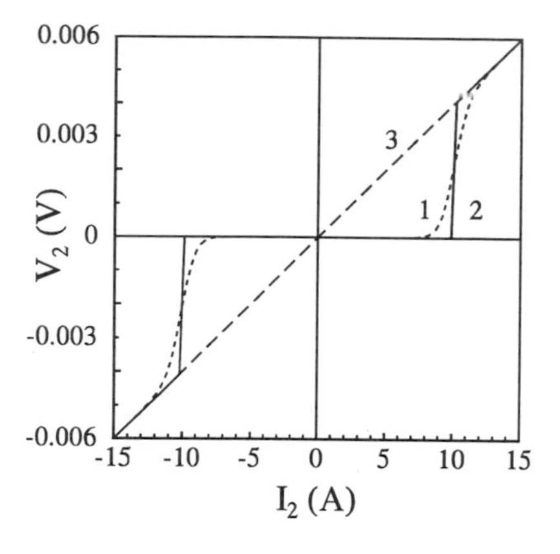

Fig. 2: Current-voltage characteristics with two different transition widths. 1-R_{2n}=0.4 mΩ, I_{2c}=10 A, ΔI_2=0.5 A, 2-R_{2n}=0.4 mΩ, I_{2c}=10 A, ΔI_2=0.015 A, 3 - Ohms' law

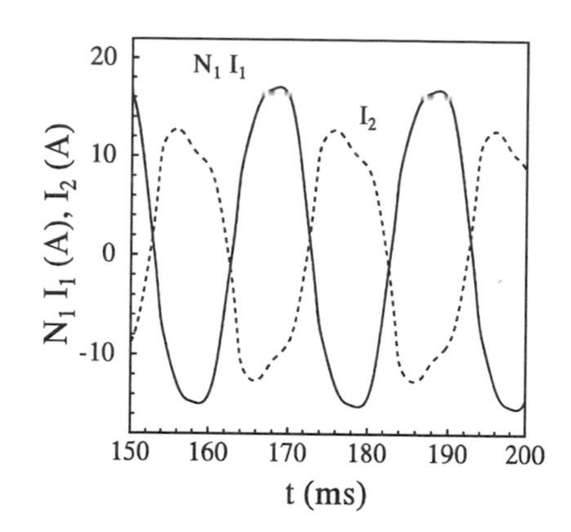

Fig. 3: Primary and secondary currents for current-voltage characteristic 1 in Fig.2. (V_o=2.5 V, f=50 Hz).

Rosenbrock-type algorithm [5]. The computational scheme was designed to garantee unconditional stability properties and to minimize at the same time the local truncation error. This scheme enables to make calculations with very steep transition of the HTS ring.

3. Results and discussion

On the basis of experimental results [2] the parameters a, b, c of eq. (4) of the iron core were evaluated (Tab. 1). The influence of two current-voltage characteristics with different transition widths (curves 1,2 in Fig. 2) on the primary and secondary currents were studied. Their normal state resistivities and critical currents were taken from experiment [2]. In Fig. 3 we show the time dependences of the primary and secondary currents in steady state, i.e. after decay of the transient effect connected with the turning on the power supply, for the current-voltage characteristic of the HTS ring with rather broad transition (curve 1 in Fig. 2). Making the transition characteristic of HTS ring more steep (curve 2 in Fig. 2) causes a blip in $I_2(t)$ dependence which is transformed into the primary current $I_1(t)$ as an abrupt decrease in form of a peak (Fig. 4). This effect is similar to onset of instabilities observed in [2] with YBaCuO melt-textured rings of the same critical current and normal state resistance. Also the position of the peak in primary current corresponds to the experiment [2] in appearing only on increasing part of the primary current when it is positive and only on decreasing part for negative values of $I_1(t)$ (compare Fig. 5b in [2]). Increasing the voltage V_o causes the shift of the peak in the primary current towards its maximum (Fig. 5). Observed instabilities in $I_1(t)$ [2] consisting of several peaks can be understood considering the current-voltage characteristic with several steep transitions, each corresponding to one peak in $I_1(t)$. Such a current-voltage characteristic can be caused by dynamic conditions when melt - textured YBCO samples with several quite large grains [2] are used. The reason can be the redistribution of the induced current related to the transition of the weak links at grain boundaries.

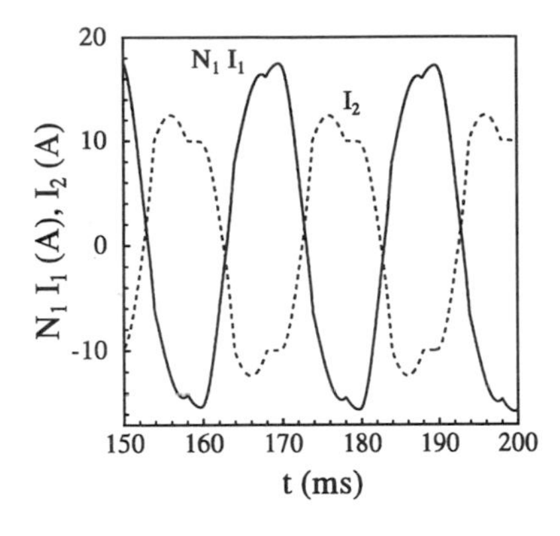

Fig. 4: Primary and secondary currents for current-voltage characteristic 2 in Fig.2 (V_o=2.5 V, f=50 Hz)

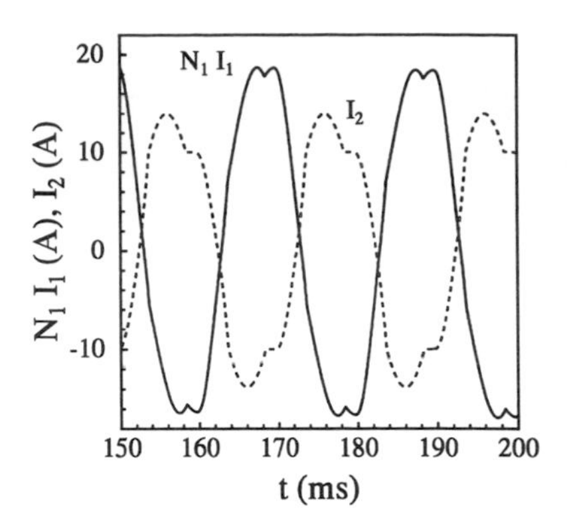

Fig. 5: Primary and secondary currents for current-voltage characteristic 2 in Fig.2 (V_o=2.8 V, f=50 Hz)

4. Conclusions

Using a mathematical model enabling calculations with very steep transition characteristics of the HTS ring, the properties of an inductive fault current limiter were analysed. A peak in the primary current appears when a fast transition characteristic of HTS ring was considered and its position corresponds to the onset of self-oscillations experimentally observed [2,3]. Increasing the power voltage the peak moves towards the maximum of the primary current. An explanation based on assumption of multiple rapid transitions of the weak links at grain boundaries was proposed as a possible mechanism of self-oscillations in circuits with HTS current limiters [2,3] (the work is in progress).

Acknowledgements

Financial support from COPERNICUS contract No ERBCIPA-CT94-0185 is gratefully acknowledged.

References

[1] Paul W, Rhyner J, Baumann Th, Platter F, *Applied Superconductivity* (IOP Publishing Ltd) (1995) 73-76

[2] Jansak L, Majoros M, Frangi F, Zannella S, *Physica C* 247 (1995) 231-238

[3] Meerovich V, Sokolovsky V, Shter G E, et al., *Physica C* 275 (1997) 119-126

[4] Majoros M, Jansak L, Sello S, Zannella S, *Transient Analysis of HTS Inductive Fault Current Limiter* Presented at ASC 1996 to be published on IEEE Transactions on Applied Superconductivity

[5] Sello S, *A semiexplicit Runge-Kutta method for the solution of ordinary differential equations* Proceedings of the 6th SAS - World Conference FEMCAD-89 Gournay sur Marne France (1989) 361-368

Inst. Phys. Conf. Ser. No 158
Paper presented at Applied Superconductivity, The Netherlands, 30 June–3 July 1997

Development of high T_c superconducting elements for a novel design of fault current limiter.

A T Rowley*, F C R Wroe*, M P Saravolac†, K Tekletsadik†, J Hancox‡, D R Watson#, J E Evetts#, A Kursomovic# and A M Campbell#

*EA Technology, Capenhurst, Chester CH1 6ES, UK.
†Rolls-Royce Peebles Electric Limited, East Pilton, Edinburgh EH5 2XT, UK.
‡Rolls-Royce Applied Science Laboratory, PO Box 31, Derby DE24 8BJ, UK.
#IRC in Superconductivity, University of Cambridge, Madingley Rd., Cambridge, CB3 OHE.

Abstract. The operation of a superconducting fault current limiter (SFCL) relies on the increase in circuit impedance produced when a superconducting element changes from its superconducting to its normal state during a fault. In this work, composite reaction textured Bi-2212 has been modified to optimise its material properties specifically for use in a SFCL. The normal state resistivity, ρ_n, and rate of change of resistivity in the flux flow region, $d\rho_f/dt$, are shown to be particularly important, together with the mechanical and thermal properties of the material. A single phase test SFCL has been constructed to characterise the performance of this material during simulated faults.

1. Background

Fault current limiters (FCLs) prevent unacceptably large surges in high power systems such as those caused by lightning strikes or short circuits. The need for FCLs is associated with the continuous growth and interconnection of modern power systems which results in a progressive increase in the number of short circuits to levels far beyond their original design capacity. The present practice for limiting fault currents, based on conventional technology, is costly and reduces the reliability and flexibility of power system operation.

The rapid transition of a superconductor from its superconducting to its normal state can be used in the design of a fault current limiter[1-3]. Superconducting FCLs are designed in such a way that a significant increase in circuit impedance occurs automatically if a fault occurs. Depending on the design, this impedance rise can be due to an increase in circuit eactance (inductive FCL) or circuit resistance (resistive FCL).

Conceptually, the resistive FCL is the simpler of the two, with a superconducting element switching from effectively zero resistance to some finite value when the critical current of the element is exceeded. In addition to providing sufficient resistance to limit the current, the superconducting element has to be able to absorb without damage the energy, $^2R\Delta t$, dissipated in it before the line circuit breaker opens.

The resistive FCL developed in this work is a novel one in which the fault current miting action uses a combination of current, magnetic field and temperature to achieve a fast ansition. The magnetic field is also used to achieve a uniform quenching of the uperconductor, which helps to solve local overheating problems and to reduce electromagnetic forces. In general, the resistivity, ρ, of the superconducting material can be expressed a functions of FCL design parameters and material properties (1)

$$\rho(T,B,J,t) = a\left(B(t)+b\right)^{m}\left(J(t)-J_{c}\right)^{n} + c\Delta T(t) \qquad (1)$$

where T, B, J and t have their usual significance and where a, b, c, m and n are constants.

Composite reaction texturing (CRT) of $Bi_2SrCa_2Cu_2O_{8+\delta}$ (Bi-2212) is a process by which the necessary texture for relatively high critical current densities, J_c, can be produced in large bulk superconducting artefacts[4-6]. Furthermore, by altering the processing conditions and/or the starting material composition, it is relatively easy to alter the superconducting and normal properties of Bi-2212 superconducting material manufactured in this way. In this work, suitably modified superconducting elements made of this material are assessed for their use in a resistive fault current limiter.

2. Experimental

2.1. *Material development*

Unlike other superconductor applications, where the focus of material development is to produce components with a high critical current in a high magnetic field, the resistive FCL has different requirements which any materials development programme must address. The normal state and flux flow resistivities, ρ_n and ρ_f, are clearly extremely important. The sensitivity of ρ_f to magnetic field is also important as this can be used to aid switching of the elements. Furthermore, since the highest resistivity is reached when the element is heated above its critical temperature, T_c, it is important to have T_c as close to that of liquid nitrogen as possible whilst still maintaining an acceptable critical current density.

Careful control of the process conditions during composite reaction texturing can affect the degree of alignment of the superconducting grains, and hence determine J_c and its sensitivity to magnetic fields. Similarly, control of a post melt-processing annealing stage can vary the oxygen concentration of Bi-2212 and the ratio of ρ_n to J_c. Furthermore, the use of different doping materials in the superconductor precursor powders also allows T_c, J_c and ρ_n to be adjusted. The object of the materials development programme is to optimise these parameters for the use of Bi-2212 within a resistive FCL.

2.2. *Laboratory scale fault current limiter*

A laboratory scale prototype fault current limiter has been developed to evaluate the performance of superconducting material during realistic fault conditions (figure 1). The power supply is a single phase 400V, 15A system which can produce a peak fault current of approximately 500A at 400V for ~100ms before a line protection circuit breaker opens.

The fault current limiter elements are made of CRT Bi-2212 rods, connected togethe in series to increase the total resistance during a fault. This assembly of elements is mounte within a magnet. To simulate a fault, the normal resistive load is temporarily shorted out for pre-determined time (typically 10-50ms). The current through the system, and the voltag across one or more elements is recorded before, during and after the fault.

As the current increases after the onset of a fault, the magnetic field applied to th superconductors also increases - this enhances the superconducting to normal transition an ensures uniform switching of the elements.

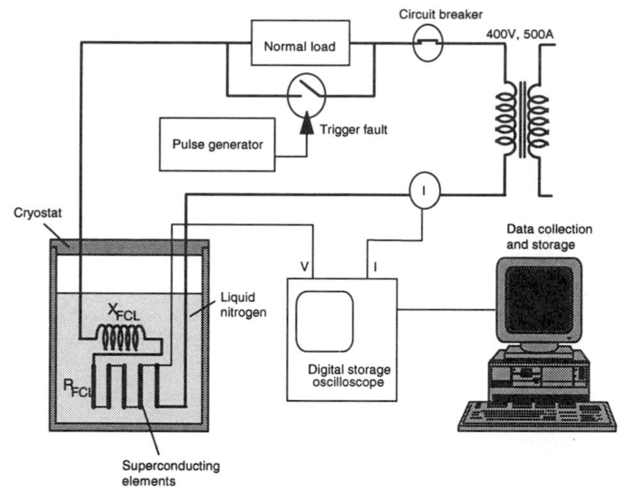

Figure 1: Schematic diagram of laboratory-scale fault current limiter

3. Results and Discussion

Figure 2 shows the variation in voltage measured across a single superconducting element when a 350A fault is applied for one half cycle (10ms). Before the fault is applied, the voltage across the element is ~zero, and rises as the current increases. About half way through the fault, heating of the material becomes significant and the voltage increases rapidly. After the fault, the temperature of the element is still above T_c, and hence a voltage remains. This voltage falls away as the element cools back down to below T_c.

The variation of element resistance corresponding to this fault is shown in figure 3. Here, the form of the resistance development is more apparent. In the first half of the fault, the resistance increases relatively slowly as the current increases. This is due to the variation of flux flow resistivity with current (and hence time, $\rho_f(t)$). However, if $d\rho_f/dt$ is large

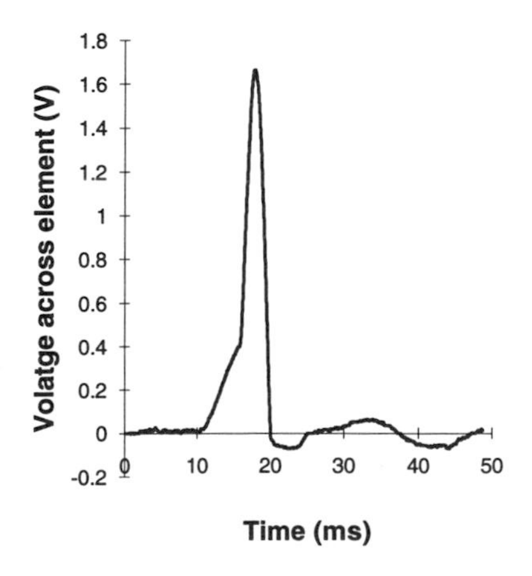

Figure 2: Variation of voltage across single FCL element with time during a 10ms fault

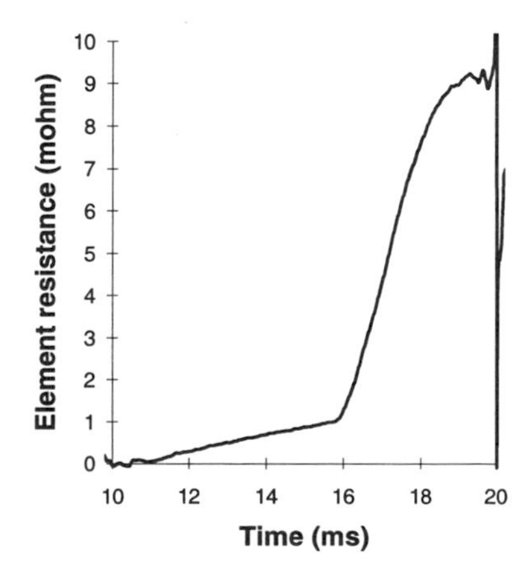

Figure 3: Variation of FCL element resistance with time during a 10ms fault

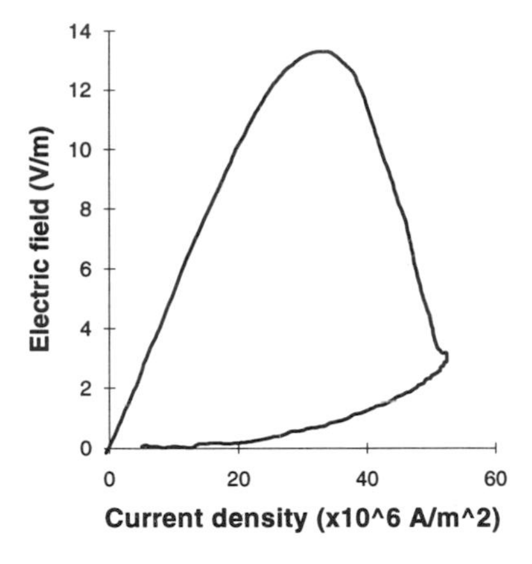

Figure 4: E-J characteristic of FCL element material during 10ms fault

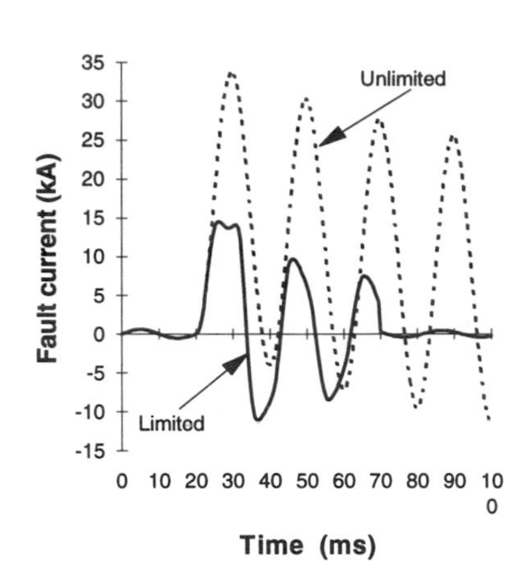

Figure 5: Projected full scale FCL current limiting performance

enough, I^2R heating will become significant, and the normal state resistivity will be reached (due to T_c being exceeded). This results in much higher resistances ($\rho_n >> \rho_f$). If ρ_f does not increase sufficiently quickly as the current increases, then not enough heat will be dissipated within the element, and the maximum resistance reached will be too small to limit the current.

In order to predict how this material will behave in a full scale device, the results are re-plotted in terms of the E-J characteristic for the material (where $\rho(t)$ is given by $E(t)/J(t)$). Such a characteristic is shown in figure 4, and can be used to model the operation of a full scale fault current limiter. A computer simulation of FCL operation is shown in figure 5.

4. Conclusions

Superconducting elements made from suitably modified (in terms of the values of J_c, ρ_n, T_c, ρ_f, and $J_c(B)$) CRT Bi-2212 material can be used in the design of a resistive fault current limiter. Such a device can limit the fault current within the first 10ms of a fault occurring. The high resistivity of the material is achieved only when the FCL elements are heated above T_c. However, before any significant heating will occur, the flux flow resistivity (due to J_c being exceeded) must increase beyond a threshold value.

Acknowledgements

This work is part funded by the DTI through the LINK Enhanced Engineering Materials programme. Two of the authors (DRW and AK) gratefully acknowledge the receipt of an EPSRC grant. The precursor superconductor powder was supplied by Merck UK Ltd..

References

[1] Paul W, Rhyner J, Baumann T and Platter F 1995 *Proc. EUCAS '95* 73-78
[2] Kleimaier M and Russo C 1995 *Proc. EUCAS '95* 615-618
[3] Cave J R, Willen D W A, Nadi R, Zhu W and Brisette Y 1995 *Proc. EUCAS '95* 623-6
[4] Watson D R, Chen M and Evetts J E 1995 *Supercon. Sci. Technol.* **8** 311-316
[5] Soylu B et al. 1992 *Appl. Phys. Lett.* **60**(25) 3183-3185
[6] Soylu B at al. 1993 *IEEE Trans. Appl. Supercon.* **3**(1) 1131-1134

Inst. Phys. Conf. Ser. No 158
Paper presented at Applied Superconductivity, The Netherlands, 30 June–3 July 1997

Development of an Inductive High-T_c Superconducting Fault Current Limiter Model

S. Zannella[1], A. Arienti[2], G. Giunchi[3], L. Jansak[4], M. Majoros[4] and V. Ottoboni[1]

[1] CISE SpA, P.O.Box 12081 - 20090 Milan, Italy; [2] Politecnico di Milano, Milan, Italy; [3] Edison SpA, Milan, Italy; [4] Institute of Electrical Engineering, SAV, Bratislava, Slovakia

Abstract. The Fault Current Limiter (FCL) is regarded as one of the first power applications of high-T_c superconductors, cooled at 77 K, that will be developed in the real size. A 4 kVA inductive FCL model, based on commercially available Bi-2212 shielding tubes, has been designed, realized and characterized. Fault tests were performed to study the limiting behavior. The results show that the impedance of the FCL suddenly changes from 0.1 Ω to ~ 0.4 Ω and then continues to increase due to heating of the superconducting tubes. A fault maximum power of 15 kVA was achieved and the limited fault current was about half of the unlimited value. A purposely developed mathematical model was used to simulate the data and the dynamic behavior of a FCL in a 15 kV network.

1. Introduction

The Fault Current Limiter (FCL) is expected to be one of the first devices incorporating High-T_c Superconductors (HTS) to be installed in the electric power sector. It is attractive because it may meet the requirements of an ideal device for current limitation at high voltages without any conventional equivalence. Several groups are working on the development of high-T_c FCL's [1-6] based on various concepts. The shielded core inductive type has been considered as suitable for commercial units in the MVA range in a near/medium term. An inductive 4 kVA FCL model has been designed, realized and tested with particular attention to the characteristics of the superconducting Bi-2212 tubes. The current limiting behavior was simulated by a mathematical model utilized also to analyze the transient response of an inductive high-T_c FCL installed in a 15 kV power system.

2. Experimental

The inductive shielded core high-T_c FCL consists of a primary copper coil and a superconducting cylindrical shield inserted coaxially between the primary coil and the central arm of a FeSi laminated core (see Fig. 1). The shield, cooled at 77 K, is constituted by two commercially available melt cast processed Bi-2212 tubes, made by Hoechst AG. The characteristics of the FCL model are reported in Table 1. The unit was tested in a 50 Hz electric circuit, including a 5 Ω load and fault/breaker switches electronically triggered, connected to the network by an autotransformer (Z_{source}= 0.6 Ω). Fault tests were performed at different nominal voltages V_n. Voltage $V_1(t)$ and current $I_1(t)$ waveforms at FCL terminals were recorded by a digital oscilloscope. In order to investigate the $V_{sc}(I_{sc})$ characteristics of the

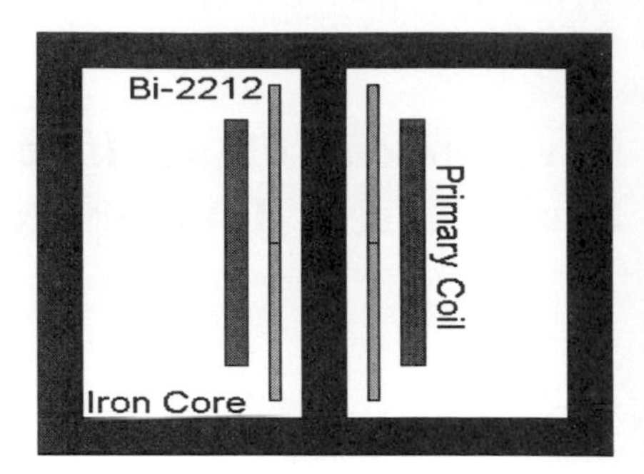

Fig. 1. Photograph of the disassembled unit and scheme of the 4 kVA high-T_c FCL model.

superconducting elements, a Rogowski and a pick-up coils were wound on the tubes; other pick-up coils were wound on the arms of the iron core to study the magnetic flux distribution during nominal and fault conditions. The dynamic $V_{sc}(I_{sc})$ curves were also deduced by the $V_1(I_1)$ traces solving the equations of the equivalent circuit [1, 7].

3. Results and discussion

The duration of the fault tests of the high-T_c FCL model were limited to a few cycles and the principal results, with V_n= 160 V and I_n= 25 A are reported in Table 1 (all AC units are expressed in rms values). Its behaviour is similar to an air core reactor during nominal operation while in fault conditions it acts like a transformer with a short circuited secondary coil having a non linear Z(I, T) characteristic.

Table 1. Characteristics of the high-T_c FCL

Primary Coil		Bi-2212 Tubes	
Number of turns	200	Total height (cm)	2 × 9.8
Internal Diameter (cm)	4.83	Thickness (cm)	0.42
Height (cm)	15	External diameter (cm)	4.75
Resistance at 77 K (mΩ)	37	J_c at 77 K (A/cm^2)	~ 1000
Air gap (mm)	0.4	**4 kVA FCL model**	
Fe-Si Core		I_n (A), V_n (V), P_n (kVA)	25, 160, 4
Cross section (cm × cm)	2.2 × 2.5	Z_n (Ω)	0.11
Magnetic permeability, μ_r	30	$I_{limited}$ (A)	94
Height (cm)	22	Z_{lim} at 1st and 5th fault cycle (Ω)	0.41 / 0.82
Magnetic reactance X_m (Ω)	0.9	$P_{fault} = V_n \times I_{limited}$ (kVA)	~ 15

Extended fault tests (15 cycles, 0.3 s) were also performed at I_n= 20 A (to limit thermal stresses) and the results are shown in Figs. 2-4. In this case, the first peak current after the fault was limited at 125 A, a factor ~ 2 with respect to the unlimited value (see Fig. 2). As the fault proceeds, the limited current continues to decrease and a deformation in its waveform appears due to the iron core saturation. It was observed that the limitation is predominantly of resistive type at the beginning and becomes mainly inductive after ~ 7-8 fault cycles when the resistance of the Bi-2212 tubes reaches the normal value (see Fig. 3). It may be explained on the basis of the equivalent circuit in which the limitation is related to the parallel between the transformed resistance N^2R_{sc} of the Bi-2212 tubes and the magnetization reactance X_m. N^2R_{sc} increases

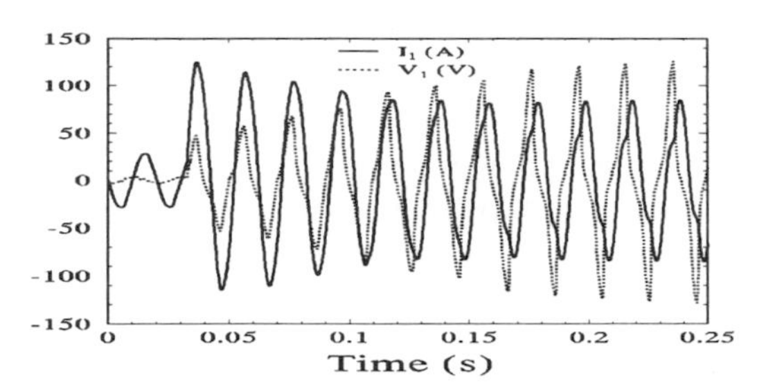

Fig. 2. Current-voltage characteristic $V_1(I_1)$ of the 4 kVA FCL model.

from 0.25 Ω at the first fault cycle to 3.4 Ω after 300 ms. Fig. 4 shows the $V_{sc}(I_{sc})$ characteristics of one Bi-2212 tube at the different fault cycles.
The change of current limitation mechanism was also confirmed by the measurement, using the Rogowski coil, of the total current circulating in the Bi-2212 tubes that decreases continuously from 16 kA at the first fault cycle to a nearly steady state value of 4.5 kA, after 12 fault cycles. In these steady conditions the heat flux to the LN_2 bath is ~ 3 W/cm^2 , i.e. below the value of 10 W/cm^2 that is the threshold to the overheating of the Bi-2212 tubes (for higher values the heat transfer mechanism changes from nucleate boiling to film boiling).

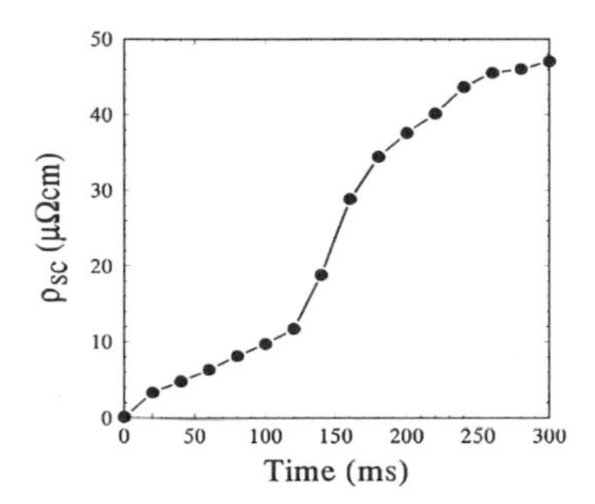

Fig. 3. Growth of the electrical resistivity ρ_{sc} of Bi-2212 tubes during the fault.

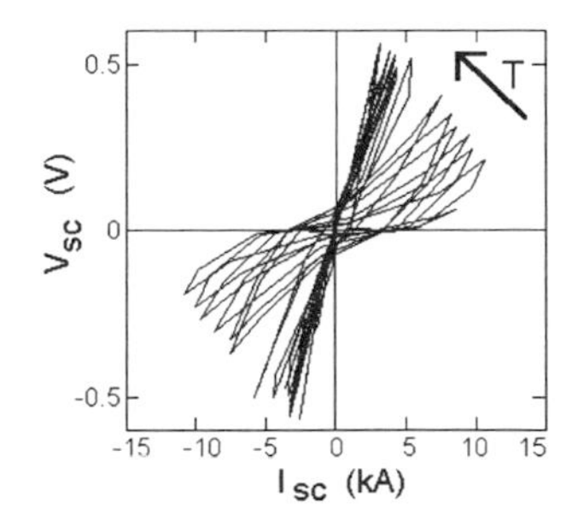

Fig. 4. $V_{sc}(I_{sc})$ curves during fault cycles for one Bi-2212 tube.

The power losses in nominal conditions are also important for economic considerations and comparison with traditional FCL. These losses are associated to the losses of the primary coil, AC loss of HTS tubes and the reactive power due to the stray magnetic field in the air gap. In our experiments, the results confirmed that the main contribution to the nominal impedance is the reactive component. The AC losses of Bi-2212 tubes (~ 12 W at I_n=25 A) were deduced from the ohmic power loss at FCL terminals (~ 30 W) by subtracting the other loss components (primary coil and iron core losses). The obtained values agree with the theoretical ones and indicate that AC losses play a significant role especially for units in the MVA range.

4. High-T_c FCL modeling

A simulation program, based on the equivalent circuit [1, 7], was developed and compared with tests results. The non-linear resistance of the Bi-2212 tubes $R_{sc}(I_{sc}, T)$ was described assuming a power law dependence from I/I_c (T) and a linear temperature dependence for $I_c(T)$.

Above T_c, a $R_n(T)$ linear function was adopted. The numerical results (see Fig. 5) are in good agreements with experimental data. Using the same numerical program and varying the dimensional parameters as number of primary coil, height and diameter of HTS rings, value of I_{sc} and J_c, we carried out the simulation of the transient response of a FCL to be inserted in a feeder line or at the interconnection of two busbars in a three-phase 15 kV distribution system. The choice of the FCL parameters influences its performances, i.e using a high number of coil turns the current limitation is more effective but the stray reactance is increased (nominal losses should be limited to 0.1% of electric power network).

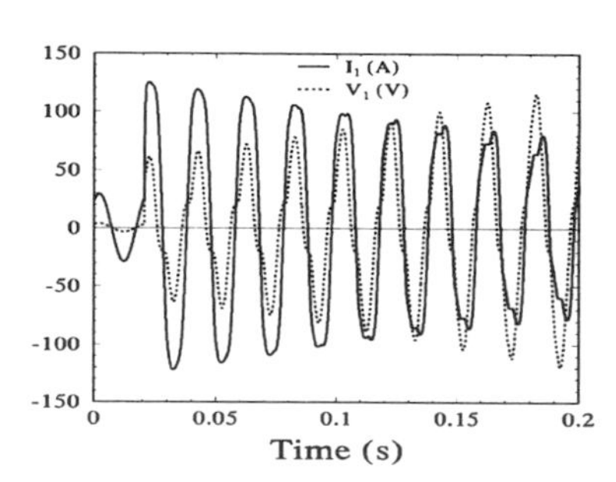

Fig. 5. Simulation of $V_1(I_1)$ oscillograms of 4 kVA FCL model

Table 2. Typical parameters for an inductive High-T_c FCL for distribution system.

Rated / single phase voltage (kV)	15 / 8.7
Nominal current (A)	200
Source impedance (Ω)	0.5
Load impedance (Ω)	43
Number of HTS rings	15
Dimensions of HTS rings (cm)	45 ϕ×10 h×0.16 t
J_c at 77 K (A/cm^2)	2500
Stray inductance (mH)	0.14
Number of primary turns	53
Nominal/Fault impedance (Ω)	0.045/1.7
Limited/Unlimited current (kA)	5.1/17.4
mQ	4.3

An illustrative result of the design and simulation for a single phase FCL is reported in Table 2. The parameter $mQ = I_{sc}/I_c$ was introduced as index for safe usage of HTS tubes to prevent high thermal local stresses and possible failure of HTS components.

Conclusions

FCL has been regarded as one of the first power applications of superconductivity. The actual characteristics of HTS components envisages a medium term commercialization of these devices for the distribution systems. Fault tests on a 4 kVA FCL model confirmed its unique properties and numerical simulations allowed to identify the critical parameters of a MVA unit in a 15 kV line during nominal and fault conditions.

Acknowledgments

Work supported by ENEL SpA, Italy. A. Arienti is grateful to Edison SpA for financial support.

References

[1] Paul W, Rhyner J, Baumann T and Platter F, *Appl. Superconductivity* (IOP Publ.) (1995) 73-78.

[2] Giese R F *Fault Current Limiters - A Second Look, IEA Report* (1995)

[3] Cave J R, Willen D W A, Nadi R, Zhu W, Paquette A, Boivin R and Brissette Y, to be published on *IEEE Trans. on Appl. Superc.*

[4] Kado H and Ichikawa M, to be published on *IEEE Trans. on Appl. Superc.*.

[5] Zannella S, Jansak L and Donadio P, to be published on *Cryogenics.*

[6] Schmidt W, Kummeth P, Schneider P, Seebacher B and Neumuller H W, *Appl. Superconductivity* (IOP Publ.) (1995) 631-634.

[7] Kajikawa K, Kaiho K, Tamada N and Onishi T *Electrical Engineering in Japan* **115** (1995) 104-110

Inst. Phys. Conf. Ser. No 158
Paper presented at Applied Superconductivity, The Netherlands, 30 June–3 July 1997

Resistive Current Limiters with YBCO Films

B Gromoll, G Ries, W Schmidt, H P Krämer, B Seebacher, P Kummeth, S Fischer* and H W Neumüller

Siemens AG, Corporate Technology, P.O. Box 3220, 91050 Erlangen, Germany
*Siemens AG, Power Transmission and Distribution, Nonnendammallee 104, 13623 Berlin, Germany

Abstract. The present status of the Siemens work on resistive current limiters is presented. It is shown, that numerical simulation can reproduce the limiting behaviour of switching elements with good agreement. This also leads to a realistic estimation of the nominal power per area. For YBaCuO films deposited on YSZ substrates with IBAD YSZ buffers nominal power densities above 300 W/cm² have been reached. On a stack of three 10 x 10 cm² switching elements a nominal power rating of 20 kVA has successfully been demonstrated.

1. Introduction

Fault Current Limiters (FCL) integrated in electrical power systems can effectively reduce the current in case of a short circuit and thereby lower mechanical and thermal loads on components such as busbars and circuit breakers. Siemens pursues the resistive concept of current limiters as an alternative to the inductive approach [1, 2]. It is based upon insertion of a HTS conductor element into the electrical circuit. In case of a fault the current exceeds the critical current and drives the superconductor into the normal and resistive state. To prevent overheating a circuit breaker disconnects the limiter from the network after some periods. After recooling the limiter can be reconnected. Superconductive FCLs therefore work as fast self restoring fuses.

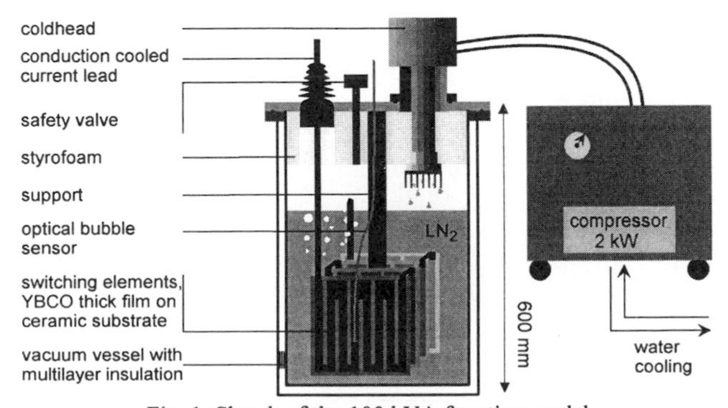

Fig. 1: Sketch of the 100 kVA function model

This paper describes latest results in the development of resistive type FCLs using YBaCuO films as switching elements. Fig. 1 shows a sketch of the planned 100 kVA functional model based on a modular concept of switching elements. It is also shown, that numerical simulation can reproduce the limiting behaviour of switching elements with good agreement.

2. Film area and nominal power

The transition from superconducting to normal conducting state due to overcurrent $I > I_c$ is crucial for fast and efficient fault current limitation. In the period of resistive state following the fault until opening the circuit breaker, the limited current continues to flow through the limiter element. In our thin films with 10^6 A/cm^2 this gives rise to efficient Joule heating and fast increase of temperature. An inevitable consequence of such thermal shocks are considerable mechanical axial and shear stresses due to nonuniform thermal expansion of the hot conductor on a cold substrate. Mechanical strength of the film and quality of bonding between HTS film and substrate will set a limit on the maximum temperature and in turn on the allowed voltage across the device.

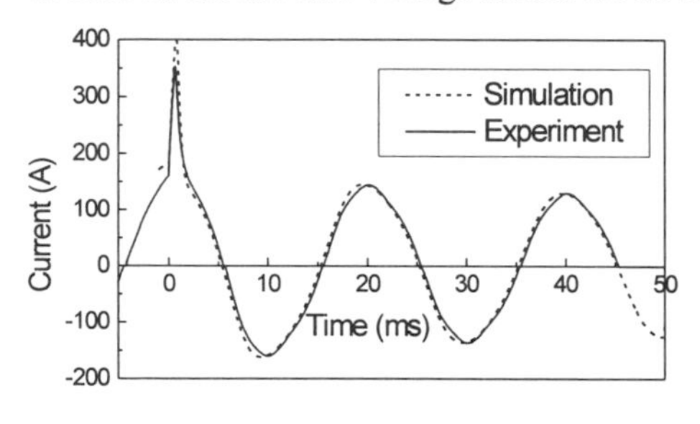

Fig. 2: Current in limiter experiment at 80 V, 127 A. Fault was at t = 0

Fig. 3: Calculated temperature and resistance in YBCO-film with Au-shunt.

As a development tool for parameter studies and to estimate the required demand of limiter area for a given nominal power, we apply numerical simulations [3]. The procedure combines consistently the evolution of resistance and dissipation in the superconductor and shunt layer, transient heat diffusion into substrate and coolant as well as reaction of the electric circuit on the time dependent resistance. Fig. 2 compares experimental and simulation results of the current in a switching experiment with a YBaCuO-meander of 29 x 1 cm² x 1.4 μm with 1 μm Au and I_c = 180 A on an IBAD YSZ substrate. The peak fault current was effectively reduced from prospective 1000 A_{pk} to 350 A_{pk}. With an operational current of 127 A_{rms} and a voltage of 80 V_{rms} on 29 cm² active film area, the limiter demonstrated reliable operation at a nominal power density of 350 VA/cm². The calculated film temperature of 320 K after 50 ms in Fig. 3 coincided nearly with the average temperature as derived from the electric resistance in the experiment.

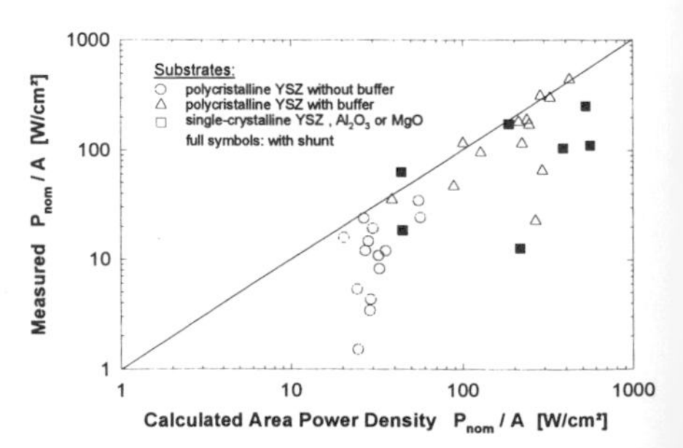

Fig. 4: Specific nominal power per area obtained in experiments and by simulations

Now three of these samples have been assembled to a stack shown in Fig. 5. The switching elements are connected in series. The switching behaviour is shown in Fig. 6. The nominal voltage and current were 200 V_{rms} and 100 A_{rms}, respectively. With a limited current near the nominal value and a peak fault current of only twice the nominal current this demonstrator shows superior limiting behaviour. The recovery time of the stack for recooling to 77 K is 1.5 s.

4. Conclusions and outlook

As shown in Fig. 7 within the past two years a significant progress in the power rating of resistive fault current limiters has been made. The results obtained so far show that the YBaCuO films on YSZ substrates are a promising concept for the modular upscaling of the resistive current limiter. The next step in the fault current limiter project will be the construction of a stack with enough 10x10 cm² switching elements to form a 100 kVA functional model. This will be realized with 5 to 10 plates. For the following 1 MVA functional model 20x20 cm² switching elements are planned. The YBaCuO deposition on 20x20 cm² substrates has already been started.

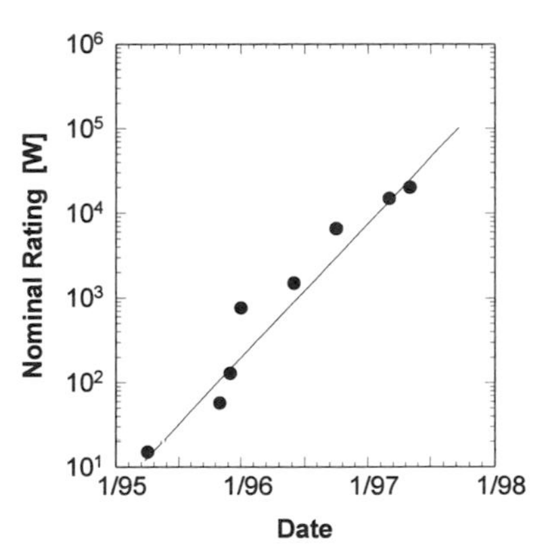

Fig. 7: Progress in nominal rating of resistive FCL demonstrators

Acknowledgements

This work has been supported by the German Federal Ministry for Education, Science, Research and Technology BMBF under grant no. 13N6842. Part of the work is done within the joint collaboration between Hydro Quebec, Canada and Siemens, Germany.

References

[1] J.R. Cave et al., "Testing and Modeling of Inductive Superconducting Fault Current Limiters," presented at *ASC 96*, Pittsburgh, August 1996.

[2] W. Paul et al., "Test of 100 kW High-T_c Superconducting Fault Current Limiter", IEEE Trans. Appl. Supercond., Vol. 5, No. 2, 1059-62 (1995)

[3] M. Lindmayer and H. Mosebach, "Quenching of High-Tc-Superconductors and Current Limitation – Numerical Simulations and Experiments", presented at *ASC 96*, Pittsburgh, August 1996

[4] P. Berberich, B. Utz, W. Prusseit and H. Kinder, "Homogeneous high quality YBa2Cu3O7 films on 3" and 4" substrates, *Physica C 219*, pp. 497-504, 1994

[5] R. Wördenweber et al., "Current limiting properties of superconducting YBCO films," *Applied Superconductivity 1995*, Inst. Phys. Conf. Ser. No. 148, pp 619-622.

[6] W. Schmidt et al., "Preparation of YBCO thick films by pulsed laser deposition for a superconducting fault current limiter," *Applied Superconductivity 1995*, Inst. Phys. Conf. Ser. No. 148, pp 631-634

[7] J. Wiesmann et al., "Biaxially textured YSZ and CeO_2 buffer layers on technical substrates for large-current HTS-applications," *Applied Superconductivity 1995*, Inst. Phys. Conf. Ser. No. 148, pp 503-506

[8] G. Ries et al., "Development of Resistive HTSC Fault Current Limiters," *Applied Superconductivity 1995*, Inst. Phys. Conf. Ser. No. 148, pp 635-638.

Fig. 4 compares specific nominal power per area obtained in experiments on various samples with values calculated by simulation for 300 K maximum temperature in the film. From this we conclude that 300 K is a reasonable upper design limit, where thermomechanical stress is still acceptable in so far as the thermal expansion of the HTS-film of about 0.3% is tolerated without destruction of the element. Failure of samples below this limit was usually due to noncomplete and nonuniform transition and subsequent burning in hot spots with deteriorated superconducting properties. Evidently improving the film homogeneity is essential for proper and effective limiter operation.

3. Demonstration of the modular FCL concept

Up to now only single switching elements have been tested. The elements consisted of YBaCuO films deposited on various substrates by thermal coevaporation [4], magnetron sputtering [5] or laser ablation [6]. Films with a thickness of 1.4 μm have been deposited on YSZ substrates with biaxially textured buffer layers of YSZ fabricated by Ion beam assisted deposition (IBAD) [7]. Critical current densities above 10^6 A/cm² are obtained routinely. A homogeneity of j_c in the range of 20% is achieved over a substrate surface of 10x10 cm². To prevent burnout at hot spots a 1 μm Au layer is deposited on top of the superconductor as an electrical shunt. A meander type current path is etched into the YBaCuO film and contacts for current and voltage are made by soldering directly on the Au shunt layer. The elements are mounted on FRP sample holders. Measurements are carried out in a LN_2 bath cryostat with two windows which allow visual observation of the vapour formation during switching. The experiments are performed using a test circuit which is described in detail in [8]. The resistive load is shorted for 45 ms by means of a thyristor switch. After this period the current is switched off completely and the limiter recovers to normal operating conditions at 77 K within about 1 s. With these elements a maximum power density above 300 W/cm² has been achieved (see Fig. 4).

Fig. 5: View of the 20 kVA demonstrator

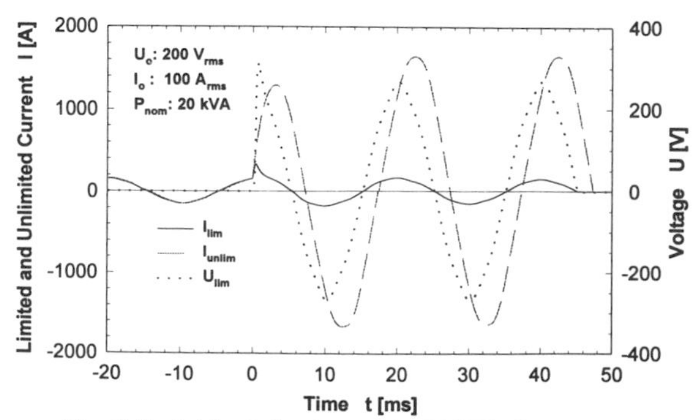

Fig. 6: Switching behaviour of the 20 kVA demonstrator

Inst. Phys. Conf. Ser. No 158
Paper presented at Applied Superconductivity, The Netherlands, 30 June–3 July 1997

Preparation of switching elements for a resistive type HTS fault current limiter

W Schmidt[1], P Kummeth[1], R Nies[1], R Schmid[2], B Seebacher[1], H-P Steinrück[2] and H-W Neumüller[1]

[1]Siemens AG, Corporate Technology, Erlangen and Munich, Germany.
[2]University of Würzburg, Experimental Physics II, Würzburg, Germany.

Abstract. The switching element of a resistive type fault current limiter consists of a shunted YBCO layer on electrical insulating substrates. Substrates of partially stabilised zirconia ceramics (PSZ) were sintered and polished with a roughness r_A of less than 1.5 nm. YBCO films were deposited on small samples and on tapes of PSZ with pulsed laser deposition (PLD). Critical current densities j_c up to 4×10^4 A/cm^2 had been achieved in films with thicknesses of about 2 μm. With an additional biaxially oriented buffer layer j_c can be increased above 10^5 A/cm^2. The distribution of j_c in large area YBCO films was measured inductively with a scanning system which has a lateral resolution of a few mm. The scanning is necessary for characterisation of the homogeneity of $j_c \times d$ before testing the switching behaviour of the samples.

1. Introduction

The change in electrical resistance during the transition from the superconducting to the normal state is considered to be suitable for the rapid limitation of fault currents in electric power transmission lines. To assess the technical feasibility of fault current limiters (FCL) a 100 kVA model will be build. The preparation of the switching elements is one of the major aspects of this function model.

Among the various superconducting materials and the large variety of known preparation techniques the deposition of YBCO on insulating substrates seems to be the most promising route for producing the required switching elements. The substrates should be insulating to avoid short-circuiting of the high resistance of the superconductor in the normal state. A sufficiently high resistance is essential to effectively limit the fault current. Due to the high power ratings (up to 10 MVA) of FCLs projected for operation in electric power grids, large substrates are required. The choice of substrate material therefore mainly depends on the availability of large sizes. Zirconia ceramics partially stabilised with yttria (PSZ) is available with dimensions up to 20×20 cm^2. It has a sufficiently matched coefficient of thermal expansion and allows the deposition of films with thicknesses above 1 μm. But to achieve j_c above 10^5 A/cm^2 an additional buffer layer of biaxially aligned zirconia [1] is necessary. The 100 kVA model e.g. would require a YBCO film with a thickness $d = 2.5$ μm covering an area of about 20×20 cm^2 and having an overall critical current density of about 10^5 A/cm^2. Physical vapour deposition techniques (PVD) like pulsed laser deposition (PLD) are appropriate to prepare YBCO films of this quality. PLD has also been proved to operate at very high growth rates of up to 1440 nm/min [2] which is essential to finish deposition within a reasonable time.

Alternative deposition methods for YBCO like thermal evaporation and sputtering as well as the preparation of the biaxially aligned buffer layer are performed within a joint research project and are described elsewhere [3-7]. This paper describes the preparation of the ceramic substrates, the attempts to upgrade PLD to areas of up to 20 × 20 cm^2, and the sample characterisation by scanning j_c.

2. Ceramic substrates from stabilised zirconia

PSZ powder with 3 mole-% Y_2O_3 was used for tape casting. Tape dimensions ranged from small samples to sheets as large as 270 × 670 mm^2. Green tape thicknesses were typically in the range of 400 to 700 μm. Debindering was performed at 1300 °C. The overall shrinkage was 20 %. In the case of the very large tapes special care had to be taken to ensure uniform shrinking. Inhomogeneous mixing of PSZ powder and organic components of the slurry or non-uniform debindering due to temperature gradients or too slow diffusion of decomposition components may cause warping and eventually scratches and fractures. The samples were sintered in air at 1500 °C with a pressure of 1 kPa applied for reducing camber. This procedure causes an additional shrinkage of 5 %. The porosity was determined by optical analysis to 0,4 to 1 %. The as-fired surface had an average roughness r_A of 50 to 100 nm. Obtaining a surface quality suitable for the deposition of well-textured films requires grinding and polishing of the substrates. Except for the largest dimensions r_A values of 1,5 nm were obtained. Measurements using a profilometer Alpha-Step of Tencor Instruments, a laser interferometer of WYCO and an atomic force microscope gave similar r_A results, but revealed different details. Measurements with the latter instrument showed particularly well the influence of a thermal treatment on the surface structure. Fig. 1 shows the effect of a heating up to 1000 °C on the free surface compared to the surface after polishing. Such high temperatures are not commonly used at the deposition of YBCO films, but a lesser grade of transformation of the surface has to be considered and will be studied.

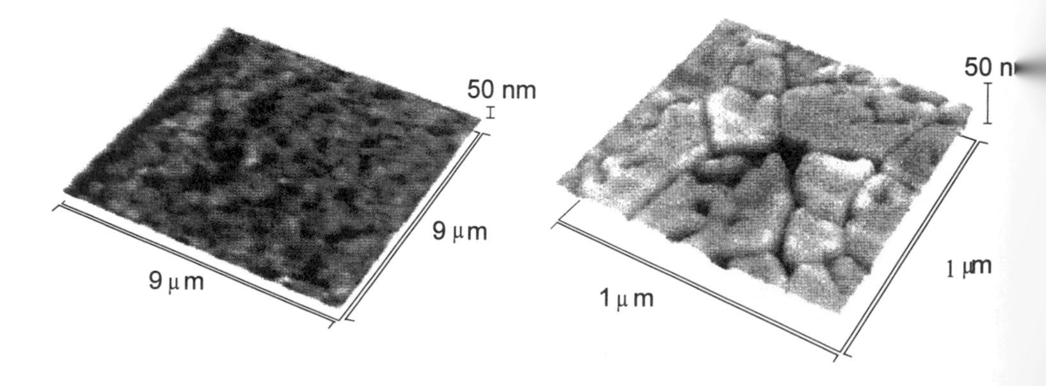

Figure 1: AFM-pictures of an as-polished PSZ sample before and after heat treatment at 1000°C. The roughness r_A before heating is below 5 nm

3. Characterisation by scanning j_c

The YBCO thick films are characterised using an inductive $j_c(x, y)$ scanning apparatus developed especially for this purpose, which allows to investigate very large samples since the scan area is up to 20×50 cm^2. Inside a closed vessel the YBCO films are cooled with liquid nitrogen to 77 K. A contactless measurement of the critical current density is performed with a single coil method by monitoring the generation of the third harmonic voltage with a digital lock-in amplifier as the current in the coil is increased [8].

The computer controlled scanning apparatus allows a spatial resolution of 1 mm while the diameter of the measuring coil is about 4 mm. The superconducting samples are mounted on a copper plate which has an electric heater on its lower side to warm up the samples after measurement inside the cryostat in a dry atmosphere. Sample characterisation is performed with a typical lateral resolution of 1 cm × 1 cm. The following figure 2 displays results on a 10×10 cm^2 sample which was provided by the Tech. Univ. of Munich [3].

Obviously there is a relatively high nearly homogeneous critical current density which reaches $j_c \approx 10^6$ A/cm^2 over the entire sample with the exception of one point,. The drop of j_c which can easily be found in the picture is due to the preparation process of the sample.

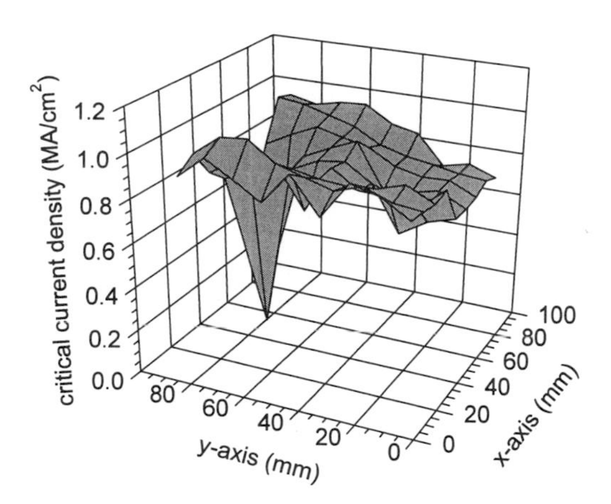

Figure 2: Results of the inductive $j_c(x, y)$ measurement on a 10×10 cm^2 sample (TUM22). The YBCO-layer was prepared by thermal evaporation (Tech. Univ. Munich [3]) and the biaxially aligned YSZ-buffer by ion-beam assisted deposition (IBAD, Univ. Göttingen [7]).

4. Preparation of YBCO films with PLD

To evaluate whether PLD can be upgraded to large areas two deposition systems were built. The first can handle tapes of 1 cm width, which are moved between the plasma plume and a stationary heater. With this system YBCO was deposited on PSZ tapes which are commercially available with length up to 20 cm.

The second deposition system was designed for substrates of 20×20 cm^2 in size and went into operation recently. The main feature of this system is a large movable substrate heater which allows scanning of the samples beneath the stationary plasma plume and parallel to the incoming laser beam. The upper part of the chamber contains the rotating YBCO-Target (∅ 4 cm) and the entrance window and is separated from the heater station in the lower part by a rotating shutter disk with a open/closed ratio of 1/9. This results in a more lower variation of the substrate temperature and is an essential feature of our experimental set-up to make large area deposition possible. The pulsed excimer laser beam (KrF 248 nm) is synchronized with the opening of the shutter, so that the substrate is just visible during the formation of the plasma plume.

The substrate heater used for the experiments presented here was built for substrate sizes up to $10 \times 10\ cm^2$. It consists of a Inconell plate into which a Thermocoax heater wire is brazed with solder melting above 1100°C.

YBCO films with a thickness up to 5 μm were produced on bare PSZ-tapes without IBAD-buffer in the smaller deposition system. The j_c reached a maximum of $4 \times 10^4\ A/cm^2$ at 2 μm with a decrease to $1 \times 10^4\ A/cm^2$ at 5 μm. The distribution of the critical transport current was measured along the tape with voltage taps spaced 1 cm apart. A variation of less than 10% could be achieved in the inner part of a 11 cm tape with a slight decrease at the outermost voltage tap. Due to the granularity of the low-j_c material a degradation of j_c of about 25% was observed during 4 cool-down cycles. Therefore further this type of material will not be considered for the current limiter. Only substrates with an additional biaxial buffer layer will be used in the future.

The first successful depositions were made with the large-area PLD-system. To test the uniformity of the deposition small samples of single crystalline $SrTiO_3$ ($1 \times 1\ cm^2$) were placed on the heater and coated with 0,5 μm of YBCO. The heater was moved over the full length of a $10 \times 10\ cm^2$ substrate holder. The inductively measured onset of the transition temperatures and the transition widths (onset - offset) ranged from 91.5 K and 1,5 K in the middle of the heater to 86 K and 5.5 K at its edge, respectively. This drop of T_c is due to a variation of the heater temperature of more than 30°C. Therefore a new heater for $20 \times 20\ cm^2$ substrates with a more homogeneous temperature profile of less than 10°C variation was designed using numerical simulations. This heater has been manufactured recently and will soon be tested.

Acknowledgements

The authors would like to thank M. Bauer and R. Semerad (TU Munich) for providing samples for testing the j_c-scanner, W. Hösler for the AFM measurements, and K. Schleicher for the numerical simulations of the heater design. The work is co-funded by the german BMBF under contract number 13N6482.

References

[1] Iijima Y, Tanabe N, Kohno O, Ikeno Y 1992 *Appl. Phys. Lett.* **60** 769-771

[2] Foltyn S R, Peterson E J, Coulter J Y , Arendt P N, Jia Q X, Dowden P C, Maley M P, Wu X D and Peterson D E 1997 to appear in *J. Mat. Res* (Nov. 1997)

[3] Bauer M, Schwachulla J, Furtner S and Kinder H 1997 *this conference*

[4] Kinder H 1997 *this conference*

[5] Utz B, Rieder-Zecha S, Kobler E and Kinder H 1997 *this conference*

[6] Schneider J, Lahl P, Königs Th, Kutzner R, and Wördenweber R, 1997 *this conference*

[7] Wiesmann J, Dzick J, Heinemann K and Freyhardt H C 1997 *this conference*

[8] Claassen J H, Reeves M E, and Soulen R J Jr 1991 *Rev. Sci. Instrum.* **62(4)** 996-1004

Inst. Phys. Conf. Ser. No 158
Paper presented at Applied Superconductivity, The Netherlands, 30 June–3 July 1997

Development of Ag sheathed Bi(2223) tapes for power applications with improved microstructure and homogeneity

Giovanni Grasso, Frank Marti, Yibing Huang, Reynald Passerini and René Flükiger

Département de Physique de la Matière Condensée, Université de Genève, 24 quai Ernest-Ansermet, CH-1211 Genève 4, Switzerland

Abstract. The development of the fabrication process of Ag sheathed Bi(2223) tapes has been carried out in order to improve their transport and mechanical properties, as required by the power applications which are so far under study. Critical current density values of 28 kA/cm^2 at 77K have been achieved on long multifilamentary Bi(2223) tapes, with a fabrication process that has been successfully employed in the fabrication of samples longer than 50 m. The microstructure and homogeneity of Ag sheathed multifilamentary Bi(2223) tapes has been markedly improved by introducing a new deformation technique, employing a prototype four-roll machine, which allows the deformation of rectangular shaped wires. The first long tapes prepared with this new technique have shown a remarkable homogeneity, both of the microstructure and of the transport properties. At present, critical current densities in excess of 25 kA/cm^2 at 77K have been achieved on long samples prepared with this new technique.

1. Introduction

Ag sheathed multifilamentary Bi(2223) tapes are intensively studied world-wide in view of industrial applications. At present, critical current densities in the range of 20-28 kA/cm^2 have been successfully achieved on long lengths of tapes at the liquid nitrogen temperature [1-3]. However, a further improvement of the transport properties of Bi(2223) tapes is still required in order to become energetically and economically favourable in comparison with normal conductors.

In this work, we present the first results we have achieved by introducing a new deformation process in the PIT method. Instead of deforming by swaging, drawing, and rolling, multifilamentary samples are reduced to the final size by means of a prototype four-roll machine. This machine allows us to work with rectangular-shaped samples, which present several advantages compared to the standard round-shaped configuration. SEM imaging, Vickers microhardness and local transport j_c measurements have revealed that a clear improvement of the homogeneity of Bi(2223) tapes has been effectively achieved.

Moreover, innovative configurations of the filaments can be assembled for some specific use. In this paper, we present Bi(2223) wires with orthogonal filaments, which show a clearly reduced anisotropy of the critical current density with respect to the magnetic field orientation.

2. Experimental

Ag sheathed Bi(2223) tapes have been prepared by the Powder-In-Tube method [1]. Calcined powders of nominal composition $Bi_{1.80}Pb_{0.40}Sr_{2.00}Ca_{2.20}Cu_{3.00}O_{10+\delta}$, and mainly composed by Bi,Pb(2212), Ca2PbO4, and CuO are filled inside pure silver tubes with an initial packing density of about 65-70%. The tubes are then swaged and drawn down to monofilamentary wires of outer diameter of about 2.0 mm.

They are then deformed with the four-roll machine in square shaped wires of about 1.0 to 1.5 mm of edge. Several pieces of them are stacked into a second square-shaped silver tube, which is deformed again with the four-roll machine down to a thickness of about 200-250 μm and a width of about 3 mm. A deformation speed of 1.5 cm/s has been employed for all the deformation processes. Then, the tapes are heat treated in order to form the Bi(2223) phase in pieces of up to 25 m in length. After a first heat treatment of about 50 hours at a temperature of 837°C in air, the tape thickness has been reduced again by about 10% by rolling. A further heat treatment at 837°C of up to 100 hours is needed to reach the highest critical current density.

3. Results

The employment of the four-roll machine instead of more conventional deformation techniques for the preparation of Bi(2223) tapes has allowed us to work with square shaped samples. The advantages of the square symmetry are evident when we look at the transverse cross section of multifilamentary tapes. In fig. 1a and fig. 1b, two different cross sections of multifilamentary tapes are presented. Both tapes have a thickness of about 250 μm, and a width of about 3 mm. The cross section of fig. 1a is relative to a standard multifilamentary tape with 61 filaments and a superconducting fraction of about 20%, while the cross section of fig. 1b is relative to a newly developed tape with 100 filaments, and a superconducting fraction of about 25%. In the standard tape of fig. 1a, the filaments near the tape centre are more compressed than those at the sides due to their non-homogeneous deformation. Moreover, as the distance between the single filaments has strong fluctuations, small bridges between them can form near the tape centre. In spite of the higher superconducting fraction, the cross section of fig. 1b is much more homogeneous, and no evidence of the formation of bridges between filaments has been detected. A confirmation of the higher homogeneity of the four-roll deformed tapes as been obtained by two different methods. Vickers microhardness measurements have been extensively performed on single filaments of both standard and four-roll deformed tapes.

Fig. 1a: Transverse cross section of a 61 filament Bi(2223) tape prepared with standard deformation techniques

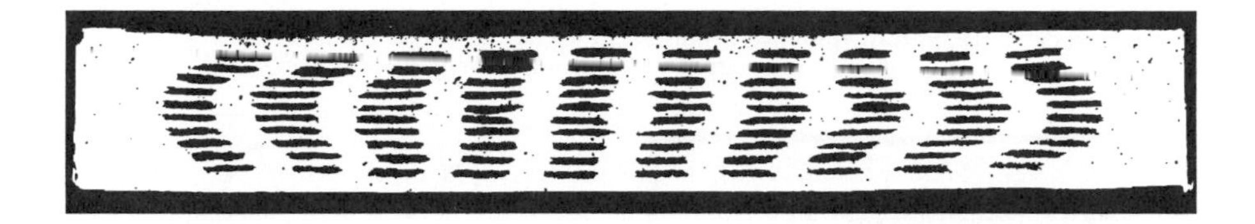

Fig. 1b: Transverse cross section of a 100 filament Bi(2223) tape prepared with the four-roll machine

The measurements are summarised in fig. 2. The Vickers microhardness of single filaments has been plotted as a function of the lateral distance from the filament centre to the tape centre.

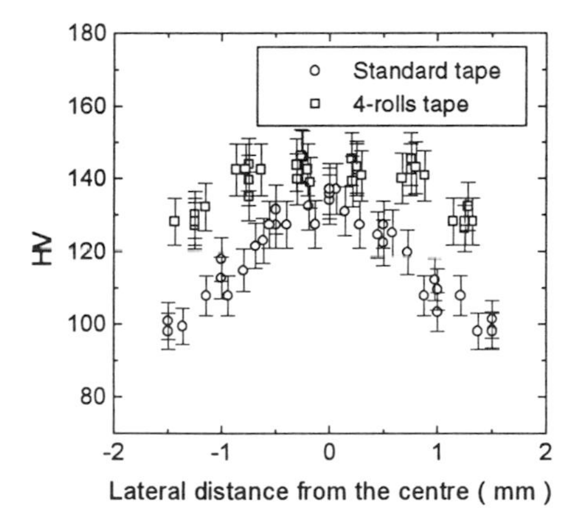

Fig. 2: Position dependence of the Vickers microhardness of single filaments

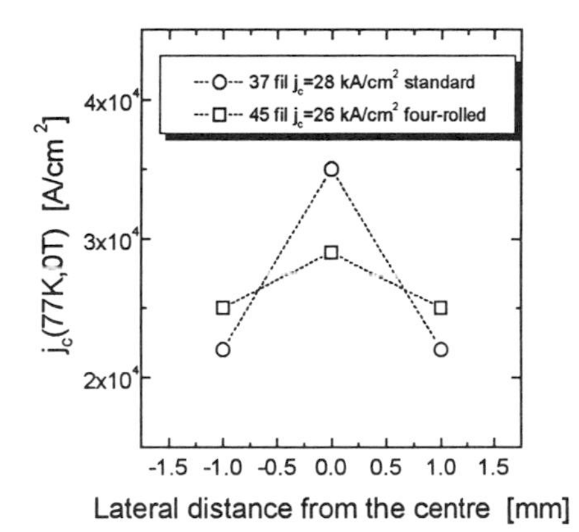

Fig. 3: Position dependence of the critical current density in multifilamentary Bi(2223) tapes

For the standard tapes, a clear variation of the microhardness has been observed between the filaments which are near the tape centre and those which are near to the sides. Typically, the filaments near to the centre presents a Vickers microhardness value of about 130-140 Hv, while near to the sides it decreases to about 90 Hv. For the four-roll deformed tapes, the Vickers microhardness is much less position dependent, going from about 145 Hv at the centre to 125 Hv near to the tape sides.

The high homogeneity of the four-roll deformed tapes has been also investigated by performing local critical current measurements. A standard and a four-roll deformed multifilamentary tape of 3 mm in width have been cut into three slides of 1 mm in width each. Then, the critical current density of each slide has been carefully measured by the four probe method. The critical current density shows a strong position dependent behaviour in the standard tapes, where the filaments near the sides carry clearly lower critical current densities due to their non optimal deformation conditions. On the contrary, the four-roll deformed tapes are much more homogeneous, the difference between the critical current density at the centre and the sides being less than 20%.

By using the four-roll machine, we also developed Bi(2223) wires with an innovative configuration of the filaments, which is particularly useful in reducing the anisotropy of the critical current density with respect to the orientation of the applied magnetic field. The transverse cross section of the wire is shown in fig. 4.

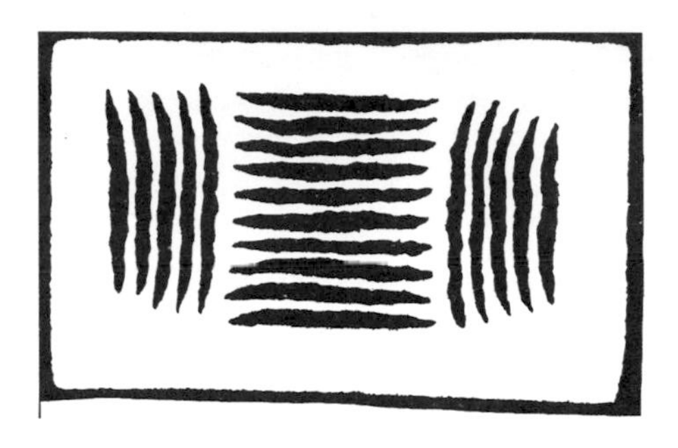

Fig. 4: Transverse cross section of the Bi(2223) wire with orthogonal filaments. Wire width is 0.9 mm, and wire thickness is 0.6 mm.

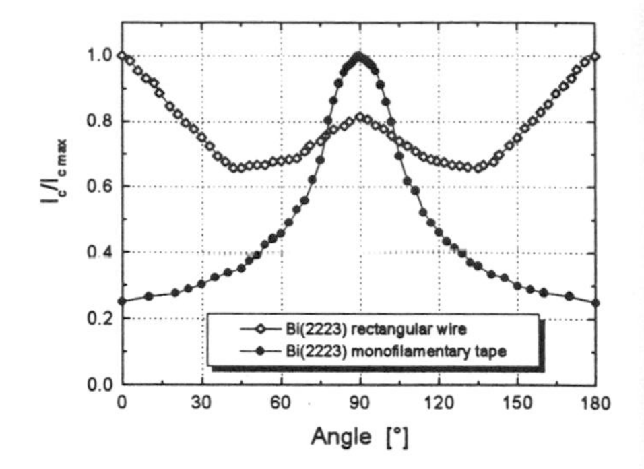

Fig. 5: Angular dependence of the critical current density at 0.2 T for a monofilamentary tape and a four-rolled wire.

So far, a critical current density of 20.5 kA/cm^2 has been reached for these wires. Complete anisotropy measurements of the transport critical current density have been measured with an applied field of fixed amplitude (0.2 T) and for $0 \leq \theta \leq 180°$, where θ is the angle between the field direction and the sample normal. The measurements are shown in fig. 5 both for a reference monofilamentary tape and for the rectangular wire. In the case of the monofilamentary tape, the critical current measured both for $\theta=0°$ and 180° is about four times lower than the critical current measured for $\theta=90°$. For the rectangular wire, the minimum I_c value for $\theta = 45°$ is only 35% lower than the maximum value measured for $\theta=0°$.

4. Conclusions

A prototype four-roll machine has been employed for the fabrication of high j_c multifilamentary Bi(2223) tapes, also with innovative filament configuration. These tapes present a higher homogeneity with respect to tapes prepared following the standard deformation route, as confirmed by Vickers microhardness measurements, as well as by local transport measurements. Further work has to be done in order to optimise the heat treatment parameters for these newly developed tapes.

Acknowledgements

The work is funded by the EC-Brite/Euram, Contract Nr. BRPR-0030, and by the Swiss Priority Program for Materials (PPM).

References

[1] Grasso G, Jeremie A, and Flükiger R 1995 *Supercond. Sci. Technol.* **8** 827.

[2] Riley G N MRS 1996 Spring Meeting, 8-12 April 1996, San Francisco (USA).

[3] Sato K, Ohkura K, Hayashi K, Hikata T, Kaneko T, Kato T, Ueyama M, Fujikami J *1995 Proc. 7th US-Japan Workshop on High Tc Superconductors*, Tsukuba (Japan), p. 15, 24-25 October 1995.

Inst. Phys. Conf. Ser. No 158
Paper presented at Applied Superconductivity, The Netherlands, 30 June–3 July 1997

The effect of silver-alloy sheaths on fabrication, microstructure and critical current density of powder-in-tube processed multifilamentary Bi-(2223) tapes

A. Hütten, M. Schubert, C. Rodig, U. Schläfer, P. Verges and K. Fischer

Institute of Solid State and Materials Research, 01069 Dresden, Germany

Abstract. Critical current densities of up to 33.8 kA/cm^2 and a matrix resistivity of 1.05 μΩcm at 77 K have been achieved in multifilamentary Bi-2223 tapes made by bundling Ag-sheathed monofilamentary wires into a $Ag_{93.86}Cu_{6.14}$ tube. A maximum resistivity of 1.2 μΩcm was measured for $Ag_{98}Pd_2$ sheathed tapes with 19.8 kA/cm^2 (77K, 0T). Homogeneous deformation of tapes with 5.8 μm thick filaments sheathed by Ag or Ag-alloys has been achieved using a rolling mill with a roll diameter < 30 mm.

1. Introduction

It has been well established [1], that the oxide-powder-in-tube-technique, OPIT, is most suitable for tape production on an industrial scale. Pure Ag, in general used as the sheath material, serves multiple purposes since it provides: (1) mechanical support during tape fabrication by cold drawing and rolling, (2) a sufficient oxygen permeability during thermo-mechanical-treatment, TMT, and (3) chemical stability to the very reactive oxide. The disadvantages of pure Silver are low strength providing insufficient mechanical support during tape fabrication and handling thereafter and a low causing current losses in ac-applications. Hence it is required to alloy the Ag sheath with secondary elements. The objective of this paper is to report on the influence of alloying elements (Au, Ca, Cu, In, Mg, Mn, Pd, Zn) on the critical current density of multifilamentary Bi-(2223) tapes.

2. Experimental details

Multifilamentary Bi-(2223) tapes with Ag and $Ag_{100-x}Y_x$ sheaths (Y= Au, Ca, Cu, In, Mg, Mn, Pd, Zn and x in wt-%) have been fabricated using the OPIT process. The starting materials for this process were monofilamentary wires, *MOFT*, sheathed either by Ag or $Ag_{100-x}Y_x$, which were previously prepared by wire or profile drawing and subsequent rolling using a variety of reduction schedules. Multifilamentary tapes, *MFT*, were then produced, bundling *MOFT* tapes into Ag or Ag-alloyed tubes, which subsequently were sealed and then drawn and rolled to tapes. After tape fabrication, the conversion of the precursor powder into the Bi-(2223) phase was achieved during a thermo-mechanical-treatment, *TMT*. This is usually a 3 step procedure, e.g., (annealing$_1$ / densification / annealing$_2$ / densification / annealing$_3$), consisting of powder densification by rolling,

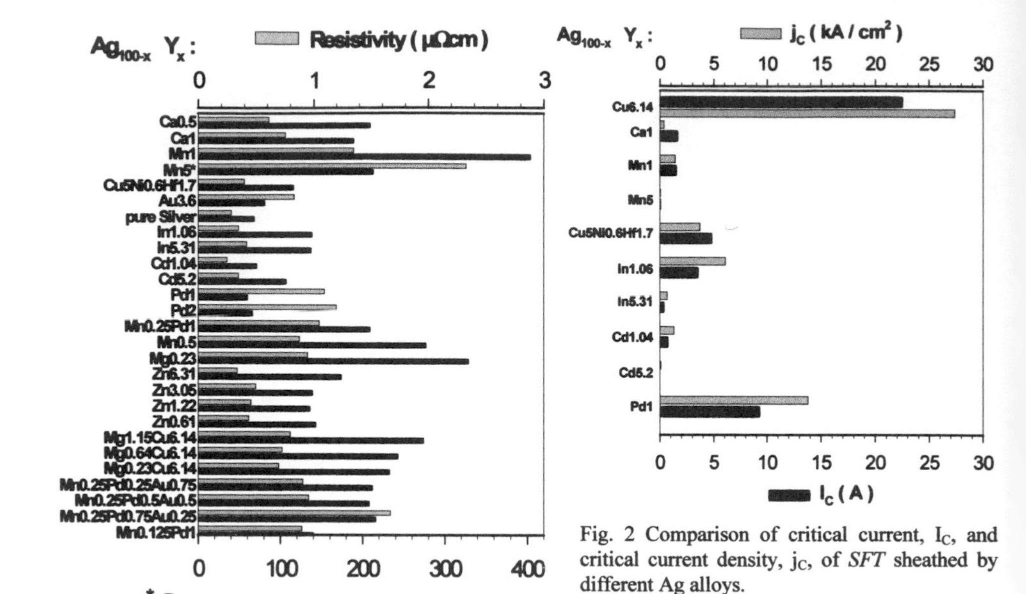

Fig. 1 Yield point ($\sigma_{0.2}$) and resistivity of Ag-alloys heat-treated at 820°C for 100h in N_2-8%O_2 atmosphere.

Fig. 2 Comparison of critical current, I_C, and critical current density, j_C, of *SFT* sheathed by different Ag alloys.

followed by subsequently sintering for several hours between 815°C to 820°C. Optical microscopy was performed to monitor the evolution of the microstructure before and after *TMT*. Short multifilamentary tapes, 5 cm long, were used to determine the critical current I_C after *TMT* by a standard four-probe technique at 77 K in a liquid nitrogen bath, employing a 1 μV/cm criterion. The corresponding J_C values were calculated, dividing I_C by the mean oxide core area which has been micro-optically measured from three transverse cross sections using a computerized image analyzer.

3. Results and discussion

Fig. 1 summarizes resistivity and yield point data of Ag-based alloys after annealing for 100 h at 820°C in N_2-8%O_2 atmosphere, so as to simulate the conditions for Ag-alloyed sheath materials present during *TMT*. Yield points above 200 MPa were achieved for Mn,

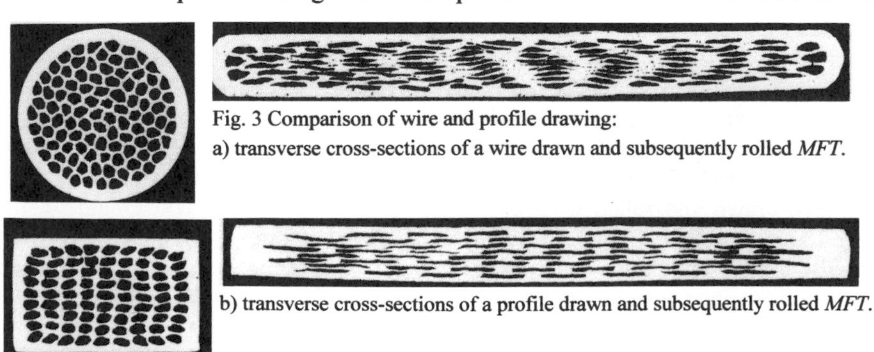

Fig. 3 Comparison of wire and profile drawing:

a) transverse cross-sections of a wire drawn and subsequently rolled *MFT*.

b) transverse cross-sections of a profile drawn and subsequently rolled *MFT*.

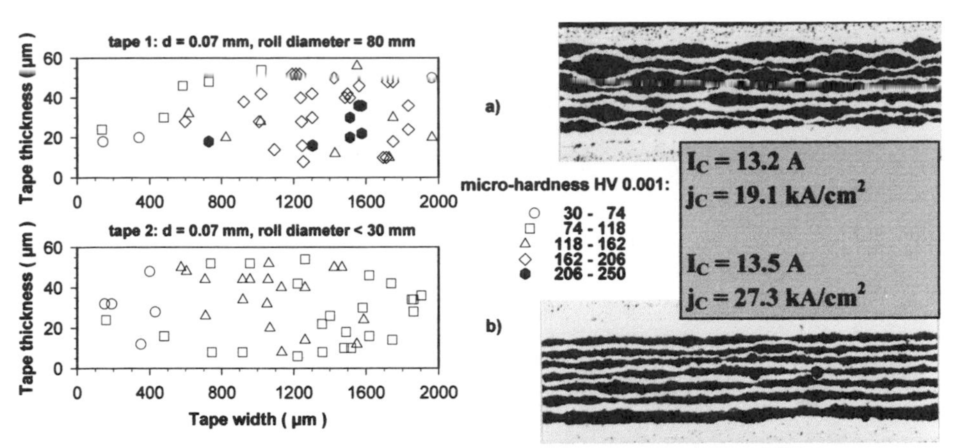

Fig. 4 Influence of the roll diameter on micro-hardness and microstructure of 64 multifilamentary tapes sheathed by $Ag_{93.86}Cu_{6.14}$. Micro-hardness of the filaments (left) across a transverse cross-section of tapes with 0.07 mm thickness, rolled using rolls with 80 mm (a) and < 30 mm diameter, respectively. Optical micrographs of the corresponding filament microstructures along longitudinal cross-sections of both tapes are given aside.

Mg or Ca containing Ag-alloys, whereas resistivity values above 1 µΩcm were reached for AgPd and AgMn alloys. To what extent the resistivity or mechanical strength of these Ag-alloys can be realized in sheath material for multifilamentary tapes together with excellent superconducting properties can be easily tested by monofilamentary tapes. As can be seen in fig. 2 reasonable current densities were only achieved for $Ag_{99}Pd_1$ and $Ag_{93.86}Cu_{6.14}$ sheathed *MOFT*. Reactions with the (2223) phase, crack formation or severe sausaging during deformation were mainly responsible for the large degradation of j_C in the other Ag alloy sheathed *MOFT*. Excellent j_C values in Ag alloy sheathed *MOFT* do not guarantee to achieve similar data in corresponding Ag alloy sheathed *MFT*. Although j_C in $Ag_{93.86}Cu_{6.14}$ sheathed *MOFT* reached up to 27 kA/cm^2 the corresponding value in $Ag_{93.86}Cu_{6.14}$ sheathed *MFT* is only about 10 kA/cm^2 at most. This can mainly be attributed to a drastic increase in the diffusion gradient causing a phase reaction between the (2223) phase and secondary elements, like Cu.

Beside the compatibility between the sheath material and the (2223) phase, high densification of all filaments [2] and avoiding severe sausaging during cold deformation and *TMT* are other major requirement so as to fabricate excellent superconducting tapes. Fig. 3 compares optical micrographs of transverse cross-sections of tapes of identical thickness but differently drawn prior to *TMT*. As a result, the filament arrangement of the profile drawn and subsequently flat rolled tape is much more homogeneous in terms of the filament aspect ratios and the overall filament alignment. In contrast, the wire drawn tape shows an inhomogeneous filament aspect ratio distribution and already sausaging of the filaments. Sausaging is not only promoted by the drawing process alone but also depending on the roll diameter of the roller mill used during *TMT* which is demonstrated in fig. 4. Comparing tapes of identical thickness which have been rolled using a 80 mm (tape 1) or < 30mm (tape 2) roll diameter, respectively, reveals that the micro-hardness of filaments across their transverse cross-section is quite different. The filament micro-hardness HV 0.001 in tape 2 varies between 30 and 162 and is more homogeneous across the transverse cross-section of this tape. In contrast, the filament micro-hardness of tape 1 is much larger and inhomogeneously distributed over the cross-section. This is related to severe sausaging of the filaments as can be seen in the optical micrograph of the corresponding longitudinal

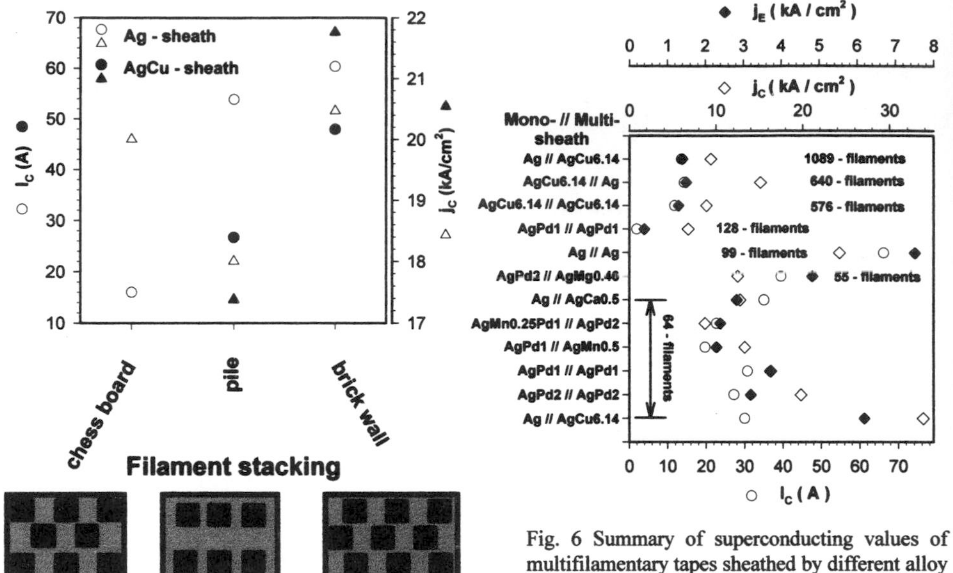

Fig. 5 Influence of different filament arrangements on critical current and critical current density comparing *MFT* sheathed by pure Ag or $Ag_{93.86}$ $Cu_{6.14}$.

Fig. 6 Summary of superconducting values of multifilamentary tapes sheathed by different alloy combinations. Indicated are the alloys used for the *MoFT* and *MFT* sheath.

cross-section. As a consequence, j_C is degraded in comparison to that of tape 2. The benefits of using rolls with small diameters has also been pointed out elsewhere [3].

In addition, there is evidence that the filament stacking influences the microstructure and the critical current density as well. Fig. 5 shows the critical current and current density of pure Ag and $Ag_{93.86}Cu_{6.14}$ sheathed *MFT* as a function of the filament stacking. In both cases, the *pile* like stacking leads to relatively small j_C values, whereas the *brick wall* like stacking of the filaments results in a maximum for j_C.

Fig. 6 summarizes critical current, critical current density and engineering current density data of short samples with different sheath materials. $Ag//Ag_{93.86}Cu_{6.14}$ sheathed 64-filamentary tapes reach up to 33.8 kA/cm^2 with a matrix resistivity of 1.05 μΩcm at 77 K. A maximum resistivity of 1.2 μΩcm was achieved for $Ag_{98}Pd_2$ sheathed tapes with 19.8 kA/cm^2. Best j_E values were measured for $Ag_{98}Pd_2//Ag_{99.54}$ $Mg_{0.46}$ with 4.83 kA/cm^2 and $Ag//Ag_{93.86}Cu_{6.14}$ with 6.2 kA/cm^2 in comparison to pure Ag//Ag sheathed 99-filamentary tapes with j_E = 7.51 kA/cm^2. Attempts to enhance j_C by increasing the number of filaments were not as promising so far and only 15.1 kA/cm^2 were achieved for 640-filamentary $Ag//Ag_{93.86}Cu_{6.14}$ tapes. A $Ag//Ag_{93.86}Cu_{6.14}$ tape, 6.6 m long, has so far been fabricated with j_C = 18.5 kA/cm^2 showing a yield strength of about 90 MPa.

References

[1] Heine K, Tenbrink J and Thoner M 1989 *Appl. Phys. Lett.* **55** 2441-2443

[2] Fischer K et al. 1995 *Proc. of EUCAS 1995* Inst. of Phys. Conf. Series Number 148 363

[3] Han Z, Skov-Hansen and Freltoft T 1997 accepted to be published in Supercon.. Science and technology

This work has been performed within the SIEMENS joint project "HTS for Power Engineering", supported by the German Federal Minister of Research under contract 13N6481.

Inst. Phys. Conf. Ser. No 158
Paper presented at Applied Superconductivity, The Netherlands, 30 June–3 July 1997

Critical transport currents of Bi(2223)/Ag tapes under axial and bending strain

Bernd Ullmann[1], Andreas Gäbler, Mario Quilitz, Wilfried Goldacker

Forschungszentrum Karlsruhe, Institut für Technische Physik, P.O.Box 3640, 76021 Karlsruhe, Germany

Abstract. The transport critical current of Bi(2223)/Ag tapes depends on the mechanical stress applied during or before j_c determination. We previously reported on our reinforced tapes with a yield strength of 300 – 350 MPa[1], which is more than 5 times the value for conventional Ag-sheathed tapes. The maximum tolerable strain is increased from about 0.3% to 0.7 – 0.8% for the modified tapes. In this contribution we extend our investigations on the variation of filament number and filament/matrix ratio and other sheath materials, e.g. AgAu. We discuss the situations for axial tensile-, hoop- and bending stress. For the special case of bending strain a carbon fibre reinforcement enhances the mechanical performance of the tapes.

1. Introduction

For technical application of high temperature superconductors (HTS) we must achieve a maximum critical current density. Furtheron, a large superconductor content in the composite conductor and small filaments for low AC loss conductors are required. From the fabrication and operation point of view there is a demand for a tensile strength of 100 – 200 MPa and a tolerable strain of 0.2%. The strategies to fullfill these demands are (1) an optimization of density, texture and phase, (2) an enhancement of the filament/matrix ratio, fmr, (3) an enhancement of the filament number and (4) new sheath materials. There is great effort on (1) which resulted in high critical current densities in short samples showing the possibilities of Bi(2223) tapes[2]. Strategies (2)-(4) have great impact on the dependence of the critical current, I_c, on the applied stress and strain. In these contributions we call a conductor which tolerates large stress and large strain before a decrease of I_c can be detected a conductor with a good mechanical performance. Bi(2223) tapes with good mechanical performance are needed for industrial fabrication, where stresses are applied during winding of cables or magnets. They are also needed during operation of most devices due to the occurence of simple axial tensile stresses, hoop stresses and more complex bending stresses.

2. Stress situations

In Fig.1 the tensile and compressive part of the stress tensor in the longitudinal direction is schematically shown for three different stress applications : (a) the axial strain experiment, (b) hoop stress during magnet operation, (c) bending stress during winding

[1] E-mail : ullmann@itphts.fzk.de, Web : http://itphts.fzk.de

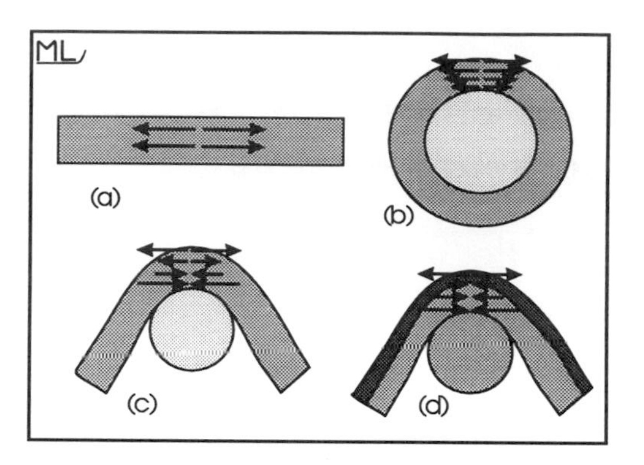

Figure 1. (a) axial strain, (b) hoop stress, (c) bending stress (d) bending of fibre reinforced tape, dark beam is a 50 micron fibre layer

and (d) bending stress during winding of a fibre reinforced tape (see section 4). For (b) to (d) the stress tensor varies over the tape thickness. Compressive stress is indicated by $\rightarrow\leftarrow$, tensile stress by $\leftarrow\rightarrow$. Note, that the very important prestress of the sheath on the filaments is not reflected in this figure. The impact of the prestress on $I_c(\epsilon)$ has been discussed previously [1] and can be added to the applied stress due to the tensor nature of σ. The hoop stress can be approximated by the axial tensile stress which is usually experimentally easier to determine. The axial strain experiment therefore gives important information on the physical properties of the tapes and their behaviour during operation, see sections 3 and 4. Bending stress, on the other hand, is described by a stress tensor with a component in tape direction which changes its sign from the bottom to the top of the tape. Therefore, it can not be approximated by the axial tensile stress experiment and the fabrication of coils and cables have to be simulated by a bending experiment, as described in section 5. The validity of a scale-up of the results obtained by the bending experiment is checked by a small single pancake coil.

3. Enhancement of the filament number and the filament content

The critical current of mono core Bi(2223) tapes decrease after a axial strain of 0.05% which corresponds to a tensile stress of about 20MPa. These values are the tolerable strain and stress, respectivly.

Table 1. Dependence of the mechanical performance on the filament number

N	ϵ_{tol} /%	σ_{tol} /MPa
1	0.05	15
7	0.15	40
19	0.25	50
37	0.20	40

Table 2. Dependence of the mechanical performance on filament content

fmr /%	ϵ_{tol} /%	σ_{tol} /MPa
8	0.5	40
15	0.2	44
28	0.17	50

From the values given in Tab.1 one can conclude that for multifilamentary tapes with similar filament-matrix ratio, fmr, the mechanical performance of the tapes is not enhanced by an increasing filament number. The significant difference between the mono-core tape and the others may be caused by the Ag-core interface which has to transfer the load into the core. In addition the matrix has two important tasks : It has to stop crack

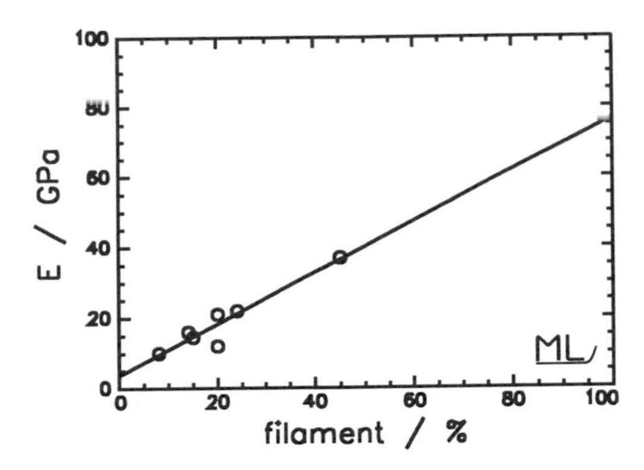

Figure 2. Elastic modulus of Bi(2223)/Ag composite conductor. From the linear fit one can extrapolate the elastic modulus of the ceramic core to 60 – 80 MPa.

formation and it has to create prestress on the filament.
Cracks correspond to a smaller cross section area, A, i.e. to a larger stress $\sigma = F/A$, thus to a larger strain ϵ. The crack widens up, it acts as a hot spot. As the crack enters the core region, it causes a decrease of I_c. Table 2 shows the tolerable stress and strain for tapes with increasing filament/matrix ratio. No significant enhancement of the mechanical performance with increasing matrix content can be observed. From Fig.2 one can conclude, that the properties of the composite conductor are determined by the soft sheath material via the rule of mixtures for elastic moduli.

4. Impact of special sheath materials on the mechanical performance

Fig.2 implies the need for stronger sheath materials, i.e. with elastic moduli nearer to, or better larger than, the value for the Bi(2223) core. Additionally, special sheath materials are requested for distinct purposes, e.g. AgAu as low thermal conductive sheath for current leads[3, 4]. Table 3 shows an enhancement of both ϵ_{tol} and σ_{tol} for alloyed sheath materials, such as AgCu and, more pronounced, for AgMg tapes. The AgAu-sheathed tapes show poorer mechanical performance than Ag-sheathed tapes, which may balance their 7 times higher resistivity and their lower heat conductivity for special application purposes. In filamentary reinforced (fi.re.) tapes some filaments are replaced by AgMg.

Table3. Comparison of the mechanical performances of BSCCO tapes with different sheath materials

	Ag	AgMg	fi.re.	AgAu	AgPd	AgCu
ϵ_{tol}	0.2–0.5	0.4–0.8	0.4	0.18	0.15	0.2
σ_{tol}	40–50	200–350	140	35	35	90

5. Bending of conventional tapes and carbon fibre reinforcement

For most technical applications the reacted tapes have to be deformed into a shape given by the device design, in most cases they have to be wound on cylindrical mandrels. Fig.3a shows a decrease of the critical current, I_c, when the tape is wound on small radii, r.

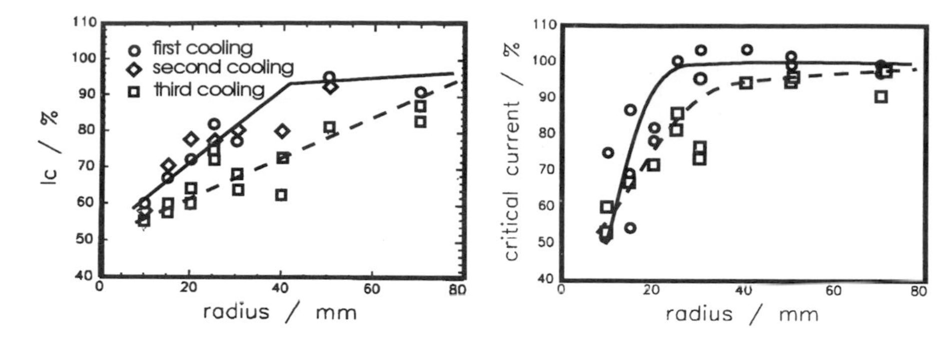

Figure 3. (a) Critical current depending on bending radius for three cooling cycles for Ag sheathed Bi(2223). Fatigue can be observed. (b) Comparison of non-reinforced (dashed line) and carbon-fibre-reinforced Bi(2223) tape.

The I_c-decrease is getting larger when the sample is warmed up and cooled down several times, i.e. thermal stress is applied in addition to the bending stress. From preliminary results we further conclude, that the wind and react method does not improve the $I_c(r)$ behaviour. The reduction of the I_c of the as-prepared tape by winding a single pancake coil with an inner radius of $r = 12$mm to 50% of the original I_c-value shows, that a scale-up from the short sample results to longer lengths is possible. Fig.3b shows that it is possible to bend Bi(2223) tapes to much smaller radii, if they are reinforced by carbon fibres on the outer surface, as shown in the schematic drawing Fig.1d.

6. Conclusions

Possible stress situations occuring during application of HTS conductors are discussed. An improved conductor behaviour under axial stress, i.e. an improved mechanical performance, is found for reinforcement by matrix hardening and filamentary reinforcement. Increasing filament number and filament content does not affect the mechanical performance in the technically interesting regime. The elastic modulus of the Bi(2223) core is estimated to 60–80 GPa. The degradation of I_c with small bending radii can be reduced by a local reinforcement applying carbon fibre bi-layer technique.

Acknowledgments

The authors acknowledge support by the European Communities under contract BRPR-CT95-0030. We would like to thank S. Zimmer and M. Sieg for technical assistance and A. Hütten and K. Fischer from IFW Dresden for AgCu and AgPd sheathed tapes.

References

[1] Ullmann B, Gäbler A, Quilitz M, Goldacker, Proc. Appl. Superc. Conf., Pittsburgh, PA, USA (1996)

[2] Goldacker W, Mossang E, Quilitz M, Rikel M, Proc. Appl. Superc. Conf., Pittsburgh, PA, USA (1996)

[3] Heller R, Friesinger G, Goldacker W, Kathol H, Ullmann B, Fuchs A, Jakob B, Pasztor G, Vécsey G, Wesche R, Proc. Appl. Superc. Conf., Pittsburgh, PA, USA (1996)

[4] Goldacker W, Ullmann B, Gäbler A, Heller R, this conference

Inst. Phys. Conf. Ser. No 158
Paper presented at Applied Superconductivity, The Netherlands, 30 June–3 July 1997

Current distributions in multi-filamentary HTS conductors

A. V. Volkozub[1], A. D. Caplin[1], H. Eckelmann[2], M. Quilitz[2], R. Flükiger[3], W. Goldacker[2], G. Grasso[3] and M. D. Johnston[1]

[1]Centre for High Temperature Superconductivity, Blackett Laboratory, Imperial College, London SW7 2BZ, UK
[2]Forschungzentrum Karlsruhe ITP, P.O. Box 3640, D-76021 Karlsruhe, Germany
[3]DPMC, Université de Genève, 24 quai Ernest-Ansermet, CH-1211 Genève 4, Switzerland

Abstract. Scanning Hall probe microscopy has been used to image in detail the current profile of high-quality multi-filamentary Bi2223 tapes with different numbers of filaments, twist pitch and sheath materials. In twisted 37-filament tapes the derived current profile evolves smoothly across the width as the transport current is increased, whereas the current profiles of the untwisted 55, 37 and in particularly 19-filament tape show considerably more variation. In all cases, the central group of filaments carries most of the current at higher levels. The derived current distribution is generally better described by the ellipsoidal solid conducting core model, rather than that of a thin flat conducting strip.

1. Introduction

In multi-filamentary superconductors intended for high-current AC applications, it is essential that the current be shared efficiently between the filaments. The manner in which this occurs depends on both the underlying microstructure and the intrinsic electrodynamics. In HTS conductors, the problems are more severe than in LTS, because of the fragility of the ceramic filaments.

In order to investigate this problem, a local probe of the current distribution is essential; further, in order to correlate it with other kinds of measurement, the probe should be non-destructive and non-invasive. We describe here the use of a scanning Hall probe to obtain the current distribution in such conductors. We aim to compare the distribution with that expected from their microstructure, and so to test the current-sharing. Discontinuities in the filaments, and current transfer between them should also be apparent.

2. Experimental approach

Our miniature scanning Hall probe maps out the normal component of the magnetic field generated above the broad face of the conductor; it was patterned to about 50 micron square from a GaAs/AlGaAs heterostructure containing a 2-D electron gas (2DEG). With lock-in detection at an excitation frequency of 2.7 kHz, the field resolution is ~ 1.5 μT.

Table 1

Tape	Number of filaments	J_c (A) at 77 K	Width (mm)	Sheath material	Twist pitch (mm)
KA1	37	12	3.26	Ag	10
KA2	37	16	3.12	Ag	20
KAM2	37	15	3.2	AgMg (2 %)	20
BRE1	19	22	3.86	Ag	untwisted
BRE2	37	>30	3.8	Ag	untwisted
BRE3	55	>30	4.3	Ag	untwisted
BRE4	37	>30	3.8	AgMg	untwisted

The deconvolution to obtain the current distribution was performed with a combination of the finite element method and an RMS minimisation technique, resulting in a spatial resolution of ~ 100 μm.

We describe measurements of seven different Ag and AgMg-sheathed Bi2223 tapes, all fabricated by the powder-in-tube method (see Table 1). The thickness of the tapes was in the range of 200 to 260 μm. Where AgMg as the sheath material is indicated, the alloy was used only for the outer tube and not for the matrix surrounding the filaments which contains only silver. The critical current J_c was measured using a 1 μV/cm criterion. All experiments were performed with sample and scanner immersed in liquid nitrogen at 77 K in zero applied magnetic field, with DC applied currents up to 20 A.

3. Results and discussion

Fig. 1 shows the normal field profile above tapes KA2 (twisted), BRE1 and BRE2 (both untwisted) as the transport current increases; the generating current profiles extracted from these scans are shown too. At low currents the distribution peaks at the superconducting core edges, the core edges being defined here as the outermost filaments. As the total transport current is increased, the peak in the distribution, or equivalently the lateral penetration of field into the sample, shifts inwards, as has been previously found for monocore and multi-filamentary tapes [1].

The three tapes show similar and smooth profiles at the lowest current, but differ at higher currents: The untwisted tapes BRE1 and BRE2 show large fluctuations in the field and current distributions, and distinct groups of current-carrying filaments can be seen. On the other hand, the profile of the twisted tape KA2 evolves smoothly, with variations that are of the order of our experimental resolution.

Fig. 2 presents extracted current profiles for the 37-filament twisted tapes KA1 (Ag sheath) and KAM2 (AgMg sheath). Comparison with predictions for current penetration as a function of the total transport current shows a resemblance to the calculations by Norris [2] for elliptical current front penetration into a superconductor. Further evidence for elliptical penetration has been advanced by Ciszek *et al.* [3], who cite intergrowths of the superconducting phase creating filament interconnections as the reason for the agreement of AC loss measurements with the Norris predictions, which were derived for a solid homogeneous ellipsoidal conductor. Previous work [1] has shown shielding of the inner filaments in tapes with untwisted filaments, with a redistribution of current to the centre

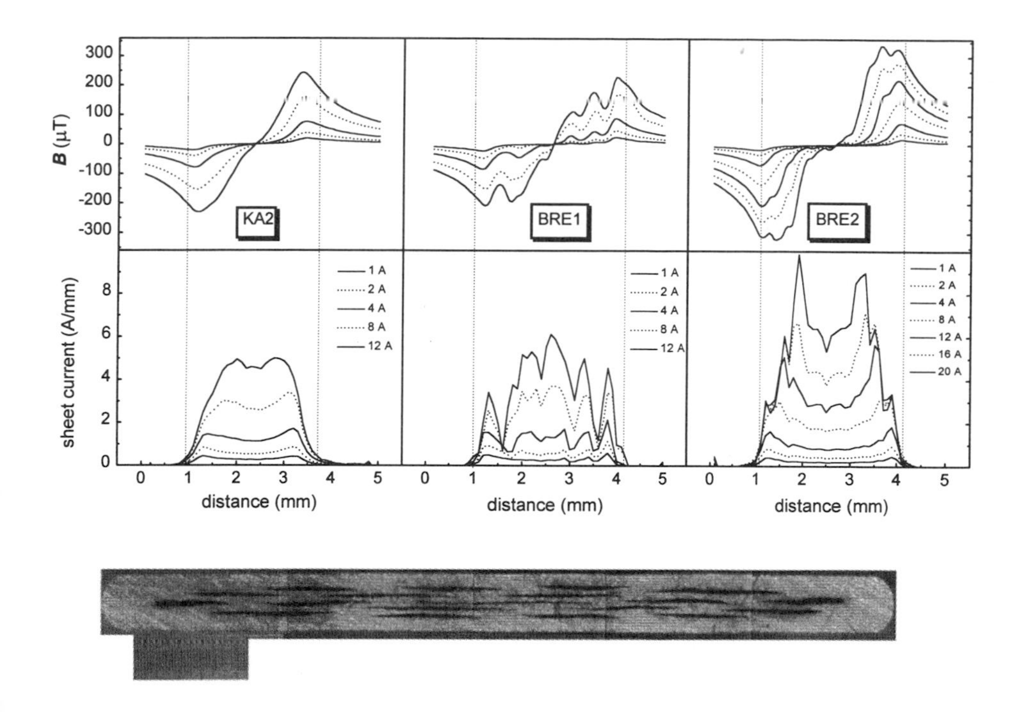

Fig. 1. Self-field and extracted current profiles for the twisted (KA2) and untwisted (BRE1, BRE2) multi-filamentary tapes, showing the penetration of the current profile as the total applied transport current increases. The dashed lines indicate the position of the outermost filaments. The micrograph shows the cross-section of the 19-filament BRE1 tape (the full length of the scale in the bottom left corner of the photograph is 500 μm).

when the current exceeds a decoupling value. This should be overcome in tapes with twisted filaments which commute between the inner and outer positions in the core — in these tapes the current should be distributed more evenly between the filaments, even at low currents.

Examination of the remanent field profiles (i.e. after a current cycle) in these tapes (data not shown here) indicates that, in addition to small-scale screening currents within the individual filaments, there are also persistent (on the scale of seconds or longer) currents circulating on a larger scale, necessarily through filament interconnections. This too is circumstantial evidence for the presence of Bi2223 intergrowths bridging adjacent filaments.

4. Conclusions

The non-destructive, non-invasive method of scanning Hall probe microscopy allows us to study the spatial distribution of both the self-field and remanent magnetic field in multi-filamentary Bi2223 tapes with spatial and current resolution sufficient for imaging groups of current-carrying filaments in detail. The extracted current profiles show smooth progressive penetration of the current front as the applied current is increased, indicating that processing

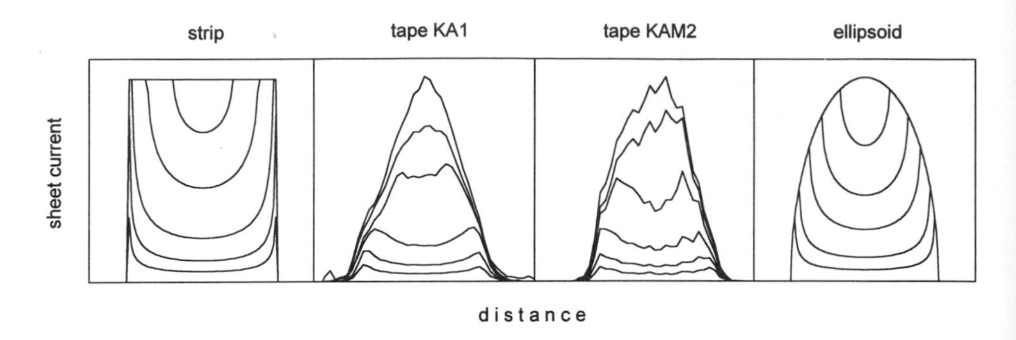

Fig. 2. Extracted current profiles for the 37-filament twisted tapes KA1 and KAM2, for the same series of increasing transport current values as in Fig. 1, compared with ellipsoidal [2] and thin-strip [4] conductor model calculations.

is now achieving tapes with a fair degree of uniformity of the filaments over the cross-section. These profiles are similar to those calculated for an elliptical superconducting core, suggesting that in these multi-filamentary tapes, there is significant filament interconnection, probably through intergrowths of the superconducting phase. Filament interconnections will degrade the conductor performance at AC, and processing of these tapes must be improved to eliminate them.

So far, we have studied only the profiles of DC transport currents; our Hall probe scanning technique is readily extendible to power-line frequencies, where we anticipate the current distributions to be rather different.

Acknowledgements

This work has been supported by UK EPSRC, and by Brite-Euram Contract No. CT920229. MDJ acknowledges the tenure of an EPSRC Research Studentship sponsored by BICC.

References

[1] Johnston M D, Everett J, Dhallé M, Caplin A D, Friend C M, LeLay L, Beales T P, Grasso G and Flükiger R 1997 to appear in *IEEE Trans. Appl. Supercond.* 7

[2] Norris W T 1970 *J. Phys. D* **3** 489-507

Yang Y, Hughes T, Beduz C, Spiller D M, Scurlock R G and Norris W T 1996 *Physica C* **256** 378-386

[3] Ciszek M, Ashworth S P, Glowacki B A, Campbell A M and Haldar P 1996 *Physica C* **272** 319-325

[4] Brandt E H and Indenbom M 1993 *Phys. Rev. B* **48** 12893-12906

Inst. Phys. Conf. Ser. No 158
Paper presented at Applied Superconductivity, The Netherlands, 30 June–3 July 1997

Progress in HTS Wire and Applications Development

R Schöttler, G Papst and J Kellers

American Superconductor Europe GmbH, Rathausstraße 7, D-41564 Kaarst, Germany

Abstract. The performance of high temperature superconductor (HTS) wire is increasing continuously and suitable for various large demonstrations in the energy sector and power applications. American Superconductor Corporation in Westborough, MA produces Bi-2223 tape in a pilot production plant with a standard wire length of 400 m and a production exceeding 120 km in 1996. The wire properties like critical current density and a.c. losses have progressed to levels which enable the realization of prototypes confirming the benefits of the HTS technology. This paper reviews the development and presents test results of American Superconductor's HTS wire and some large HTS power application prototypes including electrical motors, generators, power transmission cables, transformers and superconducting magnetic energy storage (SMES) devices. Also applications of HTS in accelerator technology and as current leads in combination with LTS magnets are presented in outline.

1. Introduction

Continuing progress in improving the properties of HTS wire at American Superconductor Corporation enabled the commercialization of current leads and laboratory magnets, and the realization of synchronous machines and SMES in demonstration sizes. These applications mainly require a high d.c. current density.

Some HTS applications for utility use, e.g. transformers, require low losses even in moderate a.c. magnetic fields. Ways to achieve this are filament twisting and increasing the inter-filament resistivity. Both techniques have shown good results in laboratory samples and will emerge to a production process within the next years [1].

2. HTS Wire Development

Significant enhancements in critical current densities j_c in rolled multifilamentary Bi-2223 HTS composite conductors have been achieved using the scalable oxide-powder-in-tube (OPIT) process. A record filament j_c of 58,000 A/cm^2 has been reached at 77 K, self field, and 1 μV/cm in short, rolled samples. Best values for the engineering current density j_e (over the total cross-section, including silver matrix) were up to 15,000 A/cm^2 [2,3]. With a reproducible, uniform j_e of more than 8,000 A/cm^2 in the pilot manufacturing process and a standard length of 400 m, American Superconductor paves the path toward more advanced prototypes and commercial development.

In addition to the standard OPIT production American Superconductor evaluates other HTS processes, like YBCO IBAD and RABiTS e.g., which have the long term potential for a better performance than BSCCO OPIT. Due to the big research effort still necessary for these processes, American Superconductor focuses on the commercialization of all applications based on OPIT.

3. HTS Applications

3.1. Motors and Generators

Synchronous machines with superconducting field windings offer reduced losses, size and weight. Latest achievements include a >200 HP motor, built and tested at Reliance Electric / Rockwell Automation in 1996 based on HTS coils from ASC. American Superconductor currently is developing and manufacturing HTS coils for a 1000 HP synchronous motor which will be delivered to Reliance Electric for demonstration in 1998. More details on this and other developments are presented in a companion paper [4].

3.2. Power Transmission Cables

Increasing global consumption of electricity creates a number of challenges for power companies. Most pressing is the need for more capacity in power transmission systems. A related problem, particularly in urban and suburban areas, will be replacing worn-out copper transmission cables in the "right-of-way", the buried steel pipes that are conduits for power and telecommunications cables.

Due to public opposition to overhead lines a number of utilities are also being forced to replace existing overhead lines with new underground cables which is disruptive, time consuming and expensive. Two or three underground copper cables are needed to achieve the capacity of one overhead transmission line, and installation costs are 20 to 30 times higher.

The solution: The HTS power transmission cable, which is expected to be commercially available in about three years. American Superconductor is manufacturing and shipping significant amounts of HTS wire to its partner Pirelli Cable, the world's second largest cable manufacturer, for developing two types of HTS cables:

- Retrofit cables that can be drawn into existing conduits as worn-out copper cables are extracted. Because HTS retrofit cables will carry two to three times as much power as conventional copper cables, the strategy addresses the fundamental need for sending more power through existing rights-of-way.
- Coaxial cables intended primarily for new underground installations. Projections indicate that HTS cables will have a four to six times greater capacity than conventional copper cables. Thus the ultimate cost will be much less.

With their higher ampere rating, HTS cables can also substitute for transmission lines at a lower voltage, thus eliminating the need for transformer substations in urban areas.

A key to the commercial success of HTS cable systems is the quality and length of individual HTS wires. By demonstrating wires more than 1,200 meters long, American Superconductor already has achieved a major step toward this goal. In addition to length requirements, the wire's mechanical strength to withstand machine stranding was a major task which has been solved successfully.

Under contract with the Electric Power Research Institute and the U.S. Department of Energy, American Superconductor and Pirelli demonstrated a 50-meter stranded cable conductor for 3,300 amps in 1996. Commercially viable, 100-meter power cable systems are expected to be tested in 1998-99 and introduced as commercial products in 2000-2001.

3.3. Transformers

In early 1997 the electric utility of Geneva, SIG, plugged the world's first HTS prototype transformer into its electric power grid. Key partners in the project are Asea Brown Boveri (ABB), the largest manufacturer of transformers, EdF, the world's largest utility, and American Superconductor, the manufacturer of the HTS wire. Supporting the project are a research consortium of the Swiss electric utility industry and the Swiss Department of Energy, Bundesamt für Energiewirtschaft.

The 630kVA transformer is the first step toward the development of larger, commercially viable transformers. Based on technical, manufacturing and marketing evaluations, the greatest benefits of HTS will be achieved in transformers larger than 40 MVA.

EdF is among a number of utilities worldwide that have expressed great interest in HTS transformers based on smaller size and weight and lower electrical losses. Economic benefits include greatly reduced expenses for transporting and installing units that are 50 % smaller and lighter, as well as lower operating costs. The elimination of dielectric oils found in conventional transformers is an additional plus. HTS transformers use liquid nitrogen, a nonflammable and environmentally friendly substance, as a coolant and a dielectric fluid.

American Superconductor is continuing to scale up its manufacturing processes to meet ABB's demand for low loss, high current wire. The company expects a number of prototypes of ever increasing size and power to be built and tested over the next three to four years. Commercial products are expected to be introduced in 2001-2002.

3.4. Superconducting Magnetic Energy Storage (SMES)

Two of the most common power quality problems encountered today by industry are voltage sags and momentary interruptions. Industrial processes today rely heavily on electronically controlled systems, e.g. adjustable speed drives. These are very sensitive to voltage disturbances which are unpredictable in time, severity and duration. Most disturbances are only a few hundred milliseconds in duration and vary from a few percent sag of the voltage to a complete outage. The SMES systems manufactured and fielded by Superconductivity Inc. (SI), a wholly-owned Subsidiary of American Superconductor Corporation, are designed specifically to correct these types of voltage disturbances for critical industrial customers.

SI's low temperature SMES systems, based on NbTi wire, are commercially available today. The devices store already 3 MJ of energy in a magnetic field created by direct current in the superconducting coil. During any disturbance of the incoming power supply line, the device senses the change and switches the load (up to 1.4 MVA) to SMES. The transfer is transparent to the customer, and the coil recharges immediately.

A large fraction of a magnet system's first and operating cost comes from the Helium liquefaction equipment, mainly necessary for cooling the current leads when conventional ones are used. To improve this situation, SI now incorporates American Superconductor's HTS current leads. This enables both, cooling of the leads and re-liquefaction of Helium boil-off, by GM cryocoolers, and eliminates the Helium liquefier, the major need for preventive maintenance.

A perspective for the future is the replacement of the LTS coil by a HTS one for a further reduction of the cooling effort. In 1996, American Superconductor delivered the first HTS SMES demonstrator worldwide to Center for Innovative Energy Conversion and Storage (E.-U.-S. GmbH) in Gelsenkirchen, Germany. This 5 kJ SMES is the most massive HTS coil manufactured so far and operates at 25-30 K, cooled by Leybold GM cryocoolers.

3.5. Laboratory Magnets

High temperature superconductors offer many advantages for laboratory magnets compared to conventional superconductors, as there are higher stability (faster ramping), less cooling effort (with respect to cooling power requirement and system complexity), and even higher achievable magnetic field (at liquid Helium temperature).

A US-New Zealand consortium comprising American Superconductor, Isys, AET, IRL, and Alphatech designed and manufactured an ion beam switching magnet. The two Bi-2223 HTS coils, conduction cooled by a single stage GM cryocooler, generate 0.72 T in the air gap of an iron yoke [5]. The system was assembled and tested at American Superconductor and has been installed successfully for operation at the Institute of Geological and Nuclear Science in New Zealand.

3.6. Current Leads

The CryoSaver™ current lead is based on American Superconductor's proprietary 85 filament composite conductor with a special alloy as matrix to reduce the thermal conductivity. Individual tapes are stacked and sintered together to form a rugged bus bar that is the central component of the CryoSaver™. This stacked structure has proven tolerance for tensile stress and thermal cycling.

CryoSaver™ current leads are available as a standard product from 50 to 2000 amps, and in custom design above 2000 amps, ready for use in any low temperature application. A 12.5 kA lead, designed, built, and successfully tested by American Superconductor, is presented in a companion paper [6].

4. Conclusions

The superconducting performance of rolled multifilamentary Bi-2223 composite conductors at American Superconductor have been improved significantly. Commercial products and development demonstrations with increasing size and in both d.c. and a.c. applications are reviewed in this paper.

References

[1] Snitchler G et al. 1996 *Applied Superconductivity Conference*

[2] Malozemoff A P, Li Q and Fleshler S 1997 *M2S Conference*

[3] Li Q et al. 1996 *Applied Superconductivity Conference*

[4] Papst G, Gamble B B, Rodenbush A J, Schöttler R 1997 *EUCAS Conference* **4Ba-1**

[5] Kalsi S S et al. 1996 *Applied Superconductivity Conference*

[6] Kellers J, Gamble B B, Rodenbush A J 1997 *EUCAS Conference* **3Ge-54**

Inst. Phys. Conf. Ser. No 158
Paper presented at Applied Superconductivity, The Netherlands, 30 June–3 July 1997

Third Round of the ITER Strand Bench Mark Test

Hennie G. Knoopers[1], Arend Nijhuis[1], Erik J. G. Krooshoop[1], Herman H. J. ten Kate[1], Pierluigi Bruzzone[2], Peter J. Lee[3] and Alexander A. Squitieri[3]

[1]Low Temperature Division, University of Twente, Enschede, The Netherlands. [2]ITER Joint Central Team, Naka-gun, Japan. [3]University of Wisconsin-Madison, Madison, Wisconsin, USA.

Abstract. The second round of the ITER strand benchmark test was completed in 1995. The two main goals were the characterization of four strand types of industrial Nb_3Sn stands for the ITER Model Coil and the comparison of the test methods and results of the four ITER Home Teams. The third ITER strand benchmark test is an extension of the characterization to all the strands manufactured for the Model Coil. This characterization includes the determination of the critical current at 12 T and 4.2 K, the hysteresis and coupling losses in fields of +/-3 T, the residual resistive ratio, twist pitch, chrome plating thickness and Cu/non-Cu ratio. In addition, from each specimen, the critical current is measured at a magnetic field of 13 T and at temperatures between 4.2 K and 8 K.

1 Introduction

In the second ITER benchmark action industrial Nb_3Sn wires of four Model Coil strand producers being EM-LMI, IGC, Bochvar and Hitachi Cable were characterized [1]. Now, this strand characterization activity is extended to the remaining manufactures i.e. Mitsubishi (MIT), Furukawa (FUR), Vacuumschmelze (VAC) and Teledyne Wah Chang (TWC). In addition, modified EM-LMI strand material is also added to this third round of the ITER Nb_3Sn strand benchmark action.

The Nb_3Sn strands applied in the ITER fusion programme can be divided in to the categories: high critical current density strands with relatively high hysteresis loss and low hysteresis loss strands exhibiting a relatively low critical current density. The qualities are indicated with HPI and HPII and will be appropriate for the TF and CS coils respectively [2]. Specimen of both categories are part of this third benchmark action.

The VAC and FUR wires are bronze process strands while the MIT, TWC and LMI wires are internal tin type of conductors. To protect the stabilising copper of the strand, the filaments (FUR) or filament bundles (VAC, MIT) are surrounded by a single layer diffusion barrier of Ta. In the TWC conductor the sub elements are surrounded by a single layer diffusion barrier of Nb-Ta alloy. The 36 sub elements in the LMI strand are equipped with a double layer barrier of Nb and Ta. All strands are Cr-coated to reduce the coupling loss in the cable manufactured with these strands by increasing the interstrand contact resistance.

All samples of each type of strand are heat treated at once under vacuum conditions according to the schedule provided by the suppliers. In order to prevent leakage of tin during heat treatment of the internal tin type of conductors, both ends of the samples are extended by a few centimetres and squeezed.

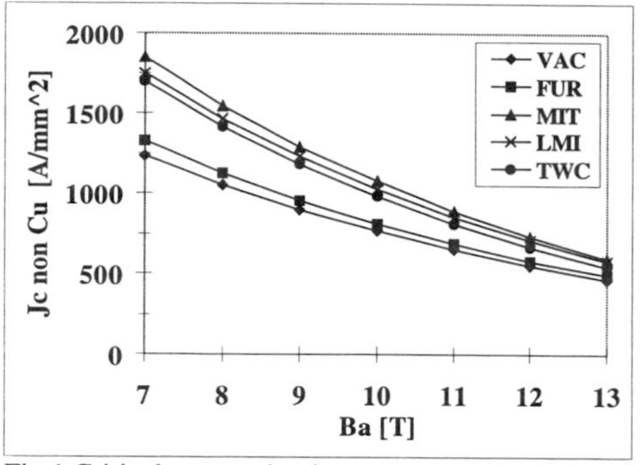

Fig 1 Critical current density versus magnetic field for T = 4.2 K and E = 0.1 μV/cm.

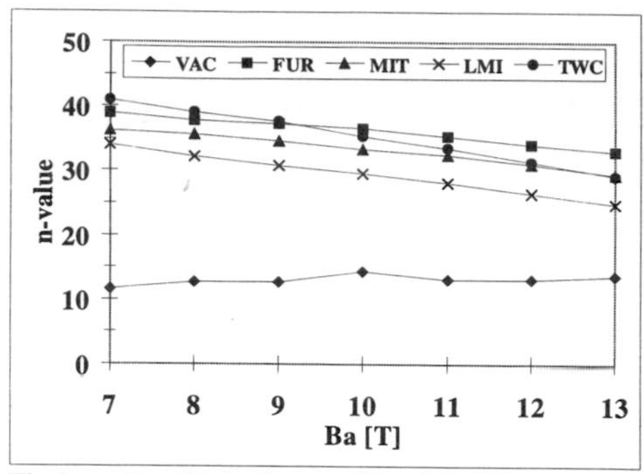

Fig 2 n-value obtained around E = 0.1 μV/cm versus the applied magnetic field

2 Critical current

In the third benchmark test the same sample holder for the I_c measurements is used as in the second benchmark action [1]. The strand on the sample holder is fixed with STYCAST 2850 FT and the voltage is measured via taps over a length of 50 cm.

From each strand type a series of four out of six specimen are measured at 4.2 K, with the criterion of 0.1 μV/cm in a background field of 12 T. In addition, for one sample out of the series the I_c-measurements are extended to a magnetic field range from 7 T up to 13 T in steps of 1 T. The results of the measurements are shown in Figure 1. The subdivision in the two ITER strand categories HPI and HPII, mentioned in the introduction, is obvious. A large scattering in I_c values is observed on the TWC specimen. Moreover, the values show a considerable lower value than the 780 A/mm^2 at 12 T provided by the manufacturer. In Figure 2 the accompanying n-values are shown. Figure 3 shows the corresponding Kramer plots of the Jc(B) curves of all strands. The upper critical field B_{c2}* is 28.3 and 27.7 T for the VAC and FUR (bronze produced) strands and 25.3, 24.9 and 24.6 T for the LMI, MIT and TWC strands respectively.

On one of the samples of each specimen the critical current is measured as function of the temperature at 13 T, starting from 4.2 K up to 8 K. A critical temperature of 10.2 K for VAC and FUR, 9.49 K for LMI and 9.31 K for the MIT and TWC strands respectively, can be deduced from the results as shown in figure 4.

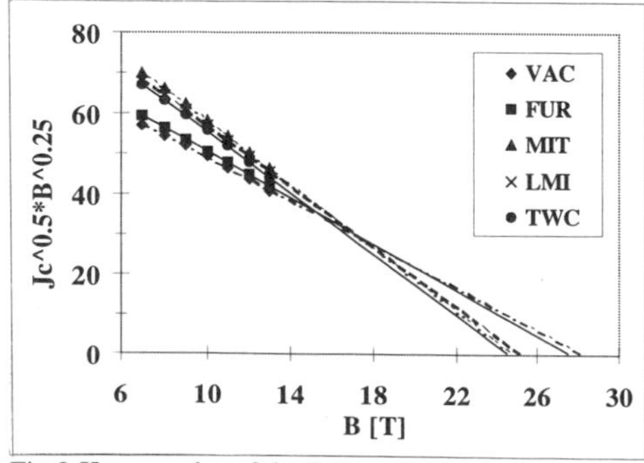

Fig.3 Kramer plot of the J_c(B) curves to determine the upper critical field.

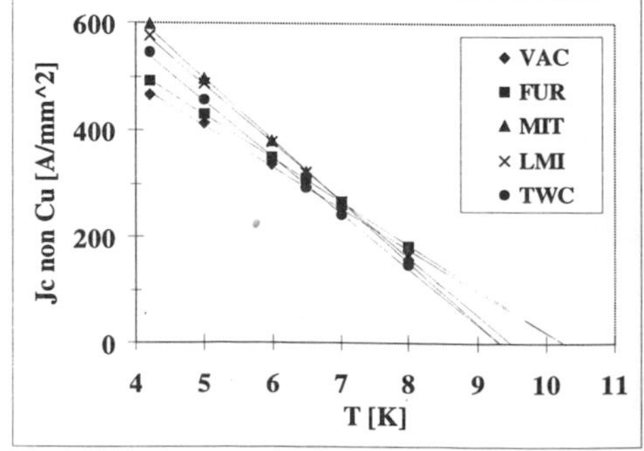

Fig 4 Critical current density versus temperature at 13T, E = 0.1μV/cm.

Table 1 Hysteresis loss and coupling current time constant

	VAC	FUR	MIT	LMI	TWC
Qhys nonCu [mJ/cm^3]	94	91	136	595	599
Coupling loss time constant [ms]	0.62	3.1	1.3	5.9	6.3

3 AC losses

The sample for the magnetization measurements is ring-shaped with an inner diameter of Ø40 mm and a height of 25 mm. The samples have two meter of strand, are insulated with glass-cloth and are wound in two layers. The loss measurements are performed in a magnetometer at liquid He temperature. The time varying magnetic field is applied perpendicular to the specimen axis. The total loss per cycle is measured as a function of the frequency at a constant amplitude of the applied AC field. The field is taken sinusoidal with a frequency up to 20 mHz and an amplitude of B_a = 3T. No DC background field is present. At low frequencies the coupling loss increases linearly with frequency. From the slope α of the linear Q_{tot} versus frequency curves, the coupling current constant $n\tau$ can be deduced using

$$n\tau = \alpha.\mu_0 /(2\pi^2.\, B_a^2) \qquad [s], \qquad (1)$$

where B_a is the applied magnetic field and n a shape factor which is 2 for circular cross sections. The calculated $n\tau$ values are collected in Table 1. Taking into account the dependence of $n\tau$ on the square of the twist pitch and the conductivity of the matrix material, the $n\tau$ values of TWC and LMI strands are roughly a factor 5 and 10 times larger than those of the low loss conductors which suggest the influence of the Nb in the diffusion barrier [3].

The hysteresis loss as shown in Table 1, obtained by extrapolation of the linear curves of the total loss down to zero frequency, is depicted as function of the critical current density in figure 5. The high critical current density of the MIT conductor in combination with the relatively low hysteresis loss is striking.

4 Other tests

The residual resistive ratio values (R273K/R20K) of the strands are determined with a direct current measuring technique. The measurements are performed on the Cr-coated specimen with a sample current of 400 mA and a distance between the voltage taps of 130 mm. The results are collected in Table 2.

To determine the Cu:non-Cu ratio two methods are used viz. digital images (2048 x 1920) analyses of pictures of the cross section taken by back scattering scanning electronic

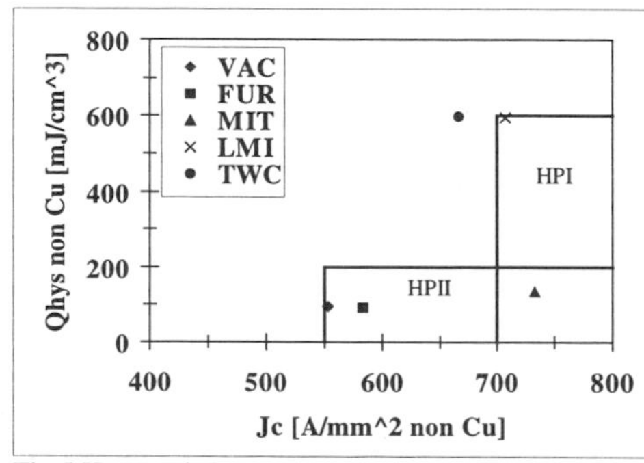

Fig 5 Hysteresis loss versus critical current density at 12 T, E = 0.1μV/cm and T = 4.2 K.

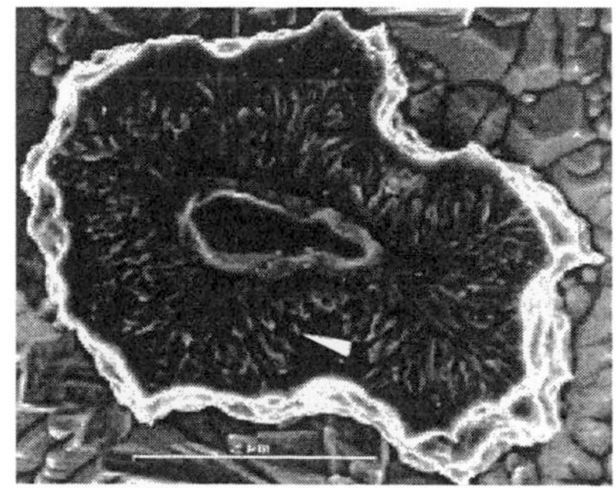

Fig 6 SEM image of etched FUR filament showing filament reaction level and columnar grain morphology.

Table 2 Summary of the strand test results

Strand manufacturer	VAC	FUR	MIT	LMI	TWC
Technique	Bronze	Bronze	Internal tin	Internal tin	Internal tin
Diameter (excl. Cr-layer)	0.803	0.802	0.801	0.806	0.802
Barrier	Ta	Ta	Ta	Ta + Nb	Ta / Nb
Thickness barrier [um]	10 - 15	6 - 8	5	1 + 2-3	3
Twist pitch [mm]	8.8	18.4	18.0	9.9	9.3
Heat treatment	570°C/220 h 650°C/175 h	650°C/240 h	200°C/6 h 350°C/18 h 450°C/28 h 580°C/180 h 650°C/240 h	220°C/175 h 340°C/96 h 650°C/180 h	185°C/120 h 340°C/72 h 650°C/200 h
Filament reacted	partially	partially	fully	fully	partially
Cu:non-Cu: Reported	1.46	1.49	1.57	1.5	1.51
Image analysis	1.56	1.45	1.56	1.35	1.51
Weighing	1.49	1.49	1.59	1.38	1.61
Critical temperature T*@ 13T	10.2	10.2	9.31	9.49	9.31
Upper critical field Bc2*	28.3	27.7	24.9	25.3	24.6
Cr thickness [μm]	2.1	3.2	2.6	1.7	2.3
RRR (R273K / R20K)	150	147	130	80	213
Overall strand density [g/cc]	9.33	9.14	9.04	8.98	9.01

microscope [4] and chemical etching and weighing. The agreement among both measuring techniques is good as can be seen in Table 2, except for the results of the TWC samples. The registered values of the later sample are obtained from multiple measurements. Nevertheless a discrepancy of 7 % is measured. The results correspond well with the data provided by the supplier except for the LMI strand were we obtained a discrepancy of 10%.

The Cr plating thickness is also determined by measuring the strand diameter in combination with chemical etching the Cr and weighing. The depicted data in Table 2 is the average of four measurements.

SEM examination of the strand material shows that in the VAC, FUR and TWC conductors almost all filaments are partially reacted. The filament reaction level of a filament in the FUR strand is shown in the SEM image depicted in Figure 6.

5 Conclusions

Except for the TWC strand, all strands are found to be within the range of the desired I_c and Q_{hys} specifications. Moreover, the Mitsubishi strand combines a high critical current density with a relatively low hysteresis loss.

A large scattering in the results of the I_c measurements on the TWC samples is observed. Despite the fact that the ends of all samples were prepared with additional length which were squeezed carefully, the TWC samples showed after the heat treatment small amounts of Sn at the ends that may explain the differences in the obtained and expected values. The influence of tin leakage on I_c and Q_{hys} is an issue for further investigation.

References

[1] Bruzzone P, ten Kate H.H.J., Nishi M , Shinkov A , Minervini J and Takayasu M 1995 Advances in Cryogenic Engineering, Vol 42, 1351 - 1358.

[2] Mitchell N 1994 Conductor designs for the ITER Toroidal and Poloidal magnet systems. IEEE trans. on magnetics Vol.30, no.4, 1602-1607.

[3] Nijhuis A, Knoopers H G and ten Kate H H J 1994 The influence of the Diffusion barrier on the AC Loss of Nb_3Sn Superconductors. ICEC 15 Proceedings Cryogenics Vol 34, 547-550.

[4] Lee P J 1997 *UW testing methods for Cu:Non-Cu Ratio.* ITER Report: 3rd ITER Benchmark-UW-ASC CU:Non-Cu.

Inst. Phys. Conf. Ser. No 158
Paper presented at Applied Superconductivity, The Netherlands, 30 June–3 July 1997

The role of vortex melting and inhomogeneities in the transport properties of Nb_3Sn superconducting wires

N. Cheggour and D. P. Hampshire.

Department of Physics, University of Durham, Durham, DH1 3LE, U.K.

Abstract. The critical current densities have been measured in a Nb_3Sn commercial superconducting multifilamentary wire from 4.2K to T_c, in magnetic fields up to 14 tesla. V(I) curves were investigated. The irreversibility line of this material is discussed in terms of a phase transition from a vortex-glass-to-liquid and high degree of disorder due to inhomogeneities.

1. Introduction

The behavior of magnetic flux lines plays an important role in the critical current densities of superconductors. When these flux lines in the form of either an ordered lattice or a glass state are pinned by various defects in the material, the supercurrent can flow through the sample. If the pinning provided is weak and the interactions between vortices are low compared to thermal energies, the vortex system can be seen as independant flux lines behaving as a liquid and the supercurrent is reduced substantially. This introduces a melting line in the B-T phase diagram, which lies well below $B_{c2}(T)$. Many investigations of the transport properties have reported the so-called phase transition from vortex-glass-to-liquid in the High-T_c superconductors [1,2]. More recently, Junod et al have given strong evidence for the melting in YBCO system using specific heat measurements [3]. In this paper, we address the question whether in case of Nb_3Sn commercial superconducting wires, the onset of the dissipative state is a consequence of a phase transition from vortex-glass-to-liquid or, alternatively, is due to disorder caused by inhomogeneities of fundamental superconducting parameters throughout the sample. We have measured the critical currents as a function of magnetic field at different temperatures, and studied the shape of the V(I) curves to address this issue.

2. Sample preparation and measurements

A Vacuumschmelze commercial multifilamentary Nb_3Sn wire, having a diameter of 0.37 mm and a total length of about 70 cm, was wound on a thin walled stainless steel sample holder and heat treated under an argon atmosphere at 700 °C for 64 hours. This heat treatment is optimum for obtaining high current densities. The wire together with the sample holder was electroplated with copper , which makes it easy to solder the wire onto its holder. This soldering protects the wire against the lorentz force exerted during measurements in high magnetic fields and provides a better shunt to prevent the wire from burning-out in the event of a quench at high currents. The sample is then mounted onto the probe, which includes an isothermal environment which contains capacitance and RhFe thermometers. The measurements were carried out under vacuum to ensure good temperature control . The uncertainty in the temperatures quoted is estimated to ±0.1K. At the bottom end of the probe where the sample is situated, the current leads are extended by Nb_3Sn wires to reduce the heat generated during the current sweep. This enables measurements up to high currents (>100A) while maintaining good temperature control. The critical current densities of the sample are measured in magnetic fields up to 14 tesla, at different temperatures ranging from 4.2 to 15 K. Measurements were taken in an increasing magnetic field. For a particular temperature, we did extend acquisition well above the critical current to study the shape of V(I) curves over a wide range of voltage (about 4 orders of magnitude).

3. Results and discussion

3.1. Critical currents and irreversibility line

In figures 1-3, we present the measured engineering critical current densities $J_c(B,T)$ defined by the standard $1\mu Vcm^{-1}$ criteria, as well as the Kramer plots from which we extrapolated the irreversibility fields $B_{irr}(T)$ at which the supercurrent vanishes. The value of T_c is about 16K, in good agreement with that given for thin film Nb_3Sn dirty samples [4]. The irreversibility line lies unambiguously below the upper critical field $B_{c2}(T)$ at which the superconductivity disappears for thin films [4]. As an example, at 11K, B_{c2} is about 14 tesla, while we found B_{irr} to be 8.9 tesla.

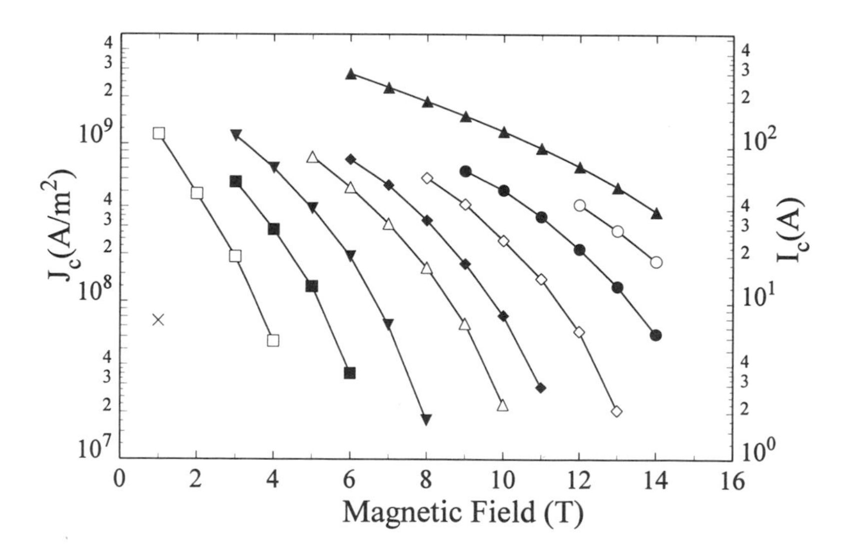

Figure 1: $J_c(B)$ at 4.2K, from 6K to 13K every 1K, and 15K for Nb_3Sn multifilamentary wires

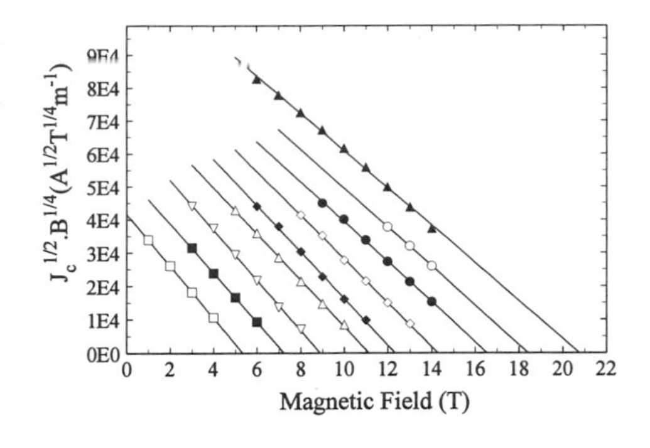

Figure 2: Kramer plots at 4.2K, and 6K to 13K every 1K, for a Nb_3Sn multifilamentary wire

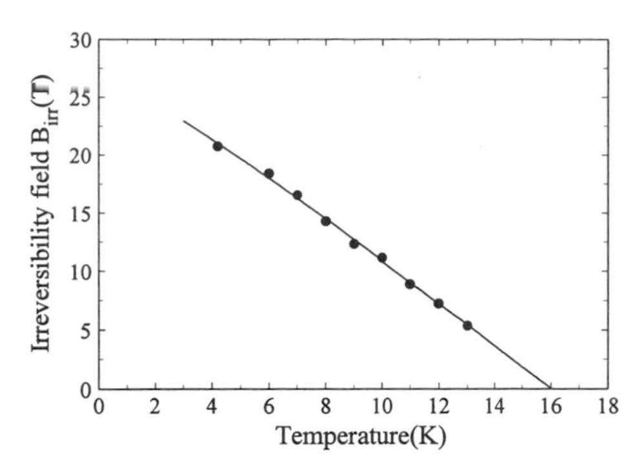

Figure 3: Irreversibility line for a Nb_3Sn multifilamentary wire

3.2. V(I) curves

At high temperatures and/or the high voltage regime, the current flowing through the shunt becomes too important to be ignored. The resistance of the shunt does not depend on temperature, within the range 6 to 20 K investigated. We have measured the shunt resistance (R_{Sh}) at 20 K for the different magnetic field values. If the shunt is considered as a parallel resistance to the superconductor, the current flowing through the superconductor is then $I_s = I - V/R_{Sh}$. Figure 4 shows V(I) curves for increasing applied magnetic fields at 11 K, after substracting the current sharing due to the shunt.

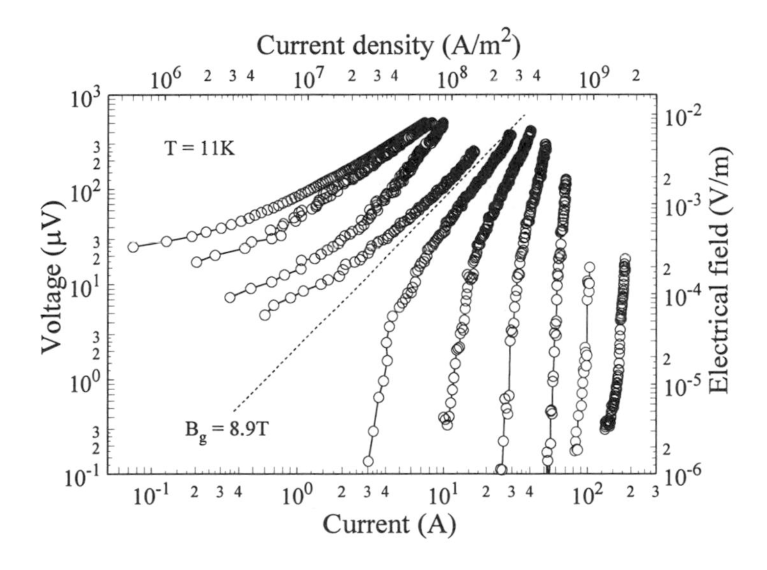

Figure 4: V(I) curves at 11K from 3T to 11T every 1T, and 13T

The dominant feature of Figure 4 is the change in curvature occuring at a particular transition magnetic field B_g. For fields ≥ 9 tesla, V(I) has a positive curvature indicating a voltage convergence towards a finite value for extremely small currents, meaning that at these fields, there is no true supercurrent flowing into the sample. For fields ≤ 8 tesla, there is a negative curvature suggesting that a true supercurrent does exist. An estimation of the irreversibility field at which the sample swiches from zero-resistance state to a dissipative one can be obtained using the Kramer extrapolation, which gives 8.9 tesla at 11K. This is consistent with the dotted line in Figure 4 which separates the two regions with different curvatures. According to that, the curvature change can be directly related to the irreversibility line. V(I) curves present similar features to those which have been attributed to a vortex-glass-to-liquid phase transition in the high-T_c superconductors. However, similar shapes can also be obtained in systems having a large degree of disorder [5]. Inhomogeneities in T_c and B_{c2} mean that superconductivity can be destroyed progressively in increasing magnetic field, within weak parts of the material. If these parts are randomly distributed throughout the sample, and the distribution of fundamental superconducting parameters is wide enough, at some particular magnetic field below the bulk B_{c2}, there is no percolative path through the superconductor to allow the flow of supercurrent. Although B_g can be seen as the field at which a transition from vortex-glass-to-liquid phase occurs, it may also be the field at which no percolative path is available through the superconductor.

4. Conclusion

Measurement of the critical current density as a function of temperature and magnetic field has shown that Nb_3Sn superconducting wire has an irreversibility line occuring below the upper critical field. Investigation of V(I) curves for increasing applied magnetic fields exhibit similar behavior to that which is widely accepted as a signature of a vortex-glass melting in case of high-T_c materials. We do not exclude this explanation in case of Nb_3Sn wires. However, inhomogeneities of superconducting properties in the sample creating a high degree of disorder could also explain the irreversibility line. At B_{irr}, a random distribution of weak regions could completely isolate the remaining superconducting parts from each other, providing no superconducting percolative path through the material. These results also suggest that the critical current densities in Nb_3Sn superconductor can eventually be enhanced by reducing inhomogeneities and/or improving the pinning to prevent the vortex system from melting.

Acknowledgements

We would like to thank H.A. Hamid for his important help during the measurements. We are also greatful to Dr. D.N. Zheng for useful discussions of the results. This work was funded by the E.P.S.R.C.

References

[1] Li Q et al 1994 *Phys. Rev. B* **50** 4256
[2] Mawatari Y, Yamasaki H, Kosaka S and Umeda M 1995 *Cryogenics* **35** 161
[3] Junod A et al 1997 *Physica C* **275** 245
[4] Orlando T P et al 1979 *Phys. Rev. B* **19** 4545
[5] Harris D C, Herbert S T, Stroud D and Garland J C 1991 *Phys. Rev. Lett.* **67** 3606

Inst. Phys. Conf. Ser. No 158
Paper presented at Applied Superconductivity, The Netherlands, 30 June–3 July 1997

Biaxially textured substrate tapes of Cu, Ni, alloyed Ni, (Ag) for YBCO films

W. Goldacker, B. Ullmann and E. Brecht, G. Linker

Forschungszentrum Karlsruhe, Institut für Technische Physik, Institut für Nukleare Festkörperphysik, Postfach 3640, D-76021 Karlsruhe, Germany

Abstract. The cubic recrystallization of the deformation texture in tapes of fcc metals and alloys like Cu, Ni, Ag, Al is applied to prepare biaxially textured substrate tapes for YBCO films. In Cu tapes of suitable purity a grain alignment within < 5 ° (FWHM) in each direction was realized. In Ni tapes different purities and rolling techniques led to significantly different cubic textures. A strong influence of side limited rolling was found. Grain misorientation of < 5° and > 10° in different sample directions were observed. For small sample sections (≈ 1 mm^2) the out of plane grain misalignment sharpens to < 3°. Cubic texture was also achieved in a mechanically tough commercial Ni65 Cu33Fe2 alloy, but still with broader grain alignment, presumably due to the non optimized treatments. In Ag tapes we found an indication for a possible cubic recrystallisation.

1. Introduction

The RABITS™ method (rolling assisted biaxially textured substrate) has demonstrated recently its potential for realizing thick YBCO films with high transport current densities of 10^6 Am^{-2} [1, 2]. A significant advantage compared with the IBAD technique [3] is the possibility to prepare readily long lengths of biaxially textured tapes. The main goal in the tape preparation is achieving completely cubic texture (100) [001] with small grain misorientation and absence of wrongly oriented grains, which cause large angle grain boundaries in the YBCO layer. The present cubically recrystallized (CR) pure Ni tapes have grain misalignment of 7 - 10° within a few mm^2 sample area in both, in plane and out of plane directions. The texture was transferred without loss of texture quality via Pd, CeO_2, YSZ layers to the YBCO film. The reported critical current density refers to 3 mm sample section. The advantages using pure Ni are the improved recrystallization and the favourable lattice parameter match with the buffer and YBCO, the disadvantage is the magnetism and the softness after annealing. For technical application, mechanical reinforcement of the tape is necessary, preferably modifying the tape material itself preserving a high engineering current density.

CR is well known to occur in heavily rolled (> 95 % deformation) fcc metals and alloys of Cu, Ni, Fe, Al, Au, Ag ... The amount and quality of the cubic texture depends sensitively on the deformation (rolling texture (110 [112]) and sample composition, since grain alignment in the rolling texture and stored dislocation energy are initial conditions determining effectiveness and quality of the CR process [4].

The aim of this paper is a first study of the influence of the rolling technique on the cubic texture, analysed for differently sized sample sections with respect to grain alignment quality. Weak alloying favours the CR but strengthening of the matrix is limited. So we investigated a higher alloyed Ni65Cu33Fe2 tape for its possibility to develop a cubic texture. Significant changes of the parameters and the kinetic of the CR are expected. Suitable alternative metals as Ag, were tried to investigated for their capability to form a cubic texture.

2. Experimental

The substrate tapes were cold rolled with 10 % reduction steps to cross sections < 5 % using either a conventional two roll mill or an innovative four roll machine (driven turk head) allowing side limitation and a true unidirectional deformation. Depending on hardness the tapes were typically 0.12 x 8-12 mm in cross section before annealing in Ar or vacuum for 2 - 100 h. Texture was analysed by X-ray powder spectra, rocking curves and pole figures on different sample areas 1 - 80 mm^2. Mechanical material data were measured on a precision strain rig. We report results on Cu, Ni, Ni alloy and Ag samples (see tab.)

Table: Specifications of the recrystallized Cu, Ni and Monel tapes

Sample	Cu 1	Cu 2	Ni 1	Ni 2	Ni 3	Monel
purity %	99.9995	99.9	99.95	99.95	99.5	---
rolling	2 R	2 R	2 R	4 R	2 R	2 R
anneal °C/hrs.	400/10	400/10	1000/10	1000/10	1000/10	1000/70
grain size μm	100-400	≈ 10	50-300	200-600	50-300	20-300
Youngs M. GPa (not annealed)	20	40 (175)	50 (237)	47 (350)	207 (240)	173
$R_p^{0.2}$ MPa	48	56	22	42	73	170
H_v Nmm^{-2}	40	56	52	71	103	123

3. Results

Copper: Especially for Cu it is well known that a small impurity content a metal or oxygen is necessary for complete CR. In the case of the extremely pure sample Cu 2, cubic texture does not form and the final Youngs modulus was only 20 GPa. In technical electrolytical copper however, sample Cu2, the cubic texture forms quantitatively as shown in the (200) and (220) pole figures (fig. 1, 2). The grains are well aligned in all three directions with < 5° FWHM. A slightly sharper alignment was observed for the rolling direction.

Nickel: The two Ni1, Ni2 tapes were rolled differently but both samples recrystallize nearly completly to the cubic texture. An exact estimation of the grain alignment is hindered through the spike like diffraction lines of the relatively large individual grains. Ni1 from the 2 roll machine had an in plane grain misalignment of 8 - 10° and < 8° (40 mm^2 sample area) out of plane, which is comparable to the results of ref [1, 2]. Varying the measured area from 40 mm^2 to 1.5 mm^2 by screening the X-ray beam, the measured out of plane misalignment reduces continuously from ≈ 8° to only ≈3°: This demonstrates that we have locally a much better texture than in average over the whole tape. The same material rolled with side limitation (4 roll machine), sample Ni2, behaves quite different (fig. 3). The grain

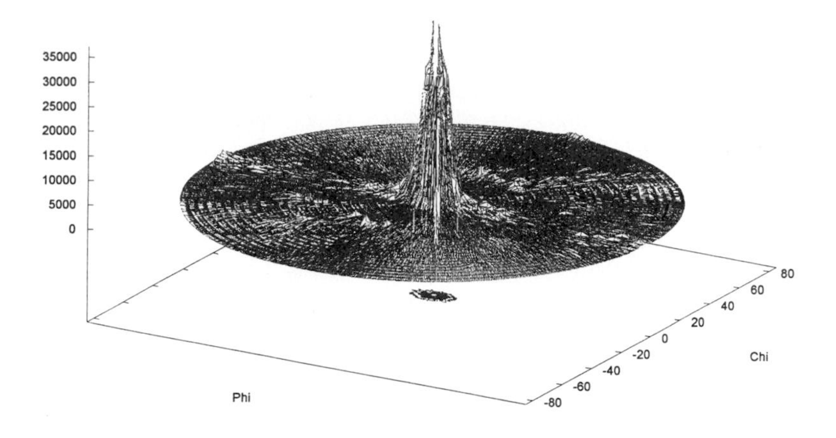

Fig. 1 Cubic texture in tape Cu2 (200) pole figure.

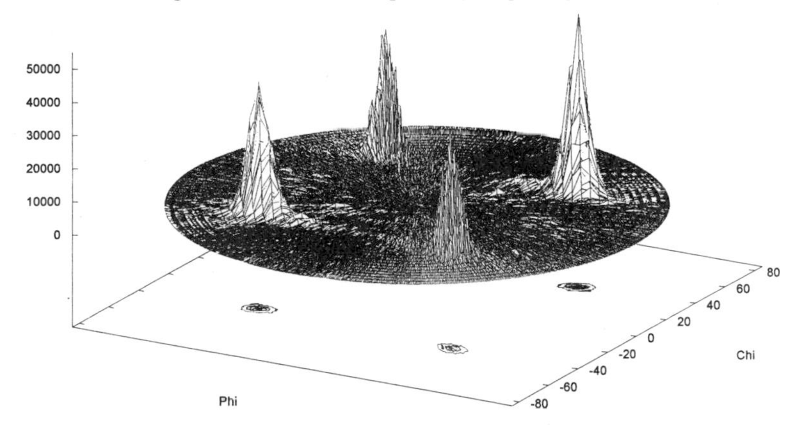

Fig. 2 Cubic texture in tape Cu2 (111) pole figure.

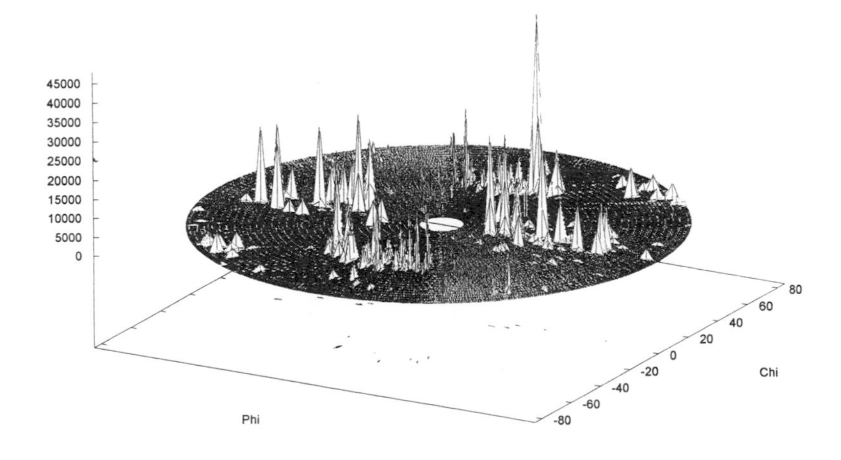

ig. 3 Cubic texture in tape Ni2 rolled in a 4 roll machine (arrow gives rolling direction) (220) pole figure.

misalignment is now anisotropic, with a sharper distribution along the rolling direction (≈ 5° for 20 mm^2 area) and a stronger scattering lateral to the rolling direction (> 10°). This proves nicely the importance of deformation strategy and the strong link between deformation texture and cubic texture. Sample Ni3 with < 0.5 % non specified impurities was treated equivalenty to sample Ni1, formed cubic texture, but with worse grain alignment between 8-15° out of plane FWHM and a significant number of non textured randomly orientated grains. The slight change of the Youngs modulus indicates an ineffective transver of dislocation energy for CR.

Ni65Cu33Fe2 (MONEL): The 3 component phase diagram Ni-Cu-Fe gives a wide range of alloys suitable for CR including MONEL [4]. Despite of a slight chemical instability (Cu loss) during heat treatment, cubic recrystallisation was obtained, but with much larger scattering in the grain orientation. Further improvement is expected for an optimized heat treatment program which is expected to differ from the Ni conditions. A very promising result, however, is the strongly improved mechanical strength compared to pure Ni (table 1) with $Rp^{0.2}$=170 Mpa. This value is more than 3 times the value of Ni, being sufficient for many future applications.

Ag tapes: Cold rolled Ag and AgMg tapes, investigated as alternatives to Ni, developed predominantly a different rolling texture which recrystallizes usually as (113) [211]. We observed a mix of texture, with an obvious partly forming cubic texture, observed via a strong increase of the (002) reflection and the decrease of (111) measured in out-of plane direction. We evaluate this as indication that CR may be achievable quantitatively for changed composition and treatment parameters.

4. Conclusions

With first investigations on Cu and Ni tapes it was demonstrated that sample composition tape deformation technique (especially side limited rolling) and annealing influence strongly the performance of the cubic tape texture. The observed narrow local grain misalignment in Ni tapes, being 3° for 1,5 mm^2 tape area, is very promising for an improved sample preparation and extension of this texture quality to long tape lengths. MONEL tapes or similar Ni alloys are possible solutions preparing mechanically stable YBCO tapes, although the quality of grain alignment needs strong improvement. But the optimization of the treatment parameters has still to be done. In the Ag system, cubic recrystallization was observed in a minority sample volume, being an indication that an improved and changed composition and treatment for a complete formation of cubic texture will be found.

References

[1] A. Goyal, D.P. Norton, J.D. Budai, M. Paranthaman, E.D. Specht, D.M. Droeger, D.K. Christen, Quing He, B. Saffian, F.A. List, D.F. Lee, P.M. Martin, C.E. Klabunde, E. Hatfield and V.K. Sikka Appl. Phys. Lett. 69 (1996) 1795

[2] Quing He, D.K. Christen, J.D. Budai, D.F. Lee, A. Goyal, D.P. Norton, M. Paranthaman, F.A. List D.M. Kroeger, Physica C 275 (1997), 155

[3] X.D. Wu, S.R. Foltyn, P. Arendt, J. Townsend, C. Adams, I.H. Campbell, P. Tiwari, Y. Coulter and D.E. Peterson, Appl. Phys. Lett. 65 (1994) 1961

[4] "Structure of Metals" Charles S. Barrett, Ed. R. F. Mehl, Mc. Graw-Hill, NY, 1952, p. 494-509 and refs therein

Inst. Phys. Conf. Ser. No 158
Paper presented at Applied Superconductivity, The Netherlands, 30 June–3 July 1997

Preparation of textured Tl(1223)/Ag superconducting tapes

Emilio Bellingeri, Roman E. Gladyshevskii and René Flükiger

Département de Physique de la Matière Condensée, Université de Genève
24 Quai Ernest Ansermet, CH-1211 Genève 4, Switzerland

Abstract. Two techniques of tape manufacturing, powder-in-tube and deposition, have been compared for substituted Tl(1223) ceramics.

Tapes prepared by the powder-in-tube method from pre-reacted (*ex-situ*) powder showed poor texture. Grain alignment and density could be slightly improved by carrying out the formation of Tl(1223) inside the tape (*in-situ*) under high isostatic pressure, or by partly substituting oxygen by fluorine in the superconducting ceramic.

A significant improvement of the texture in the superconducting layer was observed for electrophoretic deposition. The degree of texture of these samples was comparable to that usually reported for Bi-based tapes. Reproducible critical current densities of 16,000 A/cm^2 at 77 K in zero field were reached.

1. Introduction

In the last years many efforts have been made to develop Tl(1223) tapes by the powder-in-tube (PIT) method and relatively high values of the critical current density in zero magnetic field have been reached [1-3]. However, a strong magnetic-field dependence is observed and the transport current remains limited by weak links between the grains, mainly due to the negligible texture [4].

In order to obtain a higher degree of texture we tried to approach the melting point of the superconducting phase. A complete melt-texture procedure of Tl(1223), cannot be realized in the conditions explored up to now, since the phase is not in thermodynamic equilibrium with a liquid, but decomposes into Tl(1212) before melting. We found that in particular conditions it is possible to achieve a partial melting and to grow the Tl(1223) grains in presence of a liquid phase. The region where a liquid is present can be reached under an isostatic pressure of 2 kbar, or by partly substituting oxygen by fluorine in the Tl(1223) phase [5,6].

Another preparation path studied here included melting of the sample at very high temperatures (up to 1100 °C) before the formation of Tl(1223). The plate-like crystallites produced this way could be oriented by combining electrophoretic deposition and uniaxial pressing [7].

2. Experimental

2.1. Powder-in-tube

A precursor of nominal composition $Sr_{1.8}Ba_{0.2}Ca_{1.9}Cu_3O_x$ was obtained by a two-step reaction of $SrCO_3$, $BaCO_3$, $CaCO_3$ and CuO, successively at 900 and 980°C, for a total reaction time of 48 h. Starting powders were prepared by mixing the calcined precursor with Tl_2O_3 and PbO (nominal composition $Tl_{0.6}Pb_{0.5}Sr_{1.8}Ba_{0.2}Ca_{1.9}Cu_3O_x$, hereafter referred to as Tl,Pb(1223)). The samples containing fluorine were prepared using a mixture TlF / $TlO_{1.5}$ (mole ratio 5 / 1) instead of Tl_2O_3 (nominal composition $Tl_{0.6}Pb_{0.5}Sr_{1.8}Ba_{0.2}Ca_{1.9}Cu_3O_xF_y$, Tl,Pb(1223)F). The mixtures were filled into Ag(Au) tubes (6 / 4 mm diameter) which were then swaged, drawn and rolled to a thickness of 80-180 μm. The Tl,Pb(1223) tapes were uniaxially pressed at 1.5 GPa and processed in sealed Ag(Au) envelops in a high pressure furnace at a helium pressure of 2 kbar (partial oxygen pressure 1 bar) at 900 °C for 3 h. The Tl,Pb(1223)F tapes were submitted to up to four cycles of pressing at 1 GPa, followed by an *in-situ* reaction at a temperature between 920 and 950 °C (50 bar He / 1 bar O_2, 0.75 h).

In none of the treatments weight losses were detected. After the reaction the F-free and the F-containing tapes were once again mechanically pressed and annealed in flowing oxygen for 5 h at 870 and 850 °C, respectively.

2.2. Electrophoretic deposition

A calcined heavy-metal free precursor, prepared as described above, was mixed with Tl_2O_3, PbO and Bi_2O_3 and pelletized (nominal composition $Tl_{0.7}Pb_{0.2}Bi_{0.2}Sr_{1.8}Ba_{0.2}Ca_{1.9}Cu_3O_x$, Tl,Pb,Bi(1223)). The pellets were melted at 1050 °C and 50 bar He / 1 bar O_2 and then quenched. The material, ground and pelletized again, was reacted at 930 °C under the same pressure for 3 h and ground again. The powders were suspended in an organic solvent with a Tl(1223) / solvent weight ratio of ~ 1 / 250. Silver substrates were used and their shape, as well as the voltage applied between the electrodes, the time of deposition and the uniaxial pressing, were optimized to get a uniform deposition. The best results were obtained for ten cycles of deposition at 100 V for 20 sec and pressing at 0.5 GPa, which produced a ~ 10 μm-thick ceramic layer. Multilayer tapes were produced by pressing together several layers of deposited powder. A vacuum treatment at 400 °C, aimed to remove organic residues, and a final annealing between 750 and 850 °C in flowing oxygen, were performed on all samples.

3. Results and discussion

X-ray diffraction diagrams of different kinds of tape are presented in Fig. 1. High purity was obtained for all samples, even those containing fluorine. Up to 60 % texture was observed for the samples prepared by the PIT method and reacted at a pressure of 2 kbar and a slightly lower value for F-containing samples submitted to several annealing-pressing cycles. A very high degree of texture (95 %) was reached for electrophoretically deposited tapes, as indicated from the predominance of the *00l* reflections.

The different treatments applied to Ag(Au)-sheathed Tl,Pb(1223)F tapes are summarized in Fig. 2a. As can be seen from Fig. 2b, the critical current (77 K, 0 T) increased during the first three cycles, reaching a critical current density of 10,000 A/cm^2, but then dropped drastically. The sharp fall can be explained by the disappearance of the liquid phase

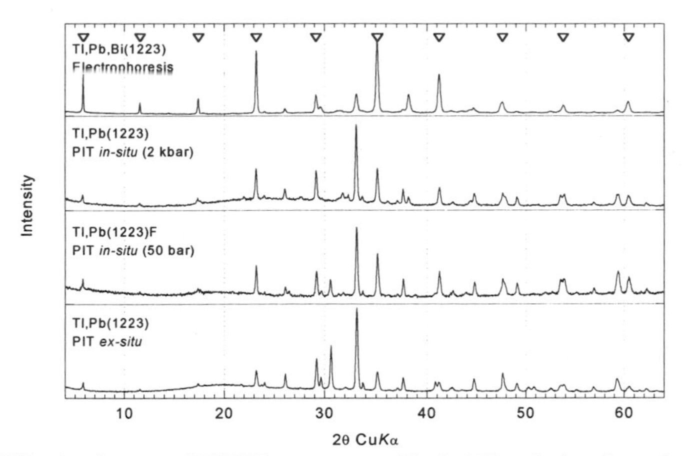

Fig. 1. X-ray diffraction diagrams of Tl(1223) tapes prepared by the PIT method or electrophoretic deposition. Symbols indicate *00l* reflections.

after a certain number of thermomechanical treatments, rendering crack healing afterpressing impossible. The drop of the critical current at 0.2 T, with respect to the value in zero field (Fig. 2c), reached a minimum factor of 16 after four cycles, whereas the ratio of the critical current with the magnetic field applied parallel and perpendicular to the tape surface reached the value 2.5 after the third cycle (Fig. 2d), indicating that some anisotropy had been introduced during the treatments.

The critical current density (77 K) *versus* the magnetic field for a Tl,Pb(1223) tape treated at 2 kbar and of a F-containing sample, with the field parallel and perpendicular to the surface of the tape, are presented in Fig. 3. The values of j_c in zero field are 6,000 and 10,000 A/cm^2, respectively. It can be seen that the field dependence could be reduced to a factor of 12 in a magnetic field of 0.2 T (field parallel to the tape surface) with respect to the value at 0 T for the sample prepared at 2 kbar. The critical current remained practically constant on increasing the field further. Both kinds of sample showed a better field

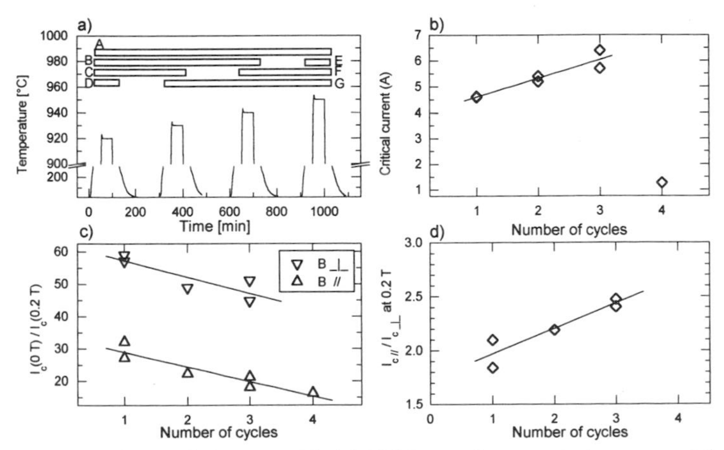

Fig. 2. Temperature profiles applied to seven Tl,Pb(1223)F tapes (*e.g.* sample A was treated four times) (**a**); critical current at 77 K and 0 T (**b**), ratio of the critical currents at 0 and 0.2 T (**c**) and ratio of the critical currents with the magnetic field (0.2 T) parallel and perpendicular to the tape surface (**d**) *versus* the number of cycles (cycles 1, 2, 3 and 4 correspond to sample E(D), C(F), B(G) and A, respectively).

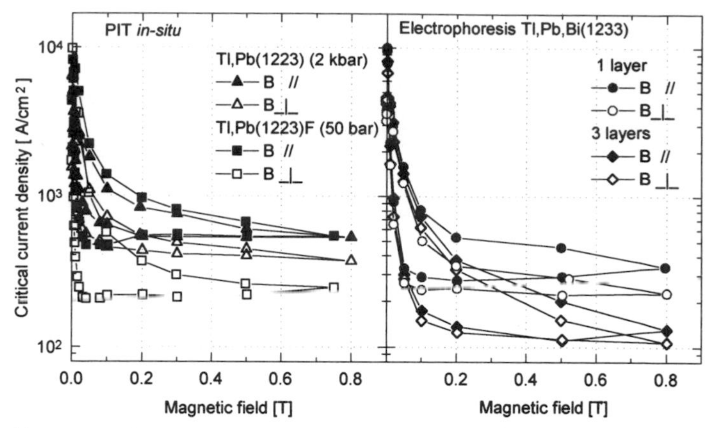

Fig. 3. Critical current density (77 K) *versus* the magnetic field applied parallel and perpendicular to the tape surface for Tl(1223) tapes prepared by the PIT method or electrophoretic deposition.

dependence than "traditional" tapes prepared by the PIT method with *ex-situ* reacted powder, where the ratio I_c(0 T) / I_c(0.2 T) is typically ~ 23 [3]. This improvement is probably a result of the higher density, rather than of the modest improvement in texture.

For samples prepared by electrophoretic deposition, reproducible critical currents of 4.5 A (77 K, 0 T), corresponding to critical current densities of 11,000 A/cm^2, were measured for 120 μm-thick 3-layer tapes with a total thickness of the superconducting core of 30 μm. Comparable values were observed in single-layer tapes. Fig. 3 shows the critical current density *versus* the magnetic field for a single- and a 3-layer tape. It was possible to preserve a high purity after annealing in flowing oxygen only up to 820 °C, *i.e.* about 50 °C below the optimal temperature required to connect the Tl(1223) grains [7]. This resulted in a strong magnetic field dependence, j_c dropping by a factor of 60 in a magnetic field of 0.2 T. Despite the high degree of *c*-axis texture, weak-link phenomena thus still limit the transport current. Treatments at temperatures higher than 820 °C did not have the desired effect on the grain connections and the annealing conditions need to be further optimized with respect to the critical current density. Experiments to improve the in-plane (*a,b*) alignment are also in progress.

Acknowledgements

This work was supported by the European Community, Brite Euram Project. No. 7055.

References

[1] Ren Z.F. *et al.*, 1995 *Physica C* **247** 163-168

[2] Gladyshevskii R.E. *et al.*, 1995 *Physica C* **255** 113–123

[3] Bellingeri E., Gladyshevskii R.E. and Flükiger R., 1996 *Il Nuovo Cimento,* accepted

[4] Everett J. *et al.*, 1996 in *High Temperature Superconductor: Synthesis, Processing, and Large Scale Applications,* Balachandran U., McGinn P.J. and Abell J.S., Eds. (TMS) pp 329-337

[5] Gladyshevskii R.E. *et al.*, *ibid.* pp 321-328

[6] Gladyshevskii R.E. *et al.*, 1997 *Proceedings of SPA'97*, accepted; *Physica C*, submitted

[7] Bellingeri E., Gladyshevskii R.E. and Flükiger R., 1997 *J. Supercond.*, submitted

Inst. Phys. Conf. Ser. No 158
Paper presented at Applied Superconductivity, The Netherlands, 30 June–3 July 1997

Critical current anisotropy minimum of Tl-1223 superconductors

B A Glowacki[1,2]

[1] IRC in Superconductivity, Madingley Road, University of Cambridge, Cambridge CB3 OHE, UK

[2] Departement of Materials Science and Metallurgy, Pembroke Street, University of Cambridge, CB2 3QZ, UK

Abstract A comparative study of the critical current anisotropy with field direction in low magnetic fields of Tl-1223 conductors manufactured by different techniques and prepared from different compositions is presented. For powder-in-tube conductors the anisotropy coefficient shows a very pronounced minimum, followed by a monotonic reduction of anisotropy with the increase of the magnetic field. This is explained in terms of: poor grain alignment with weak inter-granular superconducting coupling which cause 3D current percolation and also by demagnetising effect of the grains and the ceramic core in the powder-in-tube Tl-1223 tapes which depends on intergranular connectivity. For highly oriented layers along the c-axis perpendicular to the substrate manufactured by spray-pyrolysis technique the critical current anisotropy monotonically decreases. However there is some indication of a minimum at which can be due to the pronounced percolation of the current in the layers containing disoriented grain colonies. The power dependence J_c~$A*H_{min}^n$ for all types of conductors is discussed in details.

. Introduction

he task of developing and optimising practical conductors suitable for technological pplications at liquid nitrogen temperatures is a major area of worldwide research. The single l-O layer compound $(Tl,Pb)Sr_2Ca_2Cu_3O_9$, (Tl-1223) has an intrinsically superior in-field ehaviour relative to that of the double Bi-O layer superconductor $(Bi,Pb)_2Sr_2Ca_2Cu_3O_{10}$, (Bi-223) and is comparable only with $YBa_2Cu_3O_{7-\delta}$, (Y-123) [1]. Therefore great effort has been nade worldwide to process polycrystalline Tl-1223 conductors which will have both high ransport critical current at 77K and high magnetic flux pinning characteristics. Several factors lay an important role in determining the transport current in bulk Tl-1223 superconductors. he most important are the state of the grain boundaries and degree of texture of the ceramic rains. In polycrystalline ceramics the transport current across most high angle grain oundaries is severely limited [2]. A number of different techniques have been used to ianufacture Tl-1223 conductors. One of the common techniques widely used for production f HT_c superconducting wires is the powder-in-tube (PIT) process [3,4]. Most tapes are ibricated using a drawing-rolling method with an Ag sheath and a superconducting single re. There are two types of preparation of the superconducting powder before packing it into e metal tube: single powder, in each the powder of the given composition is partially or fully acted, [5] or two powder where it is a mixture of fully or partially reacted powders with fferent partial melting temperatures [6]. The two-powder method is aimed to improve the Tl-223 grain connectivity, rather than the grain orientation in the tape.

On the basis of the results obtained for Bi-2223, Y-123 and Tl-1223 tapes anufactured by the powder-in-tube technique, one can conclude that the superconductor grain geometry plays an important role in determining the degree of grain alignment. The Tl^{3+} ions in Tl-based cuprate materials have no lone $6s^2$ pair and the bonding between Tl-O layers is relatively strong. Therefore, from the point of view of solid state chemistry, the grain shape of Tl-based materials can be quite different from that of Bi-based cuprates but similar to Y-123. In Bi-2223 tapes, for example, grain alignment may be accomplished by different mechanical rolling or pressing procedures [7] because of the thin plate-like morphology of Bi-2223 grains. In contrast, irregular spherical grains in the Y-123 and Tl-1223 tapes are more difficult to align by the same procedure [5,8,9]. Encouraging results have been reported for Tl-1223 layers on non-metallic substrates [10,11] and also on silver textured ribbons [12] formed by the spray pyrolysis (SP) technique. In the present paper a comparative study of the critical current anisotropy in low magnetic fields of Tl-1223 conductors manufactured by different techniques, prepared from different compositions and originating from different research laboratories is presented and discussed.

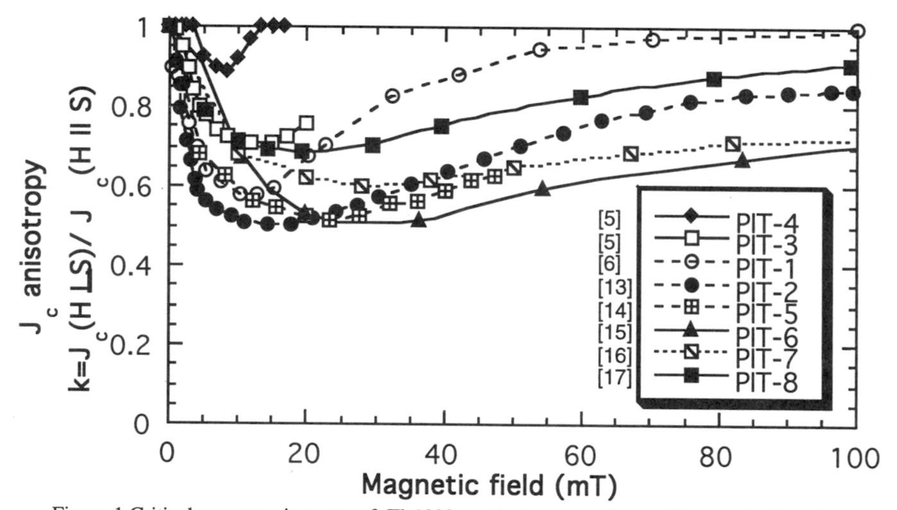

Figure 1 Critical current anisotropy of Tl-1223 conductors manufactured by powder-in-tube.

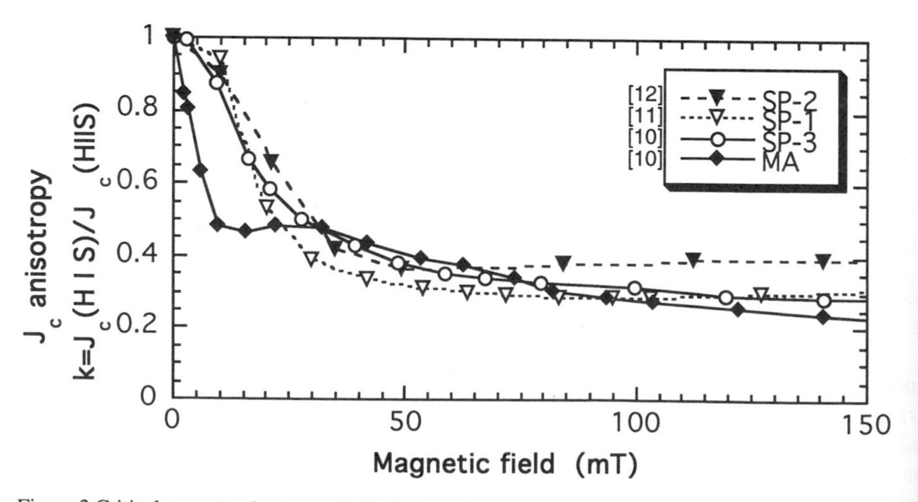

Figure 2 Critical current anisotropy of Tl-1223 conductors manufactured by spray-pyrolysis (SP) and magnetically alignment (MA) techniqe.

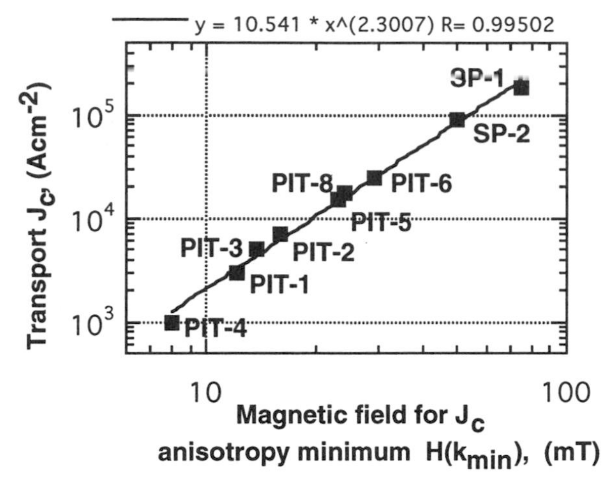

Figure 3 Transport J_c versus magnetic field of J_c anisotropy minimum.

2. Critical current density anisotropy

In comparison with other measurements techniques, the critical current density anisotropy in external magnetic field, defined in our case as $k=J_c(H \perp S)/J_c(H \parallel S)$, is probably the most sensitive indication of the grain alignment in the tapes. Changes of the anisotropy coefficient for the PIT conductors are presented in figure 1 while the data for external superconductor techniques are presented in figure 2. For all powder-in-tube conductors the anisotropy coefficient shows a very pronounced minimum at fields followed by monotonic decrease of anisotropy with increase of the magnetic field. In some cases such as PIT-1, PIT-3 and PIT-4 after reaching minimum k value the the anisotropy effect completely diminished. For the highly textured conductors manufactured by external coating techniques the k minimum is less pronounced and even for SP-3 there is no J_c anisotropy coefficient minimum, but only monotonic reduction, see fig. 2, proving that the sample has the best alignment and strong superconducting connectivity. The one exception is the magnetically aligned sample for which there is a pronounced k minimum at 15mT, due to the brick type microstructure of the sample.

It was established that there is a general trend: the higher the value of the magnetic field of anisotropy coefficient minimum $H(k_{min})$, the higher the value of the critical current density of the conductor see fig. 3. The points representing PIT samples and SP tapes in fig.3 are fitting to a power law $J_c = A[H(k_{min})]^n$ which reflects the correlation between the increasing transport J_c value and reduction of grain misalignment of the Tl-1223 layer.

. Discussion and conclusions

dependent of the processing routes. tapes produced by the powder-in-tube process by many research centres have demonstrated that weak links and random orientation of the grains is the major limitation of the critical current versus external magnetic field for PIT Tl-1223 conductors. There is evidence that the formation of the Tl-1223 is similar to YBCO on a smooth silver surface and is along the ab plane, where even small roughness of the silver surface causes nucleation of the growth along other orientations. Therefore because the perfect planar surface finish inside a PIT conductor can not be achieved during conductor deformation continous texured superconductor at the Tl-1223/Ag interface can not be developed. If one assumes, according to the microstructural data provided in the literature, that there is no alignment, (PIT-4), or little alignment along the tape surface, (PIT-6) (excluding one misleading publication [18]), of the Tl-1223 crystallites in the superconducting core for the PIT conductors then the external magnetic field and self-field affects the connectivity between the grains, decoupling some of the the inter-granular current paths and highlighting the dramatic reduction of the demagnetisation coefficient of grain agglomerates with increasing field. The poor grain alignment in plane of the tape and also magnetic division of the whole tape to the grain agglomerates in the powder-in-tube Tl-1223 tapes is responsible for the pronounced $H(k_{min})$. Further development of the PIT Tl-1223 tapes, especially using partial melting procedures or seeding procedures in analogy to composite reaction texturing technique [19], very successfully used for texturing Bi-base superconductors, may in the future provide a new source of conductors competitive with $YBa_2Cu_3O_{7-\delta}$ for application at 77K. However up to now only externally coated conductors are less affected by the 3D percolative current flow effects [20] and they present lack of the k minimum which is due to strong intrinsic current anisotropy. The J_c anisotropy coefficient appears to be a very sensitive tool for defining the granularity and texture development in Tl-1223 conductors and can be of further use for calculation of the actual transport J_c value of the superconducting core.

References

[1] Kung P J, Wahlbeck P G, McHenry M E, Maley P M and Peterson D E 1994 *Physica C* **220** 310-21

[2] Field M B, Cai, X Y, Babcock S E and Larbalestier D C 1993 *IEEE Trans. Appl. Supercond.* **3** 1479-82

[3] Glowacki B A and Evetts J E Proceedings 1987 *First European Workshop, High Tc Superconductors and Potential Applications*, Genova, Italy, 1-3 July , p.447-8

[4] Glowacki B A and Evetts J E 1988 *Mat.Res.Soc.Symp.Proc.* **99** 741-4

[5] Glowacki B A and Ashworth S A 1992 *Physica C* **200** 140-6

[6] Ciszek M, Campbell A M, Glowacki B A and Liang Y, paper presented during LTLC'96, 9-11 July 1996, Southampton, UK; Cryogenics (in press)

[7] Glowacki B A and Jackiewicz J 1994 *J.Appl..Phys.* **75** 2992-7

[8] Evetts J E and Glowacki B A, Sampson P L, Blamire, M G, Alford N McN and Harmer M A 1989 *IEEE Trans.Mag.* **25** 2041- 4

[9] Glowacki B A 1991 *Progress in High Temperature Superconductivity* **30** 216- 30 World Scientific Singapore-New Jersy-London-Hong Kong

[10] Tkaczyk J E, Lay K W, Jones B A, Bednarczyk B J and DeLuca J A 1995 *IEEE Trans. Appl.Supecond.* 2029-31

[11] DeLuca J A, Karas P L, Tkaczyk J E, Bednarczyk P J, Garbauskas M F, Briant C L and Sorensen D B 1993 *Physica C* **205** 21-31

[12] Doi J T, Yuasa Y T, Ozawa T and Higashiyama K *Proceedings of the 7th International Symposium on Superconductivity*, November 8-11 1994, Kitakyushu, Japan

[13] Ciszek M, Glowacki B A, S. P. Ashworth S P, Campbell A M, Liang W Y, Flükiger R and Gladyshevskii R E, 1996 *Physica C* **260** 93-103

[14] M. Ciszek M, Glowacki B A, Campbell A M, Ashworth S P, Haldar P and Selvamanikam V 1997 *Proc. ASC'96,* to be published in *IEEE Trans. Appl. Superconductivity* **7**

[15] Matsuda S P, Soeta A, Doi T, Okada M, Aihara K, Kamo T, Sasaoka T and Seido M 1992 , (ICMC suplement)*Cryogenics* **32** 248- 51

[16] Moore J C, Hyland D M C, Salter C J, Eastell C J, Fox S, Grovenor C R M and Goringe M J 1997 *Physica C* **276** 202-17

[17] Riepl T, Low R, Zachmayer S, Pauli C and Renk K F 1997 *Journal of Superconductivity* (in press)

[18] Richardson K A, Wu S, Bracanovic D, de Groot P A J, Al-Mosowi M K, Ogborne D M and Weller M T 1995 *Supercond. Sci.Technol.* **8** 238-44

[19] Chen M, Glowacka D M, Soylu B, Watson D, Christiansen J K, Yan Y, Glowacki B A and Evetts J E 1995*IEEE Trans. Appl. Superconductivity* **5** 1467- 70

[20] Glowacki B A 1997 *Cryogenics* (in press)

Inst. Phys. Conf. Ser. No 158
Paper presented at Applied Superconductivity, The Netherlands, 30 June–3 July 1997

Mechanical endurance of 2223-BSCCO tapes under tensile stress and strain

G.C. Montanari[1], I. Ghinello[1], L. Gherardi[2], R. Mele[2], P. Caracino[2]

[1]Dipartimento di Ingegneria Elettrica, Univ. Bologna, V.le Risorgimento 2, 40136 Bologna, Italy.
[2]Pirelli Cavi S.p.a., V.le Sarca 222, 20126 Milano, Italy

Abstract The increasing feasible length of HTSC tapes, as well as improved mechanical and electrical performances, has already led to realization of superconducting power-cable prototypes. In the view of transmission cable manufacturing, the evaluation of endurance of the HTSC tapes under mechanical stress will become more and more important. In this paper, methodologies for the endurance characterization of HTSC tapes subjected to tensile stress and strain are investigated. Diagnostic properties as critical current and quantities associated with hysteresis loops are considered for the purpose of life line inference. Data refer to tests performed on 2223-BSCCO tapes. The endurance coefficient values derived from the life lines obtained under either tensile stress or strain are compared, and the relevant degradation mechanisms discussed.

1. Introduction

The recent developments in the high-temperature superconductor (HTSC) tape echnology have allowed the manufacture of tapes hundreds meters long. This achievement, ogether with the improved electrical and mechanical performances of tapes, has led to the ealization of HTSC power-cable prototypes [1]. Consequently, manufacture of high-power uperconductor cables, which could provide a solution for ac or dc energy transport, seems a ossible target in a near future. Unfortunately, very little information is available on time ehavior of tapes under mechanical stresses, such as those that can affect tapes used as cable onductors. In particular, the most likely mechanical aging factors in energy applications are xpected to be tensile stress and strain. Therefore, this paper focuses on the evaluation of ndurance of HTSC tapes under tensile stress and strain. The diagnostic properties measured to valuate tape performance as a function of aging time are critical-current density and hysteresis oop. Multifilamentary BSCCO tapes are considered.

. Test procedures and Measurements techniques

he experiments were performed on specimens of multifilamentary 2223-BSCCO tape (which eems one of most promising HTSC for energy application), obtained with powder-in-tube rocess using a silver sheath [2]. Specimens contain 61 HTSC filaments; thickness, width and ngth are about 300 μm, 3.3 mm and 5 cm, respectively.

In order to determine the endurance lines, accelerated life tests (i.e. tests at stresses gher than those expected in service) were carried out at room temperature, at different levels of ress and strain, 4 and 3 respectively. Samples consisting of 5 specimens were used for each st. At fixed aging times, specimens were extracted from the aging cell and subjected to easurements of diagnostic properties, i.e., critical-current density and hysteresis loop.

The critical current density, J_C, was measured in liquid nitrogen by a standard four-wire procedure [3]. The estimation of J_C was made according to the corrected critical-field method with $E_c = 1\mu V / cm$, dividing the critical current, I_C, by the HTSC nominal specimen cross section [4, 5]. The results reported in the following are the mean values obtained on the samples, with their 95% confidence bounds.

The hysteresis loop was measured at 77K by means of a circuit consisting of a magnetizing coil, which provides magnetic field *H* in a wide range of amplitude and frequency, and two pick-up coils. The latter are connected in series with opposite phases. At the ends of the two coils voltage *V* is measured. The integration of *V*, obtained via software, gives the value of the flux, Φ, in the specimen which is placed inside one of two pick-up coils. The value of magnetization *M* is calculated by dividing the flux by the section. A computer program allows the dynamic hysteresis cycle of the examined material to be plotted and the relevant parameters (e.g. cycle area, proportional to AC losses, W, coercive field, H_C, residual magnetization, M_R, maximum field and magnetization, H_M and M_M) to be determined [4, 5].

The tensile-stress aging tests were performed by applying a constant force to a sliding clamp to which one side of the specimens was connected (the other side was constrained to a rigid bond). In this way, the specimens were purely subjected to tensile stress. Tests with tensile force, Ts, equal to 45, 50, 55 and 60 N, corresponding to stresses P of 45, 50, 55 and 60 N/mm2 were realized. The strain aging tests were carried out by applying a constant elongation to the specimens using a sliding clamp as described above. Tests with lengthening, ΔL, equal to 0.07, 0.08 and 0.1 mm, corresponding to strains ΔL/L of .14, .16 and .20 % were performed. Suc values were chosen so as to accelerate the life tests, and in fact they correspond to strai conditions which are significantly heavier than those which are expected to be applied to th tapes at the same (room) temperature, during fabrication and handling. Actually, the applicatio of these elongation conditions did immediately degrade the samples to some extent: it mus therefore be borne in mind that the mechanical regime explored by these tests is the one wher there are relatively large “initial” flaws in the material, which most likely brings about agin mechanisms that are more effective than should be expected to take place in “original” samples During these tests, the measurements of JCL were performed without removing the specimen from the sliding clamps, because the specimen-sliding clamp reciprocal position was extremel critical for the test results and not easily reproducible. For this reason, hysteresis loo measurements were not made on specimens under strain aging.

3. Experimental results and Discussion

A summary of the results of tensile-stress aging are reported in Figs. 1, 2, and 3. Figure shows the time behavior of critical current density (referred to the initial value, relevant unaged specimens) at the four stress levels. The chosen end point (decrease of 40% with respe to the initial value) is marked. The four values of time-to-end-point thus derived are plotte corresponding to the relevant stress values, in Fig.2, thus enabling derivation of the enduran line. As can be seen in the figure, a straight line is obtained in log-log plot, which can fit t inverse power life model:

$$L = K / S^{EC} \qquad (1)$$

where L is the time-to-failure (life) according to Fig. 1, S is the applied stress, K is a coefficie establishing life line location and EC is the endurance coefficient (the estimates of K and EC a reported in Fig. 2).

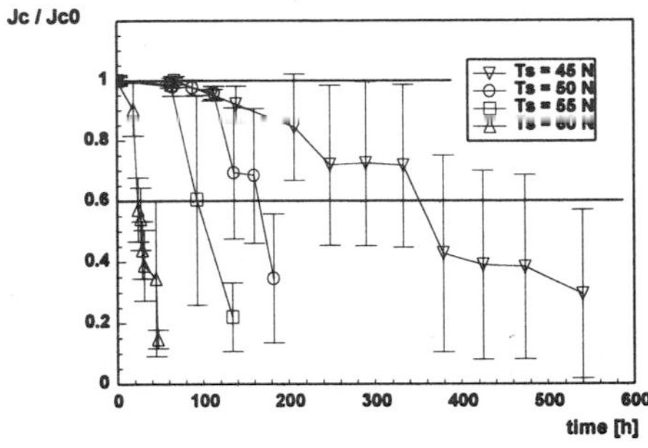

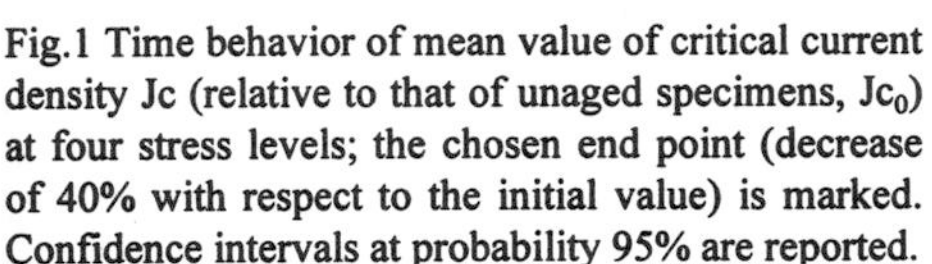
Fig.1 Time behavior of mean value of critical current density Jc (relative to that of unaged specimens, Jc_0) at four stress levels; the chosen end point (decrease of 40% with respect to the initial value) is marked. Confidence intervals at probability 95% are reported.

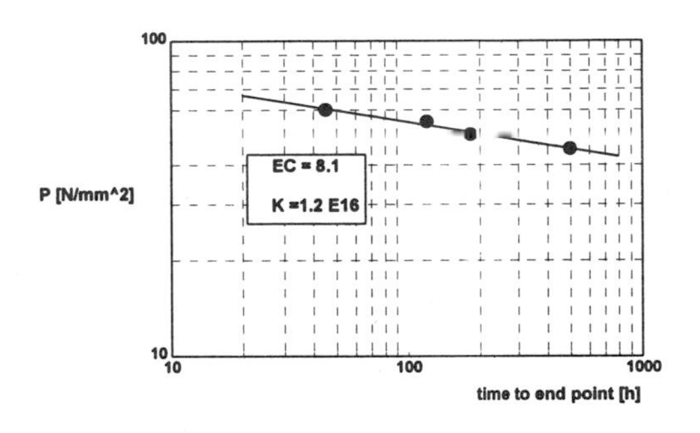

Fig.2 Life plot: experimental point and endurance line for tensile-stress. The parameters relevant to the endurance line (eq.(1)) are also reported.

An example of results relevant the hysteresis-loop measurements are depicted in Figs. 3 where the time behavior of energy density, W, (referred to the value measured on unaged specimens, Wo) for P = 45 N/mm^2 is reported for different values of applied magnetic field, H.

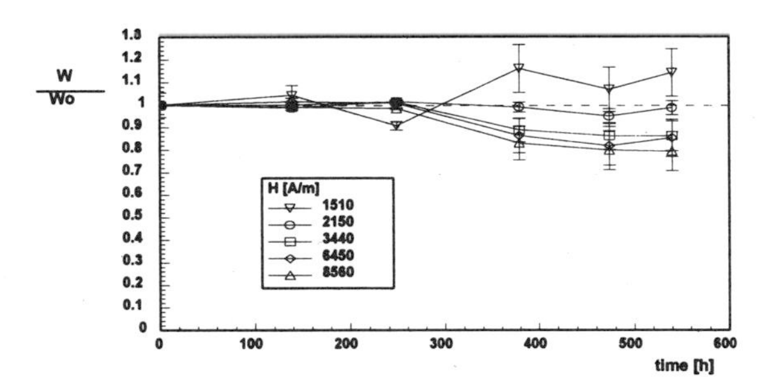

Fig.3 Time behavior of mean values of energy density during tensile-stress aging with P= 45 N/mm^2. The confidence intervals at probability 95% are reported.

The behavior of the other quantities associated to the hysteresis loop was found not to significantly differ from that of W/Wo. Hence, it comes out that the time behavior of hysteresis-loop quantities noticeably depends on H. In fact, for the lowest value of H the quantity W/Wo increases with aging time, while the same quantity decreases as H increases. This behavior can not be easily explained, but, likely, it can be correlated to the different amount of flux vortexes at different values of H [4, 5]. Comparing Fig.3 with Fig.1, it is evident that the "magnetic" quantities are much less affected by tensile-stress than the critical current density. This is presumably due to the fact that the main effect of uniaxial stress is the formation of transverse cracks which affect the transport current more than they alter the quality of single grains.

Results relevant to life tests under tensile-strain are reported in the Fig.s 4 and 5. Fig. 4 shows the time behavior of critical current density (referred to the initial value, relevant to unaged specimens) at the three test strain levels. The chosen end point (decrease of 40% with respect to the initial value, such as for tensile stress) is marked. The three values of time-to-end-point thus derived, as well as the best-fitting endurance line, are reported in Fig. 5. As in the case of tensile stress, a straight life line is obtained in log-log plot, which can be described by eq. (1) (the values of the estimates of location and endurance coefficients are given in Fig.5):

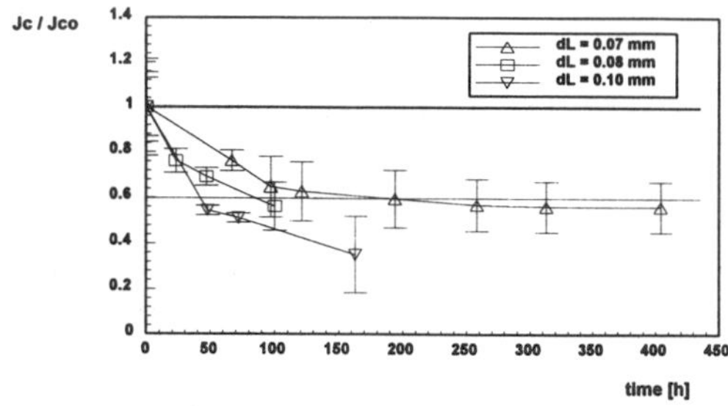

Fig.4 Time behavior of mean values of Jc (relative to that of unaged specimens, Jc_0) at three levels of strain; the chosen end point is marked. The confidence intervals at probability 95% are reported.

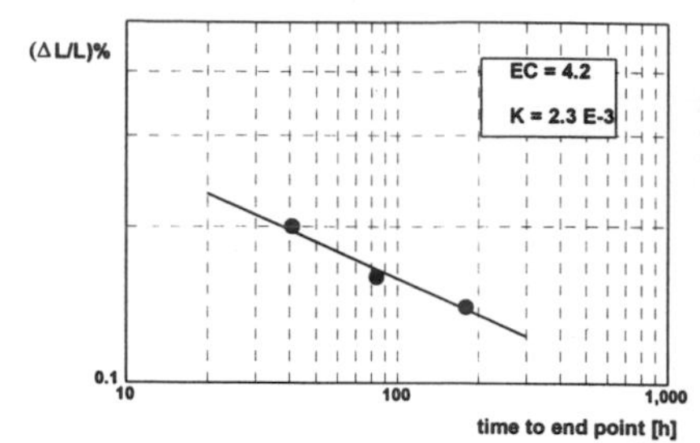

Fig.5 Life plot: experimental point and endurance line for tensile-strain aging tests. The parameters relevant to the endurance line (eq.(1)) are also reported.

The "life" data reported in Fig.2 and Fig.5 provide a very significant result, that is, the detection of a linear relationship in log-log coordinates, whose slope provides a direct figure for the capability of the tested specimens to withstand static tensile stress and strain. The difference between the two values of EC detected for the different aging conditions (EC = 9.6 for tensile stress and EC = 4.2 for strain) shows that under the presently chosen conditions the tested specimens had a better endurance under tensile stress rather than strain.

4. Conclusions

The results here reported and discussed give insight into the endurance properties of the tested 2223-BSCCO tapes under tensile stress and strain. The critical current density is, as expected, strongly affected by tensile stress and strain. The values of endurance coefficients derived from the life lines, which are straight lines in log-log coordinate, indicate a better behaviour under tensile stress rather than tensile strain. Given the high level of strains used here in order to accelerate the tests, however, the conclusions from life tests under strain should be checked by further strain testing at lower strains, such as to produce no or little "initial" degradation. Testing at lower strains will also allow to analyse the role of the silver sheath, not discussed here; in fact, although the tests are carried out at room temperature, where creep is usually assumed to be negligible, such silver sheath is very ductile, as a consequence of long annealing, which could affect the long term performance of the composite tapes to a significant extent. Different from that of the transport critical current, the behavior of the quantities associated with the magnetic hysteresis loop seem weakly influenced by tensile stress, likely because the latter does not affect the quality of grains, but their links mainly.

References

[1] A. Bolza, P. Metra, M. Nassi, M.M. Rahman, *1996 Applied Superconductivity Conference*, Pittsburgh, Pennsylvania, U.S.A., 25-30 August, 1996.

[2] Y.Yamada, M.Satou, S.Murase, T.Kitamura, Y.Kamisada, *5th Int. Symp. on Superconductivity*, Kobe, Japan, 1992.

[3] G.L. Dorofeev, A.B. Imenitov, E.Y. Klimenko *Cryogenics*, **20**, N. 6, 307-312, June 1980.

[4] G.C.Montanari, I.Ghinello, L.Gherardi, P.Caracino, *Supercond. Sci. and Technol.* **9**, N. 5, 385-392, May 1996.

[5] G.C. Montanari, I. Ghinello, L. Gherardi and P. Caracino, *IEEE Trans. on Appl. Superconductivity*, Vol. 6, n. 3, September 1996.

Inst. Phys. Conf. Ser. No 158
Paper presented at Applied Superconductivity, The Netherlands, 30 June–3 July 1997

Transverse pressure induced reduction of the critical current in BSCCO/Ag tapes

B ten Haken, A Beuink and H H J ten Kate

University of Twente, Low Temperature Division, P.O. Box 217, 7500AE Enschede, The Netherlands.

Abstract. The transport critical current of different BSCCO/Ag superconductors can be reduced severely when a certain deformation is applied. The influence of a pressure acting transversely on the wide side of the tape is investigated at 77 K. An irreversible reduction in the critical current occurs above 10 MPa. The strain induced in the superconductor is considered in a two phase mechanical model for a multi-filamentary BSCCO tape. Because of the low yielding strength of the matrix material, and the small fraction of pressurised area besides the BSCCO filaments, only a relatively small pressure can be absorbed by the matrix. Based on this analysis it is concluded that the properties of the BSCCO itself mainly determine the transverse pressure dependence of these tapes at pressures above 30 MPa. Transverse pressure experiments are made on different types of BSCCO/Ag and AgMg tapes up to 150 MPa. The relative current change due to a pressure in a AgMg tape is large –3 GPa^{-1}, compared to the value of –2 GPa^{-1} that is measured in Ag tapes. This observation supports the model that predicts a very limited role for the matrix strength in the case when a large transverse pressure is applied.

1. Introduction

Modern $(Bi,Pb)_2Sr_2Ca_2Ca_3O_x$ conductors that are being produced with the powder in a Ag tube process are sensitive for mechanical deformations. The critical current (I_c) in BSCCO/Ag tapes is reduced severely when these conductors are deformed along the current axis. A comparison of the axial strain dependence (ε_a) of I_c is described for the axial tensile and compressive strain regime [1,2]. A mechanical deformation induced by a transverse mechanical force acting on the tape surface also reduces the I_c of the tape. In important property of these BSCCO conductors is that the I_c reductions are in general irreversible. An example of this irreversible behaviour is presented in fig. 1, where the I_c is measured in a tape conductor that is deformed by a transverse pressure [2].

In general the mechanical weakness of the Ag and in particular its low yield strength is considered to be an important factor that limits the mechanical robustness of BSCCO/Ag tape conductors. Alloying the Ag increases the mechanical strength of the matrix and it is shown that an improvement can be obtained in the axial tensile stress and strain tolerance [3]. This improvement in the strain tolerance is attributed to an enhancement of the thermal compression in the BSCCO filaments and can be explained with the earlier presented model for strain induced I_c reductions [1,2]. The role of the matrix material in a transversely compressed tape conductor is considered here in a mechanical model and a first attempt is made for an experimental verification of this description.

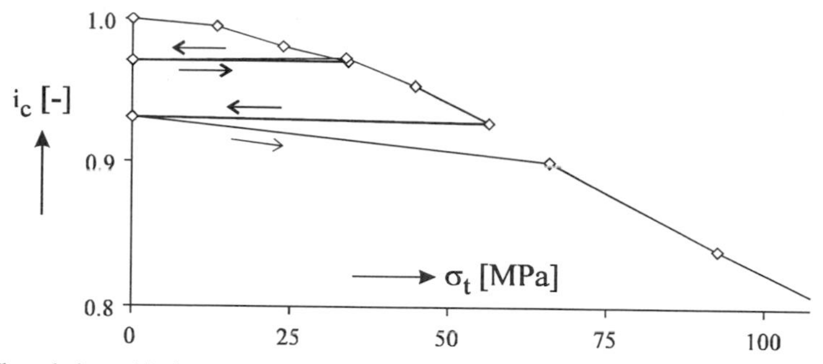

Figure 1: The relative critical current measured in a transversely pressed BSCCO/Ag tape conductor, showing the irreversible nature of the I_c reduction [2].

2. Mechanical behaviour of a multifilamentary tape

The cross-section of a multifilamentary tape consists of Ag (or a Ag-alloy) that embeds a varying number of flat-shaped BSCCO filaments. The mechanical behaviour of this structure is therefore determined by a combination of both materials. This structure can be described with a mechanical model, for instance with a detailed finite element model. Due to the uncertainties in the mechanical properties of the materials involved, the accuracy of a calculation with any a model is very limited at present. Because of these limitations in accuracy, the mechanical behaviour of the cross-section is considered here in an analytical model that includes only the most essential features of the mechanical behaviour of a real tape conductor.

The superconducting properties of a compressed tape conductor are determined by the deformation of the BSCCO filaments. In the mechanical model the cross-section of a superconductor (Fig. 2A) is represented with a series of Ag-matrix columns with BSSCO columns in between (Fig 2B). The most realistic value for the area ratio between the compressed columns of superconductor and normal matrix (A_{sc} / A_n) is much larger then the superconductor to silver ratio of the entire tape. This difference is determined by the fraction of the silver around the filaments (marked grey in figure 2A) that does not balance its stress with the superconductor. In this description this part of the matrix is considered as a thin flexible layer between the filaments.

The pressure inside the superconductor (σ_{sc}) and the matrix (σ_n) is averaged over the tape cross-section:

$$\sigma = (A_{sc}\, \sigma_{sc} + A_n\, \sigma_n) / (A_{sc} + A_n) = f\, \sigma_{sc} + (1 - f)\, \sigma_n . \tag{1}$$

Because the strain variations in the additional silver layers between the filaments are neglected there occurs an equal and uni-axial strain (ε) in both the superconductor and the silver, which in the case of an elastic deformation in a stress variation ($d\sigma$) leads to a deformation change:

$$d\varepsilon = d\sigma f E_{sc} + d\sigma (1 - f) E_n . \tag{2}$$

The Young's modulus of the BSCCO will strongly depend on the filling factor and the exac composition of the polycrystalline material. Reported values range from $E_{sc} = 2.6$ [4] t 100 GPa [5]. A reported value for the Young's modulus of Ag is 80 GPa, but this value de pends on the thermal and mechanical history of the material [6]. The plastic behaviour of A and AgMg in BSCCO tapes is investigated in experimentally. With Ag the yielding limit oc curs at $\sigma_y = 30$ MPa, by alloying the Ag with Mg the yield strength is increased to a value o about 60 MPa in these tapes [7].

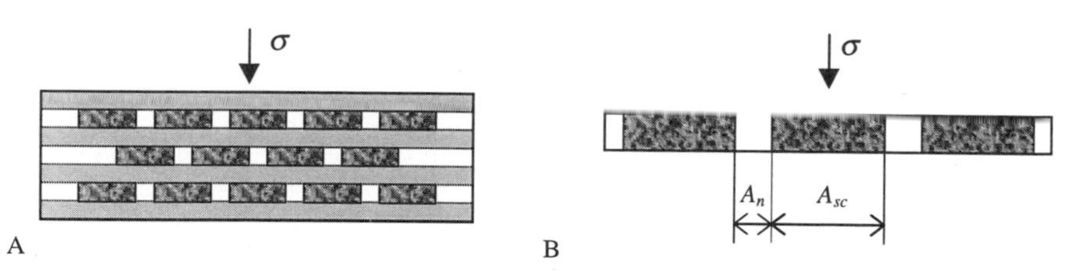

Figure 2: The cross-section of a multifilamentary tape conductor (A) and its representation in a mechanical model that considers mechanical properties of the BSCCO filaments and the Ag matrix around the filaments (B). The layered matrix area between the filaments (marked grey in A) is neglected in this description.

The mechanical description presented here shows that for realistic values of the area factor ($f = 0.2$ to 0.4) and typical mechanical properties of Ag and BSCCO, only a relatively small part of the pressure is absorbed in the matrix material and that the largest part is absorbed in the superconductor. As a consequence the role of the matrix in the mechanical behaviour is important only for small pressures. For a large transverse pressure well above $\sigma = f \cdot \sigma_y$ (typically 10 to 30 MPa) it is the BSCCO itself that determines the strain-state inside the filaments.

3. BSCCO tape conductors

A comparison is made here on the transverse pressure dependence of three different types of superconductors as obtained from various suppliers. The tapes differ in geometry, number of filaments and the type of matrix material. The critical current density in superconductor and the S.C : Ag ratio (1:4) are comparable in these conductors Table 1.

Table 1: Summary of condutor poperties:

A:	19 filaments	$J_c = 65$ A/mm^2	0.20×2.7 mm^2	Ag matrix,
B:	9 filaments	$J_c = 80$ A/mm^2	0.30×4.1 mm^2	Ag matrix,
C:	37 filaments	$J_c = 75$ A/mm^2	0.25×3.3 mm^2	Ag / 2% Mg matrix.

4. Transverse pressure

The transverse pressure is produced with a cryogenic press that applies a mechanical load on the broad side of the tape. The pressure is applied along 10 mm of sample length, and the voltage is measured just outside this pressurised region, with small voltage taps. The critical current is determined at a constant voltage criterion of 10^{-4} V/m.

The measured reduction of the critical current is depicted in Fig. 3, again only irreversible reductions of the I_c are observed, similar as reported in previous experiments samples that are compressed in the axial or transverse direction [1,2]. Above a certain pressure of typically 20 MPa, the normalised critical current ($i_c = I_c/I_c(0)$) reduces nearly proportional with the increased pressure. This relative current reduction is described well with a characteristic slope of $di_c / d\sigma = 2.0 \pm 0.4$ Gpa^{-1}, for the samples with the Ag matrix (A and B). The sample from the AgMg-tape (C) shows a more pronounced relative I_c reduction with a characteristic slope of 3.0 Gpa^{-1}. In these experiments the improved yield strength of the AgMg matrix, compared to the Ag tapes is not reflected in an improved stability for transverse pressures. On the contrary a more pronounced I_c reduction occurs in the compressed AgMg tape.

These experimental results support the idea that improving the yield strength of the matrix material by alloying the Silver, will not improve the transverse pressure dependence for large pressures above $f \cdot \sigma_y$ (typically 10 to 30 MPa). The difference that is measured in the pressure dependence between these samples has to be attributed to other mechanisms. The most probable source for these variations is the (micro-)structure of BSCCO, with its typical polycrystalline nature. The combination of mechanical and superconducting properties of BSCCO that is being monitored in a stress experiment, may very well differ between the manufacturing processes. Therefor a further improvement in the transverse pressure stability of this type of superconductors has to be realised inside the BSCCO filaments.

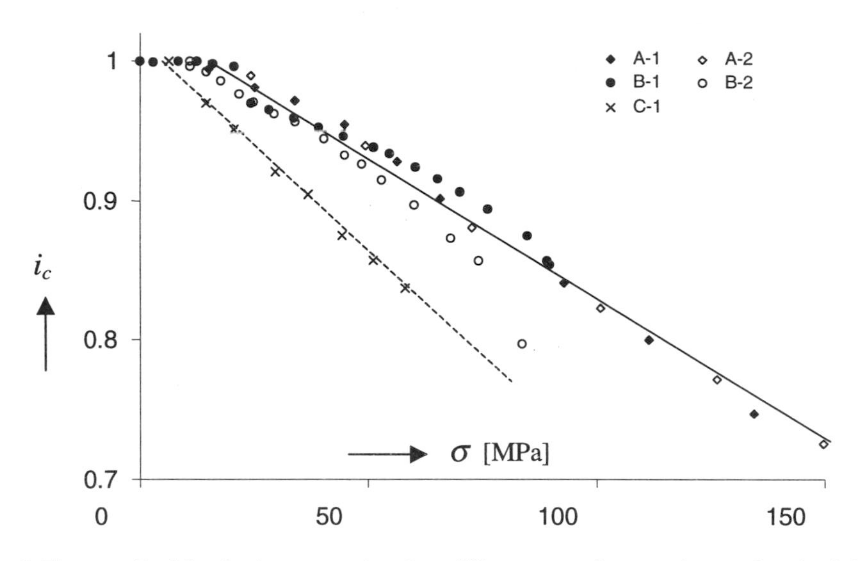

Figure 3: The normalised I_c reduction measured on three different types of tape conductors. Samples A & B are compared with a relative I_c reduction of –2 GPa^{-1} (solid line) and sample C with -3 GPa^{-1} (dotted line)

5. Conclusions

1. The I_c reduction induced by a transverse pressure is irreversible, similar to the I_c reductions that occurs in an axial compression or tension experiment on a BSCCO/Ag conductor.
2. The role of the matrix material in a transverse pressure experiment is minimal when the yield strength of the matrix material between the filaments is surpassed. In a typical BSSCO/Ag tape this occurs pressure above 10 to 30 MPa.
3. The mechanical stability for transverse pressures is not improved in the Ag-alloyed samples that are investigated here. On the contrary a reduction is observed supporting the hypothesis that the transverse pressure dependence of these tapes is mainly determined by the properties inside the BSSCO filaments.
4. Increase the yield strength of the matrix by alloying the Ag of BSCCO conductors, will not necessarily improve the stability for transverse pressures, although it will improve the strain and stress limit in the axial direction.

References

[1] B. ten Haken et al, *Adv. of Cry. Eng.* **42**, pp 651-658, 1997.
[2] B. ten Haken et al, *Appl. Supercond. Conf. Pittsburg USA*, 1996.
[3] B. Ullman et al, *Appl. Supercond. Conf. Pittsburg USA*, 1996.
[4] J.K. Yau et al, *Physica C* **243**, pp 359, 1995.
[5] L. Gherardi et al, *Cryogenics (ICEC supl.)* **34**, pp 781, 1994
[6] S. Ochiai et al, *Cryogenics* **31**, pp 954-961 1991.
[7] T. Hasegawa et al, *Appl. Supercond. Conf. Pittsburg USA* 1996.

Inst. Phys. Conf. Ser. No 158
Paper presented at Applied Superconductivity, The Netherlands, 30 June–3 July 1997

Zero field critical current of a 150 m superconducting tape measured in a pancake coil configuration

P. Bodin, Z. Han, P. Vase, M. D. Bentzon, P. Skov-Hansen, R. Bruun, and J. Goul

Nordic Superconductor Technologies, Priorparken 878, DK-2605 Brondby, Denmark.

Abstract. The zero field critical current (I_c) of a silver sheathed 150 m long superconducting $(BiPb)_2Sr_2Ca_2Cu_3O_{10+x}$ tape wound as a pancake coil was determined at 77K. The measured I_c values were expectedly lower than their individual tape zero field values due to the pancake coil self field. By computing the pancake coil self field, the lowered I_c values could be corrected for the coil self field.

1. Introduction

Today, high temperature superconducting $(BiPb)_2Sr_2Ca_2Cu_3O_{10+x}$ tapes fabricated by the oxide powder in tube (OPIT) method can be made in very long length as compared to few years ago. Superconducting tape lengths have progressed from the laboratory scale of a few centimeters to the impressive industrial scale of kilometers. The upscaling of tape lengths has set the demand to find viable routes for determining the superconducting properties of the tapes in their whole length. Basically, there are two ways of characterising a very long tape. The first and best method is to recoil the tape and let it pass through a liquid nitrogen bath with a measuring gauge. In this way detailed information on every part of the tape is obtained. The drawback is that it requires a recoiling of the tape and sliding contacts or magnetic measurements[1]. The other method described here is a less accurate but faster technique by which the whole tape is measured wound as a pancake coil. The tape coil is immersed in liquid nitrogen in a roundshaped flat dewar. The method requires no recoiling but the superconducting tape windings must be electrically insulated to force the current to run along the tape direction.

2. Experimental technique

In the here described measurement a 3.0 mm thick pancake coil of 390 mm outer diameter and 115 mm inner diameter was used. The total tape lenghts was 150 m and the tape produced has an insulating layer. Current was fed into the coil at the endpoints. 16 voltage contacts were made to the tape by needle contacts equidistantly positioned on the side of the coil. During measurement, the current was slowly increased in order to let the voltages stabilise. The I_c values (1 μV/cm criterion) found in this way are lower than the zero field values due to the field dependency of I_c to an external magnetic field. Secondly, short pieces of 5 cm length were cut off from each end of the tape and their zero field critical current values measured. The zero field critical current is denoted I_{c0}. Finally, I_c for one of the short tape pieces was measured versus applied magnetic field (B) for the field directions parallel and perpendicular to the tape. These curves were used to calculate compensation factors needed to determine the zero field current for the 150m tape that was measured as a pancake coil.

3. Results

The I_{c0} values from 5 cm pieces from each end of the 150 m tape were found to be 20.5A and 19.8A respectively. The I_c values measured for the 150 m tape pancake coil are seen in fig. 1.

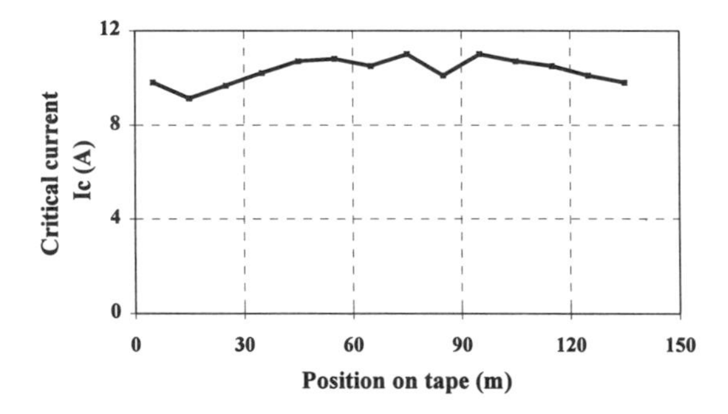

Figure 1. The measured critical current (1 μV/cm criterion) vs. position along the length for a 150 m long tape. The critical current values are lowered in the pancake coil configuration.

From fig. 1 we see that the measured I_c values are approximately 50% of the zero field values determined from the 5 cm end pieces.

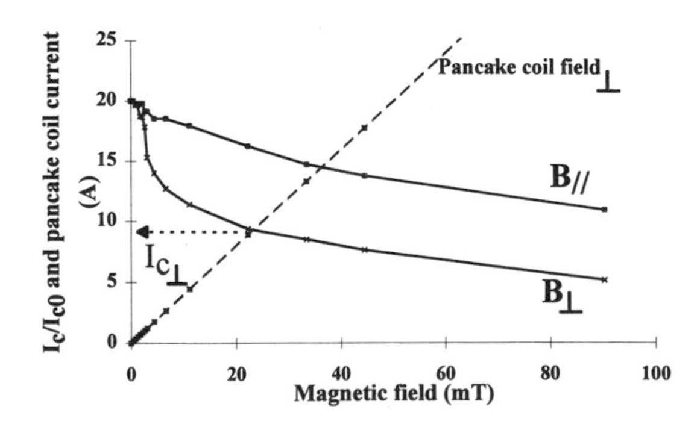

Figure 2. The critical current I_c of a 5 cm tape piece from the end of the 150 m tape vs. applied magnetic field The dashed line shows the computed self field perpendicular to the coil tape in one point of the coil.

In fig. 2 the I_c vs. B for a short tape piece is shown. On the same figure the dashed line represents the computed pancake coil self field perpendicular to the tape as a function of the pancake coil current. The dashed line in this case represents the magnetic field perpendicular to the tape at a distance 250 mm from the center. At the point where the dashed line and I_c vs. $B_\perp$ curve cross, the coil self field at that feeding current will suppress the I_c value to exacly the same value as the feeding current. In this case 9.5A. In other words, if I_{c0} is 20.0 A, we expect to measure an I_c of 9.5A in the here described pancake coil configuration only regarding the pancake coil field perpendicular to the tape in the coil.

To a first approximation, compensating the I_c values for coil perpendicular magnetic field $B_\perp$, the zero field value can be found by

$$I'_{c0} = I_c/k_\perp$$

$$k_\perp = I_{c\perp}/I_{c0}(\text{5cm piece})$$

We call $k_\perp$ the perpendicular field compensation factor. I_{c0} was found for the 5 cm piece. $I_{c\perp}$ is the crosspoint value in fig.2 for the critical current in a field perpendicular to the tape. The I'_{c0} value is quite close to the I_{c0} value because the biggest correction that needs to be performed is due to the pancake coil field perpendicular to the tape.

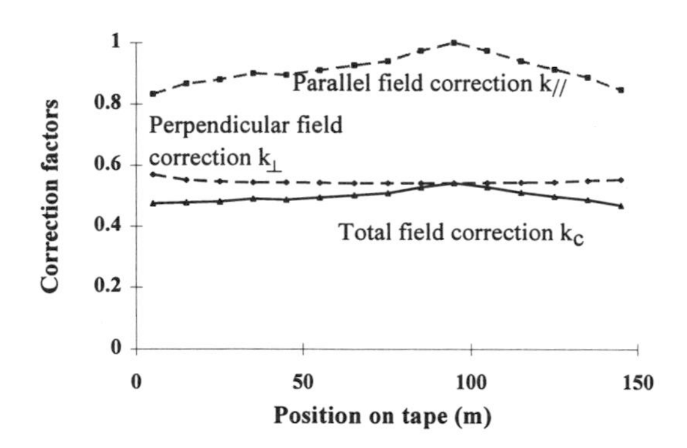

Figure 3. The computed correction factors for the measured critical current vs. position along the length for the 150 m long tape.

To find the fully corrected I_{c0} value it is further necessary to correct for the I_c suppression due to the field parallel to the tape. Correspondingly we get

$$I_{c0} = I'_{c0}/k_{//}$$

$k_{//}$ is defined similarly to $k_\perp$. Finally we define the combined compensation factor k_c as the product of $k_\perp$ and $k_{//}$

$$k_c = k_\perp\, k_{//}$$

Fig. 3 shows the field correction coefficients calculated along the tape length. It is clear that the correction needed to account for the magnetic field perpendicular to the tape is the largerst due to the higher sensitivity of the tape to the magnetic field in this direction. The correction is very constant over the whole length, reflecting that the pancake coil field is quite constant over the whole length. The field parallel to the tape direction, however, is smaller and quite dependent on position. The peak value of 1 in fig. 3 of $k_{//}$ reflects the fact that in the coil configuration the field parrallel to the tape is zero in that point.

Finally fig. 4 shows the obtained corrected I_{c0} values found from the data in fig. 1 and the total correction factor from fig. 3. It is seen that the tape shows good homegeniety. Subsequent measurements of tape pieces from the 150 m showed good agreement (± 1A) with the compensation determined I_{c0} values.

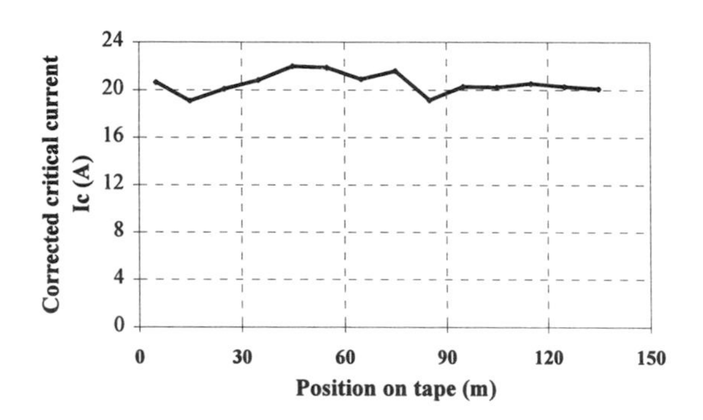

Figure 4. The zero field critical current vs. position along the length for the 150 m tape. The measured critical current values from fig. 1 have been corrected using the correction factors from fig. 3.

It should be noticed that the mathematical description used here is grossly simplified in that we assume that the total magnetic field dependency can be derived by simply determining the perpendicular and the parallel field suppression indenpendently. The other assumption we have made is that the magnetic field dependency scales directly with the local critical current value. This can only be said to be true if the texturing of the tape does not differ throughout different regions of the tape. Despite the simplifications the method has showed to be viable.

4. Conclusion

It has been demonstrated that for long superconducting tapes the zero field critical current in a pancake coil configuration can be found. The method depends on reliable computations of the magnetic field around a tape coil and a subsequent mathematical compensation for the suppressed critical current. The zero field critical current values found this way were within ± 5% of the values found from short pieces cut directly from the long tape at a later stage.

References

[1] Paasi J A J and Lahtinen M J 1993 *Physica C* **216** 382-390.

Inst. Phys. Conf. Ser. No 158
Paper presented at Applied Superconductivity, The Netherlands, 30 June–3 July 1997

Limits of application of the BSCCO conductors

Beate Lehndorff†§, Michael Hortig‡, Hans-Gerd Kürschner‡, Milan Polak§, Rainer Wilberg†, Dorothea Wehler‡ and Helmut Piel†‡

† Bergische Universität Wuppertal, FB08, Institut für Materialwissenschaften, Gaußstr. 20, D-42097 Wuppertal, Germany
‡ Cryoelectra GmbH, Wettinerstr. 6H, D-42218 Wuppertal, Germany
§ Applied Superconductivity Center, University of Wisconsin-Madison, 1500 Engineering Drive, Madison, WI 53706, U.S.A.

Abstract. One important limitation of the critical current density in BSCCO-based superconducting wires and tapes is caused by the inhomogeneity of the ceramic core. Analysis of current-voltage-characteristics of BSCCO-2212 wires and 2223 tapes exhibit scaling behavior only below T_g. Furthermore the I-V-curves behave very different for BSCCO-2212 wires with and without silver sheath. This is caused by current sharing in the metal cladding due to nonuniform cross sectional area and poor grain connectivity. Hall probe scans are used to demonstrate bad areas in BSCCO-2223 tapes. In addition ac-susceptibility experiments as a function of temperature reveal the quality of grain connectivity.

1. Introduction

Despite the improving quality of the BSCCO-based conductors their critical current density still has serious limitations especially in high magnetic fields. These limitations have at least three different origins. At first the current flow is hindered by the poor connectivity of the ceramic grains. The second reason is the porosity and inhomogeneity of cross section within the superconducting core. Finally the flux pinning within the grains influences the critical current density. To distinguish between the different sources of limitation the application of various experimental methods are necessary like transport measurements, magneto-optical imaging [1] or magnetic measurements. In this paper three basic experimental techniques are used to determine the quality of the wires and tapes of BSCCO-2212 and -2223. These methods are transport measurements of current-voltage-characteristics, scanning Hall-probe imaging and ac-susceptibility experiments.

2. Experimental details

2.1. Sample preparation

The samples were prepared by the powder in tube (PIT) method. A special combination of profile rolling and drawing enables us to produce conductors with more than 50%

filling factor [2, 3]. Short samples of Bi-2223 tapes and Bi-2212 wires and tapes have been produced and heat treated with known methods [4, 5, 6]. For the experiments without silver cladding the metal sheath was removed by chemical etching using a mixture of two parts H_2O_2 and five parts NH_4OH.

2.2. Experimental techniques

Current-voltage characteristics were measured in magnetic fields up to 8 T for temperatures between 4.2 K and 100 K. For a faster determination of j_c the samples of Bi-2212 were immersed in liquid helium and I-V-curves were measured.

The scanning Hall-probe imaging was performed with a 3D positioning system using a Hall sensor with 200x200 μm^2 sensitive area. This system has been described earlier [7]. The Bi-2223 tapes were immersed in liquid nitrogen. During cooling a magnetic field was applied by means of a permanent magnet. The remanent flux distribution was measured after removal of the magnet.

A self built ac-susceptibility setup [8] was used to measure the temperature dependent ac-susceptibility and its third harmonic. For this purpose the samples were cooled in liquid nitrogen, too.

3. Results

I-V-curves for different temperatures and magnetic fields up to 8 T have been taken for both sample types. The highest critical current densities were $j_c > 100$ kA/cm^2 for Bi-2212 at 4.2 K in self field and $j_c = 39.4$ kA/cm^2 for Bi-2223 at 77 K in self field. From the log-log plot of these curves a transition temperature T_g was determined from the change of their curvature.

In fig. 1 the Hall-probe image of a "good" and a "bad" tape of Bi-2223 is shown. It can be clearly seen, that the tape has one bad area over the whole cross section. The spatial resolution is 0.5 mm.

The temperature dependent ac-susceptibility results are shown in fig. 2 for a Bi-2223 tape. After the onset temperature $T_{c,intra}$ there is a second step $T_{c,inter}$ at a slightly lower temperature. The third harmonic susceptibility exhibits a small peak at the higher temperature and a large peak at $T_{c,inter}$. At this temperature the grains decouple and only the intrinsic losses are measured. This is supported by the fact that the higher the critical current density the closer $T_{c,inter}$ approaches $T_{c,intra}$

4. Discussion

A scaling analysis was made on the I-V-curves. The details are given elsewhere [9]. It turned out that there is no scaling above the transition temperature T_g. However, this temperature is directly correlated with the anisotropy of the critical current density for different orientations of the magnetic field. This is depicted in fig. 3 where $j_n = j_{c||c}/j_{c||ab}$ is plotted as isolines in the $T - B$-plane. This implies that the transition temperature seems to be conected rather with the decoupling of superconducting layers than with a phase transition in the flux line lattice. Nevertheless T_g or B_g might give an intrinsic

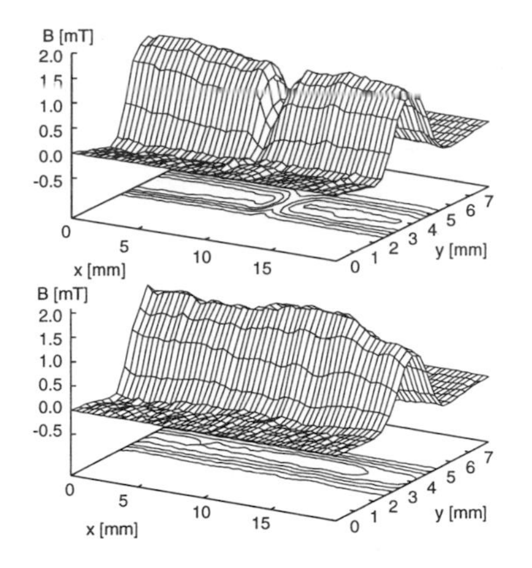

Figure 1. Scanning Hall-probe imaging of a defect in a Bi-2223 tape (top) compared to a tape with no major defect (bottom).

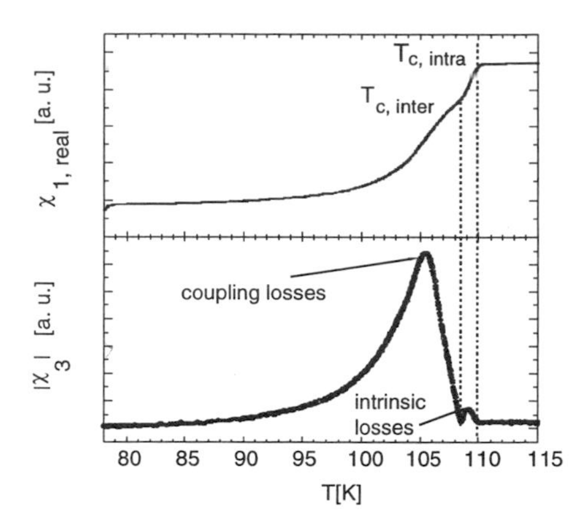

Figure 2. Fundamental (top) and 3rd harmonic (bottom) susceptibility as a function of temperature for a Bi-2223 tape.

limit to the application of these conductors as they separate the truely superconducting region from the dissipative state.

However, the interpretation of these results in any picture of flux phenomena has to be made very carefully. The I-V-curve of a superconducting transition should exhibit no resistivity up to j_c, and than rise steeply. Measured I-V-characteristics often show linear section in any part of the curve. Also the transition region mostly is not as sharp as expected. This could be connected with the inhomogeneity of the conductors and consequently current sharing through the silver cladding. As a test Bi-2212 wire was measured in liquid helium with and without silver sheath and the I-V-curves were compared. The result is given in fig. 4 in a linear scale. After etching the silver sheath linear sections vanish and the curve becomes steeper. This strongly indicates that the silver has a severe influence on the overall shape of the I-V-curve.

Also the results of the Hall-probe imaging support that bad areas are present where the current has to be shunted into the silver. In addition the ac-susceptibility results also show that the improvement of critical current is mainly a question of better connectivity of the ceramic core. A detailed description of the current sharing model is given in [10].

5. Conclusion

Results of current-voltage-characteristics imply a transition temperature T_g and an irreversibility field B_g which limit the applicability of Bi-based HTS conductors. A correlation between T_g and the anisotropy of j_c gives evidence that this limitation is connected with the decoupling of superconducting planes. However, a more detailed analysis of the shape of the I-V-curve for a conductor with and without silver sheath shows that there is current sharing through the silver cladding which makes the interpretation in terms of flux dynamics ambiguous. There are areas with defects present in these conductors which could account for this effect as demonstrated by Hall-probe imaging. Furthermore

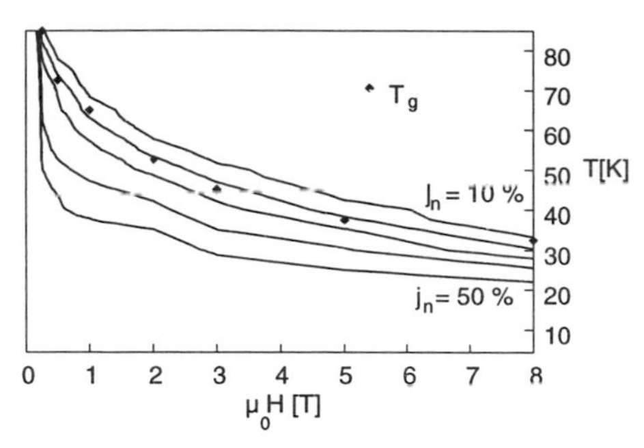

Figure 3. Different values of anisotropy plotted as isolines in the $B-T$-plane.

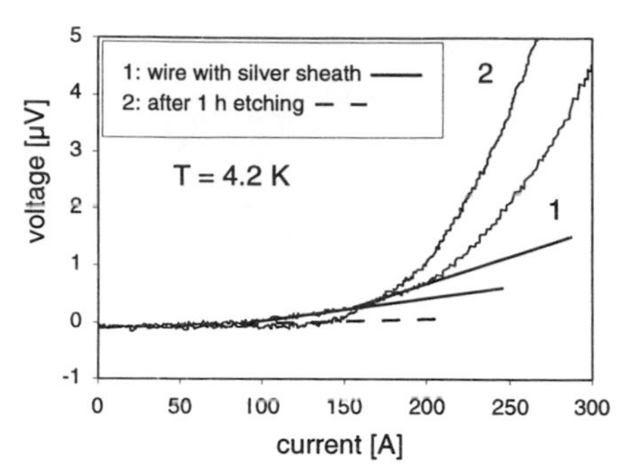

Figure 4. Comparison of I-V curves for a Bi-2212 19-filament wire with and without silver.

from the ac-susceptibility results it is clear that the grain connectivity within the ceramic core is still not satisfying.

Acknowledgment

Technical support by Rainer Theisejans, Karl Zeitzs, Christian Tilly and Huan Tran Ho is greatly acknowledged. One of us (B. L.) is grateful for the possibility of doing part of the work at the Applied Superconductivity Center at UW-Madison. This work is partly supported by the BMBF under contract number 13N6634/2.

References

[1] Parrell J A *et al* 1996 *Supercond. Sci. Technol.* **9** 393

[2] Application for a German patent

[3] Hortig M *et al* 1997 *Adv. Supercond.* **IX** (Tokyo: Springer) in press

[4] Lehndorff B *et al* 1995 *Inst. Phys. Conf. Ser.* **148** 411

[5] Zhang W *et al* 1997 ; *IEEE Trans. Appl. Supercond.* in press

[6] Reeves J *et al* 1997; *IEEE Trans.Appl. Supercond.* in press

[7] Lehndorff B *et al* 1995 Physica C **247** 280

[8] Piel H *et al* 1995 *Adv. Supercond.* **VII** (Tokyo: Springer) 757

[9] Lehndorff B *et al* submitted to Physica C

[10] Polak M *et al* 1997 submitted to *Supercond. Sci. Technol.*

Inst. Phys. Conf. Ser. No 158
Paper presented at Applied Superconductivity, The Netherlands, 30 June–3 July 1997

Observation of self-field distribution due to transport currents in Ag-sheathed Bi2223 monofilamentary and multifilamentary tapes

K.Kawano, and A.Oota

Toyohashi University of Technology, Tempaku-cho, Toyohashi, Aichi 441, Japan

Abstract. Using a scanning Hall sensor with an active area 50μm X 50μm, we measured the self-fields due to DC transport current on the surfaces of Ag-sheathed $(Bi,Pb)_2Sr_2Ca_2Cu_3O_x$ (Bi2223) monofilamentary and multifilamentary tapes at 77K and zero external field. We discuss the current distribution in the superconductor through the field slope which is obtained from the measured magnetic profile. There is a resemblance in the self-field distribution and accordingly the current distribution between the monofilamentary tape and multifilamentary tape. The multifilaments behave as a solid superconductor under DC transport currents so that the current distribution cannot be equalized in the filaments.

1. Introduction

The Ag-Bi2223 tape is one of the most promising materials for applications to electric power devices such as power cables and transformers operating at 77K under AC current transmission because of high critical current densities J_c exceeding 100 A/mm^2 (77K,0T) in long lengths of tape-form conductors on the order of 1km length [1]. Therefore, the understanding of the mechanism which generates the AC transport losses in self-fields due to AC transport current is crucial for design and construction of power devices. It is well known that the key factor dominating the AC transport losses in self-fields is the critical current I_c of the tape, and nearly independent of the number of filaments in the tape sample [2,3]. This result suggests the superconductor multifilaments are likely to behave as solid superconductor for movement of magnetic flux, so that the distribution of transport currents in the filaments is not equalized in the tape-form Ag-Bi2223 multifilamentary tape.

In this paper, we present a systematic study on the self-fields at 77K due to DC transport currents subjected to the programmed current sequence in zero external fields on the Ag-Bi2223 monofilamentary and multifilamentary tapes, and discuss the current distribution in both tapes, in order to understand the behaviors of AC transport losses.

2. Experimental

The Ag-Bi2223 monofilamentary tape (sample A) and multifilamentary untwisted tape (sample B) with 61 filaments were fabricated by powder-in-tube method [2]. Geometrical factors of the tapes, together with I_c and J_c values at 77K and 0T determined from a DC four-probe method with a criterion 0.1μV/mm, are summarized in Table 1.

Table 1 : Summary of size factors and transport properties at 77K and 0T for the Ag-Bi2223 tapes

Sample Name	Filament number	Length (mm)	Width (mm)	Thickness (mm)	I_c (A)	J_c (A/mm^2)
A	1	30	3.0	0.17	11.2	71
B	61	28	4.0	0.23	22.8	160

Using the scanning Hall sensor magnetometry with an active area 50μm X 50μm [4], the self-field distribution due to DC current was measured at 77K under zero external fields. The sample was cooled in LiqN_2 in 0T, and the Hall sensor, at a fixed distance (~1mm) from the sample surface, measured a transverse component of self-field on a fixed line along the width direction at the interval of 0.1mm. Fig.1 shows a typical current sequence for the measurements on samples A and B. All states are hereafter called their name given in Fig.1. Note that the sample goes to a normal state and becomes magnetically transparent after the virgin and third states.

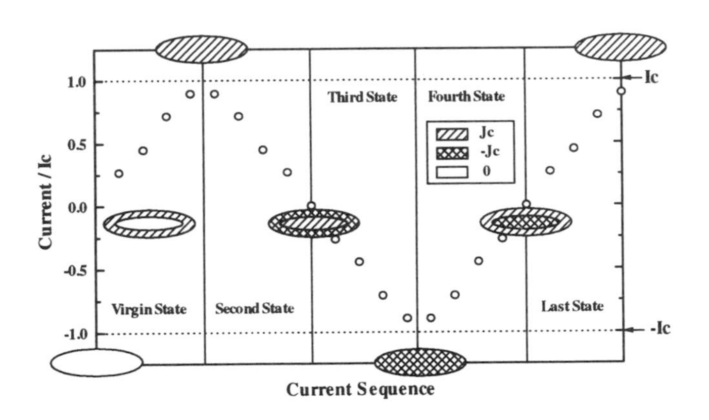

Fig.1. A schematic diagram for the current sequence to measure the self-field distribution at 77K under DC transport currents. Also shown is the definition of the state name which will be used throughout the study.

3. Results and Discussion

3.1 Self-field and current distributions for sample A

As mentioned in our previous paper [4], our samples are homogeneous without the macroscopic cracks which disturb the transport current path inside of superconductor core. Fig.2 shows the line profiles due to 3A (i.e., $0.273I_c$) of DC transport current in the virgin, second and the last states on the surface of sample A. In spite of the same current level, the results are strongly dependent on the measuring in which the current level is set up. Furthermore, the self-field distributions were measured even under no current transmission (i.e., $I = 0$A) after measurements in the second and fourth state, as shown in Fig.3.

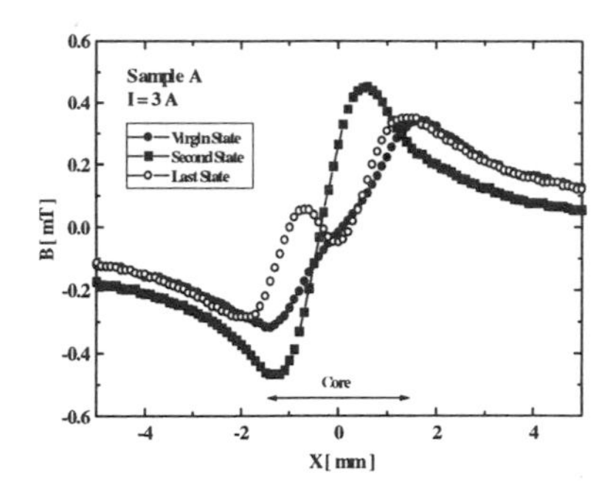

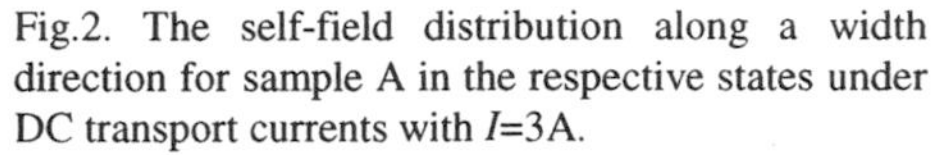
Fig.2. The self-field distribution along a width direction for sample A in the respective states under DC transport currents with I=3A.

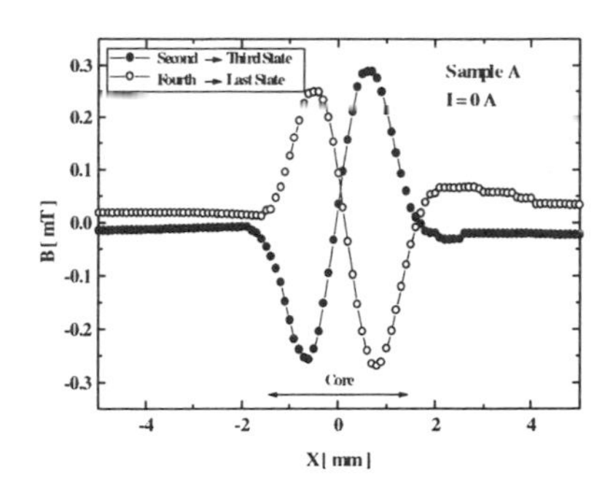

Fig.3. The self-field distribution for sample A under zero transport currents, caused by finishing the experiments in the second and fourth states.

In an attempt to discuss the current distribution, we deal with the field slope dB/dx derived from self-field distribution. Under consideration based on the electromagnetism, the value of dB/dx should be proportional to the current fragment δI at 0.1mm (a step interval of Hall sensor) along the width direction. Fig.4 shows the field slope dB/dx of sample A for I=3A (i.e., 0.273I_c) at the respective states. The broad minimum of dB/dx in virgin state indicates that the current is distributed mainly at the surface, and no-current transmission inside of core because there is no magnetic flux in the superconductor core before measurements. Next, the sharp peak of dB/dx in second state indicates that the current with +J_c dominate almost whole part of core with slightly distribution of -J_c around the surface. In the last state, the negative peak of dB/dx at center shows the existence of current transmission with -J_c inside of core. These results are completely different from the well-known conduction mechanism in a normal metal, and suggests that the current distribution in the core can be explained by the critical state model [5].

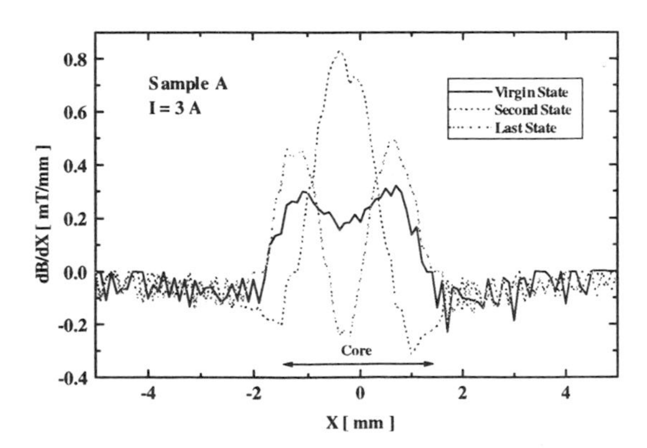

Fig.4. The field slope dB/dx derived from the field distribution for sample A in the respective states under DC transport currents with I=3A.

3.2 A comparison of self-fields and current distribution between samples A and B.

Fig.5 shows a comparison of field slope dB/dx for I=0.273I_c in virgin state (a) and I=-0.273I_c in third state (b) between samples A and B, in order to examine how the multifilaments behave under DC current transmission. Note that because of the difference in the core geometry between both samples, the horizontal axis in this figure is normalized to the width of superconductor core by regarding multifilaments as a single solid superconductor. As can be seen in Fig.5, there is strong similarity in the both shape and magnitude of the slope dB/dx between samples A and B. We note here that the similarity exists in the respective states independent of current level I. This result means that the multifilaments in Ag-Bi2223 tape behave as a solid superconductor under DC current transmission and accordingly the current distribution cannot be equalized in the filaments.

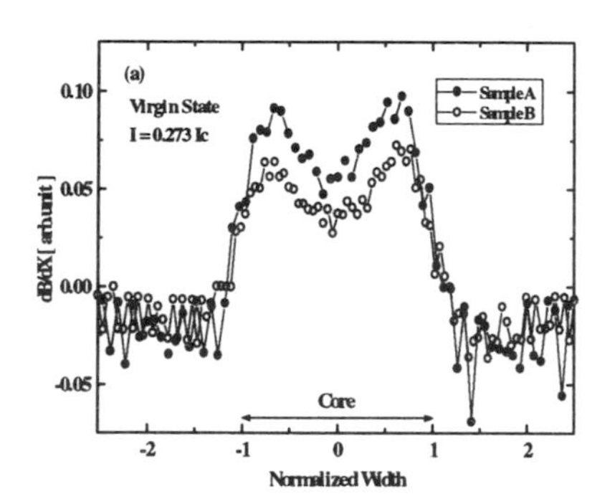

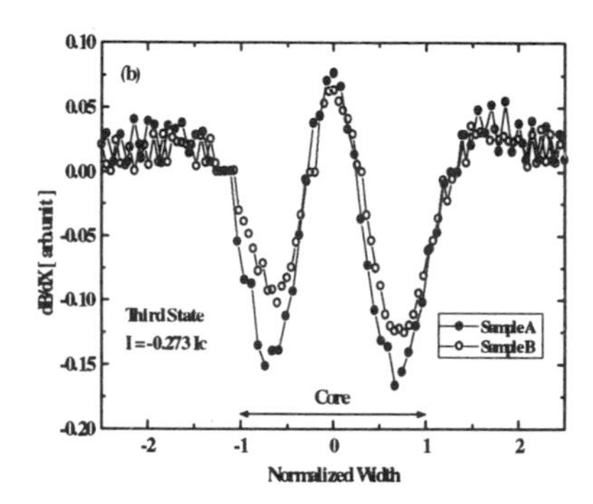

Fig.5. A comparison of the field slope dB/dx between sample A and sample B under DC transport currents: (a) in the virgin state with I = 0.273I_c, (b) in the third state with I = -0.273I_c.

4. Summary

Using a scanning Hall sensor with an active area 50μm X 50μm, the self-field distribution due to DC transport currents, programmed by the current sequence, has been investigated on the surfaces of Ag-Bi2223 monofilamentary tape and multifilamentary untwisted tape with 61 filaments. The result for monofilamentary tape can be explained by the critical state model [5]. There is no notable difference in the self-field distribution and accordingly the current distribution between the monofilamentary and multifilamentary tapes. The multifilaments behave as a solid superconductor under DC current transmission, so that the current distribution cannot be equalized in the multifilaments. The present study is consistent with our previous studies on the AC transport losses of Ag-Bi2223 tapes [2,3].

References

[1] K.Sato, K.Ohkura, K.Hayashi, M.Ueyama, J.Fujikami and T.Kato, Physica B216 (1996) 258.
[2] A.Oota, T.Fukunaga, M.Matsui, S.Yuhya and M.Hiraoka, Physica C249 (1995) 157.
[3] A.Oota, T.Fukunaga and T.Ito, Physica C270 (1996) 107.
[4] K.Kawano and A.Oota, Physica C275 (1997) 1.
[5] C.P.Bean, Phys.Rev.Lett. 8 (1962) 250; Rev.Mod.Phys.36 (1964) 31.

Inst. Phys. Conf. Ser. No 158
Paper presented at Applied Superconductivity, The Netherlands, 30 June–3 July 1997

Testing the homogeneity of Bi(2223)/Ag tapes by a Hall probe array

P. Kováč, V. Cambel, D. Gregušová, P. Eliáš, I. Hušek, R. Kúdela, S. Hasenöhrl and M.Ďurica

Institute of Electrical Engineering, Slovak Academy of Sciences, Bratislava, Slovakia

Abstract. The deforming process of a BSCCO/Ag composite is complicated resulting in inhomogeneous density and texture across and/or along the ceramic fibers and finaly in irregular distribution of transport currents in Bi(2223)/Ag tapes. To investigate such distributions, a sensitive system was developed to map the magnetic field at the surface of superconducting tapes. It is based on an integrated Hall probe array containing 16 in-line probes and PC-compatible electronics with software. The active layer of the array is madeof an InGaAs/InP heterostructure with a two-dimensional electron gas. The supporting electronics controls the Hall probe array and collects, stores, and processes measured data. A sensitive Hall probe array was employed to test the homogeneity of Bi(2223)/Ag tapes under the self-field condition. The results show the possibility to use this measurement technique for monitoring the quality of long-length Bi(2223)/Ag tapes before their concrete application.

1. Introduction

The homogeneity of Bi(2223)/Ag tapes, produced by an oxygen powder-in-tube technique (OPIT), is a very important parameter for their possible application in coils and transport cables. The mechanical deformation process of a BSCCO/Ag composite is complicated to allow for a homogeneous deformation of individual filaments. It has been demonstrated that the density, texture, and phase purity of the superconducting core correlates with J_c values [1-2]. Inhomogeneities in the aforementioned properties of the core bring about a complicated transport current distribution inside the tape. It has been shown using a Hall probe measurement near the surface of a short Bi(2223)/Ag tape that inhomogeneities related to critical current translate themselves into a magnetic field profile [3]. This work presents the measurement of such profiles using a sensitive Hall probe array. The sensor was placed close to a multicore Bi(2223)/Ag tape. This technique can be potentially used for monitoring the quality of long tapes.

2. Experimental

The multicore BSCCO/Ag tape was made by a standard powder-in-tube method which was

Fig. 1. Top view of a Hall probe array

followed by a roll-sinter process. The tape contained 36 filaments in the Ag matrix (0.2mm x 3.5mm). It was first measured in self field at liquid nitrogen temperature using a four-probe method. The tape was then monitored with a Hall probe array at zero and various transport currents. The InGaAs/InP Hall probe array is based on a semiconductor heterostructure with a two-dimensional electron gas (2DEG) as the active layer. Such arrays are being developed for use in the study of the magnetic field and its relaxation near the surface of high-temperature superconductors and other magnetic materials at 4.2 - 100 K [4]. An Ni-AuGe-Ni-Ti-Au system is used for the ohmic and contact metallization of the arrays. The topology of an array is defined in the form of a ladder-like shape (Fig. 1). The array consists of 16 probes each with an active area of 50 x 50 μm^2. The centre-to-centre distance between the probes is 200 μm. The magnetic sensitivity of the Hall probe array used for this experiment is 1100 V A^{-1} T^{-1}. The array is fixed on a ceramic holder and bonded with an Au wire. The Hall probe array was placed 0.2mm under the surface of a moved tape carrying transport current. The measuring system for mapping the magnetic field of superconducting tape consists of a semiconductor Hall probe array, supporting electronics with software, and a standard IBM-compatible PC. The purpose of the electronics and software is to control the operation of a Hall sensor. It allows collecting, storing, processing, and depicting measured data. An important role of the electronics and software is the elimination of offset voltages of the individual probes of a sensor during calibration procedures and measurement on real samples. High input sensitivity for dc signals is achieved by on-board voltage stabilization, topology of the printed-circuit board, and shielding. Further enhancement is accomplished by data accumulation. Repeating the measurement on a selected channel hundreds of times reduces the noise below 1 μV. The high-speed ADC allows scanning all 16 input channels in the short time (0.01 - 0.1s). The measurement is also possible at 50 Hz.

3. Results

The current-voltage characteristics are shown in figure 2. The values were taken from four positions of potential taps pressed to the tape surface and placed 5mm apart. There is an apparent difference between position 1 where exponent n = 60 ($U \approx I^n$) and all others (2-4)

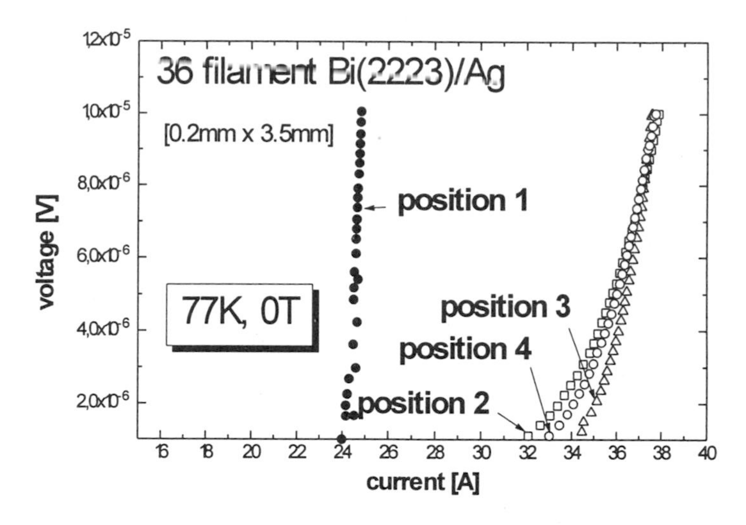

Figure 2. Current-voltage characteristics of 36 filament tape at four different positions

with n = 15 - 25. The sharp voltage rise and lower transport current reflect a large inhomogeneity of the tape at position 1.

The magnetic field profile was measured using the Hall probe array at positions 1 and 2 for transport current varying from zero up to the value which corresponds to the state of current-voltage characteristics E > 5μV/cm. The field profile was measured at both polarities of transport current because of the existence of frozen-in currents. Figure 3 shows 3D field profiles scanned from tape positions 1 and 2 for currents ranging from 26A to -26A(see Fig. 2). Shapes of the field profiles varied as a consequence of the inhomogeneity of the filaments and an irregular current distribution inside the tape. It should be noted that the field profile is strongly dependent on the current flow history (volume and direction of transport current

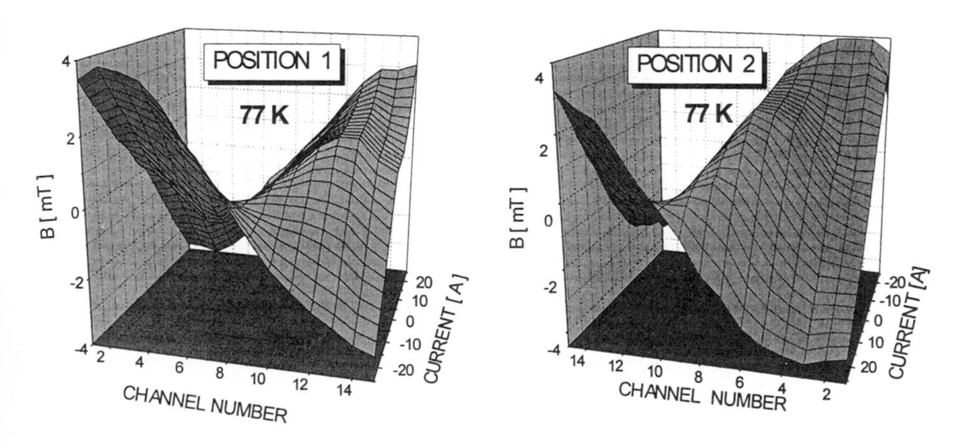

Figure 3. Magnetic field profiles as a function of transport current measured at 77K and two different positions corresponding to contact measurement in Fig. 2

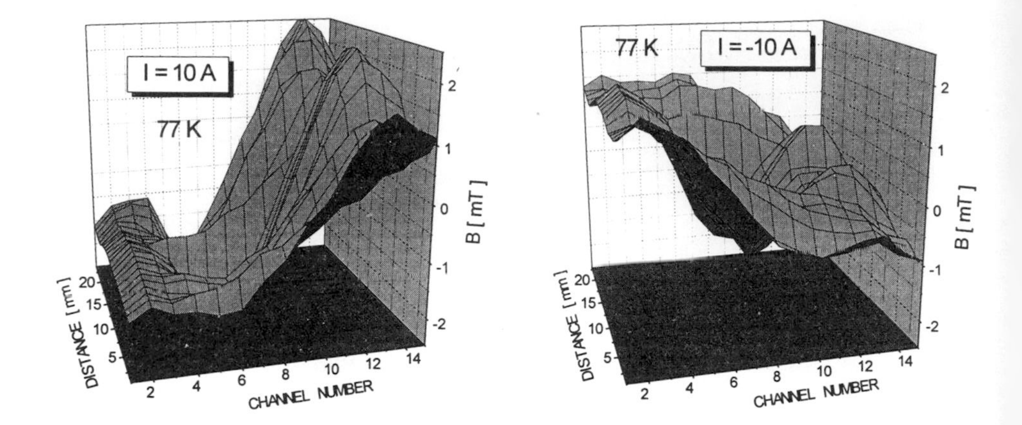

Figure 4. Magnetic field profile as a function of tape's position scanned at frozen-in currents and transport current I = 10A and I = -10A (oposite to frozen-in currents)

and direction of frozen-in currents). Therefore, mapping the field profile along the tape's length should be done at the same current state. An element of the aforementioned tape was scanned at constant current (0, 10A, and -10A) taking 1mm-long steps from position 1 to a distance of 25mm (distance equal to zero corresponds to position 1). The field profiles reflect the current distribution among individual filaments, as well as a correlation of the transport filament's current with frozen-in current (see Fig. 4). Though the length step for each measurement is only 1mm, the field profile changed markedly along the tested element of the tape. It shows that the quality of 2223 filaments along the tape was inhomogenous.

4. Conclusions

It was shown that magnetic field near a Bi(2223)/Ag multicore tape is affected by the current distribution among the filaments. The Hall probe array with supporting electronics allows for a very sensitive monitoring and short-time storing of field profile data at any position along the tape surface. The system can be used for continual testing of long Bi(2223)/Ag tapes.

Acknowledgements

This work was supported by Copernicus project CIPA CT 94-0115 and by the Grant Agency of the Slovak Academy of Sciences VEGA 4059/97 and VEGA 95/5305/108.

References

[1] Kováč P, Hušek I and Pachla W *Applied Superconductivity Conference, Pittsburg* 1996, will apear in *IEEE Transactions*

[2] Kováč P, Hušek I, Kopera L and Pachla W contribution at this conference

[3] Polak M, Majoroš M, Kvitkovič J, Kottman P, Kováč P and Melišek T *Cryogenics* **34** ICEC Supplement (1994) 805

[4] Cambel V, Eliáš P, Kúdela R,Novák J,Olejníková B, Mozolová Ž, Majoroš M, Kvitkovič J, and P. Hudek *Sol. State Electron.*, (1997) to be published.

Inst. Phys. Conf. Ser. No 158
Paper presented at Applied Superconductivity, The Netherlands, 30 June–3 July 1997

Hall probe measurements of the magnetic field above current carrying Bi-2223/Ag tapes

J Kvitkovic[1], E C L Chesneau[2,3], M Majoros[1], B A Glowacki[2,3], S Ashworth[3], M Ciszek[3] A M Campbell[4,2], J E Evetts[2,3]

[1]Institute of Electrical Engineering, Slovak Academy of Sciences, Slovak Republic.
[2]Department of Materials Science, Cambridge, UK
[3]IRC in Superconductivity, Cambridge, UK
[4]Department of Engineering, Cambridge, UK

Abstract. Here we present results of measurements made of the magnetic field above current carrying tapes of different architectures: multi-filamentary, multi-tube and single-core. The measurements were made by a high accuracy Hall probe assembly whilst the tapes were carrying a range of DC currents in a mimicked AC cycle. It is known that the AC losses in silver coated BSCCO tapes depend qualitatively on the architecture of the measured tape. Magnetic field measurements provide an insight into this behaviour.

1.Introduction

It has been found that the ac loss behaviour when carrying a transport current of many superconducting tapes [1,2] follows the equations derived by Norris [3] for superconducting wires of elliptical and slab-like cross-section. For an elliptical wire carrying the critical current, I_c, the vertical component of magnetic field, B_z, outside the wire at a position x,z, is given by [4];

$$B_z = \mathrm{Re}\left[\frac{\mu_o . I_c}{\pi \cdot (x + i.z + \sqrt{(x+i.z)^2 - a^2 + b^2})}\right]$$

where a and b are the semi-major and semi-minor axes. This results in a transport loss per cycle of;

$$L_c = \frac{I_c^2 \mu_o}{\pi}\{(1-F)\ln(1-F) + (2-F)F/2\}$$

where F is the ratio of the maximum current, I_{max}, to I_c. However, many of the tapes which fit the equations have cross-section morphologies that are neither elliptical nor slab-like.

To investigate this behaviour further we have made measurements of the vertical component of the magnetic field induced by a transport current above several Ag/Bi-2223 tapes during a slow, stepped ac cycle. One tape, which is known to closely follow the ellipse loss equation, is considered in more detail.

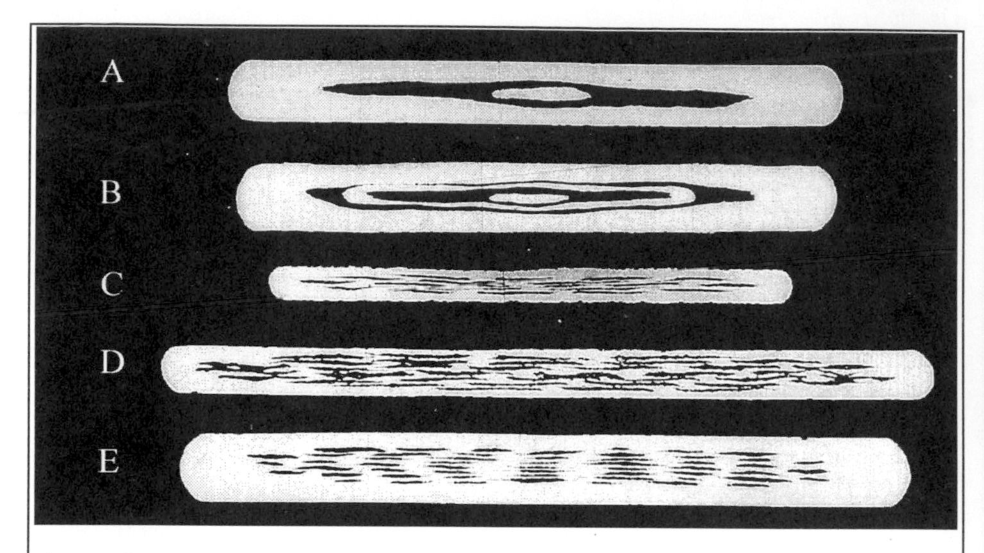

Figure 1 Cross-sections of investigated tapes; A single tube I_c = 2.5A; B double tube I_c = 4.4A; C 19 filament I_c = 14A: D 37 filament I_c = 18.8; E 55 filament I_c = 6A. The total width of tape E is 4.0mm

2.Experimental

The measurements were made by a scanning Hall probe approximately 140μm above the tape surface, with a step size of 50μm, across the tape, perpendicular to the current flow. The measurements of the vertical component of magnetic field, B_z, were made during a slow stepped ac cycle , a full scan being made at each current increment. The current was stepped from between plus and minus I_{max} which was close to I_c in steps of approximately $0.2I_{max}$. After each step the current was held constant for 16 to 20 minutes while the tape was scanned, thus the measurements were made under dc conditions. Before starting the tapes were first passed through several slow ac cycles in order to establish the same current distribution as in ac loading. Further details of the experimental arrangement are given in [5]. Optical micrographs of the cross-sections of the five tapes tested are shown in Figure 1.

3.Results and discussion

Figure 2 shows the magnetic field above the tapes while carrying I_{max}. This has been normalised with respect to the width of the superconductor within the tapes. There is variation between the tapes, in particular tape E shows dips in the field between the minimum and maximum field peaks.

Measurements made at 1mm separation along the tape showed little difference which suggested that the lateral currents in the tapes are considerably smaller than those along the tape. This, along with the high aspect ratios of the tapes (between 14.6 and 20.9) means that the gradient of the magnetic field across the tape can be taken as an approximation to the local density of current along the tape [6]. This is a reasonable approximation close to the centre of the tape. However , away from the centre, the probe is further away from the superconductor than at the centre since the superconductor becomes thinner; this will have a

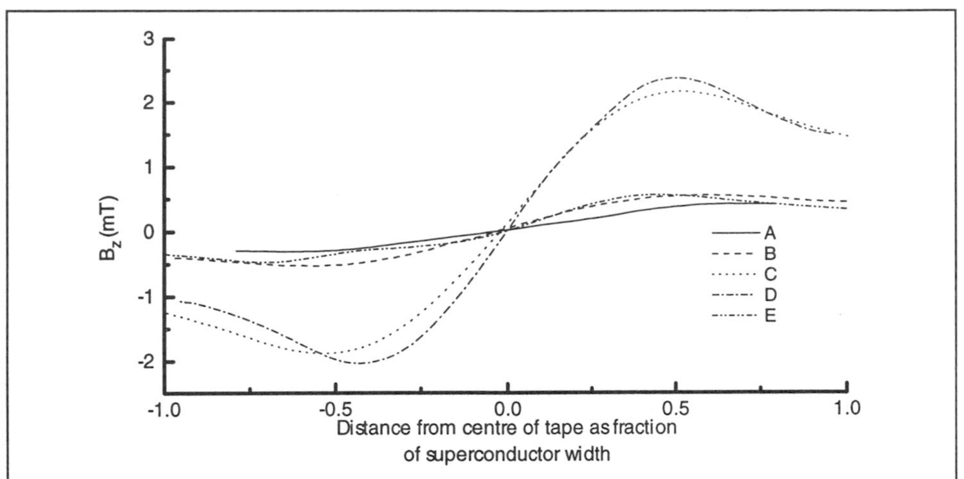

Figure 2 Magnetic field above Ag/Bi-2223 tapes at peak current during a slow stepped ac cycle. The data are normalised to the supercondutor width in each tape but, for clarity, the field is not normalised for the I_c of each tape. The curves are labelled as in Figure 1

significant effect on the field measured and the current density obtained, especially near the edges of the superconductor. Outside the width of the superconductor, the method is no longer valid.

Locally high field gradients can therefore imply regions of particularly good superconductor. They may also indicate where the current flow is closer to the tape surface, and the probe, than average. That this is possible is evident in tape E where the filaments are stacked so that it may be possible to have positions where the superconductor forms only a small fraction of the tape below the probe. Tapes, such as C and D, which have overlapping filaments or significant amount of crosslinking, do not show such dips in the field.

The multitube tapes; A and B show no sign of a reduced current density in the central regions where one might expect the silver core to have an effect. However, the introduction of a silver core has been found to increase the relative J_c of the thin layers of superconductor above and below it [7]. From our measurements, this increase in J_c compensates for the reduction in superconductor near the probe.

We now turn to the magnetic field above tape D as the current is reduced from the maximum value, 17A, to zero as shown in Figure 3. The Norris ellipse, as represented by the dotted lines, shows characteristic undulations; these extend even to I_{max} since this was lower than I_c. These undulations indicate the presence of inner regions carrying a current in the reverse direction to the outer regions. In comparison the tape D shows a magnetic field which is much more linear across the tape. This is more indicative of a uniform current distribution rather than a Bean distribution. However, the movement of the peaks of field towards the centre as the current is decreased suggests that the current is tending to occupy a central region that decreases in size as the transport current drops. Again this does not fit the Norris equations. These observations may be due to magnetic relaxation; measurements by Paasi et al [8] suggest that this may be significant over the time scale of our measurements. This could mean that dc or low frequency ac measurements are not representative of power frequencies. In the future we hope to expand our measurements to cover the gap between the regimes. It is also possible that the Norris loss equations are more general than thought such that an elliptical cross-section is not required to produce what is known as elliptical type losses.

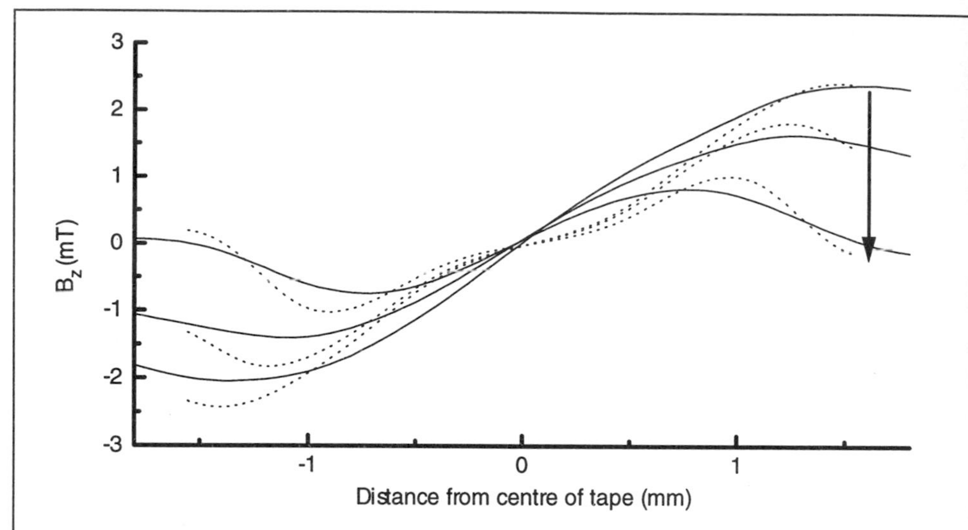

Figure 3 The field, B_z, above a tape as the transport current decreases from $+I_{max}$ to 0.6Imax to 0A. The solid lines are the experimental results for tape D, the dotted lines are the theoretical results for a wire of elliptical cross-section with identical I_c, I_{max}, minor and major axes. The total width of superconductor is 3.14mm. The arrow indicates the direction of change with decreasing current.

4. Conclusions

In conclusion, we have measured the magnetic field above several tapes which follow Norris's elliptical ac loss equations although this might not be expected from the cross sections. Many of the tapes have anomalous features that imply that the Norris model for current flow is not followed. More work is required to see if these are a result of the low frequency measurements used or if the loss equations can apply to other cross-sections.

Acknowledgements

The authors would like to thank Europa Metalli, IGC and NKT for samples. One of us (ECL Chesneau) would also like to gratefully acknowledge the financial support of Oxford Instruments, in the form of a CASE award.

References

[1] Ciszek M, Glowacki B A, Campbell B A, Ashworth S P, Liang W Y, Haldar P and Selvanamickam V *IEEE Trans. Appl. Superconductivity7 (in press)*

[2] Glowacki B A, Senderman K G, Chesneau E C L, Ciszek M, Ashworth S P, Campbell A M and Evetts J E *ibid*

[3 Norris W T 1970 *J. Phys. D* **3** 489- 507I

[4] Beth R A 1967 J. Appl. Phys. **38** 4689-4692

[5] Kvitkovic J and Majoros M 1996 *J. Magnetism and Magnetic Materials* **158** 440-441

[6] Chesneau E, Kvitkovic J, Glowacki B A, Majoros M and Haldar P *ibid*

[7] Vasanthamohan N and Singh J P 1997 *Supercond. Sci. Technol.* **10** 113-118

[8] Paasi J, Polak M, Lahtinen M, Plechacek V and Soderlund L 1992 *Cryogenics* **32** 1076-1083

Inst. Phys. Conf. Ser. No 158
Paper presented at Applied Superconductivity, The Netherlands, 30 June–3 July 1997

Magnetic flux mapping and current distribution in BSCCO tapes

H-G Kürschner‡, M Hortig‡, B Lehndorff†§, and H Piel†‡

† Bergische Universität Wuppertal, FB08, Institut für Materialwissenschaften, Gaußstr. 20, D-42097 Wuppertal, Germany
‡ Cryoelectra GmbH, Wettinerstr. 6H, D-42218 Wuppertal, Germany
§ Applied Superconductivity Center, University of Wisconsin-Madison, 1500 Engineering Drive, Madison, WI 53706, U.S.A.

Abstract. A 3D scanning Hall-probe system was used to measure the remanent flux of a multi-filamentary as well as the flux produced by a transport current. Transport currents were varied up to the critical current. From this data the local magnetic moment can be determined with a iteration procedure. The magnetisation was used to calculate the current distribution in the superconductor. The results in the transport current state deviates from the theoretical expected one at all cracks or weakly superconducting inclusions in the superconductor. Thus this gives us a good non destructive method to control the quality of tapes.

1. Introduction

The silver sheathed BSCCO-2223 tapes fabricated by the powder-in-tube process are the most promising materials for electrical power applications like magnets and transmission cables. For applicating the BSCCO-2223 tapes several 100 m of conductor with a well defined quality are needed. As there is still a large reduction of the critical current density of long tapes compared with short samples [1], the origin of the decrease has o be investigated. While the determination of j_c as a function of the tape length is a ntegral method, magnetic flux mapping allows a local, non destructive characterisation. This method is suitable for controlling the tapes quality if all j_c reducing defects as racks or non superconducting areas can be detected with high reliability.

. Experimental Setup

To get a rough overview of the quality of a long tape (0.25 mm thick, 3.2 mm wide, 9 filaments), we measure j_c over sections of about 1 cm. This distance can be increased p to 20 cm for longer tapes. Sections with strongly reduced j_c (below 70 % of the verage) are than cut out of the tape and examined further with a magnetic scanning ethod to get more information about the origin of the defects.

The magnetic field above the tape is scanned with a Hall-probe attached to a computer driven three axial moving system. The Hall-probe is 0.2 mm squared and parallel to the surface of the superconductor. Thus only the magnetic field component perpendicular to the surface (normal-component) can be detected. The spatial resolution is limited by the distance between tape and Hall-probe to better then 0.5 mm.

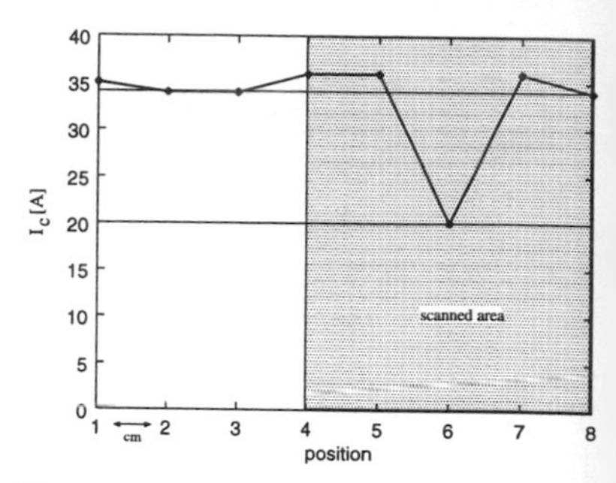

Figure 1. Critical current distribution of the tape used for scanning.

Three different current distributions in the tape are used. The first one is the remnant field after cooling in an external field perpendicular to the plane of the tape. The induced super currents are flowing in the plane than. The second one is the self induced field of the tape carrying a current $I < I_c$. The resulting magnetic field is circular. The third one is the remnant field of the transport current which remains after the current is turned off. The self induced currents are flowing in the plane of the tape but with opposite directions on both halves. The main difference between these modes is that the remanent field needs not to be produced by a macroscopic current while in the second mode there is a macroscopic transport current present. From the measured magnetic field component the corresponding magnetisation and current distribution is computed with a iteration procedure similar to the one shown by Xing et al. [2].

3. Experimental results

In figure 1 the j_c for each 1 cm-segment of a 10 cm long tape is shown. The segment 6 reduces the total current density to 60% of the average of the other positions due to a defect. The marked segment is further examined by scanning Hall-probe measurements shown in figure 2 to 4.

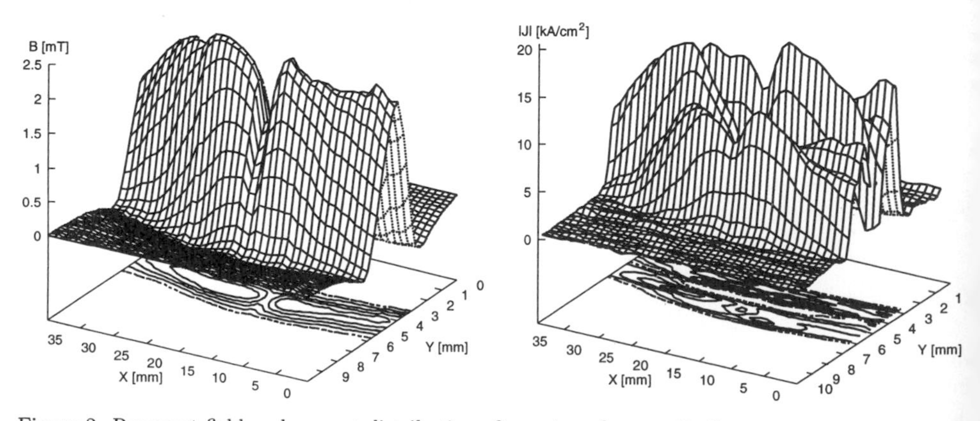

Figure 2. Remnant field and current distribution after external magnetisation

In the left part of figure 2 the remanent magnetic field $B_z(x,y)$ after the magnetisation with an external field of 25 mT is plotted over the area surround the defect. The right part is the corresponding current distribution $|j_c(x,y)|$. The magnetic field has the typical Bean like shape as predicted. The defect can be found easily due to the reduction of the magnetic remnant in the middle of the tape. The local critical current drops from 20 kA/cm^2 to 10 kA/cm^2.

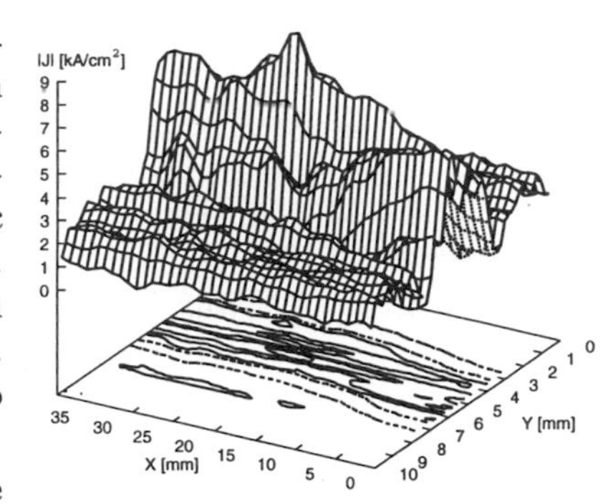

Figure 3. Transport current distribution

Figure 3 shows the absolute value of the transport current distribution $|j(x,y)|$ with an absolute current $I = 20$ A which is just below the critical current at the defect. Compared to the remnant case the current density is halved because it is limited by the critical current density of the defect. In the remnant case all flowing currents are of the local critical value.

In figure 4 the remnant field with its current distribution is depicted after switching off of the transport current. Due to the current distribution resulting from the self induced field there is no magnetic field perpendicular to the plane in the middle of the tape. All currents are flowing near the edges of the tape with opposite orientations for both halves. The defect can be easily found in the middle of the front half of the magnetic field distribution. So this mode is very sensitive to defects at the outer part of the tapes.

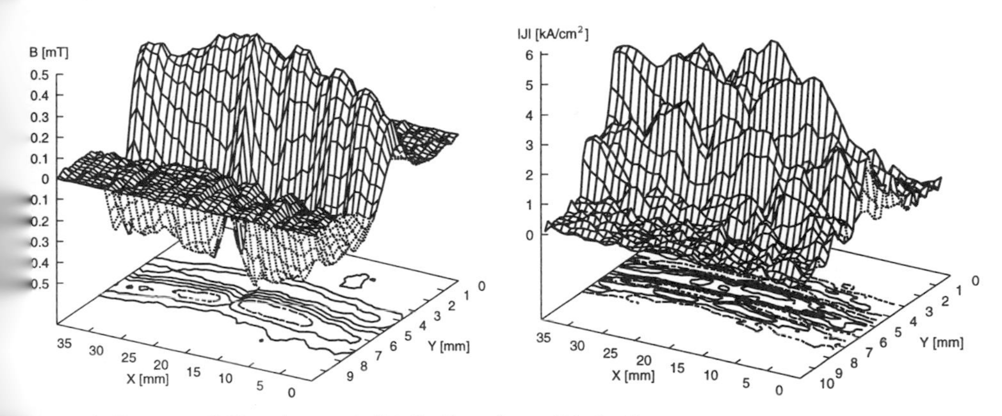

igure 4. Remnant field and current distribution after self induction

. Discussion

When using an external field for magnetising the sample the measured remanent field distributions perpendicular to the tapes length are symmetric and Bean-like (see fig. 2). The slope $\Delta M \ / \ \Delta x$ is a measure for the critical current. A sketch of the magnetisation vs. the place for a tape with and without imperfection is shown in fig. 5. Any imperfections, non superconducting areas or cracks reduce the maximum value measured in the middle of the tape independent of the localisation of the defect.

The trapped field due to transport currents is insensitive in the middle of the tape because there is no normal component of the magnetic field. This mode provides the possibility to localise the defect. Using the trapped field mode non superconduction areas are reliably detected because there is no remanent field. The situation is similar for badly superconducting areas. In both trapped flux modes the microscopic current distribution, intra or inter grain currents, is uncertain. Only in the transport current mode a macroscopic current is flowing for sure. A defect changes the local current distribution which effects the magnetic field configuration. The change can be detected by calculating the local current distribution but is not as clearly visible as in the remanent modes.

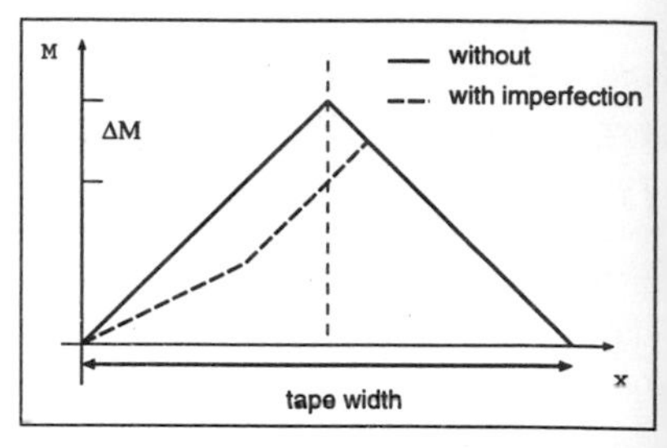

Figure 5. Magnetisation with and without imperfection vs. position

5. Conclusions

The normal component of the remanent field in the middle of BSCCO-2223 tapes is sensitive to any imperfection in the tape. This mode is most suitable for a long length quality control of the tapes. Defects can also be localised using the remanent self field or the self field of a transport current. In these modes 2 dimensional scans are necessary and the speed of measuring (3 cm/h) is limited.

Acknowledgement

We would like to thank Peter Hardenbicker and Karl Zeitzs for supplying the tapes. This work was supported by the BMBF under Contract No. 13N6634.

References

[1] Q Li et al., Appl. Supercond. Conf. (ASC'96), Pittsburgh, PA, Aug. 25-30, 1996

[2] W Xing et al. J. Appl. Phys. 76 (7), 1 Oct. 1994

[3] K Kawano and A Oota, Physica C 275 (1997) p. 1-11

Inst. Phys. Conf. Ser. No 158
Paper presented at Applied Superconductivity, The Netherlands, 30 June–3 July 1997

Influence of processing and composition on the formation of (Bi,Pb)-2223 tapes

P.Haug, D.Göhring, M.Vogt, A.Trautner, W.Wischert and S.Kemmler-Sack

Institut für Anorganische Chemie, Auf der Morgenstelle 18, D-72076 Tübingen, Germany

Abstract. The PIT process for 2223 tapes starts with a rapid formation of large plates of well textured 2212 material. However, the 2212 → 2223 transformation during the subsequent thermomechanical treatments is not accompanied by grain growth. High critical current densities were obtained for a medium 2223 content after the first heating step, whereas a prolonged final heating period destroyed the 2223 phases in favor of 2212. The employed dopants can be subdivided into promotors (Cu, Mn ($x \leq 0.1$) and Rh) and inhibitors (Mg, Al, Sn, Co, Y, Pd and high substtutions levels of Mn and Ni) for the transformation of the precursor into (Bi,Pb)-2223.

1. Introduction

ne of the most interesting applications of high-T_c oxide superconductors (Bi,Pb)-2223 is in orm of long lengths of flexible conductors for transmission cables, high field magnets, notors and generators. Ag-clad (Bi,Pb)-2223 made by using the powder-in-tube method PIT) is currently the most promising wire technology [1]. A further challenge for the abrication process is the brittle nature of (Bi,Pb)-2223 in combination with the high ductility f Ag. To solve these problems dispersion hardening or alloying of Ag with several metals ave been examined by various authors (e.g. [2] and Refs. therein). Moreover, the interaction f the resulting oxides with the core was studied by several groups (e.g. [3,4] and Refs. herein). According to a study of Tanaka et al. [5] Ag/Cu alloyed sheaths doped with Ti yield igher j_c values than for the undoped sample, because amorphous regions are likely to rovide effective flux pinning centres.

The aim of the present study is to follow certain steps of the PIT process and to select om possible alloying metals of the Ag clad those elements that will not deteriorate the uperconducting properties of (Bi,Pb)-2223 tapes.

. Experimental

he employed precursor powder has the nominal cation ratio Bi:Pb:Sr:Ca:Cu = 8:0.4:2:2.1:3. It was prepared either via the solution of nitrates or by solid state reaction cording to Ref. [6]. It was used for the standard process of tape production as well as for e doping studies with different amounts of M according to the general formula $_{1.8}Pb_{0.4}Sr_2Ca_{2.1}Cu_3M_XO_Z$ and compared with a reference ($x = 0$), heat-treated under identical conditions as indicated in [6]. The materials were investigated by XRD (CuKα radiation). The 2223 content was evaluated via the intensity I according to 2223 (%) = I[00<u>10</u>(2223)]/I {[00<u>10</u>(2223)] + I [008(2212)] + I[006(2201)]}. For all doping experiments this value was normalized to the value of the reference. Critical current density measurements were performed on the 2223 tapes at 77K and self field using the 1μV/cm voltage criterium. The dc magnetic susceptibility was measured with a SQUID magnetometer (Quantum Design).

3. Results and discussion

3.1. (Bi,Pb)-2223 formation in the PIT process

3.1.1. Development of 2212 texture in the tapes

XRD of the precursor powder from the green tape indicate the presence of some crystalline material, consisting mainly of (Bi,Pb)-2212 and a mixture of Ca_2PbO_4/3321 phase (Fig. 1). Astonishingly, after a very short heating time of 20 min at the relatively low temperature of 790°C/air a texture of the 2212 phase begins to develop. This effect is much more pronounced after increasing the firing temperature to 850°C (Fig. 1(c)). Moreover, the Pb^{4+}-containing phases (Ca_2PbO_4/3321) rapidly vanish due to an incorporation into 2212; thus helping to develop the pronounced texture.

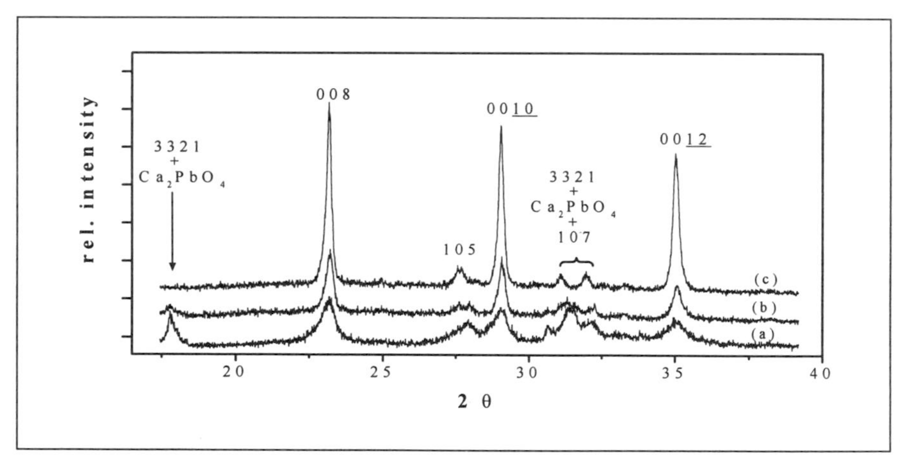

Fig. 1 Section of the XRD pattern from the surface of the core of monocore tapes (a) green tape, (b) after heating at 790°C/20min in air and (c) 850°C/20min in air.

In Fig. 2 SEM micrographs of the core are compared after several heating periods i N_2/O_2(8%). After the first furnace period the 2212 content amounts still about 50% (wi ≈50% of 2223). One observes large, extended plates with an incomplete developed textur The critical current density j_c is low (≈1.4 kA/cm^2; 77K; 0T). After two addition thermomechanical treatments the amount of 2223 increases to above 80% with $j_c \approx$ 19 kA/cm^2; 77K; 0T. The SEM micrographs indicate an improved texture. However, there is r grain growth, but the extension of the plates is reduced due to the intermediate mechanic treatment. So, the once formed large 2212 plates were destroyed during the mechanic treatment and do not recover after its transformation into 2223.

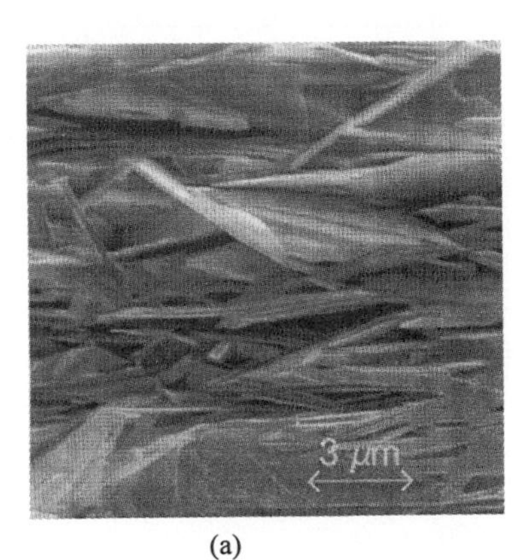

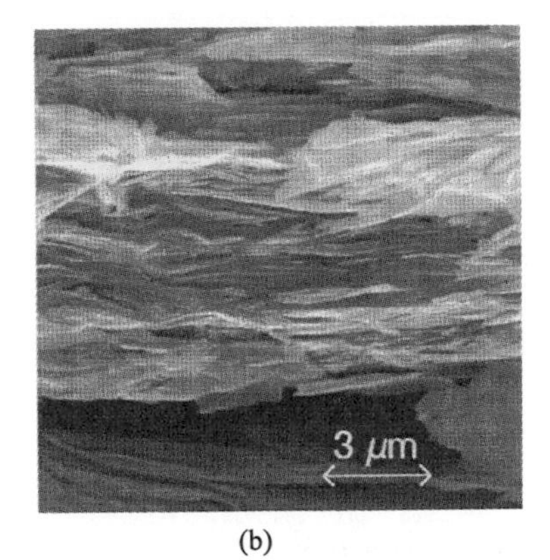

(a) (b)

Fig. 2 SEM images of the surface of the core from an Ag tape after (a) the first and (b) the third furnace period.

3.1.2. Influence of heating temperature and time

For a monocore Ag tape the influence of the heating temperature on the 2223 formation after 10h in N_2/O_2(8%) was studied. As indicated in Fig. 3 the 2223 amounts as well as the critical current increases rapidly with increasing temperature. However, it is not suitable to start with both a very low or high 2223 content. The best results were obtained in case of about 50% amount of 2223 after the first heating period. In this case even after the second step remains sufficient precursor material in the core for the transformation of 2212 into 2223 and simultaneously improving the connectivity between its grains.

Since it is well known that the velocity of the 2223 growth is low and an optimal grain/grain contact is desired we varied in a series of experiments the time of the 3. heating period. We started with 30% 2223; I_c = 0.1A after the first heating period and 88% 2223; I_c = 5.3A after the second step. The third step of 16h yield an improvement to 96% 2223; I_c = 5.5A, whereas a prolongation to 12 days results in a dramatical decrease of the 2223 content to 40% in favor of the 2212 phase. So, very long heating periods should be omitted.

3.2. Doping of (Bi,Pb)-2223

The influence of doping was studied for various elements from different groups of the periodic table as main group elements: Mg, Al, Sn, In; 3d elements: Ti, Mn, Co, Ni, Cu and 4d elements: Y, Rh, Pd. The present investigation enables a subdivision of the employed dopants into promotors (Cu, Mn at low doping level ($x \leq 0.1$) and Rh) and inhibitors (Mg, Al, Sn, Co, Y, Pd, and high substitution levels of Mn and Ni) of the transformation of the precursor into (Bi,Pb)-2223.

XRD studies indicate that the incorporation of the dopant into 2223 is uniformely restricted to a very low doping level. Thus, in most cases doping is accompanied by the formation of foreign phases. Several authors have demonstrated that the segregation of secondary phases improves the critical current density in case of Y-123. However, for the 2223 matrix we did not observe a similar effect.

From our χ vs. T data it follows that for most of the dopants the positions of T_c for the 2223 phase remains unaffected near about 110K. However, Co, Ni and Pd doping depresses T_c of 2223. As example Fig. 4 indicates the data for the Ni doped system.

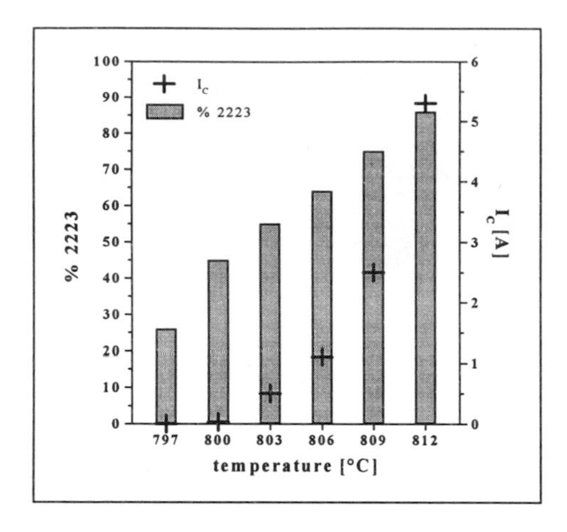

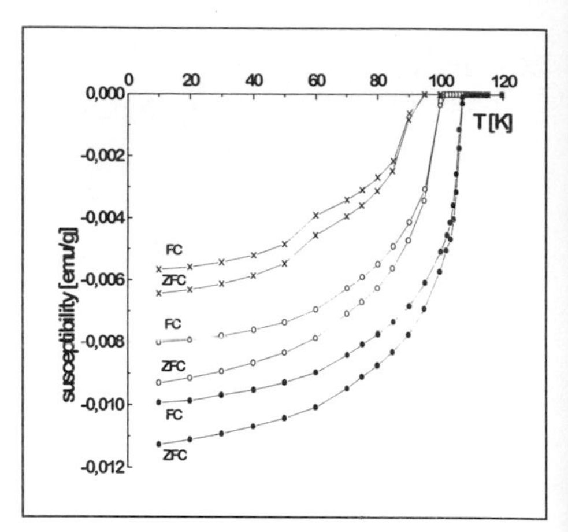

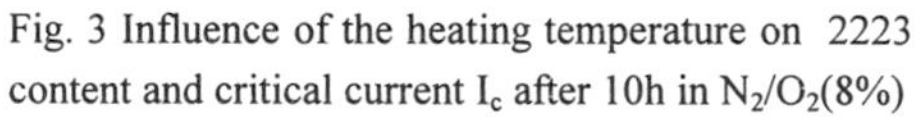

Fig. 3 Influence of the heating temperature on 2223 content and critical current I_c after 10h in N_2/O_2(8%)

Fig. 4 χ vs. T for $Bi_{1.8}Pb_{0.4}Sr_2Ca_{2.1}Cu_3Ni_xO_z$ measured at 0.001T with (x) x = 1.0; (o) x = 0.1 and (●) x = 0. FC: field cooled; ZFC: zero field cooled.

Acknowledgements

The authors are indebted to the Bundesministerium für Bildung, Wissenschaft, Forschung und Technologie and the Siemens AG (FKZ 13N6481) for financial support. This work was supported by the Verband der Chemischen Industrie. We wish to thank Prof. Dr. R. P. Huebener and T.Nissel for the SEM micrographs.

References

[1] Li S, Goa W, Hu Q Y, Liu H K and Dou S X 1997 Physica C276 229

[2] Ishizuka M, Tanaka Y and Maeda H 1995 Physica C252 339

[3] Grivel J-C and Flükiger R 1995 Supercond.Sci.Technol. 8 751

[4] Grivel J-C and Flükiger R 1996 Physica C256 283

[5] Tanaka Y, Ishizuka H, He L L, Horiuchi S and Maeda H 1996 Physica C268 133

[6] Göhring D, Vogt M, Wischert W and Kemmler-Sack S 1997 Mater.Sci.Engin., in press

Inst. Phys. Conf. Ser. No 158
Paper presented at Applied Superconductivity, The Netherlands, 30 June–3 July 1997

Multifilamentary Composite Tapes and Round Wires Based on BiPbSrCaCuO

A.D.Nikulin, A.K.Shikov, I.I.Akimov, F.V.Popov, D.N.Rakov, D.A.Filichev, N.I.Kozlenkova.

All-Russia Scientific Research Institute of Inorganic Materials, Moscow, Russia

Abstract. The architecture and critical properties of composite multifilamentary tapes as well as round wires based on Bi-system ceramics are presented, and with twisted filaments among them. The optimization of design, number and sizes of filaments as well as deformation and heat treatment regimes made it possible to achieve J_c-values higher than $3x10^4$ A/cm^2 (77 K, 0 T) and 10^5 A/cm^2 (4,2 K, 4 T) for 61-filaments Bi-2223/Ag and 361-filaments Bi-2212/Ag conductors, respectively. The procedure is proposed for making by the "powder-in-tube" technique the round wire Bi-2223/Ag having ring multilayer architecture of filaments. In short samples of this conductor processed with intermediate groove rolling between annealing stages the J_c-value as high as 10^4 A/cm^2 (77 K, 0 T) is achieved. The conductors mentioned above are used to making some pancake coils by the "wind-and-react" as well as "react-and-wind" techniques.

1. Introduction

In recent time a quite uniform approach has been formed concerning the design of composite superconductors based on high-Tc BiPbSrCaCuO compounds. This so called "multifilamentary" approach made it possible to achieve high critical current density and simultaneously high transport current values in Bi-system composite superconductors produced by the "powder-in-tube" technique [1-2]. It is necessary to note that such high critical properties are reached reproducibly in "rolled" samples which are the analogous of long length conductors. Multifilamentary conductors have higher mechanical and thermal stability than monofilamentary ones, so their high current application is more preferable.

Additionally the multifilamentary design of HTS-conductors permits to produce twisted ones for possible AC-applications [3]. On the view-point of magnet application it is very interesting to realize high current carrying capacity in round wires, in particular based on Bi-2223. In our opinion the multifilamentary approach makes it possible to perform it. Present study illustrates the possibilities of the "powder-in-tube" technique in fabrication process of multifilamentary as well as multilayer composite HTc-superconductors, Bi-2223/Ag and Bi-2212/Ag round wires.

2. Experimental procedure

To fabricate high-Tc composite conductors up to 200 m in length the "powder-in-tube" technique was used. Silver or silver alloy tubes were vibrofilled with fill factor 45-50 % by the "freeze-dried" precursors Bi-2223 as well as Bi-2212. Monofilamentary wires and tapes were made by drawing and tape rolling. To create multifilamentary billets bundles of round monofilamentary wires were placed into the tubes having round cross section. Using drawing and tape rolling multifilamentary tapes were made having up to 84 filaments. 361 and 703-filamentary conductors were produced by the re-bundling of 19 or 37-filamentary round wires, respectively. In order to produce 14-layers tape 14 monocore tapes stacked together were placed into a profiled tube, and such billet was subjected to drawing and rolling.

Bi-2223/Ag round wires have been fabricated by the filling of complex construction formed from tubes with different diameters and placed co axially. 19 and 61-filamentary twisted conductors have been made by using of torsion deformation performed simultaneously with drawing, and followed by tape rolling. Superconducting properties were realized by the "sinter-roll-sinter" method for Bi-2223 based conductors, and by multistage heat treatment in O_2 and Ar for Bi-2212 ones. Round Bi-2223 wires were heat treated with intermediate groove rolling which was performed in specially made grooved rolls. Critical current measurements on short samples were performed by the standard four-probe technique by 1 μV/cm and 0,1 μV/cm criteria at 77 K and 4,2 K, respectively.

3. Results and discussion

Longitudinal and transversal sections of composite conductors produced in present work are shown in Fig. 1-4. These conductors have different design namely number and size of filaments, geometrical size of cross sections and were made by different routes.

(a) 19-filamentary Bi-2223/Ag tape. Cross section 0.2x4.3 mm^2. J_c=29 kA/cm^2 (77 K, 0 T)

(b) 61-filamentary Bi-2223/Ag tape. Cross section 0.3x4.4 mm^2. J_c=30 kA/cm^2 (77 K, 0 T)

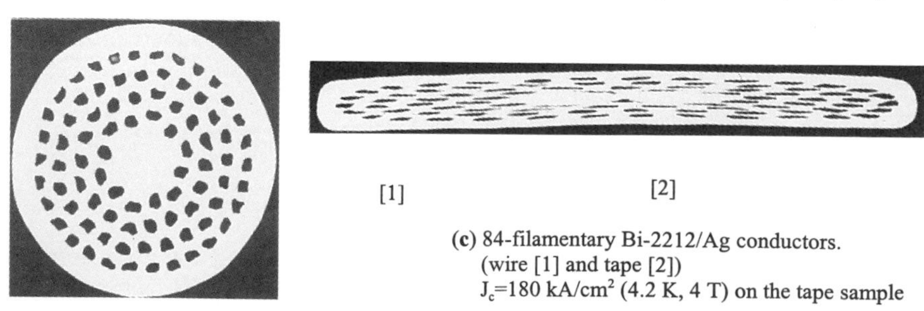

(c) 84-filamentary Bi-2212/Ag conductors.
(wire [1] and tape [2])
J_c=180 kA/cm^2 (4.2 K, 4 T) on the tape sample

Fig. 1. Multifilamentary conductors based on Bi-system.

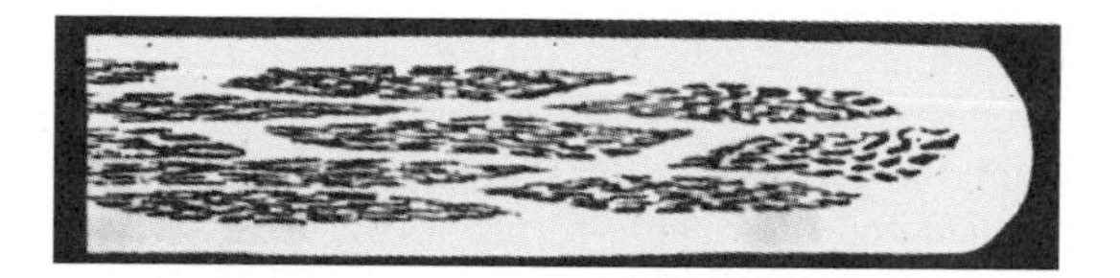

(a) 703-filamentary Bi-2223/Ag tape, part of cross section.
Cross section 0.25x4.3 mm^2. J_c=25 kA/cm^2 (77 K, 0 T)

(b) 14-layers Bi-2212/Ag tape. Cross section 0.4x5.5 mm^2.
J_c= 200 kA/cm^2 (4.2 K, 0 T)

Fig.2. Conductors based on Bi-system, having multifilamentary or multilayer design.

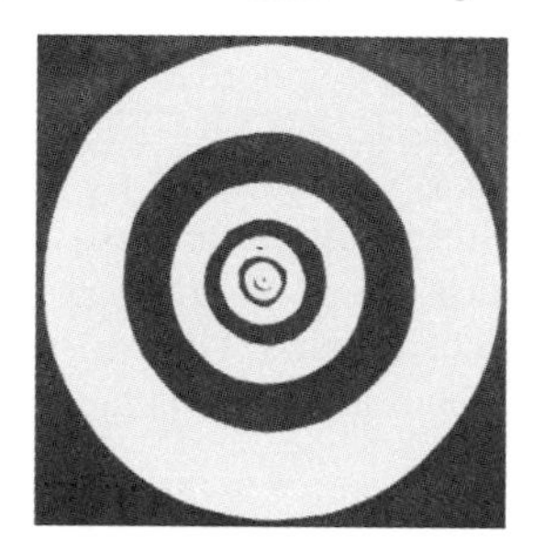

Fig.3. Bi-2223/Ag round wires 1 mm in diameter. J_c=10 kA/cm^2 (77 K, 0 T)

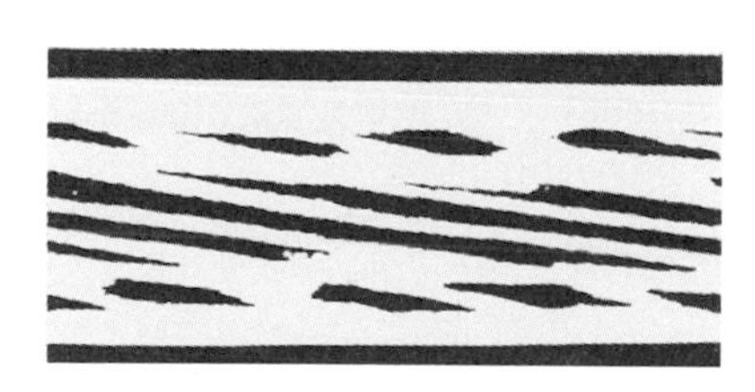

(a) round 61-filamentary Bi-2223/Ag wire 0.8 mm in diameter. Twist pitch 0,8 mm.

(b) round 19-filamentary Bi-2223/AgNiY wire 1.0 mm in diameter. Twist pitch 1.2 mm.

Fig.4. Longitudinal sections of twisted round Bi-2223 conductors.

The results of metallographic cross-sections shown in Fig. 1-4 as well as critical current measurements made it possible to suppose that there is some optimal architectures of HTS-conductors, i.e. optimal number of filaments, and initial fill factor for initial single filament at first, and optimal distribution of filaments over the cross section at second. At this view-point it is considered reasonable to use silver filaments in central part of initial multifilamentary billet, because in this case more uniform distribution of filaments on thickness over cross section is observed. Simultaneously by route above the degradation of geometrical form of central filaments (see Fig. 16, 61-filamentary tape) is excluded as in case of 84-filamentary conductor (Fig. 1c).

By increasing the filaments number in multifilamentary billet, especially by "double-billet" method, the quality of longitudinal filaments geometry is sufficiently aggravated, with the respective decreasing of critical properties (Fig. 2a, 703-filamentary conductor as compared with 19 and 61-filamentary tape). In present work the attempt was undertaken to achieve sufficient high critical current density in round Bi-2223 conductor (Fig.3). We have used the filling procedure of co axially placed tubes which made it possible to create ring multilayer architecture of HTS-core. In short samples of these conductors produced by intermediate groove rolling j_c-values as high as 10^4 A/cm^2 (77 K, 0 T) were reached. To further increase the critical properties the optimization of design is needed, however it seems the proposed approach is perspective.

It was established that is not observed any degradation of critical properties in twisted Bi-2223 conductors (Fig.4) as compared with untwisted ones. For example, in short "rolled" samples of twisted Bi-2223/AgNiY tape, at twist pitch not less than 11 mm, the j_c-value was the same with one for untwisted tape - 12 kA/cm^2 (77 K, 0 T), by analogy with results presented in [3]. Lower j_c-value in our case may be explained by the unfavourable influence of doping elements. Critical current measurements performed in same double-pancake coils made by the "react-and-wind" method using 61-filamentary and twisted 19-filamentary tapes almost not revealed the degradation of I_c-value. Thus the multifilamentary design makes it possible to use the "react-and-wind" technique for coils fabrication.

4. Conclusions

It is revealed the influence of the design as well as technological routes on the critical properties of composite HTS-conductors. The possibility of achieving high j_c-values is shown for round Bi-2223 wires and twisted conductors.

References

[1] T.Kato et.al., Submitted to the ICMC/CEM Conference held in Columbus, OH on July 17-22, 1995.
[2] Q.Li, G.N.Riley, Jr., et.al., IEEE Transactions of Applied Superconductivity, Vol.7, N 2, pp. 2026-2029, 1997.
[3] C.J.Christopherson and G.N. Riley Jr., Submitted to the ICMC/CEM Conference held in Columbus, OH on July 17-22, 1995.
[4] A.D.Nikulin et.al., IEEE Transactions of Applied Superconductivity, Vol.7, N 2, pp. 2094-2097, 1997.

Inst. Phys. Conf. Ser. No 158
Paper presented at Applied Superconductivity, The Netherlands, 30 June–3 July 1997

Processing and superconducting properties of Ag-Ti-Cu alloy-sheathed BiSrCaCuO tapes

H. Miao, M. Artigas, G. F. de la Fuente, F. Iriarte and R. Navarro

Departamento de Ciencia y Tecnología de Materiales y Fluidos, ICMA, CSIC-Universidad de Zaragoza, María de Luna, 3 50015-Zaragoza, Spain

Abstract. Monofilamentary BSCCO 2223 tapes have been produced by the powder in tube method using Ag 10 wt% (Cu-Ti) alloy sheaths. There is an increase of hardness which improve the uniformity of the tapes, but also produces microstructural changes by the diffusion of Ti and Cu. The influence of the alloy composition on the critical current density of the final tapes has been analysed. Critical temperatures higher than 108 K (onset) and J_c values up to 6.2 kA/cm^2 at 77 K in the self field have been obtained for 5 wt% Ti, 5 wt% Cu silver alloyed tapes

1. Introduction

The fabrication of uniform metallic sheathed BSCCO superconducting tapes capable of carrying high currents at liquid nitrogen temperature is a determinant factor towards the development of long cables for ac and dc power applications. Remarkable high values of the critical current density J_c have been obtained at 77 K on long Bi-2223 tapes fabricated by the powder in tube (PIT) process. The combination of ductile silver sheaths and brittle and harder BSCCO filaments gives unfavourable cold deformation behaviour for the composite wires and tapes, which are dominated by the mechanical properties of the harder materials to be textured. The enhancement of the sheath hardness would reduce the roughness of the BSCCO metal interface playing a determinant role in the final uniformity of the tapes and consequently in the achieved J_c values on long length tapes. However, some of the properties of the silver must be maintained reducing the material election mainly to binary silver alloys of Ni, Mg, Mn, Au, Cu [1, 2] and others metals [3] as well as ternary alloys [3] like Ag-Cu doped alloys [2].

This contribution, deals with recent results on BSCCO 2223 monofilamentary tapes using Ag-Cu-Ti alloy sheaths where reinforcement of the Ag-Cu alloys by means of dispersion hardening of Ti rich precipitates took place. The changes in the microstructure (SEM and EDX analysis), mechanical (microhardness and deformation uniformity) and electrical properties J_c at 77 K) of BSCCO tapes with silver rich Ag-Ti-Cu alloy sheaths are herewith reported.

2. Experimental

Unreacted powders with nominal composition $B_{1.84}Pb_{0.34}Sr_{1.91}Ca_{2.03}Cu_{3.06}O_{10}$ were obtained from wet-mixing of precursor powders containing BSCCO 2212 phase grains and $CaCuO_2$ particles, as described elsewhere [4]. The final mixture contained large 2212 grains 5 μm), dispersed with a smaller grain size of $CaCuO_2$ (0.3 μm).

Silver alloy ingots with three compositions; 3 wt% Ti, 7 wt% Cu; 5 wt% Ti, 5 wt% Cu and 7 wt% Ti, 3 wt% Cu were prepared. To avoid the differences of the Ti and Ag melting temperatures a two step process was used in the alloy fabrication. First the corresponding amounts of high purity (99.9%) metallic pieces of Ti and Cu were melted in a levitating cold crucible induction furnace, under Ar atmosphere to avoid Ti oxidation. Silver was later added to the Ti-Cu alloy and melted in a graphite crucible also under Ar. In order to homogenise the material the last melting step was repeated several times. The surface, which accumulates defects and Ti-rich segregations, was removed mechanically to yield uniform cylinders and the central hole was drilled. Finally to reach the appropriate alloy tube dimensions drawing with a harder steel anima was done.

The same PIT procedure, used on BSCCO/Ag sheathed tapes [4] has been applied for the alloy composites, involving: *i)* The use of isostatically pressed BSCCO powders, after calcination at 800 °C during 4 h, conformed to long cylinders of 2.4 mm diameter with average densities of 65 % the nominal. *ii)* The insertion of this cylinders inside 4.0 mm outer and 2.5 mm inner diameters alloy tubes. *iii)* The mechanical deformation of the wires by small step reductions (≈8 %) of round drawing down to 1.2-1.5 mm of diameter and rolling to tapes with final thickness of 0.12-0.18 mm and 2.5-2.7 mm in width. Annealing at 400 °C during 30 minutes to soften the sheath was performed several times during mechanical deformation.

A main problem has been identified with inhomogeneous precipitate distribution within the alloy. For example, in the 3 wt% Ti alloy, the differential hardening of the sheath induces local cracks during the drawing process and for this reason this composition has been discarded. For this study tapes up to 10 cm in length were optimised using three sintered steps (in 10% O_2, 90% Ar at 820-830 °C for 48-200 h) and two intermediate rolling processes. All the tapes were always heat treated without bending.

The electric resistivity of the sheaths and the transport J_c(77 K) in the self field was measured by the standard four-probe method using a dc power supply and the usual 1 μV/cm criterion. The superconducting area of the tapes determined from optical micrographs of polished transversal cross sections usually represents about 30%. The same samples were also used to measure the Vickers microhardness of the sheaths. SEM observations and semi-quantitative analyses were performed by means of a JEOL 6400 Scanning Electron Microscope, equipped with an EDX detector.

3. Results and discussion

The microstructure of the alloys with 3 and 5 wt% Ti is formed mainly by an homogeneous dispersion of Ti-rich precipitates in a matrix composed exclusively of Ag and Cu whose composition correlates with the nominal proportions of these elements. In the alloys with 7 wt% Ti, an additional Ti-rich secondary phase is observed (see Table 1). The precipitates have dimensions of the order of 1 μm and less, and in some regions they form isolated clusters with dimensions of about 40 μm. All the alloys show resistivities higher than pure silver (about four times at 77 K) and also higher microhardness (almost twice), suggesting a dispersion hardening of the Ag-Cu matrix caused by the Ti precipitates.

The improvements induced by the sheath may be observed in the SEM micrograph of Fig. 1, corresponding to polished longitudinal cross-sections of tapes with pure and alloyed silver sheaths after annealing at 825 °C for 180 h. The harder sheaths allow a more uniform deformation, denser cores and an improvement in the BSCCO grain texture. Previous studies [2] on Ag-Cu alloys with small content of Ti, Zr, or Hf have suggested the motion of this ions from the sheath to the core. In the result here presented, for much higher content of Ti precipitates, this effect is clearly observed with the presence of a gradient of precipitates along

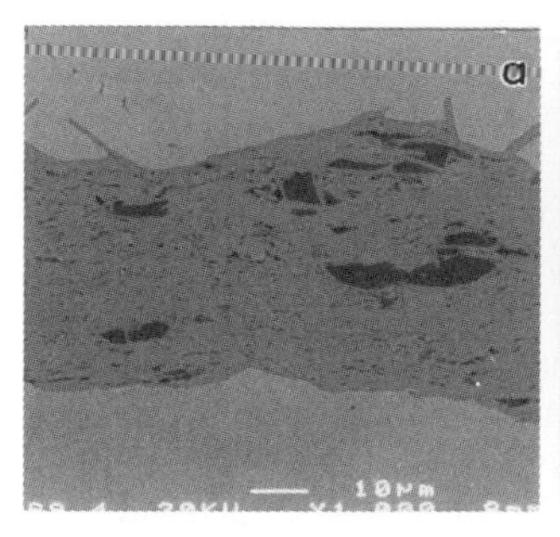

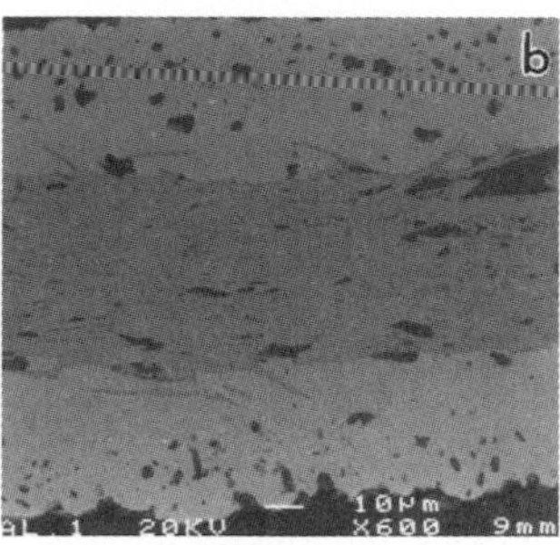

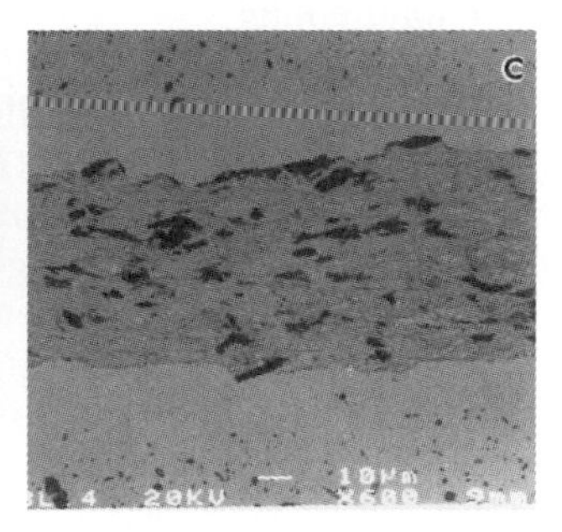

Figure 1: SEM photographs of the polished longitudinal cross-sections of mono-filamentary BSCCO-2223 tapes sheathed with pure silver (a), and 3 wt% Cu, 7 wt% Ti (b) 5 wt% Cu, 5 wt% Ti (c) silver alloys, after annealing at 825 °C for 180 h.

the sheath depth of the BSCCO/alloy composite tapes. Zones free of precipitates are observed on the sheath closed to the oxide core, as well as with less Cu. This has been confirmed by EDX analysis of the matrix, indicating an active diffusion process of both Cu and Ti from the sheath into the oxide core, which is also further evidenced by the distribution of secondary phases within the core. In pure Ag sheaths, the interfaces of the superconducting core are always between Ag and BSCCO grains [4], while secondary phases tend to be at the inner part of the core. The diffusion of Cu changes the local BSCCO composition and induces the appearance of secondary phases at the interface with the metallic sheath, as observed in Fig 1-b and 1-c. The Ti also diffuses into the core, but cannot be detected with the EDX analysis so far performed. On BSCCO tapes shielded with AgCu alloys doped with small amounts of Ti, Zr and Hf [2] however, amorphous disk-like regions of 5 nm thickness parallel to the (001) plane of BSCCO crystals have been observed using HRTEM. These have been correlated with the improvement of superconducting properties.

Optimum values for the 5 wt% Ti-5 wt% Cu sheathed tapes are obtained for the following processing conditions: 825 °C x 48 h, 825°C x 48 h and 820 °C x 100 h. The J_c value of 6.2 kA/cm^2 obtained on the tape with the 5 wt% Ti - 5 wt% Cu sheath is in contrast with the 2.7 kA/cm^2 obtained for a tape with the Ag-3 wt% Ti-7 wt% Cu sheath. There are differences in the E-J(77 K) curves for $J > J_c$ (resistive regime), which are steeper for the tapes with lower J_c adscribed to the microstructural differences.

Table 1: Microstructural composition in wt% of the Ag-Ti-Cu alloy sheaths used and physical properties including microhardness (HV) and resistivity of annealed tapes.

Alloy Ag-Cu-Ti	Matrix Ag-Cu	Precipitate Ag-Cu-Ti	Precipitate Ag-Cu-Ti	HV kg/mm^2	ρ(77 K) μΩcm	ρ(300 K) μΩcm
90-7-3	95-5	21-53-26	none	–	0.93	1.91
90-5-5	97-3	25-41-34	none	40-60	1.03	2.38
90-3-7	98-2	62-01-37	05-01-94	50-60	1.11	2.04

4. Conclusions

Monofilamentary Ag-Ti-Cu alloy sheathed BSCCO 2223 superconducting tapes have been prepared by the PIT method. Three Ag 10 wt% (CuTi) alloys with a high excess of Ti have been considered being essentially solid solutions of Ag and Cu with a large number of Ti-rich precipitates. The use of this Ag-Cu alloy has demonstrated improved workability of the wires and tapes, leading to a higher uniformity of the tapes core. During sintering, an active diffusion process of Ti and Cu from the sheaths to the core has been found. Consequently, their chemical characteristics may affect the final microstructure of the core mainly at the interface. This also may introduce effective pinning centres into the BSCCO grains that improve the J_c performance of the tapes, demonstrating a metallurgical technique to produce these effects. Two types of PIT tapes have been obtained using alloy sheaths. The tapes prepared using 5 wt% Ti- 5 wt% Cu alloyed sheath have exhibited best zero field critical current, with $J_c(77\ K) \approx 6.2\ kA/cm^2$.

Acknowledgments

This research work has been supported by the Spanish MIDAS Program (94/2442) and the SACPA (BRITE-Euram programme BRPR-CT96-0167/BE-1985) project. The technical work of J.A. Gómez and C. Estepa is acknowledged.

References

[1] Tenbrink J, Wilhelm M, Heine K and Krauth H 1993 *IEEE Trans. Appl. Supercond.* **5** 1123; Fujishiro H, Ikebe M, Noto K, Matsukawa M, Sasaoka T, Nomura K, Sato J and Kuma S 1994 *IEEE Trans. Magn.* **30** 1645;Nomura K, Sasaoka T, Sato J, Kuma S, Kumakura H, Togano K and Tomita N 1994 *Appl. Phys. Lett.* **64** 112 and Ahn J H, Ha K H, Lee S Y, Ko J W, Kim H D and Chung H 1994 *Japan. J. Appl. Phys.* **33** L1298

[2] Tanaka Y, Ishizuka M, Matsumoto F and Maeda H 1994 *Adv. Cryo. Eng.* **40** 153; Ishizuka M,Tanaka Y and Maeda H 1995 *Physica C* **252** 339 and Tanaka Y, Ishizuka M, He L L , Horiuchi S and Maeda H 1996 *Physica C* **268** 133

[3] Hubert B N, Zhou R, Holesinger T G, Hults W L, Lacerda A, Murray A S, Ray II R D, Buford C M, Phillips L G, Kebede A and Smith J L 1995 *J. Electronic Materials*

[4] Miao M, Lera F, Larrea A, De la Fuente G F and Navarro R 1997 *IEEE Trans. Appl. Supercond.* **7**, 1833.

Inst. Phys. Conf. Ser. No 158
Paper presented at Applied Superconductivity, The Netherlands, 30 June–3 July 1997

Fabrication and Properties of Extruded Ag-Mg-Ni Sheathed Bi-2223 Tapes

L. Martini, L. Bigoni, F. Curcio, R. Flukiger*, G. Grasso*, M. Migliazza, E. Varesi, S. Zannella

CISE SpA, P.O. Box 12081 - I20134 Milan, Italy
*Département de Physique de la Matière Condensée, Universite de Genève, Switzerland

Abstract. In this work we report on the fabrication process of silver-alloy (Ag-Mg-Ni) sheathed tapes by hydrostatic hot extrusion of composite billets with different layout. The ceramic powder density as well as the silver alloy matrix and BSCCO microhardness evolutions, at different stages of the mechanical deformation process, are reported and analysed. Preliminary critical current density J_c results on long (L>1 meter) Bi-2223 composite tapes are also presented.

1. Introduction

The electrical properties of Bi-2223/Ag long tapes, fabricated by the Powder-in-Tube (PIT) method, can be improved through an optimized mechanical deformation process and repeated rolling and sintering processes by achieving the most favourable microstructure of the HTS material. High-J_c Bi-2223 long tapes are attractive for the realization of superconducting electric power devices. However, for applications of Bi-2223/Ag tapes to manufacture superconducting coils or cables, the tape conductor should have an appropriate unit-length and should be able to withstand the minimum stress necessary during the assembly procedures. The combination of hot extrusion process and mechanically reinforced alloy as sheath material appears to be a very promising way to obtain long and sufficiently robust Bi-2223 composite conductors. In this work we report on the fabrication process of silver-alloy (Ag-Mg-Ni) sheathed tapes by hydrostatic hot extrusion of composite billets with different layout. The ceramic density evolution during the whole mechanical deformation process is reported for monofilamentary inserts and 55-filament wires and tapes. Silver alloy matrix and BSCCO microhardness indentations on wires and tapes at different stages of the cold-working are reported along with preliminary critical current density J_c results on long (L>1 meter) Bi-2223 composite tapes.

2. Hot extrusion experiments

2.1. Multifilamentary billet assembly

Composite Bi-2223/Ag wires have been fabricated according to the Powder-in-Tube (PIT) method by filling pure-silver tubes (14 mm outer diameter) with precursor powder having a nominal composition of $Bi_{1.8}Pb_{0.4}Sr_2Ca_2Cu_3O_x$ [1]. After sealing, the filled tubes underwent subsequent cold-drawing steps through circular dies, providing an area reduction ratio of about

10% per step. A final drawing through an hexagonal die (apothem equal to 1.75 mm) was performed in order to obtain inserts giving an optimal packing of the billet. 55 hexagonal wire inserts have been cut from the above wires and introduced in a silver-alloy billet having 45 mm o.d.. A rear end plug, provided with an evacuation nozzle has been welded to the billet central body. The assembled billet has been evacuated at 150°C for 6 hours under vacuum and then sealed by flattening the evacuation nozzle.

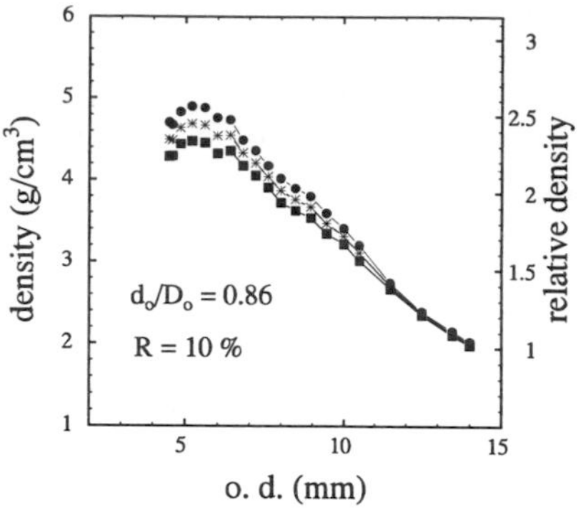

Figure 1 Powder density evolution during the cold-drawing of three single-core wires.

2.2. Concentric billet assembly

The concentric arrangement of this billet has been obtained as already presented in previous works [1, 2] and in this specific case as follows: a 21 mm o.d. silver tube has been coaxially positioned inside a 45 mm o.d. silver billet and the precursor BSCCO powder has been introduced in the hollow spaces left between the different silver walls. An external Ag-Mg-Ni tube has been placed as outer sheath of the billet. As for the previous multifilamentary billet a rear end plug was welded and the billet sealed under vacuum.

2.3. Hydrostatic hot extrusion process

The hydrostatic hot extrusion experiments have been performed by using the extrusion press available at the University of Geneva (CH). After a preheating stage at about 500°C for 1 hour, both composite billets have been hydrostatically extruded at 450°C by using a reduction ratio $R=A_0/A_f=9$ for the multifilamentary billet and R=14 for the concentric one, being A_0 the initial billet cross-sectional area and A_f the area of the extruded rod.

3. Experimental details

The extruded rods have been drawn through circular dies with an area reduction ratio of 10% per step being the initial drawing steps performed using a straight bench-drawing machine. When the composite wire length approached the maximum available length of the drawing-bench (L=5.5 meter), it was necessary to use a bull-block drawing machine (capstan diameter of about 900 mm). However for comparison, a 700 mm long section of this BSCCO composite wire has been cut and drawn on the straight drawing bench to 2.2 mm o.d.. The wires were then rolled on a ϕ=70 mm rolling mill with a thickness reduction ratio of 10% per step being the load applied to the tapes during the rolling process monitored by means of quarts ICP® force sensors. The final dimensions of the tapes were 0.28 mm x 3.8 mm. After rolling, multifilamentary Bi-2223/AgMgNi tape specimens (L>1 meter) have been wound on ceramic mandrels and sintered both in Ar/O_2 , 8% mol. O_2 , and in air atmospheres at 824-838°C for 40-80 hours. A repeated rolling and sintering process has been applied to improve the electrical properties of tapes. The BSCCO bulk density evolution was estimated by experimental elongation data (l/l_o) taken after each drawing step [3] and verified on a few wire and tape specimens. Vickers microhardness indentations (100 g , 15 seconds), HV_{100} have been performed on both the ceramic and silver alloy matrix of composite wires and tapes at different stages of the mechanical deformation process [4] as well as after the sintering heat treatment. Critical current measurements have been carried out at 77 K by the four-probe method, I_c being determined by the electric field criterion of 1 μV/cm on short and long (L>1 meter) Bi-2223 multifilamentary tapes.

3. Results and discussion

Fig. 1 shows the calculated bulk density evolution during the cold drawing process for three of the composite BSCCO/silver tubes from which the hexagonal inserts for the assembly of the multifilamentary billet have been obtained. A monotonic increase of the powder density throughout the drawing process with a relative density (ρ/ρ_o) growth of about 2.5 is observed. A straight bench-drawing machine has been initially utilised for the cold-drawing of the extruded rods while the final drawing steps have been performed by using a bull-block drawing machine. For comparison, a 700 mm long section of this Bi-2223 composite wire have been cut and drawn on the same bench-drawing. However, even when we used the same experimental conditions (area reduction ratio of 10%, dies, lubrication and drawing speed) the composite wire has broken several times during the bull-block drawing. The maximum bending strain experienced by the wire, calculated as ε=t/2R where R is the capstan radius, was ε=0.66% only; nevertheless the combination of the tensile stress due to the front load and bending strain appear to be quite severe compared to the actual mechanical properties of BSCCO wire. Composite wires, 2.2 mm o.d., have been then turned into thin tapes by rolling. Fig. 2 shows the powder density evolution for the 55-filament extruded rod through the whole cold-working process. It can be noticed the strong ρ/ρ_o decrease (of about 30%) during the very first steps of the drawing, then followed by a density increase for which the powder density come nearly back to its original value ρ_o. Finally, during the subsequent rolling process the powder density slightly increases ($\rho_f \leq 1.1\rho_o$). The observed powder density evolution could be explained as follows: the high density of the ceramic yielded by the hot extrusion process cannot be retained during the drawing because of the softness of the silver alloy sheath. However, because of grain fragmentation on drawing, the crushed small ceramic grains fill small voids, generated by tensile stresses during the repeated drawing steps, giving rise to a density increase. As soon as the "upper critical density" value ρ_c is reached, ρ_c depending on the mechanical properties of the sheath material, no powder density increase can be obtained by drawing. On the contrary, the compression force component applied to roll the tape allows

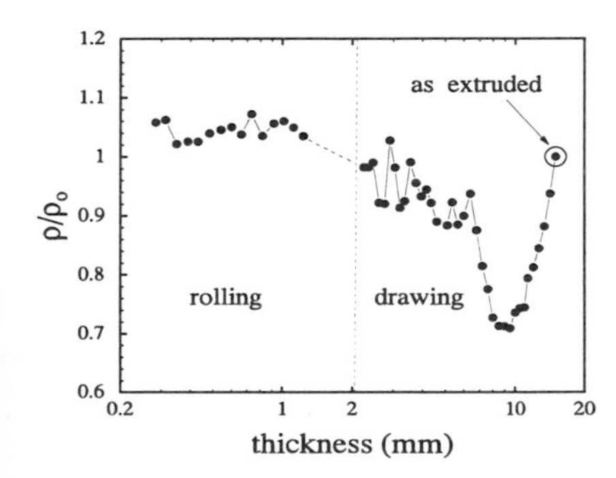

Figure 2 BSCCO powder density evolution during the cold-working of the multifilamentary extruded rod, ρ_o being the density value after extrusion.

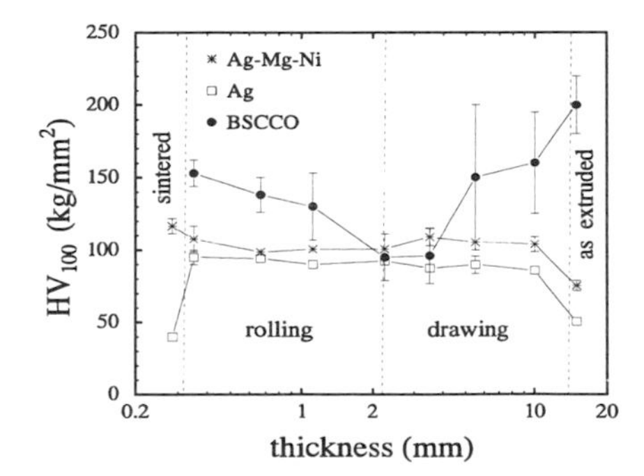

Figure 3 Microhardness results measured on the ceramic and silver matrix at different stages of the cold-working.

the slight density increase observed during the rolling process. Fig. 3 shows the results of microhardness indentations HV_{100} taken on the BSCCO powder regions as well as on the silver and Ag-Mg-Ni outer sheath of 55-filament specimens cut from the extruded rod, wires and tapes at different stages of the cold-working. HV_{100} values of metallic sheaths, i.e. pure-silver barriers around each ceramic filament and outer Ag-Mg-Ni sheath, are nearly constant during the whole deformation process, but being the $HV_{Ag\text{-}Mg\text{-}Ni}$ slightly higher than for pure-silver ($HV_{Ag\text{-}Mg\text{-}Ni} \approx 110$ Kg/mm^2 versus $HV_{Ag} \approx 100$ Kg/mm^2). A strong difference between HV_{100} values of the two distinct metals can be observed after the long sintering at high temperature because of the consequent hardening of Ag-Mg-Ni alloy and softening of pure silver: $HV_{Ag\text{-}Mg\text{-}Ni} \approx 115$ Kg/mm^2 versus $HV_{Ag} \approx 40$ Kg/mm^2. In our experiments (see Fig. 3), at the beginning and at the end of the cold-working the green compact microhardness, HV_{BSCCO} is much greater than the metallic sheath microhardness; the difference between the mechanical properties of the ceramic and the metallic sheaths is the main cause for roughness at the silver/BSCCO interface [5], i.e. sausaging. In the case of using pure silver this situation becomes even worse after the thermal treatment of composite tapes because of silver softening and therefore result in the need for the Ag-Mg-Ni alloy sheath. Composite 55-filament wires had been rolled with R= 10% per step, to a final overall thickness of 0.28 mm. The applied pressure on rolling was $250 < P_{Roll.} < 470$ MPa, the highest values of pressure being applied to the thinnest tapes. Multifilamentary BSCCO tapes with L>1 meter, sintered at 824-838°C for 60 hours, exhibited critical current values $1.8 < I_c < 3.5$A at 77 K and $\chi(T)$ measurements revealed for all specimens a high Bi-2212 fraction. After the first sintering the tapes have been rolled and then sintered in the same conditions, but for a longer time (80 hours). The critical current of the above multifilamentary tapes strongly increased after the second treatment, approaching I_c=14.8 A for the best sample, corresponding to J_c of about 11000 A/cm^2 at 77 K. For this tape, $\chi(T)$ measurements revealed an almost single phase Bi-2223 ceramic core, with the Bi-2212 phase hardly detectable.

4. Conclusions

Multifilamentary and concentric BSCCO rods have been successfully fabricated by hydrostatic hot-extrusion of Ag-Mg-Ni alloy billets. The ceramic density evolution during the whole mechanical deformation process for the 55-filament extruded rod has been presented and discussed. Moreover, the microhardness of the silver alloy matrix and BSCCO cores at different stages of cold-working are reported and compared. Preliminary and promising critical current density values above 10^4 A/cm^2 on long (L>1 meter) Bi-2223 composite tapes have been achieved after the second sintering step. Further optimization of the thermo-mechanical process is ongoing and improvement of J_c is expected.

Acknowledgments

This work was supported by ENEL SpA, Italy.

References

[1] Martini L, Ottoboni V, Zannella S, Caracino P, Gherardi L, Gandini A, (1994) *Proc. 7th IWCC*, Alpbach (A).
[2] Martini L, Bonazzi S, Majoros M, Ottoboni V, Zannella S, (1993) *IEEE Trans. Appl. Superc.* **3** 961.
[3] Bigoni L, Curcio F, Martini L, Miraglia F, Zannella S, (1995) *Proc. 4th Euro Ceramics* **7** 183.
[4] Husek I, Kovac P, Pachla W, (1995) Superc. Sci. Technol. **8** 617.
[5] Han Z, Skov-Hansen P, Freltoft T, (1997) *Superc. Sci. Technol.* **10** 1.

Inst. Phys. Conf. Ser. No 158
Paper presented at Applied Superconductivity, The Netherlands, 30 June–3 July 1997

Bi-2223/Ag- and /Ag-alloy tapes: fabrication and physical properties

Thomas Arndt†[1], Bernhard Fischer‡, Helmut Krauth†, Martin Munz†, Bernhard Roas‡, Andreas Szulczyk†

† Vacuumschmelze GmbH, Dept. HT-SE, Grüner Weg 37, D-63450 Hanau, Germany

‡ Siemens ZT MF1, Paul-Gossen-Straße 100, D-91050 Erlangen, Germany

Abstract. In this work we present some results from efforts to improve the properties of the tapes with respect to AC-losses (matrix resistivity) and mechanical load. Critical currents of multifilamentary tapes differing in matrix, sheath material and filament number have been measured at $T = 4$ K in magnetic fields up to 10 T and the critical current of long lengths has been measured at $T = 77$ K.

1. Introduction

With the Bi-2223/Ag tapes HTSC-conductors are available showing a considerable potential for usage in moderate magnetic fields at $T = 77$ K and at high magnetic field at $T = 4$ K. For applications it is necessary to reduce AC-losses and to improve the mechanical properties of the composite conductors. In this work we present some results on multifilamentary tapes (lengths ≈ 10 cm $\rightarrow \approx 600$ m) of varying number of filaments and different alloy combinations. As the measurement of the self-field critical current of long lengths (here about 200 m) is difficult to perform from a technical point of view, these lengths have been measured on the deliver reel in pancake configuration and the results have been projected to the self field case.

2. Preparation

Cold isostatic pressed pellets of $(Bi,Pb)_2Sr_2Ca_2Cu_3O_y$ (Bi-2223) are filled into an Ag- or AgPd-tube (powder-in-tube method). This tube is cold worked and the resultant monofilamentary wire is cut to be bundeled and build up a second stage multifilamentary conductor with Ag- or $AgMg_{2\,\%}$ – sheath[2]. The number of filaments is 45, 55 or 85. This

[1] E-mail :Thomas.Arndt@hau.siemens.de or Thomas-Joachim.Arndt@P1.HAU1.Siemens.net
[2] Throughout this work the given percent-values represent at.-%.

second stage conductor is cold worked by conventional drawing and rolling to a cross section of ≈ 3.7 mm$\times \approx 0.25$ mm. The final heating and rolling steps (≈ 830 °C; ≈ 100 h) textures the platelets of HTSC-ceramic and builds up the macroscopic superconducting current paths.

3. Increasing matrix resistivity using alloyed silver

Due to the high electrical conductivity of silver at temperatures of 77 K or even lower, the superconducting filaments may be coupled to a certain degree depending on the conditions of AC-application [1]. A first estimate of the importance of the coupling may be given by the relation of the actual twist length to the critical twist length L_c [2]. For round wires the actual twist pitch has to be significantly smaller than L_c and this may be taken as a first approximation for rectangular conductors, too. Increasing L_c may be possible by increasing j_c or by using a barrier-technique or alloys with higher resistivity. Alloying the silver with noble metals may raise its resistivity significantly. In this work we have focussed on the alloy AgPd with different Pd-content (see Tab.1). It

Table 1. Increasing the resistivity ρ of the matrix by adding Pd

	$\rho(T = 77\ \mathrm{K})(\mu\Omega\mathrm{cm})$
Ag	0.3
$AgPd_{1\ \%}$	0.8
$AgPd_{2\ \%}$	1.2
$AgPd_{20\ \%}$	12.0

is important that this gain in resistivity is not reduced by the final thermo-mechanical treatment of the composite conductors. Fig.1 shows that the resistivity of the $AgPd_{1\ \%}$- and the $AgPd_{2\ \%}$-alloy is insensitive to the final treatments of the conductors.

4. Improving the mechanical properties using alloyed silver

The dispersion hardening of the AgMg-alloy used for the outer sheath (second stage conductor) [3] leads to a reduced degration of I_c when the tapes are exposed to mechanical load[3] (see Tab.2).

5. Critical currents at low temperatures

Fig.2 shows the critical currents of a virgin 55-filament tape, a 55-filament tape after bending in both directions and of a 45-filament tape. Despite the high critical current at 77 K (43 A or 18 kA/cm^2; self field) the 55 filament tape shows a remarkable dependence

[3] For details refer to [4].

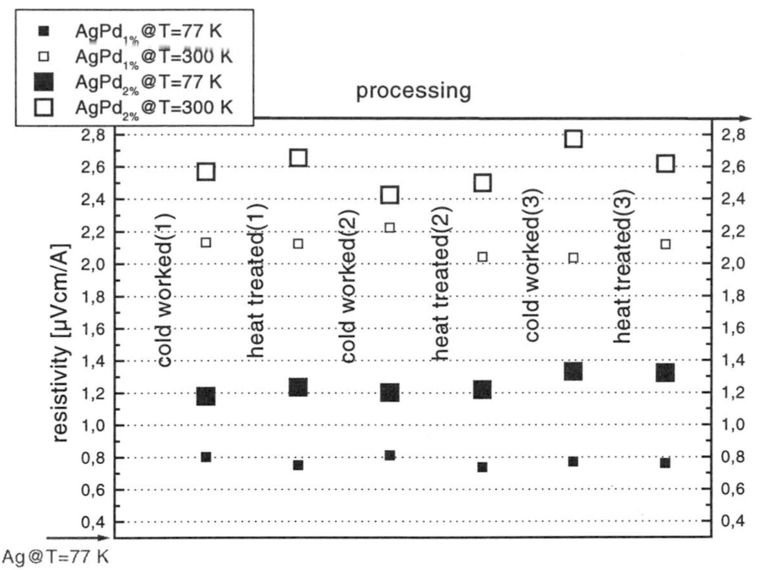

Figure 1. Resistivity of $AgPd_{1\,\%}$- and $AgPd_{2\,\%}$-alloy at different states of the thermo-mechanical treatment

Table 2. Critical strain and stress of 55-filament tapes with Ag- and $AgMg_{2\,\%}$ – sheath at $T = 300$ K

sheath material	$\epsilon(I_c/I_{c,0} = 0.9)$ (%)	$\sigma(I_c/I_{c,0} = 0.9)$ (MPa)
Ag	0.12	40
$AgMg_{2\,\%}$	0.20	110

of I_c on magnetic field at $T = 4.2$ K. The similar $I_c(B)$ dependence for tapes after bending in different directions and virgin tapes indicates that the grain-grain contact of the tapes has to be improved. The $I_c(B)$ dependence of the 45-filament tape is much less pronounced and other samples of these tapes showed higher $I_c(B = 0)$-values, too. The 55-filament tapes show the maximum of the critical current at $B > 0T$ — as expected when measuring in decreasing fields, whereas the 45-filament tape reaches maximum critical current at $B = 0$.

6. Critical currents of long lengths

The critical current of long lengths have been measured in pancake configuration on deliver reels. For example a pancake having an inner radius $R_i = 0.1$ m, an outer radius $R_o = 0.18$ m, 240 turns shows a critical current $I_{c,pancake} = 31.9$ A @ 1 μV/cm. This current produces a maximum magnetic field parallel to the tape $B_{||} \approx 0.16$ T. Considering the field dependence $I_c(B)$ one gets a self-field critical current $I_c(B = 0, T =$

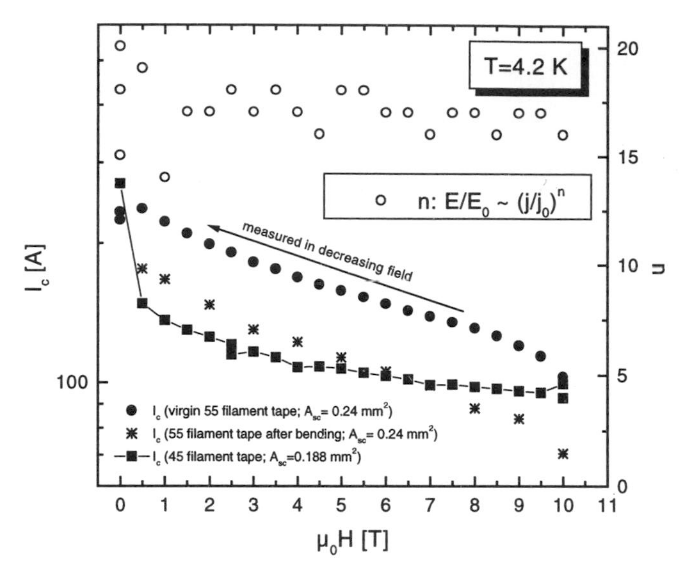

Figure 2. $I_c(\mu_0 H, T = 4\text{ K})$ of different Bi-2223/Ag tapes.(n corresponds to the virgin 55-filament tape.)

77 K) $\geq$ 47.1 A for this length of 199 m. This corresponds to 19.8 kA/cm^2 ($\lambda = 0.24$, $A = 0.9694$ mm^2). These values are only lower limits because the perpendicular field components $B_\perp$, which are much more effective in reducing I_c, are neglected. Short samples from the same batch showed I_c-values up to 56.7 A (23.8 kA/cm^2).

Acknowledgements

We wish to acknowledge our co-workers within the BMBF-programme "Superconductivity for Power Engineering" (grant no. 13N6481). A special thank is directed to R. Nanke (Universität Erlangen) and M. Leghissa (Siemens AG) for supplying us with some of the data concerning the mechanical properties.

References

[1] Oomen M P, Rieger J, Leghissa M, ten Kate H H J 1997 *Appl.Phys.Lett.* **70** 3038

[2] Wilson M N 1983 *Superconducting Magnets* (Oxford: Clarendon)

[3] Tenbrink J, Wilhelm M, Heine K, Krauth H 1993 *IEEE Trans. on Supercond.* **3** 1123
Keßler J, Blüm S, Wildgruber U, Goldacker W 1993 *JALCOM* **295** 511

[4] Nanke R, diploma thesis (in preparation), Universität Erlangen, Germany

Inst. Phys. Conf. Ser. No 158
Paper presented at Applied Superconductivity, The Netherlands, 30 June–3 July 1997

Optimization of rolling process for multicore Bi(2223)/Ag tapes made by OPIT technique

P. Kováč, I. Hušek, L. Kopera, and W. Pachla*

Institute of Electrical Engineering, Slovak Academy of Sciences, Bratislava, Slovakia
*High Pressure Research Center, Polish Academy of Sciences, Warsaw, Poland

Abstract. Usually, the bulk density of rolled BSCCO/Ag composite is varied over the cross section of the oxide core which has the direct effect on transport currents after sintering treatment. Therefore, the rolling process has been optimized with the aim to reach the homogeneous deformation of individual filaments. The effect of rolling reduction, roller diameter and rolling technique on the bulk density distribution in BSCCO core is shown and refereed to the transport current densities at 77K in self field.

1. Introduction

The mechanical deformation process plays an important role in the OPIT method for producing high quality Bi-2223/Ag composites. The homogeneity is a very important parameter for judging the quality of the composite tape [1]. It has been demonstrated that the density of the superconducting core may be correlated to J_c of the Bi-2223 tapes: a higher core density results in a higher J_c [2, 3]. The texture of the core has also been shown to influence the current carrying capabilities [4]. The alignment of the Bi-2212 powder influences on the formation of the texture of 2223 phase during later annealing [5]. It is clear that it is preferable to achieve a homogeneous and high density textured core at final mechanical deformation. Therefore, the rolling process has been optimized with the aim to reach the homogeneous deformation of individual filaments in multicore BSCCO/Ag tapes. In this paper we demonstrate the effect of rolling parameters on single core tapes and show how the final rolling should be used to obtain good quality BSCCO filaments in Ag matrix.

2. Experimental

Single core BSCCO/Ag wire has been made by the standard powder-in-tube method. Precursor powder with a nominal composition of $Bi_{1.6}Pb_{0.4}Sr_2Ca_2Cu_3O_x$ and pure Ag tube with 7mm of outer and 5.5mm inner diameter were used as starting materials. Optimization of rolling process has been performed on this composite wire of 0.9mm in diameter. Multicore tapes were fabricated by two deforming sequences: drawing - flat rolling and by two-axial rolling by Turk's head. The first tape was built by 19 round wires inserted into a circular tube and a second one by 36 rectangular wires into rectangular tube.

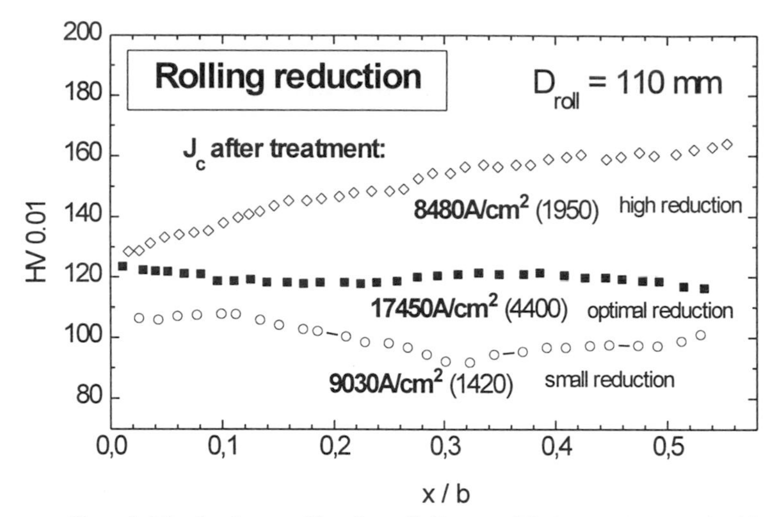

Figure 1. Microhardness profiles of as-rolled tapes and the transport current densities after treatment at 841°C/3x2GPa/250h and (841°C/150h)

Vickers microhardness HV 0.01 across the BSCCO core was measured and HV versus x/b (x - distance of measured point from core edge, b - core width) plots were done to show the core density distribution after rolling. Optical microscopy was used for filament's regularity observation and its cross-sections measurement. Single core tapes (~25mm) were sintered and press-sintered (3 x 2GPa) at 841°C in air. Multicore tapes have been annealed 50h then one time rolled and followed by 841°C/100h/air annealing. The transport currents (I_c at 1μV/cm) were measured by the four probe method in self field at 77K and J_c values were calculated by dividing I_c with BSCCO core cross-section.

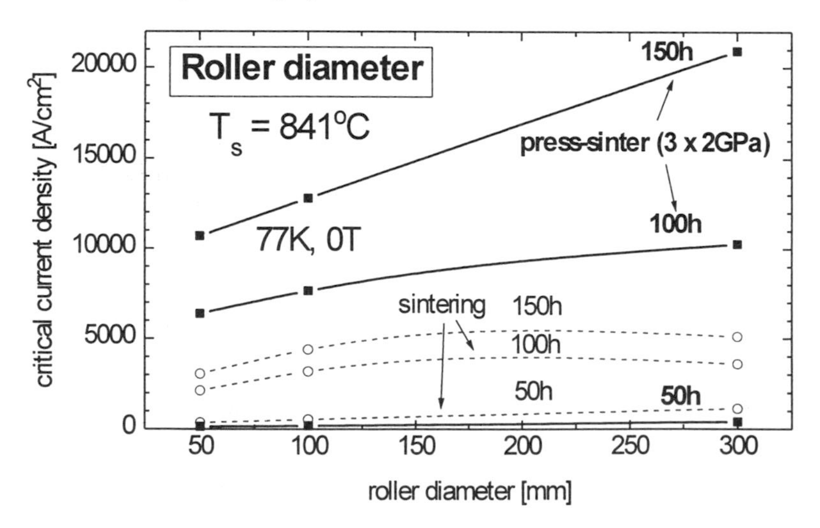

Figure 2. The effect of roller diameter on the transport current density of sintered and press-sintered tapes

3. Results and discussion

The importance of the BSCCO-core density and homogeneity on the final J_c is shown in Figure 1. Three presented microhardness profiles are the result of various reduction at rolling of single core wires 0.9mm by flat rolls of 110mm. The small reduction (open circles)leads to low density and not very homogeneous BSCCO core. High reduction (open diamonds) produces highly dense ceramics especially at the core center (x/b = 0.5) with large difference between the edge and the center. The highest J_c (nearly two times higher) were obtained for optimal rolling reduction (filled squares) of 0.075mm in this case. It is shown that simple correlation of core-density to J_c , the higher core density the higher J_c is not valid there.

Table 1. Critical current density ratios for roll diameters 300mm and 50mm

J_c(300mm)/J_c(50mm)	841°C/50h	841°C/100h	841°C/150h
sintered in air	3.16	1.70	1.68
press-sintered (3x2GPa)	2.99	1.60	1.96

Rolling by three rollers (50mm, 110mm, and 300mm) has been optimized. Figure 2 shows the effect of roll diameter on J_c of monocore tape subjected to sinter and press-sinter processes. All the presented curves show that increase of roll diameter is effective for final J_c as a result of higher core density and the better texture of 2212 crystals [5]. The highest J_c(300)/J_c(50) ratio was observed at the first sinter steps (see in Table 1). The reason is in 2223 grains creation according to 2212 texture affected by roll diameter. Though cyclic press-sinter steps were applied after final rolling, the J_c(300)/J_c(50) ratio after 150h is still positive. As it was shown in Figure 1, the high rolling reduction leads to high core density difference between the edge (x/b = 0) and the center (x/b = 0.5) of BSCCO-core. Application of two-axial rolling (by active Turk's head roller) allows to improve core homogeneity apparently. This is shown in Figure 3 where microhardness profiles of tapes obtained from identical wire by flat and two-axial rolling are presented. Two-axial rolling

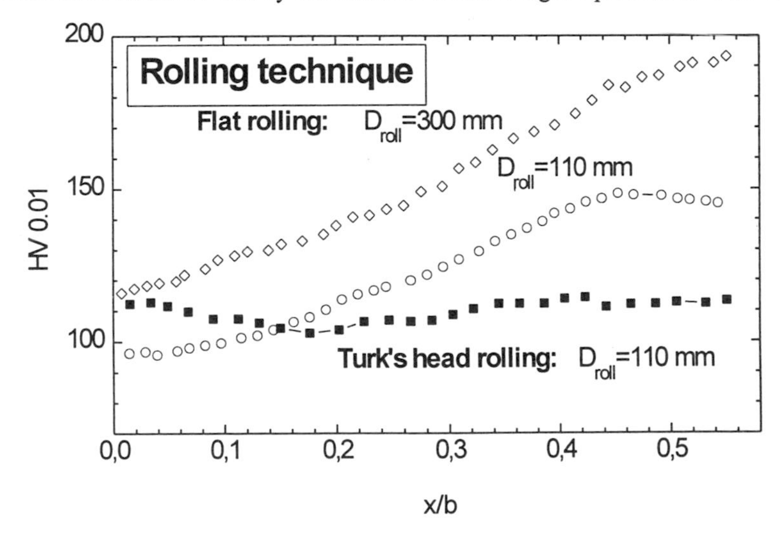

Figure 3. The effect of roller diameter and rolling technique on the density distribution inside the BSCCO core

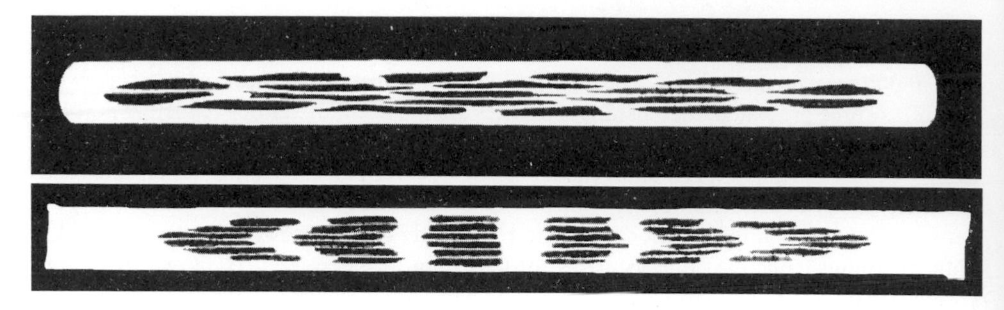

Figure 4. Flat rolled 19 filament (0.14 x 2.5mm) and two-axially rolled 36 filament (0.19 x 2.5mm) tape

allows to reach nearly the same core density at the edge as in the center. Multicore tapes made by drawing - flat rolling sequence and by two-axial rolling have been compared (see Figure 4). The same roller diameters (110mm) have been applied for both rolling techniques. It is apparent that filaments of two-axially rolled tape are more regular then those deformed by drawing and rolling. Several pieces of 36 filament wire with different sizes were rolled to reach a various rate of total tape widening and to see its effect on J_c. The tape having no widening (rolling of rectangular wire 2.5mm x 2.5mm to a tape of the same width) has reached only 5500A/cm^2 after 841°C/150h sintering. The total widening was estimated as w = (b_f - b_0 / b_0)100 (where b_f is the final tape's width and b_0 the initial wire dimension). Widening of 40%, 120%, 220%, and 280% have lead to J_c values 8000A/cm^2, 10000A/cm^2, 12000A/cm^2, and 4000A/cm^2 after the same sintering treatment. J_c = 10000 A/cm^2 was measured for optimally flat rolled 19 filament tape. This is by 20% lower in comparison to 36 filament and two-axially rolled tape.

4. Conclusions

The rolling parameters influence the density distribution inside the BSCCO-core very remarkably. The higher core density the higher J_c is not always valid. Core homogeneity has apparent effect on final J_c values. The two-axial rolling by active Turk's head has been applied for multicore BSCCO/Ag composites which allows to increase ceramic density and improve its homogeneity in comparison to drawing - flat rolling sequence.

Acknowledgements

This work was supported by Copernicus project CIPA CT 94-0115 and by the Grant Agency of the Slovak Academy of Sciences VEGA 95/5305/108.

References

[1] Kováč P, Hušek I and Pachla W *IEEE Transactions on Applied Superconductivity Vol 7 No. 2* (1997) 2098
[2] Yamada Y, Sato M, Murase S, Kitamura T and Kamisada Y *Advances in Superconductivity V* (Springer-Verlag. Tokyo) (1993) 717
[3] Parrell J A, Dorris S E and Larbalestier D C *Physica C* **231** (1994) 137
[4] Grasso G, Perin A and Flukiger R *Physica C* **250** (1995) 43
[5] Wang Y L, Bian W, Zhu Y, Cai Y X, Welch D O, Sabatini R L, Suenaga M and Thurston T R *Appl Phys Lett* **69** (1996) 580

Inst. Phys. Conf. Ser. No 158
Paper presented at Applied Superconductivity, The Netherlands, 30 June–3 July 1997

Bi 2212 tapes prepared by alternating electro-deposition and heat treatments

Fabrice Legendre[1], Philippe Gendre[2], Lelia Schmirgeld-Mignot [1] and Pierre Régnier[1]

[1] Section de Recherches de Métallurgie Physique, CEA Saclay 91 191 Gif sur Yvette CEDEX FRANCE
[2] Laboratoires SORAPEC 192 rue Carnot 94 194 Fontenay sous Bois FRANCE

Abstract.
Bi 2212 tapes have been prepared by oxidation reaction of 3.5 micrometer thick precursor layers deposited electrolytically in a sequential way on both sides of a 50 micrometer thick Ag tape. Intermediate heat treatments were performed to improve the adhesion of the deposited precursors and to benefit of the homogenisation brought about by the presence of liquid phases. After the last deposition sequence, a 2 step heat treatment was carried out to allow for Bi 2212 to synthesis and texture via a partial melt growth process, modified in such a way that the specimens were held for a few minutes only above the peritectic melting point of the Bi 2212 phase.
It was found that Bi 2212 crystallises into platelet like grains giving an excellent texture, but in some places the perfect alignment of these platelets with the substrate was hindered by the presence of cuprates. However, since our layers are typically one order of magnitude thinner than those usually prepared with the conventional partial-melt growth technique, cuprates are smaller leading to transport Jc values among the best. Moreover, the total preparation time is only 4 h. Our best transport Jc measured on a short length is $3{,}5\times10^4$ A/cm^2 at 77K and 3×10^5 A/cm^2 at 4.2 K which correspond to respectively 7.35 and 63 A for a specimen of 3 mm in width .

1. Introduction

Short lengths of Bi base high Tc superconducting wires and tapes carrying current densities, Jc, matching those required for practical applications have been produced by several groups around the world. But keeping such high Jc values over kilometric lengths is still a challenge. If there is no doubt this goal will be achieved with the powder in tube technique, it is also clear that the improvements required will make this processing route more tricky and consequently more expensive. Hence it is of a major importance to look for cheaper preparation techniques. According to that line, we have investigated the possibility of preparing Bi 2212 tapes by alternating sequential electrolytic deposition of precursors and thermal treatments [1-4].

2. Sample preparation

The 50 μm thick silver backings were cleaned with ethanol and acetone and annealed for 15 min at 860 C. The 4 precursors were electro-deposited with a current density of 12.6 A/cm^2 in the following baths: an aqueous solution of SO_4Cu for Cu, $Bi(NO_3)_3$ dissolved into dymethylsulfoxyde (DMSO) for Bi, $SrCl_2$ dissolved into a mixture of DMSO + Acétonitrile for Sr and Ca Cl_2 dissolved into DMSO for Ca. To improve the adhesion of these precursors and to benefit of the homogenisation brought about by the presence of liquid phases, we have alternate electro-deposition with short heat treatments. Accounting for electro-deposition ratios determined by ICP analysis, the 4 precursors were deposited in such a way to get the average cationic content Bi_2, $Sr_{1.7}$, Ca_1, Cu_2 and a final thickness of 3.5 µm for each 2212 layers. At this stage, the specimens were annealed for 1 min at 860 C (above the peritectic melting point of the Bi 2212 phase), quenched down to room temperature and re-annealed at a temperature just under the peritectic melting point of the Bi 2212 phase. The microstructure depends on several parameters like the temperature ramp rates used for heating and cooling the samples, the temperature and the duration of both the flash and the final synthesis treatments [3]. The optimum temperature for synthesising the Bi 2212 phase was found to be 840 C. At this temperature, according to DTA shown as figure 1 and X-ray diffraction diagrams, the Bi 2201 phase formed during the quench following the flash annealing progressively dissolves in the melt and transforms into the Bi 2212 phase. After 1 hour of annealing at 840 C, the 2201 phase is no longer detected and the texture of the 2212 phase is extremely marked (fig. 2).

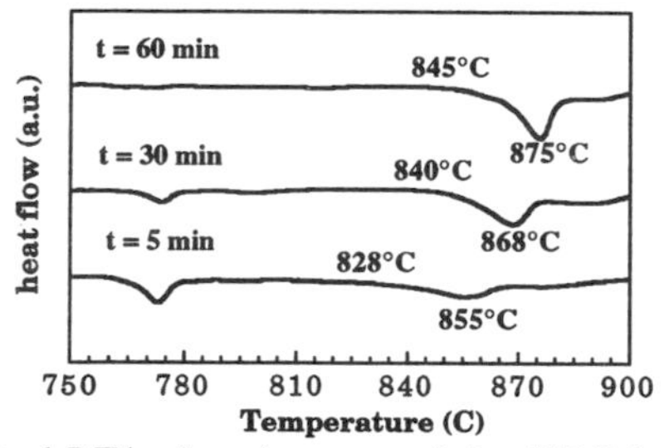

Fig. 1 DTA of specimens annealed at 840 C for 5, 30 and 60 minutes showing the fusion peaks of the Bi 2201 and the Bi 2212 phases.

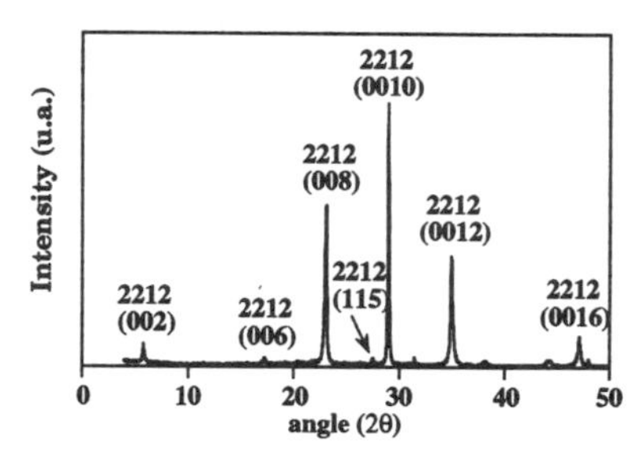

Fig. 2 X-ray diffraction diagram of a Bi 2212 layer synthesised at 840 C during 1 hour.

The texture is so strong that it masks the detection of minor phases eventually present. Hence to get some information on them, the surface and the transverse section of the specimens were observed in optical and scanning electron microscopy and also analysed with a CAMECA SX 50 electron microprobe. It was determined that the microstructure of our samples is sandwich like, with the upper and lower most layers very textured and poor in Sr-Ca cuprates, whereas the central part is rather rich in cuprates which prevent the Bi 2212 platelets to align correctly with the Ag substrate. However, since our superconducting layers are 3.5 µm thick only, our cuprates are smaller than those found in thicker specimens prepared according to the regular partial-melt growth technique, which explains why the texture of our specimens is so strong. It was found that longer annealing do not permit to dissolve further the cuprates because there is no more liquid phase after 1 hour of annealing at 840 C. What happens at longer annealing is just a coarsening of the grains which makes them more brittle during the final cooling down generating fissures.

3. Electrical characterisation

The electrical properties of our samples have been determined by transport measurements using the standard for point method. The tested specimens were 3 mm in width and 20 mm in length with a distance of 8 mm between the two potential pads. The transport critical current density Jc was determined on the basis of a voltage drop of 1 μV/cm. The performances of our samples depends on their microstructure, hence on the heat treatments which generate it. As shown in figure 3, the critical temperature, both the onset and the offset ones decrease with synthesis duration. In particular, it is remarkable that these temperatures decrease so quickly during the first hour of annealing. Though the possibility of some fast evolution of the cationic content of the superconducting phase during this period cannot be rejected, it is more probable that this effect is due to an evolution in the oxygen stoichiometry of the Bi 2212 phase.

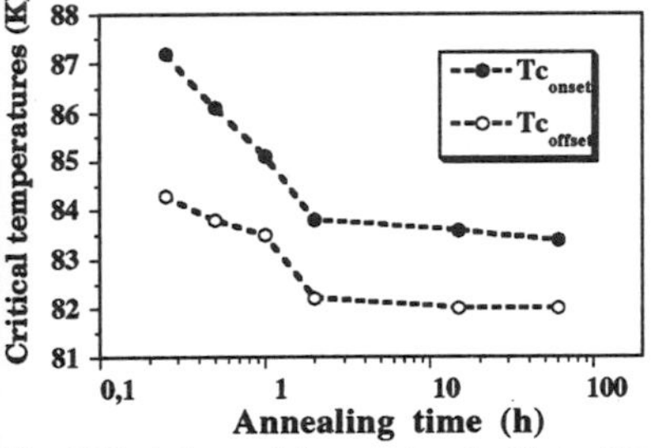

Fig. 3 Variations of the onset and offset critical temperatures as a function of the annealing time at 840 C.

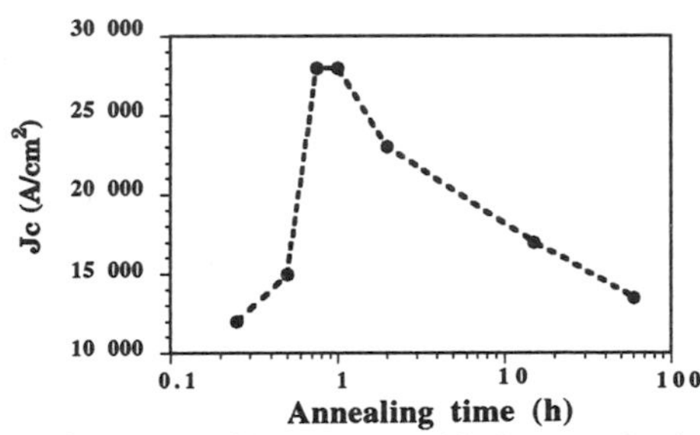

Fig. 4 Variation of the transport critical current density at 77 K as a function of the duration of the annealing time at 840 C.

The evolution of Jc as depicted in figure 4 is quite different: starting from a low value, it increases quickly with the duration of the synthesis treatment for reaching a sharp maximum at 1 hour, next it decreases slowly but significantly. This behaviour is the result of two conflicting trends: the increase of the volume fraction occupies by the Bi 2212 which is beneficial for Jc and the degradation of this phase due to its overdoping in oxygen and also to the increase of its grain size leading to a more pronounced fissuration, the last two effects being detrimental for Jc. It is worth noting that if both Tc and Jc depend on the percolation of the Bi 2212 grains, Jc is in addition sensitive to their texture and to their volume fraction, consequently, it is not too surprising that the variations of Jc and Tc with the duration of the synthesis are not correlated.

Once again, for a question of oxygen stoichiometry, since the faster the cooling rate the lower the oxygen doping, the higher the onset and offset critical temperatures (cf. Fig. 5). Presumably because Tc is getting closer and closer from 77K as the cooling rate is reduced, Jc decreases also with the cooling rate (cf. Fig. 6).

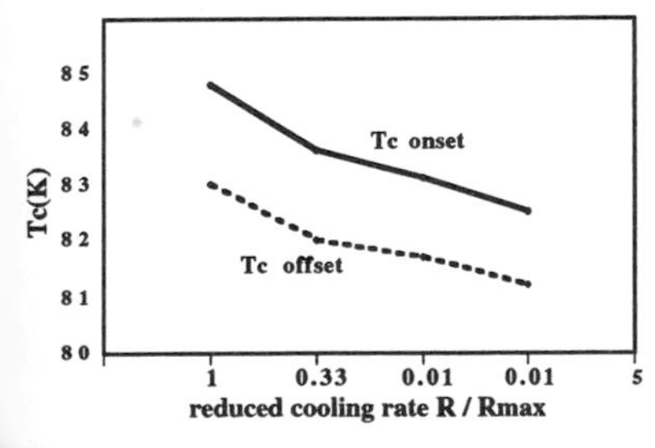

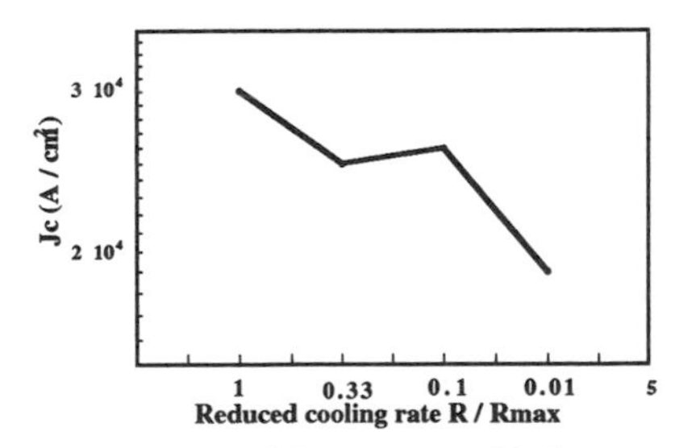

ig. 5 Variations of onset and offset temperatures as a unction of the cooling rate.

Fig. 6 Variation of the transport critical current density as a function of the cooling rate.

The Jc behaviour of our samples in applied magnetic field is quite representative of ood grade Bi 2212 materials. In particular at 4.2 k our tapes are very resistant to magnetic field: at this temperature we have measured on our best sample a Jc of 300 000 A/cm^2 under no applied field and more than 100 000 A/cm^2 in a 7 Tesla field applied normally to its plane. Figures 7 and 8 show the behaviour of the Jc of a regular grade sample as a function of magnetic field for various temperatures as determined by H. Safar et al. [5] (Los Alamos). At 75 and 64 K there is a very strong anisotropy in the resistance of Jc to magnetic field which is highlighted by the comparison of the results obtained with H parallel or perpendicular to the average "c" direction of our tapes (Fig. 8).

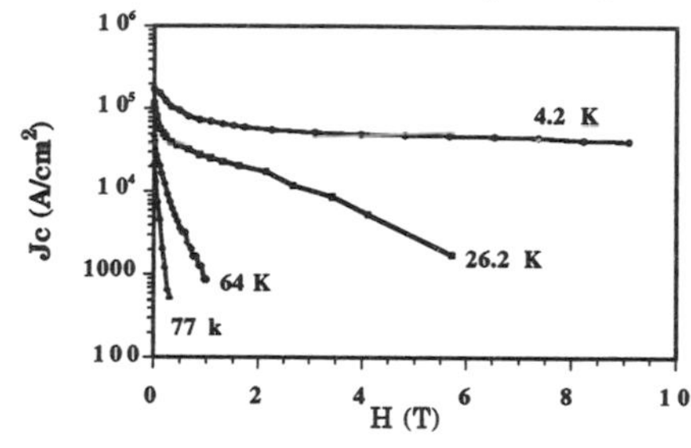

Fig. 7 Variations of Jc as a function of magnetic field applied parallel to the average "c" direction of the tape at 4.2, 26, 64 et 75 K.

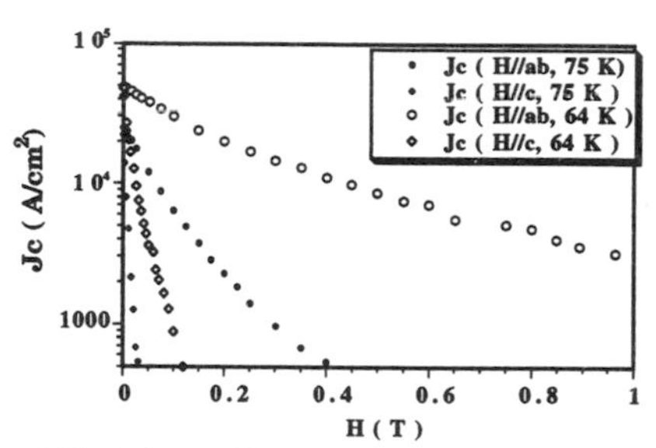

Fig. 8 Variations of Jc as a function of magnetic field applied either parallel or normal to the average "c" direction of the tape at 77 and 63 K.

4. Conclusion

We have shown that it is possible to prepare Bi 2212 tapes in alternating electrolytic deposition of precursors and heat treatments. In the present state of optimisation of our process, we currently prepare superconducting tapes which exhibit superconducting performances among the best reported for Bi 2212 tapes. In particular, the texture of our tapes is more pronounced than that of other products because the cuprates disseminated in our tapes are smaller in size. Further improvement should permit to reduce the volume fraction of non-superconducting phases which is still around 20 % and hence lead to a significant increase in Jc. The major advantages of our process are its simplicity (all the sequence is performed in open air) and its fastness (1 hour of annealing is enough to synthesise the Bi 2212 phase). As a demonstration we have electro-deposited the 4 precursors in a continuous way over 30 m on a 50 μm thick silver tape. Several short samples were cut from this tape and annealed, they all exhibit a critical current density between 20 000 and 25 000 A/cm2 at 77 K.

References

[1] Brevet Français n°94 02710 du 9 mars 1994
[2] P. Gendre, Thesis, University Paris VI, 22 April 1994
[3] F. Legendre, thesis University of Lille 7 April 1997
[4] F. Legendre, L. Schmirgeld-Mignot, P. Régnier. Proc of EUCAS 95, Edimbourg 3-6 July 1995, IOP publishing Ltd p 339-342, (1995)
[5] F. Legendre, L. Schmirgeld-Mignot, P. Régnier, H. Safar, J. Y. Coulter and P. Maley. Proc. of the 10 th anniv. HTS Worshop on Physics Materials and Applications Houston . World Scientific p 202, 203 (1996

Inst. Phys. Conf. Ser. No 158
Paper presented at Applied Superconductivity, The Netherlands, 30 June–3 July 1997

Fabrication of thin Ag/Bi-2223 mono-core tapes with an overall tape thickness of 40 μm suitable for dissipative power applications

Dag W. A. Willén, Wen Zhu and Julian R. Cave

IREQ Hydro-Québec, 1800, boul. Lionel-Boulet, Varennes (Québec) J3X-1S1 Canada

Abstract: The processing and properties of 5 mm wide rolled mono-filamentary Oxide-Powder-In-Tube tapes with an overall thickness of down to 40 μm is presented. These tapes display engineering critical current densities of 5-6 $kAcm^{-2}$ (77K, self-field) and core critical current densities of 30-40 $kAcm^{-2}$. The current transport properties and the distribution of critical-current values have been characterized at different thickness using the onset (1μVcm^{-1}) and the offset criteria of the critical current, and the second differential of the E-J curve.

1. Introduction

In industrial applications of high-temperature superconductors, AC losses and momentary high levels of power dissipation are present. The latter occur in, for example, transformers, fault-current limiters and power cables where in-rush currents and fault currents can exceed the critical current, I_c, of the superconductor. These applications may benefit from superconducting tapes with an increased surface area in contact with the cooling medium [1]. This can be obtained by reducing the overall thickness of Ag/Bi-2223 tapes while maintaining a high critical current density, J_c. Other possible advantages of thinner Ag/Bi-2223 sheathed tapes are: (1) the increase in the Ag/Bi-2223 interface area, which may increase the critical current density; (2) an increase in the permissible rolling pressure for the rolling between heat treatment steps [2, 3]; (3) the possibility of near-to-final-shape flat-form filament manufacturing followed by stacking into multifilamentary tapes [4]. Here, results on 40 μm thick Ag/Bi-2223 tapes (Fig. 1) with high J_c values (Fig. 2) are presented.

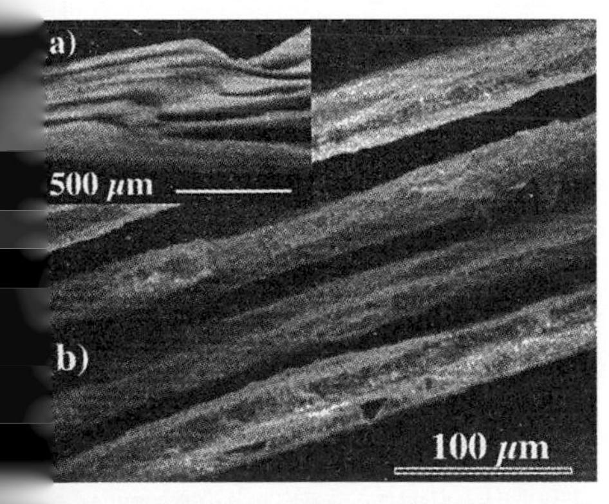

Fig. 1. (a) Six 40 μm thick Ag/Bi-2223 tapes (b) four as-rolled tape cores (Ag sheath etched off).

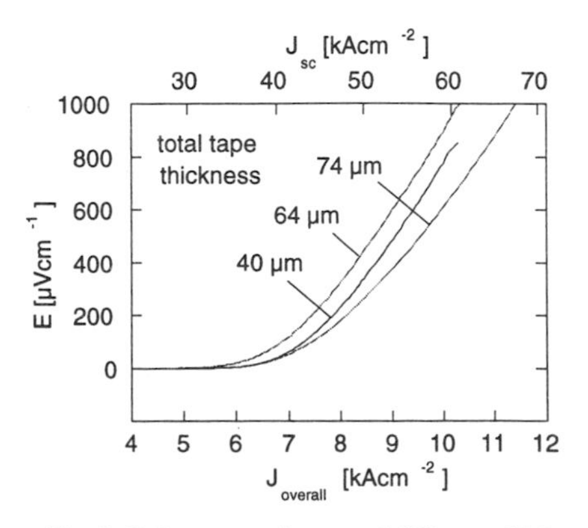

Fig. 2. E-J curves of tapes of different thickness. The ratio β of tape-to-core surface area is 5.9.

2. Fabrication procedure

There are difficulties involved in making tapes with reduced total thickness. For a given flat rolling procedure, there is normally a critical thickness associated with a given roller diameter, reduction rate, lubrication and tensioning parameters [4]. Below this critical thickness, J_c decreases due to the formation of severe non-uniformity (sausaging) in the oxide core [5].

This is explained by the pressure-dependent mechanical properties of the ceramic precursor powder. The pressure (or hydrostatic stress component) depends on the build-up of confinement forces from the contact with the tool surfaces. Therefore, the aspect ratio of the deformation zone (height divided by width or length) is of utmost importance for the uniformity of the Bi-2223 core [3, 4].

During the reduction in size of the tape (shaping), the pressure was kept low in order to avoid hardening and embrittlement of the ceramic powder [3, 6]. This was achieved by using a high aspect ratio of the deformation zone, i.e. small roller diameter (D=2 cm) and low reduction per pass (5-10 % area).

In contrast, the densification of the core between heat treatments requires a high pressure with limited plastic deformation of the tape, i.e. flat pressing, large-diameter rolling, or "sandwich rolling" [7].

An additional factor to the tool geometry is the geometry of the tape itself. This is illustrated in Figure 3 (a)-(c), where samples of various ratio, γ, between the thickness, t, and the width (γ=t/w) have been uniaxially pressed at three different pressures (a). The resulting values of J_{cov}(offset) show the inability of thick and narrow tapes (high γ) to withstand high pressures. This is explained by the large sideways plastic deformation indicated in Fig. 3 (c). In consequence, the tapes presented here were made 5 mm wide and pressed at 3 GPa.

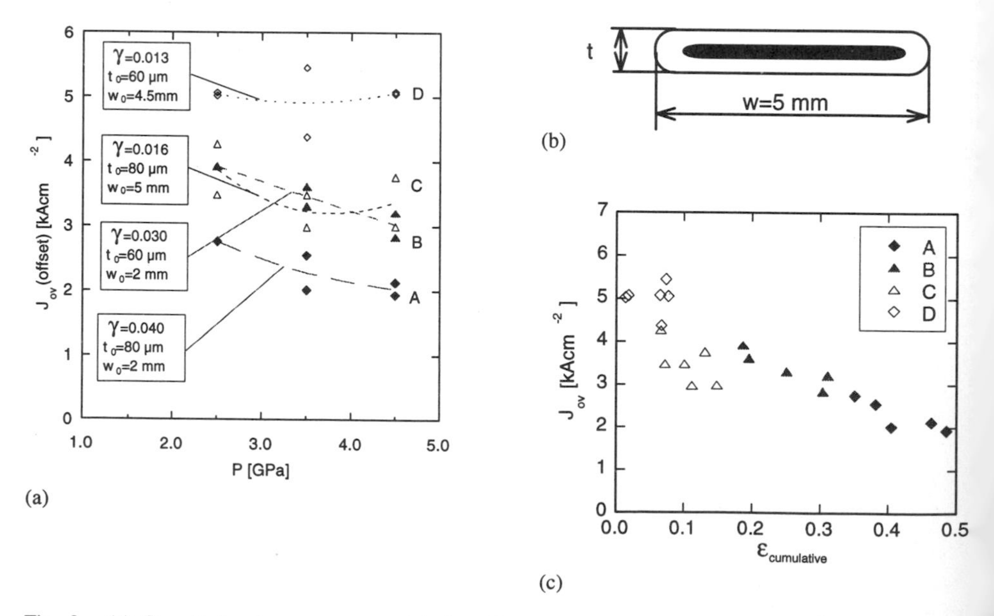

Fig. 3. (a) Overall J_c values measured with the offset criterion (E_c(offset)=250 µVcm^{-1}) for samples of different aspect ratio, γ=t/w, plotted against compaction pressure, P. (b) Schematic of the tape geometry. (c) The results are explained by the large sideways deformation of the narrow and thin tapes.

3. E-J characteristics at various thickness

The three samples of thickness 74 μm, 64 μm and 40 μm featured in Fig. 2 were heat treated using a variable-temperature procedure described in [6], with a maximum temperature of 825°C (T1) early in the treatment, and a gradual decrease of temperature down to 805°C following [8]. A second set of samples, with the same three thickness values, was treated with a maximum temperature of 835°C (T2).

The E-J characteristics of these samples are shown in Fig. 4 (a)-(c). The second differential, which can be interpreted as a distribution of J_c values, has been calculated according to [9]. It was calculated in terms of J_{ov}=I/A_{ov}, where A_{ov} is the overall cross-sectional area, in order to enable the comparison of samples of different geometry. It should be noted that the d^2E/dJ^2 curves are asymmetric primarily due to the shunting of current into the silver sheath and that the intrinsic distribution of the superconductor can be more symmetric [9].

The heat treatment T1 generally gives higher "onset" and "offset" values of the critical current, but the spread in J_c is similar for T1 and T2. The peak value of d^2E/dJ^2 is generally higher for T2. Given that the spread is similar, this indicates a higher flux-flow resistivity, ρ_f, or a higher sheath resistivity, ρ_{Ag}, or both [9].

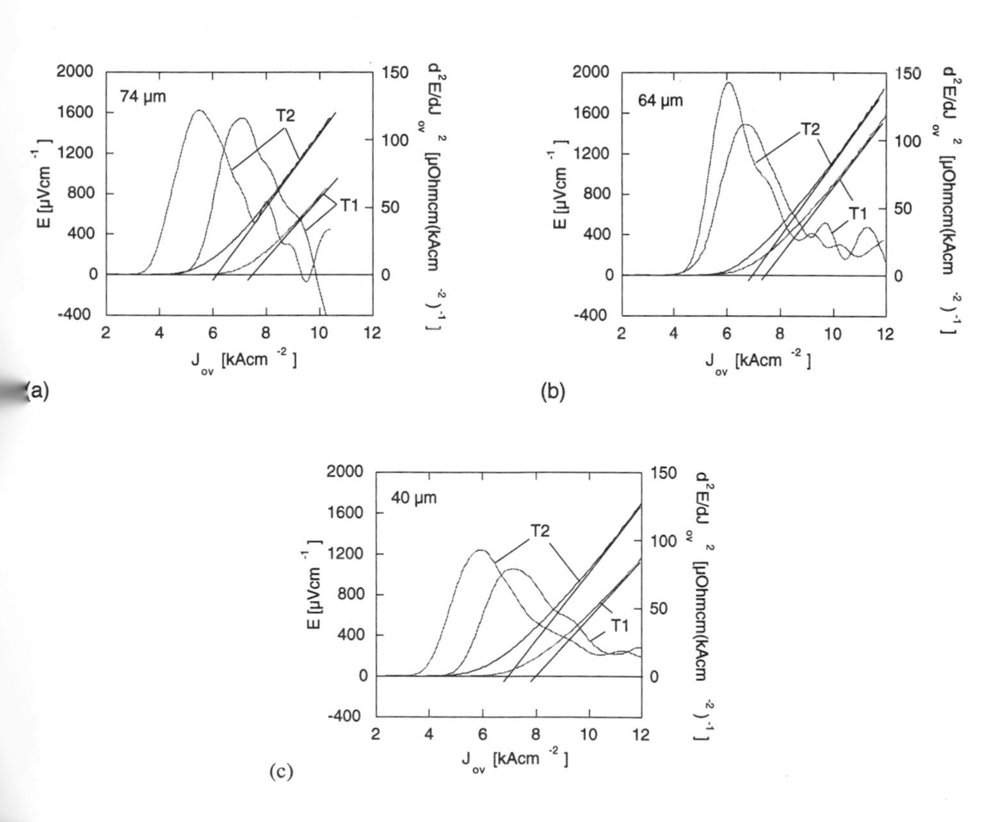

ig. 4. E-J characteristics and second differentials for Ag/Bi-2223 samples of thickness (a) 74 μm, (b) 64 m and (c) 40 μm, each with two different heat treatments, T1 and T2. The intercept of the tangent to the E-curve with the E=0 line indicates the value of J_c(offset). The ratio β of tape-to-core surface area is the same in Fig. 2.

With regards to thickness, the 64 μm thick samples have the narrowest J_{cov} distributions suggesting that the morphology and the microstructure of the superconductor core are the most homogeneous for this thickness. A marked reduction of the peak value of d^2E/dJ^2 in combination with an increase in the spread for the 40 μm thick tapes signifies a less uniform core. The maintained high J_c (Fig. 2) must then be due to an increase in the *local* J_c in the thinnest samples.

4. Conclusion

The fabrication of thin tapes has the following advantages: (1) increased surface for heat conduction; (2) increased Ag/Bi-2223 interface area; (3) allows for a higher compaction pressure in the densification step between heat treatments; (4) possibility of stacking near-to-final-shape flat-form monofilaments into multifilamentary conductors.

Ag/Bi-2223 monocore tapes with a total tape thickness of down to 40 μm and with a maintained high critical current density have been produced. The second differential of the E-J curve was used to characterize samples of different thickness and for two different heat treatments. This analysis suggests that the superconductor core is the most uniform for a tape thickness of 64 μm, while a less uniform core is compensated for by an increase in the local J_c for the 40 μm thick tapes.

References:

[1] L S Fleishman, Yu A Bashkirov, V A Aresteanu, Y Brissette and J R Cave 1993 *IEEE Trans. Appl. Supercond.* **3** p570

[2] D W A Willén, C Breau, W Zhu, D Asselin, R Nadi and J R Cave 1995 *Inst. Phys. Conf.* Ser. *no 148* p451 and references therein

[3] A recent review is: Z Han, P Skov-Hansen and T Freltolft 1997 *Supercond. Sci. Technol.* **10** p371

[4] See for example: U Syamaprasad M S Sarma P Guruswamy S G K Pillai K G K Warrier and A D Damodaran 1997 *Supercind. Sci. Technol.* **10** p100

[5] G Grasso, A Jeremie and R Flükiger 1995 *Supercond. Sci. Technol.* **8** p827

[6] D W A Willén, W Zhu, R Nadi, A Paquette and J R Cave (ASC 1996) to be published in *IEEE Trans. Appl. Supercond.*

[7] W G Wang, H K Liu, Y C Guo, P Bain and S X Dou 1995 *Applied Superconductivity* **3** p599

[8] Y B Huang, G F de la Fuente, A Larrea, and R Navarro 1994 *Supercond. Sci. Technol.* **7** p759

[9] D W A Willén, W Zhu and J R Cave, (8 Ge 33) this proceeding and references therein

Inst. Phys. Conf. Ser. No 158
Paper presented at Applied Superconductivity, The Netherlands, 30 June–3 July 1997

Evolution of core texturing in the process of Ag/BiSCCO tapes and wires fabrication by OPIT

A. Goldgirsh, V. Beilin, E. Yashchin, and M. Schieber.

Graduate School of Applied Science, the Hebrew University of Jerusalem, Israel

Abstract. c-axis texturing in ceramic core of Ag/BiSCCO tapes during cold deformation prior to thermal processing was studied. Three schemes of cold deformation used in the paper were drawing and flat rolling sequence, cold pressing of powder-filled tube down to flat tape and groove rolling followed by flat rolling. Flat rolling was demonstrated to be the most effective way of core texturing which could be connected with relatively long-range powder transport including rotation-induced particle orientation. The kinetics of 2223- phase formation from 2212 precursor and its texturing during sintering stage were also studied in order to compare deformation-induced and silver-assisted tape texturing.

1.Introducion.

It is generally accepted that c-axis texturing of the ceramic core in Ag/BiSCCO tapes is one of the main factors controlling grain connectivity and thus critical current density, J_c, in these tapes. There are strong evidences of generic relationship between texturing of 2223-phase in reacted core and that of 2212-phase in deformed non-reacted one [1][2]. This texturing is considered as resulting from shear strain of highly anisotropic 2212 phase along (ab)-planes normal to c-axis [3]. However to our knowledge there are no reported data directly supporting this idea both on experimental or theoretical basis.

This work is devoted to the study of c-axis texturing of Ag/BiSCCO tapes during flat and groove rolling and uniaxial pressing (the processes with different stress states) in order to reach better understanding of underlying grain alignment processes. Texturing after subsequent tape sintering was also studied to elucidate the Ag role in the case of different texture degree of original 2212 phase.

2.Experiment

Ag/Bi(Pb)SCCO composite tapes were fabricated by the conventional oxide owder-in-tube (OPIT) method. The details of tape fabrication and texture degree alculation have been reported elsewhere[4][2]. Three deformation schemes used for tape reparation were: a) a conventional one including drawing of powder-filled silver tube of mm OD down to a round wire of 1.6-1 mm, flat rolling and subsequent thermo-mechanical rocessing (TP) of the rolled tape, b) rolling-free scheme including drawing and uniaxial ulti-step pressing of the round wire in order to fabricate a flat tape, followed by TP, c) rawing-free scheme including groove (square) rolling of Ag/powder composite followed by flat rolling and TP. Core density before rolling or pressing in the above schemes was about 50-60%.

3. Results and discussion

Shown in Fig.1 is the dependence of texture degree, D, on relative thickness reduction Δ ($\Delta = d_0/t_i$ where d_0 and t_i are the diameter of the original round wire and thickness at the i-th deformation step, respectively) in rolling and pressing processes.

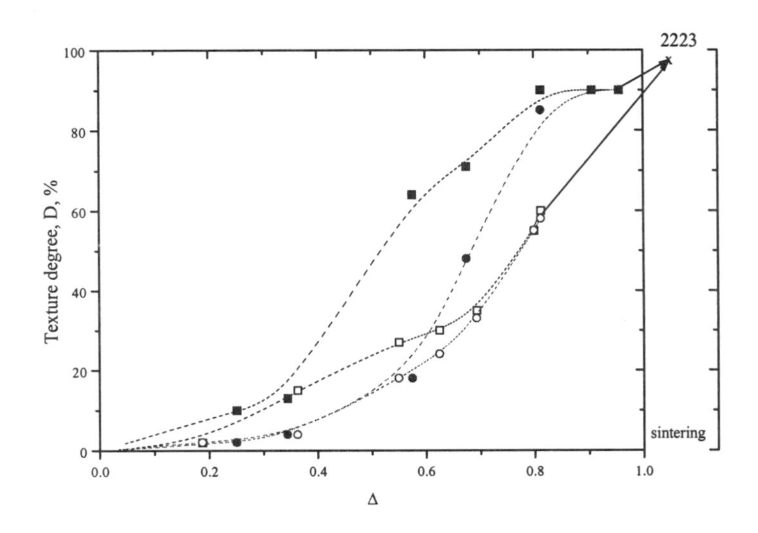

Fig. 1. Texture degree vs. thickness reduction of non-reacted Ag/BSCCO tape during rolling and uniaxial pressing processes: solid symbols- rolling, empty symbols-pressing; □, ■-core surface; ○, ●-core midplane; x - core surface after sintering.

Comparing the curves in Fig.1, one can see a more effective D increase during rolling (R) operation as compared to pressing (P). D value is saturated in the first case a high level of about 90%. Maximal D we could reach by pressing was 60% at maxima compressive stress of about 15GPa. Further texture enhancement required very large stres (increasingly growing) unattainable in our experiment. The lag of internal texturing from th surface one is significantly higher for R case. XRD study of groove-rolled composite having square-shaped cross section showed that flat core sides in rolled samples are left t be non-textured down to the ultimate composite size of $1.2 \cdot 1.2 mm^2$. This result is quit unexpected taking into account fast texturing of core surface under flat rolling.

Data on texture growth during sintering process show that in the case of 2212 cor previously highly textured by deformation (D of about 90%), the formed 2223 phase ha approximately the same high D already in a very short time period since the sintering sta (Fig.2). In the case of poorly-textured original 2212 core, new-formed 2223-phase far fro Ag sheath has also low D (Fig.3). D enhancement for residual 2212 phase, observed at co surface (Fig.3) after sintering, could be connected with the recristallization of strai hardened 2212 grains. Low-textured 2212 phase in as-pressed tape near Ag sheath transformed into highly textured 2223 phase (x-point, Fig.1), and the tapes fabricated by on

P alone (rolling-free way) show rather good J_c of about $2 \cdot 10^4$ A/cm^2 at 77K after TP. Thus, the role of Ag sheath in 2223-phase texturing looks prominent only for low-textured original 2212-core. Higher texturing capability of R comparing to that of P processes could be the result of larger contribution of shear strain(not proven though) responsible for core texturing [3], to core deformation in the R case. Meanwhile data on compressive and channel/shear strains in the channel die- and rolling experiments [5] indicate that it may be not the case. Texture distribution over core thickness would be correlated with plastic deformation zones, where the intensity of shear stress reaches the yield stress These zones having a conical shape propagate from tool surface into a core depth (see also [6]).

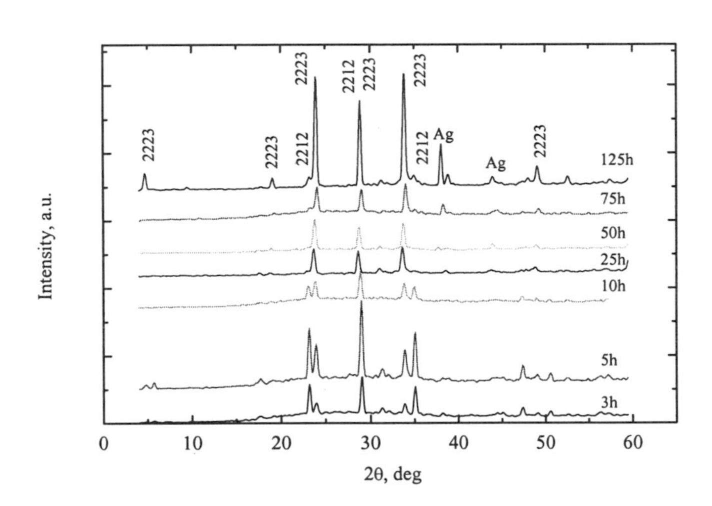

Fig. 2. XRD pattern of core surface in as-sintered state; numbers at diagrams indicate sintering duration

Thus the second possible explanation of effective rolling-induced texturing may be the rotation of anisotropic powder particles in as-ground state during their transport over long distances in rolling, which several times more than appropriate transport distances in pressing. In the presence of compression, c-axes of individual particles are known to tend to be oriented along the compression direction by rotation [5]. It is worth mentioning here that the data of X-ray ω-scan in two directions - the rolling direction and the one perpendicular to it, revealed the improvement of filament c-texture (significant suppression of texture scattering) by intermediate rolling. As to intermediate pressing, it did not result in any meaningful improvement as compared to as-sintered state [2].

4. Conclusions

c-axis texture formation in Ag/BiSCCO 2223 tapes during three deformation schemes of OPIT processes-conventional scheme including drawing and flat rolling, rolling-free scheme including drawing and uniaxial pressing and drawing-free one including only groove-and flat rolling was studied.

Flat rolling has been shown to be the most effective way of c-axis texturing resulting in 90% texture degree, D, in 2212 non-reacted core, while maximal D attainable by pressing was about 60%. This difference is considered to be connected with relatively long-range transport of ceramic particles including their rotation during rolling process and c-axis orientation along the appropriate compression direction.

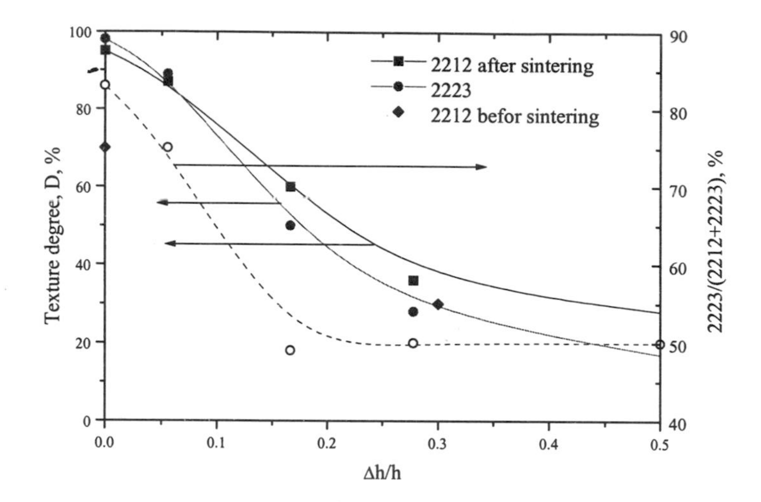

Fig. 3. Depth dependence of texture degree for non-reacted and reacted tapes: h-core thickness, Δh-distance from core surface

Groove rolling itself does not give rise to any texturing at the flat sides of as-deformed ceramic core, subsequent flat rolling results in texturing as in as-drawn composite.

The role of Ag sheath as a texture-promoting factor in the process of 2212- to 2223-phase transformation during sintering is prominent only in the case of preliminary low texture of 2212-phase in the as-deformed composite.

The enhancement of texture degree for residual 2212 phase, observed at core surface after sintering, could be connected with the recrystallization of strain-hardened 2212 grains.

Acknowledgments

This research was supported by the grants of the Ministry of National Infrastructures of Israel and of the Belfer Center for Energy Research. One of the authors (A.G.) was supported by the Ministry of Science and Arts of Israel.

References

[1] Wang Y-L, Bian W, Zhu Y , Cai Z-X, Welch D O, Sabatini R L, Suenaga N and Thurston T R 1996 *Appl. Phys. Lett* **69** 580-583

[2] Beilin V, Goldgirsh A and Schieber M, 1997 *IEEE Trans. Appl. Supercond.* to be published

[3] Wai Lo and Glowacki B A 1991 *Supercon Sci Technol* **4** S361-363

[4] Beilin V, Goldgirsh A ,Schieber M and Harel H1996 *Supercond. Sci. Technol* **9** 549-554

[5] Schoenfeld S E, Ahzi S, Asaro R J, Blumenthal W R 1996 *Phil. Mag* **6** 1565-1590

Schoenfeld S E, Ahzi S, Asaro R J, Blumenthal W R 1996 *Phil. Mag* **6** 1591-1620

[6] Korber F, 1932 *J. Inst. Met.* **48**3 17-322

Inst. Phys. Conf. Ser. No 158
Paper presented at Applied Superconductivity, The Netherlands, 30 June–3 July 1997

The effect of heating rate on Bi-2223 phase formation in composite conductors

A.D.Nikulin, A.K.Shikov, D.N.Rakov, Yu.N.Belotelova, V.E.Klepatsky, I.I.Akimov

All-Russia Scientific Research Institute of Inorganic Materials, Moscow, Russia

Abstract. The effect of heating rate in the range 0,06 °C/min and 4 °C/min is investigated on Bi-2223 phase content in the ceramic core of Bi-2223/Ag tapes during heat treatment. The sufficient influence of heating rate in the range of 810 - 840 °C is revealed by XRD-analysis on the kinetics of Bi-2223 phase growth. It is established that there is the some critical value of heating rate which strongly decreases the intensity of Bi-2223 phase growth. On base of microprobe analysis data it seems to the authors that such influence of heating rate may be connected with different distribution of Pb between different phases in the ceramic core.

1. Introduction

Despite on the sufficient progress in the increase of j_c-values for HTc-composites achieved by the some research groups and firms [1-2], there are a lot of technological problems are either not solved finally or unsufficiently described in literature.

In particular, one of this problems is the optimization of a «time-temperature» sheme for heat treatment of Bi-2223 composite conductor, namely the choice of heating rate value as well as the temperatures and duration times at different annealing stages, cooling rate value in different temperature ranges. It is considered specially interesting to reveal the optimal heating rate value before different annealing stages and intermediate deformation.

Present study was initiated both by the deficiency of published works [3-4] and in our opinion by the absence of the same approach to this problem. Our study was directed on the revealing of the kinetic features of phase formation in ceramic core of Bi-2223/Ag conductors at different heating rate values.

2. Experimental procedure

As the object for this study monofilamentary composite tapes were used on base of $Bi_{1,77}Pb_{0,38}Sr_{2,07}Ca_{2,25}Cu_{3,01}O_x$ "freeze-dried" precursor. Short samples with 3,2x0,2 mm^2 cross section and 40 mm in length were heated in air at heating rate values in the range of 0,06–4 °C/min up to 840 °C and annealed at this temperature during 1–50 hours.

To fix the ceramic core structures formed at different heat treatment regimes the samples were quickly extracted from heated furnace. To investigation of the cores phase composition XRD method was used analogically [5], as well as SEM and microprobe analysis to reveal cores microstructure features. XRD study was carried out on composite tapes after removing of silver sheathes, and SEM investigation on the tape cross sections prepared by conventional metallographic technique.

3. Results and discussion

In Figure 1 the influence of heating rate before heat treatment on Bi-2223 content into ceramic core after annealing 840 ^{0}C, 50 h is shown. These data made it possible to conclude than there is the some critical value of heating rate approximately 1 ^{0}C/min.

Bi-2223 content sufficiently decreased if the heating rate was greater than this critical value. It was established that such heating rate influence (Fig.1) is observed at varying of heating rate in narrow temperature interval 810-840 ^{0}C only. For example, at heating rate 4 ^{0}C/min from room temperature to 810 ^{0}C with followed heating to 840 ^{0}C at heating rate 1 ^{0}C/min phase composition of ceramic core was the same with the case of heating from room temperature to 840 ^{0}C at heating rate 1 ^{0}C/min.

To reveal the mechanism of heating rate influence mentioned above two types of samples were prepared by heating from 500 ^{0}C to 840 ^{0}C with hate values 0,5 ^{0}C/min ("slow" heating, "SH") and 4 ^{0}C/min ("quick" heating, "QH") The samples were extracted from furnace either at concret temperature achieving or after definite duration time at 840 ^{0}C.

Initial phase composition for ceramic core was 30-35 % Bi-2212, 35-40 % $(Ca,Sr)_5Bi_6O_{14}$, 20 % Ca_2PbO_4 and 10 % CuO. In cases of both QH and SH-regimes $(Ca,Sr)_5Bi_6O_{14}$, CuO quickly interact and Bi-2212 forms at temperature range of 780-800 ^{0}C. Bi-2212 is the main phase (75–80 %) at temperatures higher than 800 ^{0}C and small amounts of $(Ca,Sr)_2CuO_3$, Bi_2CuO_4, CaO, Bi-2201 are observed. It is revealed the sufficient distinction in Ca_2PbO_4-content in the ceramic cores of SH and QH-samples. In the temperature range 780-830 ^{0}C Ca_2PbO_4 amount decreases from 15-20 % to 5-6 % for SH-samples and remains almost constant (15-20%) for QH-samples.

Also it was observed the sufficient distinction in phase compositions as well as the kinetic of Bi-2223 phase formation for QH and SH-samples annealed at 840 ^{0}C during different time. In SH-samples the quick growth of Bi-2223 took place with the respective decreasing of other phases amount (Fig.2, curve 1). In case of QH-regime Bi-2223 growth took place in another way. After first hours of annealing Bi-2223 phase was not observed and small amount of it was revealed after 10 hours of heat treatment only. Despite on further increasing of this phase amount its content after 50 hours (Fig.2, curve 2) was less than for SH-samples. It is necessary to note different Ca_2PbO_4 phase behaviour for SH and QH-samples. In SH-core Ca_2PbO_4 amount slow decreases at heating to 840 ^{0}C, and after 20 hours at 840 ^{0}C is not observed . In case of QH-samples Ca_2PbO_4 content strongly dropped at heating to 840 ^{0}C and after 1 hour of annealing already is not revealed. But after 10 hours about 5 % of this phase is observed simultaneously with the beginning of Bi-2223 phase formation.

By XRD method among all phases Ca_2PbO_4 is revealed as the sole Pb-contained phase, but by microprobe analysis it was established Pb-presence in Bi-2212 phase. Thus in QH-samples after first hours of 840 ^{0}C - annealing when Ca_2PbO_4 is absent Bi-2212 phase is as the sole Pb-contained. Data presented above make in possible to assume that the influence of heating rate mentioned above is connected with the presence or absence of Ca_2PbO_4. At melting of Ca_2PbO_4 (827 ^{0}C, [6]) liquid phase appears and thus Bi-2223 growth is accelerated. In QH-samples in the beginning of annealing Ca_2PbO_4 is absent and Bi-2223 formation is difficult. The growth of Bi-2223 begins simultaneously with Ca_2PbO_4 appearance may be in connection with "PbO - Bi-2212" solution decomposition.

In this case Ca_2PbO_4-content not increases because Bi-2223 is Pb-contained phase and the growth of Bi-2223 phase takes place more slowly than in SH-samples. However

Ca_2PbO_4-behaviour mentioned above at different heating rate value remains unclear and the further investigation is needed.

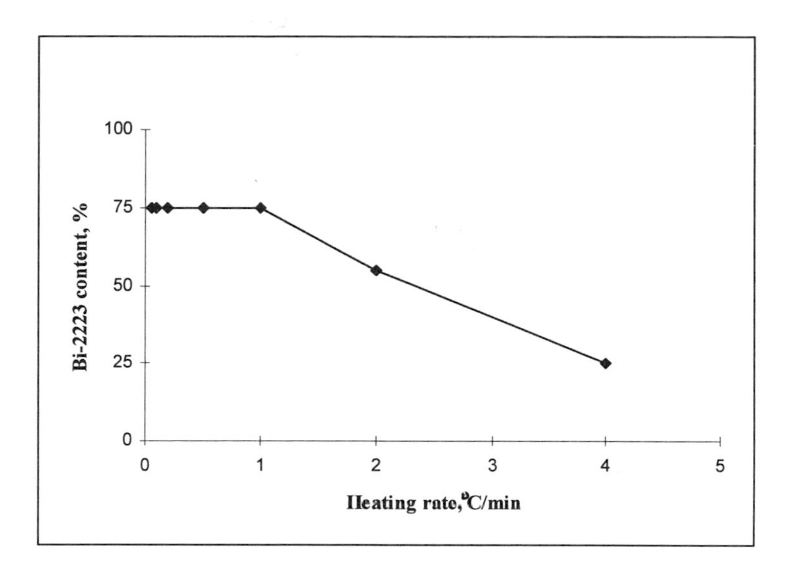

Fig.1 Bi-2223 content into ceramic core after 840 °C, 50 h VS heating rate.

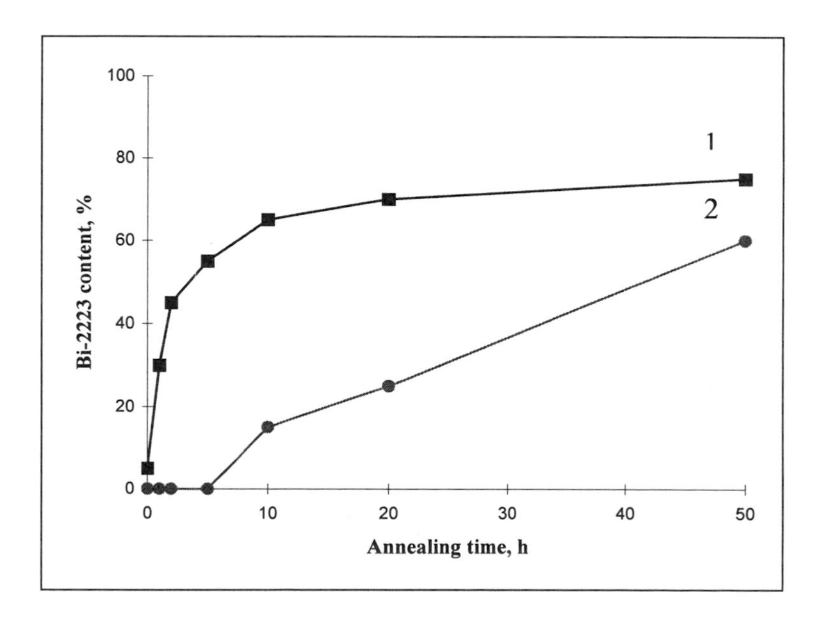

Fig.2 Bi-2223 content into ceramic core VS annealing time, 840 °C
1 - heating rate 0.5 °C/min
2 - heating rate 4 °C/min

4. Conclusions

It is established the sufficient influence of heating rate on the kinetics of phase formation in Bi-2223/Ag ceramic core. It is consider reasonable using of such influence to choice of the optimal "time-temperature" regimes of heat treatment.

References

[1] Q.Li, G.N.Riley, Jr., et.al., Progress in Superconducting Performance of Rolled Multifilamentary Bi-2223 HTS Composite Conductors, Applied Superconductivity Conference, Pittsburgh, PA, USA, Aug.26-30, 1996.
[2] J.M.Seuntjens and G.Shitchler, Practical High Temperature Superconductor Composites for High Energy Physics Applications, Applied Superconductivity Conference, Pittsburgh, PA, USA, Aug.26-30, 1996.
[3] D.M.Spiller, M.K.Al-Mosawi, Y.Yang, C.Beduz and R.Riddle, The Effect Of Heating Rate On The 2223 Phase formation and Core Morphology of (Pb,Bi)2223 Superconducting Tapes, EUCAS'95, Edinburg, Scotland, UK.
[4] Jeffrey A. Parrell, David C. Larbalestier, Steve E. Dorris, Cooling Rate Effects on the Microstructure, Critical Current Density, and T_c Transition of One -and Two-Powder BSCCO-2223 Ag-Sheathed Tapes, Paper MEB-10, presented at the 1994 Applied Superconductivity Conference, Submitted to IEEE Transactions on Applied Superconductivity.
[5] A. Jeremie, G. Grasso, R. Flukiger, Phisica C 255 (1995) 53-60.
[6] Yan Ling Chen, Ronald Stevens, Journal of the American Ceramic Society, 75 (5) 1150-1159 (1992).

Inst. Phys. Conf. Ser. No 158
Paper presented at Applied Superconductivity, The Netherlands, 30 June–3 July 1997

Annealing of Ag-clad BiSCCO Tapes Studied in-situ by High-Energy Synchrotron X-ray Diffraction

T Frello†[1], H F Poulsen†, N H Andersen†, A Abrahamsen†, S Garbe†, M D Bentzon‡ and M von Zimmermann§

† Risø National Laboratory, DK-4000 Roskilde, DENMARK
‡ Nordic Superconductor Technologies, Priorparken 878, DK-2605 Brøndby, DENMARK
§ HASYLAB at DESY, Notkestraße 85, D-22603 Hamburg, GERMANY

Abstract. Using high-energy synchrotron X-ray diffraction with a photon energy of 100 keV we have made *in-situ* studies of the solid state transformation and texture development of the Bi-2212 and Bi-2223 phases *inside* the silver-clad tape during annealing. We find that the texture improvement takes place almost exclusively for Bi-2212, and that it is independent of the phase transformation. We also observe that the phase transformation from Bi-2212 to Bi-2223 runs faster for the multifilamentary tapes than for the monofilamentary tapes.

1. Introduction

Much effort has been put into the development of superconducting tapes for power transmission. A key point in obtaining a high critical current density J_c is the optimization of the heat-treatment of the tapes. During annealing $(Bi,Pb)_2Sr_2Ca_1Cu_2O_{8+x}$ (Bi-2212, T_c=85 K) is transformed into $(Bi,Pb)_2Sr_2Ca_2Cu_3O_{10+x}$ (Bi-2223, T_c=110 K). Furthermore, the annealing procedure also improves the texture of the superconducting grains. The mechanisms governing the phase transformation and texture improvement are not yet fully understood.

The critical current density in the final tape can be severely reduced by bad grain alignment (texture), incomplete phase transformation and the presence of non-superconducting secondary phases such as Ca_2PbO_4 and CuO. The silver cladding affects both the solid state chemistry and the texture development of the superconductor material, and it is therefore highly desirable to investigate these properties *in-situ* inside the silver cladding.

High-energy synchrotron X-ray diffraction provides a unique tool for *in-situ* studies of materials in extreme environments, in particular as a completely non-destructive technique to study grain orientations [1]. With a photon energy of 100 keV the penetration length is of the order of millimeters in BiSCCO and Ag. Due to the high absorption of

[1] E-mail: FRELLO@RISOE.DK.

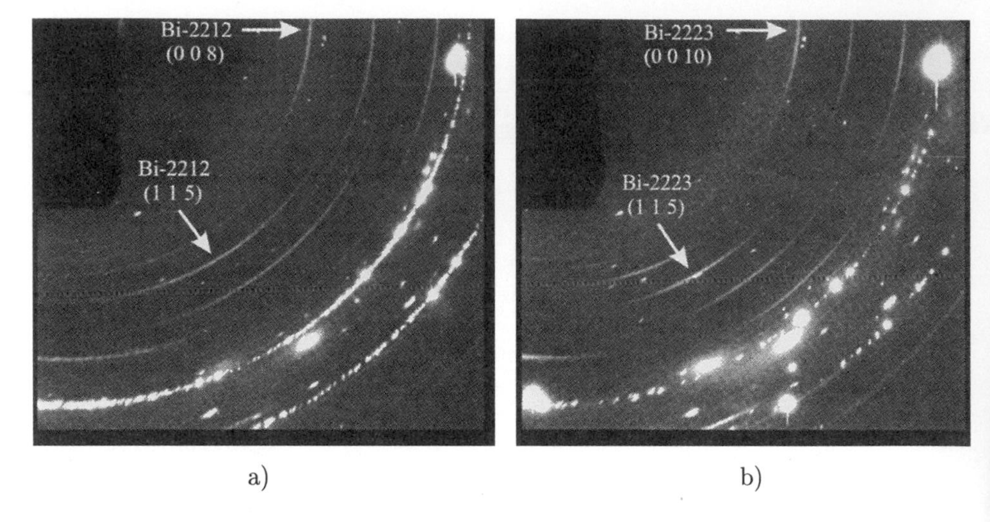

Figure 1. Diffraction images for a monofilamentary tape annealed at 835° for a) 0 hours and for b) 26 hours. The ring segments indicated by arrows on the figure are diffracted signals from the BiSCCO. Full rings and bright spots originate from the silver cladding

neutrons in Ag and the low flux from the present neutron sources, neutron diffraction is not a suitable technique for these studies.

2. Experimental

The samples used in the study were mono- and multifilamentary tapes manufactured by NKT Research Center A/S. They were produced with the powder-in-tube method by drawing and subsequent rolling. They were not exposed to any heat-treatment prior to the experiment.

The experiment was performed at the dedicated high-energy wiggler beamline BW5 at HASYLAB in Hamburg. The tapes were mounted in an Aluminium furnace, where the temperature could be controlled to within ± 0.1 degree of the desired setpoint. The normal to the tape surface could be either parallel to, or have an inclination of 75° with respect to the incoming beam. In this way either the *ab*-plane or the *c*-axis was brought into scattering condition. The experiment was done in transmission geometry with the 100 keV X-rays penetrating both furnace and tape. In this way the true bulk of the superconductor core was probed. The oxygen partial pressure could be varied during the annealing, but all experiments reported here were done in an oxygen partial pressure of 0.17 bar.

The diffraction patterns were recorded on a 2D CCD detector with an active area of 8×8 cm^2. The time to record a diffraction pattern was 1 to 2 minutes.

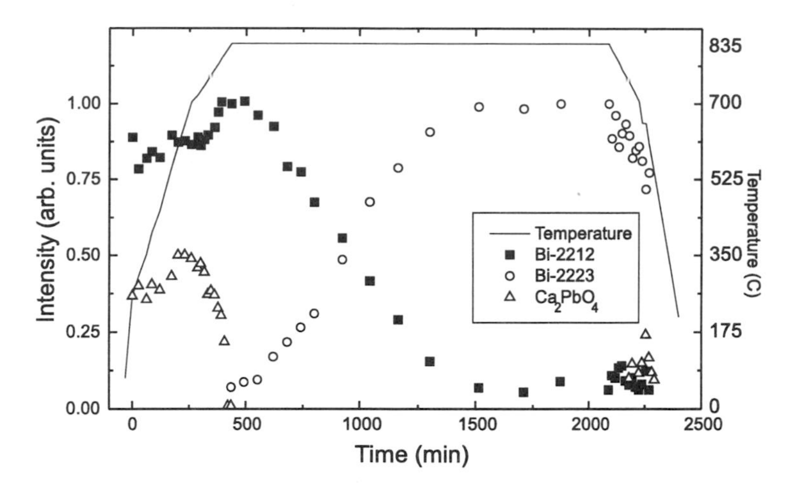

Figure 2. Phase development in a monofilamentary tape during annealing at 835°C

3. Results

An example of the diffraction patterns is seen in Fig. 1. The images show how the Bi-2212 phase is transformed into the Bi-2223 phase after ≈ 26 hours of annealing at 835° for a monofilamentary tape. The annealing also leads to a recrystallization of the silver.

A fast heating of the tapes from room temperature 835°C within 15 minutes turned out to be absolutely detrimental; the diffracted signals from BiSCCO simply disappeared, indicating that the superconductor had melted. A slow temperature ramp was used with good results, as seen in Fig. 2. The concentration of the structural phases was determined by the integrated intensity of the (0 0 8) and (0 0 10) reflection from Bi-2212 and Bi-2223, respectively (shown with horizontal arrows in Fig. 1), and from the (1 1 0) reflection of Ca_2PbO_4. The intensities were normalized to avoid overlap of the datapoints in the figure. No attempt has been made to determine the relative concentration of the different phases resulting from this figure.

The angular intensity distribution giving the *c*-axis texture was determined by analyzing azimuthally along the Debye-Scherrer cones of the (1 1 5) reflections of Bi-2212 and Bi-2223 (shown with tilted arrows in Fig. 1). The angular intensity distribution was fitted with a Gaussian and the FWHM found from the fit is plotted in Fig. 3.

It is worth noticing that an increase in the Bi-2212 texturing rate and the beginning of Bi-2223 nucleation are coinciding whith the melting of Ca_2PbO_4 ≈ 815° C.

Some of our other findings were: The multifilamentary tapes generally had better texture than the monofilamentary tapes, and they also transformed considerably faster (within ≈ 16 hours). They seemed to have a lower optimum annealing temperature, probably because of the larger silver/superconductor ratio.

The tapes all had fibre-symmetry. At no point in the annealing procedure did the *ab*-plane show any sign of texturing.

Although the tape was almost phase-pure Bi-2223 at the end of the annealing, the

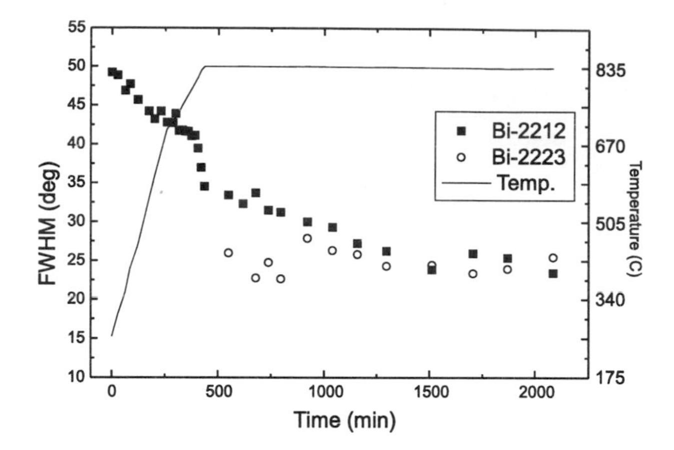

Figure 3. Development of the *c*-axis texture of a monofilamentary tape during annealing at 835°C

Bi-2212 and Ca_2PbO_4 reappeared as soon as the tape was cooled. The intensity of the Bi-2223 also decreased strongly upon cooling (see Fig. 2).

4. Discussion

From Fig. 3 it is seen that the *c*-axis texture development takes place almost exclusively for the Bi-2212 phase, with an improvement from 50° to 25° FWHM. The Bi-2223 texture is essentially constant, and from the moment the Bi-2223 phase starts to nucleate its texture is as good as the final Bi-2212 texture. We suggest that the texturing might be due to an Ostwald ripening process, where the larger Bi-2212 grains, which are expected to have the highest degree of texture from the mechanical deformation, grow on expense of the smaller grains. The Bi-2223 phase nucleates from the Bi-2212 grains, most likely via an intercalation process [2] [3], and if the growth starts from the largest grains we expect that it will also have the same texture as these.

Acknowledgements

This work was supported by The Danish Energy Research Programme, ELKRAFT, ELSAM and Dansync.

References

[1] Poulsen H F, Garbe S, Lorentzen T, Juul Jensen D, Poulsen F W, Andersen N H, Frello T, Feidenhans'l R and Graafsma H *J. Synchrotron Rad.* **4** 147–154
[2] Bian W, Zhu Y, Wang Y L and Suenaga M 1995 *Physica C* **248** 119–126
[3] Frello T and Poulsen H F *To be published*

Inst. Phys. Conf. Ser. No 158
Paper presented at Applied Superconductivity, The Netherlands, 30 June–3 July 1997

Comparative studies on Bi(2223) phase at the Ag interface and inside the ceramic core in Ag-sheathed composite tapes

W Pachla[1], H Odelius[2], U Södervall[2], P Kováč[3], I Hušek[3], H Marciniak[1] and M Wróblewski[1]

[1] High Pressure Research Center, Polish Academy of Sciences, Warszawa, Poland
[2] SIMS laboratory, Dept. of Physics, Chalmers University of Technology, Göteborg, Sweden
[3] Institute of Electrical Engineering, Slovak Academy of Sciences, Bratislava, Slovakia

Abstract. The comparative studies of grains morphology by SEM, chemical composition by SIMS, and texturing ratio and phase composition by XRD between region adjacent to Ag/oxide interface and the center of the core has been performed. Results have shown, that qualitative transverse inhomogeneities exist between these two regions. They strongly depend on the applied tape preparation technology. Observations of an increase in Bi(2223) phase content and alignment at Ag interface have confirmed commonly reported results. Secondary phases were preferentially located around the center of the oxide core. Ceramic and silver grains interpenetrate making transition zone at Ag interface of the several micrometers in width.

1. Introduction

In the past many experiments were done to determine the properties and processes that occur at the Ag/oxide core interface [1-2]. The current carrying capacity fluctuates across tape thickness [3] and width [4]. It is estimated that up to 90% of I_c is carried through the thin layer ~10µm of BSCCO at close proximity to Ag interface [4]. This 'proximity' zone is always disturbed by one or combination of instabilities such as, 'sausaging' effect [5], ceramic intrusions into soft Ag [1,6], impurities and secondary phases precipitates [6], core cracks and porosity [7], and others. In this paper the sites of 2223-phase texturization in tapes made by roll-sinter and roll-press-sinter procedures are indicated and visualized and differences between the core center and its circumference are demonstrated.

2. Experiment

The Ag/Bi(2223) tapes were fabricated by PIT technique from Bi(2212) precursor powder (Praxair, formerly SSC Inc.) of nominal stoichiometry $(Bi_{1.7}Pb_{0.3})Sr_2Ca_2Cu_3O_x$. Precursor was heat treated at 840°C for 10h in air followed by vacuum sintering at 500°C for 20h, with major phase 2212 up to 90%. The powder was wire drawn followed by flat rolling as described in [8]. Short tapes (20mm-30mm) were then sintered or press-sintered to final tape

Table 1. Thermomechanical treatment of Ag-sheathed Bi(2223) tapes

Tape Type (sintering follows each uniaxial pressing)	Uniaxial Pressing [GPa]	Total Sintering time [h]	Overall Final Tape Thickness [μm]	Critical Current Density J_c[Acm^{-2}]
Roll-Sinter	no	150	140	1,960
Roll-Press-Sinter	3x1	300	140	9,570
Roll-Press-Sinter	3-2-1-2-2	250	70	12,820

thickness 70μm and 140μm, see Table 1. Sintering temperatures were 841°C and time intervals between intermediate pressings 50h or 100h.

X-ray diffraction (XRD) scans were performed on longitudinally peeled broad face of the core. The percentage of 2223 phase was estimated from the intensity of 2223 and 2212 peaks, and texturization ratio from the full width at half maximum (FWHM) of the rocking curves of the (0014) peak. To obtain XRD information for different layers across the thickness of the core it was gently scraped with doctor blade. The microstructure was examined using scanning electron microscope (SEM). Secondary ion mass spectrometry (SIMS) measurements were applied for transverse core profiling [9]. In SIMS method the core material was continuously eroded 'in-depth' and elements profile across the thickness of the core to approx. 60μm below the Ag/oxide interface was made.

3. Results

Figure 1 presents the XRD patterns of $(0014)_{2223}$ and $(0012)_{2212}$ peaks for the roll-sinter tape corresponding to three layers across ceramic core: close to Ag/oxide interface and deeper towards the core center. Tape was rolled and three times post-sintered of total 150h. The 2223 phase content is higher for layer adjacent to Ag/oxide interface (~88%) than closer to the core center (~70%). This tendency is also true for other set of peaks $(008)_{2212}$ and $(0010)_{2223}$. Figure 2 shows the rocking curves of this same tape for four consecutive layers across the core thickness. The lowest value of FWHM=11.6° relates to Ag/oxide interface revealed by Ag etching, while the highest value of 24.8° characterizes the deepest layer towards the core center. Figure 3 compares two roll-press-sinter tapes of different overall thickness, see Table 1. Twice a thinner tape shows the more steady element distribution across the core and J_c higher by 30%. The presence of Ag throughout the entire core thickness in each tape should be noted. Figures 4 and 5 show SEM images of the Ag/oxide interface and of the secondary phase inclusions within the core, respectively. In figure 4 better aligned and dense packed grains appear within the ~5μm at the layer adjacent to Ag/oxide interface (from the top of the image). This phenomena have been observed for single cored as well as multifilamentary tapes. In figure 5 with tape pressing direction vertical the higher grain

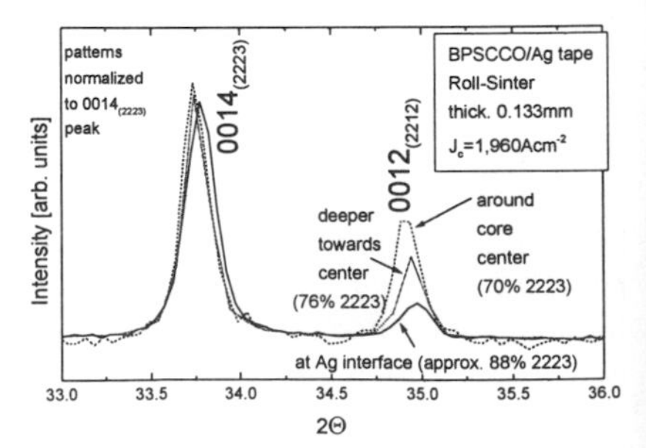

Figure 1. The phase composition by XRD K_αFe scan for various layers across the BPSCCO tape core made by Roll-Sinter procedure.

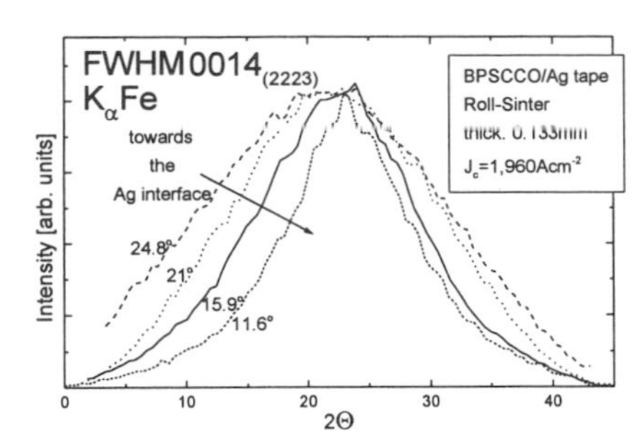

Figure 2. Rocking curves of $(0014)_{2223}$ peak using $K_{\alpha}Fe$ for four consecutive layers towards the Ag/oxide for Ag-sheathed BPSCCO tape made by Roll-Sinter procedure.

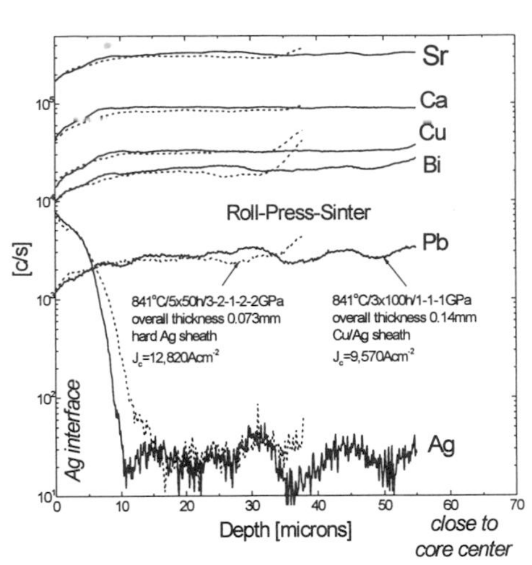

Figure 3. SIMS 'in-depth' profiles for Ag-sheathed BPSCCO tapes made by Roll-Press-Sinter procedure.

alignment of ceramic platelets is seen at horizontally oriented inclusion walls. These preferably oriented layers are ~2μm in thickness.

4. Discussion and Conclusions

It was shown in [1] that 2223 forms at Ag interface approx. 20°C below that at which 2223 forms in the tape interior. It explains higher degree of 2223 content detected at Ag/oxide in-

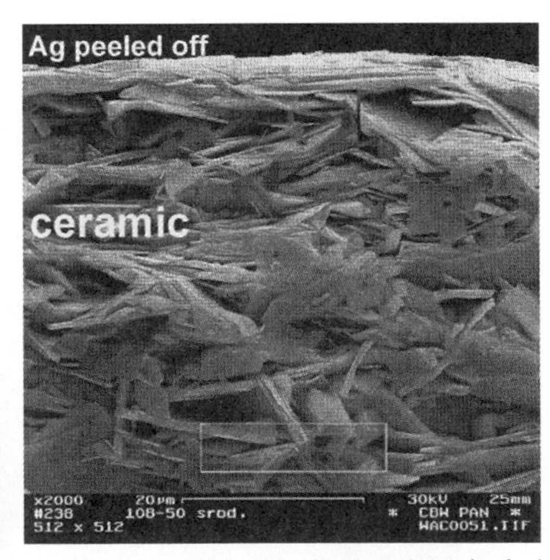

Figure 4. Variation in grain alignment (texturization) across the tape thickness made by Roll-Sinter procedure; overall tape thickness 0.14mm, J_c=3,500Acm^{-2} (77K, 0T); *Note*: better aligned grains at the Ag/BSCCO interface (*~5μm layer from the top edge of the image*).

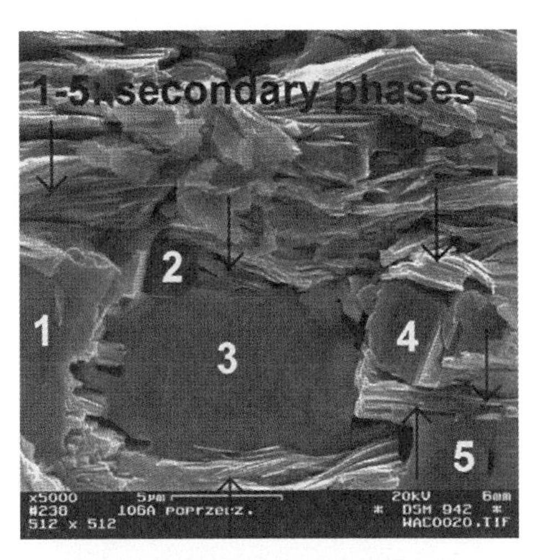

Figure 5. Secondary phases (hard and brittle inclusions) within the ceramic core of Ag-sheathed tape made by Roll-Press-Sinter procedure; overall tape thickness 0.14mm, J_c=9,570Acm^{-2} (77K, 0T); *Note*: hard inclusions act as pressing anvils on the surrounding ceramic matrix causing better grain alignment of the layers adjacent to them.

terface, figure 1. The percentage difference in 2223 content between interface and interior approaches 18%. On figure 4 the substantial higher grain alignment at the layer of ~5μm in the vicinity of Ag interface is seen. In this layer the average thickness of the disc-like grains is a few tenth of micrometers with ~20μm of an average diameter. This layer consists ~10% of the total core thickness and is commonly respected as the actual current currying region [3]. Besides high density such layer exhibits also significant resistance to bending. Below it the grain alignment is disturbed showing many voids and porosity. Grains perpendicular to broad tape face or even perpendicular to direction of rolling are observed. The coarse-grained secondary phases of sizes 2-10μm are mostly located around the core center, figure 5. Since their intrinsic brittleness the secondary phases do not elongate but crack during deformation and disturb the grain alignment continuity. They act as the hard, flat anvils to grains being in nearest neighborhood to them promoting additional texturization of the ceramics. The FWHM data confirm higher grain alignment closer to Ag/oxide interface, figure 2. However, narrower FWHM does not necessarily means higher J_c. For example, J_c<2,000Acm^{-2} was measured for FWHM=12.4° (roll-sinter tape), while J_c=3,000Acm^{-2} for FWHM=16.5° (roll-press-sinter tape). It may results from more bulk texturization during pressing in comparison to rolling, covering deeper layers from the Ag/oxide interface. The FWHM angles are lower for rolled tapes since XRD diffraction is limited to relative thinner outer layer and it does not reflect the entire core quality. From figure 3 it can be seen that ~10μm transition zone between bulk Ag and bulk ceramic exists. In this zone the Ag content drops down by three orders of magnitude but is present throughout the entire core thickness. The Ag presence was already observed by EDS maps [3]. Deeper into the core the Pb content oscillation with period of ~10μm begins. Since the grain's thickness is less than 1μm, this oscillation does not reflect switching from grain to grain. It rather reflects switching from one dense packed layer of grains to another disturbed by internal porosity between them. Pb oscillation is more pronounced than the Bi one. For thinner tape gauge, thus better homogeneity and higher density of the core, oscillations from the nominal elements content are lower. For twice thinner tape oscillations are almost not observed (figure 3, dotted line). The depth where oscillations begin denotes the depth of the pressing penetration zone and may give an information on the optimal core thickness for given tape geometry and preparation technology.

Acknowledgments

This work was performed under the Copernicus project CIPA CT 94-0115. The authors acknowledge the support of Polish Foundation of Sciences.

References

[1] Feng Y, High Y E, Larbalestier D C, Sung Y S and Hellstrom E E 1993 *Appl. Phys. Lett.* **62** 1553-1555
[2] Perin A, Grasso G, Däumling M, Hensel B, Walker E and Flükiger R 1993 *Physica C* **216** 339-344
[3] Lelovic M, Krishnaraj P, Eror N G and Balachandran U 1995 *Physica C* **242** 246-250
[4] Grasso G, Hensel B, Jeremie A and Flükiger R 1995 *Physica C* **241** 45-52
[5] Willén D W A, Zhu W, Nadi R, Paquette A and Cave J R 1997 *IEEE Trans.Appl. Supercond.* **7** 2079-2082
[6] Parrell J A, Feng Y, Dorris S E and Larbalestier D C 1996 *J. Mater. Res.*, **11** (3)
[7] Kovác P, Hušek I, Pachla W, Marciniak H and Melišek T 1996 *Physica C* **261** 131-136
[8] Hušek I, Kovác P and Pachla W 1995 *Supercond. Sci. Technol.* **8** 617-625
[9] Södervall U, Lodding A and Odelius H 1988 *Surf. Interf. Anal.* **11** 529-532

Inst. Phys. Conf. Ser. No 158
Paper presented at Applied Superconductivity, The Netherlands, 30 June–3 July 1997

A.C. losses in Bi-2223 tapes for power applications

Laura Gherardi[1], Fedor Gömöry[1,2], Renata Mele[1], Giacomo Coletta[1]

[1]Pirelli Cavi Spa, Milano, Italy. [2]Permanent address: Slovak Academy of Science, Bratislava, Slovak Republic

Abstract. Improved measuring procedures and interpretation efforts have shown that the critical state model is fully applicable to describe the behaviour of HTS tapes for power applications. By this means, we could consistently analyse magnetic and transport measurements, and study possible inhomogeneities and geometrical peculiarities in BSCCO-2223 tapes. Examples of such analysis are given in this contribution.
Losses in multistrand conductors require a more complex modellization and further development of the experimental procedures; the results of ac transport measurements on a 13m long, multilayer, prototype conductor sample, are also discussed here.

1. Introduction

Ever since it was proposed by Bean [1], the critical state model has been extensively applied to describe the a.c. behaviour of Type II, metallic, superconductors. The discovery of HTS cuprates made available a new class of materials, with strongly Type II features. Such materials were, however, generally characterised, especially at high temperatures around 77K, also by other peculiarities, namely, very high anisotropy and strong magnetic field dependence of critical current density, which made it difficult to a priori predict to what extent a simple critical state model could be successfully applied to such systems.
It is now widely recognised that the Bean model in its simplest form is an extremely powerful tool also for modelling these materials; however, special care is required in some cases in the treatment of experimental results.

Examples of application of the critical state model to the analysis of losses in tapes having a standard, or rather a peculiar geometry, or with possibly non-uniform Jc, are discussed in the following. The results of a.c. transport characterization of a multistrand, high Ic conductor, made with multifilamentary tapes will also be shown and discussed.

2. Measurements on multifilamentary tapes: transport vs magnetic

The a.c. behaviour of Ag-sheathed multifilamentary tapes can, in principle, be expected to depend upon how the filaments are magnetically coupled together.

Actually, a large amount of experimental data from a.c. transport measurements on different tapes points towards a strong (if possibly not always complete) magnetic coupling among filaments. In fact, multifilamentary tapes with several tens of filaments do show losses that are more comparable to the ones predicted by the Norris equations [2] for a monocore conductor with the same critical current than to those, lower by a factor of n equal to the number of filaments, expected if the individual filaments, carrying *1/n* of the total current each, did behave as independent. In practice, the interpretation of the results is made more complicated by the fact that calculated losses are different according to whether the geometry of the cross-section of the superconducting region is schematised as elliptical or laminar (made of closely packed ribbon-like filaments, it is actually neither).

As well known, the losses of an "elliptical" conductor are expected to be higher, and to follow a $\propto I^3$ dependence, whereas "laminar" conductors should show lower losses, increasing with current as $\propto I^4$, in a wide range of currents below Ic. What we have most commonly found measuring a large number of multifilamentary tapes is that the experimental data points fall generally in between the two curves defined by the Norris equations for ellipse and for lamina, respectively. These losses are, however, definitely higher, by at least one order of magnitude, than those expected in the case of "independent filaments (typical multifilamentary tapes have nowadays a number of filaments in the range 60-120, which should mean losses two orders of magnitude lower).

It is worth mentioning that a different approach was found to be possibly needed when analysing the results of magnetic measurements [3] which are typically carried out on much shorter specimens. In those cases, in fact, experimental losses were found to be in good agreement with those predicted for "independent filaments". This apparent discrepancy was consistently explained taking into account the coupling length calculated for typical tapes: such length is in fact generally shorter than the length of the transport specimens, and longer than that of the specimens used in magnetic measurements.

3. Transport behaviour of different tapes - "double ring" tapes

To investigate into the effect of the geometry of the superconducting core the transport losses of concentric, single- or double- ring tapes fabricated by CISE have been measured and analysed [4]. As mentioned, Norris equations allow to analytically calculate the losses for a conductor carrying transport current, provided that its cross-section can be approximated by an ellipse or by a strip. In the case of circular symmetry, straightforward derivation allows to extend the model to the case of a hollow conductor (a "tube): this is what is commonly called "monoblock" model. Losses measured on single-core concentric tapes, in fact, did show a good agreement with Norris predictions for an "elliptical" conductor [5].

Qualitative anomalies were instead found when analysing the results of measurements on double-core tapes (see fig. 1). The experimental curve shows a peculiar behaviour: losses have the same slope, but are lower than predicted for the case of an ellipse (which is apparently the shape of the core of this tape) up to a certain current level, where an upward inflexion is observed; beyond this, the slope of the curve is again the same, but the absolute values are higher than those extrapolated from the previous regime.

To qualitatively interpret this behaviour, we had previously proposed [5] a simple explanation based on two calculations, for the regime where only the outer ring was penetrated by the self field, and for highest penetration, respectively. Jc was supposed to be the same for the two rings. The result were two fits for low- and high- currents, respectively, of which the former was very satisfactory, the latter gave a slight underestimate compared to the experimental points.

We present here the fit to the whole curve (including the intermediate region) obtained by following the approach detailed in [6] developed for the calculation of losses in

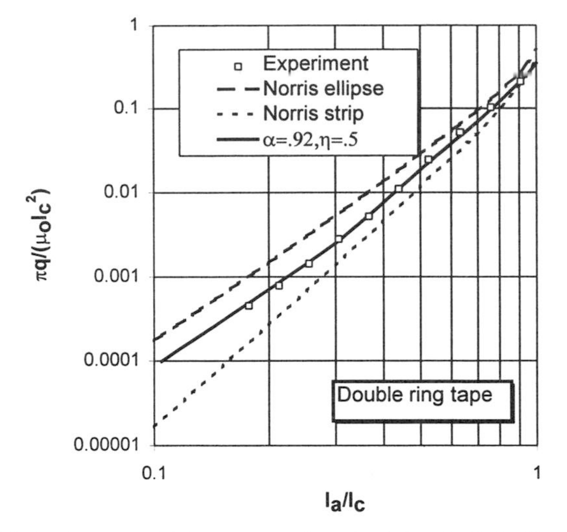

Fig.1: Transport AC losses per cycle and unit length divided by $\mu_0 I_c^2/\pi$ for the double ring tape at 77K, 35 Hz.

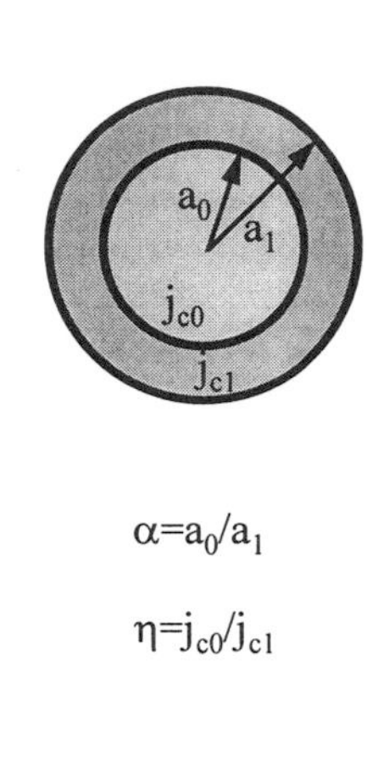

Fig. 2: Round wire with stepwise change in j_c.

a cylindrical concentric structure made of two parts (an inner core and an outer shell) with arbitrary radii and arbitrary Jc, as shown in fig. 2.

As explained in [6] the model allows by this way to take into account a possible radial non-homogeneity of Jc, and applies to elliptical as well as to circular cross-sections.

Shown in figure 1 is the fit to the experimental points obtained by assuming that the Jc of the material was the same in both rings, and modelling the tape as made of an outer shell having the inner and outer radii of the outer ring, and an inner core having an outer radius equal to the inner radius of the shell, all these dimensions being estimated from the micrographs. This means that the inner part of the tape is approximated to a homogeneous single core. The other parameter used for the fit ($\eta = Jc_0/Jc_1 = .5$) was determined empirically, and found to match rather well that determined by "scaling" the Jc of the ring according to the ratio of ring-to-core areas. As can be seen from the figure, the fit obtained by this procedure is very good, and allows to explain the observed anomaly of the loss curve in terms of the geometry of the tapes, only.

4. Multifilamentary tapes: loss curves with "inflexions"

Recently, loss curves showing "inflexions" similar to the one shown above were also obtained from measurements on multifilamentary tapes. See, as an example, fig. 3, where the losses of a multifilamentary tape fabricated by American Superconductor Co. are reported. Shown in the figure is the fit to the experimental points obtained by applying the "shell-and-core" model described above [6].

Since the model applies to a tape made of two concentric regions having different Jc, its application to the case of a multifilamentary tape implies two possible assumptions, namely,

a) that the Jc of the outer filaments is different compared to the one of the inner, or

b) that the distribution of filaments inside the s.c region is non-uniform, so that the "effective", or "average", Jc is different.

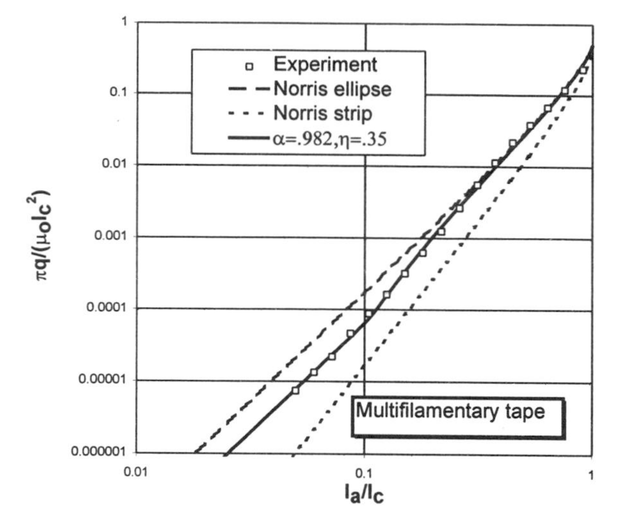

Fig.3: Transport AC losses per cycle and unit length divided by $\mu_0 I_c^2/\pi$ for the multifilamentary tape at 77K, 90 Hz.

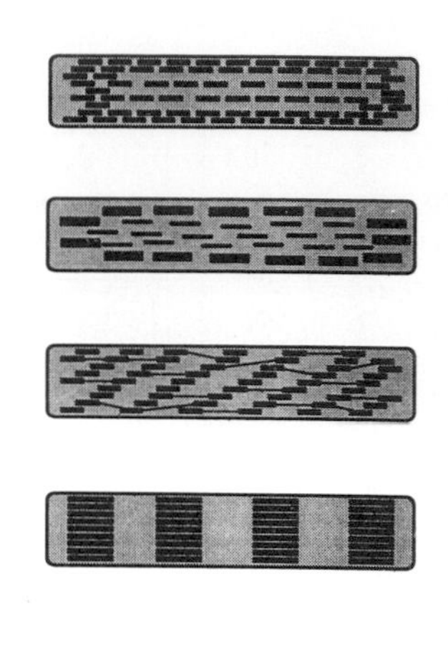

Fig. 4: Several examples of possible nonuniformity in filament distribution leading to nonuniform "effective" j_c.

In particular, the explanation of an "upward" inflexion like the one shown in the figure requires that the outer region is better than the inner (the reverse would give higher losses at low currents and a downward inflexion).

A "geometrical" interpretation as b), similar to the one proposed above for the 2-ring tape, is in principle neither straightforward nor impossible. Examples of possible peculiar filament arrangements that could result from imperfect control of the powder-in-tube technology are sketched, of course over-simplified, in fig. 4. All those situations would be expected to give rise to loss curves qualitatively similar to the one in fig.3. Hypothesis a), on the other hand, could be made from analogy to the case of monofilamentary tapes, taking into account possible non-uniform texturing and oxygenation of the superconductor throughout the cross-section of the tape.

In the specific case of fig. 3, the shape and distribution of filaments in the cross-section of the tape was apparently (from the micrograph) rather uniform, so that the most likely interpretation appears to be a). The values of α and η parameters giving the best fit to the experimental points are indicated in the figure. In the model, such high α and low η suggest the presence of a thin "shell" with significantly higher Jc than the interior: of course, further investigation, including microstructural and electromagnetic, would be required to support such a conclusion. It must be observed, for the sake of completeness, that all the above discussions are based on a standard critical state approach, that is, assume that Jc is independent of magnetic field B. An alternative hypothesis to explain the "inflexions", which is worth mentioning but will not be discussed here, could be quite based on the dependence of Jc upon B: a very non-monotonical dependence, and in particular, the presence of weak initial dependence, followed by a drastic decay of Jc, could in fact be expected to give rise to behaviours qualitatively similar to the ones discussed here.

5. AC Loss mesurements of 13m HTS cable conductor

AC loss mesurements have been performed on a 13 m HTS cable conductor prototype. This is a sample of the prototype manufactured by Pirelli in the frame of an EPRI/DOE contract (1994-1998), as the first phase of the development of an HTS "pipe" type prototype cable, which has been studied for the retrofit of a 115 kV existing transmission line, to be upgraded by this means from 220 to 400 MVA [8]. It has been fabricated in a continuous length, using industrial equipment. The conductor consists of several layers of HTS tapes spirally wound onto a flexible former. The Ag-2223 BSCCO tapes were supplied by American Superconductor Co. in 100 m lengths, according to electromechanical specifications jointly developed.

A DC critical current I_c=3300 A had been measured in liquid nitrogen using the 1 μV/cm criterion; uniformity of Ic along the conductor had also been found very good.

AC losses were measured on 13 m long specimen by an electrical, "transport", method, utilising the lock-in amplifier to single out the component of the cable voltage in-phase with transport current. To minimize the influence of the magnetic field produced by the return current, the return conductor was split into two branches, placed in the same plane as the sample (Fig.5).

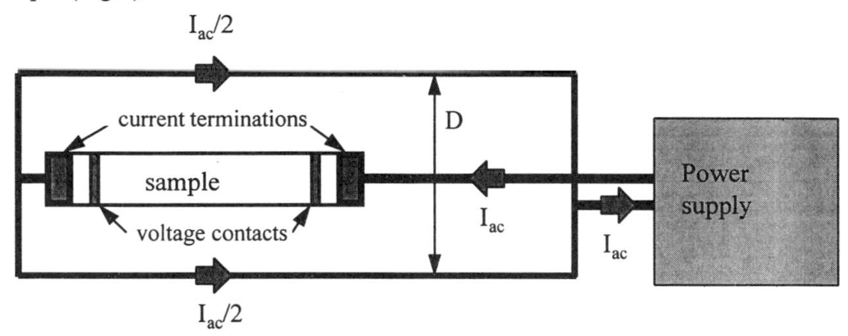

Fig. 5: Splitting of the return conductor into two branches

The loss data reported here were derived in basically two ways, i.e., from either the voltage measured between the current terminations, or from the one between ring-type voltage contacts created by soldering together the tapes of the two outermost layers. In the former case, the raw data were corrected by subtracting the ohmic losses in terminals, which were below 3 μΩ. The losses per cycle per unit length determined in both ways at 40 and 60 Hz were found to agree satisfactorily well, as shown Fig. 6.

This comparison shows that the losses per cycle and unit length depend only on the value of the AC current, in agreement with the critical state model, and confirms that at least in this range of currents the main contribution to a.c. dissipation is hysteretic loss.

These results are compared to the losses of a massive superconducting tube with the same dimensions as the conductor, calculated according to the so-called monoblock model [7]. As can be seen from the figure, measured losses are actually lower than those predicted from the above approximation. The main reason for this more favourable behaviour is supposed to be the counterwise winding of adjacent layers, which presumably promotes a more uniform sharing of current among individual layers. Rather puzzling feature is the observed power p in the $Q \sim I_{eff}^{p}$ dependence. The calculations based on the critical state model would lead to $p = 3$ because of circular symmetry of the cable. Our data show,

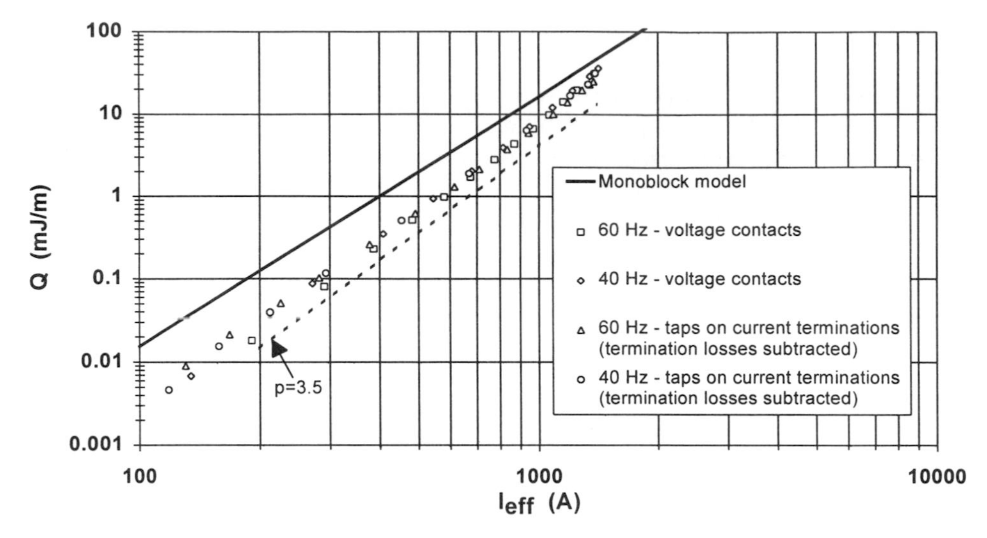

Fig. 6: Transport AC losses in a 13 m sample of 8-layer cable.

however, $p = 3.5$ at low currents, and perhaps even a slightly steeper slope at higher currents. A qualitative explanation could be given in terms of the dependence of I_c on magnetic field [9]. Further characterization in an extended range of currents is however required to better understand this behaviour.

6. Conclusions

Detailed analysis of the a.c. loss measurements carried out on different BSCCO-2223 tapes allowed to identify slightly anomalous behaviours compared to the predictions based on Norris equations. Explanation of such behaviours was shown to be possible in the frame of the standard critical state model, taking into account possible inhomogeneity of Jc or geometrical non-regularity of the superconducting core.

A.c. transport measurements carried out on a multistrand prototype conductor were also shown; experimental losses were found to increase with current with a power law with exponent close to 3.5, and to be significantly lower than predicted by a “monoblock” approach. Further testing and modellisation effort are however required to achieve a more complete picture of the behaviour of such complex structures as multilayer conductors.

References

[1] C. P. Bean, Phys. Rev. Letters 8 (1962) 250
[2] W. T. Norris, J. Phys. D. 3 (1970) 489
[3] F. Gömöry, L. Gherardi, R. Mele, D. Morin and G. Crotti, Physica C 279 (1997) 39
[4] L. Martini, S. Bonazzi, M. Majoros, V. Ottoboni and S. Zanella, IEEE Trans. Appl. Sup. 3 (1993) 961
[5] R. Mele, G. Crotti, L. Gherardi. D. Morin, L. Bigoni, L. Martini and S. Zannella, presented at Applied Superconductivity Conference 1996, Boston, August 1996
[6] F. Gömöry and L. Gherardi, to be published in Physica C
[7] D.R Salmon and J. A. Catterall, J. Phys. D: Appl. Phys. 3 (1970) 1023
[8] A. Bolza, P. Metra, M. Nassi and M. M. Rahman, presented at Applied Superconductivity Conference 1996, Boston, August 1996
[9] L. Dresner, Applied Superconductivity 4 (1966) 167

Inst. Phys. Conf. Ser. No 158
Paper presented at Applied Superconductivity, The Netherlands, 30 June–3 July 1997

The null calorimetric ac losses measurement method: present state and results

Patricia Dolez[1], Marcel Aubin[1], Dag Willén[2], Wen Zhu[2] and Julian Cave[2]

1. Département de Physique, Université de Sherbrooke, Sherbrooke (Québec), Canada J1K 2R1
2. DPRD, Hydro-Québec, 1800 montée Ste-Julie, Varennes (Québec) Canada J3X 1S1

Abstract. The principle of the null calorimetric method for ac losses measurements is based on the comparison of the temperature elevation due to the ac current in a superconducting sample with that due to a dc current in a symmetric reference. We have optimised the method and we now report on the complete automation of the measurement process leading to a reduction in the acquisition time. The addition of a new generation nanovoltmeter allows an increase in the accuracy resulting from the data analysis. We used the null calorimetric method to measure the ac losses generated by a transport current in a silver-gold alloy sheathed Bi-2223 superconducting tape. Variations in current and in frequency are studied, and the distribution of critical current in the sample is obtained by the second derivative of the V-I characteristic.

1. Introduction

Work done since the discovery of high-Tc superconductors has approached us considerably to a widespread industrial use of these materials. Possible domains of application include power cables, energy storage, current limiters and current leads. In particular, recent studies [1,2] report that long length Bi-2223 based superconducting cables are already being fabricated with reproducible properties and with mechanical and electric characteristics approaching industrial requirements. The manufacturing costs of conventional and superconducting cables are almost equivalent and the replacement of some existing power transmission lines by superconducting cables could lead to important savings, if certain lingering problems were solved.

Among the characteristics required for the superconducting cable applications are low ac losses in the relevant temperature and current ranges. This concern explains the renewed interest in the development of an accurate and flexible method for measuring all losses that could appear in a superconducting cable (Ref. [3] and references therein). In particular, the null calorimetric method is reaching the end of its development phase and is now optimised to measure the transport ac losses of various materials in various environments. The latest improvements concerning this technique and the results of the more recent measurements will be presented here.

2. End of the development phase

Since the beginning of the null calorimetric method project [4], successive improvements [5] including a new version of the vacuum chamber [3] have allowed us to optimise the measurement of the ac losses, in terms of both speed and accuracy. Nevertheless, the principle remains the same: comparing the temperature increase caused by an ac current in a superconducting sample with that caused by a dc current in a resistive reference, symmetrically situated with respect to the sample. When the temperature increase is the same, then the dissipated power must be the same in both tapes, and can be easily obtained by the V-I product of the reference. More details on the principles of the technique as well as schematics of the sample holder and of the electrical circuit are given in Ref. [3].

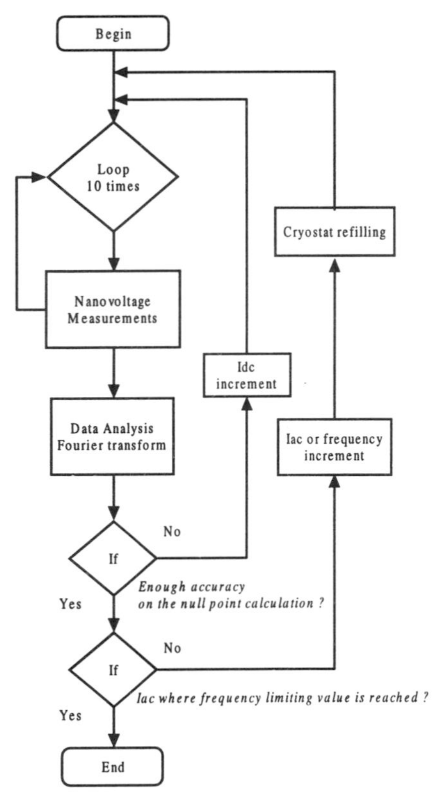

Fig 1: Flow-chart of the computer program

A major step was taken recently with the introduction of a new generation nanovoltmeter. The improvements obtained concern not only the reduced noise in the measured signal, but also the increased resolution in the delay between successive measurements, leading to an increased efficiency of the Fourier transform used in the data analysis for the noise elimination. With this advance, only about 12 loops of Idc increments are now necessary to obtain a 10% accuracy, as compared to more than 40 previously.

A further improvement has been the complete automation of the measurement process. Fig. 1 gives a flow-chart of the computer program. One chooses the range in ac current or in frequency to be studied, and the computer varies the dc current in the reference for the determination of the losses with the required accuracy for each of the selected conditions in the superconductor. The liquid nitrogen filling of the cryostat is also computer controlled. The experiment can run continuously for several days, accumulating data for a whole variation of ac current or frequency.

These last improvements complete the development phase and the null calorimetric method can now be used to routinely measure ac losses of superconducting tapes in various environments.

3. AC loss measurement with a silver-gold Bi2223 tape

A silver-gold alloy sheathed Bi2223 superconducting tape has been produced in the Hydro-Québec research laboratories by the powder-in-tube technique. The use of a silver-gold alloy instead of the usual pure silver is motivated by the desired reduction of eddy current losses which occur in the resistive sheath around the superconductor core. In this case, an introduction of 18% wt. of gold leads to a increase of the resistivity at 77 K from 0.5 to 5 μΩ.cm, and consequently to a reduction of the eddy current loss of the same ratio. The critical current density of the tape is around 13 kA/cm^2.

The variation of the ac losses as a function of the ac current is given in Fig. 2 for 559 Hz, along with the variation in frequency in the inset. The proportionality of the losses with the frequency as well as their dependence in current close to the power 4 of Norris' calculations [6] for a thin strip show the predominance of the hysteretic losses with no detectable contribution of eddy current losses (the latter varies as the square of the frequency).

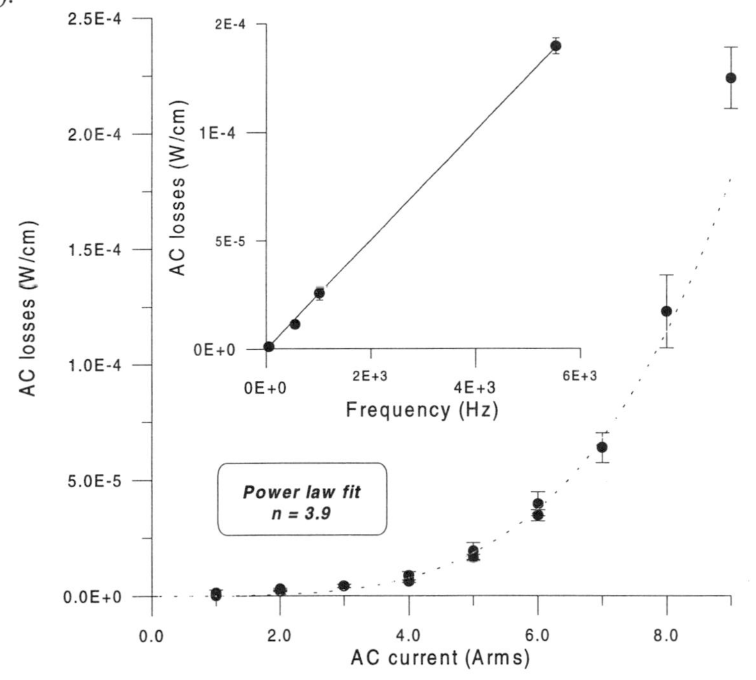

Fig. 2: ac losses variation with current (at 559 Hz) and with frequency (at 5 Arms) of Ag/Au-Bi2223 tape

From the measurement of the V-I characteristic shown in Fig. 3 can be extracted the distribution of the critical current in the superconductor core along its length. The V-I characteristic exhibits the expected behaviour for a superconductor: a power law below the critical current, followed by a linear variation in the normal state. As Jones et al. [7] demonstrate, the second derivative of the voltage versus current is proportional to the fraction of the wire having that critical current value. We fitted [8] the V-I characteristic by the double integration of the sum of two gaussian distributions, yielding the current distribution represented by a solid line (double peaked curve) in Fig. 3. The quality of the fit can be appreciated by observing the solid line passing through the experimental points. We can attribute the higher peak at 12.7 A to the bulk critical current, including the core and the well aligned and well connected Bi-2223 grains (probably at the silver alloy interface). The lower peak at 8.8 A may be due to some defects reducing locally the critical current by reducing the effective superconducting cross-section.

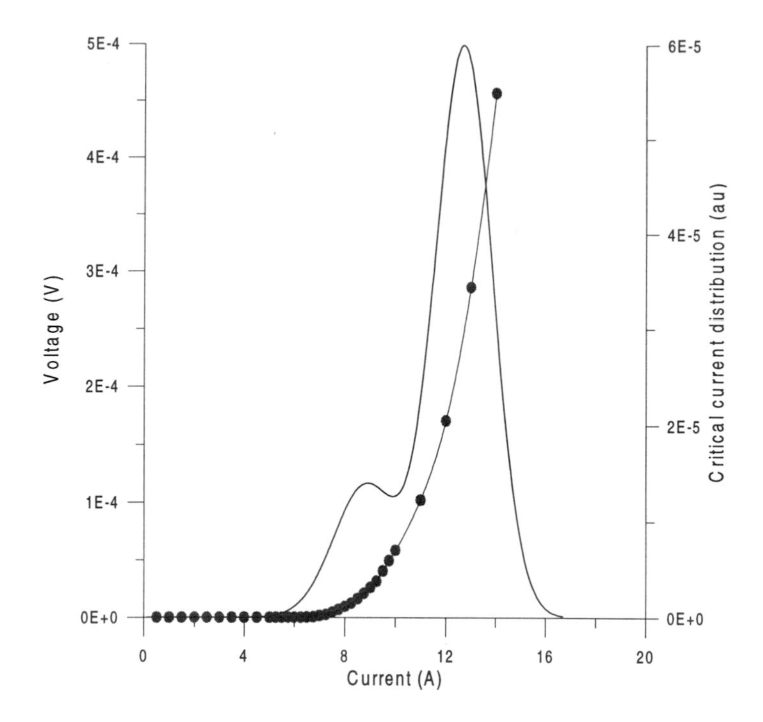

Fig. 3: V-I characteristic (experimental points), its fit by the double integration of the sum of two gaussian distributions (solid line) and critical current distribution (solid line) for the Ag/Au-Bi2223 tape

4. Conclusion

Following a period of development, the null calorimetric method is now destined to participate in the research on power cable optimisation. This will include experiments as a function of temperature and applied magnetic field. The measurements on silver alloy Bi2223 tape showed promising results upon increasing the matrix resistivity and optimising the fabrication process. Experiments have shown that Norris' predictions are in the correct range (3 and 4 for wires and tapes respectively), but in general, the correspondance is not unambiguous. We propose that a more rigorous analysis should take into account the distribution of critical current (Fig. 3) along the length in addition to the conductor cross-section.

References

[1] J. Oestergaard 1997 *IEEE Trans. on Appl. Supercond.* in press
[2] D.M. Buczek et al. 1996 *Proceedings of the Applied Superconductivity Conference* Pittsburgh
[3] P. Dolez, J. Cave, D. Willén, W. Zhu and M. Aubin, submitted to *Cryogenics*
[4] P. Dolez, M. Aubin, D.W.A. Willén, R. Nadi and J. Cave 1996 *Supercond. Sci. Technol.* **9** 374
[5] J. Cave, P. Dolez, M. Aubin, D.W.A Willén and R. Nadi 1996 *Proceedings of ICMC 16/ICEC* Kitakyushu
[6] W.T. Norris 1970 *J. Phys. D: Apply. Phys.***3** 489
[7] R.G. Jones, E.H. Rhoderick and A.C. Rose-Innes 1967 *Phys. Lett.* **24A** (6) 318
[8] The authors wish to thank Martin Parenteau (Université de Sherbrooke) for his help in this calculation.

ıst. Phys. Conf. Ser. No 158
aper presented at Applied Superconductivity, The Netherlands, 30 June–3 July 1997
) 1997 IOP Publishing Ltd

A new approach to AC loss measurement in HTS conductors

N. Chakraborty, A. V.Volkozub and A. D. Caplin

Centre for High Temperature Superconductivity, Blackett Laboratory, Imperial College, London SW7 2BZ, UK.

Abstract. AC loss measurements on superconducting tapes are now very important, but are non-trivial. It is well-known that the AC voltage contains a large reactive component, and that its magnitude depends upon the current distribution and the position of the voltage contacts; these are particularly uncertain in multi-filamentary conductors. Bolometric measurements of AC loss are intrinsically more reliable, but previous approaches have required rather long lengths, which introduces other problems. We describe a novel bolometric technique to be used with samples only a few cm. in length, and in prototype form having a sensitivity of a few μW/cm. We report here loss data for a BSCCO2223 tape obtained with this technique.

. Introduction

Development of high J_C HTS tapes (>10^4 Acm^{-2}) has stimulated great interest in their ngineering exploitation. Applications of HTS tapes in alternating fields (self or external) re limited by the AC losses at the operating conditions. These losses are mostly hysteretic 1] and so controlled primarily by J_C and the geometry of the conductor.

Three methods of AC loss measurement, inductive, transport, and calorimetric are ommon, of which calorimetric measurements are intrinsically more reliable. Dolez et al., 2] Mahedi et al.[3] and Aized et al.[4] have reported calorimetric studies on HTS tapes; ong lengths (~1 m) have to be used, causing radiation and gas conduction losses to become ignificant at higher temperatures. In addition, bath superheating as well as contact loss roblems introduce extra complications.

There are many advantages to being able to measure the AC losses on short onductor samples, not least the limited length of prototype HTS conductor that is often all hat is available. Also, it is well-known that long HTS tapes are inhomogenous along their ength. Although voltage measurements are sufficiently sensitive for short samples, the resence of a very large inductive component causes severe complications; furthermore, in ulti-filamentary HTS conductors, where the current distribution is uncertain, the usual pproach [5] of eliminating the inductive component cannot be relied on.

We report here on a new bolometric technique to determine the AC loss by neasuring the induced temperature gradient in a short (~3 cm.) sample.

2. Experimental Design

Consider a length L of conductor that is efficiently heat-sunk at its ends (Fig. 1). If the power dissipation per unit length is $P(I)$ at rms current I, the temperature difference between the centre and ends is:

$$\Delta T = \frac{P(I)L^2}{2\kappa(T)A}$$

where $\kappa(T)$ is the thermal conductivity and A is the cross-sectional area.

In superconducting Ag- (or alloy-) sheathed BSCCO tapes, the dissipation is primarily within the superconducting core, but the thermal conduction is dominated totally by the sheath. For a tape of typical cross-section, a few mm. wide and 100 μm thick, with a length of a few cm. and mounted as in Fig. 1, ΔT is of the order of 1 mK at a power dissipation of order 10 μW. Under these conditions, thermal radiation and gas conduction are negligible compared with thermal conduction through the sheath, in contrast to the situation for a long sample.

The key to measuring this small temperature difference accurately is to use a thermocouple in which the Ag sheath itself forms one leg. Constantan wires, whose thermopower against Ag is ~16 μV/K at 77 K, are soldered directly to the sheath (Fig. 1); consequently the thermocouple junctions are in excellent thermal contact with the tape.

To ensure that extraneous heating is unimportant, all wires are carefully heat-sunk, and the thermal and electrical circuits are made symmetrical. The HTS tape is indium-soldered to massive current contacts; we routinely achieve contact resistances of ~30 nΩ at 77 K. Heat generated here is absorbed primarily in the current contact, and so does not contribute to the thermal gradient within the tape. Currents up to ~50 A rms at 50 Hz are supplied *via* a centre-tapped and grounded transformer, so as to minimise the common-

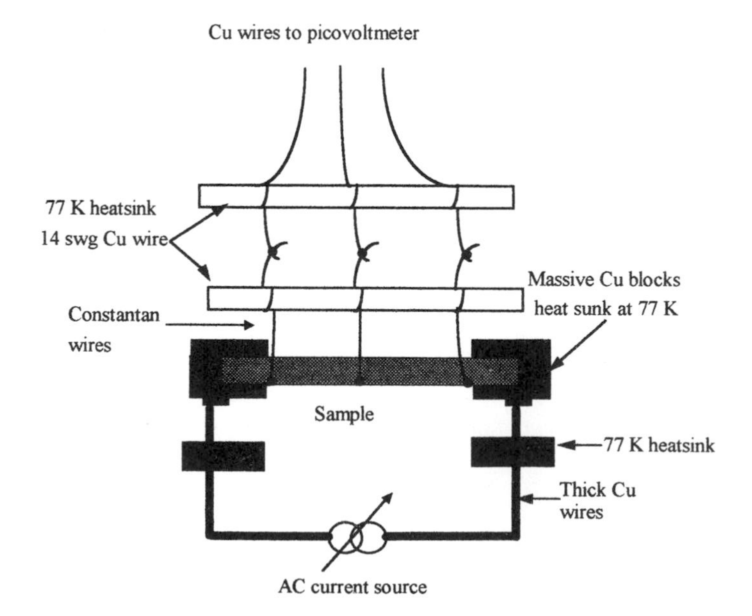

Fig. 1 Schematic diagram of the sample holder

mode voltage seen by the picovoltmeter.

As a final precaution, we monitor *two* thermocouples attached to the sample, left and right. Any asymmetry in the thermal or electrical circuits induces a voltage that has the same sign in the two thermocouples, whereas the dissipation within the sample generates voltages of opposite sign.

In its present prototype form, the sample holder is surrounded by a copper can, and simply immersed up to half its height in liquid nitrogen in a standard storage dewar. Thermal fluctuations from convection currents in the vapour are inhibited by wrapping the sample and the thermocouple junctions in cotton wool.

Calibration of the thermocouple voltage against power dissipation is easily performed *in situ*, at a temperature just above the critical temperature of the tape, where the current is carried entirely by the sheath, which has known electrical and thermal conductivities

A further feature is that the thermocouple connections to the sample can be used also for AC voltage measurement, enabling direct and *in situ* comparison of the losses measured bolometrically and by the potential tap method. This is particularly valuable for filamentary HTS tapes, because the filaments are fragile and sometimes interconnected; so that the current distribution within the tape is very uncertain.

3. Results

The bolometric technique was first tested with a Cu strip of dimensions similar to typical BSCCO-2223 tape samples. As shown in Fig. 2, the thermocouple output is accurately quadratic in I, and has the anticipated magnitude. Most importantly it can be seen that the noise level, even with simple detection, approaches 1 μW/cm.

AC loss data on a 19-filament BSCCO 2223 tape (sample no. BRE-1) with a pure Ag sheath are shown in Fig. 3; its I_C at 77 K and a 1μV/cm criterion is 22 A. The AC losses are predicted to have an I^3 dependence in a conductor of elliptical cross-section, and an I^4 dependence for a thin strip. Our data at currents below I_C are consistent with these dependences, and similar to those seen by AC voltage measurements [6,7] in other such tapes. The steep increase in losses at around 20 A may well be associated with current redistribution between filaments.[8]

Even with simple manual operation, the resolution of the rig is a few μW. It can be enhanced by about an order of magnitude by modulating the AC current at some low frequency Ω, such that $1/\Omega$ is long compared with the thermal time constant of the superconducting tape sample (of order seconds), but short compared with the thermal time constant of the massive current contacts and heat sinks (of order 10^3 seconds), and then monitoring the thermocouple voltage with a synchronous detector.

4. Conclusions

We have developed a novel and simple bolometric technique, capable of measuring AC losses in short samples of HTS tapes at a level of ~μW/cm; complementary voltage measurements of the AC loss can be made simultaneously. The comparison between the

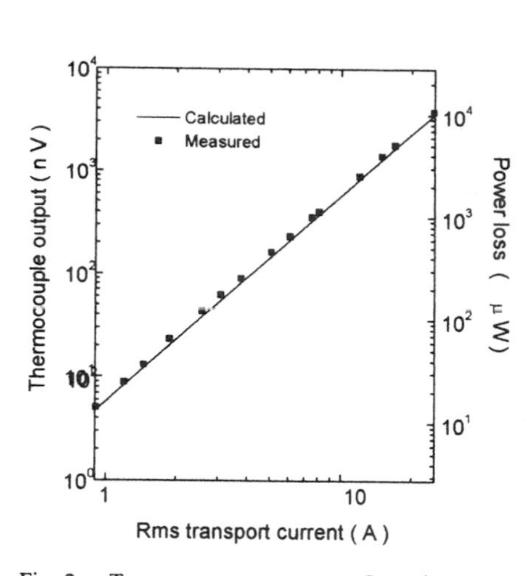

Fig. 2 Test measurements on a Cu strip at 77 K; the line is calculated from known electrical and thermal conductivities.

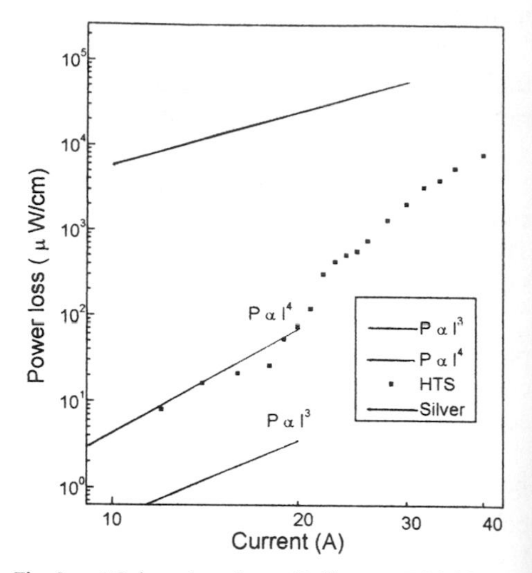

Fig. 3 AC loss data for a 19-filament BSCCO-2223 tape at 77 K and 50 Hz. The line indicates data taken at ~110 K, where the losses are entirely in the Ag sheath, and so provide a calibration.

two will allow the uncertainties of current distribution in multi-filamentary conductors to be gauged.

Acknowledgements

This work has been supported by UK Engineering and Physical Sciences Research Council, and by Brite-Euram Contract No. CT920229. NC acknowledges the award of a Commonwealth Scholarship.

References

[1] Kamper R A 1962 *Phys. Lett.* **2** 290-296

[2] Dolez P, Aubin M, Willen D, Nadi R and Cave J 1996 *Supercond. Sci. & Technol.* **9** 374-378

[3] Mahdi A E, Hughes T, Beduz C, Yang Y, Stoll R L, Sykulski J K, Haldar P and Sokolowski R S 1996 Applied Superconductivity Conference, Pittsburgh, USA

[4] Aized D, Jones E C, Snitcher G, Campbell J, Malozemoff A P, and Schwall R E 1995 European Conference on Applied Superconductivity , Edinburgh, Scotland

[5] Fleshler S, Cronis L T, Conway G E, Malozemoff A P, Pe T, McDonald J, Clem J R, Vellego G and Metra 1995 *Appl. Phys. Lett.* **67** (21) 3189-3191

[6] Passi J, Polák M, Kottman P, Suchon D, Lahtinen M, and Kokavec J 1995 *IEEE Trans. on Appl. Superconductivity* **5**, no 2, 713-716

[7] Ciszek M, Glowacki B A, Ashworth S P, Campbell A M, and Evetts J E 1995 *IEEE Trans. on Appl. Superconductivity* **5**, no 2, 709-712

[8] Volkozub A V, Caplin A D, Eckelmann H, Flükiger R, Goldacker W, Grasso G and Johnston M D 1997 European Conference on Applied Superconductivity, Veldhoven, Netherlands

Inst. Phys. Conf. Ser. No 158
Paper presented at Applied Superconductivity, The Netherlands, 30 June–3 July 1997

Low ac losses in Bi(2223) tapes with oxide barrier

Y. B. Huang,* G. Grasso,*+ F. Marti,+ M. Dhallé,+ G. Witz,*
S. Clerc,# K. Kwasnitza# and R. Flükiger*+

*Group Appl. Phys., +Dept. Phys. Mat. Cond., University of Geneva,
20 quai Ernest-Ansermet, 1211 Genève 4, Switzerland
#Paul Scherrer Institute, 5232 Villigen, Switzerland

Abstract A significant reduction of ac coupling losses has been achieved in Bi(2223) multifilamentary tapes by introducing a thin layer of high resistive $BaZrO_3$ barrier around each filament, thus increasing the transverse resistivity by a factor of 10. The electric decoupling effect of these barrier layers is clearly demonstrated by 'magnetic length scale'- and Hall sensor magnetization experiments. These tapes can also be twisted with a twist pitch of 2 cm, which yields an even higher decoupling effect without deterioration of the transport properties. The uniformity of filament size and barrier layer thickness have been improved by a modified Four Roll PIT method. With this new technique j_c values of the tapes with barriers reaches so far 15000 A/cm^2 (77K, 0T).

1. Introduction

For low T_c materials, a drastic reduction of ac losses can be obtained on multifilamentary wires by reducing the filament size and avoiding inter-filamentary coupling. Twisting the multifilamentary (MF) wire with a small pitch and increasing the transverse resistivity between filaments are two of the most efficient methods to avoid the coupling. For high T_c materials, however, these two methods are much more difficult to realize, since these materials are too brittle to be twisted with a small pitch and there is only a limited choice of suitable sheath materials, which have a much higher resistivity than Ag and no significant detrimental effects on the tape transport properties. Recently, we have successfully introduced high resistivity oxide layers between the filaments of Bi(2223) tapes which resulted in a strong reduction of the coupling [1]. $BaZrO_3$ turns out to be very well suited for this purpose. However, due to the presence of oxide barrier, the achievement of high j_c value meets with two new problems: increased brittleness and reduced oxygen diffusion. In this paper, we report our results on increasing j_c values by improving the preparation conditions. The effects of oxide barrier layers on the microstructure, transport properties as well as ac losses of the Bi2223 MF tapes will be discussed.

2. Experimental

$BaZrO_3$ has been used to prepare Ag composite sheath. The thickness of the barrier layers was controlled by the amount of $BaZrO_3$ introduced into the composite. The composite MF Bi(2223) tapes were prepared by the PIT method [2]. As an alternative route, four roll deforming (FRD) (driving turks head) technique was applied to fabricate the wires [3].

j_c values at 77 K were measured by the four-probe method with an applied field up to 0.8 T using the criterion of 1 μV/cm. The double Hall sensor ac loss measurements (f=1-130 Hz) were performed at 77 K on 3 cm long tapes in sinusoidal external field (H//c) of varying amplitude [4]. The electrical decoupling of the filaments by the $BaZrO_3$ barriers was further investigated using the magnetic length scale technique [5, 6]. The latter is a quasi-DC measurements ($dB/dt \sim 10^{-4}$ Ts^{-1}) performed in a vibrating sample magnetometer which allows the determination of the geometry of screening current flow as a function of temperature (10-60 K) and external field (0-7 T).

3. Results and discussions

3.1 Results from conventional PIT method

Recently, Ag/$BaZrO_3$ composite sheathed Bi(2223) MF tapes with different numbers of filaments (7, 19 and 37), have been fabricated -for the first time- in our group. Uniform barriers with thickness in the range 0.3 to 2 μm were successfully obtained. Fig. 1 presents a cross-sectional view of a 19-filament Bi(2223) tape with a barrier layer thickness of 1 μm. In this picture, it can be seen that each filament is well encapsulated by the high resistive $BaZrO_3$ barrier layers. The resistive layer is nearly everywhere continuous, indicating that the barrier is very stable whilst treating the tapes at high temperature. It was still possible to twist this tape with a pitch of 2 cm, which resulted in a further enhancement of the decoupling effect. j_c values of this tape under self- or external fields are very similar to that without twisting. No trace of $BaZrO_3$/Bi(2223) reaction can be found in the tapes, neither by DTA/TG nor by SEM/EDX analysis. The T_c values of the tapes is always close to 110 K, regardless of the thickness of the barrier layers. However, this type of MF tapes yield a wide range of j_c values, 0.5-1.3 x 10^4 A/cm^2 (77 K, 0 T), depending on the barrier layer thickness. Thicker barriers normally lead to lower j_c values, e.g. when the thickness is over 1 μm, j_c value is lower than 7 kA/cm^2 (77 K, 0 T). A strong sausaging of the core has been observed in the tapes with thick barriers, as shown in Fig. 1, where fluctuations in filament size are clearly noticeable. Moreover, according to measurements of I_c as a function of the angle between applied field and tape normal [7], due to the strong sausaging the misalignment angle of 2223 grains of this tape is as high as 8°, whereas this angle is about 5° for tapes sheathed by pure Ag. The sausaging appears mainly during the wire drawing and laminating processes. Since the hard $BaZrO_3$ particles within the tapes are difficult to deform with the soft Ag and mica-like BSCCO powder, these particles cause stress concentration and introduce cracking during deformation. Therefore, the presence of high hardness BaZrO barriers not only increases the deformation force, but also reduces the strength of sheath. Sausaging appears during the wire drawing process due to an easier propagation of crack under tensile stress. When a round wire is laminated into a flat tape the strongest sausaging can be found in the central part since the filaments in this part are deformed much more heavily than those on the edge.

3.2 Results from FRD method

Fig. 2 is a cross-sectional view of a 30-filament tape with a barrier layer of 1μm thickness made by the four roller deforming technique It clearly shows out the regular size of filaments which are neatly stacked and have continuous barrier layers between them. The advantage of this method is that no tensile force is applied during wire forming and every

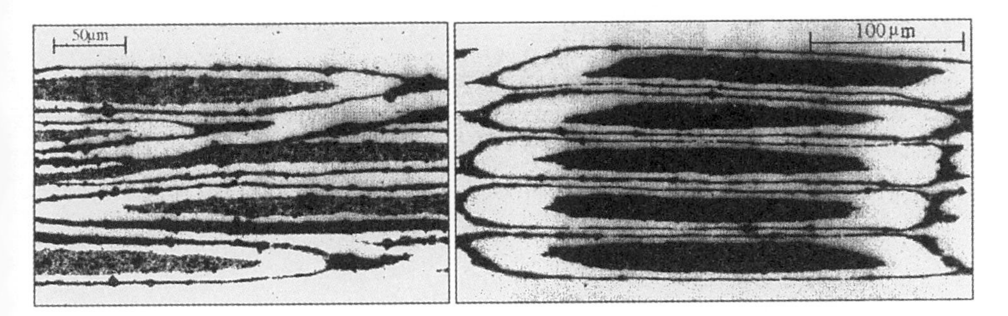

Fig. 1 A detail of cross section of a 19-filament Bi(2223) tape with oxide barrier made by the PIT method.

Fig. 2 A detail of cross section of a 30-filament Bi(2223) tape with oxide barrier made by FRD method.

filament is deformed equally when tape laminating. Consequently, the sausaging problems are diminished and the j_c values are increased by a factor of 2, up to 1.5 x 10^4 A/cm^2 (77K, 0T), comparing with that made by the normal PIT method. The j_c value field dependence of these two types of tapes is presented in Fig. 3. The tapes made by FRD method are less sensitive to the applied field especially when H//c. However, the misalignment angle is still relatively high (~ 7°), so that the new tapes with oxide barrier have lower j_c values and stronger field dependence than the standard Ag sheathed MF Bi(2223) tapes prepared in our group (j_c=28000 A/cm^2, 77 K, 0 T) [3]. Further optimization of thermal-mechanical treatment conditions in term of improving texture and j_c values is under work.

3.3 Measurements of electrical coupling

Fig.4 shows the efficiency of the $BaZrO_3$ barriers to increase the transverse electrical resistivity ρ_e in the matrix. To get low coupling current losses in externally applied time-varying magnetic fields, ρ_e must be as large as possible. In Fig.4 the total losses per cycle of twisted 19-filament tapes with and without the $BaZrO_3$ barriers are compared in a ΔB range where the coupling losses dominate. The losses are normalized by the value of the loss maximum. For the tape without barriers and peak-to-peak amplitude ΔB =9.3 mT, q_{max}=7.4 J/m^3 while for the tape with barriers and ΔB =4.6 mT, q_{max} =1.7 J/m^3. Due to the presence of the barriers the frequency at the loss maximum, which is proportional to the

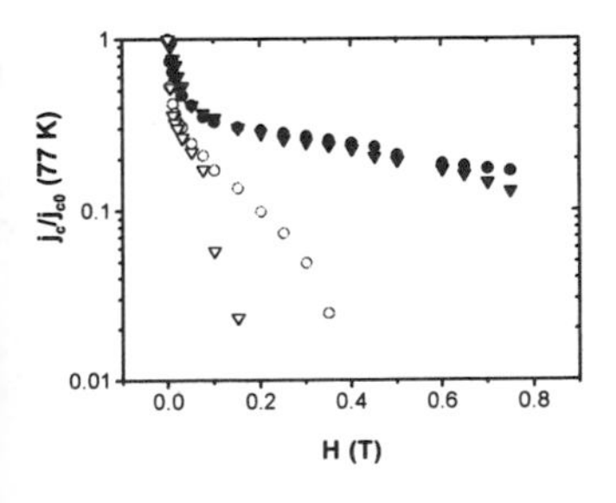

Fig. 3 j_c values versus applied field parallel (solid) and perpendicular (open) to tape plane at 77K for the tapes made by PIT method (triangles) and by FRD method (circles).

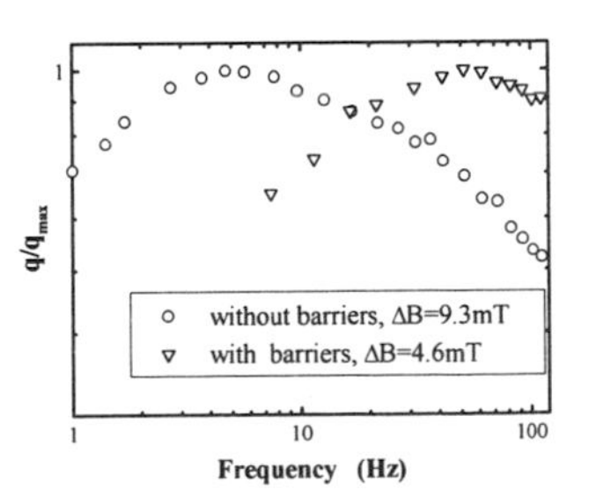

Fig. 4 Frequency dependence of normalized total losses per cycle for two 19-filament tapes twisted with a pitch of 2 cm sheathed by pure Ag (circles) and Ag/$BaZrO_3$ (triangles).

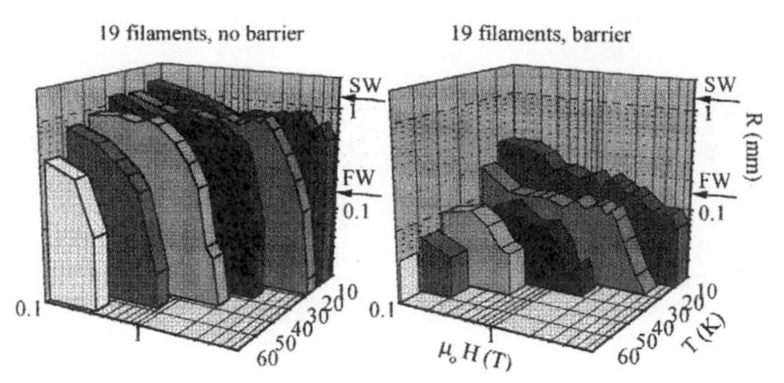

Fig. 5 Magnetic screening current lengthscale R (in logarithmic scale) plotted against temperature T and external magnetic field H. The arrows correspond to half the sample - ('SW') and filament width ('FW').

q_{max}=7.4 J/m^3 while for the tape with barriers and ΔB =4.6 mT, q_{max} =1.7 J/m^3. Due to the presence of the barriers the frequency at the loss maximum, which is proportional to the transverse resistivity, increases approximately by a factor of ten. This is in qualitative agreement with our results in untwisted samples [1, 8].

Fig. 5 shows the magnetic length scale R as a function of temperature T and external field $\mu_o H$, measured on two 19-filament tapes. In the reference sample (no barriers) the screening current loops span about half the sample width, indicating non-negligible electrical coupling between the filaments. Insertion of 1 μm thick barrier layers reduces $R(T,H)$ to essentially the filament width (500 μm), which clearly demonstrates the effect of these barriers on the filament coupling.

4. Conclusions

A remarkable reduction of electrical coupling ac losses has been achieved on Bi(2223) multifilamentary tapes -for the first time- by introducing highly resistive $BaZrO_3$ barriers around each filament. The maximum of the ac coupling losses as a function of the frequency can be shifted up by a factor of 10 using the barriers. j_c values of our new tapes are increased by a factor of 2 by using a modified Powder-In-Tube method using Four-Roller-Deforming technique which is very efficient to diminish sausaging.

Acknowledgements

This work was supported by the Swiss Priority Project of Materials (PPM) and by the European Brite/Euram project No. BRPR-CT95-0030.

References

[1] Huang Y B, Grasso G, Marti F, Erb A, Clerc St, Kwasnitza K and Flükiger R 1997 Presented at SPA'9' March, 6-8, Xi'an (China).
[2] Yamada Y, Obst B and Flükiger R 1991 *Supercond. Sci. Technol.*, **4**, 165.
[3] Grasso G and Flükiger R 1996 Presented at ISS'96 Oct., 21-24, Sapporo (Japan).
[4] Kwasnitza K and Clerc St 1994 *Physica C* **233**, 423.
[5] Angadi M A, Caplin A D, Laverty J R and Shen Z X 1991 *Physica C* **177**, 479.
[6] Dhallé M, Marti F, Grasso G, Huang Y B, Caplin A D and Flükiger R 1997 Presented at this Conference.
[7] Hensel B, Grivel J C, Jeremie A, Perin A, Pollini A and Flükiger R 1993 *Physica C* **205**, 329.
[8] Kwasnitza K, Clerc St, Flükiger R and Huang Y B 1997 *Physica C* (submitted).

Inst. Phys. Conf. Ser. No 158
Paper presented at Applied Superconductivity, The Netherlands, 30 June–3 July 1997

AC losses of twisted high-T_c superconducting multifilament Bi2223 tapes with a mixed matrix of Ag and $BaZrO_3$

K. Kwasnitza[1], St. Clerc[1], R. Flükiger[2], Y.B. Huang[2] and G. Grasso[2]

[1]Paul Scherrer Institut, Villigen, Switzerland,
[2]University of Geneva, DPMC, Switzerland

Abstract. In twisted multifilament Bi2223/Ag tapes the introduction of $BaZrO_3$ barriers around the filaments increases the transverse matrix resistivity at 77 K by a factor of 10 and shifts the coupling loss maximum in alternating magnetic fields from 4.5 to about 45 Hz. For 50 Hz applications the twist length should be further reduced and the barrier thickness increased.

1. Introduction

The development of high-T_c multifilament Bi2223/Ag tapes with low losses at 77K in alternating magnetic fields is of large technical interest. The losses are, a partly nonlinear, superposition of hysteresis loss in the filaments Q_h (per cycle) and of the interfilament coupling loss Q_c (also per cycle) in the conductor matrix. Q_c can be reduced by a small twistpitch l_t of the tape and by a large transverse electrical resisitvity ρ_e.

Lately the co-authors of the university of Geneva have succeeded to fabricate a 19 filament Bi2223/Ag tape conductor having high resistive $BaZrO_3$ barriers around the filaments in the Ag matrix [1]. On *untwisted* samples of this mixed matrix high-T_c superconductor we have lately measured at 77 K the losses and have confirmed the efficiency of the $BaZrO_3$ barriers to increase transverse ρ_e [2]. For literature on alternating field losses in the high T_c superconductors see for instance [3] - [6]. In [6] we have compiled the main aspects of Q_h and Q_c. Q_c is a function of generalized frequency $\omega\tau$ with $\omega = 2\pi\nu$ and coupling current decay time constant

$$\tau \propto l_t^2 / \rho_e \quad (1)$$

For smaller ΔB values, Q_c has a maximum at $\omega\tau = 1$. For large ΔB values the whole composite becomes saturated with increasing frequency by the coupling currents and the total losses are again of hysteretic nature. The critical rate of magnetic field change for saturation is in analogy to the multislab model calculations [7]

$$\dot{B}_c = K \cdot \rho_e \cdot j_c \cdot a / l_t^2 \quad (2)$$

j_c = critical current density, a is here the total thickness of the superconducting material in the direction of magnetic flux penetration (for applied B perpendicular to the tape width, this is in B direction). Geometry factor K has still to be calculated for flat tape geometry.

2. Experimental

Two different 19 filament Bi2223 tape conductors were used for the ac loss measurements. The first (AgMO4tF) has a pure Ag matrix, the second one (BM22tF) is with a mixed $BaZrO_3$/Ag matrix. Fig. 1 shows the high-resistive $BaZrO_3$ barriers around the filaments. The still rather inhomogeneous barriers are roughly 1 μm thick. Both tapes have the same overall dimensions 3 x 0.18 mm^2 and the same twistpitch l_t = 2 cm. For the conductor without barriers is $I_c \approx 7$ A and $j_c \approx 1.10^8$ A/m^2. The corresponding values for the barrier conductor are I_c = 2.5 - 3 A and j_c = 0.5 - 0.6 · 10^8 A/m^2. This different j_c makes a quantitative comparison of the losses in the case of composite saturation somewhat more difficult but has only marginal influence on the frequency position of the coupling loss maximum (for low ΔB) and on the derivation of ρ_e from it. All loss measurements were performed of 77 K. A sinusoidal magnetic field was applied perpendicularly to the wide side of the tapes. Using field cycles symmetrical to B_s = 0T, the losses were derived from the magnetization loops as function of frequency and peak-to-peak ΔB.

The magnetization was measured by a double Hallsensor technique [6] using a digital storage oszillascope for data acquisition.

To get larger magnetization signals the samples for the measurements consisted for each conductor type of a stack of seven isolated tape pieces, each piece having the length $l = 2\, l_t = 4$ cm. In the figures the losses are normalized by the superconducting volume.

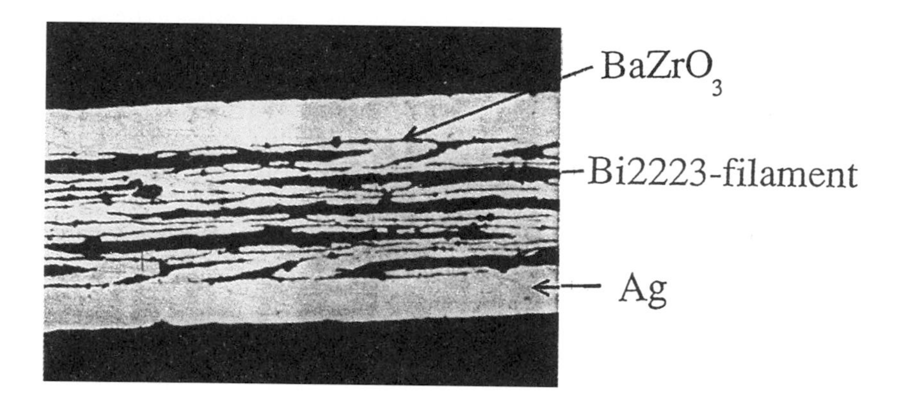

Fig. 1: Magnified part of the tape cross section showing the $BaZrO_3$ barriers

3. Results and discussion

3.1 Frequency dependence of the losses at low ΔB-values

Fig. 2 shows normalized loss $q/q_m = f(\nu)$ at small ΔB for both conductor types with and without barriers. The tape without barriers reaches for ΔB = 9.2 mT its loss maximum, for which the coupling losses are responsible, at $\nu_m \approx 4.5$ Hz (with q_m = 7.4 J/m^3). For ΔB = 3.1 mT nearly the same ν_m value is found. Using $\omega\tau = 1$, we get from this $\tau \approx 35$ ms. The second loss curve in Fig. 2 for the conductor with barriers has for ΔB = 4.6 mT its loss maximum at about 50 Hz (with q_m = 1.7 J/m^3). For ΔB = 3.1 mT, we find $\nu_m \approx 45$ Hz. Using this somewhat lower value, we get $\tau \approx 3.5$ ms and with eq (1) enhancement of ρ_e by the $BaZrO_3$ barriers by a factor of ten! This is the same order of magnitude of ρ_e increase, we have measured in the untwisted samples [2]. For 50 Hz applications, we should shift ν_m to still larger values, clearly beyond 50 Hz. Already feasible l_t = 1 cm will give $\nu_m \approx 180$ Hz. Also the barrier thickness has to be increased. The coupling losses in flat, twisted multifilament tapes are in principle large [6], as Q_c is proportional to $(c/d)^2$, with c = tape width and d = tape thickness.

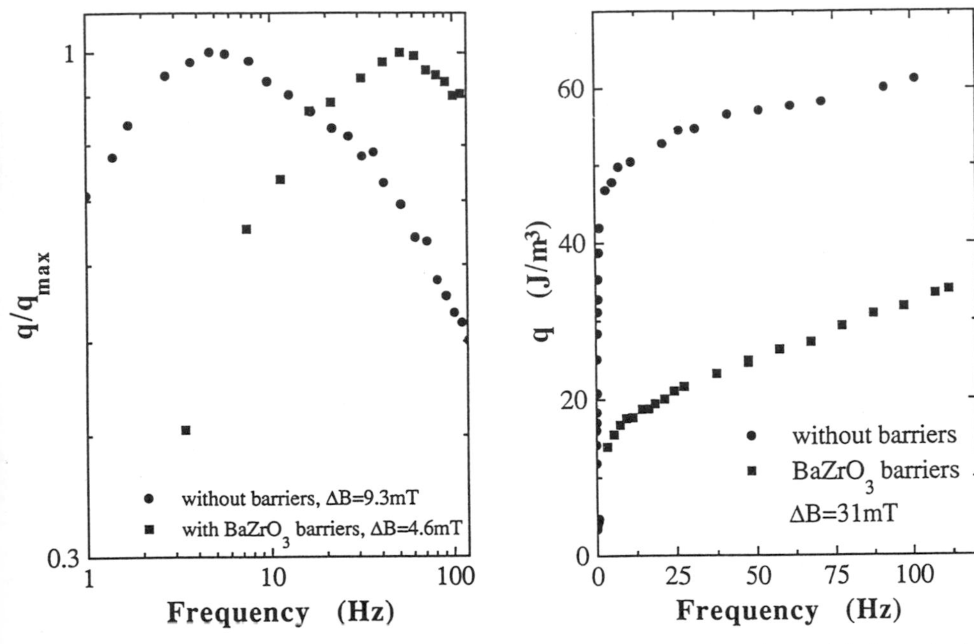

Fig. 2: Frequency shift of the loss maximum by the $BaZrO_3$ barriers

Fig. 3: Frequency dependence of the losses in the case of composite saturation

3.2 Frequency dependency of the losses at larger ΔB values

In Fig. 3 we show for $\Delta B = 31$ mT, being larger than the composite penetration field ΔB_{ps}, the total losses without and with $BaZrO_3$ barriers as function of frequency. Both conductors exhibit with increasing ν a transition from individual filament hysteresis loss behavior to full composite saturation losses of again hysteretic nature. These losses increase further with ν due to flux flow dependence of j_c [6]. The barrier conductor reaches saturation at larger frequency. But the main reason of its lower saturation losses is, at the largest ΔB-values, for the moment still its smaller I_c value.

3.3 Amplitude dependence of the losses at 50 Hz

For both conductor types the losses were measured at 50 Hz as function of ΔB. See Fig. 4a and 4b. Also displayed is the loss factor $\Gamma = q/(\Delta B^2/2\ \mu_0)$. With the pure Ag matrix (Fig. 4a) the losses are practically hysteretic over the whole ΔB-range, due to full or partial composite saturation, as here $\nu_m \ll 50$ Hz. From the position of the maximum of the loss factor we get $\Delta B_{ps} \approx 16$ mT. For $\Delta B < \Delta B_{ps}$ we have $Q \propto \Delta B^{3.2}$ and for the largest ΔB values $Q \propto \Delta B^{1.1}$, close to the predictions of Bean's model. So this multifilament conductor acts at 50 Hz like a single core one. The 50 Hz loss behavior of the barrier conductor in Fig. 4b is different for small ΔB. Here we find $Q \propto \Delta B^2$, as expected from the dominating coupling loss behavior in Fig. 2 for small ΔB. For large ΔB we get also composite saturation. Due to the smaller I_c value, the value of ΔB_{ps} is correspondingly smaller.

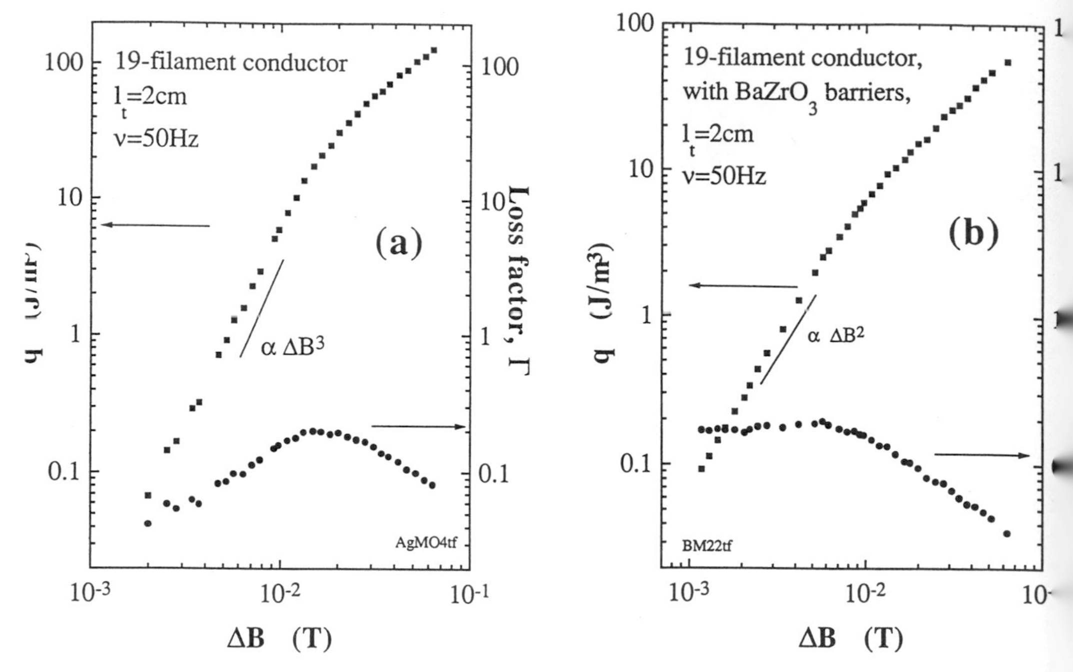

Fig. 4: Amplitude dependence of 50 Hz losses (a) without barriers, (b) with $BaZrO_3$ barriers.

4. Conclusions

Due to the use of a mixed Ag-$BaZrO_3$ matrix the transverse ρ_e has been increased in the twisted multifilament Bi2223 tape conductor by a factor of ten! This leads for larger ΔB already to losses lower in the range of $\nu \leq 45$ Hz. For 50 Hz applications both the twistlength has still to be reduced and barrier thickness increased. Also the j_c value has still be optimized for this conductor type.

Acknowledgements

The authors thank W. Fischer, PSI, for his support of this work. The financial aid of the Swiss Government in the frame of the PPM program is gratefully acknowledged.

References

[1] Huang, G. Grasso, F. Marti, A. Erb, St. Clerc, K. Kwasnitza and R. Flükiger, Presented at *SPA'97* March 6-8 (1997) Xi'an (China)
[2] K. Kwasnitza., St. Clerc, R. Flükiger and Y.B. Huang, submitted for publication to *Physica C*
[3] J.J. Gannon Jr., A.P. Malozemoff, M.J. Minot, F. Barenghi, P. Metra, G. Vellego, J. Orehotsky and M. Suenaga, *Adv. Cryog. Eng.* **40** (1994) 45
[4] Y. Fukumoto, H.J. Wiesmann, M. Garber, M. Suenaga and P. Haldar, *Appl. Phys. Lett.* **67** (1995) 3180
[5] A. Oota, T. Fukunaga, T. Abe, S. Yuhya and M. Hiraoka, *Appl. Phys. Lett.* **66** (1995) 1551
[6] K. Kwasnitza and St. Clerc, *Physica C* **233** (1994) 423
[7] M.N. Wilson, "Superconducting Magnets" *Oxford University Press*, Oxford 1980

Inst. Phys. Conf. Ser. No 158
Paper presented at Applied Superconductivity, The Netherlands, 30 June–3 July 1997

Self-field ac losses in dc external fields and non-uniform transverse Jc distribution in Ag sheathed multifilamentary PbBi-2223 tapes

Timothy J. Hughes, Yifeng Yang, Carlo Beduz and Pradeep Haldar[1]

Institute of Cryogenics, University of Southampton, Southampton SO17 1BJ, UK
[1]Intermagnetics General Corperation, Latham, NY 12110, USA

Abstract Measurements of self-field ac losses in dc external fields were carried out on Ag sheathed multifilamentary PbBi-2223 tapes at different stages of thermal processing. Using a novel method as described in [1,2], the transverse distributions of the local critical current density *Jc* were deduced from the measured losses as a function of the ac transport current. Compared with data obtained from mechanical sectioning [3] of a sample, our results showed in greater detail the *Jc* distribution in the outer region of the core, which carries most of the critical current. The non-destructive nature of the method allows the monitoring of the evolution of local *Jc* at different processing times. For example, the *Jc* in the outer region exhibits a faster deterioration when the tape is processed beyond the optimal time. It is also observed that the local critical current density in the outer area of the core has a stronger dependence on the applied magnetic field in both parallel and perpendicular directions, compared with the overall *Ic-B* dependence obtained from the dc measurement.

1. Introduction

Measurements of the self-field a.c. losses, of Ag sheathed PbBi2223 tapes, have shown that the characteristics of these losses can vary from tape to tape [4]. This variation can be interpreted in terms of different cross-section geometry of the superconducting core. However recent work [1,2] has shown that such a variation in loss is probably the result of a non-uniform distribution in the local *Jc* through the core. As well as providing an explanation of the different loss behaviour between different tapes it is possible to obtain the local *Jc* distribution from measurements of the self-field losses of the tape [1,2]. This allows a non-destructive determination of the local $Jc(x)$ which is particularly useful in monitoring the evolution of *Jc* during thermal processing. In this paper we have used the measurements of the losses to determine the local current distribution within a tape during the final sintering stages. Measurements of the losses in various dc magnetic fields is used to examine the field dependence of the local critical current density.

2. Experimental procedure.

A length of partially processed, Ag-sheathed 37 core PbBi2223 tape was supplied by IGC. To prevent degradation during thermal cycling the tape was cut into two samples, one to be used for the measurements in d.c. fields perpendicular to the tape surface and the other to be used for the parallel field. In order to monitor the evolution of the *Jc* within the core, the same piece of tape is used through the processing programme for each field direction. The voltage taps are attached with silver paint, which can easily be removed after measurement of the losses leaving no contamination during further processing. Four different sintering times were used, 0, 17.5, 35 and 102 hours. After each heat treatment the losses and critical current were measured. A similar procedure was used for the samples measured in a number of d.c. magnetic fields but with only three processing times (0, 25 and 50 hours). From the loss characteristics the local current distribution $Jc(\xi)$ can be determined [1,2], where ξ defines an ellipse $x^2+(y/b)^2=\xi^2$ within the core which is homologous to the core periphery,. An average critical current over the tape thickness $Jc(x)$ can be found as a function of the tape width x as follows

$$Jc(x) = \frac{1}{b\sqrt{a^2-x^2}} \int_0^{b\sqrt{a^2-x^2}} Jc\left(\sqrt{x^2+(y/b)^2}\right)dy$$

See insets to Figure 1 for description of axes. This average critical current density $Jc(x)$ is equivalent to the local current density obtained by mechanical sectioning of the core [3].

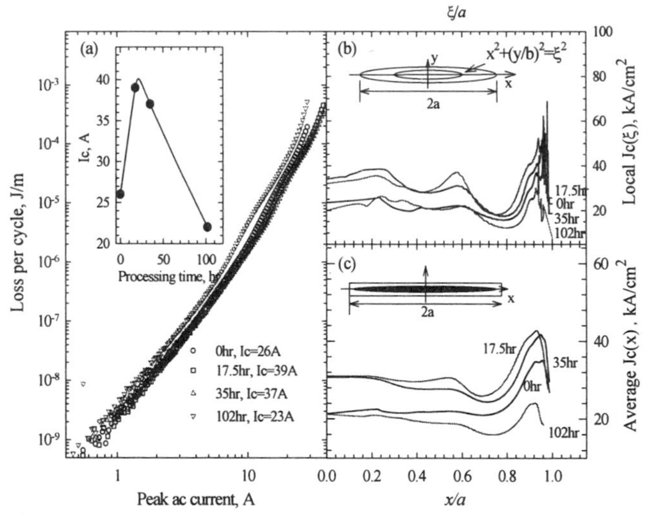

Figure 1a-c. The loss per cycle as a function of the a.c. transport current for each sintering time is shown in (a). The corresponding local $Jc(\xi)$ distributions and the average $Jc(x)$ distributions are shown in (b) and (c) plotted for several cooking times (0, 17.5, 35 and 102 hours). The inset shows the critical current as a function of cooking time.

3. Results

3.1. Evolution of the local Jc during processing.

The effect of the heat treatment of the partially processed tapes can be seen in the inset to Figure 1a with a maximum critical current seen for around 25 hours processing time. The self-field loss characteristic, for each processing time, is shown in Figure 1a. The corresponding local $Jc(\xi)$ distribution and the average $Jc(x)$ are shown in Figure 1b-c respectively. Examination of the graph reveals that after 17 hours further processing there is a general increase in the *Jc* at all points in the core.. Continued cooking (35 hours and 102 hours) acts to degrade the *Jc* throughout most of the tape with the largest percentage decrease seen near the edge. It is noted that the highest current in this tape is near but not immediate to the silver/core interface.

3.2. The local Jc-B dependence.

For three different processing times (0hrs, 25hrs and 50hrs) the losses were measured in various d.c. magnetic fields (both perpendicular and parallel orientation). At each magnetic field the local $Jc(\xi)$ distributions and the corresponding average $Jc(x)$ distribution were calculated from the loss characteristics. The data for the parallel field can be seen in Figure 2a-f, similar data was obtained for the perpendicular direction but is not shown here. It is evident that there is a greater reduction in *Jc*, for the same d.c. field, near the silver/core interface than inside the core. This feature remained unchanged by further thermal processing.

From the data shown in Figure 2 and those for the perpendicular field, the averaged critical current density at x/a=0, 0.5, 0.8, 0.9 are shown as a function of the applied d.c. field in Figure 3a (parallel) and Figure 3b (perpendicular). While the points show the actual data, the shaded region illustrates the spread in the reduced local *Jc*(*B*) across the core. It should be noted that such a significant difference in local *Jc*-B dependence has not previously been seen in measurements by mechanical sectioning. It is not clear why there is a stronger field dependence near the Ag/core interface, where the *Jc* is higher. However the fact that this

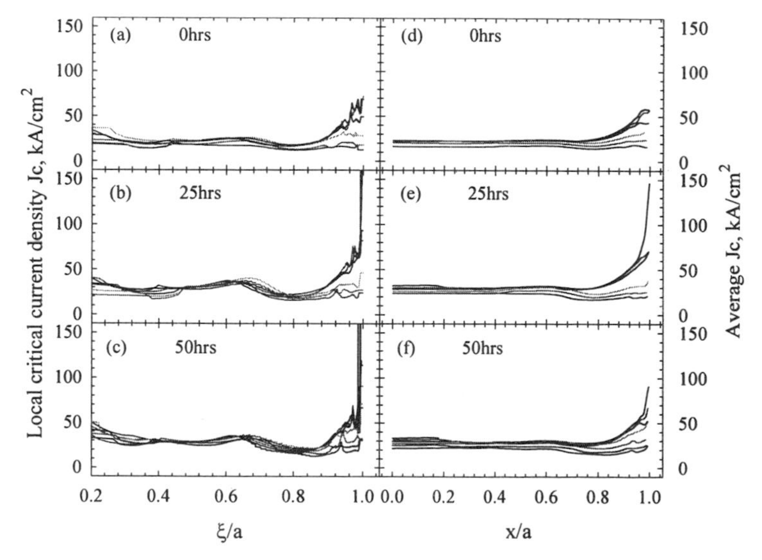

Figure 2a-f. The distributions for the local $Jc(\xi)$ (a-c) and the average $Jc(x)$ (d-f) are shown for several parallel d.c. magnetic fields for three different sintering times (a,d) 0 hours, (b,e) 25 hours and (c,f) 50 hours.

effect is present in both the parallel and perpendicular field directions suggests that it is unlikely to be due to the perpendicular component of the self-field near the edges. Instead it points to an intrinsic mechanism related to the nature of the grain boundaries.

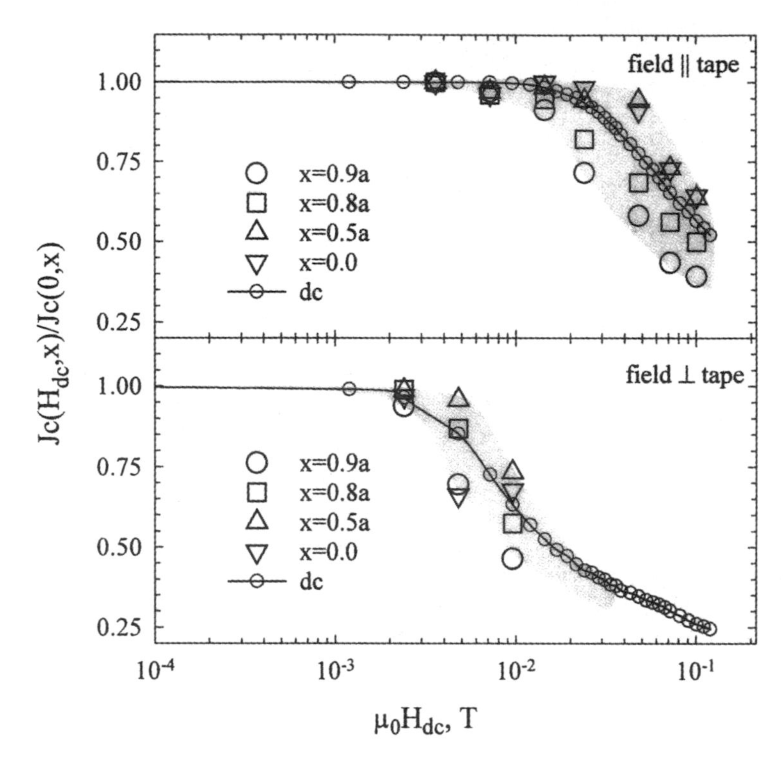

Figure 3a-b. The local values of the reduced *Jc-B* dependence *Jc*(*H*)/*Jc*(0) are shown for positions x=0, 0.5, 0.8 and 0.9 for both the parallel (a) and perpendicular (b) field directions. The reduced d.c. *Ic-B* dependence is also shown for reference.

4. Conclusions

We have demonstrated that the local critical current distribution across the core of Ag sheathed 2223 tapes can be obtained non-destructively from self-field loss measurements. Using this technique we have shown the evolution of the *Jc* distribution during processing and the different *Jc-B* dependence between the Ag/core interface and the inner core. Although the exact mechanisms responsible for such a difference have not been determined, they are likely to be linked to the nature of grain boundaries.

References

[1] Yang Y., T. Hughes and C. Beduz, *Physica C* (1997) in press
[2] Y. Yang, T, Hughes and C. Beduz, Czech J. Phys. 46 (1996) 1802
[3] Hansel B, Grasso G and Flukiger R, Phys. Rev. B 51 (1995) 15456
[4] Y. Yang, T. Hughes, S.M. Spiller, C. Beduz, M. Penny, R.G. Scurlock, P. Halder and R. Sokolowski, *Supercon. Sci. Technol.* **9** (1997) 801

Inst. Phys. Conf. Ser. No 158
Paper presented at Applied Superconductivity, The Netherlands, 30 June–3 July 1997

AC loss in Y-123/hastelloy superconducting tapes

Karl-Heinz Müller[1]

CSIRO, Division of Telecommunications and Industrial Physics, PO Box 218, Lindfield NSW 2070, Australia

Abstract. Hysteresis loss in the superconductor and eddy current loss in the metal of a Y-123/hastelloy flexible tape are calculated analytically in the low frequency limit. Cases studied are (i) where an ac transport current is passed through the tape and (ii) where an external ac magnetic field is applied perpendicular to the tape.

1. Introduction

A significant step towards the fabrication of flexible superconducting tapes made of Y-123/hastelloy with high critical current density J_c was made by Iijima *et al.* [1] and Reade *et al.* [2]. Here, a YBaCuO film is deposited onto a biaxially textured yttria stabilized zirconia buffer layer grown by ion-beam-assisted deposition on a piece of hastelloy (Ni-alloy) strip. This method was later refined by Wu *et al.* [3]. The J_c's achieved are as high as 10^6 $\mathrm{Acm^{-2}}$ at 77 K and the high J_c stays intact in applied fields up to several tesla. These tapes also show excellent bending strain tolerances. The ac losses in these Y-123/hastelloy tapes will be a key consideration in the design of ac power cables, motors and generators and other ac applications. In this paper formulae for the hysteresis loss and the eddy current loss in a Y-123/hastelloy tape are presented.

2. Hysteresis loss and eddy current ac loss in a Y-123/hastelloy tape

Let us consider a metal-superconductor composite made of a metal substrate in the shape of a very long strip of width $2s$ and thickness d_m, and centred on top, a superconducting film of width $2a$ and thickness d_{sc}, where $d_{sc} < d_m << a < s$.

The total ac loss P_{tot} of such a composite is given by $P_{tot} = P_{hy} + P_{ed}$. Here, P_{hy} is the hysteresis loss in the superconductor and P_{ed} the eddy current loss in the metal. The hysteresis loss P_{hy} is given by

$$P_{hy} = f\ J_c \oint dt \int_{V_{sc}} \mid \vec{E}(\vec{r},t) \mid \ d^3r \ , \tag{1}$$

where f is the cycle frequency, J_c the critical current density of the superconducting film and $\vec{E}(\vec{r},t)$ the time-varying electric field distribution inside the superconductor volume V_{sc}. The symbol $\oint$ indicates time integration over an ac cycle.

[1] E-mail: karl@dap.csiro.au.

The eddy current loss is given by

$$P_{ed} = \frac{f}{\rho} \oint dt \int_{V_m} E^2(\vec{r},t)\ d^3r \ , \tag{2}$$

where ρ is the resistivity of the metal and $\vec{E}(\vec{r},t)$ the time varying electric field distribution inside the metal volume V_m.

2.1. Self-field ac loss of a normal-metal-superconductor strip

Let us consider the case where an ac transport current, $I(t) = I_m \cos(2\pi f t)$, is passed through the normal-metal-superconductor strip with $I_m < I_c = 2ad_{sc}J_c$. We assume that the ac frequency f is small so that the magnetic field originating from induced eddy currents is insignificant. This is the case if

$$2\mu_o\ f\ s\ d_m\ /\ \rho << 1 \ . \tag{3}$$

Under the constraint of Eq. (3), the hysteresis loss is not affected by the metal and one obtains for the hysteresis loss per unit length [4]

$$\frac{P_{hy}}{l} = \frac{\mu_o}{\pi}\ f\ I_c^2\ [\ (1-\eta)\ln(1-\eta) + (1+\eta)\ln(1+\eta) - \eta^2\ ]\ ,\quad \eta = \frac{I_m}{I_c} \tag{4}$$

To derive the eddy current loss P_{ed}, the electric field $\vec{E}(\vec{r},t)$ in the metal has to be determined. Because of Eq. (3) and the extreme flatness of the metal-superconductor strip, one obtains from Faraday's law for the electric field the expression $E_y(x,t) = -\mu_o \frac{d}{dt} \int_0^x H_z(x,t)dx$. Here, $H_z(x,t)$ is the perpendicular component of the magnetic field at the surface of an isolated superconducting strip. An analytical expression for $H_z(x,t)$ has been derived in Refs. [5,6].

Applying Eq. (2) and the Eq. (2.6) of Ref. [5] , one derives for the eddy current loss P_{ed}/l per unit length [7]

$$\frac{P_{ed}}{l} = \frac{16\ \mu_o^2}{\pi}\ \frac{d_m\ a\ f^2}{\rho}\ I_c^2\ S(\eta,\frac{s}{a})\ , \tag{5}$$

where

$$S(\eta,\frac{s}{a}) = \int_0^\eta \sqrt{\eta u - u^2}\ [\ \frac{s}{a} - \sqrt{\frac{s^2}{a^2} - u^2} - \frac{u}{2}\ln\frac{s/a+u}{s/a-u} + \frac{s}{8a}(\ln\frac{s/a+u}{s/a-u})^2\ ]\ du\ . \tag{6}$$

Fig. 1 shows the self-field loss $P_{tot}/(lf)$ versus current amplitude I_m at different frequencies f. The parameters used are $\rho = 1.24 \times 10^{-6}\Omega$m, $s = 0.7$ cm, $a = 0.5$ cm, $d_m = 100$ μm and $I_c = 100$ A. The eddy current loss becomes important for $f > 1$ kHz. The curve labelled 0 Hz represents the hysteresis loss.

2.2. AC loss of a metal-superconductor strip in a perpendicular ac magnetic field

Let us now consider the case where an ac magnetic field, $H_a(t) = H_m \cos(2\pi f t)$, is applied perpendicular to the metal-superconductor strip. Assuming the condition of Eq. (3), the hysteresis loss is not affected by the presence of the metal and one obtains [5]

$$\frac{P_{hy}}{l} = 4\pi\ \mu_o\ f\ a^2\ H_m^2\ [\ \frac{2}{\xi}\ \ln(\cosh(\xi)) - \tanh(\xi)\ ]\ /\ \xi\ , \tag{7}$$

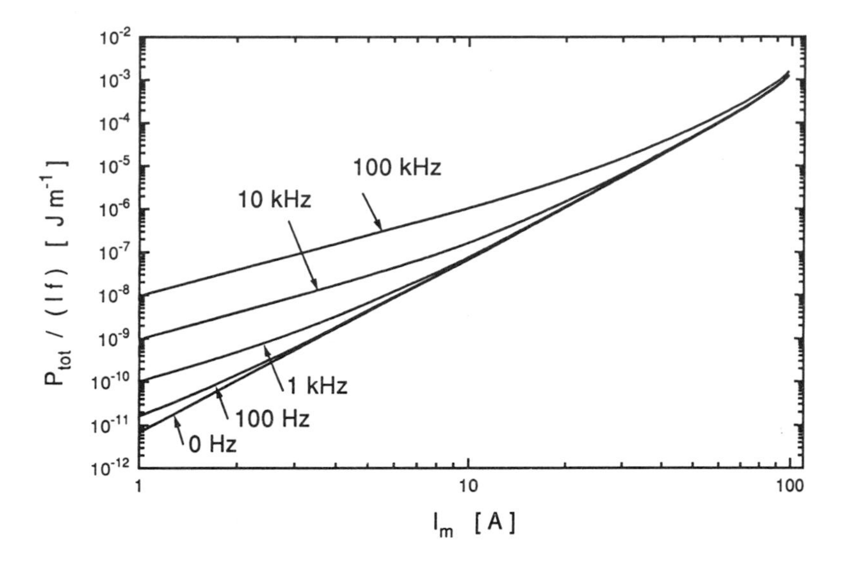

Figure 1. Total self-field loss $P_{tot}/(lf)$ per unit length per current cycle versus current amplitude I_m at different frequencies f where $s/a = 1.4$.

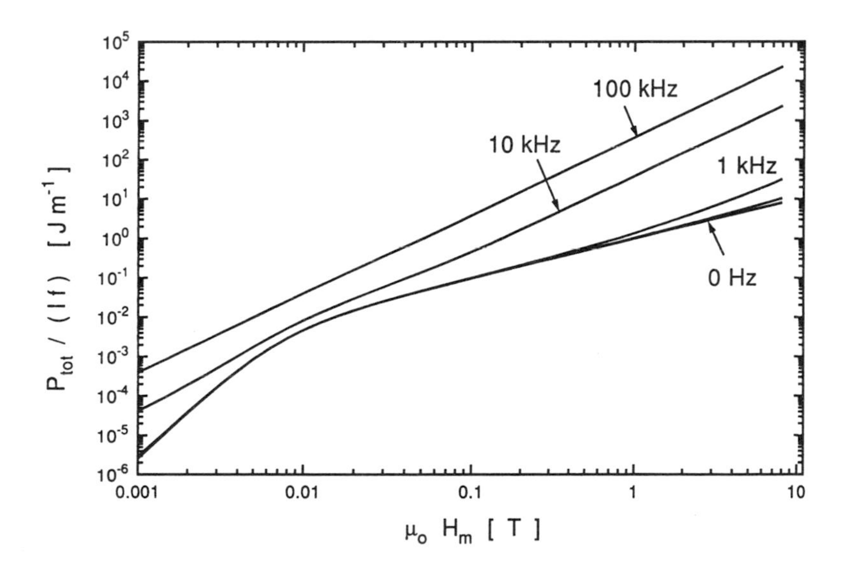

Figure 2. Total loss $P_{tot}/(lf)$ per unit length per field cycle versus the applied field amplitude H_m at different frequencies f. The field is applied perpendicular to the tape and s/a =1.4 .

where $\xi = H_m/H_d$ and $H_d = J_c d_{sc}/\pi = I_c/(2\pi a)$.

Applying Eq. (2) and Eq. (3.9) of Ref. [5], one obtains [7]

$$\frac{P_{ed}}{l} = \frac{4\pi^2\ \mu_o^2}{3}\ \frac{d_m\ a^3}{\rho}\ f^2\ H_m^2\ Q(\xi,\frac{s}{a})\ , \tag{8}$$

where

$$Q(\xi,\frac{s}{a}) = \frac{8}{\pi\xi^2}\int_0^{\xi}(\xi u - u^2)^{1/2}\ (\frac{s^3}{a^3} + \frac{3s/a - 6}{\cosh^2 u} + \frac{2}{\cosh^3 u})\ du\ . \tag{9}$$

Fig. 2 shows $P_{tot}/(lf)$ versus field amplitude H_m at different frequencies f. The parameters used are those of Fig. 1. The curve labelled 0 Hz represents the hysteresis loss.

3. Conclusions

Because of the large resistivity of hastelloy, the eddy current loss can be neglected at ac frequencies below 100 Hz and only the hysteresis loss contributes to the total ac loss. The eddy current loss decreases with decreasing s and is smallest when $s = a$.

References

[1] Iijima Y, Tanabe N, Kohno O and Ikeno Y 1992 *Appl. Phys. Lett.* **60** 769-771; Iijima Y, Onabe K, Futaki N, Tanabe N, Sadakata N, Kohno O and Ikeno Y 1993 *J. Appl. Phys.* **74** 1905-1911
[2] Reade R P, Berdahl P, Russo R E and Garrison S M 1992 *Appl. Phys. Lett.* **61** 2231-2233
[3] Wu X D, Foltyn S R, Arendt P N, Blumenthal W R, Campbell I H, Cotton J D, Coulter J Y, Hults W L, Maley M P, Safar H F and Smith J L 1995 *Appl. Phys. Lett.* **67** 2397-2399
[4] Norris W T 1970 *J. Phys. D* **3** 489-507
[5] Brandt E H and Indenbom M 1993 *Phys. Rev. B* **48** 12893-12906
[6] Zeldov E, Clem J R, McElfresh M and Darwin M 1994 *Phys. Rev. B* **49** 9802-9822
[7] Müller K-H 1997 *Physica C* (in press)

Inst. Phys. Conf. Ser. No 158
Paper presented at Applied Superconductivity, The Netherlands, 30 June–3 July 1997

Computational study of magnetisation losses in HTS composite conductors having different twist-pitch lengths

M Lahtinen, J Paasi, R Mikkonen, J-T Eriksson

Laboratory of Electricity and Magnetism, Tampere University of Technology, P.O. Box 692, FIN-33101 Tampere, Finland

Abstract. We have computed magnetisation losses of HTS conductors having different twist-pitch lengths. The computation is based on Maxwell's equations and a non-linear constitutive relation between the current density and the electric field. The computational approach allows us to take into account matrix resistivity, twist-pitch length and the gradual resistive transition of a real composite conductor. The losses are studied as a function of the ramp rate of the external flux density in the longitudinal midsection of a superconducting composite tape.

1. Introduction

Evaluation of AC losses is essential for the proper design of superconducting magnets and their cooling systems. In this paper we present magnetisation losses computed using a 2D magnetic diffusion model, which enables us to account for the gradual resistive transition of HTS materials. We will point out how the computed loss characteristics differ from standard results, which are based on the assumptions of the Bean critical state model.

2. Computation model

The low frequency electrodynamics of a composite superconductor can be described by Maxwell's equations and constitutive laws

$$\nabla \times \mathbf{E} = -\frac{\partial \mathbf{B}}{\partial t}, \quad (1a) \qquad \mathbf{E} = \rho_m \mathbf{J}, \quad (1b)$$

$$\nabla \times \mathbf{H} = \mathbf{J}, \quad (1c) \qquad \mathbf{E} = \rho_S(J)\mathbf{J}, \quad (1d)$$

where **E** is the electric field, **J** is the current density, ρ_m is the resistivity of the matrix metal and $\rho_s(J)$ is the resistivity of the superconducting filaments. When the reversible surface current is neglected, the magnetic field **H** is related to the magnetic flux density according to $\mathbf{B}=\mu_0\mathbf{H}$, where μ_0 is the vacuum permeability. For $\rho_s(J)$ we use a power-law approximation

$$\rho_S(J) = E_C \frac{J^{n-1}}{J_C{}^n}, \quad (2)$$

where n is the index number of the superconductor and the critical current density of the material is defined at the point (J_C,E_C). We call the material described by the $E(J)$-curve given by Eqs. 1d and 2 a power-law superconductor. We neglect the possible granularity of the superconductor as well as the contact resistance between the matrix metal and the superconducting filaments.

Figure 1(a) shows a cutaway view of a multifilamentary superconducting tape in external magnetic field applied parallel to the flat surface of the tape. We have formulated a 2D magnetic diffusion model for the electromagnetic fields in the longitudinal midsection of the tape [1]. In the model the current flows in the xy-plane and the magnetic field is parallel to the z-axis. Taking into account the field directions and Eqs. 1(a)-(d) we obtain a 2D diffusion equation for the z-component of the magnetic field H_Z

$$\nabla\cdot(\rho(J)\,\nabla H_Z)=\mu_0\frac{\partial H_Z}{\partial t}, \tag{3}$$

where $\rho(J)$ is defined by $\rho_m=2.5\times10^{-9}$ Ωm in the matrix and by Eq. 2 with n=10, J_C=30 kA/cm^2 and E_C=1 μV/cm in the superconducting filaments. The parameters are chosen to describe the operation of a Bi-2223/Ag tape at 77 K.

Figure 1(b) shows the midsection, where the model is applied. Equation 3 is solved in the midsection using a commercial finite element code for electromagnetic field computation [2]. A mesh consisting of 4800 linear elements is used for the discretisation of one quadrant of the midsection. We present results for four model conductors labelled A-D: one with straight filament configuration (A) and three twisted conductors with twist-pitch lengths l_{tp}=4 mm (B), l_{tp}=12 mm (C) and l_{tp}=40 mm (D). We have assumed that in the model all the crossing-points of the twisted filaments are on the boundaries parallel to the x-axis and the magnetisation current loops are confined to one half of the twist-pitch length. Therefore l_{tp} is defined by setting $H_Z=H(t)$ on the boundaries of the model conductor, Fig. 1(c). The function $\mu_0 H(t)$ is a ramp during which the external flux density is increased from zero to B_m=1 T as shown in the inset of Fig. 2. For $l_{tp}=\infty$ we set $\partial H_Z/\partial\mathbf{n}=0$ on the boundaries parallel to the x-axis, $\mathbf{n}$ being a unit vector normal to the boundary, Fig. 1(d).

We present the magnetisation losses as the energy density dissipated in a unit volume per one ramp of external flux density. The loss density is obtained from the integral

$$Q=\frac{1}{A}\int_0^T\int_A \mathbf{E}\cdot\mathbf{J}\;da\,dt, \tag{4}$$

where A is the composite cross-section in the xy-plane and T is the duration of the ramp. In the computation we will assume that the composite is always in the state of full penetration.

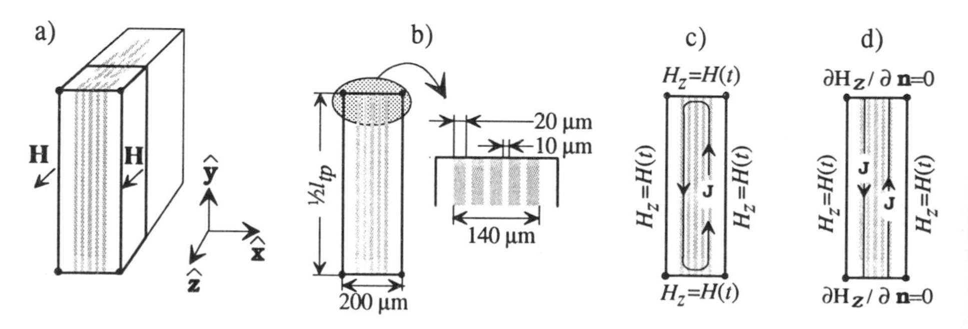

Figure 1 Composite geometry and field directions in the 2D magnetic diffusion model. A cutaway view of the composite (a), the filament configuration (b), the boundary conditions defining the magnetisation current in a conductor with a twist-pitch length l_{tp} (c), and in a conductor with straight filament configuration (d).

3. Results and discussion

Figure 2 shows the computed energy loss density characteristics as a function of the ramp rate of the external flux density dB/dt. At dB/dt=1 mT/s the filaments of the twisted conductors B-D are electrically decoupled whereas conductor A with straight filament configuration behaves like a monofilamentary conductor having an effective filament thickness of 140 μm. The magnetisation losses of conductors B-D are clearly lower than the losses of conductor A. Frequently the losses in decoupled filaments are estimated by assuming that the magnetic flux and current density distributions can be described by the Bean critical state. For a trapezoidal excitation field such an assumption leads to hysteresis losses per cycle $Q_h = B_p B_m/\mu_0$, where B_p is the field of full penetration and B_m is the maximum flux density [3]. For the present purpose we write

$$Q_h = \frac{aJ_C B_m}{4}, \tag{5}$$

where a is one half of the filament thickness and the factor ¼ accounts for the filling factor of the model conductor (50 %) and for the fact that we are considering a ramp of external field instead of a complete cycle. Let us now calculate the losses in the model conductor at dB/dt=1 mT/s using Eq. 5. Values a=10 μm, J_C=30 kA/cm^2 and B_m=1 T give Q_h=750 J/m^3, which is larger than the computed value $Q \approx 290$ J/m^3 for conductors B-D at dB/dt=1 mT/s. The explanation for the apparent contradiction is related to the electric field acting on the filaments. For dB/dt=1 mT/s the electric field on the filament surface can be approximated by

$$E = a\frac{dB}{dt}, \tag{6}$$

which gives $E = 1\cdot 10^{-4}$ μV/cm, a value essentially lower than E_C=1 μV/cm used for the definition of J_C. Inside the filaments E is even smaller. Because E and J are related through Eqs. 1(d) and 2, the current density close to the filament surface is lower than J_C by a factor of $(E/E_C)^{1/n}$ which explains the difference between the computed Q and Q_h given by Eq. (5).

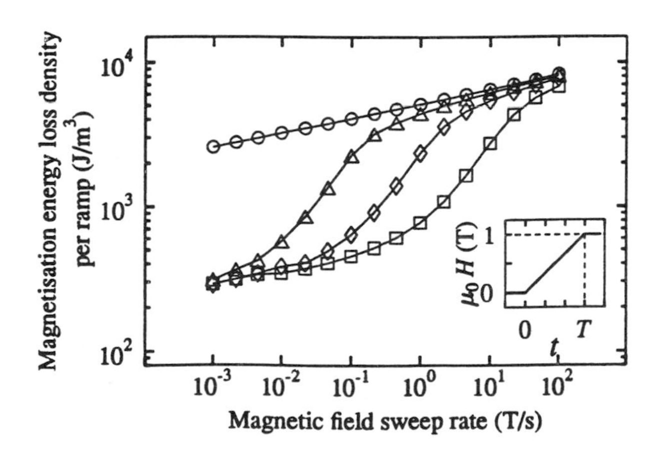

Figure 2 Magnetisation energy loss density per one ramp of magnetic field for conductor A $l_{tp}=\infty$ (O), conductor B l_{tp}=4 mm (□), conductor C l_{tp}=12 mm (◊) and conductor D l_{tp}=40 mm (Δ). The magnetic field ramp $\mu_0 H(t)$ is shown in the inset. The ramp rates 0.001 T/s<dB/dt<100 T/s are obtained by using ramp durations 0.01 s<T<1000 s.

As dB/dt increases the magnetisation losses of all the model conductors increase. In conductor A with straight filament configuration the increase is related to the increase of the current density in the filaments with increasing dB/dt. However, in the twisted conductors B-D the increase of losses is mainly due to gradually increasing filament coupling. As expected, the onset of the coupling will take place at progressively larger values of dB/dt as the twist pitch length decreases. For l_{tp}=40 mm the filament coupling becomes significant at $dB/dt\approx$ 10 mT/s, whereas for l_{tp}=12 mm and l_{tp}=4 mm the coupling starts at $dB/dt\approx$100 mT/s and $dB/dt\approx$1 T/s respectively. When 10 mT<dB/dt<10 T/s the computed loss characteristics of conductors B-D are difficult to interpret using analytical expressions for coupling losses.

For $dB/dt\approx$10-100 T/s filament coupling is strong in all conductors independent of the filament configuration. When the filaments are completely coupled magnetisation losses are high ~7 kJ/m^3 which emphasises the importance of a short twist-pitch length and high transverse resistivity of composites designed for low loss operation in AC magnetic fields.

The magnetisation losses are an important factor in the operation of a SMES magnet, especially if the magnet is cooled by a cryocooler. In applications requiring rapid charging and discharging cycles, such as flicker mitigation, the magnetisation losses produce a considerable amount of heat. If the heat load exceeds the available cooling power, the operating temperature of the magnet starts to rise. Therefore the minimum allowable cycle time together with the maximum stored energy of the magnet set an upper limit to the electric power that can be handled by a SMES system under continuous operation. The computed results will help us to estimate when the magnetisation losses give rise to a thermal instability in a given system. In the present work we have assumed that the flat face of the tape is always parallel to the external field. If the magnetic field perpendicular to the flat face of the tape becomes significant, the magnetisation losses of the tape must be re-evaluated.

4. Conclusions

We have computed magnetisation losses of composite superconductors described by a power-law $E(J)$-curve and having different twist-pitch lengths. Conventionally the losses in multifilamentary composites have been divided into hysteresis, coupling and eddy-current losses. In a power-law superconductor there is an additional loss mechanism, which arises due to the gradual resistive transition of the material. Namely in the filaments both the electric field and the current density increase with increasing dB/dt, which leads to frequency dependent hysteresis losses. This reasoning also leads to the conclusion that in a power-law superconductor the hysteresis losses are not exactly proportional to the filament thickness.

References

[1] Lahtinen M, Paasi J and Kettunen L 1996 *Cryogenics* **36** 951-956; Paasi J and Lahtinen M Applied Superconductivity Conference (ASC'96) August 25-30, 1996, Pittsburgh, PA, *IEEE Trans. Appl. Supercond.* in press

[2] Vector Fields Limited, 24 Bankside, Kidlington, Oxford, UK

[3] Wilson M N 1983 *Superconducting Magnets*, Clarendon Press, Oxford, Chap. 8; Iwasa Y 1994 *Case Studies in Superconducting Magnets*, Plenum Press, New York, Chap. 7

Inst. Phys. Conf. Ser. No 158
Paper presented at Applied Superconductivity, The Netherlands, 30 June–3 July 1997

Superconducting cables in spatially changing magnetic fields near the end portions of magnetic systems

Silvester Takács

Institute of Electrical Engineering, Slovak Academy of Sciences, 842 39 Bratislava, Slovakia

Abstract. From basic considerations for the electromagnetic coupling of different current loops in superconducting cables, the general diffusion equation for the magnetic field penetration is derived. It is shown that the model is identical to the network model used in computer calculations for the current distribution in the strands of the cable. The obtained equations enables us to consider also the effects connected with the spatial inhomogeneities in the applied magnetic field, in the contact resistance between different strands, as well as in geometrical properties along and across the cable. We focus then our attention to the effects caused by the inhomogeneous magnetic fields, especially at the ends of the magnet winding where the field changes considerably in most magnet systems. Some general results are given for the additional AC losses, the induced currents in the strands and their time constants. These results are compared with the properties of cables in homogeneous applied magnetic field.

1. Introduction

There are many examples in physics where small spatial and/or temporal changes of external parameters or of internal properties can cause considerable effects. In microscopic properties, these are the fluctuations close to phase transitions. In superconductors the influence of the spatial inhomogeneities on current-voltage characteristics is known, and in superconducting cables those are the spatial changes of applied field or the inhomogeneous current distribution along superconducting cables. The first calculations of coupling losses in spatially changing AC magnetic fields were performed for flat and round cables a long time ago [1,2]. Both for cables being partially in magnetic field [1] and for cables in periodically changing fields along the cable [2], they lead to very surprising results, mainly in quantitative sense. The loss generation in the field region is increased compared with homogeneous applied field, the loss density being a strongly oscillating function of the length of the field region. There is an extreme increase in losses if the characteristic length of the field change (or some of its Fourier components) is an odd multiple of the cabling pitch. Long-living currents in the latter case are generated [2], which are even non-decaying for an infinite cable. The inhomogeneities in the current distribution and in material parameters are analogous [3,4].

However, it is expected that the long-living supercurrents play the most important role on the stability of the conductors and magnets due to two effects: the local increase of AC losses, as the additional loss is generated in a small volume, as well as by the induced currents in the strands, decreasing their critical current.

A well-known example of spatial field change is for cables close to the magnet ends, where the applied field is reduced essentially to zero on a small length. After formulating the problem, we solve the corresponding diffusion equation for this case here.

2. The flux diffusion equation and solutions

At first, we would like to show the link between our continuum-like model [1,2,5] and the circuit-like models. For the induced currents between neighboring strands, the complicated system of coupled integro-differential equations can be simplified to one diffusion equation with effective inductance for the current loops [1,2]. Analogous picture can be used for transverse losses between strands on the opposite side of cable [5-7]. Based on the scheme of Morgan [8], the Faraday's equations for all loops between crossing strands are applied in the circuit models with four resistance points for circuits in the inner part and three at the edges (figure 1). The resulting system of equations is very complex and usually solved by computer [3,9,10]. However, one can consider only the crossing points of larger loops between any two strands (figure 1). The flux change $\dot{\Phi}_{i,k}$ is always with respect to the left corner of the loops [9], the induced voltages being $E_{i,k} = \rho_{i,k} j_{i,k}$. Then, $\dot{\Phi}_{1,k} = E_{1,k} + E_{1,k+1} - E_{2,k}$, $\dot{\Phi}_{2,k} = E_{2,k} + E_{2,k+1} - E_{3,k+1} - E_{1,k+1}$, $\dot{\Phi}_{3,k} = E_{3,k+1} + E_{3,k+2} - E_{4,k+1} - E_{2,k+1}$, $\dot{\Phi}_{4,k} = E_{4,k+1} + E_{4,k+2} - E_{5,k+2} - E_{3,k+2}$, $\dot{\Phi}_{5,k} = E_{5,k+2} + E_{5,k+3} - E_{4,k+2}$. The flux change in the loop formed by two neighboring strands (marked with squares) is $\dot{\Phi}_5 = E_{1,k} + E_{5,k+3}$. The induced voltage is therefore given by the area of loops between two strands (large "diamonds") and the resistivities at their crossing points only [11-14]. Hence, one can develop a continuum like model for the flux diffusion into strands [1,2], including even the end effects for finite cables [4].

We consider a flat superconductor cable with one-layer strands on the upper and lower side [14]. The layer between them (thickness c) may contain the surface layers on the strands only (like the helical conductor for LHD [15], or many cable-in-conduit cables) or some additional normal and/or insulating layer. We are dealing with the induced currents J in the strands due to the induced component. The transport current is supposed to be decoupled from these currents which are supposed to be essentially surface currents. At the crossing points, they are closed through the resistive layer $c << b$, the cable width b ($-b/2 \leq x \leq b/2$) is assumed as $b << l_0$ (the cabling pitch). The corresponding transverse current through the resistive layer is $j_t = \partial J/\partial y$ and the induced voltage $U = \rho c j_t$, where ρ is the resistivity of layer. The cable of length $l = \lambda l_0/\pi$ ($-l/2 \leq y \leq l/2$) is exposed to harmonically changing magnetic field $B_e = h \exp(i\omega t)$, spatially constant in the central part (figure 2) with length $l' = K l_0/\pi$ ($-l'/2 \leq y \leq l'/2$). The transition length is $a = k l_0/2\pi$. The individual strands are numerated by the parameter $\alpha_n = 2\pi(n-1)/N$. The strand $n = 1$ is at cable centre for $y = 0$.

The diffusion equation for the strand currents $J \sim \exp(i\omega t)$ in the field as shown in figure 2 is given by

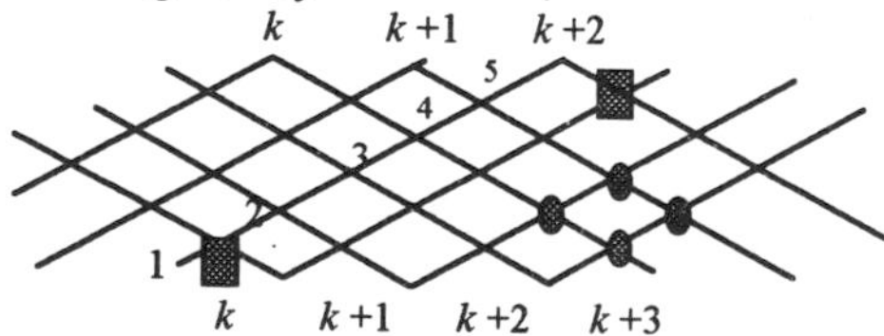

Figure 1. The circuit model and crossing points of strands. Instead of all crossing points in elementary loops (marked with small circles) within the circuit model, we consider in our continuum like-model the currents flowing in the strands superconductively, until they cross the same strand for the next time (marked with squares for two neighboring strands). As shown in the text, both models are equivalent.

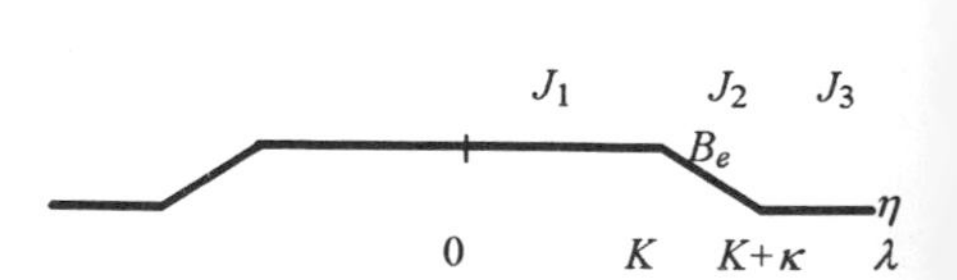

Figure 2. The field distribution along the cable in magnet. The central part is the field generated by the magnet, the linear decrease at the ends represents the conductor in the stray field and the zero level the leads. All length are related to the cabling pitch l_0. The currents J_1, J_2, J_3 are the currents in the strands (assumed to be surface currents) in the individual regions.

$$\frac{\partial^2 J}{\partial \eta^2} = \mathrm{i} m J + \beta \frac{\partial U}{\partial \eta}$$

where $\eta = 2\pi y / l_0$, $\beta = \mathrm{i}\, h\, \omega\, b\, (l_0/2\pi)^2/\rho\, c$, $m = \mu_0 l (l_0 / 2\pi)^2 \omega L_{eff} / 2\pi \rho c$ with $L_{eff} \approx 2\pi$ [16] and $U = \cos(\eta - \alpha_n)$, $(1 + K/\kappa - \eta/\kappa)\cos(\eta - \alpha_n)$ and 0 in regions 1, 2, 3, respectively. As in the case of other spatially changing magnetic fields [14], we take only the first term of the step function

$$\frac{4}{\pi} \sum_{k=1}^{\infty} \frac{(-1)^{k+1} \cos[(\eta - \alpha_n)(2k+1)]}{2k+1}$$

for the flat cable. The solutions for the strand currents in different regions are then

$$J_1 = A_1 \sin \alpha\eta + \beta_1 \sin(\eta - \alpha_n)$$

$$J_2 = A_2 \sin \alpha\eta + B_2 \cos \alpha\eta + \beta_1 (K/\kappa + 1 - \eta/\kappa) \sin(\eta - \alpha_n) - \beta_2 \cos(\eta - \alpha_n)$$

$$J_3 = A_3 \sin(\alpha - \eta)$$

where the constant A_i, B_2 are determined by the continuity of currents and $\beta_1 = \beta/(1 + \mathrm{i}\, m)$, $\beta_2 = \beta_1(1 - \mathrm{i}\, m)/(1 + \mathrm{i}\, m)$, $\alpha = (1 - \mathrm{i})\, (m/2)^{1/2}$. We suppose for this solutions an ideal transposition of all strands and constant currents coming from the leads.

3. Results

The constants determining the solutions for J, j_t and thus the coupling losses, as well as other parameters of the superconducting cables, are a bit complicated. We give here some results in the low frequency limit only. Due to the spatial inhomogeneity, the coupling losses increase [14]. These additional losses differ only slightly from the mean squared field value in the region 2. The additional losses in the regions 1 and 3 are approximately given by

$$\frac{P_1}{P_\infty} = 1 + \frac{2}{\lambda} \{\sin K[\cos(K + \kappa) - \cos K] + (\cos K - 1)[\sin(K + \kappa) - \sin K)]\}$$

$$\frac{P_3}{P_\infty} = \frac{2(\lambda - K - \kappa)}{\kappa^2} \sin^2 \frac{\kappa}{2}$$

where P_∞ is the loss density in homogeneous applied field of amplitude h. Whereas in the first region (with constant applied field) the additional losses are negligible, the losses in the "field-free" region cannot be fully neglected, having maxima at $a = l_0\ (n + 1/2)$. It is interesting to note that the loss density in the field free region of the cable is constant and depends on the value of a (or k in the dimensionless units), on which the field is changing. All these results are valid strongly speaking in the zero frequency limit only, where the field diffusion length $l_d = l_0 / \pi (2\omega\tau)^{1/2}$ is very large. Due to the large value of the time constant in inhomogeneous fields [14],

$$\tau_1 = \frac{\mu_0 l\, l_0^2}{4\pi^2 \rho c}$$

this limit means a strong restriction for many magnets. If we take e.g. the magnet length of about 1 km, with no additional layer between the strands (only the Cu sheath on them with approximate thickness of 0.2 mm), then the low frequency limit means ω below 0.01 or pulse period much larger than 600 s.

As already mentioned, the largest influence of the additional currents should be on the stability due to the decrease of the critical current in some strands. The critical point is expected to be $\eta = K$, where the magnetic field is high (the critical current decreases with magnetic field) and the field inhomogeneity also strong. Additional to the current in the homogeneous field, given by $\beta_1 \sin(K - \alpha_n)$, we obtain in the low frequency limit

$$J(K) = \beta_1 \{ \sin(K - \alpha_n) - [\cos(K - \alpha_n) - \cos(K + \kappa - \alpha_n)]/\kappa \}$$

Evidently, the maximum additional contribution for $a \geq l_0$ is at $\kappa = \pi$ (i. e. $a = l_0$) and $K - \alpha_n = (n + 3/4)\,\pi$ (nearly doubling the induced current). On the other side, no additional currents are induced at $\eta = K$ by taking $\kappa = 2\,\pi\, n$ or $a = n\, l_0$ (n - integer).

4. Conclusions

The additional losses in cables at the ends of magnets seems to cause no big problem, if the ramp rates are not too high. However, the length of the leads in the field free region should not be too large, as the loss density is constant in them.

Unfortunately, the results for the general case cannot be obtained easily. Due to the large total length of cables in the magnet, one is always not in the low frequency limit. The additional induced currents due to the field inhomogeneity penetrate then to a length comparable with the diffusion length l_d, i. e. they are more localized near the field inhomogeneity.

Acknowledgment

This work was partially supported by the Slovak Grant Agency for Science VEGA. I would like to thank B Turck for suggesting the importance of the topics presented in the paper.

References

[1] Ries G and Takács S *IEEE Trans. Magn.* **7** (1981) 2281
[2] Takács S *Cryogenics* **32** (1982) 261; **34** (1984) 234
[3] Verweij A P *Electrodynamics of Superconducting Cables in Accelerator Magnets* Thesis (Universiteit Twente, The Netherlands) 1995
[4] Faivre D and Turck B *IEEE Trans. Magn.* **17** (1981) 1048
[5] Takács S, Yanagi N and Yamamoto J *IEEE Trans. Appl. Supercond.* **5** (1995) 2
[6] Sumption M D and Collings E W *Adv. Cryog. Eng. Mater.* **40A** (1994) 579
[7] Carr W J Jr and Kovachev V T *Cryogenics* **35** (1995) 529
[8] Morgan G H *J. Appl. Phys.* **44** (1973) 3319
[9] Akhmetov A A, Kuroda K, Ono K and Takeo M *Cryogenics* **35** (1995) 495
[10] Verweij A P, ten Kate H H J, Leroy D, Oberli L and Siemko A *IEEE Trans. Appl. Supercond.* **5** (1995) 1020
[11] Kwasnitza K *Cryogenics* **17** (1977) 616
[12] Campbell A M *Cryogenics* **22** (1982) 3
[13] Wilson M *Superconducting Magnets* (Oxford: Clarendon) (1983)
[14] Takács S *Supercond. Sci. Technol.* **9** (1992) 137
[15] Yamamoto J and LHD Group *Proc. ICEC16/ICMC* (Oxford: Elsevier Science) (1997) p 725
[16] Takács S and Yamamoto J *Cryogenics* **34** (1994) 571

Inst. Phys. Conf. Ser. No 158
Paper presented at Applied Superconductivity, The Netherlands, 30 June–3 July 1997

Preisach-type hysteresis modelling in Bi-2223 tapes

Dejan Djukic, Mårten Sjöström and Bertrand Dutoit

EPFL, DE-CIRC, Lausanne, Switzerland

Abstract. An identification method for Preisach-type models of hysteresis is presented. By using a phenomenological Preisach-type model for hysteresis we can obtain an exact model for the magnetisation hysteresis in the cases of tapes with strip and with elliptical cross section geometries. With a parametrisation by Maclaurin series, it is possible to identify such a model for any other geometry. An additional advantage of this method is that the parameters are identified from the measurements of energy losses per cycle. Finally the results of application of the modelling method to the case of a cuprate ceramic (Bi–2223) superconductor strip is presented.

1. Introduction

In the recent years we have witnessed an increasing interest to apply ceramic superconductors in power transmission and distribution. However, there is a need for an accurate macroscopic model of electro-magnetic behaviour of these materials, without which a reliable design cannot be achieved. Due to the magnetic hysteresis in high temperature superconductors, the creation of such a model is fraught with difficulties.

We propose here to apply a Preisach-type model of hysteretic behaviour to Bi-2223 superconductor tapes.

2. Model description

From the aspect of magnetisation, the most prominent characteristics of type-II superconductors are very high critical field, partial flux penetration and flux pinning. A direct consequence of these is that the total flux possesses a hysteretic behaviour. A rather complete analysis for a case of an ideal superconductor strip is given in [1], from which we can conclude that properties of this hysteresis correspond exactly to a model from the class of Preisach-models. Although the analytical expressions in [1] are only given for the case of a strip, we can, in concept, extend this conclusion to the superconductors of any shape. A similar argument is given in [2].

The original Preisach-type model of hysteresis appeared as an attempt to model magnetisation of ferromagnetic materials. The models of this type are, however, not based on any physical mechanism, and can simulate hysteretic behaviour of systems of any physical nature. The Preisach-model is adequate for modelling systems with a static hysteresis with non-local memory.

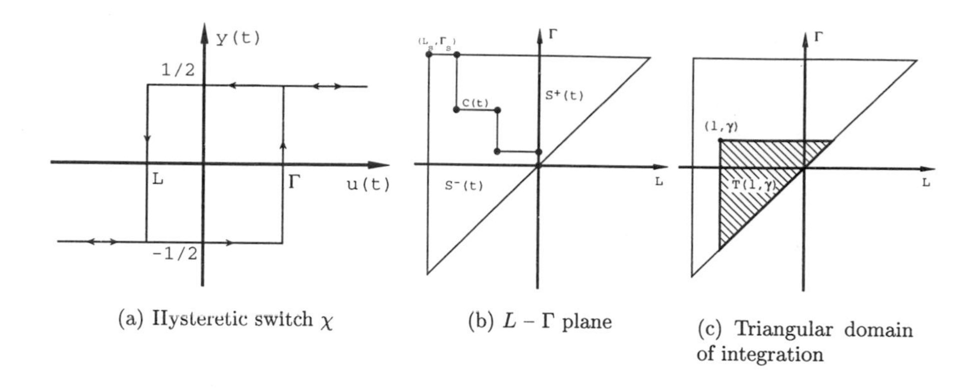

(a) Hysteretic switch χ

(b) $L - \Gamma$ plane

(c) Triangular domain of integration

Figure 1. (a) Input – output characteristics of a hysteretic switch (b) Domain of the weighting function $w(L,\Gamma)$ with the set of dominant extrema (c) Triangle in the $L - \Gamma$ plane swept by input

The output of the Preisach model can be viewed as a weighted integral of elementary hysteretic transfer functions $\chi[u]$ that works as a hysteretic switch, see Figure 1(a).

$$y = \int\int_{\mathcal{C}} w(L,\Gamma)\chi[u]\, dL\, d\Gamma \tag{1}$$

The hysteresis is described completely by $w(L,\Gamma)$ and by the set of dominant extrema of the input, $\mathcal{C}$. For an exhaustive work on the Preisach model, see [2].

Instead of the weighting function $w(L,\Gamma)$, we could also use the function $F(L,\Gamma)$ which is the integral of $w(L,\Gamma)$ over a triangular domain as the one presented in Figure 1(c). There are certain advantages for the usage of $F(L,\Gamma)$, as the calculation of double integrals can be replaced by the calculation of finite sums.

Preisach-type models allow for the energy dissipation per cycle to be calculated from the weighting function $w(L,\Gamma)$ and the extremal values of the input.

$$Q_c(l,\gamma) = \int\int_{T(l,\gamma)} w(L,\Gamma)(\Gamma - L)\, dL\, d\Gamma \tag{2}$$

Here, $T(l,\gamma)$ is the triangular surface in the $L - \Gamma$ plane swept by the input during one cycle, see Figure 1(c). From (1), an inverse formula is derived by which we can calculate $w(L,\Gamma)$ from a known energy dissipation per cycle.

$$w(l,\gamma) = -\frac{1}{\gamma - l}\frac{\partial^2}{\partial l \partial \gamma} Q_c(l,\gamma) \tag{3}$$

Since energy losses can be measured relatively easily and accurately, and since the hysteresis is described completely by its weighting function for an input of any form, we believe this formula to be of great practical importance.

The analytical expressions for the energy losses per cycle in the case of a sinusoidal transport current exist for superconductors with strip geometry (index s) and with elliptical cross sections (index e) [3]. By (3), we can calculate the corresponding weighting functions analytically.

$$w_s(L,\Gamma) = \frac{\mu_0}{4\pi I_c}\frac{x}{1-x^2}, \qquad w_e(L,\Gamma) = \frac{\mu_0}{8\pi I_c}\frac{1}{1-x}, \tag{4}$$

$$F_s(L,\Gamma) = \frac{\mu_0 I_c}{2\pi}[(1-x)\ln(1-x)-(1+x)\ln(1+x)+2x], \tag{5}$$

$$F_e(L,\Gamma) = \frac{\mu_0 I_c}{2\pi}[(1-x)\ln(1-x)+x], \quad \text{where} \quad x = \frac{\Gamma - L}{2I_c} \tag{6}$$

3. Parametrisation and Identification

The weighting functions $F(L,\Gamma)$ and $w(L,\Gamma)$ in (4) to (6) can be expressed in Maclaurin series with the variable $x = \frac{\Gamma-L}{2I_c}$. An input current $i(t) = I\sin(2\pi f_0 t)$ smaller than the critical current I_c, implies an x inferior to one so that a truncation of the series gives small errors. An approach to parametrise the weighting functions then becomes

$$F(L,\Gamma,\theta) = I_c\sum_{k=1}^{K} a_k\Big(\frac{\Gamma-L}{2I_c}\Big)^k, \quad \theta = [a_1, a_2, \ldots, a_K]^T, \tag{7}$$

$$w(L,\Gamma,\theta) = \frac{1}{4I_c}\sum_{k=2}^{K} a_k\, k(k-1)\Big(\frac{\Gamma-L}{2I_c}\Big)^{k-2}. \tag{8}$$

Note that $w(L,\Gamma,\theta)$ excludes a_1, and therefore $F(L,\Gamma,\theta)$ is a more general description. Below it is derived how a_1 contributes to reactive power but not to hysteretic losses.

An advantage with the parametrisation in (7) and (8) is its linearity in the parameters θ, which enables an identification applying a (possibly weighted) Least Square Estimate (LSE) technique. Equation (7) can be directly applied, since a raw estimate of the weighting function can be retrieved from sinusoidal time-series measurement. A good raw estimate is given by considering many measurements of different input amplitude I. Measurements have, however, shown a strong linear part a_1, such that there is no significance in the estimates of higher order parameters.

Another approach is to estimate θ from loss measurements using lock-in technique of high quality. The losses per cycle with the parametrisation (7) is given by the expression

$$Q_c(I,\theta) = 2I_c^2\sum_{k=1}^{K} a_k\frac{k-1}{k+1}\Big(\frac{I}{I_c}\Big)^{k+1}. \tag{9}$$

Again the expression is linear in θ and an identification with the LSE technique can be applied. The quality of the estimation results depend on the number of parameters and on the weighting used, see Figure 2(a). Note that a_1 has no contribution to the hysteretic losses and therefore cannot be estimated from the loss measurements.

The parameter a_1 can also be identified from lock-in measurements. The technique is to consider the reactive part of the output voltage v_r which has a phase $\pi/2$ after the input current. The chosen parametrisation (7) leads to the expression

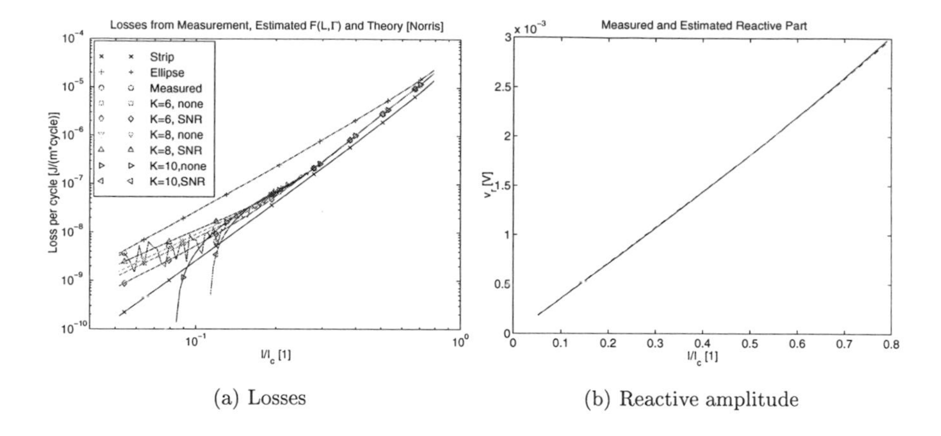

(a) Losses

(b) Reactive amplitude

Figure 2. Estimation results. (a) Measured and estimated losses per cycle. Results depend on parametrisation (K=6,8,10) and on weighting (none or signal to noise ratio, SNR). (b) Measured and estimated reactive amplitude The suggested method gives accurate results (K=6, no weighting)

$$v_r = \pi f_0 I \sum_{k=1}^{K} a_k \, k \left(\frac{I}{2I_c}\right)^{k-1}, \tag{10}$$

and it is realised that all parameters θ contribute to the reactive amplitude. The modelled losses are kept correct if the parameters estimated from losses, $\hat{a}_k, k > 1$, are inserted in (10) and a_1 thereafter is identified as the LSE. In this way, a_1 can be used to adjust the modelled output to have a correct reactive part, see Figure 2(b).

4. Conclusions

The Preisach-type model, that correspond to the hysteresis in type-II superconductors, has been presented. We have derived exact models for the special cases of strip and elliptic cross-section superconductors. Further, an identification method has been developed for the Preisach model from loss measurements independent of geometry, where one parameter can be utilised to comply the reactive part of the model output to measurements. Finally, results of identification have been given for a Bi-2223 tape.

References

[1] Brandt E H and Indenbom M 1993 *Phys. Rev.* B **48-17** 12893–12906

[2] Mayergoyz I D 1991 *Mathematical Models of Hysteresis* (New York: Springer Verlag)

[3] Norris W T 1970 *J. Phys. D: Appl. Phys.* **3** 489–507

Inst. Phys. Conf. Ser. No 158
Paper presented at Applied Superconductivity, The Netherlands, 30 June–3 July 1997

Numerical calculation of ac losses, field and current distributions for specimens with demagnetisation effects

M.Däumling

Forschungszentrum Karlsruhe, Institut für Technische Physik, D-76021 Karlsruhe, Germany

Abstract. Ac hysteresis losses P are calculated numerically for tapes in a perpendicular magnetic field H. Hereby a true three-dimensional (3D) representation of the cross section was used, enabling to model arbitrary shapes of the cross section for tapes. In particular it is found that the $P \propto H^4$ dependence expected for a 2D thin strip is only approached for extreme width to thickness ratios >> 100, leading to an exponent of around 3.5 for technically important Bi2223 tapes with aspect ratios of around 20-30.

1. Introduction

Ac losses in tape like geometries have been the subject of many recent investigations [1-8]. Most of these studies have focused on transport currents, where for thin 2D strip the hysteresis loss P is expected to vary as $P \propto I^4$ [9]. However, in a 3 phase power cable there is also a significant ac magnetic field present, which has to be included for loss considerations. For thin strips Brandt and Indenbom [10] have given the solution to the magnetic field distributions, leading to a power loss $P \propto H^n$ with n=4, just the analogue to the transport case. In this work tapes with finite thickness will be examined with respect to the ac loss in a magnetic field. The response to transport currents and self field effects in case of a field dependent current density will be reported elsewhere [11].

2. Method

If a magnetic field is applied parallel to the z axis perpendicular to tape of width w and thickness w, screening currents are induced in the tape parallel to the y direction. For fields smaller than the full penetration field the tape is only partially filled with current, shielding the applied magnetic field. The problem is now to compute the shape of the penetrating flux front, leaving j=0 and B=0 in the centre region. For j_c=const. this problem has been solved numerically by Brandt[12]. However, ac losses were not computed.

In order be able to model arbitrary shapes as well as allowing for $j_c(H)$ another method was used. For the calculation the cross section of the tape was subdivided into nxm elements (here n=m=49). The total magnetic field at the location of the element i,j is calculated as the sum of the magnetic field of all the other elements. The shape of the flux front was determined by minimising the following error function

$$f = \Sigma\, |H_z(\text{front}) - H_z(\text{centre})|$$

This method has been applied successfully to wires and ellipses [13,14]. Since we do not solve the complete set of equations (like Brandt [12] does), we don't know a priori what the external field is - due to the demagnetisation field this is *not* the field at the specimen edge. However, due to the nature of the problem this is just $-H_z$(centre) . Thus the applied field *and* the current distribution are calculated. The ac loss is determined by integrating the hysteresis in the magnetic moment when cycling the applied field. It should be noted that the magnetisation curve has to be determined very precisely in order to arrive at accurate ac losses.

3. Results and Discussion

The magnetisation curves for 5 strips of different width to thickness ratio are shown in figure 1. The magnetisation is normalised to the saturation value 1/4 j_c w, and the applied field is normalised by $j_c t$, which is something like the full penetration field for thin strips (exact values see [12]).

The corresponding ac loss is shown in figure 2. The thinnest tape has an aspect ratio of 3000, corresponding to a thin film, while the thickest strip has a square cross section. A $j_c=5\times10^8$ A/m^2 was used for the calculation. It is found that the power loss exponent n approaches 4 for the two thinnest strips, but is significantly less for thicker strips. For the Bean case (infinite slab in parallel field) one expects n=3 [15].

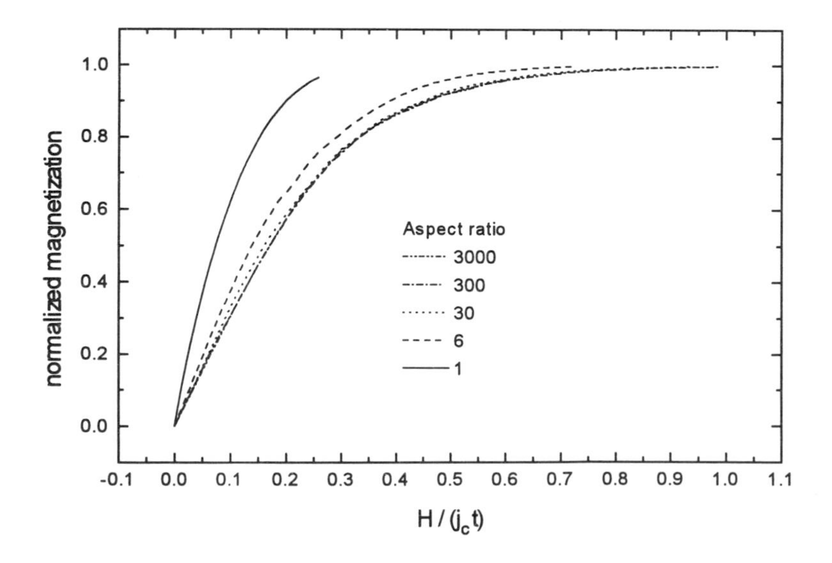

Figure 1. Magnetisation curves for strips of varying aspect ratios.

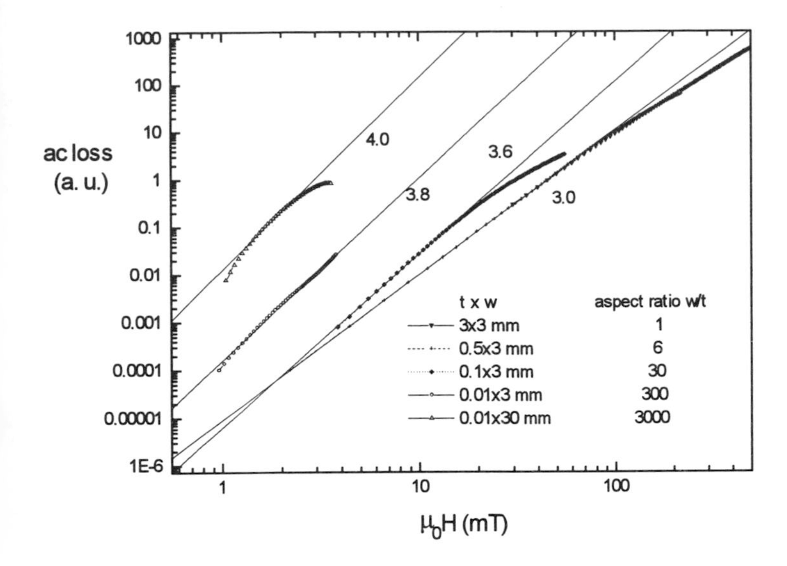

Figure 2. Ac loss as a function of applied ac peak field for strips of varying dimensions. The lines are fits to the apparent linear part of the data in log-log form, the numbers representing the slopes.

However, the square strip and the strip with w/t=6 also show n=3 already. This is a result of the more Bean like flux distribution in the thicker strips [11]. Thus for technical tapes with aspect ratios of around 30 one expects to measure an exponent of around 3.5, which is what is indeed often found.

Acknowledgments

I would like to thank the members of the high T_c wire group for discussions, and the ITP for financial support.

References

[1] Fukumoto Y, Wiesmann H J, Graber M and Suenaga M 1995 J. Appl. Phys. **78** (7) 4584-90
[2] Ishii H, Hirano S, Hara T, Fujikami J and Sato K Cryogenics **36** 697-703
[3] Fukomoto Y, Wiesmann H J, Garber M and Suenaga M 1995 App. Phys. Lett. **67** (21) 3180-82
[4] Eckelmann H, Däumling M, Quilitz M and Goldacker W, this conference
[5] Fleshler S, Cronis L T, Conway G E and Malezemoff A P 1995 Appl. Phys. Lett. **67** (21) 3189- 91
[6] Mele R, Crotti G, Gheradi L and Morin D 1996 presented at the ASC, Pittburgh
[7] Friend C M, Awan S A , Le Lay L, Sali S and Beales T P1997 to be published in Physica C
[8] Ciszek M, Ashworth S P, James M P, Glowacki B A, Campbell A M , Garre R and Conti S 1996 Supercond. Sci. Technol. **9** 379-84
[9] Norris W T 1970 J. Phys. D. **3** 489-507
[10] Brandt E H and Indenbom M 1993 Phys. Rev. **B48** 12893-905

[11] Däumling M to be published
[12] Brandt E H 1996 Phys. Rev. **B54** 4246-64
[13] Ashkin M 1970 J. Appl. Phys. 50 7060-66
[14] Navarro R and Campbell L J 1991 Phys. Rev. **B44** 10146-52
[15] Bean C B 1962 Phys. Rev. Lett. **8** 250-53

Inst. Phys. Conf. Ser. No 158
Paper presented at Applied Superconductivity, The Netherlands, 30 June–3 July 1997

Self field AC loss measurements of multifilamentary Bi(2223) tapes

Bertrand Dutoit, Nadia Nibbio, Giovanni Grasso, René Flükiger

Swiss Federal Institute of Technology EPFL-DE-CIRC, Lausanne, Switzerland. University of Geneva, DPMC, Geneva, Switzerland.

Abstract We present electrical AC loss measurements of multifilamentary Ag sheathed Bi(2223) samples in self field and under a small external DC magnetic field. The I-V measurements are performed using the AC transport method. We use a dual channel high stability lock-in amplifier in order to achieve the highest phase resolution. Tape samples of 1, 19, 37 and 55 filaments are compared. We use current amplitudes under and above the critical current at power frequency. The AC losses are analyzed as functions of the number of filaments. This effect is studied in self field and also under a small DC magnetic field of 10 mT, applied perpendicular to the tape surface. The AC losses of those multifilamentary tapes are comparable to Norris predictions of hysteretic losses. The evolution of the AC losses brings into consideration the sc/Ag ratio and the twisting of the filaments.
The Ag sheathed multifilamentary Bi(2223) tapes are prepared by Professor Flükiger's team at the University of Geneva and the measurements were performed at the EPFL.

1. Introduction

In future AC power systems, superconductive material will be used as a main component for power cables or transformers. To improve the capacity of superconductive Bi(2223), multifilamentary tapes have been developped in order to replace the monofilamentary one. The AC transport technique is used to perform the I-V measurements. We use a dual channel high stability lock-in in order to achieve the highest phase resolution. The AC transport losses of each sample is analyzed by considering the number of filaments, the twisting and the ratio between Ag and Bi(2223) surfaces, in self field or under a small external DC magnetic field. The results are then compared to the predictions of Norris based on the critical-state model of Bean, the critical current measured in self and applied field. The results of each type of sample, precisely 1, 19, 37 and 55 filaments, have shown a particular behaviour of the tape : the AC losses are higher when the number of filaments increases due to a possible bridging between the filaments. The evolution of the AC losses follows two different behaviours when the injected current increases especially for the 37 and 55 filaments tapes. In order to eliminate the AC losses of the tape, the twisting filaments and the ratio between the superconductive material and Ag have been considered.

2. Samples fabrication

Ag sheathed monofilamentary Bi(2223) tapes have been prepared by the so-called Powder-In-Tube method. Calcined precursor powders of nominal composition Bi1.8Pb0.4Sr2Ca2.2Cu3O10 have been filled inside pure Ag tubes of 5 mm of inner diameter and with wall thickness between 1.5 and 2.5 mm. The tubes are swaged and drawn down to a diameter of about 1.80 mm, after which they are drawn in a hexagonal shape. All the wires have been cut into 19, 37 or 55 pieces which are then stacked again inside Ag tubes of 10 x 8 mm. These tubes are also deformed by swaging and drawing down to a typical outer diameter of 1.3-1.8 mm. The tape-like shape is given by cold rolling, the final tape thickness being about 200-220 μm [2]. The Bi(2223) phase is formed during the heat-treatment process, which occurs at a temperature of 837°C for a total time of about 160 hours. An intermediate cold deformation process is applied after about 60 hours of heat treatment, in order to re-densify the superconducting filament, and so increase the critical current density. A typical cross section of a 37 filament Ag sheathed Bi(2223) tape is shown in fig. 1.

Figure 1 : Typical transverse cross section of a 37 filament Bi(2223) tape with thickness of 200 μm

3. Measurement set-up

The AC transport technique is used to perform the electrical measurements. The transverse magnetic field is generated by an electro-magnet at room temperature. DC field can be applied in a normal direction to the sample longitudinal axe up to 0.5 T. The cryostat is on a rotating support, position is adjusted with a computer controlled stepper motor (see fig. 2).
Time series of the current and voltage data are acquired with an analog to digital converter board installed in the computer. Lock-in calculation is performed in this computer with a software adapted to the particular conditions of our experiment. We simultaneously measure active and reactive parts of the current and of the harmonics 1 and 3 of the voltage. This layout allows us to reach a high phase stability better than 0.1 mRad necessary for the accuracy of losses measurements.

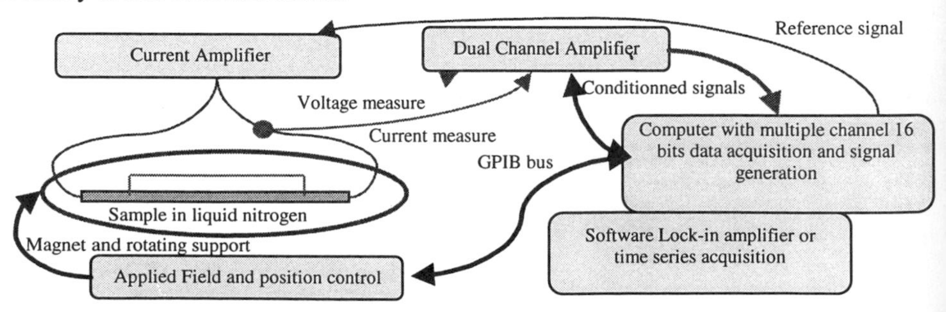

Figure 2: Experimental setup

4. AC losses results and discussion

The AC transport losses of the multifilamentary tape are expressed in terms of energy loss, compared to the Norris predictions [1], or voltage loss to compare samples when they have different characteristics. The frequency of all the measurements is 59 Hz. Table 1 reports the parameters of the samples, i.e. the DC critical current I_c, the number of filaments, the sc/Ag (superconductive material/Ag section) ratio and the twisting.

Table 1 : Measured samples

Sample	I_c [A]	sc/Ag ratio	Nb of fils.	Twisted
mono	10	20%	1	no
19 fils.	22	20%	19	no
37 fils.	32.5	20%	37	no
55 fils.	33	20%	55	no
GGS1nt	12.4	20%	37	no
GGS2nt	9.3	16%	37	no
GGS3nt	9.6	13%	37	no
GGS1tw	13.3	20%	37	yes
GGS2tw	9.3	16%	37	yes
GGS3tw	6.3	13%	37	yes

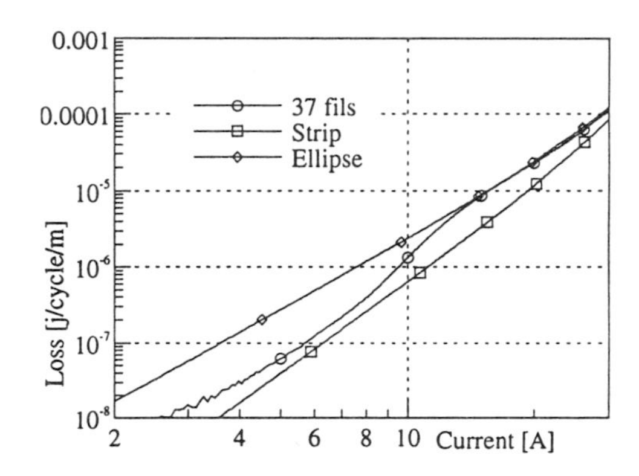

Figure 3 : Energy losses, 37 filament sample

Fig. 3 presents the losses of the 37 filaments sample and the Norris loss curves which were estimated for a strip and elliptical shape of the tape, as seen in the following formulas :

$$W_s = \frac{I^2\mu_0}{\pi}((1-F)\ln(1-F)+(1+F)\ln(1+F)-F^2) \quad W_e = \frac{I^2\mu_0}{\pi}((1-F)\ln(1-F)+(2-F)F/2)$$

with I the injected peak current and F the I/ I_c ratio.

Two different dynamical behaviours are observed : below I=0.3 I_c, the losses are similar to the strip estimation and then, the losses increase to reach the elliptical curve for current amplitudes above I_c. Fig. 4 shows the results of the monocore, 19, 37 and 55 filaments tapes. The above mentioned behaviour appears at 0.1 I_c for the 55 filaments, 0.3 I_c for the 37 filaments and 0.6 I_c for the 19 filaments.

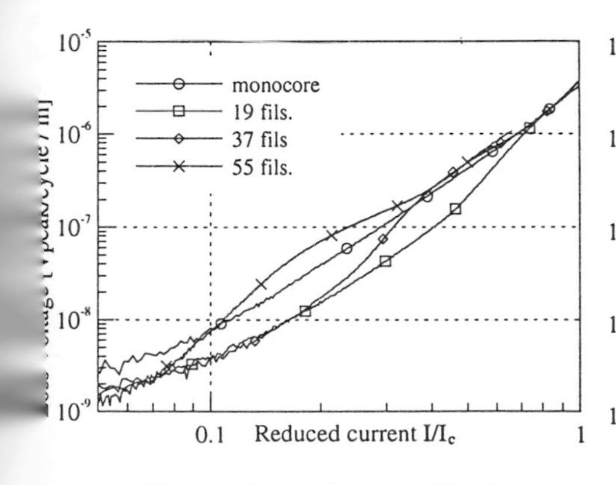

Figure 4 : Loss voltage in self field

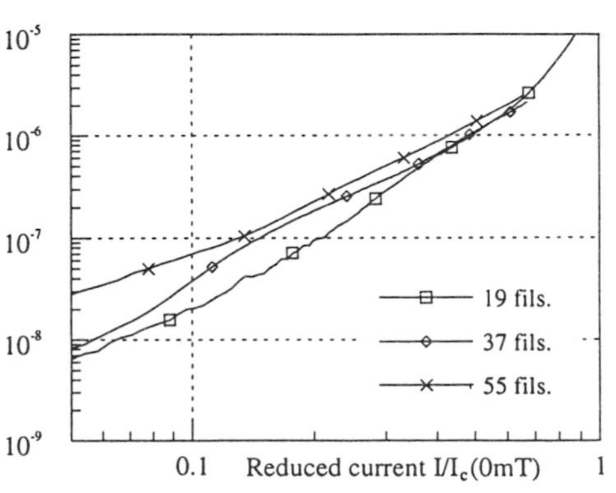

Figure 5 : Loss voltage in applied DC transverse magnetic field, 10mT

For multifilamentary tapes we expect a vanish of the losses with an increase of the number of filaments. Our measurements confirm this assumption for weak currents. For higher currents an opposite evolution is noted as seen in fig. 4 in self field : the 55 filaments tape dissipate more than the 19 and 37 filaments between 0.08 and 0.6 I_c while the 37 filaments itself dissipate more than the 19 between 0.2 and 0.7 I_c. However the 19 and 37 fils losses are weaker than those of the monocore, which confirms partly our assumptions about the evolution of the multifilamentary losses. When a DC transverse magnetic field is applied, a similar result is obtained with a transition appearing at lower current. In addition and more generally the losses increase in an applied transverse field (see fig. 5 and [2]). These results can be explained in terms of bridging between the filaments.

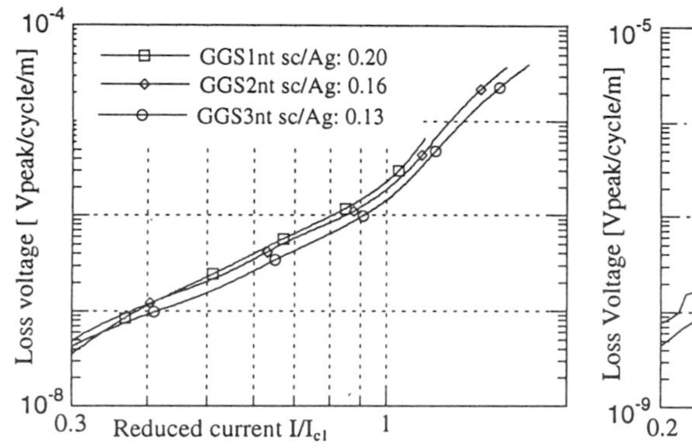

Figure 6 : Losses for different sc/Ag ratios

Figure 7 : Losses, twisted and not twisted Samples

As a conclusion, to improve the performances of the multifilamentary tape, two parameters of fabrication have been modified : the sc/Ag ratio and the twisting of the filaments. The chosen number of filaments is 37 filaments because of the two visible changes of behaviour. The measurements of the not twisted samples are shown in fig. 6. The variation of the ratio sc/Ag lead us to consider these losses as a function of the I/I_{c1} ratio. I_{c1} is not equal to the DC critical current but is determined on the curve losses vs peak current : it corresponds to the transition between flux creep and flux flow regimes. The losses of the GGS3nt are weak which maybe due to the larger distance between the filaments, avoiding coupling. In fig. 7, the twisting of the filaments diminishes the losses for injected current under $0.5I_c$ and then twisted and not twisted tapes exhibit similar losses close to I_c.

Acknowledgements

Research support acknowledgement to SWISS NEFF and CREE-RDP.

References

[1] Norris W T 1970 *J. Phys. D* **3** 489-507
[2] Grasso G, Jeremie A and Flükiger R 1995 *Supercond. Sci. Technol.* **8** 827
[3] Ishii H, Hirano S, Hara t, Fujikami J, and Sato K 1996 *Cryogenics.* **36** 697

Inst. Phys. Conf. Ser. No 158
Paper presented at Applied Superconductivity, The Netherlands, 30 June–3 July 1997

Factors affecting the accuracy of transport ac loss measurements on HTSC tapes

H.Understrup[1,2] and P.Vase[2]
[1]Department of Electric Power Engineering, Technical University of Denmark. [2]Nordic Superconductor Technologies.

Abstract. A set-up for transport ac loss measurements on HTSC tapes has been established. The loss results are one order of magnitude higher than reported by others. Therefore factors affecting the accuracy of the transport ac loss measurement are investigated. Factors including phase error and voltage loop size cannot explain the difference observed. The common mode signal on the lock-in amplifier is suggested to be a more important factor affecting the accuracy of the measurements.

1. Introduction

Within a Danish Research project Bi-2223 tapes have been developed for producing HTSC test conductors as models for future superconducting power transmission cables. It is important for the development of such cable conductor prototypes and also for the quality control of the tape production that the overall ac losses can be measured, modelled and minimised. In order to make reliable models for these conductors it is imperative that the ac losses of the individual tapes are well known. For these reasons a set-up for transport ac loss measurements has been constructed.

2. Experimental set-up

An overview of the constructed set-up for the transport ac loss measurements is shown in figure 1a. An ac current is sent through the sample placed in liquid nitrogen and the resulting voltage drop in-phase with the current is measured using a lock-in amplifier. A frequency generator together with a power amplifier delivers the current. The current through the sample is measured with a multimeter using a 0.1Ω resistor in series with the sample. The reference signal into the lock-in amplifier is the voltage across the resistor. The measured loss per cycle, Q, in J/m is the in-phase resistive voltage times the rms current divided by the frequency and distance between the voltage taps on the tape. All leads in the set-up were carefully twisted to avoid inductive pick-up. The measured samples are all taken from a 180m

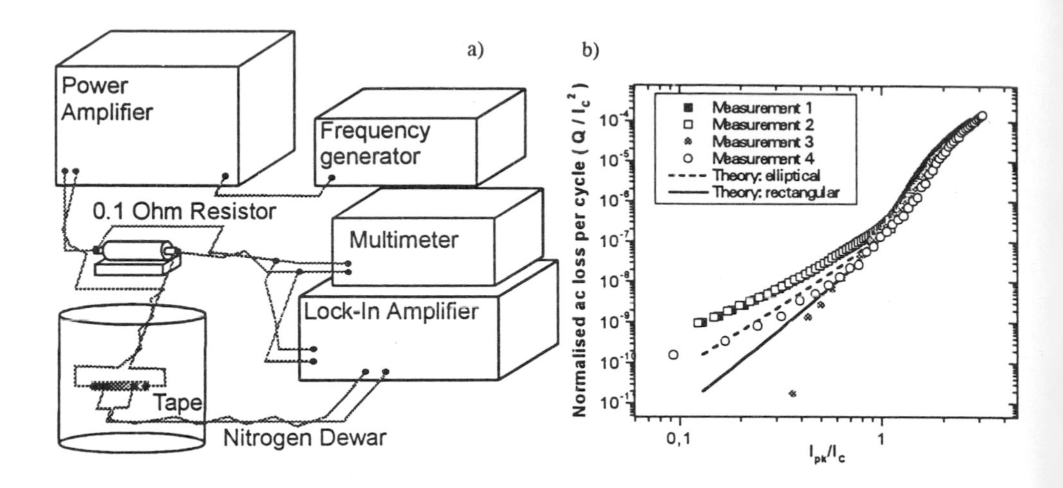

Figure 1 a) An overview of the set-up for transport ac loss measurements. b) Results of four measurements on samples from the same tape. For comparison the Norris[2] theory is plotted also.

long Bi-2223 silver clad tape fabricated using the powder in tube method[1]. The tape was fabricated by Nordic Superconductor Technologies. The total tape width is 3.54 mm and the thickness 0.21 mm. This includes a thin layer of ceramic coating which is put on during production of the tape. The critical current of the tape pieces is typically $I_C = 14$ A using the 10^{-4} V/m electric field criterion.

3. Results and discussion

Figure 1b shows four ac loss measurements on samples from the same tape versus the peak current I_{pk} divided by the critical current. The result is compared to the Norris[2] model for the self-field power loss, Q, in J/m per cycle for an elliptical and rectangular (thin strip) cross section of the superconductor given by

$$Q = \frac{\mu_0 I_C^2}{\pi}\left\{\frac{I_{pk}}{I_C} - 0.5\left(\frac{I_{pk}}{I_C}\right)^2 + \left(1-\frac{I_{pk}}{I_C}\right)\ln\left(1-\frac{I_{pk}}{I_C}\right)\right\} \qquad \text{elliptical}$$

$$Q = \frac{\mu_0 I_C^2}{\pi}\left\{-\left(\frac{I_{pk}}{I_C}\right)^2 + \left(1-\frac{I_{pk}}{I_C}\right)\ln\left(1-\frac{I_{pk}}{I_C}\right) + \left(1+\frac{I_{pk}}{I_C}\right)\ln\left(1+\frac{I_{pk}}{I_C}\right)\right\} \qquad \text{rectangular}$$

Transport ac loss measurements on high temperature tapes are reported[3][4][5] to be in good agreement with the Norris theory. The reported loss results are in the range from the Norris rectangular cross section values to the Norris elliptical cross section values. In figure 1b measurement 1 and 2 are one order of magnitude higher than the Norris values. The results from measurement 3 and 4 follow the rectangular cross section theory for currents between $0.6I_C$ and I_C. For lower currents the set-up has a large deviation resulting in loss

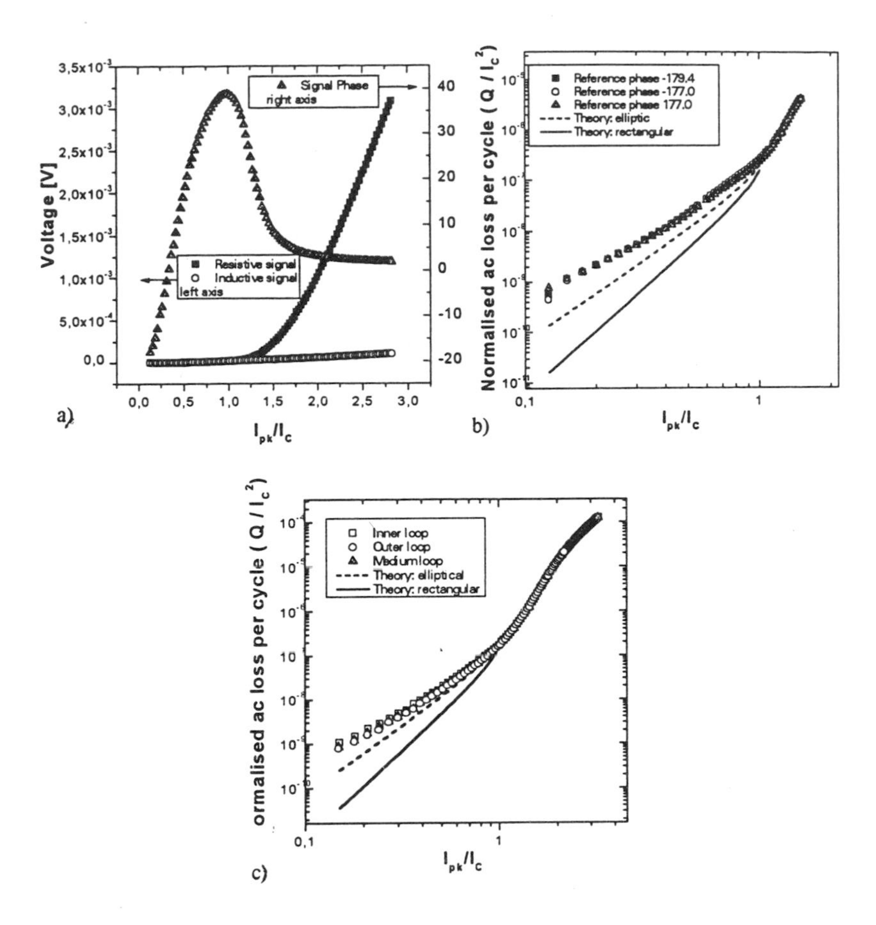

Figure 2. a) In the set-up the signal phase is kept below 40 degrees. b) A 3 degree change in reference phase cannot change the loss by one order of magnitude. c) Changing the size of the voltage pick-up loop does not change the loss by one order of magnitude.

values far from the Norris values. The only reproducible results are measurement 1 and 2. Consequently the set-up gives loss results one order of magnitude higher than reported by others. Therefore factors affecting the accuracy of the transport ac loss measurement are investigated. Three factors have been considered: the phase between the reference signal and the voltage signal from the tape, a lock-in phase error and the voltage pick-up loop size.

In the following the signal phase means the phase between the reference signal and the tape voltage measured by the lock-in. The voltage signal from the tape is not in phase with the current. Instead the signal consists of a loss signal in phase with the current and an inductive signal 90 degrees out of phase with the current. The inductive signal is a result of pick-up from the leads and the sample. The size of the inductive signal can be controlled by moving the sample voltage leads with respect to the current leads. This also controls the size of the signal phase. Figure 2.a illustrates that the signal phase is kept less than 40 degrees.

Consequently a phase error in the lock-in of 0.1° can only change the resistive voltage with 0.14%. Therefore a signal phase error is not the reason for the high loss values measured.

The resistor can introduce a phase shift. Consequently the reference phase is different from the phase of the current through the tape. By changing the reference signal phase shift on the lock-in this effect can be studied. Figure 2.b displays how a change of 3 degrees affect the measured loss very little. A reference phase error of 3 degrees can only change the resistive voltage with 4 % when the signal phase is lower than 40 degrees. Therefore a reference phase error is not the reason for the high loss values measured.

As reported by others[6] the size of the voltage pick-up loop affects the measured loss value. Figure 2.c shows how the measured loss values are changed when the loop size is changed. The largest loop is three times the width of the tape wide. This is reported to be a distance from where the resistive loss is not significantly affected by increasing the loop size further[6]. The figure shows that an error in the loop size is not the reason for the high loss values measured.

One factor which was not investigated is the common mode signal from the tape voltage. This signal can affect the result with orders of magnitude. Grounding the sample between the voltage leads is one way to reduce this effect[7]. This will be studied in the near future.

4. Conclusion

A set-up for transport ac loss measurements on HTSC tapes has been established. The loss results are one order of magnitude higher than reported by others. The investigated factors affecting the accuracy of the transport ac loss measurement are not the reasons for the high loss values measured.

References

[1] Li Q, Brodersen K, Hjuler H.A., Freltoft T. 1993 *Phys. C* **217** 360-366

[2] Norris W.T. 1970 *J. Phys. D: Appl. Phys* **3** 489-507

[3] Friend C.M. 1997 to be published in *Phys C*

[4] Yang Y. 1996 *Supercond. Sci. Technol.* **9** 801-804

[5] H.-H. Müller ASC conference in Pittsburgh Aug 25-30 1996

[6] Yang Y. 1996 *Phys C* **256** 378-386

[7] Rabbers J.J. 1996 *Measuring self-field loss of BSCCO/Ag tapes* (Graduation report:University of Twente)

Inst. Phys. Conf. Ser. No 158
Paper presented at Applied Superconductivity, The Netherlands, 30 June–3 July 1997

AC magnetic losses in multifilamentary Ag/Bi-2223 tape carrying DC transport current

M Ciszek[1)], B A Glowacki[1,2)], S P Ashworth[1)], E Chesneau[1,2)], A M Campbell[1)], and J E Evetts[1,2)]

[1)]IRC in Superconductivity, University of Cambridge, Cambridge CB3 OHE, UK
[2)]Department of Science and Metallurgy, University of Cambridge, Cambridge CB2 3QZ, UK

Abstract. In the paper we present measurements of alternating field losses (magnetic losses) in multifilamentary Bi-2223 silver sheathed tape carrying a DC transport current. An external sinusoidally varying magnetic field at a frequency of 90 Hz, with amplitudes up to 100 mT was applied parallel to the plane of the tapes and perpendicular to the current flow direction. The magnetic loss data as a function of peak applied field and transport current are given for partial and full penetration of the magnetic flux into a tape. The experimental results obtained are qualitatively compared to the theoretical predictions of the critical state model.

1. Introduction

Hysteresis losses due to an external changing magnetic field are determined mainly by the screening ability of the superconductor, which in turn depends on a critical current density and geometry of the superconductor [1]. In the majority of technical applications superconducting wires, in addition to the currents induced by the magnetic field, carry also a transport current supplied from external current sources. For partial magnetic flux penetrations the central volume of the wire is fully screened, so the transport current below its critical value does not affect magnetic losses. This situation is changed when external oscillating field exceeds the full penetration threshold; a DC voltage then appears along the wire which contribute considerably to the total loss (the so called dynamic loss). This problem was a subject of wide and intense investigations in low temperature superconductors [2-4]. Existing theoretical calculations, based on the critical state model [5-8], and developed for classical superconductors, qualitatively describe some of experimental results obtained for high temperature mono and multifilamentary wires [9-12]. We present here experimental data obtained from measurements of magnetic loss in a typical high temperature superconducting multifilamentary tape carrying DC transport current.

2. Experimental

Magnetic loss measurements were performed on a 48 filamentary Ag/Bi-2223 tape. The tape was additionally pressed with a force of 8 tonnes and annealed at 832°C for 100 hours in a reduced oxygen atmosphere (10% O_2 in Ar). The ratio of the ceramic core to the silver matrix in the cross-section of the tape was about 27%. The sample used was 50 mm long, with the overall cross section of 5×0.45mm^2. The self-field critical current I_c of the sample was 20.8A, defined using a 1μV/cm electric field criterion. This value corresponds to the

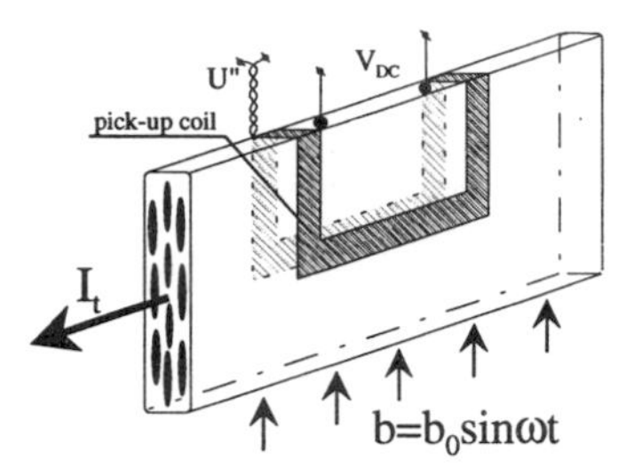

Fig.1. The arrangements of the saddle pick-up coil, potential leads (V_{DC}-critical current determination) and AC external magnetic field (b) for hysteretic loss measurements.

critical current density of approximately 4400 A/cm^2. More recently critical current densities of 16000A/cm^2 have been reached in conductors fabricated by the same technology [13]. In the central part of the tape a saddle shaped, single layer pick-up coil was wound. The average length of the coil was 1.7 cm. It is very important to place the saddle coil exactly symmetrically on both sides of the tape. The configuration of the pick-up coil, potential terminals, and external AC magnetic field is shown schematically in Fig.1. The loss component U''_{rms} of the voltage signal at the fundamental frequency induced in the pick-up coil was measured by means of a Lock-in voltmeter. Two kinds of measurements were performed: a)- keeping a DC transport current at a constant value the magnetic loss data were taken, and b)- keeping a fixed amplitude of AC magnetic field, losses were measured as a function of the DC transport current. All measurements were carried out at 77.3 K.

3. Results and Discussion

Fig.2a shows the measured losses Q_m per cycle per unit surface area, as a function of the peak external magnetic field b_0, for different DC transport current I_t flowing along the tape ($i=I_t/I_c$). The losses at amplitudes below 0.3 mT increase approximately according to the third power dependence $Q_m \propto b_0^3$. Increasing b_0 this dependence changes to the power of four, and for b_0 greater than 10 mT (well above the full penetration field) the losses change linearly with the field. When a DC transport current is flowing through the tape, the loss behaviour changes drastically, i.e. for amplitudes below full penetration losses increase with transport current values, whereas they are considerably reduced at higher amplitudes. This is typical "lossy" behaviour of a superconductor when its screening ability is reduced. Fig.2.(b) shows the imaginary part of the AC complex permeability μ'' which is related to losses *via* $Q = \mu'' \pi b_0^2/\mu_0$. The μ'' values were obtained from the loss data in Fig.2(a), taking into account the geometry and filling factor of the tape. The peaks in $\mu''(b_0)$ occur when magnetic flux fully penetrates the superconductor and are governed by the critical current density, its magnetic field dependence, and the geometry of a superconductor [1].

The presence of a DC transport current changes significantly the $\mu''(b_0)$ dependence; the maximum at permeability μ'' increases and shifts towards lower amplitudes. For higher values of i (>0.3) a kind of double stage transition is observed, the first one at very low amplitudes well below 0.1 mT. According to theoretical models based on slab geometry departure from the $Q \propto b_0^3$ dependence for partial penetration ($b_0 < b_{0p}$, where b_{0p} the full penetration field) should occur only at amplitudes above $b_{0p}(1-i)$. For our tape, for i=0.4, taking b_{0p}=6mT, loss data should be the same as for the i=0 case unless the amplitude exceeds value of 3.6 mT. In fact a large loss enhancement is observed at very low amplitudes b_0. Similar anomalies are reported in [12]. For a transport current closer to its critical value the peak in μ'' takes the form of wide plateau. The amplitudes corresponding to the maximum on $\mu''(b_0)$ curves versus transport current ratio are shown in Fig.3. The solid line represents the theoretical dependence of the form $b_{0p}(i)/b_{0p}(i=0)=(1-i^2)$. Similar dependence has been found for classical superconductors [2-3], as well for high temperature

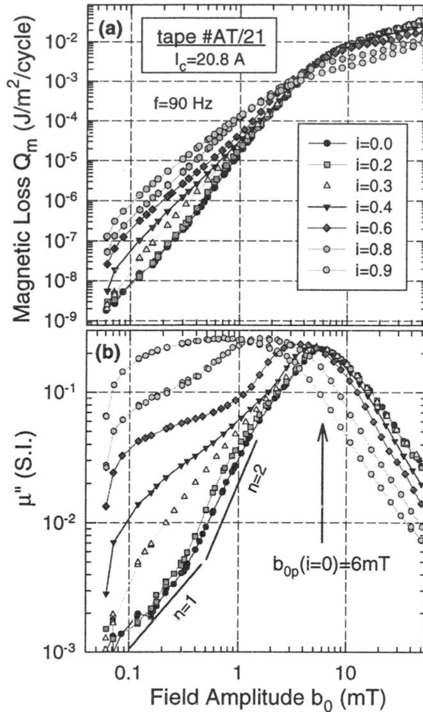

Fig.2. (a)-Magnetic loss Q_m *vs.* external field amplitude b_0, for different transport DC current I_t flowing along the tape ($i=I_t/I_c$); (b)- imaginary part of the complex magnetic permeability μ" calculated from the loss data; n=1 and n=2 correspond to $Q \propto b_0^3$ and $Q \propto b_0^4$ relations, respectively.

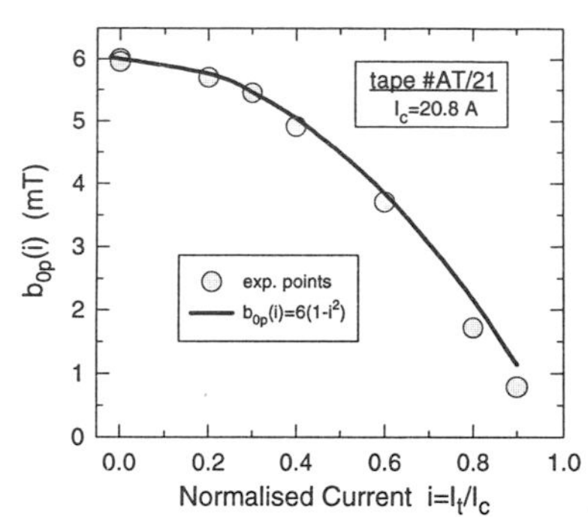

Fig.3. The dependence of the full penetration amplitude b_{0p} on the normalised transport current $i=I_t/I_c$. Data are taken from Fig.2(b). Solid curve corresponds to $b_{0p}(i)=b_{0p}(0)(1-i^2)$ function.

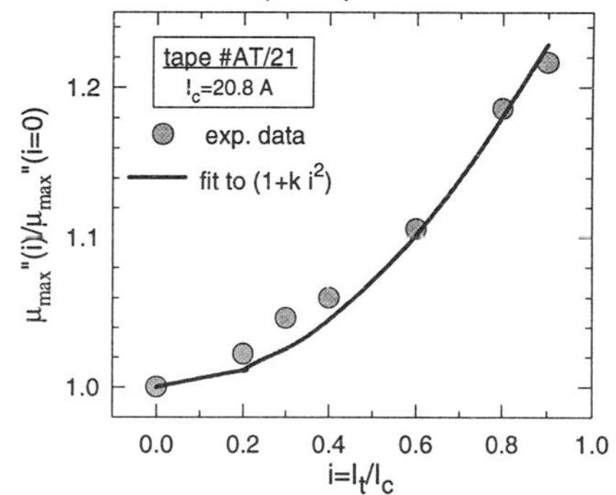

Fig.4. The dependence of μ" at the peak value on the transport current ratio i. The solid line is the theoretical value according to Ref.[7].

ceramics [9-12]. The presence of the DC transport current in the tape increases considerably the value of μ" at its maximum on μ"(b_0) plot. This enhancement in loss factor is shown in Fig.4, where the ratio of peak values of μ_{max}"(i) for the sample with transport current to the value of those without current μ_{max}"(i=0) are plotted as a function of the normalised transport current i. Qualitatively very good agreement with the theoretical model was found (solid line); the losses increase with current as $(1+ki^2)$, where k is a numerical factor less than or equal to one, which depends on the shape and whether the current is constant or oscillates with the field [7]. From our data the value obtained is about 0.3.

Losses Q_m as a function of the normalised transport current i through the tape, for different fixed AC external field amplitudes b_0 are plotted in Fig.5. The loss curves exhibit very pronounced maxima at $i=i_{max}$. These peaks shift towards lower values of i when the amplitude b_0 is increased. At an amplitude around 6 mT (the peak in μ"(b_0) function), the maximum in loss dependence disappears, and for higher b_0 the losses fall with current. For incomplete penetration the magnetic loss occurs in the regions close to the surface of the superconductor, leaving the centre of the sample fluxoid free. In this case the transport current can flow through this cross-section in a loss-free manner. Such transport current creates its self magnetic field, which locally influences the magnetic flux distribution, causing the b_p to be a decreasing function of the current. This leads to further penetration of the magnetic field towards the middle of the sample and thus magnetic loss increases.

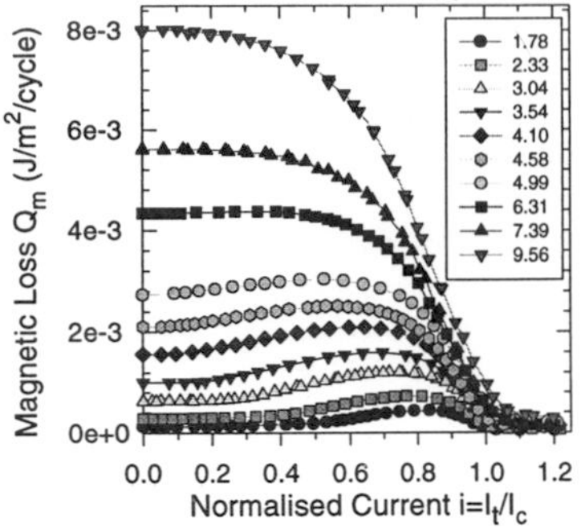

Fig.5. The set of magnetic loss data Q_m as a function of the normalised transport current i, for fixed AC field amplitudes b_0 (see also Fig.6).

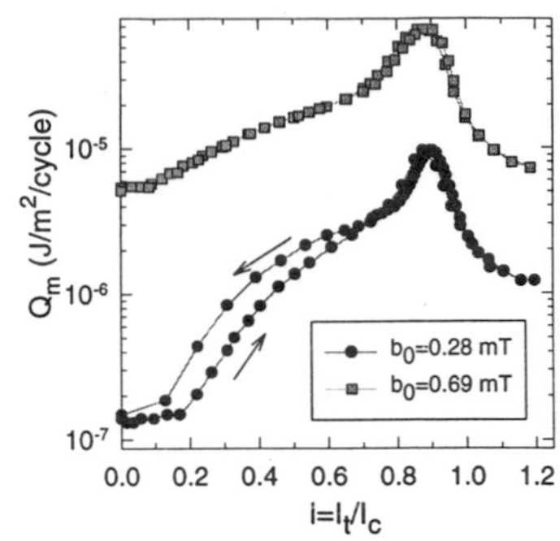

Fig.6. Loss Q_m dependence on transport current i for very low magnetic field amplitudes b_0, for increasing and decreasing transport current

Fig.6 shows the loss dependence on transport current for very small amplitudes, A history effect is here observed, probably due to trapped flux inside the filaments and redistribution of currents inside the composite. This effect is not visible at higher amplitudes. A possible explanation of this phenomenon has been given in [14], however further study on this subject is required.

4. Conclusions

The experimental data presented here are qualitatively in a fairly good agreement with the existing theoretical models, although pronounced anomalies are observed at low magnetic field amplitudes. This is due to the very inhomogenous structure of the multifilamentary ceramic tapes. Existing models are mostly based on slab geometry and deviations from this are likely to be due to penetration of the current from the edges of the tape.

Acknowledgements

We would like to thank Alcatel Alsthom Recherche for having supplied the PIT conductors which have been used for this work

References

[1] Wilson M N, 1993 *Superconducting Magnets*, Oxford University Press, London
[2] Ogasawara T, Takahashi Y, Kanbara K et al., 1979 *Cryogenics* **19** 736-740
[3] Dragomirecki M et al., 1986 *Proc. of the 11th Internat. Cryogenic Conf.* Berlin, 22-25 April 746-50
[4] de Reuver J L, 1985, Ph.D. Thesis, University of Twente, The Netherlands.
[5] Hancox R, 1966 *Proc. IEE* **113** 1221-8
[6] Carr W J, 1979 *IEEE Trans. Magn.* **15** 240-3
[7] Campbell A M, 1982 *Cryogenics* **22** .3-16
[8] Minervini J V, 1992 *Adv. in Cryogenic Engineering* **8**, Eds. R. P. Reed and A. F. Clark, 587-599
[9] Collings E W et al., 1992 *Adv. in Cryogenics Eng.* **38** Eds. Fickett F R and Reed R P, New York, 883-91
[10] Hemmes H, Woudstra M J, ten Kate H J, Tenbrink J, 1994 *Cryogenics* **34** (ICEC Suplement) 567-70
[11] Ciszek M, Glowacki B A et al., 1995 *IEEE Trans. on Appl. Superconductivity* **5** 2709-12
[12] Suenaga M, Fukumoto Y et al., 1997 *IEEE Trans. on Appl. Superconductivity* **6** 1674-8
[13] Herrmann P F, Beghin E et al., 1997 *IEEE Trans. on Appl. Superconductivity* **6** 1679-82
[14] Glowacki B A, Noji H and Oota A, 1996 *J. Appl Phys.* **80** 2311-6

Inst. Phys. Conf. Ser. No 158
Paper presented at Applied Superconductivity, The Netherlands, 30 June–3 July 1997

Critical current and self-field loss of BSCCO-2223/Ag tape in bifilar geometry

Y K Huang, J J Rabbers, O A Shevchenko, B ten Haken and H H J ten Kate

Low Temperature Division, Faculty of Applied Physics, University of Twente, P. O. Box 217, 7500 AE Enschede, The Netherlands

Abstract. The critical current and self-filed loss of BSCCO-2223/Ag tapes in bifilar geometry are investigated and compared to those in a single tape. The critical current in a bifilar tape at 77 K and zero external field is found to be 15 ~ 20% higher than that of a single tape. The increase of critical current is attributed to the substantial reduction of the self-field component perpendicular to the tape. The self-field loss in a bifilar geometry is much lower due to the changes of self-field magnitude and distribution in the tape.

1. Introduction

The progress in fabricating BSCCO-2223/Ag tapes with high critical current density and long lengths has cast a promise for large scale applications such as power transmission cables. The critical current of BSCCO-2223/Ag tapes is very sensitive to a magnetic field perpendicular to the tape broad face. A tape carrying a transport current will suffer from a self-field generated by the transport current. In zero or very low external magnetic field, the critical current of a tape is influenced by the self-field component perpendicular to the tape. In practical applications such as in a power transmission cable, certain types of tape geometry are applied, in which the self-field distribution and magnitude are different from those in a single tape. In order to evaluate the characteristics of tapes for real applications it is important to investigate various tape geometries and their effects on the critical current and self-field loss of tapes [1, 2]. In this paper, we present the results of critical current and self-field loss of the tapes in a bifilar geometry.

In a bifilar geometry, two tapes, face to face placed, carry an anti-parallel transport current. The self-field distributions in a bifilar tape and in a single tape are different. A simulation result is shown in Figure 1, assuming that the transport current is uniformly distributed in the cross section of the superconducting core. The perpendicular component of the self-field in a bifilar tape is reduced substantially, compared to that of a single tape.

2. Experimental

The samples used in this investigation are 85 filaments Ag-sheathed Bi-2223 tapes. The overall cross section is 4.2×0.25 mm^2 and the superconductor fraction is about 25%. The sample length is about 12 cm.

The two tapes in the bifilar sample are separated with a thin (30 μm) insulating layer and soldered together at one end. The current leads are soldered to the other ends of the tapes. The single tape sample used for comparison is the tape from the bifilar sample after removing the second one.

The voltage-current curves of the samples are measured at liquid nitrogen temperature (77 K) in a four-points configuration. The length between the voltage taps is 64 mm. The critical current is determined using an electric field criterion of 1 μV/cm.

The magnetic field dependence of the critical current is measured with a DC magnetic field up to 100 mT. The magnetic field is applied either perpendicular or parallel to the broad face of the tape and always perpendicular to the transport current through the sample.

The self-field loss measurements are performed using the transport method at various current frequencies (30 to 190 Hz) [3].

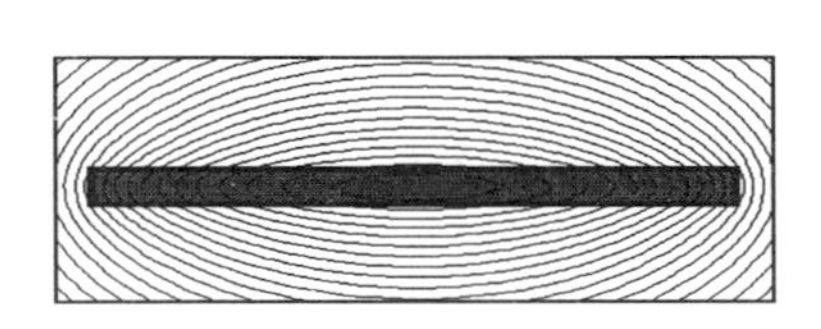

a. Single tape

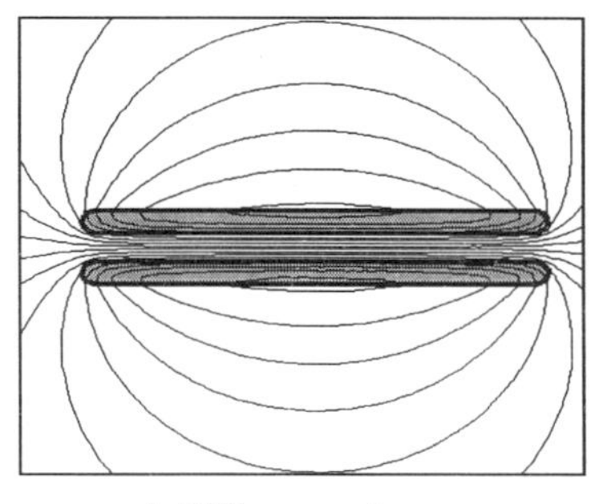

b. Bifilar geometry

Figure 1. Self-field distribution in a single tape and in a bifilar geometry, assuming the transport current is uniformly distributed.

3. Results and discussion

3.1 Critical current

The critical current as a function of the applied magnetic field for a bifilar tape geometry is presented in Figure 2 and compared with that of a single tape. The results show that the critical current of the tape in a bifilar geometry and zero external field is about 20% higher than that of a single tape (43 A and 36 A, respectively). The increase of the critical current is attributed to the reduction of the self-field component perpendicular to the tape.

The critical current of the bifilar tape decreases quicker than that of the single tape as the external field increases. The $I_C(B_\perp)$ curves of both become the same when the external field perpendicular to the tape is larger than the self-field. While the $I_C(B_{//})$ curves for B parallel to the tapes seem more different. However, this can be explained by considering the following two points.

First, the variation of critical current with the magnetic field applied parallel to the tape is related to the mis-oriented grains in the tape. If the mis-orientation in the tape is represented by an average misalignment angle θ, then applying a field parallel to the tape ($B_{//}$) is considered to be equal to applying a field perpendicular to the tape with a magnitude of $B_{//}\sin\theta$, and the scaled $I_C(B_{//}\sin\theta)$ curve is identical to the $I_C(B_\perp)$ curve [4].
Second, when the field parallel to the bifilar tape is rotated 180 degrees, the resulting two $I_C(B_{//})$ curves are different (Figure 2, Bifilar $B0°$ and $B180°$). The maximum in the $I_C(B_{//})$ curve of the bifilar tape does not occur in the zero external field. The shift is about 5 mT.

This can be explained by the fact that the self-field component parallel to the tape is not the same along the thickness of the tape. The magnitude of this field component on the inner side is much larger than that on the outer side. The net effect of this asymmetric self field on the critical current is contributed dominantly from that on the inner side. When the external parallel field $B_{//}$ is applied in the opposite direction as that of the self-field on the inner side ($B180°$), the resulting field is reduced, leading to a maximum in the $I_C(B)$ curve. When $B_{//}$ is applied in the same direction as that of the self-filed on the inner side ($B0°$), the resulting field is enhanced and the $I_C(B)$ decreases monotonously.

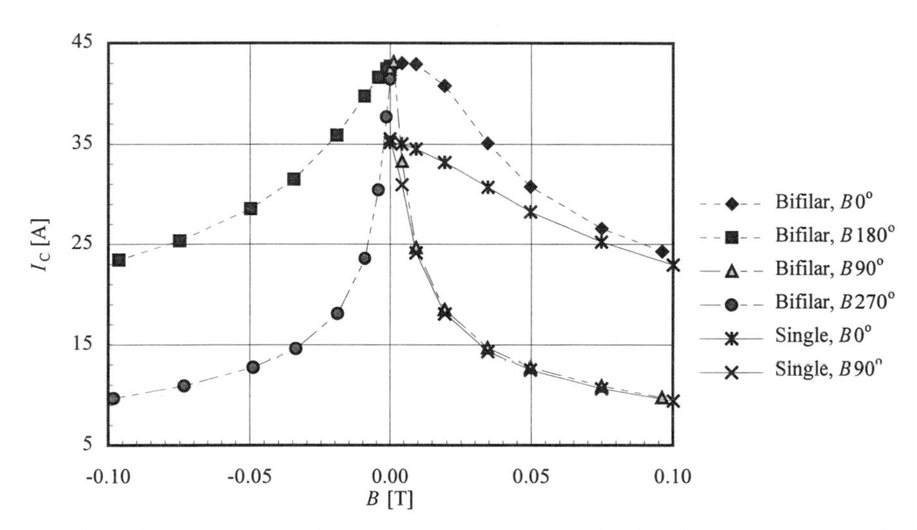

Figure 2. Critical current of the tape versus the magnetic field in the bifilar geometry compared to that in a single tape.

3.2 Self-field loss

The self-field losses of the bifilar tape and the single tape as a function of AC transport current amplitude at different frequencies (30 to 190 Hz) have been measured. The results show that the AC loss in J/m per cycle is frequency independent in both cases when the AC current amplitude is smaller than the critical current. This indicates the loss is dominated by a hysteresis loss. The curves at the frequency of 90 Hz are shown as examples in Figure 3.

It should be noticed that the AC loss measurement for a bifilar sample has encountered some uncertainty, as shown in Figure 3. The loss value strongly depends on the position and size of the voltage loop. When the voltage taps is attached at the center and closed to the tape surface, a much lower loss value is measured. This is closer to the value estimated from the Norris calculation for anti-parallel strips [5]. However, when the voltage loop is closed at the edge of the tape or extended about 10 mm from the tape, either at the center or at the edge of the tape, a much higher loss value is obtained.

Usually, if a bifilar sample consists of a continuous tape, one can measure the AC loss by attaching the voltage taps at the edges of the two branches and closing the loop. In our bifilar sample two tapes are soldered together at one end. We measured the AC loss of the bifilar sample two times using the voltage tap arrangement as described above. The tape length enclosed between the voltage taps are different in these two measurements. By subtracting the voltage of the shorter one from that of the longer one, we can eliminate the resistance effect of the soldering part. The resulting loss value is the same as that obtained from the measurement using open voltage loop at the edge (Figure 3) which has the highest loss value compared with the other situations.

The bifilar tape has a lower loss than the single tape. This is attributed to the reduction of the perpendicular component of the self-field (higher I_C) and the change in the self-field distribution in bifilar geometry.

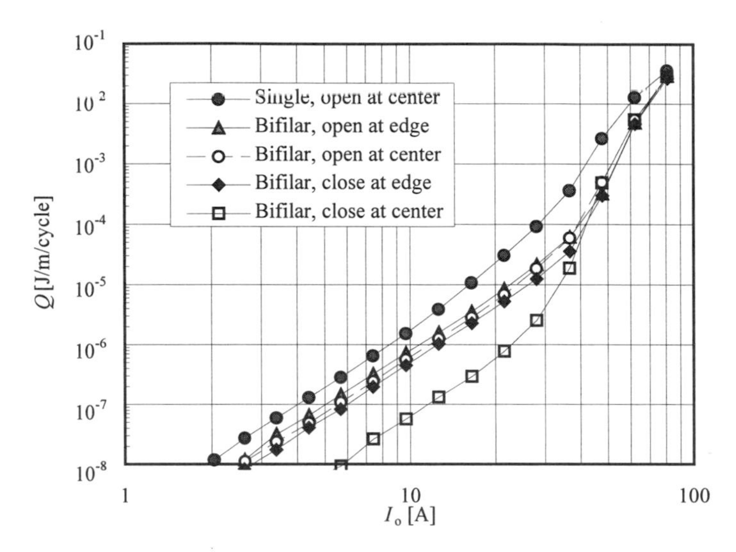

Figure 3. Self-field loss as a function of AC transport current amplitude at the frequency of 95 Hz in the bifilar tape and in the single tape.

4. Conclusion

Experimental results on critical current and self-field loss of the BSCCO-2223 tapes in a bifilar geometry and in a single tape geometry are compared. The critical current in a bifilar tape at 77 K and zero external field is increased due to the reduction of the self-field component perpendicular to the tape. The self-field loss is smaller than that of a single tape. The results indicate that the distribution and magnitude of the self-field remarkably influence the critical current and self-field loss of the tape. For accurate characterisation of the superconducting properties of a tape with regard to practical applications, it is necessary to consider the tape geometry and experimental set up.

References

[1] Grasso G, and Flükiger R 1995 *Physica C* **253** 292-296

[2] Gherardi L, Caracino P, Coletta G and Spreafico S 1996 *Materials Science and Eng. B* **39** 66-70

[3] Huang Y K, Rabbers J J, ten Haken B and ten Kate H H J, 1997 Int. Cryogenic Materials Conf., Portland, USA, July 28-Aug. 1 to be published in *Advances in Cryogenic Engineering - Materials*

[4] Kobayashi S, Kaneko T, Kato T, Fujikami J and Sato K 1996 *Physica C* **258** 336-340

[5] Norris W T 1970 *J. Phys. D* **3** 489-507

Inst. Phys. Conf. Ser. No 158
Paper presented at Applied Superconductivity, The Netherlands, 30 June–3 July 1997

Investigations of the AC current loss of twisted and untwisted multifilamentary Ag/ AgMg/ and AgAu/ Bi(2223) tapes.

H. Eckelmann, M. Däumling, M. Quilitz and W. Goldacker

Forschungszentrum Karlsruhe, Institut für Technische Physik, P.O.Box 3640 76021 Karlsruhe, Germany

Abstract: We have measured the AC transport current loss of 9 different Bi(2223) tape with varying geometry of the filament arrangement in the frequence range from 13 Hz to 770 Hz at 77 K. Tapes under investigation have an Ag-sheath (i.e. 37 or 703 filaments) or a modified sheath using Ag/AgMg or AgAu. In addition we have investigated twisted tapes with a twist-pitch of 1.4 cm for the 37 filamentary Ag and the 85 filamentary AgMg tapes. Furthermore tapes with quite novel geometries such as mulifilamentary wire-in-tube (WIT) tapes and jelly-roll tapes have been studied. We compared all tapes to find out what filament arrangement and geometry is best suited for AC power applications. We observed that the loss for the most tapes fall within the limit given by Norris [1] for rectangles and ellipses, with the exception of the jelly-roll and AgAu tapes, which have a lower loss fac tor.

1. Introduction

For the application of multifilamentary Bi(2223) tapes as power transmission cables it is necessary to reduce the AC losses in these tapes as much as possible. These losses consist of three different parts [2]: the eddy current, the coupling current and the hysteresis losses. As shown in [3] the eddy current loss at normal frequencies (50 Hz) is less than 10% of the total loss and so it will by neglected in this work. To reduce the interfilamentary coupling losses it is necessary to apply a twist-pitch much shorter than the critical coupling length. For Ag as matrix material this critical coupling length is in the order of 5 mm [4]. Mulitfilamentary tapes which are untwisted or which have a twist-pitch longer than the critical coupling length behave as a monofilamentary tape [3,4]. Thus the ac loss for the investigated tapes can be described as hysteresis losses in monofilamentray tape where all the multifilamentary tapes act as one filament.

2. Experimental

In table 1 we have summarised the main data of the tapes. The Ag-37 filament and the AgMg 85 filament tapes are tapes which have been produced by the standard PIT-technique, with sheaths og Ag and AgMg 2 at %, respectively. They were twisted at a diameter of 1.5 mm after drawing, before being rolled flat. The cross-section of the AgMg sheathed tapes resembles more an ellipse than the cross-section of the Ag sheathed tape does. The twisted tapes showed smaller j_c values because the twisting reduces the critical current [5]. The Ag-703 filament tapes were produced by the same method as the Ag-37 filament tapes with an additional intermediate bundling step. The wire-in-tube tapes were produced in a differend way which will be described in a forthcoming publication [6]. The difference between the two wire-in-tube tapes WIT 1 and tape WIT 2 is that the tape WIT 2 is 10 % thinner. As a result of this WIT 2 has a greater aspect ratio, thinner filaments and a slightly greater critical current. The jelly-roll tapes were produced using a technique which is known from low T_c-wires. The filaments consist of large ellipses extending through the width of the tape [7], the outer ones shielding the inner ones. The AgAu tapes have AgAu 8 at % as matrix material which has a greater resistivity than Ag [8]. The exact process of the production of the AgAu tapes will by published in the near future [9].

Table 1: Characteristic data for the different tapes

tape	code	fill factor (%)	I_c (A)	j_c (kA/cm^2)
Ag 37 filament	Ag 1	23	20.75	11.6
Ag 37 filament, twisted l_p = 1.4 cm	Ag 2	23	16.5	9.2
AgMg 85 filament	AgMg 1	19	18.94	11.8
AgMg 85 filament, twisted l_p = 1.4 cm	AgMg 2	19	16.08	10
AgAu 7 filament	AgAu	18.4	7.5	6.5
Ag 703 filament	Ag 703	7.9	16.38	23.7
Wire-in-tube 37 filament	WIT 1	12.7	12.06	12.2
Wire-in-tube 37 filament	WIT 2	12.7	13.57	15.6
Jelly-roll tape	Jelly 1	5.4	5	11.3

The loss was measured using the transport technique. For this an sine-wave ac current was applied to the tape, and the voltage drop was measured over a length of 10 cm. The voltage and the current in the circuit are measured with two lock-in-amplifiers. The correct phase shift setting between the current and the voltage is estimated by analyzing the U versus I signal as shown in [10]. For small currents the inductive signal of the voltage is 100 times greater than the resistive voltage. The voltage taps are soldered to the edges of the tapes and led 5 mm (nearly 3 times the half tape wide) away from the tape before they are twisted [11]. Before the ac loss measurements we have measured the dc critical current of the tapes at a length of 1.5 cm using the $1\mu V/cm$ criterion. For the ac loss measurement we used only homogeneous regions (homogeneity of I_c for the most tapes less than 10 %) of the tapes. Over the whole measurement the tapes are stored in liquid nitrogen and no degradation occured during the measurement.

3. Results and Discussion

We found a linear frequence dependence for the Power loss versus the reduced critical current (I_m/I_c), this indicating that our are hysteresis losses. These can be described using the critical state model [12], used by Norris [1] for different geometries. The hysteresis losses can be described by:

$$W_{Hyst} = \frac{I_c^{\,2} \cdot \mu_0}{\pi} \cdot L(F) \tag{1}$$

In this formula L(F) is a function which, for a given geometry, depends only on the fraction of I_m/I_c, were I_m is the peak current and I_c the critical current. For an elliptical conductor we have.

$$L(F) = (1-F)\ln(1-F) + (2-F)F/2 \tag{2}$$

while for a strip like conductor :

$$L(F) = (1-F)\ln(1-F) + (1+F)\ln(1+F) - F^2 \tag{3}$$

is valied. This loss function is plotted in the figure 1 and compared with the measured loss factors of the different tapes. The measured data fits rather well to the theoretical functions. Most of the tapes showed loss factors which are between those expected for an elliptical or a strip like conductor.

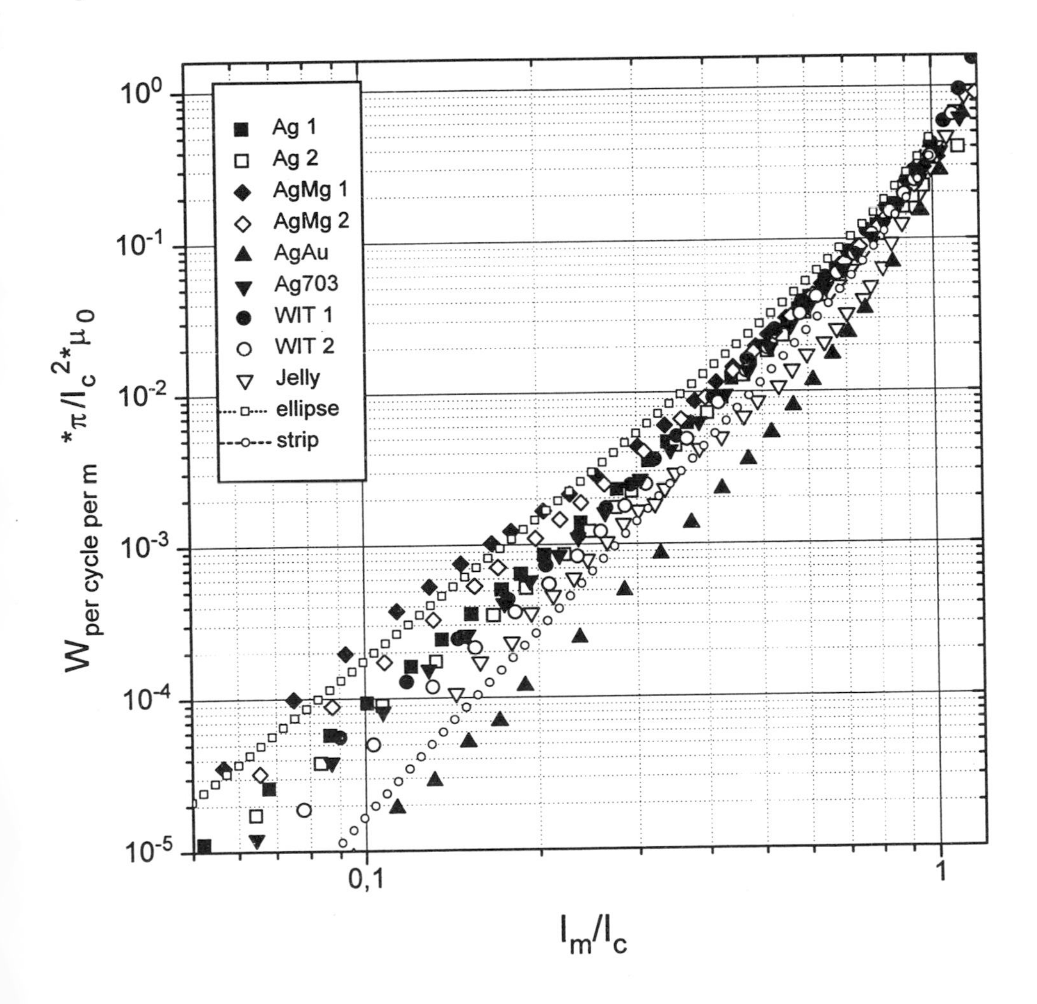

Fig. 1 Normalized loss factor per m vs the reduced the current. The lines present the loss calculated with the equations (1-3)

These results confirmed the findings of other authors e. g. [13 - 14]. If we compare the loss factors of the Ag and the AgMg sheathed tapes we find that the AgMg tape has a greater loss factor. This is consistent with the fact that the cross-section of the AgMg tapes looks more elliptical [16]. Both twisted tapes show a lower loss factor than the untwisted tapes. This is not expected because twisting is ineffective to the coupling of the filaments with respect to the self-field [17]. The fact that twisting changes the field and current distribution could serve as a tentative explanation. If we look at the WIT tapes we see that the tape WIT 2 has the greater aspect ratio, and thus looks more like a strip which has a smaller loss factor. The WIT- and the Ag 703 tapes show a smaller loss factor than the Ag- or the AgMg-tapes. An explanation for this could be found in the smaller filament thickness or in the higher j_c value. The results of the jelly-roll and AgAu tapes fit not very well to the predictions of the Norris equation. The reason for this could be found in the other geometry of the filaments which consists not of single filaments but of concentric elliptical filaments. The diagram indicates that the loss for the jell-roll tape is smaller than for the other tapes. The fact that concentric cores had a lower loss than other tapes is in agreement with [15] but in contrast with [13]. The AgAu tapes show another loss behaviour than the tapes with the Ag matrix material. The reason for this may be discussed in the termes of the greater resistivity of the matrix material which is currently studied [18].

Acknowledgments

This work was supported by the Brite/Euram project No BRPR-CT95-0030. We would like to thank S. Zimmer for technical assistance.

References

[1] Norris W T 1970 J. Phys. D. **3** 489-507
[2] Fukumoto Y, Wiesmann H J, Graber M and Suenaga M 1995 J. Appl. Phys. **78** (7) 4584-90
[3] Ishii H, Hirano S, Hara T, Fujikami J and Sato K Cryogenics **36** 697-703
[4] Fukomoto Y, Wiesmann H J, Graber M and Suenaga M 1995 App. Phys. Lett. **67** (21) 3180-82
[5] Goldacker W, Eckelmann H, Quilitz M and Ullmann B 1996 presented at the ASC held in Pittsburgh on August 25-30
[6] Quilitz M and Goldacker W to be published
[7] Däumling M to be published
[8] Hiroyuki F, Ikebe M, Noto K and Matsukawa M 1993 presented at the MTB, Victoria Canada
[9] Goldacker W to be published
[10] Müller K -H and Leslie K E 1996 presented at the ASC, Pittsburgh
[11] Fleshler S, Cronis L T, Conway G E and Malezemoff A P 1995 Appl. Phys. Lett. **67** (21) 3189- 91
[12] Bean C B 1962 Phys. Rev. Lett. **8** 250-53
[13] Mele R, Crotti G, Gheradi L and Morin D 1996 presented at the ASC, Pittburgh
[14] Friend C M, Awan S A , Le Lay L, Sali S and Beales T P1997 to be published in Physica C
[15] Ciszek M, Ashworth S P, James M P, Glowacki B A, Campbell A M , Garre R and Conti S 1996 Supercond. Sci. Technol. **9** 379-84
[16] Eckelmann H, Däumling M., Quilitz M and Goldacker W, to be published
[17] M. N. Wilson Superconducting Magnets, Oxford Science Publication 1983 Chapter 8
[18] Goldacker W, Ullmann B, Gäbler A, and Heller R presented at this conference.

Inst. Phys. Conf. Ser. No 158
Paper presented at Applied Superconductivity, The Netherlands, 30 June–3 July 1997

The influence of the silver sheathed Bi-2223 conductor architecture on the transport ac losses

B A Glowacki[1,2], K G Sandeman[3], E C L Chesneau[1,2], M Ciszek[1], S P Ashworth[1], A M Campbell[1] and J E Evetts[1,2]

[1] IRC in Superconductivity, Madingley Road, University of Cambridge, Cambridge, CB3 OHE, UK

[2] Departement of Materials Science and Metallurgy, University of Cambridge, Cambridge, CB2 3QZ, UK

[3] Churchill College, University of Cambridge, Cambridge, CB3 ODS, UK

Abstract. The influence of the architecture of silver-sheathed Bi-2223 tapes on the ac transport losses is the subject of this investigation. This work is an extension of measurements conducted on monofilamentary and multifilamentary tapes originating from different suppliers, and our own in-house material. Experimental results were selected which utilised the now standard electrical measurement technique using voltage taps. The energy loss data were normalised with respect to energy units and the critical current and compared with theoretical predictions from the Norris equations for elliptical and thin strip cross-sectional geometry. There is a wide variation in the monofilamentary tape behaviour, which, due to the complex microstructural sample specifications was difficult to attribute in general to geometrical differences in the cross-sections. The multifilamentary tapes generally obeyed the higher loss elliptical prediction, thus showing a high degree of shielding of the inner filaments by the outer ones.

1. Introduction

The discovery of high temperature superconductors in 1987 heightened interest in the potential applications of these new materials; it seemed as though research was now not very far away from developing superconducting wires and magnets for commercial application, capable of carrying larger currents at 77K, with lower losses than their metallic counterparts. London, in 1963 [1] and Bean, in 1964 [2] has developed the critical state model for type II superconductors, and used it to calculate the expected dissipative power losses in a round wire. Bean's analysis was extended by Norris [3] in 1970 to wires with elliptical cross-section, thin strips and to a gap between strips. This analysis is also applicable to the new high-T_C superconductors. Unfortunately for the YBaCuO and Tl-1223 superconductors characterised by the highest irreversibility line the powder-in-tube (PIT) technique proved not to be a suitable method to produce a reproducable conductor with even moderate value of the intergranular transport current at magnetic fields of few mT [4,5]. Therefore the only cocnductor that is currently produced that is suitable for power application at liquid nitrogen temperature is Bi-2223 PIT. Analysis of published data and our own measurements on silver-sheathed Bi-2223 tapes carrying ac transport current at frequencies of about 40 to 90 Hz (power frequencies) forms the subject of this paper. Here such tapes are divided into two categories; monofilamentary tapes which have a single body of superconducting ceramic running along the length of the sample, and multifilamentary tapes which contain more than one filament, usually randomly interconnected, including double coaxial tube superconductors.

There are four engineering benefits to multifilamentary tapes; firstly the possibility of higher current densities of the superconductor, secondly better mechanical tolerance, thirdly better thermal stability and fourthly the possibility of filament twisting to minimise the ac losses. It still remains to weigh these advantages against any disadvantages in power loss characteristics and reduction of the overall current density when the comparison is made between mono and multifilamentary conductors. In both tape types, the silver sheathing acts to bind the ceramic together, as well as being a good thermo-electric stabiliser.

2. Experimental details

Current research is poised at an interesting stage. There have been a number of papers published within the last few years, all measuring energy losses due to transport current. The

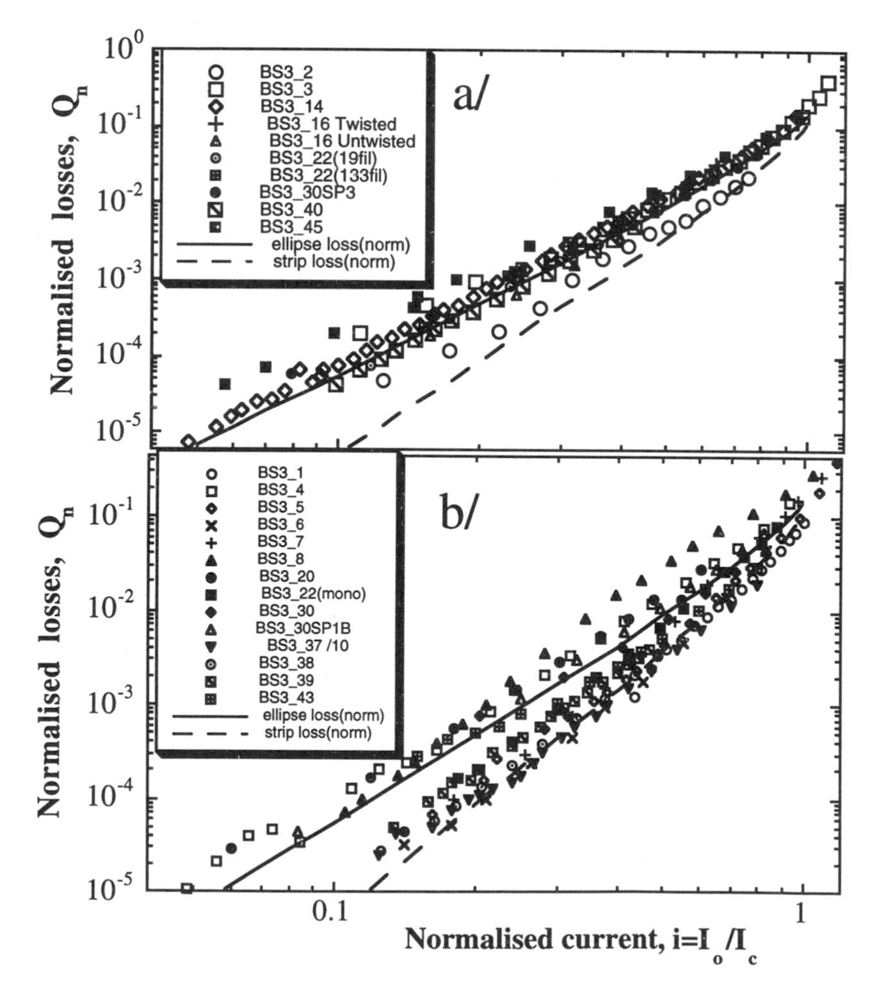

Fig.1 Normalised energy loss per cycle, Q_n vs. normalised transport current amplitude i for Bi-2223 conductors, a) multifilamentary tapes, b) monofilamentary tapes; the numbers in sqare brackets represent corresponding literature sourcs.

technique used in all the papers cited here is direct measurement of the electric field, using terminals attached to the exterior of the sample, as was described previously [6,7]. As has been pointed out [8,9], different tapes of apparently similar architecture obey different loss models some the elliptical model, others the strip model. Here an attempt is made to survey past results and collate them, in order to examine the extent of this variation and to determine possible connections between cross-sectional geometry and the dependence of energy loss on current amplitude. It has already been shown [10] that a non-uniform current density distribution could account for the variation in dependencies outlined above, with this non-uniformity arising from the geometry of the sample concerned. In this review, an alternative approach is taken, starting from geometrical properties, the aim being to compare results of our own measurements and also from the literature. For an accurate comparison, a standardised method of presenting data is required. Authors vary significantly in their choice of ordinate and abscissa and the units in which they are expressed; for the best choice, it is necessary to look at the two Norris equations, (1) and (2).

$$Q = \frac{\mu_0 I_c^2}{\pi}\left[(1-i)\ln(1-i) + \frac{(2-i)i}{2}\right] \qquad \text{elliptical approximation} \qquad (1)$$

$$Q = \frac{\mu_0 I_c^2}{\pi}\left[(1-i)\ln(1-i) + (1+i)\ln(1+i) - i^2\right] \qquad \text{strip approximation} \qquad (2)$$

In both cases, the losses are in units of Joule per metre length of wire per cycle; $i=I_0/I_c$. Since the losses are hysteresic, it is imperative to normalise for their linear variation with frequency, f, which is not contained in Norris' equations. Furthermore, a view independent of the particular critical current may be obtained if $Q_n=\frac{Q}{\mu_0 I_c^2}$ is plotted against i; from equation (1) and (2) the resulting theoretical "normalised energy losses", Q_n only contains the geometrical factors associated with ellipse and thin strip cross-sectional geometry respectively. The normalised losses, Q_n as a function of i are shown in figure 1 for multifilamentary tapes [11-17], and in figure 2 for monofilamentary tapes [8,9,14-16,18-20], along with the theoretical predictions from the normalised Norris equations.

3. Results and discussion

3.1. Multifilamentary tapes

Taking the multifilamentary tapes first, it seems as though the majority of conductors tested conform to the elliptical loss model; the only tape with significantly lower loss is BS3_2 which has only 7 filaments. The current and field distribution results presented for the 7 filament conductor appeared to be very dependent on twisting where the relative position of the filaments in untwisted wire influence the magnetic response of the conductor [21] affecting the ac loss measurements in the whole range of current due to the flux penetration pattern. This would tend to verify an assertion already made by some authors [7,16] that the critical state penetrates the tape in a series of concentric ellipses as the current is increased, indicating either that there is good connectivity between filaments or that there is a high degree of coupling between filaments due to eddy currents induced by the action of the self field of one filament on another. Either situation effectively results in the shielding of the inner filaments by the outer ones, giving bulk penetration of the critical state, as described. Further measurements are required on samples with similar filament numbers to find out whether this is a general phenomenon.

3.2. Monofilamentary tape

The behaviour of the monofilamentary tapes is more complicated to explain. The degree of variation between the samples presented here is just as pronounced as previously cited [8,9]. It is difficult to make any general statement from this subset of the plotted data as even single tube tapes of identical cross-section show different power loss characteristics [8]. Even the planar single superconducting layer in a silver matrix BS3_30SP1B with width to thickness ratio of 10^3 which should "topologically" fulfil the strip model, follows the ellipse approximation [15]. This is due to the highly textured uniform layer enabling uniform penetration of the concentric flux lines into the volume of the superconductor. Therefore it is believe that important factor defining the type of ac losses of the monofilamentary tape is the microstructure related to the local discontinuity of the superconducting core such as micro-cracks and cross-sectional non uniformity [22,23]. Such non uniformity causes macro-percolation of the transport current and as a consequence induces extensive non uniform field distribution in superconducting core. Future work is required to correlate the microstructure and architecture of the superconductors with the two main loss models, for the elliptical and strip geometry.

References

[1] London H 1963 *Phys Letters* **6** 162-5
[2] Bean C P 1964 *Phys.Rev.Lett.* **8** 250-3
[3] Norris W T 1970 *J.Phys.D* **3** 489-507
[4] N.Ozkan N, Glowacki B A, Robinson E A and Freeman P A 1991 *J.Materials Research* **6** 1829-37
[5] Glowacki B A 1997 *Cryogenics* ; Proceedings od LTEC, 1996 Southampton (in press)
[6] Ciszek M, Campbell A M and Glowacki B A 1994 *Physica C* **233** 203-8
[7] Campbell A M 1995 *IEEE Trans. on Appl. Superconductivity* **5** 682-7
[8] M Ciszek M, Ashworth S P, James M P, Glowacki B A, Campbell A M, Garré R and Conti S, 1996 *Supercond.Sci.Technol.* **9** 379-84
[9] Yang Y, Hughes T, Beduz C, Spiller D M, Scurlock R.G and Norris W T 1996 *Physica C* **256** 378-86
[10] Yang Y, Hughes T and Beduz C 1996 *Czech.J.Phys.* **46** 1803-4
[11] Awan S A, Sali S, Friend C M and Beales T P, unpublished data
[12] Fukumoto Y, Wiseman H J, Suenaga M, Haldar P 1996 *Physica C* **269** 349-53
[13] Oota A, Fukunaga T and Ito T 1996 *Physica C* **270** 107-13
[14] Gannon Jnr. J J, Malozemoff A P, Minot M J, Barenghi F, Metra P, Vellego G, Orehotsky J and Suenaga M 1994 *Adv.in Cryogenic Engineering* **40** 45-52
[15] Oota A, Fukunaga T and Matsui M 1996 *Physica C* **265** 40-4
[16] Yang Y, Hughes T, Spiller D M, Beduz C, Penny M, Scurlock R G, Haldar P and Sokolowski R S 1996 *Supercond. Sci. Technol.* **9** 801-4
[17] Ciszek M, Glowacki B A, Campbell A M, Ashworth S P and Liang W Y, 1997 *IEEE Transactions on Applied Superconductivity* **7** 314--7
[18] M Ciszek M, Campbell A M, Ashworth S P and Glowacki B A, 1996 *Appl. Supercond.* **3** 509-20
[19] Fukunaga T, Abe T, Yuhya S, Hiraoka M and Oota A 1995 *Appl. Phys. Lett.* **16** 2128-30
[20] Ashworth S P 1994 *Physica C* **229** 355-60
[21] Johnston M D, Evertt J, Dhalle M, Caplin A D, Friend C M, LeLay L, Beales T P, Grasso G and Flukiger R 1996 *IEEE Tans. Appl Supercond.* (in press)
[22] Glowacki B A and Jackiewicz J 1994 *J.Appl.Phys.* **75**(6) 2992-7
[23] Pashitski A E, Polyanskii A, Gurevich A, Parrell J A and Larbalestier D C 1995 *Physica C* **246** 133-44

Inst. Phys. Conf. Ser. No 158
Paper presented at Applied Superconductivity, The Netherlands, 30 June–3 July 1997

AC-loss measurements on HTSC cable conductor with transposed BSCCO-tapes

C. Rasmussen and S. K. Olsen

Department of Electric Power Engineering, Technical University of Denmark

Abstract. In order to minimise the AC-losses of HTSC cable conductors for power transmission lines operating at 77K, three 1.0 m long BSCCO cable conductors using different designs have been constructed and tested. Besides presenting results from a cable conductor with transposed tapes, we discuss the experimental set-up for AC-loss measurements in general. By varying the losses both as function of applied current and frequency, we have been able to separate the individual loss components.

1. Introduction

The availability of long length of high quality BSCCO-tapes promises a good opportunity to produce high capacity power cables with very low losses in the future. Before a HTSC cable becomes a reality there are, however, several technical matters to investigate. On one hand much work is currently performed in order to optimise the tape production. On the other hand there is a large step from a short length of tape to a finished cable. Before a HTSC cable becomes a reality further understanding and development of single tapes are required [1]. In this paper we present measurements performed on a cable conductor with transposed tapes and discuss the experimental set-up for the determination of AC-losses using the transport current method.

2. Cable conductors

As a part of an ongoing R&D project in Denmark concerning superconducting HTSC cables for power transmission [2], we have constructed cable conductors using several designs in order to perform a detailed investigation of conductor characteristics [3].

The material used for the 1.0 m long cable conductors is Ag-sheathed BSCCO-2223 tape, fabricated by the powder in tube (PIT) [4] method combined with a continuous pressing process. The tape consists of 19 filaments with a cross section of 3 mm × 0.18 mm. The ratio of BSCCO-2223 and Ag is 1:3.9. For all conductors 80 HTSC tapes are used. The critical current density of the tapes is usually $J_C \cong$ 12-14 kA/cm^2.

The former constitute a rigid, yet elastic, basis for the HTSC tapes. It has an outer diameter of 25 mm and consists of a thick phosphorous bronze wire wound in a helix, upon which thin Cu wires are interlaced to obtain a smooth surface. The HTSC tapes are terminated with low melting point solder in a groove cut in a brass cylinder.

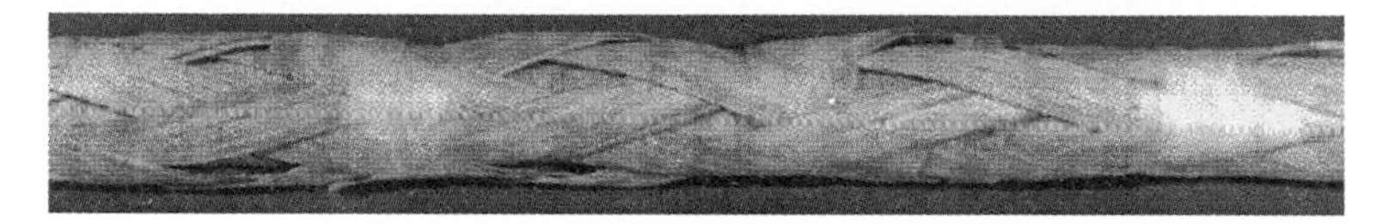

Figure 1 *Photograph of the HTSC cable conductor with transposed layers.*

The differences of the three cable conductors lies in the manner the tapes are wound. Two cable conductors have tapes which are helically wound in 4 layers in opposite directions; one cable conductor has transposed layers of HTSC tape as shown in Figure 1.
In the helically wound cable conductors a variation in the contact resistance to the individual layers is expected and in combination with Beans model [5], this results in an uneven current distribution in the 4 layers of HTSC tape.

The cable conductor with transposed layers of HTSC tapes was designed in an attempt to construct a conductor with similar impedances in all HTSC tapes. This is expected to result in an uniform current distribution. An alternative approach to this was made by Mukoyama et. al. [6], where resistances were connected to the individual layers.

3. Experimental set-up

In order to measure AC-losses, a test system (Figure 2) has been designed to measure these losses at various currents (0-600 A_{rms}) and frequencies (4-150 Hz).

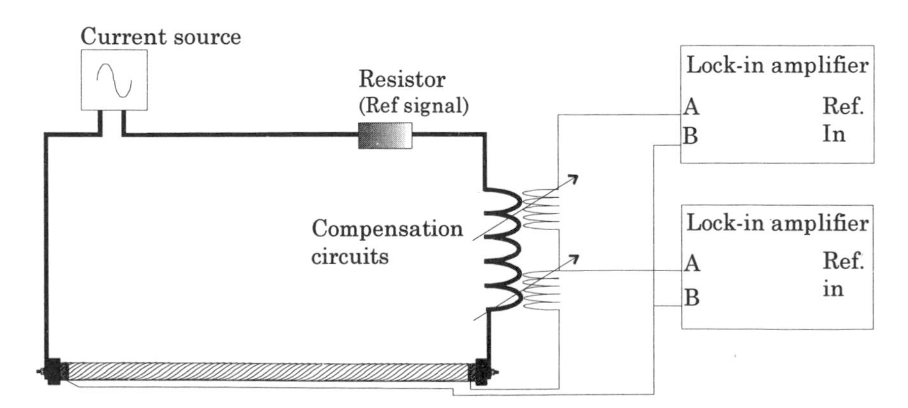

Figure 2 Test circuit for AC loss measurement.

The AC loss measurements are based on the four-terminal principle. An AC current, I, is applied to the conductor. The voltage, U, across the conductor and the voltage-current phase difference, φ, is measured with a digital lock-in amplifier.

In order to minimise errors, two lock-in amplifiers are used. This allows an easy phase calibration and thereby eliminating errors introduced by the non ideal resistor.

To lower the uncertainty of the measurement in conjunction with the lock-in amplifier, a compensation circuit is used. This compensation circuit is made as a part of the voltage leads formed as a coil and introduced in the magnetic field generated by the transport current.

Another essential issue is the position of the voltage taps. It is here discussed for two designs of cable conductors.

In a cable conductor with transposed layers the current distribution is expected to be uniform among the individual layers [6]. This implies that the total losses, P, can be found by multiplying the current with the in-phase voltage component.

In multilayer cable conductors the current distribution is not expected to be uniform. In order to measure the losses it is necessary to measure the voltage drop across and the applied current through the individual layers, respectively. The total losses, P, are then found by addition.

4. Separation of loss components

The AC-losses in a HTSC superconductor consists of three loss components. Eddy current losses (including coupling losses), P_{Eddy}, hysteresis losses, P_{Hyst}, and resistive losses, P_{Res}. Their frequency dependency are f^2, f^1 and f^0 respectively. This makes it possible to distinguish between the individual loss components if the total AC-loss is measured as a function of current and frequency. The total AC-losses are then given by Eqn. 1.

$$P_{Total} = P_{Hyst} + P_{Eddy} + P_{Res} = \alpha \cdot f + \beta \cdot f^2 + \delta \quad \textbf{(1)}$$

where the relationship between the individual loss components are given for constant current level. α, β and δ are functions of I, but independent of f.

5. Results and discussion

The cable conductor with transposed layers has a critical current of 640 A (1 μV/cm criterion). The total AC-loss (50 Hz) is measured and the hysteresis loss calculated by means of the UCD-model [6]. Figure 3.a. shows the measured AC-losses (total loss) and calculated hysteresis loss. The measured values show a good agreement with the calculated values. The result of the separation of loss components are given in Figure 3.b. It is seen that the hysteresis loss is dominant at all current levels. Eddy current losses have a little significance at low current levels and resistive losses become important at high currents.

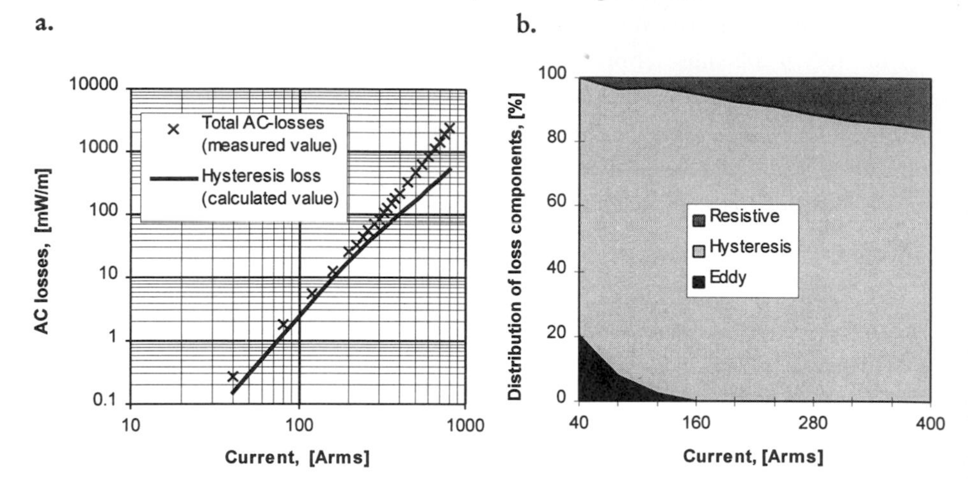

Figure 3.a. Comparison of measured and calculated losses. **b.** Separation of the individual loss components.

6. Conclusion

We have constructed 3 superconducting cable conductors and performed several test with a set-up made for this purpose. A good agreement between measured and calculated losses was found for the cable conductor with transposed layers. By varying the frequency of the current signal we were able to separate the various loss components.

Due to the uneven current distribution in the multilayer helically wound cable conductors, the used method of measuring the AC-losses does not provide us with the correct results. In the future helically wound cable conductors (for test purposes) will have separate current supply to each layer, permitting valid loss measurements.

References

[1] Hensel B, Grasso G, Flükiger 1995, *Limits to the critical transport current in superconducting $(Pi,Pb)_2Sr_2Ca_2Cu_3O_{10}$ silver-sheated tapes: The railway-switch model, Phys. Rev. B* **51** 15456-15473

[2] Oestergaard J 1996, *Superconducting Power Cables in Denmark - A Case Study*, Presented at the 1996 Applied Superconductivity Conference, Pittsburgh, USA.

[3] Rasmussen C, Olsen S K April 1997, Graduation report (in Danish), Unpublished

[4] Li Q, Brodersen K, Hjuler H.A., Freltoft T. 1993, *Critical density enhancement in Ag-sheated Bi-2223 superconducting tapes, Phys. C* **217** 360-366

[5] Bean C P 1964, *Magnetization of High-Field Superconductors*, *Reviews of Modern Physics* 31-39

[6] Mukoyama S, Miyoshi K, Tsubouti H, Mimura M, Uno N, Ichiyanagi N, Tanaka Y, Ikeda M, Ishii H, Honjo S, Sato Y, Hara T, Iwata Y 1996, *50-m Long HTSC Conductor for Power Cable, IEEE Transactions on Applied Superconductivity, Vol. 7, No. 2, June 1997* 1069-1072

Inst. Phys. Conf. Ser. No 158
Paper presented at Applied Superconductivity, The Netherlands, 30 June–3 July 1997

Hysteresis Phenomena in the I-V Characteristics of Bi(2223)/Ag Tapes at 4.2 K

P Kováč*, L Cesnak*, T Melišek*, H Kirchmayr and H Fikis****

*Institute of Electrical Engineering SAS, Dúbravská cesta 9, 84239 Bratislava, Slovakia
** Institute for Experimental Physics TU Wien, Wiedner Hauptstr. 8, 1040 Wien, Austria

Abstract. The hysteresis in the magnetic field-, angular-, and self field dependence of the I - V characteristics and, respectively, of critical currents was entirely examined at 4.2 K. These phenomena are commonly attributed to the effect of screening currents induced and then frozen-in by the changing magnetic field. The following conclusions can be drawn: 1. The self-field hysteresis manifests itself by the difference between the first and all following cycles at increasing and decreasing the transport current, namely when the external magnetic field is decreased from a higher value. 2. The angular dependence of critical currents should be followed exclusively after rotating the tape in zero transversal magnetic field. 3. The hysteresis at increasing and decreasing the external magnetic field can be overcome by adequate cycling the magnetic field around the adjusted value.

1. Introduction

The magnetic field (B) hysteresis of transport critical currents (I_c, defined at 1 μV/cm) in high T_c superconductors was firstly discovered and, as confirmed by experience, also correctly explained in 1987 investigating the $YBa_2Cu_3O_7$ material [1,2]. The $I_c(B)$ hysteresis is observable also in Bi(2223)/Ag tapes namely at low temperatures (4.2 K) and up to enough high magnetic fields (12 T) [3], while in Bi(2212)/Ag tapes it is very small. At 77 K the hysteresis vanishes. The $I_c(B)$ hysteresis is attributed to the effect of trapped flux in a network of frozen-in currents that affect electron transport in the intergrain region. In [4] the opinion was advanced that screening currents are frozen-in rather in grain colonies than in single grains and this opinion can be deduced also from our experiments [5].

In the presented contribution we refer to three problems concerning Bi(2223)/Ag tapes at 4.2 K, as follows:

1. to the hysteresis in the $I(V)$ characteristics at increasing and decreasing the transport current in a constant external magnetic field (self-field hysteresis),
2. to the correct interpretation of the critical current dependence on the tape's surface orientation in transversal magnetic field and
3. to the possibility to overcome the hysteresis effect by cycling the external magnetic field.

2. Experimental

The experiments have been carried out with four untwisted and differently treated 19-filament Bi(2223)/Ag tape samples prepared by the PIT method [6].

In Figure 1, a series of $I(V)$ characteristics is shown for the parallel orientation of the tape's surface to the magnetic field, at magnetic fields set up and down.

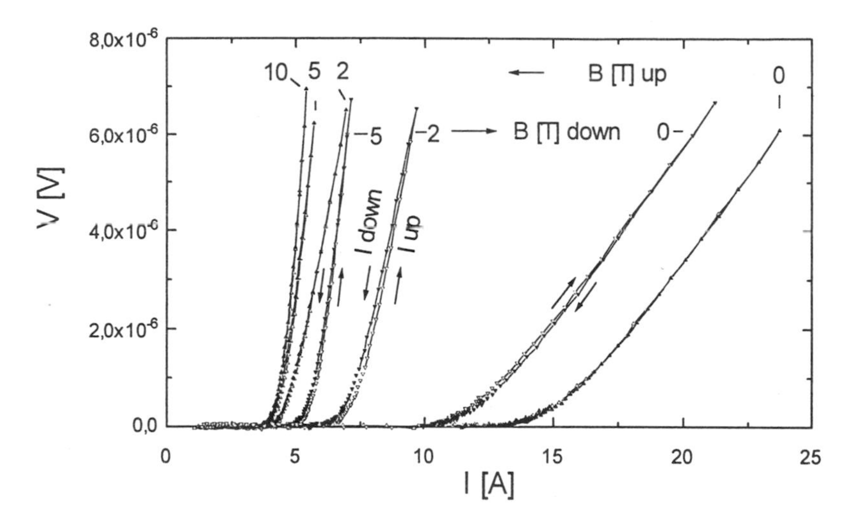

Figure 1. Self-field hysteresis

The characteristics were recorded at very slowly increasing (open symbols) as well at decreasing (full symbols) transport currents. While in zero field (omitting the remnant field of the superconducting magnet producing the external field) different narrow hysteresis loops were registered, in magnetic fields going up any loop can be observed. In contrary, at magnetic fields going down clear narrow loops can be seen. The currents are higher at the first rising the transport current. At each following cycle the hysteresis more or less vanishes and the characteristics follows the decreasing branch of the first loop. In the perpendicular orientation of the tape in the magnetic field the same behaviour was observed.

Figure 2 shows critical currents in dependence on the rotation of the tape's surface with respect to the external magnetic field at 0.5 T. Zero angle means the tape's surface being perpendicular to the magnetic field. Circle symbols are critical currents in a 0.5 T field going up from zero, while square symbols mark critical currents in a 0.5 T field going down from 1 T. The open circle (or open square) and cross symbols connected by a full line show critical currents measured at a rotation sequence from 95° to -5° and backwards when the sample was rotated in zero field and the field 0.5 T was set up from zero after each change of orientation (or set down from 1 T, respectively). In contrast to this, the full circle symbols connected by a dotted line show critical currents measured at rotating the sample backwards in the fixed magnetic field of 0.5 T. Measured values in this case are affected by screening currents induced by rotation in the magnetic field. The conclusion from this experiment is that in order to see the proper I_c hysteresis effect in a constant magnetic field at different orientations of the sample's surface with respect to the magnetic field, it is necessary to rotate the sample in zero magnetic field.

The effect of induced and frozen-in screening currents in superconductors at changing the magnetic field can be suppressed by splitting the screening current loop into patterns with oppositely flowing currents [7]. The splitting can be induced by cycling the magnetic field around the adjusted value with subsequently decreasing amplitudes. We applied this method at examining the $I(V)$ characteristics of our samples. An example of series of characteristics resulting after some cycling steps is shown in Figure 3 in $\log V$ - log

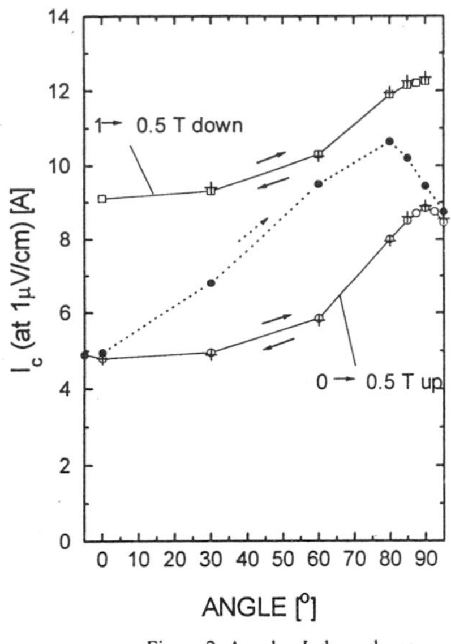

Figure 2. Angular I_c dependence

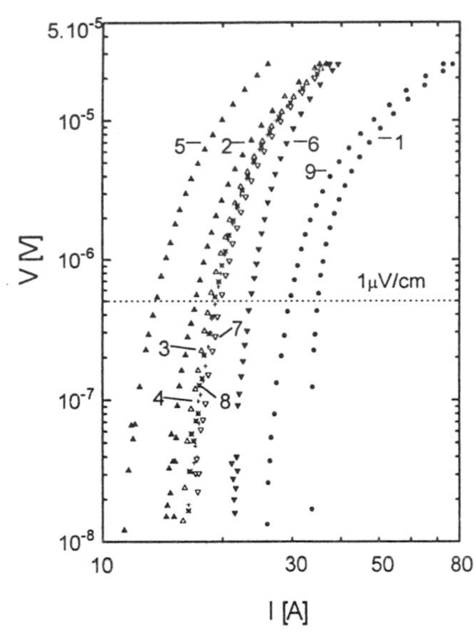

Figure 3. *V(I)* characteristics during cycling

co-ordinates. The cycling took place after adjusting the magnetic field 2 T in the parallel orientation of the tape's surface with the external magnetic field either from zero (0 to 2 T up, scatter full circle symbols (1) to full open triangle symbols (2)) or from 5 T (5 to 2 T down, scatter full triangle up symbols (5) to full triangle down symbols (6)). The characteristics 3 (scatter open triangle up symbols) resulted from characteristics 2 after the following cycling tour:2-2.5-2-1.5-2-2.4-2-1.6-2-2.3-2-1.7-2-2.2-2-1.8-2-2.1-2-1.9-2 T. The next characteristics 7 (open triangle down symbols) was recorded after a small cycle 2-2.08-2 T, followed by characteristics 4 (cross symbols) after cycling 2-1.94-2 T and characteristics 8 (star symbols) after cycling 2-2.04-2 T. The last two characteristics were almost identical indicating that the splitting of screening currents was closed. These characteristics can be seen as the "neutral characteristics" of the sample. They are set free from the effect of induced frozen-in screening currents. A similar sequence of *I*(*V*) characteristics is obtained by cycling from characteristics 6 beginning with the first cycle 2-1.5-2 T and so on. Characteristics 9 (full circle symbols) was measured after finishing the cycling procedure and decreasing the magnetic field from 2 T to 0 T. The described cycling procedure can be objectively presented in the I_c - B plane as shown in Figure 4. The open circle symbols are critical currents measured after the individual cycling steps at 2 T, while the full circle symbols show critical currents at the positive or negative ΔB amplitudes in the cycling tour.

In principle, the same situation holds in the perpendicular orientation of the tape's surface with respect to the magnetic field and a similar cycling procedure can be applied to obtain the "neutral" $I_c(B)$ characteristics.

3. Conclusions

From the above described experiments following conclusions can be drawn:

1. The *self field hysteresis* is pronounced when the magnetic field has been set down from a higher value.
2. The *proper angular dependence* of I_c can be obtained but if the tape would be rotated in zero magnetic field.
3. The *neutral $I_c(B)$ characteristics* can be obtained by cycling the external magnetic field with successively decreasing amplitudes. By this way it is possible to overcome the effect of frozen-in screening currents in grains or grain colonies.

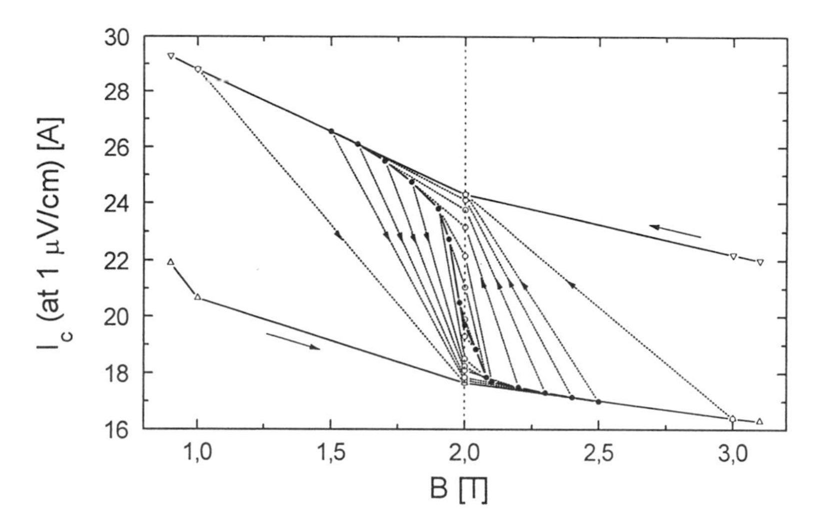

Figure 4. Cycling procedure in the $I_c(B)$ characteristics

Acknowledgements

The experiments have been carried out in the frame of the "Action Austria - Slovak Republic for scientific co-operation" (No. 15-SR 14) and within Association Euratom-ÖAW (GB6-MAG-M16/02) and were supported by the EU Fusion Project "Materials for superconducting fusion magnets" and by the Slovak Academy of Sciences project VEGA 95/5305/108.

References

[1] Evetts J E and Glowacki B A *Cryogenics* **28** (1988) 641
[2] Kwasnitza K and Widmer Ch *Cryogenics* **29** (1989) 1035

[3] Murakami Y, Itoh K, Yuyama M and Wada H *CEC/ICMC Conf.* (1995) Columbus USA
[4] Hu Q Y, Schalk R M, Weber H W, Liu H K, Wang R K, Czurda C and Dou S X *J.Appl.Phys.* **78** (2) (1995 1123
[5] Cesnak L, Melišek T, Kováč P and Hušek I will be published in *Cryogenics*
[6] Kováč P, Hušek I, Pachla W, Melišek T and Kliment V *Sup.Sci.Technol.* **8** (1995) 341
[7] Cesnak L and Kokavec J *Cryogenics* **17** (1977) 107

Inst. Phys. Conf. Ser. No 158
Paper presented at Applied Superconductivity, The Netherlands, 30 June–3 July 1997

Magnetic AC Loss in Multi-filamentary Bi-2223/Ag Tapes

M.P. Oomen, J. Rieger, M. Leghissa

Siemens AG, Corporate Technology, 91050 Erlangen, Germany

Abstract. AC losses in Bi-2223/Ag tapes with twisted and non-twisted filaments were measured by a magnetic method. Knowledge of the AC loss in fields parallel and perpendicular to the tape plane is sufficient to predict the AC loss for intermediate field orientations. Application of DC transport current together with parallel AC field made it possible to distinguish between the loss components, arising from intra-grain and from filament currents. In tapes with non-twisted filaments, the magnitude of the filament loss component indicates that the filaments are fully coupled, which agrees with theory. In tapes with short twist lengths, the AC loss at low field amplitudes is caused by hysteresis in the separate filaments.

1. Experimental method

When superconducting tapes are used in cables, transformers and other power devices, they are subjected to an AC magnetic field which is stronger than the self-field caused by the transport current in one tape. It is necessary to know the power dissipation (AC loss) which will be caused in the tapes by this oscillating magnetic field. The AC loss in multi-filamentary tapes is due to hysteretic screening currents inside the superconducting filaments, to coupling currents between the filaments and to eddy currents in the normal-conducting sheath. The interaction between these loss components is still difficult to predict and therefore high-quality measurements are essential to estimate the AC loss which will occur in applications.

An experimental set-up has been constructed in which tape samples can be subjected to an external field of 0 - 0.7 T amplitude and 50 - 1000 Hz frequency. Samples up to 10 cm long can be inserted at any orientation with respect to the field. DC transport current can be applied to samples up to 2 m long, oriented parallel to the field. The magnetisation of the sample is detected with pick-up coils, is multiplied by the field strength and integrated over several field cycles in order to obtain the AC power loss.

2. Dependence of AC loss on field orientation

The AC loss measured in a twisted-filament Bi-2223/Ag tape is displayed in Figure 1 for various orientations of the tape in the external magnetic field. In this figure the commonly used loss function [1] $\Gamma = \mu_0 Q / 2B_a^2$ (where Q is the energy dissipation per field cycle, per m^3 of tape) has been plotted against the external field amplitude B_a. The loss in perpendicular field (solid dots in Figure 1) is larger by about a factor 20 than the loss in parallel field (solid triangles), which will be studied more closely in the following sections. Both these losses have been fitted by suitable mathematical functions $Q_\perp$ resp. $Q_{//}$ (solid lines).

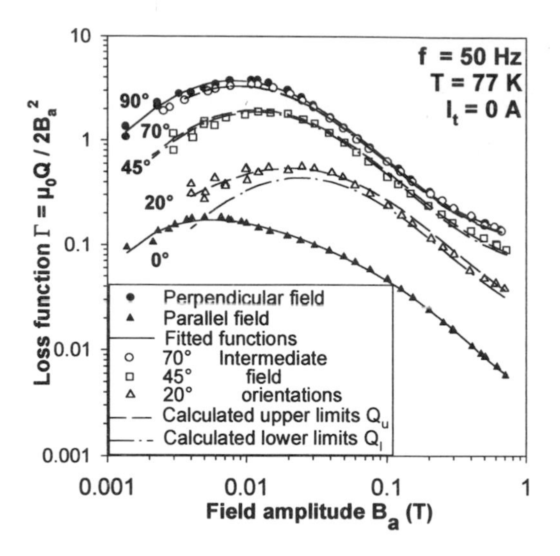

Figure 1 AC loss measured for various field angles, and calculated limits

In order to predict the AC loss at intermediate field orientations, two loss limits Q_u and Q_l have been calculated. They are displayed in Figure 1 by their respective loss functions:

dashed lines: upper limit $Q_u(B_a,\alpha) = Q_\perp(B_{a\perp}) + Q_{//}(B_{a//}) = Q_\perp(B_a \sin\alpha) + Q_{//}(B_a \cos\alpha)$

dash-dot lines: lower limit $Q_l(B_a,\alpha) \;=\; Q_\perp(B_{a\perp}) \qquad = Q_\perp(B_a \sin\alpha)$

where α is the angle between the external field direction and a normal to the tape plane, and $B_{a//}$ and $B_{a\perp}$ are the amplitudes of the field components parallel and perpendicular to the tape, respectively. The theoretical background to this calculation and its limitations are discussed in [2]. The predicted limits Q_u and Q_l form a good approximation to the measured loss at intermediate α (open dots in Figure 1). They describe the shifting of the maximum in Γ towards larger field amplitudes B_a. Furthermore, for most angles α the difference between Q_u and Q_l is relatively small. The calculation procedure leads to a loss prediction which is accurate to within 10% for all examined tapes.

3. Influence of DC transport current

The loss functions measured in a non-twisted filament tape carrying DC transport current have been plotted against field amplitude in Figure 2. For increasing transport current I_t, the loss function first shifts to the left, then decreases greatly and finally stays almost constant for $I_t > 60$A. The increasing transport current takes up more and more space in the filaments suppressing the screening and coupling currents which cause a large part of the AC loss. The filaments are fully saturated at $I_t = 60$A, which is larger than this tape's critical current of 40A. This agrees with current-sharing measurements which will be reported in [3]. The remaining AC loss at $I_t > 60$A is caused by small-scale screening currents inside the grains of Bi-2223, which have a much higher critical current density than the filament as a whole.

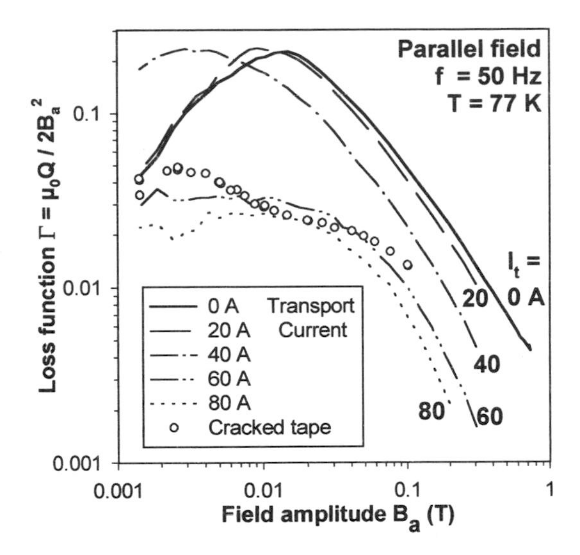

Figure 2 AC loss measured with various DC transport currents

Approximately the same grain loss is measured when the tape is bent over a small radius, so that many cracks are introduced in the filaments and most of the links between the grains are destroyed. This cracked-tape AC loss is displayed also in Figure 2.

If the transport current does not influence the screening currents inside the grains, the AC loss in the grains is independent of I_t. Then the grain loss found at $I_t > 60A$ can be subtracted from the total AC loss at $I_t = 0A$, in order to obtain the AC loss caused by screening and coupling currents following the filaments. For the above-mentioned non-twisted filament tape, the filament loss function Γ_n found in this way has been plotted against field amplitude in Figure 3 as open dots.

4. Decrease of AC loss by filament twist

For another Bi-2223/Ag tape with twist pitch $L_p = 6.5mm$ and critical current $I_c = 28A$, the AC loss was measured and the loss in the filaments was found by the procedure described above. This loss function Γ_t is displayed as solid triangles in Figure 3: its maximum is lower than that of Γ_n (for the non-twisted tape) and lies at a lower field amplitude B_a. The grain loss in both tapes was similar, making the total AC loss in the twisted-filament tape evidently smaller than in the non-twisted tape for field amplitudes larger than 0.005T.

Hysteresis loss in superconducting filaments can be described by the Bean critical-state model [1, 3, 4]. The filaments in a tape may be modelled as infinite superconducting slabs parallel to the field. The loss function for such a slab is

$$\Gamma = B_a / 3B_p \qquad \textbf{for } B_a < B_p,$$

$$\Gamma = B_p / B_a - 2B_p^2 / 3B_a^2 \qquad \textbf{for } B_a > B_p.$$

where B_a is the external field amplitude. The penetration field $B_p = \mu_0 J_c d / 2$, where J_c is the critical current density and d is the slab thickness.

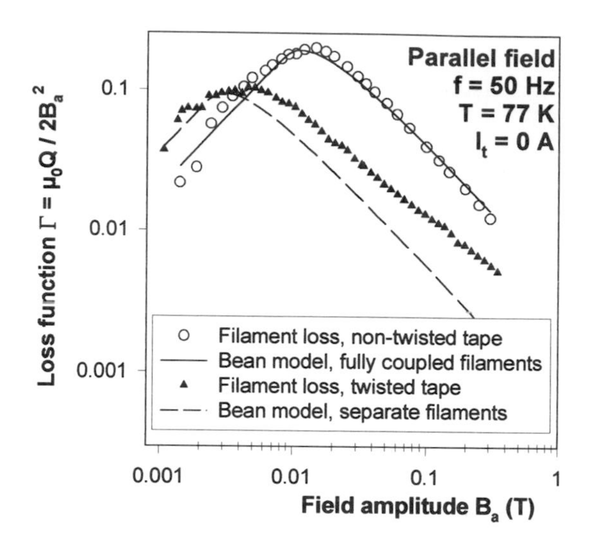

Figure 3 AC loss in filaments compared to the Bean model

This loss function has been plotted as a dashed line in Figure 3, accounting for the fact that the superconducting filaments fill about 25% of the tape volume. A penetration field of 0.0023T has been assumed, giving a reasonable filament thickness d = 19μm. This loss function agrees with the filament loss measured in the twisted-filament tape at low field amplitudes. At higher B_a the AC loss is higher due to inter-filament coupling currents.

In the non-twisted tape the maximum loop length of the coupling currents is equal to the sample length of about 1m. In that case the filaments are expected to be fully coupled: the coupling current in the filaments equals their critical current. Then the entire filamentary core of the tape (which occupies about 50% of the tape volume) behaves like one large filament, to which the Bean model can once again be applied. The solid line in Figure 3 is the loss function calculated with this model for the entire tape core, with B_p = 0.0085T, giving a reasonable core thickness d = 162μm. It describes very well the filament loss measured in the non-twisted tape at all field amplitudes. This indicates that in the non-twisted tape the filaments are indeed fully coupled [3], while in the twisted tape coupling currents play a role only at the higher field amplitudes.

This conclusion is supported by measurements of the AC loss in these tapes at higher field frequencies. The measured loss increases with frequency, but in the twisted-filament tapes this increase is much faster than in the non-twisted tapes. In the twisted tapes the magnitude of the inter-filament coupling currents can increase with frequency. In the non-twisted tapes this does not happen because at 50Hz the filaments are already saturated by the coupling currents.

Financial support by the German BMBF under contract nr. 13N6481 is gratefully acknowledged.

References

[1] Wilson M 1983 *Superconducting magnets* (Clarendon Press, Oxford)
[2] Oomen M P, Rieger J, Leghissa M and ten Kate H H J 1997 Appl. Phys. Lett. **70** 3038-3040
[3] Oomen M P, Rieger J, Leghissa M and ten Kate H H J 1997 to be published in Physica C
[4] Bean C P 1962 Phys. Rev. Lett. **8** 250-253

Inst. Phys. Conf. Ser. No 158
Paper presented at Applied Superconductivity, The Netherlands, 30 June–3 July 1997

A novel way of measuring hysteretic losses in multi-filamentary superconductors

M Dhallé[1], F Marti[1], G Grasso[1], Y B Huang[1], A D Caplin[2] and R Flükiger[1]

[1]Département de Physique de la Matière Condensée, Université de Genève, 24 quai Ernest-Ansermet, 1211 Genève 4, Switzerland. [2]Centre for High Temperature Superconductivity, Imperial College, London, UK

Abstract. At present, magnetic hysteresis proves to be a dominant contribution to AC losses in high T_c superconducting tapes. The irreversible magnetisation of a superconductor is proportional to the physical size of the screening current loops times the critical current density. In order to reduce hysteretic losses, it is clear one should decrease the loop size whilst leaving the current density unchanged. Several existing and novel methods to do so, - multi-filament configurations, twisting, electrical de-coupling - are being adapted to or pioneered for high Tc conductors. It is desirable that the success of these efforts can be tested by a relatively straightforward experiment. The 'magnetic length-scale' technique, previously used mostly to assess the degree of granularity of high Tc samples, provides such a measurement. It allows independent determination of both current density and current flow distribution in a single experiment. In this paper we present vibrating sample magnetometer data on a variety of multi-filamentary Bi(2223) tapes, discussing the experimental method and demonstrating its usefulness when applied to hysteretic losses.

1. Introduction

Hysteretic AC loss in superconductors can be understood as originating from Faradays law : a time varying magnetic flux induces an electric field, which in turn generates a current response. When the changing flux is due to a fully penetrating AC transport current in the conductor itself, the corresponding self field loss per cycle can be shown to be proportional to $I_c^{\,2}$, with I_c the critical current of the conductor [1]. When one distributes the same total current over N *independent* filaments the loss reduces by a factor $1/N$. Similar arguments show losses due to an external AC field (e.g. generated by neighbouring windings in a coil) to be proportional to j_cW, with j_c the critical current density and W the characteristic size of the conductor perpendicular to the magnetic field [2]. Also here N *independent* filaments will display a lower hysteretic loss $\propto J_cw$, where w is the filament size (see figure 1). In practice, however, the filaments are not independent: induced coupling currents flow through the metallic matrix from filament to filament, resulting in a loss $\propto j_cR$ with $w < R < W$.

In this paper we demonstrate the use of the non-destructive «length-scale» technique [3] to measure R, thus gauging the contribution of coupling currents to the AC loss.

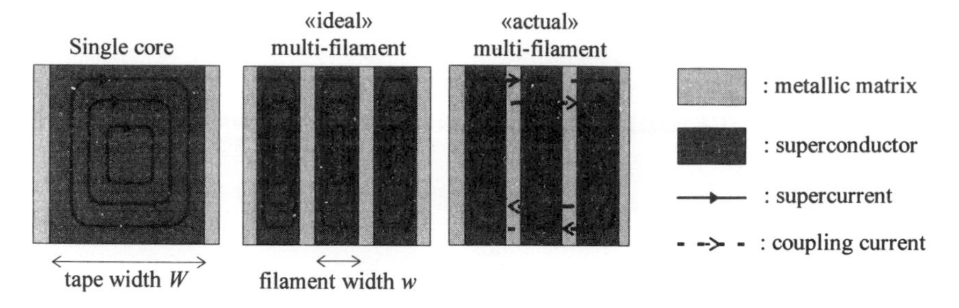

Figure 1 : schematic representation of the coupling current contribution to the AC losses

2. Experimental

The length-scale technique was originally developed to determine the degree of granularity in ceramic HTS samples [3]. The experiment consists of precise magnetisation measurements upon external field sweep reversal, i.e. during the initial re-penetration of field and current profiles just after dH/dt changed sign. Essentially, the current response in this regime is such that it counteracts the change in flux due to the external field. This screening effect is achieved by current reversal in a thin layer of width δr and thickness t around the periphery of coherently screened regions (figure 2). The flux generated by the current $I = j t \delta r$ in this surface layer is LI, where L is the self inductance of the peripheral current loop. A Bean type argument shows that $\delta r \propto 1/j$, so that the current I in this surface layer *does not depend* on the current density j. Consequently, the initial magnetic response to a field sweep reversal is solely determined by L, i.e. solely by the geometry of coherently screened regions.

A more quantitative analysis requires some model which expresses L in terms of relevant lengths. Two plausible models are a thin disk [3] or a thin strip [4], which lead to the following expressions for the initial differential susceptibility dM/dH of the re-penetrating magnetisation M :

Disk: $$\frac{dM}{dH} = \frac{\pi}{\Theta(R/t)} \frac{R}{t} \quad \text{with} \quad \Theta(x) = \ln(8x) - 0.5, \text{ radius } R \text{ and thickness } t$$

Strip: $$\frac{dM}{dH} = \pi \frac{a}{t} \quad \text{with} \quad \text{half width } a \text{ and thickness } t$$

In practice, the differences between such models are limited to constant factors of order

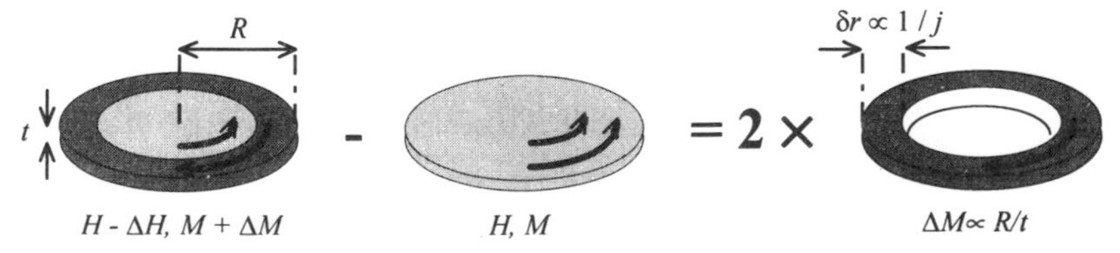

Figure 2 : schematic representation of the screening current flow upon field sweep reversal. The current direction changes initially only at the periphery of coherently screened regions, in a layer of width inversely proportional to j. The resulting change in magnetisation - corresponding to twice the flux generated by this layer - is only determined by the screening geometry and not by the value of the current density itself.

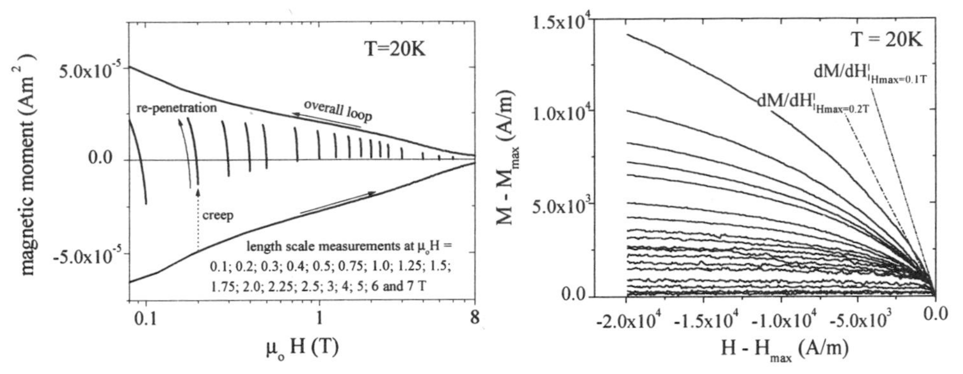

Figure 3 : Example of a length-scale measurement for a 19-filament Bi(2223) tape at $T = 20$K. At the left the raw data are shown (both the overall *mH* loop and the various re-penetration profiles), while the right figure shows the re-penetration curves, shifted by the values of external field H and magnetisation M at field sweep reversal. The length-scale R is calculated from the initial differential susceptibility dM/dH upon reversal.

unity. In what follows we will use the disk model, which has been successfully used before to study the behaviour of mono-filamentary Bi(2223)tapes [5].

The data shown in this paper were obtained in a standard vibrating sample magnetometer, ramping the field up relatively fast (~ 30 mT/s) to a desired value, pausing ~ 100s and slowly (~ 0.1mT/s) ramping back down 25mT, after which the whole cycle is repeated at the next target field. The pause ensures that the electric field induced by the creep-related relaxation of the magnetisation has sufficiently decayed to be negligible compared to the fields induced during re-penetration ($E \sim 10^{-7}$V/m). Figure 3 illustrates such a set of measurements taken at $T = 20$K on a 19 filament Bi(2223) tape.

As an example of the data obtainable with this technique, we will compare three PIT Bi(2223)/Ag tapes. The first is a 3 mm wide single core sample, with overall and oxide thickness of 100 and 45μm, resp., and a self-field DC transport j_c at 77K of 25kA/cm^2. The second is a standard 19 filament tape, 3.6mm wide and 260μm thick, with average filament section 500×15μm^2 and j_c 21kA/cm^2 [6]. The third is again a 19 filament tape (3×0.2mm^2, filaments 300×13μm^2 and j_c 14kA/cm^2) in which each individual filament is surrounded by a 1μm thick $BaZrO_3$ insulating barrier layer in an attempt to reduce coupling-current flow between the filaments [7]. Cross-sectional micrographs of the samples are shown in figure 4.

3. Results and discussion

Figure 4 presents the magnetic length-scale R for the three Bi(2223) tapes as a function of temperature (T=10-60 K) and external field ($\mu_0 H$=0.1-7 T). The data was calculated from the raw *MH* behaviour using the disk formula given above. In the case of multi-filamentary samples the question arises whether to take the tape - or filament thickness whilst extracting R. Preliminary calculations show that a stack of N disks with a separation between them which is much smaller than their radius have essentially the same magnetic response to an external field parallel to the stacking direction as a single disk of identical radius but with a N times larger thickness. The thickness t used in the length-scale formula therefore was the overall tape thickness multiplied by the superconducting fraction of each multi-filament (12 and 16% for the tapes with and without oxide barrier, resp.).

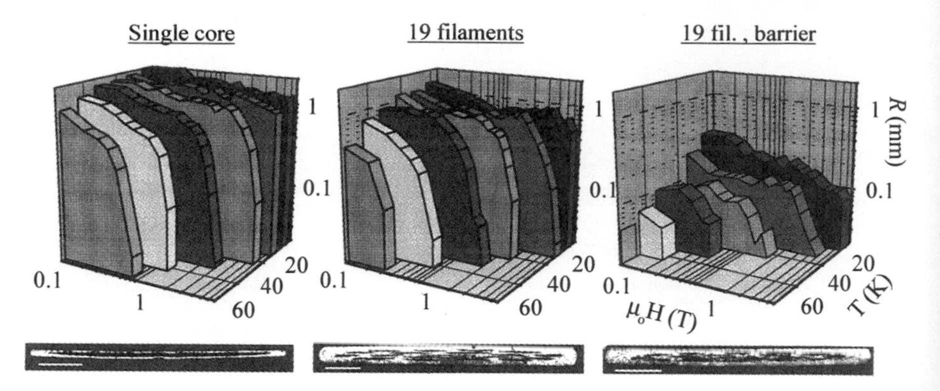

Figure 4: Magnetic screening current length-scale R plotted as a function of temperature and external field. Cross-sectional micro-graphs of the 3 samples - a mono-core Bi(2223) tape, a standard 19-filamentary Bi tape and a 19 filament tape with resistive barriers between the filaments - are shown below the magnetic data. The white bars in the micro-graphs correspond to 0.5 mm.

All three tapes behave qualitatively the same. At low external field and temperature, the length-scale R has the tendency to decrease slightly with H or T, reflecting the residual granularity in the polycrystalline Bi(2223) material [5]. However, after this initial decrease, R reaches a constant (T and H independent) plateau value persisting all the way up to the irreversibility line, where the length-scale drops steeply to zero. The plateau value of R reveals the difference between these three tapes. In the single core sample R $\approx$ 1.5mm, corresponding to the tape half-width. Such behaviour is generic for mono-filamentary Bi(2223) tapes and indicates coherent current flow throughout the core. The plateau value of R in the standard 19 filament tape (R $\approx$ 800μm) is significantly lower than the overall half-width, but still much larger than the filament width. Thus the filaments are not screened independently, but interact through non-negligible lateral coupling currents. In the third tape, with electrically insulating barriers between the filaments, these currents are effectively suppressed, leading to a reduction of R to 100-150μm, i.e. the filament half width.

4. Conclusions

We demonstrated the use of the magnetic length-scale technique in assessing the importance of filament coupling in multi-filamentary superconductors. This relatively simple magnetic experiment clearly shows the differences in current flow geometry between samples with varying degree of filament interaction. As such it provides a useful tool in further optimisation of AC loss behaviour of superconducting tapes and wires.

References

[1] Norris W T 1970 *J. Phys. D: Appl. Phys.* **3** 489-507
[2] Bean C P 1964 *Rev. Mod. Phys* **36** 31-39
[3] Angadi M A, Caplin A D, Laverty J R and Chen Z X 1991 *Physica C* **177** 479-486
[4] Brandt E H and Indenbom M 1993 *Phys. Rev. B* **48** 12893-12906
[5] Dhallé M, Cuthbert M, Johnston M D, Everett J, Flükiger R, Dou S X, Goldacker W, Beales T and Capli A D 1997 *Supercond. Sci. Technol.* **10** 21-31
[6] Grasso G, Marti F, Huang Y B and Flükiger R, *Proceedings ISS96, Sapporo, Japan, 30Oct-2Nov, 1996*
[7] Huang Y B, Grasso G, Marti F, Dhallé M, Clerc S, Kwasnitza K and Flükiger R, *this volume.*

Inst. Phys. Conf. Ser. No 158
Paper presented at Applied Superconductivity, The Netherlands, 30 June–3 July 1997

Measurements of AC-field losses (50 Hz) on Ag-sheathed $(Bi,Pb)_2Sr_2Ca_2Cu_3O_{10}$ tapes

Andrea Kasztler, Hannes Fikis, Pierre Bauer, Hans Kirchmayr

Institute for Experimental Physics, University of Technology Vienna, Austria

Abstract. For measurements of the total AC-field losses of high-T_C superconductors an inductive measurement facility has been developed. It operates at 77 K and 50 Hz in fields up to 150 mT. Apart from a description of the measurement facility first results of two special Ag-sheathed $(Bi,Pb)_2Sr_2Ca_2Cu_3O_{10}$ tapes are reported. Such conductors are among the most promising candidates for applications such as superconducting power lines, current limiters and transformers. The total field losses being proportional to the area enclosed by the magnetisation loop, the AC-losses of the samples at background fields of 50 Hz up to 150 mT have been determined.

1. Introduction

At the moment the most promising conductor material for HTS-power lines are Bi-2223/Ag-sheathed tapes. It is possible to produce such tapes in length of several hundred meters making it possible to use them for transport currents with current densities of more than 10^8 A/m^2. This is achieved by using the "powder-in-tube" production method [1].
Before using this kind of tapes in superconducting power lines it is necessary to investigate their AC-losses under low magnetic fields at 50 Hz. During the past few years a number of AC-loss measurements have been performed on high-T_c superconductors [2, 3]. But only a few on Bi-2223 multifilament tapes [4].

2. Experimental

2.1. Measurement facility

Using the inductive method for measurements of the AC-losses of BPSCCO samples there was constructed a system of two concentric coils (the pickup and the field coil) which are embedded in liquid nitrogen during the measurements.
The outer coil which provides the magnetic field consists of four layers of a 1 mm copper wire. At each side of the coil there are several additional windings which increase the homogeneity of the magnetic field up to 0,3 % in a region of ±4 cm in the middle of the coil and in the direction of its length axis. The dimensions of this coil are listed in table 1. Using

the mains it is possible to achieve magnetic fields up to 0,15 T at a frequency of 50 Hz with this coil.

Table 1: Dimensions of the field-coil

length	l = 233 mm
inner bore	a = 75 mm
diameter of the wire	d = 1 mm
number of windings	n = 998

There was also constructed a pickup coil with an inner bore of 57 mm. When using the inductive method the sample is put into the coil-system (at the position of maximum signal) and the voltage induced in the pickup coil due to magnetisation of the sample is amplified and integrated.

2.2. Samples

The two samples which are called MK7 and MK15 consist of superconducting BPSCCO-filaments embedded in pure silver as seen in fig. 1. The mean chemical structure of the untwisted superconducting filaments is $Bi_{1,8}Pb_{0,4}Sr_2Ca_{2,1}Cu_3O_{10}$. Which means that the oxides of the elements were mixed with this molar ratio when the samples were fabricated by the "powder in tube method" [1]. The dimensions of these tapes are listed in tab. 2.

For determination of the AC-losses of one tape it was brought to the form of a spiral (with windings insulated from each other) and put into the coil system which was described at 2.1. The vector of the magnetic field is then parallel to the side of the tape and perpendicular to the tape axis.

Table 2: Dimensions of the samples

	MK7:	MK15:
cross-section:	2,7 x 0,2 mm^2	3,4 x 0,25 mm^2
length:	4057 mm	2447 mm
number of filaments	19	49
ratio superconductor/silver	23%	20%
critical current (77 K)	7,5 A	12,7 A

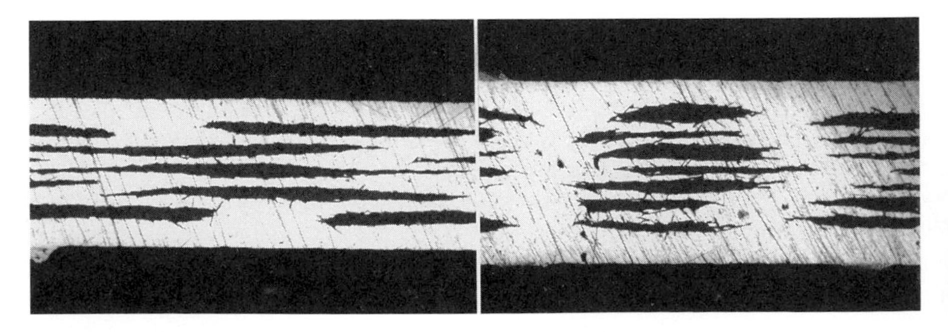

Fig. 1: Parts of the cross-sections of MK7 (left) and MK15 (right). The thickness of the tape MK7 is 0.2 mm and of MK15 0.25 mm. The total number of filaments is 19 (MK7) and 49 (MK15).

3. Results

The resulting magnetisation loops at 50 Hz and at 5 - 120 mT of the sample MK7 are shown in figure 2 and those of sample MK15 at 5 - 100 mT in figure 3. The loops show the loss of energy due to coupling currents and hysteresis. The total AC-losses of one cycle are determined by the area enclosed by the magnetization loop.

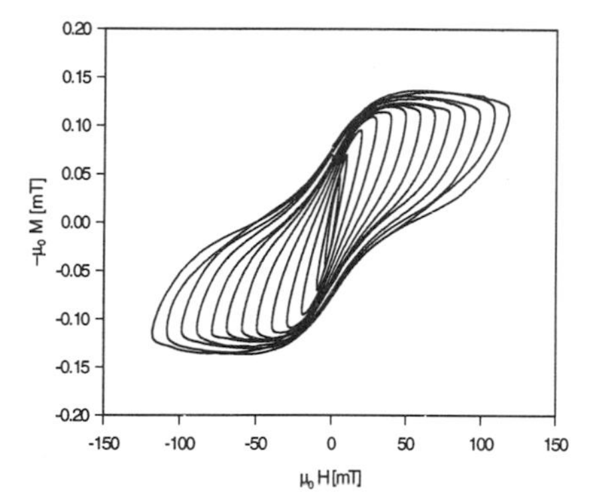

Fig. 2: Magnetisation loops of MK7 at external fields from 5 - 120 mT and 50 Hz

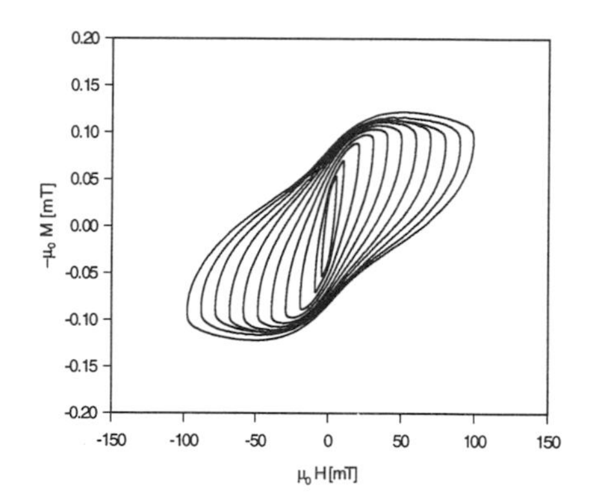

Fig. 3: Magnetisation loops of MK15 at external fields from 5 - 100 mT and 50 Hz

Fig. 4 and 5 show the mean loss values per cycle of three measurements at external fields from 5 to 150 mT. There is seen a non-linear relationship between losses and low external fields. After about 10 mT however the losses are rising linear with the external magnetic field, which is not only corresponding with 50 Hz loss-measurements on a Bi-2223/Ag-sheathed tape [3] but also on a rod [2] (both of them were monofilamentary) where the non-linear part of the curves is said to be proportional to H^3.

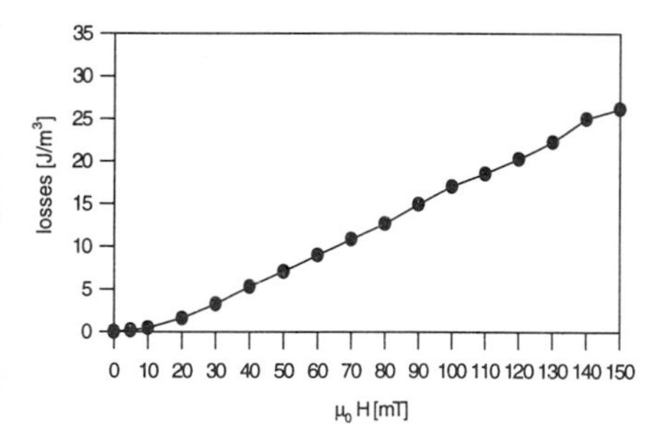

Fig. 4: Total AC-losses of MK7 at external fields from 5 - 150 mT and 50 Hz

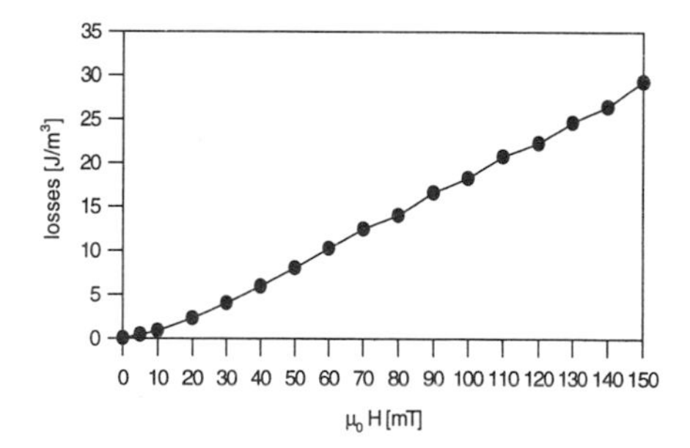

Fig. 5: Total AC-losses of MK15 at external fields from 5 - 150 mT and 50 Hz

Acknowledgement

This work has been conducted within Association EURATOM-ÖAW (Austrian Academy of Science) Task No. GB6-MAG-M16/02)

References

[1] Leghissa M., Fischer B., Roas B., Jenovelis A., Wiezoreck J., Kautz S., Neumüller H.-W. 1997 IEEE *Transactions on Applied Superconductivity (paper LE-4/ASC Pittsburgh 1996)*

[2] Fukunaga T., Maruyama S., Abe T. and Oota A. *Physica C 235-240* (1994) 3231-3232

[3] Ishii H., Hirano S., Hara T., Fujikami J., Shibuta N. and Sato K. 1994 *Advances in Superconductivity VII. Proceedings of the 7th International Symposium on Superconductivity.* 733-736 vol. 2

[4] Kwasnitza K. and Clerc St. *Physica C 233* (1994) 423-435

Inst. Phys. Conf. Ser. No 158
Paper presented at Applied Superconductivity, The Netherlands, 30 June–3 July 1997

AC losses in multifilamentary BiSrCaCuO-2223/Ag tapes studied by transport and magnetic measurements.

F. Gömöry[1,2] , L. Gherardi[1] , R. Mele[1] , D. Morin[1] and G. Crotti[1]

[1]Pirelli Cavi SpA, c.2714, Viale Sarca 222, 20126 Milano, Italy
[2]Inst. of Electrical Engineering, Slovak Acad. of Sciences, 842 39 Bratislava, Slovakia

Abstract. The possibility to relate AC losses determined by transport and magnetic measurement was studied analyzing the experimental data of two multifilamentary BiSrCaCuO-2223/Ag tapes that proved absence of bridging between filaments. In magnetic measurements, the AC field was directed perpendicularly to the tape wide side and the model of a strip was used in the analysis. We noticed that, in the tape carrying a transport AC current, a nonuniform distribution of current among individual filaments appears. In contrary, the results of magnetic measurements were explained only if we supposed an independent response of individual filaments. This difference in current distribution could explain the commonly observed discrepancy between AC transport and magnetic loss measurements.

1. Introduction

For BiSrCaCuO based tapes consisting of single or multiple superconducting core in silver matrix, the AC loss per cycle was found to be mostly frequency independent [1,5]. Then, it is natural to model the electromagnetic behavior following the assumptions of the critical state model [2] where the only parameter characterizing the superconducting material is the critical current density j_c . Combined with the sample geometry it allows to calculate the critical current I_c used in the formulas for transport AC loss [3] as well as the penetration field B_p used for the magnetic AC loss calculation [2]. In the loss measurements performed on Bi-2223 tapes the discrepancy between j_c's determined from the two types of experiments is observed mounting to 30-120 % [4]. The only geometry reported yet to give consistent transport and magnetic losses was a rod-shaped monocore sample [5].

Because of known strong anisotropy of BiSrCaCuO, it is quite expected that disagreeing j_c's are derived in experiments probing different current paths. Therefore we performed our study comparing the self-field transport losses with the magnetic losses induced by the field perpendicular to the wide face of the specimen [6]. The slab geometry seems to be no more appropriate in analyzing the AC losses measured in this arrangement. Thus we utilized the models for behavior of a superconducting disk and strip, respectively, in perpendicular magnetic field.

Table 1: Tape characteristics

Tape	Number of filaments	Filament width [μm]	Filament thickness [μm]	$I_{c,DC}$(77 K, 0 T) [A]
A	61	205	19.1	33
B	85	176	14.5	35

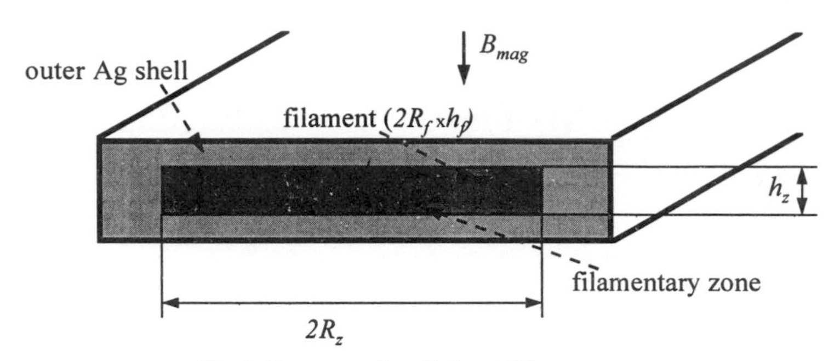

Fig. 1: Geometry of studied multifilamentary tapes

2. Samples and experiments

The tapes were manufactured by American Superconductor Corporation using the OPIT method [7]. Strip-like filaments embedded in a silver matrix form the central filamentary zone that is covered by an external Ag shell (Fig.1). Two samples with the identical dimensions of the filamentary zone (width $2R_z$= 3.3 mm, thickness h_z= 0.175 mm) but different number of filaments were measured. Optical microscopy resulted in the typical filament dimensions given in Table 1 along with the tape I_c determined at 1 μV/cm in DC transport measurements. There was no bridging encountered between filaments.

Transport AC losses were measured on 25 cm long specimens using lock-in amplifier [6]. The loss per cycle exhibited negligible frequency dependence indicating the formulas of Norris [3] can be used in theoretical modeling. Two regimes were considered: i/ uniform distribution of transport current in all the filaments that in this case are independent and ii/ strongly nonuniform current distribution governed by mutual magnetic coupling of filaments. In Fig. 2 the 35 Hz data are compared with these two models. In agreement with other works, we observed that in self-field conditions our multifilamentary tapes behave like bulk wires. However, the theoretical prediction neither for elliptical nor stripe-like section corresponds exactly to these data. Reasons for this remain unexplained yet. The I_c's yielding best fit around $I_t \approx I_c$ agree reasonably with those determined in DC conditions at 1 μV/cm.

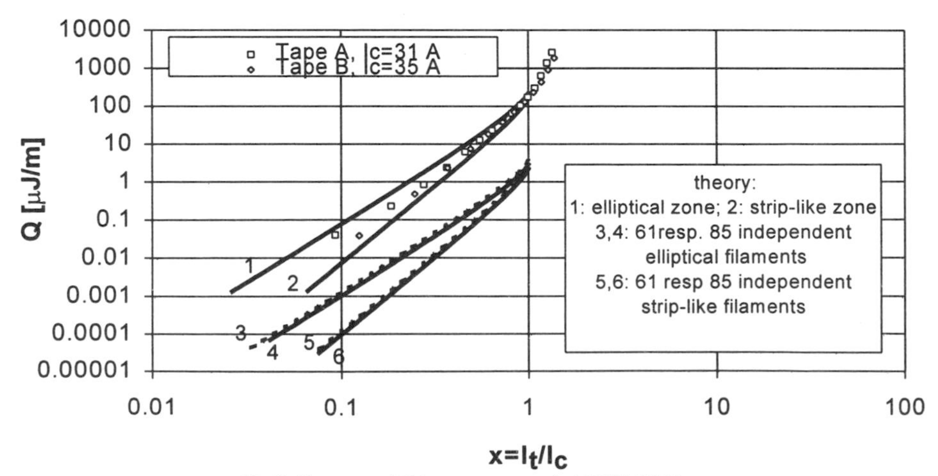

Fig.2: Transport AC losses measured at 77 K, 35 Hz.
Q are losses per cycle and unit length, I_t is the amplitude value of AC current

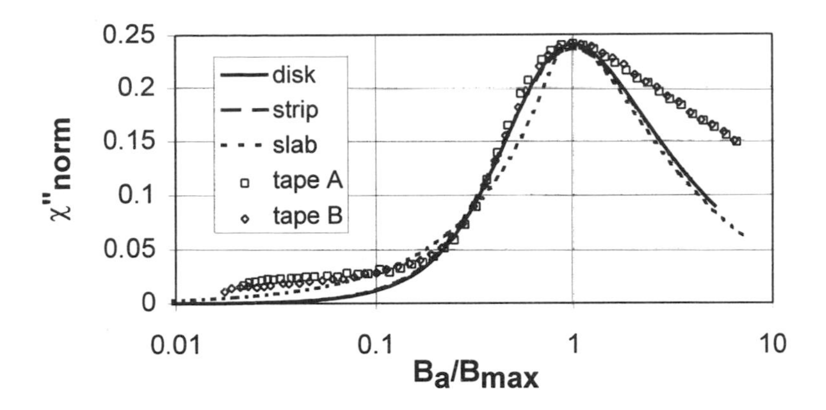

Fig.3 : Comparison of susceptibilities measured at T= 77 K and f = 35 Hz with theoretical predictions for different shapes (lines for disk and strip are nearly identical). Curves are normalised to identify better the shapes rather than the absolute values: At B_{max} the maximum value for given curve, χ''_{max}, is reached.

From χ'', the imaginary part of the internal AC susceptibility [8] measured on 6 mm long piece of the tape, the loss volume density was obtained as $q_m = \pi\chi'' \chi_0 B_a^2 / \mu_0$ where $\mu_0 = 4\pi \times 10^{-7}$ Vs/Am, B_a is the AC field amplitude and χ_0 is the absolute value of external magnetic susceptibility in conditions of perfect screening[9]. Instead of trying to calibrate the apparatus for strips in perpendicular field (denoted B_{mag} in Fig.1) we took the signal at 77 K and $B_a = 0.1$ mT to determine χ_0. The measured internal susceptibilities are given in Fig.3 together with theoretical predictions for a disk [9], strip [10] and slab [8]. One can see that the curve for slab does not agree in any portion with the experimental data, while those for disk and strip describe well the important part before the maximum. This is reached at $B_m = 0.784\mu_0 j_c h$ for strip and $B_m = 0.972\mu_0 j_c h$ for disk. For the strip geometry that corresponds better the shape of our samples, we then used the experimental B_m to calculate the critical current densities, considering again two possible regimes of current distribution between filaments. Supposing the filaments are coupled forming the filamentary zone one puts zone thickness (h_f in Fig.1) into the calculation and obtains the "zone" j_c value that should be responsible for the observed behaviour. In the hypothesis of independent filaments, filament thickness is inserted in B_m calculation and the filament j_c is calculated yielding the observed susceptibility. The latter consideration was found more plausible, as can be seen from Table 2. The difference in filament j_c found between the transport and the magnetic measurement is satisfactory: 3% (tape A) and 13% (tape B). On the other hand, one would obtain quite contradicting values for the critical current density supposing the whole filamentary zone acts as a bulk in magnetic measurement.

Table 2: j_c's calculated from transport and magnetic AC loss data

	Transport AC loss measurement			Magnetic AC loss measurement		
Sample	I_c [A]	equivalent zone j_c [A/mm^2]	equivalent filament j_c [A/mm^2]	B_m [mT]	hypothetical zone j_c [A/mm^2]	hypothetical filament j_c [A/mm^2]
A	31	53.6	130	2.53	14.7	134
B	35	61.6	161	2.6	15	182

3. Conclusions

Comparing the transport and magnetic AC losses measured on two multifilamentary BiSrCaCuO-2223/Ag tapes exhibiting no interfilament bridging we found that magnetic coupling could be quite different in these two cases. In transport measurement, filaments behaved as perfectly coupled, while the observed behaviour in magnetic measurements was explainable only if supposing the independent magnetisation of individual uncoupled filaments. Probable reason is different length of the sample, that in the case of typical magnetic measurement in perpendicular field does not induce the cross-matrix current high enough to shield the filamentary zone as a whole [6,11].

In quantitative analysis of the AC losses induced by a magnetic field directed perpendicular to the tape face, it is necessary to employ the models of disk [9] or strip [10]. These models give the loss increase with fourth power of AC field amplitude under the penetration, in contrast to the third power predicted by the slab model.

The maximum of imaginary part of AC susceptibility signifies in essence the saturation of filaments. Extensive flow of the current across the Ag matrix is thus expected for AC fields beyond B_m. Then it is not surprising that the observed -1/3 power in $\chi''(B_a)$ in this part of the curve does not correspond to any of the mentioned models derived for a superconductor in critical state.

Acknowledgement

Part of this work was supported by Slovak Grant Agency for Science VEGA.

References

[1] J. J. Gannon Jr., A. P. Malozemoff, M. J. Minot, F. Barenghi, P. Metra, G. Vellego, J. Orehotsky and M. Suenaga, Advances in Cryogenic Engineering 40 (1994) 45
M. Ciszek, B. A. Glowacki, S. P. Ashworth, A. M. Campbell and J. E. Evetts, IEEE Trans. Appl. Supercond. 5 (1995) 709
A. Oota, T. Fukunaga and T. Ito, Physica C 270 (1996) 107
P. Dolez, M. Aubin, D. Willén, R. Nadi and J. Cave, Supercond. Sci. Technol. 9 (1996) 374
[2] C. P. Bean, Phys. Rev. Letters 8 (1962) 250
[3] W. T. Norris, J. Phys. D. 3 (1970) 489
[4] Y. Fukumoto, H. J. Wiesman, M. Garber and M. Suenaga, J. Appl. Phys. 78 (1995) 4584
M. Ciszek, B. A. Glowacki, S. P. Ashworth, A. M. Campbell, W. Y. Liang, R. Flukiger and R. E Gladyshevskii Physica C 260 (1996) 93
[5] T. Fukunaga, T. Abe, A. Oota, S. Yuhya and M. Hiraoka, Appl. Phys. Lett. 66 (1995) 2128
[6] F. Gömöry, L. Gherardi, R. Mele, D. Morin and G. Crotti, Physica C 279 (1997) 39
[7] J. Cristopherson and G. N. Riley Jr., Appl. Phys. Lett. 66 (1995) 2277
[8] R. B. Goldfarb, M. Lelental, C. A. Thompson in *Magnetic Susceptibility of Superconductors and Other Spin Systems,* ed. R. A. Hein, T. L. Francavilla, D. H. Liebenberg (New York: Plenum 1991) p.49
[9] J. R. Clem and A. Sanchez, Phys. Rev. B 50 (1994) 9355
[10] E. H. Brandt, Phys. Rev. B 49 (1994) 9024
[11] Y. Fukumoto, H. J. Wiesmann, M. Garber, M. Suenaga, O.P. Haldar, Appl. Phys. Lett. 67 (1995) 3180

Inst. Phys. Conf. Ser. No 158
Paper presented at Applied Superconductivity, The Netherlands, 30 June–3 July 1997

Comparison of Transport and Magnetic AC Losses in HTS Tapes

S.P.Ashworth, M.Ciszek, B.A.Glowacki (*), A.M.Campbell, J.E.Evetts (*)

IRC in Superconductivity, Cambridge University, UK CB3 OHE
* also at Department of Materials Science and Metallurgy, Cambridge University

Abstract Losses in hts tapes due to transport currents are usually measured by a direct electrical contact method, losses due to applied magnetic fields are measured by winding 'pick-up' coils around the tape. In published data the contact method yields results up to two orders of magnitude lower for the same field. In this paper we show that the different "losses" are simply due to the differences in magnetic field penetration into the superconductor and the way in which its changes are recorded by the measurement circuit. We present data on losses due to applied magnetic fields in which the magnetic field penetration can be made to more closely reproduce that produced by a transport current. In this case the magnetic and transport losses are in good agreement.

1.Introduction

When carrying a varying current or in the presence of a varying magnetic field type II superconductors exhibit energy losses. These fields may be generated by the transport currents in the tape, may be applied externally, or may be a combination of both. Previous measurements on losses in powder in tube hts conductors have brought to light a number of interesting points. It has seemed that the two main techniques for measuring losses, the 'transport' and 'magnetic' methods yield very different results. Losses due to transport currents are usually measured by a direct electrical contact method [1], losses due to applied magnetic fields are measured by winding 'pick-up' coils around the tape [2][3]. In published data on transport losses the contact method yields results in good agreement with a simple theory whilst the coil technique suggests losses up to two orders of magnitude lower for the same field [4]. The aim of this paper is to compare the two techniques and show that the different "losses" are simply due to the differences in magnetic field penetration into the superconductor and the way in which its changes are recorded by the measurement circuit.

The hysteresis losses of the superconductor are determined by the entry and exit of magnetic field in to the superconductor. If it assumed that the critical state model (CSM) is valid in the material and that the critical current density is independent of magnetic field then the AC losses can be calculated with only a knowledge of the field distribution in the superconductor at the peak current or field [5]. Unfortunately in the general case this is very difficult to calculate and in all but the simplest geometries must be determined numerically. With the assumptions of the CSM mentioned above the AC losses have been obtained analytically for an circular superconductor carrying an AC transport current (London [6]), for elliptical and strip cross sections with transport currents (Norris [7]), for a slab in applied field (Bean [5]) and for strip perpendicular to an applied field (Brandt [8]). In the remainder of this paper we will describe the common type of experiments used to determine AC losses in hts conductors. We will then give typical data for BSCCO-223 PIT tape derived from these types of measurements.

2. Experiments

Losses due to AC transport currents are conceptually straightforward to measure. The voltage generated when a current passes along a conductor is determined, the power dissipation is then the product of voltage and current. The hysteresis loss of a superconductor is not the same as a resistive loss in a normal conductor though, the in-phase voltage generated is due to the changes in the magnetic field produced by the superconductor in ***and around*** the conductor. As pointed out by Ciszek et al [9] the measuring circuit must encompass all of the 'perturbed' magnetic field and must, consequently, extend some way from the conductor.

Losses due to AC applied magnetic fields are commonly measured using a pick up coil wound onto the sample of superconductor. The coil only returns the effect of losses due to field components perpendicular to the plane of the coil (for example [2]). This presents some limitations for typical hts PIT tapes which have typical dimension 3mm x 0.1mm, in that whilst the effects of a field parallel to the tape face can be readily measured it is difficult to wind the coil in the other plane. Consequently the losses due to perpendicular fields have not proven as easy to measure.

3. Results and Discussion

Figure 1 shows typical data for the losses due to an AC transport current along a multi-filamentary BSCCO-223 PIT tape at 77.4K at 60Hz along with the predictions of Norris [7], with which there is very good agreement. It has been shown previously that the losses per cycle are independent of frequency in the range 30-500Hz for currents $0.02<I/Ic<0.95$. In general the Norris 'ellipse' equation is a good predictor for losses in multi-filamentary BSCCO-2223 tapes and in Tl based tape, there are exceptions to this (for a survey see Glowacki [10]).

It is worth noting the assumptions built in to the Norris equation; (a) magnetic field distribution obeys the critical state model; (b) J_c does not change with position or magnetic field; (c) the superconductor has an elliptical cross section.

These assumptions ensure that the magnetic field contours within the superconductor are themselves ellipses with the same eccentricity as the conductor (this is in fact what makes the loss calculation amenable to analytic solution). As mentioned previously the losses are wholly determined by the field distribution at peak current, the distribution for an ellipse major axis *a*, minor axis *b* is sketched in Figure 2a. In Figure 3 we show the AC losses for a similar tape generated by an AC magnetic field applied *parallel* to the face of the tape. The losses due to currents and to fields can be compared by calculating the self-field due to the AC transport current, this is also shown in Figure 3. We should actually use the superconductor surface field as our parameter, the demagnetisation factor of the superconductor will lead to an error factor of order unity in our estimate of the field. We have also taken a particularly straightforward calculation of the self field due to the current ($B_{self} \approx \mu_o I/2 \times \text{width}$), this again will lead to errors, the effect of these errors is somewhat mitigated by the fact that in both cases we underestimate the surface field. It is apparent that at comparable fields the losses due to a transport current are nearly two orders of magnitude higher than those due to a parallel applied field. The reason for this becomes clear if we consider the magnetic field profile at peak field due to a field applied to an elliptical superconductor, sketched in Figure 2b, the two distributions are very different. To make a rough estimate in the difference between these losses, consider the state of penetration at peak currents/fields into an elliptical conductor (a,b) as shown in Figure 4. A transport current produces an elliptical penetration profile of the same eccentricity as the conductor (a_o,b_o).

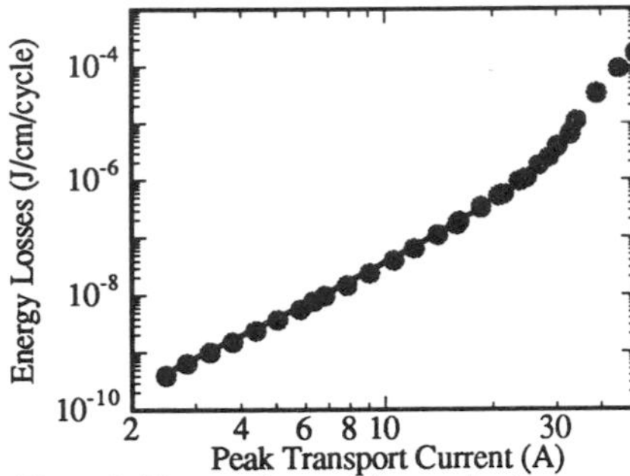

Figure 1: Measured energy losses due to an ac transport current at 77.4K as a function of peak current. The line shows the predictions due to Norris [7].

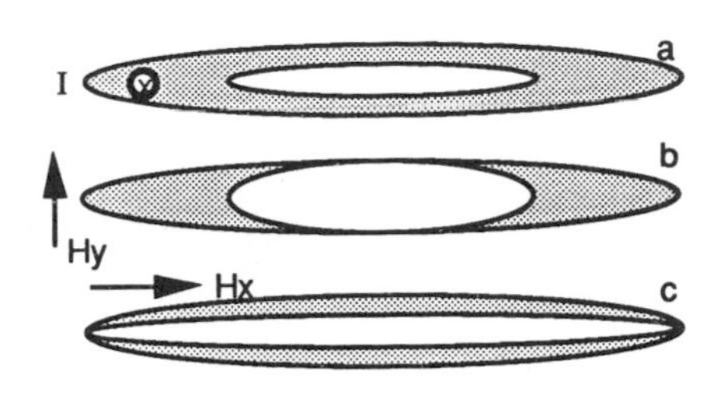

Figure 2: the magnetic field profiles at peak current or field for an elliptical superconductor; (a) carrying a transport current <Ic, (b) and (c) in applied magnetic fields as shown.

We assume that the applied field in the x direction causes the same maximum penetration into the superconductor and produces an elliptical profile (a,b_o). Proceed by estimating the difference between the transport and field generated losses, these are the losses in the cross shaded region. These are difficult to calculate in detail, but we can obtain an estimate by considering a small section of this area along the y axis. Consider the losses in the region dy at y from the common centre. This carries a current (wdyJc) and has energy losses in a half cycle;

$$dq_y = dxdyJ_c\int_{bo}^{y} B(r)dr \quad (1)$$

B(r) can be obtained from $B(y) = \frac{(y-bo)}{(b-bo)}B_s$, where B_S is the surface field. Note that $\partial B/\partial r$ is not $\mu_o J_c$ as the field has curvature at this point. The loss in the section along the y axis (0->b) is therefore;

$$q_y = \frac{1}{6}dxJ_cB_s(b-b_0)^2 \quad (2)$$

Taking into account the width of the superconductor, 2a, yields (correct up to a numerical factor of order unity). So;

$$Q^{\perp} \approx \frac{a}{dx}q_y = aJ_cB_s(b-b_0)^2 \quad (3)$$

and similarly for $Q^{//}$.

$$Q^{//} \approx \frac{b}{dx}q_x = bJ_cB_s(a-a_0)^2 \quad (4)$$

This is consistent with Campbell [11] who showed that the losses of a superconductor in an applied field are proportional to the width of the superconductor perpendicular to the applied field and also to the critical current density. Therefore;

$$\frac{Q^{\perp}}{Q^{//}} \approx \frac{adxJ_{//}B_s(b-b_0)^2}{bdxJ_{\perp}B_s(a-a_0)^2} = \alpha\frac{J_{//}}{J_{\perp}} \quad (4)$$

where α is the eccentricity of the ellipse. If the J_C of the material is anisotropic, then Jc in equation (1) is different for the Hx and Hy versions ($J_{\perp}$ and $J_{//}$) this introduces the Jc anisotropy ratio in eq. (5) giving $\alpha J_{//}/J_{\perp}$ as the scaling ratio. For a typical BSCCO-2223 PIT tape α~30 and $J_{//}/J_{\perp}$ in magnetic fields is in the range 1-5 depending on the field value, the ratio of losses will therefore be of order 100 as observed. From Figure 2 we would expect the losses for the perpendicular field configuration to more closely approximate the transport losses. Taking values as measured for a PIT tapeo fr BSCCO-2223 MF tape for

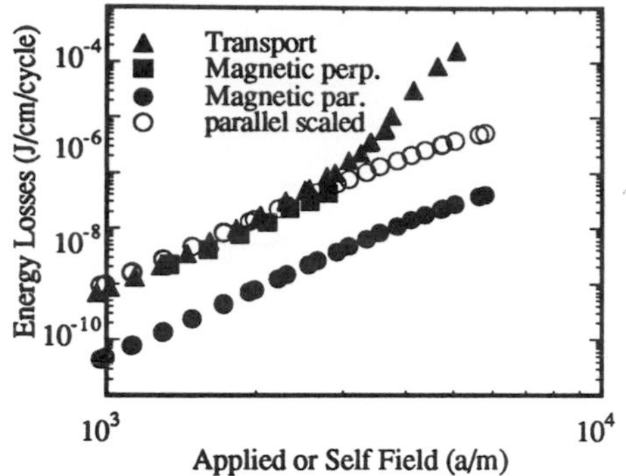

Figure 3: Losses due to applied ac magnetic fields applied parallel or perpendicular to sample compared with transport losses. Also shown are the scaled parallel field losses (see text).

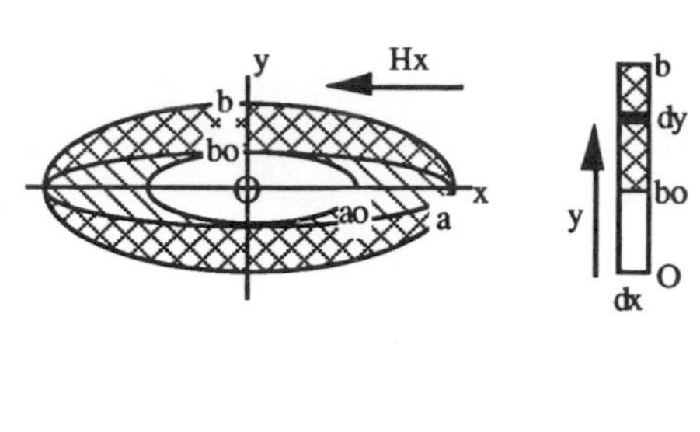

Figure 4: Field penetration profiles at peak current or field assumed for estimating ac losses. Field free region. Field due to transport current only. Field due to an applied magnetic field in x direction (Hx) and transport current

α,$J_\perp$ and $J_{//}$ and multiplying the $Q^{||}$ data by the ratio in eq.(5) we see that the scaled $Q^{||}$ data in Figure 3 now agrees well with the transport loss data. This has been achieved without consideration of superconductor microstructure, granularity etc. but simply the geometry of the experiment

It is now obvious to extend this by measuring the losses due to a perpendicular field, using a vibrating sample magnetometer. Care has to be taken to minimise remanent fields in the magnet and to warm the sample above T_C between measurements. The losses at a given peak field can then be obtained from the area enclosed within the m-H loop. This data is also shown in Figure 3, and agrees very well with the transport loss data as would be expected from the arguments above.

4. Conclusions

The values of AC loss of hts tapes measured by various techniques are shown to be consistent and to depend only on the geometry of magnetic field penetration into the superconductor. New data is presented for loss due to fields perpendicular to the tape face and it is demonstrated that if the eccentricity of the cross section and anisotropy of J_C are taken into account the field parallel losses agree well with the transport and field perpendicular losses.

5.References

[1] SP Ashworth, Physica C 229 p1838 1994
[2] M Ciszek et al IEEE Trans. Appl. Supercond. 5 p1051 1995
[3] Y Fukomoto et al, Physica C 269, p349 (1996)
[4] M Ciszek et al Physical C 260 p93 1996
[5] CP Bean Rev. Mod.Phys. 36 p31 1964
[6] H London Phys. Lett. 6 p162 1963
[7] WT Norris J.Phys. D 3 p489 1970
[8] EH Brandt, M Indenbohm Phys. Rev. B 48 p12893 1993
[9] M Ciszek, AM Campbell, BA Glowacki Physica C 233 p203 1995
[10] BA Glowacki et al EUCAS 97
[11] AM Campbell Cryogenics 22 p3 1978

Inst. Phys. Conf. Ser. No 158
Paper presented at Applied Superconductivity, The Netherlands, 30 June–3 July 1997

Calorimetric AC Loss Measuring System for HTS Carrying AC Transport Currents

N Magnusson[1] and S Hörnfeldt[2]

[1]Royal Institute of Technology, Department of Electric Power Engineering, SE-100 44 Stockholm, Sweden.
[2]ABB Corporate Research, SE-721 78 Västerås, Sweden.

Abstract For the usage of HTS in electric power systems, they need to be characterized with respect to their AC losses under application-like conditions. In this paper we present a calorimetric measuring system that provides AC loss data as a function of temperature, magnetic field, transport current and frequency, for long-length HTS in the form of tapes and wires.
The sample forms a coil and is placed in vacuum. AC transport current can either be induced in the sample or provided by an external source. To determine the AC losses, the temperature rise of the sample is measured and compared to the temperature rise caused by a heater wound together with the sample.
Preliminary results of AC loss measurements on a Bi-2223 conductor for DC applications delivered by the American Superconductor Corporation in 1995 are also presented.

1. Introduction

The development of high temperature superconductors (HTS) is presently intensive. When HTS are to be used in electric power applications the AC losses will be important. The losses (P) depend on the temperature (T), the magnetic field (B), the transport current (I) and the frequency (f). One question is how large losses can be tolerated. At 77 K the cooling penalty factor is about 10-20. If the losses in the HTS are 1% of the corresponding losses in the copper at the working temperature of an oil-filled transformer or reactor, the overall losses will be reduced by 80-90%.

Since the AC losses and the cooling penalty factor both depend on the temperature, any HTS will have an optimal operating temperature at which the AC losses multiplied by the cooling penalty factor yield a minimum. This temperature may not necessarily be 77 K. To utilize the HTS efficiently, it has to operate close to its critical current. In power applications, the HTS will be exposed to a magnetic field proportional to its transport current. To develop a semi-empirical model for the losses, they need to be determined under these conditions. The model can be of the form $P_{AC} = P_{AC}(T, B, I, f)$.

Different techniques have been used to measure AC losses. Inductive methods [1,2] do not reveal interruptions in a long conductor. When using the transport current method [3,4], in which a voltage drop in phase with the current is determined, it is hard to compensate for the unavoidable inductive voltages. Calorimetric methods [5,6,7], though less accurate, overcome these problems.

In section 2 the experimental set-up is described and the results are reported and discussed in section 3.

2. Experimental

The superconducting sample is wound in the form of a coil with a few turns (Figure 1). A diameter of 300 mm is chosen for the coil to avoid damage due to strain. The sample, which is placed in vacuum, is in thermal contact with the cold head of a cooling machine. An external alternating magnetic field is supplied by four copper coils generating a field parallel to the face of the tape. The magnetic field profile is calculated from the geometry and the currents of the copper coils. Transport currents can either be induced in the sample with the sample short-circuited, or supplied by an external source. In the case with the sample shorted, the induced transport current is determined from the voltage of a pick-up coil. The number of turns of the inner coils can be changed to vary the induced current and the external magnetic field independently. Consequently the radial field gradients over the sample coil will be moderated. The system operates at: $T \geq 50K$, $I \leq 50A$ and $B \leq 125mT$ at $f = 50Hz$, decreasing to $\leq 15mT$ at $500Hz$.

A number of thermocouples, which are attached to the sample, measure the temperature rise caused by the AC losses during 5-60 s long current pulses. The results are calibrated against the temperature rise caused by a heater wound together with the sample.

The cryostat is mainly built of fibre glass reinforced epoxy, to avoid induced currents in the cryostat walls. Liquid nitrogen is used to reduce the resistance in the copper coils and to limit the radiation to the cold region.

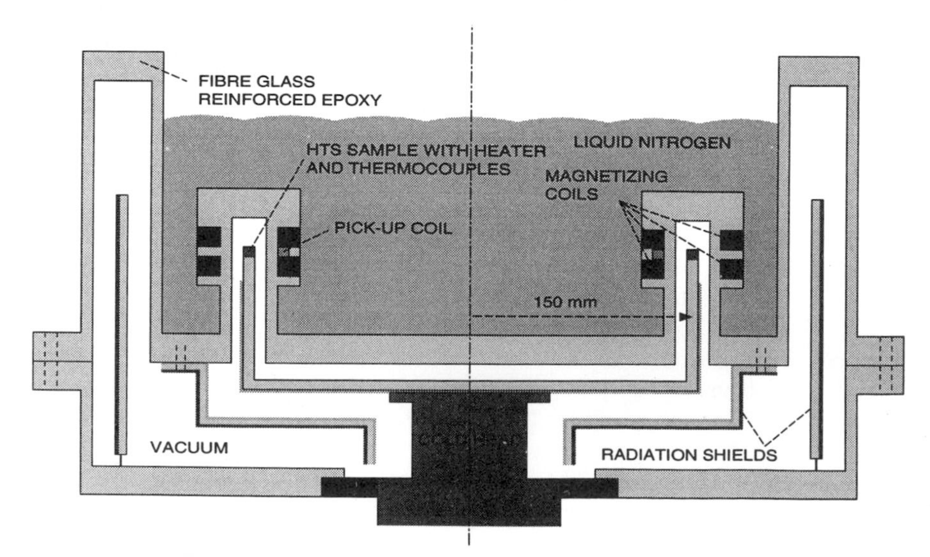

Figure 1: Layout of the cryogenic system with the HTS sample.

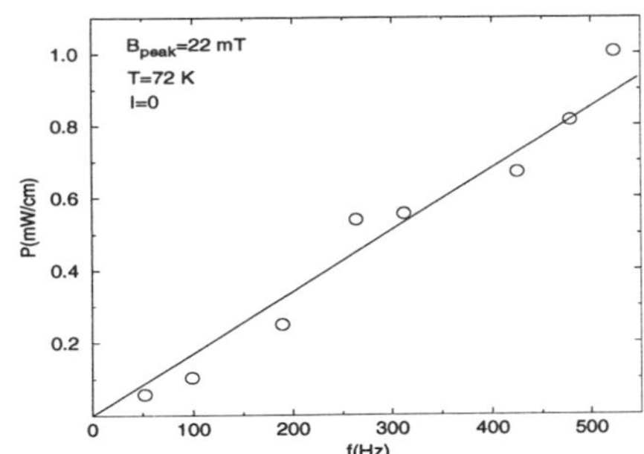

Figure 2: Frequency dependence of the losses.

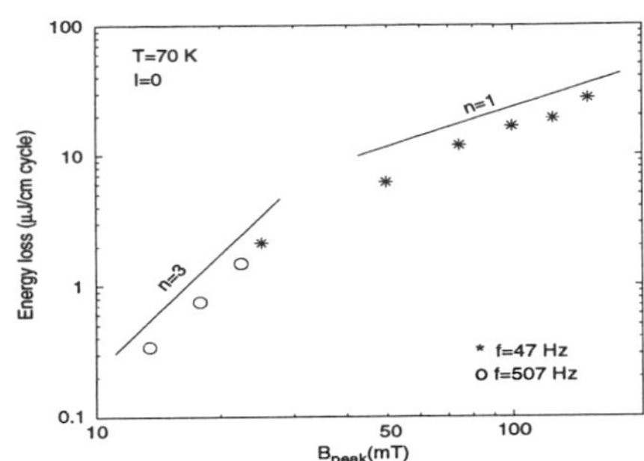

Figure 3: Magnetic field dependence of the energy losses per cycle. The lines indicate the predicted slopes according to the Bean model, $P \sim B^n$.

3. Results and discussion

We present results from AC loss measurements in external magnetic fields with the sample coil open and with the sample coil shorted. The sample consists of a Bi-2223 silver clad filamentary tape-formed (2.5 times 0.25 mm) conductor delivered by the American Superconductor Corporation in 1995.

3.1. Losses due to external magnetic fields and no transport current

In Figure 2 the frequency dependence of the AC losses of the sample is given. The result shows that the hysteresis losses are dominant for frequencies less then 500 Hz. Figure 3 shows the AC losses per cycle for different applied magnetic fields. The power law dependence predicted by Bean [8], $P \sim B^3$ for B less than the full penetration field and $P \sim B$ above the full penetration field, is given for comparison. The full penetration field can be estimated at 30-40 mT at 70 K. The temperature dependence of the losses is given in Figure 4. As expected from the Bean model for fields above the full penetration field, the losses are decreasing with temperature because of the decrease of the critical current with temperature.

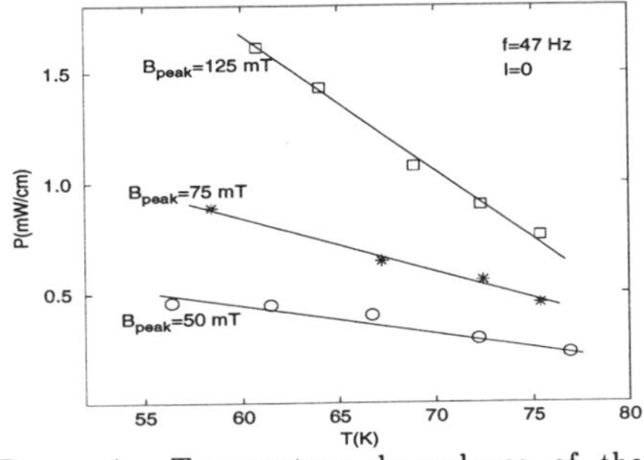

Figure 4: Temperature dependence of the losses.

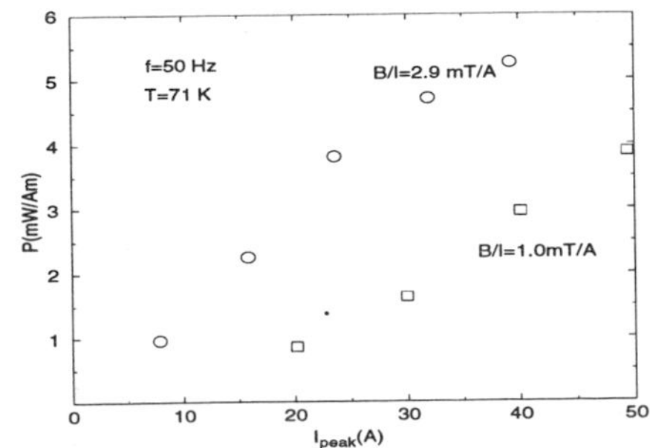

Figure 5: Transport current dependence of the losses per ampere and meter.

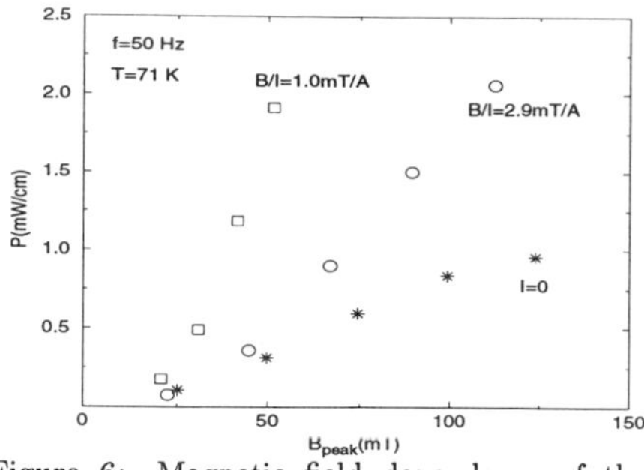

Figure 6: Magnetic field dependence of the losses.

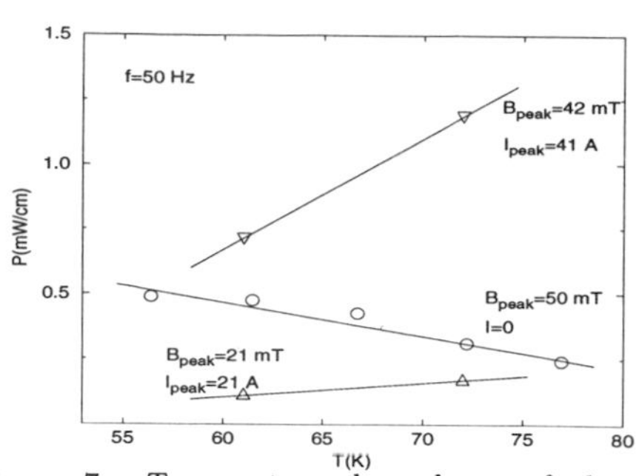

Figure 7: Temperature dependence of the losses.

3.2. Losses due to combined external magnetic fields and induced transport currents

The AC losses as functions of the transport current at different magnetic fields are shown in Figure 5. Figure 6 shows the AC losses for different magnetic fields using the field-current ratios in Figure 5. For comparison, the AC losses at zero transport current are also given. For large transport currents, the AC losses increase with temperature as shown in Figure 7. The increase of the AC losses with temperature is due to a decrease of critical current, which leads to higher flux creep and flux flow losses. The figure also includes the AC losses at zero transport current.

4. Conclusions

The presented calorimetric measuring system for AC loss measurements at application-like conditions is capable of producing useful data, which can be the basis for the development of a semi-empirical model of the form $P_{AC} = P_{AC}(T, B, I, f)$ of the AC losses in HTS. The preliminary results obtained from the Bi-2223 silver clad filamentary sample are in agreement with the critical state model.

References

[1] Fietz W A 1965 Rev. Sci. Instrum. **36** 1621 6
[2] Paasi J and Lahtinen M 1993 Physica C **216** 382 90
[3] Hughes T, Yang Y, Beduz C, Yi Z, Jansak L, Mahdi A E, Stoll R L, Sykulski J K, Harris M R and Arnold R J 1994 Physica C **235-240** 3423 24
[4] Ciszek M, Campbell A M, Ashworth S P and Glowacki B A 1995 Appl. Supercond. **3** 509 20
[5] Schmidt C and Specht E 1990 Rev. Sci. Instrum. **61** 988 92
[6] Dolez P, Aubin M, Willén D, Nadi R and Cave J 1996 Supercond. Sci. Techn. **9** 374 8
[7] Snitchler G, Campbell J, Aized D, Sidi-Yekhlef A, Fleshler S, Kalsi S and Schwall R 1997 IEEE Trans. Appl. Supercond. **7** 290 3
[8] Bean C P 1964 Rev. Mod. Phys. **36** 31 9

Inst. Phys. Conf. Ser. No 158
Paper presented at Applied Superconductivity, The Netherlands, 30 June–3 July 1997

Current redistribution and quench development in a multistrand superconducting cable for AC applications at commercial frequency

V.S. Vysotsky, K. Funaki, H. Tomiya, M. Nakamura and M. Takeo

Research Institute of Superconductivity and Graduate School of ISEEE, Kyushu University, Fukuoka, 812, Japan

Abstract. We developed the special test set-up to measure currents in each of six separate strands or subcables of a multistrand cable at frequencies from 0 to 120 Hz and at currents up to thousand amperes. At the first experiments we studied the stimulated quench test of the 36 strand NbTi cable with insulated strands. One of subcables was deliberately quenched by the heater. It was found that depending on the current level there may be a total quench of the cable, quench of the heated subcable only or quench and recovery of the heated subcable or subcables. The last, quench—recovery mode, was not observed before. The threshold current between stable and unstable regions was estimated from "single-strand" stability criteria.

1. Introduction

Multistrand superconducting cables designed for non—steady state applications have a common problem - degradation of their quench current in comparison with the sum of critical currents of all strands. It is widely believed that the reason of such behavior is a non—uniform current distribution among strands. To study the current distribution in multistrand superconducting cables we developed a special test set-up [1]. This test set-up permits us to measure currents in each of six separate strands or subcables in two ends of a sample cable. The pick-up coils and Hall sensors were employed as current probes. Numbers of potential taps and heaters could be attached to the strands or subcables. The heaters permit us to initiate a quench of one of the strands by our will to perform a stimulated quench test. Such test is necessary to study "single-strand stability" (SS-stability), i.e. stability of the cable in relation to the possible quench of one single strand [2].

The quench of one single strand of a multistrand superconducting cable may happen due to many reasons, say, a weak point in this strand, an excess by the current in the certain strand of the average level [3] in the cable, etc. The cable should be stable to this event, i.e. a quench of one strand or subcable should not lead to the quench of other strands and to an entire cable quench. Especially it is important for the cables with insulated strands or with strands with highly resistive matrix and special covering, like chromium covering of strands in the Nb_3Sn based CICC. In such cables direct current sharing from the quenched strand to other strands is difficult and the current redistribution to other strands has an electrodynamic origin. Thus, SS-stability may be referred to as "electrodynamic stability" [2].

Table 1. Parameters of the sample tested

Strand diameter, mm	Filament diameter μm	CuNi/NbTi	First stage twist, mm, (direction)	Second stage twist, mm, (direction)
0.112/0.135*	0.42	2/1	3.3 (S)	9.3 (S)

* Strand diameter with insulation

At the first experiments with the set-up developed we studied a 36 strand NbTi cable with insulated strands. Among other tests, that are described in details in [1], we performed few stimulated quench tests. In this paper we present the experimental results from stimulated quench of the multistrand sample with insulated strand and discuss them from the point of view of the SS-stability.

2. Experimental results.

The sample tested is a two stage (6+1)×(6+1) twisted cable. Its parameters are listed in Table 1. The strands are NbTi-CuNi fine filament wires with 23749 filaments. No pure copper is in the matrix. Each sextet was twisted around the central CuNi wire that has the same diameter as strands have. The first stage subcables were twisted around the central CuNi wire with the same diameter as the first stage subcable. We studied the current redistribution among the first stage subcables of this cable.

The stimulated quench test for this sample was performed at a frequency 4 Hz only, due to the limitations from our primary power supply [1]. Before the stimulated test, we measured the current distribution among the first stage subcables at different currents and frequencies. It was found that the subcables' current scattering is about 25%. We found that this current non-uniformity is determined by the difference in the joint resistances with current leads [1].

The results of the stimulated quench test are shown in Figures 1,2 and 3 for different levels of the total cable current. In all cases the quench was initiated in subcable #5.

In Figure 1 the quench has been initiated near a current amplitude of ~405 A. The heated subcable # 5 quenched and its current has been redistributed to two nearest subcables #4 and #6 only. Other subcable currents remain unchanged and the total current is not changed either. This so called "redistribution" process was observed before in [4]. At this current level the cable is stable in relation to the quench of one single subcable or it is in the region of the SS-stability.

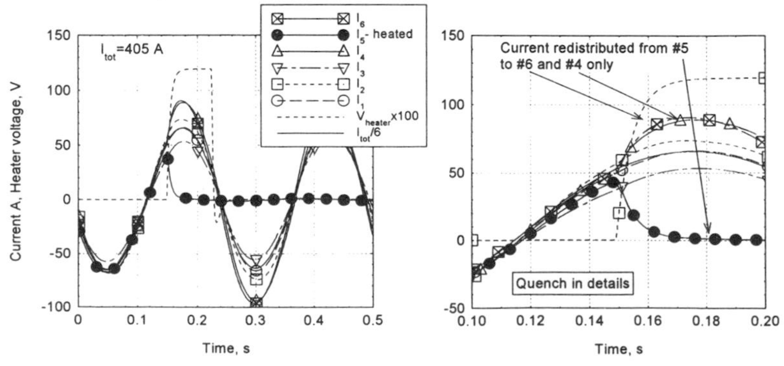

Figure 1. Redistribution quench development process. Quench of the single subcable does not lead to the quench of the entire cable. At this current level the sample is SS-stable.

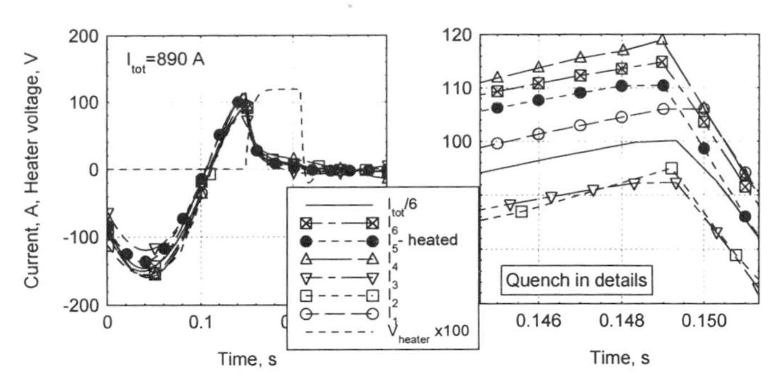

Figure 2. Total quench process. The quench of one single subcable leads to a total cable's quench. At this current level the sample is SS-unstable.

In Figure 2 the process is shown when a quench has been initiated at a much higher current of ~890 A. The quench initiated in the single subcable #5 leads to the total quench of the entire cable. It is the region at which the cable is unstable in relation to the quench of the single subcable or SS-unstable region. For this sample the threshold stability current is about 640 A. Above this current level the total cable always quenches happens when a single subcable quenches.

The quench development processes as shown in Figures 1 and 2 were studied and described in detail in [4]. This study was performed at DC current and no recovery for quenched strands was observed. Nevertheless, it was suggested in [5] that in the redistribution region the quenched strand may recover in AC mode, while passing through zero current. In this experiment we observed such quench development behavior.

In Figure 3 a and b the quench process is shown at two current levels: 600 A and 525 A. During the test shown in Figure 3a, the heated subcable #5 quenched and its current was redistributed in two adjacent subcables as usual. But unlike the process shown in Figure 1, the current in the subcable #5 recovered after passing zero level. Current distribution among subcables was slightly changed but the total cable current remained unchanged.

In Figure 3 b the heater has been initiated near the maximum of the total current. At first instance it led to a total cable quench. But after passing zero level all subcables except #5 recovered and the total cable current has been recovered also.

3. Discussion

The quench development processes shown in Figure 3 have not been observed before. We found that there is an extra quench development mode inside the "redistribution" region observed before in [4]. In some cases, the quenched subcable or even the entire cable may recover at AC current. The recovery of quenched subcables demonstrates that multistrand superconducting cables may be more stable in relation to the single strand quench at AC mode that at DC one.

In [2] we proposed the criteria of a single-strand stability as: $I_t / I_c \leq \frac{\alpha \cdot m}{1+m}$, where I_t and I_c are the threshold and critical currents correspondingly, m - is number of strand in which current from the quenched strand may redistribute and α - is a coefficient of a quench current reduction due to fast dI/dt changes. In our case we may evaluate dI/dt ~$1\text{-}3 \cdot 10^4$ A/s from Figure 2. The value of $\alpha \approx 0.78$ for such current change rates we got from experiments

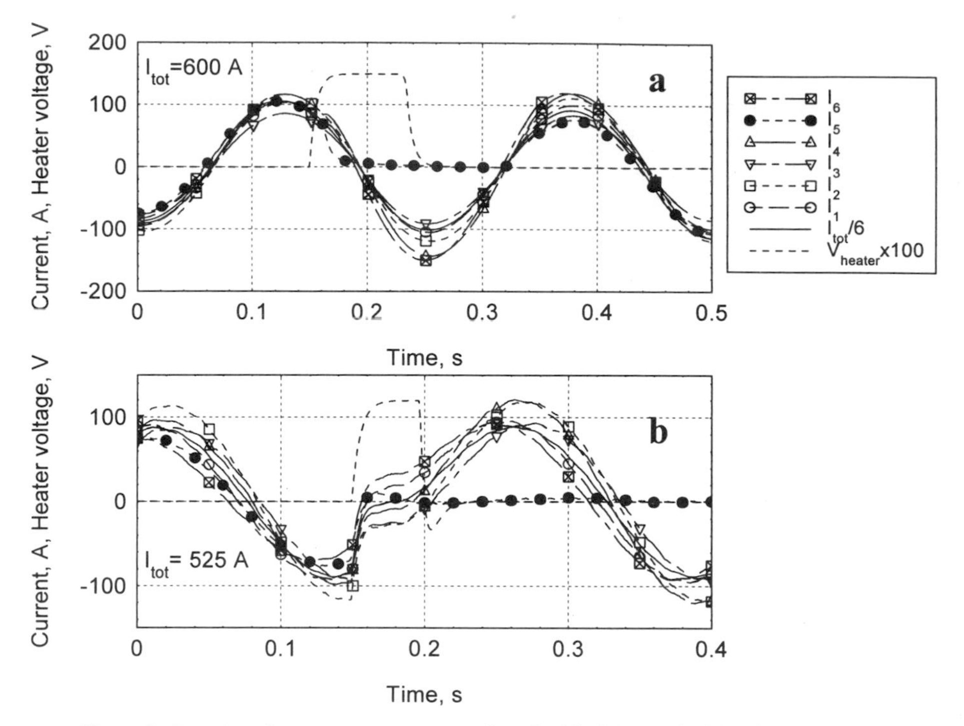

Figure 3. Quench and recovery processes. a - the subcable # 5 quenched but fully recovered after the heater was turned off. b - entire cable quench has been initiated by the heater. Nevertheless, all but #5 subcables recovered after the heater has been turned off.

in [6], where similar wires were tested. For our sample cable m=2 and the expected critical current is ~1200 A. From the expression above, one can get I_t~630 A that is close to our experimental data.

4. Conclusions

The multistrand superconducting sample cable with insulated strands has been underwent stimulated quench tests. In addition to "redistribution" and "slow quench" modes observed before, it was found that inside the "redistribution" region the quenched subcables may recover at AC current during passing zero current. Thus, in "redistribution" region multistrand cable appears to be more stable than it was suggested before. The threshold current between stable and unstable region may be estimated by use of criteria of the single-strand stability.

References

[1] Vysotsky V S Funaki K et al Paper CP-4 to be presented at CEC/ICMC '97 in Portland, Oregon, USA
[2] Vysotsky V S et al Influence of a cable design on stability 1997 *Cryogenics* **37** *in press*
[3] Vysotsky V S et al Measurements of current distribution in a 12 strand CICC 1997 *Cryogenics* **37** *in press*
[4] Vysotsky V S et al 1992 *IEEE Trans. Magn.* **28** 735-746
[5] Vysotsky V S 1991 *Proc. of International workshop on AC superconductors* Smolanice 110-115
[6] Tsuda M Takamatsu H Ishiyama A 1993 *IEEE Trans. Appl. Superconductivity* **3** 503-506

Inst. Phys. Conf. Ser. No 158
Paper presented at Applied Superconductivity, The Netherlands, 30 June–3 July 1997

Superconductors for Fusion Magnets: Realizations, Prototypes and Running Projects

Pierluigi Bruzzone

ITER JCT, 801-1 Mukouyama, Naka-machi, Naka-gun, Ibaraki-ken 311-01 Japan.

Abstract. This paper gives a review of the application of superconductivity to nuclear fusion magnets. For each machine, a short description emphasizes the special features of the winding technology and conductor design. A selection of demonstration magnets is presented, highlighting the relevant achievements. The evolution of the conductor design in the last decades and the open issues for future R&D activities are discussed .

1. Introduction

The most important progress in the field of the magnetic confinement of plasma has been achieved by experiments operating with pulsed copper magnets. However, a fusion power plant calls for long burning time and high magnetic field which can be reasonably maintained only by superconducting coils. Unlike other applications of superconductivity, in the field of fusion there is no "normal conducting" alternative: whenever a magnetic confinement fusion power plant will operate, it will have superconducting magnets.

Three main kinds of fusion devices have been studied for plasma confinement using superconducting coils: the mirror machine, the stellarator and the tokamak. The efforts of the last ten years concentrated on the last two options. The size of the superconducting fusion experiments grew by over one order of magnitude in forty years: the magnetic stored energy was 20 MJ in the first superconducting tokamak, T-7, and will be 130 GJ in ITER.

The increasing size of the superconductors, see Fig.1, and the demanding operating conditions (high field and high rate of field) have the consequence that limited R&D can be carried out on full size specimens within a reasonable time scale and budget. The design of the current and future fusion magnets is less and less based on the results of full scale components tests. Analytical models and numeric codes are extensively used to overcome the broad gap between the small scale experiments and the full scale coils. The obvious result is a tendency to take generous engineering margins and to produce a conservative design.

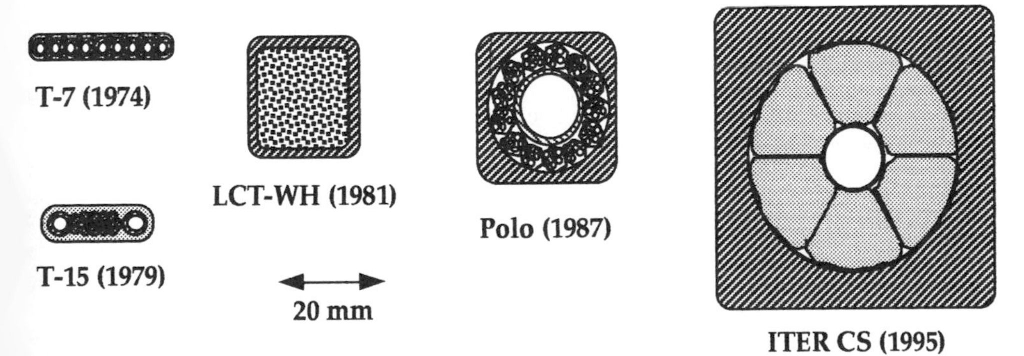

Fig.1 Size evolution of forced flow superconductors for fusion magnets.

2. Review of Superconducting Fusion Devices

2.1 Tokamak T-7

The first superconducting Tokamak ever built was assembled at the Kurchatov Institute, Moscow, in 1977 after testing the eight segments of the torus [1, 2]. The winding system consists of 48 double pancakes, with 60 turns in each coil. The average coil diameter is 1 m, the cold mass 12 t and the stored energy 20 MJ at the nominal operation point.

The forced flow conductor, see Fig.1, is made from of nine copper pipes, 16 multifilamentary and 32 single core NbTi strands, sitting in the grooves between the pipes. The components are assembled by electroplating a 0.6 mm copper layer up to the final size of 28 x 4.5 mm. The cooling is by two phase Helium @ 4.5 K. The conductor is not transposed and suffered from severe flux jumps, triggering a quench, during ramp up and down. However, 80% of the design current (6 kA @ 5 T peak field) was achieved. Operation of T-7 was discontinued at the Kurchatov Institute in 1987. Later, T-7 was transferred to the Chinese Institute of Plasma Physics, where it is now operating.

2.2 Tokamak T-15.

Ten years after the first operation in 1988 at the Kurchatov Institute in Moscow, the toroidal field coils of T-15 [3] are still the largest worldwide application of Nb_3Sn conductors, with 25 t of multifilament strand. Each of the 24 circular coils, average ø = 2.4 m, consists of 12 pancakes assembled in two vacuum impregnated coil halves with individual case, eventually bolted together. Turn insulation is obtained by wet winding. The forced flow conductor, see Fig.1, is a flat cable of 11 non stabilized Nb_3Sn strands, assembled after heat treatment on ø = 1.5 m, with two copper pipes by Cu electroplating, 1.2 mm thick. Over 100 km of conductor have been manufactured. Each coil was tested before assembly and achieved the specification, although a steady state voltage was observed, by far larger than the joint voltage, in the range of 2.5 - 10 mV/coil.

The design current in the tokamak is 5.6 kA @ 9.3 T peak field. The limited size of the cryoplant (only 0.38 g/s·conductor) and the large radiation loss limited the operating temperature to the range of 9-10 K. The highest operation point was 3.9 kA @ 6.5 T, 7-8 K, in agreement with the single coil test and in excess of the original strand specification. The field transients due to plasma disruption, up to 40 T/s, were withstood without quench, with an increase of the outlet temperature by 0.25 K. The design ground voltage is 1.5 kV. For fast discharge, 250 V was applied at the terminals, with a time constant of 104 s. The operation of T-15, discontinued in 94, is scheduled to be resumed in late 97.

2.3 Mirror Fusion Test Facility (MFTF)

MFTF is the largest set of superconducting magnets for fusion, in terms of cold mass and amount of strand [4, 5]. It consists of 8 C-type coils, 12 low field and 4 high field solenoids, including the A2 coils with a Nb_3Sn insert. All magnets are pool cooled @ 4.5 K by natural convection, with two-phase coolant outlet (<5% gas). The coils are all wound in the coil case, which acts as a cryostat, with ground insulation applied before winding. In the Yin-Yang coils, the winding form is fitted into the thick case by a copper bladder filled with urethane. The turn and pancake insulation is provided by G10 spacers.

All the conductors (two types of NbTi and one Nb_3Sn) are made as a thick multifilament composite soldered to the copper stabilizer, see Fig.2. The stabilizer of the Yin-Yang conductor is a perforated Cu strip wrapped around and soldered to the square NbTi composite, to increase the wetted surface. For the Nb_3Sn conductor, the composite is soldered in the Cu housing after heat treatment (react&wind). All the conductors are designed to be cryostable. The fraction of I_{op}/I_c is always smaller than 2/3.

The MFTF coils were manufactured in less than five years and assembled at the Lawrence Livermore National Laboratory in 1985. After cool down and successfully commissioning the magnet system, the project was discontinued. In 1990, the A2 coils were extracted and re-assembled in the FENIX conductor test facility, which operated for three years at field levels as high as 13 T [6].

Table 1. Summary of Superconducting Magnet Systems for Fusion Devices

	Strand Weight, t	Conductor/ Cooling*	Stored Energy, MJ	Peak Field, T	Operating Current, kA
Tokamak T-7	1	NbTi / FF	20	5	6
Tokamak T-15	25	Nb_3Sn / FF	795	9.3	5.6
MFTF	74	Nb_3Sn+NbTi / Pool	1 000	2 - 12.75	1.5 - 5.9
TRIAM	2	Nb_3Sn / Pool	76	11	6.2
Tore Supra	43	NbTi / Pool 1.8 K	600	9	1.4
LHD-Helical coils	10	NbTi / Pool 4.5 (1.8) K	930 (1 650)	6.9 (9.2)	13 (17.3)
LHD-Poloidal coils	23	NbTi / FF	990	5 - 6.5	20.8 - 31.25
Wendelstein	32	NbTi / FF	600	6	16

* FF = forced flow

2.4 TRIAM

The Tokamak at the University of Kyushu, Fukuoka (Japan), was first operated in 1986 [7]. The 16 D-shaped toroidal field coils, with ~3 m average perimeter, are wound as double pancakes with three conductor grades. The pressurized liquid Helium bath is at 4.5 K. The conductor, see Fig.2, consists of a large Nb_3Sn composite (10.5 x 3.3 mm for high grade, over half million filaments) soldered after heat treatment in a copper housing with roughened side surfaces to improve the heat exchange. Beside the copper housing (RRR = 90), a Cu clad high purity Al profile (RRR = 3000) is used as stabilizer.

The conductor is designed to be cryostable. At 6.2 kA, 11 T, the ratio I_{op}/I_c is 0.6. No quench event has been reported after three years of operation, including plasma disruptions.

2.5 Tore Supra

The 18 circular TF coils of the tokamak Tore Supra (Cadarache, France) are the largest magnet mass, 160 t, cooled at 1.8 K [8]. Each coil is made out of 26 double pancake, with an average diameter of 2.6 m. A co-wound prepreg tape, 0.15 mm thick, is used for the turn insulation. The pancake spacers, 2.2 mm, are built by a perforated prepreg thin plate with glued glass-epoxy cubes. The ground insulation is obtained by overlapped prepreg plates. A 2 mm thick steel case is shrink fitted to the winding and contains the atmospheric, 1.8 K He bath. A thick steel case, with thermal insulation, is shrink fitted and cooled at 4.2 K.

The conductor is a rectangular NbTi/Cu/CuNi composite, see Fig.2, wound on the short edge. The temperature margin is ~ 2.5 K, with T_{cs} = 4.25 K @1400 A, 9 T peak field. Little copper in used in the conductor: for stability, the He bath enthalpy up to the λ point is available, due to the thermal properties of He II. In case of quench, a He gas pressure builds on the top of the cryostat, open only at the bottom, and expels the He volume within 3 s, providing a very fast quench propagation and limiting the hot spot temperature below 80 K.

In 1988, about six months after first operation, an interpancake short occurred at one coil during a fast discharge [9], with 1.5 kV across the coil and ~60 V across pancakes. The damaged coil was later replaced with a spare coil and the dump voltage was decreased to 500 V in order to limit the pancake voltage to ~20 V. The pulses of the poloidal field coils and the plasma disruption result in a temperature increase in the He II bath as small as 0.01 K.

2.6 Large Helical Device (LHD)

The LHD coils are being assembled at NIFS, Toki (Japan) [10] and will be first operated in January 98. The superconducting magnet system of the stellarator consists of a double spiral wound around the toroidal vacuum vessel (Helical Coils) and three circular poloidal coils.

The DC operated Helical coils are pool cooled: initial operation will be at 4.2 K, 6.9 T, 13 kA and, at a later stage, at 1.8 K, 9.2 T, 17.3 kA. The turn and layer spacers are graded across the winding pack to provide the best mechanical support in the stressed area and the largest wet conductor surface at the peak field. The conductor, see Fig.2, is a NbTi flat cable soldered with a CuNi cladded Al stabilizer into a copper housing, eventually sealed by two electron beam welds. The conductor is designed to be cryostable, with $I_{op}/I_c = 0.55$.

The forced flow conductor for the poloidal coils is a NbTi cable-in-conduit with 486 strands (ø = 0.76 or 0.89 mm), 38% void fraction. The strands are not coated, with a coupling loss constant of 300 ms. The temperature margin is 1.2 to 1.6 K, $I_{op}/I_c = 0.33$. Conductor joints are realized by filament joining, resulting in 0.14 nΩ resistance at full current. The OV coil, ø = 11.5 m, has been wound on the site, with prepreg turn insulation.

2.7 Wendelstein W7-X

The magnet system of the stellarator Wendelstein W7-X consists of 50 non-planar and 20 planar coils assembled in five segments [11]. Completion of the machine is scheduled by 2002. The forced flow conductor, to be used for all the coils, is a NbTi cable-in-conduit with 243 strands, ø = 0.57 mm, 37% void fraction. The square jacket is made of an hardenable Al alloy, co-extruded around the cable: it is soft after extrusion and during the winding process. After hardening at 170°C, it provides the required stiffness to the winding pack. The temperature margin is >1 K and $I_{op}/I_c = 0.5$.

3. Selection of Demonstration Coils

The demonstration coils are valuable opportunities to learn about magnet and conductor technology. In such projects, the pressure for a conservative design is less strong and the performance margins can be better explored compared to a real fusion device.

In the IEA Large Coil Task at the Oak Ridge National Laboratory [12], six large D-shaped magnets have been built to the same common specification using substantially different design approaches. All the coils operated (1984-1985) at the same design point (8 T peak field) with margins ranging from 120 to 140%. For the first time, a cable-in-conduit Nb_3Sn conductor (react & wind) was used in a large scale application and, despite the broad transition observed in selected coil sections (similar to T-15 behavior), reached 8 T with T_{cs} = 8 K. The cryostable conductors for the bath cooled coils (GD, GE, JA) could be easily graded (both layer and pancake windings) and, using soldered copper profiles as stabilizer, achieved an impressive strand amount (and cost) effectiveness: from 1.4 t in the GE conductor to 8.2 t in the forced flow conductors of EU (NbTi) and WH (Nb_3Sn). However, two out of three pool cooled coils could not be dumped to the design voltage of 1 kV.

Five years after the LCT, the Demonstration Poloidal Coil (DPC) facility at JAERI,

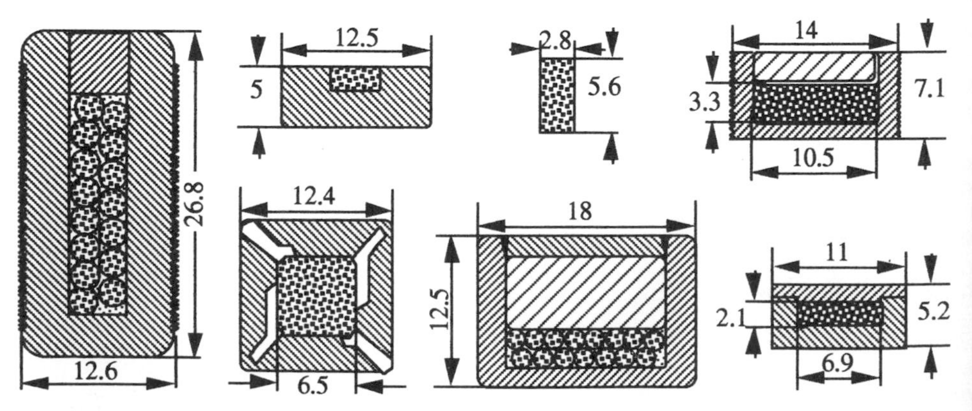

Fig.2 Monolithic superconductors for pool cooled magnets. From left to right: LCT-JA, MFTF for solenoid (upper) and Yin-Yang coils (lower), Tore Supra (upper) and LHD (lower), Nb_3Sn for TRIAM (upper) and Nb_3Sn for insert of A2 coil in MFTF

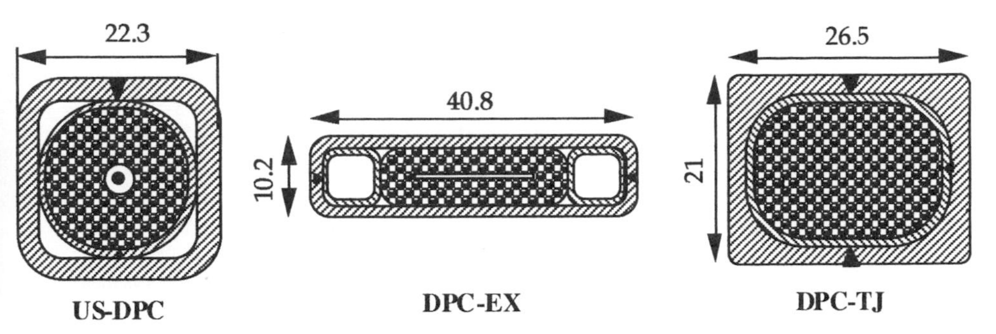

Fig.3 The Nb3Sn cable-in-conduit conductors of the Demonstration Poloidal Coils

Naka (Japan) aimed to compare design options for pulsed field coils. Three Nb3Sn react&wind winding models with $ø_{av}$ = 1.3 m, DPC-EX [13], US-DPC [14] and DPC-TJ [15] were tested in the DPC facility in pulsed mode up to about 7-8 T, sandwiched between two pulsed NbTi solenoids, connected in series. The conductors are all forced flow cable-in-conduit, see Fig.3. The jacket material is Incoloy for the US-DPC (used for the first time). The DPC-TJ had a double jacket: the outer one is 3D machined without bending and fitted by spot welding to the conductor after the heat treatment. The coupling loss is 1000 times lower in the US-DPC and DPC-EX conductors ,with Cr plated strands (2 ms). Ramp rate limitation was observed in the US-DPC, probably due to transposition errors in the cable.

The Polo coil, a NbTi circular winding with ø = 3 m, has been tested in 1994 at the Forschungszentrum Karlsruhe (Germany) [16]. The cable-in-conduit conductor, see Fig.1, has two separate hydraulic circuits: stagnant, supercritical He @ 4 bar in the annular cable region and forced flow, 2 g/s, two-phase He @ 4.5 K in the central pipe. This design allows a homogeneous temperature along the conductor with a small pressure drop. The strand is a NbTi/Cu/CuNi composite and the subcables have CuNi or insulating barriers, resulting in very low coupling loss, τ = 210 μs. Four stainless steel corner profiles are laser welded around the cable. Phase Resolved Partial Discharge was first used at 4K to assess the integrity of the glass-epoxy insulation . A midpoint electrical connection in the winding enables to create very high field transients in a half coil by a fast discharge of the other half coil. The coil has been tested up to 15 kA, 3.6 T. A steady state degradation of I_c by 30% has been observed compared to the strand performance. Polo does not have ramp rate limitation, the stability criterion being the only limiting criterion. Very fast field transient, up to 1000 T/s are withstood without quench. High voltage operation, up to 23 kV has been demonstrated.

In the scope of the ITER R&D program, two reduced scale coils are being manufactured and are scheduled to be tested in 1999. The CS Model Coil [17] is a layer wound, two in hand solenoid to be operated at 46 kA, 13 T with 0.4 T/s field rate. The stored energy is 641 MJ, compared with 13 GJ in the full size Central Solenoid. The TF Model Coil [18] is a pancake wound, race track coil to be assembled with the LCT-EU coil and tested up

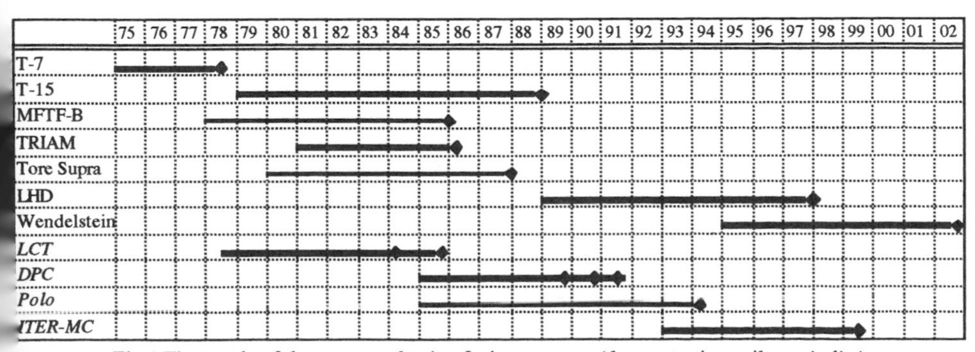

Fig.4 Time scale of the superconducting fusion magnets (demonstration coils are *italics*)

to 70 kA, 8.8 T. The weight of the TF Model Coil is 31 t, compared with 700 t of a full size TF coil. Both coils use the Nb_3Sn cable-in-conduit conductors designed for the full size coils, see Fig.1. The goals of the Model Coil program are: to test under relevant conditions the ITER conductors and to demonstrate some of the crucial technology choices, including the use of extruded Incoloy tubing as jacket material (CS), the wind & react & transfer technique for coil assembly, the grooved radial plates (TF), the electrical insulation and the joints.

4. Summary of Conductor Design

The pool cooled conductors offer a number of attractive advantages, including cryostability, low risk manufacturing and joining techniques, and effective use of the superconducting strand, i.e. cost effectiveness. However, the size and stored energy of future fusion magnets call for large mechanical loads in the winding pack and high voltage operation is required to obtain fast field pulses and to dump the energy in case of quench. The forced flow conductors, with the potted winding pack and high voltage insulation, seem presently the only option for large magnets with stored energy » 1 GJ.

The choice between NbTi and Nb_3Sn conductors is dictated by the operating field. For pool cooled coils, with the possibility to decrease the temperature down to 1.8 K, the upper limit for effective operation of NbTi conductors is ~9 T. In forced flow conductors, the limit is ~7 T: the increasing confidence with Nb_3Sn technology tends to decrease the field threshold for the NbTi vs. Nb_3Sn choice. Conductors based on Nb_3Al are in a developmental stage and may become an alternative to Nb_3Sn for selected react&wind magnets.

Among the forced flow conductors, a large variety of layouts is possible. The option to assemble the cable around a Cu pipe (T-7, T-15, LCT-CH) became obsolete when it was demonstrated that a leak tight conduit can be formed around the cable and provides a much larger wetted perimeter. Inside the conduit, several options are available: a dual channel may be required to reduce the pressure drop with large mass flow rate. The cooling layout of Polo is especially attractive for NbTi cable-in-conduit. The distribution of the copper cross section, the cable configuration, the coatings, the wrappings and other substructures depend on the specific operating requirements and are, to a large extent, the result of the designer's preference and/or supplier's experience.

A fusion power plant will need to be commercially competitive and reliable in operation. In the long term, the issues of manufacturing cost and risk will have an increasing importance in the conductor design and magnet technology.

References

[1] Ivanov, D.P. et al. 1977 *IEEE Mag* **13**, 694
[2] Ivanov, D.P. et al. 1979 *IEEE Mag* **15**, 550
[3] Bondarchuk E.N. et al. 1992 *Plasma Device and Operation* **2**, 1
[4] Wang, S.T. et al. 1983 *Proc. of 9th ICEC, Kobe,* 424 (Butterworth)
[5] Kozman, T.A. et al. 1983 *IEEE Mag* **19**, 859
[6] Slack, D.S., Patrick R.E., Miller J.R., 1991 *IEEE Mag* **27**, 1835
[7] Nakamura, Y. et al. 1990 *Proc. of MT-11, Tsukuba,* 767 (Elsevier)
[8] Turck, B., Torossian, A. 1994 *Proc. of 15th IEEE/NPSS SOFE, Hyanis,* 393
[9] Bessette, D. et al. 1991 *Proc. of 16th SOFT, London,* 1659 (North Holland)
[10] Imagawa, S. et al. 1996 Takahata et al. 1996 *Proc. Symp. Cryogenic Systems for Large Scale Superconducting Applications, Toki, NIFS-PROC-* **28**, 112, 116
[11] Sapper, J. 1997 *Proc. of annual meeting of American Nuclear Society, ANS*
[12] The IEA Large Coil Task 1988 *Fusion Engineering and Design,* **7**, 1&2
[13] Takahashi, Y. et al. 1991 *Cryogenics* **31**, 640
[14] Steeves, M.M. et al. 1992 *Adv, Cryog. Eng.* **37A**, 345
[15] Ono, M. et al. 1993 *IEEE Appl. Superconductivity* **3**, 480
[16] Darweschsad, R. et al. 1996 *IEEE Mag* **32**, 2256
[17] Mitchell, N. et al. 1997 *Proc. of 16th ICEC/ICMC, Kitakyushu,* 763 (Elsevier)
[18] Salpietro, E. et al. 1997 *Proc. of 16th IAEA Fusion Energy Conference, Montreal*

Inst. Phys. Conf. Ser. No 158
Paper presented at Applied Superconductivity, The Netherlands, 30 June–3 July 1997

Design considerations of a HTS μ-SMES

R Mikkonen, M Lahtinen, J Lehtonen and J Paasi

Tampere University of Technology, Tampere, Finland

B Connor and S S Kalsi

American Superconductor Corporation, Westborough, MA, USA

Abstract. A 5 kJ HTS SMES system has been designed at Tampere University of Technology with a coil manufactured by American Superconductor (ASC). The outer diameter of the pancake type coil is 317 mm and the axial length is 66 mm. The operating current is 160 A. The magnet will be cooled to the operating temperature of 20 K with a two stage Gifford-McMahon type cryocooler with a cooling power of 60 W at 77 K and 8 W at 20 K. The cryogenic design is determined by the application of the system and its specific AC requirements. The system will be demonstrated in occasion of a short term loss of power.

1. Introduction

HTS technology is matured to enabling different kind of prototype applications including μ-SMES. Nowadays, when speaking about HTS systems, attention is focused on the operating temperature of 20-30 K, where the critical current and flux density are fairly close to 4.2 K values. In addition by defining the ratio of the energy content of a novel HTS magnetic system and the required power to keep the system at the desired temperature, the optimum settles to the above mentioned temperature range.

2. HTS coil and conductor performance

The HTS magnet has inside and outside diameters of 252 mm and 317 mm, respectively, and axial length of 66 mm. It operates at 160 A and carries a total of 160 kA-turns to store required 5 kJ of magnetic energy. The effective magnet inductance is 0.4 H. The peak axial field is 1.7 T and it occurs at the mid-plane of the magnet bore. The peak radial component is 1.3 T and it occurs at mid-point of winding radial built at each side of the coil. The radial component of field determines the current carrying capability of the HTS wire and is responsible for generating most losses in the magnet. The magnet consists of 11 double pancakes. Pancakes employ HTS conductor of different J_e in order to minimize wire consumption. Pancakes with higher J_e conductors were employed in coil end regions. This magnet employs wire with J_e (at B = 0 and 77 K) of over 8.500 A/cm^2. The conductor was nominally 0.18 mm × 2.6 mm. Three conductors were wound in parallel in order to achieve the operating current of 160 A at 20 K. Fig. 1 shows the performance of the coil measured at 77 K in liquid nitrogen. I_c value of 27 A at 77 K should translate to 160 A at 27 K with total dissipation of 0.3-0.4 W.

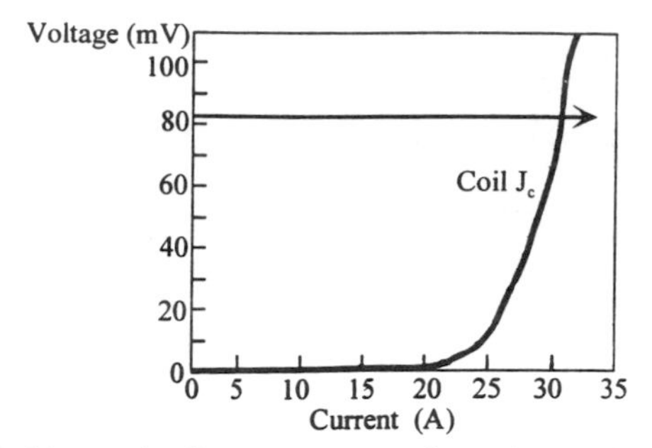

Fig. 1 Measured coil current versus voltage characteristics at 77 K.

3. Cryogenic design

The cryostat system of the HTS solenoid is presented in Fig. 2. The system is cooled with a two stage Gifford-McMahon cryocooler. The device consists of one nitrogen heat pipe (first-stage, 77 K) which is thermally anchored to the copper radiation shield and the HTS current leads and four hydrogen heat pipes (second-stage, 20 K) which are attached to the cryogenic interface at top of the magnet.

The HTS coil is installed inside the radiation shield and attached to the copper interface which is fixed to the second stage of the cryocooler. In order to diminish the eddy current losses when ramping the magnet, the copper plate is slotted into four equal segments, each of which has one hydrogen pipe. In the similar manner the radiation shield around the coil has a segmented configuration.

The HTS current leads are, like the coil, made of flexible multifilamentary composite BSCCO-2223 conductor (CryoSaver^{TM}). These leads will be able to carry 160 A up to a temperature of 77 K in self field. The contact resistance at the cold end is about 3×10^{-7} Ω per lead [1].

The calculated heat load into the first and second stage of the cryocooler is shown in Table 1. The heat loads of the system can be divided into five categories. (1) *Cryostat construction*. This heat load is due to the heat radiation from the outer vessel to the radiation shield and from the shield to the magnet, and due to the thermal conduction along the heat pipes and supporting rods. (2) *Current leads*. Between the temperature range of 300-70 K, the heat load is due to the thermal conduction along the Cu leads, Joule heating

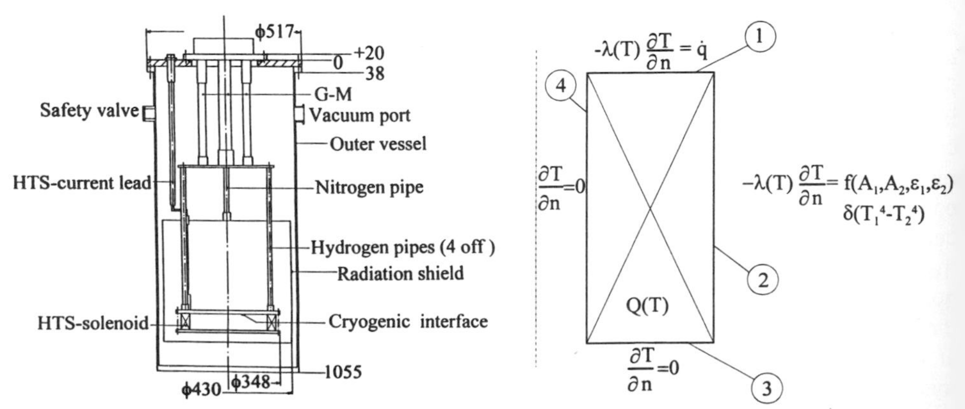

Fig. 2 The cryostat system and the coil geometry used in numerical calculation.

Table 1 Estimated heat load summary

Loss component	1st stage [W]	2nd stage [W]
Cryostat system	25	0.5
Current leads	15	0.2
Cryogenic interface	-	0.3
AC losses	-	0.1
Dissipation losses	-	0.4
Total	40	1.5

and contact resistances. The heat load into the 2nd stage is due to the heat conduction of the Bi(2223) leads and Joule heating of the contact resistance of the cold end. (3) *Cryogenic interface*. Eddy current losses are induced under a time-varying magnetic field in the interface at top of the magnet and in the radiation shield. It is possible to reduce this power dissipation to an arbitrary small value by dividing the plate and the shield into several narrow strips and by keeping the magnet ramping time at a relative moderate value, say 30 sec, which equals a flux change of a 0.05 T/sec at top of the magnet and 0.02 T/sec in the radiation shield. (4) *AC losses*. A time varying magnetic field generates AC losses in the winding which are proportional to the area of the hysteresis loop. At the inner surface of the solenoid the flux change is 0.06 T/sec (ramping time 30 sec.). (5) *Dissipation losses in the magnet*. These include the heat generation due to the joints between the separate double pancakes (resistance of a joint is typically 1 nΩ) and due to the thermal activated flux creep.

4. Stability considerations

Higher operating temperatures with increased specific heat make a HTS winding less sensitive to mechanical disturbances which are crucial for LTS magnets. Typically a disturbance energy sufficient to raise a HTS coil temperature from 30 K to 40 K is three orders of magnitude higher than what is needed to quench a Nb-based magnet [2].

The geometry under consideration is presented in Fig. 2. The purpose is to calculate the temperature distribution of the coil in different operating modes, ie. one should solve the non-linear heat balance equation

$$\Delta \cdot [\lambda(T)\, \Delta T] + Q(T) = C_p(T)\, \frac{\partial T}{\partial t}$$

with appropriate boundary conditions, see Fig. 2. In the equation above C_p is the volumetric specific heat, λ thermal conductivity and Q volumetric heat generation in the coil. The boundary conditions are all of Neumann type. $\partial T/\partial n$ represents the normal derivative of temperature. At boundary 1 the condition is determined by the heat flux density of 300 W/m^2 (equals to the cooling power of 8 W) at 20 K. At boundary 2 the heat balance is determined by the heat radiation from the radiation shield to the coil surface. The situation is modelled as two concentric cylinders. A_1, A_2, ε_1 and ε_2 are the surface areas and emissivities of the radiation shield and the coil, σ is the Stefann-Boltzmann coefficient. The boundary 3 is adiabatic and 4 is due to the symmetry.

The temperature behaviour is simulated in two different cases. In Fig. 3 the temperature rise is presented in an adiabatic case (no cooling) with four different ramping rates. The temperature has been calculated as an average of one tape which is located in the middle of

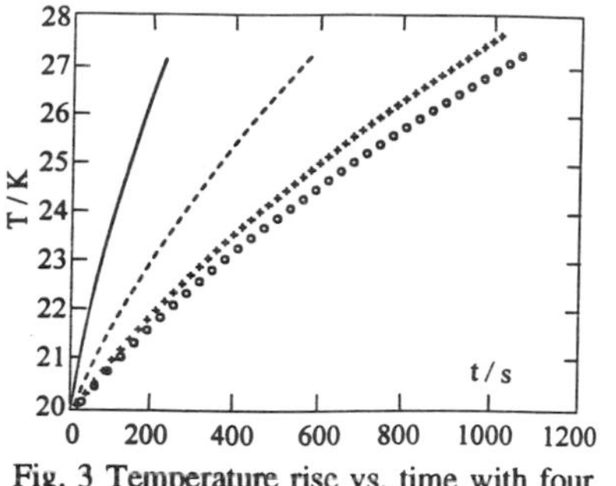

Fig. 3 Temperature risc vs. time with four different ramping rates: ___, dB/dt = 0.67 Ts^{-1}, ---, dB/dt = 0.5 Ts^{-1}, +++, dB/dt = 0.1 Ts^{-1}, 000, dB/dt = 0.033 Ts^{-1}.

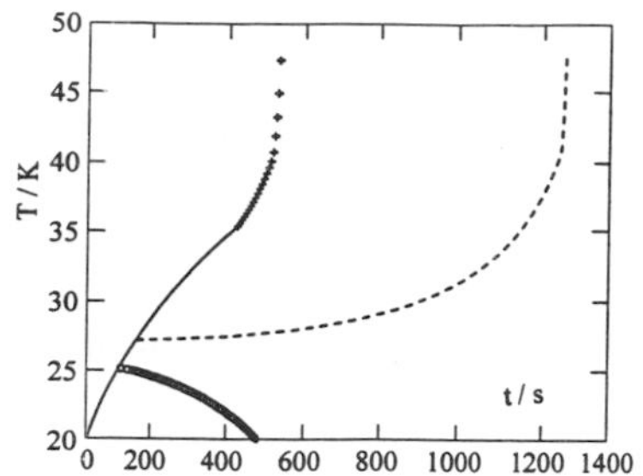

Fig. 4 Temperature rise with combined ramping and dc mode; ___, ramping phase, 000 initial temperature 25 K, --- initial temperature 27 K, +++ initial temperature 35 K.

the coil, on the lower surface of the solenoid. The average magnetic flux density in the magnet has been fixed to 1 T. The hysteresis losses with the 30 second charging time (dB/dt = 0.033 Ts^{-1}) are 0.1 W. All the other loss sources in Table 1 except the cryogenic interface has also included in the calculation. The maximum temperature of the coil is fixed to 27 K. The other simulation, Fig. 4, is a combined ramping and dc mode with a constant cooling power of 8 W at top of the magnet. The coil is first ramped with a ramping rate of 1 Ts^{-1}. After the temperature has attained a certain value the coil is fed with a dc current of 160 A. During this mode the hysteresis losses and the eddy current losses in the cryogenic interface are replaced by the Joule losses $Q_J = \rho(J)J^2$. In the vicinity of the critical current density J_c, which has a linear temperature dependency from 40 kA/cm² at 20 K to 0 A/cm² at 80 K, resistivity is assumed to obey the widely used power law $\rho(J) = E_c(J^{n-1}/J_c^n)$. In calculations n has been set to 8, which is quite typical value for HTS materials [3, 4]. During ramping the hysteresis losses are assumed to be distributed uniformly in the winding. In the dc mode the Joule losses are equal all over the coil. It can be seen that the temperature of little less than 27 K represents a value which results in stable operation.

5. Conclusions

A prototype HTS μ-SMES system has been designed and constructed in order to compensate a loss of power of a micro computer. The coil has been manufactured at ASC from multifilamentary composite BSCCO-2223 conductor in a react and wind fashion. The optimum operation temperature of the cryogen free magnet is 20-30 K. With the aid of a proper cryogenic integration and thermal analysis it is possible to determine a safe level for the operating current and magnet ramping rate before an irreversible thermal runaway.

References

[1] *Standard CryoSaver Current Leads* by American Superconductor.
[2] Iwasa Y 1994 *Case Studies in Superconducting Magnets* (New York: Plenum Press)
[3] Polak M, Zhang W, Hellstrom E E and Larbalestier D C 1995 *Inst. Phys. Conf. Ser.* **148** 427-430
[4] Wolsky A 1997, IEA Report: *The status of progress toward high-amperage conductors incorporating high-temperature superconductors* Implementing Agreement for a Cooperative Programme for Assessing the Impacts of High-Temperature Superconductivity on the Electric Power Sector.

Inst. Phys. Conf. Ser. No 158
Paper presented at Applied Superconductivity, The Netherlands, 30 June–3 July 1997

Operation of a Small SMES Power Compensator

K.P. Juengst[1], H. Salbert[1], and O. Simon[2]

[1]Forschungszentrum Karlsruhe, Institut für Technische Physik (ITP), D-76021 Karlsruhe
[2]Elektrotechnisches Institut (ETI), Universität Karlsruhe, D-76128 Karlsruhe

Abstract. A system for compensation of fluctuating loads has been designed, constructed, and laboratory operated. It consists of an IGBT based current converter and a NbTi mixed matrix conductor s.c. magnet system. Power and energy ratings are 80 kVA and 200 kJ, respectively. The dynamic converter has a very short response time and the magnet can withstand high rates of current/ field change. Operation on the laboratory grid was successful in compensating the varying power demand of a periodically stressed induction machine. For a field demonstration, the mobile system has been shipped to a saw-mill near Karlsruhe.

1. Introduction

A prerequisite for troublefree operation of automated production processes is a high quality electric power supply. Voltage sags of only 20 % and few tenths of seconds duration can lead to costly disturbances of sensitive production processes. Even much smaller cyclic variations of the voltage can lead to disturbing flicker of lamps and require countermeasures. Especially, the fringes of networks, weaker networks, and the 400 V network are affected. Industrial loads with rapidly changing power demand (e.g., hammer forges, rolling mills, electric arc furnaces) can cause power system disturbances. Instead of shielding sensitive consumers from disturbances [1], one could compensate these locally.

Therefore, a demonstration system of a SMES-based power compensator (and active filter) has been built by the ITP/ Research Center Karlsruhe, the ETI/ University of Karlsruhe, and the IAEW/ RWTH Aachen. The four utilities Badenwerk, EVS, TWS, and Neckarwerke have contributed to this project since 1993 by their knowledge of needs of networks, site selection, and funding through the Foundation for Energy Research of Baden-Württemberg.

In the past, SMES model systems for a variety of conceivable applications have been constructed and tested mainly in Japan [2]. Tests begun on a very small 5 kJ cryogen-free SMES model with HTS conductor [3] show future promise for such systems.

2. System Design

The SMES power compensator [4,5,6] consists of a power conditioning system (PCS) and a superconducting magnet system. Both systems need to be fast to achieve high quality compensation of fluctuating loads. The compensator must be on line continuously for very frequent corrective actions. The capabilities of the system should be demonstrated in an industrial environment.

Fig. 1: Six-coil superconducting solenoid system as magnetic energy storage for rated 200 kJ (Outer coil diameter 0.4 m, total height of the solenoid 0.5 m)

All of these requirements, together with the constraint of available funds, led us to design a small, mobile, and fast response system.

2.1 PCS

The PCS has been designed as an IGBT based voltage source inverter [5]. The output power of the net side inverter is 80 kVA. The SMES-side inverter is a two-quadrant chopper with a maximum power of 320 kVA. The two inverters are connected by a 700 V DC link. For the required dynamic response, a fast control has been developed and implemented. A special prediction filter [6] was designed to approximately eliminate higher harmonics.

2.2 Magnet System

The SC magnet system was designed for: mobility, modularity, continuous fast power change, and reliability. The magnet system is being constructed in two steps. First, a solenoid system was assembled from six coil modules (Fig. 1) and later on, a 10 coil torus will be built up. The AC losses in the SC have been taken into account by choosing a helium-transparent winding and a NbTi-based wire with 8600 filaments in mixed Cu/CuNi matrix. Diameters of wire and filaments are 1.2 mm and 7 μm, resp. The inductance is 4.38 H, giving 200 kJ at 302 A.

3. Results

PCS and magnet system were first tested separately. At the Research Center Karlsruhe, the magnet reached 340 A/ 5.3 T and a stored energy of 253 kJ with a rate of change of field of 2.5 T/s. Hence, the required safe operation at the rated 300 A is guaranteed. A system response time of 0.3 ms has been achieved.

Last year, a first demonstration experiment was run in the laboratory with an experimental load, i.e., an induction machine with mechanical coupling to a DC machine operated in turn by a converter. Load currents similar to real conditions in industry were

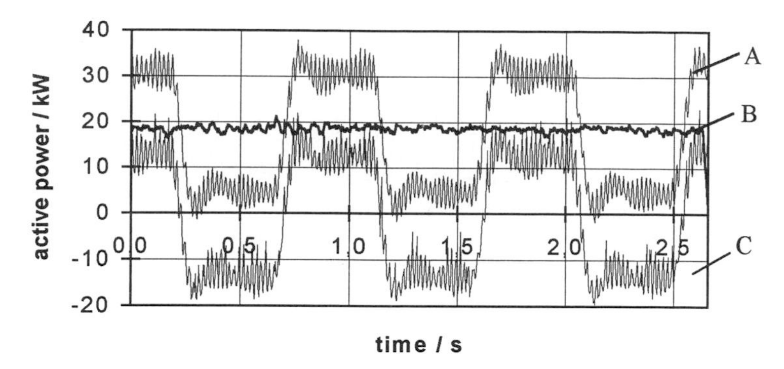

Fig. 2: Time behavior of load power, line power, and compensator power
(A: load power, B: line power, C: compensator power)

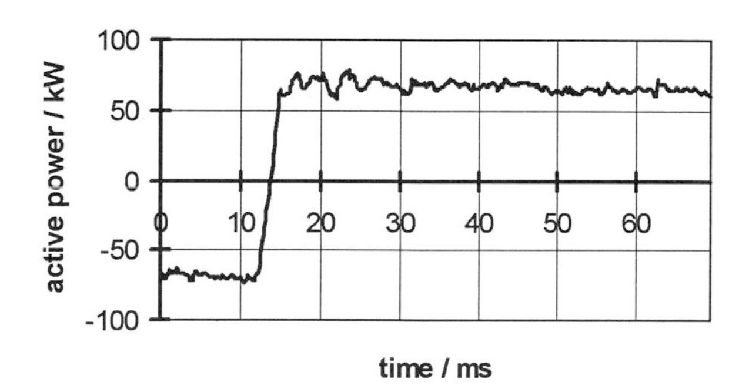

Fig. 3: Fast power swing of the compensator with 27 kW/ms

generated. The load power varied between zero and 30 kW at a rate of about 600 kW/s with a repetition rate of one cycle per second. The time behavior of the power of the load, the compensator, and the line are shown in Fig. 2. The desired condition of approximately constant power demand, as "seen" by the grid, was reached.

In further experiments, the dynamic behavior was tested. In Fig. 3, a very fast change of delivered power from - 70 kW to + 70 kW with 27 kW/ms is shown. The test of the compensator as an active filter for the current harmonics was very successful too [5].

The next step is a test period in an industrial environment. Given the power and energy capabilities of the SMES power compensator, a saw-mill has been chosen as the industrial test site. Measured data from the mill were used for lab tests. Successful smoothing of power variations has been achieved (Fig. 4). The system has been shipped to the mill for tests in July/ August 1997 (Fig. 5).

Acknowledgment

The encouraging support of Prof. Komarek and Prof. Späth of FZK and ETI, resp., the contributions of the Badenwerk and partner utilities, the funding by the Stiftung Energieforschung, Stuttgart, and the magnetic field optimization by W. Maurer are gratefully acknowledged.

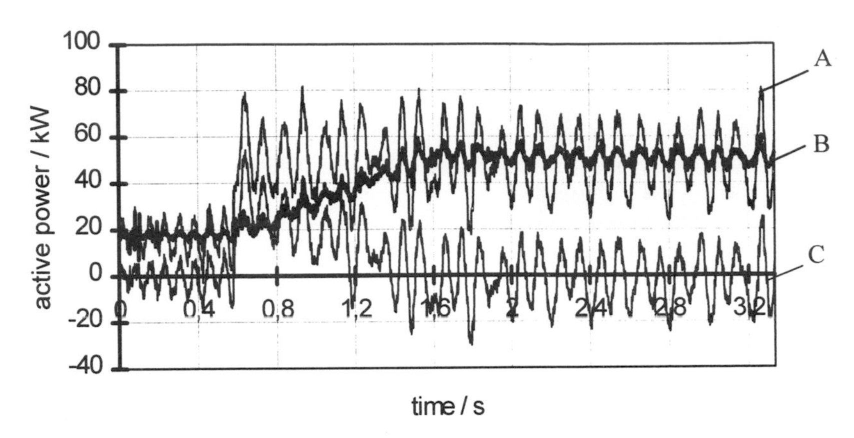

Fig. 4: Smoothing of sudden power change at a saw-mill (measured load used for simulation in the lab) (A: load power, B: line power, C: compensator power)

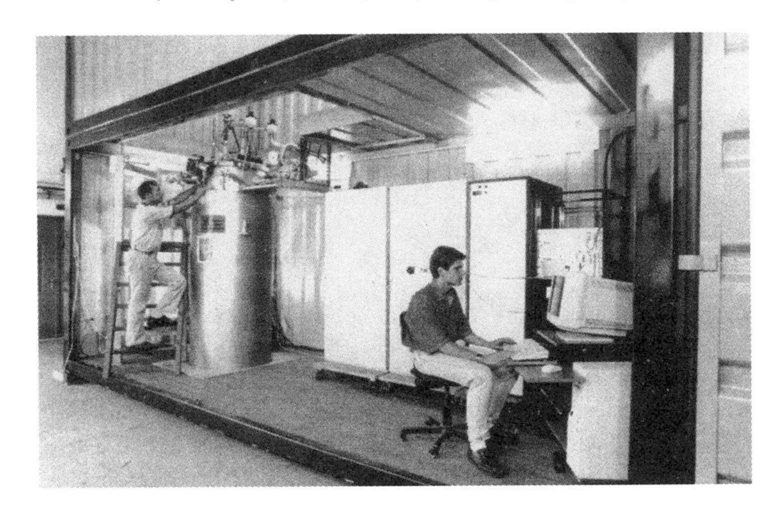

Fig. 5: Last tests of the SMES Compensator in its container before shipment to a saw mill (from left to right: cryostat with superconducting magnet, power conditioning system, magnet protection unit, and control unit)

References

[1] W.E. Buckles, M.A. Daugherty, B.R. Weber, and E.L. Kostecki; The SSD: a commercial application of magnetic energy storage, *IEEE Trans. Appl. Supercond., 3 (1993) 328-331*

[2] cf. *Superconductivity in Energy Storage*, Editors K.P. Juengst, W. Maurer, P. Komarek, *World Scientific Publishing Co., Singapore, 1995*

[3] T. Stephanblome, EUS GmbH, Gelsenkirchen, Germany, private communication

[4] H. Späth, H. Steinhart, P. Komarek, K.P. Juengst, W. Maurer, H.-J.Haubrich, T.Tischbein; 100 kJ/ 100 kW SMES for Compensation of Fast Pulsing Loads, *Ref. [1], p. 131-140*

[5] K.P. Juengst, H. Salbert, and O. Simon; Compensator for Fluctuating Loads with a SMES System, *Proc. 4th European Power Quality Conf., June 10-12,1997, Nürnberg, Germany*

[6] H. Steinhart, „Leistungselektronisches Stellglied und digitale Signalverarbeitung für einen schnellen Kompensator mit SMES“ *Thesis, Univ. Karlsruhe, 1996*

Inst. Phys. Conf. Ser. No 158
Paper presented at Applied Superconductivity, The Netherlands, 30 June–3 July 1997

Superconducting Magnet Design Issues for a Muon Collider Ring

M. A. Green

Lawrence Berkeley National Laboratory, Berkeley CA 94720, USA

Abstract Two factors govern the design of the collider ring magnets for a muon collider. These factors are the decay of the muons within the ring and the required luminosity needed to do the physics in a muon collider. The decay of the muons within the ring can deposit up to 2 kW per meter of beam energy into the magnet bore in the form of high energy electrons, positrons and gamma rays. One must keep the muon decay products from depositing their energy into the superconductor of the collider ring magnets. The luminosity at the collision point is inversely proportional to the ring circumference squared and it is inversely proportional to the beam size function squared. In order to get the circumference of the ring as small as possible, dipoles with a high central induction are desirable. The strength of the cell quadrupoles should in some way be proportional to the collider ring dipole central induction. A small beam size means that the beam beta function at the collision point must be very small. As a result, the focusing quadrupoles around the collision point must be strong with a large aperture (The beam beta function in these quadrupoles is often very large in at least one direction.) The final focus quadrupoles around the collision point of the muon storage ring should be treated as one of a kind magnets. The collider standard cell magnets would be mass produced, so cost is an issue. This report presents a number of superconducting magnet design issues that would apply to a muon collider storage ring.

1. Background

Today, two types of colliders are in use; hadron (proton and antiproton) colliders and lepton (electron-positron) colliders. Lepton colliders do a different kind of physics than is done using hadron colliders. Hadron colliders can go to much higher energies than the type of lepton colliders that are now in use. The maximum energy of an electron-positron collider ring is limited by the synchrotron radiation that is produced as the electrons and positrons are bent by the magnetic field as they go around a storage ring. The synchrotron radiation power is inversely proportional to the particle rest mass to the fourth power for a particle of a given energy in a given magnetic field. Muons, which are also leptons, have a rest mass that is 207 times larger than an electron or positron.

Muon collider rings can theoretically go to much higher energies than electron-positron collider rings without encountering the worst effects of synchrotron radiation[1,2]. Because the synchrotron radiation produced in a muon collider is much less energetic than that produced by and electron-positron collider, it appears that a muon collider can be an interesting device for probing the structure of subatomic particles. Of particular interest is the confirmation of the existence of a Higgs boson.

2. Muon decay Heating and Magnet Radiation Shielding

An important driver for the use of superconducting magnets[3] in a muon collider is muon decay. Because muons have a limited life time (2.2 μs when the muon is at rest) one must cool the muons, accelerate the muons, and collide them with muons of opposite charge before they cease to exist. When muons decay, two neutrinos and an electron or positron (depending on the charge of the muon) are formed. Thirty-five percent of the muon power ends up in the electrons and positrons[4]. The remainder of the muon power is transported to the universe by the decay neutrinos. The decay power that is of concern is in the electrons and positrons.

There is a basic incompatibility between any superconducting device and decaying muons because the secondary electrons and positrons that result from muon decay can deposit their energy within the superconducting device. Table 1 shows the muon decay rate, muon decay power and the muon decay power per unit length for various 2 TeV muon accelerator components and the 2 TeV collider ring with a luminosity of 10^{35} cm^{-2} s^{-1}. Because all of the muons decay in the collider ring, it is particularly vulnerable to the effects of muon decay. From table 1 one can see that over 1.8 kW per meter of muon beam power can be deposited inside the bending magnets.

Table 1 Muon Decay Parameters in Various Components of a 2 TeV Muon Collider

Component	Peak Energy (GeV)	No. of Turns	L_T (km)	Decay Rate (μs^{-1}) x 10^{13}	Decay Power (kW)	Decay Power per unit L (W m^{-1})
Linac	1.0	-NA-	0.12	1.9	0.6	-NA-
First Ring	9.6	9	2.17	1.2	3.6	1.64
Second Ring	79	12	11.27	0.8	19.7	1.75
Third Ring	250	18	29.24	0.4	36.8	1.26
Fourth Ring	2000	18	227.44	0.6	378	1.66
Collider Ring	2000	1000	7.93	13.1	14600	1840

In the collider ring, there are three sources of beam heating[4]: 1) the synchrotron radiation from the circulating muons (about 7.7 W per meter at a critical energy of 2.7 keV when the muon energy is 2 TeV), 2) the muon decay electrons and positrons at an average energy of 700 GeV (about 1450 W per meter), and 3) the synchrotron radiation from the decay electrons and positrons (about 400 W per meter for the small aperture dipoles in the muon storage ring at a critical energy of 2.1 GeV). The decay electrons and positrons are highly collimated along the mid-plane of the accelerator magnets, but they and their synchrotron radiation will produce numerous secondary particles when they interact with material in the walls of the vacuum chamber. The size of the region where the decay products and synchrotron radiation strike the wall is quite small (a band about 3 mm wide), but energy from the secondary particles that comes from the decay product collisions will be more evenly spread within the magnet coils.

In order to protect the superconducting magnets within the collider ring from the muon decay products, one has to absorb much of the decay product energy in a room temperature (300 K) liner within the magnets. A high specific mass density, high z material, such as tungsten, is needed for the liner in order to minimize liner thickness within the magnet bore. The power from the decay products that is allowed to be absorbed in the cold region of the superconducting magnets is two to three orders of magnitude less than the decay power of the muons in the collider ring. The amount of decay power allowed to be absorbed in the superconducting magnet coils is determined by the type of superconductor, its operating temperature and the amount of refrigeration available. The thickness of the tungsten liner needed in the dipoles is from 42 mm (for a factor of 100 attenuation) to 65 mm (for a factor of 1000 attenuation)[5]. The split coil dipole option described in Reference [6] will probably not be satisfactory because of the magnetic field reversal in the decay product channel.

3. The Need for High Magnetic Induction in the Collider Ring

Luminosity of the collider is the one factor that governs the quality of physics done on a muon collider. Higher luminosity means more Higgs bosons (or other interesting particles) are produced and analyzed. In order maximize the luminosity of the collider ring one can do the following: 1) One can store more muons in the ring. More muons stored means more muon are produced. There are limits on the number of muons that can be produced. 2) The emittance of the stored muons should be minimized. The basic emittance of the stored muons is a function of the muon cooling efficiency. (Cooling is the process that reduces the transverse momentum of the muon beam with respect to the longitudinal momentum.) A well-cooled beam also has a low momentum spread. Efficient cooling means a lower emittance, which translates directly to higher luminosity. The muon collider project is defined by the muon cooling system. 3) The muons can go around the storage ring more times before they decay. The number of turns the muons make before decaying is inversely proportional to the circumference of the ring. In order to get the ring circumference smaller one has to increase the ring dipole strength and the ring quadrupole gradient. The machine luminosity increase inversely with the ring circumference squared. 4) The beta star of the muon beam at the collision point should also be minimized. A reduction of beta star at the collision point translates to strong quadrupoles with very large apertures.

4. Dipoles and Quadrupoles for the Muon Collider Ring

A high storage ring luminosity translates to high field magnets. Most of these magnets will have a large superconducting coil aperture, because of the shielding required to reduce the heating from muon decay. Table 2 presents a number of parameters for various dipoles and quadrupoles in a 2 TeV on 2 TeV collider ring[7]. The magnets that might be found in lower energy muon collider rings are comparable, but, in general, the pole fields are not as high.

Table 2 Parameters for Selected Magnets for a 2 TeV Collider Ring

parameter	Ring Dipole	Ring Quad	QF1 Quad	QF2 Quad	QF3 Quad	QD1 Quad
5 Sigma Beam R (mm)	~7.0	~7.0	33.0	61.0	81.0	110.
Good Field R (mm)	10.0	10.0	33.0	61.0	81.0	110.
Type of Magnet	Nb-Ti	S/C Iron	Nb_3Sn	Nb_3Sn	Nb_3Sn	Nb_3Sn
Tungsten Thick (mm)	60.0	none	10.0	15.0	15.0	15.0
Pole (Coil) Radius (mm)	77.5	~12	55.0	85.0	105.0	135.0
Magnet Length (m)	~6.0	~3.0	0.8	2.0	2.0	10.6
Central Induction (T)	9.0	---	---	---	---	---
Field Gradient (T m)	---	120 to 180	191	133	119	92.6
Pole Induction (T)	9.0	1.44 to 2.16	10.5	11.3	12.5	12.5
dB/B (parts in 10000)	<3	~10	<3	<5	<5	<5
Number of Magnets	~1000	~160	2	2	2	2

The collider ring dipole could be made from either Nb-Ti at 1.8 K or it could be made from Nb_3Sn at 4.4 K. Because their aperture is large, the ring dipoles will have large forces. In order to reduce the magnetic stress, one can reduce the current density within the coils without greatly affecting the amount of superconductor needed to generate the field. The ring dipole shown in Table 2 is only about 35 percent larger than the outer coil package of the D20 dipole built at the Lawrence Berkeley National Laboratory (LBNL). It has been demonstrated that this coil alone produces a central induction that is greater than 10 tesla[8]. The collider ring dipoles can be made using either a cosine theta design or a block design with tilted ends.

Figure 1 shows a cosine theta design for the magnet that is based on the outer coil package for the LBNL D20 dipole. The magnet shown in Figure 1 could use two layers of cable with 47 strands that are 0.65 mm in diameter. The cable shown in Figure 1 is about 1.75 mm thick by 16.2 mm wide. Like the LBNL magnet, this magnet would have hard aluminum bronze wedges and poles pieces to carry the forces. The niobium tin version of this magnet would be potted, like the LBNL coils. It is evident from the recent D20 tests at Berkeley that this type of coil is quite tolerant to heating. It is not known if the Nb-Ti version of the magnet would be as heat tolerant as the niobium tin coils in the D20 dipole.

The ring quadrupole shown in Table 2 is assumed to be an iron dominated quadrupole with superconducting coils buried up inside the warm iron. The radiation attenuation of the iron and the material around the muon beam is equivalent to at least 60 mm of tungsten. The lower gradient and pole field given in Table 2 would apply to a quadrupole with pure iron pole pieces. The higher values of gradient and pole induction would apply for a magnet with holmium pole pieces and an iron flux return. The superconducting coils for the quadrupole shown in Table 2 would operate at low fields and would be made from Nb-Ti. Comparable cosine theta quadrupoles would have pole fields of 9.3 T and 13.9 T respectively.

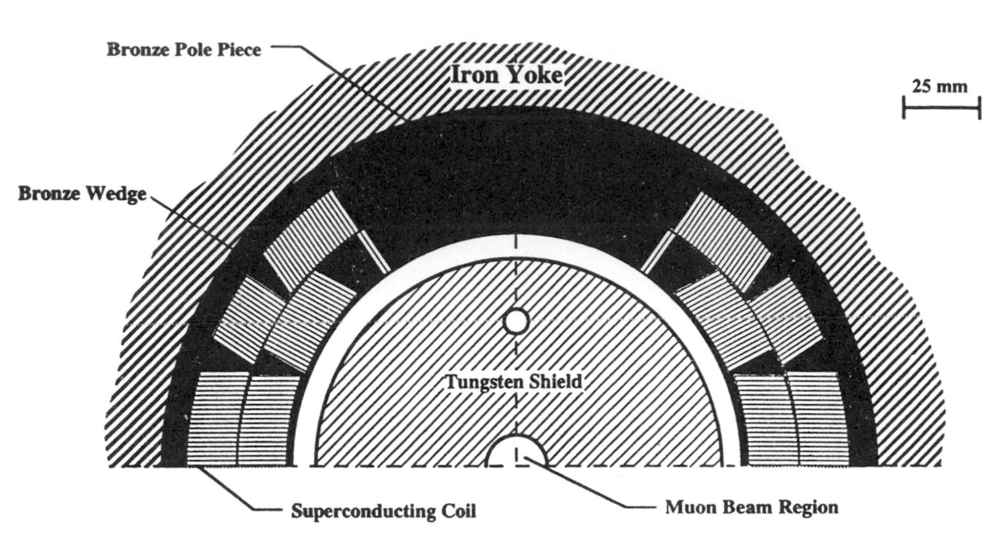

Fig. 1 A 9 Tesla Collider Ring Dipole Surrounding a 60 mm Thick Shield

The other four quadrupoles shown in Table 2 are the first four quadrupoles around the collider interaction region. There are two more interaction region quadrupole pairs that are not shown in Table 2. These quadrupoles are identical in aperture, gradient and pole field to the QD1 quadrupole shown in Table 2. These quadrupoles have the following lengths; QFM is 8.4 meters long and QD2 is 1.6 meters long. The interaction region quadrupoles are almost one of a kind in their design. They will require considerable development of both the magnet design and the conductor. The 2 TeV muon collider ring interaction region will have a total of eight quadrupoles with a pole field of 12.5 T. The other four interaction quadrupoles, which have a pole induction greater than 10 T, are almost as difficult to build as the 12.5 T pole field quadrupoles. Both cosine theta and block coil designs are being considered for the interation region quadrupoles.

Acknowledgments

The author thanks Bob Palmer, and Eric willen of the Brookhaven National Laboratory for comments concerning magnet design and other aspects of the design of a muon collider. This work was performed with the support of the Office of High Energy and Nuclear Physics, United States Department of Energy under contract number DE-AC03-76SF0098.

References

[1] The $\mu^+\mu^-$ Collider Collaboration, "μ^+ μ^- Collider, A Feasibility Study," BNL-52503, July 1996

[2] R. B Palmer, A. Sessler, A. Skrinsky, A. Tollestrup et al., "Muon Colliders," Proceedings of the Ninth ICFA Beam Dynamics Workshop, Montauk NY, 15-20 Oct. 1995

[3] M. A. Green, "Superconducting Magnets for a Muon Collider," Proceedings of the Third Workshop on the Physics Potential and Development of Muon-Muon Colliders, San Francisco, CA, 13-15 Dec. 1995

[4] *Techniques of High Energy Physics*, p 28, Edited by David M. Ritson, Interscience Publishers (1961)

[5] N. V. Mokhov & S. I. Strganov "Simulation of Backgrounds in Detectors and Energy Deposition in Superconducting Magnets at μ^+ μ^- Colliders," Proceedings of the Ninth ICFA Beam Dynamics Workshop, Montauk NY, 15-20 Oct. 1995

[6] M. A. Green and E. Willen, "Superconducting Dipoles and Quadrupoles for a 2 TeV Muon Collider," IEEE Trans. Appl. Supercon. **7**, No 2 (1997)

[7] C. Johnstone and A. Garren, "An IR and Chromatic Correction Design for a 2 TeV Muon Collider," to be published in the 1997 Particle Accelerator Conference Proceedings, (1997)

[8] R. M. Scanlan et al., "Preliminary Test Results of a 13 Tesla Niobium Tin Dipole," Proceedings of the 1997 EUCAS Conference, LBNL-40463, (1997)

Inst. Phys. Conf. Ser. No 158
Paper presented at Applied Superconductivity, The Netherlands, 30 June–3 July 1997

Status of the EU Conductor Manufacture for the ITER CS and TF Model Coils

S. Conti*, R. Garrè*, S. Rossi*, A. Della Corte, M. Ricci**, M. Spadoni**, G. Bevilacqua***, R. Maix***, E. Salpietro***, H. Krauth****, A. Szulczyk****, M. Thoener****, A. Laurenti*****, C. Siano*******

* Europa Metalli, Fornaci di Barga (Lu), Italy. ** Associazione ENEA-Euratom, Frascati Italy, *** The NET TEAM, Garching, Germany, **** Vacuumschmelze, Hanau, Germany, ***** Ansaldo Energia, Genova, Italy,

Abstract. The main R&D task of the ITER programme is the manufacture of two Model Coils, a Central Solenoid (CS) and a Toroidal Field (TF) Model Coil. The European Union is currently putting a significant effort into the production of the Nb_3Sn superconducting strands, full-size cable and jacketing for the conductors that will be used for the production of the two Model Coils. As far as the CS Model Coil is concerned, the contribution of the European Union consists in the production of 6.4 Tons of Nb_3Sn strand, 1200 m of full-size cable and 5787 m of jacketing of the conductor. The TF conductor, on the other hand, will be completely manufactured in Europe with the production of 4 Tons of internal tin Nb_3Sn strand and the cabling and jacketing of 1 Km of conductor. This paper presents the industrial activities developed and currently in progress for the realization of the different conductors.

1. Introduction

The ITER EDA stage of the ITER project foresees the realization of two Model Coils both for the Central Solenoid and for the Toroidal Field Magnet. The cables used for the winding of the two types of magnet are of the Cable in Conduit (CIC) type in which forced flow helium cooled cables are formed from Nb_3Sn superconducting strands.

2. Strand Manufacture

2.1. CSMC Conductor

For the manufacture of the CSMC two different typologies of Nb_3Sn superconducting strand were foreseen respectively denominated HP1 and HP2 for a total of 23 Tons. The technical specifications for the electromagnetic performances are reported in Table 1.

The contribution from the European Union consists in 6.4 Tons of type HP2 chromium coated strand, produced according to the bronze method by Vacuumschmelze (VAC) Company. The layout of the strand consists in a single bundle, composed of 55 sub-groups of 83 filaments of Nb, surrounded by a Ta antidiffusion barrier and assembled in a high purity copper matrix. The requested specifications have been completely satisfied. In particular the hysteretic losses are typically below 100 kJ/m^3.

Table 1 - HP1 and HP2 NET ITER Specifications

	HP1	HP2
Jc (12 T, 4.2K)	>700 A/mm^2	> 550 A/mm^2
Q (± 3T)	< 600 mJ/cm^3	< 200 mJ/cm^3

2.2. TFMC Conductor

For the TFMC all the strand of the HP1 type corresponding to a total of 4 Tons was produced in the European Union by the Company Europa Metalli. The Nb_3Sn strand is produced using the internal tin method and has a layout, defined following subsequent optimization studies [1], that consists of 36 bundles each containing 150 Nb filaments. Each single bundle is surrounded by a Ta-Nb antidiffusive barrier which protects the pure copper matrix. The strands have demonstrated an excellent workability with yields greater than 90%. The characterization tests have shown that the strand satisfies the HP1 characteristics and, in particular, the critical current is on average greater than 730 A/mm^2 at 12 T, 4.2 K.

3. Cabling Activities

The European cabling line for the two model coil cables [2], has been designed and developed in the framework of a NET Contract assigned to Europa Metalli. The line is installed at the Europa Metalli SpA Fornaci di Barga Works. Minimum unit lengths of cable which may be produced with this equipment is 1.1 Km, approx. 6 Tons in weight. The two types of cable which are produced are MC1 and MC2.

The MC1 type cable is composed of 1152 superconducting strands. The cabling sequence starts with a triplet configuration followed by intermediate steps (4 x 4 x 4) which led to the realization of sub-unit cables referred to as last but one, (see fig. 1). The sub-cables are then wrapped with Inconel 600 type tape to ensure mechanical stability and electrical decoupling of the sub-cables in the full size cables. Six of these sub-cables are then cabled around a helical spiral in AISI 304 tape to obtain a full-size cable, 39.2 mm in diameter. The full-size cable is also wrapped with Inconel 600 type tape. The MC2 type cable, made up of 1080 strands, is shown in fig. 2. The cabling sequence for this cable foresees an initial triplet formed by one copper and two superconducting strands, and the cabling steps to obtain the last but one cable are 3 x 5 x 4. Again six of these sub-cables are cabled to produce a full size cable, 38 mm in diameter. This cable is also wrapped with Inconel 600 type tape. Both MC1 and MC2 cables are used for the realization of the CSMC conductor. As far as TFMC conductors are concerned only MC2 type cables are used. In the framework of CSMC productions about 800 m of MC1 type cables, 400 m of MC2 cables and the entire production of the full size cable (about 1 Km) for the TFMC are to be produced by the European Union.

All CSMC cables have been manufactured using the Nb_3Sn bronze type method adopted by Vacuumschmelze whilst the Nb_3Sn internal tin method, developed by Europa Metalli, will be used for TFMC conductors. At present 1200 m of cable for CSMC and the first 170m of cables for TFMC have been successfully produced.

4. Jacketing

All CSMC and TFMC manufacturing activities for the entire quantity of the final conductor, which includes the jacketing and compaction activities, are performed in Europe. The CS

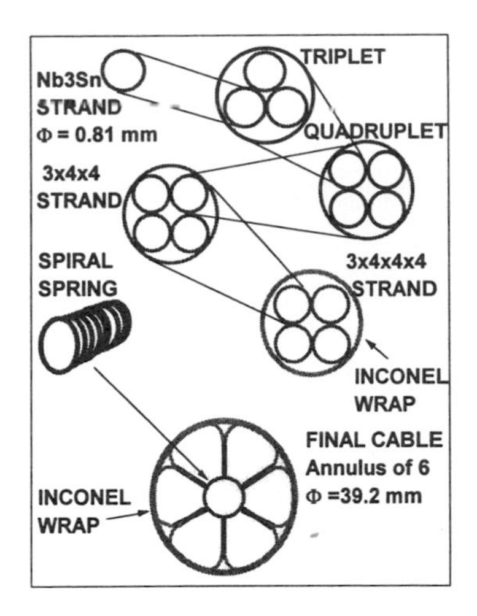

Fig. 1 Cabling Sequence of MC1 Cable.

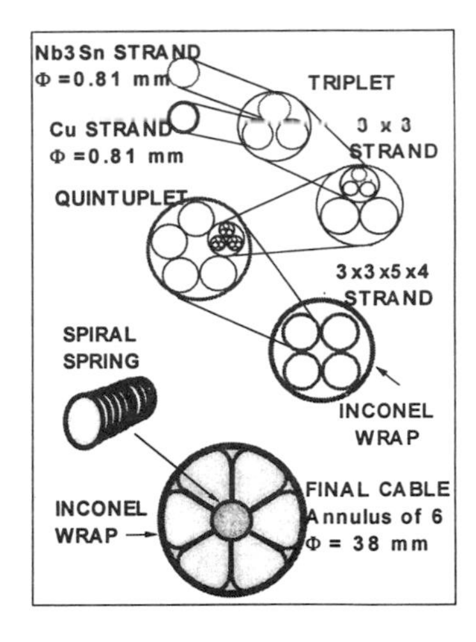

Fig. 2 Cabling Sequence of MC2 Cable.

jacketing line has been developed by Ansaldo and is installed and in function at Genova, whilst Europa Metalli has developed the jacketing line for the TF cables, which is installed at the EM Fornaci di Barga Works. The sequence for both the jacketing lines is the following: butt welding of straight jacket tube pieces; pull through of the cable inside the jacket; compaction of the straight jacketed cable to the specified dimensions; calandering of the final conductor on a suitable collecting spool.

4.1. CSMC Jacketing line

The technical characteristics of this line may be found in a previous paper [3]. The material which is used for jacketing the cable is Incoloy 908 shaped in a square jacket tube. After the initial realization of the dummy cable, 29 lengths over a total of 37 have been produced by Ansaldo. Completion of the activities is foreseen for August '97.

Since March '97 a second off line welding unit, identical to the first line equipped with bunker and radiographic machine has been in function, and this enables the welding and control of 2 - 3 tubes that enter the line, in lengths up to 20 m, substantially improving production rates which have increased from 400 to 600 m/months. The quality of the welding is considered optimum, in view of the fact that there is no waste material, although the importance of repairing surface defects, which may occur especially during the compaction and final calandering operations, must not be ignored. It has been demonstrated that these defects are therefore more likely to be attributed to the characteristics of the materials used (basic materials, welding material) rather than to the welding procedures themselves. As far as the cable insertion is concerned, pulling values have been registered which vary in the region of 200 and 500 Kg per lengths between 100 and 200 m and gaps between the cable and the conduit greater than 1.5 mm.

4.2. TFMC Line

The line which has been realized by Europa Metalli [4] in Fornaci di Barga for the jacketing of the TF cable is currently in full operation. The line is divided into 3 sections: the first section is where the conduit is welded, compacted and wound, the second section consists in a brass line where the welded conduit is located, the third is the section where the spooling device is placed, providing the pulling force for insertion. The line enables the manufacture of 400 m of jackedted cable.The material used for the jacketing is made up of steel tubes AISI 316 LN. The entire cable, after insertion of the superconducting rope, is then compacted to final diameter of 40.7 mm see fig. 3.

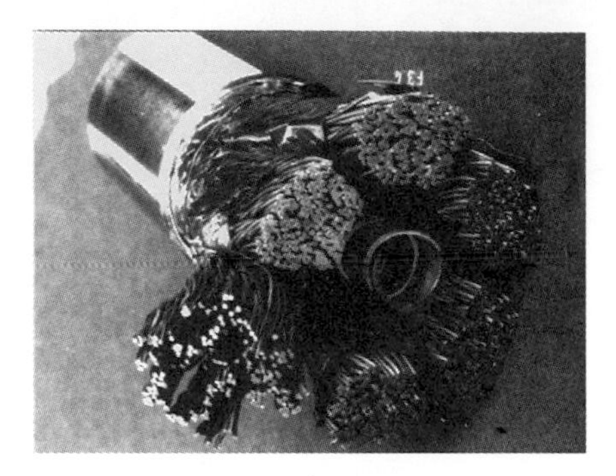

Fig. 3 Full-size TF Conductor.

Considerable preliminary test work carried out on the empty conduit and on dummy cables has enabled the validation not only of the actual working procedures but also of all control procedures defined in the framework of the quality control plan specifically developed. In particular, leak checks and X-ray controls are made on each weld; the insertion pulling force of the cable in conduit is continuously measured by a strain gauge fixed on the pulling rope; the final dimensional controls are carried out using a twin laser gauge micrometer system. Finally the cable is submitted to leak tests, pressure tests and mass flow tests. At present, besides a total of 650 m of dummy superconducting cable, the first unit length of 170 m of superconducting cable has been produced with complete success.

5. Conclusions

In the framework of the ITER-EDA phase of the ITER project the realization of two Model Coils for the Central Solenoid and the Toroidal Field Magnet is foreseen. The main results obtained can be seen in the development of appropriate industrial processes for the manufacture of Nb_3Sn superconducting strands (Europa Metalli and Vacuumschmelze), cables (Europa Metalli) and CIC type conductors (Ansaldo, Europa Metalli) which have completely satisfied the technical specifications foreseen by ITER.

References

[1] R. Garrè, S. Conti, S. Rossi, "INDUSTRIAL DEVELOPMENTS OF INTERNAL TIN Nb_3SN STRANDS FOR HIGH FIELD APPLICATIONS" Proc. of the 18th Symposium on Fusion Technology, Vol. 2, 965, (1994).

[2] A. Della Corte et al. "EU CONDUCTOR DEVELOPMENT FOR ITER CS AND TF MODEL COILS " Proc. of ASC '96, Pittsburgh (1996).

[3] A. Della Corte et al "CONDUCTOR FABRICATION FOR ITER MODEL COILS. STATUS OF THE EU CABLING AND JACKETING ACTIVITIES" Proc. of the 18th Symposium on Fusion Technology, 885-888, (August 1994).

[4] S. Conti, R. Garrè, S. Rossi et al "FABRICATION OF PROTOTYPE CONDUCTORS FOR ITER TF MODEL COIL" Proc. of the 19th Symposium on Fusion Technology, Lisbon (1996).

Inst. Phys. Conf. Ser. No 158
Paper presented at Applied Superconductivity, The Netherlands, 30 June–3 July 1997

Ramp Rate Experiments on an ITER Relevant Pulsed Coil

E. P. Balsamo, O. Cicchelli, M. Cuomo, P. Gislon, G. Pasotti, M. V. Ricci, and M. Spadoni

ASSOCIAZIONE EURATOM-ENEA SULLA FUSIONE
Via E. Fermi 45, 00044 Frascati (Rome) ITALY

Abstract Ramp rate experiments in self-field conditions up to 3.9 T and 3.5 T/s have been carried out on an ITER relevant Cable-in-Conduit-Conductor (CICC) Nb_3Sn wind-and-react solenoid. The most striking effect is that the ramp rate producing a quench is increasing with the number of pulses. Ramp rate investigations have been started after about 90 pulses necessary to optimise the shape of the ramps. To account for the experimental results, the coupling losses time constant $n\tau$ should decrease after 150 pulses to 35% of the value showed at pulse #90. Additional ramp rate experiments at different temperatures and the measurement of the AC losses over the entire magnet should allow to confirm the significant change in the transverse resistivity of the cable inside the CIC conductor.

1. Introduction

Experiments of operation in pulsed field regimes are underway at ENEA Frascati on an ITER relevant Nb_3Sn Cable-in-Conduit-Conductor solenoid. The test plant and the superconducting coil have been already described elsewhere [1,2]. Main objectives of the testing programme are the ramp rate behaviour of the magnet, the direct measurement of AC losses on the coil and the experimental study of the current distribution inside the cable during a ramp.
This paper will report about the results obtained in the ramp rate experiments.

2. Experimental results

The pulsed field experiments have been carried out in the self-field mode. The test coil was charged at constant rate to a pre-set value of the maximum current. The ramp was followed by a plateau, to detect the occurrance of a quench. A stepwise increase of the ramp rate was done until a quench occurred. The tests have been performed at maximum currents of 5, 6 and 7kA.
The coupling losses generated during a ramp can be calculated:

$$E = \frac{n\tau V}{\mu_0}\int\left(\frac{dB}{dt}\right)^2 dt \qquad (1)$$

where V is the volume of the cable metal.
Coupling losses measurements on a straigh sample of the CIC conductor used in the test coil were carried out at the University of Twente; an $n\tau$=120ms was measured.
The high-voltage generator used for the current pulses was tuned to optimize the pulse shape. A fairly good linear increase of the current, followed by a plateau, was obtained at current variations in the range 1-2kA/s (Fig.1a). However, for higher ramp rates, the slope of the pulse was found not to be linear (dI/dt is not constant) and the set plateau current was reached with some overshooting (Fig.1b).

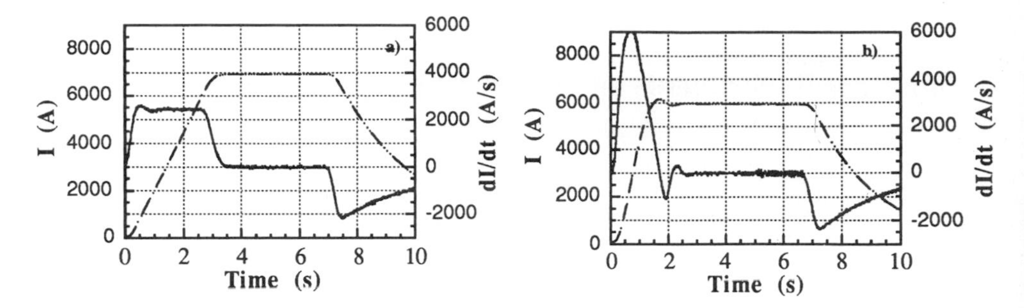

Fig.1 I(t) (dashed line) and dI/dt for a typical low and high rate current pulse. Fig 1a) refer to pulse #141, while Fig. 1b refer to pulse #135.

We define an effective dB/dt as:

$$\left(\frac{dB}{dt}\right)_{eff} = \int\left(\frac{dB}{dt}\right)^2 dt \Big/ B_{plateau} . \qquad (2)$$

The $(dB/dt)_{eff}$ is the constant ramp value producing the same coupling losses as the real pulse. A quench is considered significant when it is produced in the highest field region of the coil, as revealed by the voltage taps located there.

Table 1- Summary of the most significant ramp rate tests at T_{hf}=7.3K

Pulse n°	Iplat (kA)	Bplat (T)	(dB/dt)eff (T/s)	Q/nQ	nτ/nτ(90)	linearity (%)
90	4,95	2,76	1,67	NQ		60
91	**4,95**	**2,76**	**2,23**	**Q**	**1,00**	**45**
99	4,96	2,77	2,08	NQ		52
100	**4,96**	**2,77**	**2,79**	**Q**	**0,80**	**31**
111	4,96	2,77	2,54	NQ		30
112	**4,96**	**2,77**	**2,85**	**Q**	**0,78**	**27**
134	4,97	2,77	2,99	NQ		23
149	4,97	2,77	2,72	NQ		21
150	4,97	2,77	2,97	NQ		21
151	4,96	2,77	3,20	NQ		21
92	5,93	3,31	0,57	NQ		93
93	**5,93**	**3,31**	**1,11**	**Q**	**1,00**	**83**
102	5,96	3,33	1,05	NQ		90
109	**5,94**	**3,31**	**1,52**	**Q**	**0,72**	**71**
119	5,93	3,31	2,10	NQ		60
121	**5,93**	**3,31**	**2,57**	**Q**	**0,43**	**36**
135	**5,94**	**3,31**	**2,70**	**Q**	**0,41**	**31**
136	5,95	3,32	2,51	NQ		42
94	**6,90**	**3,85**	**0,58**	**Q**	**1,00**	**94**
141	6,94	3,87	1,27	NQ		75
142	**6,95**	**3,88**	**1,67**	**Q**	**0,34**	**66**
145	6,94	3,87	1,46	NQ		67
146	**6,92**	**3,86**	**1,70**	**Q**	**0,33**	**64**

Pressure and temperature of the supercritical helium are measured at the coolant inlet and calculated in the high field zone of the magnet (P_{hf}, T_{hf}), assuming an isenthalpic expansion. The steady state high field temperature T_{hf} is the main reference parameter for a given pulse. The most significant data for all the ramp rate experiments carried out at T_{hf}=7.3K are reported in Table 1. The runs refer to plateau currents of about 5kA, 6kA and 7kA. Pulses producing a quench are referred as "Q"; no quench is indicated as "NQ". The linearity is a rough indication of the fraction of the rise time in which the current increase could be considered linear.
The most striking feature is that, for a given plateau current, the quench is obtained at a ramp rate which increases with the number of pulses. We attribute this behaviour to a variation of the coupling losses time constant $n\tau$ of the conductor, induced by the force cycles on the cable. The ratio of $n\tau$ of a given pulse to that of the first meaningful test (pulse #90) can be calculated:

$$\frac{n\tau}{(n\tau)_{90}} = \int\left(\frac{dB}{dt}\right)^2 dt \Big/ \int\left(\frac{dB}{dt}\right)^2_{90} dt = \left(\frac{dB}{dt}\right)_{eff} \Big/ \left(\frac{dB}{dt}\right)_{eff(90)} \qquad (3)$$

and is reported in Table 1 and in Fig.2.
A similar behaviour was observed on short straight samples of CIC conductors [3], but the variation of the coupling losses saturated after a few cycles. In our case, we are still observing appreciable variations after 150 pulses.
By comparison, the 81 strand CICC of Ref. 3 showed a reduction of $n\tau$ from 9ms in the virgin state to 2ms after 4 pulses, this value remaining then constant for all the following pulses, up to 10.
Few tests on the magnet have been done also at T_{hf}=8.3K, the pulse number being >160.
As expected, the quenches occur at ramp rates lower than the previous pulses (from #130 to #150) done at T_{hf}=7.3K but definitely higher than those obtained with the pulses number 90-100. This is another confirmation that coupling losses are decreasing by increasing the number of pulses.

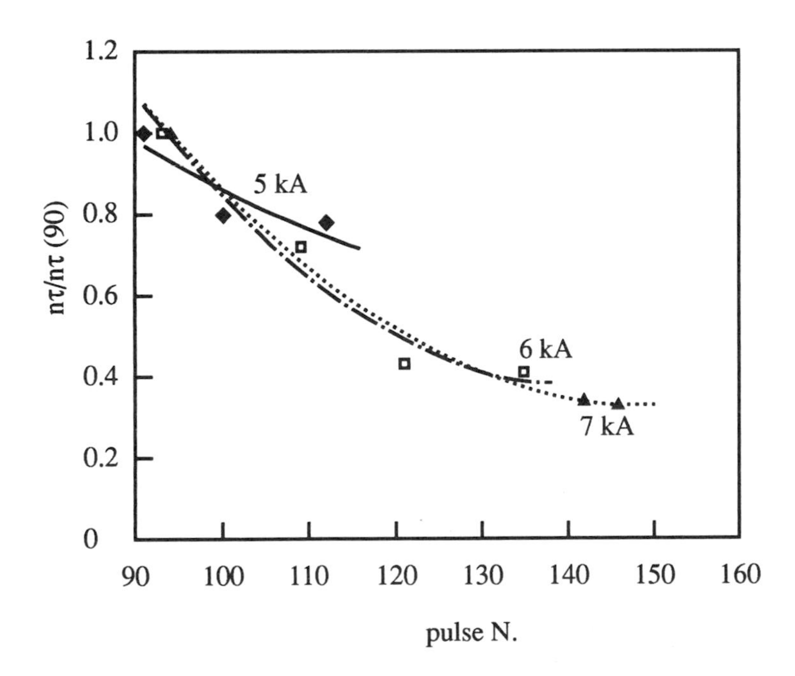

Fig.2 Change of $n\tau$ versus pulses number; the values are normalized to the pulse number #90.

We are planning in the next future to alternate ramp rate pulses at 6.3K, 7.3K and 8.3K. This should give a confirmation about the continuation of the coupling losses decrease, as well as a relative comparison at 3 different temperatures.
The direct measurement of the AC losses should also allow an independent estimate of the actual $n\tau$ in the coil.
The AC loss measurement through the coolant temperature rise is not suitable since the transit time of the coolant is typically 30-40 minutes and the average coolant temperature increase is only of the order of 0.1K.
The integration of VxI over a symmetrical excitation cycle 0-B_{max}-0 is in fact not possible, since the generator used for the ramp rate experiments doesn't allow a controlled discharge.
The method we intend to apply makes use of a power amplifier, ($\pm$200V, 100A) capable of delivering triangular or trapezoidal pulse trains, to be put in parallel to the fly-wheel generator. After the main current has reached the plateau, the amplifier generates an additional variable field (ΔB<0.05T, dB/dt<2T/s). The resulting voltage and current signals will be memorized for off-line numerical processing.
The experimental testing programme will be concluded with a direct measurement of the current distribution inside the conductor during a ramp. To this purpose, a set of miniaturized Hall probes and pick-up coils, has been provided by MIT and installed near the two interlayer joints of the magnet [4].

3. Conclusions

A significant number of pulses has allowed to determine that the ramp rate limit leading to a quench shows a continuous increase with the number of pulses. This behaviour can be explained with a decrease of the coupling losses time constant; the crossover strand contacts inside the cable lose their engagement and the transverse resistivity increases.
Owing to the significant number of pulses necessary to tune up the high voltage generator (90) we don't have data referring to a virgin state.
However, $n\tau$ appear to decrease by more than 60% going from pulse 90 to pulse 150. A direct measurement of ac losses on the magnet will be carried out, to derive the present coupling loss time constant and to compare it with the value measured on straight virgin conductor samples.
Simulations with the code SARUMAN will then be carried out to compare the predicted critical energies with the values derived from the experimental data.

Acknowledgment

The authors wish to thank Ing. F. Valente, Mr. F. Gravanti and Mr. E. Di Ferdinando for their technical support. The authors are particulary grateful to Mr. C. Mastacchini for his assistance during the tests.

References

[1] E. P. Balsamo et al. "Results of simulations of stability and quench behaviour for pulsed tests of the ENEA 12TNb_3Sn CICC coil", Journal of Fusion Energy vol.14, n°1, 1995

[2] E. P. Balsamo et al. "Ramp Rate Experiments on a Nb3Sn CICC Wind-and-React Magnet", proceeding of the ASC 1996, Pittsburgh, August 1996

[3] A. Nijhuis, H.J. Ten Kate, P. Bruzzone "The influence of Lorentz Force on the AC loss in sub-size Cable-in-Conduit Conductors for ITER", proceeding of the ASC 1996, Pittsburgh, August 1996

[4] V. S. Vysotsky, M. Takayasu, M. Ferri, J. V. Minervini and S. Shen "New method of current distribution studies for ramp rate stability of multistrand superconducting cables", IEEE Trans. on Appl. Superc. vol.5, no.2, June 1995

Inst. Phys. Conf. Ser. No 158
Paper presented at Applied Superconductivity, The Netherlands, 30 June–3 July 1997

Preliminary Test Results of a 13 Tesla Niobium Tin Dipole

R. M. Scanlan, R. J. Benjegerdes, P. A. Bish, S. Caspi, K. Chow, D. Dell'Orco, D. R. Diederich, M. A. Green, R. Hannaford, W. Harnden, H. C. Higley, A. F. Lietzke, A. D. McInturff, L. Morrison, M. E. Morrison, C. E. Taylor, and J. M. Van Oort

Lawrence Berkeley National Laboratory, Berkeley CA 94720, USA

Abstract During March and April 1997, the Lawrence Berkeley National Laboratory (LBNL) tested its four layer 13 T niobium-tin dipole. The LBNL test dipole has a coil bore of 50 mm and a length of about 1 m. This magnet was trained to a central induction of 13.5 T at a temperature of 1.8 K. The peak central induction reached at 4.5 K was 12.8 T. This report describes the dipole and presents its design parameters. Preliminary training results from the magnet test will be presented.

1. Introduction

The Lawrence Berkeley National Laboratory (LBNL) "Advancement of Accelerator Magnet Technology" program has concentrated for the last several years on the development of magnet construction techniques applicable to brittle superconductors[1-4]. Niobium tin was chosen because of its more extensive data base, and because presently it is the only superconductor with practical current density in the 11T to 16T range. This paper presents the parameters of the D20 LBNL dipole with experimental measurements taken during the spring of 1997.

The tests of the dipole have produced several surprises that are summarized as follows: 1) The dipole exhibited excellent ramp rate quench performance; 2) the dipole had excellent thermal stability (>20 watts;12T); and 3) the magnet trained up to much higher fields as compared with earlier Nb3Sn test dipoles[5-7]. The dipole achieved a central induction of 12.8T at 4.4K and 13.5T at 1.8K. The 1.8K performance was clearly not limited by the critical current of the superconductor. The highest central induction reached by the D20 dipole eclipsed the previous high field dipole record by Twente University (11.03T reached by an LHC model "MSUT"[6] magnet) by almost 2.5 T.

The highly interdependent coil fabrication steps of a wind and react Nb_3Sn magnet require a more integrated approach to cabling, insulating, stepped multi-phased heat treatment, similar expansion and contraction materials, protection heaters, epoxy impregnation, assembly, and pre-loading due to the larger temperature range that the winding must operate compared to Nb-Ti. There is a large body of Nb_3Sn data that indicates a substantial J_c loss with increasing perpendicular strain, which up to fields of 13.5 T, did not appear to be the limit in the present configuration of D20.

2. Magnet Design

D20 is a four layer graded Rutherford cable cosine theta dipole with a 50 mm coil bore. The inner two layers (1 and 2) used one size cable (37 strands 0.75 mm diameter) with a 1.57 mm (1.65 mm measured) thickness and a 14.66 mm (14.66 mm measured) width under a load of 34.5 MPa. These cable strands were made by two different manufacturers-Teledyne Wah

Chang (bottom inner coils) and Intermagnetics General (top inner coils). The outer two layers (3 and 4) use another size cable (47 strands 0.48 mm diameter) with a 1.12 mm (1.30 mm measured) thickness and a 11.89 mm (11.91 mm measured) width under a load of 34.5 MPa. These cable strands were manufactured by Teledyne Wah Chang. All of the conductor cabling was done at the LBNL cabling facility. Table 1 presents the nominal D20 dipole parameters. Figure 1 shows the nominal coil cross-section for the D20 dipole.

Table 1 Nominal Parameters for the LBNL 13 Tesla D-20 Dipole

Layer Number	1	2	3	4
Number of Blocks per Quadrant	3	4	3	3
Number of Turns per Quadrant	16	26	40	56
Packing Fraction	0.841	0.841	0.838	0.838
Copper to Non-copper Ratio	0.67 to 1.08*	0.67 to 1.08*	1.06	1.06
Nominal Layer Inner Radius (mm)	24.75	40.15	55.82	67.95
Nominal Layer Outer Radius (mm)	39.41	54.81	67.73	79.86

Yoke Inner Radius (mm)	89.57
Yoke Outer Radius (mm)	355.6
Central Iron Yoke Length (mm)	531.5
Total Stainless Yoke Length (mm)	749.8
Length of the Longest Coil (mm)	1249.4
Overall Length of the Dipole (mm)	1633.9
Overall Magnet Cold Mass (kg)	~7000
Dipole Design Current at 4.4 K (A)	6400
Central Induction at Design Current (T)	13.0
Design Field Uniformity at 10 mm (parts in 10000)	<1
Dipole Stored Energy at Design Current (kJ)	802
Peak Induction in Inner S/C at Design Current (T)	~13.3
Peak Induction in Outer S/C at Design Current (T)	~10.5
Peak Current Reached at 4.4 K (A)	6300
Central Induction Reached at 4.4 K (T)	12.8
Peak Current Reached at 1.8 K (A)	6712
Central Induction Reached at 1.8 K (T)	13.5

* The lower copper ratio is for the IGC conductor in the upper two inner layers; the higher figure is for the TWC conductor in the lower two inner layers.

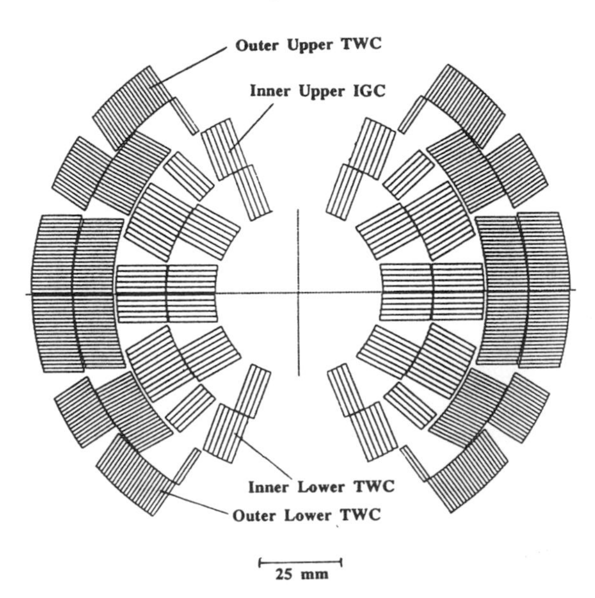

Figure 1 The Coil Cross-section for the LBNL D-20 Niobium Tin Dipole

The cable is insulated with a single glass sleeve 0.12 mm thick in the straight sections. The ends have an extra 0.28 mm thick glass tape to prevent shorts. The aluminum-bronze wedges, pole pieces and end pieces are coated with alumina (0.13 mm thick) for extra insulation. There is a layer of fiberglass tape between the bore and the inner coil heater composite structure. There is a second layer of glass tape between coil layers 1 and 2 and the coil heater composite structure. The final layer of the double winding package is a layer of fiberglass cloth. During the heat treatment, the heater, voltage tap. composite heater space was occupied by a stainless Steel shim. The glass cloth sizing was evaporated early in the bronze to Nb_3Sn formation. This process was repeated for each double coil package. D20 was protected by quench heaters with a photo-etched geometry so that coil voltages would be minimized, and so that temperature increases in the coil would be as uniform as possible.

After potting, the four double layer coils were placed back into the potting fixture with a pressure sensitive film replacing one or more insulation layers on the appropriate loading surfaces. These fixtures were then loaded to moderate levels (~10 MPa) to determine the surface profiles as well as the size of the packages. This sizing information along with design numbers and component measurements were checked with a pre-assembly of the entire magnet, exclusive of the iron and stainless steel yokes shells. Pressure sensitive film was used to measure the radial azimuthal pressure uniformity and cross calibrate the strain gauges in the poles. The bronze collared coil structure was pre-stressed with stainless steel wire up to about half the desired pre-load. The final assembly used 18 layers of 304 stainless steel 3 mm x 1 mm round edge wire at a tension of 850N for an azimuthal pre-stress of 70 MPa. The magnet end plates held the magnet winding in axial compression up to 30% of the total Lorentz force (820 kN).

3. Experimental Results

The principal goals of this four layer Nb3Sn dipole were; 1) to explore the 12 to 14 T range, 2) to validate the technology required at these field levels, and 3) to provide a test facility to test insertion coils made with advanced conductors or alternative winding geometries.

Figure 2 shows the quench history of D20. Unlike previous Nb_3Sn dipoles[5-7], D20 trained to a current 25 percent higher than the initial quench. The rate at which D20 trained was temperature dependent. The training at 1.8 K improved the performance of the magnet at 4.4 K. A full warm up and cool down cycle has yet to be completed to room temperature. D20 displayed reasonably good heat input tolerance for a completely impregnated structure. When the inner protection heater was used as a source of heat, more than 20 watts were needed to quench the magnet at 12T (at 1.9 K).

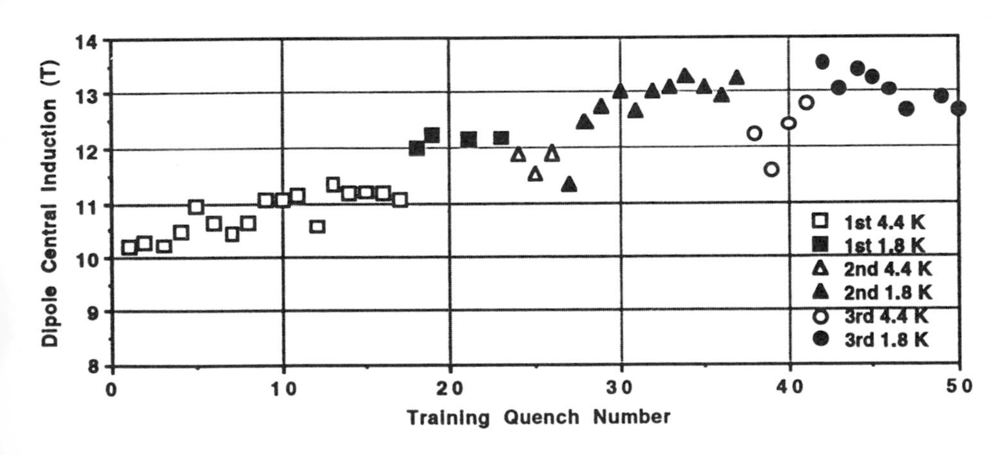

Figure 2. Dipole D20 Quench History

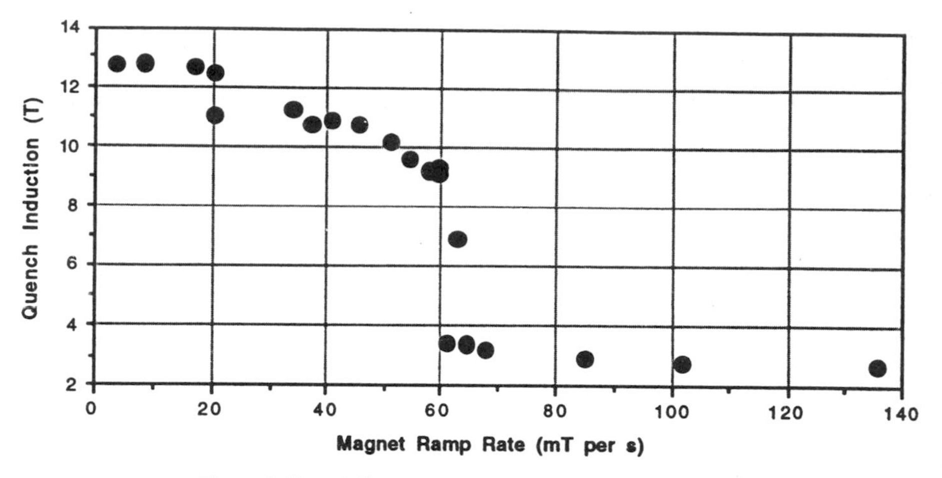

Figure 3 Quench Current versus Magnet Ramp Rate at 4.4 K

D20 appears to be operating near its short-sample at 4.4 K. At 4.4 K, the magnet quenches in the ramp between the pole turns of the inner coil pair. The ramp-rate dependence of the quench current at 4.4K (shown in Figure 3) was excellent compared with existing or proposed accelerators. At 1.8 K, the dipole is not at its short-sample current and degradation due to strain does not appear to be a factor in the magnet performance. Further comments on the test may be found in Reference 8. Magnetic measurement of the D20 dipole will be made during a future test.

Acknowledgments

This project would not have been completed without the skills and contributions of the following people: D. Byford (magnet test electronics), S. Dardin (coil reaction), Ron Oort (coil winding), and J. Remenarich (coil impregnation). The work was supported under contract # DE-AD03-76SF00098 by Director, Office of Energy Research, Office of High Energy Physics, US Dept. of Energy.

References

[1] D. Dell'Orco et. al. " Design of the Nb_3Sn Dipole D20," IEEE Trans. Appl. Supercon **3** p. 82, (1993).

[2] D. Dell'Orco et al, "Fabrication and Component Testing Results for a Nb_3Sn Dipole Magnet," IEEE Trans Appl. Supercon. **5**, No. 2, p 1000, (1995)

[3] S. Caspi et al, "Design and Fabrication of End Spacers for a 13 T Nb_3Sn Dipole Magnet," IEEE Trans Appl. Supercon. 5, No. 2, p 1004, (1995)

[4] A. F.. Lietzke et al. "Test Results for a Nb_3Sn dipole Magnet," to be published in IEEE Trans. Appl. Supercon. **7**, No 2, (1997).

[5] A. Asner, R. Perin, S. Wienger and F. Zerobin "First Nb_3Sn 1m-long Superconducting Dipole Model Magnet for LHC Breaks the 9.5T Threshold," Proc. of MT-11, Tsukuba, 1992, p.36.

[6] A. den Ouden, S. Wessel, E. Krooshoop, and H. ten Kate, "Application of Nb_3Sn Superconductors in High-Field Accelerator Magnets," IEEE Trans. on Appl. Supercon. **7**, No. 2, p , (1997).

[7] C. Taylor et al, "A Nb_3Sn Dipole Magnet Reacted after Winding," IEEE Trans. on Mag., **MAG-21**, No. 2, p 967, (1985).

[8] A. D. McInturff et al, "Test Results for a High Field (13 T) Nb_3Sn Dipole," to be published in the Proceedings of the 1997 Particle Accelerator Conference, Vancouver, Canada, 12-16 May 1997

Inst. Phys. Conf. Ser. No 158
Paper presented at Applied Superconductivity, The Netherlands, 30 June–3 July 1997

Development of Synchronous Motors and Generators with HTS Field Windings

G Papst[1], B B Gamble[2], A J Rodenbush[2] and R Schöttler[1]

[1] American Superconductor Europe GmbH, Rathausstraße 7, D-41564 Kaarst, Germany
[2] American Superconductor Corporation, 2 Technology Drive, Westborough, MA 01581, USA

Abstract. Synchronous machines with superconducting field windings offer reduced losses, size and weight. Continuous advances in the HTS conductor development at American Superconductor Corporation ASC enable the demonstration of machines with increasing power. Latest achievements include a >200 HP (150 kW) motor operated on an inverter, built and tested at Reliance Electric Corporation, and coils for a 1 MW generator for airborne applications. The HTS coils provide MMF's on the order of 70,000 amp-turns per pole and operate between 20 to 27 K. At this temperature, gaseous Helium or Hydrogen is used as coolant. The cooling is more efficient by a factor of 10 compared to cooling systems required for LTS materials operating at liquid Helium temperatures. The elevated operating temperature offers significant benefits for the overall efficiency as the coefficient of performance for removing the parasitic heat load (cryostat and current lead loss) is improved.

On this path to commercially viable machine sizes, American Superconductor currently is developing and manufacturing HTS coils for a 1000 HP (750 kW) synchronous motor which will be delivered to Reliance Electric for demonstration in 1998.

1. Introduction

Electric drives with power requirements of 1000 hp (750 kW) and more benefit by the use of synchronous motors with superconducting field windings. Compared to conventional induction machines available today these motors offer higher efficiency, and reduced size and weight. The advantage of high temperature superconducting (HTS) field windings relative to low temperature superconductors (LTS) is the higher operating temperature: In contrast to LTS which usually need liquid helium as coolant, HTS material in higher fields works in the 25-35 K range. This results in a higher coefficient of performance of the refrigeration system and leads to better overall efficiency.

The US Department of Energy funds a project to develop and demonstrate high power HTS motors within the "Superconductivity Partnership Initiative" (SPI). This project is lead by Reliance Electric / Rockwell Automation. American Superconductor Corporation, the worldwide leader in commercializing HTS applications, teamed in to develop and build the HTS field coils.

2. Progress in HTS Wire Development

The BSCCO-2223 wire production at American Superconductor Corporation is leading worldwide: The 1996 annual output of more than 120 km exceeds the combined production of all other manufacturers. The wire is very robust to withstand machine stranding and the forces experienced in high fields and rotating machines. A uniform engineering current density j_e of 8,000 A/cm^2 (77 K, self field, 1 μV/cm criterion) in long length (> 100 m) is achieved in the standard production process. Short R&D samples reach critical current densities j_c up to 58,000 A/cm^2 in tapes with high fill factor [1]. In the past, it usually took two years to develop results from short samples in R&D in long length production. Thus an HTS wire with a commercially viable j_e of 20,000 A/cm^2 can be expected for the year 1999.

3. Status in the Development of HTS Motors and Generators

3.1. U.S. Navy Homopolar Motor Demonstration

The U.S. Navy is developing d.c. homopolar motors and generators for ship propulsion. The primary reason is that these machines develop high torque at low speeds and are easy to control. They are also electrically and acoustically quiet since they do not require mechanical gears, and there is no alternating electromagnetic field during operation [2]. A homopolar motor as the only strict d.c. machine benefits greatly from the application of superconducting magnets due to the absence of a.c. losses. It is more efficient, and also smaller and lighter than conventional a.c. or d.c. machines.

After a successful at-sea demonstration of the first motor in the early 1980s, using NbTi wire for the magnet, now the same motor is used as a test bed for HTS magnets [2]. Recent experiments with HTS magnets provided by ASC and operating at 4.2 K revealed a motor power of more than 300 hp (225 kW), exceeding the design value. Future tests will be at 28 K to reduce the cooling power requirement.

3.2. U.S. Airforce Generator Demonstration

A 1 MW superconducting generator has been developed for an lightweight power system. Its field windings are manufactured by ASC from HTS wire and provide 72,000 Amp-turns at 1.5 T and 20 K. The project's perspectives are to reduce further the weight and size of the generator by increasing the magnetic flux. As the magnetic induction greatly exceeds the saturation of magnetic materials the high density iron is replaced by structures of aluminum and fiber reinforced composite [3].

3.3. 200 hp Synchronous Motor

Since 1990, Reliance Electric / Rockwell Automation and American Superconductor Corporation work together in developing synchronous motors with HTS field windings. Figure 1 shows the progress in the past and the proposed commercial size machines to be demonstrated beginning 1998.

The latest achievement in this collaboration is a 200 hp (150 kW), 4 pole, air core, synchronous motor. The HTS field windings were manufactured using the "wind-and-react" method due to the minimum bend radius of 1 cm. They operate at 27 K and produce 56,500 Amp-turns per pole [4,5]. The actual combined losses of all windings at the rated current

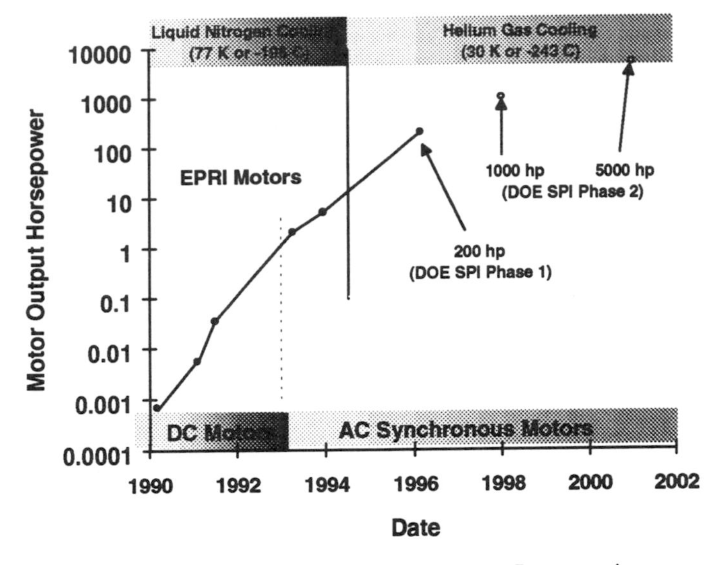

Figure 1. ASC – Reliance / Rockwell HTS Motor Demonstrations

are only 10% of the design value and represent the continuing progress in the HTS wire development.

4. Coil Design for a 1000 hp (750 kW) Synchronous Motor

The successful demonstration of the HTS technology bears significant implications for the future large motor development. Under the SPI Phase II program, the design and manufacture for a 1000 hp/750 kW motor are presently underway. This machine is scheduled for testing in 1998 (Figure 1). As shown in Figure 2, the coil analysis and design, and the wire

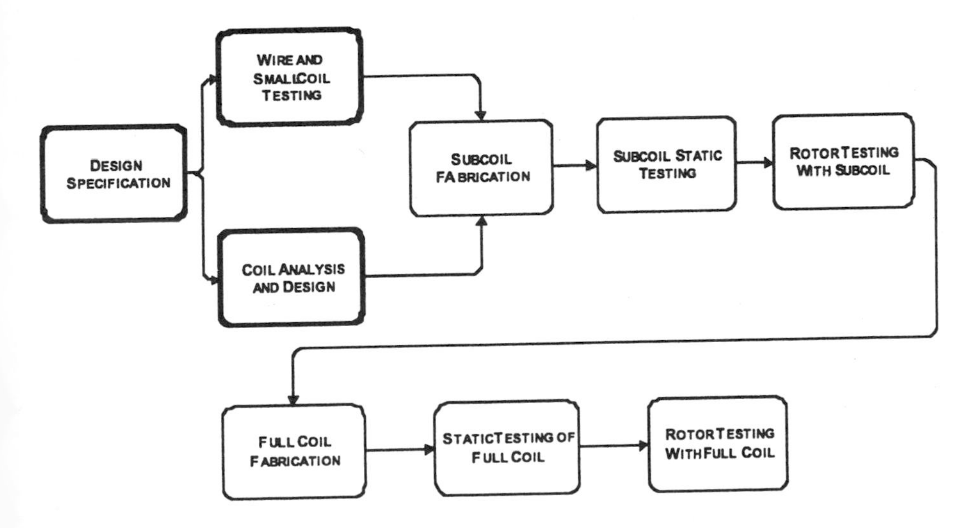

Figure 2. Flow Chart of Rotor Manufacture for 1000 hp (750 kW) Motor

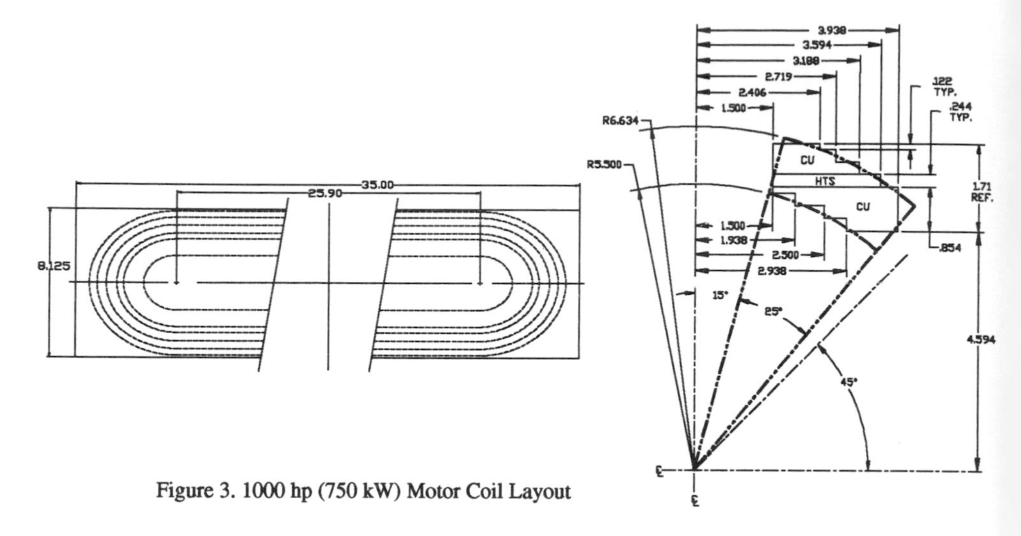

Figure 3. 1000 hp (750 kW) Motor Coil Layout

testing have been finished successfully. A subcoil assembly (Figure 3) is currently being built using the "react-and-wind" method and will be tested in the near future. After that a fully superconducting coil set will be manufactured and tested as a 1000 hp motor.

5. Conclusions

The successful development of HTS field coils for various motor and generator demonstrations and the subsequent testing at power levels is an important step toward the commercialization of HTS applications.

A 1000 hp synchronous motor is designed, its HTS field windings are currently in the first testing phase. This motor to be in operation in 1998 will be the first HTS synchronous machine in a commercially viable size. It will lead to motors which will be about half the volume and half the losses of conventional motors built today of the same rating.

Superconducting synchronous machines can be built without iron in the rotor and iron teeth in the armature windings. This results in a significantly larger effective air gap which improves the operating characteristics. Superconducting generators for utility power generation are more stable against power oscillations and can supply more reactive power than a conventional generator. Furthermore, it is possible to build superconducting generators with high nominal voltage (110 kV, e.g.). In combination with other superconducting equipment, like cables, this makes step up transformers obsolete [6].

References

[1] Malozemoff A P, Li Q and Fleshler S 1997 *M2S Conference*

[2] Gubser D U 1996 *JOM* **48** (10) 30-34

[3] Oberly C E et al. 1994 *Adv. Cryo. Eng.* **39** 949-956

[4] Schiferl R et al. 1996 *American Power Conference*

[5] Voccio J P, Gamble B B, Prum C B and Picard H 1996 *Applied Superconductivity Conference*

[6] Oswald B R 1997 *etz* **3** 52-53

Inst. Phys. Conf. Ser. No 158
Paper presented at Applied Superconductivity, The Netherlands, 30 June–3 July 1997

Experimental Investigation of a Model HTSC Alternator

Igor I. Akimov[1], Lidia I. Chubraeva[2], Denis A. Filichev[1], Leonid M. Fisher[3], Alexey V. Kalinov[3], Dmitry V. Sirotko[2], Alexander K. Shikov[1], Vladimir M. Soukhov[4], Igor F. Voloshin[3]

[1] Bochvar RIIM, Moscow, Russia. [2] Research Institute of Electrical Machinery, St.-Petersburg, Russia. [3] All-Russian Electrical Engineering Institute, Moscow, Russia. [4] State Committee of Science and Technologies, Moscow, Russia

Abstract. A new version of the disc model synchronous alternator with HTSC rotor winding and the stator winding of high-purity aluminum was developed and tested. The coils of the excitation winding were manufactured of *Bi-2223* tape of 5×0.2 mm size with 19 filaments in silver matrix. The critical currents of the coils have been measured in LHe and are in the range of 70-100 A. Excitation winding of the 8-pole alternator comprises 8 coils, the 3-phase armature winding has 12 coils. After the alternator has been assembled and undergone cryogenic tests with no-load and short-circuit curves the main parameters have been determined. The alternator rating at operating temperature will be around 1-3 kVA.

1. Introduction

Beginning from 1990 several variants of the disc-type alternators were manufactured and tested. They comprise a LTSC model machine with rotor and stator winding coils manufactured of Nb_3Sn [1], a cryogenic version with rotor and stator windings of high-purity aluminum, a rotor with rare-earth *Nd-Fe-B* circular magnets [2] and now a rotor with HTSC coils of *Bi-2223*. The combination of HTSC rotor and high-purity aluminum stator were assembled together and investigated as a synchronous generator.

2. Parameters of HTSC wire and pancake coils

The basic wire for the rotor winding represents a high-Tc superconducting $(BiPb)_2Sr_2Ca_2Cu_3O_x$ rectangular tape 0.2×4.9 mm [3]. The matrix is 99.99% silver. The number of HTSC filaments is 19. The total length of the wire used for the excitation system manufacturing is 100 m, the tested short sample critical current was in the range 25-30 A (77 K, 0 T). Prior to the coil manufacturing the tape was covered by a specially developed ceramic electrical insulation, based on the zirconium dioxide. The wire manufacturing process is based on powder-in-tube technology. Each pancake coil has the following dimensions: OD - 45 mm, ID - 15 mm, thickness - 5 mm. The stainless steel outer bandage and cheeks are 0.5 mm thick. The number of turns in each coil equals to 70. The coils are epoxy impregnated. The wire and coils were manufactured by the Bochvar RIIM.

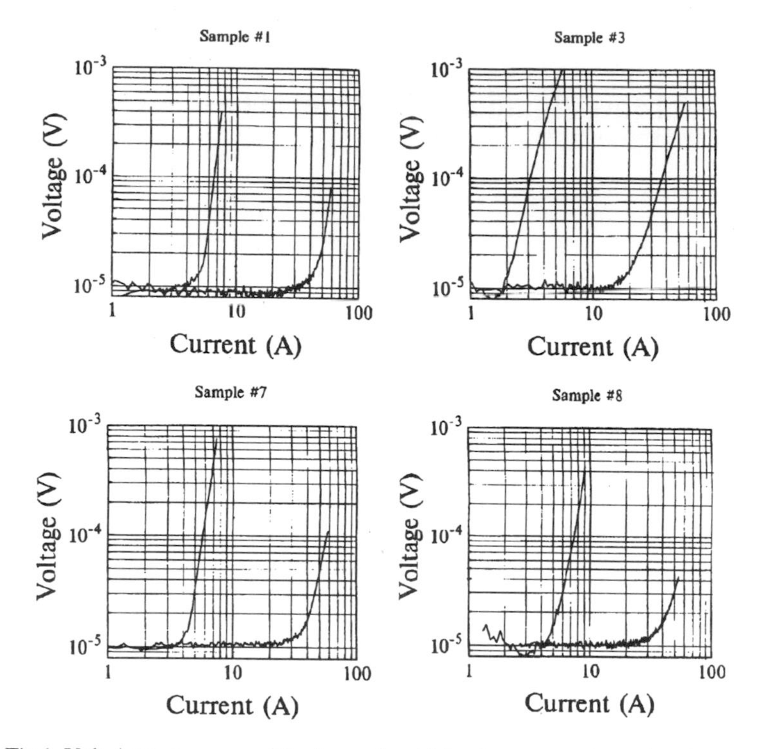

Fig.1. Volt-Ampere curves of the four solenoidal coils, measured at 4.2 and 77 K (maximum field in the coil axis equals 0.24 T)

3. Testing of HTSC coils

After manufacturing the coils have undergone experimental investigation with determination of their individual parameters. The 3 of 8 coils have potential ends. As 5 coils had no potential ends special potential wires were soldered 1 cm from the fixed tape ends. The current was supplied to all coils via the HTSC tape ends. The critical parameters of the coil were evaluated at 77 and 4.2 K. The cooling was performed by liquid nitrogen and liquid helium at normal pressure. The DC current was supplied from a controlled current stabilizer. Its time variation was a linear one, the rate of linear variation was chosen so as not to exceed several mV of self-EMF of the coil ends. The constant speed of current variation during measurements allowed to compensate the self-EMF with high accuracy. The voltage of the potential ends was amplified by nanovoltmeter KEITHLEY 181. Experimental results are presented in Fig.1. The voltage drop along the non-superconducting elements was subtracted from the measured signal value. The coils were investigated in All-Russian Electrical Engineering Institute.

4. Synchronous generator test results

The assembled disc-type alternator comprises a HTSC rotor with 8 excitation coils *1* and a high-purity aluminum stator with a 3-phase winding consisting of 12 coils *4* (Fig.2). The rotor and stator supporting discs *2* and *5* are non-metallic ones.

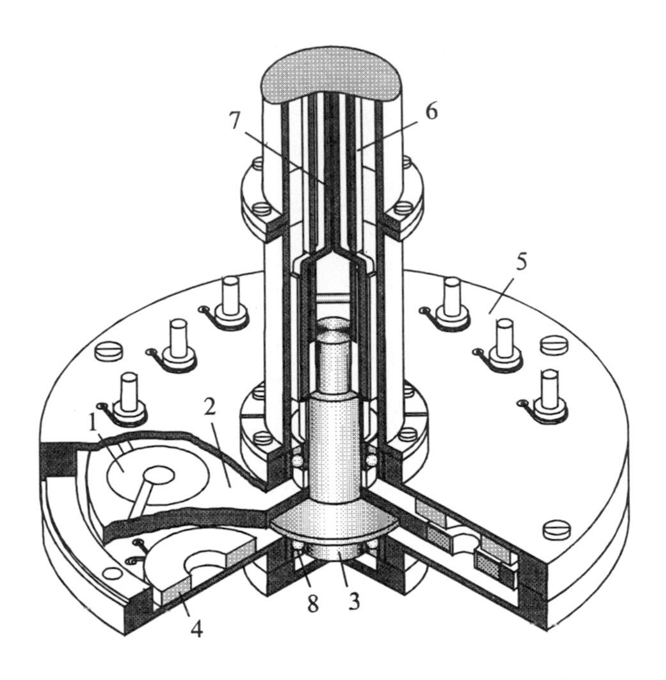

Fig.2. Model disc alternator with HTSC field winding and armature winding of high-purity aluminum

Table 1. Parameters of the model disc-type synchronous alternators

Parameter\ rotor type	LTSC	Perm. magnets	HTSC
OD of stator disc, mm	230	230	230
Alternator thickness, mm	40	45	43
Diameter of coil centres, mm	140	140	140
Frequency of rotation, rpm	750	750	750
Pole number / phase number	8 / 3	8 / 3	8 / 3
Rotor coil OD / ID, mm	50 / 18	30 / 9	45 / 15
Rotor coil thickness, mm	20	5	6
Rotor coil material	*Nb-Ti*	*Nd-Fe-B*	*Bi2223*
Number of turns in each rotor coil	1137	-	70
Rotor wire size, mm^2	0.2	-	0.2×5.0
Packing factor of rotor wire, %	80	-	20
Number of filaments in rotor wire	210	-	19
Rated excitation current, A	78	-	50
Stator coil OD / ID, mm	64 / 32	48 / 28	48 / 28
Stator coil thickness, mm	2.5	6	6
Stator coil material	Nb_3Sn	*high-purity Al*	*high-purity Al*
Number of turns in each stator coil	50	175	175
Wire diameter of stator coil, mm	0.5	0.8	0.8
Packing factor of stator wire, %	20	-	-
Number of filaments in stator wire	14641	28	28
Excitation winding inductance, H	0.229	-	0.2×10^{-3}
Mutual inductance of rotor and stator windings, H	2.46×10^{-3}	-	4.5×10^{-4}
Synchronous reactance, Ω	0.205	1.1	1.1
Sub-synchronous reactance, Ω	0.191	0.73	0.73

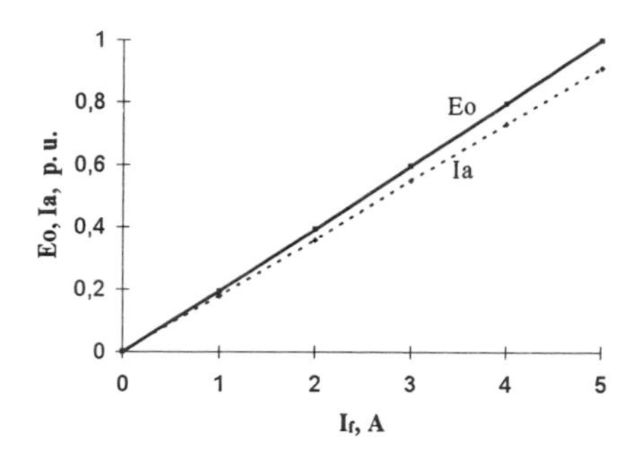

Fig.3. No-load and short-circuit curves of the alternator at LN_2 temperature

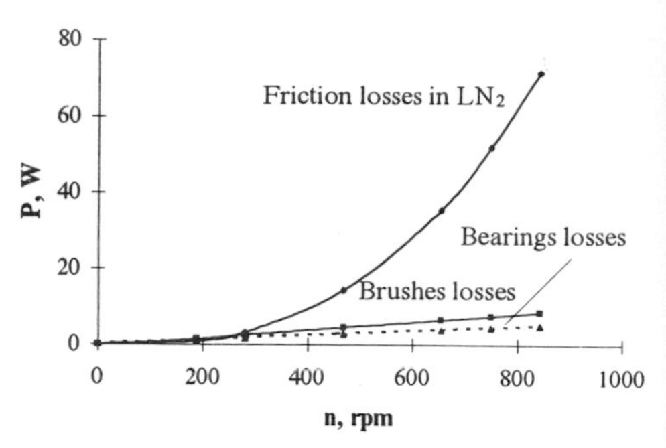

Fig.4. Dependence of components of mechanical losses of the alternator in LN_2 medium

The alternator was positioned vertically in a cryostat with a drive motor fixed in its upper part. The drive motor was connected with the alternator via a hollow shaft *6*, with current leads *7* inside of it. The final variant of bearings *8* is a fluoroplastic one.

The experimental investigations of the alternator were carried out in liquid nitrogen. There were measured the cool-down curve with 40 min of cooling period, the rotor magnetic flux distribution along the diameter occupied by coil axises, no-load and short-circuit curves (Fig.3), the curve of mechanical losses. The latter was divided by components, caused by rotor body friction and mechanical losses in brushes and bearings (Fig.4). The measured moment of inertia of the rotor against the axis of rotation equals 0.9×10^{-4} kg·m^2. The operating temperature of the model is around 25 K, the rating of the model is to be 1-3 kVA. Comparison of HTSC model with previous versions is given in Table 1.

5. Conclusions

1. The manufactured HTSC wire has quite an acceptable level of spread in critical parameters, i. e. it is uniform.
2. The average current density at liquid helium and liquid hydrogen temperature is around 50-75 A/mm^2. At this current density level it is already reasonable to apply HTSC for the excitation windings of synchronous alternators.
3. Experimental results will be used for the development of the main generator in the independent source of power supply and for an electromechanical energy storage designing.

Acknowledgement.

The researches are supported by the State Program, Direction "Superconductivity" and by the Ministry of Science and Technical Policy of Russia.

References

[1] Klimenko E Ju et al 1991 *Superconducting field and armature winding turbogenerator mode* (Leningrad: VNIIelectromash) (in Russian)
[2] Chubraeva L I et al 1995 *EUCAS '95* Le1.29
[3] Nikulin S A et al 1995 *EUCAS '95* Ld2.41

Inst. Phys. Conf. Ser. No 158
Paper presented at Applied Superconductivity, The Netherlands, 30 June–3 July 1997

Hysteresis electrical motors with bulk melt textured YBCO

L.K.Kovalev*, K.V.Ilushin*, V.T.Penkin*, K.L.Kovalev*, V.S.Semenikhin*, V.N.Poltavets*, A.E.Larionoff*, W.Gawalek, T.Habisreuther**, T.Strasser**, A.K.Shikov***, E.G.Kazakov***, V.V.Alexandrov******

* - Moscow State Aviation Institute (MAI), Department 310, 4, Volokolamskoe shosse, Moscow, 125871, Russia

** - Institut für Phisikalische Hoch Technologie e.V.(IPHT), Helmholtzweg 4, D-07743 Jena, Germany

*** - All-Russian Scientific-Research Institute of Inorganic Materials, Moscow, Russia

**** - All-Russian Electrotechnical Institute, Moscow, Russia

Abstract. New types of electrical motors based on bulk high temperature superconductors is presented. Theoretical and experimental research of these motors are given. Results for series of 100 W, 300 W, 500 W, 1000 W HTS motors with cylindrical and disk rotors are given. It is shown, that at temperature level of liquid nitrogen the specific mass-dimension parameters of hysteresis HTS machines are in 5-7 times better then the similar one for the conventional hysteresis machines.

1. Introduction

Since 1989 MAI investigates a possibility of the bulk high temperature superconductors (HTS) application for electromachinery [1-6]. Since 1994 this work was continued in co-operation with IPHT (Jena, Germany), ARSRIIM, VEI (Russia). The theory describing the electrodynamics of such patterns under influence of the alternative magnetic field has been elaborated for the different values of in-granular and intra-granular currents. On the base of developed theory the series of motors with HTS rotors of various shape were developed, fabricated and tested [3]. These motors operate due to the remagnetisation process in the bulk polycristal HTS rotor elements. These process deal with AC losses due to intra- and in-granular currents and provide the rotating mechanical torque on the shaft of a rotor. The useful torque magnitude is defined by interaction of in-granular and intra-granular currents in the bulk HTS element with the rotating magnetic field. It was shown that the torque is in the linear proportion to total hysteresis losses in HTS rotor and does not depend on the rotor angular velocity. The schematics of motor with HTS cylindrical rotors are shown in Fig. 1.

2. Theory and experiment

The theoretical approach for calculation of hysteresis losses as well as for electrical, magnetic and current fields distributions were developed for the bulk HTS elements of different shapes (spheroids, cylinders, plates). The influence of external alternative polarisation which takes place in the various models of electrical machines has been taken into account [1].

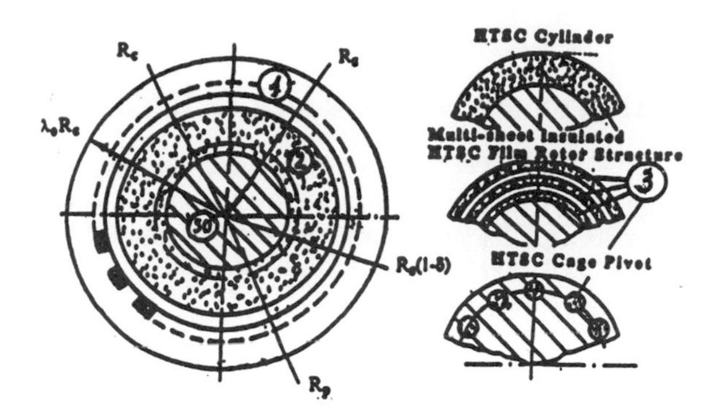

Fig. 1. HTS hysteresis motor model for analysis and rotor design variants. 1 - multi-phase stator, 2 - HTS rotor element, 3 - rotor steel core.

The analytical dependencies were obtained on the base of the solution of a set of two-dimensional and three-dimensional Maxwell equations. The Bean-London models were used for the critical current description when HTS elements have high current leading ability ($\xi = J_t\Delta/J_s a >> 1$, where J_t and J_s are intra- and in-granular current densities, Δ and "a" - characterising dimensions of HTS element and HTS grain). The spherical grain model (average sphere approach) was used for the bulk HTS elements with relatively low current leading ability $\xi << 1$ (e.g. MELT textured YBCO containing large crystals). The mutual influence of magnetic moment of HTS grains was realised with Lorenz approach [2]. The calculation examples for distributions of magnetic and current fields in HTS cylindrical rotors of $\xi >> 1$ and $\xi << 1$ placed in rotating magnetic fields are shown in Fig. 2.

On the base of developed theoretical models for electromagnetic and hysteresis phenomena in the bulk HTS elements the applied theory was carried out for calculation of hysteresis electrical HTS machines parameters with wide output power range $100 \div 1000$ W. The applied theory for hysteresis machines with the bulk HTS patterns from MELT textured YBCO ceramics were compared with the experimental results and used for elaboration of experimental series of hysteresis machines of 100, 500 W and 1000 W output power rating [3].

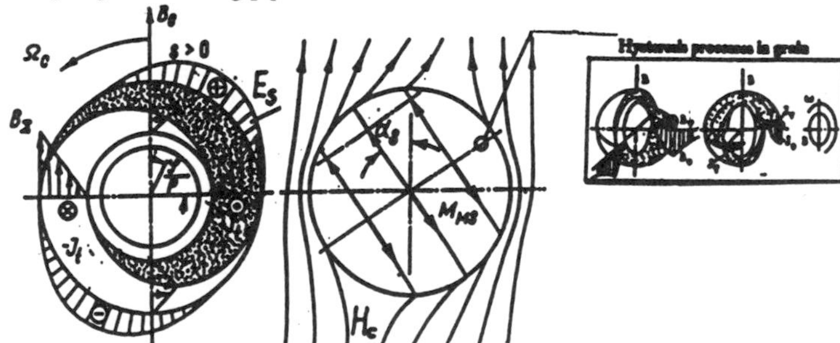

Fig. 2. The calculation results of the magnetic fields, currents and magnetic moment M_{ms} in cylindrical HTS elements for motor regimes under different values of ξ (a – $\xi >> 1$, b – $\xi << 1$).

3. Experimental HTS hysteresis motors

The experimental series consisting of 4 hysteresis immersed 100 W HTS motors containing cylindrical and disk HTS rotor operating in liquid nitrogen was fabricated and tested.

Theoretical and experimental data for 100 W hysteresis motors are presented in Fig. 3.

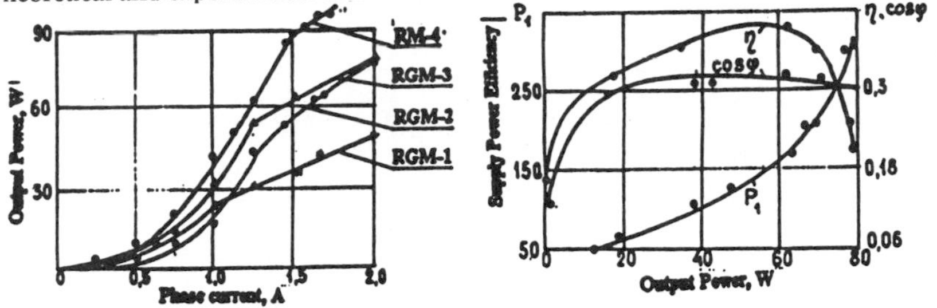

Fig. 3. 100 W HTS motors, theoretical and test results, —— theory, o. test.

Two-phase motor with assembled disk HTS rotor of 300 W output power rating was developed, fabricated and tested at the liquid nitrogen temperature level. Parameters of this motor are given in Fig. 4.

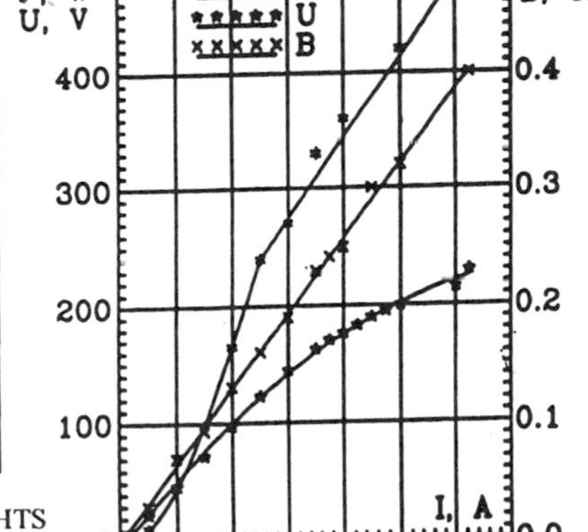

	300 W HTS disk motor	500 W HTS cylindrical motor
Current frequency, Hz	50	
Phase voltage, V	220	
Dimensions of the YBCO rotor material, mm	Ø 160 × 8	Ø 42 × 70
Phase/pole number	2/4	3/1

Fig. 4. Main parameters of 300W HTS disk motor, 500W HTS cylindrical motor, theoretical and test results for 500W HTS motor

In Fig. 4 typical parameters, experimental and theoretical results for series of 500 W HTS motors are also presented. It was shown that the specific output power parameters for HTS motor excel the similar one for traditional hysteresis motor in 7÷8 times and are at the level of traditional asynchronous motors (see Fig. 5).

The tests has been carried out for hysteresis HTS motor of output power rating about 1 kW. It was experimentally shown that this HTS motor operates in stable state and provides 1 kW output power rating for 20 A phase current. Test results are presented in Fig. 6.

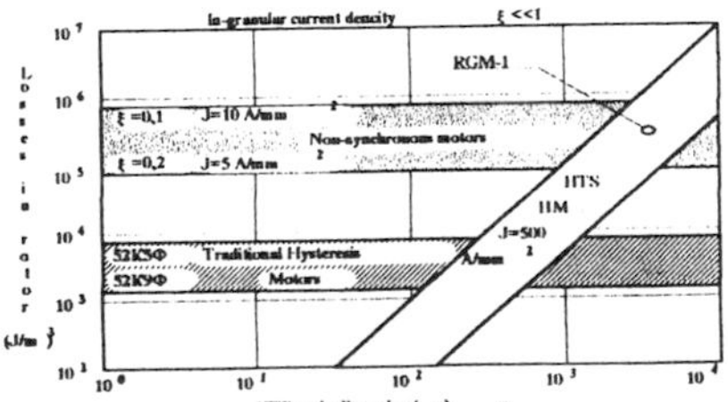

Fig. 5. Comparison of specific volume power Q of hysteresis HTS motors and electrical motors of conventional design (where $Q = N/(Vf)$, N – power on shaft, V – volume of active rotor elements, f – synchronous frequency of the magnetic field rotation.).

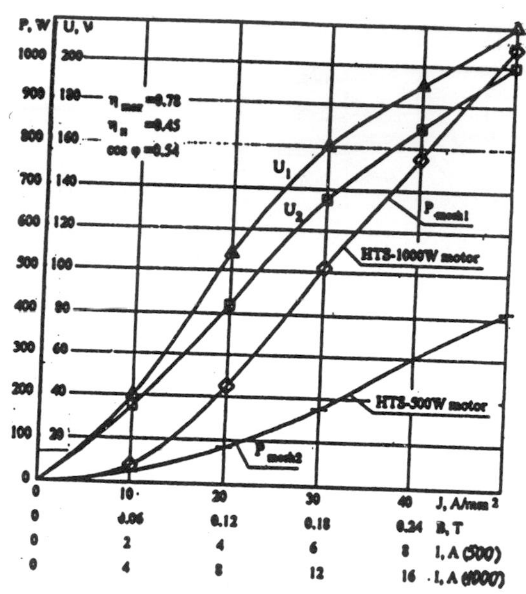

Fig. 6. Test results for 1 kW HTS motor.

5. Conclusion

The experimental investigations of the series of motors with the cylindrical HTS rotors showed that polycristal bulk YBCO HTS samples with grain size 2...4 mm provide the high values of specific hysteresis torque (M > 0.2 N m) at the liquid nitrogen temperature level and magnetic induction B ~ 0.5 T. The attained hysteresis torque are sufficed for the elaboration of new types of hysteresis HTS motors with the specific output power comparable or higher than for traditional electrical motors.

The obtained experimental and theoretical data allow to start the development of hysteresis HTS motors with output power up to 10 kW. New types of developed hysteresis HTS motors can find various applications in cryogenic and aerospace technology (e. g. they can be used as a driving motor for the submerged cryogenic pumps).

Acknowledgements

The authors gratefully acknowledge to Dr. N.A.Chernoplekov, Dr. L.M.Fisher, Dr. V.E.Keilin, Dr. V.M.Suhov and Dr. D.A.Bout for their attention to our work.

References

[1] L.Kovalev, K.Ilushin, V.Penkin, K.Kovalev, Hysteresis Electrical Machines with High Temperature Superconducting Rotors, Electrical Technology, N 2, 1994, pp. 145-170.

[2] L.Kovalev et al. The bulk high temperature superconductors implication for electromachinery. J. Superconductivity - Application & Development, N 5,6, 1995, pp. 17-20.

[3] L.Kovalev, W.Gawalek et al. Theoretical and experimental study of magnetisation and hysteresis process in single grain YBCO sphere and bulk melt textured YBCO ceramics. IX Trilateral German-Russian-Ukrainian Seminar on High Temperature Superconductivity. Jena, Germany, September 22-25, 1996.

Inst. Phys. Conf. Ser. No 158
Paper presented at Applied Superconductivity, The Netherlands, 30 June–3 July 1997

Prospects for Brushless AC Machines with HTS Rotors

M D McCulloch, K Jim, Y Kawai D Dew-Hughes,

Dept of Engineering, University of Oxford, OX1 3PJ

Abstract: The rotors of conventional brushless ac machines may be replaced by superconducting equivalents, with the objective of increasing the power densities and reducing the machine losses. The behaviour at 77K, of superconducting rotors, fabricated in the form of (a) squirrel cages made up from silver tape coated with meltprocessed Bi-2212, (b) small disks of seeded and melt textured YBCO has been studied in rotating magnetic fields produced by conventional three phase motor coils. Measurements of dynamic torque, deduced from the angular acceleration, have been used to characterise the potential performance of the embryonic machines. Two broad types of behaviour have been observed. In the Bi-2212 rotors the torque decreases with increasing speed: this behaviour is believed to be due to flux creep. By contrast the strong pinning YBCO rotors maintain a constant torque up to synchronous speed. Mathematical modelling is able to reproduce both types of observed behaviour, although only those for YBCO are presented here. Power densities some 5-10 times that of conventional machines are predicted to be achievable in optimised prototype machines, if the flux density can be increased in the airgap.

1 Introduction

The advent of High Temperature Superconductors (HTS) in 1986 caused a flourish of remarks expounding the new miracle material. However, reality soon sunk in and it is only recently that practicable applications are becoming viable. Some examples of work using HTS materials are described in [1-5]. This paper outlines the use of bulk HTS materials used in the rotor of electrical machines, and more specifically the measurement of the dynamic torque speed characteristics of various configurations.
In induction machine theory, it is well known that the torque-speed curve reveals a lot about the operation of the motor. Kovalev [1] had predicted a flat torque speed curve, but had also found that the motors always ran with some slip. This paper explores the use of BiSCO and YBCO superconductors and their different characteristics. The measured results are also compared to computed curves.

Three different geometry's of motor were tested, two made from BiSCO deposited on silver and one from YBCO. A constraint imposed by the test setup was that they should all be cylindrical and about the same size of diameter 25mm and height 10mm The first, shown in Figure 1(a) was that of BiSCO/Ag tape, wound in a spiral and then processed. The second, Figure 1(b) was that of a simple cage, made by cutting BiSCO/Ag sheets into appropriate shapes, fixing them together and then re-processing the final artefact. The third rotor was that of a puck of seeded, single domain YBCO, shown in Figure 1c).

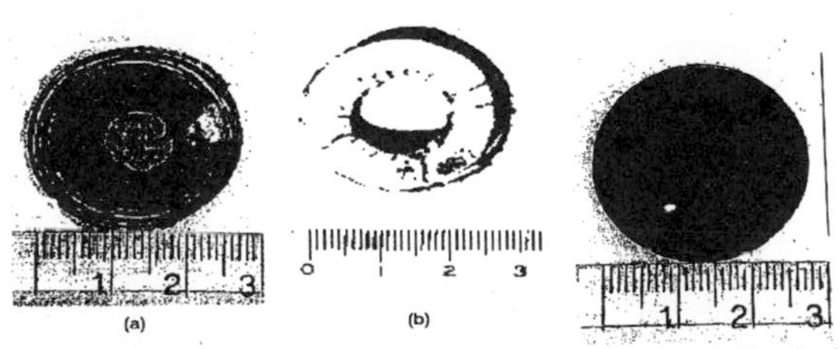

Figure 1 a) Spiral and b) cage rotor made from BiSCO deposited onto Ag c) Single domain YBCO rotor

2 Torque-speed measurements

The accurate measurement of torque is one of the more challenging aspects of working with electrical machines. In this case it is exaggerated by the fact that the rotors are small and that they have to be kept at 77K.

The method adopted was that of a non-contact technique. Secondly, as there was no mechanical load connected to the rotor the torque was derived from the acceleration. A simple variation on the digital shaft encoder was used. The encoder had to be kept small, and have near zero frictional resistance. Also the optical pickup had to be kept away from the cold rotor.

Once the samples were collected, the data was adjusted to account for the small variations in the segment size. From the time taken for each segment, the speed and then the acceleration can be determined. Due care must be taken because the samples are not equally spaced in time.

3 Measured results

Kovalev [1] predicted that the quasi steady state torque speed curve would be flat. This characteristic is typical of conventional hysteresis motors, owing to the fact that the losses in the rotor are proportional to the slip speed. Comparing to the model of the induction motor, and representing the hysteresis loss component as a resistor which varies with the slip, it can be shown that the torque speed characteristic should be flat. However, Kovalev noted that the machines using an inferior YBCO material ran with some slip when producing maximum torque.

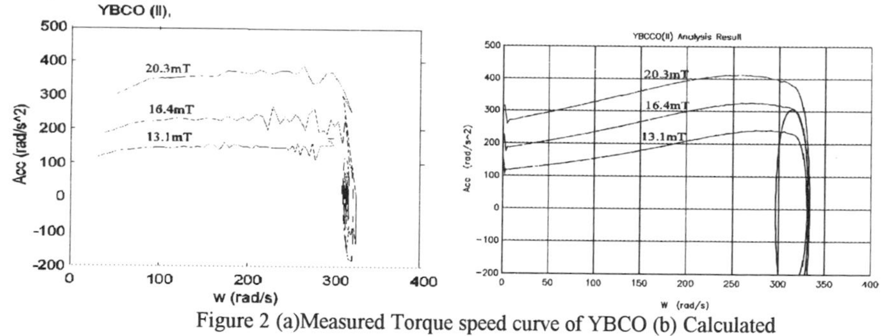

Figure 2 (a)Measured Torque speed curve of YBCO (b) Calculated

Figure 2 shows the measured accelerating torque speed curve for the single domain YBCO rotating in various strengths of magnetic fields. (Note that because the superconductor material has an effective permeability close to free space, it is difficult to generate large fields using a conventional stator winding.)

It is clear that once the field is established, the torque is flat with respect to speed. The speed overshoots and then spirals in to the synchronous speed. When observed with a strobe light synchronised to the supply, it was observed that the rotor spun at synchronous speed. When a load torque was applied, the speed remained synchronous, although the angle changed. When the load was removed, the angle moved towards its original position, but did not return to the original position, implying that the magnetic structure on the rotor had changed.

Kovalev predicts that the torque will increase with the cube root of the field strength. These results predict that the torque is proportional to $B^{2.5}$.

When the same experiments were repeated with BiSCO artefacts, a different characteristic was observed. Figure 3(a) shows the performance of the spiral rotor and Figure 3(b) shows the performance for the cage type rotor.

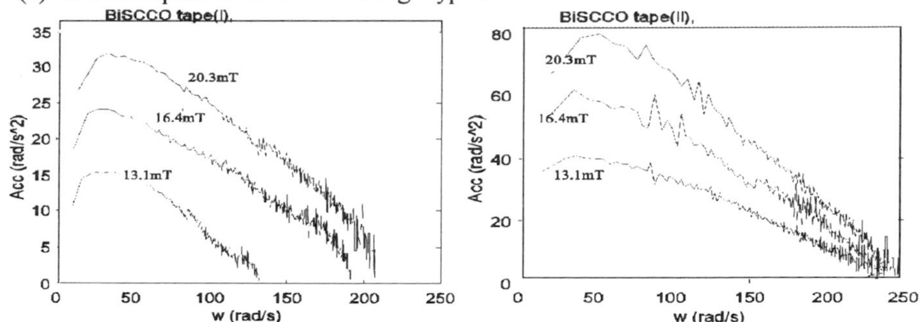

Figure 3 Measured Torque Speed curve of (a) BiSCO spiral rotor(b) BiSCO cage rotor

The most striking characteristic is that both these motors do not reach synchronous speed. Secondly, the torque *decreases* with speed. Thirdly the cage motor produces about 2.5 as much torque, despite containing the same amount of superconducting material.

The behaviour of the BiSCO material can be explained in terms of the fact that once the field increases above about 1mT at 77K, the material moves from the flux creep regime to the flux flow regime. Thus the pinning forces are small, and the torque that can be developed is very much lower. The material also appears to be much more lossy in the presence of the alternating field.

The reason that the cage machine could produce more torque, is that there were clear paths for the supercurrent to travel on the faces parallel to the field, whereas for the spiral motor, due to the poor connection between adjacent wires, a radial component of current could not readily flow, restricting the size of the magnetic moment.

4 Numerical Modelling

The phenomenological critical state model gives a very useful means to model the macroscopic magnetic properties of HTS materials. The critical state phenomenon was first hinted at by Bean [6], where he stated that the supercurrents flow to the full amount J_c for a depth necessary to reduce the field to H_{C1}. This model can be extended to account for the fact that the field penetrates the material in a finite time period, by introducing the concept of flux line viscosity.

This model can be derived using the following argument. Consider the case of a sample of superconducting material, which is placed in an applied magnetic field. At the boundaries, a supercurrent will start to flow, whose density is given by Ampere's law, (ignoring the effect of displacement currents) as $J = \frac{1}{\mu} \nabla \wedge B$. This supercurrent, which flows around the vortices, interacts with the magnetic field and a force is exerted on the vortices. The magnitude and direction of the force per unit volume is given by Lorentz force equation: $F = J \wedge B$

However, due to the action of pinning, this effect of the Lorentz force is reduced by the pinning force, Fp. Various models of pinning have been put forward: Bean, Kim, generalised Kim and logarithmic [8]. Note that this pinning force acts in the opposite direction to the Lorentz force, much like friction does in mechanical systems. The net force then acts to accelerate the vortices in the direction of the force. Various models exist to describe this motion. The one selected in this paper makes the assumptions that the vortices have no mass, ie the acceleration is infinite and that the force to move the vortices is proportional to the velocity ie there is a viscous flow motion of the vortices. This motion is described by $F-Fp = v.\eta$, where η is the viscosity. Solving for the velocity, $v= (F-Fp)/ \eta$. Now, when there is a movement of charge in a magnetic field, a voltage is generated, and is given by $E = v \wedge B$. However, from Maxwell's equations, we also have that the time rate of change of a magnetic field is proportional to the curl of the electric field, thus $\frac{\partial B}{\partial t} = \nabla \wedge E \frac{\partial B}{\partial t}$, which implies that as the vortices move, so the magnetic field redistributes itself, however while it is doing so an emf appears, which implies that a loss occurs during the motion of the field.

The results are given the calculation for the case of the YBCO sample. Figure 3(b) shows the predicted results of the accelerating torque vs speed characteristics, for a value of $Jc= 10^8$ and for an applied field of 20 mT. Although the basic shape of the curve is similar to that which was measured, the predicted magnitude is less than the measured acceleration. This is probably due to the difficulty in measuring the applied field in the actual system.

5 Conclusions

It has been over ten years since the discovery of high temperature superconductors. However, much effort has been needed for the materials processing procedures of these materials to mature. Even at this point in time there is still a rapid advancement in the preparation of ceramic superconductors. Because of the brittle nature of the material, tapes and wires have only recently become available, and mainly using BiSCO, however bulk material can also be used.

This paper has described the use of superconductors in electrical machines, which exploit the hysteresis phenomenon, rather than just the high current densities with near zero resistance. In particular, the torque speed characteristics of three topologies of rotor were measured. From these measurements it is clear that the single domain YBCO crystal material provides the best characteristics. However the large effective airgap means that it is difficult to develop a large applied magnetic field using conventional stators.

In order to better understand the processes involved, a numerical method was developed to simulate the behaviour of the magnetic field and the super currents. However, further refinement is required before these methods provide an accurate picture of the processes.

References

[1] Kovalev,LK, Ilyushin, K V, Penkin, V T, Kovalev K L, Electrical Technology, **2** pp 145-170, 1994.
[2] Spooner E, Haines N, Bucknall R,IEE Proceedings Electric Power Applications.**143**,pp.443-8, 1996
[3] Schiff-N; Schiferl-R ,Power Transmission Design, **37**, pp.55-8, 1995
[4] Mikkonen R, Soderlund L, Eriksson J T, IEEE Transactions on Magnetics,**32**,1996, pp2377-80
[5] Waltman D J, Superczynski M J,IEEE Transactions on Applied Superconductivity,**5**,1995 p.3532-5
[6] Bean C P, Physical Review Letters,**8**, pp 250-253, 1962
[7] Weinstein R et al, Proceedings of the Fourth International Conference and Exhibition: (NASA Conf. publ. 3290), Houston,USA;**2** pp.158-66,1994
[8] Poole CP et al, "Copper Oxide Superconductors" John Wiley & Sons, New York, 1988.

Inst. Phys. Conf. Ser. No 158
Paper presented at Applied Superconductivity, The Netherlands, 30 June–3 July 1997

Dynamics of a Rotor Levitated above a High-T_c Superconductor

Toshihiko Sugiura[1], Yoshitaka Uematsu and Takaaki Aoyagi

Department of Mechanical Engineering, Faculty of Science and Technology, Keio University, 3-14-1 Hiyoshi, Kohoku-ku, Yokohama, 223 Japan

Abstract. This research deals with dynamic behavior of a spinning permanent magnet freely levitated above a high-T_c superconductor. The magnetic forces acting on the magnet have been measured and modelled as nonlinear functions of both the vertical and lateral displacement of the magnet. They cause cross-coupling and cubic nonlinearities of rotational and vertical motions. Dynamics analysis predicts that the levitation gap, as well as the whirl amplitude, depend on the rotational speed, and that the steady-state whirling can have divergence instability, leading to the gap increase or decay.

1. Introduction

High-T_c superconducting magnetic bearings are expected to be one of the most promising applications of high-T_c superconducting bulk materials. So far there has been some research on static forces and 1-d.o.f. vibrations[1]-[4]. For the design of rotary bearings, it is important to analyze rotor dynamics coupled with magnetic interaction between a magnet and a superconductor.

This research deals with dynamic behavior of a spinning permanent magnet freely levitated above a high-T_c superconductor. Rotational and vertical motions of a magnet can be coupled by magnetic forces due to the superconductor. Therefore, from measurements of forces, we first approximate the magnetic forces as nonlinear functions of both the vertical and radial (lateral) displacement of the magnet. Then, using that modelling, we theoretically discuss rotor dynamics, including cross-coupling of translational motions, their nonlinearities, steady-state solutions, their instability, etc.

2. Measurements of magnetic forces acting on a magnet

Experiments were carried out for measuring magnetic forces acting on a magnet above a high-T_c superconductor. A ceramic YBCO sample of melt-quench type was used in our

[1] E-mail : sugiura@mech.keio.ac.jp.

experiments. It had a hexangular prism shape, 22mm in one side of its base and 15mm in height. A permanent magnet, consisting of Nd, Fe and B, was a cylindrical disc in shape, 22mm in diameter and 10mm in height. Its mass was 24g and its residual magnetization was 1.37T. The magnet, placed above the center of the superconductor, was vertically or laterally moved by a translation stage at very small velocity. The lateral and vertical force was measured by a load cell attached to the magnet. After the field-cooling at the levitation height, 13mm, the magnet was first vertically moved down a little, and then laterally moved while the magnetic force acting on it was measured.

Figure 1 shows dependence of the magnetic force on the lateral displacement of the magnet. As shown in Fig.1(a), the lateral force acts as a restoring force linear to the lateral displacement. The results are compared with those for no vertical shift down of the magnet before its lateral movement. They show slight increase in the lateral stiffness with the shift down of the magnet. Fig.1(b) shows increment of the levitation force from that at the start of the lateral movement. It can be seen that the levitation force increases with the lateral displacement and that its increase is proportional to the square of the lateral displacement.

3. Dynamics analysis of a spinning levitated magnet

A system to be considered here consists of a cylindrical permanent magnet and a high-T_c superconductor. Freely levitated above the superconductor by the field-cooling, the magnet is spinning around its cylindrical axis at angular velocity Ω. It is supposed to have small eccentricity ε defined by the distance between the mass center and the geometric center. For simplicity, we neglect damping terms and assume that the magnet has no deformation and that its cylindrical axis has no tilt and that its angular velocity Ω is constant, that is, its rotation is in steady state. Then motions of the magnet to be determined are only translational ones. Equations describing these three-degree-of-freedom motions are given by

$$m\ddot{r} - f_r(r,z) = m\varepsilon\Omega^2 e^{j(\Omega t+\gamma)} \quad , \quad m\ddot{z} - f_z(r,z) = -mg, \tag{1}$$

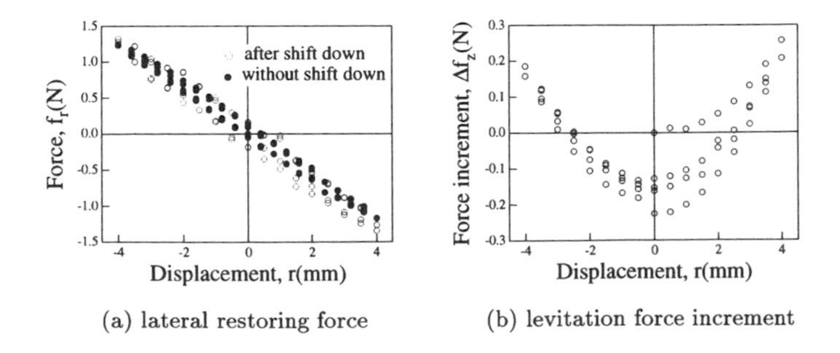

(a) lateral restoring force (b) levitation force increment

Figure 1. Magnetic force acting on the permanent magnet versus its lateral displacement

where r and z denote the lateral and vertical displacement from the static equilibrium. Here the complex number representation is used as $r = x + jy$. m is the mass of the magnet, g is the gravitational acceleration, γ is the initial phase of the rotation. f_r and f_z are the magnetic forces in the respective directions. They generally depend on both the lateral and vertical displacements of the magnet. Thus, its translational motions in each direction can be coupled with each other.

According to our experimental results stated above, these force terms can be modelled approximately by nonlinear functions with respect to the displacement, as follows:

$$f_r(r,z) = -(k_{r0} - k_{r1}z)r \quad , \quad f_z(r,z) = k_{zr}|r|^2 - k_{zz}z - k_{z0}. \tag{2}$$

By using this expression, with some shift of the origin of z, and by substituting $r = \varepsilon r^*, z = \varepsilon z^*$ and $t = \sqrt{m/k_{r0}}t^*$, the equations of motion can be nondimensionalized as,

$$\ddot{r} + (1 - \varepsilon\alpha z)r = \nu^2 e^{j(\nu t+\gamma)} \quad , \quad \ddot{z} - \varepsilon\beta_1|r|^2 + \beta_2 z = 0, \tag{3}$$

where $\alpha = k_{r1}/k_{r0}, \beta_1 = k_{zr}/k_{r0}, \beta_2 = k_{zz}/k_{r0}, \nu = \Omega\sqrt{m/k_{r0}}$, and the asterisks are omitted.

In the rotating coordinate system the above equations become

$$\ddot{\xi} - 2\nu\dot{\eta} - \nu^2\xi + (1 - \varepsilon\alpha z)\xi = \nu^2 \cos\gamma, \tag{4}$$
$$\ddot{\eta} + 2\nu\dot{\xi} - \nu^2\eta + (1 - \varepsilon\alpha z)\eta = \nu^2 \sin\gamma, \tag{5}$$
$$\ddot{z} - \varepsilon\beta_1(\xi^2 + \eta^2) + \beta_2 z = 0, \tag{6}$$

where $r = \rho e^{j\nu t}$ and $\rho = \xi + j\eta$. The mode of vibration at angular frequency ν in the steady-state solutions in terms of r can be found as a static solution in terms of ρ:

$$z_{ss} = \frac{\varepsilon\beta_1|\rho_{ss}|^2}{\beta_2} \quad , \quad \frac{-\varepsilon^2\alpha\beta_1|\rho_{ss}|^3 + \beta_2|\rho_{ss}|}{\beta_2(|\rho_{ss}|+1)} = \nu^2. \tag{7}$$

This gives the whirling at angular velocity ν with its constant amplitude ρ_{ss} and its constant vertical displacement z_{ss}. The vertical displacement is proportional to the square of the whirling amplitude. Figure 2 shows a frequency-response curve for the solution obtained by using our experimental parameters: $m = 0.024\text{kg}, k_{r0} = 3.25 \times 10^2\text{N/m}, k_{zr} = 2.50 \times 10^4\text{N/m}^2, k_{r1} = 1.25 \times 10^4\text{N/m}^2, k_{zz} = 8.00 \times 10^2\text{N/m}$. Here

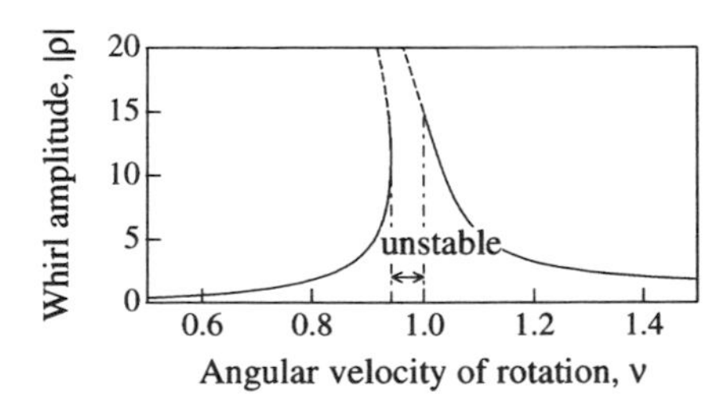

Figure 2. Frequency-response curve for primary resonance ($\alpha = 38.5, \beta_1 = 76.9, \beta_2 = 2.46, \varepsilon = 0.5\text{mm}$)

ε = 0.5mm is used as the eccentricity. For ν closer to 1.0 (the critical speed), the levitation gap, as well as the whirl amplitude, are larger. Furthermore, the peak of the curve bends to lower velocities. This tendency is due to the cubic nonlinearity of the force terms, which comes from the cross-coupling of the lateral and vertical motions.

Local stability of the steady-state solution can be determined by superposing a perturbation $(\Delta\xi, \Delta\eta, \Delta z)$ on the steady-state solution. Since both the perturbed and steady-state solutions satisfy Eqs.(7)-(9), their difference gives a set of 2nd-order differential equations for $(\Delta\xi, \Delta\eta, \Delta z)$. After its linearization, following substitutions $\Delta\xi = a_\xi e^{\lambda t}, \Delta\eta = a_\eta e^{\lambda t}, \Delta z = a_z e^{\lambda t}$ lead to a set of homogeneous linear equations of $(\Delta\xi, \Delta\eta, \Delta z)$. Thus, the eigenvalues λ can be obtained from the corresponding characterisistic equation. If the real part of λ is positive, the steady-state motions are unstable.

In Fig.2, on the dotted parts of the response curve, one of the six eigenvalues is positive, real number, which gives divergence instability. Therefore, when ν is close to and less than 1.0 (the critical speed), a perturbation can change exponentially with time, and the gap, as well as the whirl amplitude, can increase or decay without vibrating. This unstable region increases with the eccentricity ε. It should be noted that the unstable region is somewhat different from that of 1-degree-of-freedom forced vibration of a soft spring, though the response curve itself are both similar. The reason may be that the present case has 3-degree-of-freedom cross-coupling motions.

According to the imaginary part of eigenvalues, perturbations have three modes of vibrations at different frequencies, which are closely related with the natural frequencies in the lateral and vertical direction. In our case the mode of the lowest frequency decreases its frequency from 1.0 to 0, with ν getting larger from 0 toward 1.0, and finally one of the corresponding two eigenvalues becomes real and positive, leading to the static instability stated above.

4. Conclusion

This research has investigated rotor dynamics in high-Tc superconducting bearings. The magnetic forces acting on the magnet have been measured and modelled as nonlinear functions of both the vertical and lateral displacements of the magnet. They cause cross-coupling and cubic nonlinearities of rotational and vertical motions of the spinning levitated magnet. Dynamics analysis of the 3-degree-of-freedom translational motions predicts that the levitation gap, as well as the whirl amplitude, depend on the rotational speed, and that the steady-state whirling can have divergence instability, which may lead to the gap increase or decay with time.

References

[1] Coombs T A and Campbell A M 1996 *Physica C* **256** 298–302

[2] Nemoshkalenko V V, Brandt E H, Kordyuk A A and Nikitin B G 1990 *Physica C* **170** 481–485

[3] Sugiura T and Fujimori H 1991 *IEEE Transaction on Magnetics* **32** 1066–1069

[4] Sugiura T, Tashiro M, Uematsu Y and Yoshizawa M 1997 *IEEE Transaction on Applied Superconductivity* **7** to be published

Inst. Phys. Conf. Ser. No 158
Paper presented at Applied Superconductivity, The Netherlands, 30 June–3 July 1997

Hysteresis Motor With Self Sustained Rotor For High Speed Applications

X. Granados[a], I. Marquez[a], J. Mora[a], J. Fontcuberta[a], X. Obradors[a], J. Pallares[b] and R. Bosch[b]

[a]Institut de Ciència de Materials de Barcelona, CSIC, Campus UAB, 08193, Bellaterra, Spain
[b]Escola Tècnica Superior d'Enginyers Industrials de Barcelona, Universitat Politècnica de Catalunya, Diagonal 647, 08028, Barcelona Spain

Abstract. Pellets of YBCO HTSC have been used to construct a cylindrically shaped rotor. A four poles/four phases stator has been wound in order to build a motor. The motor behaves as an hysteresis motor when it starts and as synchronous in the steady state. The levitation forces associated to the flux inside the rotor region and the flux dispersed at the basal surfaces contribute to maintain the rotor anchored to its cavity in a self-sustained configuration. Power and torque in the synchronous regime, centering forces, static excitation field mapping and field distribution at the HTSC surface have been investigated.

1. Introduction

The development of high critical current HTSC has allowed the fabrication of devices in several areas of electromechanical engineering. In particular, HTSC have a large potential for the improvement of the efficiency of electrical motors either to substitute the copper winding[1] or the rotors[2,3] and for the realization of new designs [4].

The peculiar hysteresis cycle of superconductors allows to improve the performances of classical hysteresis motors [3]. Even more, the diamagnetic behavior of superconductors allows to levitate the rotor and thus to eliminate any mechanical contact between rotor and stator. This property allows to reach a high speed rotation of short diameter rotors and it seems to be a good solution for some applications as pumping of cryogenics liquids [3].

In this work we present our progress on the study of the behavior of a superconducting rotor powered by a classical copper wound armature. Magnetic field distributions around the rotor, magnetic interaction between inductor and superconducting rotor and static torque measurements are the main items investigated in this study.

2. The Motor

2.1. Designing and operation

The motor is built by the HTSC cylinder shaped rotor and a conventional copper wound stator. The stator core has an external diameter of 40 mm, the internal hollow is of 16 mm in

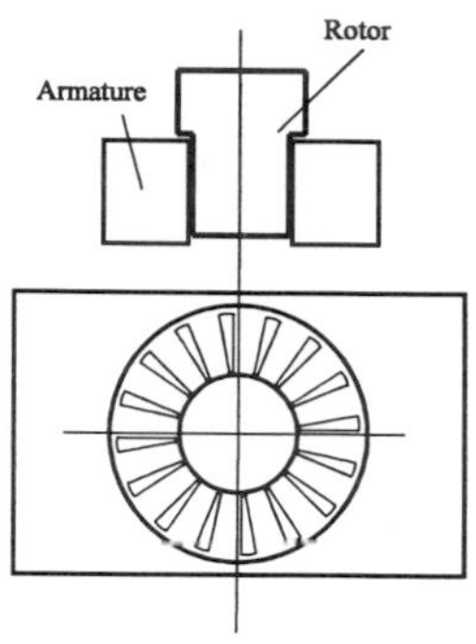

Figure 1 Scheme of the stator and rotor of the hysteresis motor

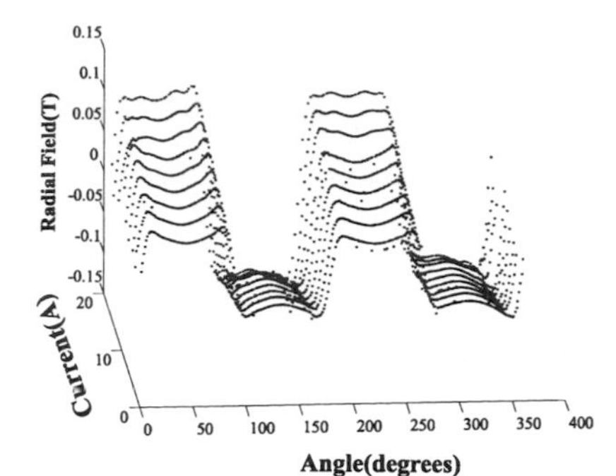

Figure 2. Profile of the magnetic field generated by the stator for applied currents from 2-16 A

diameter and the highness is 16 mm. The stator has been wound in a four poles/four phases configuration which allows the adequate symmetry for the rotor self-centering and sustentation. No mechanical bearing has been used.

The rotor, of 15.7 mm in diameter and 16 mm high, has been constructed from pieces of melt textured YBCO [5]. In order to gain homogeneity around the external area of the cylinders we have chosen multidomain pellets. In order to improve the axial stability, a disk of melt textured YBCO has been stuck on the upper face of the HTSC rotor.

The motor works in a liquid nitrogen bath, cooling both the superconducting rotor and the armature. No effects associated with the heating of the rotor has been observed. When the armature is energized the rotor levitates and it is centered by the lateral magnetic pressure. It spins driven by the applied magnetic field. The rotor reaches a steady state synchronous with the field rotation. The motor has been tested at several frequencies up to a rotation speed of 6500 tpm . This limit is determined by the saturation of the source. By stroboscopic lighting, the movement of the rotor has been observed showing its synchronous behavior. When mechanical charge is applied, the phase between the field and the rotor changes reversibly as in conventional synchronous motors. When the charge exceeds a critical limit the displacement becomes hysteretic.

The magnetic field generated by the stator, when only one phase is powered, has been measured by using a Hall miniprobe. It provides the radial component of the magnetic flux density. Figure 2 shows the measured field (at 0.2 mm from the pole to mimic the rotor surface). We note that the magnetic strength is maximum at each pole border. It can be seen in Fig. 2, that the radial field tends to be homogeneous when saturation is reached. The maximum value of the radial component of the field generated at 16A is about 0.12 T. Actually, this is a relatively low value in comparison with that usually obtained in motors with iron rotors.

3. Interaction with the rotor

3.1. Magnetization

The principle in which the motor is based is the ability of the superconductors to be magnetized. After a zero-field cooling process, the external magnetic field penetrates in the rotor, and it is pinned. Displacement of the trapped flux on the edge of the magnetized zone induces a torque Γ. Torque depends on the magnetization of the superconductor **M** and on the applied field **H**. Consequently, Γ can be either reversible or irreversible depending on the field **H** excursion and on the **H-M** angle. The displacement of the trapped flux on the limits of the magnetized zone is reached through a change of magnetization which follows either minor or large loops of an hysteresis cycle. Therefore understanding of the characteristics of the motor starts by the analysis of the magnetization of the rotor.

To this purpose we have used Hall probes. The probe is placed at the center of a slot of the stator, where the radial component of field created by the poles vanishes. In a first experiment, we have applied an excitation current I (2-16A) to the four poles. After zeroing the field, the remanence has been measured while turning the rotor along a cycle. The modulation of the remanence, with the corresponding sign changes, clearly reflects the magnetic pole structure. That is, the superconducting blocks have been correctly magnetized. The fine details of the magnetic remanence profile clearly mimics that of the poles. Notice, for instance the typical U-shape, reflecting the higher flux density close to the pole limits (see also Fig. 2). In a second series of experiments, the magnetization of the rotor is measured by the Hall probe while the stator is energized. In Fig. 4 we show the measured field for a complete turn of the rotor. We note that in this experimental set up we record the flux pinned at the superconductor after the explored zone is exposed to a peak of field generated by the nearest pole. Data in this figure clearly reflects the inhomegeneous flux pinning properties of the cylinder.

3.2. Torque and axial levitation forces

We have performed measurements of the torque under static conditions, i.e., when only one phase is powered by a constant current. As mentioned above, this may be considered as an approximation to the synchronous behavior.

The torque has been measured as a function of the applied current and the angle of misalignment (Θ) from the original position at which the rotor was magnetized. As general trends, we can observe that the torque increases with the current as expected from the remanence curves. It results form the increasing of both **M** and **B**(I). The angular dependence of Γ is more interesting. At small angles (Θ<5°), Γ increases almost linearly with Θ whereas for larger angular excursions the torque tends to saturate. The low angle region is believed to originate from small magnetization loops induced in the

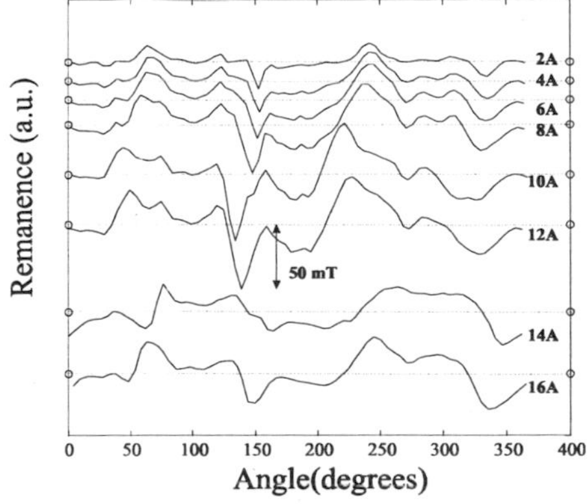

Figure 3 Remanence in the surface of the rotor after being magnetized by a phase.

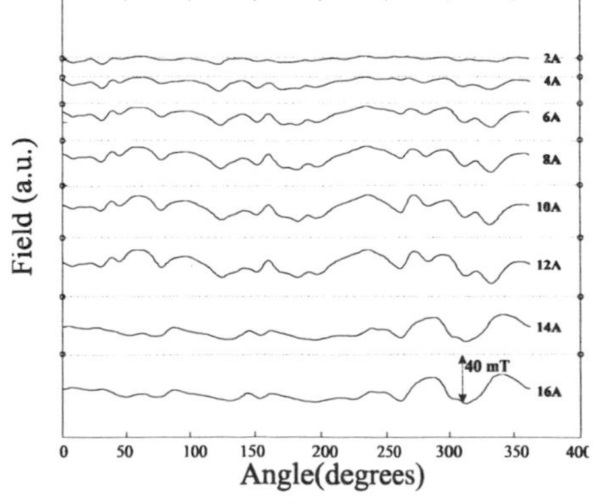

Figure 4 Induction in the gap, associated to the remanence of the rotor.

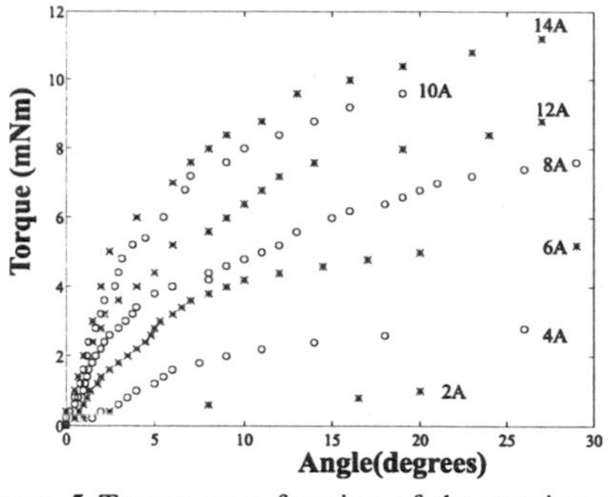

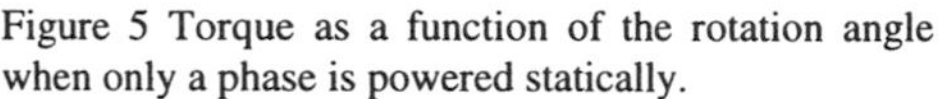

Figure 5 Torque as a function of the rotation angle when only a phase is powered statically.

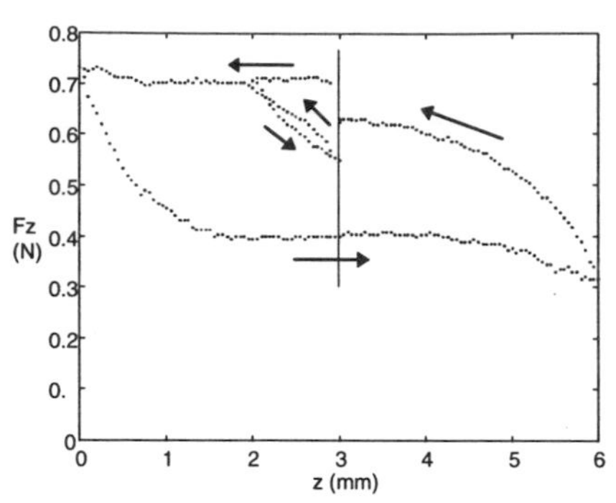

Fig 6 Vertical force loop when a current of 16 A is applied. The vertical line signals the starting position.

superconductor when displaced from its equilibrium position at the pole face. Due to the fact that $\Gamma = |\mathbf{M} \wedge \mathbf{B}|$ the observed increase of torque with Θ should be expected. As observed in Fig. 2, there is region of about ±5° at both sides of each pole where the field changes markedly. Consequently a torque appears when turning the rotor. It is important to mention that whereas in the Θ<5° region the torque brings the sample back to its original position, i.e. it is reversible, for Θ>5° the equilibrium final position is not longer the initial one. Therefore, in this region the magnetization changes are not longer reversible. Comparison with data of Fig. 2 suggests that the stronger irreversibility arises when the field polarity changes, i.e. when part of the superconducting block are forced to describe a complete hysteresis loop.

For some applications, the axial force is an important parameter of the motor. Figure 6 shows the measured vertical stabilization force when a current of 16 A is applied to the stator. The cycles starts at 3 mm over the limit of the stator (vertical line in Fig. 6). The vertical force experienced by the rotor is of 0.7N. From this position, displacements of the rotor towards the stator are considered. In Fig. 6 a minor loop is shown; the vertical stiffness is of about 137 Nm^{-1}.

In summary, we have developed a demonstrator of a bearingless motor using a superconducting rotor formed by melt-textured YBCO cylinders. Although more work is required to fully characterize this device, the present preliminary results are suggestive of interesting future applications

We would like to acknowledge finantial support by the CICYT (MAT 96-1052) and MAT (94-1924), EC-EURAM(BRE2CT94-011) and the Generalitat de Catalunya (GRQ95-8029).

References

[1] Y.Itoh et. al., “A Construction of High Temperature Superconducting...”, Physsica C, 235-240(1994)

[2] P.Tixador et al., “Electrical Motor With Bulk YBCO pellets”, ASC , Pittsburgh, (1996), to be published.

[3] T.Habisreuther et al ,“Magnetic Processes in Hysteresis Motors...”,ASC, Pittsburgh,(1996),to be published

[4] V.V. Nemoshkalenko et al., “190000 rpm Low-Power Magnetic...”,Phys. Metals, Vol. 11(9),(1993)

[5] R.Yu et al. This conference

Inst. Phys. Conf. Ser. No 158
Paper presented at Applied Superconductivity, The Netherlands, 30 June–3 July 1997

Superconducting Bearings in High-Speed Rotating Machinery

T.A. Coombs, A.M. Campbell, I.Ganney, W. Lo,T. Twardowski*, B Dawson**

IRC in Superconductivity, Cambridge University, Cambridge, UK
*International Energy Systems, Chester High Road, Neston, South Wirral, UK
**British Nuclear Fuels, Capenhurst, South Wirral, UK

Abstract - The characteristics of superconducting magnetic bearings designed for high load, high speed applications is being investigated. In work being carried out at Cambridge we have investigated various different configurations of superconducting bearings both passive and active. These configurations are designed to overcome the various problems associated with the use of YBa2Cu3O7 as an active component in a magnetic bearing. For example although the load bearing capacity is high and increases with the square of the magnetic field trapped, the stiffness is low. Both the stiffness and the levitation height are a function of the loading history of the bearing and, since a superconducting bearing has little inherent damping, it is vulnerable to large excursions due to for example cyclic loads applied at or near its natural frequency. The information being gathered at Cambridge will be used to enable these effects to be mitigated in the bearing design process.

1. Introduction

The use of bulk YBCO as a high field magnet has many potential applications such as motors, generators, bearings etc. The most obvious of these applications is in bearings in which the superconductor can not only provide a high magnetic field and therefore a high restoring force but can in addition act as a stabilising component. A magnetic bearing cannot be composed of entirely passive components [1] either an active or a diamagnetic component is required. This may be in the form of an electromagnet employing feedback control or it may be in the form of a superconductor which since it is diamagnetic reacts to changes in magnetic field and hence stabilises the bearing. We are investigating this aspect of superconductors in a specific application, that of energy storage flywheels.

This paper describes the measurement rig which we have constructed at the IRC in Cambridge. This rig is intended to provide data for the design of a full scale flywheel demonstrator currently under construction at International Energy Systems (IES). This is part of an ongoing programme of work in progress at the IRC [2-4].

2. Equipment

The rig (figure 1) is built around a 40 kg IES rotor. This rotor is composed of glass fibre and carbon fibre reinforced polymers and is approximately 300 mm tall by 330 mm in diameter. The rotor has been modified by the addition of a loading plate to the top and a magnet mounting plate to the bottom. These three elements together form the "flywheel".

The flywheel is supported in an Evershed[5] type arrangement by two magnetic bearings. The principle behind an Evershed type system is that two bearings are used, one, a magnetic bearing, is used to bear the load and the second which may in principle be any type of bearing is used to stabilise the magnetic bearing. In our arrangement the second bearing is a superconducting magnetic bearing.

The Loading bearing is formed by two cylindrical magnets acting in attraction and the stabilising bearing is formed by the interaction between a cylindrical permanent magnet and sections of superconductor. The superconductors are mounted separately from the flywheel and thus none of the moving parts are cooled.

There are three motors built into the rig. The first, the set-up motor, adjusts the separation between the magnet and the superconductors, the second, the load motor, adjusts the gap between the magnets and the third drives the flywheel round once it is levitated.
The motor which adjusts the separation between the magnets does this by raising and lowering an assembly which contains three load cells. These load cells measure the residual force between the two permanent magnets (the levitation magnets).

The position of the rotor and the separation of the magnets are measured by two separate methods. The first method involves the use of tachos fitted to the drive motors. These send a pulse for every revolution to a counter board which can then be accessed by the computer which is logging the experiment. The motion of the set-up and load motors is translated into linear motion via a 500:1 gearbox and a 1 mm pitch thread. This means that nominally at least the resolution of the position measurement is 20 microns (i.e 1mm / 500) the positional accuracy is however less than this due to backlash in the gear-box and the motor drive belt.

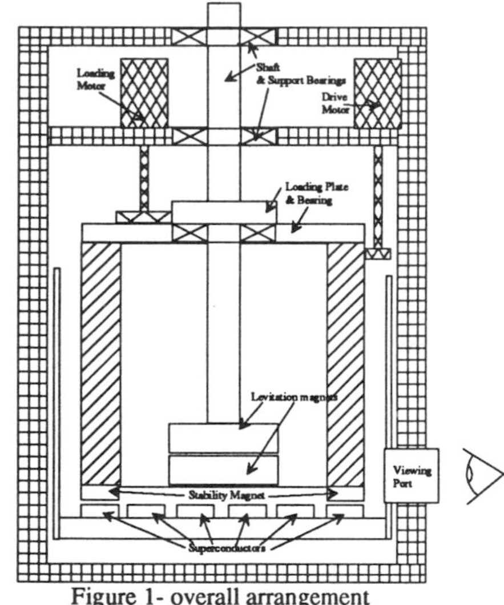

Figure 1- overall arrangement

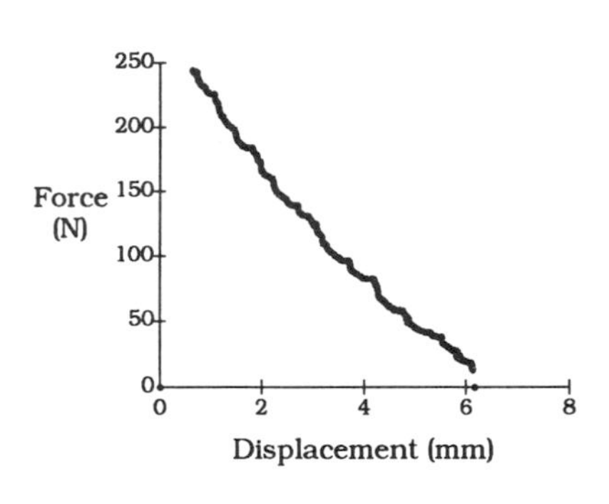

Figure 2 - Load Bearing Magnetic Bearing : Total Force V. Levitation Magnet Separation

The second measurement uses capacitive sensors. These work by measuring the changes in capacitance of a bridge circuit when one leg of it is brought up to a conducting plane. These devices are low cost and highly non-linear but they may be calibrated by comparing the measurements with those from the motor tachos.

3. Results

In order to stabilise the arrangement the positive stiffness of the superconducting bearing must be greater than the negative stiffness of the magnetic bearing. The initial test of this is to determine whether stable levitation may be achieved. In order to plot the actual stiffness of the bearing and thus to determine over what range the bearing will be stable it is necessary to provide an out of balance force. This can be done using the arrangement shown in Fig. 1 by bringing the loading plate down onto the rotor. The out of balance force is then measured by the load cells and will be a function of the mass of the rotor, the force provided by the loading bearing (Fig. 2) and the restoring force provided by the superconducting bearing (Fig. 3).

Both figure 3 and figure 4 show the force developed as the superconducting bearing gap is first reduced and then increased again. Figure 4 shows how the characteristics of the overall bearing system (superconducting + levitation) vary with the levitation bearing gap. A series of force-displacement curves are shown. They show the behaviour of the system as the bearing gap is reduced from 3.0 mm to 1.0 mm and then increased again - i.e. load is applied and then removed. Curve 1 shows the characteristics with a levitation magnet gap of 6 mm, curve 2 with a gap of 7 mm, curve 3 with an 8 mm gap and so on. Levitation occurs when the net force falls to zero and this is shown in the return trace on curve 8.

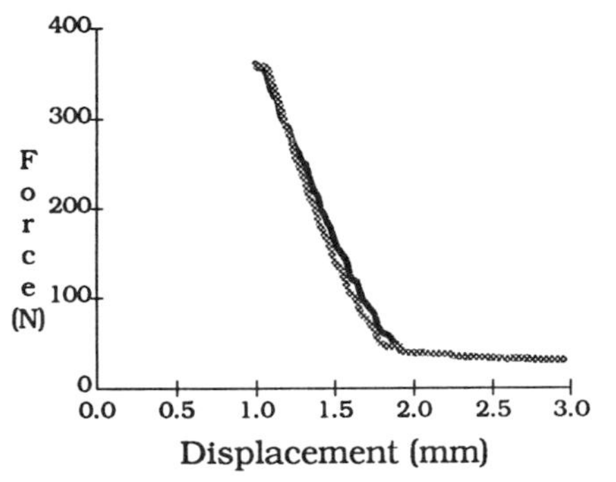

Figure 3 - Superconducting Bearing Stiffness

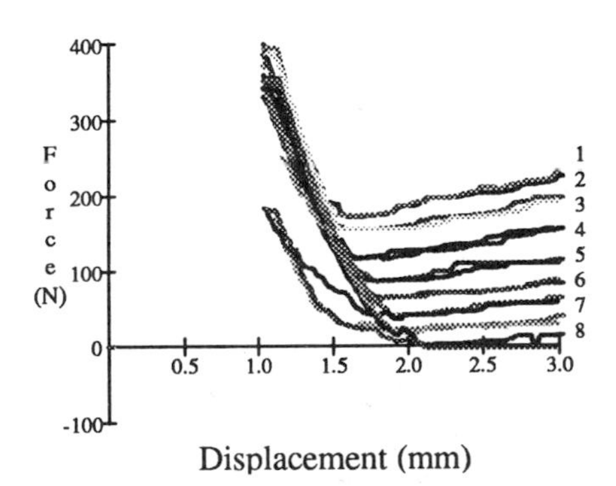

Figure 4 - Overall Bearing System Stiffness

In addition to performing pseudo-static tests it is possible to perform dynamic tests by spinning the rotor. Since the overall stiffness of the bearing is fairly low and the damping is small while the mass of the rotor is large the natural frequency will also be low. The natural frequency is given by equation 1, where 's' is the stiffness and m is the rotor mass.

$$\omega_n = \sqrt{s/m} \qquad (1)$$

The rig is designed for low-speed rotation in order to examine the behaviour of the bearing when stimulated at at or near one of the rotor's natural frequency. This enables the behaviour of the high-speed (50,000 rpm) device under construction at IES to be predicted. By taking the speed of the rotor to just below the natural frequency an approximate coefficient of friction μ may be calculated from equation 2, where k is the radius of gyration, m is the mass of the rotor and r the radius of the bearing.

$$mk^2 \frac{d\omega}{dt} = \mu mgr \qquad (2)$$

In tests the rotor took 30 seconds to spin down from a speed of 1 Hz. Taking k as approximately 2/3 of the radius of the rotor r then we have, $\mu = 1.4 * 10^{-3}$.

4. Conclusions

We have demonstrated stable levitation using an Evershed-type bearing system of a flywheel rotor weighing in excess of 40 kg. This arrangement has been chosen to minimise the problems associated with creep and vibrations. In addition in this arrangement the superconductor is field cooled and since it is providing stabilisation rather than a levitation force and therefore does not require a displacement to develop this force the system may be cooled in its operating position. The characteristics of the bearing system are determined by the relative stiffnesses of the levitation and stabilisation bearings and these in turn are determined by the bearing gaps. Thus at least in principle the stiffness of the bearing system may be varied to suit different operating conditions. Finally we have achieved a reasonably low coefficient of friction which we expect to be able to reduce still further by working in vacuum and by detailed attention to magnet design.

Acknowledgements

This project is supported by an EPSRC grant and by contributions from British Nuclear Fuels. I am also indebted to R. Storey, C. Clementsen, R. Weller, P. Hunneyball and to S. Fasham

References

[1] S. Earnshaw, Trans. Camb. Phil. Soc. 1842 7 97
[2] T.A. Coombs, D.A. Cardwell and A.M. Campbell "Development of an Active Superconducting Bearing", IEEE Transactions on Applied Superconductivity 1995 5 2 630-633
[3] T.A. Coombs and A.M. Campbell "Gap decay in superconducting magnetic bearings under the influence of vibrations", Physica C 256 1996 298-302
[4] T.A. Coombs, D.A. Cardwell and A.M. Campbell "Dynamic Properties of Superconducting Magnetic Bearings", IEEE Transactions on Applied Superconductivity, In Print.
[5] S. Evershed, J. Inst. Electron. Engr. 1900 29 743

Inst. Phys. Conf. Ser. No 158
Paper presented at Applied Superconductivity, The Netherlands, 30 June–3 July 1997

Superconducting Magnetic Bearings in a High Speed Motor

P Stoye[1], W Gawalek[2], P Görnert[3], A Gladun[4], G Fuchs[1]

[1] Institut für Festkörper- und Werkstofforschung Dresden, D-01171 Dresden, Germany
[2] Institut für Physikalische Hochtechnologie, D-07743 Jena, Germany
[3] SurATech GmbH, D-07743 Jena, Germany
[4] Technische Universität Dresden, D-01062 Dresden, Germany

Abstract. The application of superconducting magnetic bearings has been tested in a high speed motor. The shaft of the asynchronous motor has been stabilized by two superconducting magnetic bearings. One superconducting magnetic bearing is built up of two concentric NdFeB ring magnets fixed to the shaft, which are supported between two YBCO rings, 70 mm in diameter. The YBCO rings consist of 6 segments of melt textured YBCO material. In the working position a static levitation force of 25 N and values of the axial and radial stiffnesses of 15 N/mm and 10 N/mm, respectively, have been achieved. At atmospheric pressure the shaft was safely rotated at speeds up to 12 000 rpm. By measuring the radial run out, we found a small average displacement of the shaft of about 40 μm from its centered position. In experiments at a reduced gas pressure of 1.5 mbar a stable rotation speed of 42 000 rpm was achieved.

1. Introduction

Passive superconducting magnetic bearings (SMB) consisting of permanent magnets and high temperature superconductors (HTSC) are one of the most promising applications of HTSC in the near future. Today there are different fields for possible applications of SMB: flywheels for energy storage [1], motors of small size which may be used in liquid gas pumps [2] and linear transport systems [3], e. g. in clean room environment. In this paper we present results of the test of a high speed motor with passive superconducting magnetic bearings designed for demonstration as liquid nitrogen pump.

2. Experimental

In order to use the flux trapping properties of YBCO in a practical device a motor with superconducting magnetic bearings (SMB) has been constructed [4]. YBCO blocks (∅ 30 mm x 10 mm) processed by a melt texturing technology [5] were shaped to form rings of 6 segments as bearing stator. One SMB (Fig. 1) is made of two such rings stabilizing the bearing rotor which consists of two concentric NdFeB ring magnets fixed to the shaft.

By optimizing the geometry of the SMB an axial load force of 25 N in the middle position of the shaft is found experimentally, depending on the initial bearing gaps before cooling. Maximum values of axial and radial stiffness of 15 N/mm and 10 N/mm, respectively, were derived from the experimental data [6].

These bearings were tested in a high speed asynchronous motor (Fig. 2) with a power of 50 W, which is designed for low losses, high frequency (1 kHz) and use at 77 K. The YBCO

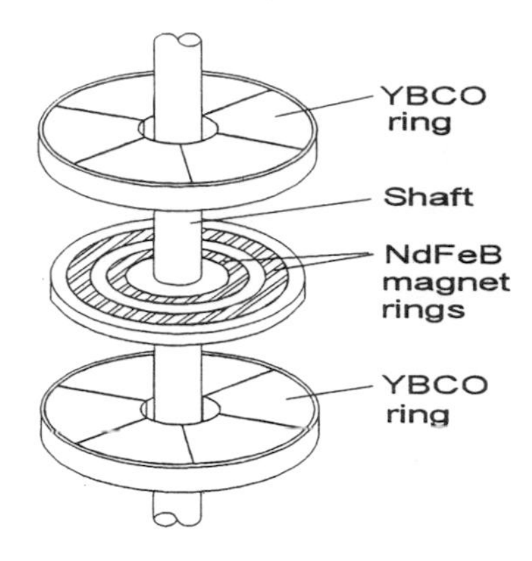

Fig. 1. Geometry of the SMB with YBaCuO rings and concentric magnet rings.

rings are cooled by heat conduction using liquid nitrogen flowing in the copper flanges of the cooling block.

Mechanical bearings serve as centering devices to hold the shaft in a fixed position during start up procedure. After the YBCO rings are cooled down to a temperature near 77 K the shaft is left free and moves to its equilibrium position. The rotation is observed by distance sensors and a sensor measuring the rotation speed. In order to investigate the behaviour of the shaft under reduced gas pressure the cooling block with motor and bearings has been built into a vacuum container (Fig. 2).

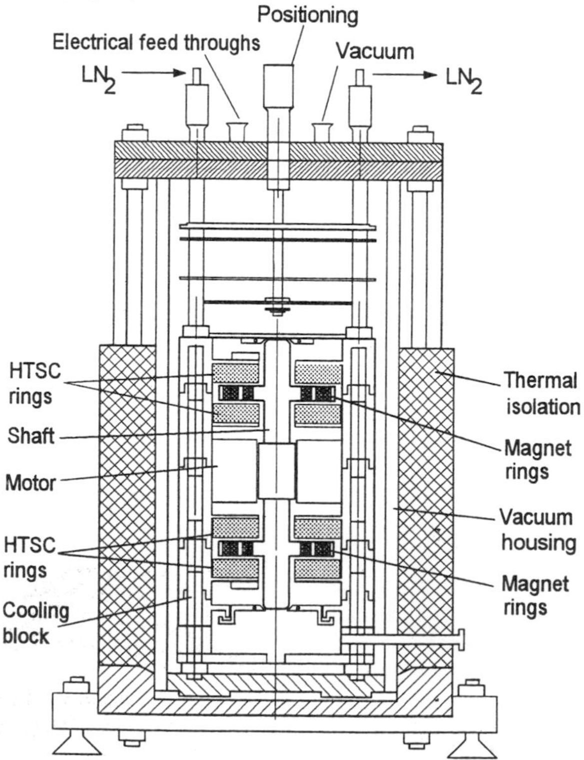

Fig. 2. Motor with superconducting magnetic bearings built into a vacuum container.

3. Results and Discussion

Testing the motor with SMB at 77 K at atmospheric pressure, the shaft rotated stable with a maximum speed of 12 000 rpm (Fig. 3). Beyond a critical frequeny of about 30 Hz the radial run out decreases with increasing frequency by self centering of the shaft. At speeds above 5000 rpm the average radial run out of the shaft is approximately 40 μm. Similar behaviour has also been observed in comparable bearing arrangements [2]. The radial run out of the shaft compared to the centered position is caused by the residual unbalance of the shaft. It has an axis of inertia which is different from its geometrical axis. Additionally the axis of symmetry of the permanent magnet field is also different from this geometrical axis.

The rotation at atmospheric pressure is limited by air drag. Compared to other losses as eddy current and hysteresis losses (proportional to speed) and the magnetic friction losses (independent of speed), the air drag losses increase with n^3 (n - rotational speed) [7]. Therefore, at higher frequencies the drastically increasing air drag losses lead to an overload of the motor, which limits the maximum rotor speed. The result is a reduction of the increase of rotor speed and a strong increase of the stator current of the motor (Fig.4).

Additional tests were performed at a gas pressure of 1.5 mbar. A maximum stable rotation of the shaft up to 42 000 rpm was achieved. Higher speeds are limited by the tensile strength of the NdFeB ring magnets [8].

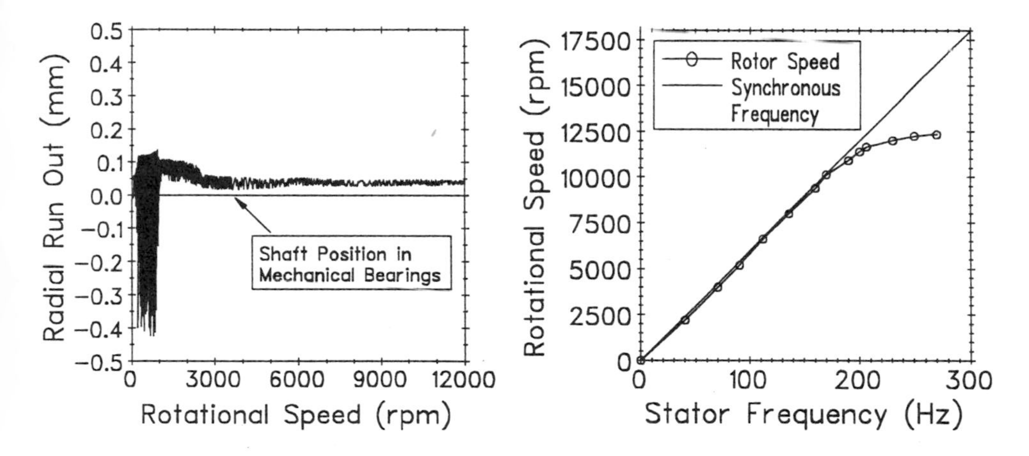

Fig. 3. Radial run out as a function of rotational speed. Straight line indicates shaft position with mechanical bearing support.

Fig. 4. Rotational speed as a function of stator frequency.

Several spin-down tests were performed. The rotor was accelerated to a distinct rotational frequency and the speed decay was measured as a function of time after the motor was switched off. The data at atmospheric pressure show an exponential decay with time (Fig. 5), which indicates frequency dependent contributions to the overall drag, such as air drag and eddy current drag. A linear decay with time was found at a reduced gas pressure of 1.5 mbar (Fig. 6). This behaviour is caused by a constant drag torque acting on the rotor [9]. Air drag as main loss contribution has been lowered significantly.

From the experimental data, the drag torque $\tau=2\pi I/(df/dt)$, where I is the moment of inertia, can be estimated to $\tau=3\cdot10^{-3}$ Nm at a pressure of 1.5 mbar and $\tau=8.8\cdot10^{-3}$ Nm for f=200 Hz at atmospheric pressure [10,11].

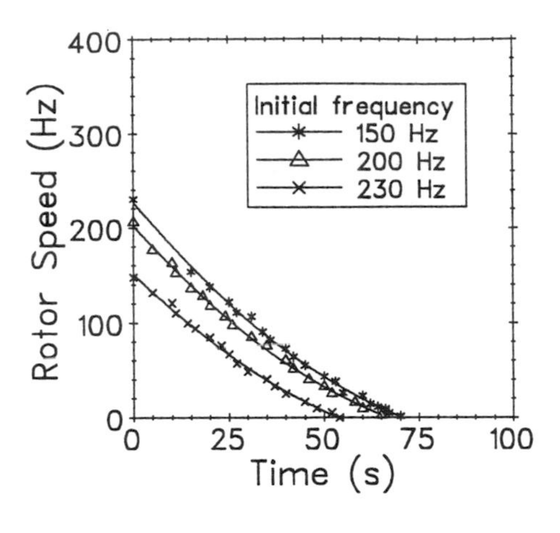

Fig. 5. Rotor spin rate decay vs. time at atmospheric pressure for various initial spin rates.

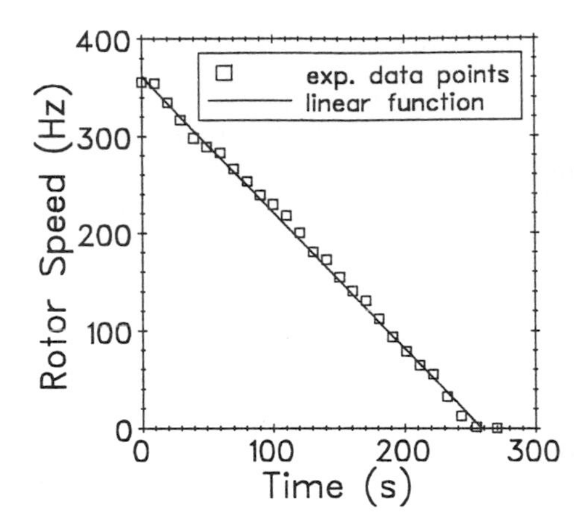

Fig. 6. Rotor spin rate decay vs. time at 1.5 mbar reduced gas pressure.

4. Conclusions

A high speed motor with superconducting magnetic bearings has been tested succesfully with carrying a load of 0.5 kg and stabilizing the shaft by two superconducting magnetic bearings. At atmospheric pressure the shaft was safely rotated at speeds up to 12 000 rpm. The radial run out of the shaft showed a small average displacement of about 40 μm from centered position of the shaft. In experiments at a reduced gas pressure of 1.5 mbar, a maximum stable rotation speed of 42 000 rpm was achieved. At this pressure, the air drag loss is reduced, but other losses as magnetic unbalance and eddy current loss still influence the rotation. A potential application of our design could be the demonstration of the operation as a pump for cryogenic liquids.

Acknowledgements

This work is supported by Bundesministerium für Bildung, Wissenschaft, Forschung und Technologie under BMBF contract no. 13N6662.

References

[1] Ma K B et al. 1995 *3rd Int. Symp. on Magnetic Suspension Technology* (Tallahassee, Florida, USA)
[2] Fukuyama H et al. 1994 *Advances in Superconductivity* **6** 1341-1344
[3] Miyamoto K et al. 1995 *Appl. Supercond.* **2** 487-497
[4] Stoye P et al. 1995 *IEEE Trans. Mag.* **31** 4220-4422
[5] Gawalek W et al. 1994 *Appl. Supercond.* **2** 465-478
[6] Stoye P et al. 1996 *Low Temp. Phys.* **105** 1457-1462
[7] Moser H 1958 *Technische Rundschau* Bern Schweiz **44** 18-23
[8] Moon F C et al. 1993 *Appl. Supercond.* **1** 1175-1184
[9] Moon F C and Chang P Z 1990 *Appl. Phys. Lett.* **56** 397-399
[10] Hull J R et al. 1995 *Appl. Supercond.* **2** 449-455
[11] Bornemann H J 1995 *3rd Int. Symp. on Magnetic Suspension Technology* (Tallahassee, Florida, USA)

Inst. Phys. Conf. Ser. No 158
Paper presented at Applied Superconductivity, The Netherlands, 30 June–3 July 1997

Electromagnetic Processes in a Disc-Type Cryoalternator

Lidia I. Chubraeva[1], Dmitry V. Tchoubraev[2]

[1] Research Institute of Electrical Machinery, St.-Petersburg, Russia. [2] Institute of Electrical Machines, ETH-Zentrum, Zurich, Switzerland

Abstract. A disc-type superconducting machine represents a convenient model to investigate experimentally electromagnetic processes in synchronous cryoalternators*. The model may be considered as well a prototype of an efficient storage system with the rotor body used as a flywheel and the field winding as a SMES. Investigation of magnetic fields and parameters of a number of disc-type alternators was carried out by the finite element method with the help of the developed program FEMAG. There are presented results of calculation of the magnetic field distribution and main principles of determination of the alternator winding sizes.**

1. Introduction

We have first developed a disc-type cryoalternator model with a simplified geometry of the rotor and stator windings, when the LTSC alternator was manufactured in 1990. The choice was determined by the Nb_3Sn rigid wire applied for both windings. Since then several other machines of similar geometry were developed, as they appeared to be a convenient and cheap model for investigation of electromagnetic processes in synchronous generators [1]. The disc configuration allows to mount and dismount rotor and stator magnetic screens without problems.

A disc-type synchronous alternator represents as well an interesting variant of an electromechanical energy storage (EMES) system. It may store mechanical energy of the rotor flywheel and magnetic energy of the superconducting field winding (if any) with a subsequent output in the form of electrical energy [2]:

$$P_{ch}t_{ch}\eta = P_d t_d \,,$$

P_{ch}, P_d - the average values of charged and discharged energy, t_{ch}, t_d - periods of charge and discharge, η - storage system efficiency. During the charge mode the alternator operates as a motor, during the discharge - as a generator. In case $t_d < t_{ch}$, $P_d > P_{ch}$ and therefore EMES operates as an amplifier of electrical energy.

2. Main peculiarities of a disc EMES system

When the rotor represents a flywheel the energy from the source is being stored in the rotor body in the form of kinetic energy $W = J\Omega^2/2$, with J - moment of inertia of the rotor body against the axis of rotation, $\Omega = 2\pi n$ - angular speed, n - frequency of the flywheel rotation [2].

The main principles of the disc flywheel development are presented in [2]. They may be co-related with the rotor coils (or magnets) development as the latter are being

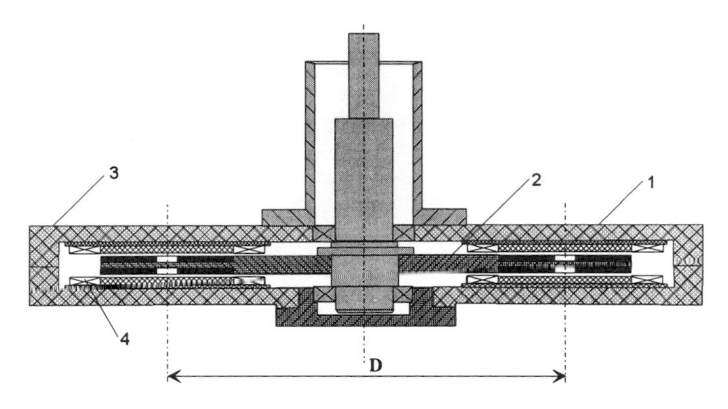

Fig.1. Cross-section of a disc-type alternator

mounted on the rotor periphery. The rotor/flywheel body may be a metallic as well as a non-metallic one.

A certain amount of energy may be stored in the coils of superconducting excitation winding. The storage will be associated with the losses in the winding only during the charge and discharge periods due to excitation current variation.
The amount of energy, stored by the field coils, representing SMES with total inductance L and excitation current I, is determined by $W = LI^2/2$.

3. Dimensions of the rotor and stator coils

The principle scheme of the alternator is presented in Fig.1. The preferable number of the rotor pole pairs is to be $p=2n$, n - odd number. In this case the 3-phase armature winding may be distributed without problems on both sides of the rotor. The outer and inner diameters of the stator coils *4* may be determined with respect to the diameter D of the circumference, occupied by the coil centers as $D_{so}=\pi D/mp$, $D_{si}=\pi D/2mp$.

The rotor coils *1* are to have the outer diameter equal to $D_{ro}=\pi D/2p$. The inner diameter of the coils is determined in accordance with the final task. In our case the preferable outer diameter was selected as $D_{ro}=\pi D/6p$.

The inductances of the rotor and stator windings and their mutual inductances may be determined in accordance with the rules accepted for the solenoidal coils.

4. Magnetic field computation

Though a disc-type alternator is very simple by geometry, calculation of the magnetic field distribution is one of the most complicated due to the presence of solenoidal coils. The computer program FEMAG was used for investigation of the magnetic fields and parameters. It allows to perform computation of magnetic field distribution for electrical machines with and without magnetic elements in the active zone. The program is based on the finite element method with application of a special built in graphic editor (AutoCAD may be used as well). The mesh dynamic optimisation was carried out. There were used computers DEC Alpha (UNIX) and DEC PC Pentium (Windows 96) to perform calculations of magnetic fields, inductances, torques, etc. The evolution of the circumference D of the 8-pole alternator with 12 armature coils is

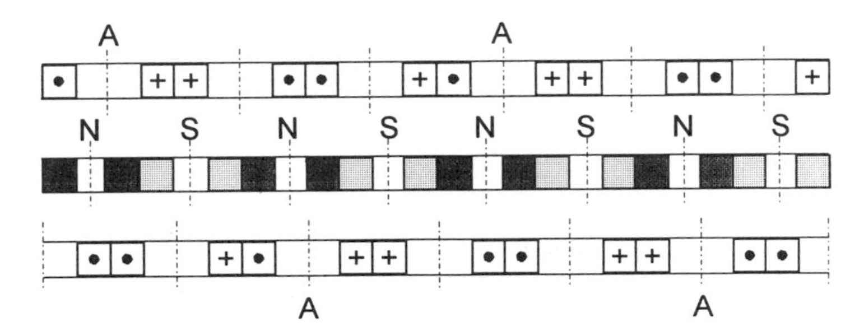

Fig.2. Evolution of the circumference with diameter D with armature coils positioned on both sides of the field coils

presented in Fig.2. Examples of magnetic field calculations with their distribution along this circumference are shown in Fig.3.

5. Transient behavior of the alternator

The energy discharge in EMES is usually a generator short-circuit mode. The peak currents of the armature winding at sudden 3-phase short-circuit in case of synchronous generator with a damper rotor screen I_{s3}'' and without it I_{s3}' co-relate as:

$$I_{a3m}''/I_{a3m}' = x_d'/x_d'',$$

where x_d' and x_d'' - correspondingly transient and sub-transient synchronous reactances. The excitation current behavior during transients is contrary to the armature current. It means the presence of a damper screen makes transients more heavy for the stator winding, but more favorable for the rotor one.

In a damperless machine the rotor winding contains extra volume of superconductor and it must have low AC losses, in an alternator with a rotor screen the armature winding is to have additional volume of superconductor or to be able to stand short-circuit currents without an avalanche heating and resistance increase, typical for the high-purity aluminum windings. The type of armature winding material is to be determined then in accordance with armature loss value.

6. Efficiency of energy storage

The EMES efficiency depends on friction losses of the rotor body and in the bearings and on additional losses in the armature during the storage mode (no-load rotation). To decrease the flywheel body friction losses it must rotate in vacuum or in gaseous medium (helium or hydrogen) [2]. Application of magnetic suspension system allows to decrease the losses in the bearings. One of the most attractive version for the configuration of the alternator being discussed is presented in [3]. In the multi-pole alternator design the role of external stator magnetic screen is negligible, therefore it can be excluded. Then there will be mainly the losses in the armature winding. In the no-load mode it means primary the eddy-currents as the circulating currents may be suppressed by the application of the multi-stage twisting or multi-stage transposing.

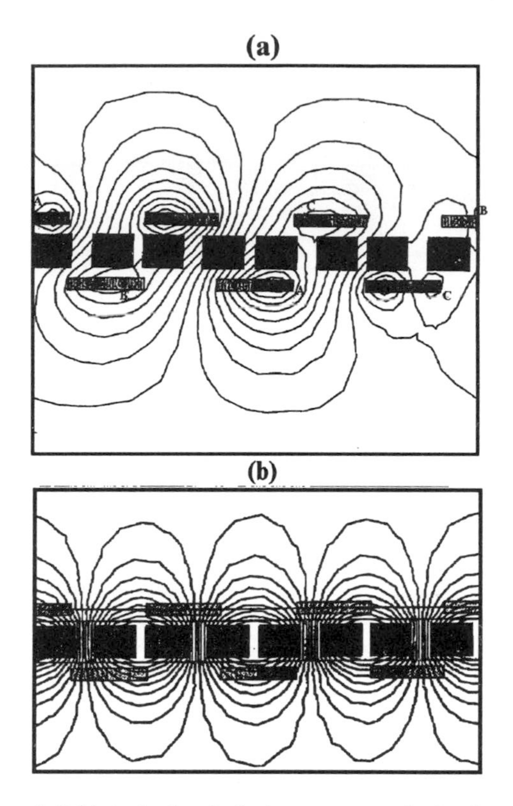

Fig.3. Magnetic fields in the 8-pole, 3-phase generator in the short-circuit (a) and no-load (b) modes

7. Conclusions

1. The disc-type electrical alternator, representing a convenient and cheap model for experimental investigation of the processes in synchronous machines, is an interesting version of a storage system as well.
2. The rotor body of the disc alternator can be designed as a flywheel to store kinetic energy.
3. The rotor superconducting field winding, representing *2p* solenoidal coils, can store magnetic energy similarly to SMES systems.

Acknowledgements.

*The researches are being supported by the State Program, Direction "Superconductivity" and by the Ministry of Science and Technical Policy of Russia. **The program FEMAG was developed under the scientific leadership of Prof. K. Reichert, Institute of Electrical Machines, Zurich, Switzerland.

References

[1] Akimov I I et al 1997 *EUCAS'97* ML3_606
[2] But D A et al 1991 *Energy Storage Systems* (Moscow: Energoatomisdat) (In Russian)
[3] El-Hamalawy A A 1995 *ISSHTS* **1/A** 81-110

Inst. Phys. Conf. Ser. No 158
Paper presented at Applied Superconductivity, The Netherlands, 30 June–3 July 1997

Fabrication and transport properties of Bi-2212/Ag multifilamentary coils for high magnetic field generation (II); high field performance test

H. Kitaguchi[1], H. Kumakura[1], K. Togano[1], M. Okada[2], K. Tanaka[2], K. Fukushima[2], K. Nomura[3], and J. Sato[3]

[1]National Research Institute for Metals, Tsukuba 305, Japan, [2]Hitachi Research Laboratory, Hitachi Ltd., Hitachi 319-12, Japan, [3]Hitachi Cable Ltd., Tsuchiura 300, Japan

Abstract. A compact superconducting magnet with 30 mm inner bore has been fabricated in a wind and react process by using Bi-2212/Ag multifilamentary tapes. The magnet is composed of 3 double pancake coils of 88 mm in outer diameter. As for a reinforcement against large magneto-electric force, the combination of Ag-Mg alloy tape co-winding and epoxy impregnation is employed. High field performance of this magnet is examined in backup fields up to 20 T at 4.2 K. Critical current density at 20 T is 516 Amm^{-2} (generating field: 1.10 T) and 112 Amm^{-2} (0.95 T) with the criteria of 1×10^{-13} Ωm for oxide part and whole cross section of the conductor, respectively. The results indicate that the reinforcement is effective against the hoop stress of 120 MPa.

1. Introduction

Extraordinary high upper critical fields B_{c2} and irreversibility fields B_{irr} of oxide high-T_c superconductors (HTSC) at low temperatures indicate that HTSC have great potential to be used in a high field superconducting magnet system operated at low temperatures. $Bi_2Sr_2CaCu_2O_x$(Bi-2212)/Ag tape is the most promising candidate for the conductor for the high field superconducting magnet which is developed in 1 GHz NMR project [1] in National Research Institute for Metals. In this system, HTSC part is designed to generate 2.4 T in the backup field of 21.1 T by the part of conventional metalic superconductors. Preliminary specification of this HTSC part is as follows; inner diameter of 75 mm, outer diameter of 130 mm, coil height of 600 mm, and coil current density over 70.5 Amm^{-2} for driven mode operation. Calculated magneto-electric stress (hoop stress) induced to the conductor in the outermost turn is $97f^{-1}$ MPa (f: packing factor). In order to be feasible against this large stress, some effective reinforcement should be developed because the mechanical strength of the Bi-2212/Ag conductor is about 40 MPa in 0.2 % yield stress [2] and this value is less than half of the required performance. Next to 50mm-class small test magnets [3-5] in our HTSC magnet development program, we planned to develop an effective reinforcement of the conductor and to fabricate several middle size magnets up to 130 mm in outer diameter and 150 mm in height. In this paper, we report the results of high field performance test in bias fields up to 20 T at 4.2 K for a magnet composed of 3 double pancake coils of 88 mm in outer diameter. Co-winding of silver based alloy tape is also examined in order to establish an effective reinforcement of the conductor.

2. Experimental

Figure 1 shows sample configuration. Detailed fabrication process is reported in the reference [6]. Two Bi-2212/Ag multifilamentary tapes with 19 filaments were bundled into a conductor. Ag-0.5wt%Mg alloy tape of 0.1 mm in thickness was co-wound as for a reinforcement member. 0.2% yield stress of this alloy tape exceeds 500 MPa after the heat treatment [2]. Melt-solidification heat treatment and subsequent post annealing were performed. The coils were impregnated with an epoxy resin to improve the mechanical strength because the alumina insulator becomes porous and has not enough mechanical strength after the heat treatment. In the coil-stacking process, the same Bi-2212/Ag tape as used in the coil fabrication was used for current leads and junctions between coils in order to reduce resistance and heat formation. All the conductors and tapes were soldered directly to the surface of the conductor of the coils except for the coil 3 at the junction between the coils 2 and 3 where the tapes were soldered to the outermost alloy tape surface. The connected coils were set in a stainless steel casing and then impregnated by the epoxy resin.

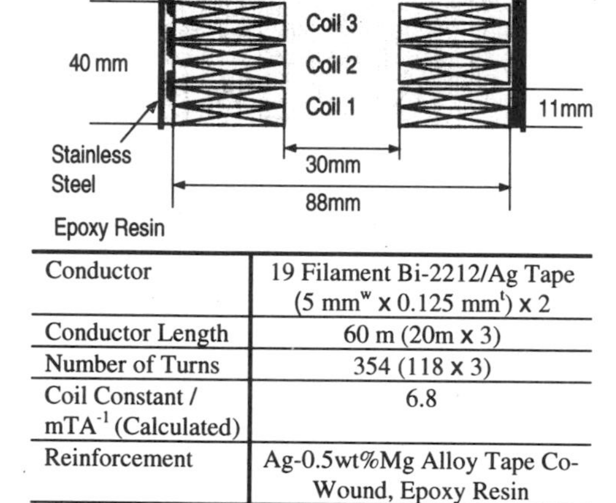

Conductor	19 Filament Bi-2212/Ag Tape (5 mm^w x 0.125 mm^t) x 2
Conductor Length	60 m (20m x 3)
Number of Turns	354 (118 x 3)
Coil Constant / mTA^{-1} (Calculated)	6.8
Reinforcement	Ag-0.5wt%Mg Alloy Tape Co-Wound, Epoxy Resin
Casing	Stainless Steel
Current Leads	Conductor x 2 / Soldering
Junction between Coils	Bi-2212/Ag Tape x 4 / Soldering

Figure 1 Configuration of the sample.

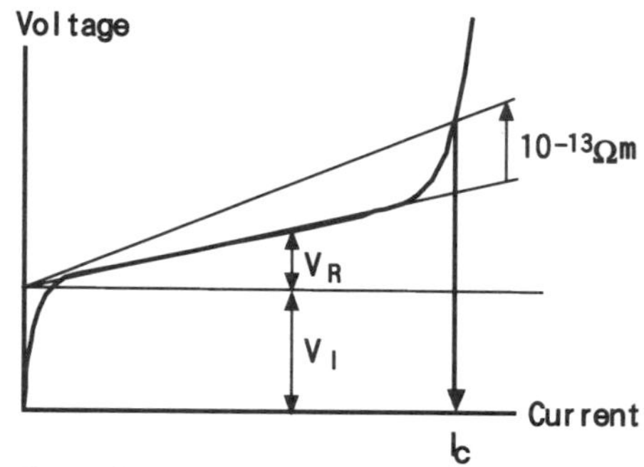

Figure 2 Typical V-I curve with continuous current sweep and I_c determination employed in this study. V_I and V_R are inductive and resistive components in V-I curve, and correspond to induction of the sample and the current ramp rate, respectively.

High field performance was examined at 4.2 K in the backup fields up to 20 T by using the hybrid magnet system installed in Grenoble High Magnetic Field Laboratory. The sample was set in 100 mm bore insert cryostat and kept at 4.2 K with liquid helium. V-I characteristic was measured for each double-pancake coil. The I_c values for the conductor (I_c-conductor) and for the Bi-2212 oxide part (I_c-oxide) were determined by using a resistivity criterion of 1×10^{-13} Ωm, which was derived from a resistance and the length between voltage terminals with cross section of the conductor and oxide part, respectively. The J_c values for the conductor (J_c-conductor) were calculated from I_c-conductor and the overall cross section of the conductor. The J_c values for the oxide part (J_c-oxide) were also calculated from I_c-oxide, the cross section of the conductor, and silver matrix/oxide core ratio. Figure 2 shows a schematic V-I curve and the way of I_c determination in this study. Voltage measured for coil shape sample with continuous current sweep contains resistive component (V_R) even in superconducting state in addition to constant inductive component (V_I) corresponding to the inductance of the sample and the current ramp rate. This resistive component is proportional to the current ramp rate [7,8] and can be explained by the self field effect [9]. The resistive component should be zero at zero ramp rate where the current is increased incrementally (step by step) with time. In this study, resistive component at zero ramp rate is about 3×10^{-15} Ωm which is enough lower than the criterion of 1×10^{-13} Ωm employed in the I_c determination. V-I curves were measured with continuous current sweep at the ramp rate of 200 $Amin^{-1}$.

3. Results and discussion

Results of high field performance test are summarized in figure 3. I_c-conductor of the coil 1 at 4.2 K, 20 T is 140 A, which corresponds to 112 Amm^{-2} in J_c-conductor, 0.95 T in generated field, and 99 MPa in hoop stress. With the criterion for the oxide part, I_c-oxide and J_c-oxide of the coil 1 are 162 A and 518 Amm^{-2}, respectively, at 4.2 K, 20 T and then the generated field is 1.10 T and the hoop stress is 114 MPa. For this coil, I_c-oxide below 18 T and I_c-conductor at 18 T and below 5 T could not be measured because of a thermal runaway of the whole sample. For each coils, no significant degradation in I_c can be confirmed after the several times of V-I measurement at 20 T with applying the current more than 20% above I_c-conductor of the coil 1 where the hoop stress reaches 120 MPa. This fact strongly supports the effectiveness of the reinforcement by the combination of alloy tape co-winding and epoxy impregnation. I_c-conductor of the coil 2 and 3 at 20 T are 100 A and 108 A, respectively, which are ~25% lower than that of the coil 1. I_c values for these 3 double pancake coils, which were measured below 7 T before the coil-stacking process, showed no significant difference. V-I relationship for the coil 3 has the resistive component of 2×10^{-13} Ωm even in the step by step current increment mode. This resistivity indicates that the coil 3 has a damaged part in the conductor itself. Figure 4 shows a simplified circuit corresponding to the sample and measured value of the resistance for each connections. Resistance lower than 0.5 μΩ is achieved even in high fields at the current lead contacts and the junction between coils 1 and 2, however the resistance at the junction between coils 2 and 3 reaches ~190 μΩ. This large resistance can be explained by the fact that the junction was soldered not directly to the conductor itself but to the Ag-Mg alloy tape at the contact to the coil 3. The soldering with an excellent conductivity could not be performed there because the alloy tape surface was oxidized in the heat treatment. These resistances in coil 3 and the junction metween coils 2 and 3 cause a large heat formation. This thermal problem gives an explanation for the facts that I_c of the coils 2 and 3 are lower

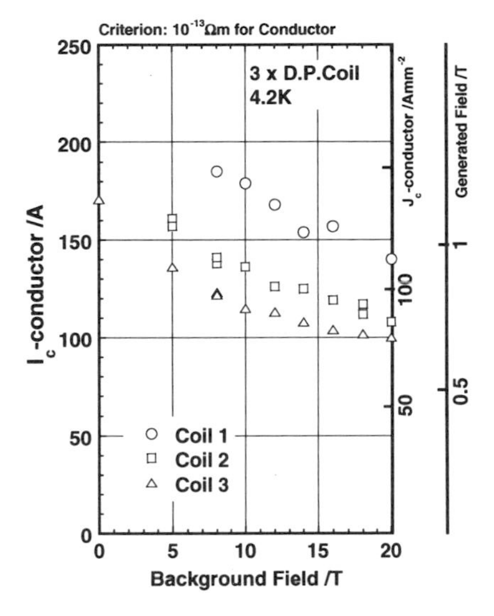

Figure 3 Results of the high field performance test. I_c-conductor, J_c-conductor, and generated field are plotted against the background fields

than the coil 1 and that I_c-B relationship for the coil 1 is anomalous (see figure 3). Without such a thermal problem, an excellent high field performance can be expected for the coils 2 and 3. The results achieved in this study show great possibilities of Bi-2212/Ag conductor for a higher field generating HTSC magnet,

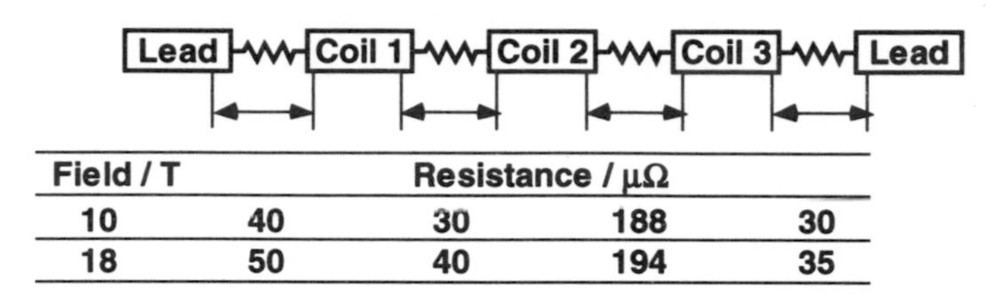

Field / T	Resistance / μΩ			
10	40	30	188	30
18	50	40	194	35

Figure 4 Simplified circuit for the sample and resistance at each connection.

4. Summary

High field performance of a HTSC magnet composed of 3 double pancake coils of 88 mm in outer diameter amd 30 mm in inner bore fabricated by using Bi-2212/Ag multifilamentary conductor is examined. At 4.2 K, in 20T, the best one of the double pancake coils carries J_c-oxide = 518 Amm^{-2} (I_c-oxide = 162 A) and J_c-conductor = 112 Amm^{-2} (I_c-conductor = 140 A) which corresponds to 114 MPa in hoop stress. At I_c-oxide of 162 A, the sample generated 1.10 T successfully in 30 mm diameter inner bore. Performance of the other two coils is ~25% lower than the best one. This degradation can be explained by the heat formation because of large resistance at the junction between these two coils and by the damage of the coil 3 suffered in the coil-stacking process. No significant degradation in I_c can be confirmed after the several times of V-I measurement at 20 T with applying the current where the hoop stress reaches 120 MPa. The reinforcement against large magneto-electric force by the combination of alloy tape co-winding and epoxy impregnation is effective.

References

[1] Kiyoshi T, Inoue K, Kosuge M, Kitaguchi H, Kumakura H, Wada H, and Maeda H 1996 *Proc. 16th Int. Cryo. Engr. Conf./ Int. Cryo. Mater. Conf., Mai 20-2/'96 Kitakyusyu* 1099-1102.

[2] Kitaguchi H, Kumakura H, Togano K, Tomita N, Takeda N, and Igata N 1995 *Proc. 4th Jpn. Int. SAMPE Sympo. Sep. 25-28/'95 Tokyo* 495-500.

[3] Kitaguchi H, Kumakura H, Togano K, Kiyoshi T, Inoue K, Maeda H, Tomita N, Kase J, Yanagisawa E, and Kato K 1995 *J. Electronic Mater.* **24** 1883-.1886

[4] Okada M, Tanaka K, Inoue N, Sato J, Kitaguchi H, Kumakura H, Kiyoshi T, Inoue K, and Togano K 1995 *Jpn. J. Appl. Phys.* **34** L981-L984.

[5] Okada M, Tanaka K, Fukusima K, Sato J, Kitaguchi H, Kumakura H, Kiyoshi T, Inoue K, and Togano K 1996 *Jpn. J. Appl. Phys.* **35** L623-L626.

[6] Kitaguchi H, Kumakura H, Togano K, Okada M, Tanaka K, and Sato J, *Cryogenics* (in press).

[7] Tomita N, Arai M, Yanagisawa E, Morimoto T, Kitaguchi H, Kumakura H, Togano K, and Nomura K 1996 *Cryogenics* **36** 485-490.

[8] Kumakura H, Kitaguchi H, Togano K, Kato K, and Sato J 1996 *Advances in Cryogenic Engr.* **42** 777-785.

[9] Duchateau J L, Turck B, Krempasky L, and Polak M *Cryogenics* **16** 97-102.

Inst. Phys. Conf. Ser. No 158
Paper presented at Applied Superconductivity, The Netherlands, 30 June–3 July 1997

Behavior of the magnetic flux within an HTS magnetic shielded cylinder for measuring the biomagnetic field

M. Itoh, K. Mori*, S. Yoshizawa, and S. Haseyama*****

Dept. of Electronic Engi., Kinki Univ., Higashi-Osaka, Osaka 577, Japan
*Div. of Sys. Sci., Kobe Univ., Kobe, Hyogo 657, Japan
**Dept. of Chemistry, Meisei Univ. Hino, Tokyo 191, Japan
***Central Res. Lab., Dowa Mining Co. Ltd., Hachioji, Tokyo 192, Japan

Abstract. The high-critical temperature superconductor (HTS) is an ideal material for use as a magnetic shielding vessel which employs perfect diamagnetism. The behavior of the magnetic flux within the HTS vessel for magnetic fields less than the value of the maximum shielded magnetic field, however, is unknown. The authors have fabricated a new magnetic shielding system, constructed from a BPSCCO cylinder and a special cooling chamber, for use in measuring the biomagnetic field associated with small animals and plants during physical stimulation. The present paper examines the magnetic behavior within the shielded system which include the magnetic noise power spectra (NPS) and the temporal change of the trapped magnetic field. The characteristic of these quantities in the presence of the earth's magnetism are measured by using the HTS dc-SQUID magnetometer. In addition, a discussion is conducted of the influence of the direction of the earth's magnetism on the magnetic NPS during cooling of the shielding system.

1. Introduction

There has been an increasing need for systems of magnetic shields capable of limiting regions to very low magnetic field exposure, such as those associated with biomagnetic measurements. The ideal magnetic shielded vessel (e.g. [1]) can be realized by use of a high-critical temperature superconductor (HTS). Little is known, however, of the characteristics and evaluation procedures for the behavior of a magnetic flux density less than the value of the maximum shielded magnetic flux density (B_s) of the shielded vessel.

New magnetic shielding system, such as those used for measuring the biomagnetic fields of small animals and plants during physical stimulation, have been fabricated from BPSCCO (Bi-Pb-Sr-Ca-Cu-O) cylinder and special cooling chamber such as that schematically illustrated in Fig. 1. In the present research, the magnetic noise power spectra (NPS) are measured with the use of the HTS dc-SQUID magnetometer and spectrum analyzer in order to evaluate the behavior of the magnetic field within the shielded system.

Experimental results reveal several characteristics of the magnetic behavior of the shielded system which include the magnetic NPS, the temporal change of the trapped magnetic flux density (B_{tf}), and the effect of the excited magnetic flux density (B_{ex}) on the magnetic flux density (B_{in}) at the sensor position in the shielded system. Also discussed is the influence of the direction of the earth's magnetism on the magnetic NPS as the magnetic shielding system undergoes cooling.

2. Experimental procedure

Figure 1 schematically illustrate the cross-sectional view of the magnetic shielding system. As was reported in Ref. [2], the BPSCCO cylinder was formed by exerting the powder to a pressure of 1-3 ton/cm^2, making use of the cold isostatic pressing (CIP) method. The material used in construction of the cooling chamber is stainless-steel plated with chromium. The shielding system is cooled over a period of two hours in the presence of the earth's magnetic field with the axial direction of the system aligned either perpendicular or horizontal to the direction of the magnetic field. The system is then placed in the vertical on a fixed platform.

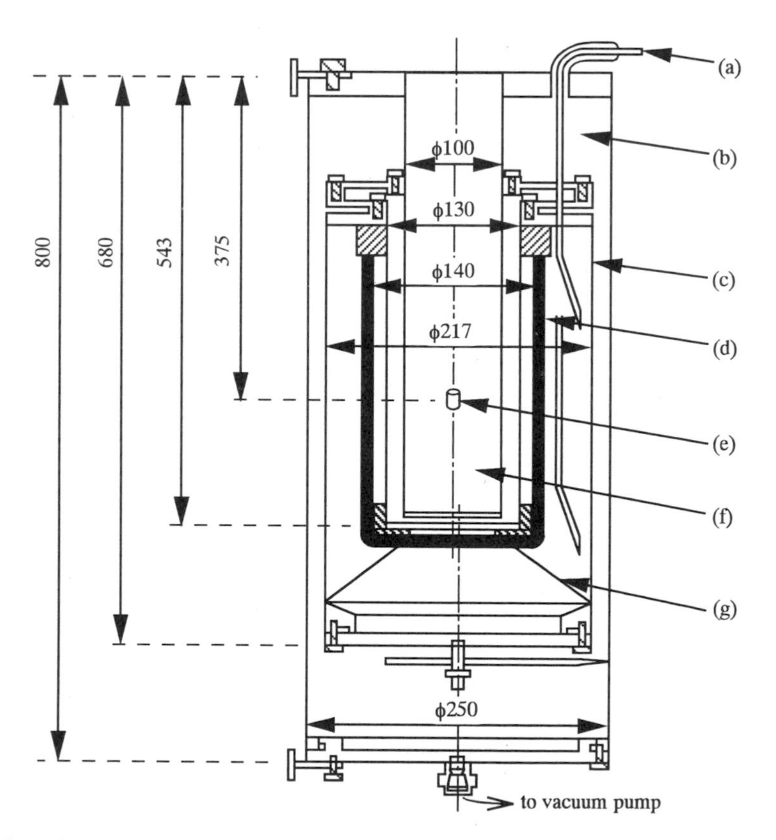

Fig. 1. Schematic cross-section of the magnetic shielding system illustrating (a) the vent for liquid nitrogen, (b) the vacuum, (c) the vessel of liquid nitrogen at 77.4 K, (d) the BPSCCO cylinder, (e) the sensor position (Hall device and SQUID sensor), (f) the shielded space (air), and (g) the holder for the BPSCCO cylinder (d).

Using a calibrated GaAs Hall device such as described in Refs [1] and [3], the characteristics of B_{in} are evaluated for the magnetic shielded system in presence of B_{ex} applied horizontal to the axial direction of the system. Figure 2 illustrates the schematic diagram [4] of the experimental arrangement used to measure the magnetic NPS at the sensor position in the shielding system such as shown in Fig. 1.

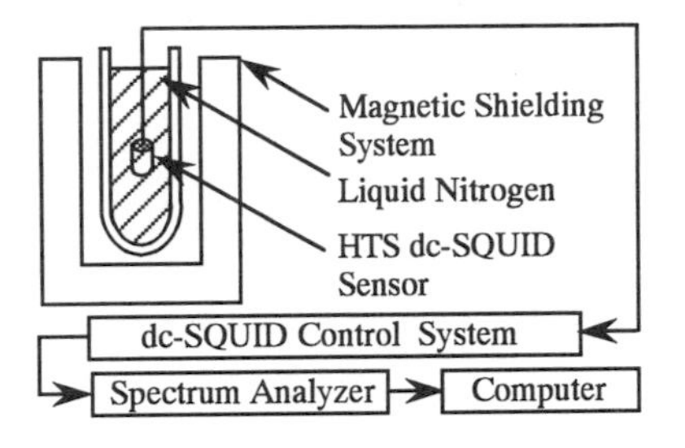

Fig. 2. Schematic diagram of the system used for measuring the magnetic NPS at the sensor position.

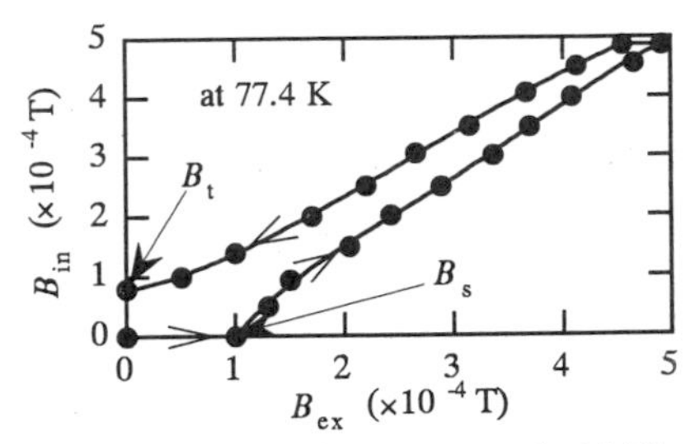

Fig. 3. Characteristics of the magnetic shielding at the sensor position.

3. Results and discussion

Figure 3 shows the characteristics of the magnetic shielding within the shielded system. The maximum shielded and trapped magnetic flux densities, B_s and B_t, were found to be 1.0 gauss and 0.8 gauss, respectively, under temperature conditions of 77.4 K. The value of B_s in the shielded vessel is an important parameter used in the design of magnetic shielding. The upgrading of the HTS vessel has been achieved by use of special techniques [1, 2], making it possible to improve the value of B_s by a factor of about 100.

Figure 4 shows the typical characteristics of the magnetic NPS in the presence of the earth's magnetic field at the sensor position, such as shown in Fig. 1. Figures (a) and (b) represent the results when the axial direction of the system is aligned perpendicular or horizontal to the direction of the earth's magnetic field, termed the perpendicular and horizontal cases, respectively, while the system undergoes cooling. Figure 4 (a) displays the magnetic

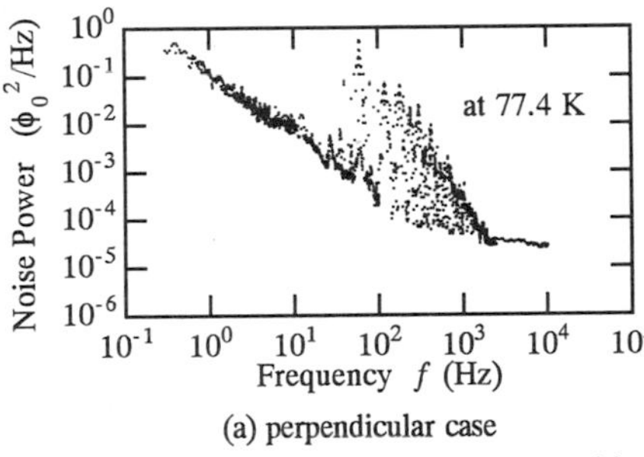

(a) perpendicular case

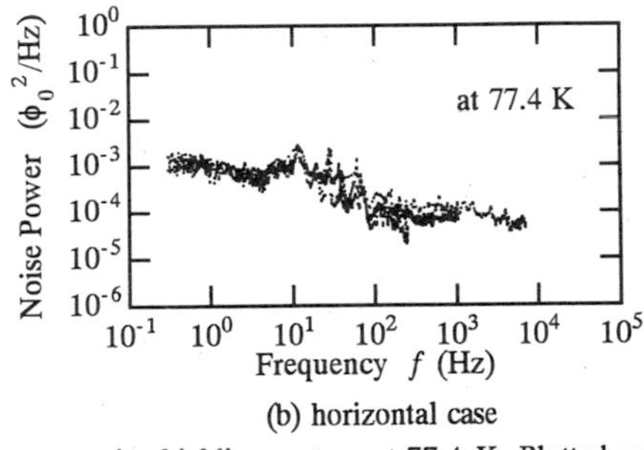

(b) horizontal case

Fig. 4. Magnetic NPS within the sensor position of the magnetic shielding system at 77.4 K. Plotted points in (a) and (b) indicate measurements when the earth's magnetism is perpendicular and horizontal to the axial direction of the magnetic shielding system, respectively, as the system undergoes cooling.

NPS exhibiting numerous high harmonic waves of commercial frequency (60 Hz) with a $1/f$ fluctuation at frequencies ranging from 20 Hz to 2 kHz. Figure 4 (b) reveals a decrease in the high harmonic waves of the magnetic NPS compared with those in Fig. 4 (a). Furthermore, the magnetic NPS at 1 Hz in Fig. 4 (b) are improved by a factor of about 10^{-2} of those in Fig. 4 (a). The magnetic NPS with an application of 0.01-0.25 gauss B_{ex}, exhibited a similar tendency as those in the results obtain in Fig. 4 (b) (not shown). In general, the demagnetization coefficients (D) for perpendicularly and horizontally applied magnetic fields to the cylinder are 1/3 and 1 (e.g. [5]), respectively. In the present results, the value of D for the horizontally applied case is also larger than that of the perpendicularly applied case. Therefore, it is concluded when the axial direction of the shielded system is horizontal to the direction of the applied magnetism, a greater degree of magnetic shielding occurs.

After 50 thermal cycles between room temperature (300 K) and the boiling point of liquid nitrogen (77.4 K), the characteristics of the shielded system, such as shown in Figs. 3, 4 (a), and 4 (b) underwent no significant change in the degree of magnetic shielding.

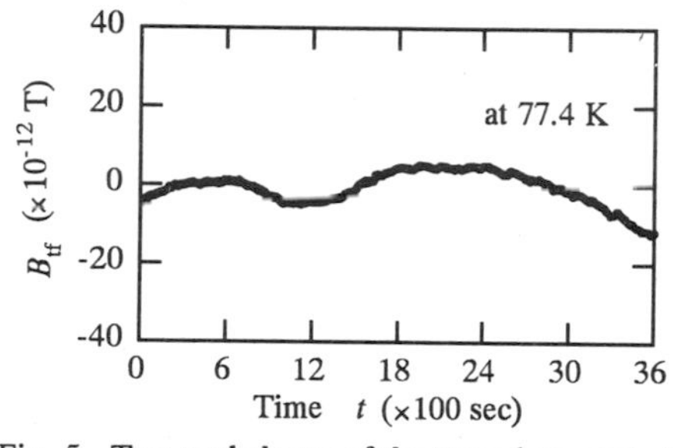

Fig. 5. Temporal change of the trapped magnetic flux density within the magnetic shielded system.

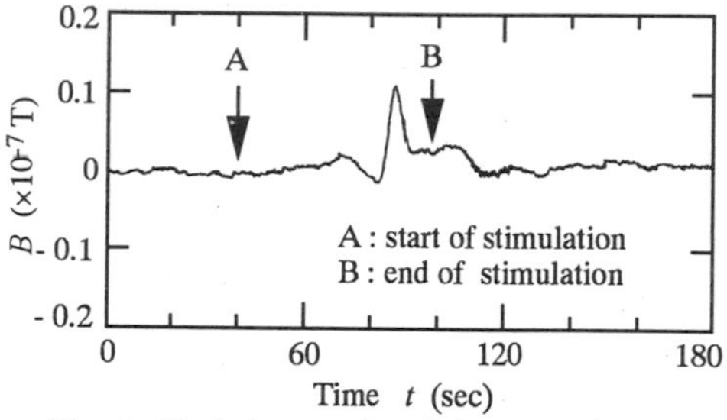

A
B
A : start of stimulation
B : end of stimulation
B ($\times 10^{-7}$ T)
0.2
0.1
0
- 0.1
- 0.2
0 60 120 180
Time t (sec)

Fig. 6. Typical example of the magnetic response of mimosa to physical stimulation.

Figure 5 displays the characteristics of the temporal change (over a period 1 sec to 10^3 sec) of the trapped magnetic flux density (B_{tf}) within the shielded system in the presence of the earth's magnetic field obtained using a HTS dc-SQUID magnetometer. The results are for the case when the axial direction of the system is horizontal to the earth's magnetic field, being known to exhibit very stable characteristics in the fluctuation of the trapped magnetic flux. As a typical example, Fig. 6 shows the magnetic response (B) for a mimosa plant to stimulation by a bamboo needle, which is measured by a flux gate meter.

4. Conclusions

One of the basic areas of research is the fabrication of large-sized HTS vessels and the improvement in their magnetic shielding effects. For these purposes, it is important to understand the behavior of the trapped magnetic field. To accomplish this, such characteristics as the magnetic noise power spectra and the temporal change of the trapped magnetic field are examined. On particular, it was found that when the axial direction of the shielding system is aligned horizontal to the earth's magnetic field during cooling of the system, a greater degree of magnetic shielding will occur. These results are important criteria, fundamental in the design of effective and reliable magnetic shielding for large-sized vessels. From the present results, the authors are investigating applicable methods for the future fabrication of large-sized vessels having better shielding characteristics.

References

[1] Itoh M Ohyama T Minemoto T Numata K and Hoshino K 1992 *J. Physics D: Appl. Phys*. **25** 1630-34

[2] Ishikawa Y Yoshizawa S Tenya K and Miyajima H 1991 Advances in Superconductivity IV, 1073-1076 (Springer-Verlag, Tokyo)

[3] Itoh M 1992 *Advances in High Temperature Superconductivity*, edited by Andreone D Connelli R S and Mezzetti E 232-241 (Singapore, World Scientific Publishing)

[4] Itoh M Iguchi S Minemoto T and Yoshizawa S 1997 *IEEE Trans. Appl. Superconductivity* **7**, in printing

[5] Lynton E A 1969 *Superconductivity* (London, Methuen)

Inst. Phys. Conf. Ser. No 158
Paper presented at Applied Superconductivity, The Netherlands, 30 June–3 July 1997

Normal zone propagation studies on a single pancake coil of multifilamentry BSCCO-2223 tape operating at 65K

M Penny*, C Beduz, Y Yang, S Manton and R Wroe+

Institute of Cryogenics, University of Southampton, Southampton, SO17 1BJ. U.K.
+E.A. Technology, Capenhurst, Chester, CH1 6ES, U.K.

Abstract. Normal-zone propagation and stability measurements have been performed on a single pancake coil made from BSCCO-2223 multifilamentry tape. The coil composite was held above a subcooled nitrogen bath at 65K. The majority of the cooling was by conduction of the copper-links which thermally connect the coil's inner and outer turns to the bath temperature. The combination of impregnated single pancake and cooling conditions simulate a long potted solenoid, where the cooling of the centre turn was dominated by the low thermal conductivity between adjacent turns ($k_{r=}$ 0.8$Wm^{-1}K^{-1}$@77K). The normal zone stability was investigated using heated regions of various lengths and power inputs. We have found that the total energy required to initiate a minimum propagation zone (E_{mpz}) reduces as the length of the heated region decreases. For example, E_{mpz}=2.8J and 4.7J for heater lengths of 2.5cm and 14.6cm respectively. The velocity of the normal zone in the transverse direction was 1.2mms^{-1} after a energy pulse of 5.7J was applied over a 12.5cm length. If the zone was undetected then the quench would have damaged the coil by ohmic heating. Therefore, careful magnet design may be required for HT_c devices with the further development of high-I_c superconductors.

1. Introduction

The normal zone propagation velocity is useful to quantify the natural ability of a superconductor to protect itself from a quench. A very slow propagation velocity can result in damage to the superconducting coil as the large stored energy is dissipated in a small volume of the coil[1]. In general, a high velocity implies that the quench volume grows rapidly which enables the stored electromagnetic energy to be dissipated over a larger volume and thus reduce the risk of damage.

The thermal behaviour of superconductors is primarily controlled by the heat generation term of the matrix, ($\rho_m(T)$). Therefore, the behaviours of low-T_c and high-T_c superconductors during a quench are different, chiefly because relevant material properties are greatly temperature dependent. For example, the specific heat capacity (C_p) of a typical matrix material increases by three orders of magnitude[2] from 4.2K to 100K. Thus, much more energy is required to raise the temperature by a given amount for a superconductor operating at 77K compared to 4.2K. For this reason many magnet designers interpret this property as an indication of higher stability of high-T_c over the low-T_c materials[3-5]. For high-T_c superconductors the propagation speed of the normal zone is slower than for low-T_c superconductors[4-6], which could cause problems of detection and protection. At

* CASE studentship supported by E.A. Technology and EPSRC

temperatures below 30K, quench studies have been possible due to the lower C_p of the sheath material and higher critical current densities. Above 30K, the study of propagating normal zones has been less frequently investigated because of the low J_c values of BSCCO tapes. The experimental work presented in this paper utilises improvements of J_c at 65K and low transverse thermal conductivity.

2. Coil fabrication and characterisation

A substrate was made from a copper plate wound and fixed onto a tufnol ring of Ø=52mm. The copper plate acts as a current lead and a thermal link to the liquid nitrogen. 5 metres of silver sheathed BSCCO-2223 tape (produced by IGC) was wound inductively with an insulating fibreglass cloth layer. After half of the tape had been wound, a constantan heater strip was placed in the coil. In addition, voltage taps were soldered to the edge of the superconducting tape and differential thermocouples were placed in between the tape and the fibre-glass at various positions. On the outer turn another copper plate was soldered to act as a current lead. The coil was then vacuum impregnated with stycast. The single pancake coil then simulates a long potted solenoid cooled by wetted surfaces. The characteristics of the coil are shown in Table 1.

Table 1. Parameters of the test coil

Conductor type	37-multifilamentry BSCCO-2223 tape
Matrix/ superconductor ratio $(1-\lambda)/\lambda$	1.9
Critical current I_c ($1\mu Vcm^{-1}$ @65K)	30.5 A
Critical Temperature T_c	110K
Resistivity of Ag ρ_m @77K	3.7×10^{-9} Ωm
Transverse thermal cond. of coil k_r @77K	$0.8 Wm^{-1}K^{-1}$

Longitudinal thermal cond. of tape k_z @77K	$224\ Wm^{-1}K^{-1}$
Insulator material	Fibreglass cloth thickness=120μm
Dimensions of tape	220μm x 0.35cm x 5m
Coil internal Ø	5.2 cm
Coil external Ø	8 cm
Fill factor of winding λ_w	0.5

Generally, superconducting multifilamentry tape carries an inhomogeneous transport current along its length which can be shown in the V-I characteristics in Fig 1. This highlights the broad current sharing region and makes the criterion of I_c for a long tape difficult to determine. The reason for the tapes characteristics is due to the thermomechanical process. The Powder-in-Tube method produces filaments which can close and/or join and results in weak regions being distributed along the conductors length. When a magnet is wound with such a conductor, a transport current determined by ($1\mu Vcm^{-1}$) would be carried by the majority of 'good' regions. However, some hot spots may appear due to some 'bad' regions produced by these weak regions.

3. Experimental details

A vacuum pump was used to reduce the pressure within the dewar and produce a bath temperature of 65K. The coil composite was suspended 2cm above the nitrogen bath and the peripheral copper links were submerged into the liquid nitrogen, see Fig 2. The current leads were connected below the nitrogen level to the copper plates. A thermometer diode was attached to the inner copper link in order to monitor the bath temperature and to provide a reference temperature for the differential thermocouples within the coil. The coil carried a 30A transport current and allowed to reach a steady-state temperature of 65.3K before any heat pulse was applied.

A pulse generator was amplified to various power settings and the electrical pulse was applied to a constantan heater. The heat pulse was adjusted to various energy levels per unit length depending the spatial distance between current leads on the heater(2.5, 4.6, 7.8, 9.6, 12.5 and 14.6cm). The minimum energy required to quench the coil (E_{mpz}) as a function of heater length was investigated. Also the propagation velocity in the transverse direction was measured for a heat pulse of 5.7J applied to a 12.5cm heater. The velocity was estimated by monitoring two differential thermocouples mounted adjacent to the heater turn and the other 5mm away in the transverse direction.

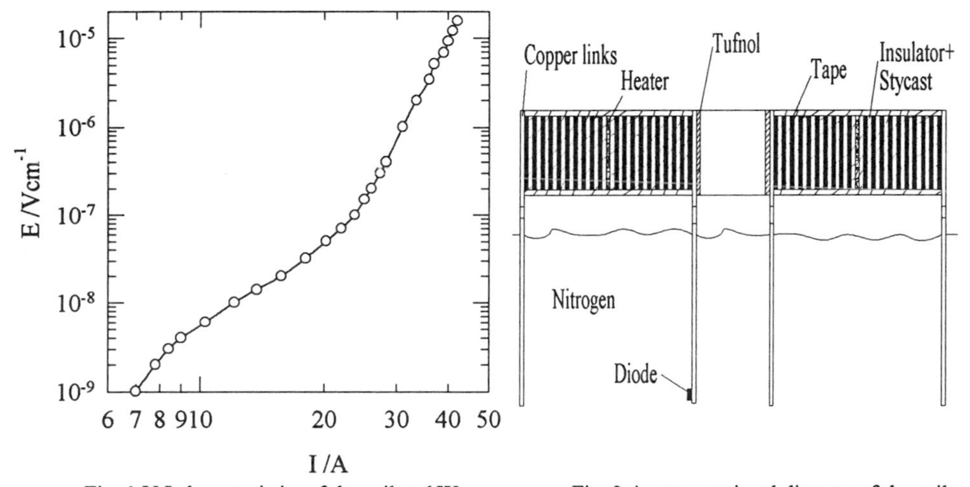

Fig. 1 V-I characteristics of the coil at 65K

Fig. 2 A cross-sectional diagram of the coil

4. Results and discussion

The minimum energy required to quench the coil for each heated region are shown in Fig. 3. It can be seen that small heated regions require less energy to reach a minimum propagation zone (MPZ) than longer regions. This trend cannot be explained by the adiabatic model proposed by Wipf[7] and Wilson[8,9] which predicts a constant energy associated with a MPZ. The adiabatic model assumes a constant heat generation above T_c which is a good approximation due to the temperature dependence of $\rho(T)$ being negligible in the operating temperatures $4.2<T_{op}<40K$. However, for HT_c materials and $T_{op}>65K$, the resistivity has a strong temperature dependence and therefore the heat generation is no longer constant but exponential above T_c. This difference in the heat generation due to $\rho(T)$ explains the results of small heater lengths requiring less total energy than longer lengths, only providing the temperature increase of the heat pulse is sufficient to benefit from $\rho(T>65K)$.

Fig. 4 shows the temperature of two thermocouples as a function of time after a 5.7J of heat was applied over 12.5cm. Using the criterion of T_c to determine a normal state, the normal zone 'appeared' at the superconducting tape adjacent to the heater 66s after the initial heat pulse. The normal zone then propagates transversely and reaches the second thermocouple 4s later, transverse normal zone propagation velocity can be estimated to be $1.2mms^{-1}$.

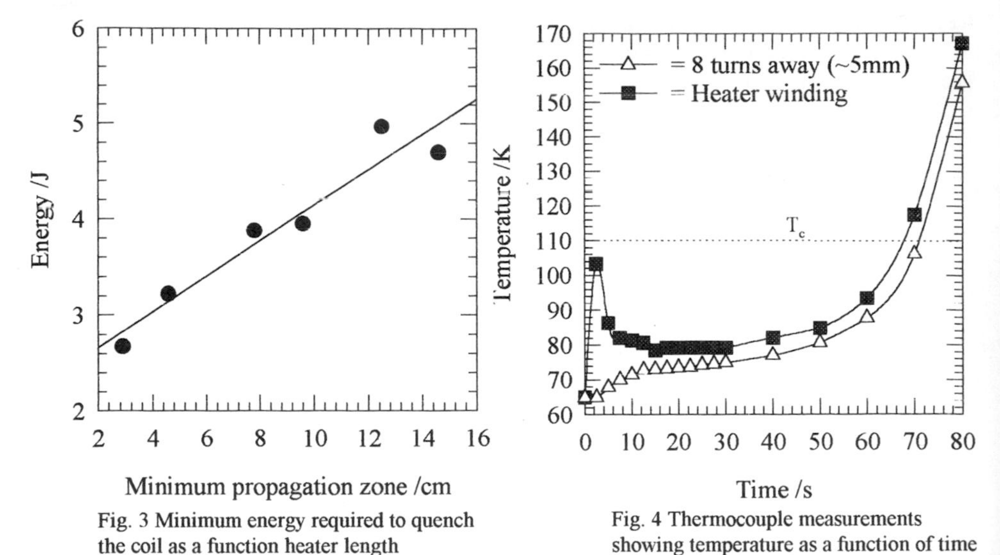

Fig. 3 Minimum energy required to quench the coil as a function heater length

Fig. 4 Thermocouple measurements showing temperature as a function of time

4. Conclusion

In this paper we have shown a normal/superconducting boundary propagating transversely through a pancake coil operating at 65K and I_{op}~I_c=30A. The slow propagation velocity of $1.2mms^{-1}$ would damage the coil if undetected. The results show that the minimum energy required to quench the coil reduces as a function of heater length. Although the energy required to quench was large E_{mpz} >2.8J, the results suggest that careful magnet design and attention to the cooling regime will be required to ensure safe operation.

Acknowledgements

M. Penny has an EPSRC CASE award supported by E.A. Technology Ltd and S. Manton has an EPSRC award.

References

[1] C Oberly et al 1992 Adv. Cryo. Eng. **38B** 479-489
[2] R Weast & M. Astle 1979 "CRC handbook of Chemistry and Physics 59th edition" D220
[3] Y Iwasa 1993 Proc 6th Conf on Superconductivity and its applications 445-454
[4] Y Iwasa 1993 Proc 5th Int Symp Superconductivity (ISS'92) Tokyo, Japan 1205-1210
[5] R Bellis and Y Iwasa 1994 Cryogenics **34** 129-144
[6] Y Iwasa 1993 IEEE Trans Magn MAG-**24** 1211-1214
[7] S Wipf and A Martinelli 1972 IEEE **72** 331-340
[8] M Wilson and Y Iwasa 1978 Cryogenics **18** 12-25
[9] M Wilson 1991 Cryogenics **31** 499-503

Inst. Phys. Conf. Ser. No 158
Paper presented at Applied Superconductivity, The Netherlands, 30 June–3 July 1997

Loss reduction in a high quality superconducting coil

O. A. Shevchenko, H. G. Knoopers and H. H. J. ten Kate

Faculty of Applied Physics, University of Twente, Enschede, the Netherlands

G. C. Damstra

Technical University Eindhoven, Eindhoven, The Netherlands

Abstract. A potential application of high temperature superconductors is in AC coils for energy related purposes. Of particular interest is to evaluate the BSCCO tape conductors now available and to find optimum coil design in order to maximise the transport properties and thus the device performance. We analysed available options to reduce the main loss components of AC coils by which the focus is on the design of a high-T_c superconducting coil for generating high voltages. Important issues are comparison and selection of optimum parameters of the single tape, arrangements of conductor in the coil, optimisation of the coil design and the operating conditions. The ratio between the impedance and the active resistance of the coil is expected to be as high as 1000 for an optimal design for use at 50-60 Hz . The results of a scale model experiment and numerical simulations are presented.

1. Introduction

Superconducting (sc.) coils made with High-T_c superconductors and operating at 77 K in liquid Nitrogen may be used in resonant circuits as it is depicted in Figure 1 to generate high voltage at 50-60 Hz [1]. The options available to reduce the loss in such HTS AC coils are reviewed. A number of coils with HTS manufactured during the last years reflects the progress made in the tape development and the growing interest of potential customers. Up to now less attention has been paid to resonant inductors. When compared to a transformer coil, an important property of the inductor is that no iron core is allowed because of mass constrains. To arrive at a quality factor of the coil exceeding 1000 is the key issue for the success of this type of application of HTS.

2. HTS tape as coil conductor

When for power applications the operating temperature has to exceed 63 K, there are currently no serious alternative superconductors to BSCCO/Ag tapes. Tapes from different vendors vary in price, size, critical current, characteristic field and loss. It is not always trivial how to compare them and to choose the best. A high price of a tape is a major obstacle preventing applications. The return of investment time of the bare tape exceeds 10 years by far [2, 3]. A cost reduction of several times is required to make HTS outside the laboratory really competitive in especially AC applications. The commercially available long length tapes are basically DC conductors. To make available a good AC conductor exhibiting many fine, de-coupled and twisted filaments is subject of ongoing research and will take years.

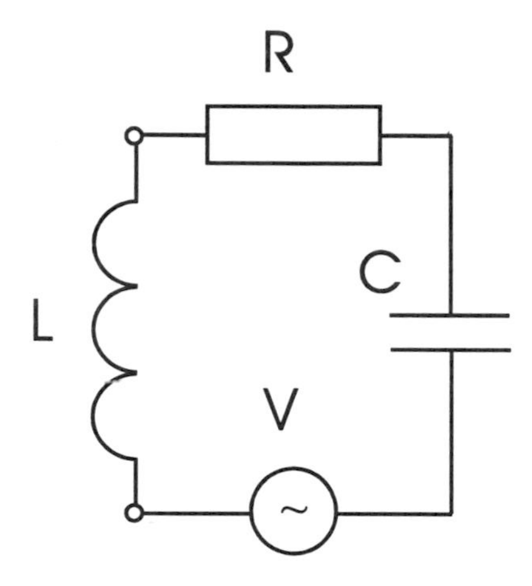

Figure 1 A resonant circuit for high voltage generation: $L = 0.3$ H, $C = 34$ μF, $R = 94$ mΩ, $V = 20$ V, $I_{max} = 100$ A.

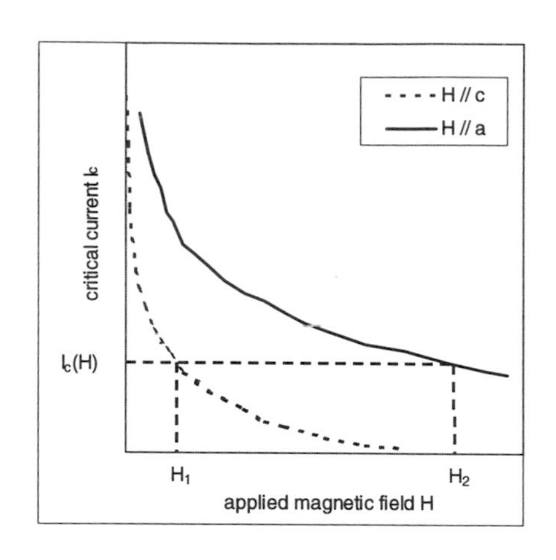

Figure 2 Typical experimental $I_c(H)$ curve of an HTS tape at a fixed temperature for two directions of the applied field H // c and H // ab

Characteristic dimensions of the tapes available are: width w ranging from 3 to 4 mm; thickness t ranging from 0.2 to 0.3 mm [4]; and an aspect ratio ranging from 10 to 20. The critical current of a typical tape depends strongly on the mutual orientation of the external magnetic field and the grain alignment as illustrated in Figure 2. For the same tape at 77 K the ratio H_2 / H_1 at the same critical current $I_c(H)$, ranges from 5 to 10 for a series of tapes examined in the field range from 0.01 to 1 T. The same ratio of the axial to radial field components in a coil must be provided to use the tape more efficiently.

Several tapes can be connected in parallel to increase the operating current. The high stability offered by HTS enables insulation of the tapes. When transposed properly it will bring the specific loss of the cable/conductor down to that one of a single tape [5].

3. Superconducting coil

A new problem arises from the anisotropy of the tape (only circular coils are considered below). The perpendicular field component i.e. the radial field in the coil must be made minimal. For DC applications an analysis was done leading to rather unpractical results [6-7]. When playing with the shape of an air-core coil, there is no escape: the ratio of the maximum axial B_z and the radial B_r field in the windings is about two at best with a strong conflict with respect to the amount of the tape required.

Ferro-magnetic material positioned at a coil edge will affect the radial field. A simple way is to consider a set of coaxial coils with equal dimensions. The application of iron shaping the B-field can reduce the AC loss to a quarter (or less) in the case of an air core coil just by using proper boundary conditions. This is illustrated in Figure 3, where the coil on the right side is replaced by the imaginary one taking Neuman condition along the boundary s. The edge problem can be analysed analytically. Options for the basic shape of the HTS coil first are the following.

Pancake. When the anisotropy is not relevant, a pancake type of coil requires somewhat less tape than a solenoid, edge cooling is almost perfect, but more conductor is located at the edge.

Solenoid. A reduction of the average magnetic field in combination with minimal cost, implies a large diameter, long and thin solenoidal coil. Both B_r and B_z rise primarily with the average field and the thickness of the coil and only slightly with the length.

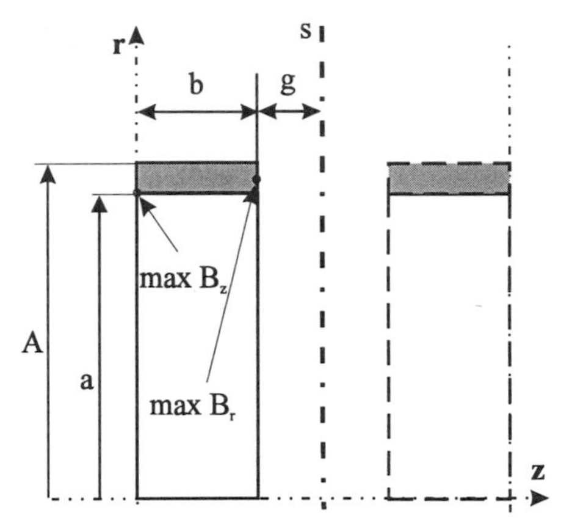

Figure 3 Model to calculate the radial field component versus the edge air gap size between coaxial solenoids of equal size.

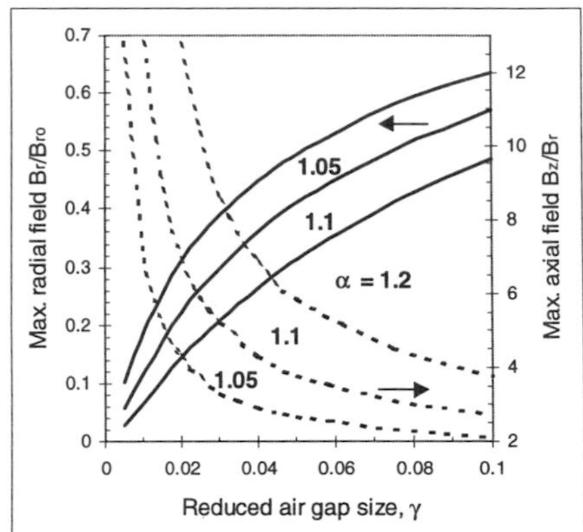

Figure 4 Radial field component in the windings versus the edge air gap size between coaxial solenoids of equal size ($\beta = 3$).

To keep the temperature within the defined limits a cooling channel per each few layers is required. *Toroid.* A toroid would require 2 - 3 times more conductor than a solenoid. It is complicated to produce and it has a higher field in the windings [8]. This leaves us with a solenoid.

A numerically analysed solution is shown in Figure 4 with: $\alpha = A/a$; $\beta = b/a$; $\gamma = g/a$; approximate locations of B_r and B_z are depicted in Figure 2; B_{r0} is the maximum field component of a single coil, $\gamma \to \infty$. The ratio B_z/B_r is plotted in dashed lines using the right axis. It is possible to reach the ratio B_z/B_r ~5...10 with an air gap size of a few mm (when $2a = 0.1$ m, $g = 5$ mm for $\gamma = 0.1$). It will not be easy to make a solenoid by stacking pancakes. Inevitably in this case air gaps of a few mm are between them and multiple joints cancel expected benefits. The layer winding technique allows to reduce the air gap between two adjacent turns in a layer to ~ 0.1 mm and may be preferable [9].

4. Practical example

The solution to use iron close to an AC coil requires a specific optimisation as eddy current loss in the iron can cancel the gain from the reduction of B_r. A 3-D lamination of iron is possible, but not easy. A simplifying solution proposed here is the following. The iron cup at the coil edges can be comprised from multiple strip wound cut C-cores (or similar). Such elements are commercially available. The arrangement is depicted in Figure 5. The concept has been verified using a small coil at 77 K. The tape (Showa [4]) used has dimensions: width 2.9 mm, thickness 0.2 mm and exhibits in self-field and 77 K a critical current of 21.4 A. The tape is wet-wound on a G-10 former using STYCAST 2850FT and thin glass-cloth, comprising a solenoid with $a = 47$ mm, $A = 48.8$ mm, $b = 9.4$ mm; $\alpha = 1.04$; $\beta = 0.2$. It has 7 layers and 44 turns in total.

The bare air-core coil has a well-expressed edge limitation, $B_z/B_{r0} = 0.9$ that limits the operating current to $I_o = 10$ A and a dominant loss contribution due to B_{r0}. An (almost) infinite sheet of iron with a thickness of 0.6 mm added to each coil edge ($g \approx 0.25$ mm) will give at the same DC resistive voltage across the coil an increase of the maximum operating current from 10 to 16 A. This solution provides $B_z/B_{r0} = 6.4$, a value very close to the calculated value. An end cup of the same total thickness, $c = 2$ mm, $d = 2.4$ mm and a shape as shown in Figure 5 will yield $B_z/B_{r0} = 8.7$, with further reduced loss. The estimates given in Table 1 agree well with the results in Figure 4.

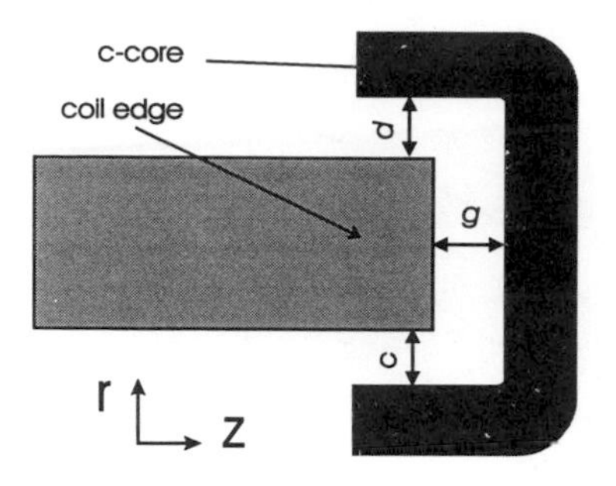

Table 1 Field and current in the coil windings

Arrange-ment of the coil	B_z [mT]	B_r [mT]	I_o [A]	I_{cmin} [A]
Air core	15.2	17.3	10	10
Iron flange	25.8	4	15	17
Iron end cup	26	3	15	18

Figure 5 A 2D model for the field calculation at the coil edge (principle arrangement, the coil former and most of the coil in z direction are not shown).

5. Concluding remarks

Often the magnetisation loss [10] dominates in a HTS coil. When an air-core coil is designed, the dominant loss is due to the radial field component causing a quality factor of 250 at best. When the edge of a solenoid is modified with iron the dominant loss component is due to the axial field with a quality factor well above 1000. A further increase of the quality factor can be achieved by de-coupling the filaments in the tape (by introducing resistive barriers and twisting).

- ✓ A strategy for a more optimum use of HTS tape in a coil is presented.
- ✓ Options to increase the quality factor of a MVA class resonator coil to values well above 1000 are discussed.
- ✓ A large diameter, long and thin solenoid with edges reinforced by ferro-magnetic cups is proposed for this purpose.

Acknowledgement

This research is supported by the Netherlands Technology Foundation (STW).

References

[1] Jin J X Dou S X Liu H K and Grantham C 1996 paper LSB-4 presented at Applied Sc. Conference

[2] Superconductor Week, 1997 Jan 13 2-3; Superconductor Industry 1997 **10** N 1 18-23

[3] Gamble B B Snitchler G L and Schwall R E 1996 IEEE Trans on Magn **32** N 4 2714-2719

[4] **1:** American Superconductor Corp. (USA); **2**: Intermagnetics General Corp. (USA); **3**: Showa Electric Wire & Cable Co Ltd (Japan); **4**: University of Geneva (Switzerland); **5**: Sumitomo Electric Industries Ltd (Japan; tape width 10; thickness 1 mm); **6**:Plastronic Inc. (USA); **7**: VACUUMSCHMELZE GMBH (Germany) references on request

[5] Iwakuma M *et al* 1997 Proc. of the 16th ICEC/ICMC Kitakyushu Japan 20-24 May 1996 **2** 1325-1328

[6] Selvaggi J A *et al* 1996 Cryogenics **36** N 7 555-558

[7] Pitel J and Kovac P Superc Sci and Techn 1997 **10** 4 7-16 and the refs within; this Conf paper 10Ge-44

[8] Mulder G B J PhD 1988 University of Twente Enschede The Netherlands

[9] Funaki K *et al* ibid. [5] 1009-1012

[10] Oomen M P Rieger J Leghissa M and ten Kate H H J submitted to Physica C

Inst. Phys. Conf. Ser. No 158
Paper presented at Applied Superconductivity, The Netherlands, 30 June–3 July 1997

3-Dimensional Levitation Force Measurements in the 15 - 85 K Range

S. Gauss, J. H. Albering

Hoechst AG, Corporate Research & Technology, D-65926 Frankfurt / M, Germany

Abstract. Levitation applications are not only limited to an operation temperature of 77 K. In the field of LH_2 (T=21 K) as a new fuel technology several levitation and bearing applications have to be considered. We established an experimental set up to determine the levitation properties in x-, y-, and z-direction simultanously in the range of 15-85 K. The 3-dimensional repulsion forces and the stiffness both in ZFC and in FC mode are presented. These results are the basis for an optimized bearing design at LH_2 temperatures.

1. Introduction

Magnetic bearings based on high temperature superconductors (HTS) have made strong progress in the last years [1-7]. This is based on the effect of autostability and very low friction in these systems. In the meantime several engineering prototypes have been constructed and tested. They are all focused on applications with high rotation speeds, low losses or abrasion freeness, like flywheels or transportation systems in clean room facilities. In addition these kinds of levitation systems may also be used as a suspension in vacuum without mechanical support. In this case the residual time of liquid gases in their storage tank can be extended. The levitation forces between a permanent magnet (PM) and the HTS bulk part are the basis for the design of all kinds of bearings and suspensions.

Most of these measurements are performed at 77 K for materials based on YBCO, but the use of cheap and reliable cryocoolers allow the operation at lower temperatures too. For the use in liquid gas systems or containers (tanks of LHe, LH_2 or LNe) the boiling temperatures of these gases are the operation temperature and the cooling is inherently integrated and no additional cooling costs or cooling equipment have to be considered. In the case of LH_2 the operation temperature will be in the range of 20 - 25 K. Little is known about the levitation properties of different HTS materials in this temperature range [6]. In this study a 3-dimensional levitation force measurement equipment operating in the range of 15-85 K is presented and basic results will be discussed.

2. Experimental Set Up

To obtain three dimensional levitation force measurements at variable temperatures a new test bench has to be installed. This consists of 3 different subsystems. The temperature variable sample holder, the mechanic translation system with the force sensors, the electronic control and data acquisition system. In difference to systems with three dimensional force measuremens at 77 K the sample is in our case fixed and the magnet including force sensors has to be moved.

The sample is mounted on the cold finger of a two stage Gifford-McMahon cryocooler (CTI) allowing levitation force measurements in the range of 15 K to T_c.

To improve the thermal conditions an additional copper shield was fixed on the second cooling stage in the insulation vacuum. Two Si-diodes at the top and the bottom of the sample continuously monitor the temperature of the HTS part. Only small differences of less than 1.0 K during a measuring cycle have been observed. This shows the uniform temperature distribution at and around the sample. During cool down the different thermal properties of the different HTS materials could be observed. For YBCO the cooling from RT to 20 K is performed within 3 hours. A quartz glass window was placed in the vacuum vessel just above the superconductor to minimise the influence on the flux distribution.
The upper surface of the HTS part has to have a fixed position to give a distinct distance to the magnet. Therefore the sample is fixed on a holder with variable height. The size of the superconductor can range up to 20 mm in height and 38 mm in diameter. In addition the samples must be fixed to withstand lateral forces. Due to the insulation (vacuum, copper shield, glass window) the minimum distance between sample surface and magnet is 4.6 mm. This shows the effectiveness of this insulation system between 15 K and room temperature. The sample temperature is computer controlled (Lake Shore) with four heaters at each stage. The magnet is mounted on a tripod fixed on three linear translation slides (see below). As a standard a $SmCo_5$ magnet (0.34 T, 25 mm Ø, 15 mm high) is used but also other magnets or magnet systems can easily be mounted. The forces between HTS and magnet are detected by three strain gauge force sensors (<0.15% accurancy) at the end of the tripod. The mechanical connection is done by a gimbal suspension to decouple the forces from the geometric axis.

The linear translation system has to withstand much higher forces than systems operated at 77 K. Three translation slides (ISEL) consisting of stepper motors with a resolution of 25 μm as they are used for CNC-machine tools have been installed. To improve the mechanic stability the y-axis is constructed like a portal with drives on each side. The maximum translation range is +/- 310 mm in the x-, +/- 160 mm in y- and 220 mm in z-direction, respectively. The speed of the translation can be varied between 0.8 and 250 m/sec, usually 0.8 mm/sec was used. Due to the high forces both in axial and radial direction the stiffness and the damping of the whole set up had to be controlled. Extended tests and further improvements of the mechanical properties led to a stiffness of 100 N/mm in radial directions and 200 N/mm in axial directions. This mechanical stiffness is limited by the tripod and the translation slides while the cold finger exhibit higher values.

3. Results and Discussion

Levitation force measurements are usually performed at liquid nitrogen temperature. Using cryocoolers or other liquid gases lower operation temperatures of magnetic bearings have to be considered. Figure 1 shows the temperature dependence between 15 K and 85 K of the repulsion force F_z at 4.6 mm distance of a single domain YBCO sample and a Melt Cast Processed (MCP) BSCCO 2212 sample in the zero field cooled (ZFC) mode. Both materials had the shape of a 30 mm diameter cylinder with 15 mm (YBCO) and 5 mm (BSCCO) thickness. The preparation and microstructure of these samples is described elsewhere [8-10]. For this measurement the standard $SmCo_5$ magnet was used. Lowering the operation temperature increased F_z (4.6 mm) only from 20.5 N (77 K) to 26.2 N (15.6 K) at YBCO. For the BSCCO sample significant levitation forces are observed for temperatures lower than 65 K. Further decreasing temperatures led to a fast increase of the forces, especially at near distances, i.e. higher fields. These results are a consequence of the temperature dependence of j_c in both materials where circular currents are not limited by grain boundaries.

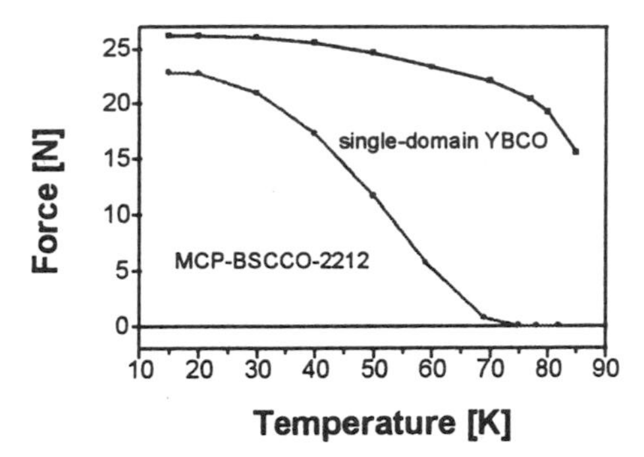

Fig.1 Temperature depence of F_Z (5mm) for two HTS materials. Both samples had 30mm Ø

Most bearing designs are based on a field cooling (FC) of the HTS material. In this case no force is observed as long as the position of the magnet is unchanged. Both an axial and a radial movement results in attractive forces to restore the cooling position. Figure 2 presents the axial force F_z in dependence of the radial x-y movement of the magnet. Here a 30 mm Ø single domain YBCO cylinder was cooled to 20 K. The magnet was then scanned at a level of 5 mm above the surface of the sample. High axial levitation forces are created by this movement.
For the design of a bearing the stiffness of the HTS-magnet set up is very important. This property has to be compared with the stiffness of conventional bearing systems and should be higher than 100 N/m. As it is known from conventional levitation force vs. distance measurements the stiffness will increase as the distance HTS-magnet be comes smaller. The stiffness for lateral movements of the same sample is shown in Figure 3. For our magnet a stiffness of up to -900 N/m was observed after FC at 20 K.

The absolute values of the lateral forces due to movements in x- and y-direction for the same YBCO sample are given in Figure 4. Again the results of a FC mode are presented. For small deviations of the HTS-PM system small forces are observed, raising up to a maximum at 6.8 Nwhen the center of the magnet is at the boundary of the cylindrical (30 mm Ø) sample and disappear when the magnet is totally outside the sample. The results presented here are mainly on single domain YBCO parts, operation at lower temperatures enables the use of various HTS materials. The same experiments have also be performed on other HTS parts, the results are given elsewhere [10].

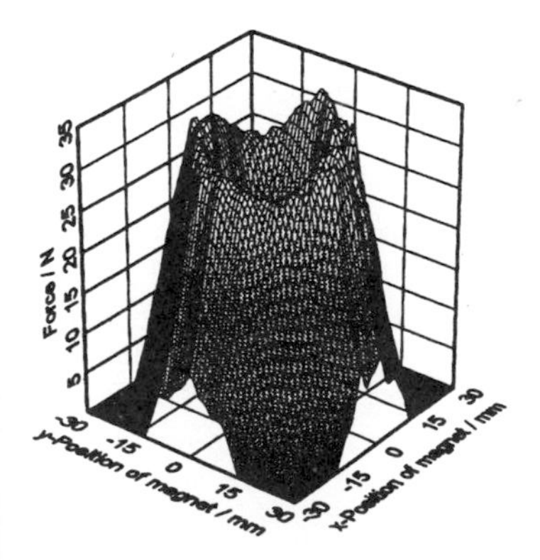

Fig.2 Distribution of Fz at 5mm distance above a 30mm Ø YBCO cylinder after field cooling (20K)

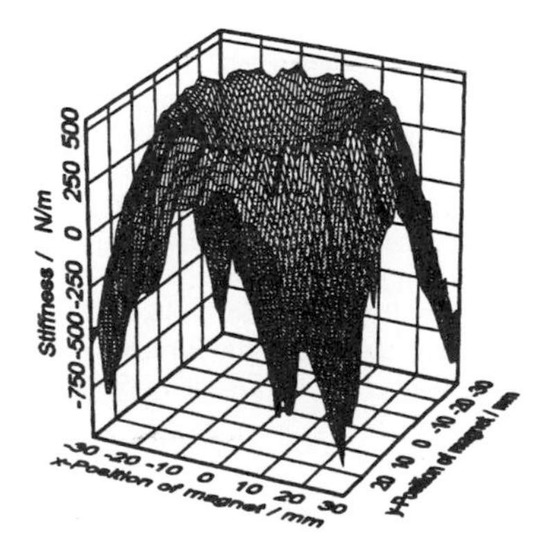

Fig.3 Lateral stiffness at 20 K for x-y movements at 5 mm distance for the same sample

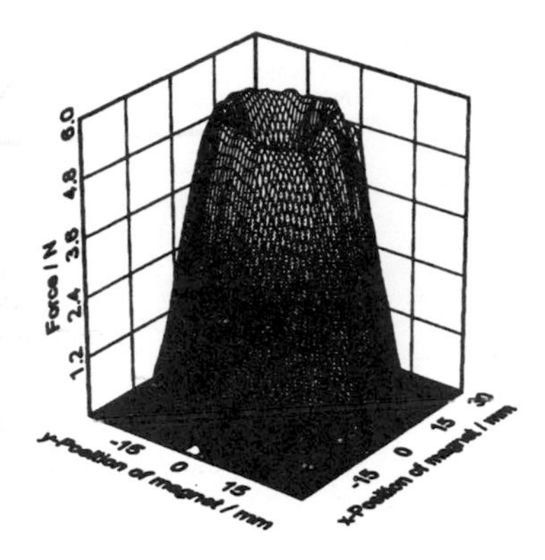

Fig. 4 Lateral forces at 20 K after field cooling at 5 mm distance HTS-magnet.

4. Conclusions

The experimental set up and some results of 3-dimensional levitation force measurements of HTS materials at temperatures from 15 K to 85 K have been reported for the first time. Further data are presented at this conference [10]. The obtained results are important for an optimum design and construction of magnetic bearings using HTS. In particular the reaction due to lateral movements and operation after FC has to be analysed. The operation at lower temperatures enables the use of various materials and simultaneously the overall basic properties of materials (weak link behaviour, j_c (T)) can be investigated.

Acknowledgements

The authors would like to thank Dr. J. Bock for supplying the BSCCO 2212 samples, H. May (TU Braunschweig) for the supply of two magnet systems. This study was partially funded by the German BMBF under contract 13N6939/9.

References

1. F.C. Moon et al., Appl. Supercond. 1, 1175 (1993)
2. M. Itoh, H, Ishigaki, IEEE Trans. Magn. 24, 732 (1989)
3. W.K. Chu et al. Appl. Supercond. 1, 1259 (1993)
4. H. Furuyama et al., Adv. Supercond. IV, Springer Verlag, p. 569 (1992)
5. H. Takaichi et al., Adv. Supercond.V, Springer Verlag, p.775 (1993)
6. S. Gauss, S. Elschner, Proc. 4th Intern. Symposium on Magn. Bearings, Zürich (1994)
7. P. Boegler et al., Appl. Supercond. 2, 315 (1994)
8. J. Bock et al., Inst. Phys. Conf. Ser., 148, P.67
9. J. Bock et al., Adv. Supercond. III, Springer Verlag, p.797 (1991)
10. J.H. Albering et al., Proc. EUCAS'97, (1997)

Inst. Phys. Conf. Ser. No 158
Paper presented at Applied Superconductivity, The Netherlands, 30 June–3 July 1997

The Design and Fabrication of a 6 Tesla EBIT Solenoid

M. A. Green[a], S. M. Dardin[a], R. E. Marrs[b], E. Magee[b], S. K. Mukhergee[a]

a Lawrence Berkeley National Laboratory, Berkeley CA 94720, USA
b Lawrence Livermore National Laboratory, Livermore CA 94551, USA

Abstract An electron-beam ion trap (EBIT) experiment allows one to strip virtually all of the electrons from a heavy element such as uranium in a device that is not much larger than a table top. Key to trapping the highly stripped ions is a superconducting magnet. The 6 T solenoid built for the Livermore EBIT experiment uses a modified Helmholtz coil design. The 6 tesla field region must be about 230 mm long and about 50 mm in diameter with a field uniformity of about one percent. Because the electron beam and the trapped ions are observed from the outside, gaps between the solenoid coils must be provided. The 6 T EBIT was fabricated from a Nb-Ti conductor with a copper to superconductor ratio of 1.8 and a critical current density of 3100 amperes per square mm at 4.2 K and 5 T. The conductor is a 55 filament conductor with a matrix diameter of 0.75 mm. The solenoid consists of four separate coils with holes for trapped ion observation between them. The solenoid will operate at a relative low current in persistent mode cooled by liquid helium in a 50 liter storage dewar. This report describes the design and fabrication of the 6 tesla EBIT solenoid for the Lawrence Livermore National Laboratory.

1. Introduction

Ions with energies of few electron volts are usually associated with the light elements such as hydrogen, helium or lithium. These atoms are part of a sequence of ions that have the same number of electrons but different nuclear charges. An ion near the end of such a sequence might be a uranium nucleus with one or two electrons. Atomic physics and other applications for highly charged ions are receiving increased attention[1]. Very highly charged ions are interesting objects, that play important roles in hot plasma, such as those in astrophysical x-ray sources, or controlled fusion devices. If the energy of these ions is low, they can interact with surfaces in surprising ways that may have applications in nano-technology.

Production of very highly charged ions is extremely difficult. One method for achieving such production is by directing a relativistic heavy ion beam from a large accelerator into a stationary foil target. This method is used at the Gesellschaft fur Schwerinenforschung (GSI) in Darmstadt Germany or at the Bevalac (before it was shut down in 1993) or the 88 inch cyclotron at the Lawrence Berkeley National Laboratory (LBNL), USA.

Another method of producing ions with very few electrons is to direct a modest energy electron beam into stationary ion target. The target of trapped ions is continuously kept stripped of electrons by the beam. This method is used in an electron beam ion trap (EBIT) such as the facility at the Lawrence Livermore National Laboratory (LLNL), USA. The EBIT approach is a much less expensive method for producing hydrogen like ions of heavy elements. The new LLNL apparatus for producing highly charged heavy nuclei ions is less than three meters long. A key to trapping the heavy ions long enough to strip off their electrons is a high field superconducting magnet. The superconducting magnet that is described in this report will be the central element for an improved EBIT experiment that produces 100 times more x-ray emission than the EBIT currently running at LLNL.

2. Solenoid Requirements

The superconducting magnet is a four coil solenoid that permits one to observe the plasma at two different levels in the device. The x-ray detectors should observe the plasma with as wide a viewing angle as possible. As a result, the detectors should be located as close to the plasma as possible and the observation ports should be as large as possible. Large viewing holes means that the field within the solenoid will not be very good. There is a compromise between viewing port size and the uniformity of the field in the solenoid. The solenoid is designed with a field uniformity of around plus or minus one percent over a region that is 20 mm in diameter and 290 mm in length. The nominal central induction for the solenoid within its good field region is 6 tesla. The inside cold bore diameter of the solenoid in the central region is 127 mm. At the ends of the solenoid the cold bore diameter is reduced to 101.6 mm. The nominal outside diameter for the solenoid helium vessel is 310 mm. This allows the x-ray detectors to be placed 160 mm from the central axis of the solenoid. The viewing angle from the detector into the central region of the plasma through any one of central eight ports is about 230 mrad. There are four more viewing ports that located axially about 95 mm from the eight ports located around the central region of the solenoid.

The inside bore of the solenoid is exposed directly to the plasma. In order for the EBIT to work efficiently, the vacuum within the cold bore must be very good (about 10^{-11} torr). Since vacuum in the magnet bore must be very good and the x-ray detectors are located close to the outside of the helium temperature vessel, a standard multilayer insulation system can not be used. The helium vessel for the magnet and the dewar above the magnet is vacuum insulated with a liquid nitrogen and a gas cooled shield. The expected helium usage for the superconducting magnet is expected to be about a half liter per hour.

3. The Superconducting Solenoid as Designed

The superconducting solenoid was wound using a round conductor made from a Nb-Ti alloy that is nominally 46.5 percent titanium. The design critical current density for the superconductor is 3100 A mm^{-2} at 4.2 K and 5 T. The bare 0.75 mm diameter conductor strand contains 54 filaments Nb-Ti in a pure copper matrix with a RRR of about 100. The copper to superconductor ratio for the strand is 1.8, and the twist pitch is about 12.7 mm. The nominal diameter of the formar insulated strand is 0.795 mm.

The superconducting solenoid consists of four coils that are level wound on a bobbin made from 6061-T4 aluminum. The four superconducting coils are hooked in series and they are inductively coupled to the aluminum bobbin. The inductive coupling to the bobbin contributes to the quench protection of the magnet.

The layers of round wire conductor were wound over a layer to layer separator of Nomex paper that was 0.125 mm thick. The conductor was wound under a tension of about 125 N. When this level of tension is applied to the conductor, the Nomex paper was supposed to be forced into the regions around the wire so that its average thickness was to be reduced to 0.075 mm. The coil ground plane insulation is 1.6 mm thick sheets of fiberglass epoxy between the coil and the aluminum bobbin. The four superconducting coils were potted in an epoxy resin that cured at a temperature of 130 C.

4. The solenoid as Built

The physical and electrical parameters of the 6 T EBIT solenoid as wound are given in Table 1. When the magnet was built, the current center for the four coils moved in the radial direction outward as compared to design for the following reasons: 1) The compressed thickness of the Nomex paper between the coil layers increased from 0.75 mm to about 0.90 mm. 2) The effective thickness of the ground plane insulation under each of the four coils is slightly greater than 1.6 mm because of extra Nomex paper. The current center radius of the two central coils (coils 2 and 3) averages about 0.77 mm larger than the design value while the average current center radius for the two end coils (coils 1 and 4) is about 0.49 mm greater than design. As a result, more coil current is needed to achieve 6 T.

Table 1 Nominal Parameters for the LLNL 6 Tesla EBIT Solenoid

Coil Number	1	2	3	4
Actual Inside Diameter at 292 K (mm)	127.31	160.66	160.61	127.36
Actual Outside Diameter at 292 K (mm)	214.21	269.85	270.08	214.26
Actual Coil Length at 292 K (mm)	66.71	53.00	53.04	66.68
Number of Layers	54	68	68	54
Design Turns per Layer	88	70	70	88
Actual Number of Turns	4752	4758	4758	4752
Current Center Radial Error (mm)	0.46	0.73	0.82	0.51
Current Center Axial Error (mm)	0.01	0.05	0.03	0.03
Bobbin Length at 292 K (mm)	472.44			
Bobbin OD at 292 K (mm)	304.80			
Bobbin ID Center at 292 K (mm)	127.00			
Bobbin ID Ends at 292 K (mm)	101.60			
Bobbin Mass (kg)	72.66			
Total Magnet Mass (kg)	125.63			
Solenoid Design Current (A)	123.12			
Central Induction at design Current (T)	6.000			
Peak Induction in Coil at Design Current (T)	~7.34			
Good Field Length (mm)	299			
Good Field Diameter (mm)	20			
Field Uniformity dB/Bo (%)	1.42			
Coil Self Inductance (H)	24.05			
Bobbin Self Inductance (μH)	0.05			
Mutual Inductance (mH)	0.934			
Coupling Coefficient	0.72			
Stored Energy at Design Current (kJ)	182.3			

There are coil to coil differences that affect field uniformity: 1) The conductor diameter in coils 1 and 2 is slightly smaller than in coils 3 and 4. 2) The two central coils (coils 2 and 3) have two fewer turns than were called for in the original design. As a result of the increased radius of coils 2 and 3 compared to coils 1 and 4 and the missing turns in coils 2 and 3, the calculated field uniformity increased from plus or minus 0.9 percent to about plus or minus 1.25 percent along the solenoid axis. Within a region 20 mm in diameter and 300 mm long, the calculated field uniformity is about plus or minus 1.4 percent. The field uniformity can be improved by adding about 200 ampere turns on either side of the eight detectors pointing to the center of the magnet. The correction coils can be room temperature coils wound with insulated copper wire on the outside of the cryostat. Figure 1 shows the measured magnetic induction at room temperature as compared to the calculated induction with the coil at 0.528 A.

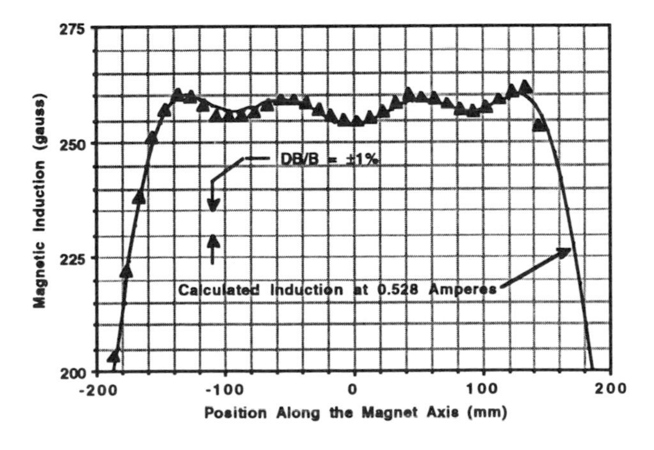

Figure 1 A comparison of Measured and Calculated Field Profiles

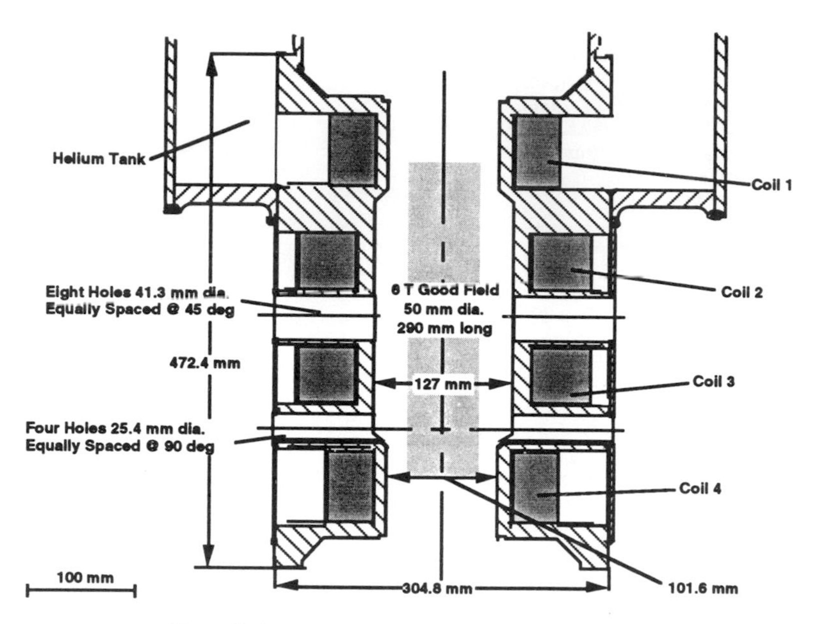

Figure 2 A cross-section of the 6 Tesla EBIT Solenoid

5. Cooling the EBIT Magnet

A cross-section of the EBIT solenoid is shown in Figure 2. Three of the coils are located below the bottom of the helium tank. The space on the outside of the coils between the coil surface and the helium vessel skin is filled with liquid helium. The helium filled spaces are connected by eight axial slots that are 12.7 mm deep by 25.4 mm wide. One of these slots contains the superconducting leads for the four coils. A second slot contains a feed pipe that carries liquid helium from the dewar feed tube to the bottom of the magnet (coil 4). The other six slots are used to circulate helium between the 50 liter helium tank and the space around the four coils. The bottom of the dewar between coils 1 and 2 serves as a platform for the superconducting splices between the four coils and the persistent switch. The bottom of the dewar liquid level gauge will be located just above top of coil one. With the use of a persistent switch, the current decay rate for the EBIT solenoid is expected to be less than 10^{-5} parts per hour. The design boil off rate for helium dewar is 0.4 liters per hour when the gas cooled electrical leads are retracted and the magnet is operated in persistent mode. It is expected that one can operate the magnet for five days without filling the dewar or recharging the magnet.

Acknowledgment

This work was performed with the support United States Department of Energy under LLNL and LBNL contract numbers W-7405-ENG-48 and DE-AC03-76SF0098.

References

[1] R. E. Marrs, P. Beiersdorfer, and D. Schneider, "The Electron Beam Ion Trap," Physics Today, p 27, October 1994

Inst. Phys. Conf. Ser. No 158
Paper presented at Applied Superconductivity, The Netherlands, 30 June–3 July 1997

Ion-Beam Switching Magnet with HTS Coils

D.M. Pooke[1], J.L. Tallon[1], S.S. Kalsi[2], A. Szczepanowski[2], G. Snitchler[2], H. Picard[2], R.E. Schwall[2], R. Neale[3], B. MacKinnon[4] and R.G. Buckley[1].

[1]N. Z. Institute for Industrial Research, PO Box 31310, Lower Hutt, New Zealand.
[2]American Superconductor Corporation, Westborough, MA 01581, U.S.A.
[3]Alphatech International Ltd., P.O. Box 33878, Takapuna, New Zealand.
[4]ISYS, 929 Lorne Way, Sunnyvale, CA 94087, U.S.A.

Abstract. An ion-beam switching magnet employing two HTS racetrack coils operating at 100 A has been installed in the beam line of a 26MeV tandem Van de Graaff accelerator at the Institute of Geological and Nuclear Science (IGNS) in New Zealand. Designed and constructed by a consortium comprising American Superconductor Corporation (ASC), ISYS, Alphatech, AET and the N.Z. Institute for Industrial Research, the magnet generates a field of 0.72 tesla in the airgap between two poles of an H-shape iron yoke and is designed to operate at a temperature of 50K. The Bi-2223 coils are conduction cooled using a single-stage cryocooler.

1. Introduction

Distributed magnets in particle accelerators represent a major operational cost whether as power dissipation in conventional electromagnets or in cooling costs for LTS magnets. The advent of high-temperature superconductor (HTS) materials promised important applications in the accelerator industry as well as wider magnet applications including MRI, minerals separation and materials processing. It has, however, taken the full 10 years since discovery for HTS wire technology to mature to the point that such applications are viable. Moreover, practical HTS wires are still confined to the Bi-2212 and Bi-2223 systems with their limitations in operating field and temperature due to poor flux pinning. Nonetheless, continued performance improvement of Bi-2223 tapes [1] has now reached the point that HTS coils provide a serious alternative to the conventional or LTS approach and, in some applications, is actually now enabling [2]. The scheduled replacement of the existing ion-beam switching magnet in the 6 MeV tandem Van de Graaff accelerator in the Accelerator Mass Spectrometry laboratory of IGNS has provided a timely opportunity for the first implementation of an accelerator magnet energised by HTS coils. Installation of this magnet for continuous use in a working environment represents a significant milestone in the development of HTS wire production. This paper describes the design, control and implementation of the HTS switching magnet.

2. AMS Facility and Magnet Requirements

Coinciding with the tenth anniversary of the discovery of HTS materials superconducting above liquid nitrogen temperature, it is also the tenth anniversary of the commissioning of the

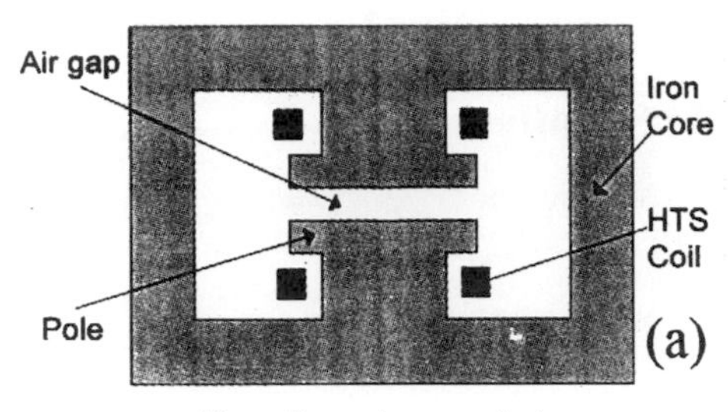

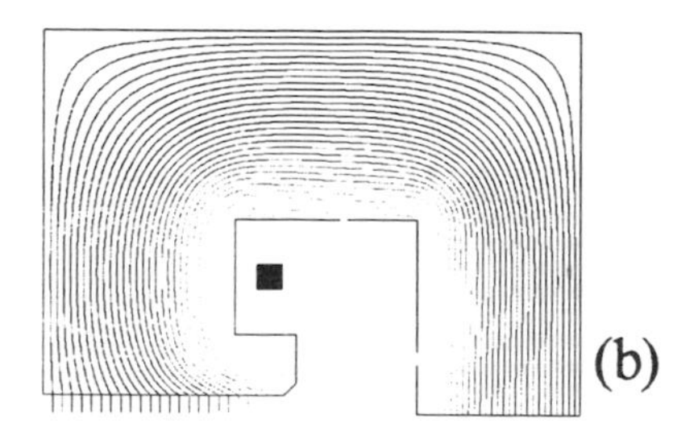

Fig. 1(a) Cross section of the magnet layout showing iron-core and HTS coils. (b) Flux distribution in core and air gap (¼ section).

New Zealand AMS system incorporating a tandem Van de Graaff accelerator. The system is used primarily for radio carbon dating (80%) and ^{10}Be for archeology and earth sciences studies (10%). Examples include dating of coral sediments, studies of coral damage to the Australian Great Barrier Reef, dating of plate-boundary earthquakes, studies of atmospheric green-house gases and dating of the earliest settlement of New Zealand from the age of bones from introduced rats [3]. The development of the new switching magnet is part of an upgrade to higher mass-energy product to extend to important cosmogenic isotopes for geological and hydrological studies, including ^{26}Al, ^{32}Si and ^{36}Cl e.g. for dating of glacial ice.

A switching magnet is used to direct the ion-beam to one of three detector beam lines, zero field selecting the straight-through target line. The existing switching magnet has a mass-energy product $ME/Q^2 = 36$ amu.MeV, and was the limiting element in the performance of the AMS facility. Design requirements for the replacement HTS magnet included a larger mass-energy product (68 amu.MeV) to allow for more versatility in operation and inclusion of the heavier nucleides, as well as more uniform field for good beam optics and a larger pole gap.

3. Magnet Design

The field is established over an air-gap in an iron yoke and is set positive or negative to deflect the beam horizontally right or left and adjusted in magnitude to deflect specific ionic species to emerge exactly on the axis of either of the exit-beam ports disposed at ±25° . The deflection radius is 1654mm. At zero field the undeflected beam emerges through the 0° port. Finite-element modelling was used in the computation of magnetic fields. The H-shaped room-temperature iron yoke is designed to a generate a uniform magnetic field of 0.72T in a 30mm air gap between the poles. The yoke plays a key role in ensuring sufficient field homogeneity for the required ion optical quality. The HTS coils and yoke cross-section are shown in Fig. 1 (a).

Larger air-gaps require larger ampere-turn coils, and in conventional water-cooled copper systems this necessitates significant increase in the capacity of both the power supply and the cooling system. On the other hand, HTS coils and cryostats can be designed to replace conventional copper coils, providing much higher ampere-turn capability without impacting on the iron-core size. In the present system each HTS coil generates 9,200 ampere-turns to produce the 0.72 Tesla field in the air-gap. Flux concentration in the iron core means that the field experienced at the coils is substantially lower, not exceeding 0.14 Tesla. The computed flux distribution within a ¼ cross-section is shown in Fig. 1(b). Low fields at the coils allow operation at higher temperatures, a key consideration with the Bi-2223 system, and also ensures lower mechanical stresses (calculated to be less than 50MPa).

4. HTS Coils

The two series-connected coils are made from high-performance $Bi_{2-x}Pb_xSr_2Ca_2Cu_3O_{10}$ multifilamentary, flexible silver-matrix tapes [1]. They were hand-wound employing a pancake configuration in the form of race-track shaped solenoids and vacuum impregnated with epoxy. Dimensional and operational parameters are as follows:

Coil cross section	20mm x 20mm
Clear-bore dimensions	346mm x 620mm
Maximum current	110A
Maximum Voltage	2V
Design temperature	50K

The HTS conductor manufacture, coil winding and cryo-integration were carried out by American Superconductor Corporation. The coil control unit consists of an instrumentation panel featuring emergency stop, a keyed system reset and a multichannel scanner for the temperature sensors, with alarm setpoints on each of the 8 channels scanned. The current supply, a Lakeshore MPS 622, is interlinked to the control panel so that the magnet is shut down if over-temperature or over-current fault conditions are detected. An RS-232 commun-ication port allows remote moinitoring and operation of the magnet.

5. Cryo-integration

The HTS coils are conduction cooled using a single-stage GM-type Leybold RGS 120T cryo-cooler providing a refrigeration capacity at 20K of 25W, which is considerably in excess of requirements and thus allows the cryo-system to cool well below the coil-design operating temperature of 50K. The coils are wrapped with superinsulation and centrally supported by G-10 fibreglass members within a series 300 stainless-steel vacuum vessel. The fibreglass supports are designed for thermal insulation and and to withstand mechanical shock during transportation. Upper and lower cryostats are connected together with a common vacuum manifold which houses a cold copper bus connected to the cold stage of the crycooler. Each HTS coil is thermally linked to the copper bus through a flexible thermal shunt. Fig. 2 shows a sketch of the positioning of the cryostat and cryocooler.

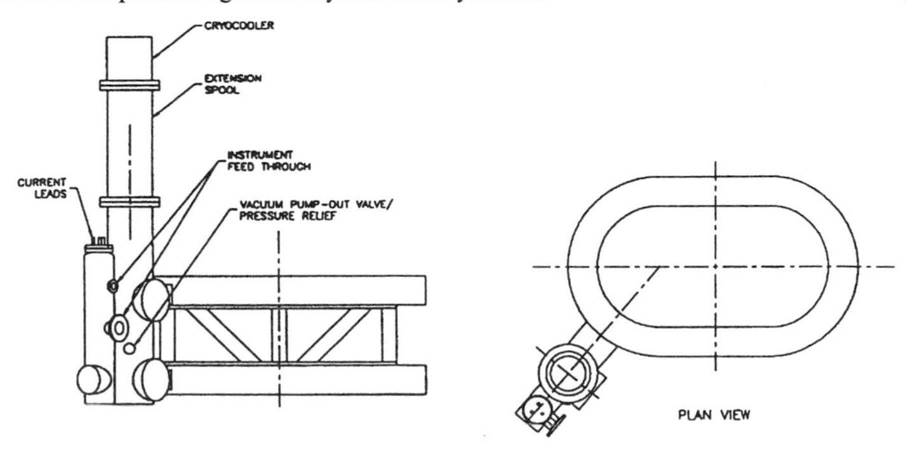

Fig. 2. Coil cryostat and cooler assembly (elevation and plan view)

A separate manifold carries twin 100A electrical feedthroughs to energise the coils. The coil ends of the feedthroughs, anchored at the cold-head temperature, are connected to the coils through conduction-cooled HTS current leads made by ASC. This allows a substantial reduction of heat load at the coils. An electrical feedthrough for eight temperature sensors mounted throughout the thermal circuit is also provided, as is a thermocouple vacuum sensor, and evacuation valve. Following evacuation to 2.10^{-6} mbar the system cool down to base temperature occurs within 9 hours. Excess capacity in the cryocooler allows operation at a base temperature of 21K, giving a wide margin of safety in operation, compatible with the long-term reliability dictated by the commercial requirements of the AMS facility. The magnet has been successfully tested, however, at temperatures up to 47K. At 47K and 100A excitation, the coil terminal voltage remained constant within 1mV over a 4-hour test period. A change of $\approx$ 20mV is expected if the coils were driven normal. System stability within the range 20-47K is thus confirmed.

6. Outlook and conclusions

The initial aims of the magnet design, namely demonstration of HTS capability and reliability, higher mass-energy product, larger pole gap for increased beam transmission, and field uniformity for good beam-optical properties were accomplished. The design field of 0.72T was achieved at an excitation current of 102A. The magnet was installed in the beam line of the AMS facility in February 1997 and commissioned in March 1997. Recently, 1000 operational hours at full excitation current were completed. Continuous use in a working environment represents a major milestone in the development of HTS wire and products. Experience with this magnet, together with continued improvements in HTS conductor performance which have occurred since the manufacture of the coils for this project, will enable less conservative designs with respect to refrigeration and size as well as push out to significantly higher fields. The design, manufacture and delivery to a customer of HTS magnets for a variety of physics research and industrial applications is now a real option. Moreover, the demonstration of long-term reliability in a working environment will also positively impact on the application of other HTS-wire-based systems, e.g. for the electrical power industry.

Acknowledgements. The financial support of the New Zealand FRST Technology for Business Growth Programme is acknowledged. We thank Alex Malozemoff (ASC) for constant encouragement and technical advice, Helder Falacho (ASC) for manufacturing the control centre, Buckley Systems (New Zealand) for manufacturing the magnet pedestal and beam-line vacuum chamber, Harry Wolters (ISYS) for mechanical engineering and P. Winn of Applied Engineering Technologies, Woburn, MA, USA for supplying the cryostat.

References

[1] Malozemoff A P, Li Q and Fleshler S 1997 Proc. M^{2}S-HTSC-V, Beijing (to be published, Physica C).
[2] Duggan G and Rogers J 1997 Proc. Particle Accelerator Conference, Vancouver , (to be published).
[3] Zondervan A and Sparks R J 1996 Nucl. Instr. and Meth. **B123**, 79.

Inst. Phys. Conf. Ser. No 158
Paper presented at Applied Superconductivity, The Netherlands, 30 June–3 July 1997

Nonlinear flux diffusion in superimposed ac magnetic field

A L Kasatkin, V M Pan, V V Vysotskii

Institute of Metal Physics, National Academy of Sciences of Ukraine,
36 Vernadsky St., Kiev, 252142, Ukraine

H C Freyhardt

Institute for Metal Physics, University of Göttingen, Göttingen 37073, Germany

Abstract. The influence of superimposed weak ac magnetic field on flux creep and relaxation phenomena in a superconducting slab (or cylinder) situated in parallel dc magnetic field is studied theoretically by solution of the appropriate nonlinear flux diffusion equation with periodical boundary conditions. The flux creep rate and magnetization decay are shown to be enhanced due to this influence in the case of rather strong nonlinearity of the flux diffusion process. The evolution of the dynamic vortex response during the relaxation process is also studied and appropriate relaxation characteristics of ac resistivity and ac magnetic susceptibility are obtained.

1. Introduction

The flux creep in high-T_c superconductors is known to be a very important factor that determines the electromagnetic properties and defines the limitations in possible applications of these materials. Usually flux creep and related phenomena (e.g. resistivity, magnetic moment relaxation, etc.) may be described on the basis of nonlinear flux diffusion equation with appropriate initial and boundary conditions and diffusion coefficient which for given material strongly depends on temperature, magnetic field and current density (see, e.g., [1,2] and references therein.)

In the present work we study the influence of weak ac magnetic field on flux creep and relaxation processes in a superconducting slab settled in dc and ac magnetic fields parallel to its surface. For this case we have obtained both numerical and approximate analytical solutions of one-dimensional nonlinear diffusion equation with periodical boundary conditions, which give the relaxation profiles of induced electric field (screening current) and magnetization decay. These results demonstrate strong effect of rather weak ac magnetic fields on the flux creep rate and the character of relaxation processes in superconductors. We have also calculated the evolution of ac vortex response (namely, ac resistivity and dynamic magnetic susceptibility) during the relaxation process, which may be used for investigation of relaxation processes in superconductors [4-6].

2. Flux creep changes caused by ac field

We consider flux creep and relaxation phenomena in a superconducting slab of thickness d along the x-axis and infinite in the yz-plane, settled in the dc $\vec{H}_0$ and weak ac $\vec{h}_{\sim} \sin \omega t$

fields that are parallel to the z-axis (Fig.1). For the case $\vec{h}_{\sim} = 0$ the relaxation processes in this geometry for different kinds of nonlinear current-voltage characteristics (CVC) $j(E)$ of the superconductor were studied in [3-6] on the basis of nonlinear diffusion equation for the induced electric field component $\vec{E} = E(x,t) \cdot \vec{n}_y$:

$$\partial E/\partial t = D(E)\, \partial^2 E/\partial x^2 ; \quad D(E) = 1/\mu_0 \;\partial E/\partial j. \quad (1)$$

The external magnetic field H_e was supposed to increase with a constant ramp rate $\dot{H}_e$ for sufficiently long time before $t = 0$, and then remain fixed at the value H_0. This completely determined the initial and boundary conditions for the diffusion problem (1):

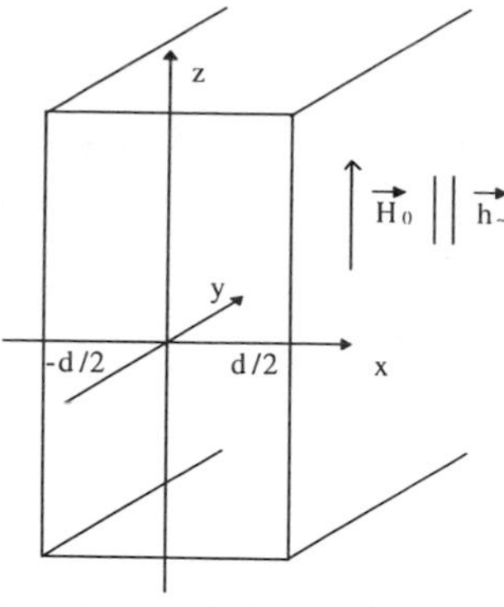

Fig. 1. Sample in parallel dc and weak ac magnetic fields.

$$E(x,0) = \dot{B}_e x ; \quad E(0,t) = 0 ; \quad \partial E/\partial x\big|_{x=\pm d/2} = 0 , \quad (2)$$

where $B_e = \mu_0 H_e$, and allowed to find the relaxation profiles $E(x,t)$ at $t > 0$ for different kinds of nonlinearity $D(E)$ corresponding to appropriate CVC $j(E)$. For instance, in case of usual (exponential) flux creep one has

$$E(j) = E_c \exp\left(\frac{j - j_c}{j_1}\right); \quad D(E) = \frac{E}{\mu_0\, j_1}, \tag{3}$$

and the asymptotic solution of (1,2) for $E(x,t)$ at $t > 0$ may be written as follows [3]:

$$E_0(x,t) = \frac{\mu_0\, j_1}{2(t+\tau)}\left(d\,x - x^2 \operatorname{sgn} x\right). \tag{4}$$

The magnetization decay of the subcritical state ($j < j_c$) in this case is given by

$$M(t) = M(0) - M_1 \ln(1 + t/\tau). \tag{5}$$

Here τ is the macroscopic diffusion time constant that is inversely proportional to $\dot{B}_e$; $M_1 = dM/d\ln t$ (at $t \gg \tau$) – the flux creep rate on the logarithmic stage of magnetization decay.

In the present work we study the role of superimposed weak ac magnetic field on flux creep and relaxation phenomena by solving the nonlinear diffusion equation (1) with non-zero periodical boundary conditions

$$\partial E/\partial x\big|_{x=\pm d/2} = h_{\sim}\omega \cos\omega t \tag{6}$$

which follow in the considered case from the Maxwell equations.

Numeric solutions of Eqs. (1), (6) for relaxation profiles $E(x,t)$ are shown of Fig. 2 in comparison with solution (4) obtained for $h_{\sim} = 0$. Drastic changes of relaxation profiles $E(x,t)$ spreading over the whole volume of the specimen and mostly prominent near the surface are obvious. These changes are caused by ac magnetic field influence. The smeared region around each relaxation profile $E(x,t)$ near the surface represents penetration of ac component $E_{ac}(x,t)$ within the skin layer. The changes in relaxation profiles $E(x,t)$ manifest themselves in magnetization decay

$$M(t) = \int_{-d/2}^{d/2} x\, j(x,t)\, dx \tag{7}$$

as it is shown in Fig. 3, which demonstrates enhancement of the magnetization decay and flux creep rate due to the influence of ac field.

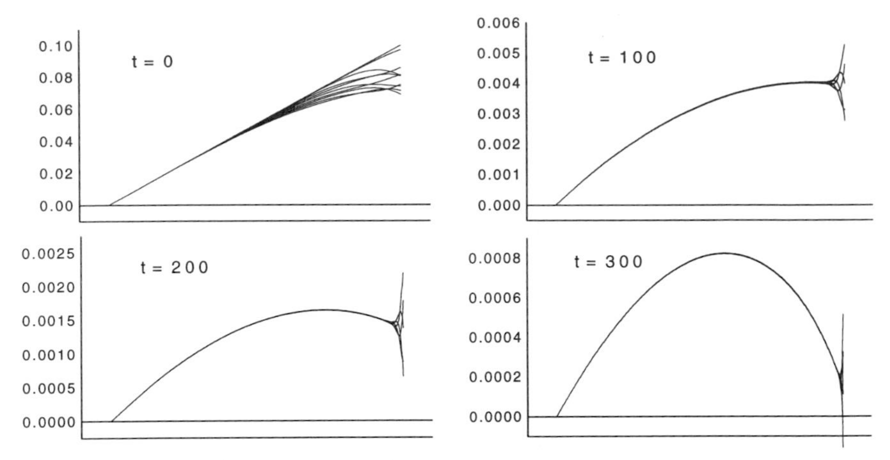

Fig. 2. Profiles of E(x, t) at different moments of time.

To analyze this effect of weak ac magnetic field influence on flux creep we consider the case of small amplitudes $h_\sim$ and not very long times when the changes in relaxation profiles $E(x,t)$ caused by ac field are small compared to $E_0(x,t)$ given by (4). Under such conditions one may look for $E(x,t)$ in the form

$$E(x,t) = E_0(x,t) + E_\sim(x,t), \quad |E_\sim(x,t)| \ll E_0(x,t). \tag{8}$$

Using the perturbation approach we expand $E_\sim(x,t)$ in series on powers of $h_\sim$

$$E_\sim(x,t) = \sum_{n\geq 1} E_\sim^{(n)}(x,t); \quad E_\sim^{(n)} \sim h_\sim^n. \tag{9}$$

By substitution of $E(x,t)$ in the form (8), (9) in Eq. (1) one obtains the system of equations for powers of $h_\sim$. The solution of this system may be found using the iteration procedure to find the consecutive functions $E_\sim^{(n)}$. Up to the second order of $h_\sim$ it may be written as follows:

$$\partial E_\sim^{(1)}/\partial t = F_0(x,t)\,\partial^2 E_\sim^{(1)}/\partial x^2 + F_1(x,t)E_\sim^{(1)};$$
$$E_\sim^{(1)}(0,t) = 0; \quad \partial E_\sim^{(1)}/\partial x\Big|_{x=\pm d/2} = h_\sim \omega \cos\omega t; \tag{10}$$

$$\partial E_\sim^{(2)}/\partial t = F_0(x,t)\,\partial^2 E_\sim^{(2)}/\partial x^2 + F_1(x,t)E_\sim^{(2)} + F_2(x,t);$$
$$E_\sim^{(2)}(0,t) = 0; \quad \partial E_\sim^{(2)}/\partial x\Big|_{x=\pm d/2} = 0. \tag{11}$$

Eqs. (11), (12) are linear diffusion type equations with coefficients $F_0(x,t)$, $F_1(x,t)$ and $F_2(x,t)$ given by

$$F_0(x,t) = D(E_0(x,t)); \tag{12}$$

$$F_1(x,t) = \partial D/\partial E\Big|_{E=E_0(x,t)} \partial^2 E_0(x,t)/\partial x^2; \tag{13}$$

$$F_2(x,t) = \frac{1}{2}\frac{\partial^2 D}{\partial E^2}\Big|_{E=E_0} \frac{\partial^2 E_0}{\partial x^2}\left(E_\sim^{(1)}(x,t)\right)^2 + \frac{\partial D}{\partial E}\Big|_{E=E_0} E_\sim^{(1)}(x,t)\frac{\partial^2 E_\sim^{(1)}}{\partial x^2}. \tag{14}$$

For the functions $D(E)$ and $E_0(x,t)$ given by (3), (4) the coefficients $F_0(x,t)$ and $F_1(x,t)$ may be found directly according to Eqs. (13), (14), while for determining the coefficient

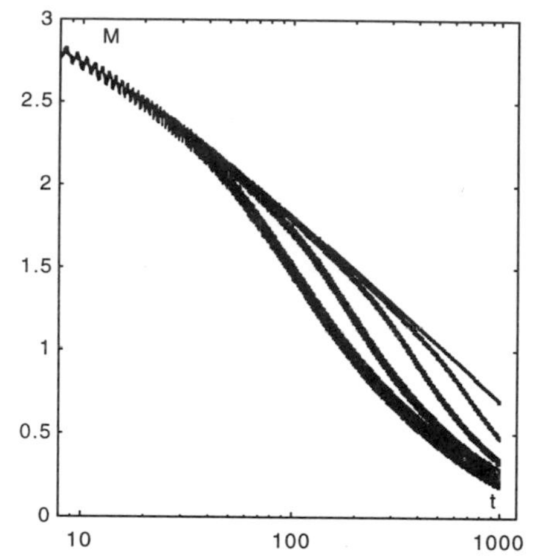

Fig. 3. Relaxation of magnetic moment at different amplitudes of applied ac magnetic field.

$F_2(x,t)$ one needs to know the solution for $E_{\sim}^{(1)}(x,t)$ given by the first step of the iteration procedure.

For $D(E)$ dependence described by Eq. (3) the approximate solution for $E_{\sim}^{(1)}(x,t)$ may be found from Eq. (11) (by using the WKB approach.) This solution contains only the oscillating component of the induced electric field:

$$E_{\sim}^{(1)} \approx \frac{h_{\sim}\omega d}{2\sqrt{2\omega(t+\tau)}} \exp\left[k_0\left(x-\frac{d}{2}\right)\right] \times$$
$$\times \cos\left[k_0\left(x-\frac{d}{2}\right)-\omega t-\frac{\pi}{4}\right]. \qquad (15)$$

Here $k_0(x,t)=2\sqrt{\omega(t+\tau)}/d$. Meanwhile the solution of Eq. (12) for $E_{\sim}^{(2)}(x,t)$ besides the oscillating component (with frequency 2ω) has also the slowly varying part $E_{\sim}^{(2)}{}_{slow}(x,t)$, which describes the changes in relaxation profiles:

$$E_{\sim}^{(2)}{}_{slow}(x,t) = -\frac{h_{\sim}^2\omega^{1/2}}{2\,ch(1)\mu_0 j_1(t+\tau)^{1/2}} sh\left(\frac{2x}{d}\right) + \frac{h_{\sim}^2 \exp[2k_0(x,t)(x-d/2)]}{4\mu_0 j_1(t+\tau)}. \qquad (16)$$

This solution describes (at least qualitatively) the changes in relaxation profiles that were obtained numerically and are shown in Fig. 2.

These changes in $E(x,t)$ profiles seem to be caused by detection of the ac field on nonlinearity of the CVC near the surface of the specimen and subsequent diffusion of the detected component of electric field inside the slab.

3. Conclusion

The results obtained in the present work indicate that application of a rather weak ac field may lead to significant changes of relaxation processes in superconductors, namely, to enhancement of flux creep and magnetization decay rates. Study of dynamic vortex response and its evolution during the relaxation may be useful for investigation of nonlinear flux diffusion and possible states of the vortex ensemble in superconductors.

References

[1] Brandt E.H., Rep. Progr. Phys., **58**, 1465, 1955
[2] Gurevich A., Int. J. Mod. Phys., B**9**, 1045, 1995
[3] Gurevich A. and Küpfer H., Phys. Rev. B, **48**, 6477, 1993
[4] Brandt E. H. and Gurevich A., Phys. Rev. Lett., **76**, 1723, 1996
[5] Vysotskii V. V., Czech J. Phys., **46**, Suppl. S2, 905, 1996
[6] Kasatkin A. L., Pan V. M., Vysotskii V. V. and Freyhardt H. C., Proc.ASC'97, IEEE Trans. Appl. Supercond., 1997

Inst. Phys. Conf. Ser. No 158
Paper presented at Applied Superconductivity, The Netherlands, 30 June–3 July 1997

Trapped Field Magnets from Melt Textured YBCO Samples

G Fuchs, S Gruss, G Krabbes, P Schätzle, K H Müller, J Fink and L Schultz

Institut für Festkörper- und Werkstofforschung Dresden, P.O. Box 270016, D-01171 Dresden, Germany

Abstract. The properties of bulk melt textured YBCO disks have been investigated with respect for use as trapped field magnets. Large single-domain grains up to 24 mm in diameter were obtained by a modified melt process using Sm-123 as seed crystal. Typical values of the maximum trapped field B_o on the surface of the disks are 0.6 T. The temperature dependence of B_o was measured in the 1 mm gap between two single-grain disks. The maximum trapped field increased from 1 T at 77 K up to 9 T at 47 K which is the highest trapped field achieved in non-irradiated samples. We observed an exponential decrease of the trapped field with increasing temperature, which can be explained by flux creep in the framework of collective pinning. The trapped field at lower temperatures was found to be limited by the mechanical strength of the samples for which a value of 25 MPa was estimated.

1. Introduction

In bulk type II superconductors with strong pinning high magnetic fields can be trapped. The typical spatial field distribution of the trapped field observed on the surface of superconductors is attributed to persistent currents circulating as volume currents in the samples. The maximum field B_o of this field distribution is determined by the critical current density j_c and the size of the persistent current loops. Therefore, high values of B_o require large grains and high j_c values.

In YBCO, large single grains can be obtained by the melt-texture process, especially by using Sm-123 as seed crystal [1]. An important advantage of the seed technique is that the grains are oriented with their a,b-planes parallel to the sample surface. The pinning effect in melt textured samples has been improved by irradiation methods [2]. In mini-magnets consisting of stacked irradiated single tiles, high trapped fields B_o of 3.1 T at 77 K and of 10.1 T at 42 K have been reported [3]. Recently, it could be shown, that at temperatures around 50 K high trapped fields up to 8.5 T can be achieved without irradiation [4].

In the present paper, we report new field trapping results on melt textured YBCO samples prepared by a top seeding method. The aim of this work was to study the temperature dependence of trapped fields and critical currents in our YBCO samples and to compare the results with that obtained for irradiated YBCO samples.

2. Experimental

Melt textured YBCO bulk samples were prepared using Sm-123 as seed crystal. The Sm-123 seeds were positioned on the top of uniaxial pressed cylinders consisting of Y-123, Y_2O_3 and

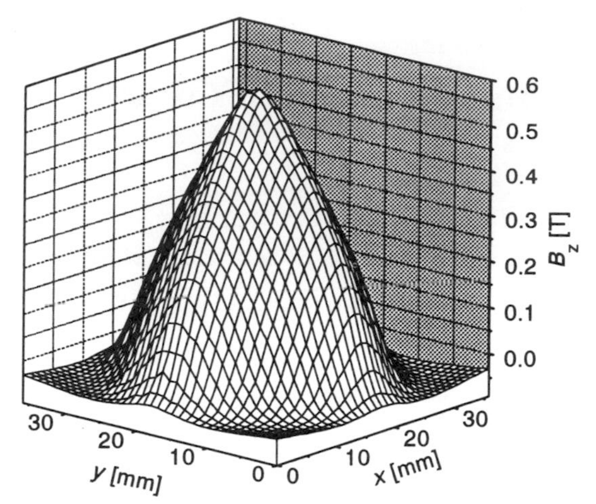

Fig. 1. Spatial distribution of trapped field in a YBCO sample at 77 K with a maximum trapped field of 0.69 T

Pt powder. Details of the sample preparation have been reported elsewhere [5,6]. After the melt process, the samples consist of one single grain with a preferred orientation of the a, b planes parallel to the surface. The microstructure revealed small, homogeneously distributed Y_2BaCuO_5 precipitates.

The disks were characterized by field mapping of the trapped field on the surface of the samples at 77 K using an axial Hall sensor with an active area of 0.2 mm^2. The temperature dependence of the maximum trapped field B_o was measured in the 1 mm gap between two disks which were mounted in a sample holder together with a transversal Hall sensor. These measurements were performed in the variable temperature cryostat of a superconducting 18 T magnet. The disks were field cooled from T=100 K to the measuring temperature. Then, the magnetic field was slowly reduced to zero at a rate of 0.1 T/min in order to avoid flux jumps due to thermal instabilities. The magnetic field at the center between the two disks was continuously recorded as a function of the applied magnetic field.

3. Results and discussion

Fig. 1 shows the field profile of a disk (diameter 26 mm, hight 12 mm) measured at 77 K. This sample consists of a large single grain of about 24mm x 24mm in size. A maximum trapped field B_o=0.69 T, was measured five minutes after activation of the sample.

In Fig. 2 the temperature dependence of B_o is shown for two pairs of disks denoted as P1 and P2. At 75 K, the maximum trapped field in the gap between two disks is $B_o \approx 1$ T in both cases. The trapped field increases strongly with decreasing temperature. The result of B_o(51.5 K) =8.5 T for disk pair P1 has been reported recently [4]. In the new disk pair P2, a trapped field of B_o= 9 T could be achieved at 47 K.

At lower temperatures, a strong reduction of the trapped field was observed during activation of the disk pairs which was found to be caused by cracking of the disks. The cracks could be revealed by visual inspection on the surface of the disks. The cracking of the disks can be explained by tensile stresses during the activation process exceeding the tensile strength of the material. According to a model proposed by Ren et al. [7], the maximum stress occurs at the center of the samples, where cracking is therefore most probable. Using this model, a value

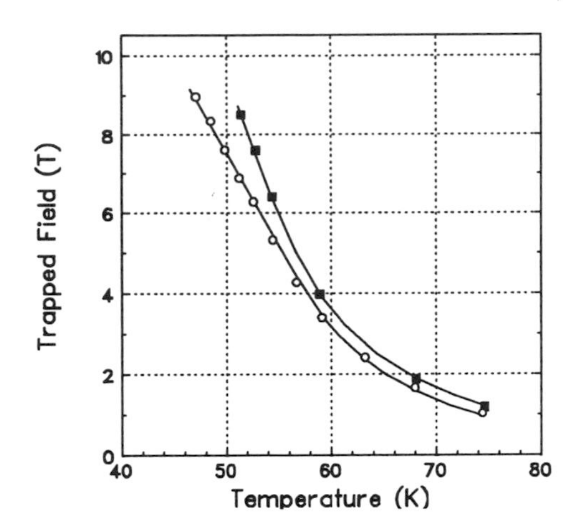

Fig. 2. Temperature dependence of the maximum trapped field B_o for disk pairs P1 (■) and P2 (o)

of 25 MPa was estimated for the tensile strength of our YBCO disks.

In the *ln* B_o vs. T plot of Fig. 3, experimental results for our disk pairs are compared with that of two mini-magnets of proton-irradiated tiles reported by Weinstein et al. [3] and Liu et al. [8]. At 77 K, the trapped field B_o of one of the mini-magnets shown in Fig. 3 is more than a factor of two higher than B_o of our disk pairs. This large difference observed in the B_o values at 77 K decreases with decreasing temperature, and at temperatures around 50 K the trapped fields observed in non-irradiated and irradiated YBCO become comparable. To the best of our knowledge, the value of $B_o = 9$ T at 47 K for disk pair P2 is the highest trapped field achieved with non-irradiated YBCO samples.

The $B_o(T)$ or $j_c(T)$ dependence for the mini-magnets can be described by a quadratic law which has been found to be valid in the intervall $20 \leq T \leq 65$ K [9]. In general, the $B_o(T)$ dependence of our samples is stronger and follows in the investigated temperature range $T > 50$ K an

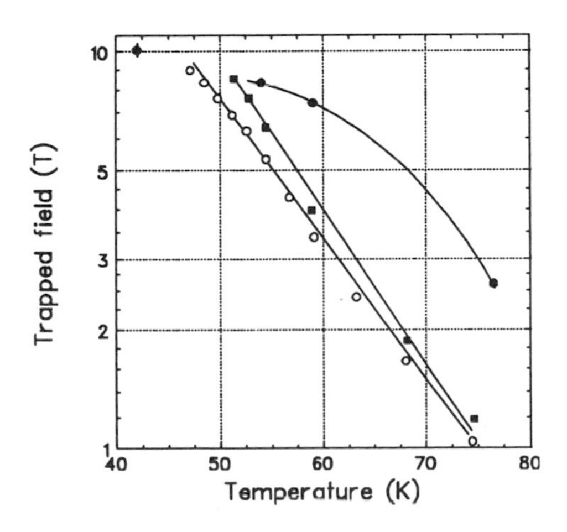

Fig. 3. Temperature dependence of the maximum trapped field B_o for disk pairs P1 (■) and P2 (o) compared with that of two mini-magnets (● - Ref. 3, ◆ - Ref. 8)

exponential law as shown in the *ln* B_o vs. T plot of Fig. 3. The different $B_o(T)$ behaviour observed in the mini-magnets mentioned above and in our samples reflects differences of the pinning mechanism.

The critical current density $j_c(T)$ of YBCO, which determines $B_o(T)$, has been extensively studied in single crystals. A quasi-exponential $j_c(T)$ dependence has been reported in many cases [10]. This behaviour can be explained by flux creep in the framework of the collective pinning model [11]. We investigated small samples prepared from a large YBCO disk by magnetization measurements. The critical current density was determined from hysteresis loops at different temperatures. By fitting the $j_c(T)$ relation of the collective pinning model to the experimental data, a value of μ=0.86 was obtained. Additionally, relaxation measurements were performed for these samples at 77 K in magnetic fields of 1 T. A non-logarithmic current decay was observed, which could be described by $j^{-\mu} \propto ln(t)$ with μ=0.75. The similar results for μ obtained from $j_c(T)$ and relaxation measurements give us confidence in the fitting procedure. The resulting value for μ corresponds to thermally activated flux creep of large flux bundles in the investigated YBCO samples.

In conclusion, a defect structure could be produced without irradiation by a modified melt texture process in YBCO samples, which leads to high critical current densities of about 10^5 A/cm^2 and trapped fields of 9 T at 47 K. The strong exponential decrease of the trapped field $B_o \propto j_c$ with increasing temperature is caused by flux creep and suggests weak pinning at higher temperatures by a too small pinning barrier height. On the other hand, the flux creep effect at 77 K is reduced considerably by stronger pinning defects produced in irradiated samples. The trapped field at temperatures below 50 K was found to be limited by the mechanical strength of the samples.

Acknowledgements

This work was supported by Bundesministerium für Bildung, Wissenschaft, Forschung und Technologie, contracts no. 13N6662 and 13N5897A.

References

[1] Lee D F, Partsinevelos C S, Presswood R G Jr. and Samala K 1994 *J. Appl. Phys.* **76** 603

[2] Weinstein R, Liu J, Ren Y, Chen I G, Obot V, Sawh R P, Foster C and Crapo A 1994 *Proc. Int. Workshop on Superconductivity,* Kyoto, Japan

[3] Weinstein R, Liu J, Ren Y, Sawh R P, Parks D, Foster C and Obot V 1996 *Proc. 10th Anniversary HTS Workshop on Physics, Materials, and Applications*, Houston

[4] Fuchs G, Krabbes G, Schätzle P, Gruß S, Stoye P, Staiger T, Müller K-H, Fink J and Schultz L 1997 *Appl. Phys. Lett.* **70** 117

[5] Schätzle P, Bieger W, Krabbes G, Klosowski J and Fuchs G 1995 *Applied Superconductivity 1995*, Vol. 1, (Bristol and Philadelphia: IOP Publishing Ltd) p. 155

[6] Krabbes G, Schätzle P, Bieger W, Wiesner U, Stöver G, Wu M, Strasser T, Köhler A, Litzkendorf D, Fischer K and Görnert P 1995 *Physica C* **244** 145

[7] Ren Y, Weinstein R, Liu J, Sawh R P and Foster C 1995 *Physica C* **251** 15

[8] Liu J, Weinstein R, Ren Y, Sawh R P, Foster C and Obot V 1995 *Proc. 1995 Int. Workshop on Superconductivity,* Maui, p. 353

[9] Weinstein R, Ren Y, Liu J, Chen I G, Sawh R, Foster C, and Obot V 1993 *Proc. Int. Symp. on Superconductivity* (Hiroshima: Springer) p. 855

[10] Thompson J R, Yang Ren Sun, Civale L, Malozemoff A P, McElfrsh M W, Marwick A D and Holtzberg F 1993 *Phys. Rev. B* **47** 14440

[11] Feigel'man M V and Vinokur V M 1990 *Phys. Rev B* **41** 8986

Inst. Phys. Conf. Ser. No 158
Paper presented at Applied Superconductivity, The Netherlands, 30 June–3 July 1997

Locally Resolved Trapped Field Measurements of Melt Textured YBCO at Low Temperatures

T Straßer, T Habisreuther, B Jung, D Litzkendorf, M Wu, W Gawalek.

Institut für Physikalische Hochtechnologie, Helmholtzweg 4, D-07743 Jena, Germany

Abstract. The trapped field of melt textured YBCO bulk cylinders (approx. 30mm ∅ x 15mm) has been mapped at a temperature range from 12K to 85K. The samples fixed on a refrigerator cryostate were put into the warm bore of an 8T superconducting magnet. Single domain, multi domain and sintered ceramic samples have been compared.
The results show no detectable intergrain currents at any temperature. A high sensitivity to flux jumps occurs below T = 20K. Mechanical strength of the superconducting material becomes important due to high magnetic forces at high fields. Despite the low temperatures the relaxation is high due to the high field gradient. 6.8T had been trapped at 12K. A trapped field of 10T can be expected if higher magnetizing fields were available.

1. Introduction

The development of YBCO ($Y_1Ba_2Cu_3O_{7-x}$) bulk high temperature superconductors has nowadays been progressed in a way that it can be used for manifold applications. The batch processed material is favourable to electric motors and generators, HTSC permanent magnets, magnetic field screens as well as for magnetic bearings and levitated transport systems.

The superconducting magnetic properties are depending on the real structure of the material (e.g. grain size) and on the critical current density j_c. The temperature dependence of these properties has been examined by trapped field mappings at a temperarture range from 12K to 85K.

Intergrain current contribution, relaxation and mechanical problems are discussed besides the indication of maximum achievable trapped fields.

2. Material preparation

The examined YBCO cylindrical blocks of 30mm in diameter and 15-17mm in height are prepared in a batch process by a precursor powder composed of $Y_{1.5}Ba_2Cu_3O_x$ and 0.5% to 1% platinum or cer addition. The powder is uniaxially pressed to cylindrical shape.

For the modified melt textured growth process the cylinders are placed in a six side heated chamber furnace. After establishing the growth starting temperature the furnace is cooled down with a cooling rate of 0.5 to 2K/h at a temperature gradient of $\nabla T < 10K/cm$. Unseeded blocks consist of several domains in the diameter of 1 to 2 cm. To obtain single domain material MgO single crystals are used as a seed on the top face of the blocks. The grains and magnetic domains are observed by polarization photography and trapped magnetic field mapping.

In a seperated step the YBCO material is oxygenated by annealing in flowing oxygen.

The intra domain critical current densities are up to $5x10^4$ A/cm^2 measured by vibration sample magnetometry.

3. Experimental setup

For trapped field measurements at liquid nitrogen the sample is field cooled magnetized in the axial homogeneous field of a superconducting magnet at 2.5T. The sample is afterwards scanned by a Hall probe attached to the robot. The distance of the Hall active area to the sample's surface is 0.25mm, the scan distance is 1mm in each horizontal direction.

Field mapping at various temperatures is provided with the refrigerator cryostate. A Hall row consisting of 20 Hall probes horizontally arranged is fixed to the feedthrough and can be rotated along the cylinder axis above the sample. The distance of the Hall row to the sample's surface is approx. 1-2mm depending on the sample's surface and accuracy of adjustment. The whole system is submerged into the warm bore of the superconducting magnet. Fields up to 8T can be applied. The field cooled magnetization was carried out at 12K. Trapped field measurements at higher temperatures were done by followed heating. At temperatures below 20K the magnetized sample is very sensitive to flux jumps. Therefore, after magnetization the external magnetizing field must be reduced at a rate of < 0.2mT/s in the final approach to zero due to the high field gradient between the sample and the low external field. Furthermore fast heating also favours flux jumps.

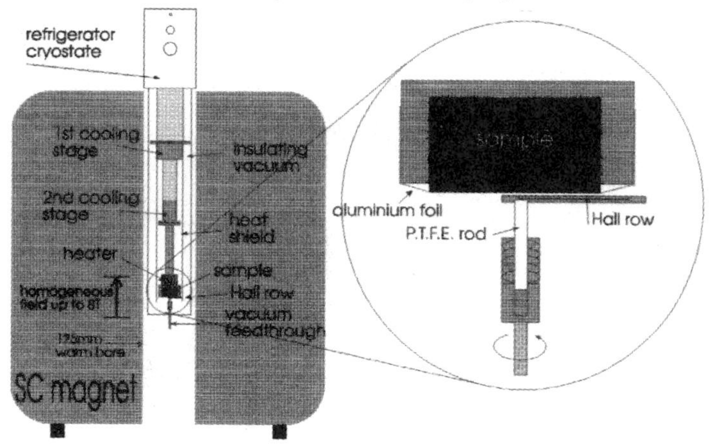

Fig 1. Experimental setup for temperature dependent field mapping

4. Results

Fig.2a and 2b display the trapped field of the three domain sample at 12K and 85K. The magnetizing field was 5T. At 12K the sample is not fully magnetized, which can be seen at the flattened peak of the Bean cone. A flattening due to a lower j_c at higher fields cannot be the reason because the ratio of $j_c(B=0)$ to $j_c(B_{max})$ remains approximately constant for all temperatures (see magnetization loops at Fig. 5 [1]) and only little flattening can be observed at higher temperatures. The lower trapped field compared to the magnetizing field is due to relaxation.

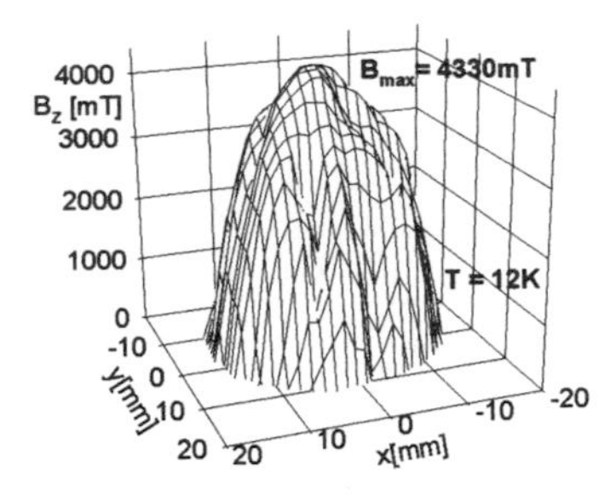

Fig 2a: Trapped field of a 3-domain sample at T = 12K. Magnetizing field = 5T

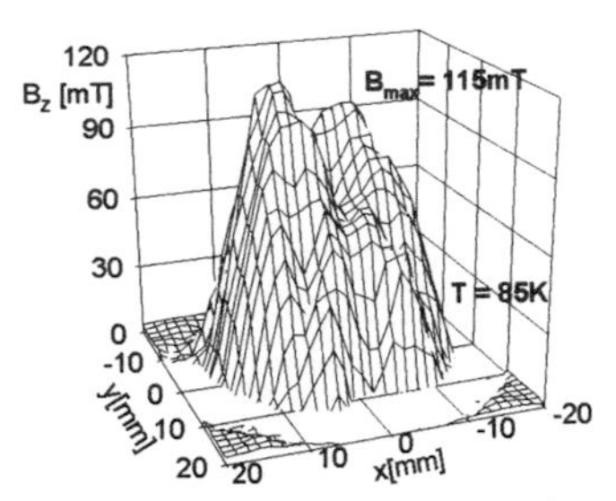

Fig 2b: Trapped field of a 3-domain sample at T = 85K. Magnetizing field = 5T

On the 85K graph the grain structure can be seen by the two peaks near the center. This is not being observed at 12K. The reason is not an increasing intergrain current with decreasing temperatures but the limited local resolution of the measurement: Fig 3 shows the trapped field of two partially overlapping domains (solid lines). The distance of the Hall sensors to the surface of the sample results in a smoothing (dashed lines) of the sharp bean cones. In the fully magnetized state (left) the horizontal distance δr is big enough to be resolved. A

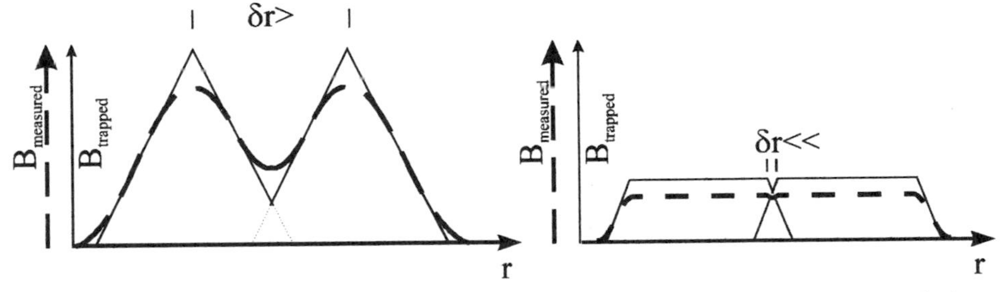

Fig 3: Trapped field of two partially overlapping domains (solid line). The measured signal is smoothed (dashed line). In the fully magnetized state (left) the domains can be resolved, but not in the partially magnetized state (right).

partially magnetized state may lead to a gap δr being too small to be resolved. The sample looks single domain like.

Fig.4a and 4b show the trapped field of a single domain sample at 12K and 85K. The magnetizing field was 8T. The distance of the Hall active area to the sample's top was 2mm so that the maximum trapped field on top of the sample can be extrapolated to 6.8T. The flattened peak at 12K compared to the pointed Bean cone at 85K shows that the sample was

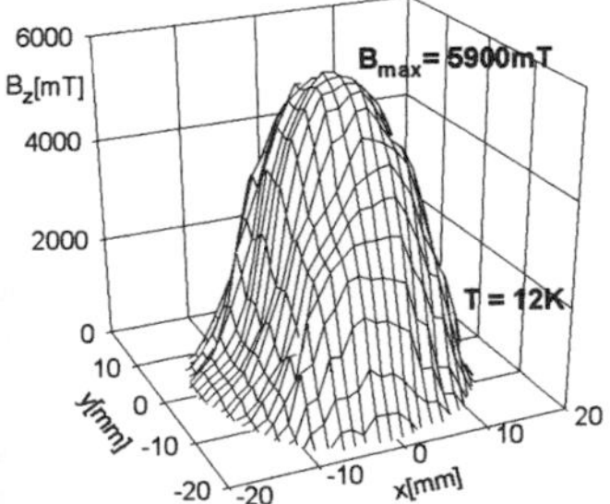

Fig 4a: Trapped field of a single domain sample at T = 12K. Magnetizing field = 8T

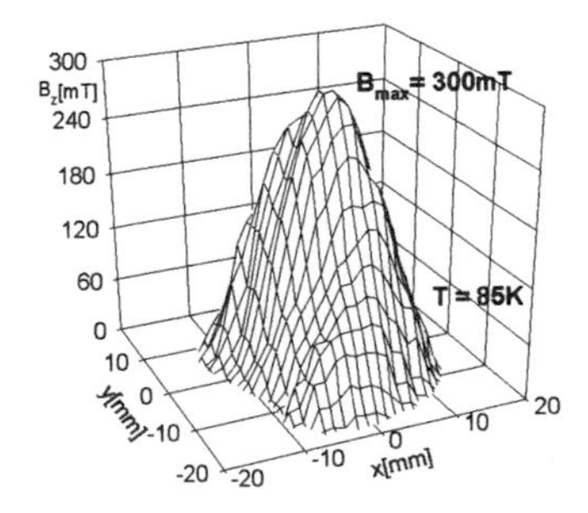

Fig 4b: Trapped field of a single domain sample at T = 85K. Magnetizing field = 8T

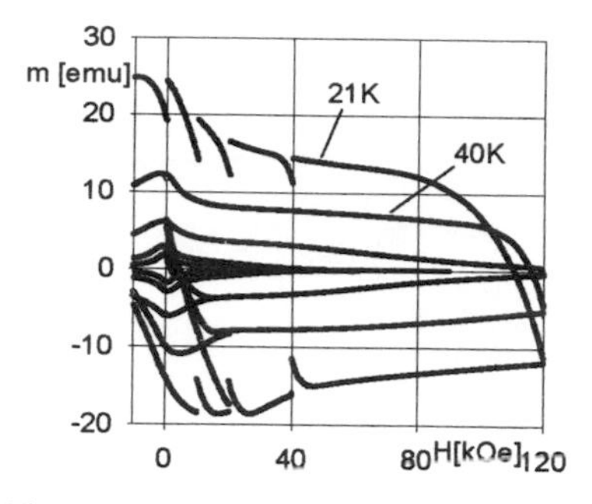

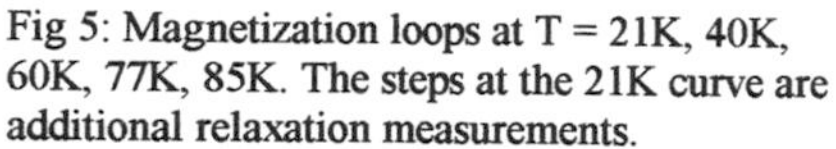
Fig 5: Magnetization loops at T = 21K, 40K, 60K, 77K, 85K. The steps at the 21K curve are additional relaxation measurements.

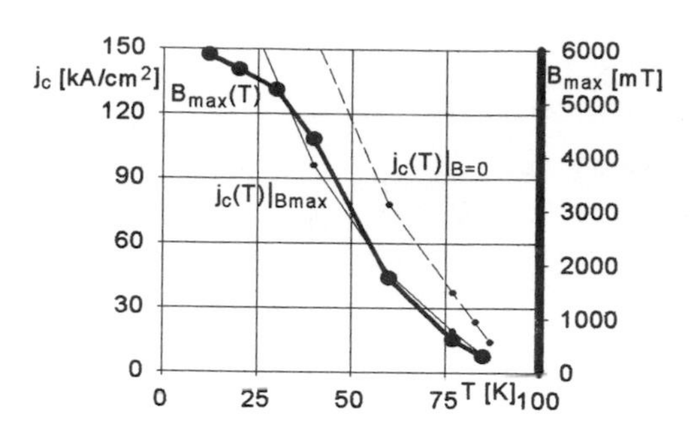

Fig 6: Comparision between the maximum trapped field B_{max} (thick line) and j_c at B=0 (dashed line) and at $B=B_{max}$ (thin solid line) vs. temperature.

not fully magnetized. The decrease from 8T to 6.8T is due to a high relaxation rate at high trapped fields: Fig 5 [1] shows VSM magnetization loops of a small single domain cylinder at various temperatures. The steps at the 21K curve display relaxation effects when the sweep of the external field was stopped for 90 minutes. The magnetic moment m decreased by 25% at an external field of 4T, which can be expected to be similar at 12K at a higher trapped field.

The accordance to the trapped field with j_c versus the temperature is shown in Fig.6. At low temperatures a flattening of the $B_{max}(T)$ curve can be observed This points out to a non fully magnetized state.

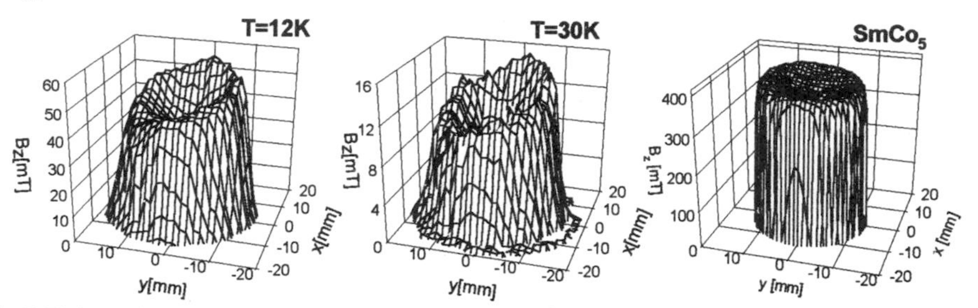

Fig 7: Field profiles of a sintered ceramic YBCO cylinder (right and center) and a permanent magnet

Fig 7 shows the magnetic field profiles of a sintered ceramic YBCO at 12K (left) and 30K (center) compared to a $SmCo_5$ permanent magnet (right). The superconductors show no Bean cone as for homogeneous and single domain melt textured material with large current loops.The ceramic YBCO field profiles look similar to the permanent magnet one. Permanent magnet fields originate from many small magnetic domains. Thus it seems to be obvious that the sintered ceramic YBCO consists of many small domains without intergrain current connection at any temperature. The deeper field dip in the center of the YBCO compared to the $SmCo_5$ one is originated in a higher aspect ratio of the superconductor.

6. Acknowledgements:

The authors wish to thank Dr. V.F. Solovyov for making the Hall line.
This work was supported by the German BMBF, No. 13N6646.

7. Reference:

[1] T. Habisreuther, T. Straßer, W. Gawalek, P. Görnert, K.V. Ilushin and L.K. Kovalev
Magnetic Processes in Hysteresis Motors Equipped with Melt Textured YBCO
Contributed paper to ASC-conference, Pittsburgh, 25th to 30th Aug. '96. In press for Transactions on Applied Superconductivity.

Inst. Phys. Conf. Ser. No 158
Paper presented at Applied Superconductivity, The Netherlands, 30 June–3 July 1997

Penetration of Magnetic Field into Melt-textured YBCO-Samples

Ch. Wenger*, G. Fuchs, G. Krabbes**,Th. Staiger** and A. Gladun***

* Institut für Tieftemperaturphysik, TU Dresden, D - 01062 Dresden
**Institut für Festkörper- und Werkstofforschung, D - 01171 Dresden

Abstract. We investigate the penetration of magnetic field into melt-textured YBCO-samples using Hall generators. Below the temperature of 15 K the formation of a Bean-profile, corresponding to the critical current density is depressed by magnetic flux jumps. The occurrence of these instabilities is well described by an isothermal critical state model.

1. Introduction

We investigated the penetration of magnetic field into zero field cooled samples. The magnetization at the centre of the sample was measured using a Hall generator. The screening of external field is governed by the critical current density. At temperatures below 12.5 K magnetic flux jumps appeared and strongly reduced the screening field. To explain the origin of these instabilities, we investigated the magnetothermal properties of the melt-textured sample.

2. Experimental details

We investigated cylindrical melt-textured YBCO-samples with a diameter of 2,5 cm and 1,2 cm height. The samples exhibit a texture along the c-axis. The samples were prepared at IFW Dresden.

The magnetization is determined using a Hall generator, placed at the centre of the sample, using the ZFC-mode and a rate of 0.5 T/min. The external field is applied along the c axis of the sample.

The experimental set-up to measure the specific heat and the thermal conductivity is described elsewhere [1].

3. Results and Discussion

To determine the magnetic behavior at lower temperature we used a Hall-generator at the top of the same sample and external fields up to 9 Tesla. In the temperature range from 13 K to 90 K we found a continuos lapse of the magnetization. Below 13 K the formation of a complete Bean-profile [2] is depressed by flux jumps as shown in Fig. 1. Similar phenomena are observed in the magnetization of several superconducting materials as NbTi [3], LaSrCuO [4] and YBCO [5,6].

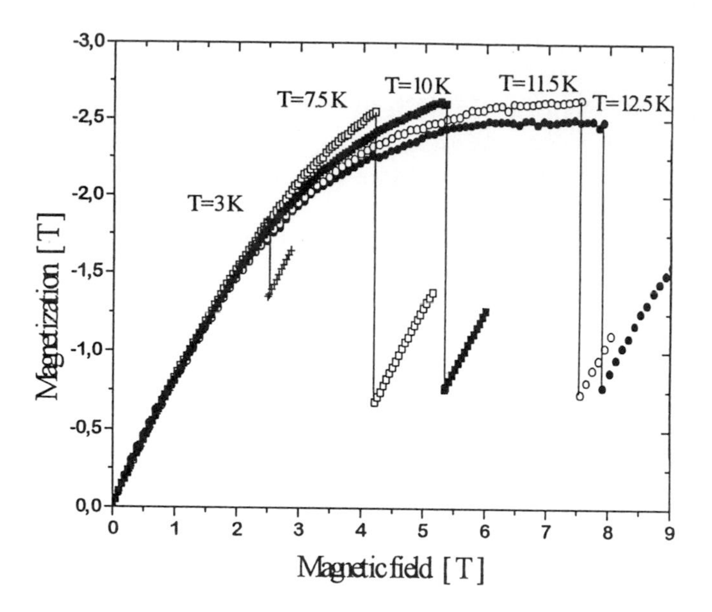

Fig. 1: Magnetization jumps at T = 3, 7.5, 10, 11.5 and 12.5 K as a function of the external field.

Swartz et al. [7] proposed an adiabatic critical state model, assuming that the thermal diffusivity is much smaller than the magnetic diffusivity. We wanted to proof this supposition and measured the thermal conductivity and the specific heat (Fig. 2).

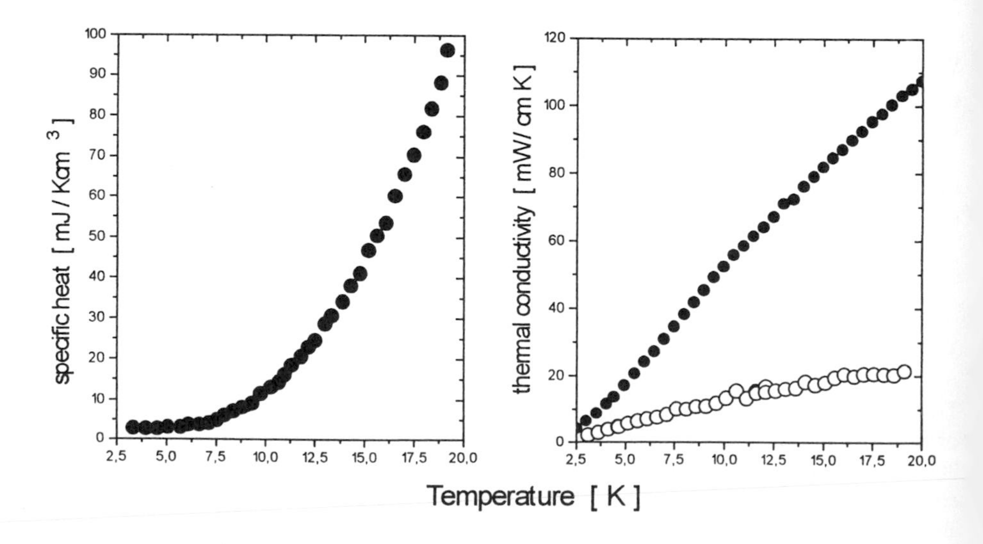

Fig. 2: The volumetric specific heat is plotted on the left picture and the thermal conductivity along the c-axis (O) and in the ab-plane (●) is plotted on the right side of the figure.

The thermal diffusivity D_{therm} is defined as the quotient of the thermal conductivity and the volumetric specific heat. The results of the calculation of D_{therm} from our measurements are shown in Fig. 3. The magnetic diffusivity can be described as $D_{mag} = \rho_{eff} / \mu_0$. Gandolfo et al. [8] studied the effective resistance of NbTi samples due to the motion of vortex lines. Neglecting the field dependence, they determined for ρ_{eff} a mean value of 10^{-11} Ωcm, corresponding to a magnetic diffusivity $D_{mag} \approx 10^{-3}$ cm^2/s. Our results indicate, that the thermal diffusivities are in both directions considerable larger then D_{mag}. Therefore the adiabatic model is not satisfied for our samples.

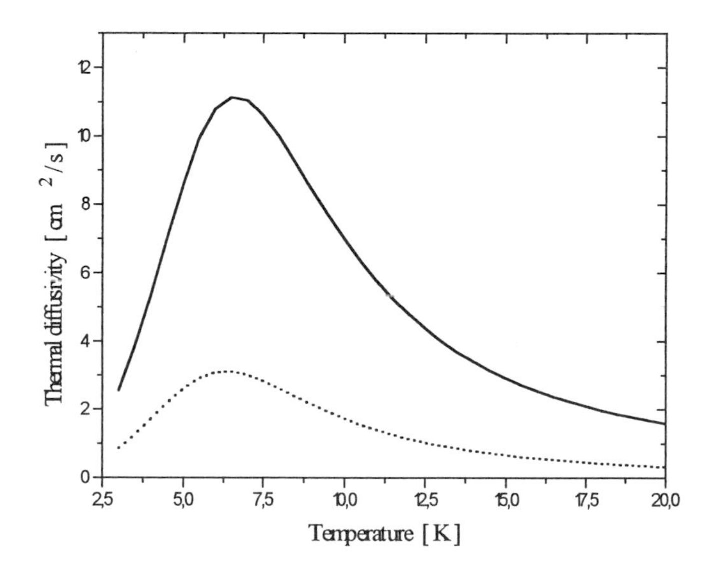

Fig. 3 : Temperature dependence of the thermal diffusivity along the c- axis (dotted line) and in the ab-plane (straight line).

The condition for instability is determined by the amount of heat developed by the motion of flux and the fraction of energy absorbed to heat up the sample. If the heat, produced by the motion of flux, is large enough to increase the temperature of the sample, the system is unstable against small disturbances. Akachi et al. [2] considered the case, that $D_{therm} >> D_{mag}$. The heat produced by the flux motion is not limited to the penetrating region and can extend over the whole sample. They obtained for the stability field H_{fj} at which the first flux jump occurred the expression:

$$H_{fj} = \left(\mu_0 C(T) H^*(T) j_c(T) \left(\partial j_c / \partial T\right)^{-1}\right)^{1/3}$$

H* is defined by the magnetic field at which the Bean-profile is completely formed. Our measurements indicate for H* a temperature dependence $H^*(T) \sim (T_c - T)^{5/2}$. C(T) is the volumetric specific heat. The temperature dependence of the critical current density is calculated from the trapped field data.

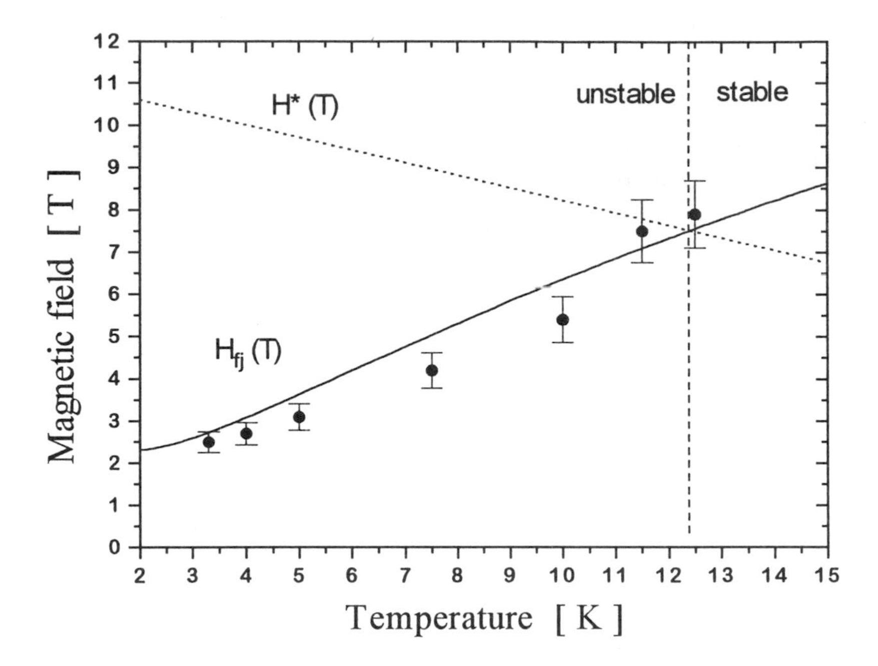

Fig. 4: Comparison between experiment and theory. The filled circles indicate the first instabilities. The solid line is the fit according to the isothermal model. H*(T) is described by the dotted line. There aren't any flux jumps possible above 12.5 K (dashed line).

As shown in Fig. 4, the isothermal critical state model is able to describe the temperature dependence of the first flux jumps. Above the H_{fj} line the magnetization is unstable. For T>12.5 K the final Bean-profile is achieved at magnetic fields lower than H_{fj} and a further increase of the external field do not produce any heating of the sample. Therefore the magnetization is stable above 12.5 K (point of intersection between H* and H_{fj}).

References

[1] Eder F 1956 *„Moderne Messmethoden der Physik, Teil II"* Berlin

[2] Bean C 1964 Rev Mod. Phys. **36** 31-39

[3] Akachi T, Ogasawara T and Yaukochi K 1981 Jap. J. Appl. Phys. **20** 1559-1571

[4] Gerber A, Tarnawski Z and Franse J 1993 Physica C **209** 147-150

[5] Chabanenko V, D'yachenko A, Szymczak H and Piechota S 1996 Physica C **273** 127-134

[6] Müller K and Andrikidis 1994 Phys. Rev. B **49** 1294-1307

[7] Swartz P and Bean C 1968 J. Appl. Phys. **39** 4991-4998

[8] Gandolfo D, Dubeck L and Rothwarf F 1968 Sol. Stat. Com. **6** 799-803

Acknowledgement

This work was supported by BMBF 13N6664.

Inst. Phys. Conf. Ser. No 158
Paper presented at Applied Superconductivity, The Netherlands, 30 June–3 July 1997

Influence of Processing Parameters on Critical Currents and Irreversibility Fields of Fast Melt Processed $YBa_2Cu_3O_7$ with Y_2BaCuO_5 Inclusions

K Rosseel*, D Dierickx, J Vanacken, L Trappeniers, W Boon, F Herlach, V V Moshchalkov, Y Bruynseraede

K.U.Leuven, Laboratorium voor Vaste-Stoffysica en Magnetisme, Celestijnenlaan 200D, B-3001, Belgium.

Abstract. The presence of a homogeneous distribution of submicron Y_2BaCuO_5 green phase inclusions strongly enhances the critical currents in melt-textured $YBa_2Cu_3O_7$. In order to minimize the mean Y_2BaCuO_5 particle size, a fast melt-processing (FMP) scheme was developed, using cooling rates of 30°C/hour, 120°/hour and 180°C/hour during the $YBa_2Cu_3O_7$ crystal growth. Detailed magnetisation measurements in pulsed fields up to 50 tesla and in the temperature interval of 5K to 77K show a clear decrease of the temperature and field dependence of the critical current J_c as well as an enhancement of the irreversibilty field H^{irr} with increasing cooling rate (e.g. H^{irr}(60K)> 50 tesla at the highest applied cooling rates).

1. Introduction

For many foreseen applications of the High-T_c superconductors (HTSC's), the ability to carry high currents in high magnetic fields is required. To achieve this, a strong pinning of the flux lines in a broad field regime is essential. Due to the complicated nature of the phase diagram for the $YBa_2Cu_3O_{7-x}$ (or 123) HTSC, several kinds of structural defects and precipitates are naturally introduced during sample growth. Many of these crystal imperfections have been shown to be effective pinning centers in melt-processed 123. Especially the addition of small non-superconducting Y_2BaCuO_5 (211 or green phase) inclusions results in a substantial increase of the critical current density $J_c(T,\mu_0H)$ over a broad range of applied fields [1]. Whether the flux is pinned by the 123/211 interface itself or by the defect structure surrounding the 211 particles remains unclear. TEM investigations show an increase of the defect density around the 211 particles with decreasing particle size and spacing [2]. Microstructural investigations using polarized light show a linear relation between the a-b microcrack spacing and the 211 particle size [3]. According to several 123 growth models, a decrease of the 211 particle size leads to a more efficient diffusion of yttrium through the melt, thus enhancing the maximum attainable growth speeds. All these results indicate that an enhancement of the critical current density J_c and of the mechanical properties of melt-processed 123 can be achieved by reducing the mean 211 particle size and spacing.

2. Sample preparation

Samples were prepared using freeze dried precursor powders consisting of Y_2O_3 and $Ba_2Cu_3O_5$ with a nominal composition of 80wt% $YBa_2Cu_3O_{7-x}$ and 20wt% Y_2BaCuO_5 [4]. These powders are isostatically cold pressed at 1200 MPa and put on a high quality $BaZrO_3$ substrate. A Fast Melt Processing scheme (see Figure 1) was developed in an attempt to minimize the 211 precipitate size during sample growth, ignoring the issue of texturation. The samples are first heated up to 1030°C, were the Y_2O_3 and $Ba_2Cu_3O_x$ react to form the 211 phase + liquid. They are kept at this temperature for 2 hours to allow the 211 formation reaction to complete. Very fast cooling of the melt through the peritectic with rates up to 180°/hour results in the growth of the 123 phase, while preventing total decomposition of submicron 211 particles and reducing the coarsening of larger 211 inclusions. This results in an average 211 particle size well below 1 μm. The applied cooling rate is the only parameter varied for the different samples presented in this work. These samples were grown using cooling rates of 30°C/h (sample A), 120°C/h (sample B) and 180°C/hour (sample C).

The as-grown samples consist of rather large but randomly ordered 123 domains of about 0.5 to 1mm in size as deduced from optical microscopy. From XRD measurements two major phases i.e. 123 and 211 were identified with some residual non-reacted liquid phases present.

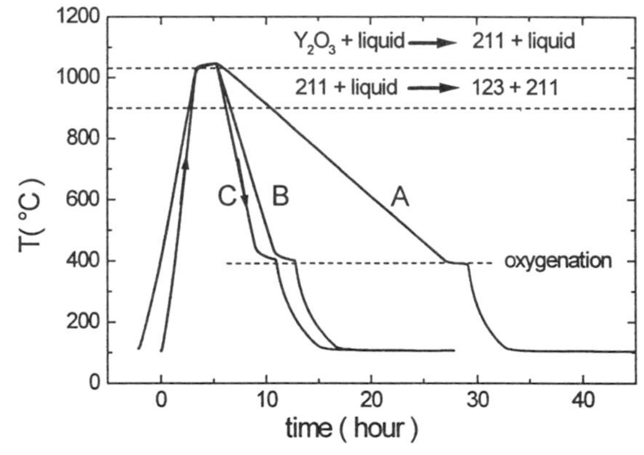

Figure 1: Heat treatment profile for 3 fast melt processed samples using a cooling rate of 30°C/hour (sample A), 120°C/hour (B) and 180°C/hour (C).

3. Results and discussion

Table 1 lists the critical temperatures T_c as determined from AC susceptibility measurements. A trend towards higher T_c and smaller ΔT_c with decreasing cooling rate can be observed. Although the transition is almost complete around $T_{c,mid}$ the magnitude of the susceptibility still increases in lowering the temperature down to 40K, where it levels off.

Table 1: The critical temperatures T_c as determined from AC susceptibility measurements.

Sample	cooling rate	$T_{c,mid}$ (K)	ΔT_c (K) @ $T_{c,mid}$
A	30°/h	93.5	2
B	120°/h	92.1	2.4
C	180°/h	90.8	4

Magnetization measurements in pulsed fields up to 50 tesla (PFMM) [5], [6] were performed to access the irreversible properties (critical current J_c, irreversibility field H^{irr}) in an as large part of the H-T phase diagram as possible. For these measurements, the samples were shaped in the form of a disk with a radius of about 0.75 mm and a thickness of 0.6mm.

The irreversibility field $H^{irr}(T)$ (see Figure 2) is estimated as the field were the width of the hysteresis loops becomes smaller than the resolution of the sensor. Very high values of H^{irr} are obtained from the pulsed fields measurements. Despite the very high fields available, an estimate of $H^{irr}(T)$ could only be made at elevated temperatures (>50K). In this temperature range a clear correlation between H^{irr} and the cooling rate applied during sample growth can be observed. The larger this cooling rate, the higher is H^{irr}.

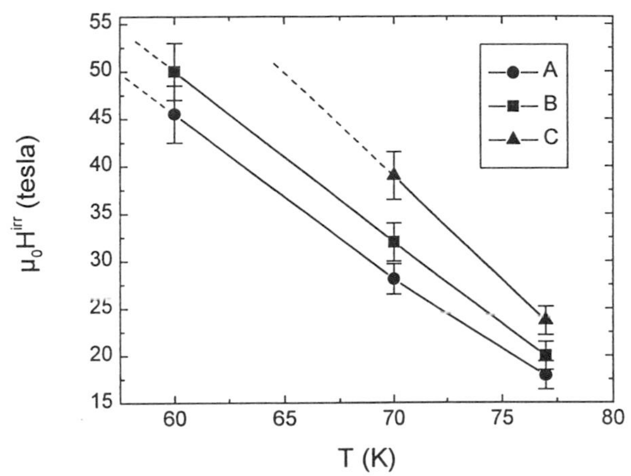

Figure 2 : Irreversibility field H^{irr} as a function of temperature measured in pulsed magnetic fields. Besides the very high absolute values, a clear increase of H^{irr} with cooling rate used for sample processing can be seen.

From the M(H) magnetization loops the critical currents were estimated using the Bean critical state model as a first approximation. Because the path over which the superconducting currents flow is unknown due to decoupling effects of the individual 123 grains, the sample radius R was taken to calculate J_c. The results are summarized in Figure 3a, 3b and 3c. All samples show very high magnetic critical currents of the order of

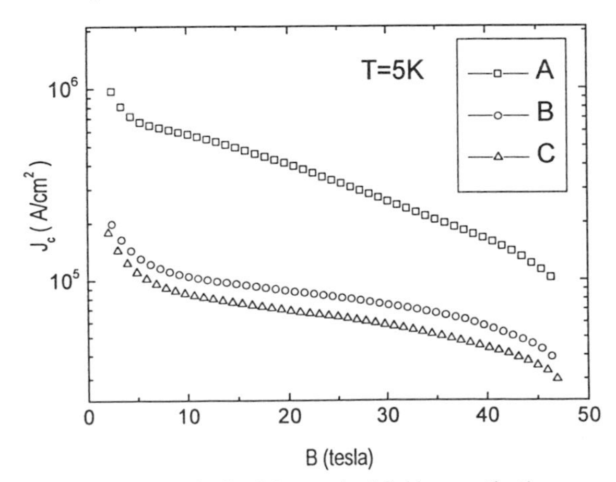

Figure 3a : Field dependence of J_c at 5K obtained from pulsed field magnetization measurements (PFMM)

10^5 A/cm^2 at 20K and 10 tesla. These results can only partially be explained by the very high voltage criterion used in PFMM [7].

At low temperatures the estimated critical currents decrease with increasing cooling rate. On the other hand, the field dependence, as well as temperature dependence, decreases with increasing cooling rate: e.g. there is only a very slight decrease in J_c when the temperature is raised from 5K to 20K, as can be seen from Figure 3a and 3b. This $J_c(H,T)$ behavior causes the J_c versus field curves to cross at elevated temperatures (above 60K), where the maximum critical current in very high fields is obtained for the sample with the fastest cooling rate (figure 3c, Δ). Even at 77K, the J_c values are well above 10^4 A/cm^2 at fields of the ordcr of 10 tesla in all the samples measured by PFMM.

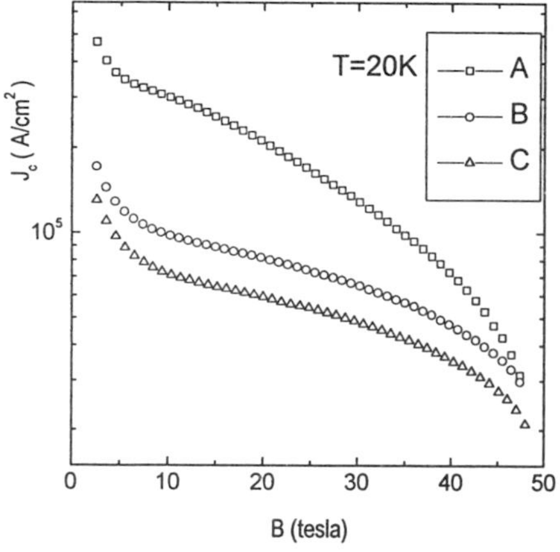

Figure 3b : Field dependence of J_c at 20K obtained from PFMM.

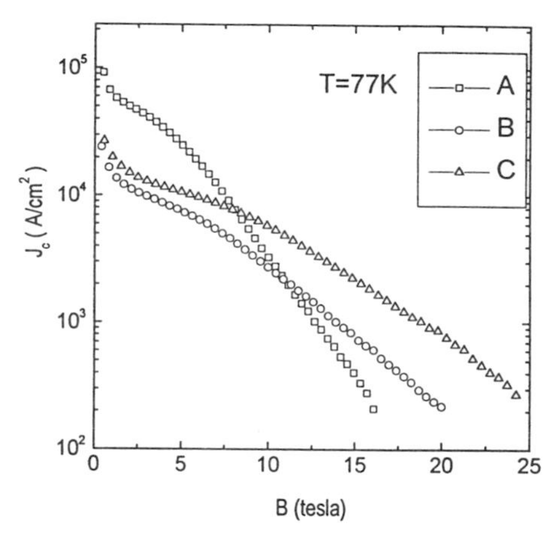

Figure 3c : Field dependence of J_c at 77K obtained from PFMM.

Acknowledgments.

This work is supported by the FWO-Vlaanderen, the Flemish GOA and the Belgian IUAP programs. J.V.and D.D. are Postdoctoral Fellows supported by the K. U. Leuven Onderzoeksraad. L. T. and K. R. are Research Fellows supported by the I.W.T.

References

[1] Kung P J, Maley M P, McHenry M E, Willis J O, Murakami M and Tanaka S 1993 *Phys. Rev. B* **48** 13922
Salama K and Lee D F 1994 *Supercond. Sci. Technol.* 7, 177
Martínez B, Obradors X, Gou A, Gomis V, Piñol S, Fontcuberta J and Van Tol H 1996 *Phys. Rev. B* **53** 2797
[2] Mironova M, Lee D F and Salama K 1993 *Physica C* **211** 188
[3] Diko P, Pelerin N and Odier P 1995 *Physica C* **247** 169
[4] Dierickx D and Van der Biest O 1995 *Eur. J. Solid State Inorg. Chem.* **32** 711
[5] Herlach F, Agosta C C, Bogaerts R, Boon, W, Deckers I, De Keyser A, Harrison N, Lagutin A S, Li L, Trappeniers L, Vanacken J, Van Bockstal L and Van Esch A 1996 *Physica B* 216 161
[6] Lagutin A S, Vanacken J, Harrison N and Herlach F 1995 *Rev Sci. Inst.* **66** 4267
[7] Vanacken J et al, proceedings of EUCAS97.

* E-mail: Kris.Rosseel@fys.kuleuven.ac.be

Inst. Phys. Conf. Ser. No 158
Paper presented at Applied Superconductivity, The Netherlands, 30 June–3 July 1997

Influence of Columnar Defects on Critical Currents and Irreversibility Fields in $(Y_xTm_{1-x})Ba_2Cu_3O_7$ Single Crystals

L Trappeniers[1], J Vanacken[1], K Rosseel[1], A Yu Didyk[2], I N Goncharov[2], L I Leonyuk[3], W Boon[1], F Herlach[1], V V Moshchalkov[1], Y Bruynseraede[1]

[1] Laboratorium voor Vaste-Stoffysica en Magnetisme, K.U.Leuven
Celestijnenlaan 200D, B-3001 Leuven, Belgium
[2] Laboratory of High Energy, Joint Institute for Nuclear Research, Dubna, Russia
[3] Faculty of Geology, M.S.U., Moscow 119899, Russia

Abstract. We have studied the influence of columnar defects, created by heavy-ion (Kr) irradiation, on the superconducting critical parameters of single crystalline $(Y_xTm_{1-x})Ba_2Cu_3O_7$. Superconductivity has not been deteriorated for irradiation fluences up to $6{\cdot}10^{11}$ ions/cm^2. Magnetisation measurements in pulsed fields up to 52 T in the temperature range 4.2 - 90 K have revealed that both critical current $J_c(H,T)$ and irreversibility field $H_{irr}(T)$ are strongly enhanced. Thus, due to the strong pinning of flux lines by these linear defects, the superconducting critical parameters exceed the ones associated with the defect structures present in the unirradiated material. The field range and magnitude of the $J_c(H,T)$ and $H_{irr}(T)$ enhancement are dependent on the irradiation dose.

1. Introduction

High temperature superconductors irradiated with high energy heavy ions form a particularly interesting physical system due to the presence of a distribution of amorphous columnar tracks. In these materials, correlated disorder associated with the tracks becomes dominant in flux line pinning and in vortex dynamics.

The most striking and, from the view of applications, most desired effect of these columnar defects is the enhancement of the irreversible properties of the irradiated material e.g. the critical current and the irreversibility field. The fact that these columnar tracks act as very efficient pinning centres can already be seen from the similarity in geometry of the linear defects and the flux lines. In this way, when the linear defects are nearly parallel to the applied magnetic field, vortices can be pinned over the entire length of the columnar defect.

At low values of the magnetic field, almost all flux lines are pinned by columnar defects. When the magnetic field is increased, at a matching field B_ϕ the flux line density equals to the density of the defects, thus enabling optimal flux pinning. This enhanced pinning at fields $B \leq B_\phi$ is expected to give a higher critical current and a shift of the irreversibility line to higher magnetic fields. At even higher fields $B > B_\phi$ the additional flux lines are subjected to the repulsion from the vortices trapped by the columnar defects and consequently they will be localized in the interstitial positions. The pinning potential associated with them is much weaker and the critical current is decreased.

2. Experimental techniques

2.1 Magnetisation measurements in pulsed magnetic fields

Since critical fields in high temperature superconductors increase sharply with decreasing temperature, it is necessary to use pulsed magnetic fields in order to characterise the superconductor below a reduced temperature $t = T/T_c = 0.8$. The K.U.Leuven pulsed fields facility [1] allows to perform magnetisation measurements in fields up to 60 tesla, at temperatures down to 350 mK. During a pulsed field magnetisation measurement (PFMM), a 20 ms magnetic field pulse of up to 60 tesla is applied parallel to the c-axis of the superconductor while at the same time, the susceptibility of the sample is measured with a calibrated susceptometer. After that, a simple integration gives the field dependence of the magnetic moment of the sample. The sensitivity of this home-made susceptometer is better than 10^{-3} emu at fields below 20 tesla and 10^{-2} emu at higher fields [1].

2.2 Sample preparation

In this study, PFMM have been performed on a series of both unirradiated and Kr-irradiated single crystals of the $(Y_{0.14}Tm_{0.85})Ba_2Cu_3O_7$ high-temperature superconductor (with $T_{c,mid} \sim 91$ K, Fig. 1), all from the same growing session. The crystals have been grown by the self-flux method in ZrO_2:Y crucibles, starting from initial powders containing Tm oxide, $BaCO_3$ and CuO in a ratio 3:25:72. The single crystals were separated by breaking apart the crucibles [2]. The Tm/Y doping is a consequence of the ZrO_2:Y crucible used in the crystal growth and has been determined by EDAX analysis.

Subsequently, these single crystals were irradiated at the JINR U-400 cyclotron in Dubna at room temperature with irradiation doses between $0.75 \cdot 10^{11}$ Kr-ion/cm^2 and $6 \cdot 10^{11}$ Kr-ion/cm^2. In this way, amorphous columnar tracks parallel to the c-axis have been created in the single crystals. The irradiation with the ^{84}Kr-ion beam of energy 2.5 MeV/amu resulted in amorphous tracks with diameter of 25 Å to 35 Å, penetrating the sample up to 13 μm [3].

3. Results and discussion

Since the irradiation process leads to a local amorphisation of the crystal lattice, it is important to make sure that the superconducting matrix between the defects retains its original superconducting properties. A direct way to check the superconducting parameters of the bulk is a low frequency AC susceptibility measurement. Figure 1 shows such a measurement for the $3 \cdot 10^{11}$ Kr ion/cm^2 irradiated crystal, the inset shows the T_c and ΔT_c values as a function of the irradiation dose. The transitions are sharp and there is no significant reduction of T_c after irradiation.

The irradiated $(Y_{0.14}Tm_{0.85})Ba_2Cu_3O_7$ single crystals thus contain random distributions of both correlated (columnar defects) and uncorrelated pinning centres (Y/Tm point defects, oxygen deficiency or local stress). The high field study of these samples makes it possible to determine the magnitude of the expected enhancement of the critical current and the irreversibility line at low temperatures. We can then conclude upon the additivity and efficiency of these two defect structures.

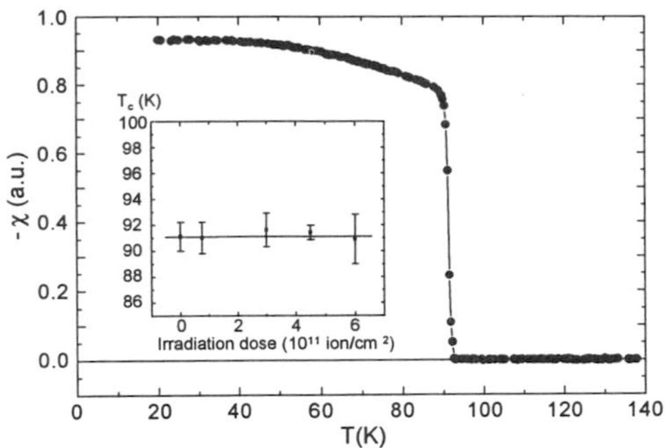

Figure 1 AC susceptibility vs. temperature for the $(Y_xTm_{1-x})Ba_2Cu_3O_7$ single crystal irradiated with $3 \cdot 10^{11}$ Kr ion/cm^2. The inset shows the T_c and ΔT_c values as a function of the irradiation dose.

Applying the simple Bean model to the PFMM gives us the magnetic critical current density $j_{c,m}$. Figure 2 shows $j_{c,m}$ versus magnetic field for the whole set of irradiated samples. It is clear that adding columnar tracks to the superconductor causes a qualitative change in the $J_{c,m}(H)$ behaviour. Instead of a broad maximum, the curves now show a pronounced peak at low fields. At 50 K (left) $j_{c,m}$ is enhanced, but the introduction of tracks does not induce a shift of the irreversibility field (the field where $j_{c,m}$ vanishes). At a higher temperature, 77 K (right), we do see a shift of the irreversibility field (see Figure 3 below).

The position and the width of the observed maximum in the critical current are clearly correlated to the dose of irradiation. This can be understood as a certain commensurability between the mean distance between the defects (derived from the irradiation dose) and the distance between the flux lines. Calculating the first matching field B_ϕ at which each flux line of the vortex lattice ideally corresponds to one defect gives us $B_\phi \sim 1.6$ tesla for the $0.75 \cdot 10^{11}$ ions/cm^2 irradiated sample and $B_\phi \sim 13$ tesla for the $6.0 \cdot 10^{11}$ ions/cm^2 irradiated sample. This does not directly correspond to the experimentally obtained values for B_{max} of resp. 0.5 T and 1.6 T but agrees with earlier observations where the ratio B_{max}/B_ϕ decreases to below 1 with increasing fluence of irradiation. This deviation is easily understood if one takes into consideration that 'matching' is a very local effect since the flux line density itself is not constant over the sample (due to pinning) and that the defects have a random spatial distribution. Moreover, while sweeping up the magnetic field, the area where the commensurability occurs moves inwards the sample.

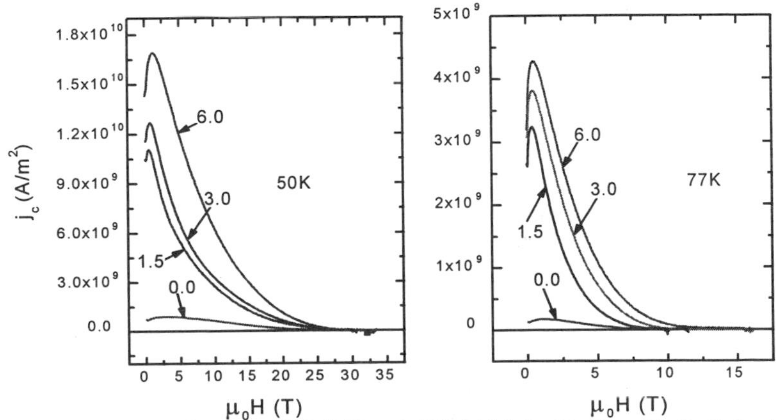

Figure 2 The critical current density at 50K (left) and 77K (right) for $Y_{0.14}Tm_{0.85}Ba_2Cu_3O_x$ irradiated with different doses of Kr-ions. Defect densities of zero, $1.5 \cdot 10^{11}$, $3.0 \cdot 10^{11}$ and $6.0 \cdot 10^{11}$ Kr ions/cm^2 were used.

Let's now focus on the effect of the columnar defects on the irreversibility field. Figure 3 shows the irreversibility line for all the samples. At low temperatures, H_{irr} does not increase with augmenting irradiation dose. It's only at higher temperatures (and hence lower fields) that a shift of the irreversibility line can be observed. This is further substantiated by the two insets: at 50 K we could not find any enhancement of the irreversibility field. At 77 K, however, a significant shift of $H_{irr}(T)$ can be observed when the crystal is irradiated with $6.0 \cdot 10^{11}$ ions/cm^2.

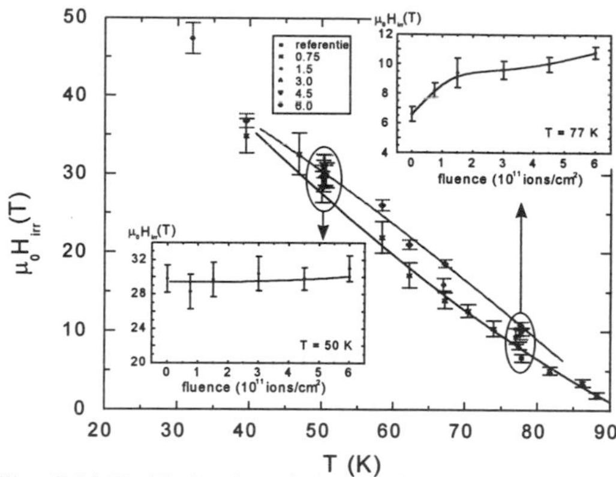

Figure 3 The irreversibility field $H_{irr}(T)$ for the whole set of $Y_{0.14}Tm_{0.85}Ba_2Cu_3O_x$ crystals; the irradiation varies from 0 to $6 \cdot 10^{11}$ ions/cm^2. The insets represent H_{irr} versus irradiation dose at fixed temperature.

4. Conclusions

It has been shown that at fields around the matching field B_ϕ the presence of linear amorphous tracks in $(Y_{0.14}Tm_{0.85})Ba_2Cu_3O_7$ single crystals results in a clear enhancement of the critical current, well above the one in the unirradiated single crystal. Also, at magnetic fields around B_ϕ the irreversibility line shifts to higher temperatures. The magnitude and the position in the H-T diagram of the enhancement of the superconducting critical parameters j_c and H_{irr} depends upon the dose of the heavy ion irradiation. The most striking conclusion however, is the fact that even at fairly high fluences of $6.0 \cdot 10^{11}$ ions/cm^2 the pinning properties of the material are improved and the limit where bulk superconductivity itself is destroyed has not yet been reached.

Acknowledgements

This work is supported by the FWO-Vlaanderen, the Flemish GOA and the Belgian IUAP programs. The cooperation with JINR and MSU was possible thanks to the European INTAS-94-3562 project. J.V. is a Postdoctoral Fellow supported by the K.U.Leuven Onderzoeksraad. L.T. and K.R. are Research Fellows supported by the Flemish Institute for Stimulation of Scientific and Technological Research in Industry (IWT).

References

[1] Herlach F et al. 1996 *Physica B* **216** 16-165.
[2] Leonyuk L, Babonas G J and Vetkin A 1994 *Supercond. Sci. Techn.* **7** 573.
[3] Toulemonde M, Bouffard S and Studer F 1994 *Nucl. Instr. Meth. Phys. Res.* **B91** 108.

Inst. Phys. Conf. Ser. No 158
Paper presented at Applied Superconductivity, The Netherlands, 30 June–3 July 1997

Observation of Local Variations of Stress in Fast Melt Processed $YBa_2Cu_3O_7$ Superconductors at Y_2BaCuO_5 Inclusions

R Provoost, K Rosseel, D Dierickx, W Boon, V V Moshchalkov, R E Silverans, and Y Bruynseraede

Laboratorium voor Vaste-Stoffysica en Magnetisme, Katholieke Universiteit Leuven, Celestijnenlaan 200 D, B-3001 Leuven, Belgium

Abstract. We have performed a detailed analysis of the micro-Raman spectra obtained around the non superconducting Y_2BaCuO_5 (Y211) inclusions embedded in the bulk $YBa_2Cu_3O_7$ (Y123) matrix of fast melt processed samples. The micro-Raman spectra have been recorded along lines crossing the Y211 inclusions with a resolution of 0.5 μm. The energies of the Raman modes of the bulk Y123 matrix are higher than reported values which can be explained by the presence of accumulated compressive stress. Around the Y211 inclusions we observe, beside the appearance of the Raman modes of Y211, a softening of the Y123 Raman modes towards the reported values. The softening occurs in a similar way for all Y123 modes. This indicates a release of mechanical stress around the inclusions. The stress variations around Y211 inclusions can enhance the formation of micro cracks and so deteriorate the crystal quality. For practical applications of the fast melt processed materials one should, besides optimizing pinning, provide also uniform mechanical properties around the pinning centers.

1. Introduction

The improvement of the critical current density (J_c) in a magnetic field is one of the major objectives in the study on the feasibility of the practical applications of high temperature superconductors. In melt processed $YBa_2Cu_3O_7$, the critical current density can be enhanced over a broad range of magnetic fields due to the presence of small non superconducting inclusions of Y_2BaCuO_5 (Y211). The Y211 inclusions of 1 - 10 μm are homogeneously distributed in the superconducting matrix and are good pinning centers for flux lines [1]. An optimisation of the size distribution of the Y211 inclusions can be achieved by using a fast melt processed preparation scheme [2]. On the other hand when transport currents are passing through the sample very high Lorentz forces are acting on the pinned flux lines and this can lead to the formation of cracks. Therefore the mechanical properties of the material should be investigated locally around the inclusions, used as pinning centers.

Micro-Raman spectroscopy has the ability to probe locally the phonon modes of the material. Out of the phonon modes in Raman spectra, one can determine variations of oxygen content [3], orientation [4] and stress [5]. To probe the mechanical properties of the fast melt processed $YBa_2Cu_3O_7$ (FMP123), we have studied the Raman spectra recorded every 0.5 μm along a line crossing an inclusion [6].

2. Experimental

Precursor powder of Y_2O_3 and Ba_2CuO_x corresponding to 80wt% of Y123 and 20wt% of Y211 nominal compositions, were cold-isostatically pressed at 1200 MPa and put on a $BaZrO_3$ substrate. The fast melt proccdure consists of a heating up to 1030°C and stabilization for 2 h to form Y211 + Liquid. A fast cooling of this melt at 60°/h results in the fast growth of the Y123 superconducting phase and prevents the formation of large Y211 inclusions. A photograph of the surface taken by an optical microscopy, with polarised light is shown in figure 1. The mean particle size of the Y211 is below

Fig. 1 Photograph of a polished surface of a fast melt processed $YBa_2Cu_3O_7$ sample.

1 μm. The randomly oriented Y123 domains are about 0.5 mm. The superconducting properties will be reported in detail elsewhere [7]. The critical temperature obtained from resistance measurements is 92 K with a transition less than 1 K. At 77 K the magnetic hysteretic M(H) behaviour is observed up to 15 T. Below 50 K, a pulsed field of 50 T was not sufficient to destroy the irreversible M(H) behaviour.

The Raman spectra were all collected at room temperature at ambient conditions. The 514.5 nm Ar^+ laser line was focused through a 100X objective of a microscope to a spot with a diameter of 1 μm. The backscattered light was analysed using a DILOR XY spectrometer equipped with a double foremonochromator and a LN cooled CCD detector. The accumulation time for each Raman spectrum is about 30 min. and the laser power is kept low enough to avoid local heating effects. Incident and scattered light was polarised parallel to the c-axis of the Y123 matrix. The penetration depth of the laser light in Y123 is about 100 nm [3].

3. Results and discussion

The Raman spectra taken every 0.5 μm along a line of 13 μm are shown in figure 2. The line is crossing two inclusions visible from the surface. The spectra far away from inclusions (the lowest and highest spectra in fig. 2) show a typical set of phonon peaks of Y123. The phonon modes are situated at 119, 152, 438, 510 cm^{-1} and are assigned to c-axis

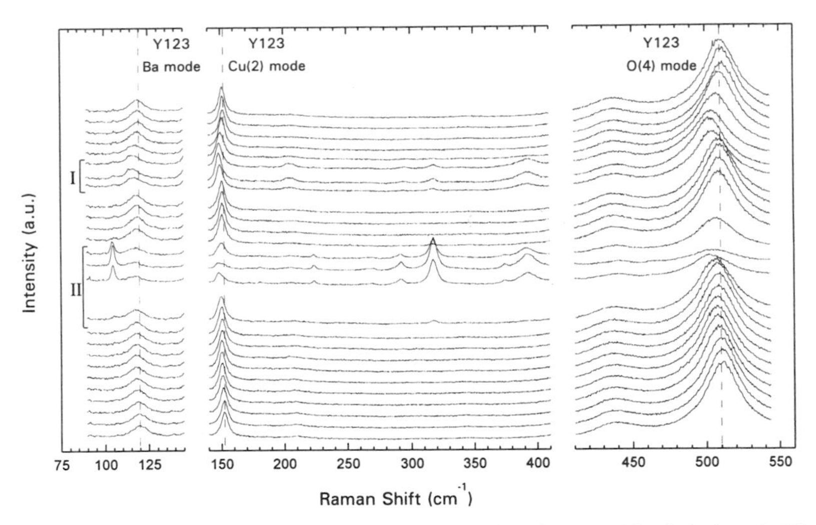

Fig. 2. Micro-Raman spectra taken every 0.5 μm along a line of 13 μm. The dashed vertical lines indicate the position of the $YBa_2Cu_3O_{7-x}$ modes far away from inclusions.

vibrations of respectively Ba, Cu(2), O(2)-O(3) in phase, and O(4). The Cu(2), O(2) and O(3) are the atoms in the CuO_2 planes. The O(4) is the apical oxygen situated between the CuO chains and CuO_2 planes. Around the Y211 inclusion, the spectra, indicated with I and II, shows additional phonon modes at 106, 205, 224, 292, 318, 375, 393, and 441 cm^{-1}, corresponding to the non superconducting Y211 phase [8]. Besides the appearance of these Y211 phonon modes around the inclusions, a remarkable shift of the Y123 phonon modes is seen.

The fitted peak positions of the Ba, Cu(2) and O(4) mode as a function of the position on the sample are plotted in respectively figure 3a, 3b and 3c. It was not possible to distinguish the O(2)-O(3) peak of Y123 and the 441 cm^{-1} peak of Y211. In figure 3 we see

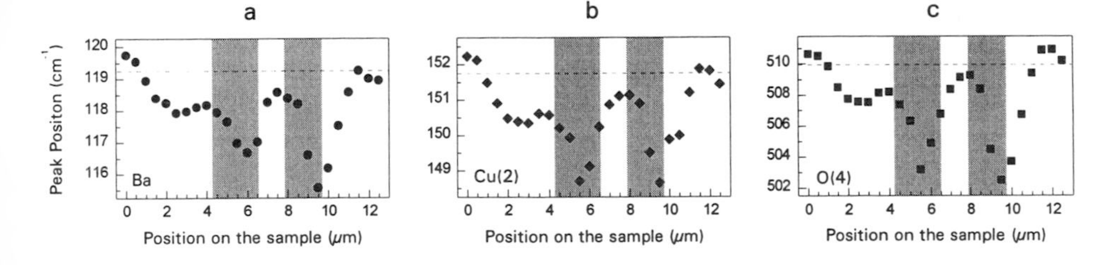

Fig. 3. The energy shifts of the Raman modes of Y123 related to the a. Ba-atoms, b. Cu(2)-atoms, and c. O(4) atoms. The shaded area indicate the places where the Y211 inclusions are located.

that all three Raman modes are softened in a similar way around the inclusion. The horizontal dashed lines indicate the energies of the Y123 modes relatively far away of an inclusion. These energies are higher than reported values for single crystal, namely 118, 150, and 504 cm^{-1} [3]. The shifts of the phonon modes can not be related to local variations of oxygen concentration. In oxygen deficient Y123 crystal only a large shift of the apical oxygen phonon mode has been observed [3]. On the other hand a shift of all Raman modes has been seen when external pressure was applied to $YBa_2Cu_3O_7$ crystals [5]. Therefore we believe that the higher energies of the phonon modes and the softening around an inclusion can be explained if we assume that mechanical compressive stress is accumulated in the bulk matrix of the Y123 during the growth and that this stress is released around the Y211 inclusions. The release of stress around the inclusion is a indication for a good intergrowth between the two phases, however the local variation of stress and the high Lorentz forces, generated by large transport currents, acting on these inclusions could intensify the formation of cracks in the samples. To obtain melt grown samples with improved mechanical properties, several processing parameters should be controlled such as the size and the distribution of the inclusions and the cooling rate.

4. Conclusion

In FMP123 samples, one can strongly enhance the critical current density due to the pinning of the flux lines by Y211 inclusions. However, by analysing the micro-Raman spectra at the interface between the superconducting matrix and the inclusions, we observed local variations of stress, which can have a negative influence on the mechanical properties of the samples. For the use of the FMP123 in practical devices, the strong pinning effect at the inclusion should be combined with improvement of the mechanical properties around the interface area.

Acknowledgements

We would like to thank for the financial support, the Fund for Scientific Research Flanders (FWO), the Concerted Action (GOA), and the Belgian Inter-University Attraction Poles (IUAP) programmes.

References

[1] P.J. Kung, M.P. Maley, M.E. McHenry, J.O. Willis, M. Murakami, and S. Tanaka, Phys. Rev. B **48**, 13922 (1993); K. Salama and D.F. Lee, Supercond. Sci. Technol. **7**, 177 (1994); B. Martínez, X. Obradors, A. Gou, V. Gomis, S. Piñol, J. Fontcuberta, and H. Van Tol, Phys. Rev. B **53**, 2797 (1996).
[2] D. Dierickx, K. Rosseel, W. Boon, V.V. Moshchalkov, Y. Bruynseraede, and O. Van der Biest, *Applied Superconductivity 1995*, Proceedings of EUCAS, 3-6 July 1995, Edinburgh, UK (Bristol, UK: IOP 1995) p. 175-8.
[3] D. Palles, N. Poulakis, E. Liarokapis, K. Conder, E. Kaldis, and K.A. Müller, Phys. Rev. B **54**, 6721 (1996).
[4] K.F. McCarty, J.Z. Liu, R.N. Shelton, and H.B. Radousky, Phys. Rev. B **41**, 8792 (1990).
[5] A.F. Goncharov, M.R. Muinov, T.G. Uvarova, and S.M. Stishov, JETP Lett. **54**, 111 (1991); K. Syassen, M. Hanfland, K. Strössner, M. Holtz, W. Kress, M. Cardona, U. Schröder, J.Prade, A.D. Kulkarni, and F.W. De Wette, Physica C **153-155**, 264 (1988).
[6] R. Provoost, K. Rosseel, V.V. Moshchalkov, R.E. Silverans, Y. Bruynseraede, D. Dierickx, and O. Van der Biest, Appl. Phys. Lett. **70**, 2897 (1997).
[7] K. Rosseel, D. Dierickx, J. Vanacken, L. Trappeniers, W. Boon, F. Herlach, V.V. Moshchalkov, and Y. Bruynseraede, Published in the proceedings of **this** EUCAS (1997), Veldhoven, Nederland.
[8] M.V. Abrashev and M.N. Iliev, Phys. Rev. B **45**, 8046 (1992).

Inst. Phys. Conf. Ser. No 158
Paper presented at Applied Superconductivity, The Netherlands, 30 June–3 July 1997

Fabrication and Transport Properties of Bi-2212/Ag Multifilamentary Tapes and Coils for High Magnetic Field Generation

M Okada[1], K Fukushima[1], J Sato[2], K Nomura[2], H Kitaguchi[3]
H Kumakura[3], T Kiyoshi[3], K Togano[3] and H Wada[3]

[1]Hitachi Research Laboratory, Hitachi Ltd., 7-1-1 Ohmika, Ibaraki, 319-12 Japan
[2]ARC, Hitachi Cable, Ltd., 3550 Kidamari, Tsuchiura, Ibaraki, 300 Japan
[3]National Research Institute for Metals, 1-2-1 Sengen, Tsukuba, Ibaraki 305 Japan

Abstract. Bi-2212/Ag insert magnets have been fabricated and tested at 4.2K and 0-20T. The magnets were made by a wind and react process and included a stack of 3-10 double pancakes. Silver sheathed multifilamentary tapes were used to fabricate the double pancake coils. A Ag-0.5mass%Mg tape was co-wound as a mechanical support member against a large electromagnetic force. After winding, the coils were partially melted at around 880C. The dimensions of the stacked coils were 40-120 mm in height, 50-130 mm in outer diameter and 15-30 mm in clear core. A small stacked pancake coil with persistent current switch (PCS) was also successfully fabricated. The relaxation measurement of the trapped magnetic field for the magnet with a closed circuit proved an excellent possibility to realize the persistent current operation of the HTS magnets.

1. Introduction

Superconducting magnets are useful for obtaining high magnetic fields with excellent stability and homogeneity without the need for a bulky power source, however, the maximum magnetic field generated by a superconducting magnet has been restricted to below around 20T because of the limitations imposed by the upper critical fields of such conventional superconductors as NbTi and Nb_3Sn [1]. The discovery of oxide superconductors opened the door to realize magnets with fields of well above 20T because of their excellent irreversible fields of far over 20T. One of the promising applications of the oxide superconductors is known to be an insert magnet for 1GHz-class NMR magnet system , which requires a quite stable magnetic filed of 23.5T [2]. In order to realize the 1GHz NMR magnet, it is considered that a persistent mode operation of an oxide magnet is required with a field of 2.4T in a backup field of 21.1 T which generated by metallic superconductors[2]. The size of insert magnet will be with 60mm bore, 130mm outer diameter, and 600mm height, and the coil is also required to carry a coil current density of $70A/mm^2$, i.e. the maximum hoop stress will be reached to be around 110MPa. Since such hard parameters seems very difficult to realize at the present stage of HTS coil development, we have tried to fabricate a stack of double pancakes in order to investigate the fabrication methodology and transport properties of the Bi-2212/Ag insert magnet.

Furthermore, we have also tried to fabricate a closed circuit with a persistent current switch.

2. Experimental

2.1. Fabrication of W&R type stacked double pancake coils

All the coils were constructed by means of a wind and react process (W&R) using silver sheathed Bi-2212 multifilamentary tapes[3,4]. A powder-in-tube (PIT) method was used to fabricate the Bi-2212/Ag 19 filament tapes with 0.15mm thick, 5mm wide 300m long. The tapes were then wound in a double pancake form, and then partially melted at around 880℃ in a flowing oxygen atmosphere. A Ag-0.5wt.%Mg alloy tape was co-wound as a mechanical support member for against a electromagnetic force in backup magnetic fields. Alumina paper 0.1mm thick and 5mm wide was used as an insulator. The double pancake coils were 11 mm high, with an outer diameter of 90-130mm and an inner diameter of 30mm. Finally, the coils were annealed at around 800℃ for 1-100h in order to optimize the carrier concentration. All the process were carefully undertaken to prevent contamination by H_2O and CO_2. After the final heat treatment, 3-10 pieces of the double pancake coils were stacked, and were then impregnated with an epoxy resin to improve the mechanical strength.

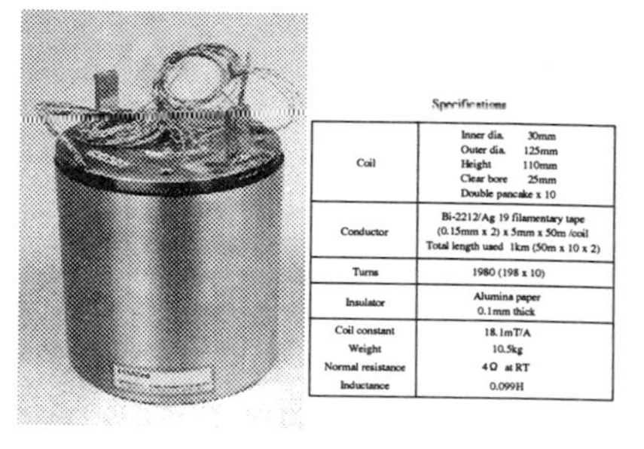

Specifications

Coil	Inner dia. 30mm Outer dia. 125mm Height 110mm Clear bore 25mm Double pancake x 10
Conductor	Bi-2212/Ag 19 filamentary tape (0.15mm x 2) x 5mm x 50m /coil Total length used 1km (50m x 10 x 2)
Turns	1980 (198 x 10)
Insulator	Alumina paper 0.1mm thick
Coil constant Weight Normal resistance Inductance	18.1mT/A 10.5kg 4Ω at RT 0.099H

Fig.1 Bi-2212/Ag insert magnet.

Backup magnetic fields up to 18-21T were applied at Tsukuba Magnet Laboratory, National Research Institute for Metals. The Bi-2212/Ag insert magnets were placed in the center of a superconducting magnet constructed from metallic superconductors. All the *V-I* measurements in the backup magnetic fields were performed in the insert dewar at 4.2K.

2.3. Fabrication of Bi-2212/Ag closed circuits with persistent current switch

The Bi-2212/Ag closed circuit with a persistent current switch [PCS] was also fabricated by a wind and react process using silver sheathed 19 filament tapes[5]. The PCS was designed as a thermal switch with non-inductive winding using a Ag-10at%Au alloy sheathed 19 filament tape. In order to obtain sufficient resistance when the PCS is off, the Ag-10at%Au alloy was used for the sheath material, since it has two orders of high resistivity compared with pure silver in low temperature regions. The magnet part consist of 4 stacks of double pancakes, and were jointed together with, using a butt-jointing method. The butt-joint enables a superconducting joint with a current carrying capacity of over 500A[6]. Alumina paper was also used as an insulator. The coils were partially melted at around 880℃ in a flowing oxygen atmosphere, and then the coil was annealed at around 800℃ for 1-100h in order to optimize the carrier concentration. After

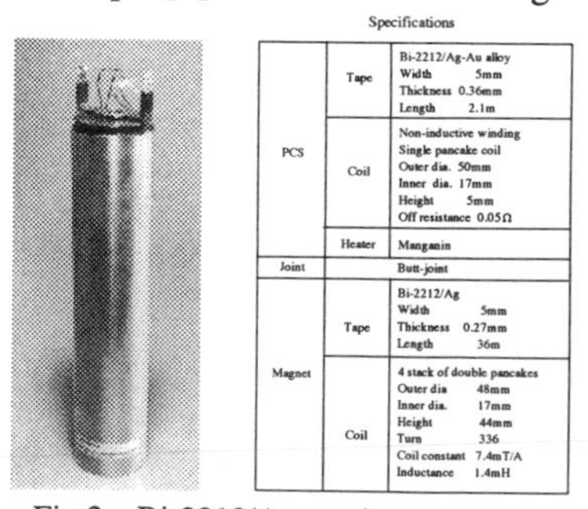

Specifications

PCS	Tape	Bi-2212/Ag-Au alloy Width 5mm Thickness 0.36mm Length 2.1m
	Coil	Non-inductive winding Single pancake coil Outer dia. 50mm Inner dia. 17mm Height 5mm Off resistance 0.05Ω
	Heater	Manganin
Joint	Butt-joint	
Magnet	Tape	Bi-2212/Ag Width 5mm Thickness 0.27mm Length 36m
	Coil	4 stack of double pancakes Outer dia 48mm Inner dia. 17mm Height 44mm Turn 336 Coil constant 7.4mT/A Inductance 1.4mH

Fig.2 Bi-2212/Ag persistent magnet.

the final heat treatment, the coil was then impregnated with an epoxy resin in a brass case as shown in Fig.2.

3. Results and discussion

3.1. Transport properties of stacked double pancake coils

Figure 3 shows the field dependence of J_c for one of the insert magnets. This field dependence is basically consistent with that for a short sample under a *B*//tape surface condition. The $J_{c,oxide}$, 300A/mm^2 was obtained at a backup field of 18T, the J_c showed only about 15% of the best value for short sample tapes. However, the magnet still generated 1.8T at a backup field of 18T, i.e. the total magnetic field reached 19.8T, carrying an I_c of 98A, which correspond to a $J_{c,coil}$ of 36A/mm^2. Thus, the maximum hoop stress was estimated to be 77MPa for conductors. The stress was just above the yield stress of the silver sheathed tapes. During the experiment, we never found marked training effect of the transport property for the Bi-2212/Ag coil, that we observed in our previous study[4]. We also did not observe marked degradation of transport properties induced by the large electromagnetic force. The coil generated over 3T, carrying a supercurrent of over 150A without backup fields just after the transport measurement in 18T backup fields.

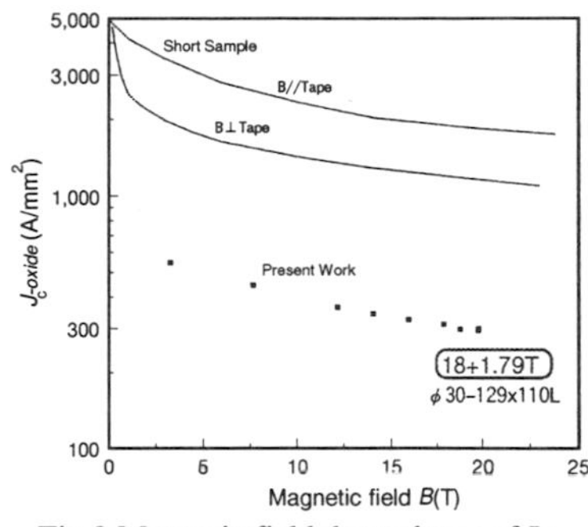

Fig.3 Magnetic field dependence of Jc .

3.2. Persistent mode operation of a Bi-2212/Ag closed circuit

Although present superconducting closed circuit includes 5 pieces of the butt-joints, the transport properties showed an excellent decay curve. Figure 4 shows the relaxation curve of the trapped magnetic field for the closed circuit. The coil trapped ~1T magnetic field with a current of 134A. Those values are almost one order larger than those of our previous work as shown in Fig.4. After 116h, the trapped field was decreased to be 0.24T, showing a resistance in the overall loop to be below 0.2nΩ.

A *V-I* curve was calculated from the decay curve, and the index number was estimated. The index number *n* is given by the following equation,

$$V \propto I^n .$$

The *n*-value estimated from a current range between 40 and 100A was 4.2, which almost corresponds to those of the coils obtained by transport measurements at a resistivity of around $10^{-13}\,\Omega$cm.

However, we found that the n-value increased rapidly over 30 when the trapped current decreased below 40A, showing the resistivity below 10^{-15} Ω cm. The observed sudden increase of n-value would be related to a change of current distribution in a superconducting tape. In a small current level $J<<J_c$, supercurrent would be able to flow through only in superconducting paths without any dissipation. Thus, it is thought that the observed n-value could be improved further with improving homogeneity of J_c of the superconducting filaments.

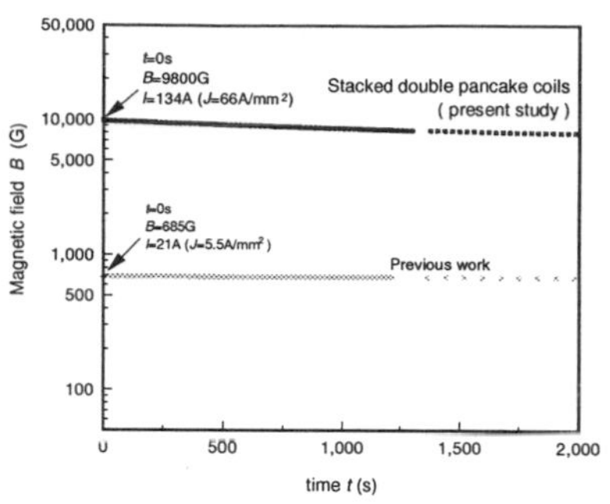

Fig.4 Relaxation of trapped magnetic field.

4. Conclusion

A Bi-2212/Ag superconducting insert magnet and a persistent magnet with PCS have successfully been fabricated and tested.

The insert magnet consisting of 10 double pancakes with 130mm outer diameter has been fabricated and tested at 4.2K with various magnetic fields. The I_c of the insert magnet was 180A with a self field of around 3.3T and 98A at a backup field of 18T, respectively. The magnetic field generated by the insert magnet at a backup field of 18T was 1.8T , i.e. 19.8T was generated by the total of the present superconducting magnet system.

The persistent magnet trapped a ~1T magnetic field with a supercurrent of 134A. The relaxation of the trapped field was measured for up to 116 hours in the operation of the persistent mode. The resistance of the overall closed circuit, including 5 superconducting butt joint, was estimated to be below 0.2nΩ.

References

[1] Kiyoshi T, Kosuge M, Inoue K, Maeda H, 1996 IEEE Trans. Magn. 32 2478-2481

[2] Kiyoshi T, Inoue K, Kosuge M, Itoh K, Yuyama M, Maeda H, 1996 *Proc. 16th Int. Cryo. Eng. Conf./ Int.Cryo. Mater. Conf. (ICMC16/ICMC), May 20-24, Kitakyushu*

[3] Okada M, Tanaka K, Sato J, Awaji S, Watanabe K 1995 *Jpn.J.Appl.Phys.* **34** 4770-4773

[4] Okada M, Tanaka K, Fukushima K, Sato J, Kitaguchi H, Kumakura H, Kiyoshi T, Inoue K, Togano K, 1996 *Jpn.J.Appl.Phys.* **35** L623-L626

[5] Okada M, Fukushima K, Tanaka,K, Kumakura H, Togano K, Kiyoshi T, Inoue K, 1996 *Jpn. J. Appl. Phys.* **35** L627-L629

[6] Okada M, Fukushima K, Tanaka K, Hirano T, Sato J, Kitaguchi H, Kumakura H, Togano K, Wada H, 1996 *J. Jpn. Inst. Metals, to be published*

Inst. Phys. Conf. Ser. No 158
Paper presented at Applied Superconductivity, The Netherlands, 30 June–3 July 1997

Fabrication and Properties of Pancake Coils Using Bi-2223 / Ag Tapes

P Verges, K Fischer, A Hütten, T Staiger, G Fuchs

Institut für Festkörper- und Werkstofforschung Dresden, P.O. Box 270016, D-01171 Dresden, Germany

Abstract. Small double pancake coils with a 15 mm bore were fabricated from Ag sheathed mulifilamentary Bi-2223 tapes by the 'wind and react' technique. By optimizing the parameters of the final heat treatment, the critical current density of the tapes could be achieved also in the winding of the coils. A typical value of the critical current density at 77 K is 18 000 A/cm^2 for a 90 m long tape. In a double pancake coil with a total tape length of 48 m, a magnetic field of 1 T could be generated at 20 K. A filling factor of 15 % was obtained for the superconductor in the winding of the coil, compared with a value of 25 % for the Bi-2223/Ag tapes. The critical current in different sections of the coils was controlled by electric field-current measurements. Another problem is the electric connection between both pancake coils. A contact resistivity of less than $2 \cdot 10^{-7}$ Ω was achieved at 20 K. Investigations of the current limiting mechanisms were performed on short tapes in the temperature range between 4.2 K and 77 K.

1. Introduction

Significant progress has been made in the development of long silver sheathed Bi-2223 tapes by the 'powder in tube' technique. Several groups reported high critical current densities in multifilamentary Bi-2223/Ag tapes with lengths of several 100 meters up to 1500 meters [1-4].

The long Bi-2223/Ag tapes have been wound into pancake-shaped coils. Test magnets have been assembled from the pancake coils and investigated at various temperatures. A record value of 4 T has been achieved at 4.2 K in a test magnet containing 1000 m of Bi-2223/Ag tape [3].

In our institute, multifilamentary Bi-2223 tapes with Ag and AgCu sheath were developed in collaboration with Siemens AG [4,5]. In the present paper, results for Bi-2223 pancake coils are presented.

2. Bi-2223/Ag tapes

Silver sheathed Bi-2223 tapes were prepared by the 'powder in tube' technology in lengths of several hundred meters [4].

In Fig. 1, the distribution of the critical current density j_c at 77 K along the tape length is shown for the tapes which were used in the pancake coil described in the following. A variation of j_c between 15 kA/cm^2 and 20 kA/cm^2 was found for these two

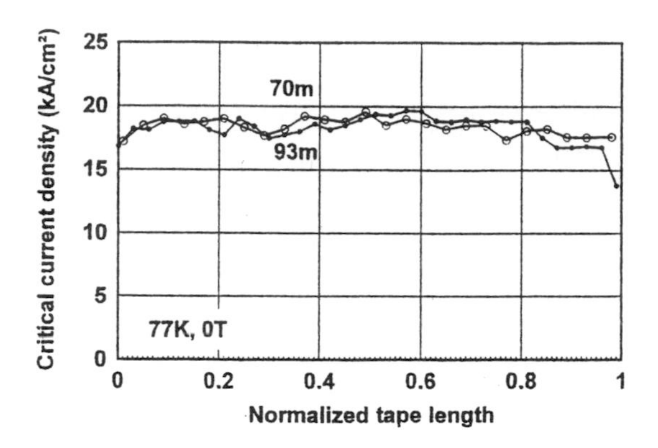

Fig. 1: Distribution of j_c values along the length of a 70m and 93m Bi-2223/Ag tape at 77K and zero field

tapes with lengths of 70 and 93 m. In the last time, longer tapes with similar j_c values could be prepared. In a tape of 550 m length, a critical current density of 25 kA/cm^2 was achieved recently.

Aiming on a future application in conductors (at 77 K) and magnets (at $T \leq 30$ K) we investigate the current limiting mechanisms in this material by performing transport experiments on short samples in the temperature range $4.2 \text{ K} \leq T \leq 77$ K and in external magnetic fields up to 16 T. The measured $j_c(H,T)$ and $E(j,H,T)$ characteristics can be quantitatively analysed [6] in the framework of a modified brick wall model [7] assuming current flow both along firmly coupled "high current paths" and weak link restricted "low current ways". At 77 K the $j_c(H)$ characteristic is governed by intragranular flux creep of pancake vortices in all current paths and can be improved only by the introduction of a pinning effective defect structure. With decreasing temperature intragranular flux creep is reduced and the intergranular coupling determines the critical current density. At 30 K only about 10% of the grains remain firmly coupled and form high current paths with a critical current density of $j_p \approx 10^6$ A/cm^2, whereas 90% are weak link limited with $j_w \approx 10^4$ A/cm^2. Comparative studies with tapes of different filament numbers and sheath materials reveal that multifilamentary tapes with hardened AgCu sheath contain the highest volume fraction of high current paths (≈10%). Therefore this tape design seems advantageous to construct coils and magnets for low temperature applications.

3. Pancake coils

Small pancake coils with an inner winding diameter of 20 mm and a bore of 15 mm were manufactured by the 'wind and react' technology. Three Bi-2223/Ag tapes were co-wound in parallel on alumina formers together with a thin AgCu tape. The top side of the AgCu tape was coated with an alumina screen print paste as insulating material separating each turn. During heating the coils up to 400° C, the organic solvent, binder and dispersant of the green tape was removed. The superconducting Bi-2223 phase develops by annealing the coils at temperatures of about 820° C in an nitrogen atmosphere containing 8% oxygen. In order to achieve the critical current density j_c of the Bi 2223/Ag-tape also in the coil winding, the parameters of the final heat treatment were optimized. More than ten small pancake test coils were built up and tested until the j_c value of the Bi 2223/Ag-tape was obtained in the coils. During the heat treatment, the

Table 1 Main parameters of a double pancake coil

Bi-2223/Ag tape				Winding				
Number of filaments	Cross section (mm^2)	Total length (m)	j_c at 77 K (A/cm^2)	Filling factor %	Number of turns	Inner diameter (mm)	Outer diameter (mm)	Length (mm)
81	3.7x 0.27	48.1	17000	24.5	2 x 61	20	64	27

clean backside of the AgCu tape and the Ag sheath of the three Bi-2223 tapes weld together by diffusion. The AgCu tape works as a mechanical support in the winding. The coils were impregnated with epoxy resin of low viscosity in order to obtain a strongly fixed winding after the heat treatment. The thickness of the AgCu tape including the insulating alumina sheath is about 80 μm and comparable with the thickness of ceramic paper used also for insulation [8]. Hence, the advantage of a strongly fixed winding is combined with a relatively high packing factor up to 67 % for the Bi-2223/Ag tape in the coil.

Double pancake coils were built up from two pancake coils by connecting them with a low resistivity contact at the central tube.

4. Test results

In this section, test results of a double pancake coil are presented. Multifilamentary Bi-2223/Ag tapes with 81 filaments were used in this coil. The main parameters of the coil are given in Table 1. Test results for this coil are summarised in Table 2. Taking into account the field dependence of the critical current density of Bi-2223 at 77 K, j_c values between 15 and 20 kA/cm^2 were found for this coil in zero magnetic field, which corresponds to the critical current density in the tape (see Table 1).

The coil was measured at different temperatures. A maximum magnetic field of 0.20 T was generated at 77 K with a current of 55.8 A. The field in the centre of the bore is by 8.5 % lower. A field of 0.39 T was measured at 64 K. At 20.4 K, a maximum magnetic field of 1.05 T was achieved with a superconducting critical current density of 40.6 kA/cm^2 and an engineering current density of 6.1 kA/cm^2. Different sections in the coil were investigated in more detail by measuring the electric field current (E-I) characteristics. Eight voltage taps were placed at each double pancake coil in order to characterise the inner, middle and outer section of the coil separately. The E-I characteristics of the coil measured at 77 K are shown in Fig. 2. The exponential law found for the E-I characteristics of the coil differs from the power law observed for Bi-2223 tapes at constant magnetic field [9]. The simple reason is that the E-j relation in the coil is influenced by the magnetic field generated by the coil itself.

Table 2 Test results of the double pancake coil

T (K)	Max. magnetic field (T)	Critical current (A)	Superconducting critical current density (kA/cm^2)
77.3	0.20	55.8	7.7
64	0.39	110.2	15.1
20.4	1.05	295.5	40.6

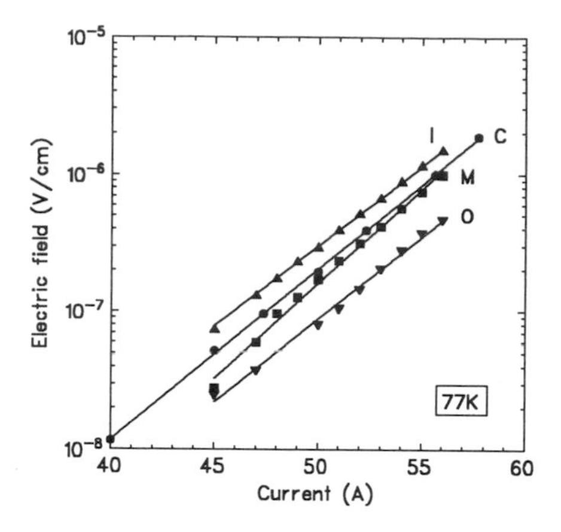

Fig. 2: Electric field - current curves of one coil (C) of the double pancake magnet at 77K. (I): inner section, (M): middle section, (O): outer section

The logE-I characteristics of Fig. 2 for the different sections have about the same slope. Considering the 10^{-6} V/cm criterion, the critical current decreases from the outer section (O) to the inner section (I), which reflects the field dependence of the critical current of the Bi-2223 tape. The critical current averaged over the whole coil (C) is similar to that of the middle section (M).

The joint between the pancake coils is very critical, because the contact between the two Bi-2223 tapes is not only non-superconducting, but it is also located in the highest magnetic field inside of the winding. The contact resistivity of the joint between the pancake coils was measured at different temperatures. Contact resistivities of $1\cdot10^{-6}\ \Omega$ at 77 K and of $2\cdot10^{-7}\ \Omega$ at 20 K were determined at the critical current of the coil resulting in power values of 3 mW at 77 K and of 18 mW at 20 K. These values are small compared with the heat generated in the current leads or with hysteresis losses during field sweeps of the magnet. Nevertheless, an improved version of the joint between pancake coils was tested in a small coil. At 77 K, a contact resistivity of 1·10-7 Ω was achieved at the critical current of 52.5 A of the coil.

Acknowledgement

This work was supported by Bundesministerium für Bildung, Wissenschaft, Forschung und Technologie under BMBF contract no. 13N6102.

References

[1] Balachandran U et al 1997 *IEEE Trans. on Appl. Supercond.* **7** 2207-2210
[2] Buczek D M et al 1997 *IEEE Trans.on Appl. Supercond.* **7** 2196-2199
[3] Sato K et al 1996 *Physica B* **216** 258-264
[4] Hütten A et al presented at EUCAS '97, Eindhoven, the Netherlands, June 30 - July 3, 1997
[5] Fischer K et al 1995 *IEEE Trans. on Appl. Supercond.* **5** 1259-1262
[6] Staiger T et al 1997 *IEEE Trans. on Appl. Supercond.* **7** 1347-1350
[7] Bulaevski L et al 1993 *Phys. Rev. B* **48** 13798-13816
[8] Okada M et al 1996 *Jpn. J. Appl. Phys.* **35** L623-L626
[9] Fuchs G et al 1996 *Advances in Superconductivity VIII* (Tokyo: Springer) 847-850

Inst. Phys. Conf. Ser. No 158
Paper presented at Applied Superconductivity, The Netherlands, 30 June–3 July 1997

High field performance of Nb_3Al multifilamentary conductors prepared by phase transformation from bcc solid solution

M Kosuge[1], T Takeuchi[1], M Yuyama[1], Y Iijima[1], K Inoue[1], H Wada[1], K Fukuda[2], G Iwaki[2], S Sakai[2], H Moriai[2], B ten Haken[3] and H H J ten Kate[3]

[1]Tsukuba Magnet Laboratory, National Research Institute for Metals, Tsukuba 305, Japan.
[2]Hitachi Cable Ltd., 3550 Kidamari, Tsuchiura 300, Japan.
[3]University of Twente, P.O.Box 217, 7500 AE Enschede, The Netherlands.

Abstract. Nb_3Al multifilamentary conductors have been fabricated via a newly developed process where supersaturated bcc-solid-solution is first formed by quenching from the melting point region and subsequently transformed to Nb_3Al. These conductors show large n-indices in the voltage-current characteristics, even at 20T, comparable to Nb_3Sn. The strain-tolerance of critical current density in these conductors at 12 T is comparable to that for conventionally heat-treated Nb_3Al conductors. Based on the test result obtained for the 1331-turn solenoid-coil, the newly developed Nb_3Al conductors are regarded as realistic alternatives to practical Nb_3Sn conductors for high-field, large-scale applications.

1. Introduction

Recently, we have developed a new fabrication technique for Nb_3Al multifilamentary superconductors with very high stoichiometry by exploiting the transformation from supersaturated bcc-solid-solution $Nb(Al)_{ss}$.[1,2] The rod-in-tube (RIT) or jelly-roll (JR) processed Nb/Al multifilamentary wires, several hundred meters in length, were continuously ohmic-heated (~2000°C) and then quenched in a molten gallium bath to retain the metastable $Nb(Al)_{ss}$ filaments embedded in a Nb matrix at room temperature. The resulting $Nb/Nb(Al)_{ss}$ composites were annealed at 700-900°C to transform the $Nb(Al)_{ss}$ to nearly-stoichiometric A15 Nb_3Al with fine grains. Both of the transformed RIT and JR conductors show nearly identical non-Nb J_c at 4.2 K which are substantially higher than non-Cu J_c values of the conventional JR processed Nb_3Al[3] and the ITER Nb_3Sn[4] strands over the whole range of magnetic fields examined. Although the volume fraction of Nb_3Al phase shall furthermore be increased by optimizing the cross-sectional structure, the overall J_c which is defined for the cross-sectional area including Nb-matrix of the transformed JR conductor is, even at present, higher than their non-Cu J_c values.[5]

In the present study, we have examined various J_c-characteristics of the transformed RIT and JR Nb_3Al conductors, which are required in designing the high-field superconducting magnet with theses conductors; the n index in the voltage-current characteristics, the degradation in J_c with mechanical strain (compression and tension), the load line performance of the small coil, and so on.

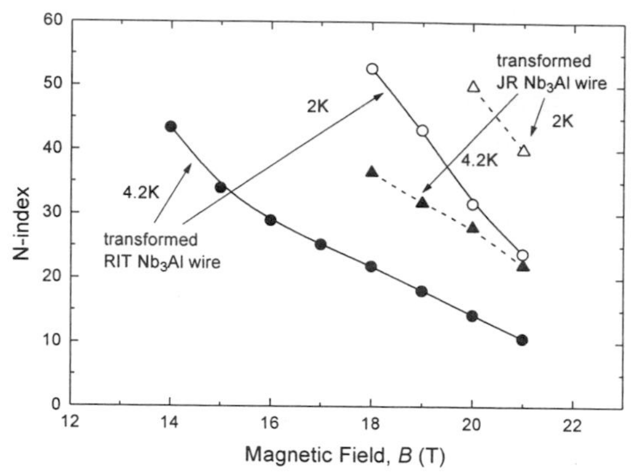

Figure 1 The n-index in the voltage-current characteristics of the transformed RIT and JR Nb_3Al conductors.

2. Voltage-current characteristics

The n-index is quite important for practical use, in particular for nuclear-magnetic-resonance (NMR) magnet operated in persistent current mode, since the small n-index restricts the operation current of the coil to much smaller than the critical current (I_c) of the conductor for ensuring a negligible ohmic loss. It is believed that inhomogeneity of filament structure, such as sausaging, compositional deviation and strained distribution, is responsible for the small n-index value. Since the heat treatment to form the $Nb(Al)_{ss}$ was carried out at a quite high temperature, the multifilamentary structure seems to be no longer uniform in the longitudinal direction and the n-index should be small accordingly.

However, as shown in Fig. 1, both the transformed RIT and JR Nb_3Al conductors show, contrary to our expectation, large n-indices even at 20 T and the remarkable field dependence, which are comparable to those of the bronze processed Nb_3Sn conductors. The n-indices at 20 T are further enhanced by reducing temperatures to 2 K. The large n-indices and hence the uniform structure may mean that the elemental Nb/Al filaments in the basic bundle almost completely reacts with each other during ohmic-heating up to temperatures close to the melting point of Nb, and each basic bundle then behaves as a single $Nb(Al)_{ss}$ filament before the transformation to Nb_3Al. The effective filament diameter of the RIT Nb_3Al conductor, estimated with magnetization, is nearly the same as the basic bundle size. The large n-index for the JR conductor corresponds to its basic-bundle-size which is larger than that of the RIT conductor. It is noted that the large n-index at high fields (>20 T) enables the application of the transformed Nb_3Al conductors to GHz class NMR analysis.

3. Strain sensitivity

The strain sensitivity of J_c of A15 phases was explained to increase dramatically with long-range atomic order parameter.[6] If this is the case, the J_c-enhancement in highly stoichiometric Nb_3Al would trade off the excellent strain tolerance. Accordingly, the strain sensitivity of the transformed RIT and JR Nb_3Al conductors were actually evaluated[5], as shown in Fig. 2, being compared with the conventional JR processed Nb_3Al[7] and the ITER Nb_3Sn[8] conductors at 12 T. The detail of the loading apparatus using the bending-substrate has been reported elsewhere.[8] There is not appreciable difference in strain sen-

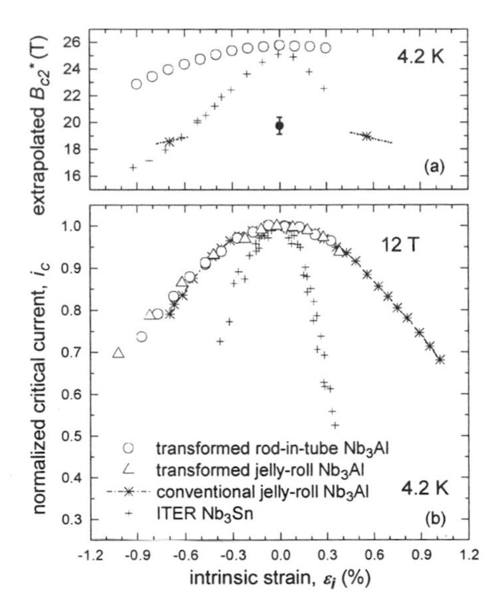

Figure 2 Strain dependence of (a) B_{c2}^* and (b) the normalized i_c at 12 T and 4.2 K. In order to compare the conventionally heat treated JR Nb_3Al conductor[7] and the ITER Nb_3Sn conductors[8], the data are plotted against the intrinsic strain.

sitivity between the transformed RIT and JR processed Nb_3Al conductors examined. The upper critical field (B_{c2}^*), obtained by extrapolating Kramer's plot, is also given in Fig. 2(a). $B_{c2}^*(0)$ of the transformed RIT Nb_3Al conductor is larger by about 5 T than that of the conventionally heat treated Nb_3Al which is known to be off-stoichiometric and of high strain-tolerance (Fig. 3(a)). The B_{c2}^* degradation with -0.7 % intrinsic-strain is 8 % for the transformed RIT Nb_3Al, and almost comparable to conventionally heat treated Nb_3Al. It is thus clear that the B_{c2}^* degradation for the both transformed and conventionally heat-treated Nb_3Al is much smaller than 30 % for the ITER Nb_3Sn. In line with the prediction [6], the high stoichiometry achieved in the present transformed-Nb_3Al-conductors seems to have slightly increased the strain-sensitivity of the B_{c2}^*. However, since J_c strongly depends on $B_{c2}^*(\varepsilon_i)$ itself as well, J_c at a given magnetic-field may look less sensitive to strain if $B_{c2}^*(\varepsilon_i)$ is larger. It should anyway have sense from the practical standpoint to compare the strain-sensitivity of J_c at 12T among different conductors, since this is a designed operation field for fusion magnets. The J_c degradation with -0.7 % intrinsic strain is only 20 % for the both transformed RIT and JR Nb_3Al conductors, almost the same in magnitude as that for the conventional JR Nb_3Al conductor that has a B_{c2}^* lower by 5T. The degradation of J_c at 12 T in these Nb_3Al conductors is much smaller than that of the ITER Nb_3Sn conductors.

4. Small coil test

A 1331-turn solenoid-coil was fabricated by a wind-and-react technique using a 137m JR $Nb/Nb(Al)_{ss}$ wire, which contains no copper as a stabilizer. The conductor was insulated with glass-fiber to withstand the subsequent heat treatment at 800°C to transform the $Nb(Al)_{ss}$ to Nb_3Al. The packing factor was 41%. Figure 3 shows the load lines of this coil in the backup fields from 19 to 21 T at 4.2 K and 2 K. As long as the operation current is less than 20A, the quench current is almost consistent with the I_c of the short sample,

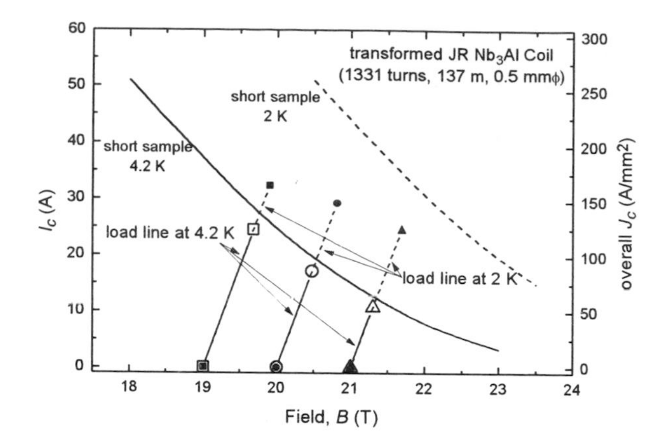

Figure 3 Load lines of the 1331-turn coil wound with the transformed JR Nb_3Al conductor 137 m in length.

indicating the characteristic uniformity in the longitudinal direction, at least, for the 137m transformed JR Nb_3Al conductor. At 2 K, this coil could generate 0.68 T in the backup field of 21 T. In order to suppress the thermal runaway of the coil, it is required to add some copper to the $Nb/Nb(Al)_{ss}$.

Since the as-quenched $Nb/Nb(Al)_{ss}$ composites are ductile enough to be twisted, cabled and flat-roll-formed before aging and transforming $Nb(Al)_{ss}$ to A15-type Nb_3Al,[9] large cryogenically stabilized conductor, for example, could be made by soldering the compacted-strand-cable into a grooved copper strip. Cable-in-conduit conductors may be also possible for the same reason. Therefore, the present transformed-Nb_3Al-conductors are regarded as a realistic alternative to Nb_3Sn conductor for high-field and large-scale applications.

Acknowledgements

Authors thank the staffs of TML for their assist in utilizing high-field facilities.

References

[1] Iijima Y, Kosuge M, Takeuchi T and Inoue K 1994 *Adv. Cryog. Eng.* **40**, 899-905

[2] Fukuda K, Iwaki G, Kimura M, Sakai S, Iijima Y, Takeuchi T, Inoue K, Kobayashi N, Watanabe K and Awaji S 1996 *Proc. 16th Int'l Cryog. Eng. Conf./Int'l Cryog. Mat. Conf.*, Eds. T. Hruyama, T. Mitsui and K. Yamafuji Kitakyushu May Elsevier Science 1669-72

[3] Yamada Y, Ayai N, Takahashi K, Sato K, Sugimoto M, Ando T, Takahashi Y and Nishi M 1994 *Adv. Cryog. Eng.* **40**, 907-14

[4] Isono T, Hosono F, Koizumi N, Sugimoto M, Hanawa H, Wadayama Y, Tsukamoto H, Sakai T, Nishi M, Yoshida K, Ando T and Tsuji H 1993 *IEEE Trans. Appl. Superconductivity* **3**, 496-9

[5] Takeuchi T, Iijima Y, Inoue K, Wada H, ten Haken B, ten Kate H H J, Fukuda K, Iwaki G, Sakai S and Moriai H 1997 *Appl. Phys. Lett.* **71**, 122-124

[6] Flukiger R, Isernhagen R, Goldacker W and Specking W 1984 *Adv. Cryog. Eng.* **30**, 851-9

[7] Specking W, Kiesel H, Nakajima H, Ando T, Tsuji H, Yamada Y and Nagata M 1993 *IEEE Trans. Appl. Superconductivity*, **3**, 1342-5

[8] ten Haken B, Godeke A, ten Kate H. H. J and Specking W 1996 *IEEE Trans. Magn.* **32**, 2739-42

[9] Takeuchi T, Iijima Y, Inoue K and Wada H 1997 *IEEE Trans. Appl. Superconductivity* **7**, 1529-1532

Inst. Phys. Conf. Ser. No 158
Paper presented at Applied Superconductivity, The Netherlands, 30 June–3 July 1997

Some factors affecting the use of internal-tin Nb_3Sn for high field dipole magnets and fusion applications

Eric Gregory, Ellina Gulko, Taeyoung Pyon

IGC Advanced Superconductors, Waterbury, CT 06704, USA

Daniel Dietderich

Lawrence Berkeley National Laboratory, Berkeley, CA 94720, USA

Abstract. The internal-tin strands described here were originally developed to meet the HP-1 specifications for The International Thermonuclear Experimental Reactor (ITER). More than four tonnes of this material has been delivered to MIT for use in the US portion of ITER Central Solenoid (CS) model coil. Recently modifications have been made to this strand to bring its properties closer to those required for the HP-2 specification for ITER. A different modification was used in the high field section of the D20 dipole magnet at the Lawrence Berkeley National Laboratory (LBNL). This recently achieved the world record field for an accelerator dipole - 13.5 T at 1.8 K. The nature of these strand modifications are indicated. While these materials show outstanding performance, a more thorough understanding of their behavior is desirable. Much of the complexity of the material results from the reaction between Sn and Cu during heat treatment. Dimensional changes resulting from these interactions are significant and must be taken into account if successful device performance is to be achieved. The importance of some of these effects is discussed.

1. Introduction

IGC Advanced Superconductors has been engaged for some years in developing and manufacturing over four tonnes of internal-tin Nb_3Sn strand to meet the ITER HP 1 specification [1]. To demonstrate that the wire produced by this process exhibits reproducible properties, an effort was made to maintain the conditions of manufacture as constant as possible, despite the fact that it was known from the start that the J_c would be below 700 A/mm^2 at 12 T. A variance in the J_c specification was made to allow material with J_c above 650 A/mm^2 at 12 T to be accepted. In a parallel investigation, and without making major changes in design, strands were produced which would meet the HP 2 specification for ITER. Another small variation of the basic ITER strand was made to produce a material which has been used successfully as one of the inner high field coils of D20, the recently reported [2], LBNL 13.5 T magnet. - the world's highest field dipole.

In this paper we report briefly on the ITER strand properties and describe the variations we have made to fully meet the HP 1 and HP 2 specifications and the D 20 requirements.

2. ITER strand results

The design of the IGC ITER strand has been described on several occasions together with the specifications for HP I and HP 2. These were summarized in a recent paper.[1]

In order to fulfill the order for 3.72 tonnes for Stage IV, the production phase of the program, the number of restacks produced and tested was 165. 13 Groups of approximately

13 restacks each, were tested for I_c, *n* value, Cu/non-Cu ratio, and RRR on both ends. The loss values were sufficiently below the limit of 600 mJ/cm^3, that, in most of the program, these measurements were only made on one in every five of the restacks. The J_c and loss results are enclosed in the ellipse shown in Figure 1 and the average J_c lies slightly below 700 A/mm^2 at 12T. There appears to be little relationship between J_c and losses.

Only two small changes were permitted by the US Home Team throughout the total production cycle. These were allowed when it was found, in the middle of the program, that the Cu/Sc ratios were centered around the low end of the specification. The amount of Cu in the stabilizer was increased to correct this. In an attempt to increase the J_c, the diameter of the Nb 7.5 wt.% Ta rods was increased slightly. This had the effect of increasing the amount of alloy by 3.15%.

The data on the ITER strand properties are divided into two categories, Groups 1 through 6 and Groups 7 through 13. The changes mentioned above were made after Group 6. The first category is shown inside the ellipse as open squares and the second as filled circles. There was a shift in the mean J_c from 685 A/mm^2 to 690 A/mm^2 between the first and second categories and the corresponding shift in the mean of the losses was from 445 to 430. The shift in the J_c was smaller than that expected from the design changes. The fact that the losses are reduced even though the J_c increases can be explained by the fact that the amount of tin was not increased when the alloy rods were enlarged. This may also account for the less than expected increase in J_c. These results emphasize the criticality of the Sn concentration and of small dimensional changes. The total number of lengths shipped was 254 from 165 restacks. Approximately half of these lengths were longer than 4000 m. RRR values were measured on each end of all the restacks shipped and the mean was 122. Only four groups had any values below 100. It was shown that the readings were highly dependent on the atmosphere in the furnaces used for the heat treatment and it is probable that these low values were due to this cause rather than to wire characteristics.

3. Changes in the J_c and loss values

The changes that were allowed in the material for Stage IV were limited as described above and therefore a parallel development program had to be launched to bring the average values of J_c and losses into the HP 1 and HP 2 specifications. The following changes were made without major alterations in the design of the strand or in the heat treatment. The latter was governed in all ITER applications by the necessity to limit chromium diffusion from the surface coating. The changes were: 1.) The diameter of the filaments was increased. 2.) Titanium was added to the Sn. 3.) The Sn concentration was varied. 4.) The barrier thickness

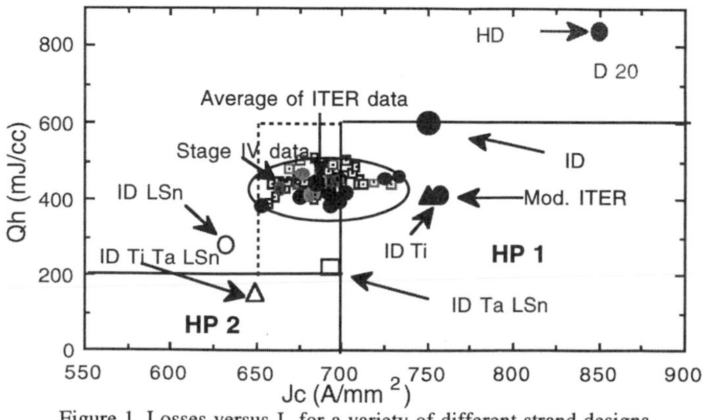

Figure 1. Losses versus J_c for a variety of different strand designs.

and composition was changed. The effects are shown in Figure 1. Increasing the filament area by 8% resulted in the properties shown by *ID*. Adding Ti to the Sn in this material gives values which are well within the HP 1 specification, *ID Ti*. Lowering the Sn by 17% instead of adding Ti results in *ID LSn*. Changing the dual Nb/Ta barrier to an all Ta one gives *ID Ta LSn*. If, in addition Ti is added, properties well within the HP 2 specification are achieved, *ID Ti Ta LSn*.

It was pointed out early[3] that the T_c and the H_{C2} of the original ITER strand were low. This has been corrected in all the materials where Ti was added to the tin core [1].

4. Material for D20 at LBNL

Accelerator dipole magnets with field capabilities above 10T have been produced recently by use of Nb_3Sn [2, 4]. LBNL has successfully produced a Nb_3Sn, one meter long dipole magnet, D20, with a 50 mm bore. One of the two inner windings was made from the IGC strand described below and the other from Teledyne Wah Chang (TWC) material. The IGC material was a variation of the ITER strand. Instead of increasing the filament area by 8% to reach the ID properties, the area was increased 17% to reach the *HD value* of 850 A/mm^2 at 12 T with 850 mJ/cm^3 in losses. This material had no Ti in the Sn and a Ta-lined Nb barrier similar to that in the regular ITER material. The Cu percentage was reduced from 60% used for ITER, to 30%. Originally, since the losses were of less importance than J_c, a 100% Nb barrier was contemplated. The idea was abandoned when it was found that the magnetization curve for such material showed considerable instability in the low field region, Figure 2. The magnetization curve of the material actually used in D20 was very similar to that of the ITER material shown previously [1] where the losses were greatly reduced and no low field instability existed.

The operation of D20 up to fields of 13.5 T indicated that the strain limitations of the Nb_3Sn conductor were not a limiting factor. Much of the success of the magnet may, however, be related to the protection system which involved an extensive array of heaters [2]. This enabled almost 50 training quenches to be induced without any degradation of the cable or windings being detected. The fact that D20 employed a cable rather than a single strand perhaps allowed it to accommodate the appreciable volume changes which take place during heat treatment. An indication of these are given in Figure 3a & b which shows dilatometer traces of strands similar to those used in D20 (3a) and the ITER Central Solenoid model coil (3b). The difference in the expansion is, in part, related to the volume of the copper in the stabilizer which in 3a is about one half that in 3b. The contraction occurring around 175°C is assumed to be due to stress relief of the filamentss. As manufactured, the Nb 7.5 wt.% Ta filaments are under tension. When the Cu surrounding these core filaments softens it allows them to contract. The Sn then melts and interactions and phase transformations take place.

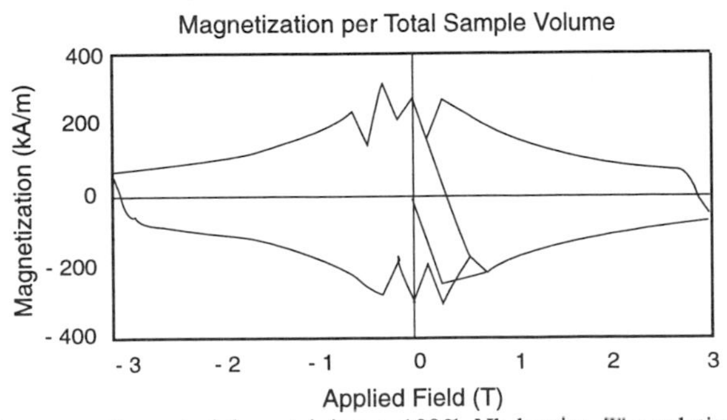

Figure 2 Magnetization curve for material containing a 100% Nb barrier. The relatively large flux jumps in the low field region are very apparent.

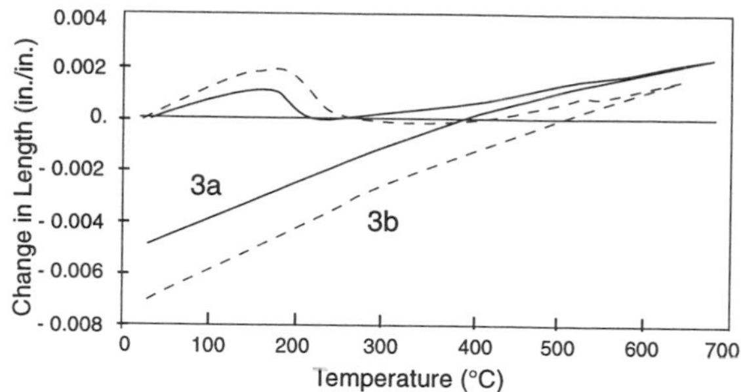

Figure 3. Dilatometer curves for D 20 and an ITER strands.

From these preliminary results, it appears that the 3a material begins to extend again earlier and to a greater extent than the 3b material. The net result of this is that the overall contraction of the ITER type material is greater. A small contraction can be seen at around 520°C in both materials due, presumably, to a phase change. The importance of these changes on the performance of the material is unknown at this time and work in this area is continuing [5].

5. Summary

The strand properties for the material for the ITER CS Model Coil are described briefly and more details will be given in another paper [6]. Recent modifications have been made meeting both the HP 1 and HP 2 specifications. The variation used in the D20 magnet at LBNL, the world's highest field dipole is also mentioned. Some of the dimensional changes occurring during heat treatment due to the interaction of Sn and Cu are described. These must be taken into account in designing devices from these materials.

Acknowledgments

This work was supported in part by the US Department of Energy under subcontracts FC-A-395276 and FT-S-560409 from MIT and a series of Small Business Innovative Research Grants. The authors thank Drs. R.B. Goldfarb and L.F. Goodrich of the National Institute of Standards and Technology (NIST) for considerable help in testing the materials for losses and J_C. The authors acknowledge the assistance of our colleagues at IGC, particularly, D. Birdsall, M. Dormody and M. Vincenti.

References

[1] Gregory E, Gulko E A and Pyon T, "Improvements in the properties of internal-tin Nb_3Sn", 1996 Paper MI-1, ASC, Pittsburgh, in press.

[2] McIntruff A D, Benjegerdes R J, Bish P A, Caspi S, Chow K, Dell'Orco D, Dietderich D R, Green M A, Hannaford R, Harnden W, Highley H C, Lietzke A F, Morrison L, Morrison M E, Scanlan R M, Taylor C E and Van Oort J M, 1997 "Preliminary Test Results of a 13 T Niobium Tin Dipole" Particle Accelerator Conference Vancouver Canada and Session LP 1 and LP 2, Eucas '97, Veldhoven Netherlands.

[3] Bruzzone P, Mitchell N, Steeves M, Spadoni M, Takahashi Y and Sytnikov V E, Conductor fabrication for the ITER model coils, (MT-14), 1996, , IEEE Trans. Vol 32, No. 4, pp. 2300-2303.

[4] den Ouden A, Wessel S, Krooshoop E and ten Kate H, "Application of Nb_3Sn superconductors in High Field Accelerator Magnets, Paper LN-2, 1996 ASC, Pittsburgh. in press.

[5] Dietderich D, Litty J R, and Scanlan R M, "Dimensional Changes of Nb_3Sn, Nb_3Al and $Bi_2Sr_2CaCu_2O_{8+d}$ conductor during heat treatment and their implications for coil design", Paper BB 5, CEC/ICMC, Portland, OR July 1997.

[6] Gregory E, Gulko E A and Pyon T, "Development of Nb_3Sn Wires made by the internal-tin process", Paper AB 6 CEC/ICMC, Portland, OR July 1997.

Inst. Phys. Conf. Ser. No 158
Paper presented at Applied Superconductivity, The Netherlands, 30 June–3 July 1997

Superconducting windings with "short–circuited" turns

A V Dudarev, A V Gavrilin, Yu A Ilyin, V E Keilin, N Ph Kopeikin, V I Shcherbakov, I O Shugaev, and V V Stepanov
Kurchatov Institute, 1 Kurchatov's sq., 123182 Moscow, Russia

Abstract. Feasibility to avoid an electrical insulation in superconducting (*sc*) windings, i.e., to "short–circuit" their turns, is attractive from several points of view. Firstly, it may be beneficial from the standpoint of average current density increase of a *sc* winding. Secondly, it improves the winding's mechanical properties. Last but not least is that for a *sc* winding with "short–circuited" turns very fast quench propagation should be typical, and hence stored energy should more or less uniformly dissipate over the winding volume, while an overvoltage is impossible "by definition". Most likely "short–circuited" *sc* windings can be used only for DC applications. The main problem is to obtain acceptable times of charging. Characteristic time of charging is evaluated by the ratio of the winding's inductance to the integrated inter–turn (transverse) resistance. Therefore, a proper increase of the latter is one of the ways to go to acceptable charging times. Test results of several "short–circuited" windings are described. Some methods to increase the inter–turn resistance are discussed and tested. A model of quench propagation in "short–circuited" windings is developed, and relevant computational results are presented.

1. Introduction

The most important problems in superconducting (*sc*) magnets development are quench protection (against overheating and overvoltage) and an increase of mechanical strength as average current density increases [1]. One of the most promising ways to ruggedize and to protect reliably the windings is abandoning an electical insulation, i.e., short–circuiting turns by soldering, welding or other methods some of which are tested and discussed in present work. Some methods suggested enable to avoid the main and probably the only pitfall of large windings with "short–circuited" turns – too long charging time. Note that the approach being discussed is presumably acceptable only for DC magnets. Our early experiments with "short–circuited" *sc* windings were treated in [2].

2. Peculiarities of windings with "short–circuited" turns

It is evident that going from inter-turn connection of "metal–insulation–metal" type to "metal–metal" one sharply increases mechanical strength of a *sc* winding and is helpful in avoiding the winding's training of mechanical origin [3]. In addition, any turn (or its any part) within a *sc* winding with "short–circuited" turns proves to be shunted by quite well thermally and electrically conducting transverse "metal–metal" resistive connection (contact, solder or/and inter–turn metallic band–spacer – see below). It leads to a fast flow of transport current from this turn, just after normal zone origination in it, to other turns and to a redistribution of the current throughout all inductively coupled turns of the winding under quench. Such a process can cause fast transfer of turns into normal state (electro–magnetic avalanche) as this takes place in conventional multi–section DC *sc* magnets; as a consequence of this process internal voltage and hot–spot temperature are very low owing to more or less uniform dissipation of stored energy even in large windings not possessing any external active protection, i.e., windings with "short–circuited" turns are practically self–protected, and, in other words, such windings are of maximum possible degree of sectioning with internal inter–turn shunt–resistances (actually continiously distributed shunting).

On the other hand, a presence of inter–turn metal connection is assumed to lead to considerable limitation of the windings charging rates, as the inductive component of a turn's resistance should be much less than transverse inter–turn ohmic resistance. For example, it is easy to show that for infinitely long one–layer solenoid with "short–circuited" turns the charging rate $\dot{I}$ obeys the relation derived from the condition not to exceed the critical temperature T_C by the coil temperature

$$\dot{I} < \frac{4d\sqrt{d}}{\pi\mu_0}\frac{\sqrt{2h(T_C - T_b)\rho_{\perp}^{eff}}}{D^2} \tag{1}$$

Here, D is the coil diameter, d is the cross–size of a superconductor used, h is the heat transfer coefficient to the coolant, T_b is the coolant (bath) temperature, $\rho_{\perp}^{eff}$ is the effective resistivity of inter-turn connection (with due regard for contact resistances). Hence, a proper choice (an increase) of transverse resistivity value can give charging rate required.

3. Examples of windings with "short–circuited" turns

3.1. One–layer coil

A test coil of 100 mm length and 20 mm diameter was made of conventional $NbTi/Cu$ composite superconductor of 1 mm diameter. The coil's inductance was $4 \cdot 10^{-5}\,H$. Resistive connection of turns had to be attained by means of high–ohmic solder usage with the resistivity about $3 \cdot 10^{-8}\,\Omega \cdot m$. So, measured resistance between a pair of turns was $R_{\perp} = \rho_{\perp}^{eff}\delta/S_c = 3 \cdot 10^{-5}\Omega$ (δ – the effective thickness of solder streak, S_c – the inter–turn contact surface area of the conductor). For a control of the current flowing in $NbTi$–superconductor, a Hall–probe was positioned into the coil center. Charging rate was equal to 2 A/s. As shown in Fig. 1, dragged transfer to quench was observed due to the current re–flow between turns through the inter–turn connections.

Such a type of inter–turn connection permits to attain very high mechanical strength, and it is especially useful for *sc* windings of a complicated shape where the problem of mechanical strength is put in the forefront, but a proper banding is hampered.

3.2. Race–track

The winding of "race–track" type with straight parts of 100 mm length was manufactured according to "wind–and–react" route from Nb_3Sn/Cu multi–strand rectangular

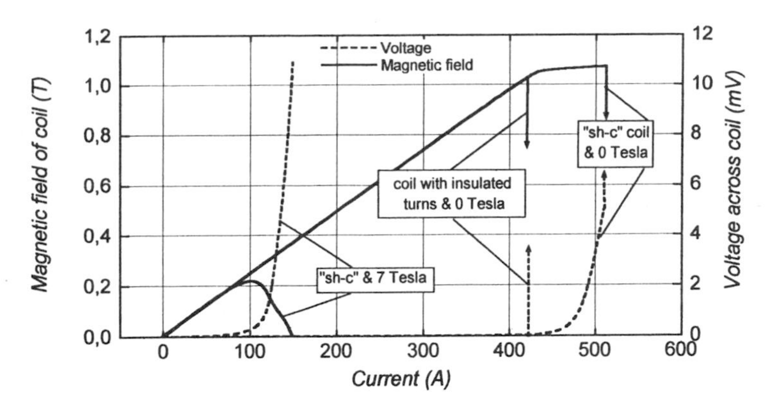

Fig. 1: Dependence of magnetic field and internal voltage on current of coil under charging.

conductor with $1 \times 1.8\,mm^2$ cross–section area. The winding consisted of 8 race–track pancakes of 32 turns. The conductor was soldered with small quaintity of tin that resulted in nice diffusive connection of turns after the heat treatment. The pancakes were separated with glass-cloth insulation of $0.3\,mm$ thickness; the winding inductance was $L \simeq 10^{-2}\,H$, and $R_\perp \simeq 10^{-5}\,\Omega$. In process of the winding testing, the short–sample current was attained at maximum field within the winding $5\,T$ without any training. The presence of inter–turn connections enabled to charge the winding with the current even somewhat higher than the critical one without heating up the coil up to T_C.

3.3. Double–pancake

It is obvious that the use of soldering or welding for resistive connection of turns causes quite low inter–turn resistance that leads to unacceptable increase of the magnets charging times as the magnets' sizes increase. To avoid this fault it is necessary to increase the value of inter–turn resistance. Really, the characteristic time τ of charging can be estimated as $\tau = L/R_\perp^\Sigma$, where $R_\perp^\Sigma$ represents the integrated inter–turn resistance. As the magnet's over–all dimensions increase, the inductance L grows faster than $R_\perp^\Sigma$.

The double–pancake of about $252\,mm$ inner diameter and of 60 (30×2) turns was made of $NbTi/Cu$ rectangular ($2 \times 3.5\,mm^2$ cross-section area) wire. For this winding, another method of high-resistive connection of the turns was used. In the winding process, $0.1\,mm$ thick and $3\,mm$ wide stainless steel strip was wound with the wire (stacked between the wire's turns) and glued to the surface of adjacent turns with special epoxy–resin. Proper preload of the wire and the strip permitted to attain the value of $R_\perp \sim 3{\cdot}10^{-4}\,\Omega$ (at room temperature) and quite uniform inter–turn contact over all length of the wire. Generally, to ensure a proper inter–turn electric connection in more large windings of such a type – through multitudinous uniformly distributed contact micro–spots, additional routes, like, for example, a metallic band–spacer pre–corrugating or a wire contact surface knurling, could be used.

The characteristic time τ of the double-pancake manufactured was estimated as $0.23\,ms$ ($R_\perp^\Sigma \simeq 80\,m\Omega$, $L \simeq 1.8\,mH$). At relatively low charging rates ($\dot{I} \sim 10\,A/s$), the magnet reached the short–sample current ($4250\,A$ at maximum field within the winding $3.5\,T$). As the rate increased, the dependence of the winding quench current on the rate $\dot{I}$ became noticeable (Fig. 2).

High-accuracy computer simulation of quench behaviour of a double–pancake of

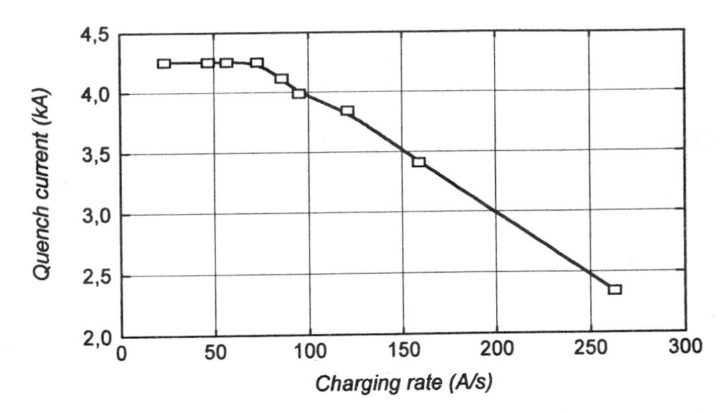

Fig. 2: Experimental dependence of test double–pancake quench current on charging rate.

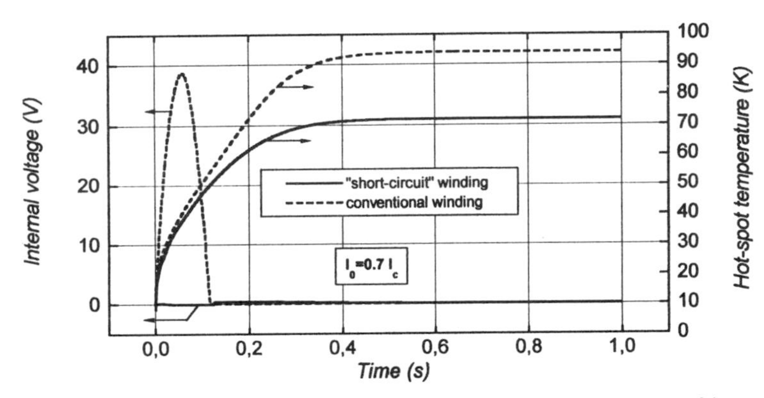

Fig. 3: Computed time dependence of large double–pancake hot-spot temperature and internal voltage. For the "short-circuited" winding, $\rho_{\perp}^{eff} = 5{\cdot}10^{-3}\,\Omega$, thermal conductivity of the strip $\kappa_{\perp}^{eff} = 2\,W/(m{\cdot}K)$; for the winding with electrically insulated turns, $\kappa_{\perp}^{eff} = 0.1\,W/(m{\cdot}K)$.

ten times more inner diameter ($2.5\,m$ and of 60 turns) that could be manufactured in the same way was carried out. Comparative computer analysis of quench behaviour of the winding with traditional insulation was also done. The results are presented in Fig. 3. It is seen that in the case of the "short–circuited" winding there is practically no internal voltage and the hot-spot temperature is considerably lower.

So, in closing, we would like to notice that the described method of short–circuiting the turns by means of thin stainless steel strip can be effectively applied to rather large *sc* DC magnets with quite acceptable charging rates.

References

[1] Smith P F, Colyer B A 1975 *Cryogenics* **15** (4) 201

[2] Barkov A V, Dudarev A V, Keilin V E 1988 *Reports of the USSR Academy of Sciences: Tech. Phys.* (*Dokl. Akad. Nauk SSSR*) **300** (6) 1370–1371 (in Russian)

[3] Anashkin O P, Keilin V E, Krivykh A V 1996 *Cryogenics* **36** (2) 107–111

Inst. Phys. Conf. Ser. No 158
Paper presented at Applied Superconductivity, The Netherlands, 30 June–3 July 1997

Modelling of current transfer in current contacts to anisotropic bulk superconductors

R P Baranowski, D R Watson and J E Evetts.

Department of Materials Science and Metallurgy, University of Cambridge, Pembroke Street, CB2 3QZ and IRC in Superconductivity, University of Cambridge, Madingley Road, Cambridge, CB3 0HE

Abstract. Many practical applications of bulk superconductors require low resistance current contacts. The optimisation of such contacts requires a thorough understanding of the nature of current transfer across the metal-superconductor interface. This work describes modelling of the spatial distribution of current in current contacts to Composite Reaction Textured (CRT) Bi-2212 material. The nonlinear nature of the E-J characteristic of the superconductor can be accounted for in a model where the resistivity of the bulk material includes a flux flow component. Thus, when the critical current of the superconductor is locally exceeded, there is a flux flow resistivity component in addition to the surface resistivity at the contact interface. The anisotropic nature of the current flow and this local flux flow condition generates a spatially dependent effective surface resistivity. This is examined in a 2D nonlinear numerical analysis.

1. Introduction

Many applications of bulk high temperature superconductors (HTSs) in power engineering systems require high quality current contacts, and for ready integration into existing power systems such contacts must be normal metal to HTS ceramic. Such applications include current leads and resistive fault current limiters. One requirement of such contacts is that they have a low contact resistance, principally to avoid excessive heating of the superconductor above the critical temperature near the contact area. A more stringent requirement in, for example, a resistive fault current limiter is that the conductor (superconductor and contact) can carry currents possibly as high as 50 times I_c (1 μVcm^{-1}). Through this regime the initially zero resistivity of the HTS increases to values well in excess of that of the contact metal, as the superconductor makes a progressive transition to the normal state [1].

Due to the nonlinear nature of the HTS the high current regime is difficult to analyse both analytically and numerically. This is further complicated by both the inherent and engineered anisotropies of the Composite Reaction Textured HTS material [2,3] modelled in this analysis. This paper describes a numerical model of a 2D *in situ* current contact, Fig. 1, to a HTS. The HTS possesses a nonlinear flux flow resistivity and current anisotropy ratio of 30 [3].

2. Current transfer model

There are two principle contributions to the effective electrical surface resistivity of a current contact, one from the interface resistance and the other from the bulk resistivity of the device material. In geometries where the direction of current injection is not perpendicular to the contact-device material interface, there is bending of the current path away from the injection direction. This results in current passing across the contact area nonuniformly, and modifying the bulk resistivity. This effect increases as the resistivities of the contacted materials become more dissimilar. The extreme case where the device material is a superconductor results in current crowding at the interface nearest the injcction point.

The CRT process results in highly textured microstructures in bulk components of arbitrary shape, in this case the grain ab-planes are parallel to the contact interface. This is similar to the preferential alignment of grains near a Ag or Au interfaces [3,4] although is not limited by the proximity to the interface. Current crowding in the *in situ* geometry, Fig. 1, produces a current flow across the interface, which includes a large c-axis component. In the c-direction of current flow the critical current, J_c, can be up to a factor of 30 less than that in the ab-plane. It is proposed that this current anisotropy and to a lesser extent current crowding can lead to the J_c of the superconductor becoming locally exceeded, and a local flux flow resistivity being generated; producing a spatially dependent bulk resistivity (flux flow resistivity).

3. Numerical solution

The numerical solution to the above model solves for a number of electrostatic state variables in the 2D geometry of the *in situ* contact, Fig. 1. The solution space is discretised into rectangular grids of solution points (nodes), where the local electric potential, resistivity and the horizontal (ab-plane) and vertical (c-direction) current densities are calculated (see technical addendum below); the discretisation limits the spatial resolution of the solution.

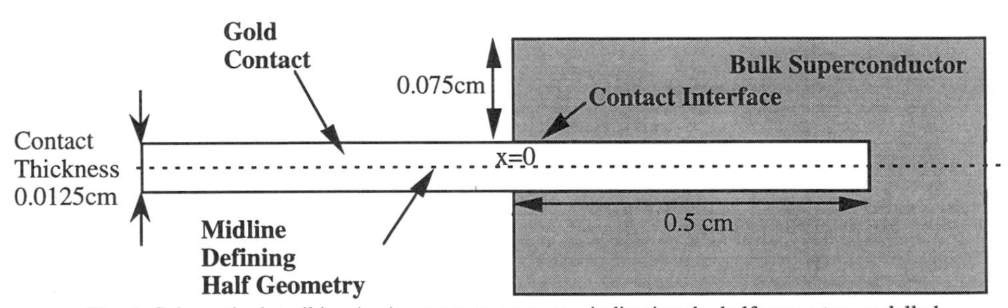

Fig. 1. Schematic describing *in situ* contact geometry, indicating the half geometry modelled.

The symmetry of the lateral midline of the contact geometry is used to halve the solution space. At low currents where the current transfer is negligible at the end of the contact, this half space solution approximates that of a lap contact. Electrical potential boundary conditions at both ends of the contact induce a current flow in the superconductor subject to nonlinear resistivities in the horizontal and vertical directions. The nonlinear resistivity in the ab-plane is derived from the E-J curve of an ab textured sample, with critical current 1000 Acm^{-2} and n-value of 10. The curve is scaled by an anisotropy ratio of 30 to provide the nonlinear resistivity in the c-direction. A contact interface resistance of 0.25 $\mu\Omega.cm^2$ is specified as a boundary condition across the contact-HTS interface, and is typical of CRT *in situ* current contacts.

4. Results and discussion

The global solution to the current contact is used to provide boundary conditions for a submodel that examined more closely the spatial variation of current density, Fig. 2, and electric potential, Fig. 3, in the superconductor region above the current contact for the *in situ* geometry. The figures show the results of the model contact with an applied current such that in the region where the current density in the ab plane becomes, across the thickness, uniform the current density is equal to the ab plane critical current. The current component perpendicular to the current contact interface, J^c, starts at a maximal value (344 Acm^{-2}) and decreases; however the absolute magnitude of current density within the superconductor region is maximal (1387 Acm^{-2}) 0.02 cm away from the point of first transfer (x=0). The maximum starts at x=0 at low current and moves along the interface for increasing current. This is principally due to the transport current anisotropy of the HTS. Due to the lower critical current in the c-direction J_c^c than the ab-direction J_c^{ab}, the current transferred near x=0 flows predominantly near and parallel to the contact interface. The local absolute magnitude of the current density becomes maximal near the contact interface as additional current transfers across the interface further down, Fig 2.

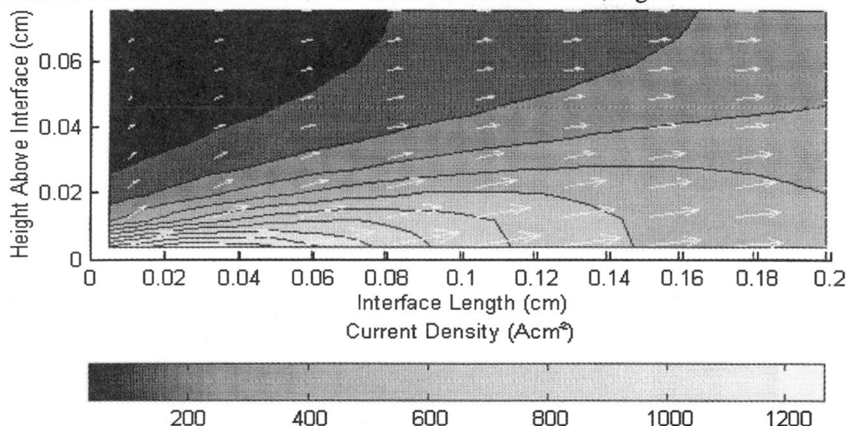

Fig. 2. Contour plot of the absolute magnitude of current density in the superconductor above the contact interface. J_c^{ab} is 1000 Acm^{-2}, and J_c^c is 30 Acm^{-2}. Current density vectors are represented by white arrows.

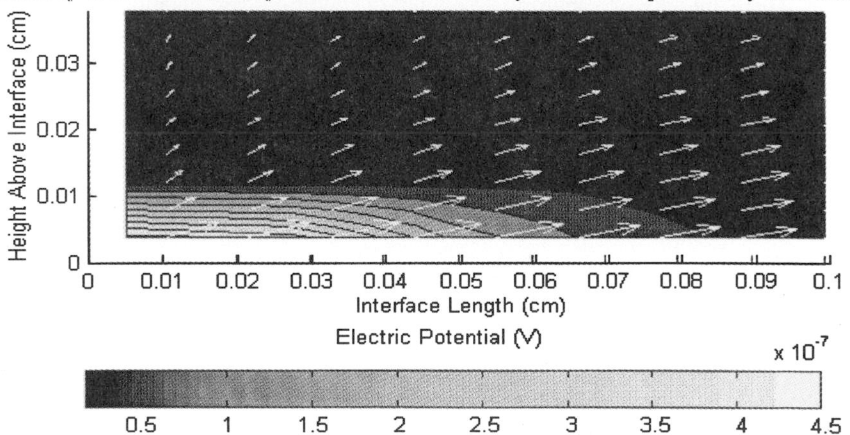

Fig. 3. Contour plot of electric potential in the superconductor above the contact interface, bottom left quarter of current density contour plot. Current density vectors are represented by white arrows.

Even though the local current density can become high, the contact is still intrinsically fail safe in the contact region at currents many times higher than the I_c of the superconductor, as the contact metal shunts current away from the superconductor. The anisotropic nature of the critical current also leads to the current density vectors pointing in a direction which is not perpendicular to the constant voltage contours, Fig. 3.
Due to the current dependence of the flux flow resistivity, Fig. 4, this is intrinsically linked to the regions of high current transfer. At the point of first transfer J^c is greatest, and the current flow exceeds J_c^c to a greater degree than J_c^{ab} and the flux flow resistivity attains a local maximum value at x=0. Further even though the ab-plane current flow is predominant in this structure, it is J_c^c which principally determines the flux flow region.

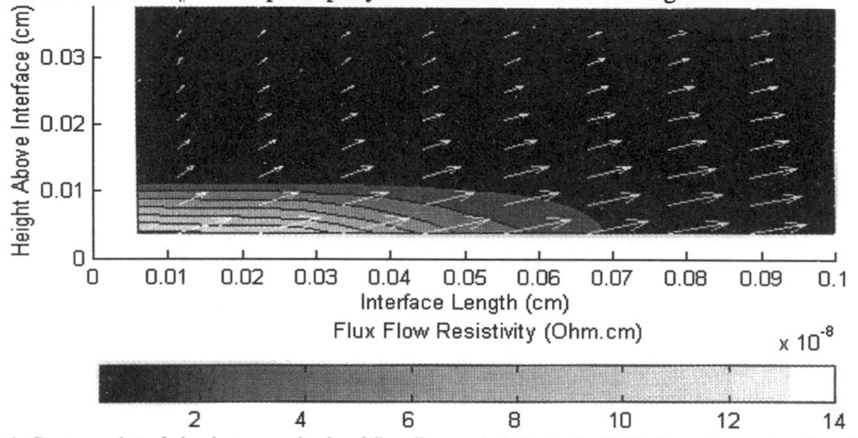

Fig. 4. Contour plot of absolute magnitude of flux flow resistivity in the HTS above the contact interface, bottom left quarter of current density contour plot. Current density vectors are represented by white arrows.

Technical Addendum

The numerical model solves the uniformly elliptic harmonic nonlinear Laplace equation and continuity equation. The nonlinearity is a current dependent conductivity. The finite difference method is used to approximate the Laplace equation, over a rectangular grid of 150 nodes. There are only two appropriate boundary conditions for this model, Dirichlet (1st type) and Neumann (2nd type). The resulting sets of nonlinear equations are solved by least squares optimisation employing the Levenburg Marquardt search direction method.

Acknowledgements

This work is supported by the Oxford Instruments (UK) Ltd and the Engineering and Physical Sciences Research Council.

References

[1] Ahmed Kursomosovic et al, EUCAS 97 Conference Proceeding W16548
[2] Chen M et al, *IEEE Trans. on Appl. Supercond.*, Vol. **5**, No. 2 (1995) 1467-1470
[3] Watson D R, and Evetts J E, *Supercond. Sci. Technology*, Vol. **9** (1996) 327-332
[4] Pashitski A E et al, *Physica C*, Vol. **246** (1995) 133-144

Inst. Phys. Conf. Ser. No 158
Paper presented at Applied Superconductivity, The Netherlands, 30 June–3 July 1997

Power dissipation at high current CRT Bi-2212 SC-Ag contacts

A Kuršumovic, J E Evetts, D R Watson, B A Glowacki, S P Ashworth and A M Campbell

IRC in Superconductivity, University of Cambridge, Madingley Road, Cambridge CB3 0HE, UK, Department of Materials Science and Metallurgy, University of Cambridge, Cambridge CB2 3QZ, UK.

Abstract. In-situ electric contacts, between silver leads and Composite Reaction Textured (CRT) Bi-2212 superconductor, were made for handling high currents at 77 K. Contact resistance and power dissipation at the metal-superconductor contacts were estimated by monitoring the voltage drop during continuous current increase in self- and applied field. A transmission line model is used to describe current transport across the junction. At currents below critical, $I<I_C$, this approach was satisfactory; the metal current lead resistance, and contact interfacial resistivity together dominated the power loss. However, at higher currents non-ohmic behaviour is observed. The apparent contact resistance increased with increasing current, and with applied magnetic field (parallel to *c*-axis). The addition contribution to the resistance arises from flux flow effects in the anisotropic superconductor and is shown to vary quadratically with both increasing current and increasing applied field.

1. Introduction

Superconductors that are connected to a normal network are interfaced by normal metal power leads. The quality of this junction is of great importance and should have a design specification adequate to assure trouble free operation (i.e. no quenching of the superconductor in the contact zone). Any non-uniformity in the contact can result in hot spot formation that can lead to localised thermal quenching and destruction of the contact due to much increased power dissipation. In addition a large contact resistance results in excessive loss of liquid nitrogen during normal operation. The issue of contact formation is especially important in the case of ceramic superconductors due to their brittle nature with no direct welding, soldering or fastening possibilities. There is a classical approach to the problem in low temperature superconductors [1] and several studies on YBCO 123, with only a few concerning Bi-2212 [2-4]. However, they usually address contact resistance at low currents ($I \sim I_C$). In applications such as fault current limiters [5] currents many times higher than the critical current are experienced. The resulting flux flow resistivity of the ceramic superconductor becomes comparable to the resistivity of the metallic current leads. Applying a magnetic field has principally the same effect. In this study, contact resistance and power dissipation at contacts in self- and applied field are investigated for d.c. currents up to 5 I_C.

2. Experimental

Ten Bi-2212 samples were prepared by the CRT method [6]. All samples were about 60 mm long, 1.25 mm thick and 5 mm wide. Current leads were in-situ processed 0.125 mm thick silver tapes of the same width. They were all embedded ~8 mm into the samples (see Fig. 1). As-processed contacts were machined in order to improve geometry. Electrical measurements were performed using a four probe method, both in self-field and in applied field up to 50 mT, an HP 6680A current source was used. Voltage contacts (25 μm Au-wires) were placed at X=0 and X=L along a lateral midline of the silver lead and the E-I behaviour of the superconductor itself was monitored by additional voltage contacts along the top midsection of the superconductor. All voltages were monitored by Keithley 181 nanovoltmeters. The critical current was defined at 10^{-6} V/cm electric field.

3. Results

Fig. 2 shows E-I characteristics for the superconductor and also the total voltage drop, U, within the superconductor-metal contact in self- and applied field. A range of field values was applied up to 50 mT, however, they are not shown here for the sake of clarity. As can be seen, at low currents and in self-field U was linear with I. At high currents and in an applied field U deviates from linear behaviour approaching nonlinearity of the superconductor itself. Results for all samples showed good reproducibility. In-situ made contacts on Bi-2212, do not suffer from surface contamination that can otherwise severely deteriorate contact quality [7].

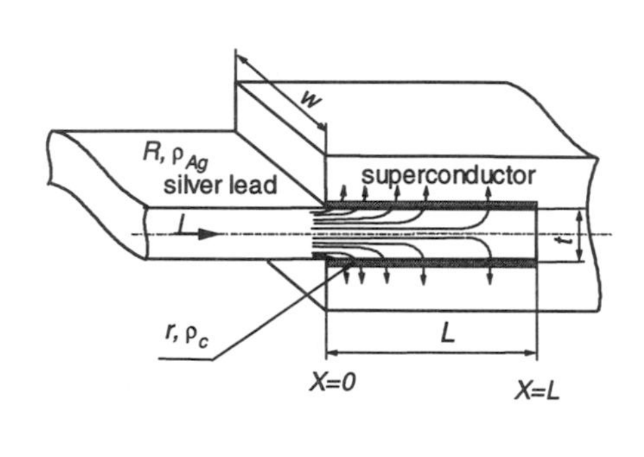

Figure 1. Current contact configuration used, and visualisation of current flow. Voltage contacts described in text.

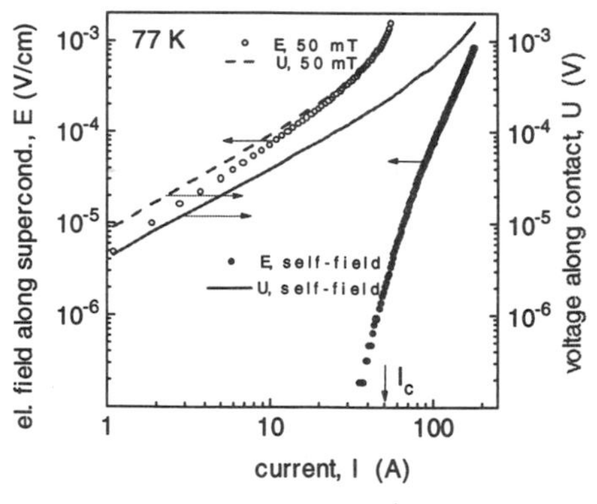

Figure 2. Voltage-current behaviour in bulk (E-I) and at contacts (U-I) of CRT Bi-2212.

Fig. 3 shows power dissipation at contacts that appears as Joule heat calculated as $P=UI=R_CI^2$. Assuming ohmic losses, R_C is the measured contact resistance. The power dissipation exhibited simple power law behaviour ($P \propto I^2$) for currents below the critical current (I_C). R_C in a self-field is simply calculated as P/I^2 or alternatively as a derivative (dU/dI), showing ohmic behaviour for $I<I_C$ with $R_C \approx 3.5$ μOhm.

4. Model and analysis of the results

Planar contacts between metals and superconductors at power engineering frequencies (50-440 Hz), are regarded as pure ohmic. A transmission line model is often used in describing metal-superconductor [1, 4] and metal-semiconductor [8] contacts. In this sense the contact can be regarded as a lossy transmission line that has an attenuation factor $\alpha=\sqrt{(R/r)}$, where R is the line resistance (Ag here) per unit length [Ω/cm] and r the contact resistivity per unit length [Ωcm].

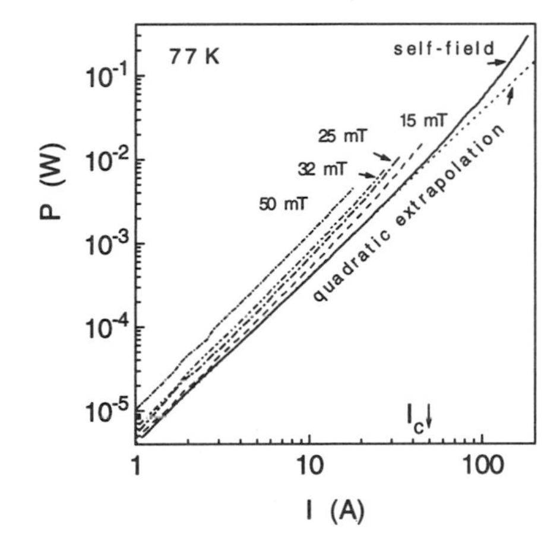

Figure 3. Power dissipation at current contact in self- and applied field. Nonlinearity is observed at $I>I_c$.

The transmission line equation (for the potential, $V(x)$) with respect to the potential of the superconductor, and current along the contact are related as [1, 8]: $V(x)=-rdI(x)/dx$ and $I(x)=-R^{-1}dV(x)/dx$. Separation of variables leads to the second order differential equation where: $d^2V/dx^2=\alpha^2V=V/(L_T)^2$. Term $\alpha^{-1}=\sqrt{(r/R)}=L_T$ is called transfer length, over which virtually all the current is transferred from the metal lead to the superconductor. Applying appropriate boundary conditions for the second order equation, rather straight forward mathematics gives the total power dissipated at the junction [1]: $P=I_0^2(Rr)^{1/2}\,ctnh(L/L_T)$.

Assuming that $L>>L_T$ is a good approximation for $I<I_C$, gives $P=I_0^2(Rr)^{1/2}$ and contact resistivity $\rho_C=rL=LR_C^2/R$ [Ωcm^2]. From the contact dimensions used and ρ_{Ag}(77K)=0.29 μΩcm follows the contact resistivity ρ_C =0.26 μΩcm^2 (Fig. 4). The resulting transfer length $L_T=\sqrt{(r/R)}$=0.7 mm fulfilling the condition $L>>L_T$ in the ohmic region.

5. Discussion

The ohmic interfacial resistivity found here, ρ_C =0.26 μΩcm^2, is in agreement with the previous results from the voltage drop across gold contacts [4] in CRT Bi-2212. This value appears to be four times smaller than other reported values for in-situ processing [3]. The higher total contact resistance, found here, is due to effectively thinner Ag leads used in this study. For small currents ($I<I_C$) ρ_C showed ohmic behaviour. Hence, the usual transmission line approach was satisfactory. It was found that ρ_C started to significantly deviate around I_C. Since at that point the resistance of the metal lead is still >100 times larger than the flux flow resistance of the superconductor in the *a-b* plane this can not be the source of the deviation from the strict applicability of the transmission line model for the case of an isotropic superconductor. However, strong anisotropy in the superconductor results in a much lower I_C value and higher flux flow resistivity in the *c*-axis direction and is believed to be the main reason for this behaviour [4]. For convenience, the effective interfacial resistivity, ρ_C, can be expressed as a sum of an ohmic (ρ_0) and flux flow part [4]. Variation of flux flow part in self-field ρ_{SF}, is shown in Fig. 4, and with applied field, ρ_{AF}, in Fig. 5. It is concluded (from Figures 4 and 5) that ρ_{SF} and ρ_{AF} have initially the same quadratic behaviour with current and applied field, respectively: $\rho_C = \rho_0 + \rho_{SF}$ (I) + ρ_{AF}(B) = ρ_0 + $f_1(I^2) + f_2(B^2)$. The simplicity of this quadratic behaviour suggests that when current transfer across the contact is dominated by anisotropy a simple analysis can be made.

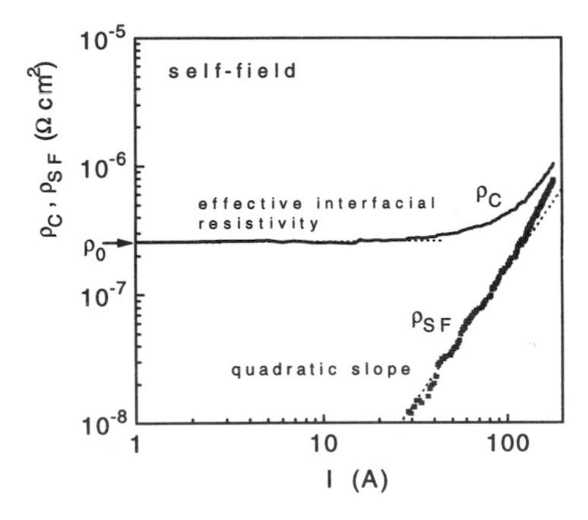

Figure 4. Interface resistivity between silver lead and Bi-2212, and apparent flux flow part (squares).

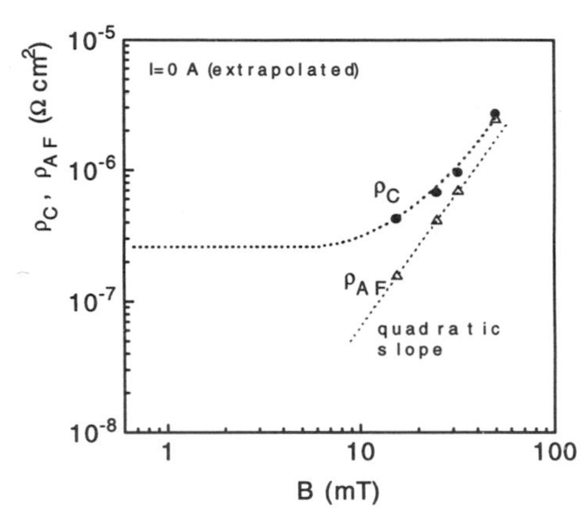

Figure 5. Apparent interface resistivity and flux flow part (triangles).

Further increase in contact resistance non-linearity, with increasing field and current (B>35 mT, $I>2I_C$) can be directly attributed to the sample *E-I* curve. Namely, in relatively strong magnetic fields (50 mT at 77 K for Bi-2212) and high currents, even the *a-b* plane resistance of the superconductor increases to a higher value than that of the contact metal, playing a mayor role in the current transfer [9].

The power dissipation at the contacts scales over six orders of magnitude for three orders of magnitude in the current density, approaching 1W per contact for ~4 I_c current. At the same time power dissipation in the superconductor itself, due to flux flow resistivity, per unit of volume is estimated to be an order of magnitude lower. Moreover, despite the increasing transfer length due to increasing r for $I>I_C$, most of the current is still transferred in the first few millimetres of the contact [9], "heating" it more than the rest of the sample.

6. References

[1] Wilson M N 1983 *Monographs on Cryogenics 2: Superconducting Magnets* (Oxford: Clarendon Press)

[2] Elschner S and Bock J 1992 *Adv. Mater.* **3** 242-244

[3] Preisler E, Bayersdörfer J, Brunner M, Bock J and Elschner S 1994 *Supercond. Sci. Technol.* **7** 389-96

[4] Watson D R and Evetts J E 1996 *Supercond. Sci. Technol.* **9** 327-32

[5] Rowley A et al., this Conference

[6] Watson D R, Chen M and Evetts J E 1995 *Supercond. Sci. Technol.* **8** 311-16

[7] Imao H, Kishida S and Tokutaka H 1996 *Jpn. J. Appl. Phys.* **35** 5299-5303

[8] Shur M 1987 *Microdevices Physics and Fabrication Technologies: GaAs Devices and Circuits* (New York, Plenum Press), ch 3

[9] Kursumovic A and Evetts J E, to be published

Inst. Phys. Conf. Ser. No 158
Paper presented at Applied Superconductivity, The Netherlands, 30 June–3 July 1997

Improvements of electric metal superconductor contacts in Bi-2212 polycrystalline textured materials

J.M. Mayor[1], L.A. Angurel[1], R. Navarro[1] and L. García-Tabarés[2]

[1]Departamento de Ciencia y Tecnología de Materiales y Fluidos, ICMA, CSIC-Universidad de Zaragoza, María de Luna, 3 50015-Zaragoza, Spain, [2]CEDEX, Alfonso XII, 3-5 28014-Madrid, Spain

Abstract. This article describes the performance of silver painted contacts on Bi-2212 textured materials. The effect of the thermal annealing on the final contact strength has been analysed determining the optimum process that leads to lower resistance without downgrading the critical current, I_c. Contacts bellow 1 μΩ and a very low reduction of I_c with time and thermal cycles have been obtained. A simple model, which accounts for the influence of the different contact parameters in its final properties has been studied.

1. Introduction

The development of electrotechnical applications based on high temperature superconductors, as current leads or current limiters, is closely related to the feasibility of low resistance contacts in the feeding current terminals. The heat leaks of a current lead has two main sources, one associated to its thermal conductivity and another due to the electrical resistance of the contacts [1]. The advantageous low thermal conductivity of Bi-2212 polycrystalline materials, introduces strong requirements in the contacts quality. The heat dissipation of poor contacts may increase the local temperature reducing the transport critical current, I_c, and even may produce a catastrophic melting [2].

In this paper, the improvements of silver painted contacts in textured Bi-2212 polycrystalline rods are described. These materials, which for instance on 2 mm of diameter samples reachI_c(77 K)≈90 A and three times more at 63 K have been fabricated using Laser Floating Zone (LFZ) methods [3]. Their transport characterization has been possible after reducing the resistance to values bellow 1 μΩ.

2. Contact preparation and performance

Cylindrical LFZ rods of 1 mm in diameter and up to 100 mm in length have been used along this study. In a previous work [4] it has been shown that annealing treatments at 845°C during 60 h generate I_c(77 K) = 45 - 50 A that are representative of the maximum current capability, while longer annealing does not change these values. To allow further comparisons, the Bi-2212 rods were initially annealed at this temperature for 60 h and after this initial process, all the contacts have been fabricated in the same way. Two current feeding contacts of 5 mm length and two voltage ones were silver painted on the rod surface and annealed again for 12 h

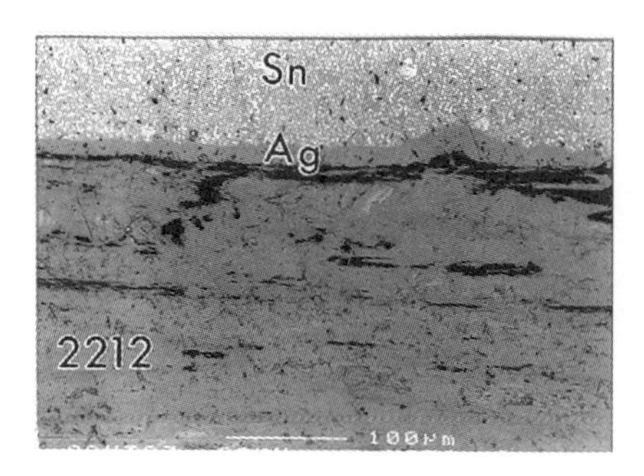

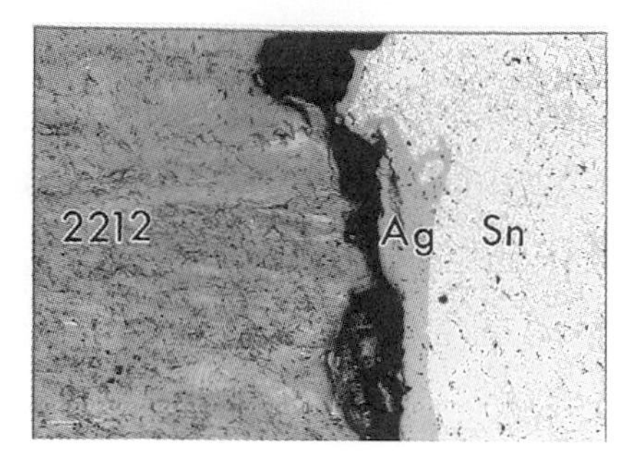

Figure 1: SEM micrographs (backscattered electrons) of a polished tin-Ag-superconductor interface of a contact: (a) Parallel and (b) perpendicular to the texture direction. Black contrast corresponds to non-superconducting phases in (a) and to a crack between the Ag and Bi-2212 phases in (b).

at different temperatures, between 750 and 860°C. In the final step, samples were inserted inside copper blocks, just after drilling a hole 2 mm of diameter and soldered with a commercial tin alloy. In this working device, contact resistance as well as I_c(77 K) measurements were performed with the usual four-probe technique.

Two different kind of contacts have been prepared. In this first case, silver has been painted over the surface of the sample and in the second case the sample was previously filed. The SEM micrographs of the Fig. 1 show the Bi-2212 /Ag/tin interfaces of a contact prepared without modifications of the superconducting rod. In its lateral surface, parallel to the texture direction, there is a near uniform layer of Ag (≈ 30 μm thick) with a good adherence to the tin and the superconductor which accumulate in its neighborhood the highest amount of non-superconducting phases (black contrast in Fig. 1(a)). The best results have been obtained when this thickness is uniform all around the sample. In the opposite, all the samples analysed in this study show that the adherence of the Ag layer in the bases of the cylinder, which are perpendicular to the Bi-2212 texture direction, is lost, remaining attached to the tin (the black contrast in Fig. 1(b) correspond to a big crack). This behaviour is believed to be related with the preferred disposition of the silver parallel to the ab planes of the Bi-2212 crystallites, as it has been observed in Ag/ Bi-2223 composite LFZ rods [5]. In consequence, the electric contact is only established along the lateral surface of the sample.

For this reason, when the surface of Bi-2212 is filed, a poorer adherence, a higher contact resistance and a worse thermal cycling behaviour were found. As an example, samples that have been annealed at 850 °C have contact resistance values of 0.5-1 μΩ without any observable degradation after more that 50 thermal cycles between 77 K and room temperature. In the same sample but with its surface filed these values increase up to 3-5 μΩ showing a fast degradation which manifests in the fact that after 10 thermal cycles the resistance has increased up to more than 15 μΩ.

Another fact that has been analysed is the influence of the annealing temperature on the final value of the contacts. It has been observed that there is two different regimes: Using temperatures lower than 835 °C the resistance value decreases if the temperature is increased, showing that the adherence of the silver with the superconductor is improving. This regime changes at higher temperatures reaching a constant value until temperatures in which a partial melting of the sample starts. This minimum value is of the order of 0.5-1 μΩ and it has been obtained in a repetitive way. These values are associated to specific resistivities in the order of $\approx 10^{-11}$ Ωm^2, which are one order of magnitude lower than the values obtained in melt cast

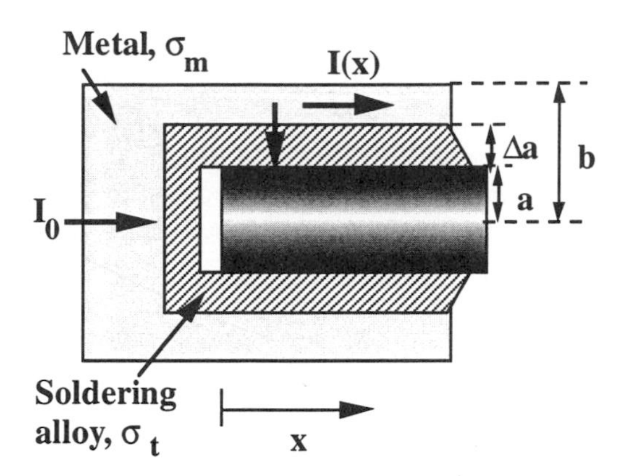

Figure 2: Squematic diagram of the contact showing the main parameters used in the model.

processed Bi-2212 [6].

In the worse conditions, the melting of the sample, due to the heat generated in the contacts, has been induced. These experiments have shown that for the sample immersed in liquid nitrogen, the maximum energy dissipated by these contacts is ≈0.2 J. With this level of energy, sinusoidal rips of 2 s, reaching maximum currents up to 400 A have been used in the characterization of these textured samples.

3. Theoretical modeling

It is believed that the limiting resistance value so far reached is not associated to the silver-superconductor interface, but to the contribution of the rest of the contact elements. A previous simple model [7] has been modified in order to determine the limiting resistance values associated to the current injection from the Cu blocks. As presented in Fig. 2, this model assumes that the sample, of radius a, is introduced inside a metallic cylinder of outer radius b and a electrical conductivity, σ_m. The contact is performed soldering with a tin layer characterized by a width Δa and conductivity, σ_t. It is postulated that the contact is long enough in order to be sure that all the current has penetrated inside the superconductor before reaching the end of the contact and that the problem has a cylindrical symmetry because only the lateral surface has to be considered. With these assumptions, the equipotential surfaces are perpendicular to the Bi-2212 rod and neglecting the resistance of the silver and of its interface with the superconductor it is obtained that the current along the metallic cylinder has an exponential dependence, $I(x)=I_0\exp(-x/x_0)$, where

$$x_0 = \sqrt{\frac{\sigma_m}{\sigma_t}\frac{\Delta a\left[b^2-(a+\Delta a)^2\right]}{2(a+\Delta a)}} \tag{1}$$

and the origin has been taken at the beginning of the superconducting sample (see Fig. 2).

The equivalent resistance can be calculated from the following equation:

$$R_{eq} = \sqrt{\frac{1}{\sigma_m\sigma_t}\frac{\Delta a}{2\,\pi^2(a+\Delta a)\left[b^2-(a+\Delta a)^2\right]}} \tag{2}$$

In this equation, the resistance of the metallic blocks before the beginning of the sample has not been considered. From the values of the electrical conductivities that have been measured in our materials, namely, σ_m=1.38 10^8 $(\Omega m)^{-1}$ and σ_t= 2.88 10^7 $(\Omega m)^{-1}$ and from the geometrical factors a=0.5 mm, b=3 mm and Δa=0.5 mm a value of R_{eq}=0.9 μΩ is obtained and x_0=3 mm. The value of R_{eq} is of the right order of magnitude, and in consequence, the original idea can be considered to be correct.

The value of x_0 seems to indicate that the contact length should be higher. The calculations show that the current at x=5 mm is less than 0.2 I_0. With a contact of 9 mm this value is reduced bellow 0.05I_0. If b is increased, i.e., the resistance of the metallic block is reduced, the contact length should be longer because there is a competence between the fact of flowing the current along the block or to be introduced to the superconductor.

The above expressions give some trends to reduce the contact resistance. Increasing b and reducing Δa the value of R_{eq} is reduced up to 0.3 μΩ using the same material to make the contact. This value cannot be reduced, even if the contact length can be increased as much as possible, there is a minimum value of the order of 0.1 μΩ that cannot be reduced making the contact longer. This value cannot be reduced neither using different materials because the most limiting factors are geometric. In consequence, different contact geometries should be used with the aim of introducing the current in a radial way, instead of doing it in a longitudinal one as it has been analysed in this study.

4. Conclusions

Resistance contacts bellow 1μΩ have been obtained in silver painted contacts on Bi-2212 textured samples which have allowed to use more that 400 A in the transport characterization of these samples. The contact is established through the lateral surface of the sample and the best results have been obtained after having annealed the sample with the contact at temperatures between 835°C and 855°C. The lowest values have been associated with the geometry of the contact and with a simple model the lowest limit of the resistance value has been obtained. In order to reduce this limit alternative geometries should be used.

Acknowledgments

The financial support of the Spanish Ministry of Education and Science (CICYT 95-0921-C02-01 and 02) and R.E.E. is acknowledged.

References

[1] Yang S, Chen B, Hellstrom E E, Stiers E and Pfotenhauer J M 1995 *IEEE Trans. on Applied Superconductivity* **5** 1471-1474

[2] Diez J C, Angurel L A, Miao H, Fernández J M and De la Fuente G F 1997 *submited to Supercond. Sci. Technol.*

[3] Angurel L A, De la Fuente G F, Badía A, Larrea A, Diez J C, Peña J I, Martínez E and Navarro R 1997 *Studies of High Temperature Superconductors* vol 21 (New York: Nova Sciece Publishers, Inc.) p 1

[4] Diez J C, Martínez E, Angurel L A, Peña J I, De la Fuente G F and Navarro R 1997 *to be submited*

[5] Larrea A, Snoeck E, Badía A, De la Fuente G F and Navarro R 1994 *Physica C* **220** 21-32

[6] Elschneer S, Bock J, Brommer G and Herrmann PF 1996 *IEEE Transactions on Magnetics* **32** 2724-2727

[7] Watson D R and Evetts J E 1996 *Supercond. Sci. Technol.* **9** 327-332

Inst. Phys. Conf. Ser. No 158
Paper presented at Applied Superconductivity, The Netherlands, 30 June–3 July 1997

A fast and non-destructive procedure to determine the transport properties of BSCCO/Ag superconducting tapes

Jan-Jaap Rabbers, William F.A. Klein Zeggelink, Bennie ten Haken and Herman H. J. ten Kate

Low Temperature division, University of Twente, P.O. Box 217 7500 AE Enschede, The Netherlands

Abstract. A fast method to determine the critical current of a short section in a long length of BSCCO/Ag tape is developed. A quick measurement of the critical current is required for an in-process quality control of long lengths of BSCCO/Ag tapes. Non-destructive temporary contacts for current transfer and voltage measurement are developed and tested. Different procedures to determine the critical current are evaluated on their speed, accuracy and correlation with a standard critical current determination. An AC-method is based on the relation between self-field loss and critical current. Offset or thermoelectric voltages do not disturb the measurement but the relation between loss and critical current is not unambiguous. With a DC-method the critical current can be measured fast and accurate because of the steep DC voltage-current relation. With an optimised DC-method the critical current of a typical BSCCO/Ag tape can be measured in 2 s with an uncertainty of about 1%.

1. Introduction

Devices based on HTS tape are expensive. Therefore an in-process quality test of the conductor is a reasonable last production step. The in-process quality control can consist of measuring the critical current of short sections along the length of a long tape. With an adequate in-process quality control the production loss can be minimised and a detailed specification of the performance of the tape can be supplied with the conductor. First, non-destructive contacts for cryogenic in-process quality control are presented, after that fast critical current measurement procedures are evaluated and experimental results are presented and evaluated.

2. Non-destructive contacts

Non-destructive contacts for current transfer and voltage probing are developed. Figure 1 shows a schematic picture of the contacts. The current contacts are copper cylinders covered with a silver layer. The resistance of the contact depends mainly on the quality of the surface of a tape. A typical contact resistance of $30\pm10\mu\Omega$ is obtained with a pressure of 3 MPa. The contact resistance is only slightly reduced when the pressure is increased to the maximal allowable value of 10 MPa. The voltage contact consists of a piece of silver tape. A line contact along the width of the tape gives better electrical results than a point contact. The spatial resolution of a point contact is better but it is not possible to get a reproducible electrical contact without damaging the tape. A line contact with a spatial resolution of 0.2 mm and a contact area of 0.5 mm^2 gives a stable electrical contact for a pressure of 5 Mpa.

A comparison with soldered contacts showed no measurable difference in voltage noise (<10 nV) and no increase of the offset voltage.

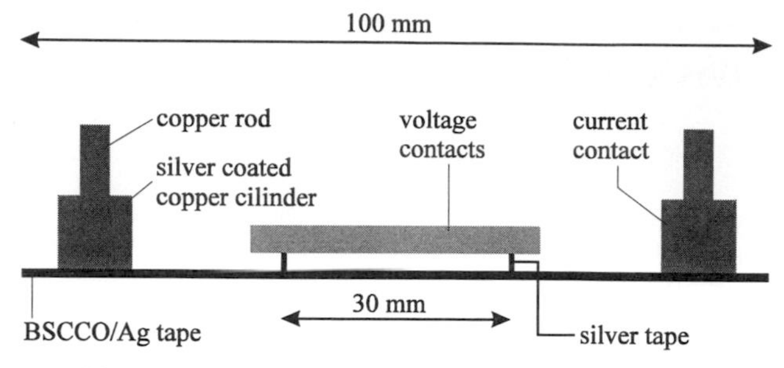

Figure 1 Schematic picture of the non-destructive contacts.

3. I_c measurement methods

When the desired result of a voltage-current measurement is not a complete VI-curve accurate to the nanovolt level, a relatively fast measurement method can be used. Two different measurement methods are investigated here. An AC-method based on the self-field loss and a DC-method that is optimised with respect to speed.

3.1. The AC loss method

When a superconductor is fed with an AC current (*I*), a voltage (*U*) across the tape can be measured due to the self-field loss. The critical current can be calculated according to:

$$U = f_1(I, I_c, f) \Rightarrow I_c = f_2(U, I, f) \tag{1}$$

The function f_2 is not unambiguous, see e.g. [1,2]. An initial gauging is necessary and the function f_2 must remain the same along the whole tested length. The relation between loss voltage and critical current, f_1, is given by:

$$U \propto \frac{1}{I_c^N} \qquad 1 < N < 2. \tag{2}$$

Because there is no strong relation between loss voltage and critical current, measuring one voltage is sufficient to characterise samples with a wide spread of critical currents. The uncertainty can be calculated from (2) by taking a partial differential:

$$\frac{\partial I_c}{I_c} = \frac{1}{N}\frac{\partial U}{U} \qquad 1 < N < 2. \tag{3}$$

The consequence of (3) is that the uncertainty of the voltage measurement is a limiting factor because the uncertainties in voltage and critical current are of the same order. A possible disadvantage of the AC-method is the complicated measurement procedure with a lock-in amplifier and phase calibration. An advantage of the AC-method is that offset and thermoelectric voltages do not disturb the measurement.

3.2. The DC-method

The DC-method is based on the DC voltage-current relation. The difference is that not the complete voltage-current curve is measured but only the current at a specified electrical field, e.g. 10^{-4} V/m. The uncertainty of the method can be evaluated with:

$$U \propto I^N \quad \Rightarrow \quad \frac{\partial I_C}{I_C} = \frac{1}{N}\frac{\partial U}{U}\,. \tag{6}$$

Formula (6) shows that the large acceptable uncertainty in the voltage measurement is directly related to the steep voltage current curve via the N-value. The relation between voltage and current uncertainty for a few realistic N-values is shown in figure 2. A disadvantage of the DC-method is thermoelectric voltage and offset voltage that is probably not constant in time. A method to solve this complication is to alternate the current or measure additional points at I_0. A fast measurement with such a DC-method is presented in the next section.

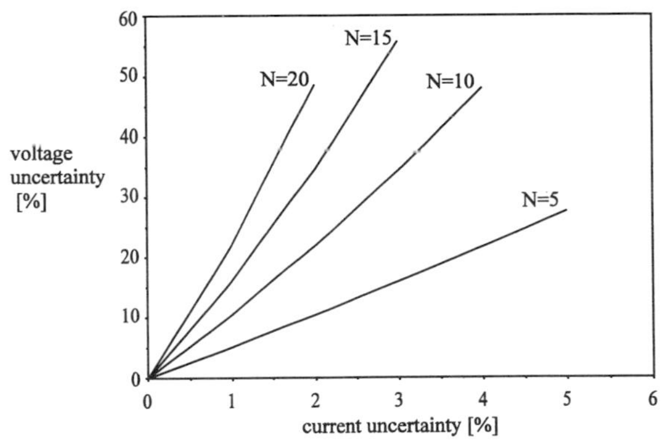

Figure 2 Uncertainty in the critical current as a function of uncertainty in voltage for different N-values.

3.3. Optimised DC-method

Not the uncertainty itself but the uncertainty in combination with measurement time is in this specific situation the interesting subject. To investigate the minimum allowable measurement time, the voltage response on a rectangular shaped current is investigated, see figure 3. After a current step a huge voltage peak is observed due to a very high *dI/dt*. The voltage decays exponentially, e.g. coupling currents in the tape, and after a time T_1 the voltage is constant within the desired uncertainty and a time T_2 is used for the voltage measurement. When the voltage is measured and averaged for a current $+I$ and $-I$, offset and thermoelectric voltages can be eliminated. Measuring this way combines the advantages of the AC-method and the DC-method. Figure 4 shows a measured response on a current step for a typical conductor (I_c=35 A, N=13). The voltage is measured with a sample frequency of 10 kHz. In order to minimise 50Hz noise the points shown in the figure are the average over 0.02 s (1 Power Line Cycle). The time T_1 for this typical sample is 180 ms, after that a time T_2 of e.g. 20 ms (1 PLC) can be used for one measurement point. This leads to $\pm 0.5\ \mu$V uncertainty in this set-up. The total time for measuring one voltage point with two directions of the current is $2(T_1+T_2)$=400 ms.

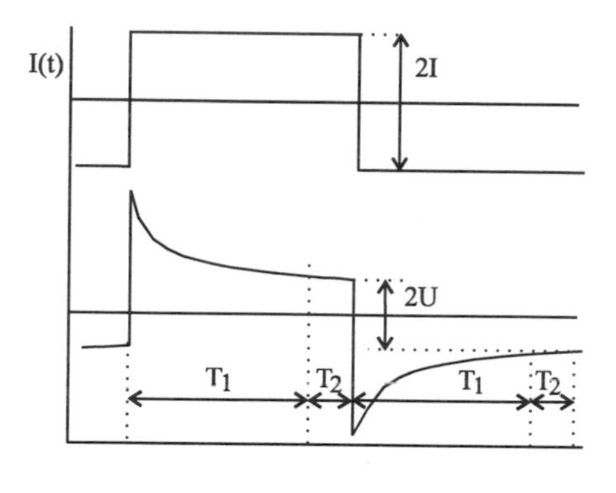

Figure 3 Principle of the optimised DC-method.

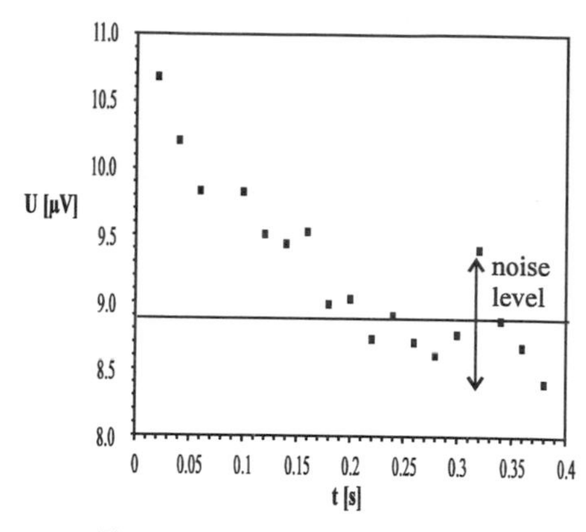

Figure 4 Voltage response on a current step voltage averaged over 1PLC.

To reach a set voltage, e.g. 10^{-4} V/m, an iteration method is used. With this method the critical current can be reached in five steps i.e. 5×400ms=2s. With a distance of 50 mm between the voltage contacts and a noise level of 1 μV (figure 4), the uncertainty is of the order of 1% for a high quality conductor with N>10 (figure 2). Even if T_1=0 and T_2=20 ms the error (1.5 μV) and the uncertainty (±0.5 μV) in the voltage leads to an uncertainty in the critical current of a few percent. The measurement time in this case is 5×20ms=100 ms. It is not necessary to switch to the AC-loss based method with all the complications of interpreting the results.

From the time constant observed in a tape as depicted in figure 4, it can be concluded that voltage current measurement methods using a constant current ramp are not possible in a time scale of seconds, because the time constant is of the same order of magnitude as the ramp time.

4. Conclusions

1. Non-destructive temporary contacts to perform a four point voltage-current measurement on a piece of BSCCO/Ag tape at 77 K are developed and tested. Compared with soldered contacts no extra contributions to the offset and noise are found.
2. A fast I_c determination can be performed best using the optimised DC-method, using a rectangular shaped current, because of the small uncertainty, the unambiguous result and the elimination of offset and thermoelectric voltages.
3. With an AC loss based method the relation between critical current and voltage is not clear and the uncertainty is a limiting factor.
4. An experiment with a typical conductor shows that with the optimised DC-method and an iteration method I_c can be measured in 2 s with an uncertainty of about 1%.

References

[1] Yang Y, Hughes T, Spiller D M, Beduz C, Penny M, Scurlock R G, Haldar P and Sokolowski R S 1996 *Supercond. Sci. Technol.* **9** 801-804

[2] Ciszek M, Ashworth S P, James M P, Glowacki B A, Camplbell A M, Garré R and Conti S 1996 *Supercond. Sci. Technol.* **9** 379-384

Inst. Phys. Conf. Ser. No 158
Paper presented at Applied Superconductivity, The Netherlands, 30 June–3 July 1997

Contact less characterisation of BSCCO tape

Michael D. Bentzon, Daniel Suchon, Peter Bodin, and Per Vase

Nordic Superconductor Technologies, Priorparken 878, DK-2605 Brondby, Denmark.

Abstract. Characterisation is important for improving and controlling the quality in the production of high temperature superconducting tape. The characterisation must be of a non-destructive kind and a contact free type is preferable, especially for long length tapes.
Magnetic characterisation is contact free and have the potential of spatial resolution of the critical current in the superconducting tape. Until now work has been focusing on single filament tapes and has not considered multifilament tapes and the coupling of filaments.
In this work we apply a Hall probe technique and measure the shielding and remanent fields induced by single and multifilament tapes. Measurements have been performed during ac and dc conditions and the results show that magnetic characterisation may be useful for measuring critical currents as well as filament coupling in multifilament tapes.

1. Introduction

The application of high temperature superconducting (HTSC) tapes in long length scale, e.g. for manufacturing of conductors and cables, makes it essential to possess a spatial resolving method capable of characterising continuos length of HTSC tape. It is important for the manufacturer for development of homogeneous HTSC tapes and for a quality control of the final HTSC tape to have a characterisation method with spatial resolution. It is also essential for the manufacturer of magnets, cables etc. to have a well documented semi-manufactured product. To fulfil demands from customers the manufacturer must be able to perform a spatially resolved characterisation on continuos lengths of HTSC tape.

Work performed until now concerning magnetic characterisation of HTSC tapes has focused on the single filament type [1,2] or have concentrated on measuring the spatial distribution of the trapped field during dc conditions [2].

The application potential of HTSC multifilament tapes is much higher but the fact that the tapes contains many and smaller filaments causes the remanent fields from these tapes to diminish. In this work we will show that by combining ac and dc techniques it is possible to use the magnetic characterisation to obtain information concerning critical current and filament coupling in HTSC multifilament tapes.

2. Experimental technique

Bi-2223 tapes are mounted inside a solenoid in a dewar with liquid nitrogen. The tape normal is parallel to the axis of the solenoid. A Hall probe is scanned across the tape axis. During dc conditions data are collected when the external field is on (shielding field) and after removing of the external field (remanent field). Tapes are cooled in zero field and the

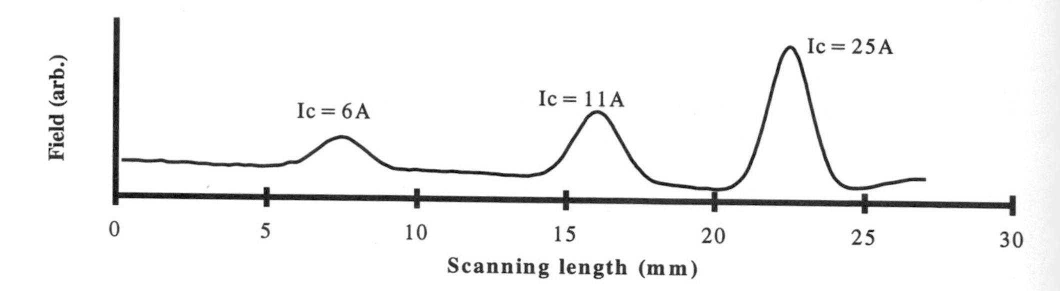

Figure 1. Hall probe scanning across 3 single filament tapes with different critical current. Remanent field, dc conditions. External field 45 mT.

remanent field is measured for increasing values of the external field. After switching on/off external fields 2 minutes are passed on before scanning of the Hall probe.

During ac measurements the solenoid field has a sine shape and the Hall probe signal is detected by a lock-in amplifier. The real part of the signal (in phase with the external field) detected by the lock-in amplifier is proportional to the shielding field of the tape and the imaginary part (out of phase) is proportional to the remanent field. A more detailed description is given in [1].

3. Results

Fig.1 shows the remanent field from dc measurements above three single filament tapes with different critical currents and the remanent field is increasing for increasing critical current.

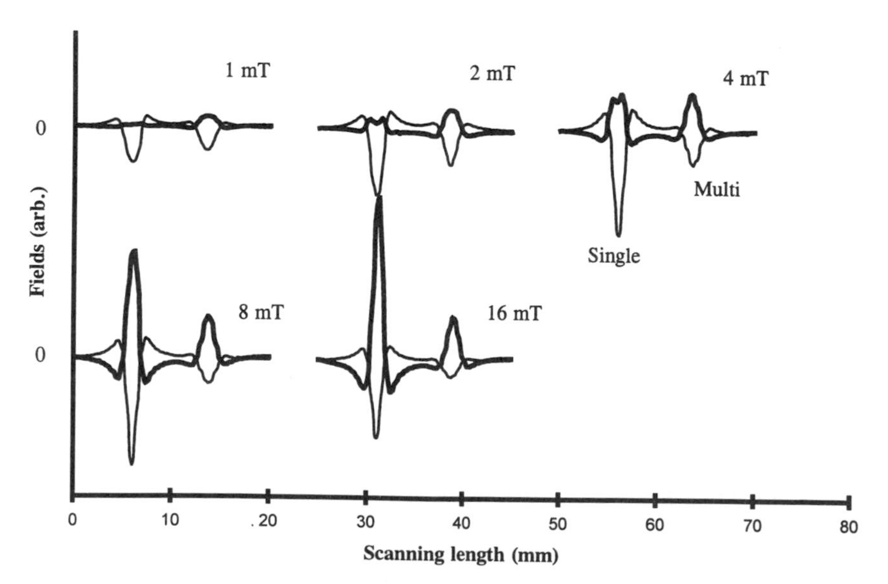

Figure 2. 5 Hall probe scannings across one single and one multifilament tape with critical currents of 23,9 A and 18,0 A, respectively. Remanent (broad curve) and shielding (narrow curve) fields are measured using 5 different external dc fields.

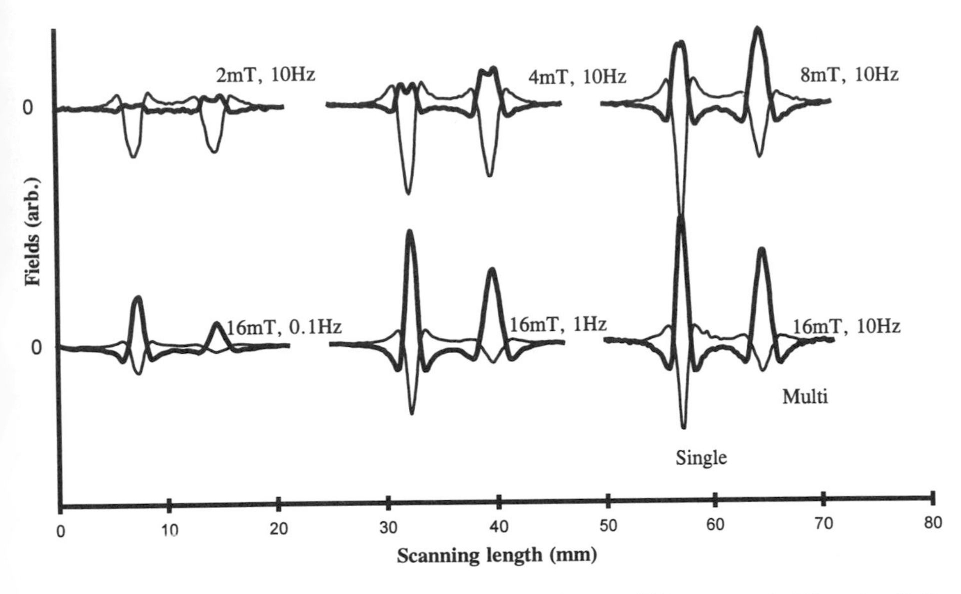

Figure 3. 6 Hall probe scannings across one single and one multifilament tape in different ac fields. The remanent fields are shown by a broad line, the shielding field with a narrow. Same tapes as in figure 2.

Fig.2 shows the results of scanning the Hall probe across one single and one multifilament tape. The left tape is a single filament tape while the other tape is a 19-filament tape. The scannings are performed at dc conditions for 5 different values of the external field. We observe a characteristic M-shape of the remanent field for the single filament tape at low fields (<4 mT). This is not observed for the multifilament tape.

Fig.3 shows the shielding and the remanent fields above the same tapes as in fig.2. Here the measurements are performed during ac conditions. At low fields the remanent field has a characteristic M-shape turning into a more simple shape for higher fields. And this is the case for the single as well as for the multifilament tape. For higher fields (8 mT and 16 mT) the shielding field of the multifilament tape is strongly reduced. The shielding field is also reduced somewhat at 16 mT for the single filament tape. The lower sets of data in figure 3 are measured at 0.1, 1, and 10 Hz and it is clear that reducing the frequency to 0.1 Hz has strong influence on the fields.

4. Discussion

The discussion will mainly concern the potential of magnetic measurements to characterise HTSC multifilament tapes. However, some considerations concerning the interaction between magnetic fields and the superconductor is performed and is based on Bean´s critical state model [3]. Considering the results in fig. 1, the trapped flux being present after removal of the external field images the capability of the superconducting material to draw a current, and we find that the remanent field is proportional to the critical current determined using standard I-V measurements.

In fig. 2 the two tapes are different in the sense that one of them is a 19-filament tape. Due to smaller current loops with smaller current the shielding and the remanent fields are expected to be much smaller for the multifilament tape when the external field is approaching the field of full penetration. This is also observed for the highest applied field,

16 mT, and is taken as a strong evidence that the filaments are not short circuited. In some cases we have observed that tapes being manufactured as multifilament may appear as single filament tapes. This is probably due to short circuiting of the superconducting filaments, and this is confirmed by microscopy.

In fig.3 the same two tapes are compared at different external ac conditions. The induced electromotive force increases with field amplitude and frequency and may cause the filaments to couple. At very low frequency (0.1 Hz) the conditions are approaching dc, and a strong difference between the single and the multi filament tape is observed. At 10 Hz and at low fields we observe the characteristic M-shape for both tapes. This shape is to be expected for low fields and single filament tapes [3] and it is a strong indication that the filaments in the multifilament tape are now coupled due to the ac field. These observations show that magnetic measurements during dc and ac conditions may be applied for detecting the filament coupling.

When increasing the frequency of the applied field from 0.1 to 1 and 10 Hz (at 16 mT) we observe a strong increase in amplitude on the measured fields and specially the remanent field for the multifilament tape increases strongly. We interpret this as a coupling of the filaments through the silver matrix due to induced electromotive force caused by a faster alternating magnetic field. This coupling of the filaments due to the ac magnetic field indicates that the magnetic measurement can be applied to measure the critical current, also for multifilament tapes. The coupling may be performed either by an ac magnetic field or by moving the tape through a field gradient.

Differences between the single and the multifilament tapes are also seen when changing the field amplitude at constant frequency (10 Hz), fig.3. We observe that the remanent field of the multifilament tape rises faster at low external fields. We also observe that the shielding field starts decreasing already at 8 mT external field for the multifilament tape, while this happens at 16 mT for the single filament tape. These informations are not analysed in this work.

Also for the single filament tape we observe an effect when reducing the frequency from 1 Hz to 0.1 Hz. The fields are much higher at 1 Hz and the difference may be explained by the relaxation of field lines. This relaxation is faster than observed in [4].

5. Conclusion

Magnetic measurements using Hall probe technique is well suited for characterisation of single and multifilament superconducting tapes. The method is sensitive to both critical current and filament coupling. The method may be applied on both fixed and moving tapes, and may be performed for basic studies as well as on-line characterisation of moving and long length tapes. Due to the non-contact nature of the method it is specially well suited for characterisation of long length and moving tapes.

References

[1] Paasi J A J and Lahtinen M J 1993 *Physica C* **216** 382-390.
[2] Tanihata A, Sakai A, Matsui M, Nonaka N and Osamura K 1996 *Supercond. Sci. Techn.* **9** 1055-1059.
[3] Bean C P 1962 *Phys. Rev. Lett.* **8** 250.
[4] Paasi J A J, Polák M, Lathinen M, Plechácek V and Söderlund L 1992 *Cryogenics* **32** 1076-1083.

Inst. Phys. Conf. Ser. No 158
Paper presented at Applied Superconductivity, The Netherlands, 30 June–3 July 1997

Computation of critical current through magnetic flux profile measurements

J. Amorós[1]*, M. Carrera[1+], X. Granados[1], J. Fontcuberta[1] and X. Obradors[1]

[1]Inst. de Ciència de Materials de Barcelona (CSIC); Campus U.A.B., 08193 Bellaterra (Spain)
* D. Matemàtica Aplicada. U.P.C. Diagonal 647. 08028 Barcelona (Spain)
+ DMACS. Universitat de Lleida. Pl. Víctor Siurana,1. 25003 Lleida (Spain)

Abstract. Hall probe magnetometry is used to measure critical currents of large melt-textured superconducting $YBa_2Cu_3O_7$ blocks. Comparison of the current densities deduced based on critical state analysis of magnetization and by solving the inverse problem from magnetic flux maps is reported for samples having a single domain and multidomain structure.

1. Introduction

Determination of critical current density distributions is a fundamental aspect of superconductor's characterization. Whenever direct measurements of critical current density J_c of superconducting samples cannot be employed other techniques should be used. This is particularly relevant for the case of levitation problems, where large superconducting blocks are needed. Commonly, inductive critical current densities are deduced under Bean model approximations with only limited success mainly because of finite sample geometry, demagnetizing effects, anisotropy and the field dependence of $J_c(B)$.

The magnetic flux maps B(x,y,z) determined using Hall probes provide a global measure of the sample magnetization, so it contains information about the current distribution $J_c(x,y,z)$ in the bulk sample. Consequently, B(x,y,z) maps can be measured and analyzed to recover $J_c(x,y,z)$ by solving the inverse problem.

In this contribution we present $J_c(x,y)$ determinations based on Hall probe magnetometry, either by using critical state analysis of magnetization or by solving the inverse problem from magnetic flux maps. Results are presented and compared for $YBa_2Cu_3O_7$ (YBCO) melt textured blocks having different microstructure.

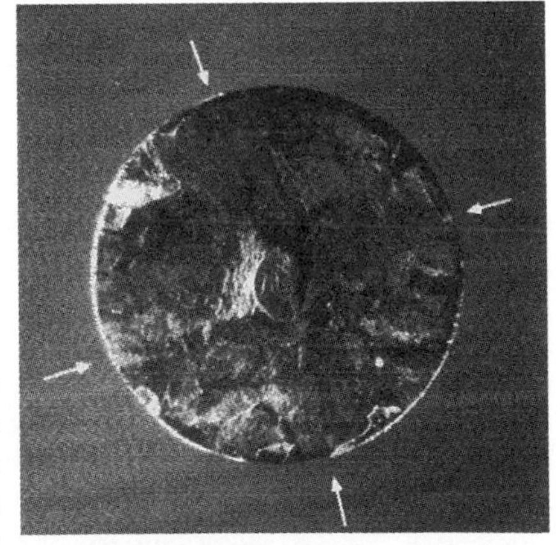

Fig. 1. Photography of cylinder top surface. Note the central semicircunference corresponding to the seed position and the X-cross section indicated by arrows.

2. Experimental

We report and analyze data corresponding to two different YBCO melt textured samples prepared by top seeding growth [1]. One of the samples, so called Rectangular, has been cut to a parallelepiped of $10\times8.5\times4mm^3$ from a long cylinder. Under the optical microscope, the largest face (ab plane) appears to be a single domain, i.e. large angle grain boundaries are absent. A smaller piece ($1\times1\times1$ mm^3) has been cut and used for magnetization measurements by a magnetometer. A second sample, so called Cylindrical, is a cylinder of 31 mm diameter and 18 mm height. It has an identical initial composition than the previous ones. Under optical microscope, the upper face of the block (ab plane) reveals (see Fig. 1) the typical X-cross generated around the seed.

Different sets of measurements have been performed at 77 K to fully characterize the magnetic properties of these samples: a) magnetic flux mapping B(x,y) using a Hall probe. The magnetic flux over the sample is scanned at a distance of 0.5 mm, using a step of 0.2 mm. The active area of the Hall probe is 0.01 mm^2. In these experiments field is applied by using either a permanent magnet (SmCo, diameter: ϕ= 25 mm and height: h = 20 mm) or an electromagnet. b) Induction measurements at a fixed position have been used to measure magnetization hysteresis M(H) loops up to 120 kOe and remanence B_{rem}(H). c) Hysteresis loops have also been measured by using a SQUID magnetometer.

3. The computation method

The starting point of our computation is a superconducting sample A×I, which is a straight cylinder over a base A of arbitrary geometry. A current **J**= (J_x,J_y,J_z) circulates through this sample. It is assumed that : (i) The current circulation is horizontal, i.e. J_z=0. (ii) The planar current in every horizontal section is the same, i.e. the functions J_x,J_y depend only on x,y. The first hypothesis is fulfilled by well crystallised samples of YBCO. The second one is more restrictive; it is reasonable in the case of samples with a thickness of few mm fully penetrated by the field. We do *not* make any hypothesis on the distribution of currents in the basal planes. Thus the algorithm may detect the existence of in-plane irregular current distributions.

Our method is an adaptation of that of [2] and based on the measurement of the vertical component B_z(x,y,z) of the magnetic field generated by the currents over the sample. Let **M**=(0,0,M) be the magnetization of the material, which is vertical by our previous assumptions. We measure the vertical magnetic field B_z in a mesh at a constant height h over the sample with a Hall scanning probe. The sample base A and its neighbourhood is divided in a fine rectangular mesh, henceforth called the *discretization lattice* and we assume that the magnetization M has a constant value M(i,j) in the prism over every rectangle Δ_{ij} of this lattice. Therefore:

$$B_z(x_m,y_n,h)=\sum_{i,j} M(i,j)\frac{\mu_0}{4\pi}\int_{\Delta ij\times I}\frac{3z^2-r^2}{r^5}dxdydz \tag{1}$$

where r=|(x_m,y_n,h)-(x,y,z)|, and the summation extends over all prisms Δ_{ij} × I in the *discretization lattice*. The integrals in equation (1) depend only on the geometry of the problem. Computing them, equation (1) becomes a linear equation on the unknowns M(i,j), and using measured B_z over a mesh of points (x_m,y_n,h) we can construct a determined or overdetermined linear system of equations. The discrete solutions M(i,j) is a good approximation to M if a fine discretization lattice is chosen. Due to the limited precision of the measures of B_z and the propagation of errors in the resolution of the linear system, it is necessary to use overdetermined systems, i.e. measures of B_z should be redundant. To solve the system, the QR algorithm has been selected. Once the magnetization **M** has been determined, the current **J**= (J_x,J_y) is obtained by computing ∇× **M**. The solver is written in C language. When applied to the present samples, computing **J** with a discretization resolution of 0.3 × 0.3 mm, involves solving overdetermined systems of about 1600 × 14000.

4. Results and discussion

4.1. Rectangular sample

Fig. 2 shows the remanent flux distribution (at 77K) after a ZFC process corresponding to an applied field $\mu_0 H_{ap}$=0.33 T. A flux distribution having a unique central peak (B_{rem} ≈60 mT) as expected for a single domain sample, is observed. Measurements of B_{rem} for various applied fields (not shown) revealed that the full penetration field (H*) value is about 175-210 mT. In order to estimate the current density from these local remanence B_{rem} measurements, it should be taken into account the fact that self-field and demagnetizing fields reduce the observed B_{rem} value. According to Ref. 3, $B_{rem}= \mu_0(1-D)\ M_{rem}\ C_B(r)$, where D is the demagnetization factor and $C_B(r)$ is a factor that depends on the position of the probe on the sample. For our experimental set up: D≈0.55 and $C_B(0)$≈0.77 and thus $\mu_0 M_{rem}$ (H) ≈0.43 T. Using the appropriate expression [3] to evaluate the critical current, it turns out that J_c ≈1.2×10^4 A/cm^2. The critical current obtained by using global magnetization (SQUID) measurements on a smaller piece (1mm^3) at zero-field and for H//c, leads to J_c^{ab} ≈3.0×10^4 A/cm^2 which is in good agreement with the above estimates obtained from local probes.

Therefore, from the above comparison it is clear that local magnetization measurements can be used to consistently estimate the critical current of large samples. However, we note that in all cases, usc has been made of the critical state model; that is, we assume that currents flow undisturbed over all the sample. This approximation seems appropriate when considering the map of Fig.2.

The induced current distribution $\mathbf{J}(J_x,J_y)$ calculated from the remanence induction map (Fig. 2.) is shown in Fig.3. The current circulates mainly in the more external part of the sample describing almost rectangular paths that are progressively circular towards the center of the sample. The maximum current density takes values around 4.0×10^3A/cm^2 with a considerable decrease towards central zone where J≈10^3-1.5×10^3A/cm^2.

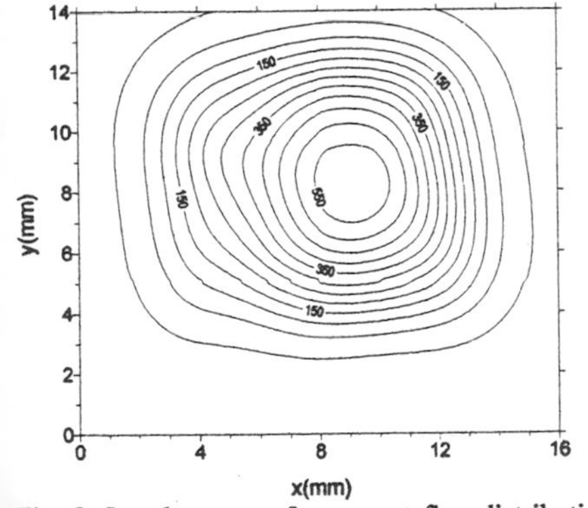

Fig. 2. Level curves of remanent flux distribution after a ZFC process with an applied field of 0.33T.

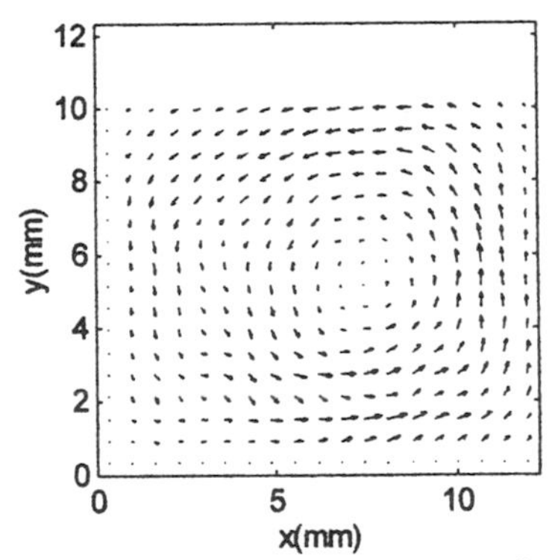

Fig. 3. Circulating current calculated from flux mapping of Fig. 2.

The currents computed by solving the inversion problem from remanent flux maps are significantly smaller than measured by magnetization techniques. We should ask ourselves about the factors that justify these differences. First, remanent flux mappings were started about 30 min after the field was removed. Measurement of magnetic relaxation indicates a decrease of about a 30% in 30 min; so relaxation is not sufficient to account for the observed differences. Probably of more significance is the fact that flux profiles were taken from a non fully penetrated initial state (H<2H*), so current loops of opposite polarity should flow at the sample center.

4.2. Cylindrical sample

This sample is a single domain. The X-cross is somehow smeared and thus may behave as subgrain boundary [4]. In Fig. 4 we show the remanent flux distribution measured after a FC process for a $\mu_0 H_{ap}$=1 T. The perturbation of the flux at the edges is apparent and this is reflected in the calculated J(x,y) shown in Fig. 5. It is a beautiful evidence of how current distribution is conditioned by the microstructure. The maximum J_c is about 9500 A/cm^2. We would like to mention that in this case, the sample was fully penetrated (FC), the values of J_c are thus more significant and so they approach to values obtained by magnetization methods.

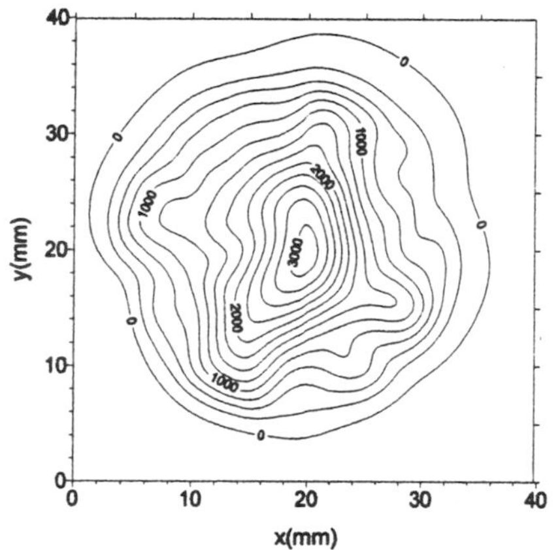

Fig. 4. Level curves of remanent flux distribution after a FC process for an applied field of 1T.

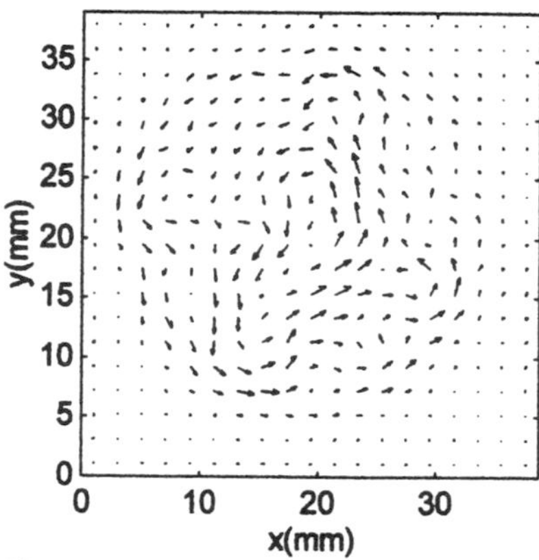

Fig. 5. Circulating current calculated from flux mapping of Fig. 4.

5. Conclusions

In summary, we have shown that Hall magnetometry is very sensitive to detect inhomogeneities of flux profile and allows determination of critical currents in large superconducting blocks. By solving the inversion problem, accurate maps of the current circulation can be obtained although precise numerical values can only be extracted for magnetically fully penetrated initial state.

Acknowledgements

We would like to acknowledge financial support by the CICYT (MAT 96-1052) and MAT (94-1924) and the Generalitat de Catalunya (GRQ95-8029).

References

[1] R. Yu, J. Mora, S. Piñol, F. Sandiumenge, N. Vilalta, V. Gomis, B. Martínez, E. Rodríguez, J. Amorós, M. Carrera, X. Granados, D. Camacho, J. Fontcuberta, X. Obradors 1997 IEEE Trans. on Appl. Supercond. (in press)
[2] W. Xing, B. Heinrich, Hu Zhou, A. Fife, A. Cragg 1994 J. Appl. Phys. **76**, 4244-4255
[3] H.P. Wiesinger, F.M. Sauerzopf, H.W. Weber. 1992 Physica C **203**, 121-128
[4] F. Sandiumenge, N. Vilalta, X. Obradors, S. Piñol, J. Bassas, Y. Mariette 1996 J. Appl. Phys. **79**, 8847

Inst. Phys. Conf. Ser. No 158
Paper presented at Applied Superconductivity, The Netherlands, 30 June–3 July 1997

Levitation of a small permanent magnet between superconducting $YBa_2Cu_3O_{7-\delta}$ thin films

R Grosser[1], A Martin, O Kus, E V Pechen[2] and W Schoepe

Institut für Experimentelle und Angewandte Physik, Universität Regensburg, D-93040 Regensburg, Germany

Abstract. A small permanent magnet (spherical shape, radius 0.1 mm) is levitating inside a superconducting parallel plate capacitor made of YBCO thin films. The nonlinear damping of the translational oscillations of the magnet is investigated as a function of amplitude and temperature. The dissipation is found to be orders of magnitude lower than in bulk YBCO samples at small amplitudes. With increasing amplitude the dissipation becomes exponentially large. We attribute our results to thermally activated flux creep in the YBCO thin films assuming a linear current dependence of the pinning barriers.

1. Introduction

Among the possible applications of high-temperature superconductors for practical use superconducting levitation, e.g. for bearings, is most promising. In particular static levitation forces, dynamical stiffness and damping of levitation systems consisting of permanent magnets and $YBa_2Cu_3O_{7-\delta}$ (YBCO) bulk samples have been widely investigated [1]. However, there are only few experiments considering the application of YBCO thin films, although a very high static force density is reached. In our present work we have investigated the elastic and dissipative forces of the oscillations of a small spherical magnet levitating between two horizontally YBCO epitaxial thin films. We focus in particular on the damping of the oscillations which for small amplitudes is found to be much lower than in bulk samples studied in earlier work [2] but becomes exponentially large with increasing amplitude.

2. Experiment

Our experimental method has been described in detail in earlier work [2, 3]. A small spherical permanent magnet ($SmCo_5$, radius 0.1 mm, mass 52 μg) is levitating inside of a superconducting parallel plate capacitor (spacing d = 1 mm, diameter 4 mm). In the present work the electrodes consist of YBCO epitaxial thin films (thickness δ = 450 nm) laser deposited on $SrTiO_3$ (lower electrode) and Y_2O_3-stabilized ZrO_2 (upper electrode) substrates [4]. The substates are cut with an inclination of 2 degrees off the

[1] E-mail: reiner.grosser@physik.uni-regensburg.de
[2] Permanent address: Lebedev Physics Institute of the Academy of Sciences, Leninskii Prospect 53, 117924 Moscow, Russia

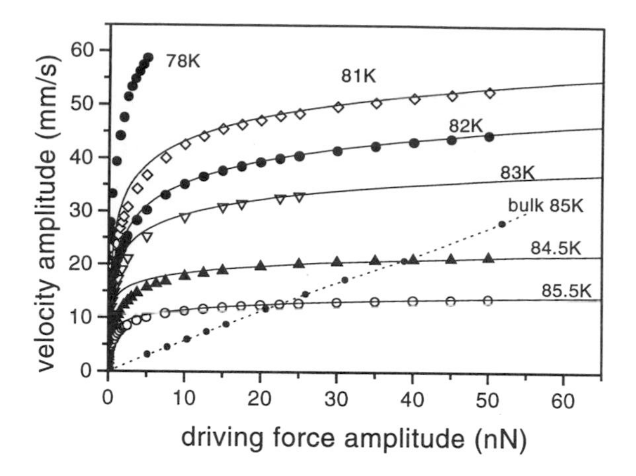

Figure 1. Resonant velocity amplitude of the oscillating magnet as a function of the driving force at various temperatures. Note the steep increase at small driving forces and a logarithmic dependence at large drives. The lines are fits of our model to the data, see text. A typical linear dependence observed with melt-textured bulk samples is shown for comparison.

(001) orientation, improving both flux pinning and, due to initiating a terrace growth, crystallinity of the films. The lower electrode is protected by a 20 nm thick $PrBa_2Cu_3O_7$ epitaxial layer. The critical temperature $T_c = 89.5$ K is measured by an ac susceptibility method showing an extremly sharp transition width of 0.1 K.

Before the capacitor is cooled through T_c a dc voltage of about 800 V is applied to the bottom electrode. The magnet then carries an electrical charge $q \sim 2$ pC when levitating and the voltage is switched off. Now, vertical oscillations of the magnet can be excited by applying an ac voltage U_{ac} ranging from 0.1 to 20 V. The moving charged magnet induces a current $q \cdot v/d$ (v is the velocity of the magnet) in the electrodes which is detected by an electrometer and a lock-in amplifier. We measure the maximum signal at a given driving force amplitude $F = q \cdot U_{ac}/d$ ranging from 10^{-13} N to 10^{-7} N while sweeping its frequency towards resonance. Then the driving force amplitude is varied at constant temperature. The velocity amplitude ranges from 0.1 mm/s to 50 mm/s, which corresponds to an oscillation amplitude $a = v/\omega$ of about 50 nm to 30 μm. The measurements are carried out in a temperature range from 4 K to 87 K.

3. Results and discussion

We investigate the damping of the oscillations by analyzing the velocity amplitude at resonance as a function of driving force amplitude, see Fig. 1. After a steep initial increase at small drives the amplitude grows only logarithmically for large $F > 10$ nN: $v = v_0 \cdot \ln(F/F_0)$, where v_0 is a characteristic velocity which increases with decreasing temperature. This implies an exponential increase of the dissipation. This logarithmic regime could not be reached at lower temperatures because the oscillations became unstable at large amplitudes. The initial increase of the amplitude is shown in Fig. 2 on a largely expanded force scale. Only for the smallest drives a linear dependence $F = \lambda \cdot v$

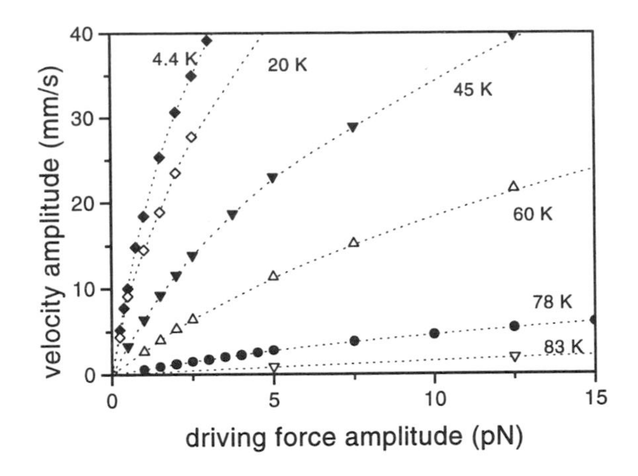

Figure 2. Initial velocity amplitude at very small driving forces indicating the linear initial increase at various temperatures. (The dashed curves are merely a guide to the eye.)

is observed, where the drag coefficient $\lambda(T)$ increases strongly with temperature. The temperature dependence of λ is given by $\lambda(T) \propto (T_c - T)^{-2}$ down to about 20 K where it becomes constant $\lambda \sim 5 \cdot 10^{-11}$kg/s. This corresponds to a Q-factor $Q = m \cdot \omega/\lambda \sim 10^6$. For such high values of Q we cannot completely rule out dissipation due to residual gas or eddy currents in normal metals (vacuum can or contact leads) but we propose that dissipation due to viscous motion of pinned flux lines dominates.

For a more quantitative analysis of our data, we first calculate the dc sheet current $J(x, y)$ on the films given by the dipolar fields of the magnet (levitating in the middle between the electrodes with its dipole moment $4 \cdot 10^{-6}$ Am^2 orientated parallel to the surfaces) and the first image dipoles [5], neglecting images further away and modifications of the field due to trapped flux. The maximum static field at the film surfaces is about 5 mT directly below or above the magnet. This is less than the lower critical field. The ac surface current caused by the oscillations of the magnet is then calculated from $\Delta J(x, y, t) = \left(\frac{\partial J(x,y,h)}{\partial h}\right)_{h_0} \cdot a \cos \omega t$, where a and ω are the oscillation amplitude and frequency, respectively. The maximum ac field is about 0.2 mT for a velocity amplitude of 10 mm/s. This ac current gives rise to an alternating Lorentz force on the vortices, disturbing the stable non-dissipative trapped flux configuration in the films. This causes flux lines to move and an electric field $E = v_f n \Phi_0$ arises, where v_f is the vortex velocity and n the moving flux line density. The dissipated energy caused by this movement is balanced by the energy gained from the external force at resonance during period τ:

$$\int_0^\tau F \cdot v \, dt = \int_0^\tau \int_{x,y} \Delta J \cdot E \, dx dy dt.$$

We note that we have assumed here a homogeneous current distribution and a homogeneous flux line velocity over the thickness of the films. For the current dependence of the vortex velocity v_f we use the model of Kim and Anderson for thermal activated flux jumps which leads to a hyperbolic-sine dependence $E \propto \sinh(j/j_0)$, with $j_0 = -kT/(\partial U/\partial j)_0$ and a linear current dependence of the pinning barrier $U(j) = U_0 + (\partial U/\partial j)_0 \cdot j$, $(\partial U/\partial j)_0 < 0$, which we approximate by a first order expansion [6]. This gives the

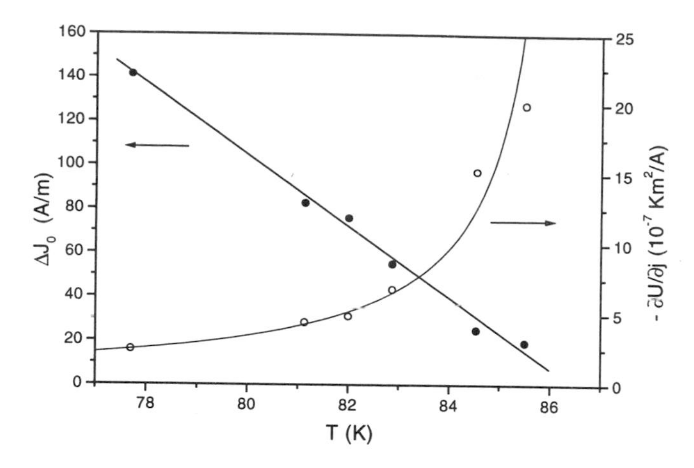

Figure 3. Characteristic sheet current ΔJ_0 obtained from a fit of the model to the data (left scale) and the first derivative $(\partial U/\partial j)_0$ of the pinning potential (right scale) as a function of temperature. (The full lines are a linear fit to the ΔJ_0 data.)

right asymptotic behaviours, both for the inital linear regime and the exponential one at large drives even after the integration is performed[3]. In Fig. 1 the evaluated $v(F)$ curves from our model are compared with the experimental data. From the measured characteristic velocity $v_0(T)$ we obtain a characteristic surface current $\Delta J_0(T) = j_0 \cdot \delta$ which is shown in Fig. 3. It decreases linearly with increasing temperature and goes to zero at $T_0 = 86.5$ K $< \mathrm{T_c} = 89.5$ K. Because of the linear decrease of $j_0 \propto (T_0 - T)$ we obtain $(\partial U/\partial j)_0 \propto t/(1-t)$, with $t = T/T_0$ (see Fig. 3). This means that at T_0 the barrier $U(j)$ vanishes for infinitesimal small currents, free flux flow sets in and the levitating magnet looses its lateral stability.

4. Conclusion

In conclusion our phenomenological model describes rather well the measured energy dissipation in the superconducting films caused by the oscillations of the levitating magnet, although some simplifying assumptions are made. The increase of dissipation from a very low and linear regime to an exponential one is in sharp contrast to bulk samples.

References

[1] Moon F C 1994 *Superconducting Levitation* (New York: John Wiley & Sons)
[2] Grosser R, Jäger J, Betz J and Schoepe W 1995 *Appl. Phys. Lett.* **67**, 2400–2
[3] Barowski H, Sattler K M and Schoepe W 1993 *J. Low Temp. Phys.* **93** 85–100
[4] Pechen E V, Varlashkin A V, Krasnosvobodtsev S I, Brunner B and Renk K F 1995 *Appl. Phys. Lett.* **66**, 2292–4
[5] Haley S B and Fink H J 1996 *Phys. Rev.* **B53**, 3506–15
[6] Blatter G et al. 1994 *Revs. Mod. Phys.* **66**, 1125–1388
Brandt E H 1995 *Rep. Prog. Phys.* **58**, 1465–1594

[3] We assume a constant density n of the trapped flux lines

Inst. Phys. Conf. Ser. No 158
Paper presented at Applied Superconductivity, The Netherlands, 30 June–3 July 1997

Modelling Of $YBa_2Cu_3O_7$ Superconductors For Magnetic Levitation Applications

E. Portabella, J. Mora, B. Martínez, J. Fontcuberta and X. Obradors

Institut de Ciència de Materials de Barcelona (C.S.I.C.), Campus U.A.B., Bellaterra 08193, Catalunya, Spain

Abstract.- A numerical method based on a current model and employing finite element analysis is developed to calculate the flux profiles and magnetisation loops for superconductor cylinder of arbitrary size. Results are compared with analytical solutions, when available, and with experimental magnetisation loops, with excellent agreement. As a further application of the model, the forces between a permanent magnet and a superconductor are computed.

1. Introduction

The development of bulk superconductors with strong pinning forces, has opened up the possibility of applications of large monolithic blocks for levitation purposes. In this context, the importance and utility of simulation tools in order to understand and predict the behaviour of these materials is obvious. The finite dimensions of the superconducting blocks and the non-uniformity of the field generated by permanent magnets pushes the computing difficulties far beyond the analytical solutions. Then finite element (FE) analysis becomes an appropriate tool. In this work, a new method, so-called *current method*, is developed [1]. Our model simulates the hysteretic behaviour of a superconducting material by using the critical state. That means that we consider all the shielding currents as volumetric and with values that are the critical current at any point of the superconductor. By imposing this restriction to the current value at any point and using as an input the precise $J_C(H)$ dependence, an iterative model has been developed which computes the current distribution and the magnetisation of the sample.

2. Implementation procedure

In the critical state model, volumetric currents are induced in the superconductor to avoid the variation of the magnetic flux density inside the material. Our method imposes elementary currents inside the sample with this purpose. A field dependent current function $J_C(H_i)$, where H_i is the field intensity at the element i, has to be chosen.

In the initial state, when no external field is applied, there are no induced currents inside the superconductor. As the magnetic field is increased to an arbitrary value ΔH, the internal magnetic field is calculated by the FE software, by solving the Maxwell's equations. In order to reduce the flux density at the superconductor a current $J_C(H_i)$, with the appropriated polarity, is fixed over the whole sample. This current assignment and field calculations are repeated iteratively until a stable solution is found, that is, until the values of

J_C verify the imposed law $J_C(H_i)$. Finally, the coherence of the solution is checked and the global magnetisation $M(H_{app})$ is calculated from the final current distribution. The process can be repeated for any value of the applied field and the complete M(H) cycle can be computed.

The algorithm described above is applicable to the first cycle of a zero-field-cooled process , that means, when the external field is increased from zero to the maximum field (H_m). In order to calculate the second (field retreating) and third (field reversing) cycles, and due to the fact that in these processes the magnetisation is dependent on the previous field history, the internal field values for each element inside the superconductor should be stored in a matrix. The elementary fixed currents should have the appropriate polarity to maintain this stored flux density inside the superconductor. The calculation method for any ZFC or FC cycle are basically the same.

In order to validate the numerical method, the magnetisation of a superconducting infinitely long cylinder, for which analytical expressions are available [2], has been computed and compared to the analytical predictions. As a convenient and general enough approximation we have used the $J_C(H)$ given by Kim's model [3], that is: $J_C(H_i) = k/(H_o+|H_i|)$, where k and H_o are parameters that define the zero-field value of the critical current and the sensitivity to field. The agreement between the results of the FE numerical model and the analytical expressions given in Ref. (2) is complete.

Therefore, we can conclude that the algorithm developed, based on the new *current method* to represent the response of the superconductor to field variations, provides an accurate description of the magnetisation loops. As it can be seen, for this infinitely long cylinder, the results are extremely accurate.

The success of this approach, has opened the possibility to explore more delicate situations, like finite geometries or the effects of a non-uniform external field. The interest of such situations is evident from many points of view

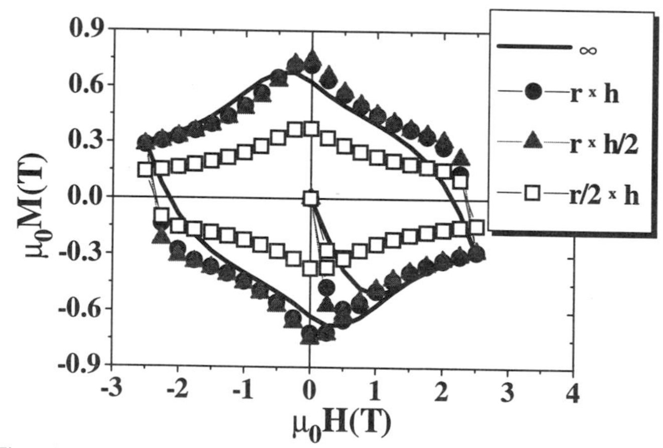

Fig. 1: Computed magnetisation curves for an infinite and finite superconducting samples (h=7.5mm , r=10mm ; h/2 x r ; h x r/2)

3. Results and discussions

To progress in that direction, the next step to be taken is to use our method to simulate superconductors with a finite geometry immersed in an homogenous magnetic field. Thus the influence of the demagnetising effects can be evaluated.

3.1. Geometry effects

Fig. 1 shows magnetisation cycles corresponding to the Kim critical current dependence $J_C(k_3=2\cdot10^{14}A^2/m^3, H_{o3}=10^6\ A/m)$ for a finite superconductor. The maximum applied field is μ_oH_m=2.5T and the field step is $\mu_o\Delta H$=0.25T. The superconductor sample has a radius r=10mm and height h=7.5mm. The computed magnetisation curve for this r×h cylinder , is compared with that obtained for the infinite cylinder. In Fig. 1 the demagnetisation effects due to the finite geometry are cleared appreciated. On the other hand, the numerical results for the different geometry relations are in agreement with the theoretical expectations [4].

The current distribution inside the superconductor sample is shown in Fig. 2. As it was expected, due to the demagnetising effects current profiles are different for any height in the sample.

3.2. Experimental validation

In order to validate the finite geometry results, the M(H) cycle of a very short sample (1.22mm×1.38mm×0.88mm) has been obtained by means of SQUID magnetisation measurements and compared with the calculations for a cylindrical equivalent geometry (r=0.65mm, h=0.9mm). The critical current deduced from the experimental hysteresis loop is represented in Fig. 3a. This curve has been fitted to an exponential law $J(H) = C_1e^{-B/C_2}+C_3$, and used as an input in our calculation method (solid line in Fig. 3a). The global magnetisation cycle is shown in Fig. 3b.

As it can be seen, good agreement is found for all the points in the curve, also for those close to the origin.

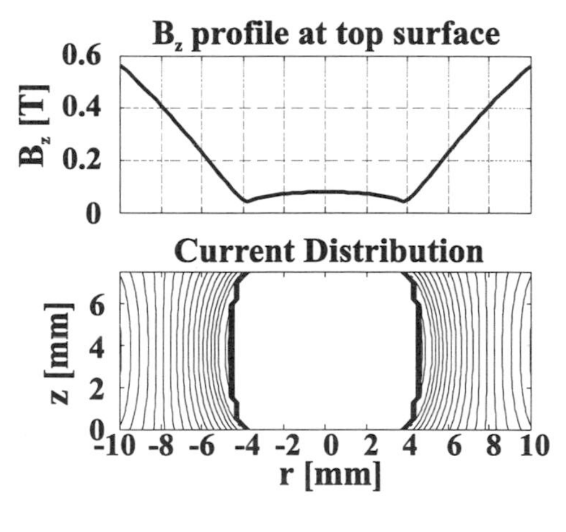

Fig. 2: Field profile and current distribution in a finite superconductor sample (h=7.5mm, r=10mm) in a ZFC process.

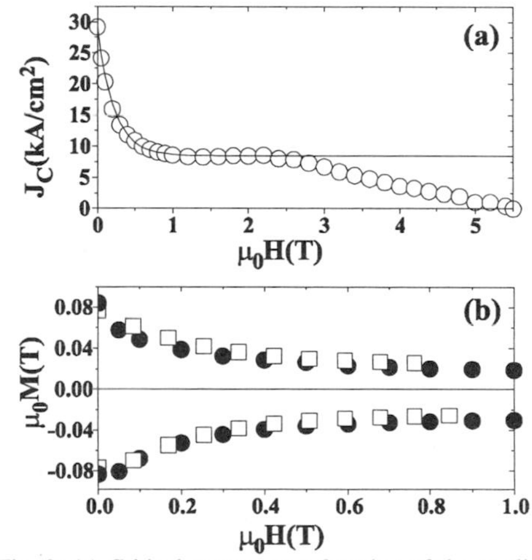

Fig. 3: (a) Critical current as a function of the applied magnetic field. The solid line through the experimental points is the exponential curve used in the FEM analysis. (b) Magnetisation cycle measured with the SQUID magnetometer (solid circles) compared with the numerical results (open squares).

3.3. Further results

Finally, the interaction between a superconductor and a permanent magnet in a FC process can be evaluated [1][7]. Assuming a Kim-like critical current $J_C(k_3=2\cdot10^{14}\ A^2/m^3,\ H_{o3}=10^6\ A/m)$. In this example, a cylindrical permanent magnet PM (r=12mm, h=20mm) axially magnetised with a remanent field 0.35T has been used. The superconductor sample has a radius 10mm and height 15mm. In Fig. 4 we show the attractive force when retreating the magnet from the superconductor and the repulsive force observed in the approaching process. The qualitative agreement with experimental levitation forces is evident [5][6].

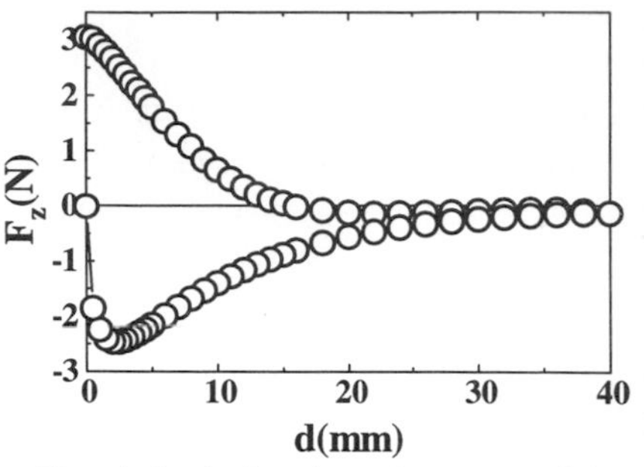

Fig. 4: Levitation force between a finite superconductor sample and a permanent magnet after a FC process.

4. Conclusions

The model here presented provides a very simple and general method to calculate the current and the magnetic field distributions inside a cylindrical superconducting sample after a ZFC or FC process. The algorithm has been successfully validated for infinite geometry, leading to excellent agreement with analytical solutions.

Experimental validation has also been performed for finite geometries, comparing the numerical results with SQUID magnetisation measurements. Thus, we have demonstrated that the use of current distributions to simulate the magnetic superconductor behaviour leads to a close agreement with the experimental results. Preliminary tests for situations where the applied filed is not uniform, a typical configuration found in levitation experiments, also seem to present very promising results. Flux penetration maps can be calculated for arbitrary axisymetrical geometries and levitation forces conveniently evaluated. Due to its extreme versatility, the method we have developed can be a very convenient tool to be used not only by designers of superconducting applications based on levitation systems, but also to help the understanding of flux penetration in real (finite) superconducting samples.

Acknowledgements

We would like to acknowledge financial support by the CICYT (MAT 96-1052) and MAT (94-1924), EC-EURAM (BRE2CT94-011) and the Generalitat de Catalunya (GRQ95-8029).

References

[1] E. Portabella, J. Mora, J. Fontcuberta and X. Obradors, Physica C, submitted
[2] X. Chen and R. B: Goldfarb, J. Appl. Phys 66, 6 (1989)
[3] Y. B. Kim, C. F. Hempstead and A. R. Strnad, Phys. Rev. Lett. 9, 306 (1962)
[4] H. P. Wiesinger, F. M. Sauerzopf and H. W. Weber, Physica C 203, 121-128 (1992)
[5] F. C. Moon, in *Superconducting Levitation,* Ed. John Wiley & Sons Inc., 1994
[6] J. Mora, X. Granados, V. Gomis, M. Carrera, F. Sandiumenge, S. Piñol, J. Foncuberta and X. Obradors, in *Applied Superconductivity*, Ed. D. Dew-Hughes, Publ. Institute of Phys., Vol 148, p. 679 (1995)
[7] D. Camacho, J. Mora, J. Fontcuberta and X. Obradors, J. Of Applied Physics, 82 (2), 15 July 1997

Inst. Phys. Conf. Ser. No 158
Paper presented at Applied Superconductivity, The Netherlands, 30 June–3 July 1997

A hybrid magnetic shield employing ferromagnetic iron and HTSC rings

V Meerovich[1], V Sokolovsky[1], S Goren[1], G Jung[1], G Shter[2] and G S Grader[2]

[1]Physics Department, Ben-Gurion University, P.O.B. 653, 84105 Beer- Sheva, Israel
[2]Chemical Engineering Department, Technion, 32000, Haifa, Israel

Abstract. HTSC shields performed in the form of close loops can be used to protect sensitive devices and human beings from stray magnetic fields produced by electric equipment. We present the investigation results of a laboratory model of the magnet shielded by a stack of HTSC rings in combination with a ferromagnetic shield. It was shown that the HTSC rings allow one to increase the operating current by 30% with keeping the same level of device field outside. As a example of application of HTSC shields in full-scale devices, we estimated the performance of magnetic shielding for an MRI magnet. The application of IITSC rings reduces the weight of the ferromagnetic shield in this magnet by about 33%.

1. Introduction

The shielding of magnetic fields is one possible application of HTSC materials. Many papers were devoted to HTSC shields for screening of extremely sensitive devices (e. v. SQUIDS's) [1]. Because the magnetic field penetrates into HTSC materials at the values less than 100 G, the shields based on Messner effect can be successfully employed only at low magnetic fields. However, the need for magnetic field shielding is also raised in power electric equipment: generators, transformers, magnets, etc. produce high stray magnetic fields influenced on sensitive equipment and live organisms. Conventional method in this case is the application of ferromagnetic shields (FS) of large weight and dimensions. Another approach, developed recently for MRI magnets, consists in the active shielding by a set of coaxial coils producing the magnetic field that compensates the main magnetic field outside the magnet [2]. It was demonstrated that the combination of superconducting coils with FS (hybrid shield) is much better from the point of view of the cost and weight of a device [2]. The further development of the method can be made by substituting compensating coils by superconducting shields shaped into close loops (rings, cylinders). In this case, the shielding ability of the HTSC loops depends on the critical current value. We present the results of the experimental study of the effectiveness of a hybrid superconductor-ferromagnetic iron shield employing HTSC rings.

2. Experimental

The laboratory model of an electric device shielded by the hybrid shield is shown in Fig. 1.

The model consists from a cylindrical coil with 400 turns and 20 mm in diameter, placed on a ferromagnetic core and surrounded by a FS of 60 mm in length performed from the transformer ferromagnetic plate of 0.35 mm in thickness. The ferromagnetic core was used for the enhancement of magnetic field allowing us the operation at rather low currents. Two YBCO HTSC rings of 35 mm in outer diameter, 26 mm in inner diameter and 5 mm in thickness were prepared by the procedure described in [3]. They were placed over the FS. The magnetic field distribution outside the model along its axis was measured by a Hall-probe connected with Bell Gaussmeter (Model 9200 with sensitivity 0.01 G). The Hall probe was placed 7 mm above the surface of the FS and moved along the system axes.

The design of the laboratory model was chosen as an example of the application of magnetic shields in transformers or magnets for MRI. In the last case, inner ferromagnetic core is not used. Two various locations of the HTSC rings were investigated: (1) at the ends of the FS; (2) at the central part of the shield. Centrally disposed HTSC rings are not effective: the magnetic field outside the device is even increased. Fig. 2 shows the basic results for AC and DC cases for the rings placed at the distance of 10 mm (middle of the ring) from the FS ends.

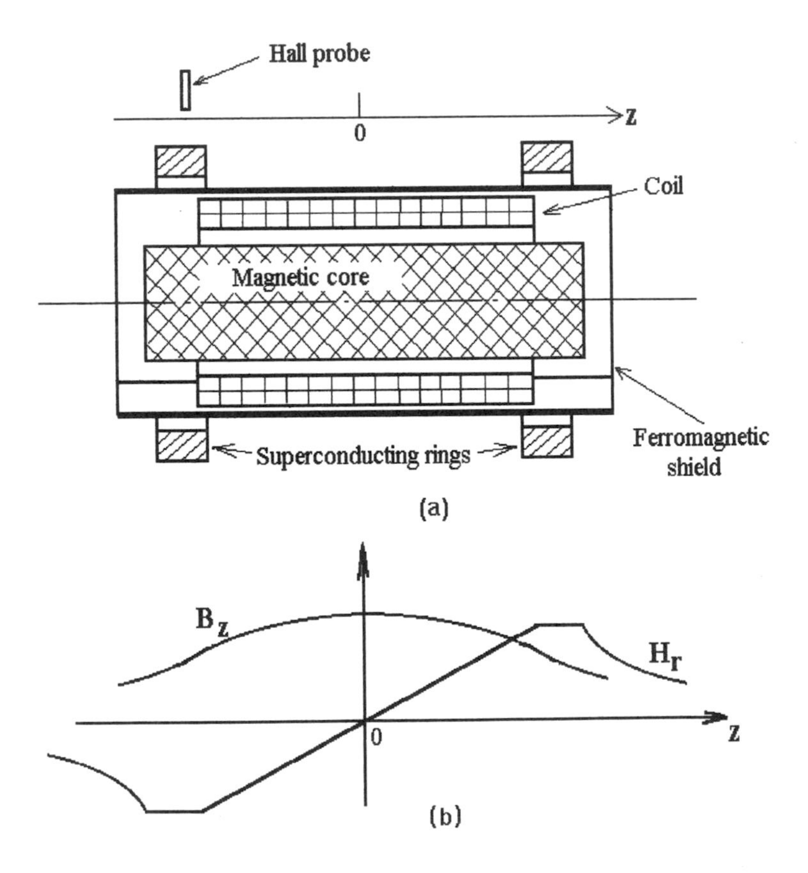

Fig. 1 (a) Schematic of the experimental model. (b) Radial component of magnetic field intensity H_r on the inner surface of the ferromagnetic shield and magnetic flux density B_z inside the shield.

3. Discussion

The main part of the magnetic flux in the device is closed through the FS. Without the rings, we observe the maximum of the field near the ends of the FS at low currents, and in the central part - at high currents. The shift of the maximum to the central part is explained by the saturation of the ferromagnetic material in the center of the shield (Fig. 1).

The HTSC rings screen only the difference between the entire magnetic flux and the its part passing in the FS. Therefore, their influence is substantially higher at high currents produced magnetic fields saturating the FS. When the FS is saturated, the flux at the ends of the model increases. This, in turn, leads to the increase of the induced current in the rings and their shielding ability. So, at the DC current of 1 A, the rings reduce the magnetic field more than 30% (Fig. 2a). At the AC current of 0.8 A, the reduction is about 20%. The HTSC rings influence sufficiently on the magnetic field distribution. In neighborhood of the rings, the DC magnetic field reverses the direction and AC magnetic field changes the phase by 180 degrees (Fig. 2b shows the effective value of the field). The difference in the magnetic field magnitudes for DC and AC cases is explained that, under AC, the magnetic flux in the FS exceeds the saturation level only a part of the AC period. At higher currents, the influence of the rings is reduced. There are two possible reasons of this reduction. First, the part of the FS near the ring is saturated due to the common influence of magnetic fields produced by the coil and ring. Note that the magnetic fields have the same direction in this region. Second, the current in the rings achieves the critical value.

In the difference from the shielding of AC magnetic fields, the DC shielding is related to the persistent current mode. Decay of persistent current is determined by the properties of an HTSC material and, especially, by the flux creep rate. The choice of the operating current for HTSC rings is governed by the condition of small current decay during the device operation. Fig. 3 illustrates the temporal dependence of the magnetic field trapped in an YBCO ring. The current was induced by increasing DC magnetic field in the ferromagnetic core passing through the ring. The current decreased to 85% of the critical value remaining constant in the limits of measurement precision during next few hours. Studying the influence of fluctuations on the stability of persistent current, we froze the magnetic flux in the ring and applied an additional 1 kHz AC magnetic field of about 10% of magnitude of the DC magnetic field corresponding to the critical current. The response of the ring is shown in the

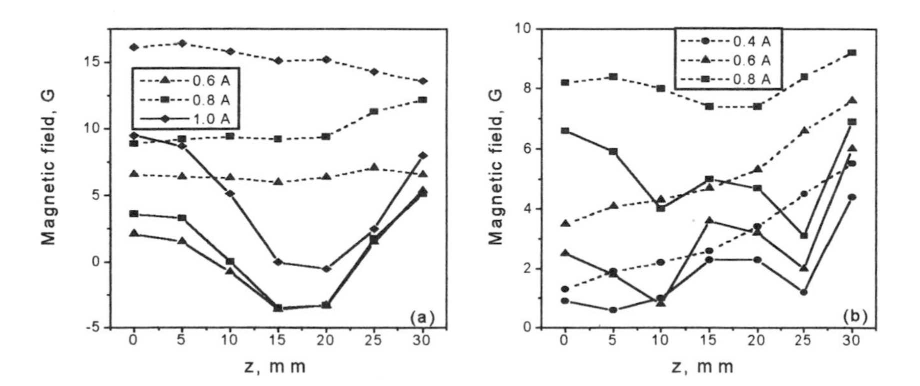

Fig. 2 Distribution of magnetic field in the model in (a) DC case and (b) AC case. Solid curves - with HTSC rings, dashed curves - without rings.

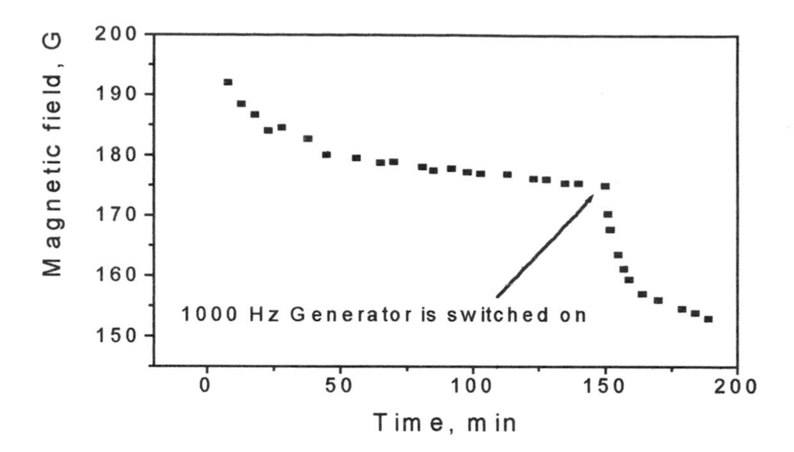

Fig. 3 Temporal dependence of trapped magnetic field in YBCO ring

same figure. In contrast from this behavior of the YBCO ring, we observed the prolonged and pronounced decay of the persistent current in a BSCCO ring.

As a example of application of HTSC shields in full-scale devices, we estimated the performance of magnetic shielding for an MRI magnet. Assuming that the coil of the magnet has the inner diameter of 0.7 m, the length of 2 m and produces the magnetic field inside the system of about 2 T, we estimated that the cross-section of HTSC ring with the critical current density 10^4A/cm^2 is 35 cm^2. The application of HTSC rings reduces the weight of the FS in this magnet by about 33%. This is appreciable because the weight of a standard FS in MRI magnets achieves 12 ton.

4. Conclusion

The application of HTSC shields allows one to reduce the weight of ferromagnetic shields employed in the electric devices. The choice of the operating current of HTSC shields in DC case is dictated by the flux creep rate and level of magnetic field fluctuations in the device. In AC case, the operating current is determined by the critical current and level of AC losses.

Acknowledgements

This work was supported by the Ministry of Science of Israel and the Technion Crown Center for Superconductivity.

References

[1] Polushkin V, Buev A and Koch H 1994 *Appl. Supercond.* **2** (9) 597
[2] Kalafala A K 1990 *IEEE Tans. on Magn.* **26** (3) 1181
[3] Shter G E and Grader G S 1994 *J. Am. Cer. Soc.* **77** 1436

Inst. Phys. Conf. Ser. No 158
Paper presented at Applied Superconductivity, The Netherlands, 30 June–3 July 1997

Possibilities to increase the critical current of solenoids made of anisotropic HTS tapes

J Pitel and P Kováč

Institute of Electrical Engineering, Slovak Academy of Sciences,
Dúbravská 9, 842 39 Bratislava, Slovakia

Abstract. The results of the theoretical analysis of the influence of external magnetic field on the critical currents of cylindrical solenoids made of BSCCO tapes with the high anisotropy in $I_c(B)$ characteristic at 77 K are presented. Critical current of such coils is determined especially by the radial component of the magnetic field close to the coil's flanges. In principle there are two ways how to increase the coil's critical current. First, applying the magnetic field parallel with the coil axis generated by the outer coil leads to decrease of the angle between the tape surface and the direction of the resulting magnetic field. However, the best results can be obtained using supplementary windings that are located close to the coil's flanges where rather the effect of the partial compensation of undesired radial component of the magnetic field is utilised. Depending on the coil's geometry, tape $I_c(B)$ characteristic and its anisotropy, the increase in the coil's critical current higher than 50 % can be expected.

1. Introduction

Measurements of the Bi(2223)Ag tapes show considerable anisotropy in the $I_c(B)$ characteristic [1,2]. When the angle α between the tape surface and the external magnetic field B_e is being changed from the parallel orientation ($\alpha = 0^o$) to the perpendicular one ($\alpha = 90^o$) the value of the transport critical current I_c decreases remarkably. As for the cylindrical coils that are wound with the anisotropic conductor (tape) it has been shown [3,4,5] that the radial component of the magnetic field close to the coil's flanges may be considered the limiting factor influencing the value I_{cmin} of the coil critical current. It can be expected that if we are able to decrease the angle α between the tape surface and the magnetic field, the load lines of the individual turns will intersect the part of the $I_c(B,\alpha)$ surface corresponding to lower value of α.

2. Two methods of decreasing the angle between the tape surface and the magnetic field

The most simple way is to apply the magnetic field parallel with the coil axis. Though the radial component of the magnetic field being unchanged, the axial component of the magnetic field will be the sum of the applied field and the axial component generated by the coil itself. However, this method of decreasing the angle α is rather disadvantageous. The decrease in angle α can be partially compensated by simultaneous increase in resulting absolute value of the axial magnetic field component. For the coil with a given geometrical dimensions the enhancement of its critical current I_{cmin} is now the compromise between the decrease of angle α and the increase of the resulting value of the axial magnetic field component. Another method is based on compensation of the radial magnetic field component close to the coil's flanges. It is evident that if the radial component of the magnetic field decreases considerably, the angle α between the tape surface and the direction of the magnetic field approaches zero automatically. Consequently, the turns load lines will intersect the $I_c(B,\alpha)$ surface in the parts with higher values of I_c, i.e. the coil's individual turns show the isotropic like behaviour.

3. Mathematical model

In order to investigate both above mentioned methods we have developed mathematical model [5] which enables to calculate the critical currents of individual coil turns. The real coil's winding is replaced by a grid of infinitely thin fictitious filaments. The critical current I_c of each filament is obtained like the numerical solution of the non-linear equation representing geometrically the point of intersection of the filament load line and the $I_c(B,\alpha)$ surface of the tape short sample in 3-D representation. It has been shown that if the anisotropy and the external field is taking into account the load lines of the individual turns are not linear, in fact they are the curvilinears with a very slight curvature. The critical current of the coil I_{cmin} represents consequently the minimum of the set of the values I_c.

4. Parameters of the model magnet and $I_c(B,\alpha)$ characteristic of the tape

In order to demonstrate the possible increase of the critical current of the coil made of anisotropic tape we have used the parameters of the magnet [6] consisting of 8 pancake coils. Each coil is wound with Bi(2223)Ag tape of width 4 mm and thickness 0.2 mm. The number of layers in each coil is 40 and the distance between the adjacent coils is 1 m. Assuming the width of the outer flanges is 0.5 mm, the length of the whole magnet is 2b = 40 mm. The inner and outer diameters of the winding are $2a_1$ = 36 mm and $2a_2$ = 72 mm, respectively. The values of the $I_c(B,\alpha)$ characteristic of the tape short sample, that were obtained from the V-I characteristic measurements at 77 K using an ordinary criterion of 1 $\mu V\ cm^{-1}$, chosen for the calculations are presented in Fig. 1.

5. Results

Figs. 2 and 3 represent the distribution of the critical currents I_c of individual infinitely thin filaments replacing the real magnet winding without the presence of external magnetic field. While Fig. 2 concerns the case like the pancake coils would be wound with an ideal isotropic tape, i.e. only the part of $I_c(B,\alpha)$ characteristic corresponding to $\alpha = 0^o$ is considered, the Fig. 3 is related to the anisotropic $I_c(B,\alpha)$ as it is shown in Fig. 1. The r-z frame represents the cross-section of the winding. We can see the difference between the position of the filament with the minimum value of I_c which is considered the critical current I_{cmin} of the magnet. In case of isotropic tape it is the turn located in the middle of the innermost layer. As for the anisotropic tape, the turns with the lowest values of I_c are situated close to the magnet flanges. Of course, the differences in absolute values are evident.

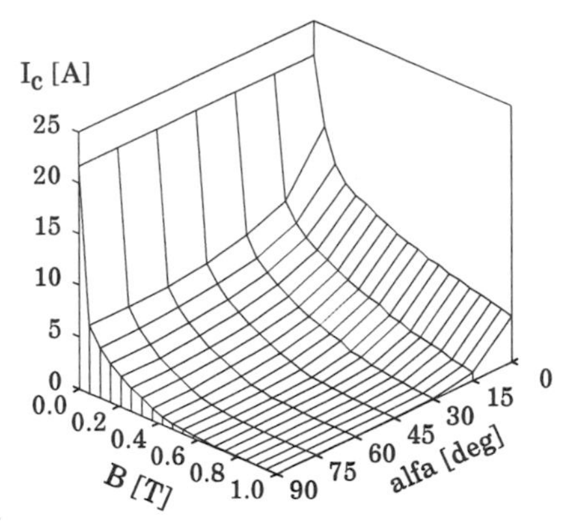

Fig.1 $I_c(B,\alpha)$ of the Bi(2223)Ag tape at 77 K

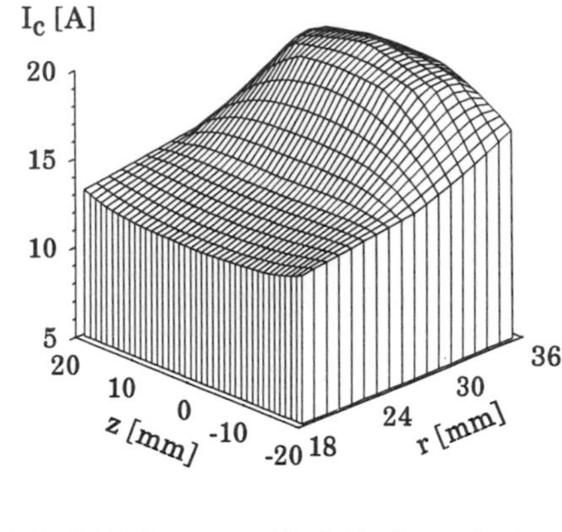

Fig.2 Critical currents of individual turns for the case of isotropic $I_c(B,\alpha)$ characteristic

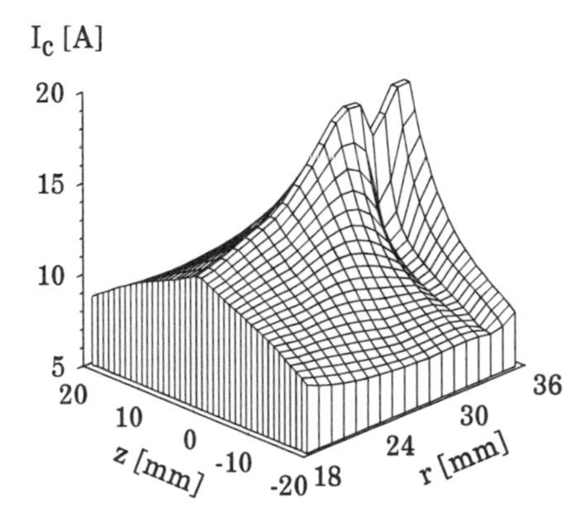

Fig.3 Critical currents of individual turns for the case of anisotropic $I_c(B,\alpha)$ characteristic

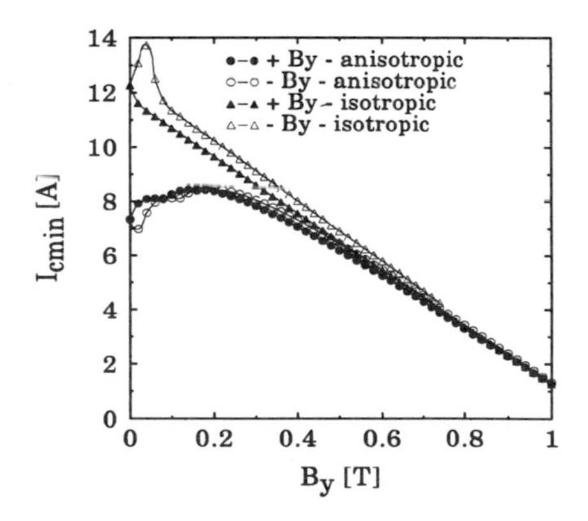

Fig.4 Critical current of the model coil as a function of an external magnetic field parallel (filled) and antiparallel (empty) with the coil axis

5.1 Influence of an external field parallel with the coil axis

Dependence of the coil's critical current I_{cmin} on the value of the homogeneous magnetic field B_y parallel with the coil axis is shown in Fig. 4. Parallel as well as antiparallel orientations are considered for both isotropic and aisotropic cases. The slight increase in the coil's critical current I_{cmin} is evident. The curves corresponding to the isotropic conductor can be considered the upper limit achievable with an anisotropic one.

5.2. Partial compensation of the radial component of the magnetic field

The most simple way how to compensate the radial component of the magnetic field generated in the winding close to the flanges is to utilise the couple of supplementary copper windings which are symmetrically situated on both sides of the original anisotropic HTS coil. Let us suppose that the supplementary coils which have the same dimensions and are made of the copper wire 1.5 mm in a diameter, will be energised by the current I_{Cu} flowing in the same direction like in the HTS magnet between them. The vortex of the magnetic field generated by the upper copper coil acts towards that of the HTS coil at the top of the common flanges area. As for the lower copper coil, the situation is the same at the bottom of the HTS coil. Due to this the radial component of the HTS magnet close to the common flanges areas decreases. Certain combination of the currents in both adjacent (copper and superconducting) coils may lead to practically entire compensation of the radial component of the magnetic field. Fig. 5 shows the results of computation of the critical current of the model magnet in combination with the couple of adjacent coils, assuming that the gap between the superconducting and copper coils is zero. The critical current of the HTS magnet increases with the increasing current I_{Cu} and reaches maximum of I_{cmin} = 11.34 A at I_{Cu} = 12 A. Higher increase in I_{Cu} leads gradually to over-compensation of the radial magnetic field component resulting in decrease of the magnet's critical current I_{cmin} . Fig. 6 shows the dependencies of the resulting (B_v) magnetic field in the magnet system center as well as that of the background field generated by the copper coils (B_{Cu}) and the HTS coil (B_{HTS}).

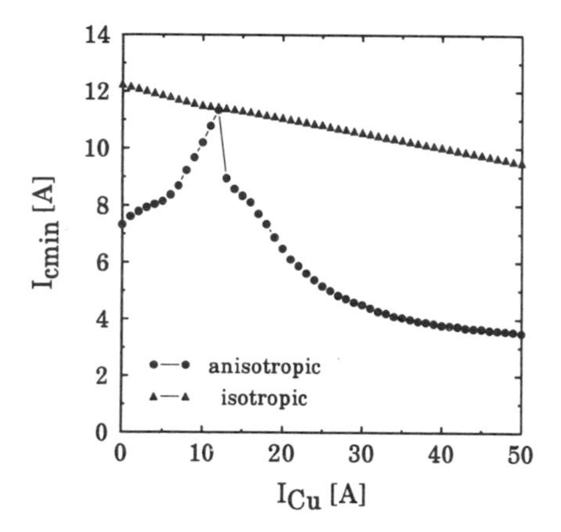

Fig.5 Critical current of the model coil as a function of the current energising the couple of adjacent copper coils

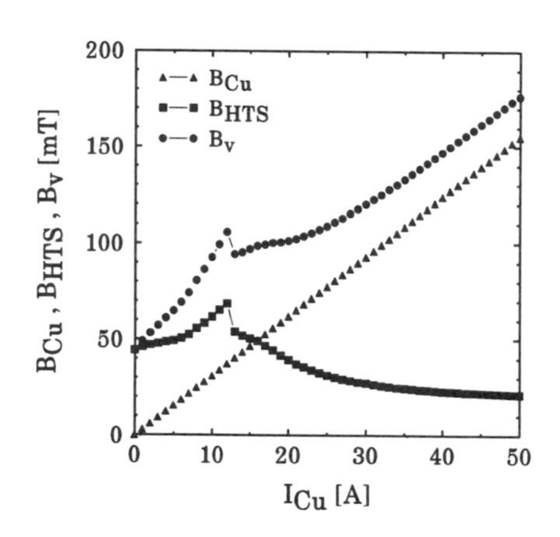

Fig.6 Magnetic field in the center of the coil's system generated by the copper coils (triangles), HTS coil (squares) as well as their sum (circles)

6. Conclusion

It has been demonstrated that applying an external magnetic field may increase the critical current of cylindrical coils which are made of anisotropic Bi(2223)Ag tapes. The method of partial compensation of the radial magnetic field component using supplementary copper windings located close to the flanges yields rather higher increase in the HTS coil's critical current than applying the homogeneous magnetic field which is parallel with the coil axis.

References

[1] Kumakura H, Togano K, Maeda H, Kase J and Morimoto T : *Appl. Phys. Lett.* **58** (1991) 2830-2832
[2] Hu Q Y, Schalk R M, Weber H W, Liu H K, Wang R K and Dou S X : *J. Appl. Phys.* **78** (1995) 1123-1130
[3] Jenkins R G and Jones H : *Inst. Phys. Conf. Se.* No **148** (1995) 79-84
[4] Daümling M and Flükiger R : *Cryogenics* **35** (1995) 867-870
[5] Pitel J and Kováč P : *Supercond. Sci. Technol.* **10** (1997) 7-16
[6] Fabbricatore P, May 1996, private communication

Inst. Phys. Conf. Ser. No 158
Paper presented at Applied Superconductivity, The Netherlands, 30 June–3 July 1997

Special features of HTS magnet design

J Paasi, M Lahtinen, J Lehtonen, R Mikkonen

Tampere University of Technology, P.O. Box 692, FIN-33101 Tampere, Finland

B Connor, S S Kalsi

American Superconductor Corporation, Two Technology Dr., Westborough, MA 01581, USA

Abstract. In this paper we consider the design of HTS magnets from the point of view where the design differs from the established principles for LTS magnet design. New principles are required because of a few intrinsic features of HTS materials very different to those of LTS conductors. Examples of such features are slanted *E(J)*-characteristics, high thermal stability, critical current anisotropy, and the $J_c(B)$-hysteresis of HTS conductors. In order to show how these issues influence the magnet design we consider a Bi-2223 magnet constructed at American Superconductor for a 5 kJ micro-SMES system at Tampere University of Technology.

1. Introduction

The design of LTS magnets is well established and many methods can be directly applied to HTS magnets as well [1]. Some questions, however, require a new approach because HTS conductors have a few intrinsic features very different to those of LTS conductors. The high stability of HTS is a great advantage. Slanted *E(J)*-characteristics (low index values) lead to considerable power dissipation even at subcritical currents. The critical current anisotropy and the $J_c(B)$-hysteresis of the conductor bring also their own impacts on the magnet design.

In this paper we show how these issues influence the design of HTS magnets. As a demonstration we consider a Bi-2223 magnet constructed at American Superconductor for a 5 kJ μ-SMES system at Tampere University of Technology. The magnet is cooled by a cryocooler and has a nominal current of 160 A determined by 1 W rated power dissipation at 20 K. The magnet was prepared by stacking 11 double pancake coils, joining the pancakes with low resistivity joints, and vacuum impregnating the magnet with epoxy. The magnet has dimensions: outer diameter 317 mm, inner diameter 252 mm, and axial length 66 mm. The design current of the coil was obtained by using a conductor bundle consisting of three Bi-2223/Ag tapes. The description of the design of the μ-SMES system is given elsewhere [2].

2. Magnetic design

Magnetic flux density within the demonstration magnet was computed for the 160 A operating current, which corresponds to an overall coil current density of 8500 A/cm^2. The results are presented in Fig. 1 for axial and radial field components. The highest flux densities of 1.7 T are at the mid-point of the inner surface of the coil. For the radial field the highest values of 1.3 T are found at the mid-point of the top and bottom surfaces of the coil. In NbTi-magnets

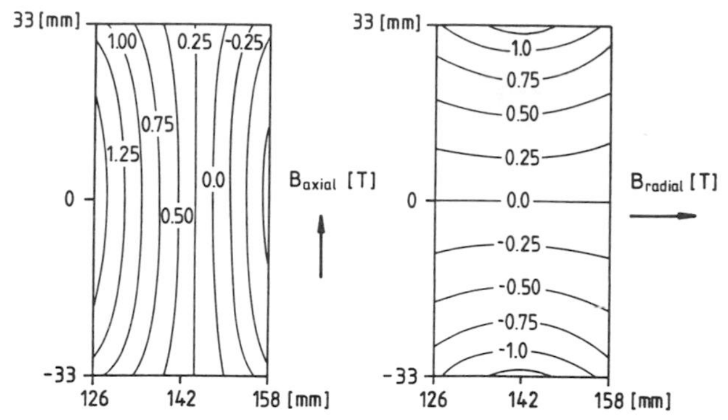

Fig. 1 Axial and radial components of the magnetic flux density in the magnet cross-section at the operating current of 160 A. The distances are from the center of the magnet.

the critical current of the coil is limited by the maximum of flux density in the winding. In HTS-magnets the situation is different. The anisotropy in the critical current density vs. magnetic flux density, i.e. $J_c(B \| \text{tape}) > J_c(B \perp \text{tape})$, results in that the radial component of the field determines the current carrying capability of the HTS magnet conductor [3].

There are several possibilities to reduce the influence of the radial field on the magnet performance: 1. The magnet can be shaped by replacing the pancakes at the ends by other pancakes of similar outer diameter but a larger inner diameter. This kind of shaping can reduce the radial field at the end sections without essentially decreasing the overall performance of the magnet; 2. The radial magnetic field inside the magnet can be reduced by altering the magnetic flux distribution by using external pieces of iron [4,5] or an additional external magnet [6]; 3. The end sections of the magnet can be designed to operate at lower current ratio, I_{op}/I_c, than other parts of the magnet [7]. This can be realised either by having an additional tape in the conductor bundle used for the winding of the end section pancakes or simply by using a tape with higher J_c in the end section coils. The last technique was used in the present magnet because wire with higher J_c was available for the end pancakes.

3. Stability

HTS windings are very stable against small mechanical or thermal disturbances. Typically a disturbance energy sufficient to raise the temperature of a HTS coil from 30 K to 40 K is three orders of magnitude higher than what is needed to quench a Nb-based magnet [1]. On the other hand, there can be considerable power dissipation in the magnet even at subcritical currents due to thermally activated flux creep and normal conducting joints between pancake coils. As a consequence, a maximum allowable power dissipation is more suitable design criterion than the critical current [8]. The type of refrigeration - liquid nitrogen, liquid helium or a cryocooler - sets further criteria for the magnet design. In some cases the temperature in the winding can increase several degrees without loosing the stability of the operation. Therefore, the allowable power dissipation depends strongly on the application, operating temperature and the available refrigeration power.

In order to study allowable losses or, more precisely, allowable operating conditions in the demonstration magnet, we computed the temperature distribution of the coil in different operating conditions by solving the non-linear heat conduction equation

$$\nabla \cdot [\,\lambda(T)\,\nabla T] + Q(T) = C_p(T)\,\frac{\partial T}{\partial t}$$

with appropriate boundary conditions (T is the temperature, λ and C_p are the averaged thermal conductivity and volumetric specific heat of Bi-2223/Ag and epoxy, respectively, and Q is the volumetric heat generation in the coil accounting for ac losses as well as Joule losses for dc current). The refrigeration power of the cryocooler, 8 W, was applied to the top surface of the magnet [2]. The resistivity of the Bi-2223 material was assumed to follow the power law $\rho(J)=E_c(J^{n-1}/J_c^{n})$ with n=10 and a linear temperature dependence of the J_c between J_c(20K)= 40 kA/cm^2 and J_c(80K)=0. More detailed describtion of the computation model together with a complete case analysis will be published elsewhere [9].

When a SMES magnet is used in a power conditioning system, the operation typically consists of a rapid discharging/charging cycle followed by a constant current mode. In the simulation of Fig. 2 the magnet current is ramped up and down with a high ramp rate of 300 A/s until the temperature at the mid-point of the bottom surface of the magnet reaches a predefined value (25 K, 27 K or 35 K for the cases of the figure). The chosen point of the magnet is assumed to be the critical one because there the radial component of the magnetic flux density is the highest leading to the highest power dissipation and because there the cooling conditions are the poorest. After the predefined temperature has been reached, the magnet is fed by the nominal dc current of 160 A. The computed Joule losses in the magnet winding corresponding to 160 A are 4.7 W, 6.7 W, and 34 W at 25 K, 27 K, and 35 K, respectively. From the results we can see that in the 25 K case the dissipated heat can be effectively conducted away resulting in stable operation. At 27 K the dissipated power is already too high with respect to the available refrigeration and the operation is unstable.

4. Hysteresis of critical current

The superconducting material in HTS-filaments includes grains or grain blocks which have essentially higher critical current density than the transport J_c of the filament. The magnetisation currents in the high-J_c regions influence the total magnetic field at weak grain interfaces: depending on the prior magnetic field cycling, the magnetisation currents can either increase or decrease the effective magnetic field at weak links resulting in history dependent transport $I_c(B)$ values [10].

In magnet operation the magnetic field affecting the superconducting grains or grain blocks inside the filaments is the self-field of the magnet. Thus the $I_c(B)$ as well as the current-voltage,$U(I)$, relation of the magnet have different values depending on whether the magnet is charged or discharged. A case-in-point is given in Fig. 3. Because the experimental data of the

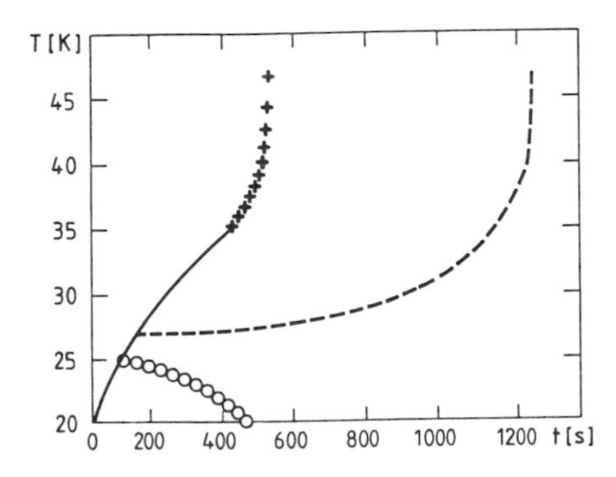

Fig. 2 Temperature at the mid-point of the bottom surface of the cryocooler refrigerated HTS magnet in a ramping mode followed by a dc mode of 160 A magnet current: ramping mode (——); dc modes with an initial temperatures of 25 K (ooo), 27 K (- - -), and 35 K (+++).

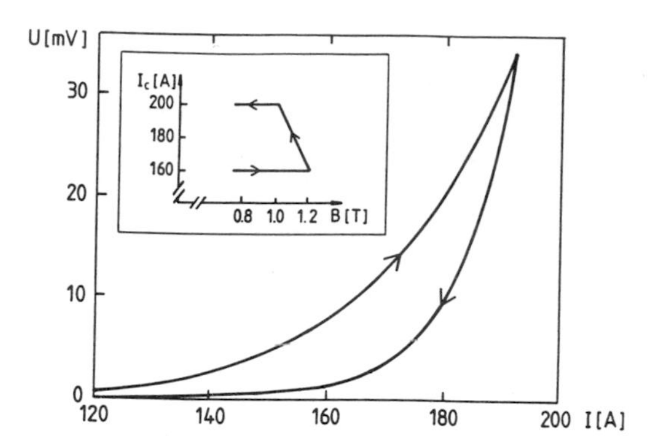

Fig. 3 Computed $U(I)$-hysteresis of a HTS magnet. The $I_c(B)$-curve of the magnet conductor used in the computation is given in the inset.

magnet conductor was not available at the time of the conference, we computed the $U(I)$-curve by using a simple model of the $I_c(B)$-dependence given in the inset of Fig. 3 together with equations $B = I/160$ (T) and $U = U_c\,(I/I_c(B))^8$ (mV), where U_c=8 mV. The computed hysteresis in the $U(I)$-curve is similar to the measured $U(I)$-curves of the excitation coils of a HTS synchronous motor [4] despite of simplifications used in the computation, which lets us assume that the real behaviour of the demonstration magnet would not differ much from the one presented here.

The hysteresis in the $U(I)$-curve of the magnet means that, if the operating current is determined by maximum allowable power dissipation, the operating current depends on the charging procedure of the magnet. A gain in operating current is obtained if the magnet current is momentarily overloaded and then decreased to a value resulting in the fixed power dissipation. In the demonstration magnet this kind of charging procedure would give approximately 10% gain in operating current if the power dissipation is fixed to 1 W.

5. Conclusions

We have considered special aspects of HTS magnet design. The negative influence of radial magnetic field on the magnet performance must be reduced by a proper design. A maximum allowable power dissipation is more suitable design criterion than the critical current. Finally, the operating current determined by a fixed power dissipation at a given temperature is not constant but dependent on the magnetic history of the magnet.

References

[1] Iwasa Y 1994 *Case Studies in Superconducting Magnets* (New York: Plenum Press)

[2] Mikkonen R, Lahtinen M, Lehtonen J, Paasi J, Kalsi S S and Connor B *EUCAS'97 Conference*, The Netherlands, 30 June - 3 July 1997

[3] Däumling M and Flükiger R 1995 *Cryogenics* **35** 867-870.

[4] Mikkonen R, Söderlund L and Eriksson J-T 1996 *IEEE Trans. Magn.* **32** 2377-2380; Eriksson J-T, Mikkonen R, Paasi J, Perälä R and Söderlund L 1997 *IEEE Trans. Appl. Supercond.* **7** 523-526

[5] US Patent allowed and pending

[6] Pitel J and Kovác P 1997 *Supercond. Sci. Technol.* **10** 7-16

[7] US Patent 5,525,583, issued on June 11, 1996

[8] Seuntjens J M and Snitchler G 1997 *IEEE Trans. Appl. Supercond.* **7** 1817-1820

[9] Lehtonen J, Lahtinen M, Mikkonen R and Paasi J, *MT-15 Conference*, Beijing, 20-24 October 1997

[10] Evetts J and Glowacki B A 1988 *Cryogenics* **28** 641-649

Inst. Phys. Conf. Ser. No 158
Paper presented at Applied Superconductivity, The Netherlands, 30 June–3 July 1997

Choice of design margins of superconducting magnets

V E Keilin

Kurchatov Institute, 1 Kurchatov's sq., 123182 Moscow, Russia

Abstract. One of the most important steps at the design stages of a superconducting (*sc*) magnet is a proper choice of current margins, i.e., of the ratio of certain critical value of a superconductor used to maximum corresponding value envisioned during normal operation of the magnet. Very often the ratio n_B of the critical current at the operational magnetic field B_0 to the transport current I_0 at the same field is put to use as the characteristic margin. However, the only meaningful margin is the energy margin that is minimum energy (per unit of the superconductor length or of the magnet's winding volume) required to induce quench at given conditions. The energy margin depends mainly on the difference ΔT between the critical temperature T_{BI} of the superconductor (at nominal values of the maximum field B_0 and the transport current I_0) and the operational temperature T_0 (usually equal to that of the coolant). In the simplest and practically important case of adiabatic windings, the interdependence between the energy margin and the temperature one is evidently single–valued. On the other hand, much-used "constant–field current margin" n_B depends not only on the temperature margin ΔT, but also on B_0/B_C ratio (B_C is the critical field). As shown in the paper, it is preferable to characterize *sc* windings by "the load–line current margin" n_{BI}, i.e., by the ratio of the critical current taken along the load line to the nominal transport current. The interdependence of ΔT and n_{BI} is practically single–valued. This statement is strict if the critical surface is approximated by a plane.

1. Introduction

When any superconducting (*sc*) magnet is designed, its operational parameters are usually chosen well below the critical surface of a superconductor used [1]. Proper margin choice is one of the most important decisions that the magnet designer should undertake (another one is the choice of average current density). If the margin is too small, there exists a risk not to attain a nominal magnetic field or, at least, to attain it in a rather undesirable way (say, after some training). On the other hand, if the margin is too large then it results in the magnet cost increase.

The goal of this paper is not to give any recommendations concerning quantitative considerations towards margins values, but to draw primary attention to the preferable format of the margins.

2. Margins choice

In principle, the necessity of the margins is due to the fact that *sc* magnets during their normal operation are subjected to some regular and/or random, steady–state and/or

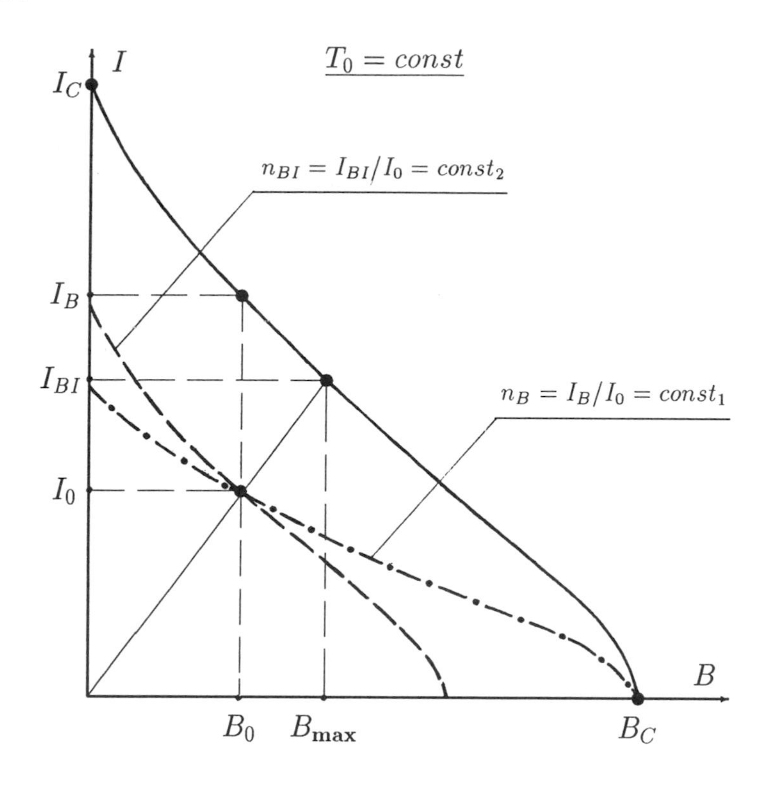

Fig. 1: Definition of different current margins.

pulsed, local and/or distributed heat depositions. So, a *sc* winding should be capable to absorb some energy without going normal, i.e., the winding should withstand some heating above normal coolant temperature, or, in other words, it should have some temperature margin. However, since exact values and even nature of heat depositions are often unknown, the temperature margin choice is arbitrary in most cases, say, 1 K above bath temperature. In practice, a *sc* magnet designer first chooses the current margin, i.e., the ratio of a superconductor critical current to the operational current of the magnet under designing and then, referring to the critical surface of the superconductor, calculates the temperature margin ΔT. It is important to point out that different values of current and/or temperature margins can be obtained keeping average current density almost constant, because the choice of the latter is determined mainly by considerations of quench protection, mechanical strength and the stability.

A common practice is to define the current margin as the ratio of the critical current I_B at the maximum field B_0 seen by a superconductor at operational conditions to the operational current I_0 (see Fig. 1).

Let us denote this "constant–field current margin" as $n_B = I_B/I_0$. Note that in this ratio both the operational current I_0 and the critical current I_B are taken at the same field B_0.

Another possible (though rarely used) definition of the current margin is the ratio of maximum theoretically possible critical current of a given magnet I_{BI} (that corresponds to the intersection point of the magnet load–line with the critical current *vs* the magnetic field curve) to the operational current I_0. Let us denote this "load–line current margin" as $n_{BI} = I_{BI}/I_0$.

It will be shown that the utilization of load–line current margin n_{BI} is preferable to the utilization of the constant–field current margin n_B.

The obvious though not of prime importance reason is that for a given magnet the situation that corresponds to the combination of I_B and B_0 values is meaningless as it can be never met. By contrast, the combination of I_{BI} and $B_{\max}$ values (Fig. 1) corresponds to theoretically possible real situation when the magnet is charged up to its "short sample critical current".

However, more important reason to prefer the load–line current margin n_{BI} to the constant–field current margin n_B is that the relationship between n_{BI} and the temperature margin ΔT does not depend practically on the fact whether the magnet should generate high or low fields, i.e., on B_0/B_C ratio ($\Delta T = T_{BI} - T_0$, where T_{BI} is the critical temperature of a superconductor at operational values of both I_0 and B_0, and T_0 – bath temperature). As mentioned above, the capability of a *sc* winding to withstand thermal disturbances depends strongly (in the case of adiabatic and distributed disturbances – entirely) on the energy or the enthalphy margin $\Delta h = h(T_{BI}) - h(T_0)$.

The relationship between load–line current margin n_{BI} and the temperature margin ΔT is especially simple if the critical surface is approximated by a plane (see Fig. 2). It can easily be shown that in this case

$$\frac{\Delta T}{\Delta T_C} = 1 - \frac{1}{n_{BI}} \ . \tag{1}$$

On the other hand, if we use the constant–field current margin n_B, the temperature margin ΔT depends not only on n_B, but also on B_0/B_C ratio. For the case of approximating the critical surface by a plane, it can be shown that

$$\frac{\Delta T}{\Delta T_C} = \left(1 - \frac{B_0}{B_C}\right) \cdot \left(1 - \frac{1}{n_B}\right) \ . \tag{2}$$

3. Example

As an example let us calculate the temperature margin $\Delta T/\Delta T_C$ value (normalized to ΔT_C), using the same value of the constant–field current margin $n_B = 2$ for "low–field" ($B_0/B_C = 0.25$) and "high–field" ($B_0/B_C = 0.75$) windings.

The results according to equations 1 & 2 are given in the Table.

TABLE. Numerical example for $n_B = 2$ (B_C & $T_0 = 4.2\,$K, T_C & $B = 0$).

B_0/B_C	$\Delta T/\Delta T_C$	n_{BI}	for $NbTi$				for Nb_3Sn			
			T_C	B_C	B_0	ΔT	T_C	B_C	B_0	ΔT
0.25	0.375	1.60	9.5K	11T	~ 2.8T	~ 2.0K	17.2K	20T	5T	~ 4.9K
0.75	0.125	1.14	9.5K	11T	~ 8.3T	~ 0.65K	17.2K	20T	15T	~ 1.6K

It is seen from the Table that depending on B_0/B_C value the same value of the constant–field current margin n_B can result in completely different temperature margins. It is not the case when we use the load–line current margin n_{BI}. In this case, the interdependence of n_B and $\Delta T/\Delta T_C$ is single–valued.

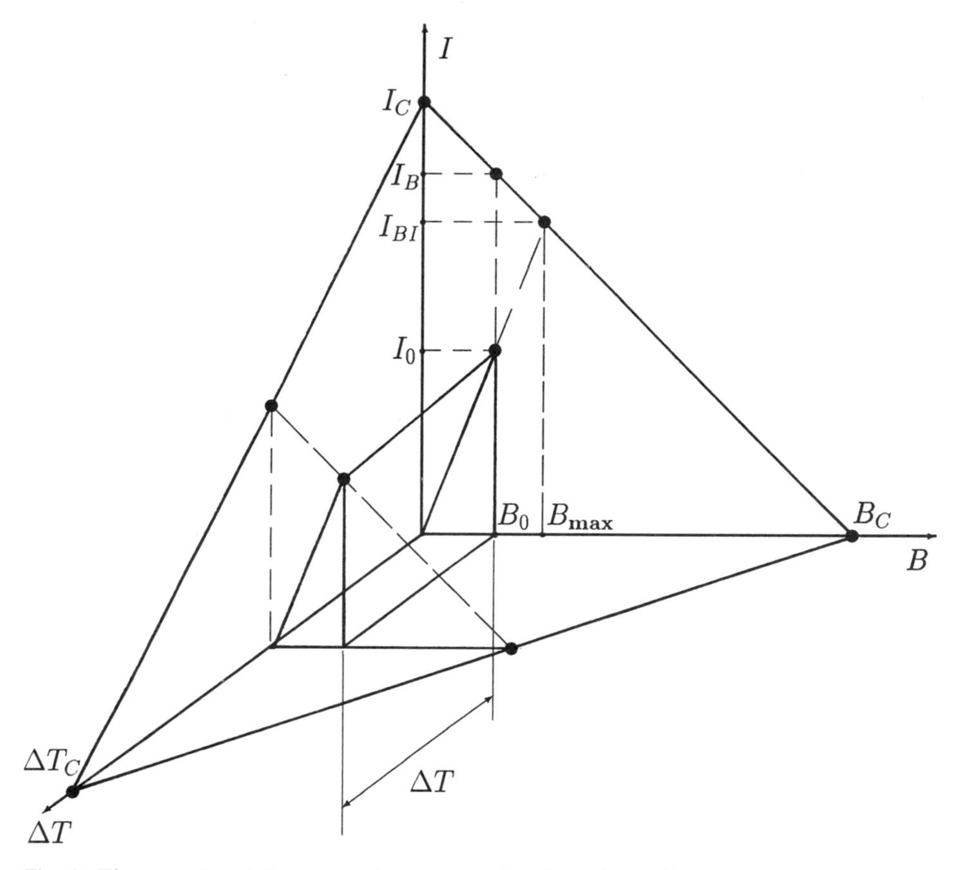

Fig. 2: The current and the temperature margins interdependence (for the case of the critical surface approximation by a plane).

It seems that this simple numerical example is rather persuasive in favour of the load–line current margin n_{BI} usage instead of the constant–field current margin n_B.

Reference

[1] Wilson M N 1983 *Superconducting Magnets* (Oxford: Clarendon Press)

Inst. Phys. Conf. Ser. No 158
Paper presented at Applied Superconductivity, The Netherlands, 30 June–3 July 1997

Simple model and optimization of HTSC bearing

V V Vysotskii and V M Pan

Institute of Metal Physics, National Academy of Sciences of Ukraine,
36 Vernadsky St., Kiev, 252142, Ukraine

Abstract. A simple model of interaction between a field-cooled extremely-hard HTSC sample and a permanent magnet is proposed. The model qualitatively describes all peculiarities of such systems and allows to investigate the dependence of forces acting in it upon the geometry of the superconductor in order to optimize the construction of HTSC bearing.

1. Introduction

Many techniques exist for computation of levitation forces and other mechanical properties of the systems consisting from permanent magnets and superconductors [1-5]. However, their application for and optimization of real magnetic levitation systems is often unfeasible due to complicated geometry of such systems. The present paper proposes a simple model of interaction between a superconductor and a permanent magnet (or for a system of superconductors and magnets), which is suitable in case of sufficiently hard superconductors. The model allows to understand qualitatively the main properties of such systems, and gives a possibility of estimating the mechanical properties and optimize the construction of HTSC-based magnetic bearing.

2. The model of ideal conductor with frozen magnetic flux

Let's regard a field-cooled HTSC sample with a stationary permanent magnet as a source of external field $H_0(\vec{r})$. If the condition $H_0 \gg H_{c1}(T)$ holds at the final cooling temperature $T \ll T_c$, we may neglect the demagnetization of the sample during the cooling, and assume that inside the sample

$$\vec{B}_i(\vec{r}) = \mu_0 \vec{H}_0(\vec{r}). \tag{1}$$

Since the field $\vec{H}_0(\vec{r})$ is potential everywhere outside the magnet, the macroscopic current density in the volume and on the surface of the sample is zero.

Let's move the magnet so that it's field becomes $\vec{H}_1(\vec{r})$. The superconductor reacts to this by appearance of induced currents screening the change of external field. In general case this change will partly penetrate into the volume of the sample, and the induced currents will flow both on the surface and in the volume of the superconductor. The exact picture of current distribution is usually very hard to calculate.

In this paper we propose a simple model of extremely hard superconductor. If the field change penetration depth $\lambda \approx \Delta H / j_c$ is very small compared to the characteristic length of the regarded system (i.e., the critical current is sufficiently large), it can be assumed that the field $\vec{B}_i$ inside superconductor remains constant and the currents flow only on its surface. In this case the superconductor is equivalent to an *ideal conductor* with certain non-zero magnetic flux frozen into it. The existence of this flux makes the difference of this model from the Meissner state, when the superconductor is not only an ideal conductor, but also an ideal diamagnetic. In our case the superconductor can be described as having the *ideal differential diamagnetism*. Our model is a limit case of Bean critical state model [2].

The surface current density $\vec{i}$ should satisfy the boundary condition

$$\left[\vec{n} \times \left(\vec{H}_e - \vec{H}_i\right)\right] = \vec{i}, \tag{2}$$

where $\vec{n}$ is the surface normal, $\vec{H}_i = \vec{H}_0(\vec{r})$ is the internal field. The external field

$$\vec{H}_e(\vec{r}) = \vec{H}_1(\vec{r}) + \vec{H}_{ind}(\vec{r}) \tag{3}$$

is defined both by the displaced magnet and induced currents:

$$\vec{H}_{ind}(\vec{r}) = \int_S \frac{\left[\vec{i}(\vec{r}\,') \times (\vec{r} - \vec{r}\,')\right]}{|\vec{r} - \vec{r}\,'|^3} dS. \tag{4}$$

From (1)-(4) follows the integral equation which defines the density of the induced current on the surface of superconductor:

$$\vec{i}(\vec{r}) + \int_S \frac{\left[\vec{i}(\vec{r}\,') \times (\vec{r} - \vec{r}\,')\right]}{|\vec{r} - \vec{r}\,'|^3} dS = \left[\vec{n} \times \left(\vec{H}_1(\vec{r}) - \vec{H}_0(\vec{r})\right)\right]. \tag{5}$$

The solution of (5) allows to find the magnetic field and force acting upon the permanent magnet. Equation (5) can be solved analytically for simple cases (e.g., in case of point dipole above a superconducting half-space it reduces to a modified mirror image model, which was proposed independently by us and the authors of [6].) However, in case of finite-size samples and magnets the analytic solution becomes impossible. We have performed numerical calculations aimed at optimization of constructed HTSC bearing.

3. Results and discussion

We regarded a model of simple HTSC bearing with a cylindrical magnet and cylindrical or cup-shaped superconductor (See Fig. 1.) The surface of the magnet was divided into small elements, and the integral equation (5) was transformed into a system of linear equations for the values $\vec{i}_k$ of current flowing on the surface element k. This system was then solved numerically for different right parts corresponding to certain initial (during cooling) and final positions of the magnet.

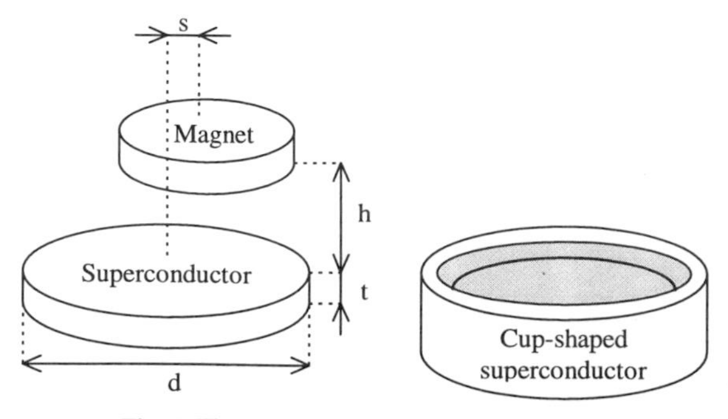

Fig. 1. The geometry of model HTSC bearing.

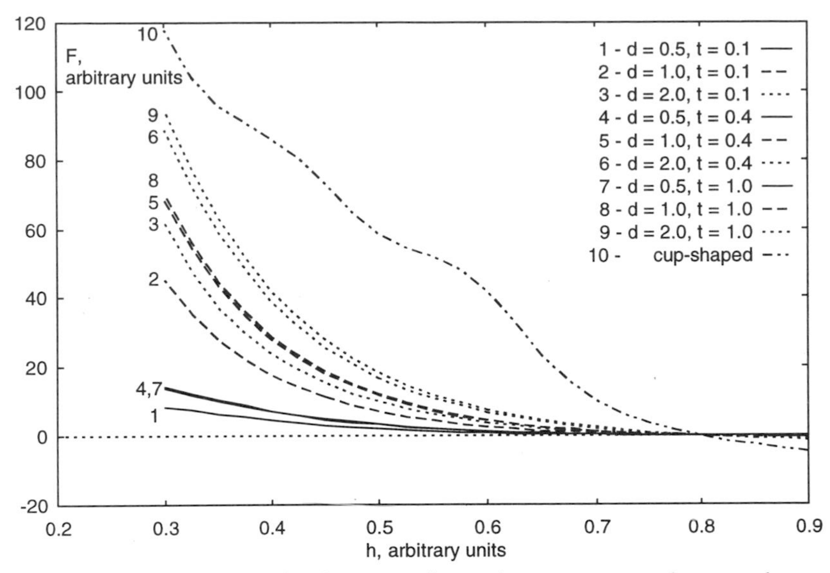

Fig. 2. Dependence $F_z(h)$ of levitation force upon distance between magnet and superconductor at different values of superconductor diameter and thickness.

In order to find out the dependence between the mechanical properties of the system and its geometry we varied the diameter and thickness of the superconductor, as well as the distance between the magnet and superconductor during cooling.

The dimensions of magnet in all presented cases were the same: diameter d=1.0, thickness t=0.4 arb. units.

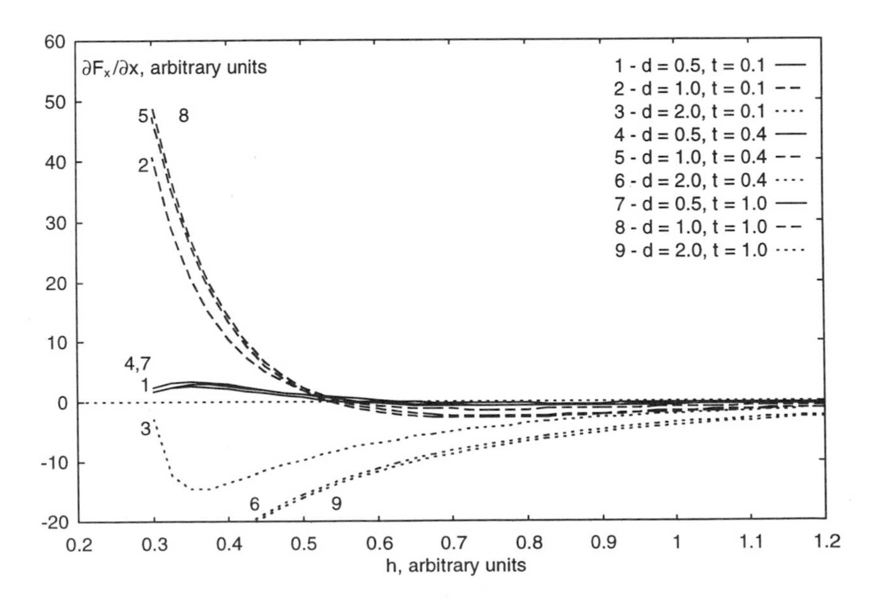

Fig. 3. Dependence of horizontal stiffness coefficient upon distance between magnet and superconductor

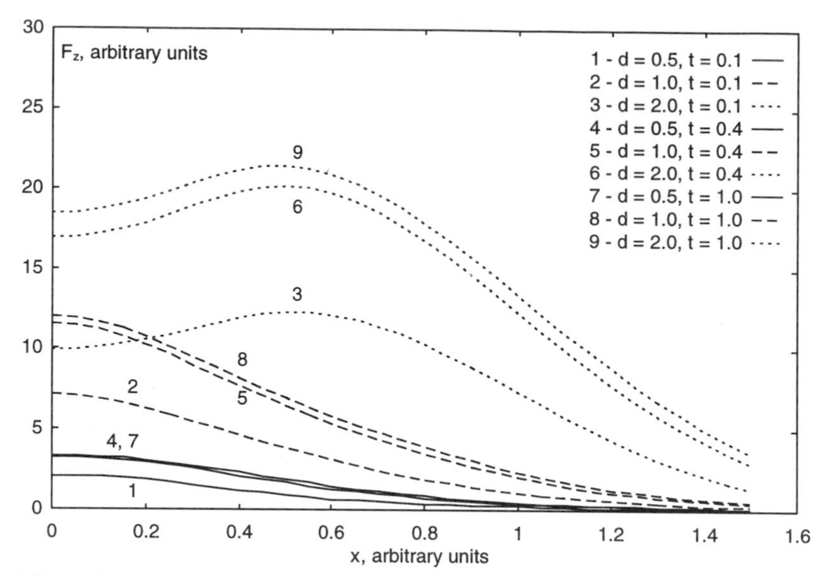

Fig. 4. Dependence $F_z(x)$ of levitation force upon horizontal displacement of the magnet from the symmetry axis of the system

The dependence $F_z(h)$ of levitation force upon distance between magnet and superconductor (with constant initial distance during cooling) is shown in Fig. 2 for different values of diameter and thickness of the superconductor, as well as for cup-shaped superconductor (with diameter d=1.0, thickness t=0.4 and height of the border equal to 0.5). The force strongly depends upon the diameter of the superconductor when it is less than the diameter of the magnet, and much less when superconductor is wider than the magnet. It can also be seen that there is an optimal thickness of the superconductor, above which the levitation force depends on thickness only slightly. In case of cup-shaped sample the levitation force was predictably much higher than for the flat one.

Fig 3 demonstrates the dependence $\partial F_x(h)/\partial x$ of horizontal elasticity coefficient upon the distance between magnet and superconductor. The negative values correspond to restoring force and stability with regard to horizontal displacements of the sample. We see that the region of stability has a lower bound for narrow superconducting samples. At very small distances the axially symmetric position of the magnet may become unstable.

Fig. 4 shows the dependence of levitation force upon the horizontal displacement of the magnet.

References

[1] Kozorez V.V., Dynamic systems of magnetically interacting free bodies (in Russian), Kiev, 1981

[2] Bean C.P., Rev. Mod. Phys. **36**, 31 (1964)

[3] Nemoshkalenko V.V., Ivanov M.A., Nikitin B.G., Pogorelov Yu.G., Hight-T_c Superconductivity, Springer-Verlag, Berlin, p. 175 (1992)

[4] Yang Z.J., Hull J.R., Mulcahy T.M., Rossing T.D., J. Appl. Phys. **78**, 2097 (1995)

[5] Yang Z.J., to appear in J. Supercond. (1997)

[6] Kordyuk A., Nemoshkalenko V.V., poster presentation 3Ge-42 at EUCAS'97

Inst. Phys. Conf. Ser. No 158
Paper presented at Applied Superconductivity, The Netherlands, 30 June–3 July 1997

Numerical Modelling of High Temperature Superconducting Bearings

F. Negrini*, P.L. Ribani*, E. Varesi, S. Zannella****

* Department of Electrical Engineering, University of Bologna, Viale Risorgimento 2, 40136 Bologna, Italy
** CISE S.p.A., Via Reggio Emilia 39, 20134 Segrate (Milano), Italy

Abstract. The levitation of a permanent magnet on a high temperature bulk superconductor is studied. An axial symmetry system is analyzed by means of the electric vector potential and the finite element method. The flux-creep and the flux-flow model is utilized. The dependence of the critical current density with the magnetic field is taken into account. A comparison of some simulation results with experimental data is presented.

1. Introduction

High temperature superconducting magnetic bearings are largely studied in the world. At the Department of Electrical Engineering of the University of Bologna a simulation research activity has been recently started. As a first step, the levitation of a permanent magnet on an YBCO bulk superconductor has been studied. An axial symmetry system is considered. A cylindrical permanent magnet levitates on a coaxial cylindrical YBCO pellet ; the symmetry axis of the system is the c-axis of the superconductor. In this work, the integral form of the Maxwell equations is utilized to simulate the system. High temperature superconducting bearings have been studied at CISE, in Milan both experimentally and theoretically [1] [2]. The results of the numerical simulations are compared with some of the experimental data.

2. Mathematical model

A screening current density is present inside the superconductor. Let $\mathbf{B}_e$ the magnetic flux density which is due to the permanent magnet and $\mathbf{B}_i$ the magnetic flux density which is due to the superconducting current. The magnetization vector into the permanent magnet is supposed to be a known constant, thus $\mathbf{B}_e$ can be calculated, by means of its integral expression [3], once the position of the magnet is known. An integral expression of $\mathbf{B}_i$, as a function of the electric vector potential $\mathbf{T}$ ($\nabla\times\mathbf{T}=\mathbf{J}$) and its divergence, is obtained [3]-[5]. It is experimental evidence that the component of the current density in the direction of the c-axis of YBCO is much lower than the component in the a-b plane [4] ; thus the z-component of the current density is neglected and only the z-component of $\mathbf{T}$ is considered: $\mathbf{T} = T\,\mathbf{k}$.

With reference to the generic values z_k of the z-coordinate, and r_i and r_j of the radial coordinate, the equation (1) is obtained where Z_1 and Z_2 are the z-coordinates of the lower and upper surfaces of the superconductor, R is the radius of the superconductor, μ_0 is the magnetic permeability of the free space, E_φ is the azimuthal component of the electric field.

$$r_j\, E_\varphi(r_j,z_k,t) - r_i\, E_\varphi(r_i,z_k,t) = -\int_{r_i}^{r_j} \frac{\partial B_{ez}}{\partial t}(\chi,z_k,t)\,\chi\, d\chi - C_k \int_{r_i}^{r_j} \frac{\partial T^*}{\partial t}(\chi,z_k,t)\,\chi\, d\chi +$$
$$\int_{r_i}^{r_j}\int_0^R\int_{Z_1}^{Z_2} \frac{\partial^2 T^*}{\partial t\,\partial \zeta}(\rho,\zeta,t)\, G(\chi,z_k,\rho,\zeta)\,\chi\, d\zeta\, d\rho\, d\chi + \sum_{h=1}^{2}(-1)^h \int_{r_i}^{r_j}\int_0^R \frac{\partial T^*}{\partial t}(\rho,Z_h,t)\, G(\chi,z_k,\rho,Z_h)\,\chi\, d\rho\, d\chi \quad (1)$$

where: $G(r,z,\rho,\zeta) = \int_0^\pi \frac{\rho\,(z-\zeta)}{2\,\pi\left[r^2+\rho^2-2\,r\,\rho\cos(\theta)+(z-\zeta)^2\right]^{\frac{3}{2}}}\, d\theta \quad ; \quad T^* = \mu_0\, T$

Eq. (1) is the induction law in integral form : the term on the left is the electric circulation along a closed line, in the plane $z=z_k$, made by two circle segments (at r_i and r_j coordinates) and two line segments in the radial direction. The term on the right is the time derivative of the magnetic flux across the surface enclosed by the closed line. The first term is the contribution of the external field $\mathbf{B}_e$, the other terms are the contribution of the internal field $\mathbf{B}_i$. C_k is a coefficient which is equal to 1, if $Z_1<z_k<Z_2$, is equal to 0.5, if $z_k=Z_1$ or $z_k=Z_2$. In deriving eq. (1) the axial symmetry of the problem has been taken into account. In order to uniquely determine T, boundary conditions are required : T is equal to zero on the lateral surface of the pellet (r=R), and the radial derivative of T is equal to zero on the z-axis.

The flux-flow and flux-creep model [6] is utilized to relate the electric field to the current density inside the superconductor. The dependence of the critical current density to the magnetic field is taken into account by means of the Kim model where only the component of **B** along the c-axis is considered :

$$\mathbf{E} = f(|\mathbf{J}|)\frac{\mathbf{J}}{|\mathbf{J}|} \; ; \; f(|\mathbf{J}|) = \begin{cases} \dfrac{E_c}{\sinh\left(\dfrac{U_0}{k\theta}\right)} \sinh\left(\dfrac{U_0}{k\theta}\dfrac{|\mathbf{J}|}{J_c}\right) & \text{if } 0 \le |\mathbf{J}| \le J_t \\ E_t + \rho_f\,(|\mathbf{J}| - J_t) & \text{if } J_t < |\mathbf{J}| \end{cases} \; ; \; J_c = J_{c0}\frac{B_0}{|B_z| + B_0} \quad (2)$$

In eq. (2) J_c is the critical current density and E_c is the corresponding value of the electric field, U_0 is the pinning potential, k is the Boltzmann constant, θ is the temperature and ρ_f is the flow resistivity. E_t and J_t are calculated in order to achieve C^1 continuity of **E** vs. **J** curve.

The 2D space domain is discretized by means of rectangular finite elements [7] with middle points in the sides parallel to the r-axis. The element shape functions are linear in the z-direction and quadratic in the r-direction. The values of T in the middle points are calculated by imposing the continuity of the radial derivative of T (the current density), on the boundaries of each finite element. From eq. (1), a system of $n_r\cdot(n_z+1)$ ordinary differential equations of the first order is obtained. Gaussian quadrature formulae are used to calculate the coefficients of the system. The system is numerically solved, by means of the Adams method [8]. The levitation force is calculated by means of numerical integration of its expression referred to the volume of the superconductor.

3. Simulation results

In order to test the described method, some of the experimental data which were obtained at CISE [2] were utilized. The NdFeB permanent magnet is 28 mm large and 20 mm high, with

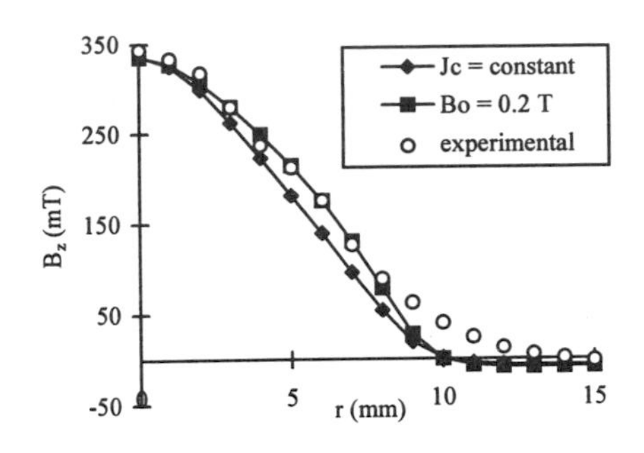

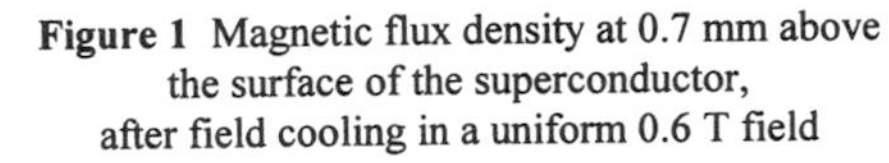

Figure 1 Magnetic flux density at 0.7 mm above the surface of the superconductor, after field cooling in a uniform 0.6 T field

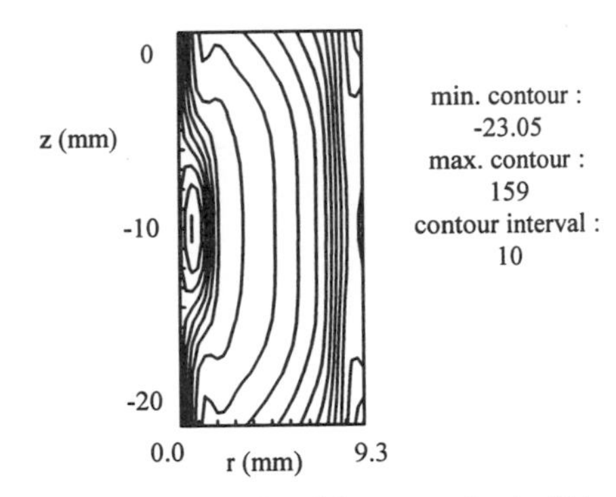

Figure 2 Contour plot of the current density (A/mm^2) distribution, after field cooling in a uniform 0.6 T field ($B_0 = 0.2$)

an equivalent surface current density of 1.2 10^6 A/m. The YBCO melt textured pellet [1] is 18.6 mm large and 20 mm high and is cooled to 77 K. No direct experimental measurements of the physical parameters in eq. (2) could be done, thus, many different values were utilized in the simulations. In the following, the results obtained for $\rho_f = 7\ 10^{-9}\ \Omega$ m, $E_c = 10^{-4}$ V/m and $U_0 = 0.2$ eV are reported. As much as the critical current density is concerned, for each value of B_0, the zero field critical current density J_{c0} was calculated in order to verify, as much as possible, the measured magnetic field distribution at the surface of the YBCO sample, after field cooling in a uniform 0.6 T magnetic field (fig. 1). The results corresponding to $B_0 = 0.2$ T and to a constant critical current density ($B_0 = \infty$) are reported in the following. In the first case, the zero field critical current density is 2.23 10^8 A/m^2, in the second case, the constant critical current density is equal to 1.05 10^8 A/m^2. The calculated current density distribution after field cooling in a uniform 0.6 T magnetic field is highly non uniform (fig. 2).

Figures 3 and 4, reports the comparison between the simulation results and the experimental data for the levitation force. In the zero field cooling (z.f.c.) case, the gap between the magnet and the superconductor is 75 mm when the superconductor is cooled, T is equal to zero and then simulation starts : the magnet approaches the superconductor at a constant speed of 3.375 cm/s till to the desired final value of the gap, then it stops. In the lower branch of the curve, the magnet, after reaching the minimum distance of 0.7 mm from the superconductor reverses its motion till to the final value of the gap.

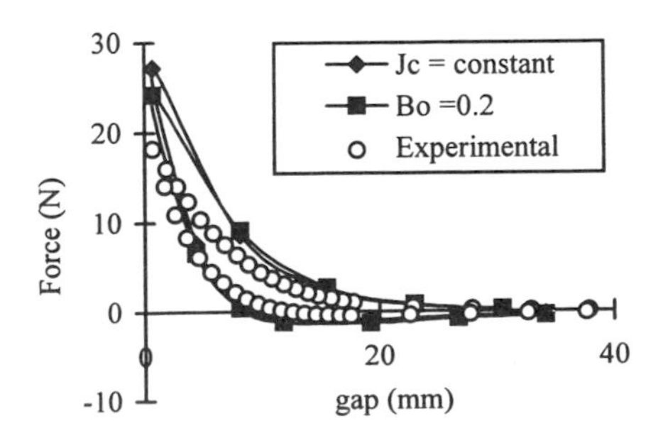

Figure 3 Levitation force vs. gap, zero field cooling case

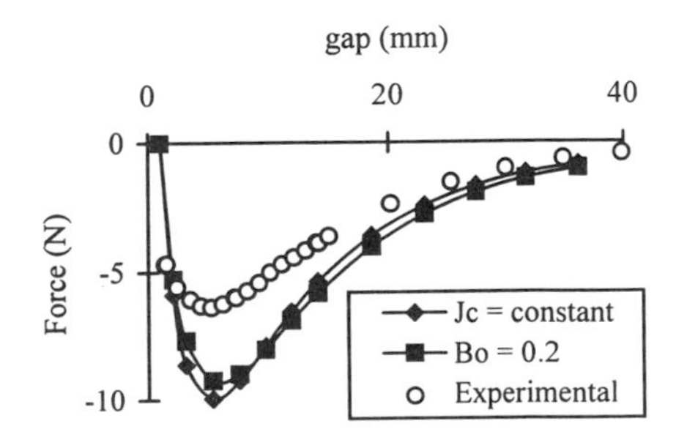

Figure 4 Levitation force vs. gap, field cooling case

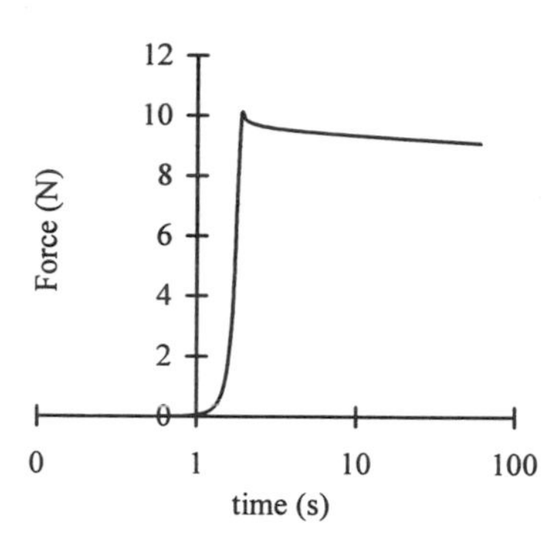

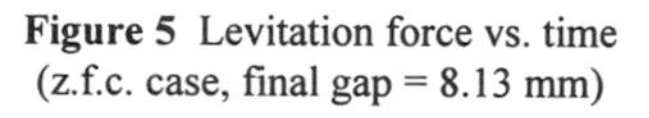

Figure 5 Levitation force vs. time (z.f.c. case, final gap = 8.13 mm)

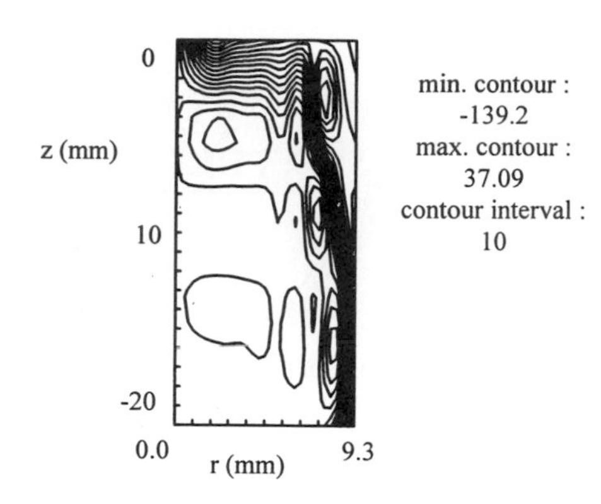

Figure 6 Contour plot of the current density distribution (z.f.c. case, final gap = 8.13 mm)

In the field cooling (f.c.) case, the initial gap between the magnet and the superconductor is 1 mm; and the velocity of the magnet is 2.2 cm/s. In figures 3 and 4, the values of the levitation force after 60 s are reported and compared with the experimental data.

The levitation force vs. time in a z.f.c. simulation ($B_0 = 0.2$) is reported in figure 5 : it can be seen that, in the flux-creep regime, the decay is logarithmic, with time constant equal to $k\theta/U_0$. The current density distribution at the end of the simulation (60 s) is shown in figure 6.

The simulation results were obtained with a 50 element mesh. Each element of the mesh is 0.93 mm large (r-direction) and 4.0 mm high (z-direction).

4. Conclusions

The most important phenomena of the system have been properly described by the simulations. The observed discrepancies can be due to the uncertainty on the value of the physical parameters of the flux-flow and flux-creep model and, or, to a non uniform structure of the YBCO pellet in the z-direction. As much as numerical aspects are concerned, improvement can be obtained by proper selection of the method for calculating the multidimensional integrals in eq. (1).

References

[1] Frangi F, Varesi E, Ripamonti G and Zannella S. 1994 *Physica c* **233** 301
[2] Varesi E, Borghi L and Zannella S 1996 to be published in *Nuovo Cimento D*
[3] Steel C W 1987 *Numerical Computation of Electric and Magnetic Fields* (New York : Van Nostrand Reinhold Company)
[4] Hashizume H, Sugiura T, Miya K and Toda S 1992 *IEEE Trans. on Magnetic* **28** 1332
[5] Yoshida Y, Uesaka M and Miya K 1994 *IEEE Trans. on Magnetic* **30** 3503
[6] Grissen R, Hagen CW, Lensik J and Groot DG 1989 *Physica c* **162** 661
[7] Zienckiewicz O C 1977 *The Finite Element Method* (New York :McGraw-Hill)
[8] Kockler N 1994 *Numerical methods and scientific computing using software libraries for problem solving* (New York : Oxford University Press)

Inst. Phys. Conf. Ser. No 158
Paper presented at Applied Superconductivity, The Netherlands, 30 June–3 July 1997

A NEW-TYPE PASSIVE MAGNETIC BEARING BASED ON HIGH-TEMPERATURE SUPERCONDUCTIVITY

A. V. Filatov, O. L. Poluschenko
Moscow N. E. Bauman State Technical University
5, 2-nd Baumanskaya str., Moscow, 107005, Russia

L. K. Kovalev
Moscow State Aviation Institute
Volokolamskoe shosse 4, Moscow, 125871, Russia

A new approach to the design of passive magnetic bearings using hard type II superconductors and, particularly high-temperature superconductors, is proposed. The advantages of this approach are drastic reduction of the force-displacement hysteresis and increase of the load capacity and stiffness per unit surface area of the bearing. A prototype of the element of the proposed bearing is produced and tested. With the superconducting ring of 0.4-gram mass, the 5.2-N load capacity was reached while the width of the hysteresis loop did not exceed 4%.

1. Introduction

Passive magnetic bearings using the interaction of a superconductor with a permanent magnet are known to have several significant advantages such as very low drag torque and zero energy consumption. However, the usage of superconducting bearings is often impracticable because of low stiffness, low load capacity and strong force-displacement hysteresis, which makes a bearing cease operation under influence of vibrations [1]. The origin of this hysteresis is well understood [2] and directly related with a remagnetization hysteresis of a hard type II superconductor being exposed to an external magnetic field changing during a displacement of the movable bearing part with respect to the stationary one.

We propose a new approach to the design of a superconducting bearing aimed to reduce the force-displacement hysteresis and increase stiffness and load capacity of the bearing.

2. A new approach to the design of passive magnetic bearings using the interaction of a hard type II superconductor with a permanent magnet

The core of our approach is to design such a system comprising superconductors and permanent magnets that the interaction between them takes place while the superconductors are exposed to magnetic field independent of rotor displacements [3]. The key idea of the new structure is explained in Fig. 1 showing a square-shaped flat shortened turn made of a hard type II superconductor, two opposite sides of which are exposed to oppositely directed magnetic fields uniform in the z-direction. If the turn were shifted in this direction from the position in which it had been cooled down, a change of the external magnetic flux through the turn would take place and a screening current would appear in the turn. Because some regions of the current-carrying turn are exposed to the magnetic fields, the Lorenz force directed oppositely to the displacement would act on the turn. It is easy to see that there is no change of the external magnetic field within the volume of the superconductor during the turn motion.

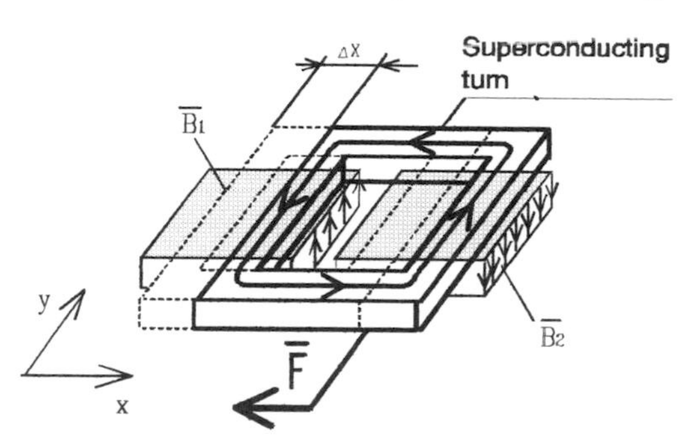

Fig.1 Principle of the low-hysteresis interaction of a hard type II superconductor with a permanent magnet.

It is to be noticed that besides of the reduction of the force-displacement hysteresis the above described system has many other advantages. Let us analyze the superconducting turn carrying the maximal current. As it can be seen from Fig.1, at any point of the superconductor volume exposed to the external magnetic field, the direction of the current flow is perpendicular to both the direction of the field and the interaction direction and, therefore, Lorenz force is directed oppositely to the turn displacement. Besides, because the external magnetic field is uniform, the Lorenz force exerted at unit volume of the superconductor is of the same value everywhere within the superconductor portion exposed to the magnetic field. Because of the above said, in the proposed system, the current-carrying capacity of the superconductor is used optimally and the required load capacity can be obtained with the minimal volume of the superconductor. The above conditions cannot be realized in previous bearing designs were a superconductor is exposed to a strongly nonhomogeneous magnetic field. Further, if high-temperature superconducting ceramics are used in the proposed system, the anisotropy of their electromagnetic characteristics can be accounted so that the maximal critical current will be reached. In such ceramics, to obtain the maximal current density, the current is required to flow in the crystallographic “ab” plane [4], while external magnetic field is recommended to be applied along the “c” axis [5]. In the system Fig.2, if the plane of the superconducting turn coincides with the “ab” plane of the material, the critical current is maximal.

It is also to be noticed that in the proposed system the magnetic field is generated in a narrow gap and, therefore, the stray flux is minimal and a high value of the field can be reached with a permanent magnets of minimal volume.

In both linear and rotational bearings, continuous frictionless motion in the direction normal to the interaction direction is required. In the proposed system, it could be provided if the areas of uniform magnetic fields were lengthened in the y-direction. Unfortunately, in this case, some parts of the turn would be exposed to magnetic fields changing during the motion of the turn in the z-direction and some hysteresis of the force-displacement characteristics would appear. Nevertheless, the force-displacement hysteresis can be reduced to the required level by means of appropriate choosing the ratio between sizes of the turn aperture and wall thickness. Speaking about rotational bearings, the continuous circular motion of the rotor can be provided if the areas of uniform magnetic fields form concentric annular regions. An example of the journal rotational bearing based on this principle is shown in Fig.2.

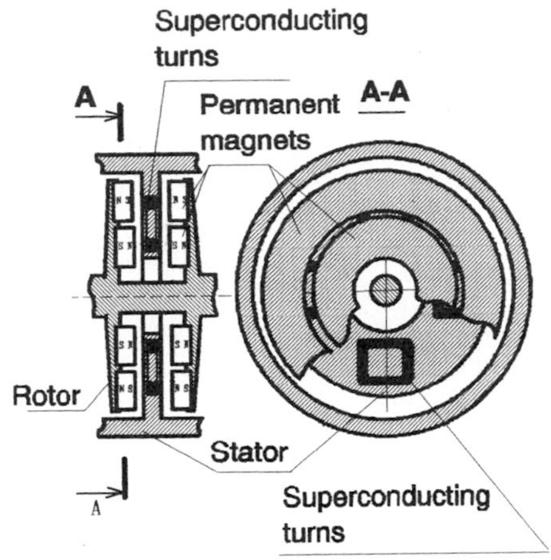

Fig.2 Journal passive magnetic bearing

3. Prototype of the magnetic bearing unit

As it can be seen from Fig.2, the proposed journal bearing comprises several (at least three) superconducting shortened turns placed circumferentially about the bearing axis. To investigate the force-displacement characteristics of the bearing we produced the prototype of the bearing unit comprising one superconducting shortened turn, two opposite sides of which are exposed to oppositely directed uniform magnetic fields normal to the turn plane (Fig.3). This turn shaped as a ring with 10-mm outer diameter, 5-mm inner diameter and 1.3-mm thickness has been cut out from a bulk melt-textured ceramic sample prepared by the top-seeding technology in Superconductivity Research Laboratory of Korea Atomic Energy Research Institute. The cutting has been carried out so that the crystallographic 'ab' plane coincides with the ring plane. The critical current value in the

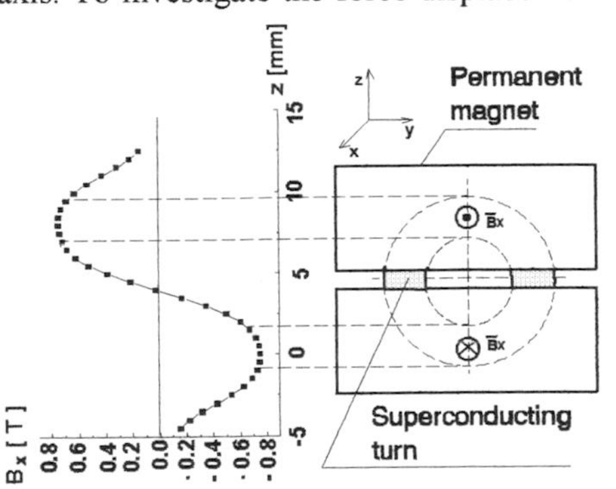

Fig.3 Prototype of the bearing unit

ring measured by an inductive method was found to be about 360 A, corresponding to the critical current density of 11000 A/cm^2. The magnetic fields are generated by four Nd-Fe-B permanents magnets 3×6×15 mm^3. To measure the force-displacement characteristics the superconducting ring was set in a stationary fixture while the magnetic circuit along with the permanent magnets was connected to a load cell driven by motorized stage in the z-

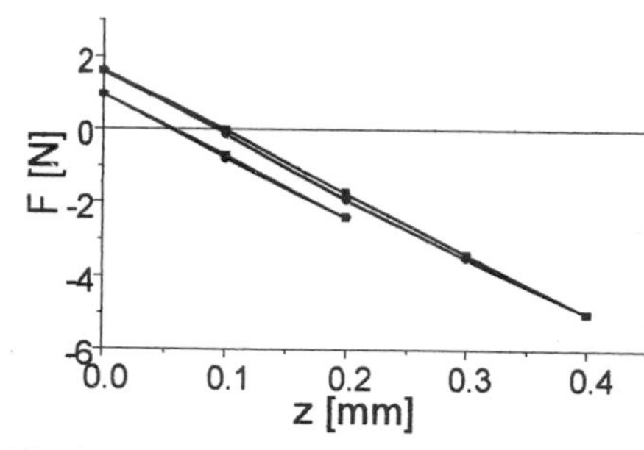

Fig.4 Force-displacement characteristics of the bearing unit

direction with 0.1-mm step. The magnetic circuit was moved up and back to the initial position with long holds at every point where force magnitude was measured. Two steady-state force-displacement characteristics corresponding to the maximal displacements 0.2 mm and 0.4 mm are shown in Fig.4. These characteristics have been obtained after several preliminary cycles of the reciprocal motion. The force-displacement graphs were observed to shift up after every increment of the displacement, what is resulted from the depinning of some portion of vortices and corresponding change of the flux linked with the ring after every increase of the screening current value. Under 0.4-mm displacement the maximal 5.2-N value of the interaction force was reached. Accounting that the mass of the superconducting ring is 0.4 g only, the ratio of the load capacity of the bearing prototype vs the prototype weight exceeded 1300. The stiffness of the suspension was measured to be about 17 N/mm and the width of the hysteresis loop was in order of 4% of the maximal interaction force. The maximal pressure (ratio of the load capacity vs bearing surface area) of the bearing prototype is evaluated to be about 3.7 N/cm^2, the specific stiffness (ratio of the stiffness vs bearing surface area) is about 12 N/mm/cm^2.

3. Conclusion

The measurements of the characteristics of the produced prototype of the proposed bearing unit have shown that the force-displacement hysteresis in such bearings can be made much lower than one exhibited by conventional superconducting bearings [6], while the load capacity and stiffness can be made much higher.

It can be noticed that critical current density of 11000 A/cm^2 measured in the superconductor used in the prototype corresponds to a moderate quality of the material. Usage of high-quality materials promises to provides a significant improvement of the bearing characteristics. Further increase of both the maximal pressure and the specific stiffness could be obtained if a square-shaped turn were used instead of ring-shaped one.

References

[1] Coombs, T.A., Campbell, A.M. 1996 *Physica C* 256 298

[2] Komori, M., Iwakuma, M., Kitamura, T., and Matsushita T. 1993 *J. Appl. Phys.* 73 2535

[3] Filatov, A. V., Konovalov, S.F., 1996 *Physica C* 271 225

[4] Selvamanickam, V., and Salama, K., 1990 *Appl. Phys. Lett.* 57 1575

[5] Salama, K., and Lee, D., 1994 *Supercond. Sci. Technol.* 7 177

[6] Komori, M., Tsuruta, A., Fukata, S., Matsushita, T., 1994 *Proceedings of the 4-th International Symposium on Magnetic Bearings*, Zurich, Switzerland, 401

Inst. Phys. Conf. Ser. No 158
Paper presented at Applied Superconductivity, The Netherlands, 30 June–3 July 1997

Magnetic shielding effects found in the superposition of a ferromagnetic cylinder over a BPSCCO cylinder : Influence of the air gap between the BPSCCO and ferromagnetic cylinders

K. Mori, M. Itoh*, and T. Minemoto

Div. of Sys. Sci., Kobe Univ., Nada, Kobe 657, Japan
* Dept. of Electronic Engi., Kinki Univ., Higashi-Osaka, Osaka 577, Japan

Abstract. The high-critical temperature superconductor (HTS) vessel is an ideal magnetic shield which employs the Meissner effect. It does not, however, generally satisfy the value of the maximum shielded magnetic flux density (B_S) required of an HTS vessel for practical use. The authors have been improving the value of B_S by the superposition of a soft-iron cylinder over a BPSCCO cylinder, termed the superimposed cylinder, and limited the evaluation of the magnetic shielding effects to the superimposed cylinder, in order to simplify theoretical analysis. The values of B_S are systematically measured by changing the air gap distance (δ) between BPSCCO and soft-iron cylinders under conditions of a constant length of the soft-iron cylinder. The values of B_S are also theoretically analyzed. It was found that the experimental values of B_S agree well with theoretical values. The results confirm that B_S is an important criterion in the design of effective magnetic shielding for large-sized vessels.

1. Introduction

It is often required that effective magnetic shielding be available in many applications, for example, the measurement of very weak magnetic fields such as biomagnetic fields. The idealized magnetic shielded vessel can be realized by use of a high-critical temperature superconductor (HTS). The HTS vessel does not, however, generally satisfy the maximum shielded magnetic flux density (B_S) required for practical use of the HTS vessel. Therefore, it is necessary to improve the B_S of the HTS vessel.

The authors have improved the value of B_S by superimposing a soft-iron cylinder over a BPSCCO (Bi-Pb-Sr-Ca-Cu-O) cylinder [1 - 7], termed the superimposed cylinder. In the present paper, the values of B_S are systematically measured when changing the air gap distance (δ) between the BPSCCO and soft-iron cylinders in which a parameter is used to control the value of the length (l) of the soft-iron cylinder. In addition, values of B_S are theoretically analyzed by using the results of Refs. [5] and [6]. It is found that the experimental values of B_S agree well with the theoretical values. Also described are the effect of the excited magnetic flux density (B_{ex}) on the magnetic flux density (B_{in}) within the superimposed cylinder, and the temporal change of the trapped magnetic flux density (B_t) within the BPSCCO cylinder.

2. Experimental details

Magnetic shielding effects have been evaluated for all cylinders, both single BPSCCO and superimposed cylinders, when placed in a homogeneous dc magnetic flux density (B_{ex}) applied parallel to the axial direction of the cylinders. The effects of B_{ex} on the magnetic flux density (B_{in}) within all cylinders were measured with the use of a GaAs Hall device, such as that described in Ref. [1].

Table 1 lists the inner radius (r_{in}) of the soft-iron (1.2 mm thickness) cylinders superimposed over the BPSCCO cylinder. In this paper, use is made of soft-iron cylinders having lengths (l) of 60 mm, 80 mm, 100 mm, 120 mm, and 140 mm. The soft-iron cylinders were degaussed by an ac magnetic field (60 Hz) at room temperature (300 K), prior to carrying out the experiments. The outside of the BPSCCO cylinder (2.9 mm inner radius, 5.5 mm outer radius, 30.4 mm in length) was wrapped with several turns of fluoroplastic (PTFE) tape, using the troidal winding method, in order to avoid sudden temperature changes [5]. After undergoing 100 thermal cycles between room temperature (300 K) and the boiling point of liquid nitrogen (77.4 K), the characteristics of the BPSCCO cylinder, such as shown in Figs. 2, 3, and 4, exhibited no significant change in the degree of magnetic shielding. In addition, all characteristics of the BPSCCO cylinder in the present experiments were obtained by the zero field cooling method. Figure 1 illustrates a schematic diagram of the superimposed BPSCCO cylinder.

Table 1. Inner radius of the soft-iron cylinder. (1.2 mm thickness)

cylinder	r_{in} (mm)
1	6.75
2	8.35
3	9.90
4	11.50
5	13.10
6	14.70
7	16.30
8	17.85

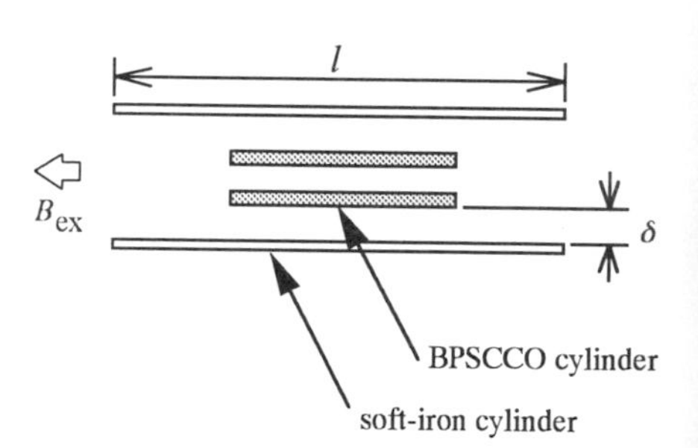

Fig. 1. Schematic illustration of the superposition of a soft-iron cylinder over a BPSCCO cylinder.

3. Results and discussion

Figure 2 shows the characteristics of the magnetic shielding at the center of the innermost BPSCCO cylinder under temperature conditions of 77.4 K. The curves (a) and (b) represent the single BPSCCO cylinder (open circle) and superimposed cylinder (solid circles), respectively. The dimensions of the soft-iron cylinder in the superimposed cylinder system are 9.9 mm inner radius, 11.1 mm outer radius, and 60 mm in length. In this figure, the points denoted as B_{sa} (=9.8 × 10^{-4} T) and B_{sb} (=155 × 10^{-4} T) express the maximum shielded magnetic flux density (B_s) for the respective curves. It is found that the value of B_{sa} associated with the single BPSCCO cylinder is improved by a factor of about 16 when superimposing a soft-iron cylinder.

Figure 3 displays the temporal change (over a period of 10 - 10000 sec) of the trapped magnetic flux density (B_t) at the center of the BPSCCO cylinder, such as shown in Fig. 2 (a) at 77.4 K. As can be seen, the trapped magnetic flux density in the BPSCCO cylinder remained stable over a long period of time. This characteristic is an important criterion in evaluating the magnetic shielding effect and the properties of superconductors.

From Ref. [5], the theoretical value of B_s for a superimposed cylinder can be written as

$$B_s = B_s' \left(|S_1|_{l/r_{in} \geq k} + |S_2|_{l/r_{in} < k} \right) . \tag{1}$$

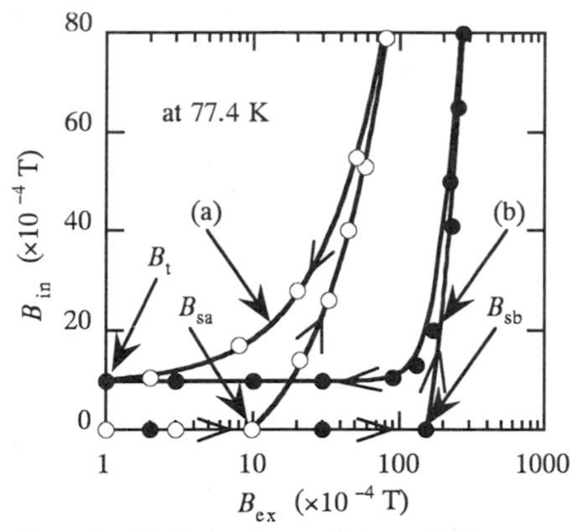

Fig. 2. Typical characteristics of the magnetic shielding at the center of the BPSCCO cylinder at 77.4 K. Curves (a) and (b) represent the single BPSCCO cylinder and the superimposed BPSCCO cylinder, respectively.

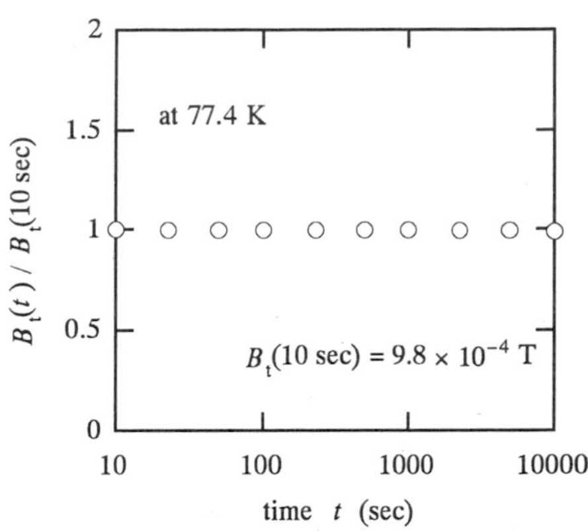

Fig. 3. Temporal change of the trapped magnetic flux density (B_t) at the center of the BPSCCO cylinder at 77.4 K.

The B_s' in Eq. (1) has been modified from Ref. [5] to include a function of the air gap (δ), which is the gap of air between the superconducting cylinder and ferromagnetic cylinder. The modified B_s' can be expressed as

$$B_s' = B_{sa}\frac{3(a+\delta)^2-(a^2+ab+b^2)}{3(a+\delta)^2}, \qquad (2)$$

where B_{sa} is the maximum shielded magnetic flux density for the single superconducting cylinder, and a and b are the inner and outer radii of the superconducting cylinder, respectively. In Eq. (1), the shielding factors ($|s_1|$ and $|s_2|$) are given, following [4 - 6], for cases in which the end effects of the superimposed cylinder are neglected and not neglected as

$$|s_1|_{l/r_{in}\geq k} = 1+4N\left[1+\frac{\mu_s}{4}\left\{1-\left(\frac{a+\delta}{r_{out}}\right)^2\right\}\right], \qquad (3)$$

and

$$|s_2|_{l/r_{in}<k} = \frac{K}{2.6\sqrt{l/2r_{out}}}\exp\left\{2.25\frac{l}{2(a+\delta)}\right\}, \qquad (4)$$

respectively. Here, l is the length of the ferromagnetic cylinder, k the value of l/r_{in} at the crossing point of $|s_1|$ and $|s_2|$, and r_{in} and r_{out} are the inner and outer radii of the ferromagnetic cylinder. In addition, μ_s is the relative permeability of the ferromagnetic cylinder, N the demagnetizing factor represented by the axis ratio ($p = l/2r_{out}$) of the ellipsoid, and K (= 0.4) the correction factor determined from numerous experimental results [5, 6].

For investigating the influence of the value of δ on the value of B_s, Fig. 4 displays the experimental values (solid circles) of B_s as a function of the air gap (δ) and the length (l) of the soft-iron cylinder. In Fig. 4, the solid curves are obtained by the use of Eq. (1). The value of μ_s for the soft-iron cylinder superimposed over the BPSCCO cylinder is determined from the characteristics of μ_s versus B_{ex} [5]. The theoretical values of B_s agree well with experimental

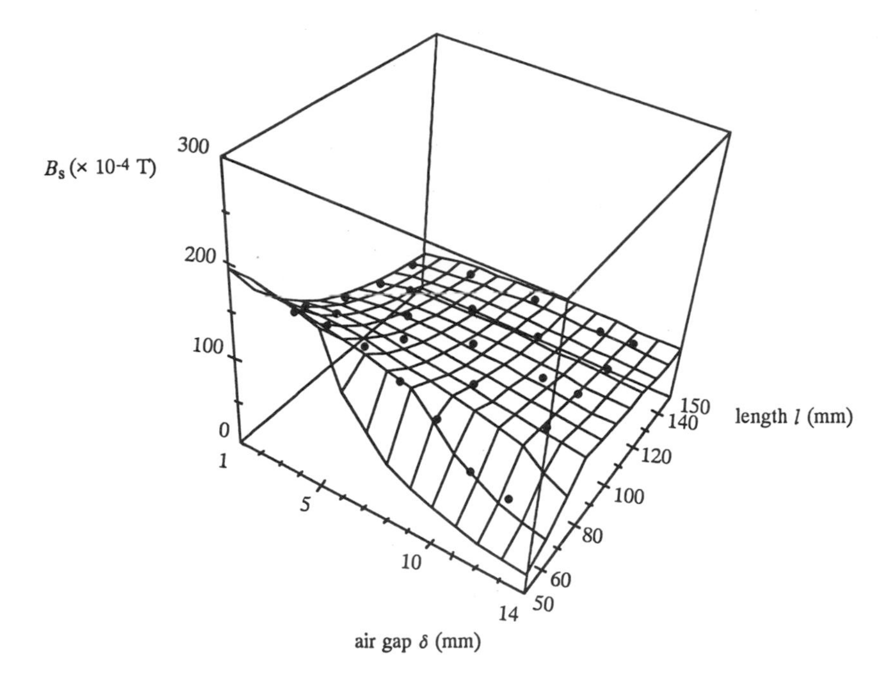

Fig. 4. Typical air gap (δ) - length (l) distribution of the maximum shielded magnetic flux density (B_s) for the superimposed cylinder. Solid circles are the experimental values at 77.4 K.

values. Therefore, as a result, on the value of δ is an important criterion fundamental in the design of an effective and reliable shield for a large-sized vessel.

4. Conclusions

The present research has examined the influence of the air gap distance (δ) on the value of B_s. It was found that the value of δ is an important parameter for design of large-sized vessels having highly effective magnetic shielding. The present paper has been directed at the design of the superposition of a ferromagnetic cylinder over a superconducting cylinder for a magnetic shielded vessel. In addition, it was found that theoretical values of B_s agree well with experimental values. The results of the present research have determined an important criterion, i.e., the air gap, as fundamental to the design of a large-sized vessel having better shielding characteristics.

References

[1] Itoh M Ohyama T Minemoto T Numata T and Hoshino K 1992 *J. Phys. D: Appl. Phys.* **25** 1630-1634

[2] Itoh M Ohyama T Hoshino K Ishigaki H and Minemoto T 1993 *IEEE Trans. on Appl. Superconductivity* **3** 181-184

[3] Ohyama T Minemoto T Itoh M and Hoshino K 1993 *IEEE Trans. Magnetics*, **29** 3583-3585

[4] Itoh M Ohyama T Mori K and Minemoto T 1994 *Cryogenics* **34** *ICEC Suppl.* 817-820

[5] Itoh M Ohyama T Mori K and Minemoto T 1995 *T IEE Japan* **115-C** 696-701

[6] Mori K Itoh M and Minemoto T 1996 *J. Mag. Japan* **20** 257-260

[7] Itoh M Mori K and Minemoto T 1996 *IEEE Trans. on Magnetics* **32** 2605-260

Inst. Phys. Conf. Ser. No 158
Paper presented at Applied Superconductivity, The Netherlands, 30 June–3 July 1997

Liquid phase sintered YBCO hollow cylinders for magnetic shielding

J. Plewa[1], W. Jaszczuk, C. Seega, C. Magerkurth, E. Kiefer[2], H. Altenburg

Fachhochschule Münster, Supraleitertechnologie und Kristalltechnik, D-48565 Steinfurt
[1]SIMa, D-48565 Steinfurt, [2]SIPERM, D-44287 Dortmund, Germany

Abstract. Shielding property measurements on various YBCO sintered bulk materials were performed at 77 K. By optimising the fabrication procedure with respect to homogeneity, density and shielding it was possible to make YBCO hollow cylinders with a relatively thin wall size. The process is only successful within a given temperature range. YBCO hollow cylinders show a critical shielding field up to 1,25 mT (BSCCO up to 6,7 mT) and YBCO plates up to 4,6 mT.

1. Introduction

Magnetic shielding is commonly used to reduce the susceptibility of a material or device to an external magnetic field. There are many possible applications, for example the shielding of the electromagnetic radiation from electrical machines and the shielding of the fluctuation of the earth's field. Ferromagnetic materials such as iron and μ-metal are usually used for shielding dc fields. Highly conductive metals such as copper or aluminium (inductive current shielding) are used for shielding ac signals of higher frequencies. Superconductors provide an alternative for shielding low and high frequencies and low amplitude magnetic fields [1-4]. Also ac screening based on the high T_c materials will be superior to normal metals. Passive superconducting shielding uses the Meissner effect where the magnetic flux does not penetrate a superconductor. For textured high T_c materials the flux-pinning effect coexists with the Meissner effect. For application in biomagnetic measurements, nondestructive testing and electrochemical reactions the three essential requirements are the field exclusion (up to mT), small flux creep and low magnetic noise. The flux creep and the magnetic noise must have low values compared to the resolution of the SQUID sensor.
The microstructure of supercondutors in the bulk monolithic form consists of a granular structure with grain boundaries and crystal defects [5-6]. This would suggest that magnetic shielding is not fully effective, while flux exclusion occurs from each grain by the induction of surface currents. These flux lines will be concentrated between the grains. However at low frequencies the high T_c materials can still shield electromagnetic radiation more effectively than other materials.

2. Sample Preparation

The tested YBCO tubes and plates were compacted from reactive mixture powders by cold isostatic pressing (with plasticiziers) and sintered by a temperature programme similar to a standard texturing method. The initial materials, the forming processes and thermal treatment of the bulk material are shown in Table 1.
The mixtures react together to form YBCO at a temperature above 700°C. The reaction is more rapid above 850°C [7]. The sintering process consists of several steps with maximum temperatures between 950 and 1050°C. During the sintering and oxidation of YBCO bulk materials a high stress level was observed. To decrease this stress the cooling must be carried out very carefully, especially between 950 and 500°C. A high stress level was also observed whibt decreasing the temperature from 500 to 200°C.

Table.1 Fabrication of the YBCO bulk materials

Synthesis of special powder	forming of pieces	thermal treatment
solid-state reaction	cold isostatic pressing with plasticiers	liquid phase sintering
• precursor materials $YBa_2Cu_3O_{7-x}$ $BaCuO_2$ • reactive mixtures Y:Ba:Cu=1:2:3 or 21:33:49 • particle size 5-15 µm	• dimension of hollow cylinders 3,5 cm inner radius 2-5 mm wall thickness 15-30 cm length • pressure 100-200 MPa • green density 2,5-2,9 g/cm^3	• rapid heating up to 950-1050°C and slow cooling to 920°C • hardness (Vickers) 7-10 GPa • density 5-6 g/cm^3

3. The Experiment

For the investigation of the shielding properties at a temperature of 77 K a new test bench had to be developed (Fig.1).
The HTSC hollow cylinders or plates were fixed in a liquid nitrogen dewar. A cryogenic Hall probe can be positioned inside the cylinder or directly onto a plate to measure the flux density as a function of the applied dc field.
If the magnetic field is weak some of the penetrating flux cannot be measured by the Hall probe because the sensitivity is too low. If the external field is increased beyond a critical value the shielding abruptly breaks down which allows the flux to penetrate the shield. The measured values for the shielding of the YBCO tube are shown in Table 2. An example of a plot of the hysteresis graph is represented by Fig. 2.

4. Results and Conclusion

The liquid phase sintered YBCO hollow cylinder and plates were tested for homogeneity and the electrical and magnetic properties were investigated.
Firstly the test probes underwent microscopic investigation as shielding behaviour is dramatically reduced by every fissure or open pore present. The critical temperature readings (electronically by the four-point-method) were always, for small sections of the test probe, successfully achieved. Indicating that test probes with some structural defects will also display the correct T_c-curve. Therefore the quality of the test probe (hollow cylinders or plates) can only be clearly determined by the deciding test of shielding properties. Fig.2 shows a hysteresis loop of slab (Tab.2 No 8) when the external field has a value of 4,6 mT the magnetic flux penetrate the materials.

Table.2 Shielding properties for various HT_c bulk materials at 77 K

No.	YBCO sample	out.diameter cm	wall thicknes, mm	length cm	max. temp °C	density, g/cm^3	shielding mT
1	cylinder (alt1)	3	5	6,5	1050	5,89	1,25
2	cylinder (alt4)	3	5	6,0	1100		0,56
3	tube (sint3)	3,5	2	12,0	950	5,04	0,17
4	tube (41196)	3	1,5	14,0	950	5,38	0,42
5	tube (21495)	4	1	4,5	1050	5,82	0,19
6	disc	2	3		1100	6,22	1,05
7	plate	2	2		1050	6,05	2,75
8	slab (R1)	2	3		1050	6,15	4,60
9	BSCCO tube with bottom	2,2	1,5	9	850	5,95	6,73

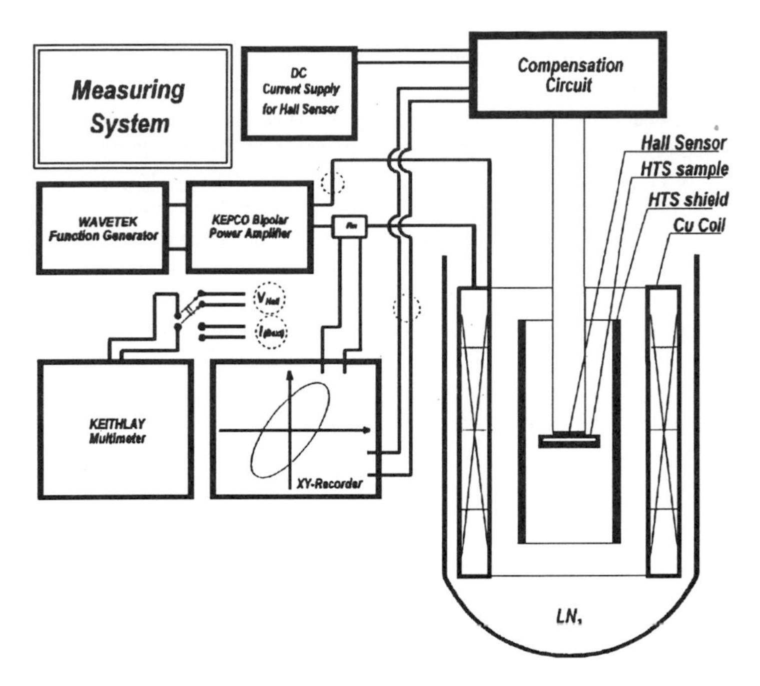

Fig.1 The test system constructed to characterize the shielding properties of YBCO bulk material

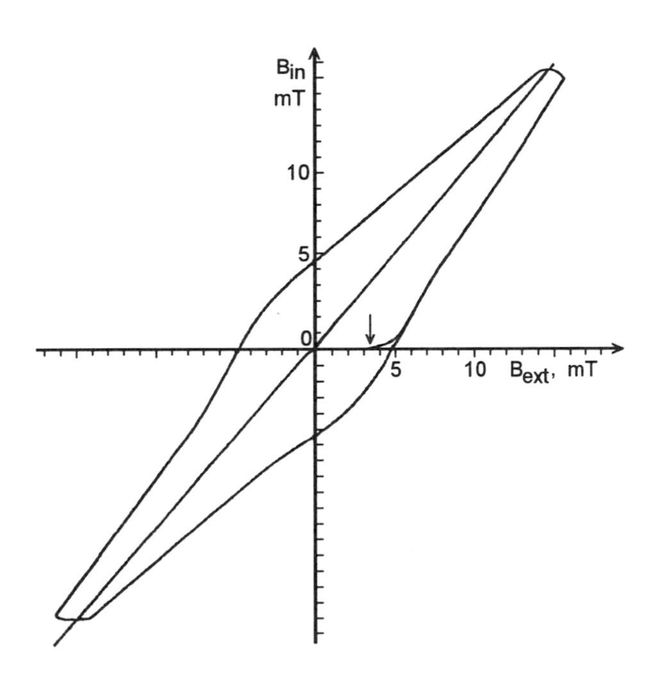

Fig.2 Magnetic hysteresis loop merasured for the YBCO slab

After some comparison of the thermal treatment, the shielding properties of the different YBCO materials were determined. Good shielding properties are correlated with high material compression. It is known [8], that geometric shielding factors play a very important role in the shielding process. Using the measuring system given (Fig.1) it was experimentally confirmed that a wider hysteresis for hollow cylinders than plates is found. Particular importance can be placed on the path of the new curve, its horizontal section agreed with a complete displacement of the field. This effect will be determined by applying forming and sinter techniques. It was found that a high sinter temperature was required to achieve a particularly good YBCO-material compression (Tab.2). To have a high maxium temperature a suitable temperature programme is required, which must also guarantee the removal of the material used to help the pressing process, as well as heating and cooling the test probes without the formation of fissures or inhomogeneities. By varying the starting compositions and changing the temperature programme [5-7] better shielding elements were produced The mechanical, electrical and magnetic properties must be considered with respect to homogeneity and density, weak links and the quantity of the Meissner phase. Good electrical properties as well as the observation of leviation effects are not sufficient for shielding with some mT. The measurement of shielding properties can indirectly, in some of these cases, indicate bad coupling of a single grain (Tab.2 No.2-5). Optimisation of fabrication methods of HTSC shields is possible by increasing the density in the pressing process and by some adjustment of the thermal treatment process. It is possible to improve the samples properties by reducing the wall thickness.
In general the pressing process of thin walled hollow cylinders gives the possibilities of reducing fissure formaton (by sintering and by design) and achieving a greater material compression. However this requires some changes to the sintering methods, in order to avoid the breaking of the hollow cylinders when sintering. The hitherto readings using dc fields allow shielding hysteresis to be measured without needing to give information about the noise level. Further investigations will be made in ac fields with SQUID sensors.

Acknowledgements

The authors would like to thank Dr. M. Itoh for supply of cryogenic Hall probe. This study was funded by the german BMBF under contract 13N5555

References

1 S. Elschner, J. Bock, G. Brommer, P.F. Herrmann, IEEE Trans. Magn. 32 (1996) 2724
2 H. Matsuzawa, J. Appl. Phys. 74 (1993) R111
3 V. Palushkin, A. Buev, H. Koch, Appl. Supercond. 2 (1994) 597
4 M. Itoh, K. Mori, T. Minemoto, F. Pavese, M. Vanalo, D. Giraudi, Y. Hotta, EUCAS'95, Edinburgh, July 3-6, 1995
5 J. Plewa, W. Jaszczuk, E. Kiefer, C. Renzing, M. Lerch, G. Farkasch, M. Barkhoff, C. Risse, H. Altenburg, 10. HTSL-Verbundtreffen, Texturierte HTSL-Materialien, 20.-22. Nov. 1996, Braunschweig
6 J. Plewa, H. Altenburg, EUCAS'95, Edinburgh, July 3-6, 1995
7 H. Altenburg, J. Plewa, S. Cherepov, J. Hauck, Statusseminar „Supraleitung und Tieftemperaturtechnik" Weimar, Juni 1994
8 Q.Y. Chen, TCSUH Report, Preprint No 93:001, Houston, 1993, 1

Inst. Phys. Conf. Ser. No 158
Paper presented at Applied Superconductivity, The Netherlands, 30 June–3 July 1997

Switching properties of high quality superconductors

O. A. Shevchenko, H. J. G. Krooshoop and H. H. J. ten Kate

University of Twente, Faculty of Applied Physics, Enschede, the Netherlands

Abstract. Switching and current limiting characteristics of several high-T_c superconductors are reviewed and compared to those for low-T_c superconductors in perspective of power applications. The comparison includes BSCCO tape as well as melt-processed tube, and YBCO film under repetitive operation. Potentially, YBCO film shows the best characteristics, but several problems prevent it from immediate application. The presence of a silver or silver alloy matrix in BSCCO tape also limits the ways of application. Here a hybrid switch is proposed employing the best properties of low- and high-T_c superconductors, of semiconductors and magnetic materials. Simulations of the devices were performed to validate the operation of the switches in pulse as well as in a repetitive mode.

1. Introduction

The proper operation of superconducting (sc.) magnets requires special properties of the powering systems. When a persistent mode is realised it provides a high quality magnetic field (no ripple, high stability) by using a rather simple power supply [1, 2]. But, in order to be really persistent the internal resistance of the conductor in the magnet must be sufficiently low. Another option is to apply a superconducting converter as it would avoid this condition and moreover, it widens the control features. A continuous improvement of sc. converters based on low-T_c superconductors was demonstrated during the last years. With high-T_c conductors some progress has been reported as well [3-5], but a better understanding of the limiting factors is required for further improvement of the performance.

In this paper a comparative study of switching properties of several superconductors is presented focusing on repetitive operation modes, essential for converters powering sc. magnets.

2. Switching a superconductor

Principle of operation. At present most of the methods employ a transition between the superconducting (**s**-) and normal metal (**n**-) states. The characteristic operating cycle consists of 4 stages including the two mentioned steady states and two transient states: **s** $\rightarrow$ **n** (activation) or **n** $\rightarrow$ **s** (recovery). There is an extended literature on the subject. Properties of several superconductors relevant to the switching process are summarised in Table 1.The basic methods to control the state of a superconductor used for repetitive operation are the following ones.

Thermal activation. A heat or an irradiation pulse triggers the switch gate. In the n-state the gate temperature exceeds the critical one. The recovery time is determined by a thermal time constant. The loss in the **s**-state is negligible compared to the loss in the **n**-state. While for LTS efficient 50 Hz operation can be achieved, for HTS, due to a much higher heat capacity, it is doubtful, especially when the efficiency is of importance.

Table 1. Properties of low- and high-T_c superconductors used for repetitive switching.

Material	Nb1%Zr [6]	NbTi [7]	BSCCO [4,5]	BSCCO/Ag [present]	YBCO [3,4]
Geometry	wire, tape	tape	rod, tube	tape	film
Operating temperature T_o, K	4.2	4.2	77	77	77
Critical properties at T_o:					
Current density j_c, A/m^2	2 10^9	5 10^9	4 10^7	4 10^7	*10^{10}
Magnetic field B_{c2}, T	1	11	0.2 (B//c)	1 (B//c)	>10 (B//c)
Temperature T_c, K	8.6	9.3	90	110	90
n-state resistivity ρ, Ω·m	2 10^{-8}	5 10^{-7}	10^{-6}	3 10^{-9}	10^{-6}
Switching properties:					
Control	magnetic (m)	thermal (t)	m, f, or t	flux flow (f)	thermal
Trigger pulse	magnetic field	heat	magnetic field	current	current
$J_c^2\rho$, W/m^3	8 10^{10}	10^{13}	2 10^9	4 10^6	10^{14}
Activation time, ms	< 10	< 1	10	< 1	10
Recovery time, ms	< 10	< 1	10	< 1	50
Electric efficiency, %	98	98	?	< 50	95*

* updated with recent data

Magnetic activation. An external pulse of magnetic field triggers the gate. In the **n**-state the field exceeds the critical one. The recovery time is determined by a magnetic time constant. The loss in the **n**-state can be made comparable to the loss in the **s**-state. Unfortunately, there is a remarkable loss present in the required coil to generate the control field as well, which together with other constrains makes this method rather unpractical. The obvious limitation for good quality HTS materials is the high critical magnetic field exceeding 100 T, in granular superconductors the low specific power $J_c^2\rho$ (above 10^{14} for a semiconductor) and the efficiency is therefore poor.

Flux flow activation. The gate is triggered by a current pulse kept below the quench current. In the **n**-state (which is the resistive state) the current exceeds the critical one. The recovery time is defined by the electro-magnetic time constant. The efficiency is poor due to the loss in the **n**-state. The method is applicable for pulse operation and where the loss is not important.
These are briefly the main methods and the reasons why, despite impressive progress shown in the quality of films, bulk materials and tapes, there is in fact no progress visible in repetitive switching of HTS since the early attempts [3-4]. The obvious high critical parameters of HTS causes serious limitation that apparently hinders further development. Therefore, at present the feasibility of efficient repetitive switching in HTS circuits has remained unanswered for power applications. Here a new concept is presented focusing on a circuit that contains a superconducting magnet as a source and/or load.

3. The concept of a hybrid switch

When the inductance L of a magnet is relatively small, a sc. transformer can be used for powering the magnet, a method that can be extended to a very high current [8]. A draw back is that the self-inductance of the transformer must be larger than the inductance of the load magnet. A (full-wave, single-phase) superconducting converter [6, 7] can be much smaller than the load magnet, but it is not very efficient due to the loss in the repetitive switches $S_{1,2}$. When cold semiconducting switches are used, a reasonable efficiency is reached, but the voltage drop across the switch limits the lower output voltage of the converter [1].

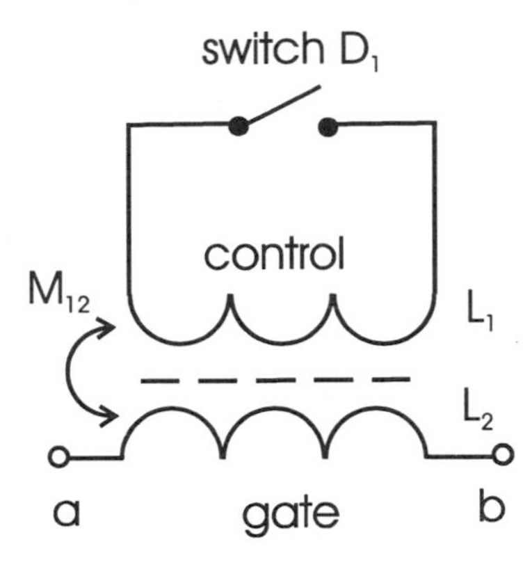

Figure 1 Schematic view of the hybrid switch. L_2 is the gate coil, L_1 and D_1 are the control coil and switch respectively; M_{12} is the mutual inductance between the coils and the core is shown by the dashed line.

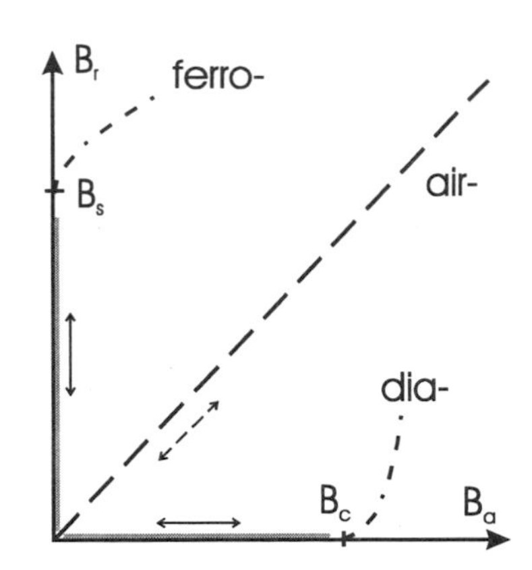

Figure 2 B(H) characteristic of the switch: B_a, B_r, B_c, B_s are applied, resulting, critical and saturation magnetic fields respectively for a superconductor (diamagnetic state) and the core (ferromagnetic state or air: linear conductor state).

The proposal presented here is to use in the converter a new hybrid switch, lacking the mentioned drawbacks and using the best features from each material involved. This paper focuses only on the switch itself. The operation of the converter based on the same principle is described elsewhere [9]. The device provides fast and efficient switching of magnetic flux between self- and mutual inductances, which results in the adequate powering of a load magnet.

The principle is sketched in Figure 1. The switch consists of two coaxial (superconducting) coils wound around an (iron) core and a semi-conducting switch D_1 placed at the room temperature. The primary and secondary coils have self-inductances L_1 and L_2 respectively and a mutual inductance M_{12}. The coil L_2 is used as the switch gate in the high current secondary circuit of the converter including a magnet as a load. The coil L_1 with the switch D_1 comprises the control circuit of the switch S_1. It operates at low current and high voltage in order to keep the loss of D_1 below an acceptable limit. When D_1 is closed, the gate shows a stray inductance L_s chosen much smaller than L and it exhibits a small impedance (coils L_1 and L_2 are sandwiched for this purpose). When D_1 is opened, the gate exhibits the self-inductance L_2 designed to be much larger than L.

When an air core is used the switch S is in general large compared to the load magnet. This mismatch can be dramatically changed when an iron core is employed (~μ times). The DC component of the voltage across the gate, present during repetitive operation, might saturate the core. This problem is avoided by using large enough core for each switch $S_{1,2}$. The functions of the switch S are separated as follows. The switch D1 operating at optimal conditions provides resistive switching. Coils L_1 and L_2 are always in the superconducting state and provide a low impedance L_s and a low loss at high currents. The iron core causes increased impedance of S when open and provides a good coupling between the coils. The *B(H)* characteristic of the switch S as depicted in Figure 2 is close to the ideal switching characteristic. Air-core coils can also be used (the dashed lines). The diamagnetic property provides low L_s while the ferromagnetic property causes a high L_2. Switching by *D1* provides a high efficiency and a fast repetition rate. The grey lines and arrows indicate the normal operation areas of the device, horizontal part: switch D_1 closed, vertical part: switch opened. The arrangement and the operation diagram of the switch are sketched in Figures 3 and 4

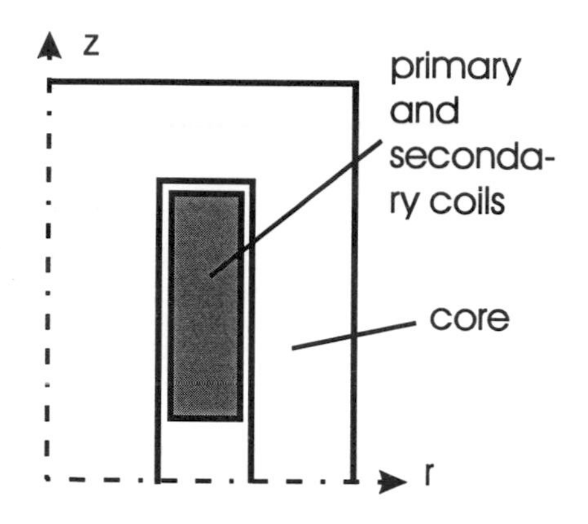

Figure 3 Cross-sectional view of the hybrid switch cold part. Only one quarter of the actual cross-section is shown due to the symmetry.

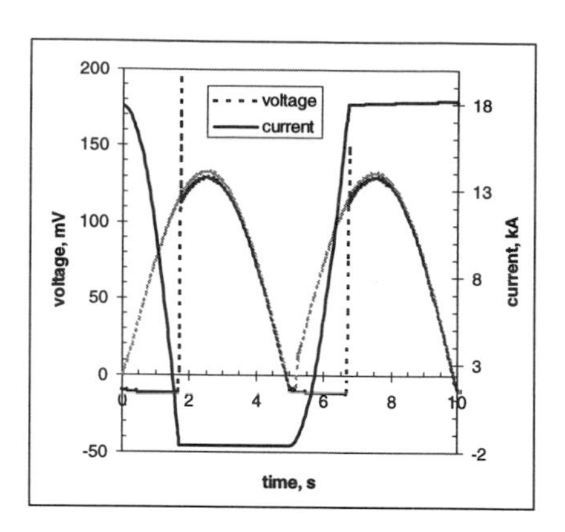

Figure 4 Operating diagram of the switch when ramping up the magnet current of 18 kA. L=2 mH, L_1=100 H, L_2=20 mH, M_{12}= 1.4141 H, D_1 = diode, power from a voltage source, frequency 0.1 Hz.

(black lines: gate voltage and current after 500 cycles; light grey lines: gate voltage at first cycle). Obviously, the superconductor limits the repetition rate. Nevertheless, both activation and recovery time constants are below a few ms and result in a reasonable efficiency of 99.5 %. This enables operation at 50-60 Hz with currents of 1 kA. At higher currents lower frequency is due to a cable conductor, see for example Figure 4 where results of numerical simulation are depicted for one switch of the single phase and full-wave superconducting converter.

4. Conclusions

1. A gap is present between classical concepts of a repetitive switching and the progress made so far with high temperature superconductors.
2. A new concept for a hybrid switch based on switching magnetic flux is proposed. The concept is applicable for any kind of a superconductor.
3. The device provides fast and efficient switching and can be used for powering a superconducting magnet.

Acknowledgement

This work is supported by the Netherlands Organisation for Scientific Research (NWO) and by EC INTAS grant N 95-0197.

References

[1] Shevchenko O A Fedorovsky M A and ten Kate H H J 1997 Adv. in Cryog. Eng., **41** 1873-1880
[2] T. Hase *et al* Cryogenics 1997 **37** N 4 201 – 206
[3] Vysotsky *et al* Supercond.: Sci&Technol. 1990 **3** 259 – 262
[4] Markovsky N V *et al* Cryogenics 1992 **32** 548 –549; 624 – 627
[5] Paul W *et al* 1995 Inst. Phys. Conf. Ser. No 148 **1** 73- 78; this conference paper 10Ge-23
[6] Mulder G B J 1988 PhD University of Twente Enschede The Netherlands
[7] Shevchenko O A *et al* ibid. [5] 647-650
[8] Knoopers H J *et al* 1997 Proc. of the ICEC/ICMC Elsevier Science **2** 803-806
[9] Shevchenko O A and ten Kate H H J paper to be presented at CEC-ICMC'97

Inst. Phys. Conf. Ser. No 158
Paper presented at Applied Superconductivity, The Netherlands, 30 June–3 July 1997

The Life and Times of Heike Kamerlingh Onnes

R J Soulen, Jr

Naval Research Laboratory, Code 6344, Washington, DC 20375-5000

Abstract. This paper reviews some of the highlights of this famous pioneer in superconductivity and low temperature physics.

1. The beginnings

Heike Kamerlingh Onnes was born September 21, 1853 in Groningen, which is in the northeastern corner of the Netherlands. He inherited the separate gifts of the arts and science from his parents. His mother, Anna G. Coers, was artistic and was the root of this family predilection. Thus a younger brother, Menso, became a highly regarded artist, and his son, Harm an even greater one. It is said that Heike dabbled in poetry when he was a boy.

His father, Harm Kamerlingh Onnes, was a successful businessman who owned a roofing-tile factory and who stressed the value of a formal education. Thus Heike was sent to the *Hoogere Burgerschool*, a public school in Groningen. There he did well [1] and in 1870 he enrolled in physics and mathematics at the University of Groningen. Onnes was an exceptional student academically. In only one year he had finished his bachelor's degree (the norm was two to three years) and at the end of the same year he won a gold medal from the University of Utrecht for writing an essay which evaluated the methods used for determination of vapor density. Onnes also showed a knack for organization and social responsibility. He was elected head of the student fraternities at the University and one of his accomplishments was to make the *Lustrum*, a special celebration given once every five years, a financial success for the first time in student memory. On another occasion, Onnes led a student delegation to the Hague to testify against a proposed merger of the University of Groningen with another university.

In 1871 Onnes won a prize which allowed him to spend three semesters at the University of Heidelberg. There he studied calorimetry with the chemist, Robert Bunsen and was inspired to write another essay which won him a silver medal from the University of Groningen. He later joined the physicist, Gustav Kirchhoff, who started him on a study of Foucault's pendulum. Onnes returned to Groningen in 1873 to complete the pendulum work under R. A. Mees. Onnes not only showed considerable skill as an experimentalist in designing and using the pendulum, but he also carried out the first exact mathematical treatment of Foucault's pendulum which is regarded as a first rate job of mathematical physics.[2] The thesis was finished in 1878, and in that year Onnes was awarded a doctorate *magna cum laude*. Although Onnes was destined to pursue a career dedicated to careful experimentation and exacting measurement, he did remain in constant and intimate contact with the relevant theoretical underpinnings and scientific issues germane to his experimental programs.

Shortly after completion of his doctorate, Onnes was appointed a teaching assistant to Johannes Bosscha at the Technical University in Delft. During this time Onnes became acquainted with Johannes Diderik van der Waals, then a professor of physics at the University of Amsterdam. Van der Waals, who as a mature man of 35, had left business to study science and had devoted his thesis effort at the University of Leiden to the development of a more accurate equation of state for real gases. In his doctor's dissertation given in 1873, van der Waals first presented the famous equation which bears his name: [3]

$$\left(P + \frac{a}{V^2}\right)(V - b) = \frac{1}{3}nmv^2 = \frac{2}{3}\left(\frac{1}{2}nmv^2\right) = \frac{2}{3}E = RT \qquad (1)$$

van der Waals realized thereafter that that Eq. (1) could be rewritten as a universal function for all gases with normalized variables, $p=P/P_c$, $v=V/V_c$ and $t=T/T_c$ (the so-called the law of corresponding states).

2. Building a laboratory and establishing a career

In 1882 P. C. Rijke retired as professor of physics at the University of Leiden and eventually Onnes was appointed his replacement at the age of 29. It was the first Dutch Chair in Experimental Physics. At his inaugural address in November 11, 1882, entitled "The Significance of Quantitative Research in Physics", Onnes made some initial philosophical statements about the societal benefits of science and then went on to state his credo that was to mark his career in science. In his own words:

> According to my views, aiming at quantitative investigations, that is at establishing relations between measurements of phenomena, should take first place in the experimental practice of physics. Through measurement to knowledge (*Door meten tot weten*) I should like to write as a motto above the entrance to every physics laboratory.

The crucial phrase is written in rhyming Dutch and likely harks back to Onnes's childhood interest in poetry.

Heike Kamerlingh Onnes had very specific plans for physics. He wanted to establish a comprehensive physical measurement laboratory which could encompass studies of gases, electrical resistance, magnetism, x-rays, and radioactivity. Inspired by van der Waals, Onnes was particularly interested in testing the law of corresponding states for gases. Indeed, Onnes had already written a book discussing several theoretical issues related to van der Waals work.[4] Since Onnes wanted to study simpler molecules which had low condensation temperatures, he was obliged to develop a cryogenics program.

Thus Onnes systematically set about building a laboratory at Leiden on a grand scale. He carried out this program with efficiency, determination, and energy. Since the city of Leiden did not provide electrical power, Onnes purchased a gas motor and generator and made his own electricity. He purchased liquefiers and built his own pumps. By 1892 his liquefaction apparatus was producing several liters of liquid nitrogen and oxygen per day. His laboratory resembled a small factory as much as it did a laboratory: noisy pumps and compressors were connected by large diameter tubes to the cryostats and storage facilities. What distinguished his laboratory from a factory, however, was the central "measurement room". The floor of this room rested on pillars sunk deep into the sandy soil and thus gave the floor sufficient mechanical stability from vibration that all the senstive galvonometers could be used there. This room served as the focus for the whole laboratory since all signals were sent there for registration and for close scrutiny by its founder. It was a living testimony to Onnes's motto.

By 1896 the laboratory was in full swing and Onnes had caught up with the three leaders in cryogenic physics (James Dewar in London, Raoul Pictet in Berlin, and Karol Olszewski in Crakow) and was ready to join them in the international competition dubbed *La Guerre du Froid.* [5] That is, all but two of the known gases had been liquefied, and there was a keen race under way to be the first to liquefy hydrogen and helium.

3. Onnes vs the citizens of Leiden

At the very time that this competition was gaining momentum, however, Onnes's laboratory was to suffer an unexpected and very unfortunate setback. The original University of Leiden had been established in the center of town in the 1500's. Onnes's laboratory was at the nearby site of the science laboratories (anatomy, physiology, chemistry and physics) which had been built more recently. That site had, in fact, a rather turbulent history. In 1807 a barge laden with gunpowder from a factory in Amsterdam had made a stop in Leiden on its way to an armory in

Delft. On January 12 the ship exploded, destroying nearly five hundred houses and leveling a few square blocks of the city. The "ruins" lay dormant until the University eventually decided to expand its facilities by building the science departments there. When it became known that a professor in Physics was conducting experiments with compressed gasses-worse yet, explosive ones-on the very site of the ruins, the citizens of Leiden became increasingly restless. A groundswell of public protest grew and, by 1896, Onnes faced the combined forces of: the municipality of Leiden, the Leiden section of the Society of Public Benefit, the University curators, and, Professor T. Zaayer, Director of the University Anatomy Theatre which was situated between the Physical and Chemical Laboratories, and who feared for his laboratory. [6] In 1896 this group had petitioned the Special Committee of the States General of South Holland (the province to which Leiden belongs) to stop all research involving compressed gasses. The matter was remanded to the Minister of Internal Affairs who formed a committee chaired by van der Waals, and staffed by Kortweg, S. Hoogewerff (professor chemistry at Delft), and C. Ley (a leading engineer in hydrology).

The committee asked Onnes to suspend his work during the course of its investigation. It also asked him to describe his own safety procedures and to inquire as to the safety procedures carried out by the other cryogenic laboratories. Onnes did so and passed the responses onto the Commission. The one from Dewar was particularly direct:

> It would be a terrible disaster for science in your country (and universal science) if the municipality of Leiden succeeded in carrying our any restrictions on your splendid cryogenic laboratory and the fine work you are doing. I cannot understand such a position... I may say that I have made all my experiments with high pressure apparatus before the Prince of Wales and the Sister of your Queen Dowager the Duchess of Albany without the slightest hesitation and no suggestions of danger were even suggested.

The committee studied these responses as well as information on the safety practices for the manufacture and storage of high pressure vessels. It also studied the records for accidents which had occurred from ruptured vessels containing compressed gases. In the course of its deliberations, it concluded that a gas cylinder, for typical pressures and volumes, had the comparatively benign explosive power of only 3 kg of gunpowder. The committee also recognized the fact that many Dutch factories had been built in densely populated areas and represented a far more serious safety hazard than the activities of the physics department at Leiden. The committee considered all this information, ground out a report favorable to Onnes and, finally in 1898, Onnes was granted permission to continue his research.

4. Hydrogen lost, helium gained

About to rejoin the cryogenics race in 1898, Onnes received yet another blow. "Hydrogen... liquefied..." read the telegram from James Dewar, announcing the fact that on May 10, 1898 Dewar had liquefied 20 cc of hydrogen. Dewar was soon liquefying large amounts of this valuable cryogen and, exploiting this accomplishment, he systematically measured several properties of materials, including the temperature dependence of the electrical resistance of metals. Meanwhile Onnes continued his program where he had left off, setting his sights for the remaining prize: the liquefaction of helium. His approach was again systematic and even on a grander scale. It was clear to Onnes by this time that the liquefaction of helium was going to take a far greater commitment of resources than before and he realized the importance of having a cadre of trained technicians available to carry out the increasingly complex experiments. Accordingly, in 1901 he established the Society for the Promotion of the Training of Instrumentation Makers, and a school, the *Leidse Instrumentmakersschool.* The school was put under the direction of Flim. He also created a glassblower's school under the directorship of the master glassblower, Kesselring.

The laboratory grew in power, scale, and personnel, relentlessly outstripping the resources of the three other laboratories. By February 1906, Onnes was liquefying large quantities of hydrogen and it seemed almost inevitable that he would capture the remaining prize. Indeed, after a great number of problems were overcome Onnes and his staff first liquefied helium on July 10, 1908. [7] They found that it boiled at 4.25 K and had a critical temperature of about 5 K. The latter temperature was near the value predicted by Onnes on the basis of his extensive study of the P-V isotherms for helium; once again he had verified the law of corresponding states.

5. The debate concerning the electrical resistance of metals

The resolution of a simmering debate had long been awaiting this momentous occasion. Studies of the electrical resistance of metals had been carried out for many years prior to 1908- motivated in part by the prospect of a new type of thermometer. The relative ease of a resistance measurement combined with the small size of the resistor meant that the resistance-temperature characteristic R(T) of a wire of stable material could be calibrated and thus used to replace the massive and cumbersome gas thermometer.[8] Accordingly, surveys of the R(T) characteristics of metals and alloys had already begun in the early 1800's. At first the interest was in the temperature range above room temperature but, as the cryogenic laboratories succeeded in liquefying various gases, the measurements were extended to ever lower temperatures. The most intriguing and common feature of the measured R(T) curves for various metals was that they all decreased as the temperature was lowered. Such data tempted several theoreticians to offer explanations for this observed behavior. An even more intriguing question arose as the gap between accessible temperatures and absolute zero shrank: What would happen to the electrical resistance of metals as absolute zero was approached? Indeed, the predictions could not be more disparate.

Lord Kelvin took one extreme side of the debate, advancing a thermal activation argument. He noted that glasses were conductive at high temperatures and became insulating as the temperature was reduced. He argued that the conduction electrons in metals should analogously become bound to the lattice atoms at low temperature and he thus predicted [9] that the electrical resistance of metals should become *infinite* at T=0. Kelvin also concluded that any significant increase in the resistance due to this mechanism should only appear at temperatures below 1 K.

The resolution of a puzzle in the specific heat of solids led others to adopt a completely opposite position. Experiments on several materials had showed that their specific heat decreased at low temperatures, thereby contradicting the prediction of Dulong and Petit. Einstein came to the rescue by postulating that Max Planck's theory for quantized oscillators for light could be applied to quantized lattice vibrations in solids. Using the simplification that the lattice vibrated at a single eigenfreqeuncy ν, and thus the energy E of the oscillator was given by Planck's theory as, $E= 3kT/(e^{h\nu/kT} -1)$, Einstein showed that his expression accounted for the observed temperature dependence of the specific heat of the few solids studied. Indeed, from the fits, he could determine the value of ν for each material. Walther Nernst then carried out an extensive survey of the specific heat of many solids and demonstrated a remarkable concordance between Einstein's prediction and all the available data. Emboldened by this success, Nernst went on to suggest that the decrease in resistance of metals had the very same physical basis as the specific heat. He therefore suggested that the resistance should follow an Einstein function of the form

$$R(T)= AE + B = A/(e^{h\nu/kT} -1) + B \quad (2)$$

and he found that this equation provided a rather good account of the temperature dependence of metals over the low temperatures accessible at that time. Onnes, using Drude's equation for the electrical conductivity and arguing that the mean free path should be inversely proportional to the square root of the energy of the quantized lattice vibrations, predicted that $R{\sim}(TE)^{1/2}$. The implication of both equations is that the amplitude of the oscillator decreases with temperature thus causing the resistance to decrease accordingly. At T=0 the scattering from the lattice has vanished, leaving the temperature-independent scattering from residual impurities [the term B in Eq. (2)]. This line of argument led to the conclusion that a pure metal (B=0) would be a perfect conductor at T=0.

What then did the experiments have to say about this issue? In the decade preceding 1908 James Dewar was best positioned to answer this question because he alone could muster the lowest possible temperature on Earth (approximately 14 K) by pumping on liquid hydrogen. The conclusion of his survey of metallic resistance was, regrettably, that 14 K was still too high to resolve the issue. Dewar did note, however, that the trend was for the resistance to approach a constant value at his lowest temperature. Onnes, once he had liquefied helium, was ready to expand his research into this new frontier and to finally answer this question with authority. He chose samples of platinum, gold, and mercury because they could be made very pure. The last element was particularly easy to purify by distillation and the study of mercury enclosed in the glass tubes made by the glassblowing shop became very popular at Leiden.

6. The discovery of superconductivity

The study of the electrical resistance in metals and the liquefaction of helium have been woven into the fabric of the story which we know as the discovery of superconductivity. Since this story has been told by several authors, including Paul H. E. Meijer [10], Rudolph de Bruyn Ouboter [11], Jacobus de Nobel [12] and by Per F. Dahl [13], I will not repeat it here. I will offer instead a few observations. My overall impression is that the phenomenon of superconductivity taunted Onnes with frustration almost as much as it rewarded him with the excitement of discovery. For instance, the shape of the R(T) curve of mercury was truly puzzling: At temperatures down to liquid hydrogen, Onnes obtained a nice fit with his equation, obtaining a fitted value for ν (0.5 mm) which was nearly coincident with a strong frequency (0.3 mm) found in the spectral lines of a mercury vapor lamp. At lower temperatures, however R(T) gradually departed from the simple equation. Furthermore, at about 4.18 K the resistance dropped abruptly to zero-too abrupt in fact for the equations suggested above to apply.

The effect of sample purity on superconductivity was also contrary to what Onnes expected. Thinking that superconductivity could only occur in very pure samples, Onnes sought to destroy it by alloying his distilled, superconductive mercury samples with gold. The superconducting transitions were barely changed. Furthermore, he found that samples of mercury amalgamated with tin had transitions even higher (4.29 K) than that of pure mercury. For many years, despite mounting evidence largely from his own laboratory to the contrary, Onnes persisted in the belief that only pure metals should be superconducting. He stated in 1913, "there is little doubt, that, if gold and platinum could be obtained absolutely pure, they would also pass into the superconducting state at helium temperatures"[14] This was a view he was to cling to as late as 1922.

7. Magnets and more frustration

Although Onnes was unable to completely understand this new phenomenon, he was quick to realize its potential for an immediate practical application. Experimentalists at that time were faced with a limit in the size of the magnetic field they could generate. The maximum flow rate of a coolant sent to remove the power dissipated in the windings of air core copper coils set a limit on the maximum current and thus the magnetic field which could be generated by them. Filling the core with iron helped appreciably, but the saturation of this material set a limit of about 2 T to the magnetic fields which could be generated by this technique. The simple empirical fact established by Onnes that a superconductor possessed zero resistance offered the prospect that an air core coil, freed from the dissipation of power, could be wound which could be used to generate magnetic fields as high as 10 T.

Onnes realized that making such a coil from mercury was not feasible because it was a liquid at room temperature and required that samples to be enclosed in tubes of glass which were not particularly amenable to being shaped into multi-turn coils. In December 1912 the prospects suddenly brightened. Onnes and his collaborators discovered superconductivity in tin (T_c=3.78 K) and in lead (T_c> 6 K). Since these materials would be readily drawn into wires, they immediately eliminated the problem posed by mercury's liquid state. Soon a small coil was wound from tin wire and another from lead wire. Unfortunately the limitation encountered in passing current through mercury appeared in these materials as well. Even more puzzling was the observation that a straight wire segment of tin had a critical current of 8 amperes whereas the coil wound from it had a disappointing I_c value of only 1.0 A. The lead coil performed even worse, possessing a critical current of only 0.8 A which generated a trifling magnetic field of 0.04 T. Again the specters of "bad places", poor thermal contact to the liquid helium (the lead wires were wrapped in a silk insulation), and magnetoresistance were revived.

8. The Nobel Prize; Back to Magnets and more frustration

In November of 1913 Onnes received a telegram from Stockholm notifying him that he was that year's Nobelist in Physics for "his investigations on the properties of matter at low temperatures which led,*inter alia*, to the production of liquid helium". Note that superconductivity is not specifically mentioned in the citation. When he returned from the Nobel Prize ceremony, Onnes revived his study of superconductors in a magnetic field. Hc placed the lead and tin coils in a new copper coil electromagnet recently installed in his laboratory. A series of experiments showed that there was a critical magnetic field between 0.05 and 0.07 T which was sufficient to destroy superconductivity in lead, whereas a magnetic field of only 0.02 T quenced the tin coil. Disappointed, Onnes turned to another aspect of superconductivity: the demonstration of a persistent current. Onnes trapped a magnetic field of about 0.04 T in a lead ring and measured the decay of the trapped field. He established that the resistance of lead in the superconducting state was less than 0.2×10^{-10} of the normal state.

9. Conclusion and summary

World War I essentially suspended low temperature research throughout the world because helium was needed for balloons. Onnes preserved his precious store of helium and was able to eke out a few experiments during this time. Onnes and Tuyn discovered superconductivity in a fourth material (Tl; T_c~2.32 K) in 1919. After the War, experiments expanded and Onnes and Tuyn reported the fifth superconductor (In; T_c~3.4 K) in 1923.

By the 1920's Onnes health was failing and he was frequently absent from the laboratory. In 1922 the Leiden physics department held its 50th anniversary and Onnes presence was clearly felt even though he was not present for most of the celebrations. In 1923 Onnes retired as director of the physics department and in 1926 he died. He left behind a legacy of students and discoveries and associated research issues which were to occupy the attention of low temperature physicists at the laboratory he started for many years to come. The Leiden laboratory was to remain preeminent for many decades following his death and, although a sign emblazoned with the phrase *Door meten tot weten* was never permanently hung above any doorway at Leiden, its spirit was always there.

Acknowledgements

The author thanks Paul H. E. Meijer and Peter Lindenfeld for generously sharing their resouces on Onnes; thanks to Paul Meijer, J. Claassen, and A. Clark for their careful reading of this manuscript.

References

[1] Cohen H 1927*Chemical Society, Journal* 1 1195
[2] Gavroglu K 1994 Phys. **15** 9
[3] Mendelsohn K 1977 *The Quest for Absolute Zero* (Taylor & Francis, Ltd, London) 43
[4] Kamerlingh Onnes H 1881*General Theory of the Fluid State*
[5] Matricon J and Waysand W 1994 *La Guerre du Froid* (Editions du Seuil, Paris)
[6] Kipnis A, Yavelov B E and Rowlinson J S 1966 *Van der Waals and Molecular Science* (Clarendon Press, Oxford) 97-99.
[7] A dramatic account of that long day may be found in Onnes's Nobel Lecture
[8] This effort succeeded, for the platinum resistance thermometer is the standard for the temperature scale.
[9] Lord Kelvin 1902 *Philos. Mag.* 3 257
[10] Meijer P H E 1994*Am. J. Phys.* **62** 1105
[11] de Bruyn Ouboter R 1997 *Scientific American* March
[12] Jacobus de Nobel J 1996 *Physics Today* Sept.
[13] Dahl P F 1992*Superconductivity* (American Institute of Physics, New York) chapter 3
[14] Onnes H K 1913 PLS **34b** 55-70

Index

Abbas F 275, 315
Abell J S 85, 185, 209, 817, 953
Abrahamsen A 1363
Abramowicz A 351
Adrian H 563, 583, 805
Aindow M 185
Akimov I I 1327, 1359, 1511
Albering J H 977, 1033, 1559
Alexandrov V V 1515
Amatuni L 599
Aminov B A 319
Andersen N H 1363
Anderson J 893
Andreev K E 755
Andreone A 177
Andreone D 343, 535
Angurel L A 1211
Aoyagi M 377
Aoyagi T 1523
Apperley M 937
Arienti A 1239
Arndt T 1339
Artigas M 1331
Aruta C 177
Arzumanov A V 559, 627
Ashworth S P 949, 1315, 1425, 1437, 1465
Attanasio C 169, 205
Aubin M 1377
Auer R 93, 153
Auguste F 1029
Auyeung R C Y 789
Azoulay J 73

Balashov D V 445, 731
Ballarino A 1203
Balsamo E P 1499
Baranowski R P 1049
Barbut J M 889
Barholz K-U 651
Barnett M 185
Barone A 457
Barowski H 193
Basset M 563
Baudenbacher F 1089
Bauer C 45
Bauer F 739
Bauer M 1077
Bauer P 1457
Bäuerle D 165
Baumann Th 1173
Baumfalk A 319
Bedard F D 429
Beduz C 1393, 1551
Beghin E 825
Behr R 631
Beilin V 1355
Beille J 889
Beljaev A V 727
Bellingeri E 1283
Belmont O 889
Belotelova Yu N 1359
Belyi D I 1199
Beňačka Š 237, 503, 523
Benjegerdes R J 1503
Bentzon M D 1299, 1363
Benzing W 441
Berberich P 1077
Betz V 1081
Beuink A 1295
Bevilacqua G 1495
Bhasale R 1165
Biegel W 133, 225, 829
Biehl M 441
Bigoni L 1335
Bish P A 1503
Blamire M G 453, 483, 519, 547
Blank D H A 49, 113, 181, 257, 385
Blondel J 373
Blumers M 563
Bock J 825, 977, 1179
Bodin P 1299
Boffa M 849

Boikov Yu A 129, 327
Bolkhovsky D 531
Bonaldi M 303
Bondarenko S I 667
Booij W E 483, 519
Boon W 985, 1587, 1591, 1595
Borgmann J 675
Bornarel A-C 925
Bornemann H J 837
Börner H 283
Borovitskii S I 587
Bosch R 1527
Bourgault D 889
Bousack H 711, 743, 775
Brabetz S 759
Bracanovic D 1113
Braginski A I 675, 743, 751, 775
Bramley A P 201, 271
Brandt C 833
Bratsberg H 1065, 1137
Brecht E 93, 1279
Bringmann B 833
Brons G C S 635
Bruchlos H 351
Bruijn M P 397, 425
Brunetti L 343
Bruun R 1299
Bruynseraede Y 985, 1587, 1591
Bruzzone P 917, 921, 1271, 1477
Buchholz F-Im 433
Buckley R G 1207, 1567
Buisson O 793
Burkhardt H 471
Bussell D M 659
Butkute R 13
Byers J M 789

Calestani G 193
Calleja A 897
Camerlingo C 81, 343, 647
Campbell A M 1235, 1315, 1425, 1437, 1465, 1531
Campbell V 1311
Campion R P 245, 249
Caplin A D 905, 1157, 1263, 1381, 1453
Caracino P 1291
Carapella G 551
Cardwell D A 865
Carelli P 679, 809
Carlsson E F 327, 339
Carolissen R 141
Carr C 747
Carrera M 845
Caspi S 1503
Cassinese A 177
Castellano M G 647, 679, 809
Castellanos A 797
Cave J R 1013, 1351, 1377
Čepelis D 393
Cerdonio M 303, 307
Cesnak L 1445
Chakalov R A 327, 331, 339
Chakraborty N 1381
Chaloupka H J 319, 323
Champion G 515
Chaudhari P 571
Cheggour N 1275
Chen C 1037
Chen H Q 197
Chen J 385, 695
Chen M 1173
Cherubini R 1153
Chesca B 671
Chesneau E C L 1009, 1315, 1425, 1437
Chew N G 185
Chmaissen O 1117
Chow K 1503
Chrisey D B 789
Chromik Š 237, 503, 523
Chubraeva L I 1511, 1539
Chudzik M P 949
Chwala A 739
Cicchelli O 1499
Čihař R 699
Cimberle M R 213
Ciszek M 1315, 1425, 1437, 1465
Claassen J H 57
Claeson T 197, 571
Claridge R 817
Clark T D 663

Clerc S 1385
Clerc St 1389
Clinton T W 789
Cloots R 1029
Coccorese C 17, 169, 205
Cohen D 29
Cohen L F 33, 53, 65
Coletta G 1371
Connor B 1483
Constantinian K Y 409, 417, 559
Conti S 1495
Coombs T A 1531
Costabile G 551
Cottevieille C 825
Cowie A L 33, 53, 65
Cristiano R 539
Crossley A L 881, 1157
Crossley A 877
Crotti G 1461
Cuomo M 1499
Curcio F 1335

Dabrowski B 1117
Daly G M 789
Damen C A J 257
Daniel I J 1169
Dardin S M 1563
Darula M 503, 599
Däumling M 1413, 1433
David P 675, 751
Davidson B A 575
Davis L E 331
Dawson B 1531
De Luca R 1021
Dechert J 755
de Graauw M W M 367
de Groot P A J 37, 1113
de Korte P A J 397, 425
Delabouglise G 97
de la Fuente G F 1211, 1331
Dell'Orco D 1503
Della Corte A 1495
Demsar J 149
Denhoff M W 515
de Nivelle M J M E 397
de Vries R 397
Devyatov I V 463
Dew-Hughes D 201, 271, 1519
Dewhurst C D 865
Dhalle M 1109
Dhallé M 1385, 1453
Di Luccio T 205, 1021
DiTrolio A 849
Diaspro A 213
Didyk A Yu 1591
Diederich D R 1503
Dieleman P 367, 377, 421
Dierickx D 985, 1587, 1595
Diete W 45, 101
Díez J C 1211
Diggins J 663
Diko P 989
Dittmann R 137
Divin Y Y 467
Djukic D 1409
Doderer T 623
Dolabdjian C 785
Dolata R 433, 445, 623
Dolesi R 303
Dolez P 1377
Dolgosheev P I 1199
Donaldson G B 747
Dörrer L 651
Dou S X 869, 937, 957, 1165, 1215
Doyle R A 1129, 1133
Doyle T B 1129, 1133, 1149
Driver R 507
Ďurica M 1311
Dutoit B 1409, 1417
Dzick J 909, 997, 1001, 1085, 1093

Eastell C J 901, 905
Eastell C 1157
Eckelmann H 1263, 1433
Eddy C R 789
Edwards D J 271
Einfeld J 221
Eisenmenger J 133
Ekstrom H 417
Eliáš P 1311
Elsner H 253
Elwenspoek M 397

Emelyanov D 1041
Engelhardt A 137
Engelmann M 583
Erb A 941, 1109
Erickson L E 515
Eriksson J-T 1401
Ermakov A B 389
Erts D 197
Eulenburg A 747
Everett J 905, 1157
Evetts J E 149, 547, 1037, 1049, 1073, 1235, 1315, 1425, 1437, 1465

Fàbrega L 897, 1045
Fahr T 961
Falco C M 169
Faley M I 137, 743, 775
Faley M 137
Falferi P 303, 307
Feld G 257
Ferdeghini C 89, 213
Fikis H 1445, 1457
Filhol P 69
Filichev D A 1327, 1511
Filippenko L V 389
Fink J 1575
Fischer B 965, 969, 1191, 1339
Fischer K 961, 1191, 1255
Fischer S 1243
Fisher B 21
Fisher L M 1511
Flacco K 679
Fletcher J R 245
Flögel-Delor U 821
Flokstra J 425, 635, 703, 707
Flükiger R 941, 1057, 1105, 1109, 1161, 1251, 1263, 1283, 1335, 1385, 1389, 1417, 1453
Foley C P 507, 715
Fontana F 177
Fontana G 307
Fontcuberta J 311, 845, 897, 1045, 1527
Forterre G 69
Fossheim K 1137
Foulds S A L 85
Fox S 905
Francke C 161, 643
Frello T 1363
Frey J 283
Frey U 563
Freyhardt H C 833, 861, 909, 933, 997, 1001, 1033, 1085, 1093
Fritzsch L 491
Frölich K 97, 233
Frunzio L 539
Fuchs G 829, 1535, 1575, 1583
Fuchs M 193
Fukushima K 1543
Funaki K 1473
Furtner S 1077

Gäbler A 1223, 1259
Gagnon R 37, 1113
Galkin A Yu 1057, 1161
Gallop J C 33, 53, 65, 275, 315, 655, 723
Gambardella U 849
Gamble B B 1507
Ganney I 1531
Gao J-R 389, 401
Gao M 515
Garbe S 1363
García López J 113
García-Moreno F 909, 1001, 1085, 1093
Garrè S 1495
Gauss S 977, 1033, 1179, 1559
Gawalek W 989, 1515, 1535, 1579
Gaži Š 237
Gelikonova V D 587
Gemeinder B 981
Gendre P 1347
Genossar J 21
Genoud J-Y 1109
Gerardino A 809
Gerbaldo R 539, 1101, 1153
Gerritsma G J 437, 463
Gerster J 813
Gevorgian S 279, 327, 339
Gherardi L 1291, 1371, 1461
Ghigo G 1101, 1153

Ghinello I 1291
Ghosh I S 267, 611
Giannini E 89, 213
Gibson G 53
Gierl J 965, 969
Ginovker M 81
Gislon P 1499
Giunchi G 1239
Gladun A 1535, 1583
Gladyshevskii R E 1105, 1283
Glowacki B A 949, 1009, 1073, 1287, 1315, 1425, 1437, 1465
Gobl N 1227
Godeke A 917
Göhring D 1323
Gol'tsman G 405
Goldacker W 1223, 1259, 1263, 1279, 1433
Goldgirsh A 1355
Goldobin E 547
Golubov A A 463
Gomis V 845, 857
Gömöry F 993, 1371, 1461
Goncharov I N 1591
Gorbenko O Y 1097
Gordeev S N 37, 1113
Goren S 1179, 1227
Goringe M J 901
Görnert P 1535
Gösele U 173
Gough C E 125, 817
Goul J 1299
Gozzelino L 539, 1101, 1153
Granados X 1527
Granata C 479, 487, 647
Grassano G 89, 213
Grasso G 941, 1251, 1263, 1335, 1385, 1389, 1417, 1453
Green M A 1491, 1503, 1563
Gregušová D 1311
Grigorashev D I 929
Grivel J-C 941
Gromoll B 1243
Grovenor C R M 201, 271, 901, 905, 1157
Grube K 981
Gruss S 829, 1575
Grüneklee M 743, 775
Gubankov V N 467
Gubina M 279
Gudoshnikov S A 755
Guérig A 1173
Guillon H 885
Guilloux-Viry M 81
Gundlach K H 373, 377, 421
Guo Y C 1165
Gustafsson M 241

Ha Y S 347, 363
Habisreuther T 1515, 1579
Halbritter J 41, 61
Haldar P 1009, 1393
Haller A 751
Hampshire D P 1005, 1169, 1275
Hamster A W 425
Han S-K 287, 347, 363
Han Z 1299
Hanayama Y 1017
Hancox J 1235
Hangyo M 595, 603
Hannaford R 1503
Hao L 275, 315, 655, 723
Hara T 1183
Harnack O 503
Harnden W 1503
Harrison S 1203
Hasegawa K 1121
Hasenöhrl S 1311
Haseyama S 1547
Haug P 1323
Hauser S 193
Heiden C 755
Heidenblut T 397
Heidinger R 61
Hein G 555
Hein M A 45, 261, 283, 319, 101
Heine G 1061
Heinemann K 833, 909, 933, 997, 1001, 1985, 1093
Heinicke K 161
Heinrich J 651
Heller R 1223

Hellstrom E 893
Hensen S 101
Herlach F 985, 1591, 1587
Herrmann K 93
Herrmann P F 825
Herzog P 969
Herzog R 1203
Hesse D 173
Heutink J 181
Hidaka M 449
Higley H C 1503
Higuchi T 1145
Hilgenkamp H 1
Hill F 319
Hinaus B M 575
Hirst P J 185
Hochmuth H 283
Hoenig H E 567
Hoffmann J 909, 997, 1001, 1085, 1093
Hohmann R 743, 775
Hojczyk R 137
Holzapfel B 1081
Hong G-W 873
Hong J S 359
Honjo S 1183
Hörnfeldt S 1469
Horng H E 157, 639, 683
Horstmann C 137
Hortig M 1303, 1319
Horvat J 869, 1165
Horwitz J S 789
Hosking M W 291
Hoste S 141
Hu Q-H 197
Huang C 189
Huang S-L 1137
Huang Y B 1385, 1389, 1453
Huang Y K 1429
Huang Y 1251
Huber S 351
Hubert-Pfalzgraf L 885
Hübner U 651, 699
Huebener R P 623
Hughes T J 1393
Huh Y 691
Hulshoff W 401
Humphreys R G 125, 185
Hüning F 121
Hušek I 1311, 1343, 1367
Hutchison J M S 659
Hütten A 961, 1255
Hyland D 271, 905

Iavarone M 177
Iguchi I 417, 527, 619
IJspeert A 1203
IJsselsteijn R P J 253, 567, 687
Il'ichev E 567
Ilushin K V 1515
Inoue M 1121
Ionescu M 869, 937
Iriarte F 1331
Irie F 1017, 1121
Isaac S P 453
Isaev A 1001
Ishibashi T 527
Issaev A 909, 1085
Itoh M 1547
Ivanov Z G 7, 129, 197, 327, 331, 339, 571
Iwashita H 381
Iwata Y 1183
Izquierdo M 703, 707

Jackson T J 1073
Jaekel C 121
Jansak L 1231, 1239
Jansman A B M 703, 707
Jarvis A L L 1149
Jegers J B M 377
Jeng J T 157
Jenkins A P 201, 271
Jeong S 1195
Jia C L 137
Jiang J 953
Jim K 1519
Jirsa M 1057, 1161
Jockel D 1085
Johansen T H 1065, 1137
Johansson L-G 7, 197
Johnson M 789

Johnston M D 905, 1263
Jorgensen J D 1117
Jou D C 639
Juengst K P 1487
Juhás P 237
Jung B 1579
Jung G 17, 81, 1179, 1227
Jutzi W 441

Kachlicki T 425
Kadowaki K 1157
Kaiser A W 837
Kaiser T 45, 101, 283, 319
Kalinov A V 1511
Kallscheuer J 45
Kalsi S S 1483, 1567
Kanda Y 299
Kang K-Y 287, 347, 363
Kaparkov D 279
Kaplan T 189
Karpov A 373
Kasatkin A L 1571
Kästner G 173
Kasztler A 1457
Kattouw H E 635
Kaul A R 1097
Kautschor L-O 861
Kautz S 965, 969
Kawai Y 1519
Kawano K 1307
Kazakov E G 1515
Keck K 797
Keck M 623
Kellers J 1267
Kemmler-Sack S 1323
Kessel W 433
Khabipov M I 433
Kidiyarova-Shevechenko A Yu 433, 445
Kiewiet F B 425
Kim D W 691
Kim I S 763
Kim J M 763
Kim J 287, 363
Kim J 789
Kim S-J 695
Kinder H 1077
King P J 245, 249
Kirchmayr H 1445, 1457
Kirichenko A 539
Kirichenko D E 727
Kiryakov N 1041
Kiss T 1017, 1121
Kitaguchi H 1543
Klapwijk T M 367, 377, 401, 421
Klarmann R 225
Klein C 1105
Klein N 69, 267, 295, 611
Klemm F 759
Klepatsky V E 1359
Klinger M 351
Klippe L 885, 1097
Klupsch Th 989
Klushin A M 587, 599
Knizhnik A 21
Knoopers H G 917, 1271, 1555
Koblischka M R 1057, 1065, 1137, 1141, 1145, 1161
Kobzev A P 523
Koch H 781
Koch R H 767
Koch R 441, 837
Koenitzer J W 1117
Kohjiro S 615
Köhler H-J 491
Köhler O 699
Kohlmann J 631
Kohlstedt H 499, 587, 599
Koishikawa S 85
Kojo H 1145
Kolesov S 319, 323
Kollberg E 279, 405
Koller D 789
Komissinski P 109
Königs Th 221
Kopera L 1343
Kopperschmidt P 173
Kordyuk A A 973
Koren G 29
Kornev V K 559, 627
Koshelets V P 389, 731
Kosmetatos P 1049

Koster G 181
Kotelyanskii I M 229, 467, 475
Kötzler J 121
Koutzarova T 145
Kováč P 1311, 1343, 1367, 1445
Kovalev K L 1515
Kovalev L K 1515
Kozlenkova N I 1327
Krabbes G 829, 1575, 1583
Krämer A 161, 643
Krämer H-P 1191, 1243
Krasnopolin I Y 631
Krauns Ch 833
Krause H J 675, 711, 743, 751, 775
Krauth H 1339, 1495
Krelaus J 861, 933
Krellmann M 885
Krooshoop E J G 1271
Kropman B L 181
Kroug M 405
Kúdela R 1311
Kühle A 217
Kuhn M 133, 225
Kuk I-H 873
Kula W 117
Kumakura H 1543
Kumar S 767
Kummeth P 1243, 1247
Küpfer H 925
Kupriyanov M Yu 463
Kurihara M 1219
Kurin V 531
Kürschner H-G 1303, 1319
Kuršumovic A 1037, 1049, 1235
Kurz H 121
Kutzner R 221
Kuznetsov M 1041
Kvitkovic J 1009, 1315
Kwasnitza K 1385, 1389
Kwon H C 763

Labusch R 1129, 1133
Lacquaniti V 413, 535, 543
Ladenberger F 933
Lagutin A S 985
Lahl P 77, 221
Lahtinen M 1401, 1483
Lakner M 1173
Lam S K H 507, 511
Lanagan M T 949
Lancaster M J 125, 359
Lang S Y 157
Lang W 117, 165, 1061
Lapsker I 73
Larbalestier D 893
Larionoff A E 1515
Larsson P 327, 339
Laurenti A 1495
Laurinavičius A 393
Leake J A 1073
Lebedev O I 49
Lee E-H 579
Lee H-G 873
Lee K 417, 527, 619
Lee M C 691
Lee P J 1271
Lee S-G 691
Lee S 1041
Lee W L 683
Lee Y H 763
Leenders A 861, 1033
Legendre F 1347
Leghissa M 1191, 1449
Lehikoinen P 389
Lehndorff B 1303, 1319
Lehtonen J 1483
Leiderer P 133
Lemaitre Y 295
Lenzner J 283
Leoni R 679, 809
Leonyuk L I 1591
Leriche A 825
Leslie K E 715
Li Y H 53, 945
Liang G-C 311
Liang J-F 311
Libera M 575
Liday J 523
Liégeois M 1029
Lietzke A F 1503
Linker G 1279
Linzen S 699

Lippold G 283
Lisauskas V 13, 393
Litskevitch P G 445
Litzkendorf D 1579
Liu H K 869, 937, 957, 1165, 1215
Lo W 865, 1531
Lomatch S 591
Lomparski D 775
Lorenz M 173, 283
Löw R 1069
Lu J H 639
Lu Z 189
Luinge W 389
Luiten O J 425
Luzanov V A 229

MacKinnon B 1567
MacManus-Driscoll J L 53, 877, 881, 945
Macfarlane J C 715, 723, 747
Machajdík D 233, 523
Macks L D 715
Maezawa M 615
Mage J C 295
Magee E 1563
Maggi S 413, 535, 543
Magnusson N 1469
Maix R 1495
Majoros M 1009, 1231, 1239, 1315
Mannhart J 1
Manton S 1551
Manzel M 351
Marciniak H 1367
Maritato L 169, 205
Mariño A 1053
Markowitsch W 117, 165
Marquez I 1527
Marré D 89, 213
Marrs R E 1563
Marti F 1109, 1251, 1385, 1453
Martínez B 857, 1045, 1211
Martinez J C 563, 583, 805
Martini L 1335
Mashtakov A D 109, 229, 475, 559
Matijasevic V 189
Matsumoto K 85
Matsushita T 1017
Mattern N 1081
Matthews R 767
Matveets L V 755
Maus M 775
Mawdsley A 1207
Mayerhöfer T 193
McCulloch M D 1519
McInturff A D 1503
McN Alford N 69, 355
Meerovich V 1179, 1227
Meersschaut J 985
Meier O 735
Melchiorri F 809
Mele R 1291, 1371, 1461
Melišek T 1445
Mercaldo L 169, 205
Mex L 161, 643
Meyer H-G 253, 491, 567, 687, 739
Mezzena R 307, 727
Mezzetti E 539, 1101, 1153
Miao H 1211, 1331
Michaels P C 1195
Michalke W 397
Migliazza M 1335
Migliori A 193
Mihailovic D 149
Mikkonen R 1401, 1483
Minervini J V 1195
Minetti B 1101, 1153
Mitchell E E 507
Miteva S 109, 145
Moerman R 437
Monaco A 647
Monaco R 81
Montanari G C 1291
Monticone E 413, 535, 543
Moore D F 483, 519
Moore J C 901, 905
Mora J 845, 1527
Morgan C 1157
Morgenroth W 253, 687
Mori K 1547
Morin D 1461
Morley S M 245, 249
Morrison L 1503

Morrison M E 1503
Moshchalkov V V 985, 1587, 1591
Moya A 453
Mozhaev P B 217, 229, 409
Mück M 755
Mukhanov O 539
Mukhergee S K 1563
Müller F 495, 555, 631
Müller G 45, 101, 283, 319
Müller J 161, 643, 965, 969
Müller K-H 715, 1397, 1575
Munz M 1339
Murakami M 85, 1145
Murashov V A 869, 937
Murashova G E 937, 957
Mygind J 409, 731
Myoren H 385, 695

Nagashima T 595
Nakajima K 385, 695
Nakamura M 1473
Nakamura T 1017, 1121
Nakielski G 121
Nam B C 691
Navarro R 1211, 1331
Neale R 1567
Nedkov I 145
Nemoshkalenko V V 973
Nesher O 29
Neuhaus M 441
Neumüller H-W 965, 969, 1105, 1191, 1243, 1247
Nevirkovets I P 547
Nibbio N 1417
Niemeyer J 433, 495, 555, 623, 631
Nies R 1247
Nietzsche S 781
Nijhuis A 917, 921, 1271
Nikulin A D 1327, 1359
Noguchi T 381
Nomura K 1543
Nomura S 1183
Nordman J E 575
Noudem J G 889
Nurgaliev T 109, 331, 801
Nálevka P 1057, 1161

O'Callaghan J 311
Obradors X 845, 853, 857, 897, 1045, 1527
Ockenfuss G 25, 137, 675, 797
Odelius H 1367
Ohkuma T 1183
Okada M 1543
Okamoto H 1017
Oleynikov N N 929
Olsen S K 1441
Olsson E 241
Ong C K 105
Ooi S 1129
Oomen M P 1449
Oota A 1307
Osofsky M S 789
Ottoboni V 1239
Ovsyannikov G A 109, 217, 229, 409, 417, 475, 559

Paasi J 1125, 1401, 1483
Pace S 849, 1021
Pachla W 1343, 1367
Pagano S 457, 539
Pallares J 1527
Pallecchi I 89, 213
Palmieri V G 479
Palomba F 177
Pan V M 1057, 1571
Papst G 1267, 1507
Parage F 793
Park G 579
Park J C 763
Park S 579
Park Y K 763
Park Y W 347
Pasotti G 1499
Passerini R 1251
Patlagan L 21
Paul W 1173
Pauza A J 483
Pavolotskij A B 727
Peden D 723
Pedersen N F 409, 551
Pedyash M V 519
Pegrum C M 747

Peña J 885
Penkin V T 1515
Penn S J 69, 355
Penny M 1551
Perkins G K 905
Perpeet M 101
Perrin A 81
Peters M J 771
Petraglia A 551
Petrizzelli M 343
Petrov P K 339
Picard H 1567
Piel H 45, 101, 319, 1303, 1319
Piñol S 845, 897, 1045
Pitschke W 961
Plathner B 377
Podobnik B 149
Poeppelmeier K R 1117
Polak M 893, 1303
Polichetti M 849, 1021
Polimeni A 245
Poltavets V N 1515
Polturak E 29
Polyakova N V 1199
Polyanskii A 893
Pooke D M 1207, 1567
Pöpel R 495, 555, 631
Popov F V 1327
Poppe U 137, 467, 743
Porcar L 889
Porch A 125
Poulsen H F 1363
Pouryamout J 101
Pous R 311
Powell J R 125
Prance H 663
Prance R J 663
Predel H 471
Pressler H 623
Price J P G 817
Prishepa S 169, 205
Pröbst F 735
Prokhorova I G 727
Prokopenko G V 731
Provoost R 1595
Prusseit W 587
Puig T 857
Pust L 1057, 1161
Putti M 213

Quilitz M 1259, 1263, 1433
Quinton W A J 1089

Rabbers J J 1429
Rajteri M 413
Rakov D N 1327, 1359
Ralph J F 663
Rasmussen C 1441
Rassau A P 37, 1113
Rastello M L 413
Reder M 933
Reed R P 655
Régnier P 1347
Reißig L 813
Reimann C 1069
Reiner H 981
Reisner G M 21
Renk K F 193, 1069
Repšas K 393
Reppel M 323
Reschauer N 193
Rhyner J 1173
Ricci M V 1495, 1499
Rice J P 785
Rieger J 1191, 1449
Riepl T 1069
Ries G 1243
Rijnders A J H M 49, 181
Rijpma A P 771
Riley A 209
Rippert E D 591
Ritzer A 165
Roas B 965, 969, 1191, 1339
Robbes D 785
Robertazzi R 539
Robertson I D 335
Rodenbush A J 1507
Rodig C 961, 1255
Rodríguez H 1053
Roesthuis F J G 703
Rogacki K 1117

Rogalla H 113, 181, 257, 385, 425, 437, 463, 635, 703, 707, 771
Rojas C E 1053
Rolandi R 213
Romans E J 723, 747
Roskos H G 121
Rosseel K 985, 1587, 1591, 1595
Rossi S 1495
Roth H 253
Rowley A T 1235
Rozen J R 767
Rüders F 711
Ruggiero B 479
Rühlich I 813
Rulmont A 1029
Russo M L 177, 479, 487, 647
Rzchowski M S 575

Sachse H 555
Saini R 817
Sakai K 595, 603
Sakai S 499
Salbert H 1487
Salluzzo M 205
Salpietro E 1495
Salvato M 169, 205
Samoilova T 279
Samoylenkov S V 1097
Sánchez S 397
Sancho D 311
Sandemann K G 1437
Sandiumenge F 845, 853, 857
Sankrithyan B 507
Sans C 311
Saravolac M P 1235
Sarnelli E 647
Sato J 1543
Sato K 527
Satoh T 449
Savo B 17
Sawada K 1145
Scanlan R M 1503
Schattke A 583
Schätzle P 829, 1575
Schauer T 193, 1069
Schauer W 981
Scheider R 93
Schemion D 611
Scherer T 441
Schey B 133, 225
Schicke M 377, 421
Schieber M 1355
Schiek B 607
Schiller H-P 981
Schilling M 471
Schläfer U 961, 1255
Schmatz U 885
Schmid R 1247
Schmidl F 651, 699, 759
Schmidt W 1081, 1243, 1247
Schmirgeld-Mignot L 1347
Schmitz G J 841
Schneider C W 437
Schneider J 221
Schneider R 153, 437
Schneidewind H 651
Schöllmann V 663
Schornstein S 267
Schöttler R 1267, 1507
Schubert J 675
Schubert M 961, 1255
Schultz J H 1195
Schultz L 1081, 1575
Schultze V 567, 687, 739
Schulz G W 1105
Schulz H 467
Schulze H 495
Schwab R 61
Schwachulla J 1077
Schwall R E 1567
Schwan C 805
Schwarz S E 335
Schwarzmann E 933
Schwierzi B 397
Seeßelberg M 841
Seebacher B 1243, 1247
Segarra M 897
Seidel P 651, 699, 759, 813
Seidel W 735
Selbmann D 885
Selders P 797
Sello S 1231

Semenikhin V S 1515
Semerad R 587
Seppenwoolde Y 771
Seton H C 659
Severloh P 61
Shablo A A 667
Shadrin P M 467
Shapiro B Ya 81
Shevchenko O A 1429, 1555
Shi S-C 381
Shi Y H 865
Shields T C 209
Shikii S 595
Shikov A K 1327, 1359, 1511, 1515
Shiles P 901
Shirotov V V 467
Shitov S V 389, 731
Shoji A 615
Siangchaew K 575
Siano C 1495
Sicking F 607
Siegel M 719
Siejka J 113
Silvestrini P 479
Simon O 1487
Sin A 897, 1045
Sing M L C 785
Singer R F 965
Sipos G 1081
Siri A S 89, 213
Sirotko D V 1511
Sjöström M 1409
Skov J L 217
Skov-Hansen P 1299
Slaughter J M 169
Sloggett G L 715
Smith K 1203
Sneary A B 1005
Snigirev O V 727, 755
Snitchler G 1567
Sobolewski R 117
Södervall U 1367
Sodtke E 587
Sokolovsky V 1179, 1227
Soltner H 743
Somekh R E 1073
Song I 579
Sonntag I 989
Šouc J 97, 233
Soukhov V M 1511
Soulen Jr R J 789
Spadoni M 1495, 1499
Španková M 237
Spasov A 109, 331
Specht H 651
Spreitzer U 193
Spörl R 61
Squitieri A A 1271
Stadel O 885, 1097
Staiger Th 1583
Steinbeiss E 397
Steinrück H-P 1247
Steni R 413, 535, 543
Stepantsov E A 571
Stockinger C 117, 165
Stolz R 739
Stoye P 1535
Straßer T 1579
Strasser T 1515
Štrbík V 503, 523
Stritzker B 133, 225
Stroud R M 789
Stöver G 829
Subke K O 121
Sugiura T 1523
Sumption M D 913
Suurling A K 401
Suzuki E 1219
Svalov G G 1199
Sytnikov V E 1199
Szalay A 1227
Szczepanowski A 1567
Szulczyk A 1339, 1495

Tahara S 449
Taillefer L 37, 1113
Takada S 377
Takács S 97, 913, 993, 1187, 1405
Takayasu M 1195
Takeo M 1017, 1121, 1473
Tallon J L 1207, 1567
Tamegai T 1129

Tanaka K 1543
Tani M 595, 603
Tanichi N 595
Tarte E J 453, 483
Taussig D A 767
Tavrin Y 719, 751
Taylor C E 1503
Tchoubraev D V 1539
Tekletsadki K 1235
Templeton A 355
Teng M 1203
ten Haken B 917, 1295, 1429
ten Kate H H J 917, 921, 1271, 1295, 1429, 1555
Terai H 449
ter Brake H J M 771
Teske C L 977, 1025
Thoener M 1495
Thrum F 351, 491
Thyssen N 499, 531
Thürk M 813
Tian Y 699
Timlin C 1203
Tixador P 889
Togano K 1543
Tomiya H 1473
Tonkin B A 291
Tonouchi M 595, 603
Torres Bruna J M 703
Torrioli G 647, 679, 809
Tournier R 889
Traeuble T 623
Trappeniers L 985, 1587, 1591
Trautner A 1323
Tretyakov Yu D 929, 1041
Trinks P 961
Trofimenko E A 929
Tsurunaga K 1183
Tuohimaa A 1125
Turner C W 335
Twardowski T 1531

Uchaikin S 735
Uematsu Y 1523
Uezono I 299
Ukhansky N N 755
Ullmann B 1223, 1259, 1279
Ullrich M 833, 841, 861, 1033
Understrup H 1421
Unternährer P 1173
Urata M 1183
Urban K 137, 743
Usoskin A 909, 1001, 1085, 1093
Ustinov A V 499, 531, 547

Vaglio R 177
Vajda I 1227
Vallet-Regi M 885
Van Bael M 985
van Dalen A J J 1145
van den Berg M L 425
van der Hart A 797
van de Stadt H 377, 401, 421
Van Driessche I 141
van Duuren M J 635
Van Oort J M 1503
Van Tendeloo G 49
Vanacken J 985, 1587, 1591
Vanhoyland G 141
Varesi E 1335
Varlashkin A 193
Vase P 1299, 1421
Vaškevičius A R 393
Vaupel M 241, 797
Vávra I 237
Vázquez-Navarro M D 1037
Vecchione A 849
Vendik I 279
Vendik O 279
Vengalis B 13
Verbist K 49
Verdyan A 73
Verges P 1255
Verhage T 825
Verhoeven M A J 49
Vilalta N 853, 857
Vitale S 303, 307, 727
Vodel W 781
Vogrinčič P 523
Vogt M 1323
Volkov O Y 467
Volkozub A V 905, 1263, 1381

Voloshin Igor F 1511
von Papen T 885
von Zimmermann M 1363
Voss M 373
Vysotskaia V V 1195
Vysotskii V V 1571
Vysotsky V S 1017, 1195, 1473

Wada N 603
Wahl G 885, 1097
Walker E 1109
Wallraff A 531
Walsh G 817
Wang C H 1073
Wang H Y 359
Wang W G 1165
Wang X L 1061
Wang X Z 1061
Wang X 355
Warburton P A 335
Watson D R 1049, 1235
Weber C 587
Weber H W 1105
Webers J M A 425
Wehler D 1303
Weidl R 759
Weimann T 555, 623
Weiss F 97, 233, 885
Wells J J 877
Wende C 989
Wende G 491
Wenger Ch 1583
Wenger F 571
Werfel F N 821
Whitecotton B R 767
Whiteman R 663
Widenhorn L 1173
Wiesmann J 909, 997, 1001, 1085, 1093
Wiezoreck J 1191
Wijnbergen J J 397
Wilberg R 1303
Wilke I 121
Wilke T 841
Wilkinson A J 201
Willén D W A 1013, 1351, 1377
Wilms Floet D 401
Wilson Y 507
Winter M 69
Wippich D 821
Wirth G 1157
Wischert W 1323
Wisniewski A 1153
Witz G 1385
Woeltgens P 767
Wolf T 925
Woodfield B F 789
Wooldridge I 335
Wördenweber R 25, 77, 221, 241, 279, 295, 797
Wormeester H 425
Wróblewski M 1367
Wroe F C R 1235
Wroe R 1551
Wu J M 639, 683
Wu M 1579
Wu Z 331
Wunderlich S 651, 759

Xu S Y 105

Yagoubov P 405
Yamafuji K 1017, 1121
Yamashita T 385, 695
Yanagi N 1187
Yang H C 157, 639, 683
Yang S Y 639, 683
Yang Y 1393, 1551
Yashchin E 1355
Yazawa T 1183
Yi H R 241, 711
Yoneda E 1183
Yoo S I 1145
Yoshida K 299
Yoshizawa S 1547
Yu H W 639
Yu I-K 287
Yu R 845, 853, 857
Yuzhelevski Y 81

Zachmayer S 1069
Zafar N 877

Zaitsev A G 25, 279, 295
Zakosarenko V 567, 687
Zamboni M 85
Zameck Glyscinski J v 781
Zannella S 1231, 1239, 1335
Zavrtanik M 149
Zeimetz B 957, 1215
Zeng X H 711
Zhang W 893
Zhang X 105
Zhang Y 711, 775
Zheng D N 1005, 1169
Zhou Y L 105
Zhu W 1013, 1351, 1377
Zuccaro C 69, 295